OTHER PH DIFFERENTIAL EQUATIONS TITI

CUSHING
Differential Equations: An Applied Approach (0-13-044930-X)

EDWARDS / PENNEY
Differential Equations: Computing and Modeling, 3/E (0-13-067337-4)
Differential Equations and Boundary Value Problems:
Computing and Modeling, 3/E (0-13-065245-8)

EDWARDS / PENNEY
Elementary Differential Equations, 5/E (0-13-145773-X)
Elementary Differential Equations with Boundary Value Problems, 5/E (0-13-145774-8)

CONRAD
Differential Equations: A Systems Approach (0-13-046026-5)
Differential Equations and Boundary Value Problems:
A Systems Approach (0-13-093419-4)

HOLLIS
Differential Equations with Boundary Value Problems (0-13-015927-1)

BANKS
Differential Equations with Graphical and Numerical Methods (0-13-084376-8)

DAVIS
Differential Equations: Modeling with MATLAB® (0-13-736539-X)

RAINVILLE / BEDIENT / BEDIENT
Elementary Differential Equations, 8/E (0-13-508011-8)

PH TEXTS IN DIFFERENTIAL EQUATIONS AND LINEAR ALGEBRA

EDWARDS / PENNEY
Differential Equations and Linear Algebra, 2/E (0-13-148146-0)

FARLOW / HALL / MCDILL / WEST
Differential Equations and Linear Algebra, 2/E (0-13-186061-5)

GOODE
Differential Equations and Linear Algebra, 2/E (0-13-263757-X)

GREENBERG
Differential Equations and Linear Algebra (0-13-011118-X)

WILLIAMSON/TROTTER
Multivariable Mathematics, 4/E (0-13-067276-9)

Contact **www.prenhall.com** to order any of these titles

Differential Equations
with Boundary
Value Problems

Differential Equations with Boundary Value Problems

Second Edition

John Polking
Rice University

Albert Boggess
Texas A&M University

David Arnold
College of the Redwoods

PEARSON

Prentice
Hall

Upper Saddle River, New Jersey 07458

Library of Congress Cataloging in Publication Data

Polking, John C.
  Differential equations with boundary value problems / John Polking,
Albert Boggess, David Arnold.—2nd ed.
     p.   cm.
  Includes index.
  ISBN 0-13-186236-7
  1. Differential equations. 2. Boundary value problems. I. Boggess,
Albert. II. Arnold, David. III. Title.

QA371.D4495   2005
515'.35--dc22   2005049284

Executive Acquisitions Editor: George Lobell
Editor-in-Chief: Sally Yagan
Production Editor: Barbara Mack
Senior Managing Editor: Linda Mihatov Behrens
Executive Managing Editor: Kathleen Schiaparelli
Manufacturing Buyer: Alan Fischer
Director of Marketing: Patrice Jones
Marketing Manager: Halee Dinsey
Marketing Assistant: Joon Won Moon
Editorial Assistant/Print Supplements Editor: Jennifer Urban
Art Director: Jonathan Boylan
Interior and Cover Design: Koala Bear Design
Art Editor: Thomas Benfatti
Director of Creative Services: Paul Belfanti
Manager, Cover Visual Research & Permissions: Karen Sanatar
Cover Photo: Hemisfèric (right) and Museo de las Ciencias Príncipe Felipe (left). City of Arts and Sciences. Valencia. Spain.
Photo: age fotostock/SuperStock
Art Studio: Laserwords

© 2006, 2002 Pearson Education, Inc.
Pearson Prentice Hall
Pearson Education, Inc.
Upper Saddle River, NJ 07458

Pearson Prentice Hall™ is a trademark of Pearson Education, Inc.

MATLAB®is a trademark of The MathWorks, Inc. and is used with permission. The MathWorks does not warrant
the accuracy of the text or exercises in this book. This book's use or discussion of MATLAB® software or related
products does not constitute endorsement or sponsorship by The MathWorks of a particular pedagogical approach or
particular use of the MATLAB® software.

Printed in the United States of America

10   9   8   7   6

ISBN 0-13-186236-7

Pearson Education LTD., *London*
Pearson Education Australia PTY, Limited, *Sydney*
Pearson Education Singapore, Pte. Ltd
Pearson Education North Asia Ltd, *Hong Kong*
Pearson Education Canada, Ltd., *Toronto*
Pearson Educación de Mexico, *S.A. de C.V.*
Pearson Education—Japan, *Tokyo*
Pearson Education Malaysia, Pte. Ltd

Contents

Preface

This book started in 1993, when the first author began to reorganize the teaching of ODEs at Rice University. It soon became apparent that a textbook was needed that brought to the students the expanded outlook that modern developments in the subject required, and the use of technology allowed. Over the ensuing years this book has evolved.

The mathematical subject matter of this book has not changed dramatically from that of many books published ten or even twenty years ago. The book strikes a balance between the traditional and the modern. It covers all of the traditional material and somewhat more. It does so in a way that makes it easily possible, but not necessary, to use modern technology, especially for the visualization of the ideas involved in ordinary differential equations. It offers flexibility of use that will allow instructors at a variety of institutions to use the book. In fact, this book could easily be used in a traditional differential equations course, provided the instructor carefully chooses the exercises assigned. However, there are changes in our students, in our world, and in our mathematics that require some changes in the ODE course, and the way we teach it.

Our students are now as likely to be majoring in the biological sciences or economics as in the physical sciences or engineering. These students are more interested in systems of equations than they are in second-order equations. They are also more interested in applications to their own areas rather than to physics or engineering.

Our world is increasingly a technological world. In academia we are struggling with the problem of adapting to this new world. The easiest way to start a spirited discussion in a group of faculty is to raise the subject of the use of technology in our teaching. Regardless of one's position on this subject, it is widely agreed that the course where the use of technology makes the most sense, and where the impact of computer visualization is the most beneficial, is in the study of ODEs. The use of computer visualization pervades this book. The degree to which the student and the instructor are involved is up to the instructor.

The subject of ordinary differential equations has progressed, as has all of mathematics. To many it is now known by the new name, dynamical systems. Much of the progress, and many of the directions in which the research has gone, have been motivated by computer experiments. Much of the work is qualitative in nature. This is beautiful mathematics. Introducing some of these ideas to students at an early point is a move in the right direction. It gives them a better idea of what mathematics is about than the standard way of discussing one solution method after another.

It should be emphasized that the introduction of qualitative methods is not, in itself, a move to less rigor.

New to This Edition

We are gratified with the success of the first edition. However, in the years since its first appearance, we have continued to teach ODEs and we have new, hopefully better ideas about the subject. We have learned from many comments from and conversations with our users. As a result many small parts have been rewritten to improve the exposition. In addition, this new edition incorporates the following more substantial changes.

- We have added a large number of new figures to help visualize the ideas in the text.
- More of the examples are now application based.
- Where appropriate we have highlighted methods of solution to make them easier to find.
- The section in Chapter 2 on exact first-order equations has been entirely rewritten.
- Chapter 7 on matrix algebra has been rewritten. There is now a section devoted to the subject in two and three dimensions. For many instructors this is all that is needed. For the rest it is a good introduction to the subject.
- The introductory discussion of linear systems in Chapter 8 has been rewritten, to improve the exposition and to incorporate more applications.
- The section in Chapter 9 on phase plane portraits was too long, so it has now been split in two, resulting in a new section on the trace-determinant plane.
- We have collected the material on complex numbers and matrices into an appendix to make it more easily found.
- We have added new exercises and eliminated old ones in many sections. Many of the new exercises are application oriented. This is especially true of the exercises on partial differential equations in Chapter 13.

The Use of Technology

The book covers the standard material with an appropriate level of rigor. However, it enables the instructor to be flexible in the use of modern technology. Available to all, without the use of any technology, is the large number of graphics in the book that display the ideas in ODEs. At the next level are a large number of exercises that require the student to compute and plot solutions. For these exercises, the student will have to have access to computer (or calculator) programs that will do this easily.

The tools needed for most of these exercises are two. The student will need a program that will plot the direction field for a single differential equation, and superimpose the solution with given initial conditions. In addition, the student will need a program that will plot the vector field for an autonomous planar system of equations, and superimpose the solution with given initial conditions. Such tools are available in MATLAB®, Maple, and *Mathematica*. For many purposes it will be useful for the students to have computer (or calculator) tools for graphing functions of a single variable.

The book can also be used to teach a course in which the students learn numerical methods early and are required to use them regularly throughout the course.

Students in such a course learn the valuable skill of solving equations and systems of equations numerically and interpreting the results using the subject matter of the course. The treatment of numerical methods is somewhat more substantial than in other books. However, just enough is covered so that readers get a sense of the complexity involved. Computational error is treated, but not so rigorously as to bog the reader down and interrupt the flow of the text. Students are encouraged to do some experimental analysis of computational error.

Modeling and Applications

It is becoming a common feature of mathematics books to include a large list of applications. Usually the students are presented with the mathematical model and they are required to apply it to a variety of cases. The derivation of the model is not done. There is some sense in this. After all, mathematics does not include all of the many application areas, and the derivation of the models is the subject of the application areas. Furthermore, the derivations are very time consuming.

However, mathematicians and mathematics are part of the modeling process. It should be a greater part of our teaching. This book takes a novel approach to the teaching of modeling. While a large number of applications are covered as examples, in some cases the applications are covered in more detail than is usual. There is a historical study of the models of motion, which demonstrates to students how models continue to evolve as knowledge increases. There is an in-depth study of several population models, including their derivation. Included are historical examples of how such models were applied both where they were appropriate and where they were not. This demonstrates to students that it is necessary to understand the assumptions that lie behind a model before using it, and that any model must be checked by experiments or observations before it is accepted.

In addition, models in personal finance are discussed. This is an area of potential interest to all students, but not one that is covered in any detail in college courses. Students majoring in almost all disciplines approach these problems on an even footing. As a result it is an area where students can be required to do some modeling on their own.

Linear Algebra and Systems

Most books at this level assume that students have an understanding of elementary matrix algebra, usually in two and three dimensions. In the experience of the authors this assumption is not valid. Accordingly, this book devotes a chapter to matrix algebra. The topics covered are carefully chosen to be those needed in the study of linear systems of ODEs. With this chapter behind them, the instructor can cover linear systems of ODEs in a more substantive way. On the other hand an instructor who is confident in the knowledge of the students can skip the matrix algebra chapter.

Projects

There are a number of projects discussed in the book. These involve students in an in-depth study of either mathematics or an application that uses ODEs. The projects provide students with the opportunity to bring together much of what they have learned, including analytical, computational, and interpretative skills. The level of difficulty of the projects varies. More projects will be made available to users of this book as they are developed.

Varied Approaches Possible

It should be noticed that the book has three authors from three very different schools. The ODE courses at these institutions are quite different. Indeed, there is no standard ODE course across the country. The authors set the understandable goal of writing a book that could be used in the ODE courses at each of their own institutions. Meeting this goal required some compromises, but the result is a book that is flexible enough to allow its use in a variety of courses at a variety of institutions.

On one hand, it is possible to use the book and teach a more or less standard course. The standard material is covered in the standard order, with or without the use of technology.

However, at Rice University, after the first three chapters the class moves to numerical methods, and then to matrix algebra. This is followed by linear systems. Once this material is covered, higher-order equations, including the second-order equations that are important in science and engineering, are covered as examples of systems. This approach allows the students to use linear algebra throughout the course, thereby gaining a working knowledge of the subject. Technology is used throughout to enhance the students' understanding of the mathematical ideas.

In another approach, used at College of the Redwoods, numerical methods is done early using *Ordinary Differential Equations Using* MATLAB®, 3/E, while covering the first four chapters. Chapters 7, 8, and 9 are studied, emphasizing the material on planar systems. The course ends with nonlinear systems in Chapter 10. The goal is to locate and classify equilibrium points using the Jacobian, to locate nullclines and determine flow on the nullclines and in the regions determined by the nullclines, and then to draw the phase portrait by hand.

Mathematical Rigor

Mathematical ideas are not dodged. Motivated by a perceived lack of understanding on the part of our students, we have added material about the nature of theorems and of mathematics, and the importance of proof. Proofs are given when the proof will add to the students' understanding of the material. Difficult proofs, or those that do not add to a student's understanding, are avoided. Suggestions of how to proceed, and examples that use these suggestions, are usually offered as motivation before one has to wade through the abstraction of a proof. The authors believe that proof is fundamental to mathematics, and that students at this level should be introduced gently to proof as an integral part of their training in mathematics. This is true for the future engineer or doctor as well as for the math major.

Additional Material

The last two chapters of this version contain the solution of boundary value problems and the material needed for that. In Chapter 12 we treat Fourier series. This is an expanded version of what appears in most books, including complex Fourier series and the discrete Fourier transform. Since this material is becoming of greater interest, some instructors might want to include it in their courses. In the final chapter, we treat boundary value problems. The way this material is taught is changing, and we have tried to make the treatment a little more modern, while not abandoning the traditional approach. We have added material on the d'Alembert solution to the wave equation, and put some emphasis on the eigenvalue problem for the Laplacian, and its importance in understanding the wave and heat equations in more than one space dimension.

Supplements

Instructors who use this book will have available a number of resources. There is an Instructor's Solutions Manual, containing the complete solutions to all of the exercises. In addition there is a Student Solutions Manual with the solutions to the odd-numbered exercises. The Student Solutions Manual is available shrinkwrapped with this book at no extra cost (ISBN 0-13-155954-0).

One way to meet the software needs of the student is to use the programs **dfield** and **pplane**, written by the first author for use with MATLAB®. These programs are described in the book *Ordinary Differential Equations Using* MATLAB®, 3/E (ISBN 0-13-145679-2), written by two of the authors of this book. However, it should be emphasized that it is not necessary to use **dfield** and **pplane** with this book. There are many other possibilities. Several software manuals are available shrinkwrapped with this book at no additional cost:

- *Ordinary Differential Equations Using* MATLAB®, 3/E (ISBN 0-13-169834-6)
- *Maple Projects for Differential Equations* by Robert Gilbert and George Hsiao (ISBN 0-13-169835-4)
- *Mathematica for Differential Equations: Projects, Insights, Syntax, and Animations* by David Calvis (ISBN 0-13-169836-2)
- *Mathematica Companion for Differential Equations* by Selwyn Hollis (ISBN 0-13-169837-0)

It is also possible to get this book bundled with the Student Solutions Manual and *Ordinary Differential Equations Using* MATLAB®, 3/E (ISBN 0-13-155972-9). Contact www.prenhall.com to order any of these titles.

The Website http://www.prenhall.com/polking is a resource that is very valuable to both instructors and students. Interactive java versions of the direction field program **dfield** and the phase plane program **pplane** are accessible from this site. It also provides animations of the examples in the book, links to other web resources involving differential equations, and true-false quizzes on the subject matter. As additional projects are developed for use with the book, they will be accessible from the Website.

Acknowledgments

The development of this book depended on the efforts of a large number of people. Not the least of these is the Prentice Hall editor, George Lobell. We would also like to thank Barbara Mack, the production editor, who so patiently worked with us. Our compositor, Dennis Kletzing, was the soul of patience and worked with us to solve the problems that inevitably arise.

The reviewers of the first drafts caused us to rethink many parts of the book and certainly deserve our thanks. They are Mark Cawood, Clemson University, Charles Li, University of Missouri at Columbia, Sophia Jang, University of Louisiana, Mohammed Saleem, San Jose State University, David Russell, Virginia Tech, Moses Glasner, Penn State, Joan McCarter, Arizona State University, Gunther Uhlmann, University of Washington at Seattle, Joseph Biello, Courant Institute, New York University, Alexander Khapalov, Washington State University, Jiyuan Tao, Loyola College, and Yan Wu, Georgia Southern University.

Finally, and perhaps most important, we would like to thank the hundreds of students at Rice University, The College of the Redwoods, and Texas A&M University who patiently worked with us on preliminary versions of the text. It was they who found many of the errors that are corrected in this edition.

John Polking
polking@rice.edu

Albert Boggess
boggess@math.tamu.edu

David Arnold
david-arnold@redwoods.edu

1

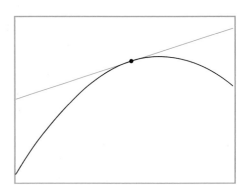

Introduction to
Differential Equations

With the systematic study of differential equations, the calculus of functions of a single variable reaches a state of completion. Modeling by differential equations greatly expands the list of possible applications. The list continues to grow as we discover more differential equation models in old and in new areas of application. The use of differential equations makes available to us the full power of the calculus.

When explicit solutions to differential equations are available, they can be used to predict a variety of phenomena. Whether explicit solutions are available or not, we can usually compute useful and very accurate approximate numerical solutions. The use of modern computer technology makes possible the visualization of the results. Furthermore, we continue to discover ways to analyze solutions without knowing the solutions explicitly.

The subject of differential equations is solving problems and making predictions. In this book, we will exhibit many examples of this—in physics, chemistry, and biology, and also in such areas as personal finance and forensics. This is the process of mathematical modeling. If it were not true that differential equations were so useful, we would not be studying them, so we will spend a lot of time on the modeling process and with specific models. In the first section of this chapter we will present some examples of the use of differential equations.

The study of differential equations, and their application, uses the derivative and the integral, the concepts that make up the calculus. We will review these ideas starting in Sections 1.2 and 1.3.

1.1 Differential Equation Models

To start our study of differential equations, we will give a number of examples. This list is meant to be indicative of the many applications of the topic. It is far from being exhaustive. In each case, our discussion will be brief. Most of the examples will be discussed later in the book in greater detail. This section should be considered as advertising for what will be done in the rest of the book.

The theme that you will see in the examples is that in every case we compute the rate of change of a variable in two different ways. First there is the mathematical way. In mathematics, the rate at which a quantity changes is the derivative of that quantity. This is the same for each example. The second way of computing the rate of change comes from the application itself and is different from one application to another. When these two ways of expressing the rate of change are equated, we get a differential equation, the subject we will be studying.

Mechanics

Isaac Newton was responsible for a large number of discoveries in physics and mathematics, but perhaps the three most important are the following:

- The systematic development of the calculus. Newton's achievement was the realization and utilization of the fact that integration and differentiation are operations inverse to each other.

Figure 1. The force F results in the acceleration of the mass m according to equation (1.1).

- The discovery of the laws of mechanics. Principal among these was Newton's second law, which says that force acting on a mass (see Figure 1) is equal to the rate of change of momentum with respect to time. Momentum is defined to be the product of mass and velocity, or mv. Thus the force is equal to the derivative of the momentum. If the mass is constant,

$$\frac{d}{dt}mv = m\frac{dv}{dt} = ma,$$

where a is the acceleration. Newton's second law says that the rate of change of momentum is equal to the force F. Expressing the equality of these two ways of looking at the rate of change, we get the equation

$$F = ma, \tag{1.1}$$

the standard expression for Newton's second law.

- The discovery of the universal law of gravitation. This law says that any body with mass M attracts any other body with mass m directly toward the mass M, with a magnitude proportional to the product of the two masses and inversely proportional to the square of the distance separating them. This means that there is a constant G, which is universal, such that the magnitude of the force is

$$\frac{GMm}{r^2}, \tag{1.2}$$

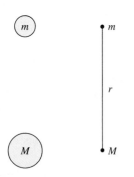

Figure 2. The magnitude of the force attracting two masses is given by equation (1.2).

where r is the distance between the centers of mass of the two bodies. See Figure 2.

All of these discoveries were made in the period between 1665 and 1671. The discoveries were presented originally in Newton's *Philosophiae Naturalis Principia Mathematica*, better known as *Principia Mathematica*, published in 1687.

Newton's development of the calculus is what makes the theory and use of differential equations possible. His laws of mechanics create a template for a model for motion in almost complete generality. It is necessary in each case to figure out what forces are acting on a body. His law of gravitation does just that in one very important case.

The simplest example is the motion of a ball thrown into the air near the surface of the earth. See Figure 3. If x measures the distance the ball is above the earth, then the velocity and acceleration of the ball are

$$v = \frac{dx}{dt} \quad \text{and} \quad a = \frac{dv}{dt} = \frac{d^2x}{dt^2}.$$

Figure 3. x is the height of the ball above the surface of the earth.

Since the ball is assumed to move only a short distance in comparison to the radius of the earth, the force given by (1.2) may be assumed to be constant. Notice that m, the mass of the ball, occurs in (1.2). We can write the force as $F = -mg$, where $g = GM/r^2$ and r is the radius of the earth. The constant g is called the earth's acceleration due to gravity. The minus sign reflects the fact that the displacement x is measured positively above the surface of the earth, and the force of gravity tends to decrease x. Newton's second law, (1.1), becomes

$$-mg = ma = m\frac{dv}{dt} = m\frac{d^2x}{dt^2}.$$

The masses cancel, and we get the differential equation

$$\frac{d^2x}{dt^2} = -g, \tag{1.3}$$

which is our mathematical model for the motion of the ball.

The equation in (1.3) is called a differential equation because it involves an unknown function $x(t)$ and at least one of its derivatives. In this case the highest derivative occurring is the second order, so this is called a differential equation of second order.

A more interesting example of the application of Newton's ideas has to do with planetary motion. For this case, we will assume that the sun with mass M is fixed and put the origin of our coordinate system at the center of the sun. We will denote by $\mathbf{x}(t)$ the vector that gives the location of a planet relative to the sun. See Figure 4. The vector $\mathbf{x}(t)$ has three components. Its derivative is

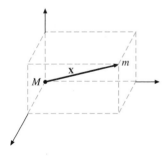

$$\mathbf{v}(t) = \frac{d\mathbf{x}}{dt},$$

Figure 4. The vector $\mathbf{x}$ gives the location of the planet m relative to the sun M.

which is the vector-valued velocity of the planet. For this example, Newton's second law and his law of gravitation become

$$m\frac{d^2\mathbf{x}}{dt^2} = -\frac{GMm}{|\mathbf{x}|^2}\frac{\mathbf{x}}{|\mathbf{x}|}.$$

This system of three second-order differential equations is Newton's model of planetary motion. Newton solved these and verified that the three laws observed by Kepler follow from his model.

Population models

Consider a population $P(t)$ that is varying with time.[1] A mathematician will say that the rate at which the population is changing with respect to time is given by the derivative

$$\frac{dP}{dt}.$$

On the other hand, a population biologist will say that the rate of change is roughly proportional to the population. This means that there is a constant r, called the reproductive rate, such that the rate of change is equal to rP. Putting together the ideas of the mathematician and the biologist, we get the equation

$$\frac{dP}{dt} = rP. \tag{1.4}$$

This is an equation for the function $P(t)$. It involves both P and its derivative, so it is a differential equation. It is not difficult to show by direct substitution into (1.4) that the exponential function

$$P(t) = P_0 e^{rt},$$

where P_0 is a constant representing the initial population, is a solution. Thus, assuming that the reproductive rate r is positive, our population will grow exponentially.

If at this point you go back to the biologist he or she will undoubtedly say that the reproductive rate is not really a constant. While that assumption works for small populations, over the long term you have to take into account the fact that resources of food and space are limited. When you do, a better model for the the reproductive rate is the function $r(1 - P/K)$, and then the rate at which the population changes is better modeled by $r(1 - P/K)P$. Here both r and K are constants.

When we equate our two ideas about the rate at which the population changes, we get the equation

$$\frac{dP}{dt} = r(1 - P/K)P. \tag{1.5}$$

This differential equation for the function $P(t)$ is called the **_logistic equation._** It is much harder to solve than (1.4), but it does a creditable job of predicting how single populations grow in isolated circumstances.

Pollution

Consider a lake that has a volume of $V = 100$ km^3. It is fed by an input river, and there is another river which is fed by the lake at a rate that keeps the volume of the lake constant. The flow of the input river varies with the season, and assuming that $t = 0$ corresponds to January 1 of the first year of the study, the input rate is

$$r(t) = 50 + 20\cos(2\pi(t - 1/4)).$$

Notice that we are measuring time in years. Thus the maximum flow into the lake occurs when $t = 1/4$, or at the beginning of April.

In addition, there is a factory on the lake that introduces a pollutant into the lake at the rate of 2 km^3/year. Let $x(t)$ denote the total amount of pollution in the lake

---

[1] For the time being, the population can be anything—humans, paramecia, butterflies, and so on. We will be more careful later.

at time t. If we make the assumption that the pollutant is rapidly mixed throughout the lake, then we can show that $x(t)$ satisfies the differential equation

$$\frac{dx}{dt} = 2 - \left(52 + 20\cos(2\pi(t - 1/4))\right)\frac{x}{100}.$$

This equation can be solved and we can then answer questions about how dangerous the pollution problem really is. For example, if we know that a concentration of less than 2% is safe, will there there be a problem? The solution will tell us.

The assumption that the pollutant is rapidly mixed into the lake is not very realistic. We know that this does not happen, especially in this situation, where there is a flow of water through the lake. This assumption can be removed, but to do so, we need to allow the concentration of the pollutant to vary with position in the lake as well as with time. Thus the concentration is a function $c(t, x, y, z)$, where (x, y, z) represents a position in the three-dimensional lake. Instead of assuming perfect mixing, we will assume that the pollutant diffuses through water at a certain rate. Once again we can construct a mathematical model. Again it will be a differential equation, but now it will involve partial derivatives with respect to the spatial coordinates x, y, and z, as well as the time t.

Personal finance

How much does a person need to save during his or her work life in order to be sure of a retirement without money worries? How much is it necessary to save each year in order to accumulate these assets? Suppose one's salary increases over time. What percent of one's salary should be saved to reach one's retirement goal?

All of these questions, and many more like them, can be modeled using differential equations. Then, assuming particular values for important parameters like return on investment and rate of increase of one's salary, answers can be found.

Other examples

We have given four examples. We could have given a hundred more. We could talk about electrical circuits, the behavior of musical instruments, the shortest paths on a complicated-looking surface, finding a family of curves that are orthogonal to a given family, discovering how two coexisting species interact, and many others.

All of these examples use ordinary differential equations. The applications of partial differential equations go much farther. We can include electricity and magnetism; quantum chromodynamics, which unifies electricity and magnetism with the weak and strong nuclear forces, the flow of heat, oscillations of many kinds, such as vibrating strings, the fair pricing of stock options, and many more.

The use of differential equations provides a way to reduce many areas of application to mathematical analysis. In this book, we will learn how to do the modeling and how to use the models after we make them.

EXERCISES

The phrase "y is proportional to x" implies that y is related to x via the equation $y = kx$, where k is a constant. In a similar manner, "y is proportional to the square of x" implies $y = kx^2$, "y is proportional to the product of x and z" implies $y = kxz$, and "y is inversely proportional to the cube of x" implies $y = k/x^3$. For example, when Newton proposed that the force of attraction of one body on another is proportional to the product of the masses and inversely proportional to the square of the distance between them, we can immediately write

$$F = \frac{GMm}{r^2},$$

where G is the constant of proportionality, usually known as the universal gravitational constant. In Exercises 1–11, use these ideas to model each application with a differential equation. All rates are assumed to be with respect to time.

1. The rate of growth of bacteria in a petri dish is proportional to the number of bacteria in the dish.

2. The rate of growth of a population of field mice is inversely proportional to the square root of the population.

3. A certain area can sustain a maximum population of 100 ferrets. The rate of growth of a population of ferrets in this area is proportional to the product of the population and the difference between the actual population and the maximum sustainable population.

4. The rate of decay of a given radioactive substance is proportional to the amount of substance remaining.

5. The rate of decay of a certain substance is inversely proportional to the amount of substance remaining.

6. A potato that has been cooking for some time is removed from a heated oven. The room temperature of the kitchen is $65°F$. The rate at which the potato cools is proportional to the difference between the room temperature and the temperature of the potato.

7. A thermometer is placed in a glass of ice water and allowed to cool for an extended period of time. The thermometer is removed from the ice water and placed in a room having temperature $77°F$. The rate at which the thermometer warms is proportional to the difference in the room temperature and the temperature of the thermometer.

8. A particle moves along the x-axis, its position from the origin at time t given by $x(t)$. A single force acts on the particle that is proportional to, but opposite the object's displacement. Use Newton's law to derive a differential equation for the object's motion.

9. Use Newton's law to develop the equation of motion for the particle in Exercise 8 if the force is proportional to, but opposite the square of the particle's velocity.

10. Use Newton's law to develop the equation of motion for the particle in Exercise 8 if the force is inversely proportional to, but opposite the square of the particle's displacement from the origin.

11. The voltage drop across an inductor is proportional to the rate at which the current is changing with respect to time.

1.2 The Derivative

Before reading this section, ask yourself, "What is the derivative?" Several answers may come to mind, but remember your first answer.

Chances are very good that your answer was one of the following five:

1. The rate of change of a function

2. The slope of the tangent line to the graph of a function

3. The best linear approximation of a function

4. The limit of difference quotients,

$$f'(x_0) = \lim_{x \to x_0} \frac{f(x) - f(x_0)}{x - x_0}$$

5. A table containing items such as we see in Table 1

Table I A table of derivatives

$f(x) =$	$f'(x) =$
C	0
x	1
x^n	nx^{n-1}
$\cos(x)$	$-\sin(x)$
$\sin(x)$	$\cos(x)$
e^x	e^x
$\ln(|x|)$	$1/x$

All of these answers are correct. Each of them provides a different way of looking at the derivative. The best answer to the question is "all of the above." Since we will be using all five ways of looking at the derivative, let's spend a little time discussing each.

The rate of change

In calculus, we learn that a function has an instantaneous rate of change, and this rate is equal to the derivative. For example, if we have a distance $x(t)$ measured from a fixed point on a line, then the rate at which x changes with respect to time is the velocity v. We know that

$$v = x' = \frac{dx}{dt}.$$

Similarly, the acceleration a is the rate of change of the velocity, so

$$a = v' = \frac{dv}{dt} = \frac{d^2x}{dt^2}.$$

These facts about linear motion are reflected in many other fields. For example, in economics, the law of supply and demand says that the price of a product is determined by the supply of that product and the demand for it. If we assume that the demand is constant, then the price P is a function of the supply S, or $P = P(S)$. The rate at which P changes with the supply is called the marginal price. In mathematical terms, the marginal price is simply the derivative $P' = dP/dS$. We can also talk about the rate of change of the mass of a radioactive material, of the size of population, of the charge on a capacitor, of the amount of money in a savings account or an investment account, or of many more quantities.[2]

We will see all of these examples and more in this book. The point is that when any quantity changes, the rate at which it changes is the derivative of that quantity. It is this fact that starts the modeling process and makes the study of differential equations so useful. For this reason we will refer to the statement that the derivative is the rate of change as the ***modeling definition*** of the derivative.

The slope of the tangent line

This provides a good way to visualize the derivative. Look at Figure 1. There you see the graph of a function f, and the tangent line to the graph of f at the point $(x_0, f(x_0))$. The equation of the tangent line is

$$y = f(x_0) + f'(x_0)(x - x_0).$$

From this formula, it is easily seen that the slope of the tangent line is $f'(x_0)$.

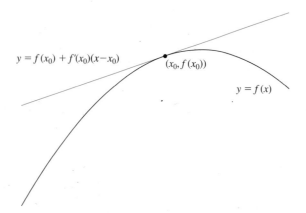

$$y = f(x_0) + f'(x_0)(x - x_0)$$

$(x_0, f(x_0))$

$y = f(x)$

Figure 1. The derivative is the slope of the tangent line to the graph of the function.

Again looking at Figure 1, we can visualize the rate at which the function f is changing as x changes near the point x_0. It is the same as the slope of the tangent line.

We will refer to this characterization of the derivative as the ***geometric definition*** of the derivative.

---

[2] In all but one of the mentioned examples, the quantity changes with respect to time. Most of the applications of ordinary differential equations involve rates of change with respect to time. For this reason, t is usually used as the independent variable. However, there are cases where things change depending on other parameters, as we will see. Where appropriate, we will use other letters to denote the independent variable. Sometimes we will do so just for practice.

The best linear approximation

Let

$$L(x) = f(x_0) + f'(x_0)(x - x_0). \tag{2.1}$$

L is a linear (or affine) function of x. Taylor's theorem says there is a remainder function $R(x)$, such that

$$f(x) = L(x) + R(x) \quad \text{and} \quad \lim_{x \to x_0} \frac{R(x)}{x - x_0} = 0. \tag{2.2}$$

The limit in (2.2) means that $R(x)$ gets small as $x \to x_0$. In fact, it gets enough smaller than $x - x_0$ that the ratio goes to 0. It turns out that the function L defined in (2.1) is the only linear function with this property. This is what we mean when we say that L is the best linear approximation to the nonlinear function f. You will also notice that the straight line in Figure 1 is the graph of L. In fact, Figure 1 provides a pictorial demonstration that $L(x)$ is a good approximation for $f(x)$ for x near x_0.

The formula in (2.1) defines $L(x)$ in terms of the derivative of f. In this sense, the derivative gives us the best linear approximation to the nonlinear function f near $x = x_0$. [Actually (2.1) contains three important pieces of data, x_0, $f(x_0)$, and $f'(x_0)$. We are perhaps stretching the point when we say that it is the derivative alone that enables us to find a linear approximation to f, but it is clear that the derivative is the most important of these three.]

Since the linear approximation is an algebraic object, we will refer to this as the **_algebraic definition_** of derivative.

The limit of difference quotients

Consider the difference quotient

$$m = \frac{f(x) - f(x_0)}{x - x_0}. \tag{2.3}$$

This is equal to the slope of the line through the two points $(x_0, f(x_0))$ and $(x, f(x))$ as illustrated in Figure 2. We will refer to this line as a secant line. As x approaches x_0, the secant line approaches the tangent line shown in Figure 1. This is reflected in the fact that

$$f'(x_0) = \lim_{x \to x_0} \frac{f(x) - f(x_0)}{x - x_0}. \tag{2.4}$$

Thus the slope of the tangent line, $f'(x_0)$, is the limit of the slopes of secant lines.

The difference quotient in (2.3) is also the average rate of change of the function f between x_0 and x. As the interval between x_0 and x is made smaller, these average rates approach the instantaneous rate of change of f. Thus we see the connection with our modeling definition.

The definition of the derivative given in (2.4) will be called the **_limit quotient definition_**. This is the definition that most mathematicians think of when asked to define the derivative. However, as we will see, it is also very useful, even when attempting to find mathematical models.

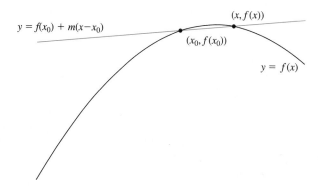

Figure 2. The secant line with slope m given by the difference quotient in (2.3).

The table of formulas

By memorizing a table of derivatives and a few formulas (especially the chain rule), we can learn the skill of differentiation. It isn't hard to be confident that you can compute the derivative of any given function. This skill is important. However, it is clear that this ***formulaic definition*** of derivative is quite different from those given previously.

A complete understanding of the formulaic definition is important, but it does not provide any information about the other definitions we have examined. Therefore, it helps us neither to apply the derivative in modeling nature nor to understand its properties. For that reason, the formulaic definition is incomplete. This is not true of the other definitions. Starting with one of them, it is possible to construct a table that will give us the formulaic finesse we need. Admittedly that is a big task. That was what was done (or should have been done) in your first calculus course.

To sum up, we have examined five definitions of the derivative. Each of these emphasizes a different aspect or property of the derivative. All of them are important. We will see this as we progress through the study of differential equations. If your answer to the question at the beginning of the section was any of these five, your answer is correct. However, a complete understanding of the derivative requires the understanding of all five definitions.

Even if your answer was not on the list of five, it may be correct. The famous mathematician William Thurston once compiled a list of over 40 "definitions" of the derivative. Of course many of these appear only in more advanced parts of mathematics, but the point is made that the derivative appears in many ways in mathematics and in its applications. It is one of the most fundamental ideas in mathematics and in its application to science and technology.

EXERCISES

You might recall the following rules of differentiation from your calculus class. Let f and g be differentiable functions of x. Then

$$(cf)' = cf' \qquad (f \pm g)' = f' \pm g'$$

$$(fg)' = f'g + fg' \qquad \left(\frac{f}{g}\right)' = \frac{f'g - fg'}{g^2}.$$

Also, the chain rule is essential.

$$(f \circ g)'(x) = f'(g(x))g'(x)$$

Use these rules, plus the table of derivatives in Table 1, to find the derivative of each of the functions in Exercises 1–12.

1. $f(x) = 3x - 5$

2. $f(x) = 5x^2 - 4x - 8$

3. $f(x) = 3 \sin 5x$

4. $f(x) = \cos 2\pi x$

5. $f(x) = e^{3x}$

6. $f(x) = 5e^{x^2}$

7. $f(x) = \ln|5x|$

8. $f(x) = \ln(\cos 2x)$

9. $f(x) = x \ln x$

10. $f(x) = e^x \sin \pi x$

11. $f(x) = \dfrac{x^2}{\ln x}$ **12.** $f(x) = \dfrac{x \ln x}{\cos x}$

13. Suppose that f is differentiable at x_0. Let L be the "best linear approximation" defined by $L(x) = f(x_0) + f'(x_0)(x - x_0)$. Given that $R(x) = f(x) - L(x)$, show that

$$\lim_{x \to x_0} \frac{R(x)}{x - x_0} = 0.$$

For each of the functions given in Exercises 14–17, sketch the function f and its linearization $L(x) = f(x_0) + f'(x_0)(x - x_0)$ at the given point x_0 on the same set of coordinate axes.

14. $f(x) = e^x$, at $x_0 = 0$

15. $f(x) = \cos x$, at $x_0 = \pi/4$

16. $f(x) = \sqrt{x}$, at $x_0 = 1$

17. $f(x) = \ln(1 + x)$, at $x_0 = 0$

In order that $R(x)/(x - x_0)$ of equation (2.2) approach zero as $x \to x_0$, the numerator $R(x)$ must approach zero at a faster rate than does the denominator $x - x_0$. For each of Exercises 18–21, sketch the graph of $y = x - x_0$ and $R(x) = f(x) - L(x)$ on the same set of coordinate axes. Do both $x - x_0$ and $R(x)$ approach zero as $x \to x_0$? Which approaches zero at a faster rate, $R(x)$ or $x - x_0$?

18. $f(x) = x^{3/2}$, at $x_0 = 1$

19. $f(x) = \sin 2x$, at $x = \pi/8$

20. $f(x) = \sqrt{x + 1}$, at $x = 0$

21. $f(x) = xe^{x-1}$, at $x = 1$

1.3 Integration

We can start once more by asking the question, "What is the integral?" This time our list of possible answers is not so long.

1. The area under the graph of a function
2. The antiderivative
3. A table containing items such as we see in Table 1

Let's look at each of them briefly.

Table I A table of integrals			
$f(x) =$	$\int f(x)\,dx =$	$f(x) =$	$\int f(x)\,dx =$
0	C	$\cos(x)$	$\sin(x) + C$
1	$x + C$	$\sin(x)$	$-\cos(x) + C$
x	$\dfrac{x^2}{2} + C$	e^x	$e^x + C$
x^n	$\dfrac{x^{n+1}}{n+1} + C$	$\dfrac{1}{x}$	$\ln(|x|) + C$

The area under the graph

The first answer emphasizes the definite integral. The definite integral

$$\int_a^b f(x)\,dx \tag{3.1}$$

is interpreted as the area under the graph of the function f between $x = a$ and $x = b$. It represents the area of the shaded region in Figure 1.

This is the most fundamental definition of the integral. The integral was invented to solve the problem of finding the area of regions that are not simple rectangles or circles. Despite its origin as a method to use in this one application, it has found numerous other applications.

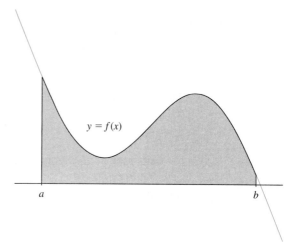

Figure 1. The area of the shaded region is the integral in (3.1).

The antiderivative

This answer emphasizes the indefinite integral. In fact, the phrase ***indefinite integral*** is a synonym for ***antiderivative.*** The definition is summed up in the following equivalence. If the function g is continuous, then

$$f' = g \quad \text{if and only if} \quad \int g(x)\, dx = f(x) + C. \tag{3.2}$$

In (3.2), C refers to the arbitrary constant of integration. Thus the process of indefinite integration involves finding antiderivatives. Given a function g, we want to find a function f such that $f' = g$.

The connection between the definite and the indefinite integral is found in the ***fundamental theorem of calculus***. This says that if $f' = g$, then

$$\int_a^b g(x)\, dx = f(b) - f(a).$$

The table of formulas

This formulaic approach to the integral has the same features and failures as the formulaic approach to the derivative. It leads to the handy skill of integration, but it does not lead to any deep understanding of the integral.

All of these approaches to the integral are important. It is very important to understand the first two and how they are connected by the fundamental theorem. However, for the elementary part of the study of ordinary differential equations, it is really the second and third approaches that are most important. In other words, it is important to be able to find antiderivatives.

Solution by integration

The solution of an important class of differential equations amounts to finding antiderivatives. A first-order differential equation can be written as

$$y' = f(t, y), \tag{3.3}$$

where the right-hand side is a function of the independent variable t and the unknown function y. Suppose that the right-hand side is a function only of t and does not depend on y. Then equation (3.3) becomes

$$y' = f(t).$$

Comparing this with (3.2), we see immediately that the solution is

$$y(t) = \int f(t)\, dt. \tag{3.4}$$

Let's look at an example.

Example 3.5   Solve the differential equation

$$y' = \cos t. \tag{3.6}$$

According to (3.4), the solution is

$$y(t) = \int \cos(t)\, dt = \sin t + C, \tag{3.7}$$

where C is an arbitrary constant. That's pretty easy. It is just the process of integration. It's old hat to you by now. Solving the more general equation in (3.3) is not so easy, as we will see.

The constant of integration C makes (3.7) a one-parameter family of solutions of (3.6) defined on $(-\infty, \infty)$. This is an example of a **general solution** to a differential equation. Some of these solutions are drawn in Figure 2.   ●

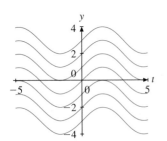

Figure 2.   Several solutions to (3.6).

It is significant that the solution curves of equation (3.6) shown in Figure 2 are vertical translates of one another. That is to say, any solution curve can be obtained from any other by a vertical translation. This is always the case for solution curves of an equation of the form $y' = f(t)$. According to (3.2), if $y(t) = F(t)$ is one solution to the equation, then all others are of the form $y(t) = F(t) + C$ for some constant C. The graphs of such functions are vertical translates of the graph of $y(t) = F(t)$.

The constant of integration allows us to put an extra condition on a solution. This is illustrated in the next example.

Example 3.8   Find the solution to $y'(t) = te^t$ that satisfies $y(0) = 2$.

This is an example of an **initial value problem**. It requires finding the particular solution that satisfies the **initial condition** $y(0) = 2$. According to (3.2), the general solution to the differential equation is given by

$$y(t) = \int te^t\, dt. \tag{3.9}$$

This integral can be evaluated using integration by parts. Since this method is so useful, we will briefly review it. In general, it says

$$\int u\, dv = uv - \int v\, du, \tag{3.10}$$

where u and v are functions. If they are functions of t, then $du = u'(t)\, dt$ and $dv = v'(t)\, dt$. For the integral in equation (3.9), we let $u(t) = t$, and $dv = v'(t)\, dt = e^t\, dt$. Then $du = dt$ and $v(t) = e^t$, and equation (3.10) gives

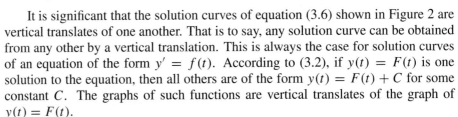
$$\int te^t\, dt = \int u\, dv = uv - \int v\, du = te^t - \int e^t\, dt.$$

After evaluating the last integral, we see that

$$y(t) = te^t - e^t + C = e^t(t - 1) + C. \tag{3.11}$$

This one-parameter family of solutions is the general solution to the equation $y' = te^t$. Each member of the family exists on the interval $(-\infty, \infty)$. The condition $y(0) = 2$ can be used to determine the constant C.

$$2 = y(0) = e^0(0 - 1) + C = -1 + C$$

Therefore, $C = 3$ and the solution of the initial value problem is

$$y(t) = e^t(t - 1) + 3. \tag{3.12}$$

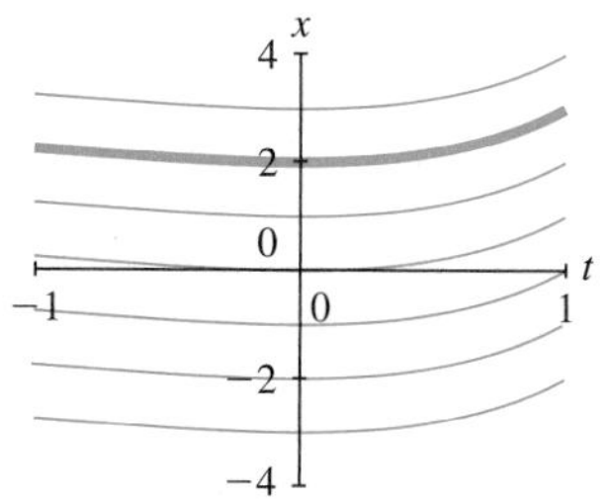

Figure 3. The solution of the initial value problem in Example 3.8 passes through the point $(0, 2)$.

It is important to note that the solution curve defined by equation (3.12) is the member of the family of solution curves defined by (3.11) that passes through the point $(0, 2)$, as shown in Figure 3.   ●

The use of initial conditions to determine a particular solution can be affected from the beginning of the solution process by using definite integrals instead of indefinite integrals. For example, in Example 3.8, we can proceed using the fundamental theorem of calculus:

$$y(t) - y(0) = \int_0^t y'(u)\, du.$$

Hence,

$$y(t) = y(0) + \int_0^t u e^u\, du$$
$$= 2 + u e^u - e^u \Big|_0^t$$
$$= e^t(t - 1) + 3.$$

We will not always use the letter t to designate the independent variable. Any letter will do, as long as we are consistent. The same is true of the dependent variable.

Example 3.13   Find the solution to the initial value problem

$$y' = \frac{1}{x} \quad \text{with} \quad y(1) = 3.$$

Here we are using x as the independent variable. By integration, we find that

$$y(x) = \ln(|x|) + C.$$

We are asked for the solution that satisfies the initial condition

$$3 = y(1) = \ln(1) + C = C.$$

Thus, $C = 3$.

A solution to a differential equation has to have a derivative at every point. Therefore, it is also continuous. However, the function $y(x) = \ln(|x|) + 3$ is not defined for $x = 0$. To get a continuous function from y, we have to limit its domain to $(0, \infty)$ or $(-\infty, 0)$. Since we want a solution that is defined at $x = 1$, we must choose $(0, \infty)$. Thus, our solution is

$$y(x) = \ln(x) + 3 \quad \text{for} \quad x > 0. \quad ●$$

The motion of a ball

In Section 1.1, we talked about the application of Newton's laws to the motion of a ball near the surface of the earth. See Figure 4. The model we derived [in equation (1.3)] was

$$\frac{d^2x}{dt^2} = -g,$$

where $x(t)$ is the height of the ball above the surface of the earth and g is the acceleration due to gravity. If we measure x in feet and time in seconds, $g = 32$ ft/s^2.

Example 3.14   Suppose a ball is thrown into the air with initial velocity $v_0 = 20$ ft/s. Assuming the ball is thrown from a height of $x_0 = 6$ feet, how long does it take for the ball to hit the ground?

We can solve this equation using the methods of this section. First we introduce the velocity to reduce the second-order equation to a system of two first-order equations:

$$\frac{dx}{dt} = v, \quad \text{and} \quad \frac{dv}{dt} = -g. \tag{3.15}$$

Solving the second equation by integration, we get

$$v(t) = -gt + C_1.$$

Figure 4. x is the height of the ball above the surface of the earth.

Evaluating this at $t = 0$, we see that the constant of integration is $C_1 = v(0) = v_0 = 20$, the initial velocity. Hence, the velocity is $v(t) = -gt + v_0 = -32t + 20$, and the first equation in (3.15) becomes

$$\frac{dx}{dt} = -gt + v_0 = -32t + 20.$$

Solving by integration, we get

$$x(t) = -\frac{1}{2}gt^2 + v_0t + C_2 = -16t^2 + 20t + C_2.$$

Once more we evaluate this at $t = 0$ to show that $C_2 = x(0) = x_0 = 6$, the initial elevation of the ball. Hence, our final solution is

$$x(t) = -\frac{1}{2}gt^2 + v_0t + x_0 = -16t^2 + 20t + 6. \tag{3.16}$$

The ball hits the ground when $x(t) = 0$. By solving the quadratic equation, and using the positive solution, we see that the ball hits the ground after 1.5 seconds. ●

EXERCISES

In Exercises 1–8, find the general solution of the given differential equation. In each case, sketch at least six members of the family of solution curves.

1. $y' = 2t + 3$

2. $y' = 3t^2 + 2t + 3$

3. $y' = \sin 2t + 2\cos 3t$

4. $y' = 2\sin 3t - \cos 5t$

5. $y' = \dfrac{t}{1+t^2}$

6. $y' = \dfrac{3t}{1+2t^2}$

7. $y' = t^2 e^{3t}$

8. $y' = t\cos 3t$

In Exercises 1–8 above, each equation has the form $y' = f(t, y)$, the goal being to find a solution $y = y(t)$; that is, find y as a function of t. Of course, you are free to choose different letters, both for the dependent and independent variables. For example, in the differential equation $s' = xe^x$, it is understood that $s' = ds/dx$, and the goal is to find a solution s as a function of x; that is, $s = s(x)$. In Exercises 9–14, find the general solution of the given differential equation. In each case, sketch at least six members of the family of solution curves.

9. $s' = e^{-2\omega} \sin \omega$ **10.** $y' = x \sin 3x$

11. $x' = s^2 e^{-s}$ **12.** $s' = e^{-u} \cos u$

13. $r' = \dfrac{1}{u(1-u)}$ **14.** $y' = \dfrac{3}{x(4-x)}$

Note: Exercises 13 and 14 require a partial fraction decomposition. If you have forgotten this technique, you can find extensive explanation in Section 5.3 of this text. In particular, see Example 3.6 in that section.

In Exercises 15–24, find the solution of each initial value problem. In each case, sketch the solution.

15. $y' = 4t - 6$, $y(0) = 1$

16. $y' = x^2 + 4$, $y(0) = -2$

17. $x' = te^{-t^2}$, $x(0) = 1$

18. $r' = t/(1+t^2)$, $r(0) = 1$

19. $s' = r^2 \cos 2r$, $s(0) = 1$

20. $P' = e^{-t} \cos 4t$, $P(0) = 1$

21. $x' = \sqrt{4-t}$, $x(0) = 1$

22. $u' = 1/(x-5)$, $u(0) = -1$

23. $y' = \dfrac{t+1}{t(t+4)}$, $y(-1) = 0$

24. $v' = \dfrac{r^2}{r+1}$, $v(0) = 0$

In Exercises 25–28, assume that the motion of a ball takes place in the absence of friction. That is, the only force acting on the ball is the force due to gravity.

25. A ball is thrown into the air from an initial height of 3 m with an initial velocity of 50 m/s. What is the position and velocity of the ball after 3 s?

26. A ball is dropped from rest from a height of 200 m. What is the velocity and position of the ball 3 seconds later?

27. A ball is thrown into the air from an initial height of 6 m with an initial velocity of 120 m/s. What will be the maximum height of the ball and at what time will this event occur?

28. A ball is propelled downward from an initial height of 1000 m with an initial speed of 25 m/s. Calculate the time that the ball hits the ground.

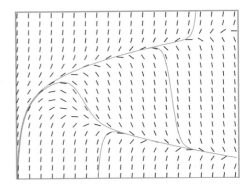

CHAPTER 2

First-Order Equations

In this chapter, we will study first-order equations. We will begin in Section 2.1 by making some definitions and presenting an overview of what we will cover in this chapter. We will then alternate between methods of finding exact solutions and some applications that can be studied using those methods. For each application, we will carefully derive the mathematical models and explore the existence of exact solutions. We will end by showing how qualitative methods can be used to derive useful information about the solutions.

2.1 Differential Equations and Solutions

In this section, we will give an overview of what we want to learn in this chapter. We will visit each topic briefly to give a flavor of what will follow in succeeding sections.

Ordinary differential equations

An *ordinary differential equation* is an equation involving an unknown function of a single variable together with one or more of its derivatives. For example, the equation

$$\frac{dy}{dt} = y - t \tag{1.1}$$

is an ordinary differential equation. Here $y = y(t)$ is the unknown function and t is the *independent variable*.

Some other examples of ordinary differential equations are

$$y' = y^2 - t \qquad\qquad ty' = y$$
$$y' + 4y = e^{-3t} \qquad\qquad y' = \cos(ty) \qquad (1.2)$$
$$yy'' + t^2 y = \cos(t) \qquad\qquad y'' = y^2.$$

The **order** of a differential equation is the order of the highest derivative that occurs in the equation. Thus the equation in (1.1) is a **first-order** equation since it involves only the first derivative of the unknown function. All of the equations listed in the first two rows of (1.2) are first order. Those in the third row are **second order** because they involve the second derivative of y.

The equation

$$\frac{\partial^2 w}{\partial t^2} = c^2 \frac{\partial^2 w}{\partial x^2} \qquad (1.3)$$

is not an ordinary differential equation, since the unknown function w is a function of the two independent variables t and x. Because it involves partial derivatives of an unknown function of more than one independent variable, equation (1.3) is called a **partial differential equation**. For the time being we are interested only in ordinary differential equations.

Normal form

Any first order equation can be put into the form

$$\phi(t, y, y') = 0, \qquad (1.4)$$

where ϕ is a function of three variables. For example, the equation in (1.1) can be written as

$$y' - y - t = 0.$$

This equation has the form in (1.4) with $\phi(t, y, z) = z - y - t$. Similarly, the general equation of order n can be written as

$$\phi(t, y, y', \ldots, y^{(n)}) = 0, \qquad (1.5)$$

where ϕ is a function of $n + 1$ variables. Notice that all of the equations in (1.2) can be put into this form.

The general forms in (1.4) and (1.5) are too general to deal with in many instances. Frequently we will find it useful to solve for the highest derivative. We will give the result a name.

DEFINITION 1.6

A first-order differential equation of the form

$$y' = f(t, y)$$

is said to be in **normal form**. Similarly, an equation of order n having the form

$$y^{(n)} = f(t, y, y', \ldots, y^{(n-1)})$$

is said to be in **normal form**.

Example 1.7 Place the differential equation $t + 4yy' = 0$ into normal form.

This is accomplished by solving the equation $t + 4yy' = 0$ for y'. We find that

$$y' = -\frac{t}{4y}. \tag{1.8}$$

Note that the right-hand side of equation (1.8) is a function of t and y, as required by the normal form $y' = f(t, y)$. ●

Solutions

A **solution** of the first-order, ordinary differential equation $\phi(t, y, y') = 0$ is a differentiable function $y(t)$ such that $\phi(t, y(t), y'(t)) = 0$ for all t in the interval[1] where $y(t)$ is defined.

To discover if a given function is a solution to a differential equation we substitute the function and its derivative(s) into the equation. For example, we can show that $y(t) = t + 1$ is a solution to equation (1.1) by substitution. It is only necessary to compute both sides of equation (1.1) and show that they are equal. We have

$$y'(t) = 1, \quad \text{and} \quad y(t) - t = t + 1 - t = 1.$$

Since the left- and right-hand sides are equal, $y(t) = t + 1$ is a solution.

The process of verifying that a given function is or is not a solution to a differential equation is a very important skill. You can use it to check that your homework solutions are correct. We will use it repeatedly for a variety of purposes, including finding solution methods. Here are two more examples.

Example 1.9 Show that $y(t) = Ce^{-t^2}$ is a solution of the first-order equation

$$y' = -2ty, \tag{1.10}$$

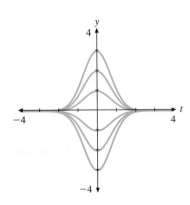

Figure 1. Several solutions to equation (1.10).

where C is an arbitrary real number.

We compute both sides of the equation and compare them. On the left, we have $y'(t) = -2tCe^{-t^2}$, and on the right, $-2ty(t) = -2tCe^{-t^2}$, so the equation is satisfied. Both $y(t)$ and $y'(t)$ are defined on the interval $(-\infty, \infty)$. Therefore, for each real number C, $y(t) = Ce^{-t^2}$ is a solution of equation (1.10) on the interval $(-\infty, \infty)$. ●

Example 1.9 illustrates the fact that a differential equation can have lots of solutions. The solution formula $y(t) = Ce^{-t^2}$ gives a different solution for very value of the constant C. We will see in Section 2.4 that every solution to equation (1.10) is of this form for some value of the constant C. For this reason the formula $y(t) = Ce^{-t^2}$ is called the **general solution** to (1.10). The graphs of these solutions are called **solution curves**, several of which are drawn in Figure 1.

Example 1.11 Is the function $y(t) = \cos t$ a solution to the differential equation $y' = 1 + y^2$?

We substitute $y(t) = \cos t$ into the equation. On the left-hand side we have $y' = -\sin t$. On the right-hand side, $1 + y^2 = 1 + \cos^2 t$. Since $-\sin t \neq 1 + \cos^2 t$ for most values of t, $y(t) = \cos t$ is not a solution. ●

---

[1] We will use the notation (a, b), $[a, b]$, $(a, b]$, $[a, b)$, (a, ∞), $[a, \infty)$, $(-\infty, b)$, $(-\infty, b]$, and $(-\infty, \infty)$ for intervals. For example, $(a, b) = \{t : a < t < b\}$, $[a, b) = \{t : a \leq t < b\}$, $(-\infty, b] = \{t : t \leq b\}$, and so on.

Initial value problems

In Example 1.9, we found a general solution, indicated by the presence of an undetermined constant in the formula. This reflects the fact that an ordinary differential equation has infinitely many solutions. In applications, it is necessary to use other information, in addition to the differential equation, to determine the value of the constant and to specify the solution completely. Such a solution is called a ***particular solution***.

Example 1.12   Given that

$$y(t) = -\frac{1}{t - C} \tag{1.13}$$

is a general solution of $y' = y^2$, find the particular solution satisfying $y(0) = 1$.

Because

$$1 = y(0) = \frac{-1}{0 - C} = \frac{1}{C},$$

$C = 1$. Substituting $C = 1$ in equation (1.13) makes

$$y(t) = -\frac{1}{t - 1}, \tag{1.14}$$

a particular solution of $y' = y^2$, satisfying $y(0) = 1$.   ●

DEFINITION 1.15

A first-order differential equation together with an initial condition,

$$y' = f(t, y), \quad y(t_0) = y_0, \tag{1.16}$$

is called an ***initial value problem***. A solution of the initial value problem is a differentiable function $y(t)$ such that

1. $y'(t) = f(t, y(t))$ for all t in an interval containing t_0 where $y(t)$ is defined, and

2. $y(t_0) = y_0$.

Thus, in Example 1.12, the function $y(t) = 1/(1 - t)$ is the solution to the initial value problem
$$y' = y^2, \quad \text{with} \quad y(0) = 1.$$

Interval of existence

The ***interval of existence*** of a solution to a differential equation is defined to be the largest interval over which the solution can be defined and remain a solution. It is important to remember that solutions to differential equations are required to be differentiable, and this implies that they are continuous. The solution to the initial value problem in Example 1.12 is revealing.

Example 1.17   Find the interval of existence for the solution to the initial value problem

$$y' = y^2 \quad \text{with} \quad y(0) = 1.$$

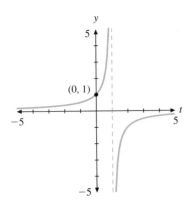

Figure 2. The graph of $y = -1/(t - 1)$.

In Example 1.12, we found that the solution is

$$y(t) = \frac{-1}{t - 1}.$$

The graph of y is a hyperbola with two branches, as shown in Figure 2. The function y has an infinite discontinuity at $t = 1$. Consequently, this function cannot be considered to be a solution to the differential equation $y' = y^2$ over the whole real line.

Note that the left branch of the hyperbola in Figure 2 passes through the point $(0, 1)$, as required by the initial condition $y(0) = 1$. Hence, the left branch of the hyperbola is the solution curve needed. This particular solution curve extends indefinitely to the left, but rises to positive infinity as it approaches the asymptote $t = 1$ from the left. Any attempt to extend this solution to the right would have to include $t = 1$, at which point the function $y(t)$ is undefined. Consequently, the maximum interval on which this solution curve is defined is the interval $(-\infty, 1)$. This is the interval of existence. ●

Using variables other than y and t

So far all of our examples have used y as the unknown function, and t as the independent variable. It is not required to use y and t. We can use any letter to designate the independent variable and any other for the unknown function. For example, the equation

$$y' = x + y$$

has the form $y' = f(x, y)$, making x the independent variable and requiring a solution y that is a function of x. This equation has general solution

$$y(x) = -1 - x + Ce^x,$$

which exists on $(-\infty, \infty)$.

Similarly, in the equation

$$s' = \sqrt{r},$$

the independent variable is r and the unknown function is s, so s must be a function of r. The general solution of this equation is

$$s(r) = \frac{2}{3}r^{3/2} + C.$$

This general solution exists on the interval $[0, \infty)$.

Example 1.18 Verify that $x(s) = 2 - Ce^{-s}$ is a solution of

$$x' = 2 - x \tag{1.19}$$

for any constant C. Find the solution that satisfies the initial condition $x(0) = 1$. What is the interval of existence of this solution?

We evaluate both sides of (1.19) for $x(s) = 2 - Ce^{-s}$.

$$x'(s) = Ce^{-s}$$
$$2 - x = 2 - (2 - Ce^{-s}) = Ce^{-s}$$

They are the same, so the differential equation is solved for all $s \in (-\infty, \infty)$. In addition,

$$x(0) = 2 - Ce^{-0} = 2 - C.$$

To satisfy the initial condition $x(0) = 1$, we must have $2 - C = 1$, or $C = 1$. Therefore, $x(s) = 2 - e^{-s}$ is a solution of the initial value problem. This solution exists for all $s \in (-\infty, \infty)$. Its graph is displayed in Figure 3.

Finally, both $x(s)$ and $x'(s)$ exist and solve the equation on $(-\infty, \infty)$. Therefore, the interval of existence is the whole real line.   ●

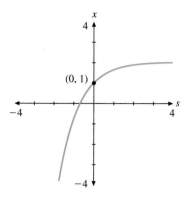

Figure 3. Solution of $x' = 2 - x$, $x(0) = 1$.

The geometric meaning of a differential equation and its solutions

Consider the differential equation

$$y' = f(t, y),$$

where the right-hand side $f(t, y)$ is defined for (t, y) in the rectangle

$$R = \{(t, y) \mid a \le t \le b \text{ and } c \le y \le d\}.$$

Let $y(t)$ be a solution of the equation $y' = f(t, y)$, and recall that the graph of the function y is called a solution curve. Because $y(t_0) = y_0$, the point (t_0, y_0) is on the solution curve. The differential equation says that $y'(t_0) = f(t_0, y_0)$. Hence $f(t_0, y_0)$ is the **slope** of any solution curve that passes through the point (t_0, y_0).

This interpretation allows us a new, geometric insight into a differential equation. Consider, if you can, a small, slanted line segment with slope $f(t, y)$ attached to every point (t, y) of the rectangle R. The result is called a **direction field**, because at each (t, y) there is assigned a direction represented by the line with slope $f(t, y)$.

Even for a simple equation like

$$y' = y, \tag{1.20}$$

it is difficult to visualize the direction field. However, a computer can calculate and plot the direction field at a large number of points—a large enough number for us to get a good understanding of the direction field. Each of the standard mathematics programs, Maple, *Mathematica*, and MATLAB®, has the capability to easily produce direction fields. Some hand-held calculators also have this capability. The student will find that the use of computer- or calculator-generated direction fields will greatly assist their understanding of differential equations. A computer-generated direction field for equation (1.20) is given in Figure 4.

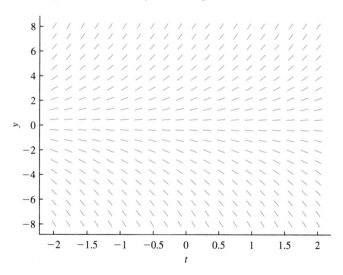

Figure 4. The direction field for $y' = y$.

The direction field is the geometric interpretation of a differential equation. However, the direction field view also gives us a new interpretation of a solution. Associated to the solution $y(t)$, we have the solution curve in the ty-plane. At each point $(t, y(t))$ on the solution curve the curve must have slope equal to $y'(t) = f(t, y(t))$. In other words, the solution curve must be tangent to the direction field at every point. Thus finding a solution to the differential equation is equivalent to the geometric problem of finding a curve in the ty-plane that is tangent to the direction field at every point.

For example, note how the solution curve of

$$y' = y, \quad y(0) = 1 \tag{1.21}$$

in Figure 5 is tangent to the direction field at each point (t, y) on the solution curve.

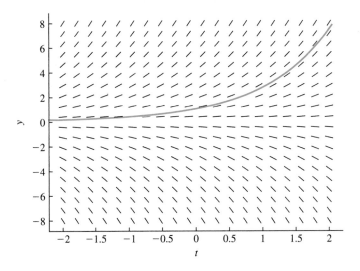

Figure 5. The solution curve is tangent to the direction field.

Approximate numerical solutions

The direction field hints at how we might produce a numerical solution of an initial value problem. To find a solution curve for the initial value problem $y' = f(t, y)$, $y(t_0) = y_0$, first plot the point $P_0 = (t_0, y_0)$. Because the slope of the solution curve at P_0 is given by $f(t_0, y_0)$, move a prescribed distance along a line with slope $f(t_0, y_0)$ to the point $P_1 = (t_1, y_1)$. Next, because the slope of the solution curve at P_1 is given by $f(t_1, y_1)$, move along a line with slope $f(t_1, y_1)$ to the point $P_2 = (t_2, y_2)$. Continue in this manner to produce an approximate solution curve of the initial value problem.

This technique is used in Figure 6 to produce an approximate solution of equation (1.21) and is the basic idea behind *Euler's method*, an algorithm used to find numerical solutions of initial value problems. Clearly, if we decrease the distance between consecutively plotted points, we should obtain an even better approximation of the actual solution curve.

Using a numerical solver

We assume that each of our readers has access to a computer. Furthermore, we assume that this computer has software designed to produce numerical solutions of initial value problems. For many purposes a hand-held graphics calculator will suffice. There is a wide variety of software packages available for the study of

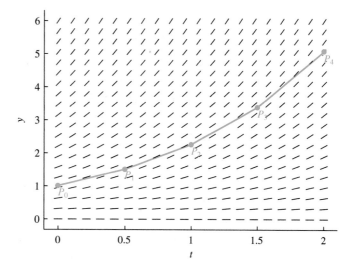

Figure 6.   An approximate solution curve of $y' = y$, $y(0) = 1$.

differential equations. Some of these packages are commercial, some are shareware, and some are even freeware. We will assume that you have access to a solver that will

- draw direction fields,
- provide numerical solutions of differential equations and systems of differential equations, and
- plot solutions of differential equations and systems of differential equations.

Test drive your solver

Let's test our solvers in order to assure ourselves that they will provide adequate support for the material in this text.

Example 1.22   Use a numerical solver to compute and plot the solution of the initial value problem

$$y' = y^2 - t, \quad y(4) = 0 \tag{1.23}$$

over the t-interval $[-2, 10]$.

Although solvers differ widely, they do share some common characteristics. First, you need to input the differential equation, and you will probably have to identify the independent variable — in this case t. Most solvers require that you specify a display window, rectangle in which the solution will be drawn. In this case we choose the bounds $-2 \le t \le 10$ and $-4 \le y \le 4$.

Finally, you need to enter the initial condition $y(4) = 0$ and plot the solution. If your solver can superimpose the solution on a direction field, then your plot should look similar to that shown in Figure 7.   ●

Qualitative methods

We are unable at this time to find analytic, closed-form solutions to the equation

$$y' = 1 - y^2. \tag{1.24}$$

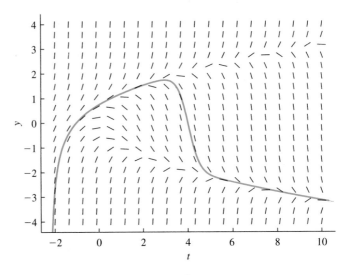

Figure 7. The solution curve for $y' = y^2 - t$, $y(4) = 0$.

This situation will be remedied in the next section. However, the lack of closed-form solutions does not prevent us from using a bit of qualitative mathematical reasoning to investigate a number of important qualities of the solutions of this equation.

Some information about the solutions can be gleaned by looking at the direction field for the equation (1.24) in Figure 8. Notice that the lines $y = 1$ and $y = -1$ seem to be tangent to the direction field. It is easy to verify directly that the constant functions

$$y_1(t) = -1 \quad \text{and} \quad y_2(t) = 1 \tag{1.25}$$

are solutions to equation (1.24).

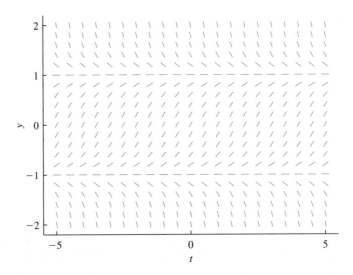

Figure 8. The direction field for the equation $y' = 1 - y^2$.

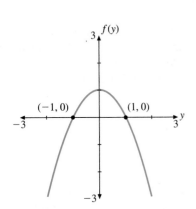

Figure 9. The graph of $f(y) = 1 - y^2$.

To see how we might find such constant solutions, consider the function $f(y) = 1 - y^2$, which is the right-hand side of (1.24). The graph of f is shown in Figure 9. Notice that $f(y) = 0$ only for $y = -1$ and $y = 1$. Each of these points (called **equilibrium points**) gives rise to one of the solutions we found in (1.25). These *equilibrium solutions* are the solutions that can be "seen" in the direction field in Figure 8. They are shown plotted in blue in Figure 10.

Next we notice that $f(y) = 1 - y^2$ is positive if $-1 < y < 1$ and negative otherwise. Thus, if $y(t)$ is a solution to equation (1.24), and $-1 < y < 1$, then

$$y' = 1 - y^2 > 0.$$

Having a positive derivative, y is an increasing function.

How large can a solution $y(t)$ get? If it gets larger than 1, then $y' = 1 - y^2 < 0$, so $y(t)$ will be decreasing. We cannot complete this line of reasoning at this point, but in Section 2.9 we will develop the argument, and we will be able to conclude that if $y(0) = y_0 > 1$, then $y(t)$ is decreasing and $y(t) \to 1$ as $t \to \infty$.

On the other hand, if $y(0) = y_0$ satisfies $-1 < y_0 < 1$, then $y' = 1 - y^2 > 0$, so $y(t)$ will be increasing. We will again conclude that $y(t)$ increases and approaches 1 as $t \to \infty$. Thus any solution to the equation $y' = 1 - y^2$ with an initial value $y_0 > -1$ approaches 1 as $t \to \infty$.

Finally, if we consider a solution $y(t)$ with $y(0) = y_0 < -1$, then a similar analysis shows that $y'(t) = 1 - y^2 < 0$, so $y(t)$ is decreasing. As $y(t)$ decreases, its derivative $y'(t) = 1 - y^2$ gets more and more negative. Hence, $y(t)$ decreases faster and faster and must approach $-\infty$ as t increases. Typical solutions to equation (1.24) are shown in Figure 11. These solutions were found with a computer, but their qualitative nature can be found simply by looking at the equation.

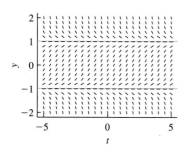

Figure 10. Equilibrium solutions to the equation $y' = 1 - y^2$.

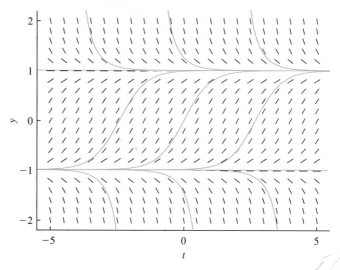

Figure 11. Typical solutions to the equation $y' = 1 - y^2$.

EXERCISES

In Exercises 1 and 2 , given the function ϕ, place the ordinary differential equation $\phi(t, y, y') = 0$ in normal form.

1. $\phi(x, y, z) = x^2 z + (1 + x)y$

2. $\phi(x, y, z) = xz - 2y - x^2$

In Exercises 3–6, show that the given solution is a general solution of the differential equation. Use a computer or calculator to sketch the solutions for the given values of the arbitrary constant. Experiment with different intervals for t until you have

a plot that shows what you consider to be the most important behavior of the family.

3. $y' = -ty$, $y(t) = Ce^{-(1/2)t^2}$, $C = -3, -2, \dots, 3$

4. $y' + y = 2t$, $y(t) = 2t - 2 + Ce^{-t}$, $C = -3, -2, \dots, 3$

5. $y' + (1/2)y = 2\cos t$, $y(t) = (4/5)\cos t + (8/5)\sin t + Ce^{-(1/2)t}$, $C = -5, -4, \dots, 5$

6. $y' = y(4 - y)$, $y(t) = 4/(1 + Ce^{-4t})$, $C = 1, 2, \dots, 5$

7. A general solution may fail to produce all solutions of a differential equation. In Exercise 6, show that $y = 0$ is a solution of the differential equation, but no value of C in the given general solution will produce this solution.

8. (a) Use implicit differentiation to show that $t^2 + y^2 = C^2$ implicitly defines solutions of the differential equation $t + yy' = 0$.

   (b) Solve $t^2 + y^2 = C^2$ for y in terms of t to provide explicit solutions. Show that these functions are also solutions of $t + yy' = 0$.

   (c) Discuss the interval of existence for each of the solutions in part (b).

   (d) Sketch the solutions in part (b) for $C = 1, 2, 3, 4$.

9. (a) Use implicit differentiation to show that $t^2 - 4y^2 = C^2$ implicitly defines solutions of the differential equation $t - 4yy' = 0$.

   (b) Solve $t^2 - 4y^2 = C^2$ for y in terms of t to provide explicit solutions. Show that these functions are also solutions of $t - 4yy' = 0$.

   (c) Discuss the interval of existence for each of the solutions in part (b).

   (d) Sketch the solutions in part (b) for $C = 1, 2, 3, 4$.

10. Show that $y(t) = 3/(6t - 11)$ is a solution of $y' = -2y^2$, $y(2) = 3$. Sketch this solution and discuss its interval of existence. Include the initial condition on your sketch.

11. Show that $y(t) = 4/(1 - 5e^{-4t})$ is a solution of the initial value problem $y' = y(4 - y)$, $y(0) = -1$. Sketch this solution and discuss its interval of existence. Include the initial condition on your sketch.

In Exercises 12–15, use the given general solution to find a solution of the differential equation having the given initial condition. Sketch the solution, the initial condition, and discuss the solution's interval of existence.

12. $y' + 4y = \cos t$, $y(t) = (4/17)\cos t + (1/17)\sin t + Ce^{-4t}$, $y(0) = -1$

13. $ty' + y = t^2$, $y(t) = (1/3)t^2 + C/t$, $y(1) = 2$

14. $ty' + (t + 1)y = 2te^{-t}$, $y(t) = e^{-t}(t + C/t)$, $y(1) = 1/e$

15. $y' = y(2 + y)$, $y(t) = 2/(-1 + Ce^{-2t})$, $y(0) = -3$

16. Maple, when asked for the solution of the initial value problem $y' = \sqrt{y}$, $y(0) = 1$, returns two solutions: $y(t) = (1/4)(t + 2)^2$ and $y(t) = (1/4)(t - 2)^2$. Present a thorough discussion of this response, including a check and a graph of each solution, interval of existence, and so on. *Hint:* Remember that $\sqrt{a^2} = |a|$.

In Exercises 17–20, plot the direction field for the differential equation by hand. Do this by drawing short lines of the appropriate slope centered at each of the integer valued coordinates (t, y), where $-2 \le t \le 2$ and $-1 \le y \le 1$.

17. $y' = y + t$

18. $y' = y^2 - t$

19. $y' = t\tan(y/2)$

20. $y' = (t^2y)/(1 + y^2)$

In Exercises 21–24, use a computer to draw a direction field for the given first-order differential equation. Use the indicated bounds for your display window. Obtain a printout and use a pencil to draw a number of possible solution trajectories on the direction field. If possible, check your solutions with a computer.

21. $y' = -ty$, $R = \{(t, y) : -3 \le t \le 3, -5 \le y \le 5\}$

22. $y' = y^2 - t$, $R = \{(t, y) : -2 \le t \le 10, -4 \le y \le 4\}$

23. $y' = t - y + 1$, $R = \{(t, y) : -6 \le t \le 6, -6 \le y \le 6\}$

24. $y' = (y + t)/(y - t)$, $R = \{(t, y) : -5 \le t \le 5, -5 \le y \le 5\}$

For each of the initial value problems in Exercises 25–28 use a numerical solver to plot the solution curve over the indicated interval. Try different display windows by experimenting with the bounds on y. *Note:* Your solver might require that you first place the differential equation in normal form.

25. $y + y' = 2$, $y(0) = 0$, $t \in [-2, 10]$

26. $y' + ty = t^2$, $y(0) = 3$, $t \in [-4, 4]$

27. $y' - 3y = \sin t$, $y(0) = -3$, $t \in [-6\pi, \pi/4]$

28. $y' + (\cos t)y = \sin t$, $y(0) = 0$, $t \in [-10, 10]$

Some solvers allow the user to choose dependent and independent variables. For example, your solver may allow the equation $r' = -2sr + e^{-s}$, but other solvers will insist that you change variables so that the equation reads $y' = -2ty + e^{-t}$, or $y' = -2xy + e^{-x}$, should your solver require t or x as the independent variable. For each of the initial value problems in Exercises 29 and 30, use your solver to plot solution curves over the indicated interval.

29. $r' + xr = \cos(2x)$, $r(0) = -3$, $x \in [-4, 4]$

30. $T' + T = s$, $T(-3) = 0$, $s \in [-5, 5]$

In Exercises 31–34, plot solution curves for each of the initial conditions on one set of axes. Experiment with the different display windows until you find one that exhibits what you feel is all of the important behavior of your solutions. *Note:* Selecting a good display window is an art, a skill developed with experience. Don't become overly frustrated in these first attempts.

31. $y' = y(3 - y)$, $y(0) = -2, -1, 0, 1, 2, 3, 4, 5$

32. $x' - x^2 = t$, $x(0) = -2, 0, 2$, $x(2) = 0$, $x(4) = -3, 0, 3$, $x(6) = 0$

33. $y' = \sin(xy)$, $y(0) = 0.5, 1.0, 1.5, 2.0, 2.5$

34. $x' = -tx$, $x(0) = -3, -2, -1, 0, 1, 2, 3$

35. Bacteria in a petri dish is growing according to the equation

$$\frac{dP}{dt} = 0.44P,$$

where P is the mass of the accumulated bacteria (measured in milligrams) after t days. Suppose that the initial mass of the bacterial sample is 1.5 mg. Use a numerical solver to estimate the amount of bacteria after 10 days.

36. A certain radioactive substance is decaying according to the equation

$$\frac{dA}{dt} = -0.25A,$$

where A is the amount of substance in milligrams remaining after t days. Suppose that the initial amount of the substance present is 400 mg. Use a numerical solver to estimate the amount of substance remaining after 4 days.

37. The concentration of pollutant in a lake is given by the equation

$$\frac{dc}{dt} = -0.055c,$$

where c is the concentration of the pollutant at t days. Suppose that the initial concentration of pollutant is 0.10. A concentration level of $c = 0.02$ is deemed safe for the fish population in the lake. If the concentration varies according to the model, how long will it be before the concentration reaches a level that is safe for the fish population?

38. An aluminum rod is heated to a temperature of 300°C. Suppose that the rate at which the rod cools is proportional to the difference between the temperature of the rod and the temperature of the surrounding air (20°C). Assume a proportionality constant $k = 0.085$ and time is measured in minutes. How long will it take the rod to cool to 100°C?

39. You're told that the "carrying capacity" for an environment populated by "critters" is 100. Further, you're also told that the rate at which the critter population is changing is proportional to the product of the number of critters and the number of critters less than the carrying capacity. Assuming a constant of proportionality $k = 0.00125$ and an initial critter population of 20, use a numerical solver to determine the size of the critter population after 30 days.

2.2 Solutions to Separable Equations

An unstable nucleus is radioactive. At any instant, it can emit a particle, transforming itself into a different nucleus in the process. For example, ^{238}U is an alpha emitter that decays spontaneously according to the scheme ^{238}U $\rightarrow$ ^{234}Th $+$ ^{4}He, where ^{4}He is the alpha particle. In a sample of ^{238}U, a certain percentage of the nuclei will decay during a given observation period. If at time t the sample contains $N(t)$ radioactive nuclei, then we expect that the number of nuclei that decay in the time interval Δt will be approximately proportional to both N and Δt. In symbols,

$$\Delta N = N(t + \Delta t) - N(t) \approx -\lambda N(t)\Delta t, \tag{2.1}$$

where $\lambda > 0$ is a constant of proportionality. The minus sign is indicative of the fact that there are fewer radioactive nuclei at time $t + \Delta t$ than there are at time t.

Dividing both sides of equation (2.1) by Δt, then taking the limit as $\Delta t \rightarrow 0$,

$$N'(t) = \lim_{\Delta t \to 0} \frac{N(t + \Delta t) - N(t)}{\Delta t} = -\lambda N(t).$$

This equation is one that arises often in applications. Because of the form of its solutions, the equation

$$N' = -\lambda N \tag{2.2}$$

is called the ***exponential equation***.

Equation (2.2) is an example of what is called a ***separable equation*** because it can be rewritten with its variables separated and then easily solved. To do this, we first write the equation using dN/dt instead of N',

$$\frac{dN}{dt} = -\lambda N. \tag{2.3}$$

Next, we separate the variables by putting every expression involving the unknown function N on the left and everything involving the independent variable t on the right. This includes dN and dt. The result is

$$\frac{1}{N} dN = -\lambda \, dt. \tag{2.4}$$

It is important to note that this step is valid only if $N \neq 0$, since we cannot divide by zero. Then we integrate both sides of equation (2.4), getting[2]

$$\int \frac{1}{N} \, dN = -\lambda \int dt, \quad \text{or}$$
$$\ln |N| = -\lambda t + C. \tag{2.5}$$

It remains to solve for N. Taking the exponential of both sides of equation (2.5), we get

$$|N(t)| = e^{-\lambda t + C} = e^C e^{-\lambda t}. \tag{2.6}$$

Since e^C and $e^{-\lambda t}$ are both positive, there are two cases

$$N(t) = \begin{cases} e^C e^{-\lambda t}, & \text{if } N > 0; \\ -e^C e^{-\lambda t}, & \text{if } N < 0. \end{cases}$$

We can simplify the solution by introducing

$$A = \begin{cases} e^C, & \text{if } N > 0; \\ -e^C, & \text{if } N < 0. \end{cases}$$

Therefore, the solution is also described by the simpler formula

$$N(t) = Ae^{-\lambda t}, \tag{2.7}$$

where A is a constant different from zero, but otherwise arbitrary.

In arriving at equation (2.4), we divided both sides of equation (2.3) by N, and this procedure is not valid when $N = 0$. We will discuss this a bit later. For now, let's notice that if we set $A = 0$ in equation (2.7), we get the constant function $N(t) = 0$, and we can verify by substitution that this is a solution of the original equation, $N' = -\lambda N$. Consequently, equation (2.7) with A completely arbitrary, gives us the solution in all cases.

Example 2.8 ^{32}P, an isotope of phosphorus, is used in leukemia therapy. After 10 hours, 615 mg of an initial 1000 mg sample remain. The **half-life** of a radioactive substance is the amount of time required for 50% of the substance to decay. Determine the half-life of ^{32}P.

The differential equation $N' = -\lambda N$ was used to model the number of remaining nuclei. However, the number of nuclei is proportional to the mass, so we will let N represent the mass of the remaining nuclei in this example. As seen earlier, this differential equation has solution

$$N = Ae^{-\lambda t}, \tag{2.9}$$

where A is an arbitrary constant. At time $t = 0$ we have $N = 1000$ mg of the isotope. Substituting these quantities in equation (2.9),

$$1000 = Ae^{-\lambda(0)} = A. \tag{2.10}$$

---

[2] Our understanding of integration first has us use two constants of integration,

$$\ln |N| + C_1 = -\lambda t + C_2.$$

We get (2.5) by setting $C = C_2 - C_1$. This combining of the two constants into one works in the solution of any separable equation.

Consequently, equation (2.9) becomes

$$N = 1000e^{-\lambda t}. \tag{2.11}$$

After $t = 10\,\text{hr}$, only $N = 615\,\text{mg}$ of the substance remains. Substituting these values into equation (2.11), we get

$$615 = 1000e^{-\lambda(10)}. \tag{2.12}$$

Using a little algebra and a calculator to compute a logarithm shows that $\lambda = 0.04861$, correct to six decimal places, and equation (2.11) becomes

$$N = 1000e^{-0.04861t}. \tag{2.13}$$

To find the half-life, we substitute $N = 500\,\text{mg}$ in equation (2.13).

$$500 = 1000e^{-0.04861t}$$

Solving for t, we find that the half-life of the isotope is approximately 14.3 hours.●

A large number of equations are separable and can be solved exactly like we solved the exponential equation. Let's look at another example.

Example 2.14 Solve the differential equation

$$y' = ty^2. \tag{2.15}$$

Again, we rewrite the equation using dy/dt instead of y', so

$$\frac{dy}{dt} = ty^2. \tag{2.16}$$

Next we separate the variables by putting every expression involving the unknown function y on the left and everything involving the independent variable t on the right, including dy and dt. The result is

$$\frac{1}{y^2}\,dy = t\,dt. \tag{2.17}$$

Notice that this step is valid only if $y \neq 0$, since we cannot divide by zero. Next we integrate both sides of equation (2.17), getting

$$\int \frac{1}{y^2}\,dy = \int t\,dt, \quad \text{or} \quad -\frac{1}{y} = \frac{1}{2}t^2 + C. \tag{2.18}$$

Finally, we solve equation (2.18) for y. The equation for the solution is

$$y(t) = \frac{-1}{\frac{1}{2}t^2 + C} = \frac{-2}{t^2 + 2C}. \tag{2.19}$$

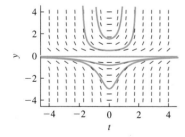

Figure 1. Several solutions to $y' = ty^2$.

Several solutions are shown in Figure 1. Included among the functions plotted in Figure 1 is the constant function $y(t) = 0$. It is easily verified by substitution that this is a solution of (2.15), although no finite value of C in equation (2.19) will yield this solution. We will have more to say about this on page 30. ●

Treating dy and dt as mathematical entities, as we did in separating the variables in equation (2.17), may be troublesome to you. If so, it is probably because you have learned your calculus very well. We will explain this step at the end of this section under the heading "Why separation of variables works."

The general method

Clearly the key step in this method is the separation of variables. This is the step going from equation (2.3) to equation (2.4) or from equation (2.16) to equation (2.17). The method of solution illustrated here will work whenever we can perform this step, and this can be done for any equation of the two equivalent forms

$$\frac{dy}{dt} = \frac{g(t)}{h(y)} \tag{2.20}$$

and

$$\frac{dy}{dt} = g(t) f(y). \tag{2.21}$$

Equations of either form are called **separable** differential equations. For both we can separate the variables.

The method we used to solve equation (2.16) will work for any separable equation.

> We can solve any separable equation of the form (2.21) using the following three steps.
>
> 1. Separate the variables: $\dfrac{dy}{f(y)} = g(t)\,dt$.
> 2. Integrate both sides: $\displaystyle\int \frac{dy}{f(y)} = \int g(t)\,dt$.
> 3. Solve for the solution $y(t)$, if possible.

Avoiding division by zero

When separating the variables we do have to worry about dividing by zero, but otherwise things work well. What about those points where $f(y) = 0$ in equation (2.21)? It turns out to be quite easy to find the solutions in such a case, since if $f(y_0) = 0$, then by substitution we see that the constant function $y(t) = y_0$ is a solution to (2.21).

In particular, the function $y(t) = 0$ is a solution to the equation $y' = ty^2$. We found in (2.19) that, under the assumption that $y \neq 0$, the general solution to the equation $y' = ty^2$ is

$$y(t) = \frac{-2}{t^2 + 2C}.$$

If we naively substitute the initial condition $y(0) = 0$ into this general solution, we get $0 = -1/C$. No finite value of the constant C solves this equation. This should not be a surprise, since (2.19) was derived on the assumption that $y \neq 0$. Nevertheless, we will want to call (2.19) a general solution to equation (2.16). We define a **general solution** to a differential equation to be a family of solutions depending on sufficiently many parameters to give all but finitely many solutions.

Thus the general solution to a differential equation does not always yield the solution to every initial value problem, and for separable equations this is related to the problem of dividing by 0. In the case of $y' = ty^2$, we can find the exceptional solution by setting $C = \infty$. This often the case, but we will not explore this further.

Using definite integration

Sometimes it is useful to use definite integrals when solving initial value problems for separable equations.

Example 2.22   A can of beer at 40°F is placed into a room where the temperature is 70°F. After 10 minutes the temperature of the beer is 50°F. What is the temperature of the beer as a function of time? What is the temperature of the beer 30 minutes after the beer was placed into the room?

According to ***Newton's law of cooling***, the rate of change of an object's temperature (T) is proportional to the difference between its temperature and the ambient temperature (A). Thus we have

$$\frac{dT}{dt} = -k(T - A). \tag{2.23}$$

We introduce the minus sign so that the proportionality constant k is positive. Notice that if $T < A$, the temperature of the object will be increasing. The equation is separable, so we separate variables to get

$$\frac{dT}{T - A} = -k\, dt.$$

The next step is to integrate both sides, but this time let's use definite integrals to bring in the initial condition $T(0) = T_0$. Since $t = 0$ corresponds to $T = T_0$, we have

$$\int_{T_0}^{T} \frac{ds}{s - A} = -k \int_{0}^{t} du.$$

Notice that we changed the variables of integration because we want the upper limits of our integrals to be T and t. Performing the integration, we get

$$\ln \frac{|T - A|}{|T_0 - A|} = \ln |T - A| - \ln |T_0 - A| = -kt.$$

We can solve for T by exponentiating, and since $T - A$ and $T_0 - A$ both have the same sign, our answer is

$$T(t) = A + (T_0 - A)e^{-kt}. \tag{2.24}$$

We first use the fact that $T(10) = 50$ in addition to the initial condition $T(0) = T_0 = 40$ and the ambient temperature $A = 70$ to evaluate k. Equation (2.24) becomes $50 = 70 - 30e^{-10k}$. Therefore, $k = \ln(3/2)/10 = 0.0405$. Thus from equation (2.24) we see that the temperature is

$$T(t) = 70 - 30e^{-0.0405t}.$$

After 30 minutes the temperature is 61.1°F. The solution is plotted in Figure 2.   ●

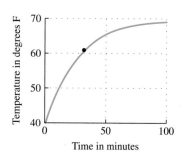

Figure 2. The temperature of the can of beer in Example 2.22.

Implicitly defined solutions

After the integration step, we need to solve for the solution. However, this is not always easy. In fact, it is not always possible. We will look at a series of examples.

Example 2.25   Find the solutions of the equation $y' = e^x/(1+y)$, having initial conditions $y(0) = 1$ and $y(0) = -4$.

Separate the variables and integrate.

$$(1 + y)\,dy = e^x\,dx$$

$$y + \frac{1}{2}y^2 = e^x + C \tag{2.26}$$

Rearrange equation (2.26) as

$$y^2 + 2y - 2(e^x + C) = 0. \tag{2.27}$$

This is an implicit equation for $y(x)$ that we can solve using the quadratic formula.

$$y(x) = \frac{1}{2}\left[-2 \pm \sqrt{4 + 8(e^x + C)}\right]$$

$$= -1 \pm \sqrt{1 + 2(e^x + C)}$$

We get two solutions from the quadratic formula, and the initial condition will dictate which solution we choose. If $y(0) = 1$, then we must use the positive square root and we find that $C = 1/2$. The solution is

$$y(x) = -1 + \sqrt{2 + 2e^x}. \tag{2.28}$$

On the other hand, if $y(0) = -4$, then we must use the negative square root and we find that $C = 3$. The solution in this case is

$$y(x) = -1 - \sqrt{7 + 2e^x}. \tag{2.29}$$

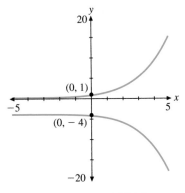

Figure 3. $y = -1 + \sqrt{2 + 2e^x}$ passes through $(0, 1)$, while $y = -1 - \sqrt{7 + 2e^x}$ passes through $(0, -4)$.

Both solutions are shown in Figure 3.

What about the interval of existence? A quick glance reveals that each solution is defined on the interval $(-\infty, \infty)$. Some calculation will reveal that $y'(x)$ is also defined on $(-\infty, \infty)$. However, for each solution to satisfy the equation $y' = e^x/(1 + y)$, y must not equal -1. Fortunately, neither solution (2.28) or (2.29) can ever equal -1. Therefore, the interval of existence is $(-\infty, \infty)$. ●

Let's be sure we know what the terminology means. An **explicit** solution is one for which we have a formula as a function of the independent variable. For example, (2.28) is an explicit solution to the equation in Example 2.25. In contrast, (2.27) is an implicit equation for the solution. In this example, the implicit equation can be solved easily to find an explicit equation, but this is not always the case.

Unfortunately, implicit solutions occur frequently. Consider again the general problem in the form $dy/dt = g(t)/h(y)$. Separating variables and integrating, we get

$$\int h(y)\,dy = \int g(t)\,dt. \tag{2.30}$$

If we let

$$H(y) = \int h(y)\,dy \quad \text{and} \quad G(t) = \int g(t)\,dt,$$

and then introduce a constant of integration, equation (2.30) can be rewritten as

$$H(y) = G(t) + C. \tag{2.31}$$

Unless $H(y) = y$, and therefore $h(y) = 1$, this is an implicit equation for $y(t)$. To find an explicit solution we must be able to compute the inverse function H^{-1}. If this is possible, then we have

$$y(t) = H^{-1}(G(t) + C).$$

Let's do one more example.

Example 2.32   Find the solutions to the differential equation

$$x' = \frac{2tx}{1+x},$$

having initial conditions $x(0) = 1$, $x(0) = -2$, and $x(0) = 0$.

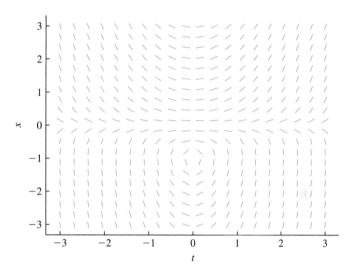

Figure 4. The direction field for $x' = 2tx(1+x)$.

The direction field for this equation is shown in Figure 4. This equation is separable since it can be written as

$$\frac{dx}{dt} = 2t\frac{x}{1+x}.$$

When we separate variables, we get

$$\left(1 + \frac{1}{x}\right) dx = 2t \, dt,$$

assuming that $x \neq 0$. Integrating, we get

$$x + \ln(|x|) = t^2 + C, \tag{2.33}$$

where C is an arbitrary constant. For the initial condition $x(0) = 1$, this becomes $1 = C$. Hence our solution is implicitly defined by $x + \ln(|x|) = t^2 + 1$. The function $\ln(|x|)$ is not defined at $x = 0$, so our solution can never be equal to 0. Since our initial condition is positive, and a solution must be continuous, our solution $x(t)$ must be positive for all t. Hence $|x| = x$ and our solution is given implicitly by

$$x + \ln(x) = t^2 + 1. \tag{2.34}$$

This is as far as we can go. We cannot solve equation (2.34) explicitly for $x(t)$, so we have to be satisfied with this as our answer. The solution x is defined implicitly by equation (2.34).

For the initial condition $x(0) = -2$, we can find the constant C in the same manner. We get $-2 + \ln(|-2|) = C$, or $C = \ln 2 - 2$. Hence the solution is defined implicitly by

$$x + \ln(|x|) = t^2 + \ln 2 - 2.$$

This time our initial condition is negative, so $|x| = -x$, and our implicit equation for the solution is

$$x + \ln(-x) = t^2 + \ln 2 - 2.$$

For the initial condition $x(0) = 0$, we cannot divide by $x/(1 + x)$ to separate variables. However, we know that this means that $x(t) = 0$ is a solution. We can easily verify that by direct substitution. Thus we do get an explicit formula for the solution with this initial condition. ●

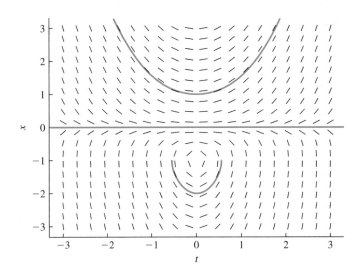

Figure 5. Solutions to $x' = 2tx(1 + x)$.

The solutions sought in the previous example were computed numerically and are plotted in Figure 5. Since the solutions are defined implicitly, it is a difficult task to visualize them without the aid of numerical methods.

Why separation of variables works

If we start with a separable equation

$$y' = g(t)/h(y), \tag{2.35}$$

then separation of variables leads to the equation

$$h(y)\,dy = g(t)\,dt. \tag{2.36}$$

However, many readers will have been taught that the terms dy and dt have no meaning and so equation (2.36) has no meaning. Yet the method works, so what is going on here?

To understand this better, let's start with (2.35) and perform legitimate steps

$$y' = g(t)/h(y) \quad \text{or} \quad h(y)y' = g(t).$$

Integrating both sides with respect to t, we get

$$\int h(y(t))y'(t)\,dt = \int g(t)\,dt.$$

The integral on the left contains the expression $y'(t) \, dt$. This is inviting us to change the variable of integration to y, since when we do that, we use the equation $dy = y'(t) \, dt$. Making the change of variables leads to

$$\int h(y) \, dy = \int g(t) \, dt. \tag{2.37}$$

Notice the similarity between (2.36) and (2.37). Equation (2.36), which has no meaning by itself, acquires a precise meaning when both sides are integrated. Since this is precisely the next step that we take when solving separable equations, we can be sure that our method is valid.

We mention in closing that the objects in (2.36), $h(y) \, dy$ and $g(t) \, dt$, can be given meaning as formal objects that can be integrated. They are called *differential forms*, and the special cases like dy and dt are called *differentials*. The basic formula connecting differentials dy and dt when y is a function of t is

$$dy = y'(t) \, dt,$$

the change-of-variables formula in integration. These techniques will assume greater importance in Section 2.6, where we will deal with exact equations. The use of differential forms is very important in the study of the calculus of functions of several variables and especially in applications to geometry and to parts of physics.

EXERCISES

In Exercises 1–12, find the general solution of the indicated differential equation. If possible, find an explicit solution.

1. $y' = xy$

2. $xy' = 2y$

3. $y' = e^{x-y}$

4. $y' = (1 + y^2)e^x$

5. $y' = xy + y$

6. $y' = ye^x - 2e^x + y - 2$

7. $y' = x/(y + 2)$

8. $y' = xy/(x - 1)$

9. $x^2 y' = y \ln y - y'$

10. $xy' - y = 2x^2 y$

11. $y^3 y' = x + 2y'$

12. $y' = (2xy + 2x)/(x^2 - 1)$

In Exercises 13–18, find the exact solution of the initial value problem. Indicate the interval of existence.

13. $y' = y/x$, $y(1) = -2$

14. $y' = -2t(1 + y^2)/y$, $y(0) = 1$

15. $y' = (\sin x)/y$, $y(\pi/2) = 1$

16. $y' = e^{x+y}$, $y(0) = 0$

17. $y' = (1 + y^2)$, $y(0) = 1$

18. $y' = x/(1 + 2y)$, $y(-1) = 0$

In Exercises 19–22, find exact solutions for each given initial condition. State the interval of existence in each case. Plot each exact solution on the interval of existence. Use a numerical solver to duplicate the solution curve for each initial value problem.

19. $y' = x/y$, $y(0) = 1$, $y(0) = -1$

20. $y' = -x/y$, $y(0) = 2$, $y(0) = -2$

21. $y' = 2 - y$, $y(0) = 3$, $y(0) = 1$

22. $y' = (y^2 + 1)/y$, $y(1) = 2$

23. Suppose that a radioactive substance decays according to the model $N' = N_0 e^{-\lambda t}$. Show that the half-life of the radioactive substance is given by the equation

$$T_{1/2} = \frac{\ln 2}{\lambda}. \tag{2.38}$$

24. The half-life of ^{238}U is 4.47×10^7 yr.

(a) Use equation (2.38) to compute the *decay constant* λ for ^{238}U.

(b) Suppose that 1000 mg of ^{238}U are present initially. Use the equation $N = N_0 e^{-\lambda t}$ and the decay constant determined in part (a) to determine the time for this sample to decay to 100 mg.

25. Tritium, ^{3}H, is an isotope of hydrogen that is sometimes used as a biochemical tracer. Suppose that 100 mg of ^{3}H decays to 80 mg in 4 hours. Determine the half-life of ^{3}H.

26. The isotope Technetium 99m is used in medical imaging. It has a half-life of about 6 hours, a useful feature for radioisotopes that are injected into humans. The Technetium, having such a short half-life, is created artificially on scene by harvesting from a more stable isotope, ^{99}Mb. If 10 g of ^{99m}Tc are "harvested" from the Molybdenum, how much of this sample remains after 9 hours?

27. The isotope Iodine 131 is used to destroy tissue in an overactive thyroid gland. It has a half-life of 8.04 days. If a hospital receives a shipment of 500 mg of ^{131}I, how much of the isotope will be left after 20 days?

28. A substance contains two Radon isotopes, ^{210}Rn $[t_{1/2} = 2.42\,\text{h}]$ and ^{211}Rn $[t_{1/2} = 15\,\text{h}]$. At first, 20% of the decays come from ^{211}Rn. How long must one wait until 80% do so?

29. Suppose that a radioactive substance decays according to the model $N = N_0 e^{-\lambda t}$.

(a) Show that after a period of $T_\lambda = 1/\lambda$, the material has decreased to e^{-1} of its original value. T_λ is called the ***time constant*** and it is defined by this property.

(b) A certain radioactive substance has a half-life of 12 hours. Compute the time constant for this substance.

(c) If there are originally 1000 mg of this radioactive substance present, plot the amount of substance remaining over four time periods T_λ.

In the laboratory, a more useful measurement is the decay rate R, usually measured in disintegrations per second, counts per minute, etc. Thus, the ***decay rate*** is defined as $R = -dN/dt$. Using the equation $dN/dt = -\lambda N$, it is easily seen that $R = \lambda N$. Furthermore, differentiating the solution $N = N_0 e^{-\lambda t}$ with respect to t reveals that

$$R = R_0 e^{-\lambda t}, \qquad (2.39)$$

in which R_0 is the decay rate at $t = 0$. That is, because R and N are proportional, they both decrease with time according to the same exponential law. Use this idea to help solve Exercises 30–31.

30. Jim, working with a sample of ^{131}I in the lab, measures the decay rate at the end of each day.

TIME (DAYS)	COUNTS (COUNTS/DAY)	TIME (DAYS)	COUNTS (COUNTS/DAY)
1	938	6	587
2	822	7	536
3	753	8	494
4	738	9	455
5	647	10	429

Like any modern scientist, Jim wants to use all of the data instead of only two points to estimate the constants R_0 and λ in equation (2.39). He will use the technique of ***regression*** to do so. Use the first method in the following list that your technology makes available to you to estimate λ (and R_0 at the same time). Use this estimate to approximate the half-life of ^{131}I.

(a) Some modern calculators and the spreadsheet Excel can do an exponential regression to directly estimate R_0 and λ.

(b) Taking the natural logarithm of both sides of equation (2.39) produces the result

$$\ln R = -\lambda t + \ln R_0.$$

Now $\ln R$ is a linear function of t. Most calculators, numerical software such as MATLAB®, and computer algebra systems such as *Mathematica* and Maple will do a linear regression, enabling you to estimate $\ln R_0$ and λ (e.g., use the MATLAB® command `polyfit`).

(c) If all else fails, plotting the natural logarithm of the decay rates versus the time will produce a curve that is almost linear. Draw the straight line that in your estimation provides the best fit. The slope of this line provides an estimate of $-\lambda$.

31. A 1.0 g sample of Radium 226 is measured to have a decay rate of 3.7×10^{10} disintegrations/s. What is the half-life of ^{226}Ra in years? *Note:* A chemical constant, called Avogadro's number, says that there are 6.02×10^{23} atoms per mole, a common unit of measurement in chemistry. Furthermore, the atomic mass of ^{226}Ra is 226 g/mol.

32. Radiocarbon dating. Carbon 14 is produced naturally in the earth's atmosphere through the interaction of cosmic rays and Nitrogen 14. A neutron comes along and strikes a ^{14}N nucleus, knocking off a proton and creating a ^{14}C atom. This atom now has an affinity for oxygen and quickly oxidizes as a ^{14}CO$_2$ molecule, which has many of the same chemical properties as regular CO$_2$. Through photosynthesis, the ^{14}CO$_2$ molecules work their way into the plant system, and from there into the food chain. The ratio of ^{14}C to regular carbon in living things is the same as the ratio of these carbon atoms in the earth's atmosphere, which is fairly constant, being in a state of equilibrium. When a living being dies, it no longer ingests ^{14}C and the existing ^{14}C in the now defunct life form begins to decay. In 1949, Willard F. Libby and his associates at the University of Chicago measured the half-life of this decay at 5568 ± 30 years, which to this day is known as the ***Libby half-life***. We now know that the half-life is closer to 5730 years, called the ***Cambridge half-life***, but radiocarbon dating labs still use the Libby half-life for technical and historical reasons. Libby was awarded the Nobel prize in chemistry for his discovery.

(a) Carbon 14 dating is a useful dating tool for organisms that lived during a specific time period. Why is that? Estimate this period.

(b) Suppose that the ratio of ^{14}C to carbon in the charcoal on a cave wall is 0.617 times a similar ratio in living wood in the area. Use the Libby half-life to estimate the age of the charcoal.

33. A murder victim is discovered at midnight and the temperature of the body is recorded at 31°C. One hour later, the temperature of the body is 29°C. Assume that the surrounding air temperature remains constant at 21°C. Use Newton's law of cooling to calculate the victim's time of death. *Note:* The "normal" temperature of a living human being is approximately 37°C.

34. Suppose a cold beer at 40°F is placed into a warm room at 70°F. Suppose 10 minutes later, the temperature of the beer is 48°F. Use Newton's law of cooling to find the temperature 25 minutes after the beer was placed into the room.

35. Referring to the previous problem, suppose a 50° bottle of beer is discovered on a kitchen counter in a 70° room. Ten minutes later, the bottle is 60°. If the refrigerator is kept at 40° how long had the bottle of beer been sitting on the counter when it was first discovered?

36. Consider the equation

$$y' = f(at + by + c),$$

where a, b, and c are constants. Show that the substitution $x = at + by + c$ changes the equation to the separable equation $x' = a + bf(x)$. Use this method to find the general solution of the equation $y' = (y + t)^2$.

37. Suppose a curve, $y = y(x)$ lies in the first quadrant and suppose that for each x the piece of the tangent line at $(x, y(x))$ which lies in the first quadrant is bisected by the point $(x, y(x))$. Find $y(x)$.

38. Suppose the projection of the part of the normal line to the graph of $y = y(x)$ from the point $(x, y(x))$ to the x-axis has length 2. Find $y(x)$.

39. Suppose a polar graph $r = r(\theta)$ has the property that θ always equals twice the angle from the radial line (i.e. the line from the origin to $(\theta, r(\theta))$) to the tangent. Find the function $r(\theta)$.

40. Suppose $y(x)$ is a continuous, nonnegative function with $y(0) = 0$. Find $y(x)$ if the area under the curve, $y = y(t)$, from 0 to x is always equal to one-fourth the area of the rectangle with vertices at $(0, 0)$ and $(x, y(x))$.

41. A football, in the shape of an ellipsoid, is lying on the ground in the rain. Its length is 8 inches and its cross section at its widest point is a circular disc of radius 2 inches. A rain drop hits the top half of the football. Find the path that it follows as it runs down the top half of the football. *Hint:* Recall that the gradient of a function $f(x, y)$ points in the (x, y)-direction of maximum increase of f.

42. From Torricelli's law, water in an open tank will flow out through a hole in the bottom at a speed equal to that it would acquire in a free-fall from the level of the water to the hole. A parabolic bowl has the shape of $y = x^2$, $0 \le x \le 1$, (units are feet) revolved around the y-axis. This bowl is initially full of water and at $t = 0$, a hole of radius a is punched at the bottom. How long will it take for the bowl to drain? *Hint:* An object dropped from height h will hit the ground at a speed of $v = \sqrt{2gh}$, where g is the gravitational constant. This formula is derived from equating the kinetic energy on impact, $(1/2)mv^2$, with the work required to raise the object, mgh.

43. Referring to the previous problem, find the shape of the bowl if the water level drops at a constant rate.

44. A destroyer is hunting a submarine in a dense fog. The fog lifts for a moment, disclosing that the submarine lies on the surface 4 miles away. The submarine immediately descends and departs in a straight line in an unknown direction. The speed of the destroyer is three times that of the submarine. What path should the destroyer follow to be certain of intercepting the submarine? *Hint:* Establish a polar coordinate system with the origin located at the point where the submarine was sighted. Look up the formula for arc length in polar coordinates.

2.3 Models of Motion

One of the most intensively studied scientific problems is the study of motion. This is true in particular for the motion of the planets. The history of the ideas involved is one of the most interesting chapters of human history. We will start by giving a brief summary of the development of models of motion.

A brief history of models of motion

The study of the stars and planets is as old as humankind. Even the most primitive people have been fascinated by the nightly display of the stars, and they soon noticed that some objects, now called planets, moved against the background of the "fixed" stars. The systematic study of planetary motion goes back at least 3000 years to the Babylonians, who made the first recorded observations.

Their interest was furthered by the Greek civilization. There were a number of explanations posed, including that of Aristarchus who put the sun at the center of the universe. However, the one that lasted was developed over hundreds of years and culminated with the work of Hipparchus and Ptolemy. It was published in Claudius Ptolemy's *Almagest* in the second century A.D. Their theory was a descriptive model of the motion of the planets. They assumed that the earth was the center of the universe and that everything revolved around it. At first they thought that the planets, the sun, and the moon moved with constant velocities in circular paths around the earth. As they grew more proficient in their measurements they realized that this was not true. They modified their theory by inventing *epicycles*. These were smaller circles, the centers of which moved with constant velocity along circular paths centered at the earth. The planets moved with constant velocity along the epicycles as the epicycles moved around the earth, as illustrated in Figure 1. When this theory proved to be inadequate in some cases, the Greeks added epicycles to the epicycles.

The theory of epicycles enabled the Greeks to compute and predict the motion of the planets. In many ways it was a highly satisfactory scientific theory, but it left many questions unanswered. Most important, why do the planets move in the complicated manner suggested by the theory of epicycles? The explanation was not causal. It was only descriptive in nature.

A major improvement on this theory came in 1543, when Nicholas Copernicus made the radical suggestion that the earth was not the center of the universe. Instead, he proposed that the sun was the center. Of course this required a major change in the thinking of all humankind in matters of religion and philosophy as well as in astronomy. It did, however, make the theory of epicycles somewhat easier, because fewer epicycles were needed to explain the motion of the planets.

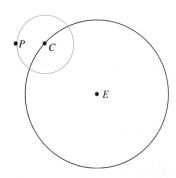

Figure 1. A planet moving on an epicycle.

Starting in 1609, and based on extensive and careful astronomical observations made by Tycho Brahe, Johannes Kepler postulated three laws of planetary motion:

1. Each planet moves in an ellipse with the sun at one focus.

2. The line between the sun and a planet sweeps out equal areas in equal times.

3. The squares of the periods of revolution of the planets are proportional to the cubes of the semimajor axis of their elliptic orbits.

Each of Kepler's laws made a major break with the past. The first abandoned circular motion and the need for epicycles. The second abandoned the uniformity of speed that had been part of the Ptolemaic theory, and replaced it by a beautiful mathematical expression of how fast a planet moved. The first two laws are illustrated in Figure 2. The planet P moves along an ellipse with the sun S at one focus. The two pie-shaped regions have equal area, so the planet will traverse the two arcs in equal times. The third law was equally dramatic, since it displays a commonality in the motion of all of the planets. Although a major accomplishment, Kepler's results remained descriptive. His three laws provided no causal explanation for the motion of the planets.

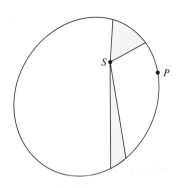

Figure 2. Kepler's second law.

A causal explanation was provided by Isaac Newton. However he did much more. He made three major advances.[3] First, he proved the fundamental theorem of calculus, and for that reason he is given credit for inventing the calculus. The fundamental theorem made possible the easy evaluation of integrals. As has been demonstrated, this made possible the solution of differential equations. Newton's second contribution was his formulation of the laws of mechanics. In particular, his second law, which says that force is equal to mass times acceleration, means that the study of motion can be reduced to a differential equation or to a system of differential equations. Finally, he discovered the universal law of gravity, which gave a mathematical description of the force of gravity. All of these results were published in 1687 in his *Philosophiae Naturalis Principia Mathematica* (*The Mathematical Principles of Natural Philosophy*), commonly referred to as the *Principia*.

Using his three discoveries, Newton was able to derive Kepler's laws of planetary motion. This means that for the first time there was a causal explanation of the motion of the planets. Newton's results were much broader in application, since they explained any kind of mechanical motion once the nature of the force was understood.

There were still difficulties with Newton's explanation. In particular, the force of gravity, as Newton described it, was a force acting at a distance. One body

---

[3] We have already discussed this briefly in Section 1.1 of Chapter 1.

acts on any other without any indication of a physical connection. Philosophers and physicists wondered how this was possible. In addition, by the end of the nineteenth century, some physical and mathematical anomalies had been observed. Although in most cases Newton's theory provided good answers, there were some situations in which the predictions of Newton's theory were not quite accurate.

These difficulties were apparently resolved in 1919, when Albert Einstein proposed his general theory of relativity. In this theory, gravity is explained as being the result of the curvature of four-dimensional space-time. This curvature in turn is caused by the masses of the bodies. The space-time itself provided the connection between the bodies and did away with problems of action at a distance. Finally, the general theory seems to have adequately explained most of the anomalies.

However, this is not the end of the story. Most physicists are convinced that all forces should be manifestations of one unified force. Early in the twentieth century they realized that there were four fundamental forces: gravity, the weak and strong nuclear forces, and electromagnetism. In the 1970s they were able to use quantum mechanics to unify the last three of these forces, but to date there is no generally accepted theory that unites gravity with the other three. There seems to be a fundamental conflict between general relativity and quantum mechanics.

A number of theories have been proposed to unify the two, but they remain unverified by experimental findings. Principal among these is *string theory*. The fundamental idea of string theory is that a particle is a tiny string that is moving in a 10-dimensional space-time. Four of these dimensions correspond to ordinary space-time. The extra six dimensions are assumed to have a tiny extent, on the order of 10^{-33} cm. This explains why these directions are not noticeable. It also gives a clue as to why string theory has no experimental verification. Nevertheless, as a theory it is very exciting. Hopefully someday it will be possible to devise an experimental test of the validity of string theory.

The modeling process

What we have described is a sequence of at least six different theories or mathematical models. The first were devised to explain the motion of the planets. Each was an improvement on the previous one, and starting with Newton they began to have more general application. With Newton's theory we have a model of all motion based on ordinary differential equations. His model was a complete departure from those that preceded it. It is his model that is used today, except when the relative velocities are so large that relativistic effects must be taken into account.

The continual improvement of the model in this case is what should take place wherever a mathematical model is used. As we learn more, we change the model to make it better. Furthermore, changes are always made on the basis of experimental findings that show faults in the existing model. The scientific theories of motion are probably the most mature of all scientific theories. Yet as our brief history shows, they are still being refined. This skepticism of the validity of existing theories is an important part of the scientific method. As good as our theories may seem, they can always be improved.

Linear motion

Let's look now at Newton's theory of motion. We will limit ourselves for the moment to motion in one dimension. Think in terms of a ball that is moving only up and down near the surface of the earth, as shown in Figure 3. Recall that we have already discussed this in Sections 1.1 and 1.3 of Chapter 1.

Figure 3. A ball near the surface of the earth.

To set the stage, we recall from Chapter 1 that the displacement x is the distance the ball is above the surface of the earth. Its derivative $v = x'$ is the velocity, and its second derivative $a = v' = x''$ is the acceleration. The mathematical model for motion is provided by Newton's second law. In our terms this is

$$F = ma, \tag{3.1}$$

where F is the force on the body and m is its mass. The gravitational force on a body moving near the surface of the earth is

$$F = -mg,$$

where g is the gravitational constant. It has value $g = 32 \text{ ft/s}^2 = 9.8 \text{ m/s}^2$. The minus sign is there because the direction of the force of gravity is always down, in the direction opposite to the positive x-direction. Thus, in this case, Newton's second law (3.1) becomes

$$m\frac{dv}{dt} = m\frac{d^2x}{dt^2} = -mg, \quad \text{or} \quad \frac{dv}{dt} = \frac{d^2x}{dt^2} = -g. \tag{3.2}$$

We solved equation (3.2) in Example 3.14 in Section 1.3, and the solution is

$$x(t) = -\frac{1}{2}gt^2 + v_0 t + x_0, \tag{3.3}$$

where v_0 is the original velocity and x_0 is the initial height.

Air resistance

In the derivation of our model in equation (3.2), we assumed that the only force acting was gravity. Now let's take into account the resistance of the air to the motion of the ball. If we think about how the resistance force acts, we come up with three simple facts. First, if there is no motion, then the velocity is zero, and there is no resistance. Second, the force always acts in the direction opposite to the motion. Thus if the ball is moving up, the resistance force is in the down direction, and if the ball is moving down, the force is in the up direction. From these considerations, we conclude that the resistance force has sign opposite to that of the velocity. We can put this mathematically by saying that the resistance force R has the form $R(x, v) = -r(x, v)v$, where r is a function that is always nonnegative. There are cases where r depends on x as well as v, such as when a ball is falling from a very high altitude so the density of the air has to be taken into account. However, in the cases we will consider r will depend only on v, so we will write

$$R(v) = -r(v)v. \tag{3.4}$$

Beyond these considerations, experiments have shown that the resistance force is somewhat complicated and there is no law that applies in all cases. Physicists use several models. We will look at two. In the first, resistance is proportional to the velocity, and in the second, the magnitude of the resistance is proportional to the square of the velocity. We will look at each of these cases in turn.

In the first case, r is a positive constant. Since forces add, our total force is the sum of the forces of gravity and air resistance,

$$F = -mg + R(v) = -mg - rv.$$

Using Newton's second law, we get

$$m\frac{dv}{dt} = -mg - rv, \quad \text{or} \quad \frac{dv}{dt} = -g - \frac{r}{m}v. \tag{3.5}$$

Notice that equation (3.5) is separable. Let's look for solutions. We separate variables to get

$$\frac{dv}{g + rv/m} = -dt.$$

When we integrate this and solve for v, we find the solution

$$v(t) = Ce^{-rt/m} - mg/r, \tag{3.6}$$

where C is a constant of integration.

We discover an interesting fact if we look at the limit of the velocity for large t. The exponential term in (3.6) decays to 0, so the velocity reaches the limit

$$\lim_{t\to\infty} v(t) = -\frac{mg}{r}.$$

Thus the velocity does not continue to increase in magnitude as the ball is falling. Instead it approaches the **terminal velocity**

$$v_{\text{term}} = -mg/r. \tag{3.7}$$

We still have to solve for the displacement and for this we use equation (3.6), which we rewrite as

$$\frac{dx}{dt} = v = Ce^{-rt/m} - mg/r.$$

This equation can be solved by integration to get

$$x = -\frac{mC}{r}e^{-rt/m} - \frac{mgt}{r} + A,$$

where A is another constant of integration.

Example 3.8 Suppose you drop a brick from the top of a building that is 250 m high. The brick has a mass of 2 kg, and the resistance force is given by $R = -4v$. How long will it take the brick to reach the ground? What will be its velocity at that time?

The equation for the velocity of the brick is given in (3.6). Since we are dropping the brick, the initial condition is $v(0) = 0$, and we can use (3.6) to find that

$$0 = v(0) = C - mg/r \quad \text{or} \quad C = mg/r = 2 \times 9.8/4 = 4.9.$$

Then

$$\frac{dx}{dt} = v(t) = 4.9\left(e^{-2t} - 1\right). \tag{3.9}$$

Integrating, we get

$$x(t) = 4.9\left(-\frac{1}{2}e^{-2t} - t\right) + A.$$

The initial condition $x(0) = 250$ enables us to compute A, since evaluating the previous equation at $t = 0$ gives

$$250 = -\frac{4.9}{2} + A \quad \text{or} \quad A = 252.45.$$

Thus the equation for the height of the brick becomes

$$x(t) = 4.9\left(-\frac{1}{2}e^{-2t} - t\right) + 252.45.$$

We want to find t such that $x(t) = 0$. This equation cannot be solved using algebra, but a hand-held calculator or a computer can find a very accurate approximate solution. In this way we obtain $t = 51.5204$ seconds.

For a time this large the exponential term in (3.9) is negligible, so the brick has reached its terminal velocity of $v_{\text{term}} = -4.9\text{m/s}$.   ●

Now let's turn to the second case, where the magnitude of the resistance force is proportional to the square of the velocity. Given the form of R in (3.4) together with the fact that $r \geq 0$, we see that the magnitude of R is

$$|R(v)| = r(v)|v| = kv^2$$

for some nonnegative constant k. Since $v^2 = |v|^2$, we conclude that $r = k|v|$, and the resistance force is $R(v) = -k|v|v$. In this case, Newton's second law becomes

$$m\frac{dv}{dt} = -mg - k|v|v, \quad \text{or} \quad \frac{dv}{dt} = -g - \frac{k}{m}|v|v. \tag{3.10}$$

Again, (3.10) is a separable equation. Let's look for solutions. Because of the absolute value, we have to consider separately the situation when the velocity is positive and the ball is moving upward and when the velocity is negative and the ball is descending. We will solve the equation for negative velocity and leave the other case to the exercises. When $v < 0$, $|v| = -v$, so (3.10) becomes

$$\frac{dv}{dt} = -g + \frac{k}{m}v^2. \tag{3.11}$$

Scaling variables to ease computation

We could solve (3.11) using separation of variables, but the constants cause things to get a little complicated. Instead, let's first introduce new variables by scaling the old ones. We introduce

$$v = \alpha w \quad \text{and} \quad t = \beta s,$$

where the constants α and β will be determined in a moment. Then

$$\frac{dv}{dt} = \frac{dv}{dw}\frac{dw}{ds}\frac{ds}{dt} = \frac{\alpha}{\beta}\frac{dw}{ds},$$

and equation (3.11) becomes

$$\frac{\alpha}{\beta}\frac{dw}{ds} = -g + \frac{k}{m}\alpha^2 w^2, \quad \text{or} \quad \frac{dw}{ds} = -\frac{g\beta}{\alpha} + \frac{k\alpha\beta}{m}w^2. \tag{3.12}$$

We choose α and β to make this equation as simple as possible. We require that

$$\frac{g\beta}{\alpha} = 1 \quad \text{and} \quad \frac{k\alpha\beta}{m} = 1.$$

Solving the first equation for α, we get $\alpha = g\beta$. Making this substitution in the second equation, it becomes $kg\beta^2/m = 1$. Solving for β and then for α, we get

$$\beta = \sqrt{\frac{m}{kg}} \quad \text{and} \quad \alpha = \sqrt{\frac{mg}{k}}.$$

As a reward for all of this, our differential equation in (3.12) simplifies to

$$\frac{dw}{ds} = -1 + w^2. \tag{3.13}$$

The separable equation (3.13) can be solved in the usual way. We first get

$$\frac{dw}{1 - w^2} = -ds.$$

Next we use partial fractions to write this as

$$\frac{1}{2}\left[\frac{dw}{1 + w} + \frac{dw}{1 - w}\right] = -ds.$$

This can be integrated to get

$$\frac{1}{2}\ln\left|\frac{1 + w}{1 - w}\right| = C - s,$$

where C is an arbitrary constant. When we exponentiate, we get

$$\left|\frac{1 + w}{1 - w}\right| = e^{2C - 2s} = Ae^{-2s}.$$

By allowing A to be negative or 0, we see that in general

$$\frac{1 + w}{1 - w} = Ae^{-2s}.$$

Solving for w, we find that

$$w(t) = \frac{Ae^{-2s} - 1}{Ae^{-2s} + 1}.$$

In terms of our original variables v and t, this becomes

$$v(t) = -\sqrt{\frac{mg}{k}}\frac{1 - Ae^{-2t\sqrt{kg/m}}}{1 + Ae^{-2t\sqrt{kg/m}}}. \tag{3.14}$$

We want to observe the limiting behavior of $v(t)$ as $t \to \infty$. The exponential terms in (3.14) decay to 0, so the velocity approaches the terminal velocity

$$v_{\text{term}} = -\sqrt{mg/k}.$$

This should be compared to equation (3.7), which gives the terminal velocity when the air resistance is proportional to the velocity instead of to its square.

Finding the displacement

Integrating equation (3.14) to find the displacement is a daunting task to say the least. For certain problems the task can be made easier by eliminating the variable t from the equation $a = dv/dt$. This is done using the chain rule to write

$$a = \frac{dv}{dt} = \frac{dv}{dx} \cdot \frac{dx}{dt} = \frac{dv}{dx} \cdot v. \tag{3.15}$$

Using this, equation (3.10) becomes

$$v\frac{dv}{dx} = -g - \frac{k}{m}|v|v. \tag{3.16}$$

Here is an example of how this can be useful.

Example 3.17 A ball of mass $m = 0.2$ kg is projected from the surface of the earth with velocity $v_0 = 50$ m/s. Assume that the force of air resistance is given by $R = -k\,|v|v$, where $k = 0.02$. What is the maximum height reached by the ball?

Since the ball is going up, the velocity is positive, so equation (3.16) becomes

$$v\frac{dv}{dx} = -g - \frac{k}{m}v^2 = -\frac{mg + kv^2}{m}.$$

When we separate variables, we get

$$\frac{v\,dv}{mg + kv^2} = -\frac{dx}{m}. \tag{3.18}$$

We will integrate this equation using the definite integral. To find what the end points of the integrations are, we notice first that at time $t = 0$ we have $x(0) = 0$, and $v(0) = v_0$. At a later time T, which is unknown and which need not be computed, the ball is at the top of its path, where $x(T) = x_{\max}$ and $v(T) = 0$. With these limits the integral of (3.18) is

$$\int_{v_0}^{0} \frac{v\,dv}{mg + kv^2} = -\int_{0}^{x_{\max}} \frac{dx}{m}.$$

Evaluating the integrals and solving for $x_{\max}$, we get

$$x_{\max} = \frac{m}{2k}\ln\left(1 + \frac{kv_0^2}{mg}\right).$$

With the data given we find that $x_{\max} = 16.4$ m.

EXERCISES

1. The acceleration due to gravity (near the earth's surface) is 9.8 m/s^2. If a rocketship in free space were able to maintain this constant acceleration indefinitely, how long would it take the ship to reach a speed equaling $(1/5)c$, where c is the speed of light? How far will the ship have traveled in this time? Ignore air resistance. *Note:* The speed of light is 3.0×10^8 m/s.

2. A balloon is ascending at a rate of 15 m/s at a height of 100 m above the ground when a package is dropped from the gondola. How long will it take the package to reach the ground? Ignore air resistance.

3. A stone is released from rest and dropped into a deep well. Eight seconds later, the sound of the stone splashing into the water at the bottom of the well returns to the ear of the person who released the stone. How long does it take the stone to drop to the bottom of the well? How deep is the

well? Ignore air resistance. *Note:* The speed of sound is 340 m/s.

4. A rocket is fired vertically and ascends with constant acceleration $a = 100$ m/s^2 for 1.0 min. At that point, the rocket motor shuts off and the rocket continues upward under the influence of gravity. Find the maximum altitude acquired by the rocket and the total time elapsed from the take-off until the rocket returns to the earth. Ignore air resistance.

5. A body is released from rest and travels the last half of the total distance fallen in precisely one second. How far did the body fall and how long did it take to fall the complete distance? Ignore air resistance.

6. A ball is projected vertically upward with initial velocity v_0 from ground level. Ignore air resistance.

   (a) What is the maximum height acquired by the ball?

   (b) How long does it take the ball to reach its maximum height? How long does it take the ball to return to the ground? Are these times identical?

   (c) What is the speed of the ball when it impacts with the ground on its return?

7. A particle moves along a line with x, v, and a representing position, velocity, and acceleration, respectively. Assuming that the acceleration a is constant, use equation (3.15) to show that
$$v^2 = v_0^2 + 2a(x - x_0),$$
where x_0 and v_0 are the position and velocity of the particle at time $t = 0$, respectively. A car's speed is reduced from 60 mi/h to 30 mi/h in a span covering 500 ft. Calculate the magnitude and direction of the constant deceleration.

8. Near the surface of the earth, a ball is released from rest and its flight through the air offers resistance that is proportional to its velocity. How long will it take the ball to reach one-half of its terminal velocity? How far will it travel during this time?

9. A ball having mass $m = 0.1$ kg falls from rest under the influence of gravity in a medium that provides a resistance that is proportional to its velocity. For a velocity of 0.2 m/s, the force due to the resistance of the medium is -1 N. [One Newton [N] is the force required to accelerate a 1kg mass at a rate of 1 m/s^2. Hence, 1 N $= 1$ kg m/s^2.] Find the terminal velocity of the ball.

10. An object having mass 70 kg falls from rest under the influence of gravity. The terminal velocity of the object is -20 m/s. Assume that the air resistance is proportional to the velocity.

    (a) Find the velocity and distance traveled at the end of 2 seconds.

    (b) How long does it take the object to reach 80% of its terminal velocity?

11. A ball is thrown vertically into the air with unknown velocity v_0 at time $t = 0$. Assume that the ball is thrown from about shoulder height, say $y_0 = 1.5$ m. If you ignore air resistance, then it is easy to show that $dv/dt = -g$, where $g = 9.8$ m/s^2 is the acceleration due to gravity. Follow the lead of exercise 7 to show that $v\,dv = -g\,dy$. Further, because the velocity of the ball is zero when it reaches its maximum height,

$$\int_{v_0}^{0} v\,dv = \int_{1.5}^{15} -g\,dy.$$

Find the initial velocity of the ball if the ball reaches a maximum height of 15 m.

Next, let's include air resistance. Suppose that $R(v) = -rv$ and show that the equation of motion becomes

$$v\,dv = \left(-g - \frac{r}{m}v\right)dy.$$

If the mass of the ball is 0.1 kg and $r = 0.02$ N/(m/s), find the initial velocity if the ball is again released from shoulder height ($y_0 = 1.5$ m) and reaches a maximum height of 15 m.

12. A mass of 0.2 kg is released from rest. As the object falls, air provides a resistance proportional to the velocity ($R(v) = -0.1v$), where the velocity is measured in m/s. If the mass is dropped from a height of 50 m, what is its velocity when it hits the ground? *Hint:* You may find equation (3.15) useful. Find v when $y = 0$.

13. An object having mass $m = 0.1$ kg is launched from ground level with an initial vertical velocity of 230 m/s. The air offers resistance proportional to the square of the object's velocity ($R(v) = -0.05v|v|$), where the velocity is measured in m/s. Find the maximum height acquired by the object.

14. One of the great discoveries in science is Newton's universal law of gravitation, which states that the magnitude of the gravitational force exerted by one point mass on another is proportional to their masses and inversely proportional to the square of the distance between them. In symbols,

$$|F| = \frac{GMm}{r^2},$$

where G is a universal gravitational constant. This constant, first measured by Lord Cavendish in 1798, has a currently accepted value approximately equal to 6.6726×10^{-11} Nm^2/kg^2. Newton also showed that the law was valid for two spherical masses. In this case, you assume that the mass is concentrated at a point at the center of each sphere.

Suppose that an object with mass m is launched from the earth's surface with initial velocity v_0. Let y represent its position above the earth's surface, as shown in Figure 4.

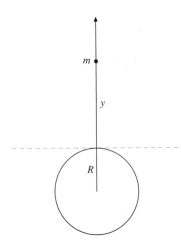

Figure 4. The object in Exercise 14.

(a) If air resistance is ignored, use the idea in equation (3.15) to help show that

$$v \frac{dv}{dy} = -\frac{GM}{(R+y)^2}. \qquad (3.19)$$

(b) Assuming that $y(0) = 0$ (the object is launched from earth's surface) and $v(0) = v_0$, solve equation (3.19) to show that

$$v^2 = v_0^2 - 2GM \left(\frac{1}{R} - \frac{1}{R+y} \right).$$

(c) Show that the maximum height reached by the object is given by

$$y = \frac{v_0^2 R}{2GM/R - v_0^2}.$$

(d) Show that the initial velocity

$$v_0 = \sqrt{\frac{2GM}{R}}$$

is the minimum required for the object to "escape" earth's gravitational field. *Hint:* If an object "escapes" earth's gravitational field, then the maximum height acquired by the object is potentially infinite.

15. Inside the earth, the surrounding mass exerts a gravitational pull in all directions. Of course, there is more mass towards the center of the earth than any other direction. The magnitude of this force is proportional to the distance from the center. (Can you prove this?) Suppose a hole is drilled to the center of the earth and a mass is dropped in the hole. Ignoring air resistance, with what velocity will the mass strike the center of the earth? As a hint, let x represent the distance of the mass from the center of the earth and note that equation (3.15) implies that the acceleration is $a = v(dv/dx)$.

16. An object with mass m is released from rest at a distance of a meters above the earth's surface (see Figure 5). Use Newton's universal law of gravitation (see Exercise 14) to show that the object impacts the earth surface with a velocity determined by

$$v = \sqrt{\frac{2agR}{a+R}},$$

where g is the acceleration due to gravity at the earth's surface and R is the radius of the earth. Ignore any affects due to the earth's rotation and atmosphere. *Hint:* On the earth's surface, explain why $mg = GMm/R^2$, where M is the mass of the earth and G is the universal gravitational constant.

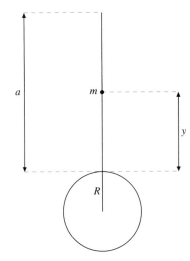

Figure 5. The object in Exercise 16.

17. A 2-foot length of a 10-foot chain hangs off the end of a high table. Neglecting friction, find the time required for the chain to slide off the table. *Hint*: Model this problem with a second-order differential equation and then solve it using the following reduction of order technique: If x is the length of the chain hanging off the table, then by equation (3.15) the acceleration is $a = v(dv/dx)$.

18. A parachutist of mass 60 kg free-falls from an airplane at an altitude of 5000 meters. He is subjected to an air resistance force that is proportional to his speed . Assume the constant of proportionality is 10 (kg/sec). Find and solve the differential equation governing the altitude of the parachuter at time t seconds after the start of his free-fall. Assuming he does not deploy his parachute, find his limiting velocity and how much time will elapse before he hits the ground.

19. In our models of air resistance the resistance force has depended only on the velocity. However, for an object that drops a considerable distance, such as the parachutist in the previous exercise, there is a dependence on the altitude as well. It is reasonable to assume that the resistance force is proportional to air pressure, as well as to the velocity. Furthermore, to a first approximation the air pressure varies exponentially with the altitude (i.e., it is proportional to e^{-ax}, where a is a constant and x is the altitude). Present a model using Newton's second law for the motion of an object in the earth's atmosphere subject to such a resistance force.

2.4 Linear Equations

A first-order *linear* equation is one of the form

$$x' = a(t)x + f(t). \tag{4.1}$$

If $f(t) = 0$, the equation has the form

$$x' = a(t)x, \tag{4.2}$$

and the linear equation is said to be **homogeneous**. Otherwise it is inhomogeneous.

The functions $a(t)$ and $f(t)$ in (4.1) are called the **coefficients** of the equation. We will sometimes consider equations of the more general form

$$b(t)x' = c(t)x + g(t). \tag{4.3}$$

These are still linear equations, and they can be put into the form (4.1) by dividing by $b(t)$—provided $b(t)$ is not zero. The important point about linear equations is that the unknown function x and its derivative x' both appear alone and only to first order. This means that we do not allow x^2, $(x')^3$, xx', e^x, $\cos(x')$, or anything more complicated than just x and x' to appear in the equation. Thus the equations

$$x' = \sin(t)x,$$
$$y' = e^{2t}y + \cos t, \quad \text{and}$$
$$x' = (3t + 2)x + t^2 - 1$$

are all linear, while

$$x' = t\sin(x),$$
$$y' = yy', \quad \text{and}$$
$$y' = 1 - y^2$$

are all nonlinear.

Solution of the homogeneous equation

Linear equations can be solved exactly, and we will show how in this section. We start with the homogeneous equation (4.2). You will notice that this is a separable equation. Separating variables and then integrating gives us

$$\frac{dx}{x} = a(t)\,dt \quad \implies \quad \ln|x| = \int a(t)\,dt + C.$$

Exponentiating, we get

$$|x| = e^{\int a(t)\,dt + C} = e^C e^{\int a(t)\,dt}.$$

The constant e^C is positive. We will replace it with the constant A and we will allow it to be positive, negative, or zero, so that we can get rid of the absolute value. Hence the general solution is

$$x(t) = A e^{\int a(t)\,dt}. \tag{4.4}$$

Example 4.5   Solve

$$x' = \sin(t)x.$$

Using the method for separable equations,

$$\frac{dx}{x} = \sin(t)\,dt$$

$$\ln|x| = -\cos(t) + C$$

$$|x(t)| = e^{-\cos t + C} = e^{C}e^{-\cos t}$$

or

$$x(t) = Ae^{-\cos t}.$$ ●

Solution of the inhomogeneous equation

We will illustrate the solution method with an example.

Example 4.6   Newton's law of cooling states that the rate at which an object losses or gains heat is proportional to the difference between the temperature of the object (T) and the temperature of the surrounding medium (A). Mathematically, this translates to

$$T' = -k(T - A),$$

where k is the proportionality constant. Solve this equation, assuming that the ambient temperature A is constant.

If we rewrite the equation as

$$T' + kT = kA, \tag{4.7}$$

then the left-hand side reminds us of the formula for the derivative of a product. In fact, if we multiply the left-hand side of equation (4.7) by e^{kt}, it becomes

$$e^{kt}(T' + kT) = e^{kt}T' + ke^{kt}T = [e^{kt}T]', \tag{4.8}$$

the derivative of a product. Using this, equation (4.7) becomes

$$[e^{kt}T]' = kAe^{kt}. \tag{4.9}$$

We can now integrate both sides of this equation to get

$$e^{kt}T(t) = Ae^{kt} + C, \quad \text{or} \quad T(t) = A + Ce^{-kt}. \tag{4.10}$$

This is the general solution to our linear equation.   ●

That worked pretty well. Can we always do this? Let's start with the general linear equation in (4.1) and go through the same steps. First we rewrite it as

$$x' - ax = f, \tag{4.11}$$

in analogy to (4.7). Next, in analogy to (4.8) and (4.9), we want to multiply equation (4.11) by a function $u(t)$, like e^{kt} in the previous example, which will turn the left-hand side into the derivative of a product. Thus we want

$$(ux)' = u(x' - ax). \tag{4.12}$$

We will call such a function an **integrating factor**.

Assume for the moment that we have found an integrating factor u. Multiplying (4.11) by u, and using (4.12), we get

$$(ux)' = u(x' - ax) = uf.$$

As we did for equation (4.9) in Example 4.6, we can integrate this directly to get

$$u(t)x(t) = \int u(t)f(t)\,dt + C,$$

or

$$x(t) = \frac{1}{u(t)} \int u(t)f(t)\,dt + \frac{C}{u(t)}, \tag{4.13}$$

which is the general solution to (4.1).

Thus, the key to the method is finding an integrating factor, a function u that satisfies equation (4.12). If we expand both sides of (4.12), this becomes

$$ux' + u'x = ux' - aux.$$

Subtracting ux' from each side, this becomes $u'x = -aux$, which will be satisfied if

$$u' = -au. \tag{4.14}$$

This is a linear homogeneous equation, and, as we saw earlier in (4.4), a particular solution is given by

$$u(t) = e^{-\int a(t)\,dt}. \tag{4.15}$$

(Notice that we do not need the constant A that appears in (4.4) because we only need one particular solution. Any solution to (4.14) will do for the present purpose.)

Summary of the method

We have found a general method of solving arbitrary linear equations.

The equation

$$x' = ax + f. \tag{4.16}$$

can be solved using the following four steps.

1. Rewrite the equation as

$$x' - ax = f.$$

2. Multiply by the integrating factor

$$u(t) = e^{-\int a(t)\,dt},$$

so that the equation becomes

$$(ux)' = u(x' - ax) = uf. \tag{4.17}$$

3. Integrate this equation to obtain

$$u(t)x(t) = \int u(t)f(t)\,dt + C.$$

4. Solve for $x(t)$.

After you have found the integrating factor u in step 2, it is always a good idea to check that equation (4.17) is satisfied.

Let's look at some examples.

Example 4.18   Find the general solution to the equation

$$x' = x + e^{-t}.$$

Let's go about this very carefully. The first thing to do is to bring the term involving x to the left-hand side,

$$x' - x = e^{-t}. \tag{4.19}$$

Next, since $a(t) = 1$, the integrating factor is

$$u(t) = e^{-\int 1\, dt} = e^{-t}.$$

Multiply equation (4.19) by the integrating factor, getting

$$e^{-t}(x' - x) = e^{-2t}. \tag{4.20}$$

Verify that the left-hand side of (4.20) is the derivative of the product $u(t)x(t) = e^{-t}x(t)$, or

$$[e^{-t}x(t)]' = e^{-t}(x' - x) = e^{-2t}. \tag{4.21}$$

We can now integrate both sides of (4.21),

$$e^{-t}x(t) = \int e^{-2t}\, dt$$

$$= -\frac{1}{2}e^{-2t} + C.$$

Finally, we solve for x by multiplying both sides by e^t, getting

$$x(t) = -\frac{1}{2}e^{-t} + Ce^t. \tag{4.22}$$

●

Example 4.23   Find the general solution of

$$x' = x \sin t + 2te^{-\cos t}$$

and the particular solution that satisfies $x(0) = 1$.

This equation is more clearly in the linear form of (4.16) if we rewrite it as $x' = (\sin t)x + 2te^{-\cos t}$. Again we start to find the solution by rewriting the equation as

$$x' - x \sin t = 2te^{-\cos t}.$$

This time $a(t) = \sin t$, so the integrating factor is

$$u(t) = e^{-\int \sin t\, dt} = e^{\cos t}.$$

Multiplying by u, we get

$$[e^{\cos t}x(t)]' = e^{\cos t}(x' - x \sin t) = 2t.$$

After integrating both sides we have

$$x(t)e^{\cos t} = 2 \int t \, dt = t^2 + C.$$

Therefore, the general solution is

$$x(t) = (t^2 + C)e^{-\cos t}. \tag{4.24}$$

The particular solution we want satisfies $x(0) = 1$, so

$$1 = Ce^{-1} \quad \text{or} \quad C = e.$$

Thus the solution to the initial value problem is

$$x(t) = (t^2 + e)e^{-\cos t}. \qquad \bullet$$

Example 4.25 Find the general solution to

$$x' = x \tan t + \sin t,$$

and find the particular solution that satisfies $x(0) = 2$.

Rewrite the equation as

$$x' - x \tan t = \sin t.$$

Then $a(t) = \tan t$, so an integrating factor is

$$u(t) = e^{-\int \tan t \, dt} = e^{\ln(\cos t)} = \cos t.$$

Multiplying by the integrating factor, we get

$$[x \cos t]' = \cos t \left(x' - x \tan t\right) = \cos t \sin t,$$

so

$$x(t) \cos t = \int \cos t \sin t \, dt = -\frac{\cos^2 t}{2} + C.$$

Finally, we divide by $\cos t$ to get

$$x(t) = -\frac{\cos t}{2} + \frac{C}{\cos t}. \tag{4.26}$$

This is the general solution. To find the particular solution with $x(0) = 2$, we substitute this into the formula for the general solution and compute that $C = 5/2$. Thus our particular solution is

$$x(t) = -\frac{\cos t}{2} + \frac{5}{2\cos t}. \qquad \bullet$$

An alternate solution method

There is another method of solving linear equations that you might find easier to remember and use. Let's begin with an example.

Example 4.27   Find the general solution of

$$y' = -2y + 3. \tag{4.28}$$

First, the solution to the associated homogeneous equation, $y'_h = -2y_h$ is $y_h = Ce^{-2t}$. Replace the constant in the homogeneous solution with $v = v(t)$, a yet to be determined function of t, so

$$y(t) = v(t)e^{-2t}. \tag{4.29}$$

Then we substitute this expression for y into the inhomogeneous equation (4.28) and solve for v.

$$(ve^{-2t})' = -2(ve^{-2t}) + 3$$
$$-2ve^{-2t} + v'e^{-2t} = -2ve^{-2t} + 3$$
$$v' = 3e^{2t} \tag{4.30}$$
$$v = \frac{3}{2}e^{2t} + C$$

Finally, substitute this last result into equation (4.29) to obtain the general solution of equation (4.28).

$$y = \left(\frac{3}{2}e^{2t} + C\right)e^{-2t} = \frac{3}{2} + Ce^{-2t} \qquad \bullet$$

Notice that the derivation in (4.30) left us with a formula for v', which we only needed to integrate to find v. It is fair to ask if this always happens. Let's look at the general case.

We want to solve the linear equation

$$y' = a(t)y + f(t). \tag{4.31}$$

We start by finding a particular solution to the associated homogeneous equation

$$y'_h = a(t)y_h. \tag{4.32}$$

According to (4.4), a solution is

$$y_h(t) = e^{\int a(t)\,dt}. \tag{4.33}$$

Notice that because of its exponential form the function $y_h(t)$ is never equal to zero. Hence we can safely divide by it. If $y(t)$ is any solution to (4.31), we can define

$$v(t) = \frac{y(t)}{y_h(t)}, \quad \text{so that} \quad y(t) = v(t)y_h(t).$$

This is the key idea. We write an arbitrary solution to (4.31) in the form $y(t) = v(t)y_h(t)$. The function v is as yet unknown. It is what is sometimes called a variable parameter, and this method is called **variation of parameters**. To solve for v we substitute $y = v(t)y_h(t)$ into the differential equation (4.31). We get

$$(vy_h)' = a(vy_h) + f \quad \text{or}$$
$$vy'_h + v'y_h = avy_h + f.$$

Remember that y_h is a solution of the homogeneous equation (4.32). Hence, $y_h' = ay_h$, and proceeding, we get

$$avy_h + v'y_h = avy_h + f,$$
$$v'y_h = f,$$
$$v' = \frac{f}{y_h}.$$

(4.34)

From this, we can compute v by integration. The solution is the product $y = v(t)y_h(t)$.

> Here is a list of the steps in the method of variation of parameters for solving
>
> $$y' = ay + f. \tag{4.35}$$
>
> **1.** A particular solution to the associated homogeneous equation $y_h' = ay_h$ is
>
> $$y_h(t) = e^{\int a(t)\,dt}$$
>
> **2.** Substitute $y = vy_h$ into equation (4.35) to find v, or remember that
>
> $$v' = \frac{f}{y_h} \tag{4.36}$$
>
> **3.** Write down the general solution $y(t) = v(t)y_h(t)$.

In step 2 you have a choice. Many people find it easier to substitute $y = vy_h$ into the equation than to remember the exact formula for v.

Example 4.37 Use variation of parameters to find the general solution of

$$x' = x \tan t + \sin t, \tag{4.38}$$

which we solved in Example 4.25.

The associated homogeneous equation is $x_h' = x_h \tan t$, which has solution

$$x_h(t) = 1/\cos t.$$

Next we look for a solution of the form $x = vx_h = v/\cos t$. Substituting x into equation (4.38) and using the homogeneous equation, we get

$$\left[\frac{v}{\cos t}\right]' = \frac{v}{\cos t}\frac{\sin t}{\cos t} + \sin t$$
$$\frac{v'\cos t + v \sin t}{\cos^2 t} = \frac{v}{\cos t}\frac{\sin t}{\cos t} + \sin t$$
$$\frac{v'}{\cos t} = \sin t$$
$$v' = \sin t \cos t.$$

The last equation can be integrated, finding that

$$v(t) = \int \sin t \cos t \, dt = -\frac{\cos^2 t}{2} + C.$$

Finally, our solution is

$$x(t) = v(t)x_h(t) = \left(-\frac{\cos^2 t}{2} + C\right) \Big/ \cos t = -\frac{\cos t}{2} + \frac{C}{\cos t},$$

which agrees with our previous answer.

Structure of the solution

The method of variation of parameters shows that an arbitrary solution to the inhomogeneous linear equation

$$y' = ay + f$$

has the form

$$y(t) = v(t)y_h(t),$$

where

$$y_h(t) = e^{\int a(t)\,dt}$$

is a particular solution to the associated homogeneous equation and where, according to (4.34),

$$v'(t) = f(t)/y_h(t) = f(t)e^{-\int a(t)\,dt}.$$

Performing the integration, we see that

$$v(t) = \int f(t)e^{-\int a(t)\,dt}\,dt + C.$$

We have added the constant C to this formula to emphasize the presence of a constant of integration.

Hence, we can write an arbitrary solution as

$$
\begin{aligned}
y(t) &= v(t)y_h(t) \\
&= y_h(t)\int f(t)e^{-\int a(t)\,dt}\,dt + Cy_h(t).
\end{aligned}
\tag{4.39}
$$

Notice how the constant of integration C appears in this formula. It is the coefficient of the solution y_h to the associated inhomogeneous equation.

If we pick a particular solution $y_p(t)$, it will be associated with a particular value of the constant C, say C_p, so that

$$y_p(t) = y_h(t)\int f(t)e^{-\int a(t)\,dt}\,dt + C_p y_h(t). \tag{4.40}$$

Comparing (4.39) and (4.40), we see that the difference of the two solutions to the inhomogeneous equation is

$$y(t) - y_p(t) = (C - C_p)y_h(t).$$

Thus, the difference $y - y_p$ is a constant multiple of y_h and is itself a solution to the homogeneous equation. Furthermore, if we set $A = C - C_p$, we see that an arbitrary solution y can be written as

$$y(t) = y_p(t) + Ay_h(t).$$

Thus, we have demonstrated the following result, showing how the constant of integration appears in the general solution to a linear equation.

THEOREM 4.41   Suppose that y_p is a particular solution to the inhomogeneous equation

$$y' = a(t)y + f(t),$$

and that y_h is a particular solution to the associated homogeneous equation. Then every solution to the inhomogeneous equation is of the form

$$y(t) = y_p(t) + Ay_h(t), \tag{4.42}$$

where A is an arbitrary constant.   ∎

EXERCISES

In Exercises 1–12, find the general solution of the first-order, linear equation.

1. $y' + y = 2$

2. $y' - 3y = 5$

3. $y' + (2/x)y = (\cos x)/x^2$

4. $y' + 2ty = 5t$

5. $x' - 2x/(t + 1) = (t + 1)^2$

6. $tx' = 4x + t^4$

7. $(1 + x)y' + y = \cos x$

8. $(1 + x^3)y' = 3x^2 y + x^2 + x^5$

9. $L(di/dt) + Ri = E$, L, R, E real constants

10. $y' = my + c_1 e^{mx}$, m, c_1 real constants

11. $y' = \cos x - y \sec x$

12. $x' - (n/t)x = e^t t^n$, n an positive integer

13. (a) The differential equation $y' + y \cos x = \cos x$ is linear. Use the integrating factor technique of this section to find the general solution.

(b) The equation $y' + y \cos x = \cos x$ is also separable. Use the separation of variables technique to solve the equation and discuss any discrepancies (if any) between this solution and the solution found in part (a).

In Exercises 14–17, find the solution of the initial value problem.

14. $y' = y + 2xe^{2x}$, $y(0) = 3$

15. $(x^2 + 1)y' + 3xy = 6x$, $y(0) = -1$

16. $(1 + t^2)y' + 4ty = (1 + t^2)^{-2}$, $y(1) = 0$

17. $x' + x \cos t = \frac{1}{2} \sin 2t$, $x(0) = 1$

In Exercises 18–21, find the solution of the initial value problem. Discuss the interval of existence and provide a sketch of your solution.

18. $xy' + 2y = \sin x$, $y(\pi/2) = 0$

19. $(2x + 3)y' = y + (2x + 3)^{1/2}$, $y(-1) = 0$

20. $y' = \cos x - y \sec x$, $y(0) = 1$

21. $(1 + t)x' + x = \cos t$, $x(-\pi/2) = 0$

22. The presence of nonlinear terms prevents us from using the technique of this section. In special cases, a change of variable will transform the nonlinear equation into one that is linear. The equation known as **Bernoulli's equation**,

$$x' = a(t)x + f(t)x^n, \quad n \neq 0, 1,$$

was proposed for solution by James Bernoulli in December 1695. In 1696, Leibniz pointed out that the equation can be reduced to a linear equation by taking x^{1-n} as the dependent variable. Show that the change of variable, $z = x^{1-n}$, will transform the nonlinear Bernoulli equation into the linear equation

$$z' = (1 - n)a(t)z + (1 - n)f(t).$$

Hint: If $z = x^{1-n}$, then $dz/dt = (dz/dx)(dx/dt) = (1 - n)x^{-n}(dx/dt)$.

In Exercises 23–26, use the technique of Exercise 22 to transform the Bernoulli equation into a linear equation. Find the general solution of the resulting linear equation.

23. $y' + x^{-1}y = xy^2$

24. $y' + y = y^2$

25. $xy' + y = x^4 y^3$

26. $P' = aP - bP^2$

27. The equation

$$\frac{dy}{dt} + \psi y^2 + \phi y + \chi = 0,$$

where ψ, ϕ, and χ are functions of t, is called the **generalized Riccati equation**. In general, the equation is not integrable by quadratures. However, suppose that one solution, say $y = y_1$, is known.

(a) Show that the substitution $y = y_1 + z$ reduces the generalized Riccati equation to

$$\frac{dz}{dt} + (2y_1\psi + \phi)z + \psi z^2 = 0,$$

which is an instance of Bernoulli's equation (see Exercise 22).

(b) Use the fact that $y_1 = 1/t$ is a particular solution of

$$\frac{dy}{dt} = -\frac{1}{t^2} - \frac{y}{t} + y^2$$

to find the equation's general solution.

28. Suppose that you have a closed system containing 1000 individuals. A flu epidemic starts. Let $N(t)$ represent the number of infected individuals in the closed system at time t. Assume that the rate at which the number of infected individuals is changing is jointly proportional to the number of infected individuals and to the number of uninfected individuals. Furthermore, suppose that when 100 individuals are infected, the rate at which individuals are becoming infected is 90 individuals per day. If 20 individuals are infected at time $t = 0$, when will 90% of the population be infected? *Hint:* The assumption here is that there are only healthy individuals and sick individuals. Furthermore, the resulting model can be solved using the technique introduced in Exercise 22.

29. In Exercise 33 of Section 2.2, the time of death of a murder victim is determined using Newton's law of cooling. In particular it was discovered that the proportionality constant in Newton's law was $k = \ln(5/4) \approx 0.223$. Suppose we discover another murder victim at midnight with a body temperature of $31°C$. However, this time the air temperature at midnight is $0°C$, and is falling at a constant rate of $1°C$ per hour. At what time did the victim die? (Remember that the normal body temperature is $37°C$.)

In Exercises 30–35, use the variation of parameters technique to find the general solution of the given differential equation.

30. $y' = -3y + 4$

31. $y' + 2y = 5$

32. $y' + (2/x)y = 8x$

33. $ty' + y = 4t^2$

34. $x' + 2x = t$

35. $y' + 2xy = 4x$

In Exercises 36–41, use the variation of parameters technique to find the general solution of the given differential equation. Then find the particular solution satisfying the given initial condition.

36. $y' - 3y = 4$, $\quad y(0) = 2$

37. $y' + (1/2)y = t$, $\quad y(0) = 1$

38. $y' + y = e^t$, $\quad y(0) = 1$

39. $y' + 2xy = 2x^3$, $\quad y(0) = -1$

40. $x' - (2/t^2)x = 1/t^2$, $\quad x(1) = 0$

41. $(t^2 + 1)x' + 4tx = t$, $\quad x(0) = 1$

42. Consider anew Newton's law of cooling, where the temperature of a body is modeled by the equation

$$T' = -k(T - A), \qquad (4.43)$$

where T is the temperature of the body and A is the temperature of the surrounding medium (*ambient temperature*). Although this equation is linear and its solution can be found by using an integrating factor, Theorem 4.41 provides a far simpler approach.

(a) Find a solution T_h of the homogenous equation $T' + kT = 0$.

(b) Find a particular solution T_p of the inhomogeneous equation (4.43). *Note:* This equation is autonomous. See Section 2.9.

(c) Form the general solution $T = T_h + T_p$.

(d) Add a source of constant heat (like a heater in a room) to the model (4.43), as in

$$T' = -k(T - A) + H. \qquad (4.44)$$

Use the technique outlined in parts (a)–(c) to find the general solution of equation (4.44).

43. Suppose that the ambient temperature of Exercise 42 varies sinusoidally with time, as in

$$T' = -k(T - A \sin \omega t). \qquad (4.45)$$

(a) Find a solution T_h of the homogeneous equation $T' + kT = 0$.

(b) The equation (4.45) is not autonomous, so finding a particular solution T_p is a bit more difficult. However, it doesn't hurt to guess.[4] As a first guess, substitute $T_p = C \cos \omega t + D \sin \omega t$ into the equation $T' + kT = kA \sin \omega t$ and show that

$$-\omega C + kD = kA \qquad \text{and} \qquad kC + \omega D = 0.$$

(c) Solve the simultaneous equations in part (b) and use Theorem 4.41 to show that the general solution of equation (4.45) is

$$T = T_h + T_p$$

$$= Fe^{-kt} + \frac{kA}{k^2 + \omega^2}[k \sin \omega t - \omega \cos \omega t],$$

where F is an arbitrary constant.

44. Suppose that the temperature T inside a mountain cabin behaves according to Newton's law of cooling, as in

$$\frac{dT}{dt} = -\frac{1}{2}(T - A), \qquad (4.46)$$

where t is measured in hours and the ambient temperature A outside the cabin varies sinusoidally with a period of 24 hours. At 6 A.M., the ambient temperature outside is at a minimum of 40°F. At 6 P.M., the ambient temperature is at a maximum of 80°F.

(a) Adjust equation (4.46) to model the sinusoidal nature of the ambient temperature.

(b) Suppose that at midnight the temperature inside the cabin is 50°F. Solve the resulting initial value problem. *Hint:* You can simplify the calculation by letting $t = 0$ represent midnight. You might also consider technique suggested in Exericse 43.

(c) Sketch the graph of the temperature inside the cabin. On the same coordinate system, superimpose the plot of the ambient temperature outside the cabin. Comment on the appearance of the plot.

2.5 Mixing Problems

Consider a lake with a factory on its shore that introduces a pollutant into the lake. The lake is fed by one river and drained by another, keeping the volume of the lake constant. Using the methods we discuss in this section, we can model how the amount of pollutant in the lake varies with time. We can then make intelligent decisions about the danger involved in this situation.

The problems we will discuss are called *mixing problems*. They employ tanks, beakers, and other receptacles that hold solutions, mixtures usually containing water and an additional element such as salt. While these examples might appear to be

---

[4] The technique of guessing presented here is commonly called the *method of undetermined coefficients*, a technique we will visit in some detail in Section 4.5.

inane, they should not be underestimated. They take on an urgency when the tanks and beakers are replaced with the heart, stomach, or gastrointestinal systems, or indeed by the lake mentioned earlier. We will return to the lake in the exercises.

We will illustrate the principles involved in a series of three examples.

Example 5.1   The tank in Figure 1 initially holds 100 gal of pure water. At time $t = 0$, a solution containing 2 lb of salt per gallon begins to enter the tank at the rate of 3 gallons per minute. At the same time a drain is opened at the bottom of the tank so that the volume of solution in the tank remains constant. How much salt is in the tank after 60 min?

Let us begin by letting $x(t)$ represent the number of pounds of salt in the tank after t min. Consequently, the derivative dx/dt is the mathematical representation of the rate at which the amount of salt is changing with respect to time. The modeling process consists of computing a physical representation of the rate of change and setting them equal. The end result is a differential equation for the function $x(t)$.

It is very helpful in the modeling process to keep track of the units used. For example, since we are measuring x in pounds and time in minutes, the derivative dx/dt is measured in pounds per minute (lb/min). In the statement of the problem we are given rates of flow into and out of the tank. These are ***volume rates***, measured in gallons per minute (gal/min). In addition we are given the ***concentration*** of the solution entering the tank, and this is measured in pounds per gallon (lb/gal).

The physical representation of the rate of change splits naturally into two parts since there is a flow into the tank and another out. Thus we can write

rate of change = rate in − rate out.

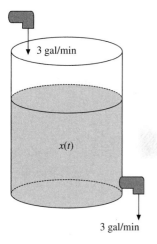

Figure 1.   The tank in Example 5.1.

3 gal/min

$x(t)$

3 gal/min

This useful separation of effects is sometimes referred to as a ***balance law***. Since the rate of change is measured in lb/min, the same must be true for the rate in and the rate out.

Let's examine the rate in. Solution enters the tank at a volume rate of 3 gal/min. The concentration of salt in this solution is 2 lb/gal. Consequently,

$$\text{rate in} = \text{volume rate} \times \text{concentration}$$
$$= 3 \text{ gal/min} \times 2 \text{ lb/gal}$$
$$= 6 \text{ lb/min}.$$

The rate at which salt leaves the tank is a little trickier. We still have

$$\text{rate out} = \text{volume rate} \times \text{concentration}.$$

Since the volume is kept constant, we know that the solution leaves through the drain at the bottom of the tank with a volume rate of 3 gal/min, but what is the concentration of salt in the water leaving the tank? We will assume that the mixture in the tank is "instantaneously mixed" at all times so that the concentration of the salt does not change from point to point. Granted, this is probably not an accurate assumption, but it is a good starting point and enables us to complete the model using ordinary differential equations. Without this assumption we can still model the system, but the concentration would depend on the position within the tank, and the model would be a partial differential equation. With our assumption of perfect mixing the concentration of salt in the tank at any time t is calculated by dividing the amount of salt in the tank at time t by the volume of solution in the tank. The concentration at time t, $c(t)$, is given by

$$c(t) = \frac{x(t)}{100} \text{ lb/gal.}$$

We can now determine the rate at which salt is leaving the tank.

$$\text{rate out} = 3 \text{ gal/min} \times \frac{x(t)}{100} \text{ lb/gal} = \frac{3x(t)}{100} \text{ lb/min}$$

Our discussion has led us to the differential equation

$$\frac{dx}{dt} = \text{rate of change}$$

$$= \text{rate in} - \text{rate out}$$

$$= 6 - \frac{3x}{100}.$$

This equation has the form $dx/dt = a(t)x + f(t)$, so it is linear, and we can use the technique of Section 2.4 to find its solution. First, we rewrite the equation as

$$\frac{dx}{dt} + \frac{3x}{100} = 6. \tag{5.2}$$

Next we compute the integrating factor,

$$u(t) = e^{-\int(-3/100)\,dt} = e^{3t/100}.$$

When we multiply both sides of equation (5.2) by u, we notice that

$$\left[e^{3t/100}x\right]' = e^{3t/100}\left(\frac{dx}{dt} + \frac{3x}{100}\right) = 6e^{3t/100}.$$

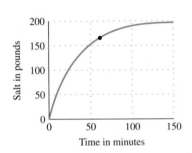

We integrate both sides of this equation to get

$$e^{3t/100}x = \int 6e^{3t/100}\,dt = \frac{600}{3}e^{3t/100} + C.$$

Finally, we get the general solution by solving for x:

$$x(t) = 200 + Ce^{-3t/100}.$$

Figure 2. The amount of salt in the tank in Example 5.1.

Since there was no salt present in the tank initially, the initial condition is $x(0) = 0$. Hence

$$0 = x(0) = 200 + Ce^{-3(0)/100} = 200 + C.$$

Consequently, $C = -200$ and our solution is

$$x(t) = 200 - 200e^{-3t/100}.$$

The solution is plotted in Figure 2. The amount of salt present in the tank after 60 minutes is

$$x(60) = 200 - 200e^{-3(60)/100} \approx 167 \text{ lb}.$$

Example 5.3

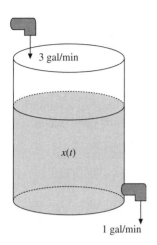

3 gal/min

$x(t)$

1 gal/min

Figure 3. The tank in Example 5.3.

The 600-gal tank in Figure 3 is filled with 300 gal of pure water. A spigot is opened above the tank and a salt solution containing 1.5 lb of salt per gallon of solution begins flowing into the tank at a rate of 3 gal/min. Simultaneously, a drain is opened at the bottom of the tank allowing the solution to leave the tank at a rate of 1 gal/min. What will be the salt content in the tank at the precise moment that the volume of solution in the tank is equal to the tank's capacity (600 gal)?

This problem differs from Example 5.1 in that the volume of solution in the tank is not constant. Because the solution enters the tank at a rate of 3 gal/min and leaves the tank at a rate of 1 gal/min, the volume increases at a net rate of 2 gal/min. Since the initial amount of solution in the tank is 300 gal, the volume of solution in the tank, at any time t, is given by $V(t) = 300 + 2t$.

With this change, the solution of this example now parallels that of Example 5.1. The rate at which the salt enters the tank is given by

$$\text{rate in} = 3 \text{ gal/min} \times 1.5 \text{ lb/gal} = 4.5 \text{ lb/min}.$$

Assuming perfect mixing, the concentration of the solution in the tank is

$$c(t) = \frac{x(t)}{V(t)} = \frac{x(t)}{300 + 2t} \text{ lb/gal}.$$

Therefore, the rate at which salt leaves through the drain at the bottom of the tank is

$$\text{rate out} = 1 \text{ gal/min} \times \frac{x(t)}{300 + 2t} \text{ lb/gal} = \frac{x(t)}{300 + 2t} \text{ lb/min}.$$

The balance law now yields

$$\frac{dx}{dt} = \text{rate in} - \text{rate out},$$

$$\frac{dx}{dt} = 4.5 - \frac{x}{300 + 2t}.$$

This equation is linear, and the technique of Example 5.1 and Section 2.4 can be used to calculate the solution

$$x(t) = 450 + 3t + \frac{C}{\sqrt{300 + 2t}}.$$

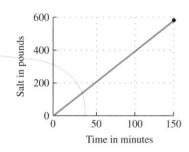

Figure 4. The amount of salt in the tank in Example 5.3.

Again, the tank is filled with pure water initially, so the initial salt content is zero. Thus, $x(0) = 0$ and

$$0 = x(0) = 450 + 3(0) + \frac{C}{\sqrt{300 + 2(0)}} = 450 + \frac{C}{\sqrt{300}}.$$

Consequently, $C = -4500\sqrt{3}$ and

$$x(t) = 450 + 3t - \frac{4500\sqrt{3}}{\sqrt{300 + 2t}}.$$

The solution is plotted in Figure 4.

We are left with the business of finding the salt content at the moment that the solution in the tank reaches the tank's capacity of 600 gal. The equation $V(t) = 300 + 2t$ will produce the time of this event.

$$600 = 300 + 2t$$

$$t = 150 \text{ min}$$

Hence the final salt content is

$$x(150) = 450 + 3(150) - \frac{4500\sqrt{3}}{\sqrt{300 + 2(150)}} \approx 582 \text{ lb.}$$

Example 5.4   Consider two tanks, labeled tank A and tank B (Figure 5). Tank A contains 100 gal of solution in which is dissolved 20 lb of salt. Tank B contains 200 gal of solution in which is dissolved 40 lb of salt. Pure water flows into tank A at a rate of 5 gal/s. There is a drain at the bottom of tank A. Solution leaves tank A via this drain at a rate of 5 gal/s and flows immediately into tank B at the same rate. A drain at the bottom of tank B allows the solution to leave tank B, also at a rate of 5 gal/s. What is the salt content in tank B after 1 minute?

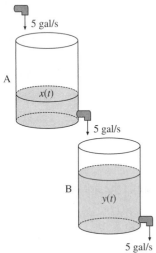

5 gal/s

A

$x(t)$

5 gal/s

B

$y(t)$

5 gal/s

Figure 5. The tanks in Example 5.4.

Let $x(t)$ represent the number of pounds of salt in tank A after t seconds. Because pure water flows into tank A, rate in $= 0$. Solution enters and leaves tank A at the same rate (5 gal/s), so the volume of solution in tank A remains constant (100 gal). Once more we assume "perfect mixing," so the concentration of the salt in tank A at time t is given by $c_A(t) = x(t)/100$ lb/gal. Consequently, the rate at which salt is leaving tank A is given by

$$\text{rate out} = 5 \text{ gal/s} \times \frac{x(t)}{100} \text{ lb/gal} = \frac{1}{20}x(t) \text{ lb/s}.$$

Substituting the rate in and the rate out into the balance law yields

$$\frac{dx}{dt} = \text{rate in} - \text{rate out} = -\frac{1}{20}x.$$

Because there is initially 20 lb of salt present in the solution in tank A, $x(0) = 20$.

Now, let's turn our attention to tank B. The rate at which salt enters tank B is equal to the rate at which salt is leaving tank A. Consequently,

$$\text{rate in} = \frac{1}{20}x \text{ lb/s}.$$

Solution enters and leaves tank B at the same rate (5 gal/s), so the volume of solution in tank B remains constant (200 gal). Assuming "perfect mixing," the concentration of salt in tank B at time t is given by $c_B(t) = y(t)/200$ lb/gal. Consequently, the rate at which salt is leaving tank B is given by

$$\text{rate out} = 5 \text{ gal/s} \times \frac{y(t)}{200} \text{ lb/gal} = \frac{1}{40}y(t) \text{ lb/s}.$$

Substituting the rate in and the rate out into the balance law yields a differential equation defining the rate at which the salt content is changing in tank B.

$$\frac{dy}{dt} = \frac{1}{20}x - \frac{1}{40}y$$

Because there is initially 40 lb of salt present in the solution in tank B, $y(0) = 40$.

Our discussion has led us to the **system** of first-order differential equations

$$\frac{dx}{dt} = -\frac{1}{20}x, \tag{5.5}$$

$$\frac{dy}{dt} = \frac{1}{20}x - \frac{1}{40}y, \tag{5.6}$$

with initial conditions $x(0) = 20$ and $y(0) = 40$. Systems of equations will be a major topic in the remainder of this book. However, because of the special nature of this particular system, we do not need any special knowledge to find a solution. We can solve equation (5.5) for x, then substitute the result into equation (5.6). This will allow us to solve (5.6) with a minimum of difficulty.

Equation (5.5) is separable, so we can separate the variables, integrate, and solve for x, finding that

$$x(t) = C_1 e^{-t/20}.$$

The initial condition $x(0) = 20$ yields $C_1 = 20$, so

$$x(t) = 20e^{-t/20}. \tag{5.7}$$

We now substitute equation (5.7) into equation (5.6) and simplify to obtain the linear equation

$$\frac{dy}{dt} = e^{-t/20} - \frac{1}{40}y. \tag{5.8}$$

Solving in the usual way, we get the general solution

$$y(t) = -40e^{-t/20} + C_2 e^{-t/40}.$$

The initial condition $y(0) = 40$ yields $C_2 = 80$ and

$$y(t) = -40e^{-t/20} + 80e^{-t/40}. \tag{5.9}$$

The solutions in (5.7) and (5.9) are plotted in Figure 6.

Finally, we can use equation (5.9) to find the salt content in tank B at $t = 1$ min $= 60$ seconds, finding that

$$y(60) = -40e^{-(60)/20} + 80e^{-(60)/40} \approx 15.9 \text{ lb.} \qquad \bullet$$

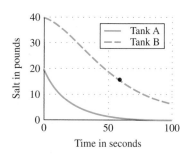

Figure 6. The amount of salt in the tanks in Example 5.4.

EXERCISES

1. A tank contains 100 gal of pure water. At time zero, a sugar-water solution containing 0.2 lb of sugar per gal enters the tank at a rate of 3 gal per minute. Simultaneously, a drain is opened at the bottom of the tank allowing the sugar solution to leave the tank at 3 gal per minute. Assume that the solution in the tank is kept perfectly mixed at all times.

   (a) What will be the sugar content in the tank after 20 minutes?

   (b) How long will it take the sugar content in the tank to reach 15 lb?

   (c) What will be the eventual sugar content in the tank?

2. A tank initially contains 50 gal of sugar water having a concentration of 2 lb of sugar for each gal of water. At time zero, pure water begins pouring into the tank at a rate of 2 gal per minute. Simultaneously, a drain is opened at the bottom of the tank so that the volume of sugar-water solution in the tank remains constant.

   (a) How much sugar is in the tank after 10 minutes?

   (b) How long will it take the sugar content in the tank to dip below 20 lb?

   (c) What will be the eventual sugar content in the tank?

3. A tank initially contains 100 gal of water in which is dissolved 2 lb of salt. The salt-water solution containing 1 lb of salt for every 4 gal of solution enters the tank at a rate of 5 gal per minute. The solution leaves the tank at the same rate, allowing for a constant solution volume in the tank.

   (a) Use an analytic method to determine the eventual salt content in the tank.

   (b) Use a numerical solver to determine the eventual salt content in the tank and compare your approximation with the analytical solution found in part (a).

4. A tank contains 500 gal of a salt-water solution containing 0.05 lb of salt per gallon of water. Pure water is poured into the tank and a drain at the bottom of the tank is adjusted so as to keep the volume of solution in the tank constant. At what rate (gal/min) should the water be poured into the tank to lower the salt concentration to 0.01 lb/gal of water in under one hour?

5. A 50-gal tank initially contains 20 gal of pure water. Salt-water solution containing 0.5 lb of salt for each gallon of water begins entering the tank at a rate of 4 gal/min. Simultaneously, a drain is opened at the bottom of the tank, allowing the salt-water solution to leave the tank at a rate of 2 gal/min. What is the salt content (lb) in the tank at the precise moment that the tank is full of salt-water solution?

6. A tank initially contains 100 gal of a salt-water solution containing 0.05 lb of salt for each gallon of water. At time zero, pure water is poured into the tank at a rate of 3 gal per minute. Simultaneously, a drain is opened at the bottom of the tank that allows the salt-water solution to leave the tank at a rate of 2 gal per minute. What will be the salt

content in the tank when precisely 50 gal of salt solution remain?

7. A tank initially contains 100 gal of pure water. Water begins entering a tank via two pipes: through pipe A at 6 gal per minute, and pipe B at 4 gal per minute. Simultaneously, a drain is opened at the bottom of the tank through which solution leaves the tank at a rate of 8 gal per minute.

   (a) To their dismay, supervisors discover that the water coming into the tank through pipe A is contaminated, containing 0.5 lb of pollutant per gallon of water. If the process had been running undetected for 10 minutes, how much pollutant is in the tank at the end of this 10-minute period?

   (b) The supervisors correct their error and shut down pipe A, allowing pipe B and the drain to function in precisely the same manner as they did before the contaminant was discovered in pipe A. How long will it take the pollutant in the tank to reach one half of the level achieved in part (a)?

8. Suppose that a solution containing a drug enters an organ at the rate a cm^3/s, with drug concentration κ g/cm^3. Solution leaves the organ at a slower rate of b cm^3/s. Further, the faster rate of infusion causes the organ's volume to increase with time according to $V(t) = V_0 + rt$, with V_0 its initial volume. If there is no initial quantity of the drug in the organ, show that the concentration of the drug in the organ is given by

$$c(t) = \frac{a\kappa}{b+r}\left[1 - \left(\frac{V_0}{V_0 + rt}\right)^{(b+r)/r}\right].$$

9. A lake, with volume $V = 100$ km^3, is fed by a river at a rate of r km^3/yr. In addition, there is a factory on the lake that introduces a pollutant into the lake at the rate of p km^3/yr. There is another river that is fed by the lake at a rate that keeps the volume of the lake constant. This means that the rate of flow from the lake into the outlet river is $(p+r)$ km^3/yr. Let $x(t)$ denote the volume of the pollutant in the lake at time t. Then $c(t) = x(t)/V$ is the concentration of the pollutant.

   (a) Show that, under the assumption of immediate and perfect mixing of the pollutant into the lake water, the concentration satisfies the differential equation

$$c' + \frac{p+r}{V}c = \frac{p}{V}.$$

   (b) It has been determined that a concentration of over 2% is hazardous for the fish in the lake. Suppose that $r = 50$ km^3/yr, $p = 2$ km^3/yr, and the initial concentration of pollutant in the lake is zero. How long will it take the lake to become hazardous to the health of the fish?

10. Suppose that the factory in Exercise 9 stops operating at time $t = 0$ and that the concentration of pollutant in the lake was 3.5% at the time. Approximately how long will it

take before the concentration falls below 2% and the lake is no longer hazardous for the fish?

11. Rivers do not flow at the same rate the year around. They tend to be full in the spring when the snow melts and to flow more slowly in the fall. To take this into account, suppose the flow of the input river in Exercise 9 is

$$r = 50 + 20\cos(2\pi(t - 1/3)).$$

Our river flows at its maximum rate one-third into the year (i.e., around the first of April) and at its minimum around the first of October.

   (a) Setting $p = 2$, and using this flow rate, use your numerical solver to plot the concentration for several choices of initial concentration between 0% and 4%. (You might have to reduce the relative error tolerance of your solver, perhaps to 5×10^{-12}.) How would you describe the behavior of the concentration for large values of time?

   (b) It might be expected that after settling into a steady state, the concentration would be greatest when the flow was smallest (i.e., around the first of October). At what time of year does it actually occur?

12. Consider two tanks, labeled tank A and tank B for reference. Tank A contains 100 gal of solution in which is dissolved 20 lb of salt. Tank B contains 200 gal of solution in which is dissolved 40 lb of salt. Pure water flows into tank A at a rate of 5 gal/s. There is a drain at the bottom of tank A. The solution leaves tank A via this drain at a rate of 5 gal/s and flows immediately into tank B at the same rate. A drain at the bottom of tank B allows the solution to leave tank B at a rate of 2.5 gal/s. What is the salt content in tank B at the precise moment that tank B contains 250 gal of solution?

13. Lake Happy Times contains 100 km^3 of pure water. It is fed by a river at a rate of 50 km^3/yr. At time zero, there is a factory on one shore of Lake Happy Times that begins introducing a pollutant to the lake at a rate of 2 km^3/yr. There is another river that is fed by Lake Happy Times at a rate that keeps the volume of Lake Happy Times constant. This means that the rate of flow from Lake Happy Times into the outlet river is 52 km^3/yr. In turn, the flow from this outlet river goes into another lake, Lake Sad Times, at an equal rate. Finally, there is an outlet river from Lake Sad Times flowing at a rate that keeps the volume of Lake Sad Times at a constant 100 km^3.

   (a) Find the amount of pollutant in Lake Sad Times at the end of 3 months.

   (b) At the end of 3 months, observers close the factory due to environmental concerns and no further pollutant enters Lake Happy Times. How long will it take for the pollutant in Lake Sad Times (found in part (a)) to be cut in half?

14. Two tanks, Tank I and Tank II, are filled with V gal of pure water. A solution containing a lb of salt per gallon of water is poured into Tank I at a rate of b gal per

minute. The solution leaves Tank I at a rate of b gal/min and enters Tank II at the same rate (b gal/min). A drain is adjusted on Tank II and solution leaves Tank II at a rate of b gal/min. This keeps the volume of solution con-

stant in both tanks (V gal). Show that the amount of salt solution in Tank II, as a function of time t, is given by $aV - abte^{-(b/V)t} - aVe^{-(b/V)t}$.

2.6 Exact Differential Equations

So far we have found methods for finding analytic solutions for differential equations which are either separable or linear. In this section we will explore further the task of finding analytic solutions. We will find some new methods of finding analytic solutions. However, we will also discover some hints as to why this task is so daunting, and often impossible.

Differential forms and differential equations

We will consider differential equations that can be written as

$$P(x, y) + Q(x, y)\frac{dy}{dx} = 0, \tag{6.1}$$

where P and Q are functions of both the independent variable x and the dependent variable y. This is a very general class of differential equations.

As usual, a solution will be a differentiable function $y(x)$ defined for x in an interval, such that equation (6.1) is satisfied at each point in the interval. We discovered in our discussion of separable equations that it is necessary to allow solutions defined implicitly by equations of the form

$$F(x, y) = C, \tag{6.2}$$

and we will do so here as well.

Notice that equation (6.2) treats the two variables x and y symmetrically. There is no distinction made between the two. It is natural and convenient to symmetrize the differential equation as well. This leads to the language of differential forms. A **differential form** in the two variables x and y is an expression of the type

$$\omega = P(x, y)\,dx + Q(x, y)\,dy, \tag{6.3}$$

where P and Q are functions of x and y. The simple forms dx and dy are called **differentials**.

Suppose that $y = y(x)$. Then $dy = y'(x)\,dx$. If we substitute this into the differential form ω in (6.3), we get

$$P(x, y)\,dx + Q(x, y)\,dy = \left(P(x, y) + Q(x, y)\frac{dy}{dx}\right) dx.$$

Thus, y is a solution to the differential equation (6.1) if and only if

$$P(x, y)\,dx + Q(x, y)\,dy = 0. \tag{6.4}$$

For this reason, we will consider (6.4) as another way of writing the differential equation in (6.1). The differential form variant of a differential equation will be used systematically in this section.

Example 6.5 Consider the differential equation

$$\omega = x\,dx + y\,dy = 0, \quad \text{or} \quad \frac{dy}{dx} = -\frac{x}{y}. \tag{6.6}$$

Show that this equation has solutions defined implicitly by the equation

$$x^2 + y^2 = C. \tag{6.7}$$

This can be verified by differentiating formula (6.7) with respect to x, getting

$$2x + 2y\frac{dy}{dx} = 0, \quad \text{or} \quad x\,dx + y\,dy = 0.$$

Of course, we can solve (6.7) for y, obtaining two solutions

$$y(x) = \pm\sqrt{C - x^2} \tag{6.8}$$

defined for $|x| \le \sqrt{C}$. ●

Solution curves and integral curves

Example 6.5 illustrates some features that we want to emphasize because they apply more generally. First, the level set defined by $x^2 + y^2 = C$ is the circle with center at the origin and radius $\sqrt{C}$. (See Figure 1.) This level set is not the graph of a function, but it contains the graphs of both of the solutions in (6.8). This means that the level set contains two solution curves, which motivates the following definition.

DEFINITION 6.9

> Suppose that solutions to the differential equation (6.1) or (6.4) are given implicitly by the equation
> $$F(x, y) = C.$$
> Then the level sets defined by $F(x, y) = C$ are called ***integral curves*** of the differential equation.

Thus, we have shown that an integral curve can contain two or more solution curves as illustrated in Figure 1.

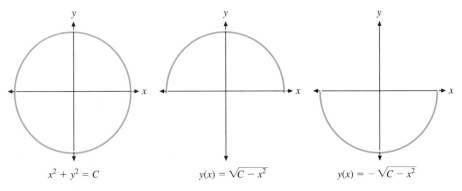

Figure 1. The integral curve defined by (6.7) and the solution curves in (6.8).

Solutions

Let's carefully examine what it means when we say that the equation

$$F(x, y) = C \tag{6.10}$$

gives the general solution to the differential equation

$$\omega = P(x, y)\,dx + Q(x, y)\,dy = 0. \tag{6.11}$$

Suppose $y = y(x)$ is defined by the implicit equation (6.10). Differentiating (6.10) with respect to x, we get

$$\frac{\partial F}{\partial x} + \frac{\partial F}{\partial y}\frac{dy}{dx} = 0, \quad \text{or} \quad \frac{dy}{dx} = -\frac{\partial F/\partial x}{\partial F/\partial y}. \tag{6.12}$$

Since $y(x)$ is a solution to (6.11), we have

$$\frac{dy}{dx} = -\frac{P(x, y)}{Q(x, y)}. \tag{6.13}$$

Comparing (6.13) with (6.12), we see that

$$\frac{\partial F/\partial x}{\partial F/\partial y} = \frac{P}{Q},$$

or

$$\frac{1}{P}\frac{\partial F}{\partial x} = \frac{1}{Q}\frac{\partial F}{\partial y}. \tag{6.14}$$

If we let $\mu = \mu(x, y)$ be defined as the function in (6.14), then

$$\frac{\partial F}{\partial x} = \mu P \quad \text{and} \quad \frac{\partial F}{\partial y} = \mu Q. \tag{6.15}$$

Thus to find the solution to the differential equation (6.11), we must find functions μ and F that satisfy (6.15). This is a daunting task.

Exact differential equations

The task is made easier if we approach it systematically. We will first look at the more tractable case when $\mu = 1$ in (6.15). In this case our differential form in (6.11) becomes

$$\omega = P(x, y)\, dx + Q(x, y)\, dy = \frac{\partial F}{\partial x}\, dx + \frac{\partial F}{\partial y}\, dy.$$

We will give forms of this type a special name.

DEFINITION 6.16

The **differential** of a continuously differentiable function F is the differential form

$$dF = \frac{\partial F}{\partial x}\, dx + \frac{\partial F}{\partial y}\, dy.$$

A differential form is said to be **exact** if it is the differential of a continuously differentiable function.

Let's point out explicitly that the differential form $P\, dx + Q\, dy$ is exact if and only if there is a continuously differentiable function $F(x, y)$ such that

$$dF = \frac{\partial F}{\partial x}\, dx + \frac{\partial F}{\partial y}\, dy = P\, dx + Q\, dy.$$

This means that the coefficients of dx and dy must be equal, or

$$\frac{\partial F}{\partial x} = P(x, y) \quad \text{and} \quad \frac{\partial F}{\partial y} = Q(x, y). \tag{6.17}$$

Furthermore, if the form $\omega = P\, dx + Q\, dy$ is exact, and is equal to the differential dF, then by the discussion leading to equation (6.15), the general solution to $dF = P\, dx + Q\, dy = 0$ is given by $F(x, y) = C$.

Example 6.18   Solve the equation $2x\,dx + 4y^3\,dy = 0$.

Using (6.17), we see that we want to find a function $F(x, y)$ satisfying

$$\frac{\partial F}{\partial x} = 2x \quad \text{and} \quad \frac{\partial F}{\partial y} = 4y^3.$$

Because the variables are separated in these equations, it is natural to look for a function F which is the sum of a function of x and a function of y. Then integration leads us to $F(x, y) = x^2 + y^4$. Consequently, the differential form $2x\,dx + 4y^3\,dy$ is exact. Furthermore, the general solution to the equation $2x\,dx + 4y^3\,dy = 0$ is given by

$$x^2 + y^4 = C.$$

Three integral curves are plotted in Figure 2.   ●

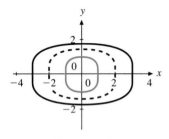

Figure 2. Three integral curves for the equation in Example 6.18.

As Example 6.18 illustrates, it is quite easy to solve a differential equation that has **separated variables**, by which we mean an equation of the form

$$P(x)\,dx + Q(y)\,dy = 0. \tag{6.19}$$

Using the method in Example 6.18 we find that the solution is given by $F(x, y) = C$, where

$$F(x, y) = \int P(x)\,dx + \int Q(y)\,dy.$$

However, two questions come to mind with respect to the general question.

1. Given a differential form $\omega = P\,dx + Q\,dy$, how do we know if it is exact?
2. If a differential form is exact, is there a way to find F such that $dF = P\,dx + Q\,dy$?

Both of these questions are answered in the next result.

THEOREM 6.20   Let $\omega = P(x, y)\,dx + Q(x, y)\,dy$ be a differential form where both P and Q are continuously differentiable.

(a) If ω is exact, then

$$\frac{\partial P}{\partial y} = \frac{\partial Q}{\partial x}. \tag{6.21}$$

(b) If equation (6.21) is true in a rectangle R, then ω is exact in R.

Proof   To prove (a), suppose that $\omega = dF$. Then

$$\frac{\partial F}{\partial x} = P \quad \text{and} \quad \frac{\partial F}{\partial y} = Q. \tag{6.22}$$

Both P and Q are continuously differentiable, so F is twice continuously differentiable. This means that the mixed second-order derivatives of F are equal. Consequently,

$$\frac{\partial P}{\partial y} = \frac{\partial^2 F}{\partial y \partial x} = \frac{\partial^2 F}{\partial x \partial y} = \frac{\partial Q}{\partial x}.$$

To prove (b), we need to find a function F satisfying both equations in (6.22). If we integrate the first equation, $\partial F/\partial x = P$, then, by the fundamental theorem of calculus, we must have

$$F(x, y) = \int P(x, y)\,dx + \phi(y), \tag{6.23}$$

where ϕ is a function of y only. In this formula $\int P(x, y) \, dx$ represents a particular indefinite integral, and $\phi(y)$ represents the constant of integration. Since we are integrating with respect to x, this "constant" can still depend on y.

To discover what ϕ is, we differentiate both sides of equation (6.23) with respect to y to get

$$\frac{\partial F}{\partial y} = \frac{\partial}{\partial y} \int P(x, y) \, dx + \phi'(y) = \int \frac{\partial P}{\partial y}(x, y) \, dx + \phi'(y).$$

The second formula in (6.22) says that $\partial F / \partial y = Q$, so ϕ must satisfy

$$\phi'(y) = Q(x, y) - \frac{\partial}{\partial y} \int P(x, y) \, dx = Q(x, y) - \int \frac{\partial P}{\partial y} \, dx.$$

This equation can be solved by integration provided the function on the right does not depend on the variable x. To see that this is true, it suffices to show that the derivative of the function on the right with respect to x is zero. We have

$$\frac{\partial}{\partial x} \left(Q(x, y) - \int \frac{\partial P}{\partial y} \, dx \right) = \frac{\partial Q}{\partial x} - \frac{\partial}{\partial x} \int \frac{\partial P}{\partial y} \, dx$$

$$= \frac{\partial Q}{\partial x} - \frac{\partial P}{\partial y}$$

$$= 0.$$

The very last step is true by the hypothesis (6.21) of our theorem. ∎

Solving exact differential equations

We now have two methods to solve exact equations. The first method is that used in Example 6.18, where we found the function F satisfying (6.17) by inspection and experimentation. The second, and more systematic method, is that used in the proof of part (b) of Theorem 6.20.

If the equation $P(x, y) \, dx + Q(x, y) \, dy = 0$ is exact, the solution is given by $F(x, y) = C$ where F is found by solving (6.17) using the steps:

1. Solve $\partial F / \partial x = P$ by integration:

$$F(x, y) = \int P(x, y) \, dx + \phi(y). \tag{6.24}$$

2. Solve $\partial F / \partial y = Q$ using (6.24) by choosing ϕ so that

$$\frac{\partial F}{\partial y} = \frac{\partial}{\partial y} \int P(x, y) \, dx + \phi'(y) = Q(x, y).$$

It is also possible to solve $\partial F/\partial y = Q$ first by setting $F(x, y) = \int Q(x, y)\,dy + \psi(x)$, and then to solve $\partial F/\partial x = P$ by choosing the "constant" of integration $\psi(x)$.

Let's look at an example.

Example 6.25 Show that the equation $\sin(x + y)\,dx + (2y + \sin(x + y))\,dy = 0$ is exact and find a general solution.

Since

$$\frac{\partial}{\partial y}\sin(x + y) = \cos(x + y) = \frac{\partial}{\partial x}(2y + \sin(x + y)),$$

we know the equation is exact. To find a general solution, we need to find a function $F(x, y)$ such that

$$\frac{\partial F}{\partial x} = \sin(x + y) \quad \text{and} \quad \frac{\partial F}{\partial y} = 2y + \sin(x + y). \tag{6.26}$$

We solve the first equation in (6.26) by integrating,

$$F(x, y) = \int \sin(x + y)\,dx + \phi(y) = -\cos(x + y) + \phi(y).$$

We solve the second equation in (6.26) by choosing ϕ so that

$$\frac{\partial F}{\partial y} = \sin(x + y) + \phi'(y) = 2y + \sin(x + y).$$

Therefore, $\phi'(y) = 2y$, which has solution $\phi(y) = y^2$. Finally, $F(x, y) = y^2 - \cos(x + y)$, and solutions are given implicitly by

$$F(x, y) = y^2 - \cos(x + y) = C. \tag{6.27}$$

Two integral curves are shown in Figure 3. The blue curve is for $C = 0$ and the black curve for $C = 2$. ●

Sometimes it is easier to solve the equations in the opposite order. Let's see how this works out in Example 6.25. We solve the second equation in (6.26) by integrating:

$$F(x, y) = \int [2y + \sin(x + y)]\,dy = y^2 - \cos(x + y) + \psi(x).$$

The we solve the first equation in (6.26) by choosing ψ so that

$$\frac{\partial F}{\partial x} = \sin(x + y) + \psi'(x) = \sin(x + y).$$

We want $\psi'(x) = 0$, so we take $\psi(x) = 0$. Again we get the solution $F(x, y) = y^2 - \cos(x + y)$.

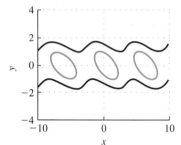

Figure 3. Two integral curves for the equation in Example 6.25.

Solutions and integrating factors

We learned earlier, in equation (6.15), that to solve the differential equation $P\,dx + Q\,dy = 0$ we must find functions μ and F that satisfy

$$\frac{\partial F}{\partial x} = \mu P \quad \text{and} \quad \frac{\partial F}{\partial y} = \mu Q. \tag{6.28}$$

Notice that as a result,

$$\mu\,[P\,dx + Q\,dy] = \mu P\,dx + \mu Q\,dy = \frac{\partial F}{\partial x}\,dx + \frac{\partial F}{\partial y}\,dy = dF,$$

so the form $\mu\,[P\,dx + Q\,dy]$ is exact. Let's make a definition.

DEFINITION 6.29

An ***integrating factor*** for the differential equation $\omega = P\,dx + Q\,dy = 0$ is a function $\mu(x, y)$ such that the form $\mu\omega = \mu(x, y)P(x, y)\,dx + \mu(x, y)Q(x, y)\,dy$ is exact.

Example 6.30

Consider the equation $(x + 2y^2)\,dx - 2xy\,dy = 0$. Show that the equation is not exact and that $1/x^3$ is an integrating factor. Find a general solution.

Since

$$\frac{\partial}{\partial y}(x + 2y^2) = 4y \quad \text{and} \quad \frac{\partial}{\partial x}(-2xy) = -2y,$$

the equation is not exact. On the other hand, after we multiply the equation by $1/x^3$, we get the equation

$$\frac{(x + 2y^2)\,dx}{x^3} - \frac{2y\,dy}{x^2} = 0.$$

For this equation, we have

$$\frac{\partial}{\partial y}\left(\frac{x + 2y^2}{x^3}\right) = \frac{4y}{x^3} = \frac{\partial}{\partial x}\left(-\frac{2y}{x^2}\right),$$

so the equation is exact and $1/x^3$ is an integrating factor. To solve it, we set

$$F(x, y) = \int \frac{(x + 2y^2)\,dx}{x^3} + \phi(y) = -\frac{1}{x} - \frac{y^2}{x^2} + \phi(y).$$

To find ϕ, we differentiate this with respect to y, using the fact that $\partial F/\partial y = -2y/x^2$. We get

$$-\frac{2y}{x^2} = -\frac{2y}{x^2} + \phi'(y),$$

so $\phi'(y) = 0$, and we can take $\phi(y) = 0$. Consequently, our general solution is

$$F(x, y) = -\frac{1}{x} - \frac{y^2}{x^2} = C.$$

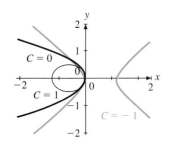

Figure 4. Three integral curves for the equation in Example 6.30.

We multiply through by x^2 and rearrange this so that the solutions are defined by

$$x + y^2 + Cx^2 = 0.$$

Thus the integral curves are all conics. Three are shown plotted in Figure 4.   ●

Finding integrating factors

As illustrated in Example 6.30, we now have a strategy for finding a general solution to the differential equation $P\,dx + Q\,dy = 0$.

1. Find an integrating factor μ, so that $\mu P\,dx + \mu Q\,dy$ is exact.
2. Find a function F such that $dF = \mu P\,dx + \mu Q\,dy$.

Then a general solution is given implicitly by $F(x, y) = C$.

    Since we already know how to solve exact equations, the remaining difficulty is the first step. Integrating factors always exist, but as we will see, it is not always easy to find one. Even in Example 6.30, the choice of $1/x^3$ is not at all obvious. Before proceeding to new situations, let's put what we already know about separable and linear equations into this new context.

Separable equations

A differential equation is said to be **separable** if there is an integrating factor that will separate the variables by putting the equation in the form

$$P(x)\,dx + Q(y)\,dy = 0$$

found in (6.19). The solution is then given by

$$F(x, y) = \int P(x)\,dx + \int Q(y)\,dy = C.$$

A special case is the separable equations we studied in Section 2.2. These are equations of the form

$$\frac{dy}{dx} = \frac{p(x)}{q(y)} \quad \text{or} \quad \frac{p(x)}{q(y)}\,dx - dy = 0. \tag{6.31}$$

Multiplication by the integrating factor $q(y)$ transforms the equation to $p(x)\,dx - q(y)\,dy = 0$, which has its variables separated. Here is a slightly different example.

Example 6.32   Solve the equation $-y^2\,dx + x^3\,dy = 0$.

    This differential equation is not exact. However, it is separable. If we multiply by the integrating factor $1/x^3 y^2$, we get

$$-\frac{dx}{x^3} + \frac{dy}{y^2} = 0, \tag{6.33}$$

an equation with separated variables. Consequently, we can write down the solution by integrating (6.33) to get

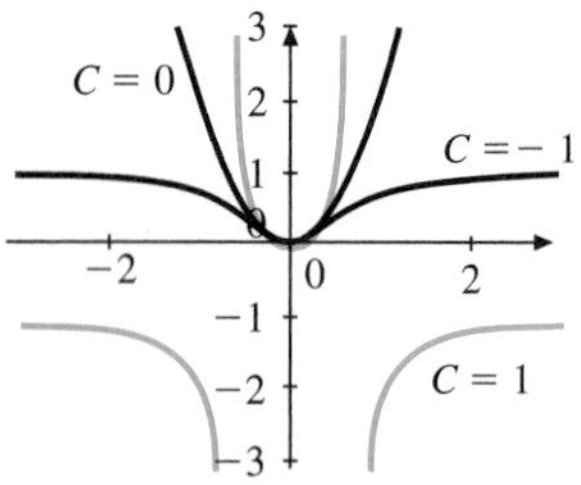

$$-\int \frac{dx}{x^3} + \int \frac{dy}{y^2} = \frac{1}{2x^2} - \frac{1}{y} = C.$$

We can solve this equation for y, getting

$$y(x) = \frac{2x^2}{1 - 2Cx^2}.$$

Some integral curves are plotted in Figure 5.

Figure 5. Three integral curves for the equation in Example 6.32.

Linear equations

Linear equations are equations of the special form

$$\frac{dy}{dx} = a(x)y + f(x) \quad \text{or} \quad [a(x)y + f(x)]\, dx - dy = 0. \tag{6.34}$$

In Section 2.4 we showed how to solve a linear equation using the integrating factor $\mu(x)$ defined by

$$\frac{d\mu}{dx} = -a(x)\mu \quad (\text{or} \quad \mu(x) = e^{-\int a(x)\, dx}). \tag{6.35}$$

We will leave it as an exercise to show that if $\mu(x)$ satisfies the differential equation in (6.35), then

$$\mu(x)\, [a(x)y + f(x)]\, dx - \mu(x)dy = 0$$

is exact. We will also leave it as an exercise to solve this exact equation and show that the general solution to the linear equation in (6.34) is given by

$$y(x) = \frac{1}{\mu(x)} \int \mu(x) f(x)\, dx + C\frac{1}{\mu(x)}.$$

Integrating factors depending on only one variable

The general procedure to search for an integrating factor starts from the criterion for exactness that we found in Theorem 6.20. Suppose $\omega = P\, dx + Q\, dy$ and we want to find μ such that $\mu\omega = \mu P\, dx + \mu Q\, dy$ is exact. According to Theorem 6.20, μ must satisfy

$$\frac{\partial}{\partial y}(\mu P) = \frac{\partial}{\partial x}(\mu Q). \tag{6.36}$$

This is a partial differential equation for μ. There is no procedure for solving this equation in general.

However, sometimes we can make assumptions about μ that make this equation simpler. Notice that the integrating factor for the separable equation in (6.31) depended only on y, while the integrating factor for the linear equation in (6.34) depended only on x. If there is an integrating factor that depends on only one variable, it is much easier to find.

Let's find out when the equation

$$P(x, y)\, dx + Q(x, y)\, dy = 0$$

has an integrating factor $\mu(x)$ depending only on x. If μ does not depend on y, equation (6.36) simplifies to

$$\mu\frac{\partial P}{\partial y} = \frac{d\mu}{dx} Q + \mu\frac{\partial Q}{\partial x}.$$

Solving for the derivative of μ, we get

$$\frac{d\mu}{dx} = \frac{1}{Q}\left(\frac{\partial P}{\partial y} - \frac{\partial Q}{\partial x}\right)\mu. \tag{6.37}$$

This differential equation for μ will have a solution that depends only on x and is independent of y only if the quantity

$$h = \frac{1}{Q}\left(\frac{\partial P}{\partial y} - \frac{\partial Q}{\partial x}\right) \tag{6.38}$$

does not depend on the variable y and is a function of x only. If this is so, by (6.37), the integrating factor $\mu(x)$, is a solution to the ordinary differential equation

$$\frac{d\mu}{dx} = h\mu.$$

This equation is separable and linear. A solution is

$$\mu(x) = e^{\int h(x)\,dx}.$$

Here's an example where this process is successful.

Example 6.39 Solve the equation

$$(xy - 2)\,dx + (x^2 - xy)\,dy = 0.$$

In this case, $\partial P/\partial y = x$ and $\partial Q/\partial x = 2x - y$, so the equation is not exact. However, the quantity in (6.38) is

$$h = \frac{1}{Q}\left(\frac{\partial P}{\partial y} - \frac{\partial Q}{\partial x}\right) = \frac{1}{x^2 - xy}[x - (2x - y)] = -\frac{1}{x}.$$

Since this is a function only of x, we know that there is an integrating factor that depends only on x, and it is a solution of the differential equation

$$\mu' = -\frac{1}{x}\mu.$$

A solution is $\mu = 1/x$. Multiplying the original equation by μ, we get the exact equation

$$\left(y - \frac{2}{x}\right)dx + (x - y)\,dy = 0.$$

Thus,

$$F(x, y) = \int \left(y - \frac{2}{x}\right)dx + \phi(y) = xy - 2\ln|x| + \phi(y).$$

To find ϕ, we differentiate with respect to y, using $\partial F/\partial y = x - y$.

$$x - y = x + \phi'(y), \quad \text{or} \quad \phi'(y) = -y$$

A solution is $\phi(y) = -y^2/2$, and the general solution is given implicitly by

$$F(x, y) = xy - 2\ln|x| - y^2/2 = C.$$

We can solve this equation for y using the quadratic formula, and we find that the solutions are

$$y(x) = x \pm \sqrt{x^2 - 4\ln|x| - 2C}.$$

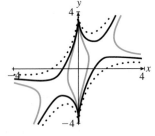

Figure 6. Three integral curves for the equation in Example 6.39.

Three integral curves are shown in Figure 6. The dotted black curve corresponds to $C = -1$, the solid black curve to $C = 0$, and the blue curve to $C = 1$.

Of course, we can explore the possibility that there is an integrating factor depending only on y. We will leave this as an exercise. Let's summarize these results.

> The form $P\,dx + Q\,dy$ has an integrating factor depending on one of the variables under the following conditions.
>
> - If
>
> $$h = \frac{1}{Q}\left(\frac{\partial P}{\partial y} - \frac{\partial Q}{\partial x}\right)$$
>
>   is a function of x only, then $\mu(x) = e^{\int h(x)\,dx}$ is an integrating factor.
> - If
>
> $$g = \frac{1}{P}\left(\frac{\partial P}{\partial y} - \frac{\partial Q}{\partial x}\right)$$
>
>   is a function of y only, then $\mu(y) = e^{-\int g(y)\,dy}$ is an integrating factor.

Homogeneous equations

A function $G(x, y)$ is **homogeneous of degree n** if

$$G(tx, ty) = t^n G(x, y)$$

for all $t > 0$ and all $x \neq 0$ and $y \neq 0$. Thus the functions

$$\frac{1}{x^2 + y^2}, \quad \ln(y/x), \quad 2x^3 - 3x^2 y + 2xy^2 - y^3, \quad \text{and} \quad \sqrt{x^2 + y^2}$$

are homogeneous of degrees -2, 0, 3, and 1, respectively. The functions

$$x + xy, \quad \sin(x), \quad \ln(x + y + 1), \quad \text{and} \quad x - y - 2$$

are not homogeneous.

A differential equation

$$P\,dx + Q\,dy = 0$$

is said to be **homogeneous**[5] if both of the coefficients P and Q are homogeneous of the same degree. Homogeneous equations can be put into a form in which they can be solved by using the substitution $y = xv$, where v is a new variable. Let's look at an example first and then we will examine the general case.

Example 6.40   Verify that $(x^2 + y^2)\,dx + xy\,dy = 0$ is homogeneous and find a solution.

Both $x^2 + y^2$ and xy are homogeneous of degree 2, so the equation is homogeneous. To solve the equation, we make the substitution $y = xv$. Then $dy = v\,dx + x\,dv$, so the equation becomes

$$(x^2 + x^2 v^2)\,dx + x^2 v(v\,dx + x\,dv) = 0.$$

After canceling out the common factor x^2 and collecting terms, this becomes

$$(1 + 2v^2)\,dx + xv\,dv = 0.$$

---

[5] We have used the term *homogeneous differential equation* with a completely different meaning in Section 2.4. Unfortunately, both usages have become standard. The meanings are sufficiently different that you should not have any difficulty, but keep your eyes open.

Although this is not immediately solvable, it is separable. The integrating factor

$$\frac{1}{x(1 + 2v^2)}$$

transforms the equation into the equation

$$\frac{dx}{x} + \frac{v\,dv}{1 + 2v^2} = 0.$$

Integrating, we get

$$\ln|x| + \ln(1 + 2v^2)^{1/4} = k,$$

where k is a constant. If we multiply by 4 and exponentiate, this becomes

$$x^4(1 + 2v^2) = e^{4k} = C.$$

Substituting $v = y/x$, we get our final answer

$$x^4 + 2x^2y^2 = C. \tag{6.41}$$

Three integral curves are shown in Figure 7. The dotted curve corresponds to $C = 1$, the solid black curve to $C = 4$, and the blue curve to $C = 9$. ●

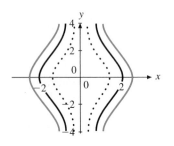

Figure 7. Three integral curves for the equation in Example 6.40.

In working this example, we did two things. First, we made the substitution $y = xv$, and then we looked for an integrating factor that will separate the variables. These two steps will serve to find the solution for any homogeneous equation. To see this, let's start with

$$P(x, y)\,dx + Q(x, y)\,dy = 0,$$

where P and Q are both homogeneous of degree n. We make the substitution $y = xv$, and we get

$$P(x, xv)\,dx + Q(x, xv)\,(v\,dx + x\,dv) = 0.$$

The homogeneity means that $P(x, xv) = x^n P(1, v)$ and $Q(x, xv) = x^n Q(1, v)$. Using this, dividing out the common term x^n, and collecting terms, our differential equation becomes

$$(P(1, v) + vQ(1, v))\,dx + xQ(1, v)\,dv = 0.$$

We recognize that the integrating factor

$$\frac{1}{x(P(1, v) + vQ(1, v))}$$

will separate the variables, leaving us with the equation

$$\frac{dx}{x} + \frac{Q(1, v)\,dv}{P(1, v) + vQ(1, v)} = 0. \tag{6.42}$$

This equation has separated variables, so it can be solved. Finally, we substitute $v = y/x$ to put the answer in terms of the original variables. This verifies that the method works in general. However, when working problems of this type, it is usually better to make the substitution $y = vx$ and then compute with the result, rather than remember the formula in (6.42).

EXERCISES

In Exercises 1–8, calculate the differential dF for the given function F.

1. $F(x, y) = 2xy + y^2$

2. $F(x, y) = x^2 - xy + y^2$

3. $F(x, y) = \sqrt{x^2 + y^2}$

4. $F(x, y) = 1/\sqrt{x^2 + y^2}$

5. $F(x, y) = xy + \tan^{-1}(y/x)$

6. $F(x, y) = \ln(xy) + x^2 y^3$

7. $F(x, y) = \ln(x^2 + y^2) + x/y$

8. $F(x, y) = \tan^{-1}(x/y) + y^4$

In Exercises 9–21, determine which of the equations are exact and solve the ones that are.

9. $(2x + y)\, dx + (x - 6y)\, dy = 0$

10. $(1 - y\sin x)\, dx + (\cos x)\, dy = 0$

11. $\left(1 + \dfrac{y}{x}\right) dx - \dfrac{1}{x}\, dy = 0$

12. $\dfrac{x}{\sqrt{x^2 + y^2}}\, dx + \dfrac{y}{\sqrt{x^2 + y^2}}\, dy = 0$

13. $\dfrac{dy}{dx} = \dfrac{3x^2 + y}{3y^2 - x}$

14. $\dfrac{dy}{dx} = \dfrac{x}{x - y}$

15. $(u + v)\, du + (u - v)\, dv = 0$

16. $\dfrac{2u}{u^2 + v^2}\, du + \dfrac{2v}{u^2 + v^2}\, dv = 0$

17. $\dfrac{dr}{ds} = \dfrac{\ln s}{r/s - 2s}$

18. $\dfrac{dy}{du} = \dfrac{2 - y/u}{\ln u}$

19. $\sin 2t\, dx + (2x \cos 2t - 2t)\, dt = 0$

20. $2xy^2 + 4x^3 + 2x^2 y \dfrac{dy}{dx} = 0$

21. $(2r + \ln y)\, dr + ry\, dy = 0$

In Exercises 22–25, the equations are not exact. However, if you multiply by the given integrating factor, then you can solve the resulting exact equation.

22. $(y^2 - xy)\, dx + x^2\, dy = 0$, $\mu(x, y) = \dfrac{1}{xy^2}$

23. $(x^2 y^2 - 1)y\, dx + (1 + x^2 y^2)x\, dy = 0$, $\mu(x, y) = \dfrac{1}{xy}$

24. $3(y + 1)\, dx - 2x\, dy = 0$, $\mu(x, y) = \dfrac{y + 1}{x^4}$

25. $(x^2 + y^2 - x)\, dx - y\, dy = 0$, $\mu(x, y) = \dfrac{1}{x^2 + y^2}$

26. Suppose that $y\, dx + (x^2 y - x)\, dy = 0$ has an integrating factor that is a function of x alone [i.e., $\mu = \mu(x)$]. Find the integrating factor and use it to solve the differential equation.

27. Suppose that $(xy - 1)\, dx + (x^2 - xy)\, dy = 0$ has an inte-

grating factor that is a function of x alone [i.e., $\mu = \mu(x)$]. Find the integrating factor and use it to solve the differential equation.

28. Suppose that $2y\, dx + (x + y)\, dy = 0$ has an integrating factor that is a function of y alone [i.e., $\mu = \mu(y)$]. Find the integrating factor and use it to solve the differential equation.

29. Suppose that $(y^2 + 2xy)\, dx - x^2\, dy = 0$ has an integrating factor that is a function of y alone [i.e., $\mu = \mu(y)$]. Find the integrating factor and use it to solve the differential equation.

30. Consider the differential equation $2y\, dx + 3x\, dx = 0$. Determine conditions on a and b so that $\mu(x, y) = x^a y^b$ is an integrating factor. Find a particular integrating factor and use it to solve the differential equation.

The equations in Exercises 31–34 each have the form $P(x, y)\, dx + Q(x, y)\, dy = 0$. In each case, show that P and Q are homogeneous of the same degree. State that degree.

31. $(x + y)\, dx + (x - y)\, dy = 0$

32. $(x^2 - xy - y^2)\, dx + 4xy\, dy = 0$

33. $\left(x - \sqrt{x^2 + y^2}\right) dx - y\, dy = 0$

34. $(\ln x - \ln y)\, dx + dy = 0$

Find the general solution of each homogeneous equation in Exercises 35–39.

35. $(x^2 + y^2)\, dx - 2xy\, dy = 0$

36. $(x + y)\, dx + (y - x)\, dy = 0$

37. $(3x + y)\, dx + x\, dy = 0$

38. $\dfrac{dy}{dx} = \dfrac{y(x^2 + y^2)}{xy^2 - 2x^3}$

39. $x^2 y' = 2y^2 - x^2$

40. $(y + 2xe^{-y/x})\, dx - x\, dy = 0$

41. In Figure 8, a goose starts in flight a miles due east of its nest. Assume that the goose maintains constant flight speed (relative to the air) so that it is always flying directly toward its nest. The wind is blowing due north at w miles per hour. Figure 8 shows a coordinate frame with the nest at $(0, 0)$ and the goose at (x, y). It is easily seen (but you should verify it yourself) that

$$\frac{dx}{dt} = -v_0 \cos\theta,$$

$$\frac{dy}{dt} = w - v_0 \sin\theta.$$

(a) Show that

$$\frac{dy}{dx} = \frac{y - k\sqrt{x^2 + y^2}}{x}, \qquad (6.43)$$

where $k = w/v_0$, the ratio of the wind speed to the speed of the goose.

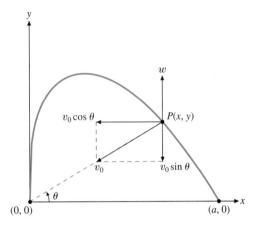

Figure 8. The geometry in Exercise 41.

(b) Solve equation (6.43) and show that

$$y(x) = \frac{a}{2}\left[\left(\frac{x}{a}\right)^{1-k} - \left(\frac{x}{a}\right)^{1+k}\right].$$

(c) Three distinctly different outcomes are possible, each depending on the value of k. Find and discuss each case and use a grapher to depict a sample flight trajectory in each case.

An equation of the form $F(x, y) = C$ defines a family of curves in the plane. Furthermore, we know these curves are the integral curves of the differential equation

$$dF = \frac{\partial F}{\partial x}\,dx + \frac{\partial F}{\partial y}\,dy = 0 \quad \text{or} \quad \frac{dy}{dx} = -\frac{\partial F}{\partial x}\bigg/\frac{\partial F}{\partial y}.$$
$$(6.44)$$

A family of curves is said to be **orthogonal** to a second family if each member of one family intersects all members of the other family at right angles. For example, the families $y = mx$ and $x^2 + y^2 = c^2$ are orthogonal. For a curve $y = y(x)$ to be everywhere orthogonal to the curves defined by $F(x, y) = C$ its derivative must be the negative reciprocal of that in (6.44), or

$$\frac{dy}{dx} = \frac{\partial F}{\partial y}\bigg/\frac{\partial F}{\partial x}.$$

The family of solutions to this differential equation are orthogonal to the family defined by $F(x, y) = C$.

42. Find the family of curves that is orthogonal to the family defined by the equation $y^2 = cx$ and provide a sketch depicting the orthogonality of the two families.

43. The equation $x^2 + y^2 = 2cx$ defines the family of circles tangent to the y-axis at the origin.

    (a) Show that the family of curves orthogonal to this family satisfies the differential equation

$$\frac{dy}{dx} = \frac{2xy}{x^2 - y^2}.$$

    (b) Find the orthogonal family and provide a sketch depicting the orthogonality of the two families.

Knowing an integrating factor exists and finding one suitable for a particular equation are two completely different things. Indeed, as stated previously, finding an integrating factor can be a genuine mathematical art. However, certain differential forms can remind us of differentiation techniques that may aid in the solution of the equation at hand. For example, seeing $x\,dy + y\,dx$ reminds us of the product rule, as in $d(xy) = x\,dx + y\,dy$, and $x\,dy - y\,dx$ might remind us of the quotient rule, $d(x/y) = (y\,dx - x\,dy)/y^2$. In the equation

$$x\,dy + y\,dx + 3xy^2\,dy = 0,$$

we are again reminded of the product rule. In fact, if you multiply the equation by $1/(xy)$, then

$$\frac{x\,dy + y\,dx}{xy} + 3y\,dy = 0,$$

$$d(\ln xy) + 3y\,dy = 0,$$

$$\ln xy + \frac{3}{2}y^2 = C.$$

In Exercises 44–49, use these ideas to find a general solution for the given differential equation. Hints are provided for some exercises.

44. $x\,dx + y\,dy = y^2(x^2 + y^2)\,dy$
    Hint: Consider $d(\ln(x^2 + y^2))$.

45. $x\,dy - y\,dx = y^3(x^2 + y^2)\,dy$
    Hint: Consider $d(\tan^{-1}(y/x))$.

46. $x\,dy + y\,dx = x^m y^n\,dx, \quad m \neq n - 1$

47. $x\,dy - y\,dx = (x^2 + y^2)^2(x\,dx + y\,dy)$
    Hint: Consider $d(x^2 + y^2)^2$.

48. $(xy + 1)(x\,dy - y\,dx) = y^2(x\,dy + y\,dx)$
    Hint: Consider $d(\ln(xy + 1))$.

49. $(x^2 - y^2)(x\,dy + y\,dx) = 2xy(x\,dy - y\,dx)$

50. A light situated at a point in a plane sends out beams of light in all directions. The beams in the plane meet a curve and are all reflected parallel to a line in the plane, as shown in Figure 9.

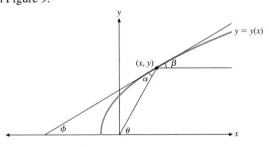

Figure 9. The reflector in Exercise 50.

The light is reflected so that the angle of incidence α equals the angle of reflection β.

    (a) Show that $\tan\theta = \tan 2\beta$; then use trigonometry to show that

$$\frac{y}{\sqrt{x^2 + y^2}} = \frac{2y'}{1 - (y')^2}. \qquad (6.45)$$

(b) Use the quadratic formula to solve equation (6.45); then solve the resulting first-order differential equation to find the equation of the reflecting curve. *Hint:*

You may want to try some of Exercises 44–49 before attempting this solution.

2.7 Existence and Uniqueness of Solutions

We have now discovered how to solve a few differential equations explicitly. We have also seen a few equations that cannot be solved explicitly, and we will find that, unfortunately, most differential equations are of this type. In this section, we begin to discover methods for studying properties of solutions when we do not know the solution explicitly. We will start with two very basic questions about an initial value problem.

- When can we be sure that a solution exists at all?
- How many different solutions are there to a given initial value problem?

These are the questions of existence and uniqueness.

Existence of solutions

We will start with an example.

Example 7.1 Consider the initial value problem

$$tx' = x + 3t^2, \quad \text{with} \quad x(0) = 1. \tag{7.2}$$

We recall that an equation of the form

$$x' = f(t, x) \tag{7.3}$$

is said to be in normal form. To put (7.2) into normal form, we divide by t to get

$$x' = \frac{1}{t}x + 3t. \tag{7.4}$$

This equation is linear, but makes no sense at $t = 0$ since the coefficient $1/t$ is undefined there. However, using Theorem 4.41, we see that, if we stay away from $t = 0$, every solution is of the form

$$x(t) = 3t^2 + Ct, \tag{7.5}$$

for some constant C. Notice that these solutions are defined for all values of t, including $t = 0$, and that $x(0) = 0$ for every solution. Furthermore, they are solutions to (7.2) even for $t = 0$. Consequently, if we want to solve the initial value problem in (7.2), we are out of luck. There is no solution to this initial value problem! (See Figure 1.) ●

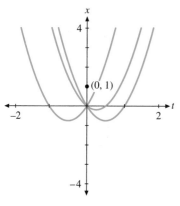

Figure 1. All solutions of $tx' = x + 3t^2$ pass through $(0, 0)$.

Initial value problems like that in Example 7.1 are anomalies. The reason for the nonexistence in Example 7.1 begins to appear if we put the differential equation into normal form (7.4). The resulting equation, makes no sense at $t = 0$. Most of the equations that we deal with are in normal form. It is extremely rare that an equation that arises in applications cannot be put into normal form. It turns out that for equations in normal form there is little problem with existence.

We will assume that the function $f(t, x)$ on the right-hand side of (7.3) is defined in a rectangle R defined by $a < t < b$, and $c < x < d$. Given a point $(t_0, x_0) \in R$, we want to know if there is a solution to (7.3) that satisfies the initial condition $x(t_0) = x_0$.

THEOREM 7.6 **Existence of solutions**

Suppose the function $f(t, x)$ is defined and continuous on the rectangle R in the tx-plane. Then given any point $(t_0, x_0) \in R$, the initial value problem

$$x' = f(t, x) \quad \text{and} \quad x(t_0) = x_0$$

has a solution $x(t)$ defined in an interval containing t_0. Furthermore, the solution will be defined at least until the solution curve $t \to (t, x(t))$ leaves the rectangle R. ■

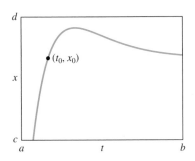

Figure 2. A solution to $x' = f(t, x)$ with $x(t_0) = x_0$ exists in both directions until it leaves the rectangle R.

The results of Theorem 7.6 are illustrated in Figure 2.

Notice that it is required that the equation be written in normal form as displayed in (7.3). Thus, in order to apply the theorem to an equation like that in (7.2), we first have to put it into normal form, as we did in (7.4). For the case in (7.4) the function on the right-hand side is

$$f(t, x) = \frac{1}{t}x + 3t.$$

Since f is discontinuous when $t = 0$, the existence theorem does not apply in any rectangle including points (t, x) with $t = 0$. Hence, the nonexistence of a solution to the initial value problem does not contradict the theorem.

On the other hand, according to the theorem the only condition on the right-hand side is that the function $f(t, x)$ be continuous. This is a very mild condition, and it is satisfied in most cases.

The interval of existence of a solution

We defined the interval of existence of a solution in Section 2.1 to be the largest interval in which the solution can be defined. Let's examine what Theorem 7.6 has to say about this concept.

Example 7.7 Consider the initial value problem

$$x' = 1 + x^2 \quad \text{with} \quad x(0) = 0. \tag{7.8}$$

Find the solution and its interval of existence.

The right-hand side is
$$f(t, x) = 1 + x^2,$$

which is continuous on the entire tx-plane. Hence we can take our rectangle to be the entire plane ($a = -\infty$, $b = \infty$, $c = -\infty$, $d = \infty$). Does this mean that the solutions are defined for $-\infty < t < \infty$?

Unfortunately it is not true, as we see when we find that the solution to the initial value problem is

$$x(t) = \tan t. \tag{7.9}$$

Notice that $x(t) = \tan t$ is discontinuous at $t = \pm\pi/2$. Hence the solution to the initial value problem given in (7.8) is defined only for $-\pi/2 < t < \pi/2$, so the interval of existence of the solution is the interval $(-\pi/2, \pi/2)$. ●

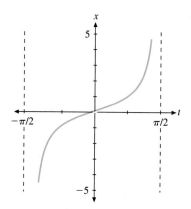

Figure 3. The solution to $x' = 1 + x^2$, $x(0) = 0$ becomes infinite at $t = \pm\pi/2$.

The last sentence in the existence theorem says that solutions exist until the solution curve leaves the rectangle R. In this case the solution curve leaves through the **top** of R as $t \to \pi/2$ from below, since $x(t) \to \infty$ there. In addition, it leaves through the **bottom** as $t \to -\pi/2$ from above, since $x(t) \to -\infty$. The theorem allows that solution curves can leave the rectangle R through any of its four sides, but that is the only thing that can happen. Nevertheless, it is important to realize that solutions to very nice equations, such as that in (7.8), can approach $\pm\infty$ in finite time. It cannot be assumed that solutions exist for all values of the independent variable. These facts are illustrated in Figure 3.

As Example 7.7 shows, the interval of existence of a solution cannot usually be found from the existence theorem. The only really reliable way to discover the interval of existence of a solution is to find an explicit formula for the solution. At best, the existence theorem gives an interval that is a subset of the interval of existence.

Existence for linear equations

Linear equations have the special form

$$x' = a(t)x + g(t).$$

This means that the right-hand side is of the special form

$$f(t, x) = a(t)x + g(t).$$

If $a(t)$ and $g(t)$ are continuous on the interval $b < t < c$, the function f is continuous on the rectangle R defined by $b < t < c$ and $-\infty < x < \infty$. In this case, a stronger existence theorem can be proved, which guarantees that solutions exist over the entire interval $b < t < c$.

Existence when the right-hand side is discontinuous

There are times when the right-hand side of the equation in (7.3) is discontinuous, yet we will want to talk about a "solution" to the initial value problem. Some of these examples are important in applications as well.

Example 7.10 Consider the initial value problem

$$y' = -2y + f(t), \qquad y(0) = 3, \tag{7.11}$$

where

$$f(t) = \begin{cases} 0, & \text{if } t < 1 \\ 5, & \text{otherwise.} \end{cases}$$

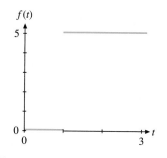

Figure 4. The discontinuous function in Example 7.10.

As illustrated in Figure 4, $f(t)$ has a discontinuity at $t = 1$. Nevertheless we will seek a "solution" to the initial value problem. For $0 \le t < 1$, the equation is $y' = -2y$ with the initial condition $y(0) = 3$. The solution in this smaller interval is $y(t) = 3e^{-2t}$. At $t = 1$, we have $y(1) = 3e^{-2}$. Having found the solution up to $t = 1$, we now are left with a new initial value problem for $t \ge 1$, namely

$$y' = -2y + 5, \qquad y(1) = 3e^{-2}.$$

This is a perfectly respectable initial value problem and can be solved easily. The solution is $y(t) = 5/2 + (3 - 5e^2/2)e^{-2t}$. Thus the initial value problem has a piecewise defined "solution"

$$y(t) = \begin{cases} 3e^{-2t}, & \text{for } t < 1, \\ 5/2 + (3 - 5e^2/2)e^{-2t}, & \text{for } t \geq 1. \end{cases} \tag{7.12}$$

The function defined in (7.12) solves the differential equation in (7.11) everywhere except at $t = 1$, and it is continuous everywhere. The solution is shown in Figure 5. As might be gathered from the sharp cusp in Figure 5, y fails to have a derivative at $t = 1$. Example 7.10 shows that there are cases when the hypothesis of Theorem 7.6 are not satisfied, yet solutions to initial value problems are desirable. In cases that arise in applications, the equation is linear,

$$x' = a(t)x + f(t),$$

and the only discontinuity is in the $f(t)$ term. In situations like this, we will agree to accept as a solution a continuous function $x(t)$ that satisfies the equation except where f is discontinuous. ●

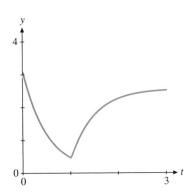

Figure 5. The solution to the initial value problem in Example 7.10.

Uniqueness of solutions

It is interesting to contemplate the existence theorem in conjunction with the physical systems that are modeled by the differential equations. The existence of a solution to an ordinary differential equation (ODE) simply reflects the fact that the physical systems change according to the relationships modeled by the equation. We would expect that solutions to equations that model physical behavior would exist. Next we turn to the question of the number of solutions to an initial value problem. If there is only one solution, then the physical system acts the same way each time it is started from the same set of initial conditions. Such a system is therefore **deterministic**. If an equation has more than one solution, then the physical response is unpredictable. Thus the uniqueness of solutions of initial value problems is equivalent to the system being deterministic. It is not too much to say that the success of science requires that solutions to initial value problems be unique.

Before we state our uniqueness theorem, we present an example that shows that we must restrict the right-hand side of the equation

$$x' = f(t, x) \tag{7.13}$$

more than we did in the existence theorem in order to have uniqueness. Consider the initial value problem

$$x' = x^{1/3}, \quad \text{with} \quad x(0) = 0. \tag{7.14}$$

This is a separable equation, and you are encouraged to find a solution. First we notice that

$$x(t) = 0$$

is a solution. Next we define

$$y(t) = \begin{cases} \left(\dfrac{2t}{3}\right)^{3/2}, & t > 0 \\ 0, & t \leq 0. \end{cases}$$

It is easily verified by direct substitution that y is also a solution to (7.14) (although technically it is necessary to use the limit quotient definition of derivative to calculate that $y'(0) = 0$).

Thus, we have two solutions to the initial value problem (7.14). (See Figure 6.) Notice that the function $f(t, x) = x^{1/3}$ is continuous and therefore satisfies the hypothesis of the existence theorem. Consequently, we will need a stronger condition on the right-hand side of (7.13) to ensure uniqueness.

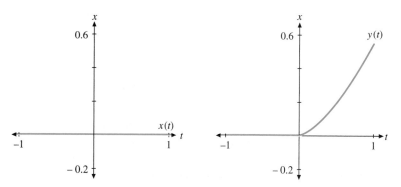

Figure 6. Two solutions to the initial value problem in (7.14).

The uniqueness theorem will follow easily from the following theorem, which we will find useful in other ways.

THEOREM 7.15   Suppose the function $f(t, x)$ and its partial derivative $\dfrac{\partial f}{\partial x}$ are both continuous on the rectangle R in the tx-plane and let

$$M = \max_{(t,x) \in R} \left| \frac{\partial f}{\partial x}(t, x) \right|.$$

Suppose (t_0, x_0) and (t_0, y_0) are in R and that

$$x'(t) = f(t, x(t)), \quad \text{and} \quad x(t_0) = x_0$$
$$y'(t) = f(t, y(t)), \quad \text{and} \quad y(t_0) = y_0.$$

Then as long as $(t, x(t))$ and $(t, y(t))$ belong to R, we have

$$|x(t) - y(t)| \leq |x_0 - y_0| e^{M|t - t_0|}. \qquad \blacksquare$$

The theorem provides an estimate of how much two solutions, $x(t)$ and $y(t)$, to the same differential equation can differ depending on how close together their initial values are—this is the $|x_0 - y_0|$ term—and on how far we are from the initial points—this is the $e^{M|t-t_0|}$ term. The special case when the initial values are equal is of most interest to us at the moment. In this case, we have $x_0 = y_0$, so $|x_0 - y_0| = 0$. Hence the theorem implies that $|x(t) - y(t)| \leq 0$ for all t. Since the absolute value is always nonnegative, we must have $x(t) - y(t) = 0$, or $x(t) = y(t)$ for all t. This is the uniqueness theorem, and we will state it separately.

THEOREM 7.16 Uniqueness of solutions

Suppose the function $f(t, x)$ and its partial derivative $\partial f/\partial x$ are both continuous on the rectangle R in the tx-plane. Suppose $(t_0, x_0) \in R$ and that the solutions

$$x' = f(t, x) \quad \text{and} \quad y' = f(t, y)$$

satisfy

$$x(t_0) = y(t_0) = x_0.$$

Then as long as $(t, x(t))$ and $(t, y(t))$ stay in R, we have

$$x(t) = y(t). \qquad \blacksquare$$

There are several ways to look at the uniqueness theorem. The simplest is just to rephrase the statement of the theorem. Roughly it says that, under suitable hypotheses, two solutions to the same equation that start together stay together. The upshot of this is that through any point $(t_0, x_0) \in R$, there is only one solution curve.

It is important to realize that any point in R can be the starting point. For example, suppose we have two solutions $x(t)$ and $y(t)$ to the same equation in our rectangle R and at some point t_1 the two agree, so $x(t_1) = y(t_1)$. We can take t_1 as our starting point (relabel it t_0 if you wish), and the uniqueness theorem says that the two solutions must agree everywhere.

Mathematics and theorems

Any theorem is a logical statement. Like all logical statements, it has hypotheses and conclusions. A simple example is the statement, "If it rains, the sidewalk gets wet." Here the hypothesis is, "If it rains," and the conclusion is, "the sidewalk gets wet." The point of a logical statement is that when the hypotheses are true, then the conclusion is true as well. It is important to realize that the implication goes only one way. In our simple example, if the sidewalk is wet, it does not follow that it has rained. There are lots of ways for a sidewalk to get wet.

Let's examine the uniqueness theorem, Theorem 7.16, and list the hypotheses and conclusions. The hypotheses are as follows:

1. The equation is in normal form $y' = f(t, y)$.
2. The right-hand side $f(t, y)$ and its derivative $\partial f/\partial y$ are both continuous in the rectangle R.
3. The initial point (t_0, y_0) is in the rectangle R.

Notice that the first hypothesis is not stated precisely in the theorem. It is implicit in the form of the equations used.

If these hypotheses are valid, then the conclusions of the theorem are true. For the uniqueness theorem the conclusions are as follows:

1. There is one and only one solution to the initial value problem.
2. The solution exists until the solution curve $t \to (t, y(t))$ leaves the rectangle R.

Most logical statements, like our example of the wet sidewalk, are based on experience. Mathematical theorems are different, however. They are based on logical proof starting from more basic assumptions and other theorems that have previously been proved. For this reason mathematical theorems have a truth value that goes well beyond the ordinary logical statement.

That said, you might well ask how the existence and uniqueness theorems are proved. In the case of the existence theorem, it is necessary to exhibit a solution to the initial value problem. This means we have to exhibit a function $y(t)$ which satisfies $y(t_0) = y_0$ and $y'(t) = f(t, y(t))$. Even in a simple case like $y' = y^2 - t$, it will not be possible to exhibit the solution by giving a formula. We need to have other ways of defining functions. Remember that a function y is a rule that assigns a value $y(t)$ to every value of t in the domain of y. The proof of the existence theorem exhibits the solution as the limit of a sequence of functions that are explicitly presented. In general, this limiting process does not allow us to give a formula for the solution.

Applying the existence and uniqueness theorems

We usually want to apply the theorems to a specific initial value problem

$$y' = f(t, y), \quad \text{with} \quad y(t_0) = y_0.$$

To be able to apply the theorems, it is necessary to find a rectangle R in which the equation satisfies the hypotheses.

Example 7.17 Consider the equation $tx' = x + 3t^2$ from Example 7.1. Is there a solution to this equation with the initial condition $x(1) = 2$? If so, is the solution unique?

To begin with, we write the equation in normal form

$$x' = \frac{1}{t}x + 3t.$$

The right-hand side, $f(t, x) = (x/t) + 3t$, is continuous except where $t = 0$. We can take R to be any rectangle which contains the initial point $(1, 2)$ and avoids $t = 0$. One good choice is to define R by $1/2 < t < 2$ and $0 < x < 4$. Then f is continuous everywhere in R, so the hypotheses of the existence theorem are satisfied. We can conclude that there is a solution to the initial value theorem. Furthermore, since $\partial f/\partial x = 1/t$ is also continuous everywhere in R, the hypotheses of the uniqueness theorem are satisfied as well. Consequently there is only one solution. ●

Example 7.18 Consider the equation

$$x' = (x - 1)\cos(xt)$$

and suppose we have a solution $x(t)$ that satisfies $x(0) = 1$. We claim that $x(t) = 1$ for all t. How do we prove it?

The right-hand side, $f(t, x) = (x - 1)\cos(xt)$, is continuous in the entire plane, as is the derivative $\partial f/\partial x = \cos(xt) - t(x - 1)\sin(tx)$. Therefore, the hypotheses of the uniqueness theorem are satisfied in any rectangle containing the initial point $(0, 1)$. The key fact is the observation that $y(t) = 1$ is also a solution to the equation as we see by direct substitution. We have $x(0) = y(0) = 1$, so the uniqueness theorem implies that $x(t) = y(t) = 1$ for all t. ●

This example illustrates a very typical use of the uniqueness theorem. The tricky part of the example was the solution $y(t) = 1$, which we apparently pulled out of a hat. This particular hat is available to everyone. The trick is to look for solutions to a differential equation that are constant functions. In this case, we looked for a constant c such that

$$(c - 1)\cos ct = 0 \quad \text{for all } t.$$

Clearly we want $c = 1$. Then the constant function $x(t) = c$ (in our case, $x(t) = 1$) is a solution to the differential equation.

Let's go over the more general case. We are looking for constant solutions $x(t) = c$ to the equation

$$x' = f(t, x). \tag{7.19}$$

On the left-hand side we have $x' = 0$, since $x(t) = c$ is a constant function. To have equality in (7.19), the right-hand side must also be equal to 0. Hence we need

$$f(t, c) = 0 \quad \text{for all } t. \tag{7.20}$$

Thus, to find constant solutions $x(t) = c$, we look for constants that satisfy (7.20).

Geometric interpretation of uniqueness

The uniqueness theorem has an important geometric interpretation. Let's look at the graphs of the solutions—the solution curves. If we have two functions $x(t)$ and $y(t)$ that satisfy $x(t_0) = y(t_0) = x_0$ at some point, then the graphs of $x(t)$ and $y(t)$ meet at the point (t_0, x_0). If in addition we know that $x(t)$ and $y(t)$ are solutions to the same differential equation, then the uniqueness theorem implies that $x(t) = y(t)$ for all t. In other words, the graphs of $x(t)$ and $y(t)$ coincide. Stated in a different way, two distinct solution curves cannot meet. This means they cannot cross each other or even touch each other.

Example 7.21 The geometric view of the uniqueness theorem illustrates how knowledge of one solution to a differential equation can give us information about another solution that we do not know as well. In Example 7.18, we discovered that $y(t) = 1$ is a solution to
$$y' = (y - 1) \cos yt.$$

Now consider the solution x of the initial value problem

$$x' = (x - 1) \cos xt, \quad x(0) = 2.$$

Is it possible that $x(2) = 0$?

If $x(2) = 0$, then since $x(0) = 2$ there must be some point t_0 between 0 and 2 where $x(t_0) = 1$. This is an application of the intermediate value theorem from calculus, but it is most easily seen by looking at the graphs of x and y in Figure 7. Thus, to get from the initial point $(0, 2)$ to $(2, 0)$, the graph of x must cross the graph of y. The uniqueness theorem says this cannot happen. Consequently, we conclude that $x(2) \neq 0$.

In fact, the same reasoning implies that we cannot have $x(t) \leq 1$ for any value of t, and we conclude that
$$x(t) > 1 \quad \text{for all } t.$$

Geometrically we see that the graph of x must lie above the graph of the solution y. Thus, knowledge of the solution $y(t) = 1$, together with the uniqueness theorem, gives us information about the solution x, or about any other solution.

In Figure 7, the numerical solution for the solution x is shown, verifying that its solution curve lies above the graph of $y(t) = 1$. ●

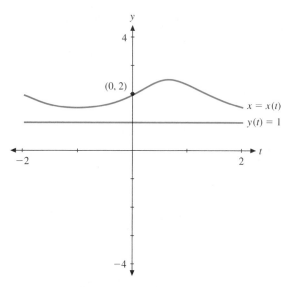

Figure 7. The solutions to the initial value problems in Example 7.21. Solution curves cannot cross, so $x(2) \neq 0$.

The geometric fact that solution curves cannot meet will be important in what follows. A curve in the plane divides the plane into two separate pieces, so any solution curve limits the space available to any other. This simple fact will be exploited in a remarkable variety of ways. When we study higher order equations and systems of more than one equation, this geometric interpretation cannot be made, simply because curves in dimensions bigger than two do not divide space into separate parts. If there is a third dimension available, it is always possible to move into that direction to get around any curve.

Computer-drawn pictures can sometimes be misleading with regard to uniqueness. Consider the solution curves on the left in Figure 8. Here we are looking at solutions of the equation $x' = x^2 - t$. It seems as though three solution curves merge in the lower right-hand part of the figure. However, they are only getting very close. In fact, they are getting exponentially close. It happens frequently that solution curves get exponentially close, but the uniqueness theorem assures us that they never actually meet. On the right in Figure 8 we show the result of zooming in

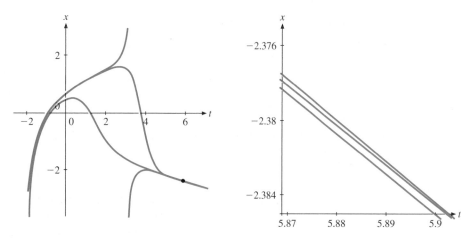

Figure 8. Sometimes solution curves seem to run together shown on the left. However, if we zoom in near the indicated point to get the graph on the right, the solution curves can be separated.

near the indicated point in the graph on the left. Notice that with this magnification, the solution curves are distinct, as uniqueness requires.

EXERCISES

Which of the initial value problems in Exercises 1–6 are guaranteed a unique solution by the hypotheses of Theorem 7.16? Justify your answer.

1. $y' = 4 + y^2$, $y(0) = 1$ **2.** $y' = \sqrt{y}$, $y(4) = 0$

3. $y' = t \tan^{-1} y$, $y(0) = 2$

4. $\omega' = \omega \sin \omega + s$, $\omega(0) = -1$

5. $x' = \dfrac{t}{x+1}$, $x(0) = 0$

6. $y' = \dfrac{1}{x} y + 2$, $y(0) = 1$

For each differential equation in Exercises 7–8, perform each of the following tasks.

(i) Find the general solution of the differential equation. Sketch several members of the family of solutions portrayed by the general solution.

(ii) Show that there is no solution satisfying the given initial condition. Explain why this lack of solution does not contradict the existence theorem.

7. $ty' - y = t^2 \cos t$, $y(0) = -3$

8. $ty' = 2y - t$, $y(0) = 2$

9. Show that $y(t) = 0$ and $y(t) = t^3$ are both solutions of the initial value problem $y' = 3y^{2/3}$, where $y(0) = 0$. Explain why this fact does not contradict Theorem 7.16.

10. Show that $y(t) = 0$ and $y(t) = (1/16)t^4$ are both solutions of the initial value problem $y' = ty^{1/2}$, where $y(0) = 0$. Explain why this fact does not contradict Theorem 7.16.

In Exercises 11–16, use a numerical solver to sketch the solution of the given initial value problem.

(i) Where does your solver experience difficulty? Why? Use the image of your solution to estimate the interval of existence.

(ii) For 11–14 only, find an explicit solution; then use your formula to determine the interval of existence. How does it compare with the approximation found in part (i)?

11. $\dfrac{dy}{dt} = \dfrac{t}{y+1}$, $y(2) = 0$

12. $\dfrac{dy}{dt} = \dfrac{t-2}{y+1}$, $y(-1) = 1$

13. $\dfrac{dy}{dt} = \dfrac{1}{(t-1)(y+1)}$, $y(0) = 1$

14. $\dfrac{dy}{dt} = \dfrac{1}{(t+2)(y-3)}$, $y(0) = 1$

15. $\dfrac{dy}{dt} = \dfrac{2t^2}{(y+3)(y-1)}$, $y(0) = 0$

16. $\dfrac{dy}{dt} = \dfrac{-t^2}{y(y-5)}$, $y(0) = 3$

An electric circuit, consisting of a capacitor, resistor, and an electromotive force can be modeled by the differential equation

$$R\frac{dq}{dt} + \frac{1}{C}q = E(t),$$

where R and C are constants (resistance and capacitance) and $q = q(t)$ is the amount of charge on the capacitor at time t. For simplicity in the following analysis, let $R = C = 1$, forming the differential equation $dq/dt + q = E(t)$. In Exercises 17–20, an electromotive force is given in piecewise form, a favorite among engineers. Assume that the initial charge on the capacitor is zero [$q(0) = 0$].

(i) Use a numerical solver to draw a graph of the charge on the capacitor during the time interval [0, 4].

(ii) Find an explicit solution and use the formula to determine the charge on the capacitor at the end of the four-second time period.

17. $E(t) = \begin{cases} 5, & \text{if } 0 < t < 2, \\ 0, & \text{if } t \ge 2, \end{cases}$

18. $E(t) = \begin{cases} 0, & \text{if } 0 < t < 2, \\ 3, & \text{if } t \ge 2, \end{cases}$

19. $E(t) = \begin{cases} 2t, & \text{if } 0 < t < 2, \\ 0, & \text{if } t \ge 2, \end{cases}$

20. $E(t) = \begin{cases} 0, & \text{if } 0 < t < 2, \\ t, & \text{if } t \ge 2, \end{cases}$

21. Consider the initial value problem

$$y' = 3y^{2/3}, \quad y(0) = 0.$$

It is not difficult to construct an infinite number of solutions. Consider

$$y(t) = \begin{cases} 0, & \text{if } t \le t_0, \\ (t - t_0)^3, & \text{if } t > t_0, \end{cases}$$

where t_0 is any positive number. It is easy to calculate the derivative of $y(t)$, when $t \ne t_0$,

$$y'(t) = \begin{cases} 0, & \text{if } t < t_0, \\ 3(t - t_0)^2, & \text{if } t > t_0, \end{cases}$$

but the derivative at t_0 remains uncertain.

(a) Evaluate both

$$y'(t_0^+) = \lim_{t \searrow t_0} \frac{y(t) - y(t_0)}{t - t_0}$$

and

$$y'(t_0^-) = \lim_{t \nearrow t_0} \frac{y(t) - y(t_0)}{t - t_0},$$

showing that

$$y'(t) = \begin{cases} 0, & \text{if } t \le t_0, \\ 3(t - t_0)^2, & \text{if } t > t_0. \end{cases}$$

(b) Finally, show that $y(t)$ is a solution of 21. Why doesn't this example contradict Theorem 7.16?

22. Consider again the "solution" of equation (7.11) in Example 7.10,

$$y(t) = \begin{cases} 3e^{-2t}, & \text{for } t < 1, \\ 5/2 + (3 - 5e^2/2)e^{-2t} & \text{for } t \ge 1. \end{cases}$$

(a) Follow the lead in Exercise 21 to calculate the derivative of $y(t)$.

(b) In the sense of the definition of solution in Section 2.1, is $y(t)$ a solution of (7.11)? Why or why not?

(c) Show that $y(t)$ satisfies equation (7.11) for all t except $t = 1$.

23. Show that

$$y(t) = \begin{cases} 0, & \text{for } t < 0, \\ t^4 & \text{for } t \ge 0 \end{cases}$$

is a solution of the initial value problem $ty' = 4y$, where $y(0) = 0$, in the sense of Definition 1.15 from Section 2.1. Find a second solution and explain why this lack of uniqueness does not contradict Theorem 7.16.

24. Uniqueness is not just an abstraction designed to please theoretical mathematicians. For example, consider a cylindrical drum filled with water. A circular drain is opened at the bottom of the drum and the water is allowed to pour out. Imagine that you come upon the scene and witness an empty drum. You have no idea how long the drum has been empty. Is it possible for you to determine when the drum was full?

(a) Using physical intuition only, sketch several possible graphs of the height of the water in the drum versus time. Be sure to mark the time that you appeared on the scene on your graph.

(b) It is reasonable to expect that the speed at which the water leaves through the drain depends upon the height of the water in the drum. Indeed, Torricelli's law predicts that this speed is related to the height by the formula $v^2 = 2gh$, where g is the acceleration due to gravity near the surface of the earth. Let A and a represent the area of a cross section of the drum and drain, respectively. Argue that $A \, \Delta h = av \, \Delta t$, and in the limit, $A \, dh/dt = av$. Show that $dh/dt = -(a/A)\sqrt{2gh}$.

(c) By introducing the dimensionless variables $\omega = \alpha h$ and $s = \beta t$ and then choosing parameters

$$\alpha = \frac{1}{h_0} \quad \text{and} \quad \beta = \left(\frac{a}{A}\right)\sqrt{\frac{2g}{h_0}},$$

where h_0 represents the height of a full tank, show that the equation $dh/dt = -(a/A)\sqrt{2gh}$ becomes $dw/ds = -\sqrt{w}$. Note that when $w = 0$, the tank is empty, and when $w = 1$, the tank is full.

(d) You come along at time $s = s_0$ and note that the tank is empty. Show that the initial value problem, $dw/ds = -\sqrt{w}$, where $w(s_0) = 0$, has an infinite number of solutions. Why doesn't this fact contradict the uniqueness theorem? *Hint:* The equation is separable and the graphs you drew in part (a) should provide the necessary hint on how to proceed.

25. Is it possible to find a function $f(t, x)$ that is continuous and has continuous partial derivatives such that the functions $x_1(t) = t$ and $x_2(t) = \sin t$ are both solutions to $x' = f(t, x)$ near $t = 0$?

26. Is it possible to find a function $f(t, x)$ that is continuous and has continuous partial derivatives such that the functions $x_1(t) = \cos t$ and $x_2(t) = 1 - \sin t$ are both solutions to $x' = f(t, x)$ near $t = \pi/2$?

27. Suppose that x is a solution to the initial value problem

$$x' = x \cos^2 t \quad \text{and} \quad x(0) = 1.$$

Show that $x(t) > 0$ for all t for which x is defined.

28. Suppose that y is a solution to the initial value problem

$$y' = (y - 3)e^{\cos(ty)} \quad \text{and} \quad y(1) = 1.$$

Show that $y(t) < 3$ for all t for which y is defined.

29. Suppose that y is a solution to the initial value problem

$$y' = (y^2 - 1)e^{ty} \quad \text{and} \quad y(1) = 0.$$

Show that $-1 < y(t) < 1$ for all t for which y is defined.

30. Suppose that x is a solution to the initial value problem

$$x' = \frac{x^3 - x}{1 + t^2 x^2} \quad \text{and} \quad x(0) = 1/2.$$

Show that $0 < x(t) < 1$ for all t for which x is defined.

31. Suppose that x is a solution to the initial value problem

$$x' = x - t^2 + 2t \quad \text{and} \quad x(0) = 1.$$

Show that $x(t) > t^2$ for all t for which x is defined.

32. Suppose that y is a solution to the initial value problem

$$y' = y^2 - \cos^2 t - \sin t \quad \text{and} \quad y(0) = 2.$$

Show that $y(t) > \cos t$ for all t for which y is defined.

2.8 Dependence of Solutions on Initial Conditions

Suppose we have two initial value problems involving the same differential equation, but with different initial conditions that are very close to each other. Do the solutions stay close to each other? This is the question we will address in this section. The question is important, since in many situations the initial condition is determined experimentally and therefore is subject to experimental error. If we use the slightly incorrect initial condition in an initial value problem, instead of the correct one, how accurate will the solution be at later times?

There are two aspects to the problem. The first question is, "Can we ensure that the solution with incorrect initial data is close enough to the real solution that we can use it to predict behavior?" This is the problem of *continuity of the solution with respect to initial data*. The second aspect looks at the problem from the other end. Given that we have an error in the initial conditions, just how far from the true situation can the solution be? This is the problem of *sensitivity to initial conditions*.

Everything we do in this section will follow from Theorem 7.15 in the previous section. We will analyze its implications when the initial conditions of the two solutions are not equal (i.e., when $x_0 \neq y_0$). In this case, Theorem 7.15 provides limits on how far apart the corresponding solutions can be as the independent variable changes. It provides an upper bound on how the initial error propagates.

Continuity with respect to initial conditions

Let's look at a specific example.

Example 8.1 Examine the behavior of solutions to

$$x' = (x - 1) \cos t. \tag{8.2}$$

In this case, $f(t, x) = (x - 1) \cos t$, and $\dfrac{\partial f}{\partial x} = \cos t$. Hence

$$M = \max_{(t,x) \in R} \left| \frac{\partial f}{\partial x} \right| \leq 1$$

regardless of which rectangle R we choose. Therefore, we may as well take $M = 1$. Suppose that we have two solutions $x(t)$ and $y(t)$ of (8.2) with initial conditions $x(t_0) = x_0$, and $y(t_0) = y_0$. According to Theorem 7.15, with $M = 1$,

$$|x(t) - y(t)| \leq |x_0 - y_0| e^{|t - t_0|} \quad \text{for all } t. \tag{8.3}$$

This is illustrated in Figures 1 and 2. The black curve in each is the solution to (8.2) with initial condition $x(0) = 0$. The colored curves in Figure 1 show the limits in (8.3) when $|x_0 - y_0| \leq 0.1$, while those in Figure 2 show the limits when $|x_0 - y_0| \leq 0.01$.

To take a concrete example, suppose $t_0 = 0$ and the t-dimension of the rectangle R is $-1 \leq t \leq 1$. (In this case, the x-dimension of R is not important.) Then if $(t, x) \in R$, we have $|t - t_0| = |t| \leq 1$, and (8.3) becomes

$$|x(t) - y(t)| \leq e|x_0 - y_0| \quad \text{if } |t| \leq 1. \tag{8.4}$$

If we wanted to ensure that $|x(t) - y(t)| \leq 0.01$ for $|t| \leq 1$, we should insist that the initial difference satisfies $|x_0 - y_0| \leq 0.01/e$. It is clear that we can ensure that $|x(t) - y(t)|$ is as small as we wish simply by making sure that $|x_0 - y_0|$ is small enough. ●

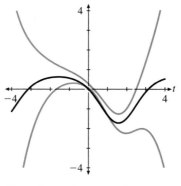

Figure 1. A solution to (8.2) with $|x(0)| \leq 0.1$ must lie between the colored curves.

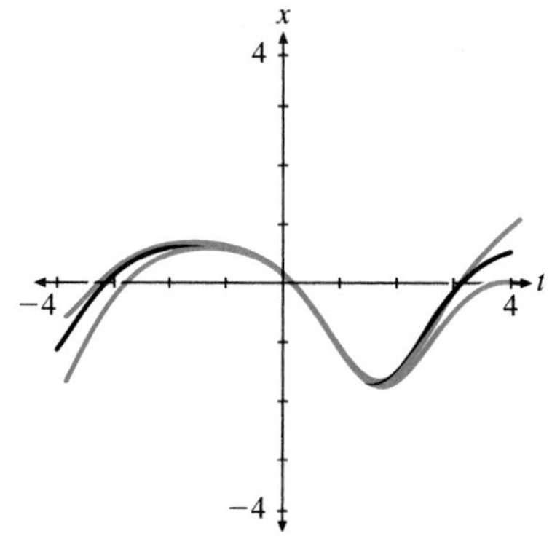

Figure 2. A solution to (8.2) with $|x(0)| \leq 0.01$ must lie between the colored curves.

Theorem 7.15 of the previous section implies that what is seen in Example 8.1 is true in general. If $R = \{(t, x) | a \leq t \leq b$ and $c \leq x \leq d\}$, and if

$$M = \max_{(t,x) \in R} \left| \frac{\partial f}{\partial x}(t, x) \right|,$$

then for two solutions $x(t)$ and $y(t)$ we have

$$|x(t) - y(t)| \leq |x_0 - y_0| e^{M|t - t_0|} \tag{8.5}$$

as long as $(t, x(t))$ and $(t, y(t))$ stay in the rectangle R. In particular, since $|t - t_0| \leq b - a$, we have

$$|x(t) - y(t)| \leq e^{M(b-a)} |x_0 - y_0|,$$

provided that $(t, x(t))$ and $(t, y(t))$ stay in R. As we did in Example 8.1, we can ensure that $|x(t) - y(t)| < \epsilon$ by taking $|x_0 - y_0| < e^{-M(b-a)}\epsilon$. Thus we can be sure that the two solutions stay very close (to be precise, within ϵ of each other) over the interval (a, b) by ensuring that the initial conditions are very close (within $e^{-M(b-a)}\epsilon$ of each other).

We sum up these thoughts by saying that the solutions to an ODE are continuous with respect to the initial conditions.

Sensitivity of solutions to the initial condition

While our deliberations in the preceding section are reassuring, the exponential term in (8.5) is a cause for concern. This term can get extremely large if $|t - t_0|$ is large. For example, if $M = 2$ and $|t - t_0| = 3$, then

$$e^{M|t - t_0|} = e^6 \sim 403.4,$$

while if $|t - t_0| = 10$, then

$$e^{M|t - t_0|} = e^{20} = 4.85 \times 10^8.$$

Thus, we see that as $|t - t_0|$ gets large, the control given by equation (8.5) on the difference of the solutions rapidly gets weaker.

Of course, equation (8.5) is an inequality and therefore provides an upper bound to the difference between the solutions. Does such "worst case" behavior actually occur? The next example shows that it does, and in some of the simplest examples.

E x a m p l e 8 . 6   Consider the exponential equation

$$x' = x.$$

The solutions with initial values $x(t_0) = x_0$ and $y(t_0) = y_0$ are

$$x(t) = x_0 e^{t-t_0} \quad \text{and} \quad y(t) = y_0 e^{t-t_0}.$$

Hence

$$x(t) - y(t) = (x_0 - y_0)e^{t-t_0}. \tag{8.7}$$

Since for the exponential equation $x' = x$ the right-hand side is

$$f(t, x) = x,$$

we have $\partial f/\partial x = 1$. Hence $M = 1$. We see therefore that the two solutions to the exponential equation give precisely the worst case behavior predicted by the inequality in equation (8.5). The difference between the two solutions becomes exponentially larger as t increases. ●

Although this example shows that the worst case behavior does occur, it does not always occur. Quite the opposite phenomenon occurs with the exponential equation if we let t decrease from t_0. Then (8.7) shows that the difference between the solutions actually decreases exponentially. If we are predicting physical phenomena from initial conditions, this is the best case scenario.

As these examples show, the sensitivity of solutions to initial conditions is limited by the inequality in Theorem 7.15, but beyond that not much can be said. It can be as bad as allowed by Theorem 7.15, but in some situations it can be much better. Let's look at a more visual example.

E x a m p l e 8 . 8   Consider the equation

$$x' = x \sin(x) + t.$$

Figure 3 shows solutions to three initial value problems with initial values differing by 2×10^{-5}. The solution curves remain very close for $0 \le t \le 2$. Nevertheless, they diverge pretty quickly after that, indicating sensitivity to initial conditions. ●

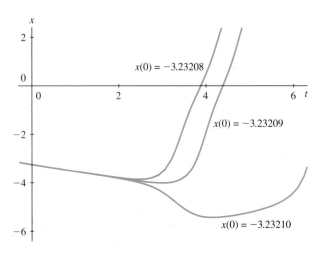

Figure 3. Sensitivity to initial conditions for solutions to $x' = x \sin(x) + t$.

Sensitivity to initial conditions is the idea behind the theory of chaos, which has developed over the past 25 years. In chaotic situations, solutions are sensitive to initial conditions for a large set of possible initial conditions. In the situations we have examined, the sensitivity occurs only at a few isolated points. Such equations do not give rise to truly chaotic behavior.

EXERCISES

Sensitivity to initial conditions is well illustrated by a little target practice with your numerical solver. In Exercises 1–12, you are given a differential equation $x' = f(t, x)$ and a "target." In each case, enter the equation into your numerical solver, and then experiment with initial conditions at the given value of t_0 until the solution of $x' = f(t, x)$, with $x(t_0, x_0)$, "hits" the given target.

We will use the simple linear equation, $x' = x - t$ in Exercises 1–4. The initial conditions are at $t_0 = 0$. The target is

1. $(3, 0)$ **2.** $(4, 0)$

3. $(5, 0)$ **4.** $(6, 0)$

In Exercises 5–8, we use the slightly more complicated nonlinear equation, $x' = x^2 - t$. Again the initial conditions are at $t_0 = 0$. The target is

5. $(3, 0)$ **6.** $(4, 0)$

7. $(5, 0)$ **8.** $(6, 0)$

For Exercises 9–12, we use the equation in Example 8.8, $x' = x \sin x + t$. Again the initial conditions are at $t_0 = 0$. The target is

9. $(3, 0)$ **10.** $(4, 0)$

11. $(5, 0)$ **12.** $(6, 0)$

13. This exercise addresses a very common instance of a motion that is sensitive to initial conditions. Flip a coin with your thumb and forefinger, and let the coin land on a pillow. The motion of the coin is governed by a system of ordinary differential equations. It is not immediately important what that system is. It is only important to realize that the motion is governed entirely by the initial conditions (i.e., the upward velocity of the coin and the rotational energy imparted to it when it is flipped). If the motion were not sensitive to initial conditions, it would be possible to learn how to flip 10 heads in a row. Try to learn how to do this, and report the longest chain of heads you are able to achieve. The flipping of a coin is often considered to have a random outcome. In fact, the result is determined by the initial conditions. It is the sensitivity of the result to the initial conditions that gives the appearance of randomness.

14. Let's plot the error bounds shown in Figure 1. First, solve $x' = (x - 1) \cos t$, $x(0) = 0$, and plot the solution over

the interval $[-4, 4]$. Next, as we saw in Example 8.1, if $y(t)$ is a second solution with $|x(0) - y(0)| \leq 0.1$, then the inequality (8.3) becomes $|x(t) - y(t)| \leq 0.1 e^{|t|}$. Solve this inequality for $x(t)$, placing your final answer in the form $e_L(t) \leq x(t) \leq e_H(t)$. Then add the graphs of $e_L(t)$ and $e_H(t)$ to your plot. How can you use Theorem 7.16 to show that no solution starting with initial condition $|x(0) - y(0)| \leq 0.1$ has any chance of rising as far as indicated by $e_H(t)$?

15. Draw the error bounds shown in Figure 2. See Exercise 14 for assistance.

16. Consider the equation $x' = (x - 1) \cos t$.

(a) Let $x(t)$ and $y(t)$ be two solutions. What is the upper bound on the separation $|x(t) - y(t)|$ predicted by Theorem 7.15?

(b) Find the solution $x(t)$ with initial value $x(0) = 0$, and the solution $y(t)$ with initial value $y(0) = 1/10$. Does the separation $x(t) - y(t)$ satisfy the inequality found in part (a)?

(c) Are there any values of t where the separation achieves the maximum predicted?

17. Consider $x' + 2x = \sin t$.

(a) Let $x(t)$ and $y(t)$ be two solutions. What is the upper bound on the separation $|x(t) - y(t)|$ predicted by Theorem 7.15?

(b) Find the solution $x(t)$ with initial value $x(0) = -1/5$, and the solution $y(t)$ with initial value $y(0) = -3/10$. Does the separation $x(t) - y(t)$ satisfy the inequality found in part (a)?

(c) Are there any values of t where the separation achieves the maximum predicted?

18. Let $x_1(t)$ and $x_2(t)$ be solutions of $x' = x^2 - t$ having initial conditions $x_1(0) = 0$ and $x_2(0) = 3/4$. Use Theorem 7.15 to determine an upper bound for $|x_1(t) - x_2(t)|$, as long as the solutions $x_1(t)$ and $x_2(t)$ remain inside the rectangle defined by $R = \{(t, x) : -1 \leq t \leq 1, -2 \leq x \leq 2\}$. Use your numerical solver to draw the solutions $x_1(t)$ and $x_2(t)$, restricted to the rectangular region R. Estimate $\max_R |x_1(t) - x_2(t)|$ and compare with the estimated upper bound.

2.9 Autonomous Equations and Stability

A first-order **autonomous** equation is an equation of the special form

$$x' = f(x). \tag{9.1}$$

Notice that the independent variable, which we have usually been denoting by t, does not appear explicitly on the right-hand side of equation (9.1). This is the defining feature of an autonomous equation.

In Section 2.3, we derived the differential equation

$$v' = -g - kv|v|/m, \tag{9.2}$$

which models the velocity v of an object near the surface of the earth with air resistance proportional to the square of the velocity. Here g is the acceleration due to gravity, m is the mass of the object, and k is a proportionality constant. This is an autonomous equation, since the independent variable t does not appear on the right-hand side. Other examples of autonomous equations are

$$x' = \sin(x), \quad y' = y^2 + 1, \quad \text{and} \quad z' = e^z.$$

The equations

$$x' = \sin(tx), \quad y' = y^2 + t, \quad z' = t^z, \quad \text{and} \quad y' = xy$$

are not autonomous, since the independent variable appears on the right-hand side of each equation.

Autonomous equations occur very frequently in applications. A differential equation model of any physical system that is evolving without external forces will almost always be autonomous. It is usually the external forces that give rise to terms that depend explicitly on time.

In this section, we will describe ways to discover the qualitative behavior of solutions to autonomous equations, without actually finding the solutions. Although autonomous equations are in principle solvable, finding the solutions explicitly may be difficult and the results may be so complicated that the formula does not reveal the behavior of the solutions. In contrast, qualitative methods are so easy that it will be useful to study the solutions qualitatively in addition to finding the solutions explicitly, when that is possible. In some cases, it might be sufficient to do the qualitative analysis without finding exact solutions.

In particular, we will learn how to discover the limiting behavior of solutions as the independent variable goes to $\pm\infty$. If an autonomous equation is modeling some scientific phenomenon, the limiting behavior tells us the ultimate outcome. In addition our qualitative methods will enable us to sketch the graphs of all solutions to the autonomous equation.

The direction field and solutions

Since the function $f(t, x)$ on the right-hand side of (9.1) does not depend on t, the slopes of the direction lines have the same feature. This is illustrated by the direction field for equation (9.2) shown in Figure 1. The slopes do not change as we move from right to left in this figure.

Because of this fact, we would expect the same behavior for the solution curves. We would expect that one solution curve translated to the left or right would be another solution curve. We can see this analytically. Suppose that $y(t)$ is a solution to the autonomous equation $y' = f(y)$. Consider the function $y_1(t) = y(t + C)$, where C is a constant. Since

$$y_1'(t) = y'(t + C) = f(y(t + C)) = f(y_1(t)),$$

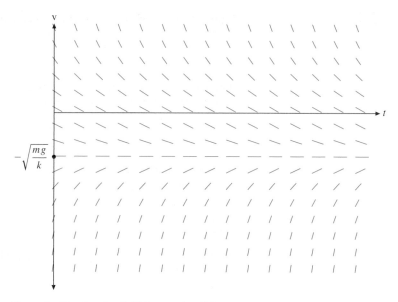

Figure 1. The direction field for equation (9.2).

we see that y_1 is also a solution. This means that we get different solution curves by translating one curve left and right. See Figure 2, which displays several solutions to equation (9.4).

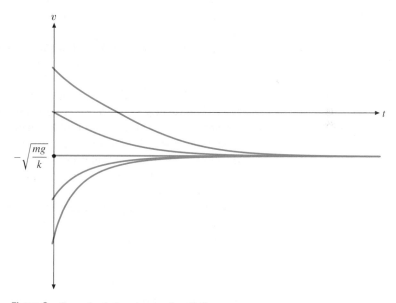

Figure 2. Several solutions to equation (9.4).

Equilibrium points and solutions

The starting point to the qualitative analysis of an autonomous equation is the discovery of some easily found particular solutions. If $f(x_0) = 0$, then the constant function $x(t) = x_0$ satisfies

$$x'(t) = 0 = f(x_0) = f(x(t)).$$

Hence this constant function is a particular solution to (9.1). We will call a point x_0 such that $f(x_0) = 0$ an **equilibrium point**. The constant function $x(t) = x_0$ is called an **equilibrium solution**.

Example 9.3   Find the equilibrium points and equilibrium solutions for the equation

$$v' = -g - kv|v|/m, \tag{9.4}$$

which models the velocity of an object subject to gravity and air resistance.

The right-hand side is the function

$$f(v) = -g - kv|v|/m = \begin{cases} -g - kv^2/m & \text{for } v \geq 0, \\ -g + kv^2/m & \text{for } v < 0. \end{cases}$$

From these equations, we see that the graph of f consists of half of a parabola curved down for $v \geq 0$, and half of a parabola curved up for $v \leq 0$ as shown in the graph of f in Figure 3. Hence, f is a decreasing function, and can have only one zero. Since $f(v) < 0$ for $v > 0$, the zero of f must occur when $v < 0$, for which $f(v) = -g + kv^2/m$. Therefore, the equilibrium point is

$$v = -\sqrt{mg/k},$$

and the corresponding equilibrium solution is

$$v_1(t) = -\sqrt{mg/k}.$$

The graph of this solution is shown in Figure 4.   ●

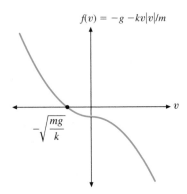

Figure 3. The graph of the right-hand side of equation (9.4).

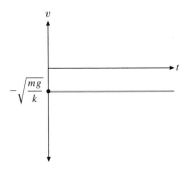

Figure 4. The equilibrium solution for equation (9.4).

Nonequilibrium solutions

Now suppose that $v(t)$ is a solution to (9.4) that satisfies $v(t_0) > -\sqrt{mg/k}$ for some t_0. Notice that $f(v) = -g - kv|v|/m$ and its derivative $df/dv = -2k|v|/m$ are both continuous for all v. Thus equation (9.4) satisfies the hypotheses of the uniqueness theorem. Therefore, the graphs of v and the equilibrium solution $v_1(t) = -\sqrt{mg/k}$ cannot cross. Consequently we must have

$$v(t) > v_1(t) = -\sqrt{mg/k} \quad \text{for all } t. \tag{9.5}$$

Since $f(v)$ is decreasing and $f(-\sqrt{mg/k}) = 0$, when $v(t)$ satisfies (9.5), we have $f(v(t)) < 0$. Hence,

$$v'(t) = -g - kv|v|/m < 0 \quad \text{for all } t.$$

Because it has a negative derivative, $v(t)$ is a monotone decreasing function. Since the decreasing function $v(t)$ is bounded below by $v(t) > -\sqrt{mg/k}$ for all t, we know that $v(t)$ approaches a limit as $t \to \infty$. It can be shown that this limit must be $-\sqrt{mg/k}$. A similar train of thought shows that $v(t) \to \infty$ as $t \to -\infty$.

Notice that without solving the initial value problem we have learned three things about the solution $v(t)$ that has an initial value $v(t_0) > -\sqrt{mg/k}$:

1. $v(t)$ is monotone decreasing
2. $v(t) \to -\sqrt{mg/k}$ as $t \to \infty$
3. $v(t) \to \infty$ as $t \to -\infty$

Thus the solution curves have the appearance shown in Figure 2. We cannot say how fast $v(t) \to -\sqrt{mg/k}$ as $t \to \infty$, or $v(t) \to \infty$ as $t \to -\infty$. For this reason, we have not included any tick marks along the t-axis in Figure 2. The same reasoning shows that if $v(0) = v_0 < -\sqrt{mg/k}$, then $v(t)$ is increasing to $-\sqrt{mg/k}$ as $t \to \infty$, and tends to $-\infty$ as $t \to -\infty$.

We have shown that as t increases, the velocity always tends to

$$v_{\text{term}} = -\sqrt{mg/k}.$$

We reached the same result at the end of Section 2.3. Because of this fact, we called v_{term} the terminal velocity. However, it is interesting to compare the amount of work involved in the two different methods used. Qualitative analysis is almost always easier when we want to discover the limiting behavior of solutions.

The phase line

An autonomous equation $y' = f(y)$ can be a mathematical model for a variety of phenomena. Regardless of the actual application, it is useful to think of y as measuring the distance from 0 along a number line, a y-axis. With this interpretation, the equation $y' = f(y)$ describes the dynamics involved in the motion of $y(t)$ along the line, which is called the ***phase line***.

For example, consider the graph of the right-hand side $f(v) = -g - kv|v|/m$ of equation (9.4) in Figure 5. The v-axis will be the phase line. We mark the equilibrium point $v_{\text{term}} = -\sqrt{mg/k}$ with a solid point. To the left of v_{term}, $f(v) > 0$, so a solution $v(t)$ is increasing, and we indicate this on the phase line by a blue arrow pointing to the right. Similarly, to the right v_{term}, $v(t)$ is decreasing, indicated by a blue arrow pointing to the left. A solution on either side of the equilibrium point must approach the equilibrium point as $t \to \infty$. Notice that the information on the phase line encapsulates everything we know about the solutions. In addition, notice that all of the information on the phase line was easily obtained from the graph of the $f(v)$, the right-hand side of the differential equation.

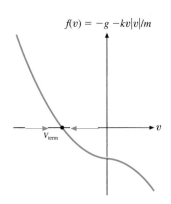

Figure 5. The phase line for equation (9.4).

Constructing and using the phase line

The analysis carried out above for equation (9.4) can be done for any autonomous equation. Let's illustrate this with another example.

Example 9.6   Discover the behavior as $t \to \infty$ of all solutions to the differential equation

$$x' = f(x) = (x^2 - 1)(x - 2). \tag{9.7}$$

We will use the phase line to do the analysis, and we start by graphing the right-hand side, $f(x) = (x^2 - 1)(x - 2)$, as shown in Figure 6. The graph can be easily plotted on a computer or a calculator. However, we do not need great precision in this graph. We need to know the zeros precisely, but in the intervals limited by the zeros we only need to know the sign of f. Our phase line will be the x-axis in Figure 6. We will consider x to be the position of a point on this line with its motion modeled by equation (9.7).

Since we can factor the right-hand side of (9.7) as $f(x) = (x-1)(x+1)(x-2)$, the zeros of f are $x_1 = -1$, $x_2 = 1$, and $x_3 = 2$. These are the equilibrium points, and we plot them on the x-axis in Figure 6, either with a dot or a small circle. Corresponding to these, we have three equilibrium solutions:

$$x(t) = -1, \quad x(t) = 1, \quad \text{and} \quad x(t) = 2.$$

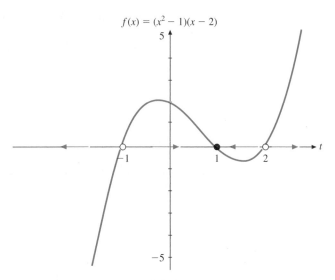

Figure 6. The phase line for the equation $x' = (x^2 - 1)(x - 2)$.

Since these are constant functions, the position of the point on the phase line modeled by them is also constant.

The equilibrium solutions are plotted in blue in Figure 7. We easily show that the uniqueness theorem applies to equation (9.7). Therefore, the solution curves of nonequilibrium solutions cannot cross those of the equilibrium solutions. There are four intervals limited by the equilibrium points. If a solution x starts in one of these intervals, then $x(t)$ is in that interval forever. Thus the motion of the solution point along the phase line is limited by the equilibrium points.

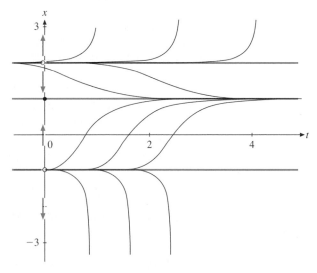

Figure 7. Solutions of the equation $x' = (x^2 - 1)(x - 2)$.

In each of the four intervals the right-hand side $f(x) = (x^2 - 1)(x - 2)$ has a constant sign:

$$
\begin{aligned}
f(x) < 0 \quad &\text{in} \quad (-\infty, -1), \\
f(x) > 0 \quad &\text{in} \quad (-1, 1), \\
f(x) < 0 \quad &\text{in} \quad (1, 2), \\
f(x) > 0 \quad &\text{in} \quad (2, \infty).
\end{aligned}
$$

This information can be obtained from the graph of f in Figure 6. If you do not have a graph handy, you can check the sign of f at one point in each of the intervals.

If $x(t)$ is a nonequilibrium solution to equation (9.7), and $x(t)$ is in one of the intervals $(-1, 1)$ and $(2, \infty)$ where $f(x) > 0$, then $x'(t) = f(x(t)) > 0$. Thus x is increasing. This means that x is moving to the right along the phase line. We indicate this on the phase line in Figure 6 by arrows pointing to the right in these intervals. Similarly, in the intervals $(-\infty, -1)$ and $(1, 2)$ where $f(x) < 0$, x is moving to the left, and we have arrows pointing to left.

Because of its monotone behavior, a nonequilibrium solution in one of the intervals $(-1, 1)$ and $(1, 2)$ must move in the direction of the arrow, and must approach the equilibrium point $x_2 = 1$ as $t \to \infty$. By similar reasoning, an equilibrium solution in $(2, \infty)$ must approach ∞, and one in $(-\infty, -1)$ must approach $-\infty$. On the other hand, as $t \to -\infty$, the motion is in the direction opposite to the arrows, and approaches the equilibrium point in that direction. With this information we can sketch examples of the nonequilibrium solutions in each interval, as shown in Figure 7. ●

Notice that we transferred all of the phase line information from the x-axis in Figure 6 to the x-axis in Figure 7 to assist us in sketching the solutions. Thus we have two examples of phase lines for equation (9.7). It should be emphasized that a phase line can be constructed on any x-axis. Another example is given in Figure 8.

Figure 8.   The phase line for the equation $x' = (x^2 - 1)(x - 2)$.

Stability

Some equilibrium points, like $x_2 = 1$ in Example 9.6, have the property that solution curves which start near them approach the equilibrium point as $t \to \infty$. These are called ***asymptotically stable*** equilibrium points. We have marked asymptotically stable equilibrium points with solid points on our phase lines. There are also equilibrium points, like $x_1 = -1$ and $x_3 = 2$ in Example 9.6, where some solutions move away. These are called ***unstable***.[6] Unstable equilibrium points are marked by open circles. If we focus our attention on the phase line near an equilibrium point, then we see that it is an asymptotically stable equilibrium point if and only if both adjacent arrows point toward the point. In fact, since each arrow can have only two directions, there are a total of four possibilities, only one of which represents an asymptotically stable equilibrium point.

These possibilities are shown in Figure 9, together with an indication of what the graph of f looks like near the associated equilibrium point. Notice that only Figure 9(b) depicts an asymptotically stable equilibrium point. In Figures 9(c) and (d), one arrow points toward the point and the other points away. Although these points might be called semistable, we will not stress that terminology. They are unstable equilibrium points.

Examining the possibilities, we see that an equilibrium point x_0 for $x' = f(x)$ is asymptotically stable if and only if f is decreasing at x_0. We can use this fact

---

[6] Consider the equation $y' = 0$. For this equation, every point is an equilibrium point, and every solution is a constant function. These solutions do not move nearer to the equilibrium points, nor do they move away. The property of "not moving away" is described by saying that the equilibrium points are ***stable***. In one dimension, the equation $y' = 0$ provides essentially the only example of stable equilibrium points that are not asymptotically stable. In higher dimensions, the concept of stability is more interesting.

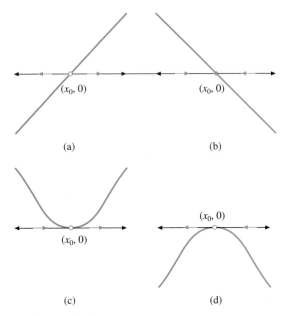

Figure 9. Possible configurations of equilibrium points.

to derive a ***first derivative test*** for stability. In the figures in this section, we have systematically indicated asymptotically stable equilibrium points with solid points, and unstable equilibrium points with open circles.

THEOREM 9.8   Suppose that x_0 is an equilibrium point for the differential equation $x' = f(x)$, where f is a differentiable function.

1. If $f'(x_0) < 0$, then f is decreasing at x_0 and x_0 is asymptotically stable.
2. If $f'(x_0) > 0$, then f is increasing at x_0 and x_0 is unstable.
3. If $f'(x_0) = 0$, no conclusion can be drawn.   ∎

E x a m p l e   9 . 9   Classify the equilibrium points for the equation

$$x' = (x^2 - 1)(x - 2)$$

from Example 9.6.

We saw in Example 9.6 that the equilibrium points are -1, 1, and 2. We can analyze these by looking at the phase lines in Figures 6, 7, or 8, and noticing that the solutions starting near -1 or near 2 are driven away from these values. Hence these are unstable points. On the other hand, the solutions starting near 1, either above or below, are drawn toward 1 as $t \to \infty$. Thus 1 is an asymptotically stable equilibrium point.

We could have also classified these equilibrium points by looking at the graph of the right-hand side in Figure 6. The right-hand side $f(x) = (x^2 - 1)(x - 2)$ is decreasing when it passes through point 1, but increasing as it passes through the other two. Hence point 1 is asymptotically stable and the others are unstable.

Finally, a third way is to use Theorem 9.8. We compute that $f'(x) = 3x^2 - 4x - 1$. At the equilibrium points, we have $f'(-1) = 6$, $f'(1) = -2$, and $f'(2) = 3$. Thus -1 and 2 are unstable and 1 is asymptotically stable.   ●

Summary of the method

We now have a method for analyzing the solutions of the autonomous equation

$$x' = f(x).$$

Let's summarize the procedure.

> 1. Graph the right-hand side $f(x)$ and add the phase line information to the x-axis. Find, mark, and classify the equilibrium points where $f(x) = 0$. In each of the intervals limited by the equilibrium points, find the sign of f and draw an arrow to the right if f is positive and to the left if f is negative.
>
> 2. Create a tx-plane, transfer the phase line information to the x-axis, draw the equilibrium solutions, and then use the phase line information to sketch nonequilibrium solutions in each interval limited by the equilibrium points.

As an example consider the equation $x' = f(x)$, where the graph of f is shown in Figure 10. Using the information in the graph we construct the phase line on the x-axis. Notice that asymptotically stable equilibrium points are marked with solid points and unstable points with small circles.

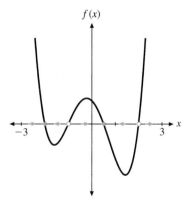

Figure 10. The graph of $f(x)$ and the associated phase line.

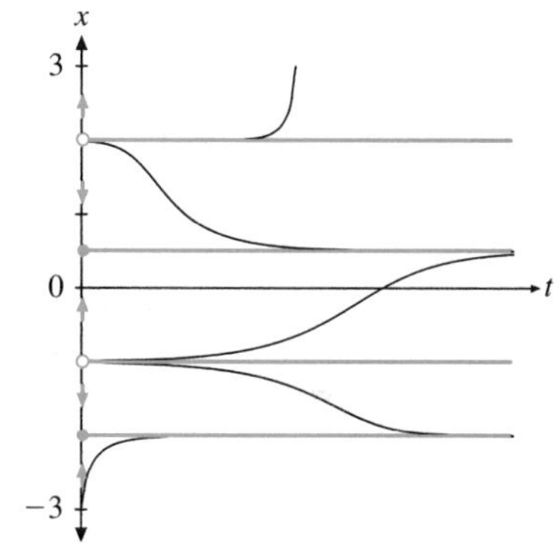

Figure 11. The phase line and graphs of solutions.

Next we transfer the phase line information from Figure 10 to the vertical x-axis in Figure 11. We drew the equilibrium solutions in blue. The arrows on the phase line show whether solutions increase or decrease, and allow us to sketch the nonequilibrium solutions.

Example 9.10 Analyze the solutions of $x' = x^3 - 2x^2 + x$.

We can factor the right-hand side as $f(x) = x(x - 1)^2$. Hence 0 and 1 are equilibrium points. Figure 12(a) shows the graph of f turned 90° counterclockwise. We see that $x^3 - 2x^2 + x$ is increasing through 0. It has a local minimum at 1, so it is not decreasing there. Hence both of these equilibrium points are unstable. We complete the phase line information on the x-axis. Figure 12(b) shows the phase line all by itself, and this information is repeated on the x-axis in Figure 12(c). There it is used to plot the equilibrium and nonequilibrium solutions.

Showing the three versions of the phase line arranged as they are in Figure 12 makes the relationships among them more obvious. ●

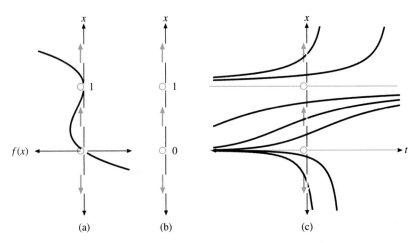

(a)   (b)   (c)

Figure 12. Three versions of the phase line for $y' = f(x) = x(x-1)^2$. In (a) the graph of f is rotated $90°$ counterclockwise, so the three phase lines are parallel. Notice that in (b) the phase line stands alone.

EXERCISES

In Exercises 1–6, if the given differential equation is autonomous, identify the equilibrium solution(s). Use a numerical solver to sketch the direction field and superimpose the plot of the equilibrium solution(s) on the direction field. Classify each equilibrium point as either unstable or asymptotically stable.

1. $P' = 0.05P - 1000$ **2.** $y' = 1 - 2y + y^2$

3. $x' = t^2 - x^2$ **4.** $P' = 0.13P(1 - P/200)$

5. $q' = (2 - q)\sin q$ **6.** $y' = (1 - y)\cos t$

In Exercises 7–10, the graph of the right-hand side of $y' = f(y)$ is shown. Identify the equilibrium points and sketch the equilibrium solutions in the ty-plane. Classify each equilibrium point as either unstable or asymptotically stable.

7.

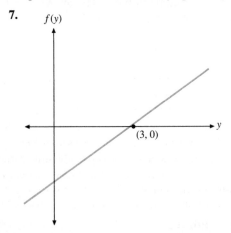

8.

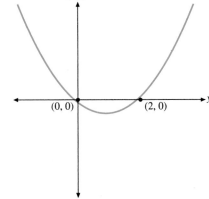

9.

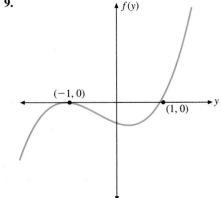

10.

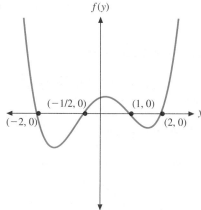

In Exercises 11–12, only a small part of the direction field for the differential equation $y' = f(y)$ is shown. Sketch the remainder of the direction field. Then superimpose the equilibrium solution(s), classifying each as either unstable or asymptotically stable.

11.

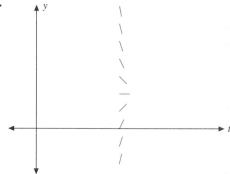

12.

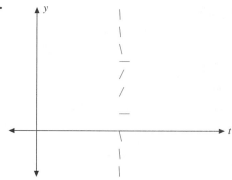

In Exercises 13–14, the sketch of $f(y)$ is given, where $f(y)$ is the right-hand side of the autonomous differential equation $y' = f(y)$. Use the sketch of $f(y)$ to help sketch the direction field for the differential equation $y' = f(y)$. Superimpose the equilibrium solution(s), classifying each as either unstable or asymptotically stable.

13.

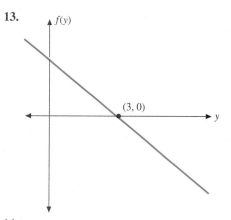

14.

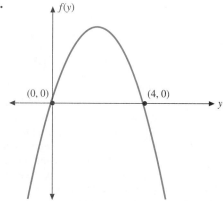

In each of Exercises 15–22, an autonomous differential equation is given in the form $y' = f(y)$. Perform each of the following tasks without the aid of technology.
  (i) Sketch a graph of $f(y)$.
 (ii) Use the graph of f to develop a phase line for the autonomous equation. Classify each equilibrium point as either unstable or asymptotically stable.
(iii) Sketch the equilibrium solutions in the ty-plane. These equilibrium solutions divide the ty-plane into regions. Sketch at least one solution trajectory in each of these regions.

15. $y' = 2 - y$ **16.** $y' = 2y - 7$
17. $y' = (y + 1)(y - 4)$ **18.** $y' = 6 + y - y^2$
19. $y' = 9y - y^3$ **20.** $y' = (y + 1)(y^2 - 9)$
21. $y' = \sin y$ **22.** $y' = \cos 2y$

For each initial value problem presented in Exercises 23–26, perform each of the following tasks.
  (i) Solve the initial value problem analytically.
 (ii) Use the analytical solution from part (i) and the theory of limits to find the behavior of the function as $t \to +\infty$.
(iii) Without the aid of technology, use the theory of qualitative analysis presented in this section to predict the long-term behavior of the solution. Does your answer agree with that found in part (ii)? Which is the easier method?

23. $y' = 6 - y$, $y(0) = 2$

24. $y' + 2y = 5$, $y(0) = 0$

25. $y' = (1 + y)(5 - y)$, $y(0) = 2$

26. $y' = (3 + y)(1 - y)$, $y(0) = 2$

In Exercises 27–28, use the calculus technique suggested in Theorem 9.8 to determine the stability of the equilibrium solutions.

27. $x' = 4 - x^2$ **28.** $x' = x(x - 1)(x + 2)$

29. In Theorem 9.8, if $f'(x_0) = 0$, no conclusion can be drawn about the equilibrium point x_0 of $x' = f(x)$. Explain this phenomenon by providing examples of equations $x' = f(x)$ where

(a) $f'(x_0) = 0$ and x_0 is unstable, and

(b) $f'(x_0) = 0$ and x_0 is asymptotically stable.

30. A skydiver jumps from a plane and opens her chute. One possible model of her velocity v is given by

$$m \frac{dv}{dt} = mg - kv,$$

where m is the combined mass of the skydiver and her parachute, g is the acceleration due to gravity, and k is a proportionality constant. Assuming that m, g, and k are all positive constants, use qualitative analysis to determine the skydiver's "terminal velocity."

31. A tank contains 100 gal of pure water. A salt solution with concentration 3 lb/gal enters the tank at a rate of 2 gal/min. Solution drains from the tank at a rate of 2 gal/min. Use qualitative analysis to find the eventual concentration of the salt solution in the tank.

PROJECT 2.10   The Daredevil Skydiver

A skydiver jumps out of an airplane at an altitude of 1200 m. The person's mass, including gear, is 75 kg. Assume that the force of air resistance is proportional to the velocity, with a proportionality constant of $k_1 = 14$ kg/s during free fall. After t_d seconds, the parachute is opened, and the proportionality constant becomes $k_2 = 160$ kg/s. Assume, for the moment, that the chute deploys instantaneously when the skydiver pulls his ripcord.

It will be helpful to review Section 2.3.

1. Use your physical intuition to sketch three graphs: the distance the skydiver falls versus time, the velocity versus time, and the acceleration versus time. *Hint*: Do not use technology, do not solve any differential equations. Simply rely on your understanding of the physical model to craft your sketches. You might find qualitative analysis useful. Keep in mind that you do not have to draw the graphs in the order listed. This is much harder than it looks. Once your drawings are complete, put them aside and save them for comparison once you've completed item 4.

2. In order to establish time limits on the problem, examine the two extreme cases. In the first, the skydiver never pulls the ripcord, and in the second, the ripcord is pulled immediately, so $t_d = 0$. An intelligent skydiver would avoid each of these strategies, but they serve to put upper and lower limits on the general problem.

For each of these cases use a numerical solver to estimate the time it takes the skydiver to impact the ground. You should verify this result analytically. (You will have to solve an implicit equation to find the time.) What is the velocity at this moment? How close is this velocity to the terminal velocity?

3. Suppose that the skydiver deploys the chute by pulling the ripcord $t_d = 20$ s after leaving the airplane. Use a numerical solver to find an estimate of the time when the person hits the ground. What is the velocity at this moment? Verify these results analytically. Compare the final velocity with the terminal velocity. *Hint*: You will get better numerical results if t_d is one of the points at which the solver computes an approximate solution.

4. Using the numerical data from item 3, plot three graphs: the distance the skydiver falls versus time, the velocity versus time, and the acceleration versus time. Compare these graphs with those you created in item 1.

5. Recall that we made the assumption that the parachute deploys instantaneously. In aviation parlance, the unit of acceleration is a "g", which is equal to the acceleration due to gravity near the surface of the earth. How many "g"s does the skydiver experience at the instant the chute opens? You can use the plot of the acceleration made in item 4 or you can compute this analytically. Do you think a skydiver could withstand such a jolt? Do some research on this question before answering.

6. Special gear allows the skydiver to land safely provided that the impact velocity is below 5.2 m/s. Do some numerical experimentation to discover approximately the last possible moment that the ripcord can be pulled to achieve a safe landing.

7. Let's change our assumption about chute deployment. Suppose that the chute actually takes $\tau = 3$ s to deploy. Moreover, suppose that during deployment, the proportionality constant varies linearly from $k_1 = 14$ kg/s to $k_2 = 160$ kg/s from the time t_d that the ripcord is pulled to the time $t_d + \tau$ when the chute is fully deployed. (See Figure 1.) Repeat the numerical parts of items 3, 4, 5, and 6 with this new assumption. (The analytical parts are not so easy with this assumption about the proportionality constant.)

There are a number of fascinating adaptations you can make to this model. For example, suppose that k varies between t_d and $t_d + \tau$ according to some cubic interpolation. Or, suppose that the skydiver started at a higher altitude and you take the density of the air into account when determining the force of resistance.

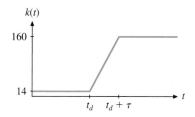

Figure 1. Linear interpolation.

Finally, there are a number of useful articles that will aid in your pursuit of this model.

- Drucker, J., Minimal Time of Descent, *The College Mathematics Journal*, 26 (1995), pp. 252–235.
- Meade, D.B., ODE Models for the Parachute Problem, *SIAM Review*, 40 (1998), pp. 327–332.
- Melka and Farrior, Exploration of the Parachute Problem with STELLA, *Newsletter for the Consortium for Ordinary Differential Equations Experiments*, Summer–Fall, 1995.

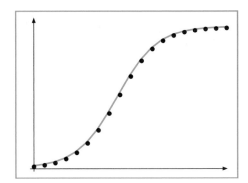

CHAPTER **3**

Modeling and Applications

The discovery of the calculus occurred at the beginning of the scientific revolution in the seventeenth century. This discovery was not a side issue in the revolution. Rather, it was the linchpin on which much of what followed was based. For the first time, humankind had a systematic way to study how things changed. In many cases, the study of change has led to a differential equation, or to a system of differential equations through the process known as ***modeling***.

We have explored a few applications, and we have constructed the corresponding models in Chapter 2. In this chapter, we will look carefully at the modeling process itself. The process will then be used in several applications. Along the way, we will also consider some examples of modeling that are faulty.

The main idea in the modeling process is easily explained. Suppose x is a quantity that varies with respect to the variable t. We want to model how it changes. From the mathematical point of view, the rate of change of x is the derivative

$$x' = \frac{dx}{dt}.$$

Building a model of the process involves finding an alternate expression for the rate of change of x as a function of t and x, say $f(t, x)$. This leads us to the differential equation

$$\frac{dx}{dt} = f(t, x).$$

This equation is the mathematical model of the process.

The problem, of course, is discovering how the rate of change varies, and this means discovering the function $f(t, x)$. Let's look at some examples.

3.1 Modeling Population Growth

The process of modeling is exemplified by the modeling of the growth of populations. We will study the growth of a population of protozoa, which are single-celled organisms such as amoebae or paramecia. We will assume that there is no lack of nutrients or of room to grow. Let $P(t)$ denote the number of cells at time t.

Protozoa multiply by cell division. We may assume that any cell is as likely to divide as any other. Let b be the probability that a cell will divide in a unit of time, where time is measured in some convenient unit such as days or hours. Then between times t and $t + \Delta t$, there will be approximately

$$bP(t)\Delta t.$$

divisions, resulting in the same number of new cells. Furthermore, this approximation gets better as Δt gets smaller. We will call b the **birth rate**.

Like any living creature, single-celled organisms die for a variety of reasons. Let d denote the probability that any individual cell will die in a unit of time. Just as in our discussion of births, the number of deaths between t and $t + \Delta t$ will be approximately

$$dP(t)\Delta t,$$

and this approximation gets better as Δt gets smaller. We will call d the **death rate**.

Putting together the results on births and deaths, we see that the change in the population between times t and $t + \Delta t$ is

$$P(t + \Delta t) - P(t) \approx bP(t)\Delta t - dP(t)\Delta t = (b - d)P(t)\Delta t.$$

Hence, by the limit quotient definition of the derivative,

$$P'(t) = \lim_{\Delta t \to 0} \frac{P(t + \Delta t) - P(t)}{\Delta t} = (b - d)P(t).$$

If we let

$$r = b - d \tag{1.1}$$

and call r the **reproductive rate**, the equation is

$$P' = rP. \tag{1.2}$$

Equation (1.2) is a first-order differential equation involving the function $P(t)$. It is a mathematical model of the growth of a population. The nature of the predictions of the model depend on the nature of the reproductive rate r.

The Malthusian model

If we assume that the population has plenty of resources like food and space, then it makes sense that the birth and death rates do not depend on time or on the size of the population. Hence both are constants, and the reproductive rate r is also a constant. Then equation (1.2) is the exponential equation, and solutions are easily found. Separation of variables leads us to the solution

$$P(t) = Ce^{rt}$$

for any constant C. Furthermore, if the population at time $t = 0$ is P_0, then we see that $C = P_0$ and our solution is

$$P(t) = P_0 e^{rt}. \tag{1.3}$$

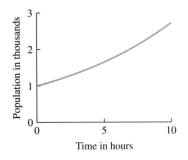

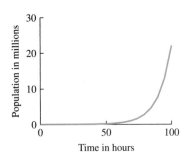

Figure 1. Exponential growth of a population.

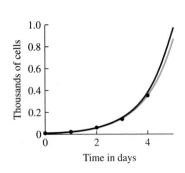

Figure 2. Exponential growth over a longer period of time.

Example 1.4

Figure 3. The black curve is the solution in Example 1.4. The blue is the result of using regression in Exercise 1.8.

We can use our model (1.2) and the solution (1.3) to predict the population of our colony of protozoa. There are two cases. If the death rate exceeds the birth rate, then $r = b - d < 0$, and the population given by (1.3) will decline. The colony will eventually disappear. On the other hand, if $r = b - d > 0$, the population will grow. In fact, it will grow exponentially.

The economist Thomas Malthus is usually given credit for the model of population growth in (1.2) and (1.3). It is referred to as the *Malthusian model*.[1]

As an example, suppose we have a population of bacteria, which has 1000 individuals at time $t = 0$, and that the reproductive rate is $r = 0.1$. Suppose also that we are measuring time in hours. Then according to (1.3), the population at later times will be given by

$$P(t) = 1000e^{0.1t}.$$

The graph of this solution over the first 10 hours is shown in Figure 1. Notice that the population almost triples in that time. Figure 2 shows the same population, but over the first 100 hours. The difference is quite striking and illustrates how rapid exponential growth really is. After 100 hours there are about 22,000,000 bacteria.

Thus, we see that the exact solution is important because it allows us to extract quantitative information about a population obeying the Malthusian law of growth.

Evaluating the parameters in the Malthusian model

The Malthusian model contains the parameter r, the reproductive rate. This constant must be chosen to reflect reality. Usually this is done on the basis of experimentation or observation. Let's look at an example.

A biologist starts with 10 cells in a culture. Exactly 24 hours later he counts 25. Assuming a Malthusian model, what is the reproductive rate? What will be the number of cells at the end of 10 days?

The model is given by equation (1.2), with a constant reproductive rate r. This is the exponential equation, and the general solution is

$$P(t) = Ce^{rt}. \tag{1.5}$$

Let's measure time in days. Then for our initial condition we have

$$10 = P(0) = C.$$

Then from the second data point we have

$$25 = P(1) = 10e^r.$$

Solving for r, we get

$$r = \ln(2.5) \approx 0.9163.$$

After 10 days, the population is predicted to be

$$P(10) = 10e^{r \times 10} \approx 95,367. \tag{1.6}$$

The solution $P(t) = 10e^{rt}$ is plotted in black for $0 \leq t \leq 5$ in Figure 3. ●

---

[1] Thomas Robert Malthus (1766–1834) was an English clergyman and political economist. In his famous treatise, *An Essay on the Principle of Population,* he postulated that population tended to increase "geometrically" (by which he meant exponentially, as we see in equation (1.3)) while resources, especially food, increased only linearly. As a result, human and other populations will increase until checked by natural limitations, principally to do with food supply. In the case of humans he predicted that population growth would be limited by wars, pestilence, and other natural disasters.

Of course the supply of food and other resources (except space) has tended to increase exponentially since the time of Malthus. Nevertheless, there are modern Malthusians who think he was on to something.

In practice, the biologist in Example 1.4 would probably count the cells at least once each day. After each count, he would reestimate r using all of his data. He will do this using a process called ***linear regression***. It is our experience that many students are able to perform the process of linear regression on calculators, even though they do not know how the calculator does the job.

The key to the method is found by taking the logarithm of equation (1.5), getting

$$\ln P(t) = \ln C + rt.$$

Thus, $\ln P(t)$ is a linear function of t with coefficients $\ln C$ and r. Suppose we make measurements at N different times $t_0, t_1, \ldots, t_{N-1}$, and the populations at these times are $P_0, P_1, \ldots, P_{N-1}$. The method of linear regression chooses $\ln C$ and r to minimize the function

$$\sum_{n=0}^{N-1} (\ln P_n - \ln P(t_n))^2 = \sum_{n=0}^{N-1} (\ln P_n - \ln C - rt_n)^2. \tag{1.7}$$

The method of linear regression is a special case of the ***method of least squares,*** since it minimizes the sum of the squares of errors. Using multivariable calculus the solutions can be found explicitly. We will leave this to Exercise 9. Computing $\ln C$ and r is not something to be done by hand unless N is small. However, computers and some handheld calculators can do the computation easily.

If you do not know how to compute a linear regression, you may skip the following example.

Example 1.8 After each of the first four days, the biologist in Example 1.4 counts 25, 61, 144, and 360 cells. What is the reproductive rate r? What does he estimate the population will be after 10 days?

We have estimated values for $t = 0, 1, 2, 3, 4$. We will choose the coefficients using linear regression. Using a calculator or a computer, this is an easy process. In doing so we find that $\ln C = 2.3141$ or $C = 10.1154$, and $r = 0.8918$.[2]

After 10 days the population is predicted to be

$$P(10) = Ce^{r \times 10} \approx 75,511. \tag{1.9}$$

The solution $P(t) = Ce^{rt}$ is plotted in blue in Figure 3. The five data points are also plotted there. We can visually verify that this solution is a closer match to the data points than is the curve in black from Example 1.4, which uses only the first two data points.   ●

The use of linear regression in the previous example allows the biologist to use all of his data to estimate the parameters. This will usually result in better estimates.

The logistic model of growth

Whenever we have a mathematical model, we should examine its implications carefully to discover to what extent the model correctly predicts what happens in the laboratory or in the real world. In the case of the Malthusian model the prediction of unlimited exponential growth is clearly impossible. If a colony of protozoa grew exponentially, it would cover the earth in times that are observable, and this just

---

[2] Some calculators will do an *exponential regression*, which approximates the data directly by $P(t) = Cb^t$ using the method we have described. This is equivalent to the form we want, $P(t) = Ce^{rt}$, if $b = e^r$, or $r = \ln b$.

does not happen. Nevertheless, laboratory experiments do verify the exponential growth of populations that are relatively small.

If we reexamine our assumptions, we quickly see the problem. We have assumed that the colony has no lack of nutrients and no lack of space in which to grow. These assumptions may be valid for relatively small populations but become untrue as the population grows. Let's look again at our derivation of (1.2) and take into account limits of growth.

Let's first take a closer look at the death rate. A lack of food means that the cells will have to compete with each other for what food there is, and some cells will die of malnutrition as well as of natural causes. Similarly, a lack of space will cause increased interaction between cells, which may increase the number of deaths. Thus, both a limited food supply and limited space increase the death rate by increasing the number of interactions between cells. The number of interactions that an individual cell will have will be proportional to the probability of an interaction. Assuming that the population is evenly distributed over the area occupied by the species, the probability will be proportional to the size of the population. Hence a more realistic formulation of the death rate would have it increase proportionally to the size of the population. This leads us to the formula

$$d + aP$$

for the death rate. The constant d is the same as before, the death rate for small populations. The new term aP measures additional deaths that are due to interactions between a cell and other cells. The constant a is a measure of the impact of interactions on the rate at which deaths occur.

Similarly, the birth rate will decrease proportionally to the size of the population. The modified birth rate will have the formula

$$b - cP.$$

Arguing as we did before, we see that the change of population between t and Δt is

$$P(t + \Delta t) - P(t) \approx (b - cP(t))P(t)\Delta t - (d + aP(t))P(t)\Delta t$$
$$= (b - d - (a + c)P(t))P(t)\Delta t.$$

Hence

$$P'(t) = \lim_{\Delta t \to 0} \frac{P(t + \Delta t) - P(t)}{\Delta t} = (b - d - (a + c)P(t))P(t).$$

We set $r_0 = b - d$, but now we call it the ***natural reproductive rate***. To put the equation into a nicer form, we set $a + c = r_0/K$, where $K = r_0/(a + c)$ is a new constant. Then

$$P' = (r_0 - r_0 P/K)P = r_0(1 - P/K)P. \tag{1.10}$$

Equation (1.10) is called the ***logistic equation***. The model of population growth embodied in the logistic equation is called the ***logistic model***.[3] Referring to equation (1.2), we see that the reproductive rate assumed in the logistic model is

$$r = r_0(1 - P/K).$$

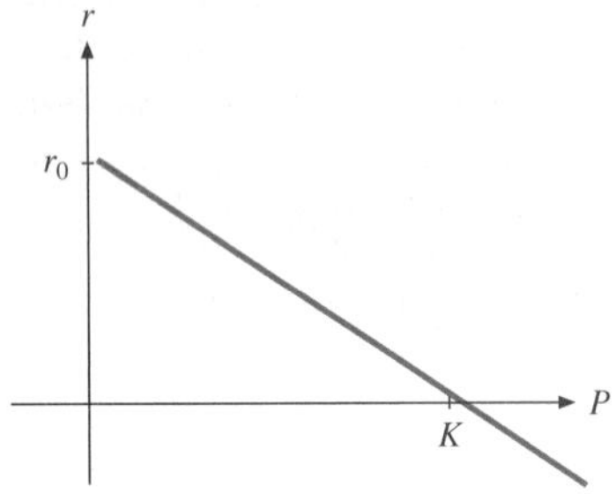

This is no longer a constant but rather depends on the population, as shown in Figure 4. Notice that the reproductive rate becomes negative for $P > K$.

The direction field for the logistic equation is shown in Figure 5.

Figure 4. The reproductive rate in the logistics model varies with the population.

---

[3] The logistic model was first constructed by Pierre-François Verhulst (1804–1849). His work was mostly ignored until around 1920, when it was published again by R. Pearl and L. J. Reed. Since then there have been many experiments that verify the logistic model in carefully controlled circumstances. In particular, we mention the Russian biologist G. F. Gause, who in 1934 published *The Struggle for Existence*, a classic text with many examples of logistic growth, especially of paramecia.

Figure 5. The direction field for the logistic equation.

Qualitative analysis of the logistic equation

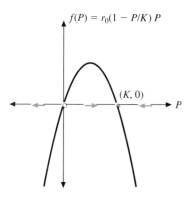

Figure 6. The graph of the right-hand side of the logistic equation.

The logistic equation in equation (1.10) is autonomous, since the right-hand side

$$f(P) = r_0 P(1 - P/K)$$

does not depend on t. The graph of f is shown in Figure 6. The equilibrium points are where $f(P) = 0$, or at $P_1 = 0$, and $P_2 = K$. Consequently,

$$P_1(t) = 0 \quad \text{and} \quad P_2(t) = K$$

are equilibrium solutions. The graphs of these solutions are shown in Figure 7.

Next we notice that $P' = f(P) > 0$ if $0 < P < K$, and $P' = f(P) < 0$ if $P > K$ or if $P < 0$. Hence $P_1 = 0$ is an unstable equilibrium point, and $P_2 = K$ is stable. In particular, we see that if $P(t)$ is any solution with a positive population, then it must stay positive, and it must tend to K as $t \to \infty$. Consequently, every positive population governed by the logistic equation tends to K as time increases. For this reason K is called the ***carrying capacity***. Of course populations are always positive, but mathematically equation (1.10) can have negative solutions. It is easily seen using qualitative analysis that these solutions tend to $-\infty$.

Hence without solving the initial value problem we know that the solution curves have the appearance shown in Figure 8. We cannot say how fast $P(t) \to K$ as $t \to \infty$. For this reason we have not included any tick marks along the t-axis in Figure 8.

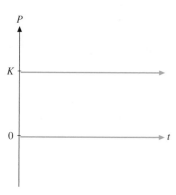

Figure 7. The equilibrium solutions for the logistic equation.

Solution of the logistic equation

To make our notation a bit simpler, let's drop the subscript in r_0 in the logistic equation. Then it becomes

$$P' = rP(1 - P/K),$$

and r is now a constant. The logistic equation, being autonomous, is separable. We can write

$$\frac{K \, dP}{P(K - P)} = r \, dt.$$

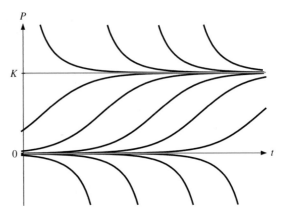

Figure 8. Several solutions to the logistic equation.

To solve the equation, we must integrate both sides. On the left, we can use partial fractions to discover that

$$\frac{K}{P(K-P)} = \frac{1}{P} + \frac{1}{K-P}.$$

Hence we want to solve

$$\left(\frac{1}{P} + \frac{1}{K-P}\right) dP = r\, dt. \tag{1.11}$$

Integrating both sides of equation (1.11), we get

$$\ln|P| - \ln|K-P| = rt + C$$

$$\ln\left|\frac{P}{K-P}\right| = rt + C,$$

or

$$\left|\frac{P}{K-P}\right| = e^{rt+C} = e^C e^{rt}.$$

The constant e^C is positive. If we replace it by A, and allow it to be positive, negative, or zero we can drop the absolute values and write

$$\frac{P}{K-P} = Ae^{rt}. \tag{1.12}$$

If we let P_0 represent the population at time $t = t_0$, then $P(t_0) = P_0$ and

$$Ae^{rt_0} = \frac{P_0}{K-P_0}. \tag{1.13}$$

Next we solve equation (1.12) for P to get

$$P(t) = \frac{KAe^{rt}}{1 + Ae^{rt}}. \tag{1.14}$$

Using (1.13) and (1.14) and a little algebra, we can eliminate A and find that

$$P(t) = \frac{KP_0}{P_0 + (K-P_0)e^{-r(t-t_0)}}. \tag{1.15}$$

We already know as a result of our qualitative analysis that $P(t) \to K$ as $t \to \infty$. This can also be seen from (1.15). As $t \to \infty$, $e^{-r(t-t_0)} \to 0$ and

$$\lim_{t \to \infty} P(t) = \lim_{t \to \infty} \frac{K P_0}{P_0 + (K - P_0)e^{-r(t-t_0)}} = \frac{K P_0}{P_0} = K.$$

It is interesting to notice that the qualitative analysis that leads to the notion and interpretation of K as the carrying capacity is so much easier than finding the exact solution.

Evaluation of the parameters in the logistic equation

However, the exact solution in (1.15) is important because it allows us to be quantitative about our conclusion in well-controlled circumstances. There are two constants in the logistic equation, the carrying capacity K and the rate r. To find a specific solution we need in addition the initial population P_0. With just three observations we ought to be able to compute these constants and use (1.15) to tell us the complete history.

Example 1.16   Suppose we start at time $t_0 = 0$ with a sample of 1000 cells. One day later we see that the population has doubled, and sometime later we notice that the population has stabilized at 100,000.

To keep the size of the numbers smaller, we will use 1000 cells as our unit. Then, from the data, we see that the initial population is $P_0 = 1$, and the carrying capacity is $K = 100$. To find r we use the solution and the fact that $P(1) = 2$. Although we could use (1.15), it is easier to combine equations (1.12) and (1.13), using $t_0 = 0$, resulting in

$$\frac{P(t)}{K - P(t)} = \frac{P_0}{K - P_0} e^{rt}.$$

With $P_0 = 1$, $K = 100$, $t = 1$, and $P(1) = 2$ this becomes

$$\frac{2}{100 - 2} = \frac{1}{100 - 1} e^{r(1)}.$$

Thus

$$e^r = 198/98, \quad \text{and} \quad r = \ln\left(\frac{198}{98}\right) \approx 0.7033.$$

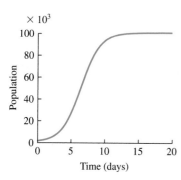

Consequently, the population model is captured by substituting the appropriate parameters into equation (1.15)

$$P(t) = \frac{(100)(1)}{1 + (100 - 1)e^{-0.7033t}} = \frac{100}{1 + 99e^{-0.7033t}}. \tag{1.17}$$

Figure 9. The population $P(t)$ from Example 1.16.

The plot of $P(t)$ is shown in Figure 9. Note that the population eventually nears its carrying capacity of 100,000 in about 15 days.

Of course, now that we have an equation modeling the population, there are a number of interesting quantitative questions that can be posed and answered. For example, the population after 5 days is projected to be

$$P(5) = \frac{100}{1 + 99e^{-0.7033(5)}} \approx 25,377 \text{ cells}.$$

In addition, if you wish to know how long it takes the population to reach a level of 50,000 cells, then the equation

$$50 = \frac{100}{1 + 99e^{-0.7033t}}$$

and a little algebra will produce the solution $t = (\ln 99)/0.7033 \approx 6.5$ days.

It is somewhat more difficult to estimate the parameters for a logistic model when we are not far enough along in the growth of the population to know the carrying capacity K. There are three constants in the solution given in equation (1.15), r, the reproductive rate for small populations, K, the carrying capacity, and P_0, the initial population. Estimates of these three can be computed from any three measurements of the population. However, the calculations are somewhat difficult.

The situation is somewhat easier if the times of the measurements are equally spaced, say at 0, h, and $2h$. Suppose the populations at these times are P_0, P_1, and P_2. If we solve equation (1.15) for K, we get

$$K = \frac{P_0 P(t)(1 - e^{-r(t-t_0)})}{P_0 - P(t)e^{-r(t-t_0)}}.$$

Let's apply this equation twice, first with $t_0 = 0$ and $t = h$, and then with $t_0 = h$ and $t = 2h$. We get two equations for K:

$$K = \frac{P_0 P_1(1 - e^{-rh})}{P_0 - P_1 e^{-rh}} \quad \text{and} \quad K = \frac{P_1 P_2(1 - e^{-rh})}{P_1 - P_2 e^{-rh}}. \tag{1.18}$$

Setting these equal and solving for r, we get

$$r = \frac{1}{h} \ln\left(\frac{P_2(P_1 - P_0)}{P_0(P_2 - P_1)}\right). \tag{1.19}$$

Then we can substitute this value of r into one of the expressions in (1.18) to find K.

Let's see how well this works.

Example 1.20 Early in the twentieth century, the German biologist T. Carlson performed experiments in which he measured the growth of a yeast culture.[4] He dropped a few yeast cells into an appropriate nutritive sugar solution, and over the next 18 days he measured carefully the size of the population of yeast cells. His data are shown in Table 1. Find a logistic curve that fits this data.

Table 1 The growth of a population of yeast cells					
DAY	**QUANTITY OF YEAST**	**DAY**	**QUANTITY OF YEAST**	**DAY**	**QUANTITY OF YEAST**
0	9.6	7	257.3	13	629.4
1	18.3	8	350.7	14	640.8
2	29.0	9	441.0	15	651.1
3	47.2	10	513.3	16	655.9
4	71.1	11	559.7	17	659.6
5	119.1	12	594.8	18	661.8
6	174.6				

This problem is not well posed. We have not defined what it means to fit the data. We will look at a number of possible answers. First we take the first three data points and use the method indicated in equations (1.18) and (1.19). Undoubtedly, the experimenter would do this computation after the second day to get some idea of what will happen to the yeast culture. However, when we compute the carrying capacity, we get $K = 48.5$. One look at the rest of data shows that this figure is

---

[4] Über Geschwindigkeit und Grösse der Hefevermehrung in Würze, *Biochem. Ztschr.,* Bd. 57, pp. 313–334, 1913. The data in Table 1 labeled "Quantity of yeast" are not actually the number of yeast cells. Instead Table 1 contains a quantity that Carlson measured that is proportional to the number of cells.

completely unrealistic. The data show that the carrying capacity should be larger than 660.

What went wrong? The answer is twofold. First, we are dealing with experimental data, and there is always some error associated with real data. Second, and perhaps more important, is the fact that a logistic curve is fairly complicated. Look at Figure 10. We are trying to get at the entire S-shaped logistic curve by looking only at its three starting points in the lower part of the S-shaped curve. To expect high accuracy is unrealistic.

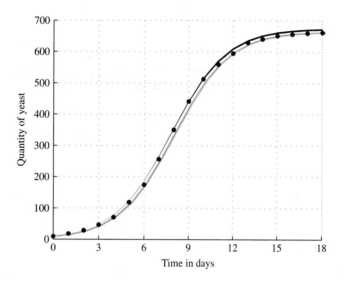

Figure 10. The growth of a yeast culture.

Next let's look at what happens if we choose three data points that are farther apart, so that they represent what is happening in more of the logistic curve. Let's use the data after 5 and 10 days in addition to the initial data. Then using (1.18) and (1.19) we find that $r = 0.5396$ and $K = 674.4$. The results are shown in the black curve in Figure 10. The black dots are the experimental data in Table 1. Clearly the match of the data and the computed curve is quite good.

A scientist who collected data such as that in Table 1 would undoubtedly want to use all of the data when looking for a matching logistic curve. It would be nice to find the best-fitting curve. Although it takes us beyond the subject matter of this book, we will explain briefly how this would be accomplished. First, let P_j be the measured population in day j. These are the numbers in Table 1. We are looking for the function $P(t)$ of the form in equation (1.15) that best matches the data. We could look for the function that minimizes the sum of the squares of the errors,

$$\sum_0^{18} \left(P(t_j) - P_j\right)^2.$$ (1.21)

(Notice that in our case $t_j = j$.)

However, since our populations vary over two orders of magnitude, from 9.6 to 665, it makes more sense to minimize the error relative to size of the individual numbers. This is done by minimizing

$$\sum_0^{18} \left(\ln P(t_j) - \ln P_j\right)^2 = \sum_0^{18} \left(\ln(P(t_j)/P_j)\right)^2.$$ (1.22)

The function P in equation (1.15) has three parameters, P_0, r, and K. The task is to find the values of these parameters that minimize (1.21) or (1.22). This is not a task to be done using a pencil and paper. It is much too difficult. However, it can be done quite easily on a computer if the required programs are available. In either case, the result is called a ***least squares approximation***. Unlike the situation in Example 2, this is a nonlinear least squares problem, which is usually an order of magnitude more difficult than the linear least squares problem found there.[5]

In the case at hand, we used a computer to minimize (1.22), our calculations revealed that the minimum occurred with $P_0 = 9.5998$, $r = 0.5391$, and $K = 665.0000$. The colored curve in Figure 10 is the logistic curve with these parameters. If you have a calculator that will do a logistics regression, you should try to replicate this result.[6]   ●

The fit shown by the colored curve in Figure 10 is quite good. Many experiments similar to the one described in Example 1 have shown that the logistic model does a very good job of describing population growth of populations under very carefully controlled conditions, such as those found in a well-run laboratory.

EXERCISES

1. A biologist starts with 100 cells in a culture. After 24 hours, he counts 300. Assuming a Malthusian model, what is the reproduction rate? What will be the number of cells at the end of 5 days?

2. A biologist prepares a culture. After 1 day of growth, the biologist counts 1000 cells. After 2 days of growth, he counts 3000. Assuming a Malthusian model, what is the reproduction rate and how many cells were present initially?

3. A population of bacteria is growing according to the Malthusian model. If the population triples in 10 hours, what is the reproduction rate? How often does the population double itself?

4. A population of bacteria, growing according to the Malthusian model, doubles itself in 10 days. If there are 1000 bacteria present initially, how long will it take the population to reach 10,000?

5. A certain bacterium, given plenty of nutrient and room, is known to grow according to the Malthusian model with reproductive rate r. Suppose that the biologist working with the culture harvests the bacteria at a constant rate of h bacteria per hour. Use qualitative analysis to discuss the fate of the culture.

6. A certain bacterium is known to grow according to the Malthusian model, doubling itself every 8 hours. If a biologist starts with a culture containing 20,000 bacteria, then harvests the culture at a constant rate of 2000 bacteria per hour, how long until the culture is depleted? What would happen in the same time span if the initial culture contained 25,000 bacteria?

7. A certain bacterium is known to grow according to the Malthusian model, doubling itself every 4 hours. If a biologist starts with a culture of 10,000 bacteria, at what minimal rate does he need to harvest the culture so that it won't overwhelm the container with bacteria?

8. A biologist grows a culture of fruit flies in a very large enclosure with substantial nutrients available. The following table contains the data on the numbers for each of the first ten days.

DAY	NUMBER OF FLIES	DAY	NUMBER OF FLIES
0	10	6	55
1	14	7	72
2	19	8	85
3	24	9	123
4	28	10	136
5	38		

---

[5] Some calculators will do a "logistic regression." However, in at least one case this routine fits a function of the form

$$f(t) = \frac{a}{1 + be^{ct}} + d$$

to the data. Such a routine will not solve the problem we are dealing with. What is needed is a routine that fits a function of the form

$$f(t) = \frac{a}{1 + be^{ct}}$$

to the data. The method used probably changes from calculator to calculator, so you should check it out.

[6] To learn more about fitting the logistic equation, see "Fitting a Logistic Curve to Data," Fabio Cavallini, *College Mathematics Journal*, Vol. 24, Num. 3, pp. 247–253.

(a) As was done in Example 1.4, use the first two data points to estimate the reproduction rate. Plot the results of the Malthusian model against the data points.

(b) Now use the number of flies after 5 days to estimate the reproduction rate. Plot the results of the Malthusian model against the data points.

(c) If you have a calculator or a computer with the right program, estimate the reproduction rate using linear regression. Plot the results of the Malthusian model against the data points.

(d) Explain the differences in the reproduction rates found in your computations.

(e) In your estimation is the Malthusian model a good one for this experiment?

9. Some readers might not like the "black box" nature of regression routines on calculators and computers and would prefer to understand regression in more depth. Suppose that we wish to "fit" a line having equation $y = mx + b$ to a set of data points $(x_1, y_1), \ldots, (x_n, y_n)$. Then the points on the line having the same x-coordinates as the given data points are $(x_1, mx_1 + b), \ldots, (x_n, mx_n + b)$. The error made at the ith data point is $e_i = y_i - (mx_i + b)$. If the ith data point lies above the line of best fit, then this error is positive. If the ith data point lies below the line of best fit, then this error is negative. To keep the sum of the errors from canceling, we square each error before summing. Then the sum of the squares of the errors is given by

$$S = \sum_{i=1}^{n} [y_i - (mx_i + b)]^2. \qquad (1.23)$$

Note that S is a function of m and b. The idea behind the "line of best fit" is to minimize this sum of the squares of the errors. That's how the process gets the name **least squares fit**. We want to pick values of m and b that minimize S. To complete this minimization, we'll need to find critical values using differentiation.

(a) Take the partial derivative of S, defined by equation (1.23), with respect to m and set it equal to zero to obtain

$$m \sum_{i=1}^{n} x_i^2 + b \sum_{i=1}^{n} x_i = \sum_{i=1}^{n} x_i y_i. \qquad (1.24)$$

If you don't know what a partial derivative is, that's fine; just differentiate with respect to m, holding b constant. Next, take the partial derivative of S with respect to b (hold m constant) and set it equal to zero to obtain

$$m \sum_{i=1}^{n} x_i + nb = \sum_{i=1}^{n} y_i. \qquad (1.25)$$

Solve equations (1.24) and (1.25) simultaneously, eliminating b, to show that

$$m = \frac{n \sum_{i=1}^{n} x_i y_i - \sum_{i=1}^{n} x_i \sum_{i=1}^{n} y_i}{n \sum_{i=1}^{n} x_i^2 - \left(\sum_{i=1}^{n} x_i\right)^2}. \qquad (1.26)$$

Finally, you can use this value of m and equation (1.24) to calculate b.

(b) Compare equation (1.7) in the narrative with equation (1.23) and then make the appropriate changes to equation (1.26) to show that

$$r = \frac{n \sum_{i=1}^{n} t_i \ln P_i - \sum_{i=1}^{n} t_i \sum_{i=1}^{n} \ln P_i}{n \sum_{i=1}^{n} t_i^2 - \left(\sum_{i=1}^{n} t_i\right)^2}. \qquad (1.27)$$

Use this result to show that

$$\ln C = \frac{\sum_{i=1}^{n} \ln P_i - r \sum_{i=1}^{n} t_i}{n}. \qquad (1.28)$$

(c) Use equations (1.27) and (1.28) to compute the values of r and C found in Example 1.8. Plot the data and superimpose the resulting exponential curve of best fit.

10. Suppose a population is growing according to the logistic equation,

$$\frac{dP}{dt} = rP\left(1 - \frac{P}{K}\right).$$

Prove that the rate at which the population is increasing is at its greatest when the population is at one-half of its carrying capacity. *Hint:* Consider the second derivative of P.

11. Consider anew the logistic equation,

$$\frac{dP}{dt} = rP\left(1 - \frac{P}{K}\right), \qquad P(t_0) = P_0. \qquad (1.29)$$

Typically, mathematical ecologists will introduce *dimensionless* variables to reduce the number of parameters in the logistic equation before proceeding with their analysis.

(a) Show that the substitutions $\omega = \alpha P$ and $s = \beta t$ transform equation (1.29) into

$$\frac{d\omega}{ds} = \frac{r}{\beta} w - \frac{r}{\alpha \beta K} \omega^2. \qquad (1.30)$$

(b) Find values of α and β that transform equation (1.30) into

$$\frac{d\omega}{ds} = \omega - \omega^2. \qquad (1.31)$$

(c) Note that equation (1.31) is a variant of Bernoulli's equation. Use the technique of Exercise 22 in Section 2.4 to show that equation (1.31) has the solution

$$\omega = \frac{1}{1 - Ce^{-s}}. \qquad (1.32)$$

(d) Finally, use the change of variables $\omega = \alpha P$ and $s = \beta t$, your parameters α and β found in part (b), and the initial condition $P(t_0) = P_0$ to show that equation (1.32) is equivalent to the solution given in equation (1.15).

12. A population, obeying the logistic equation, begins with 1000 bacteria, then doubles itself in 10 hours. The population is observed eventually to stabilize at 20,000 bacteria. Find the number of bacteria present after 25 hours and the time it takes the population to reach one-half of its carrying capacity.

13. A population is observed to obey the logistic equation with eventual population 20,000. The initial population is 1000, and 8 hours later, the observed population is 1200. Find the reproductive rate and the time required for the population to reach 75% of its carrying capacity.

14. In *The Biology of Population Growth*, published in 1925, the biologist Raymond Pearl reported the data shown in the following table for the growth of a population of fruit flies.

DAY	NUMBER OF FLIES	DAY	NUMBER OF FLIES
0	6	18	163
3	10	21	226
7	21	24	265
9	52	28	282
12	67	32	319
15	104		

(a) Notice that data were not collected systematically. However, data were collected on days 9 and 18. Use the method of Example 1.20 to estimate the natural reproductive rate and the carrying capacity for a logistic model.

(b) If you have a computer program with a least squares program, use it with all of the data in the table to estimate the natural reproductive rate and the carrying capacity for a logistic model.

(c) Is the logistic model a good one for this data?

15. G. F. Gause, in his *Struggle for Existence*, simply estimated the carrying capacity of a population from the graph of his data. Plot the data shown in the following table, and use your plot to obtain an estimate of the carrying capacity.

DAY	QUANTITY	DAY	QUANTITY
0	100	80	8587
20	476	100	9679
40	1986	120	9933
60	5510	140	9986

You now know the carrying capacity and the initial population. You can use any other point in the table to determine the reproduction rate r. Do so, and then superimpose the resulting logistic curve on your data plot for comparison. Can you see any problems that could occur with this method?

16. Consider a lake that is stocked with walleye pike and that the population of pike is governed by the logistic equation

$$P' = 0.1P(1 - P/10),$$

where time is measured in days and P in thousands of fish. Suppose that fishing is started in this lake and that 100 fish are removed each day.

(a) Modify the logistic model to account for the fishing.

(b) Find and classify the equilibrium points for your model.

(c) Use qualitative analysis to completely discuss the fate of the fish population with this model. In particular, if the initial fish population is 1000, what happens to the fish as time passes? What will happen to an initial population having 2000 fish?

17. A biologist develops a culture that obeys the modified logistic equation

$$P' = 0.38p\left(1 - \frac{P}{1000}\right) - h(t),$$

where the "harvesting" is defined by the piecewise function

$$h(t) = \begin{cases} 200, & \text{if } t < 3, \\ 0, & \text{otherwise.} \end{cases}$$

(a) Use a numerical solver to plot solution trajectories for initial bacterial populations ranging between 0 and 1000. You'll note that in some cases, the population "recovers," but in others, the bacterial count goes to zero. Determine experimentally the critical initial population that separates these two behaviors.

(b) Use an analytic method to determine the exact value of the "critical" initial population found in part (a). Justify your answer.

18. A population, left alone, obeys the logistic law with an initial population of 1000 doubling itself in about 2.3 hours. It is known that the environment can sustain approximately 10,000 individuals. Harvesting is introduced into this environment, with 1500 individuals removed per hour, but only during the last 4 hours of a 24-hour day. Suppose that the population numbers 6000 at the beginning of the day. Use a numerical solver to sketch a graph of the population over the course of the next three days. Find approximately the size of the population at the end of each day.

19. Consider the same lake as in the Exercise 16, but suppose that the fishing is done for a fixed time every day, with the result that 1% of the fish are caught each day.

(a) Modify the logistic model to account for the fishing.

(b) Find and classify the equilibrium points for your model.

(c) If the initial fish population is 1000, what happens to the fish as time passes?

20. Adjust the "standard" logistic equation

$$\frac{dP}{dt} = rP\left(1 - \frac{P}{K}\right)$$

to reflect the fact that a fixed percentage γ of the population is harvested per unit time. Use qualitative analysis to discuss the fate of the population. In your analysis, discuss two particular cases: (1) $\gamma < r$ and (2) $\gamma > r$.

21. In Exercise 20, examine the units of the term γP and explain why the function $Y(\gamma) = \gamma P$ is called the *yield*. Suppose that the harvesting strategy defined in Exercise 20 is kept in place long enough for the population to adjust to its new equilibrium level. What value of γ will maximize the yield at this level? What will be the yield and the new equilibrium level of the population for this value of γ?

3.2 Models and the Real World

Mathematical models are meant to explain what is happening in the real world. It is not enough to derive models from theoretical considerations. It is necessary to check the predictions of our models with what is happening in reality. We did this in the previous section when we looked at the implications of the Malthusian model of population growth. We realized that its prediction of unlimited exponential growth was unrealistic and is only good if the assumption of unlimited resources is satisfied. While this might be true for small populations, it is certainly not true in the long run. We then went back to the drawing board and came up with the logistic model. Our analysis of the experimental data in Example 1.20 showed that in that case, under controlled circumstances, the logistic model worked very well.

We should do the same kind of analysis for the logistic model. If we want to apply the logistic model in new circumstances, it is important to know if these circumstances fit the assumptions behind the logistic model. Let's recall what those assumptions are. First we assumed that the population changed due to births and deaths. We then allowed the birth and death rates to vary with the population in a way that reflected the competition between individuals for limited resources. We did not allow any change of the reproductive rate with respect to time. Now let's look at some other situations.

A Malthusian model of early U.S. population

A number of attempts have been made to model the population of the United States. Pierre-François Verhulst, a Belgian mathematician, argued that the Malthusian model could be used to model rapidly growing populations in environments containing seemingly unlimited resources.

> The United States (in the late eighteenth and early nineteenth centuries) offers just such an example of a rapidly growing population that is expanding as if it had unlimited resources. [Verhulst, Pierre-François. 1845. Recherches mathematiques sur la loi d'accroissement de la population. *Noveaux Mémoires de l'Académie Royale des Sciences et Belles-Letters de Bruxelles* 18: 1–38]

Verhulst used United States census data for the years 1790–1840, provided in Table 1. He used the arithmetic mean to estimate the population in intercensal years, then proceeded to show that the population grew in the geometric progression predicted by the Malthusian model.

If we use regression on the data points in Table 1 as we did in Example 1.8, we get the equation

$$P = 3,966,000\, e^{0.0294(t-1790)}, \tag{2.1}$$

where the time t is measured in years. The plot of equation (2.1) is superimposed on Verhulst's data in Figure 1. Note that the graph of equation (2.1) is a good fit with the census data.

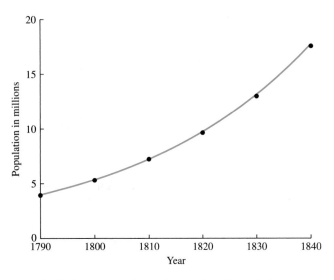

Table I	Early population of the United States	
YEAR	TIME	POPULATION
1790	0	3,929,827
1800	10	5,305,925
1810	20	7,239,814
1820	30	9,638,151
1830	40	12,866,020
1840	50	17,062,566

Figure 1. Fitting a Malthusian model to early U.S. population.

Now we come to the main question of this section. Is Verhulst's assertion that this period of U.S. population growth is a good example of "a rapidly growing population that is expanding as if it had unlimited resources" valid? It might be argued that, in addition to births and deaths, there was a third factor affecting the growth of the population, namely immigration. However, in the period between 1790 and 1820, immigration to the United States was rather small in comparison to the population increases that were occurring. Consequently, we can say that Verhulst's use of the Malthusian model was a valid approximation. Indeed, the agreement of (2.1) with the data in Table 1 as shown in Figure 1 is impressive.

Logistic models of U.S. population growth

We have just seen that the early growth of the U.S. population was Malthusian in nature. What about the growth since then?

In 1920, Pearl and Reed used the logistic equation to model the United States population. Their census data, taken from their report to the National Academy of Sciences,[7] came from the Bureau of Census figures. It included the data in Table 2, up to the year 1910. Table 2 comes from the *Statistical Abstract of the United States, 2002*. It shows the population of the United States in thousands.[8] These data are plotted in Figure 2.

By selecting three data points at equally spaced time intervals and using the technique in equations (1.18) and (1.19) of Section 1, Pearl and Reed were able to fit the logistic model to the population data in Table 2 up to 1910. Using the same techniques as before and the populations for 1790, 1850, and 1910, we fit the data

---

[7] Pearl and Reed, *On the Rate of Growth of the Population of the United States Since 1790 and Its Mathematical Representation*, Proceedings of the National Academy of Sciences, Volume 6, June 15, 1920, Number 6.

[8] The observant reader will notice that the data in Table 2 differs somewhat from that in Table 1. This is simply because the data is from different sources. It frequently happens that different sources report conflicting data. This is sometimes very disconcerting.

Table 2 Population of the United States (in thousands)

YEAR	POPULATION	YEAR	POPULATION	YEAR	POPULATION
1790	3,929	1870	39,818	1940	131,699
1800	5,308	1880	50,156	1950	151,326
1810	7,240	1890	62,948	1960	179,323
1820	9,638	1900	75,995	1970	203,302
1830	12,866	1910	91,972	1980	226,542
1840	17,069	1920	105,711	1990	248,718
1850	23,192	1930	122,755	2000	281,422
1860	31,443				

in Table 2 with the logistic model. The estimates of the parameters are[9]

$$r = 0.0313 \quad \text{and} \quad K = 197{,}274. \tag{2.2}$$

These data, together with $P_0 = 3{,}929{,}000$, are used in equation (1.15), which is

$$P(t) = \frac{KP_0}{P_0 + (K - P_0)e^{-r(t-t_0)}}. \tag{2.3}$$

The plot of this function is superimposed on the plot of the U.S. population data in Figure 2. Note the excellent fit.

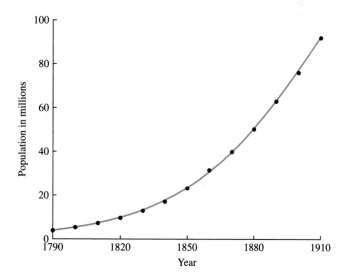

Figure 2. Fitting the logistic model to U.S. population.

Pearl and Reed then used their model to extrapolate the eventual behavior of the U.S. population. Let's check how well their model predicted the population of the United States. In Figure 3, we extend the plot of equation (2.3) with the parameters in (2.2) to the present, and we show the actual data from Table 2.

The model of Pearl and Reed predicted a carrying capacity of 197,274,000 people. However, in 1990 the U.S. population was 248,718,301 people, far beyond that predicted by Pearl and Reed. Thus, we see that using the logistic model to predict the population of the United States failed spectacularly after about 1950. It appears that the logistic model is not very good at predicting the population of the United States.

---

[9] These computations are a little sensitive, especially to the value of r used to compute K. It is important to keep the actual computed value of r when computing K. Even then, a calculator may not get the value found here due to lack of precision.

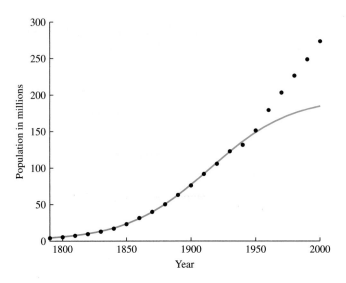

Figure 3. Logistic model projection of U.S. population based on data from 1790, 1850, and 1910, measured against the actual data.

Why is this true? If we think about how the population has grown in the United States, we come to the conclusion that the assumptions of the logistic model are simply not satisfied. For one thing, since about 1830 immigration has been a very important factor. For example, in the first decade of the twentieth century there were 8,795,386 immigrants, while the population increased by 15,977,691. Clearly immigration cannot be ignored. In addition, great improvements in public hygiene and in health care have had a big impact on the birth and death rates. Agricultural technology has increased the production of food by an incredible amount. Birth rates have decreased over the past century because of individual choice—a factor that is unrelated to the availability of resources or to the size of the population.

For these reasons, it should not be a surprise that the logistic model gives such a poor fit with the U.S. population curve. Indeed, it would be highly surprising if the logistic model were a good model for any segment of human population over a very long period of time.

For the total human population of the earth, such a model might be accurate for the period up to about 10,000 years ago, while humans were all hunter/gatherers. However, with the introduction of agriculture, the carrying capacity of the earth increased dramatically. The period from then until the start of the industrial revolution might be another period when a logistic model would be accurate. However, since the start of the industrial revolution changes have occurred at a very high rate, all of which have served to allow the earth to support more people. In addition, there have been changes in the attitudes of people that affect the reproductive rate. It is not likely that a logistic model would be useful for the human population of the earth since that time.

EXERCISES

1. Use the method of Example 1.4 of Section 3.1 and the first two data points in Table 1 to derive equation (2.1).

2. Use equations (1.18) and (1.19) in Section 3.1 together with the population data for 1790, 1850, and 1910 in Table 2 to verify the estimates of the parameters in (2.2).

3. Find in your library or on the Internet the historical census data for one of the United States, or for some other country. Attempt to model the data using the Malthusian model and the logistic model. Critique the effectiveness of the models.

3.3 Personal Finance

There are a number of problems involving personal finance that can be modeled using differential equations. We will start by considering what happens to the balance of a savings account, or of money invested in a portfolio of stocks and bonds.

Let $P(t)$ be the balance at time t, and suppose the account pays interest at a rate of r percent per year, compounded continuously. This means that the increase in the balance between times t and $t + \Delta t$ is

$$P(t + \Delta t) - P(t) = \text{interest earned in time } \Delta t. \tag{3.1}$$

Since r is the interest rate over a year, the interest earned in time Δt is approximately

$$\text{interest earned in time } \Delta t \approx r P \Delta t. \tag{3.2}$$

Hence, our model is

$$
\begin{aligned}
P'(t) &= \lim_{\Delta t \to 0} \frac{P(t + \Delta t) - P(t)}{\Delta t} \\
&= \lim_{\Delta t \to 0} \frac{r P \Delta t}{\Delta t} \\
&= r P.
\end{aligned}
\tag{3.3}
$$

This is once more the exponential equation, which we have already seen in a variety of applications. We know that the solutions have the form

$$P(t) = C e^{rt}.$$

If the initial balance is $P(0) = P_0$, then we see that $C = P_0$ and

$$P(t) = P_0 e^{rt}.$$

Example 3.4   Suppose you put \$1000 in a savings account with a continuously compounded interest rate of 5% per year. What will be the balance after 40 years?

In this case, $P_0 = 1000$, and $r = 0.05$. Hence

$$P(40) = 1000 e^{0.05 \times 40} = 1000 e^2 \approx 7389.$$ ●

Under the assumptions we have made, the balance in our savings account will grow exponentially. The basic assumption is that the interest rate r is a constant. Of course, in practice this is not true. Interest rates change constantly in reaction to a variety of economic and political events and are unpredictable. This limits the effectiveness of our model. To allow for this unpredictability, we should do the analysis for a variety of interest rates ranging from the lowest to the highest that we expect. This kind of "what if" analysis will allow us to bracket the real outcome.

An interest-bearing account with steady withdrawals

Next let's look at the balance $P(t)$ in a savings or an investment account from which the amount W is withdrawn every year and that pays interest at the rate of r percent per year, compounded continuously. When we add the effect of the withdrawals, equation (3.1) becomes

$$P(t + \Delta t) - P(t) = \text{interest earned in time } \Delta t - \text{withdrawal in time } \Delta t. \tag{3.5}$$

Once more, the interest earned is given by (3.2). On the other hand, if W is the amount withdrawn per unit time, the amount withdrawn in time Δt is

$$W \Delta t. \tag{3.6}$$

Thus,

$$P(t + \Delta t) - P(t) \approx r P(t) \Delta t - W \Delta t.$$

We can compute the derivative using the limit quotient definition:

$$\frac{dP}{dt} = \lim_{\Delta t \to 0} \frac{P(t + \Delta t) - P(t)}{\Delta t}$$

$$= \lim_{\Delta t \to 0} \frac{rP(t)\Delta t - W\Delta t}{\Delta t}$$

$$= rP - W.$$

Hence our model is

$$\frac{dP}{dt} = rP - W. \tag{3.7}$$

On the other hand, if we deposit D dollars per year, the equation becomes

$$P' = rP + D. \tag{3.8}$$

Systematic use of dimensions and units

Another factor that can often be used to keep confusion at bay during the modeling process is the careful and systematic use of dimensions and units. For example, the difference between equations (3.2) and (3.6) can be misleading. In particular, why doesn't (3.6) read $WP\Delta t$ in complete analogy to (3.2)? The answer is that r and W have different dimensions. The interest rate r is usually referred to as a percentage. It measures the fraction of a dollar earned per unit time. Since a fraction has no dimension and we are measuring time in years, the dimensions of r are simply years^{-1}. On the other hand, the withdrawal rate W is expressed in terms of dollars per year. Hence the units of W are dollars $\times$ years^{-1}. The left-hand side of (3.5) is measured in dollars. With the dimensions of r and W as found earlier, the expressions of both (3.2) and (3.6) are also measured in dollars. On the other hand, the expression $WP\Delta t$ is measured in (dollars)^2, which does not match.

The point is that by keeping careful track of the dimensions of the quantities being used, it is possible to keep from making errors.

Systematic savings

Our next problem is to start planning for retirement. The same process models savings towards any financial goal, such as the saving for the education of a child or the purchase of a house.

Example 3.9   Suppose you are just starting to work and you decide to save $2000 each year. Assuming that you have no savings to begin with and that your savings will earn 5% per year, compounded continuously, how much will you accumulate after 30 years?

We developed the model for this in equation (3.8). The equation in this case is

$$P' = 0.05\,P + 2,$$

where P is the principal balance in thousands of dollars. This is a linear equation. Let's work through the solution process carefully.

We first look for an integrating factor, and we know that one is given by

$$u(t) = e^{-\int 0.05\,dt} = e^{-0.05t}.$$

Hence

$$\left[e^{-0.05t}P\right]' = e^{-0.05t}\left[P' - 0.05P\right] = 2e^{-0.05t}. \qquad (3.10)$$

Integrating both sides yields

$$e^{-0.05t}P(t) = -40e^{-0.05t} + C,$$

or

$$P(t) = -40 + Ce^{0.05t}.$$

The initial condition becomes

$$0 = P(0) = -40 + C,$$

so $C = 40$ and the solution is

$$P(t) = 40(e^{0.05t} - 1).$$

To answer the question, we evaluate this when $t = 30$ to get

$$P(30) = 40(e^{1.5} - 1) = 139.2676.$$

Our balance is measured in thousands of dollars, so we learn that our balance after 30 years is $139,268. ●

Planning for retirement

That's not a bad start for retirement. After all, the amount of money put into savings was $2000 per year for 30 years, for a total of $60,000. Because of the compounded interest, this has grown to more than double that amount.

Now, however, let's turn the question around. How much money do you need to retire on?

Example 3.11 After some thought, you have decided that you will need $50,000 each year to live on after you retire, and that you should plan on living 30 years after your retirement. Assuming that your retirement account will earn 5% interest while you are taking out $50,000 each year, how much money must be in the retirement account when you retire?

Let $P(t)$ be the balance in your retirement account at time t after retirement. Let P_0 denote the balance in your retirement account when you retire. Then $P(0) = P_0$. The problem is to find P_0 so that $P(30) \geq 0$. Once more we will use a thousand dollars as our unit.

According to the model developed in equation (3.7),

$$P' = 0.05P - 50.$$

The integrating factor we used in (3.10) will also work here. Thus we have

$$\left[e^{-0.05t}P\right]' = -50e^{-0.05t}.$$

Integrating, we get

$$e^{-0.05t}P(t) = 1000e^{-0.05t} + C,$$

or

$$P(t) = 1000 + Ce^{0.05t}.$$

If we let $P(0) = P_0$ denote the balance at the time you retire, then we can evaluate the constant C,

$$P_0 = P(0) = 1000 + C.$$

Therefore, $C = P_0 - 1000$, and the solution is

$$P(t) = 1000 + (P_0 - 1000)e^{0.05t}.$$

Since you want to have that $50,000 each year until you die 30 years after retiring, you will want $P(30) \geq 0$. If you spend your last cent the day you die, you will want

$$0 = P(30) = 1000 + (P_0 - 1000)e^{1.5}.$$

We can solve this equation for P_0, getting

$$P_0 = 1000(1 - e^{-1.5}) = 776.8698.$$

Thus, you will need to have saved $776,870 before you retire in order to have the retirement you want. ●

Saving for retirement

Well, that $139,000 we saved in Example 3.9 doesn't seem so great any more. You are going to have to do much better than that. But just how are you going to accumulate the funds needed to finance a comfortable retirement?

Example 3.12 After some more thought prompted by the previous example, you decide that you should put a fixed percentage ρ of your salary into your retirement account. The question is, what value of ρ will achieve our goal?

First, you realize that your current salary of $35,000 per year will not stay at that level forever, hopefully for no more than one year. We need a model of how your salary will grow over time. Let's assume that your salary will grow at 4% per year. That's only a little more than the inflation rate. This thought leads to the differential equation $S' = 0.04S$, where $S(t)$ is your annual salary in thousands of dollars. We are very familiar with the exponential equation, so we easily solve this equation to get

$$S(t) = 35e^{0.04t}.$$

Now you notice that with this model, your salary in year 40 will be over $173,000. This seems excessive, but you are assured by your financial advisors that this type of increase over a lifetime is not at all unusual. Remember, we are including inflation in our forecasts, and the 4% per year increase we are projecting hardly covers the historical inflation rate. However, there is another concern. If your salary at retirement is this large, what kind of income should you plan for in your retirement years? Clearly the $50,000 in the previous example is too little. You decide that $100,000 is a more reasonable figure. This means that the size of your retirement account at retirement has to be double that found in the previous example. (Why?) You decide to be cautious and aim for a retirement fund of $1,600,000.

We will assume once more that your retirement account will earn an interest rate of 5%. Let $P(t)$ denote the balance in thousands of dollars in your retirement account at time t. The balance P will grow between times t and $t + \Delta t$ from two sources, the interest on the balance, which is $0.05P(t)\Delta t$, and from your investment, which is $\rho S(t)\Delta t$. Hence we have

$$P(t + \Delta t) - P(t) \approx 0.05P(t)\Delta t + \rho S(t)\Delta t.$$

Therefore,

$$P'(t) = \lim_{\Delta t \to 0} \frac{P(t + \Delta t) - P(t)}{\Delta t}$$

$$= 0.05P(t) + \rho S(t)$$

$$= 0.05P(t) + 35\rho e^{0.04t}.$$

We solve this linear equation as before. Again, the integrating factor $u(t) = e^{-0.05t}$ used in (3.10) will work here. We get

$$\left[e^{-0.05t} P\right]' = 35\rho e^{-0.01t}.$$

Integrating, we get

$$e^{-0.05t} P(t) = 35\rho \int e^{-0.01t} \, dt = -3500\rho e^{-0.01t} + C,$$

or

$$P(t) = -3500\rho e^{0.04t} + Ce^{0.05t}.$$

We will again assume that you are starting with no money in your retirement account, so this initial condition says

$$0 = P(0) = -3500\rho + C,$$

which allows us to conclude that $C = 3500\rho$, and

$$P(t) = 3500\rho \left(e^{0.05t} - e^{0.04t}\right).$$

We can now compute what ρ has to be by comparing this with our goal, which is that $P(40) = 1600$. Completing the computation, you find that you must save 18.77% of your salary to ensure a comfortable retirement. ●

Saving almost 19% of your salary every year seems like a lot. You decide to explore other strategies.

Example 3.13 When your salary is low, at the beginning of your career, saving this much of your salary will be hard. Perhaps even while you are young you should enjoy more of the fruits of your labor. What would happen if you were to start saving at a more modest rate and slowly increase your savings rate over time? You decide that the percent to save in year t is

$$\rho(t) = \frac{Rt}{40}.$$

With this choice, R will be the savings rate just before you retire after 40 years of work. What does R have to be in order to achieve your retirement goal?

The model in this case is similar to what it was in the previous example, but it has to change to accommodate the variability of your savings rate:

$$P' = 0.05P + \rho S$$

$$= 0.05P + \frac{Rt}{40} \times 35e^{0.04t}$$

$$= 0.05P + 0.875Rte^{0.04t}.$$

Again this equation is linear, and again the integrating factor $u(t) = e^{-0.05t}$ used in (3.10) will work. Integrating, we get

$$\left[e^{-0.05t} P\right]' = 0.875Rte^{-0.01t},$$

to get

$$e^{-0.05t} P(t) = 0.875R \int te^{-0.01t} \, dt$$
$$= -87.5R(t + 100)e^{-0.01t} + C,$$

or

$$P(t) = -87.5R(t + 100)e^{0.04t} + Ce^{0.05t}.$$

Using $P(0) = 0$ to compute that $C = 8750R$, we end up with

$$P(t) = 87.5R \left(100e^{0.05t} - (t + 100)e^{0.04t}\right).$$

Finally, we use our goal $P(40) = 1600$ to compute that $R = 0.4021$. This means that with this plan, although your savings rate in your early career will be small, in the last year before your retirement you will have to save over 40% of your salary. ●

Of course other strategies are possible. In particular, you might think about exploring ways to increase the income on your investments. Didn't you hear somewhere that over the long term, stocks return more than 8%? We will explore some other cases and strategies in the exercises.

EXERCISES

1. Suppose that $1200 is invested at a yearly rate of 5%, compounded continuously.

   (a) Assuming no additional withdrawals or deposits, how much will be in the account after 10 years?

   (b) How long will it take the balance to reach $5000?

2. Jamal wishes to invest an unknown sum in an account where interest is compounded continuously. Assuming that Jamal makes no additional deposits or withdrawals, what annual interest rate will allow his initial investment to double in exactly five years?

3. Alicia opens an account that pays an annual rate of 6% compounded continuously with an initial investment of $5000. After that, she deposits an additional $1200 per year. Assuming that no withdrawals are made and she continues with the same yearly deposit over a period of 10 years, how much will be in the account at the end of the 10-year period?

4. On the day of his birth, Jason's grandmother pledges to make available $50,000 on his eighteenth birthday for his college education. She negotiates an account paying 6.25% annual interest, compounded continuously, with no initial deposit, but agrees to deposit a fixed amount each year. What annual deposit should be made to reach her goal?

5. Andre inherits $50,000 from his grandfather's estate. The money is in an account that pays 5% annual interest, compounded continuously. The terms of the inheritance require that $8000 be withdrawn each year for Andre's educational expenses. If no additional deposits are made, how long will the inheritance last before the funds are completely gone?

6. Clarissa wants to buy a new car. Her loan officer tells her that her annual rate is 8%, compounded continuously, over a four-year term. Clarissa informs her loan officer that she can make equal monthly payments of $225. How much can Clarissa afford to borrow?

7. David and Mary would like to purchase a new home. They borrow $100,000 at 8% annual interest, compounded continuously. The term of the loan is 30 years. What fixed, annual payment will satisfy the terms of their loan?

8. Don and Heidi would like to buy a home. They've examined their budget and determined that they can afford monthly payments of $1000. If the annual interest is 7.25% and the term of the loan is 30 years, what amount can they afford to borrow?

9. José is 25 years old. His current annual salary is $28,000. Over the next 20 years, he expects his salary to increase continuously at a rate of 1% per year. He establishes a fund paying 6% annual interest, compounded continuously, with an initial deposit of $2500 and a promise

to deposit a fixed percentage of his annual income each year. Find that fixed percentage if José wants his balance to reach $50,000 at the end of the 20-year period.

10. Adriana opens a savings account with an initial deposit of $1000. The annual rate is 6%, compounded continuously. Adriana pledges that each year her annual deposit will exceed that of the previous year by $500. How much will be in the account at the end of the tenth year?

Discrete versus continuous You may have cast a somewhat skeptical eye at our financial models involving continuous compounding of interest. After all, no one pays off their loans continuously. It would be difficult to imagine how that could be accomplished. Payments are made at regular intervals, perhaps yearly, but more likely monthly. Let's examine how accurately our continuous models reflect the real world of finance.

11. The *recursive* definition,

$$a(n + 1) = ra(n), \quad a(0) = a_0,$$

is called a *first-order difference equation* and generates the sequence

$$a_0, ra_0, r(ra_0), r(r(ra_0)), \ldots.$$

A little simplification shows that the nth term of this sequence is

$$a(n) = a_0 r^n.$$

Now suppose that I represents the annual interest rate, but the interest is awarded in discrete packets, m times per year. Then the rate awarded during each compounding period is I/m. Consequently, if the initial investment is P_0, the balance is $P_0(1 + I/m)$ at the end of the first compounding period, $P_0(1 + I/m)^2$ at the end of the second compounding period, and so on.

(a) Give a first-order difference equation with an initial condition that generates a sequence describing the balance in the account at the end of each compounding period.

(b) Find a formula for the nth term of the sequence generated by the first-order difference equation created in part (a).

12. Arkady invests $2000 in an account paying 6% annual interest.

(a) If the interest is compounded continuously and no additional deposits or withdrawals are made, how much will be in the account at the end of 10 years?

(b) If the interest is awarded in discrete annual packets, then you want to use the discrete formula generated in Exercise 11. In this case, $m = 1$, so

$$P(10) = 2000 \left(1 + \frac{0.06}{1}\right)^{10} \approx \$3,581.70.$$

Calculate the balance in the account at the end of the 10-year period if the interest is compounded

- semiannually (twice per year)
- monthly (12 times per year)
- daily (365 times per year)

(c) Write a short paragraph explaining the point of this problem.

13. The first-order difference equation

$$a(n + 1) = ra(n) + b, \quad a(0) = a_0$$

generates the sequence

$$a_0, ra_0 + b, r(ra_0 + b) + b, r(r(ra_0 + b) + b) + b, \ldots.$$

A little simplification shows that the nth term of this sequence is

$$a(n) = a_0 r^n + b\left(1 + r + r^2 + \cdots + r^{n-1}\right).$$

(a) Show that

$$a(n) = \left(a_0 - \frac{b}{1 - r}\right) r^n + \frac{b}{1 - r}.$$

(b) Let I represent the annual interest rate, m the number of compounding periods in a year, P_0 the initial investment, and d the fixed deposit at the end of each compounding period. Then the balance at the end of each compounding period is generated by the first order difference equation

$$P(n + 1) = \left(1 + \frac{I}{m}\right) P(n) + d, \quad P(0) = P_0.$$

Use the result of part (a) to show that the balance at the end of n compounding periods is given by

$$P(n) = \left(P_0 + \frac{md}{I}\right)\left(1 + \frac{I}{m}\right)^n - \frac{md}{I}.$$

14. Demetrios opens an account with an initial investment of $2000. The annual interest rate is 5%.

(a) If the interest is compounded continuously and Demetrios makes an additional $1000 deposit every year, what will be the balance at the end of 10 years?

(b) If the interest is compounded quarterly (four times per year) and $250 is deposited at the end of each compounding period, what will be the balance after 10 years?

(c) What happens if the interest is compounded daily?

15. Chieh-Hsien purchases a new car and finances $12,000 at an annual rate of 8% for 5 years.

(a) If the interest is compounded continuously, what are his monthly payments?

(b) If the interest is compounded monthly, what are his monthly payments?

3.4 Electrical Circuits

A typical electric circuit involves a voltage source, a resistor, a capacitor, and an inductor. A simple diagram of an electric circuit is given in Figure 1. The voltage source is denoted by E, the resistor by R, the capacitor by C, and the inductor by L.

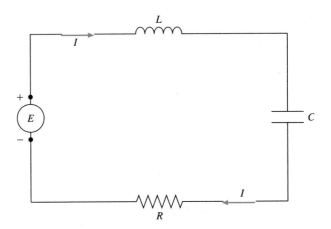

Figure 1. An *RLC* circuit.

The voltage source is sometimes referred to as the electromotive force and is denoted by emf. It could be a battery or a generator. A battery provides a steady, constant voltage, while a generator produces a variable voltage, usually sinusoidal in nature. The voltage causes electrons, and hence electric charge, to flow through the circuit. The rate at which charge flows is called the **current**. We will denote the source voltage by $E = E(t)$. All voltages are measured in volts, denoted by V, and the currents are measured in amperes, denoted by A.

Notice that in Figure 1 we have chosen a direction in which the current flows. This choice is essentially arbitrary, but once chosen it is fixed. An alternating current by definition flows both ways at different times. When a current flows in the direction opposite to the chosen direction, it is given negative values.

The derivation of the differential equation model of the circuit depends on the following experimental laws. The first three laws concern the voltage drop that occurs across various components of a typical circuit. We will refer to these as the **component laws**.

1. **Ohm's law.** A resistor impedes the flow of electrical charge. Every electrical device, such as a heater, a toaster, or an electrical stove, has some resistance to the flow of electrical charge. The voltage drop E_R across a resistor is proportional to the current I,
    $$E_R = RI,$$
    where R is the **resistance**. The resistance is measured in ohms, denoted by Ω.

2. **Faraday's law.** A typical inductor is a coil of wire. Current flowing through the coil produces a magnetic field. The magnetic field opposes any change in the current. The voltage drop E_L across an inductor is proportional to rate of change of the current,
    $$E_L = L\frac{dI}{dt},$$
    where L is the **inductance**. The inductance is measured in henrys, denoted by H.

3. **Capacitance law.** A capacitor, or condenser, is a device that stores electric charge. It usually consists of two parallel plates separated by an insulator. Consequently, there is little or no current passing through a capacitor. As current

flows up to a capacitor, the charge builds up there. Eventually the charge will be so large that the current reverses the direction of its flow. The voltage drop E_C across a capacitor is proportional to the charge Q on the capacitor,

$$E_C = \frac{1}{C}Q,$$

where C is the **capacitance** and Q is the **charge** on the capacitor. Capacitance is measured in farads, denoted by F. Charge is measured in coulombs, denoted by C.

In addition, there are two laws discovered by Gustav Kirchhoff that are very important for understanding circuits. They are not independent and often either can be used to analyze a circuit.

4. **Kirchhoff's voltage law.** The sum of the voltage drops around any closed loop in a circuit must be zero.

5. **Kirchhoff's current law.** The sum of currents flowing into any junction is zero.

To derive the differential equation that governs the circuit in Figure 1, we start with Kirchhoff's voltage law, which states that the sum of the voltage drops across the resistor, the inductor, the capacitor, and the voltage source is zero,

$$E_L + E_C + E_R - E = 0.$$

In the preceding equation, we have started and ended at the positive terminal of the voltage source. The voltage drop is positive in the direction of the current and negative in the opposite direction. Notice that the signs on the voltage source in Figure 1 indicate a voltage gain in the direction of the current. We can rearrange this equation to read

$$E_L + E_C + E_R = E. \tag{4.1}$$

Using the component laws listed on page 128 we obtain

$$L\frac{dI}{dt} + \frac{1}{C}Q + RI = E. \tag{4.2}$$

The current I represents moving charge. Therefore,

$$I = \frac{dQ}{dt}. \tag{4.3}$$

We differentiate equation (4.2) and use (4.3) to eliminate Q from the result to obtain

$$L\frac{d^2I}{dt^2} + \frac{1}{C}\frac{dQ}{dt} + R\frac{dI}{dt} = \frac{dE}{dt},$$

or

$$L\frac{d^2I}{dt^2} + R\frac{dI}{dt} + \frac{1}{C}I = \frac{dE}{dt}. \tag{4.4}$$

Initial conditions are usually imposed on the initial charge on the capacitor $Q(0)$ and the initial current $I(0)$. To solve the second-order equation (4.4), we must know both $I(0)$ and $I'(0)$. The derivative $I'(0)$ is found by inserting $I(0)$ and $Q(0)$ into (4.2) and solving for $I'(0)$ to obtain

$$I'(0) = \frac{1}{L}\left(E(0) - RI(0) - \frac{1}{C}Q(0)\right).$$

First-order systems and second-order equations will be the topic of much of the rest of this book. However, there are special cases that can be solved using techniques learned in Chapter 2.

Suppose there is no capacitor in the circuit. Then equation (4.2) becomes

$$RI + L\frac{dI}{dt} = E. \tag{4.5}$$

This is a linear first-order equation for the current I.

Example 4.6   Suppose the electrical circuit has a resistor of $R = 1/2$ Ω, and an inductor of $L = 1$ H. Assume the voltage source is a constant $E = 1$ V. If the initial current is 0 A, find the resulting current.

With the given data (4.5) becomes

$$\frac{dI}{dt} + \frac{1}{2}I = 1, \quad I(0) = 0.$$

This equation is both linear and separable. Solving it, we find that

$$I(t) = 2(1 - e^{-t/2}).$$

The solution is shown in Figure 2.   ●

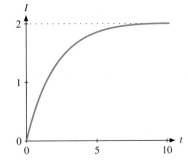

Figure 2.   The current in Example 4.6.

Next suppose there is no inductor in the circuit. Then equation (4.2) becomes $RI + Q/C = E$. Using $I = dQ/dt$, this becomes

$$R\frac{dQ}{dt} + \frac{1}{C}Q = E. \tag{4.7}$$

This time we have a first-order equation for the charge on the capacitor.

Example 4.8   Suppose the electrical circuit has a resistor of $R = 2$ Ω and a capacitor of $C = 1/5$ F. Assume the voltage source is $E = \cos t$ V. If the initial current is 0 A, find the resulting current.

With the given data, equation (4.7) becomes

$$2\frac{dQ}{dt} + 5Q = \cos t.$$

This is a linear equation for Q. We find that $e^{5t/2}$ is an integrating factor. Multiplying by it, we get

$$\left(e^{5t/2}Q\right)' = e^{5t/2}\left(\frac{dQ}{dt} + \frac{5}{2}Q\right) = \frac{1}{2}e^{5t/2}\cos t.$$

Integrating, this becomes

$$e^{5t/2}Q(t) = \frac{1}{2}\int e^{5t/2}\cos t\, dt$$

$$= \frac{1}{29}e^{5t/2}(2\sin t + 5\cos t) + C.$$

Solving for Q, we get

$$Q(t) = \frac{1}{29}(2\sin t + 5\cos t) + Ce^{-5t/2}.$$

Differentiating, we find the formula for the current:

$$I(t) = \frac{1}{29}(2\cos t - 5\sin t) - \frac{5}{2}Ce^{-5t/2}.$$

We evaluate the constant C using the initial condition $I(0) = 0$, and our final answer is

$$I(t) = \frac{1}{29}\left(2\cos t - 5\sin t - 2e^{-5t/2}\right).$$

The current is shown in Figure 3.

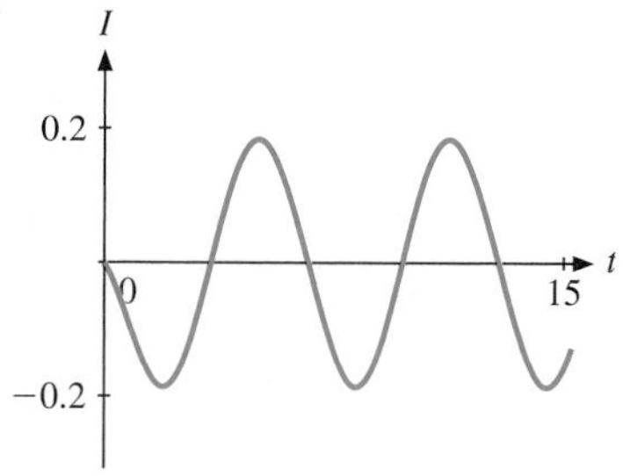

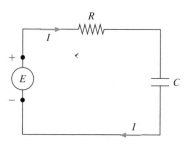

0.2

15

−0.2

Figure 3. The current in Example 4.8.

EXERCISES

A resistor (20 Ω) and capacitor (0.1 F) are joined in series with an electromotive force (emf) $E = E(t)$, as shown in Figure 4. If there is no charge on the capacitor at time $t = 0$, find the ensuing charge on the capacitor at time t for the given emf in each of Exercises 1–6.

1. $E(t) = 100$ V

2. $E(t) = 100e^{-0.1t}$ V

3. $E(t) = 100\sin 2t$ V

4. $E(t) = 100\cos 3t$ V

5. $E(t) = 100 - t$ V

6. $E(t) = 100(1 - e^{-0.1t})$ V

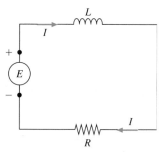

Figure 4. An RC circuit.

An inductor (1 H) and resistor (0.1 Ω) are joined in series with an electromotive force (emf) $E = E(t)$, as shown in Figure 5. If there is no current in the circuit at time $t = 0$, find the ensuing current in the circuit at time t for the given emf in each of Exercises 7–12.

7. $E(t) = 1$ V

8. $E(t) = e^{-0.1t}$ V

9. $E(t) = 5\sin 2\pi t$ V

10. $E(t) = 4\cos 3t$ V

11. $E(t) = 10 - 2t$ V

12. $E(t) = 10(1 - e^{-0.05t})$ V

13. Solve the general initial value problem modeling the RC circuit,

$$R\frac{dQ}{dt} + \frac{1}{C}Q = E, \quad Q(0) = 0,$$

where E is a constant source of emf.

14. Solve the general initial value problem modeling the LR circuit,

$$L\frac{dI}{dt} + RI = E, \quad I(0) = I_0,$$

where E is a constant source of emf.

15. A resistor (20 Ω) and capacitor (1 F) are linked in series with an electromotive force (emf) $E = E(t)$ in an RC circuit (see Figure 4). If the emf is given as $E(t) = 10e^{-0.01t}$ and the charge on the capacitor is zero at time $t = 0$, find the maximum charge on the capacitor and the time that it will occur.

16. An inductor (10 H) and resistor (1 Ω) are linked in series with an electromotive force (emf) $E = E(t)$ in an LR circuit (see Figure 5). If the emf is given as $E(t) = 5e^{-0.05t}$ and the current is zero at time $t = 0$, find the maximum current in the circuit and the time that it will occur.

17. A resistor (10 Ω) and capacitor (1 F) are linked in series with an electromotive force (see Figure 4). Initially, there is no charge on the capacitor. The emf produces a constant voltage difference of 100 V for the first five seconds, after which it is switched off.

(a) Use a numerical solver to approximate the charge on the capacitor at the end of 10 seconds.

(b) Set up and solve a differential equation modeling this circuit, then use the resulting formula to calculate the charge on the capacitor at the end of 10 seconds.

Figure 5. An LR circuit.

18. An RC circuit (see Figure 4) contains a resistor (10 Ω) and a capacitor (0.2 F) linked in series with an electromotive force $E(t) = 10 \sin 2\pi t$.

(a) Use a numerical solver to sketch the graph of the charge on the capacitor versus time for each of the initial conditions $Q(0) = -1, -0.75, -0.5, \ldots, 1$ on the time interval $[0, 15]$. What appears to happen to each of the solutions with the passage of time? Explain what might be meant by a **steady-state response**.

(b) What is the frequency of the emf? What appears to be the frequency of the steady-state response? *Hint:* Recall that in the function $f(t) = \sin \omega t$, ω is called the **angular frequency**, the **period** is given by $T = 2\pi/\omega$, and the **frequency** is the reciprocal of the period. Think of the frequency as the number of cycles per second, a unit known as Hertz (Hz).

19. An LR circuit (see Figure 5) contains an inductor (1 H) and a resistor (0.5 Ω) linked in series with an electromotive force $E(t) = \cos \pi t$. Use a numerical solver to sketch the graph of the current in the circuit versus time for each of the initial conditions $I(0) = -1, -0.75, -0.5, \ldots, 1$ on the time interval $[0, 20]$. What is the frequency of the emf? What appears to be the frequency of the steady-state response (see Exercise 18)?

20. An LR circuit, modeled by the initial value problem

$$I' + I = \cos 2t, \quad I(0) = 0,$$

has the solution

$$I = \frac{2}{5} \sin 2t + \frac{1}{5} \cos 2t - \frac{1}{5} e^{-t},$$

which is made up of a transient response, $-(1/5)e^{-t}$, and a steady-state response, $(2/5) \sin 2t + (1/5) \cos 2t$.

(a) On one plot, sketch the graph of the solution in black, the transient response in red, and the steady-state response in blue.

(b) Create a second plot that contains the steady-state response in blue. Use your numerical solver to superimpose the solutions of $I' + I = \cos 2t$ on your plot for

the initial conditions $I(0) = -1, -0.5, \ldots, 1$. Explain the behavior of the solutions in terms of the steady-state and transient response.

21. In the circuit modeled by the differential equation

$$RI + \frac{1}{C} Q = E \sin \omega t,$$

show that, regardless of initial condition, the eventual charge on the capacitor is given by the expression

$$\frac{EC}{1 + R^2 C^2 \omega^2} (\sin \omega t - RC\omega \cos \omega t).$$

22. In the lab, measuring the voltage drop across the capacitor of an RC circuit in Figure 4 is easier than measuring the charge buildup on the capacitor. Show that

$$R\frac{dQ}{dt} + \frac{1}{C} Q = E \cos \omega t$$

is equivalent to the differential equation

$$RC\frac{dV}{dt} + V = E \cos \omega t,$$

where V is the voltage drop across the capacitor. Show that, regardless of the initial voltage drop across the capacitor, the eventual voltage is given by

$$V = \frac{E}{1 + R^2 C^2 \omega^2} (RC\omega \sin \omega t + \cos \omega t).$$

23. Consider the circuit modeled by the differential equation

$$LI' + \frac{1}{C} Q = 0, \qquad (4.9)$$

with initial conditions $Q(0) = Q_0$ and $I(0) = 0$. Note that the substitution $I' = Q''$ is not much use, because we have not yet learned how to solve second-order differential equations. However, you can use the chain rule to show that $I' = I(dI/dQ)$, and this substitution will make equation (4.9) separable. Solve the resulting equation for I in terms of Q; then solve this result for Q in terms of t. Finally, find the current as a function of t.

PROJECT 3.5 **The Spruce Budworm**

In 1978, Ludwig, Jones, and Holling published a paper detailing the interaction between the spruce budworm and the balsam fir forests in eastern Canada. They began by assuming that the budworm population B obeys the logistic model defined by

$$\frac{dB}{dt} = r_B B \left(1 - \frac{B}{K_B} \right),$$

where r_B is the intrinsic growth rate of the budworm and K_B is the carrying capacity, which is assumed to depend on the amount of foliage available.

The growth of the budworm is affected by predation,

mainly in the form of birds and parasites. Thus, the equation governing the budworm population becomes

$$\frac{dB}{dt} = r_B B \left(1 - \frac{B}{K_B} \right) - p(B),$$

where the rate of predation $p(B)$ is assumed to depend on the budworm population.

1. Ludwig et al. made the assumption that

$$p(B) = \frac{\beta B^2}{\alpha^2 + B^2},$$

where α and β are positive. Sketch the graph of $p(B)$ versus B, assuming a positive budworm population. What is the limiting rate of predation? At what value of B does $p(B)$ reach one-half of its limiting rate of predation? Explain how the consumption of budworms by avian predators is limited by "saturation" and is nicely modeled by the function $p(B)$.

The analysis of the model

$$\frac{dB}{dt} = r_B B \left(1 - \frac{B}{K_B} \right) - \frac{\beta B^2}{\alpha^2 + B^2} \qquad (5.1)$$

is complicated by the presence of four different parameters, r_B, K_B, α, and β. Ludwig introduces *dimensionless* variables at this point, one immediate benefit being the reduction of the number of parameters.

2. Argue that B and α must have the same units and that $\mu = B/\alpha$ is a dimensionless quantity. Divide both sides of equation (5.1) by β; then substitute the dimensionless variable $\mu = B/\alpha$ to show that

$$\frac{\alpha}{\beta} \frac{d\mu}{dt} = \frac{\alpha r_B}{\beta} \mu \left(1 - \frac{\alpha \mu}{K_B} \right) - \frac{\mu^2}{1 + \mu^2}. \qquad (5.2)$$

Argue that

$$\tau = \frac{\beta}{\alpha} t, \quad R = \frac{\alpha r_B}{\beta}, \quad \text{and} \quad Q = \frac{K_B}{\alpha}$$

are dimensionless variables and their insertion into equation (5.2) produces the equation

$$\frac{d\mu}{d\tau} = R\mu \left(1 - \frac{\mu}{Q} \right) - \frac{\mu^2}{1 + \mu^2}. \qquad (5.3)$$

3. Ludwig continues his qualitative analysis of the budworm model by calculating the equilibrium points. Of course, this is accomplished by setting the right-hand side of equation (5.3) equal to zero. Factoring,

$$\mu \left[R \left(1 - \frac{\mu}{Q} \right) - \frac{\mu}{1 + \mu^2} \right] = 0, \qquad (5.4)$$

it is easily seen that $\mu = 0$ is an equilibrium point. Argue that $\mu = 0$ is an *unstable* equilibrium.

4. The remaining equilibrium points are calculated by setting the remaining factor of equation (5.4) equal to zero. This eventually leads to a rather intractable solution of a cubic polynomial, so Ludwig adopted an alternate strategy, that of determining where

$$R \left(1 - \frac{\mu}{Q} \right) = \frac{\mu}{1 + \mu^2}. \qquad (5.5)$$

The analysis continues by plotting graphs of the left- and right-hand sides of equation (5.5) versus μ and looking for points of intersection. Fix R. If Q is sufficiently small, there is only one point of intersection. However, for large values of Q, you can have one, two, or even three points of intersection, depending on the value of R, as shown in Figure 1. If you gradually decrease R in Figure 1, two points of intersection will coalesce into one, then disappear, leaving only one point of intersection. The same thing will happen if you increase R. Further, because $\mu > 0$, the sign of $d\mu/dt$ is determined solely by the sign of

$$R \left(1 - \frac{\mu}{Q} \right) - \frac{\mu}{1 + \mu^2},$$

which in turn is easily determined by noting where the graph of $R(1 - \mu/Q)$ lies above or below the graph of $\mu/(1 + \mu^2)$.

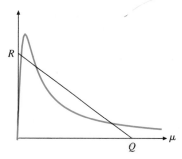

Figure 1. Three equilibrium points.

Sketch all possible ways that the graphs of $R(1 - \mu/Q)$ and $\mu/(1 + \mu^2)$ can intersect. In each case,

- create a phase line below the graph where equilibrium points and direction of flow are indicated, and
- write a description of the ecological significance of the given phase line. Why are some scenarios more/less dangerous than others?

You've seen that the positions where two equilibrium points coalesce into one, then disappear, are located where the curves in Figure 1 intersect *tangentially*. Therefore, at these positions of **bifurcation**,

$$R \left(1 - \frac{\mu}{Q} \right) = \frac{\mu}{1 + \mu^2} \qquad (5.6)$$

and

$$\frac{d}{d\mu} \left\{ R \left(1 - \frac{\mu}{Q} \right) \right\} = \frac{d}{d\mu} \left\{ \frac{\mu}{1 + \mu^2} \right\}. \qquad (5.7)$$

5. After performing the differentiation in equation (5.7), substitute the result into equation (5.6) to show that

$$R = \frac{2\mu^3}{(1 + \mu^2)^2}. \qquad (5.8)$$

After that, show that

$$Q = \frac{2\mu^3}{\mu^2 - 1}. \qquad (5.9)$$

Taken together, these equations define parametrically (in terms of μ) the bifurcation curve in the (RQ) parameter space. Sketch the bifurcation curve in (RQ) parameter space, then label the regions that have one, two, and three equilibrium points (not including $\mu = 0$ in your count). Discuss the ecological significance of these regions. Which region leads to a stable, manageable budworm population? Which region leads to a serious outbreak of the budworm population?

6. Pick an (R, Q) in each region determined by the bifurcation curve in Exercise 5 and use your numerical solver to sketch the fate of μ versus τ for various initial population sizes. [Use equation (5.3).]

If you wish to pursue this model in greater depth, here are some resources you will find useful.

- Ludwig, Jones, and Holling. "Qualitative Analysis of Insect Outbreak Systems: The Spruce Budworm and Forest," *Journal of Animal Ecology* (1978), **47**, 315–332.
- Murray, J. *Mathematical Biology* (1993), Springer-Verlag (New York).
- Strogatz, S. *Nonlinear Dynamics and Chaos* (1994), Addison-Wesley (Reading, Mass.).

PROJECT 3.6 **Social Security, Now or Later**

In the spring of 2000, a change in the Social Security law allowed individuals over their full retirement age[10] to continue to work full time and still receive full social security retirement benefits. However, they could choose to delay the start of their retirement benefits until a later time, in which case the monthly benefit would be larger. In May 2000, all working individuals over full retirement age had an important decision to make: Should I take social security now or later? Anyone who continues to work after their full retirement age will face the same question.

Many people faced with this decision find it an easy choice to make. If the government will give you money, take it. That is not bad logic, but others will want to examine the situation a little more deeply. They ask the question, Which way maximizes my financial situation?

It is not easy to see what this question means, but here is one way to approach the problem. It is based on the idea of setting up an account in which the Social Security benefits are deposited, which pays a certain fixed rate of return and from which taxes are paid on the benefits and on the return. Of course, such an account would be purely hypothetical. No one would actually do this. However, if we postulate two such accounts, one with balance $P_1(t)$ set up at full retirement age and another with balance $P_2(t)$ set up at the true retirement age, say 70, we can compare the balances over time.

Of course, $P_2(t) = 0$ for $t < 70$ while $P_1(t)$ is increasing. Then starting at $t = 70$, P_2 will start increasing at a faster rate than P_1 because of the larger benefits. The question is, At what age will P_2 catch up with P_1? This "catch-up" age might help someone make the decision on when to start receiving Social Security. If the catch-up age is low, then it might be a good idea to delay taking Social Security. On the other hand, if the catch-up age is high, then the opposite strategy might be best. Of course, it is up to each person to say what is high and what is low.

To carry out the modeling of the account balances, it is necessary to accumulate some data.

- The most important data is the monthly benefit amounts for retirement at the full retirement age and 70. The cor-

rect figures depend in a fairly complicated way on one's entire income history. However, it is possible to make estimates. Use your Internet browser to visit the Social Security Administration (http://www.ssa.gov/). There you will find calculators that will compute the monthly benefit for you, once you provide it with income data. The point is to find the monthly payment amounts starting at both full retirement age and at 70 based on the same income data. Perhaps the easiest thing to do is to find the maximum benefits for each age. Let A_1 be the annual benefit starting at full retirement age, and A_2 that starting at 70. Furthermore, let T_d be the time in years between full retirement age and 70.

- It will be necessary to deduct the taxes paid on each of the accounts. Consult with IRS documents to discover what portion of the benefits is taxable. In addition, taxes will have to be paid on the income into the accounts. Make an assumption about the marginal tax rate, which we will denote by ρ.
- The only other item of information needed is r, the rate of return on the investment accounts. Since this is uncertain, the computation should be done for a range of return rates varying from 0 to a maximum that will be computed.

With the data you have collected, perform the following tasks.

1. Along the lines of the derivations in Section 3.3, construct a differential equation model for the growth of the two accounts. Assume that all payments are made continuously. Be sure to account for taxes and return on investments.

2. Solve the differential equations exactly.

3. Show that if

$$r > r_{\max} = \frac{1}{(1 - \rho)T_d} \ln(A_2/A_1), \qquad (6.1)$$

then $P_1(t) > P_2(t)$ for all t. In other words, if the return on investment is large, the account started at the full retirement age is always larger than the one started at 70.

---

[10] As a result of the same change in the Social Security law, the full retirement age was also changed. It used to be 65 for everyone, but now it varies depending on a person's birth date. For individuals born in 1937 or earlier it is still 65, but for younger people it increases gradually to the point where for people born in 1967 or later it is 67.

4. Show that for $0 < r < r_{\max}$ the catch-up time T is

$$T = \frac{1}{(1 - \rho)r} \ln\left(\frac{A_2 - A_1}{A_2 e^{-(1-\rho)rT_d} - A_1}\right).$$

Plot the catch-up time T versus the investment rate r with the data you got from the Social Security Administration.

5. Redo the whole exercise with $r = 0$. In other words, assume that there is no return on investment. Show that the catch-up time is

$$T = \frac{T_d A_2}{A_2 - A_1}.$$

6. What would you advise an older relative or friend who asks you if he or she should delay receiving Social Security?

4

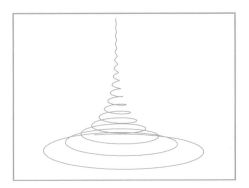

Second-Order Equations

We are ready to move on to differential equations of higher order. We will start with those of second order. These are especially important since so many of the equations that arise in science and engineering are of second order.

In Section 4.3 we will completely solve homogeneous, second-order, linear equations with constant coefficients, and in Section 4.4 we will apply this to the analysis of harmonic motion. We then examine methods of solution for inhomogeneous equations and apply these methods to the study of forced harmonic motion.

4.1 Definitions and Examples

A *second-order differential equation* is an equation involving the independent variable t and an unknown function y along with its first and second derivatives. We will assume it is possible to solve for the second derivative, in which case the equation has the form

$$y'' = f(t, y, y').$$

A *solution* to such an equation is a twice continuously differentiable function $y(t)$ such that

$$y''(t) = f(t, y(t), y'(t)).$$

Many problems in physics give rise to models that are second-order equations. For example, in the study of motion almost all models start with Newton's second law,

$$F = ma.$$

Here we are modeling the displacement $y(t)$ of a body from some reference point. The derivative of y is the velocity v and the second derivative is the acceleration a. The force F acting on the body is usually a function of time t, the displacement

y, and the velocity v, or $F = F(t, y, v)$. Thus Newton's second law gives us the differential equation

$$m\frac{d^2y}{dt^2} = F(t, y, dy/dt),$$

an equation of second order. We will discuss an important example in this section.

Linear equations

We will spend most of our time discussing *linear equations*. These have the special form

$$y'' + p(t)y' + q(t)y = g(t). \tag{1.1}$$

The *coefficients* p, q, and g can be arbitrary functions of the independent variable t, but y, y', and y'' must all appear to first order. This means we do not allow products of these to occur, nor any powers higher than 1, nor any complicated functions like $\cos y'$. For example, when the coefficients p and q are positive constants, equation (1.1) is the equation for the harmonic oscillator, which we will derive later in this section. On the other hand, the equation for the angular displacement of a pendulum bob is

$$\theta'' + k\sin\theta = 0.$$

Because of the $\sin\theta$ term this equation is nonlinear.

The function $g(t)$ on the right side of equation (1.1) is called the *forcing term* since it often arises from an external force. For example, such is the case in the equation of the harmonic oscillator. If the forcing term is equal to 0, the resulting equation is said to be *homogeneous*. Thus the equation

$$y'' + p(t)y' + q(t)y = 0 \tag{1.2}$$

will be called the homogeneous equation associated to (1.1).

An example—the vibrating spring

An important example of a second-order differential equation occurs in the model of the motion of a vibrating spring. The mathematical principles behind the vibrating spring appear in many areas of science and engineering. The differential equation that we derive here is the paradigm of oscillatory behavior.

The situation is illustrated in Figure 1. We consider the spring suspended from a beam. In Figure 1(a) we see the spring with no mass attached. It is assumed to be in equilibrium, so there is no motion. This is called the *spring equilibrium*. The position of the bottom of the spring is the reference point from which we measure displacement, so it corresponds to $x = 0$. We will orient our measurements by making x positive below the spring equilibrium.

In Figure 1(b) we have attached a weight of mass m to the spring. This weight has stretched the spring until it is once more in equilibrium at $x = x_0$. This is called the *spring-mass equilibrium*. At this point there are two forces acting on the mass. There is the force of gravity mg, and there is the restoring force of the spring, which we denote by $R(x)$ since it depends on the distance x that the spring is stretched. The fact that we have equilibrium at $x = x_0$ means that the total force on the weight is 0,

$$R(x_0) + mg = 0. \tag{1.3}$$

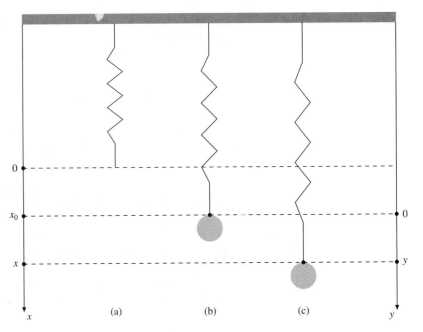

$$0 \qquad x_0 \qquad x$$

(a)   (b)   (c)

Figure 1. Analysis of a vibrating spring.

In Figure 1(c) we have stretched the spring further. The weight is no longer in equilibrium so it is most likely moving. Its velocity is

$$v = x'. \tag{1.4}$$

Now, in addition to gravity and the restoring force, there is a damping force D, which is the resistance to the motion of the weight due to the medium (air?) through which the weight is moving and perhaps to something internal to the spring. The damping force depends on a lot of factors, such as the shape of the body, but the major dependence is on the velocity. Hence we will write it as $D(v)$. To be complete, we will allow for an external force $F(t)$ as well.

Let $a = v' = x''$ denote the acceleration of the weight. According to Newton's second law,

$$ma = \text{total force acting on the weight}$$
$$= R(x) + mg + D(v) + F(t).$$

Since $a = x''$, and $v = x'$, we get the second-order differential equation

$$mx'' = R(x) + mg + D(x') + F(t). \tag{1.5}$$

To discover the form of the restoring force we resort to experimentation. For many springs, it turns out that the restoring force is proportional to the displacement. This experimental fact is referred to as Hooke's law. It says that

$$R(x) = -kx. \tag{1.6}$$

We use the minus sign because the restoring force is acting to decrease the displacement, and this allows us to say that the **spring constant** k is positive. It is important to realize that Hooke's law is an experimental fact. There are some springs for which it is not true, even for small displacements. It is not true for a bungee cord, for example. Furthermore, for any spring Hooke's law is valid only for small displacements. Assuming Hooke's law, (1.5) becomes

$$mx'' = -kx + mg + D(x') + F(t). \tag{1.7}$$

Assuming, for the moment, that there is no external force and that the weight is at spring-mass equilibrium where $x = x_0$, and $x' = x'' = 0$, then the damping force is $D = 0$, and we have (see (1.3))

$$0 = R(x_0) + mg = -kx_0 + mg \quad \text{or} \quad mg = kx_0. \tag{1.8}$$

Example 1.9 In an experiment, a 4-kg weight is suspended from a spring. The displacement of the spring-mass equilibrium from the spring equilibrium is measured to be 49 cm. What is the spring constant?

This example will give us an opportunity to discuss units. We will use the **International System**, in which the unit of length is the meter (abbreviated m), that of time is the second (abbreviated s), and the unit of mass is the kilogram (abbreviated kg). Other units are derived from these. For example, the unit for velocity is m/s, and that for acceleration is m/s^2. Thus, the acceleration due to gravity near the earth is $g = 9.8$ m/s^2. According to this system, the unit for force is kg·m/s^2, but this is called a newton (abbreviated N).

We can determine the spring constant in our example by solving (1.8) for k,

$$k = \frac{mg}{x_0}.$$

According to our data, the mass $m = 4$ kg, and $x_0 = 49$ cm $= 0.49$ m. Using $g = 9.8$ m/s^2, we find that the spring constant is

$$k = \frac{4 \text{ kg} \times 9.8 \text{ m/s}^2}{0.49 \text{ m}} = 80 \text{ N/m}.$$

Using (1.8) to substitute $mg = kx_0$, equation (1.7) becomes

$$mx'' = -k(x - x_0) + D(x') + F(t).$$

This motivates us to introduce the new variable $y = x - x_0$. Notice that y is the displacement of the weight from the spring-mass equilibrium (see Figure 1). Since $y' = x'$ and $y'' = x''$, our equation becomes

$$my'' = -ky + D(y') + F(t). \tag{1.10}$$

The damping force $D(v)$ always acts against the velocity. Hence we can write it as

$$D(v) = -\mu v, \tag{1.11}$$

where $\mu = \mu(v)$ is a nonnegative function of the velocity. Again it is an experimental fact that for objects of many shapes and for small velocities, the damping force is proportional to the velocity. In such cases μ is a nonnegative constant, called the **damping constant**. Again there are examples when the dependence of D on v is more complicated.

If we use (1.11) in equation (1.10) it becomes

$$my'' = -ky - \mu y' + F(t) \tag{1.12}$$

or

$$my'' + \mu y' + ky = F(t). \tag{1.13}$$

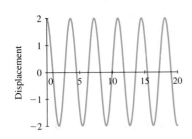

Figure 2. A vibrating spring with no damping.

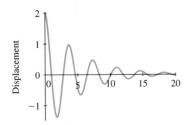

Figure 3. A vibrating spring with small damping.

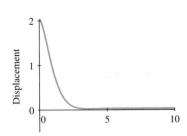

Figure 4. A vibrating spring with large damping.

If μ is a constant, this is a second-order, linear differential equation for the displacement $y(t)$. We can compute the solution numerically. With $k = 3$, $m = 1$, no damping ($\mu = 0$), and no external force ($F(t) = 0$), a solution is plotted in Figure 2. In Figures 3 and 4, we see the results with damping—in Figure 3, $\mu = 0.4$, and in Figure 4, $\mu = 4$.

Let's look at the case when the spring is undamped ($\mu = 0$) and unforced ($F(t) = 0$). Then our equation reduces to

$$y'' = -\frac{k}{m}y. \tag{1.14}$$

Can we solve this equation? We will develop systematic methods to solve such equations later, but for now we can only use our knowledge of calculus. It might help to look at the solution plotted in Figure 2. To solve equation (1.14) we must find a function whose second derivative is a negative multiple of itself. When we think in those terms (and look at Figure 2) we are led to consider the sine and cosine. With a little experimentation we can discover that

$$\cos\left(\sqrt{k/m}\,t\right) \quad \text{and} \quad \sin\left(\sqrt{k/m}\,t\right)$$

are solutions to (1.14). In fact, direct substitution shows that any function of the form

$$y(t) = a\cos\left(\sqrt{k/m}\,t\right) + b\sin\left(\sqrt{k/m}\,t\right), \tag{1.15}$$

where a and b are constants, is a solution to (1.14).

From (1.15) we see that our solutions are periodic functions. If we introduce the **natural frequency** $\omega_0 = \sqrt{k/m}$, the solution can be written as

$$y(t) = a\cos\omega_0 t + b\sin\omega_0 t. \tag{1.16}$$

If $T = 2\pi/\omega_0 = 2\pi\sqrt{m/k}$, then

$$\cos\omega_0(t + T) = \cos(\omega_0 t + \omega_0 T) = \cos(\omega_0 t + 2\pi) = \cos(\omega_0 t)$$

and the same is true for $\sin\omega_0(t + T)$. Hence $y(t + T) = y(t)$, so y is periodic and T is the **period**. Notice that ω_0 has units radians per second. It is an example of an **angular frequency**. Since there are 2π radians in a circle, the number $\nu = \omega_0/2\pi$ has units cycles per second. It is also called the frequency in many applications. To differentiate it from an angular frequency, it might be called a **numerical frequency**. When a frequency is present it is extremely important to notice whether it is an angular frequency or a numerical frequency. In this book, all frequencies will be angular frequencies.

Existence and uniqueness

The existence and uniqueness results for second-order equations are very similar to those for first-order equations. We will state a result for linear equations, which we will find quite useful.

THEOREM 1.17 Suppose the functions $p(t)$, $q(t)$, and $g(t)$ are continuous on the interval (α, β). Let t_0 be any point in (α, β). Then for any real numbers y_0 and y_1 there is one and only one function $y(t)$ defined on (α, β), which is a solution to

$$y'' + p(t)y' + q(t)y = g(t) \quad \text{for } \alpha < t < \beta$$

and satisfies the initial conditions $y(t_0) = y_0$ and $y'(t_0) = y_1$. ∎

The major difference between this result and the corresponding theorem for first-order linear equations in Section 2.7 of Chapter 2 is that an initial condition is needed not only for the function y, but also for its derivative y'. It is important to notice that we can be sure that a solution exists, and furthermore that it exists over the entire interval where the coefficients are defined and continuous.

Structure of the general solution

We will use Theorem 1.17 to find the form of the general solution to a homogeneous linear equation. It is based on the following result, which is the defining feature of linearity.

PROPOSITION 1.18 Suppose that y_1 and y_2 are both solutions to the homogeneous, linear equation

$$y'' + p(t)y' + q(t)y = 0. \tag{1.19}$$

Then the function

$$y = C_1 y_1 + C_2 y_2 \tag{1.20}$$

is also a solution to (1.19) for any constants C_1 and C_2.

Proof We notice that $y' = C_1 y_1' + C_2 y_2'$ and $y'' = C_1 y_1'' + C_2 y_2''$. Consequently, by simply rearranging the terms we get

$$
\begin{aligned}
y'' + py' + qy &= \left(C_1 y_1'' + C_2 y_2''\right) + p\left(C_1 y_1' + C_2 y_2'\right) + q\left(C_1 y_1 + C_2 y_2\right) \\
&= C_1\left(y_1'' + py_1' + qy_1\right) + C_2\left(y_2'' + py_2' + qy_2\right) \\
&= 0.
\end{aligned}
$$

The type of expression occurring in equation (1.20) will appear often. Let's give it a name.

DEFINITION 1.21

> A *linear combination* of the two functions u and v is any function of the form
>
> $$w = Au + Bv,$$
>
> where A and B are constants.

With this definition we can express Proposition 1.18 by saying that a linear combination of two solutions is also a solution. As an example, direct substitution will show that $y_1(t) = e^{-t}$ and $y_2(t) = e^{2t}$ are solutions to the linear equation $y'' - y' - 2y = 0$. In light of Proposition 1.18, we know that any linear combination $y(t) = C_1 e^{-t} + C_2 e^{2t}$ is also a solution. Again this can be checked by direct substitution.

The general linear combination, $y = C_1 y_1 + C_2 y_2$, of two solutions y_1 and y_2, contains two arbitrary constants. One might be led to expect that this is the general solution. This is often the case, but not always. This will be the content of our main theorem. However, to state it we need some terminology.

DEFINITION 1.22

> Two functions u and v are said to be *linearly independent* on the interval (α, β) if neither is a constant multiple of the other on that interval. If one is a constant multiple of the other on (α, β) they are said to be *linearly dependent* there.

Thus, the functions $u(t) = t$ and $v(t) = t^2$ are linearly independent on $\mathbf{R}$. It is true that $v(t) = tu(t)$, but the factor t is not a constant. On the other hand, $u(t) = \sin t$ and $v(t) = -4\sin t$ are obviously linearly dependent on $\mathbf{R}$.

We are going to use Theorem 1.17 to prove the following result, which will provide us with our solution strategy for homogeneous equations.

THEOREM 1.23   Suppose that y_1 and y_2 are linearly independent solutions to the equation

$$y'' + p(t)y' + q(t)y = 0. \tag{1.24}$$

Then the general solution to (1.24) is

$$y = C_1 y_1 + C_2 y_2,$$

where C_1 and C_2 are arbitrary constants.   ∎

We will prove Theorem 1.23 after some discussion of its result. Theorem 1.23 says that the general solution is the general linear combination of the solutions y_1 and y_2, provided that y_1 and y_2 are linearly independent. Because of this result we will say that two linearly independent solutions form a ***fundamental set of solutions***.

Notice that Theorem 1.23 defines a strategy to be used in solving homogeneous equations. It says that it is only necessary to find two linearly independent solutions to find the general solution. That is what we will do in what follows.

Example 1.25   Find a fundamental set of solutions to the equation

$$x'' + \omega^2 x = 0,$$

which is the equation for simple harmonic motion. (See Section 4.4.)

It can be shown by substitution that

$$x_1(t) = \cos \omega t \quad \text{and} \quad x_2(t) = \sin \omega t$$

are solutions. (See equation (1.14) and what follows.) It is clear that these functions are not multiples of each other, so they are linearly independent. It follows from Theorem 1.23 that x_1 and x_2 are a fundamental set of solutions. Therefore, every solution to the equation for simple harmonic motion is a linear combination of x_1 and x_2.   ●

To prove Theorem 1.23, we need to know a little more about the impact of linear independence. The best way to determine if two given functions are linearly independent is by simple observation. For example, in Example 1.25 it is pretty obvious that $\cos \omega t$ and $\sin \omega t$ are not multiples of each other and therefore are linearly independent. However, we will need a way of making this determination in more difficult cases. The ***Wronskian*** of two functions u and v is defined to be

$$W(t) = \det \begin{pmatrix} u(t) & v(t) \\ u'(t) & v'(t) \end{pmatrix} = u(t)v'(t) - v(t)u'(t).$$

The relationship of the Wronskian to linear independence is summed up in the next two propositions.

PROPOSITION 1.26   Suppose the functions u and v are solutions to the linear, homogeneous equation

$$y'' + p(t)y' + q(t)y = 0$$

in the interval (α, β). Then the Wronskian of u and v is either identically equal to zero on (α, β) or it is never equal to zero there.

Proof   To prove this result, we differentiate the Wronskian $W = uv' - vu'$. We get

$$W' = u'v' + uv'' - v'u' - vu'' = uv'' - vu''.$$

Since u and v are solutions to $y'' + py' + qy = 0$, we can solve for their second derivatives and substitute. We get

$$\begin{aligned} W' &= u\left(-pv' - qv\right) - v\left(-pu' - qu\right) \\ &= -p(uv' - vu') \\ &= -pW. \end{aligned}$$

This is a separable first-order equation for W. If t_0 is a point in (α, β), the solution is

$$W(t) = W(t_0)e^{-\int_{t_0}^{t} p(s)\,ds} \quad \text{for } \alpha < t < \beta.$$

If $W(t_0) = 0$, then $W(t) = 0$ for $\alpha < t < \beta$. On the other hand, if $W(t_0) \neq 0$, then $W(t) \neq 0$, since the exponential term is never zero.   ●

Consider the solutions $x_1(t) = \cos \omega t$ and $x_2(t) = \sin \omega t$ we found in Example 1.25. The Wronskian of x_1 and x_2 is

$$W(t) = x_1(t)x_2'(t) - x_1'(t)x_2(t) = \omega \cos^2 \omega t + \omega \sin^2 \omega t = \omega.$$

Thus for these two solutions the Wronskian is never equal to zero. This is always the case for a fundamental set of solutions, as we will prove in the next result.

PROPOSITION 1.27   Suppose the functions u and v are solutions to the linear, homogeneous equation

$$y'' + p(t)y' + q(t)y = 0 \tag{1.28}$$

in the interval (α, β). Then u and v are linearly dependent if and only if their Wronskian is identically zero in (α, β).

Proof   Suppose first that u and v are linearly dependent in (α, β). Then one of them is a constant multiple of the other. Suppose $u = Cv$. Then $u' = Cv'$ as well, so

$$W(t) = u(t)v'(t) - v(t)u'(t) = Cv(t)v'(t) - v(t)Cv'(t) = 0.$$

Conversely, suppose that $W(t) = 0$ for $\alpha < t < \beta$. It remains to show that u and v are linearly dependent. First, if $v(t) = 0$ for $\alpha < t < \beta$, then $v = 0u$, so u and v are linearly dependent. Suppose, therefore, that v is not identically equal to 0 on (α, β). Suppose that $v(t_1) \neq 0$ for some $t_1 \in (\alpha, \beta)$. Since v is continuous, there is an interval $(c, d) \subset (\alpha, \beta)$ containing t_1 on which $v \neq 0$. On this interval we have

$$\frac{d}{dt} \frac{u}{v} = \frac{u'v - uv'}{v^2} = \frac{-W}{v^2} = 0.$$

Hence, on the interval (c, d), u/v is equal to a constant C, or $u = Cv$. In particular, at t_1 we have $u(t_1) = Cv(t_1)$ and $u'(t_1) = Cv'(t_1)$. By Proposition 1.18 both u and Cv are solutions to the differential equation $y'' + py' + qy = 0$. Since they have the same initial conditions at t_1, it follows from Theorem 1.17 that $u = Cv$ everywhere in (α, β). Consequently, u and v are linearly dependent.   ●

Let's restate the results of Propositions 1.26 and 1.27 to highlight the points we will need.

PROPOSITION 1.29 Suppose the functions u and v are solutions to the linear, homogeneous equation

$$y'' + p(t)y' + q(t)y = 0$$

in the interval (α, β). If $W(t_0) \neq 0$ for some t_0 in the interval (α, β), then u and v are linearly independent in (α, β). On the other hand, if u and v are linearly independent in (α, β), then $W(t)$ never vanishes in (α, β).

Proof   If $W(t_0) \neq 0$ for some t_0 in the interval (α, β), then by Proposition 1.27, u and v cannot be linearly dependent. Hence they are linearly independent.

On the other hand, if u and v are linearly independent in (α, β), then by Proposition 1.27 the Wronskian is not identically equal to 0. By Proposition 1.26 it is never equal to 0 in (α, β).   ●

Now we are ready to prove Theorem 1.23.

Proof of Theorem 1.23   Suppose that y_1 and y_2 are linearly independent solutions to the equation $y'' + py' + qy = 0$, and suppose that $y(t)$ is any solution. We need to find the constants C_1 and C_2 such that $y = C_1 y_1 + C_2 y_2$. Let t_0 be any point in (α, β). We choose the constants so that

$$\begin{aligned} y(t_0) &= C_1 y_1(t_0) + C_2 y_2(t_0), \quad \text{and} \\ y'(t_0) &= C_1 y_1'(t_0) + C_2 y_2'(t_0). \end{aligned} \tag{1.30}$$

This system is solvable provided the determinant

$$\det \begin{pmatrix} y_1(t_0) & y_2(t_0) \\ y_1'(t_0) & y_2'(t_0) \end{pmatrix} \neq 0.$$

However, this determinant will be recognized as the Wronskian W of y_1 and y_2. Since y_1 and y_2 are linearly independent, by Proposition 1.29 we know that $W(t_0) \neq 0$. Therefore, we can find C_1 and C_2 solving (1.30).

We know that y and $C_1 y_1 + C_2 y_2$ are both solutions to the differential equation $y'' + p(t)y' + q(t)y = 0$, and by (1.30) they have the same initial conditions at t_0. By the uniqueness part of Theorem 1.17, we have

$$y(t) = C_1 y_1(t) + C_2 y_2(t) \quad \text{for } \alpha < t < \beta.$$

Initial value problems

Let's give some thought to formulating the initial value problem for the second-order equation $y'' = F(t, y, y')$. We see in Theorem 1.17 that to determine a solution y uniquely it is necessary to specify both $y(t_0)$ and $y'(t_0)$. While the theorem applies only to linear equations, this is true for all second-order equations.

Example 1.31   Find the solution to the equation for simple harmonic motion $x'' + 4x = 0$, with initial conditions $x(0) = 4$ and $x'(0) = 2$.

We know from Example 1.25 that the general solution has the form

$$x(t) = a \cos 2t + b \sin 2t,$$

where a and b are arbitrary constants. Substituting the initial conditions, we get

$$4 = x(0) = a, \quad \text{and} \quad 2 = x'(0) = 2b.$$

Thus, $a = 4$ and $b = 1$ and our solution is

$$x(t) = 4\cos 2t + \sin 2t.$$

EXERCISES

For each of the second-order differential equations in Exercises 1–8, decide whether the equation is linear or nonlinear. If the equation in linear, state whether the equation is homogeneous or inhomogeneous.

1. $y'' + 3y' + 5y = 3\cos 2t$

2. $t^2 y'' = 4y' - \sin t$

3. $t^2 y'' + (1 - y)y' = \cos 2t$

4. $t y'' + (\sin t)y' = 4y - \cos 5t$

5. $t^2 y'' + 4yy' = 0$

6. $y'' + 4y' + 7y = 3e^{-t}\sin t$

7. $y'' + 3y' + 4\sin y = 0$

8. $(1 - t^2)y'' = 3y$

9. In an experiment, a 2-kg mass is suspended from a spring. The displacement of the spring-mass equilibrium from the spring equilibrium is measured to be 50 cm. If the mass is then displaced 12 cm downward from its spring-mass equilibrium and released from rest, set up (but do not solve) the initial value problem that models this experiment. Assume no damping is present.

10. In an experiment, a 5-kg mass is suspended from a spring. The displacement of the spring-mass equilibrium from the spring equilibrium is measured to be 75 cm. The mass is then displaced 36 cm upward from its spring-mass equilibrium and then given a sharp downward tap, imparting an instantaneous downward velocity of 0.45 m/s. Set up (but do not solve) the initial value problem that models this experiment. Assume no damping is present.

11. Suppose that the spring mass system of Exercise 9 is suspended in a viscous solution that dampens the motion according to $R(v) = -0.05v$. Furthermore, a machine is attached to the top of the spring that shakes the spring up and down harmonically with period $T = 2$ s and amplitude $A = 0.5$ m. Assume that the spring is initially displaced 0.5 m upward from the spring-mass equilibrium by this driving force. Adjust the model in Exercise 9 to satisfy these additional constraints.

12. Suppose that the spring mass system of Exercise 10 is suspended in a viscous solution that dampens the motion according to $R(v) = -0.125v$. Furthermore, a machine is attached to the top of the spring that shakes the spring up and down harmonically with period $T = 4$ s and amplitude $A = 0.25$ m. Assume that the spring is initially

displaced 0.25 m downward from the spring-mass equilibrium by this driving force. Adjust the model in Exercise 10 to satisfy these additional constraints.

In Exercises 13–16, show, by direct substitution, that the given functions $y_1(t)$ and $y_2(t)$ are solutions of the given differential equation. Then verify, again by direct substitution, that any linear combination $C_y y_1(t) + C_2 y_2(t)$ of the two given solutions is also a solution.

13. $y'' - y' - 6y = 0$, $y_1(t) = e^{3t}$, $y_2(t) = e^{-2t}$

14. $y'' + 4y = 0$, $y_1(t) = \cos 2t$, $y_2(t) = \sin 2t$

15. $y'' - 2y' + 2y = 0$, $y_1(t) = e^t \cos t$, $y_2(t) = e^t \sin t$

16. $y'' + 4y' + 4y = 0$, $y_1(t) = e^{-2t}$, $y_2(t) = te^{-2t}$

In Exercises 17–20, use Definition 1.22 to explain why $y_1(t)$ and $y_2(t)$ are linearly independent solutions of the given differential equation. In addition, calculate the Wronskian and use it to explain the independence of the given solutions.

17. $y'' - y' - 2y = 0$, $y_1(t) = e^{-t}$, $y_2(t) = e^{2t}$

18. $y'' + 9y = 0$, $y_1(t) = \cos 3t$, $y_2(t) = \sin 3t$

19. $y'' + 4y' + 13y = 0$, $y_1(t) = e^{-2t}\cos 3t$, $y_2(t) = e^{-2t}\sin 3t$

20. $y'' + 6y' + 9y = 0$, $y_1(t) = e^{-3t}$, $y_2(t) = te^{-3t}$

21. Show that the functions

$$y_1(t) = t^2 \quad \text{and} \quad y_2(t) = t|t|$$

are linearly independent on $(-\infty, +\infty)$. Next, show that the Wronskian of the two functions is identically zero on the interval $(-\infty, +\infty)$. Why doesn't this result contradict Proposition 1.27?

22. Show that $y_1(t) = e^t$ and $y_2(t) = e^{-3t}$ form a fundamental set of solutions for $y'' + 2y' - 3y = 0$, then find a solution satisfying $y(0) = 1$ and $y'(0) = -2$.

23. Show that $y_1(t) = \cos 4t$ and $y_2(t) = \sin 4t$ form a fundamental set of solutions for $y'' + 16y = 0$, then find a solution satisfying $y(0) = 2$ and $y'(0) = -1$.

24. Show that $y_1(t) = e^{-t}\cos 2t$ and $y_2(t) = e^{-t}\sin 2t$ form a fundamental set of solutions for $y'' + 2y' + 5y = 0$, then find a solution satisfying $y(0) = -1$ and $y'(0) = 0$.

25. Show that $y_1(t) = e^{-4t}$ and $y_2(t) = te^{-4t}$ form a fundamental set of solutions for $y'' + 8y' + 16y = 0$, then find a solution satisfying $y(0) = 2$ and $y'(0) = -1$.

26. Unfortunately, Theorem 1.23 does not show us how to find two independent solutions. However, there is a technique that can be used to find a second solution when one solution is known.

(a) Show that $y_1(t) = t^2$ is a solution of

$$t^2 y'' + t y' - 4y = 0. \qquad (1.32)$$

(b) Let $y_2(t) = v y_1(t) = v t^2$, where v is a yet to be determined function of t. Note that if $y_2/y_1 = v$ and v is nonconstant, then y_1 and y_2 are independent. Show that the substitution $y_2 = v t^2$ reduces equation (1.32) to the separable equation

$$5v' + t v'' = 0. \qquad (1.33)$$

Solve equation (1.33) for v, form the solution $y_2 = v t^2$, and then state the general solution of equation (1.32).

Use the technique shown in Exercise 26 to find the general solution of the second-order equations in Exercises 27–30.

27. $t^2 y'' - 2t y' + 2y = 0, \quad y_1(t) = t$

28. $t^2 y'' + t y' - y = 0, \quad y_1(t) = t$

29. $t^2 y'' - 3t y' + 3y = 0, \quad y_1(t) = t$

30. $t^2 y'' + 4t y' + 2y = 0, \quad y_1(t) = 1/t$

4.2 Second-Order Equations and Systems

A *planar system* of first-order equations is a set of two first-order differential equations involving two unknown functions. It might be written as

$$x' = f(t, x, y)$$
$$y' = g(t, x, y),$$

where f and g are functions of the independent variable t and the two unknowns x and y.

There is a close connection between higher-order equations and first-order systems. In this section, we will begin to explore that connection. Then we will explore ways to visualize solutions to second-order equations. One of those ways involves the use of the phase plane, which is really a way of visualizing solutions to planar systems of equations. We will begin a systematic study of first-order systems in Chapter 8.

Second-order equations and planar systems

Let's look again at the second-order equation

$$y'' = F(t, y, y'). \qquad (2.1)$$

If we introduce a new variable $v = y'$, then, in terms of v, equation (2.1) can be written $v' = F(t, y, v)$. We see that the functions y and v are related by the system of first-order equations

$$y' = v$$
$$v' = F(t, y, v). \qquad (2.2)$$

If y is a solution to the second-order equation (2.1), and we set $v = y'$, then the pair of functions y and v solve the first-order system (2.2). The converse is also true. If the pair of functions y and v solve the first-order system (2.2), then y is a solution to the second-order equation (2.1). To see this, notice that by the first equation in (2.2), $y' = v$. Differentiating this and using the second equation in (2.2) we get

$$y'' = v' = F(t, y, v) = F(t, y, y'),$$

which is equation (2.1).

Thus, the second-order equation (2.1) and the first-order system (2.2) are equivalent in the sense that a solution of either leads to a solution of the other. This fact is important for two reasons. First, when solving higher-order equations numerically, it is often necessary to solve the equivalent first-order system, since numerical methods are typically designed only to solve first-order systems. Second, the equivalence allows us to study first-order systems and deduce from them results about higher-order equations. This procedure will be illustrated when we begin the study of first-order systems in Chapter 8.

Example 2.3 Let's look first at one of the most important examples, the single, second-order, linear equation. The general such equation has the form

$$y'' + p(t)y' + q(t)y = F(t). \tag{2.4}$$

To find an equivalent system, we introduce the new variable $v = y'$. Then, solving (2.4) for y'', we see that $v' = y'' = F(t) - p(t)y' - q(t)y = F(t) - p(t)v - q(t)y$. Hence, y and v solve the system

$$y' = v$$
$$v' = F(t) - p(t)v - q(t)y.$$

⬤

Visualization of solutions

The simplest and most obvious way to visualize a solution to a second-order differential equation is to graph it. We have already seen examples of this in Figures 2, 3, and 4 in Section 4.1.

Sometimes it is of interest to graph both the solution y and its first derivative, which is especially true if the derivative has physical significance, which is often the case in applications. For example, if we are solving for a displacement y, then $y' = v$ is the velocity. If we are solving for the charge Q on a condenser, then the derivative $Q' = I$ is the current. In both of these cases it might be of interest to see graphs of both the solution and its derivative.

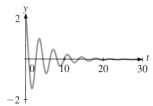

Figure 1. The graph of the displacement.

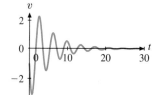

Figure 2. The graph of the velocity.

Let's use as an example a damped unforced spring. According to (1.12), the equation is

$$my'' + \mu y' + ky = 0.$$

Let's use $m = 1$, $\mu = 0.4$, and $k = 3$. We will look at the solution y which satisfies the initial conditions $y(0) = 2$ and $v(0) = y'(0) = -1$. The graph of the displacement y alone is shown in Figure 1, and the graph of the velocity alone is in Figure 2. Sometimes it is useful to see both plotted together. This is shown in Figure 3.

Another type of figure that is often useful is the plot of the curve

$$t \to (y(t), v(t))$$

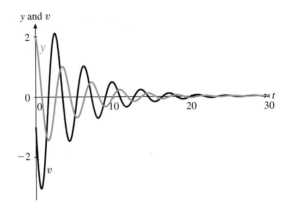

Figure 3. The displacement y and the velocity v.

in the yv-plane. The yv-plane is called the ***phase plane***, and this is called a ***phase plane plot***. For our example the phase plane plot is shown in Figure 4. Notice how this curve spirals into the origin. This gives an interesting visual interpretation of the effect of the damping. Missing from a phase plane plot is any indication of the dependence on t. However, it does show nicely the interplay between the displacement and the velocity.

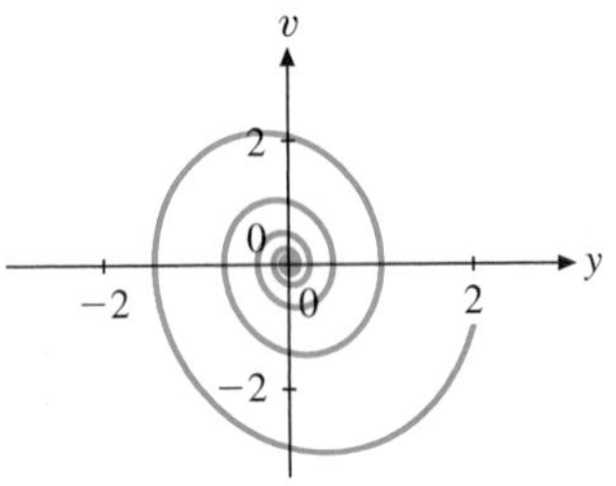

Also illuminating is a three-dimensional (3D) plot of the three variables t, y, and v. These can be plotted in either of the orders

$$t \to (t, y(t), v(t)) \quad \text{or} \quad t \to (y(t), v(t), t).$$

Which is more effective is often a matter of taste. The latter is shown in Figure 5. The 3D plots show the interplay of all three variables. However, some people find them difficult to understand. Often they become clearer simply by changing the orientation slightly.

Finally, there is the ***composite plot***, shown in Figure 6. In this one figure we see plots of y and v versus t, the phase plane plot, and the 3D plot. The relationships among all of these become clearer by looking at this figure. The 3D plot is central to the figure, and it is shown plotted in blue. This curve will be seen to be the same as the curve in Figure 5. The plot of the displacement y versus t is shown on the back. It is the projection of the 3D plot onto the back panel. Similarly, the plot of v versus t is on the right panel, and it too is a projection of the 3D plot. Finally, the phase plane plot is shown on the bottom, and, again, it is the projection of the 3D plot onto the bottom panel.

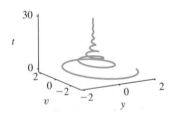

Figure 4. The phase plane plot of y and v.

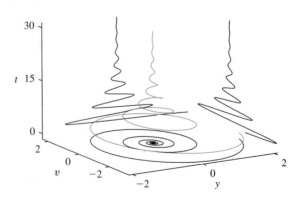

Figure 5. A 3D plot of $t \to (y(t), v(t), t)$.

Figure 6. The composite plot of the solution.

We see that there are many possible ways to represent solutions. In particular situations, one of these might be much better than the others. There might also be ad hoc representations that are better than any of those shown here.

EXERCISES

In Exercises 1–6, use the substitution $v = y'$ to write each second-order equation as a system of two first-order differential equations (planar system).

1. $y'' + 2y' - 3y = 0$

2. $4y'' + 4y' + y = 0$

3. $y'' + 3y' + 4y = 2\cos 2t$

4. $y'' + 2y' + 2y = \sin 2\pi t$

5. $y'' + \mu(t^2 - 1)y' + y = 0$

6. $y'' + cy' - ay + by^3 = A\cos \omega t$

7. Sometimes the physical situation aids in the selection of substitution variables. For example, in an LRC circuit, the current is the derivative of the charge. In the LRC circuit governed by the equation

$$LQ'' + RQ' + \frac{1}{C}Q = E(t),$$

show that the substitution $I = Q'$ transforms the equation into the planar system

$$Q' = I,$$
$$I' = -\frac{R}{L}I - \frac{1}{LC}Q + \frac{1}{L}E(t).$$

8. In general, when changing a second-order equation to a planar system, the choice of variables for substitution is arbitrary. If

$$y'' = 2y' - 3y + 2\cos 3t,$$

show that the substitutions $x_1 = y$ and $x_2 = y'$ lead to the planar system

$$x_1' = x_2,$$
$$x_2' = 2x_2 - 3x_1 + 2\cos 3t.$$

In Exercises 9–16, you are given the mass, damping, and spring constants of an undriven spring-mass system

$$my'' + \mu y' + ky = 0.$$

You are also given initial conditions. Use a numerical solver to

(i) provide separate plots of the position versus time (y vs. t) and the velocity versus time (v vs. t),

(ii) provide a combined plot of both position and velocity versus time, and

(iii) provide a plot of the velocity versus position (v vs. y) in the yv phase plane.

In each exercise, choose a viewing window that highlights the important features of the solutions.

9. $m = 1\,\text{kg}$, $\mu = 0\,\text{kg/s}$, $k = 4\,\text{kg/s}^2$, $y(0) = -2\,\text{m}$, $y'(0) = -2\,\text{m/s}$

10. $m = 1\,\text{kg}$, $\mu = 0\,\text{kg/s}$, $k = 9\,\text{kg/s}^2$, $y(0) = 3\,\text{m}$, $y'(0) = 2\,\text{m/s}$

11. $m = 1\,\text{kg}$, $\mu = 2\,\text{kg/s}$, $k = 1\,\text{kg/s}^2$, $y(0) = -3\,\text{m}$, $y'(0) = -2\,\text{m/s}$

12. $m = 4\,\text{kg}$, $\mu = 4\,\text{kg/s}$, $k = 1\,\text{kg/s}^2$, $y(0) = 3\,\text{m}$, $y'(0) = 1\,\text{m/s}$

13. $m = 1\,\text{kg}$, $\mu = 0.5\,\text{kg/s}$, $k = 4\,\text{kg/s}^2$, $y(0) = 2\,\text{m}$, $y'(0) = 0\,\text{m/s}$

14. $m = 1\,\text{kg}$, $\mu = 2\,\text{kg/s}$, $k = 1\,\text{kg/s}^2$, $y(0) = -1\,\text{m}$, $y'(0) = -5\,\text{m/s}$

15. $m = 1\,\text{kg}$, $\mu = 3\,\text{kg/s}$, $k = 1\,\text{kg/s}^2$, $y(0) = -1\,\text{m}$, $y'(0) = -5\,\text{m/s}$

16. $m = 1\,\text{kg}$, $\mu = 0.2\,\text{kg/s}$, $k = 1\,\text{kg/s}^2$, $y(0) = -3\,\text{m}$, $y'(0) = -2\,\text{m/s}$

17. Consider carefully the graphs of v and y versus t, shown in Figure 3.

(a) Why do the peaks of the of the curve $t \to y(t)$ occur where the curve $t \to v(t)$ crosses the t-axis? What is the physical significance of this fact?

(b) Do the peaks of the of the curve $t \to v(t)$ occur where the curve $t \to y(t)$ crosses the t-axis?

18. Consider carefully the phase plane plot of the spring-mass system given in Figure 4.

(a) What physical configuration of the spring-mass system is represented by the points where the solution curve $t \to (y(t), v(t))$ crosses the y-axis?

(b) What physical configuration of the spring-mass system is represented by the points where the solution curve $t \to (y(t), v(t))$ crosses the v-axis?

(c) What physical significance can be attached to the fact that the solution curve $t \to (y(t), v(t))$ spirals toward the origin with the passage of time?

If your software supports 3D capability, sketch the solution curve $t \to (y(t), v(t), t)$ for the spring-mass system with constants and initial conditions given in the indicated exercise.

19. Exercise 9   **20.** Exercise 10

21. Exercise 15   **22.** Exercise 16

In Exercises 23–28, you are given the inductance, resistance, and capacitance of a driven LRC circuit

$$LQ'' + RQ' + \frac{1}{C}Q = 2\cos 2t.$$

You are also given initial conditions. Use a numerical solver to

(i) provide separate plots of the charge on the capacitor versus time (Q vs. t) and the current in the circuit versus time (I vs. t),

(ii) provide a combined plot of both charge and the current versus time, and

(iii) provide a plot of the current versus the charge (I vs. Q) in the QI phase plane.

See Exercise 7 for aid in setting up the system. In each exercise, choose a viewing window that highlights the important features of the solutions.

23. $L = 1\,\mathrm{H}$, $R = 0\,\Omega$, $C = 1\,\mathrm{F}$, $Q(0) = -3\,\mathrm{C}$, $I(0) = -2\,\mathrm{A}$
24. $L = 1\,\mathrm{H}$, $R = 0\,\Omega$, $C = 1/4\,\mathrm{F}$, $Q(0) = 1\,\mathrm{C}$, $I(0) = 2\,\mathrm{A}$
25. $L = 1\,\mathrm{H}$, $R = 5\,\Omega$, $C = 1\,\mathrm{F}$, $Q(0) = 1\,\mathrm{C}$, $I(0) = 2\,\mathrm{A}$
26. $L = 2\,\mathrm{H}$, $R = 4\,\Omega$, $C = 1\,\mathrm{F}$, $Q(0) = -1\,\mathrm{C}$, $I(0) = -2\,\mathrm{A}$
27. $L = 1\,\mathrm{H}$, $R = 0.5\,\Omega$, $C = 1\,\mathrm{F}$, $Q(0) = 1\,\mathrm{C}$, $I(0) = 2\,\mathrm{A}$
28. $L = 1\,\mathrm{H}$, $R = 0.2\,\Omega$, $C = 1\,\mathrm{F}$, $Q(0) = -1\,\mathrm{C}$, $I(0) = -1\,\mathrm{A}$

If your software supports 3D capability, sketch the solution curve $t \rightarrow (Q(t), I(t), t)$ for the LRC circuit with constants and initial conditions given in the indicated exercise.

29. Exercise 23   30. Exercise 24
31. Exercise 27   32. Exercises 28

4.3 Linear, Homogeneous Equations with Constant Coefficients

This is a class of equations that we can solve easily. They are equations of the form

$$y'' + py' + qy = 0, \tag{3.1}$$

where p and q are constants. If $p \geq 0$ and $q > 0$, this is the equation for unforced harmonic motion, which we will discuss in the next section. We show there that it includes the equation for the unforced motion of a vibrating spring, and the equation for the behavior of an RLC circuit.

Remember the solution strategy we devised in Section 4.1 using Theorem 1.23. It is only necessary to find two linearly independent solutions, which we call a fundamental set of solutions. The general solution is the general linear combination of these.

The key idea

The analogous first-order, linear, homogeneous equation with constant coefficients is the equation

$$y' + py = 0.$$

This is the exponential equation. It is separable and easily solved. Its general solution is

$$y(t) = Ce^{-pt},$$

where C is an arbitrary constant.

Motivated by the fact that the first-order equation has an exponential solution, let's see if we can find an exponential solution to the second-order equation (3.1). We will look for a solution of the type

$$y(t) = e^{\lambda t},$$

where λ is a constant, as yet unknown. Inserting this function into our differential equation, we obtain

$$y'' + py' + qy = \lambda^2 e^{\lambda t} + p\lambda e^{\lambda t} + q e^{\lambda t}$$
$$= (\lambda^2 + p\lambda + q)e^{\lambda t}.$$

Since $e^{\lambda t} \neq 0$, we will have solution to (3.1) if and only if

$$\lambda^2 + p\lambda + q = 0. \tag{3.2}$$

This is called the ***characteristic equation*** for the differential equation in (3.1). The polynomial $\lambda^2 + p\lambda + q$ is called the ***characteristic polynomial*** for the equation. A root of the characteristic equation is called a ***characteristic root***. If λ is a characteristic root, then $y = e^{\lambda t}$ is a solution to the differential equation.

It is illuminating to write the differential equation and its characteristic equation in close proximity,

$$y'' + py' + qy = 0,$$
$$\lambda^2 + p\lambda + q = 0.$$

This indicates clearly how to pass from the differential equation to its characteristic equation.

Since the characteristic equation is a quadratic equation, its roots are given by the quadratic formula

$$\lambda = \frac{-p \pm \sqrt{p^2 - 4q}}{2}.$$

Looking at the discriminant $p^2 - 4q$, we see that there are three cases to consider:

1. two distinct real roots if $p^2 - 4q > 0$;
2. two distinct complex roots if $p^2 - 4q < 0$;
3. one repeated real root if $p^2 - 4q = 0$.

We will look at each of these in what follows. Our way will be guided by Theorem 1.23 in Section 4.1. We know that if we find a fundamental set of solutions, then the general solution is the general linear combination of these. Thus we need to find two linearly independent solutions.

Distinct real roots

If λ_1 and λ_2 are distinct real roots of the characteristic equation, then $y_1 = e^{\lambda_1 t}$ and $y_2 = e^{\lambda_2 t}$ are both solutions. Since the roots are not equal, the solutions are not constant multiples of each other. Hence, they are linearly independent, and by Theorem 1.23 we have the following result.

PROPOSITION 3.3   If the characteristic equation $\lambda^2 + p\lambda + q = 0$ has two distinct real roots λ_1 and λ_2, then the general solution to $y'' + py' + qy = 0$ is

$$y(t) = C_1 e^{\lambda_1 t} + C_2 e^{\lambda_2 t},$$

where C_1 and C_2 are arbitrary constants.   ●

The particular solution for an initial value problem can be found by evaluating the constants C_1 and C_2 using the initial conditions.

Example 3.4   Find the general solution to the equation

$$y'' - 3y' + 2y = 0.$$

Find the unique solution corresponding to the initial conditions $y(0) = 2$ and $y'(0) = 1$.

Letting $y = e^{\lambda t}$ and inserting this into our differential equation, we obtain

$$0 = y'' - 3y' + 2y = (\lambda^2 - 3\lambda + 2)e^{\lambda t}.$$

Thus, the characteristic equation is

$$0 = \lambda^2 - 3\lambda + 2 = (\lambda - 2)(\lambda - 1),$$

with solutions $\lambda_1 = 2$ and $\lambda_2 = 1$. According to Proposition 3.3, the general solution is

$$y(t) = C_1 e^{2t} + C_2 e^t. \qquad (3.5)$$

To find the particular solution for the initial conditions $y(0) = 2$ and $y'(0) = 1$, we differentiate the general solution,

$$y'(t) = 2C_1 e^{2t} + C_2 e^t. \qquad (3.6)$$

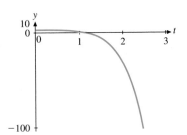

Then we substitute $t = 0$ into (3.5) and (3.6) to obtain the system of equations

$$2 = y(0) = C_1 + C_2$$
$$1 = y'(0) = 2C_1 + C_2.$$

The solutions are $C_1 = -1$ and $C_2 = 3$, so the solution to our initial value problem is

$$y(t) = -e^{2t} + 3e^t.$$

Figure 1. The solution in Example 3.4 with initial conditions $y(0) = 2$ and $y'(0) = 0$.

The solution is shown in Figure 1.

Complex roots

If the roots to the characteristic equation (3.2) are complex,[1] then, since the coefficients of the characteristic polynomial are real, the roots are complex conjugates. They have the form $\lambda = a + ib$ and $\bar{\lambda} = a - ib$. The corresponding solutions are $z(t) = e^{(a+ib)t}$ and $\bar{z}(t) = e^{(a-ib)t}$. Using Euler's formula,[2] these solutions become

$$z(t) = e^{(a+ib)t} = e^{at}[\cos(bt) + i\sin(bt)], \quad \text{and}$$
$$\bar{z}(t) = e^{(a-ib)t} = e^{at}[\cos(bt) - i\sin(bt)]. \qquad (3.7)$$

Since $\bar{z}(t) = e^{-2ibt} z(t)$, the solutions are not constant multiples of each other, so they are linearly independent. Hence using Theorem 1.23, we see that the general solution is

$$y(t) = C_1 z(t) + C_2 \bar{z}(t). \qquad (3.8)$$

The solutions z and $\bar{z}$ are complex valued. Such solutions are often preferred (for example, in circuit analysis). However, we are aiming for real valued solutions.

Notice from (3.7) that z and $\bar{z}$ are complex conjugates. Written in terms of their real and imaginary parts, we have

$$z(t) = y_1(t) + i y_2(t) \quad \text{and} \quad \bar{z}(t) = y_1(t) - i y_2(t), \qquad (3.9)$$

where

$$y_1(t) = \operatorname{Re} z(t) = e^{at}\cos(bt) \quad \text{and} \quad y_2(t) = \operatorname{Im} z(t) = e^{at}\sin(bt). \qquad (3.10)$$

---

[1] Complex numbers are discussed in detail in the Appendix near the end of this book. We refer you there for all of the facts about complex numbers used here.

[2] Euler's formula tells us that $e^{i\theta} = \cos\theta + i\sin\theta$. See the Appendix for more details.

Thus we have

$$y_1(t) = \frac{1}{2}\left(z(t) + \bar{z}(t)\right) \quad \text{and} \quad y_2(t) = \frac{1}{2i}\left(z(t) - \bar{z}(t)\right).$$

Therefore, by Proposition 1.18, $y_1(t)$ and $y_2(t)$, the real and imaginary parts of $z(t)$, are solutions. Furthermore, these are real valued solutions. Since $y_2(t) = \tan(bt) \cdot y_1(t)$ they are not constant multiples of each other, so they are linearly independent. Hence by Theorem 1.23, they form a fundamental set of solutions, and the general solution can be written as

$$y(t) = A_1 y_1(t) + A_2 y_2(t) = A_1 e^{at} \cos(bt) + A_2 e^{at} \sin(bt),$$

where A_1 and A_2 are arbitrary constants.

We summarize our discussion in the following proposition.

PROPOSITION 3.11 Suppose the characteristic equation $\lambda^2 + p\lambda + q = 0$ has two complex conjugate roots, $\lambda = a + ib$ and $\bar{\lambda} = a - ib$.

1. The functions

$$z(t) = e^{(a+ib)t} \quad \text{and} \quad \bar{z}(t) = e^{(a-ib)t}$$

form a complex-valued fundamental set of solutions, so the general solution is

$$w(t) = C_1 e^{(a+ib)t} + C_2 e^{(a-ib)t},$$

where C_1 and C_2 are arbitrary complex constants.

2. The functions

$$y_1(t) = e^{at} \cos(bt) \quad \text{and} \quad y_2(t) = e^{at} \sin(bt)$$

form a real-valued fundamental set of solutions, so the general solution is

$$y(t) = e^{at}(A_1 \cos bt + A_2 \sin bt),$$

where A_1 and A_2 are constants. ●

In either case, the constants can be determined in the usual way to find the solution to an initial value problem.

Example 3.12 Find the general solution to the system

$$y'' + 2y' + 2y = 0.$$

Find the solution corresponding to the initial conditions $y(0) = 2$ and $y'(0) = 3$.

The characteristic equation is $\lambda^2 + 2\lambda + 2 = 0$. Its roots are $\lambda = -1 \pm i$. The corresponding solutions (given in (3.10) with $a = -1$ and $b = 1$) are

$$y_1(t) = e^{-t} \cos t \quad \text{and} \quad y_2(t) = e^{-t} \sin t.$$

By Proposition 3.11, the general solution is given by

$$y(t) = C_1 y_1(t) + C_2 y_2(t) = e^{-t}(C_1 \cos t + C_2 \sin t). \tag{3.13}$$

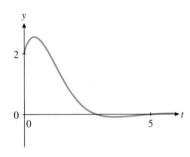

Figure 2. The solution in Example 3.12 with initial conditions $y(0) = 2$ and $y'(0) = 3$.

To find the solution with the initial conditions $y(0) = 2$ and $y'(0) = 3$, we differentiate (3.13),

$$y'(t) = -e^{-t}(C_1 \cos t + C_2 \sin t) + e^{-t}(-C_1 \sin t + C_2 \cos t). \tag{3.14}$$

At $t = 0$, equations (3.13) and (3.14) become

$$2 = y(0) = C_1$$
$$3 = y'(0) = -C_1 + C_2.$$

This system has solutions $C_1 = 2$ and $C_2 = 5$. Therefore, the solution to the initial value problem is

$$y(t) = e^{-t}(2 \cos t + 5 \sin t).$$

The solution is shown in Figure 2. ●

Repeated roots

If the roots of the characteristic equation (3.2) are repeated, then it becomes

$$0 = \lambda^2 + p\lambda + q = (\lambda - \lambda_1)^2,$$

where λ_1 is the repeated root. By the quadratic formula,

$$\lambda_1 = \left(-p \pm \sqrt{p^2 - 4q}\right)/2 = -p/2. \tag{3.15}$$

In order for λ_1 to be a double root, we must have $p^2 - 4q = 0$, or $q = p^2/4$. This value of λ gives one solution to the differential equation, namely

$$y_1 = e^{\lambda_1 t} = e^{-pt/2}.$$

To use Theorem 1.23, we need to find another solution that is not a constant multiple of this one. We will use a method of finding the second solution that works whenever we already have one solution to a second-order, linear equation.[3] We will look for a nonconstant function $v(t)$ such that

$$y_2(t) = v(t)y_1(t) = v(t)e^{-pt/2} \tag{3.16}$$

is a solution to

$$y'' + py' + qy = y'' + py' + p^2 y/4 = 0. \tag{3.17}$$

If we can find a nonconstant function v for which this is true, then y_1 and $y_2 = vy_1$ will be linearly independent and we will have a fundamental set of solutions.

Differentiating (3.16), we get

$$y_2' = e^{-pt/2}[v' - pv/2] \quad \text{and} \quad y_2'' = e^{-pt/2}[v'' - pv' + p^2 v/4].$$

Substituting into (3.17), we get

$$0 = y'' + py' + p^2 y/4$$
$$= e^{-pt/2}[v'' - pv' + p^2 v/4 + p(v' - pv/2) + p^2 v/4]$$
$$= e^{-pt/2}v''.$$

Hence for $y_2 = ve^{-pt/2}$ to be solution, we must have $v'' = 0$, or $v(t) = At + B$ for any constants A and B. Since we want v to be nonconstant, we will choose $B = 0$ and $A = 1$. Thus, our second solution is

$$y_2(t) = te^{\lambda_1 t} = te^{-pt/2}.$$

Theorem 1.23 now gives us the form of the general solution, and we summarize our discussion in the following proposition.

---

[3] See Exercise 26 in Section 4.1.

PROPOSITION 3.18   If the characteristic equation $\lambda^2 + p\lambda + q = 0$ has only one double root λ_1, then the general solution to $y'' + py' + qy = 0$ is

$$y(t) = C_1 e^{\lambda_1 t} + C_2 t e^{\lambda_1 t} = (C_1 + C_2 t)\, e^{\lambda_1 t},$$

where C_1 and C_2 are arbitrary constants.   ●

The constants C_1 and C_2 can be found from initial conditions to solve initial value problems.

E x a m p l e 3 . 1 9   Find the general solution to

$$y'' - 2y' + y = 0.$$

Find the solution corresponding to the initial conditions $y(0) = 2$ and $y'(0) = -1$.

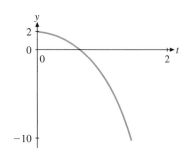

The characteristic equation is

$$0 = \lambda^2 - 2\lambda + 1 = (\lambda - 1)^2,$$

so $\lambda = 1$ is a double root. Hence, e^t and te^t form a fundamental set of solutions to this differential equation, and the general solution is

$$y(t) = C_1 e^t + C_2 t e^t. \tag{3.20}$$

Figure 3. The solution in Example 3.19 with initial conditions $y(0) = 2$ and $y'(0) = 0$.

To find the solution corresponding to the initial conditions $y(0) = 2$ and $y'(0) = -1$, we differentiate y,

$$y'(t) = C_1 e^t + C_2 t e^t + C_2 e^t. \tag{3.21}$$

At $t = 0$, equations (3.20) and (3.21) become

$$2 = y(0) = C_1$$
$$-1 = y'(0) = C_1 + C_2.$$

This system has solutions $C_1 = 2$ and $C_2 = -3$, so the solution to the initial value problem is

$$y = 2e^t - 3te^t.$$

The solution is shown in Figure 3.   ●

Summary

The equation

$$y'' + py' + qy = 0 \tag{3.22}$$

has characteristic equation

$$\lambda^2 + p\lambda + q = 0. \tag{3.23}$$

The solution is discussed in Propositions 3.3, 3.11, and 3.18.

This information can be summarized as follows:

- If $p^2 - 4q > 0$, the characteristic equation has two distinct, real roots λ_1 and λ_2. A fundamental set of solutions is

$$y_1(t) = e^{\lambda_1 t} \quad \text{and} \quad y_2(t) = e^{\lambda_2 t}.$$

- If $p^2 - 4q = 0$, the characteristic equation has one repeated real root λ. A fundamental set of solutions is

$$y_1(t) = e^{\lambda t} \quad \text{and} \quad y_2(t) = te^{\lambda t}.$$

- If $p^2 - 4q < 0$, the characteristic equation has two complex conjugate roots $a \pm ib$. A fundamental set of solutions is

$$y_1(t) = e^{at} \cos bt \quad \text{and} \quad y_2(t) = e^{at} \sin bt.$$

EXERCISES

The equations in Exercises 1–8 have distinct, real, characteristic roots. Find the general solution in each case.

1. $y'' - y' - 2y = 0$
2. $2y'' - 3y' - 2y = 0$
3. $y'' + 5y' + 6y = 0$
4. $y'' + y' - 12y = 0$
5. $2y'' - y' - y = 0$
6. $6y'' + y' - y = 0$
7. $3y'' - 2y' - y = 0$
8. $6y'' + 5y' - 6y = 0$

The equations in Exercises 9–16 have complex characteristic roots. Find the general solution in each case.

9. $y'' + y = 0$
10. $y'' + 4y = 0$
11. $y'' + 4y' + 5y = 0$
12. $y'' + 2y' + 17y = 0$
13. $y'' + 2y = 0$
14. $y'' + 2y' + 3y = 0$
15. $y'' - 2y' + 4y = 0$
16. $y'' + 2y' + 2y = 0$

The equations in Exercises 17–24 have repeated, real, characteristic roots. Find the general solution in each case.

17. $y'' - 4y' + 4y = 0$
18. $y'' - 6y' + 9y = 0$
19. $4y'' + 4y' + y = 0$
20. $4y'' + 12y' + 9y = 0$
21. $16y'' + 8y' + y = 0$
22. $y'' + 4y' + 4y = 0$
23. $16y'' + 24y' + 9y = 0$
24. $y'' + 8y' + 16y = 0$

In Exercises 25–36, find the solution of the given initial value problem.

25. $y'' - y' - 2y = 0$, $y(0) = -1$, $y'(0) = 2$
26. $10y'' - y' - 3y = 0$, $y(0) = 1$, $y'(0) = 0$
27. $y'' - 2y' + 17y = 0$, $y(0) = -2$, $y'(0) = 3$
28. $y'' + 25y = 0$, $y(0) = 1$, $y'(0) = -1$
29. $y'' + 10y' + 25y = 0$, $y(0) = 2$, $y'(0) = -1$
30. $y'' - 2y' - 3y = 0$, $y(0) = 2$, $y'(0) = -3$
31. $y'' + 2y' + 3y = 0$, $y(0) = 1$, $y'(0) = 0$
32. $y'' - 4y' - 5y = 0$, $y(1) = -1$, $y'(1) = -1$
33. $8y'' + 2y' - y = 0$, $y(-1) = 1$, $y'(-1) = 0$
34. $4y'' + y = 0$, $y(1) = 0$, $y'(1) = -2$
35. $y'' + 12y' + 36y = 0$, $y(1) = 0$, $y'(1) = -1$
36. $y'' - 4y' + 13y = 0$, $y(0) = 4$, $y'(0) = 0$

37. Suppose that λ_1 and λ_2 are characteristic roots of the characteristic equation $\lambda^2 + p\lambda + q = 0$, where p and q are real constants.
   (a) Prove that $\lambda_1 \lambda_2 = q$.
   (b) Prove that $\lambda_1 + \lambda_2 = -p$.

38. Given that the characteristic equation $\lambda^2 + p\lambda + q = 0$ has a double root, $\lambda = \lambda_1$, show, by direct substitution, that $y = te^{\lambda_1 t}$ is a solution of $y'' + py' + qy = 0$.

4.4 Harmonic Motion

In Section 4.1 we derived the equation for the motion of a vibrating spring. It is (see equation (1.13))

$$my'' + \mu y' + ky = F(t), \tag{4.1}$$

where the constant coefficients are the mass m, the damping constant μ, and the spring constant k, and the function $F(t)$ is an external force.

In Section 3.4 of Chapter 3 we derived differential equations that modeled simple *RLC* circuits. If the circuit consists of a resistor of resistance R, a condenser of capacitance C, and an inductor of inductance L in series with a source voltage $E(t)$, then the current I satisfies the equation

$$L\frac{d^2 I}{dt^2} + R\frac{dI}{dt} + \frac{1}{C}I = \frac{dE}{dt}. \qquad (4.2)$$

The coefficients R, C, and L are all constants.

It is interesting to compare equations (4.1) and (4.2). They are almost identical, differing only in the letters chosen to represent the coefficients and the unknown functions. If we compare the coefficients, we see that the inductance L acts like the mass m, the resistance R like the damping constant, and the reciprocal of the capacitance $1/C$ like the spring constant. Finally, the derivative of the source voltage acts like the external force on the spring.

It is important to keep these analogies in mind. Physically it means that the two phenomena have similar behavior. Mathematically it means that when we discover facts about solutions to one of these equations, we also have related facts about the other. For example, if we have a circuit without resistance ($R = 0$) and no source voltage, then equation (4.2) simplifies to

$$L\frac{d^2 I}{dt^2} + \frac{1}{C}I = 0. \qquad (4.3)$$

This corresponds to the equation for an unforced, undamped spring. Earlier we discovered solutions to that equation (see equation (1.15)). Replacing the mass m with the inductance L, and the spring constant k with the reciprocal of the capacitance $1/C$, we see that any function of the form

$$I(t) = a\cos(t/\sqrt{LC}) + b\sin(t/\sqrt{LC})$$

is a solution to (4.3).

If we divide equations (4.1) and (4.2) by their leading coefficient (L or m), they become

$$\frac{d^2 y}{dt^2} + \frac{\mu}{m}\frac{dy}{dt} + \frac{k}{m}y = \frac{1}{m}F(t), \quad \text{and}$$

$$\frac{d^2 I}{dt^2} + \frac{R}{L}\frac{dI}{dt} + \frac{1}{LC}I = \frac{1}{L}\frac{dE}{dt}.$$

If we make the identifications $c = \mu/2m$, $\omega_0 = \sqrt{k/m}$, $f(t) = F(t)/m$, and $x = y$ in the first equation, or $c = R/2L$, $\omega_0 = \sqrt{1/LC}$, $f(t) = (dE/dt)/L$, and $x = I$ in the second, we get the equation

$$x'' + 2cx' + \omega_0^2 x = f(t), \qquad (4.4)$$

where $c \geq 0$ and $\omega_0 > 0$ are constants. We will refer to equation (4.4) as the equation for **harmonic motion**. It includes the vibrating spring and the arbitrary *RLC* circuit, but there are many other phenomena that lead to this equation.

It is common to use the terminology of the vibrating spring when discussing harmonic motion. In particular, c is called the **damping** constant, and f is the **forcing term**.

In this section we will study unforced harmonic motion. This means that $f(t) = 0$ in (4.4), so we will be studying the homogeneous equation

$$x'' + 2cx' + \omega_0^2 x = 0, \qquad (4.5)$$

where $c \geq 0$ and $\omega_0 > 0$ are constants. This is the type of equation we considered in the previous section, so we are in a position to analyze unforced harmonic motion.

Simple harmonic motion

In the special case when there is no damping (so $c = 0$) the motion is called *simple harmonic motion*. Equation (4.5) simplifies to

$$x'' + \omega_0^2 x = 0. \tag{4.6}$$

The characteristic equation is

$$\lambda^2 + \omega_0^2 = 0.$$

The characteristic roots are $\lambda = \pm i\omega_0$. According to Proposition 3.11, the general solution is

$$x(t) = a \cos \omega_0 t + b \sin \omega_0 t, \tag{4.7}$$

where a and b are constants.

If we define $T = 2\pi/\omega_0$, so that $\omega_0 T = 2\pi$, then the periodicity of the trigonometric functions implies that $x(t + T) = x(t)$ for all t. Thus, the solution x is itself periodic with period T. For this reason, ω_0 is called the *natural frequency* of the spring. We will continue with this terminology even in the damped case.

Example 4.8   Suppose we have simple harmonic motion with a natural frequency $\omega_0 = 4$. Find the solution with initial values $x(0) = 1$ and $x'(0) = 0$.

From (4.7) we see that the general solution is

$$x = a \cos 4t + b \sin 4t.$$

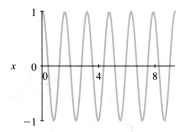

The initial condition $x(0) = 1$ becomes $a = 1$, and $x'(0) = 0$ becomes $4b = 0$. Hence the solution to the initial value problem is $x(t) = \cos 4t$. The graph of x is given in Figure 1.   ●

In the case of a vibrating spring without friction, we see that the mass on the spring oscillates up and down with the natural frequency $\omega_0 = \sqrt{k/m}$. Note that the natural frequency increases as the spring constant increases, and it decreases as the mass increases.

Figure 1. The undamped oscillation in Example 4.8.

Amplitude and phase angle

It is frequently convenient to put the solutions in (4.7) into another form that is more convenient and more revealing of the nature of the solution.

Consider the vector (a, b) in the plane (see Figure 2). We will write this in polar coordinates. Assuming that $(a, b) \neq (0, 0)$, there is a positive number A, which is the *length* of (a, b), and an angle ϕ in the interval $(-\pi, \pi]$, called the *polar angle*, such that

$$a = A \cos \phi \quad \text{and} \quad b = A \sin \phi. \tag{4.9}$$

Examples are shown in Figure 5.

If we substitute these equations into (4.7), it becomes

$$\begin{aligned}
x(t) &= a \cos(\omega_0 t) + b \sin(\omega_0 t) \\
&= A \cos \phi \cos(\omega_0 t) + A \sin \phi \sin(\omega_0 t) \tag{4.10} \\
&= A \cos(\omega_0 t - \phi).
\end{aligned}$$

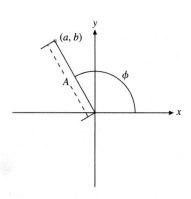

Figure 2. Polar coordinates.

Thus, we see that the general solution to the second-order equation (4.6) can be written as

$$x(t) = A \cos(\omega_0 t - \phi). \tag{4.11}$$

This expression for the solution makes it clear that undamped harmonic motion is a pure sinusoidal oscillation. The parameter A is the **amplitude** of the oscillation. Since the cosine term oscillates between ± 1, the limits of the oscillation in (4.11) are $\pm A$. The parameter ϕ represents the **phase** of the oscillation. A positive phase shifts the graph of the cosine to the right. This effect is shown in Figure 3. It can best be understood by writing (4.11) as

$$x(t) = A \cos\left(\omega_0 \left(t - \frac{\phi}{\omega_0}\right)\right).$$

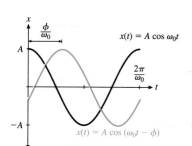

Notice that there are still two undetermined constants in (4.11). Now they are the amplitude A and the phase ϕ.

Figure 3. The phase angle shifts the graph of the cosine.

This is all very well and good, but it is not very useful unless we can find the constants A and ϕ when we know a and b. Equations (4.9) tell us how to compute a and b if we know A and ϕ. They also can be solved for A and ϕ. First, if we square each equation and add, we get

$$A^2 = a^2 + b^2 \quad \text{or} \quad A = \sqrt{a^2 + b^2}. \tag{4.12}$$

Next, if we divide the second equation by the first, we get

$$\tan \phi = \frac{b}{a}. \tag{4.13}$$

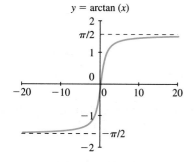

While it is tempting to "solve" this last equation for ϕ and write $\phi = \arctan(b/a)$, this would be a mistake. As Figure 4 shows, the arctan takes values between $-\pi/2$ and $\pi/2$, whereas we know that ϕ can be any angle between $-\pi$ and π. The range $-\pi/2 < \phi < \pi/2$ corresponds to the points (a, b) where $a = A \cos \phi > 0$. These are points (a, b) in the right half-plane. How do we compute ϕ when $a < 0$?

Figure 4. The arctangent takes values between $-\pi/2$ and $\pi/2$.

When (a, b) is in the left half-plane, then $(-a, -b)$ is in the right half-plane, and clearly $\arctan(b/a) = \arctan(-b/-a)$. Thus $\arctan(b/a)$ is measuring the polar angle of $(-a, -b)$, which differs by $\pm \pi$ from the polar angle of (a, b) (see Figure 5). There are three cases depending on the signs of a and b, as shown in Figure 5. Following the angles in Figure 5, we see that the polar angle is given by

$$\phi = \begin{cases} \arctan(b/a), & \text{if } a > 0; \\ \arctan(b/a) + \pi, & \text{if } a < 0 \text{ and } b > 0; \\ \arctan(b/a) - \pi, & \text{if } a < 0 \text{ and } b < 0. \end{cases}$$

Example 4.14 A mass of 4 kg is attached to a spring with a spring constant of $k = 169 \, \text{kg/s}^2$. It is then stretched 10 cm from the spring-mass equilibrium and set to oscillating with an initial velocity of 130 cm/s. Assuming it oscillates without damping, find the frequency, amplitude, and phase of the vibration.

The differential equation is

$$4y'' + 169y = 0 \quad \text{or} \quad y'' + 42.25y = 0.$$

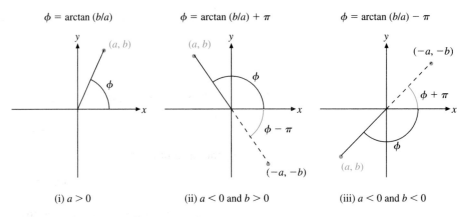

$\phi = \arctan (b/a)$ $\phi = \arctan (b/a) + \pi$ $\phi = \arctan (b/a) - \pi$

(i) $a > 0$ (ii) $a < 0$ and $b > 0$ (iii) $a < 0$ and $b < 0$

Figure 5. Polar coordinates. There are three cases for finding the angle.

The natural frequency is $\omega_0 = \sqrt{42.25} = 6.5$ and the general solution is

$$y(t) = C_1 \cos 6.5t + C_2 \sin 6.5t.$$

The initial conditions are $y(0) = 0.1$ m and $y'(0) = 1.3$ m/s. The specific solution satisfying these initial conditions is

$$y = 0.1 \cos 6.5t + 0.2 \sin 6.5t.$$

The amplitude of vibration is

$$A = \sqrt{0.01 + 0.04} = \frac{\sqrt{5}}{10} \approx 0.2236 \text{ m}.$$

The phase is $\phi = \arctan(2) \approx 1.1071$. Hence we can write the solution as $y(t) = \sqrt{5}/10 \cos(6.5t - \phi)$.

The solution is plotted in Figure 6. ●

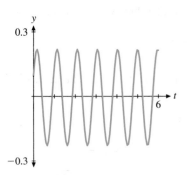

Figure 6. The undamped motion in Example 4.14.

Damped harmonic motion

Now $c > 0$. The differential equation

$$x'' + 2cx' + \omega_0^2 x = 0$$

has the characteristic equation

$$\lambda^2 + 2c\lambda + \omega_0^2 = 0.$$

The roots are

$$\lambda_1 = -c - \sqrt{c^2 - \omega_0^2} \quad \text{and} \quad \lambda_2 = -c + \sqrt{c^2 - \omega_0^2}. \tag{4.15}$$

We have three cases to consider depending on the sign of the discriminant $c^2 - \omega_0^2$.

1. $c < \omega_0$. This is the **underdamped** case. The roots in (4.15) are distinct complex numbers. Hence, the general solution is

$$x(t) = e^{-ct} [C_1 \cos \omega t + C_2 \sin \omega t],$$

where

$$\omega = \sqrt{\omega_0^2 - c^2}.$$

2. $c > \omega_0$. This is the ***overdamped case***. Now the roots in (4.15) are distinct and real. Further, since $\sqrt{c^2 - \omega_0^2} < \sqrt{c^2} = c$, we have $\lambda_1 < \lambda_2 < 0$. The general solution is

$$x(t) = C_1 e^{\lambda_1 t} + C_2 e^{\lambda_2 t}.$$

3. $c = \omega_0$. This is the ***critically damped*** case, and in this case, the root in (4.15) is a double root,

$$\lambda = -c.$$

The general solution is

$$x(t) = C_1 e^{-ct} + C_2 t e^{-ct}.$$

In each of the cases the solution decays to zero as $t \to \infty$ due to the exponential term in the solution, and the fact that $c > 0$. In the critically damped case, this follows since, for $c > 0$, $\lim_{t \to \infty} t/e^{ct} = 0$ by l'Hôpital's rule.

In the underdamped case the cosine and sine terms cause the solution to oscillate with frequency ω as it converges to zero. Notice that this frequency is always smaller than the natural frequency of the spring. In the other two cases there is no oscillation.

Let's look at specific examples of damping phenomena.

Example 4.16   Consider the spring in Example 4.14 with damping constant $\mu = 12.8 \, \text{kg/s}$. Find the solution with initial conditions $y(0) = 0.1 \, \text{m}$ and $y'(0) = 1.3 \, \text{m/s}$.

Since $m = 4$, $k = 169$ and $\mu = 12.8$, the differential equation is

$$4y'' + 12.8y' + 169y = 0 \quad \text{or} \quad y'' + 3.2y' + 42.25y = 0.$$

The characteristic polynomial $\lambda^2 + 3.2\lambda + 42.25$ has roots $-1.6 \pm i\sqrt{42.25 - 1.6^2}$ $= -1.6 \pm 6.3i$. Thus, $\omega = 6.3$ and the general solution is

$$y(t) = e^{-1.6t}(C_1 \cos 6.3t + C_2 \sin 6.3t).$$

For the initial conditions we have

$$0.1 = y(0) = C_1$$
$$1.3 = y'(0) = -1.6\,C_1 + 6.3\,C_2.$$

Hence $C_1 = 0.1$ and $C_2 = 1.46/6.3 \approx 0.2317$. The solution is

$$y(t) = e^{-1.6t}(0.1 \cos 6.3t + 0.2317 \sin 6.3t).$$

This can also be written as

$$y(t) = 0.2524 e^{-1.6t} \cos(6.3t - 1.1634).$$

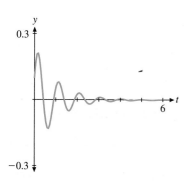

Figure 7. The underdamped motion in Example 4.16.

The underdamped motion in Example 4.16 is plotted in Figure 7. The decaying oscillatory motion that is seen there is typical of underdamped motion. As an example, consider the springs of a car, which are damped by shock absorbers. If the shocks are very old, when you press the front of the car down and release it, the car will bounce up and down for a while with decreasing amplitude until it reaches equilibrium. This is because as the shocks wear out, the damping constant decreases. When it decreases to the point that the motion is underdamped, the front end will bounce.

Example 4.17

Consider the spring in Example 4.14 with damping constant $\mu = 77.6\,\text{kg/s}$. Find the solution with initial conditions $y(0) = 0.1\,\text{m}$ and $y'(0) = 1.3\,\text{m/s}$.

Since $m = 4$, $k = 169$ and $\mu = 77.6$, the differential equation is

$$4y'' + 77.6y' + 169y = 0 \quad \text{or} \quad y'' + 19.4y' + 42.25y = 0.$$

The characteristic polynomial $\lambda^2 + 19.4\lambda + 42.25$ has roots

$$-9.7 \pm \sqrt{9.7^2 - 42.25} = -9.7 \pm 7.2.$$

Set $\lambda_1 = -16.9$ and $\lambda_2 = -2.5$. This is the overdamped case. The general solution is

$$y(t) = C_1 e^{-16.9t} + C_2 e^{-2.5t}.$$

For the initial conditions we have

$$0.1 = y(0) = C_1 + C_2$$
$$1.3 = y'(0) = -16.9C_1 - 2.5C_2.$$

Hence $C_1 = -31/288$, and $C_2 = 299/1440$. The solution is given by

$$y(t) = -\frac{31}{288}e^{-16.9t} + \frac{299}{1440}e^{-2.5t}.$$

The overdamped motion in Example 4.17 is plotted in Figure 8. Typical of such behavior, the displacement decreases to its equilibrium without the oscillation we see in Example 4.16. It is possible with properly chosen initial conditions that the displacement passes through the equilibrium just once before decaying to it. An example of overdamped motion is provided by the springs of a car with brand new shock absorbers. When you press the front of the car down and release it, the car simply rises to its original position without any oscillation.   ●

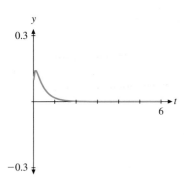

Figure 8. The overdamped motion in Example 4.17.

Example 4.18

For the spring in Example 4.14, find the value of the damping constant μ for which there is critical damping. Find the solution with initial conditions $y(0) = 0.1\,\text{m}$ and $y'(0) = 1.3\,\text{m/s}$.

Critical damping occurs when $c = \omega_0$. Since $c = \mu/2m$ and $\omega_0 = \sqrt{k/m}$, we need $\mu = 2m\sqrt{k/m} = 2\sqrt{mk} = 52\,\text{kg/s}$ With this value of μ, the equation becomes

$$4y'' + 52y' + 169y = 0 \quad \text{or} \quad y'' + 13y' + 42.25y = 0.$$

The characteristic polynomial is $\lambda^2 + 13\lambda + 42.25 = (\lambda + 6.5)^2$. The general solution is

$$y(t) = C_1 e^{-6.5t} + C_2 t e^{-6.5t}.$$

For the initial conditions we have

$$0.1 = y(0) = C_1$$
$$1.3 = y'(0) = -6.5\,C_1 + C_2.$$

Hence $C_1 = 0.1$, and $C_2 = 1.95$. The solution is

$$y(t) = 0.1e^{-6.5t} + 1.95te^{-6.5t}.$$

The critically damped motion in Example 4.18 is plotted in Figure 9. By its definition, the damping constant for critical damping is the dividing line that separates underdamping from overdamping. The motion itself looks very similar to overdamping. It simply decays to its equilibrium. Once more it is possible that it will cross the equilibrium just once.   ●

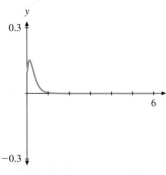

Figure 9. The critically damped motion in Example 4.18.

Designers of shock absorbers aim for a damping constant that is just a little larger than critical damping. If the damping constant is too large, the shocks prevent the springs from absorbing bounces, and the ride becomes bumpy. If the damping constant gets below critical damping, the car oscillates. So just a little above critical damping is the right goal. This prevents the oscillations of the underdamped case, and it allows the springs to absorb the bounces. In addition it allows for the damping constant to decrease a little before it becomes underdamped.

EXERCISES

In Exercises 1–6,

(i) use a computer or calculator to plot the graph of the given function, and

(ii) place the solution in the form $y = A\cos(\omega t - \phi)$ and compare the graph of your answer with the plot found in part (i).

1. $y = \cos 2t + \sin 2t$
2. $y = \cos t - \sin t$
3. $y = \cos 4t + \sqrt{3}\sin 4t$
4. $y = -\sqrt{3}\cos 2t + \sin 2t$
5. $y = 0.2\cos 2.5t - 0.1\sin 2.5t$
6. $y = 0.2\cos 6.3t - 0.5\sin 6.3t$

In Exercises 7–10, place each equation in the form $y = Ae^{-ct}\cos(\omega t - \phi)$. Then, on one plot, place the graphs of $y = Ae^{-ct}\cos(\omega t - \phi)$, $y = Ae^{-ct}$, and $y = -Ae^{-ct}$. For the last two, use a different line style and/or color than for the first.

7. $y = e^{-t/2}(\cos 5t + \sin 5t)$
8. $y = e^{-t/4}(\sqrt{3}\cos 4t - \sin 4t)$
9. $y = e^{-0.1t}(0.2\cos 2t + 0.1\sin 2t)$
10. $y = e^{-0.2t}(\cos 4.2t - 1.2\sin 4.2t)$

11. A 0.2-kg mass is attached to a spring having a spring constant $5\,\text{kg/s}^2$. The system is displaced 0.5 m from its equilibrium position and released from rest. If there is no damping present, find the amplitude, frequency, and phase of the resulting motion. Plot the solution.

12. A 0.1-kg mass is attached to a spring having a spring constant $3.6\,\text{kg/s}^2$. The system is allowed to come to rest. Then the mass is given a sharp tap, imparting an instantaneous downward velocity of 0.4 m/s. If there is no damping present, find the amplitude, frequency, and phase of the resulting motion. Plot the solution.

13. The undamped system

$$\frac{2}{5}x'' + kx = 0, \quad x(0) = 2, \quad x'(0) = v_0$$

is observed to have period $\pi/2$ and amplitude 2. Find k and v_0.

14. Consider the undamped oscillator

$$mx'' + kx = 0, \quad x(0) = x_0, \quad x'(0) = v_0.$$

Show that the amplitude of the resulting motion is $\sqrt{x_0^2 + mv_0^2/k}$.

15. A 2-μF capacitor ($1\,\mu\text{F} = 1 \times 10^{-6}\,\text{F}$) is charged to 20 V and then connected across a 6-μH inductor ($1\,\mu\text{H} = 1 \times 10^{-6}\,\text{H}$), forming an LC circuit.

(a) Find the initial charge on the capacitor.

(b) At the time of connection, the initial current is zero. Assuming no resistance, find the amplitude, frequency, and phase of the current. Plot the graph of the current versus time. (Use equations (4.2) and (4.3) in Section 3.4.)

16. A 1-kg mass, when attached to a large spring, stretches the spring a distance of 4.9 m.

(a) Calculate the spring constant.

(b) The system is placed in a viscous medium that supplies a damping constant $\mu = 3\,\text{kg/s}$. The system is allowed to come to rest. Then the mass is displaced 1 m in the downward direction and given a sharp tap, imparting an instantaneous velocity of 1 m/s in the downward direction. Find the position of the mass as a function of time and plot the solution.

17. Prove that an overdamped solution of $my'' + \mu y' + ky = 0$ can cross the time axis no more than once, regardless of the initial conditions. Use a numerical solver to create a plot of an overdamped system that crosses the time axis one time and a second plot where the plot does not cross the time axis.

18. A 50-g mass ($1\,\text{kg} = 1000\,\text{g}$) stretches a spring 20 cm ($1\,\text{m} = 100\,\text{cm}$). Find a damping constant μ so that the system is critically damped. If the mass is displaced 15 cm from its equilibrium position and released from rest, find the position of the mass as a function of time and plot the solution.

19. Prove that a critically damped solution of $my'' + \mu y' + ky = 0$ can cross the time axis no more than once, regardless of the initial conditions. Use a numerical solver to create a plot of an overdamped system that crosses the time axis one time and a second plot where the plot does not cross the time axis.

20. A spring-mass system is modeled by the equation

$$x'' + \mu x' + 4x = 0.$$

(a) Show that the system is critically damped when $\mu = 4\,\text{kg/s}$.

(b) Suppose that the mass is displaced upward 2 m and given an initial velocity of 1 m/s. Use a numerical solver to compute the solution for $\mu = 4, 4.2, 4.4, 4.6, 4.8, 5$. Plot all of the solution curves on one figure. What is special about the critically damped solution in comparison to the other solutions? Why would you want to adjust the spring on a screen door so that it was critically damped?

21. If $\mu > 2\sqrt{km}$, the system $mx'' + \mu x' + kx = 0$ is overdamped. The system is allowed to come to equilibrium. Then the mass is given a sharp tap, imparting an instantaneous downward velocity v_0.

(a) Show that the position of the mass is given by

$$x(t) = \frac{v_0}{\gamma} e^{-\mu t/(2m)} \sinh \gamma t,$$

where

$$\gamma = \frac{\sqrt{\mu^2 - 4mk}}{2m}.$$

(b) Show that the mass reaches its lowest point at

$$t = \frac{1}{\gamma} \tanh^{-1} \frac{2m\gamma}{\mu},$$

a time independent of the initial conditions.

(c) Show that, in the critically damped case, the time it takes the mass to reach its lowest point is given by $t = 2m/\mu$.

22. A 100-g mass (1 kg = 1000 g) is hung from a spring having spring constant 9.8 kg/s². The system is placed in a viscous medium that imparts a force of 0.3 N when the mass is moving at 0.2 m/s. Assume that the force applied by the medium is proportional, but opposite, to the mass's velocity. The mass is displaced 10 cm from its equilibrium position and released from rest. Find the amplitude, frequency, and phase of the resulting motion. Plot the solution.

23. A capacitor (0.008 F) is charged to 50 V and then connected in series with an inductor (4 H) and a resistor (20 Ω). Initially, there is no current in the circuit. Find the amplitude, frequency, and phase of the current and plot its graph.

24. A 10-kg mass stretches a spring 1 m. The system is placed in a viscous medium that provides a damping constant $\mu = 20$ kg/s. The system is allowed to attain equilibrium. Then a sharp tap to the mass imparts an instantaneous downward velocity of 1.2 m/s. Find the amplitude, frequency, and phase of the resulting motion. Plot the solution.

25. A capacitor (0.02 F) is charged to 1 V and then connected in series with an inductor (10 H) and a resistor (40 Ω). Initially, there is no current in the circuit. Find the amplitude, frequency, and phase of the charge on the capacitor and plot its graph.

4.5 Inhomogeneous Equations; the Method of Undetermined Coefficients

We now turn to the solution of inhomogeneous linear equations. These are equations of the form

$$y'' + py' + qy = f, \tag{5.1}$$

where $p = p(t), q = q(t)$, and $f = f(t)$ are functions of the independent variable. Remember that f is called the inhomogeneous term, or the forcing term.

Our solution strategy comes from the understanding of the structure of the general solution, which is contained in the next theorem. We refer you to a similar result for first-order linear equations in Theorem 4.41 in Section 2.4 of Chapter 2.

THEOREM 5.2 Suppose that y_p is a particular solution to the inhomogeneous equation (5.1), and that y_1 and y_2 form a fundamental set of solutions to the associated homogeneous equation

$$y'' + py' + qy = 0. \tag{5.3}$$

Then the general solution to the inhomogeneous equation (5.1) is given by

$$y = y_p + C_1 y_1 + C_2 y_2, \tag{5.4}$$

where C_1 and C_2 are arbitrary constants. ∎

Notice that the general solution can be written as $y = y_p + y_h$, where $y_h = C_1 y_1 + C_2 y_2$ is the general solution to the corresponding homogeneous equation. Thus, to find the general solution to the inhomogeneous equation (5.1) we first find the general solution, y_h, to the corresponding homogeneous equation (5.3). Next, we find a particular solution, y_p, to the inhomogeneous equation. The general solution to the inhomogeneous equation is then $y = y_p + y_h$.

Proof   Suppose that y is a solution to (5.1). We are given that y_p is also a solution, so we have the two equations

$$y'' + py' + qy = f, \quad \text{and}$$
$$y_p'' + py_p' + qy_p = f.$$

Subtracting, we get

$$(y - y_p)'' + p(y - y_p)' + q(y - y_p) = 0.$$

Therefore, $y - y_p$ is a solution to the associated homogeneous equation (5.3). Since y_1 and y_2 form a fundamental set of solutions, there are constants C_1 and C_2 such that

$$y - y_p = C_1 y_1 + C_2 y_2.$$

Consequently, $y = y_p + C_1 y_1 + C_2 y_2$, as promised in (5.4).

For equations with constant coefficients, we already know how to solve the homogeneous equation, so in this section we will concentrate on finding one particular solution to the inhomogeneous equation.

The method of undetermined coefficients

We will be looking at the inhomogeneous equation

$$y'' + py' + qy = f, \tag{5.5}$$

where p and q are constants and $f = f(t)$ is a function of the independent variable. The method of undetermined coefficients only works if the coefficients are constants. Second-order equations are the most important applications of the method, but the method works for higher-order equations in exactly the same way that it does for second-order equations. For simplicity and convenience we will emphasize the second-order case.

The method of undetermined coefficients is based on the fact that there are some situations where the form of the forcing term in (5.5) allows us to almost guess the form of a particular solution. Let's highlight the key idea.

> If the forcing term f has a form that is replicated under differentiation, then look for a solution with the same general form as the forcing term.

Exponential forcing terms

The easiest example of such a forcing term is an exponential function $f(t) = e^{at}$. Then $f'(t) = ae^{at}$, which is also an exponential function. The method is illustrated by our first example.

Example 5.6 Find a particular solution to the equation

$$y'' - y' - 2y = 2e^{-2t}. \tag{5.7}$$

The forcing term is $f(t) = 2e^{-2t}$. We look for a particular solution with the same form as f, or

$$y(t) = ae^{-2t},$$

where a is an as yet undetermined coefficient. The derivatives of y are

$$y'(t) = -2ae^{-2t} \quad \text{and} \quad y''(t) = 4ae^{-2t}.$$

If we insert these expressions into the left-hand side of (5.7), we get

$$y'' - y' - 2y = 4ae^{-2t} - (-2ae^{-2t}) - 2(ae^{-2t}) = 4ae^{-2t}.$$

This must be equal to the right-hand side of (5.7) in order for y to be a solution. Thus we must have

$$4ae^{-2t} = 2e^{-2t}.$$

Equating the coefficients of e^{-2t} gives $4a = 2$, or $a = 1/2$. Consequently,

$$y(t) = e^{-2t}/2$$

is a particular solution to (5.7). We encourage the readers to check by direct substitution that y is a solution to (5.7). ●

Trigonometric forcing terms

Next, consider a forcing term of the form

$$f(t) = A \cos \omega t + B \sin \omega t.$$

The derivative of f has the same general form, so we will look for solutions of the form

$$y(t) = a \cos \omega t + b \sin \omega t,$$

where a and b are as yet undetermined coefficients.

Example 5.8 Compute a particular solution to the equation

$$y'' + 2y' - 3y = 5 \sin 3t. \tag{5.9}$$

We look for a particular solution of the form

$$y = a \cos 3t + b \sin 3t.$$

As we will see, we will need both the cosine and the sine terms, even though only a sine term appears in the forcing term.

The derivatives of y are

$$y' = -3a \sin 3t + 3b \cos 3t \quad \text{and}$$
$$y'' = -9a \cos 3t - 9b \sin 3t.$$

Inserting these expressions into the left-hand side of (5.9), we get

$$y'' + 2y' - 3y = (-9a \cos 3t - 9b \sin 3t) + 2(-3a \sin 3t + 3b \cos 3t)$$
$$- 3(a \cos 3t + b \sin 3t)$$
$$= (-12a + 6b) \cos 3t + (-6a - 12b) \sin 3t. \tag{5.10}$$

This result must be equal to the right-hand side of (5.9) in order that y be a solution. Equating the coefficients of the sine and cosine terms in the right-hand side of (5.9) and (5.10) gives two equations for a and b,

$$-12a + 6b = 0$$
$$-6a - 12b = 5,$$

with solutions $a = -1/6$ and $b = -1/3$. Hence, a particular solution is

$$y = -\frac{1}{6}\cos 3t - \frac{1}{3}\sin 3t. \tag{5.11}$$

●

The complex method

There is another way to find a particular solution in situations where the forcing function contains a trigonometric term. We will illustrate the method by solving the same equation as we did in Example 5.8.

Example 5.12   Use the complex method to find a particular solution to the equation

$$y'' + 2y' - 3y = 5\sin 3t. \tag{5.13}$$

Notice that $5\sin 3t$, the right-hand side of (5.13), is the imaginary part of $5e^{3it} = 5\cos 3t + 5i\sin 3t$. Instead of solving (5.13) directly as we did in Example 5.8, we will look for a solution to

$$z'' + 2z' - 3z = 5e^{3it} \tag{5.14}$$

using the techniques of Example 5.6. If $z(t) = x(t) + iy(t)$ is that solution, then formally we have

$$\begin{aligned} z'' + 2z' - 3z &= (x + iy)'' + 2(x + iy)' - 3(x + iy) \\ &= (x'' + 2x' - 3x) + i(y'' + 2y' - 3y). \end{aligned} \tag{5.15}$$

On the other hand, expanding the right-hand side of (5.14) using Euler's formula, we get

$$z'' + 2z' - 3z = 5e^{3it} = 5[\cos 3t + i\sin 3t]. \tag{5.16}$$

Equating the imaginary parts of (5.15) and (5.16), we see that $y(t) = \operatorname{Im} z(t)$ is a solution to (5.13). We notice in passing that $x(t) = \operatorname{Re} z(t)$ is a solution to the equation

$$x'' + 2x' - 3x = 5\cos 3t.$$

Our solution to (5.14) should have the same form as the forcing term, so we try $z(t) = ae^{3it}$. Substituting this into the left-hand side of (5.14), we get

$$z'' + 2z' - 3z = (3i)^2 ae^{3it} + 2(3i)ae^{3it} - 3ae^{3it} = -6(2 - i)ae^{3it}.$$

For z to be a solution of (5.14) we must have $-6(2 - i)a = 5$. Therefore,

$$a = -\frac{1}{6}\frac{5}{2 - i} = -\frac{1}{6}\frac{5}{2 - i}\frac{2 + i}{2 + i} = -\frac{1}{6}\frac{10 + 5i}{5} = -\frac{2 + i}{6}.$$

Hence the complex solution is

$$z(t) = -\frac{2+i}{6}e^{3it}$$

$$= -\frac{1}{6}(2+i)(\cos 3t + i\sin 3t)$$

$$= -\frac{1}{6}\{[2\cos 3t - \sin 3t] + i[\cos 3t + 2\sin 3t]\}.$$

The solution y to (5.13) is the imaginary part of z, or

$$y(t) = -\frac{1}{6}[\cos 3t + 2\sin 3t],$$

which agrees with (5.11). ●

It is up to the individual to decide which of these methods is preferable. If you are comfortable with complex arithmetic, then the complex method is probably quicker. The complex method is preferred by some engineers and physicists because of the insight the answers provide.

Polynomial forcing terms

The derivative of a polynomial is another polynomial of lower degree. Consequently, a polynomial forcing term

$$f(t) = a_0 t^n + a_1 t^{n-1} + \cdots + a_{n-1}t + a_n$$

has a form that is replicated under differentiation. We can find a particular solution by the method of undetermined coefficients.

Example 5.17 Find a particular solution to the differential equation

$$y'' + 2y' - 3y = 3t + 4. \tag{5.18}$$

The right-hand side is a polynomial of degree 1, so we look for a particular solution of the same form. In this case that means a polynomial of the same degree, or

$$y = at + b,$$

where a and b are constants to be determined. The derivatives of y are

$$y' = a \quad \text{and} \quad y'' = 0.$$

Inserting these into our differential equation gives

$$y'' + 2y' - 3y = 0 + 2a - 3(at + b) = -3at + (2a - 3b). \tag{5.19}$$

Equating the coefficient of t and the constant term in (5.18) and (5.19) gives two equations for a and b:

$$-3a = 3,$$
$$2a - 3b = 4.$$

The solutions are $a = -1$ and $b = -2$, and our particular solution is

$$y = -t - 2.$$

We encourage the reader to check that y is indeed a solution. ●

Exceptional cases

The method of undetermined coefficients looks straightforward. There are, however, some exceptional cases to look out for. Suppose we change the forcing term in equation (5.7) in Example 5.6 to $3e^{-t}$, getting the equation

$$y'' - y' - 2y = 3e^{-t}.$$

If we look for a solution of the indicated form, $y(t) = ae^{-t}$, we run into trouble. To see this, let's insert y into the left-hand side of the differential equation. We get

$$y'' - y' - 2y = ae^{-t} + ae^{-t} - 2ae^{-t} = 0.$$

There is no choice of the constant a, which will make this equal to $3e^{-t}$.

The problem arises because the forcing term, and hence the proposed solution, is a solution to the associated homogeneous equation. This is the source of the difficulty in all exceptional cases. However, even if the solution to the homogeneous equation is only a part of the forcing term, we can have an exceptional case. As we will see, exceptional cases do arise in applications, and they are often the most interesting cases.

What do we do? In this case we look for a solution of the form $y(t) = ate^{-t}$. We multiply the usual general form by t. This is the way to find a solution whenever the usual form does not work. If this method does not work, multiply by t once more and try again.

Example 5.20 Find a particular solution to the equation

$$y'' - y' - 2y = 3e^{-t}. \tag{5.21}$$

We know from the preceding discussion that the forcing term is a solution to the homogeneous equation, so we look for a solution of the form $y(t) = ate^{-t}$. Substituting into the left-hand side of the equation, we get

$$y'' - y' - 2y = a(t - 2)e^{-t} - a(1 - t)e^{-t} - 2ate^{-t} = -3ae^{-t}.$$

This will give a solution to (5.21) provided that

$$-3ae^{-t} = 3e^{-t}.$$

Therefore, we need $a = -1$, and our particular solution is $y(t) = -te^{-t}$. ●

Typically, we do not notice that the forcing function is exceptional. The first indication that arises is when we are presented with equations for the undetermined coefficients that are inconsistent. At this point apply the remedy—multiply the trial solution by t and try again. If that still leads to inconsistent equations, multiply by t^2.

Combination forcing terms

The method of undetermined coefficients can be used whenever the forcing term is a linear combination of expressions of the forms we have already handled. This is a result of the linearity of the differential equation.

THEOREM 5.22 Suppose that $y_f(t)$ is a solution to the linear equation

$$y_f'' + py_f' + qy_f = f(t)$$

and $y_g(t)$ is a solution to

$$y_g'' + py_g' + qy_g = g(t).$$

Then $y(t) = \alpha y_f(t) + \beta y_g(t)$ is a solution to

$$y'' + py' + qy = \alpha f(t) + \beta g(t). \qquad \blacksquare$$

The proof of this result will be left to Exercise 48. Theorem 5.22 will be used in our examples.

Example 5.23 Find a particular solution to the equation

$$y'' - y' - 2y = e^{-2t} - 3e^{-t}. \qquad (5.24)$$

The forcing term is a sum, $f(t) = f_1(t) + f_2(t)$, where $f_1(t) = e^{-2t}$, and $f_2(t) = -3e^{-t}$. Suppose we break up the equation and solve the equations separately for each part of the forcing term. This means that we find functions y_1 and y_2 such that

$$\begin{aligned} y_1'' - y_1' - 2y_1 &= e^{-2t}, \quad \text{and} \\ y_2'' - y_2' - 2y_2 &= -3e^{-t}. \end{aligned} \qquad (5.25)$$

Because the equation in (5.24) is linear, Theorem 5.22 says that $y_1 + y_2$ will be a solution to (5.24).

It remains to solve the two equations in (5.25). The first is solved using the method in Example 5.6. We find that $y_1(t) = e^{-2t}/4$. In the second equation we have an exceptional case, so we use the method of Example 5.20. The solution is $y_2(t) = te^{-t}$. Hence the solution to (5.24) is

$$y(t) = y_1(t) + y_2(t) = \frac{1}{4}e^{-2t} + te^{-t}. \qquad \bullet$$

Let's look at one more example where the forcing term is a linear combination.

Example 5.26 Find a particular solution to the equation

$$y'' + 4y = \cos 2t - 2\sin 2t. \qquad (5.27)$$

We will use the complex method. It is easiest to treat the two summands in the forcing term separately. This means that we look for solutions y_1 and y_2 to the equations

$$\begin{aligned} y_1'' + 4y_1 &= \cos 2t \quad \text{and} \\ y_2'' + 4y_2 &= -2\sin 2t. \end{aligned} \qquad (5.28)$$

Then it is easily seen that $y = y_1 + y_2$ is a solution to (5.27).

To solve the first equation in (5.28), we notice that $\cos 2t = \text{Re}(e^{2it})$, so we solve the complex equation

$$z'' + 4z = e^{2it}. \qquad (5.29)$$

The forcing term is a solution to the homogeneous equation, so we look for a solution of the form $z(t) = ate^{2it}$. In the usual way we find that the solution is

$$z(t) = -\frac{it}{4}e^{2it}.$$

The solution to the first equation in (5.28) is

$$y_1(t) = \text{Re } z(t) = \text{Re}\left(-\frac{it}{4}e^{2it}\right) = \frac{t\sin 2t}{4}.$$

Notice that the imaginary part of z,

$$u(t) = \text{Im } z(t) = \text{Im}\left(-\frac{it}{4}e^{2it}\right) = -\frac{t\cos 2t}{4},$$

is a solution to the equation

$$u'' + 4u = \text{Im } e^{2it} = \sin 2t. \tag{5.30}$$

Since the right-hand side of the second equation in (5.28) is -2 times the right-hand side in (5.30), Theorem 5.22 allows us to conclude that

$$y_2(t) = -2u(t) = \frac{t\cos 2t}{2}$$

is a solution to the second equation in (5.28), and that a solution to (5.27) is

$$y(t) = y_1(t) + y_2(t) = \frac{t\sin 2t}{4} + \frac{t\cos 2t}{2}. \qquad \bullet$$

More complicated forcing terms

In addition to the three cases we have considered so far, we could have forcing terms which are products of two of these, or of all three. The most general situation would be a forcing term of the form

$$f(t) = e^{rt}P(t)\cos\omega t + e^{rt}Q(t)\sin\omega t,$$

where $P(t)$ and $Q(t)$ are polynomials. Again $f'(t)$ has the same form, so we can use undetermined coefficients.

The method of undetermined coefficients is summarized in Table 1, which shows the allowed forcing functions and the type of particular solution to be used. Even for the most general case the method works in the way we have indicated in our examples.

Table 1 The method of undetermined coefficients		
FORCING FUNCTION $f(t)$	TRIAL SOLUTION $y_p(t)$	COMMENTS
e^{rt}	ae^{rt}	
$\cos \omega t$ or $\sin \omega t$	$a \cos \omega t + b \sin \omega t$	
$P(t)$	$p(t)$	P is a polynomial; p is a polynomial of the same degree
$P(t) \cos \omega t$ or $P(t) \sin \omega t$	$p(t) \cos \omega t +$ $q(t) \sin \omega t$	P is a polynomial; p & q are polynomials of the same degree
$e^{rt} \cos \omega t$ or $e^{rt} \sin \omega t$	$e^{rt}[a \cos \omega t +$ $b \sin \omega t]$	
$e^{rt} P(t) \cos \omega t$ or $e^{rt} P(t) \sin \omega t$	$e^{rt}[p(t) \cos \omega t +$ $q(t) \sin \omega t]$	P is a polynomial; p & q are polynomials of the same degree

EXERCISES

In Exercises 1–4, use the technique demonstrated in Example 5.6 to find a particular solution for the given differential equation.

1. $y'' + 3y' + 2y = 4e^{-3t}$ **2.** $y'' + 6y' + 8y = -3e^{-t}$

3. $y'' + 2y' + 5y = 12e^{-t}$ **4.** $y'' + 3y' - 18y = 18e^{2t}$

In Exercises 5–8, use the form $y_p = a \cos \omega t + b \sin \omega t$, as in Example 5.8, to help find a particular solution for the given differential equation.

5. $y'' + 4y = \cos 3t$ **6.** $y'' + 9y = \sin 2t$

7. $y'' + 7y' + 6y = 3 \sin 2t$

8. $y'' + 7y' + 10y = -4 \sin 3t$

9. Suppose that $z(t) = x(t) + iy(t)$ is a solution of

$$z'' + pz' + qz = Ae^{i\omega t}. \qquad (5.31)$$

Substitute $z(t)$ into equation (5.31). Then compare the real and imaginary parts of each side of the resulting equation to prove two facts:

$$x'' + px' + qx = A \cos \omega t,$$
$$y'' + py' + qy = A \sin \omega t.$$

Write a short paragraph summarizing the significance of this result.

In Exercises 10–13, use the complex method, as in Example 5.12, to find a particular solution for the differential equation.

10. $y'' + 4y = \cos 3t$ **11.** $y'' + 9y = \sin 2t$

12. $y'' + 7y' + 6y = 3 \sin 2t$

13. $y'' + 7y' + 10y = -4 \sin 3t$

In Exercises 14–17, use the technique shown in Example 5.17 to find a particular solution for the given differential equation.

14. $y'' + 5y' + 4y = 2 + 3t$ **15.** $y'' + 6y' + 8y = 2t - 3$

16. $y'' + 5y' + 6y = 4 - t^2$ **17.** $y'' + 3y' + 4y = t^3$

In Exercises 18–23, use the technique of Section 4.3 to find a solution of the associated homogeneous equation; then use the technique of this section to find a particular solution. Use Theorem 5.2 to form the general solution. Then find the solution satisfying the given initial conditions.

18. $y'' + 3y' + 2y = 3e^{-4t}$, $y(0) = 1$, $y'(0) = 0$

19. $y'' - 4y' - 5y = 4e^{-2t}$, $y(0) = 0$, $y'(0) = -1$

20. $y'' + 2y' + 2y = 2 \cos 2t$, $y(0) = -2$, $y'(0) = 0$

21. $y'' - 2y' + 5y = 3 \cos t$, $y(0) = 0$, $y'(0) = -2$

22. $y'' + 4y' + 4y = 4 - t$, $y(0) = -1$, $y'(0) = 0$

23. $y'' - 2y' + y = t^3$, $y(0) = 1$, $y'(0) = 0$

In Exercises 24–29, the forcing term is also a solution of the associated homogeneous solution. Use the technique of Example 5.20 to find a particular solution.

24. $y'' - 3y' - 10y = 3e^{-2t}$ **25.** $y'' - y' - 2y = 2e^{-t}$

26. $y'' + 4y = 4 \cos 2t$ **27.** $y'' + 9y = \sin 3t$

28. $y'' + 4y' + 4y = 2e^{-2t}$ **29.** $y'' + 6y' + 9y = 5e^{-3t}$

30. If $y_f(t)$ is a solution of

$$y'' + py' + qy = f(t)$$

and $y_g(t)$ is a solution of

$$y'' + py' + qy = g(t),$$

show that $z(t) = \alpha y_f(t) + \beta y_g(t)$ is a solution of

$$y'' + py' + qy = \alpha f(t) + \beta g(t),$$

where α and β are any real numbers.

Use the technique suggested by Examples 5.23 and 5.26, as well as Exercise 30, to help find particular solutions for the differential equations in Exercises 31–38.

31. $y'' + 2y' + 2y = 2 + \cos 2t$

32. $y'' - y = t - e^{-t}$

33. $y'' + 25y = 2 + 3t + \cos 5t$

34. $y'' + 2y' + y = 3 - e^{-t}$

35. $y'' + 4y' + 3y = \cos 2t + 3 \sin 2t$

36. $y'' + 2y' + 2y = 3 \cos t - \sin t$

37. $y'' + 4y' + 4y = e^{-2t} + \sin 2t$

38. $y'' + 16y = e^{-4t} + 3 \sin 4t$

39. Use the form $y_p(t) = (at + b)e^{-4t}$ in an attempt to find a particular solution of the equation $y'' + 3y' + 2y = te^{-4t}$.

Use an approach similar to that in Exercise 39 to find particular solutions of the equations in Exercises 40–43.

40. $y'' - 3y' + 2y = te^{-3t}$ **41.** $y'' + 2y' + y = t^2 e^{-2t}$

42. $y'' + 5y' + 4y = te^{-t}$ **43.** $y'' + 3y' + 2y = t^2 e^{-2t}$

44. Use the form $y_p = e^{-2t}(a \cos t + b \sin t)$ in an attempt to find a particular solution of $y'' + 2y' + 2y = e^{-2t} \sin t$.

45. If $z(t) = x(t) + iy(t)$ is a solution of

$$z'' + pz' + qz = Ae^{(a+bi)t},$$

show that $x(t)$ and $y(t)$ are solutions of

$$x'' + px' + qx = Ae^{at} \cos bt$$

and

$$y'' + py' + qy = Ae^{at} \sin bt,$$

respectively.

46. Use the technique suggested by Exercise 45 to find a particular solution of the equation in Exercise 44.

47. Prove that the imaginary part of the solution of $z'' + z' + z = te^{it}$ is a solution of $y'' + y' + y = t \sin t$. Use this idea to find a particular solution of $y'' + y' + y = t \sin t$.

48. Prove Theorem 5.22.

4.6 Variation of Parameters

In this section we introduce a technique called *variation of parameters*. This technique is used to find a particular solution to more general higher-order equations, provided we know a fundamental set of solutions to the associated homogeneous equation. As we did in the previous section, we will illustrate the method for second-order equations. The method also works for higher-order equations, but it is usually more efficient to solve the associated first-order system using variation of parameters. This will be discussed in a later chapter.

We are interested in solving the equation

$$y'' + p(t)y' + q(t)y = g(t). \tag{6.1}$$

Notice that we are allowing the coefficients $p(t)$ and $q(t)$ to be functions of t. In particular, we are not restricting them to be constants. This might seem to be a great increase in generality, but there is a rather strong constraint. We will have to assume that we have computed a fundamental set of solutions y_1 and y_2 to the associated homogeneous equation

$$y'' + p(t)y' + q(t)y = 0. \tag{6.2}$$

Then the general solution to the homogeneous equation is

$$y_h = C_1 y_1 + C_2 y_2, \tag{6.3}$$

where C_1 and C_2 are arbitrary constants.

The idea behind variation of parameters is to replace the constants C_1 and C_2 in (6.3) by unknown functions $v_1(t)$ and $v_2(t)$ and look for a particular solution to the inhomogeneous equation (6.1) of the form

$$y_p = v_1 y_1 + v_2 y_2. \tag{6.4}$$

You will notice the similarity with the method of variation of parameters as it was used to solve first-order linear equations in Chapter 2.

Example 6.5   Find a particular solution to the equation

$$y'' + y = \tan t. \tag{6.6}$$

The right side, $\tan t$, is not one of the forms that can be handled with undetermined coefficients. We will find a solution by the method of variation of parameters. First, note that $y_1(t) = \cos t$ and $y_2(t) = \sin t$ are a fundamental set of solutions to the associated homogeneous equation $y'' + y = 0$. So we look for a particular solution of the form

$$y_p = v_1 \cos t + v_2 \sin t. \tag{6.7}$$

We start by computing the first derivative of y_p.

$$\begin{aligned} y_p' &= v_1' \cos t - v_1 \sin t + v_2' \sin t + v_2 \cos t \\ &= (v_1' \cos t + v_2' \sin t) - v_1 \sin t + v_2 \cos t \end{aligned} \tag{6.8}$$

The differential equation (6.6) puts only one constraint on the two functions v_1 and v_2. We can impose one more constraint. Notice that if we set the term in parentheses on the right in (6.8) equal to zero, then the expression simplifies, and we will eliminate the first derivatives of v_1 and v_2. Then when we compute y_p'' no second derivatives of v_1 and v_2 will appear. Hence we set

$$v_1' \cos t + v_2' \sin t = 0. \tag{6.9}$$

Now (6.8) simplifies to

$$y_p' = -v_1 \sin t + v_2 \cos t.$$

Differentiating this equation gives

$$\begin{aligned} y_p'' &= -v_1' \sin t - v_1 \cos t + v_2' \cos t - v_2 \sin t \\ &= -v_1' \sin t + v_2' \cos t - v_1 \cos t - v_2 \sin t. \end{aligned}$$

Inserting these expressions for y_p and y_p'' into the left-hand side of our differential equation, we obtain

$$\begin{aligned} y_p'' + y_p &= \left(-v_1' \sin t + v_2' \cos t - v_1 \cos t - v_2 \sin t \right) + (v_1 \cos t + v_2 \sin t) \\ &= -v_1' \sin t + v_2' \cos t. \end{aligned}$$

Comparing this with (6.6), we see that y_p is a solution to (6.6) provided $-v_1' \sin t + v_2' \cos t = \tan t$. This is our second equation for v_1' and v_2'. Let's restate it along with (6.9).

$$v_1' \cos t + v_2' \sin t = 0$$
$$-v_1' \sin t + v_2' \cos t = \tan t$$

This is a system of two linear equations for the unknowns v_1' and v_2'. The coefficient matrix is

$$\begin{pmatrix} \cos t & \sin t \\ -\sin t & \cos t \end{pmatrix}.$$

Since the determinant is $\cos^2 t + \sin^2 t = 1 \neq 0$, the system has a solution. Multiply the first equation by $\sin t$ and the second by $\cos t$ to get

$$\sin t (v_1' \cos t + v_2' \sin t) = 0$$
$$\cos t (-v_1' \sin t + v_2' \cos t) = \sin t.$$

Adding these equations, we get

$$v_2'(\sin^2 t + \cos^2 t) = \sin t \quad \text{or} \quad v_2' = \sin t.$$

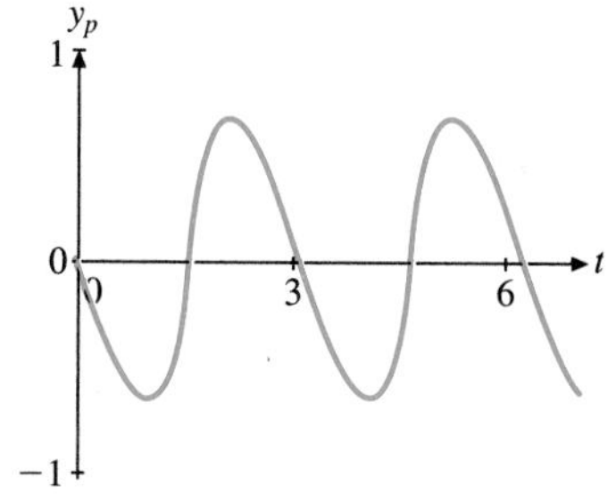

Similarly, if we multiply the first equation by $\cos t$ and the second by $\sin t$ and then subtract the second from the first, we get

$$v_1'(\cos^2 t + \sin^2 t) = v_1' = -\sin t \tan t \quad \text{or} \quad v_1' = -\frac{\sin^2 t}{\cos t}.$$

When we integrate the equations for v_1' and v_2', we get

$$v_1(t) = -\ln|\sec t + \tan t| + \sin t \quad \text{and} \quad v_2(t) = -\cos t. \tag{6.10}$$

Notice that we have omitted the constants of integration. Since we are after one particular solution, we can set the constants equal to zero.

Substituting the formulas for v_1 and v_2 into 6.7 gives our particular solution

$$\begin{aligned} y_p(t) &= (-\ln|\sec t + \tan t| + \sin t)\cos t - \cos t \sin t \\ &= -(\cos t)\ln|\sec t + \tan t|. \end{aligned}$$

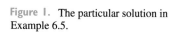

y_p

Figure 1. The particular solution in Example 6.5.

The general case

Does this procedure always work? Let's examine the general case and see what happens. To find a particular solution to

$$y'' + p(t)y' + q(t)y = g(t), \tag{6.11}$$

we look for a solution of the form

$$y_p = v_1 y_1 + v_2 y_2, \tag{6.12}$$

where v_1 and v_2 are functions that are yet to be determined, and y_1 and y_2 are a fundamental set of solutions to the associated homogeneous equation

$$y'' + p(t)y' + q(t)y = 0. \tag{6.13}$$

We compute the derivative of y_p,

$$y_p' = v_1' y_1 + v_1 y_1' + v_2' y_2 + v_2 y_2' = [v_1' y_1 + v_2' y_2] + v_1 y_1' + v_2 y_2'.$$

As in Example 6.5, we set the expression in the square brackets equal to zero, or

$$v_1' y_1 + v_2' y_2 = 0. \tag{6.14}$$

The first derivative then simplifies to

$$y_p' = v_1 y_1' + v_2 y_2'.$$

Differentiating again, we obtain

$$y_p'' = v_1' y_1' + v_2' y_2' + v_1 y_1'' + v_2 y_2''.$$

Inserting y_p, y_p', and y_p'' into the left-hand side of equation (6.11), we get

$$\begin{aligned} y_p'' + p(t)y_p' + q(t)y_p &= \left(v_1' y_1' + v_2' y_2' + v_1 y_1'' + v_2 y_2''\right) \\ &\quad + p(t)\left(v_1 y_1' + v_2 y_2'\right) + q(t)(v_1 y_1 + v_2 y_2) \\ &= v_1(y_1'' + p(t)y_1' + q(t)y_1) \\ &\quad + v_2(y_2'' + p(t)y_2' + q(t)y_2) \\ &\quad + v_1' y_1' + v_2' y_2'. \end{aligned}$$

Since y_1 and y_2 are solutions to the homogeneous equation (6.13), this simplifies to

$$y_p'' + p(t)y_p' + q(t)y_p = v_1'y_1' + v_2'y_2'.$$

Now we see that y_p is a solution to (6.11) provided that

$$v_1'y_1' + v_2'y_2' = g(t).$$

This is our second equation for v_1 and v_2. We restate it along with equation (6.14):

$$\begin{aligned} v_1'y_1 + v_2'y_2 &= 0 \\ v_1'y_1' + v_2'y_2' &= g(t). \end{aligned} \tag{6.15}$$

This is a system of two linear equations for v_1' and v_2'. The coefficient matrix is

$$\begin{pmatrix} y_1 & y_2 \\ y_1' & y_2' \end{pmatrix}.$$

The system can be solved provided that the determinant of this matrix is nonzero. The determinant will be recognized as the Wronskian of y_1 and y_2,

$$W = y_1 y_2' - y_2 y_1'.$$

Since y_1 and y_2 form a fundamental set of solutions to the homogeneous equation, they are linearly independent and we know by Proposition 1.27 that the Wronskian $W(t) \neq 0$.

The equations in (6.15) can be solved by elimination to obtain

$$v_2' = \frac{y_1 g}{y_1 y_2' - y_1' y_2} \quad \text{and} \quad v_1' = \frac{-y_2 g}{y_1 y_2' - y_1' y_2}.$$

Notice that the denominator in the above expressions is the Wronskian of y_1 and y_2 and is always nonzero.

We can integrate these equations:

$$v_1(t) = \int \frac{-y_2(t)g(t)\,dt}{y_1(t)y_2'(t) - y_1'(t)y_2(t)}, \quad \text{and}$$

$$v_2(t) = \int \frac{y_1(t)g(t)\,dt}{y_1(t)y_2'(t) - y_1'(t)y_2(t)}. \tag{6.16}$$

When they are inserted into (6.12), we get a particular solution.

Summary

To find a particular solution to the inhomogeneous equation $y'' + py' + qy = g$, follow these steps.

1. Find a fundamental set of solutions y_1, y_2 to the associated homogeneous equation $y'' + py' + qy = 0$.
2. Form $y_p = v_1 y_1 + v_2 y_2$, where v_1 and v_2 are functions to be determined.
3. Find v_1 and v_2. Here there are two possible ways to proceed. The first way is to use the formulas in (6.16). This requires that you remember these formulas or that you have the use of a book that contains the formulas.

   The second way is to follow the procedure of the method of variation of parameters. The procedure has the following steps:

(a) Differentiate y_p:

$$y_p' = (v_1'y_1 + v_2'y_2) + v_1y_1' + v_2y_2'$$

and set the first term on the right equal to zero.

$$v_1'y_1 + v_2'y_2 = 0 \tag{6.17}$$

(b) Take the second derivative of y_p and insert y_p, y_p', and y_p'' into the differential equation. After simplifying, a second equation will appear:

$$v_1'y_1' + v_2'y_2' = g(t). \tag{6.18}$$

(c) Solve (6.17) and (6.18) for v_1' and v_2' by elimination.

(d) Integrate to find v_1 and v_2.

4. Substitute v_1 and v_2 into $y_p = v_1y_1 + v_2y_2$.

It is up to the reader to decide which of the methods to use in step 3.

Although variation of parameters always works theoretically, its success requires a fundamental set of solutions to the homogeneous equation and the ability to compute the integrals in (6.16). The method is of significant theoretical importance, however.

EXERCISES

For Exercises 1–12, find a particular solution to the given second-order differential equation.

1. $y'' + 9y = \tan 3t$

2. $y'' + 4y = \sec 2t$

3. $y'' - y = t + 3$

4. $x'' - 2x' - 3x = 4e^{3t}$

5. $y'' - 2y' + y = e^t$

6. $x'' - 4x' + 4x = e^{2t}$

7. $x'' + x = \tan^2 t$

8. $x'' + x = \sec^2 t$

9. $x'' + x = \sin^2 t$

10. $y'' + 2y' + y = t^5 e^{-t}$

11. $y'' + y = \tan t + \sin t + 1$

12. $y'' + y = \sec t + \cos t - 1$

13. Verify that $y_1(t) = t$ and $y_2(t) = t^{-3}$ are solutions to the homogeneous equation

$$t^2 y''(t) + 3ty'(t) - 3y(t) = 0.$$

Use variation of parameters to find the general solution to

$$t^2 y''(t) + 3ty'(t) - 3y(t) = \frac{1}{t}.$$

14. Verify that $y_1(t) = t^{-1}$ and $y_2(t) = t^{-1} \ln t$ are solutions to the homogeneous equation

$$t^2 y''(t) + 3ty'(t) + y(t) = 0.$$

Use variation of parameters to find the general solution to

$$t^2 y''(t) + 3ty'(t) + y(t) = \frac{1}{t}.$$

4.7 Forced Harmonic Motion

In this section, we apply the technique of undetermined coefficients to analyze harmonic motion with an external sinusoidal forcing term. We derived the model in Section 4.4, and with a sinusoidal forcing term, equation (4.4) becomes

$$x'' + 2cx' + \omega_0^2 x = A \cos \omega t. \tag{7.1}$$

The constant A is the amplitude of the driving force, and ω is the **driving frequency**. Remember that c is the damping constant and ω_0 is the natural frequency.

To focus our thinking, we may suppose that we have an iron mass m suspended on a spring, with the top of the spring attached to a motor that moves the top of the spring up and down. We can also consider an *RLC* circuit in which the source voltage is sinusoidal.

We will first treat the case with no damping.

Forced undamped harmonic motion

The equation we will deal with here comes from (7.1) with $c = 0$ or

$$x'' + \omega_0^2 x = A \cos \omega t. \tag{7.2}$$

The associated homogeneous equation is

$$x'' + \omega_0^2 x = 0, \tag{7.3}$$

with general solution

$$x_h = C_1 \cos \omega_0 t + C_2 \sin \omega_0 t.$$

We have to consider separately the cases when the driving frequency ω is equal to the natural frequency and when it is not.

Case 1: $\omega \neq \omega_0$. If the driving frequency is not equal to the natural frequency, we look for a particular solution of the form

$$x_p = a \cos \omega t + b \sin \omega t, \tag{7.4}$$

where a and b are undetermined constants. Substituting x_p into the left-hand side of the inhomogeneous differential equation (7.2) gives

$$x_p'' + \omega_0^2 x_p = a(\omega_0^2 - \omega^2) \cos \omega t + b(\omega_0^2 - \omega^2) \sin \omega t.$$

Comparing this with (7.2), we see that we have a solution provided that

$$a = \frac{A}{\omega_0^2 - \omega^2} \quad \text{and} \quad b = 0.$$

So our particular solution is

$$x_p(t) = \frac{A}{\omega_0^2 - \omega^2} \cos \omega t.$$

Notice that x_p is oscillatory, with the same frequency as the driving force. The amplitude of this oscillation also depends on the driving frequency and gets larger as the driving frequency approaches the natural frequency of the spring. This is our first indication of resonance. We will have more to say about that shortly.

The general solution to the inhomogeneous equation is

$$x(t) = x_h(t) + x_p(t) = C_1 \cos \omega_0 t + C_2 \sin \omega_0 t + \frac{A}{\omega_0^2 - \omega^2} \cos \omega t.$$

Let's look at the solution where the motion starts at equilibrium. This means the initial conditions are $x(0) = x'(0) = 0$. It is easily seen that $C_1 = -A/(\omega_0^2 - \omega^2)$, and $C_2 = 0$. Hence the solution is

$$x(t) = \frac{A}{\omega_0^2 - \omega^2} (\cos \omega t - \cos \omega_0 t). \tag{7.5}$$

This is a superposition of two oscillations with different frequencies. The result is interesting.

Example 7.6 Suppose $A = 23$, $\omega_0 = 11$, and $\omega = 12$. With these values of the parameters the solution (7.5) becomes

$$x(t) = \cos 11t - \cos 12t. \qquad (7.7)$$

The graph is shown in Figure 1. This figure shows the phenomenon called **beats**. It occurs whenever two frequencies that are almost equal interfere with each other as we see in (7.5) and (7.7).

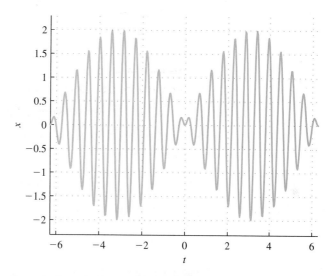

Figure 1. Beats in forced, undamped, harmonic motion.

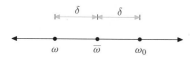

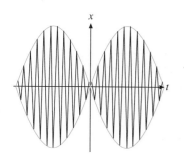

Figure 2. The relationship of ω and ω_0 to $\overline{\omega}$ and δ.

We can understand the phenomenon of beats better after a little algebra. We introduce the **mean frequency** $\overline{\omega} = (\omega_0 + \omega)/2$ and the **half difference** $\delta = (\omega_0 - \omega)/2$. Then

$$\omega = \overline{\omega} - \delta \quad \text{and} \quad \omega_0 = \overline{\omega} + \delta.$$

The relationship between these variables is shown in Figure 2 under the assumption that $\omega < \omega_0$. If we substitute these equations into (7.5) and use the addition law for the cosine, we get

$$x(t) = \frac{A \sin \delta t}{2\overline{\omega}\delta} \sin \overline{\omega} t. \qquad (7.8)$$

In Example 7.6, $\delta = -1/2$ and $\overline{\omega} = 23/2$. In that case the factor $\sin \delta t = \sin(-t/2)$ oscillates very slowly, especially in comparison with the faster oscillation with frequency $\overline{\omega} = 23/2$. Thus the solution in (7.8) can be seen to be a fast oscillation with a frequency $\overline{\omega}$ and an amplitude

$$\left| \frac{A \sin \delta t}{2\overline{\omega}\delta} \right|, \qquad (7.9)$$

Figure 3. The envelope of the solution in Example 7.6 is plotted in blue.

which oscillates much more slowly. In Figure 3 we superimpose a graph of this slow oscillating amplitude on the oscillation in Figure 1. It is the blue curve through the maxima of the faster oscillation. This is called the **envelope** of the faster oscillation. (See Exercise 2.)

The phenomenon of beats illustrated in Figure 1 occurs in many situations. For example, it is used by a piano tuner to be sure that particular piano strings are properly tuned. The tuner strikes a tuning fork, which vibrates at the correct frequency.

He next hits the poorly tuned piano key, which vibrates at a slightly different frequency. The two sounds interfere in a way modeled by (7.5), and therefore by (7.8). The tuner hears the mean frequency with an amplitude that is modulated by the difference frequency, as shown in Figure 1. This modulation gives rise to beats in the tone that are readily audible. When the string is properly tuned, the beats go away.

Case 2: $\omega = \omega_0$. In this case, the particular solution given in (7.4) is a solution to the homogeneous equation. Therefore, we look for a particular solution of the form

$$x_p = t(a \cos \omega_0 t + b \sin \omega_0 t).$$

Inserting x_p into the left-hand side of the differential equation in (7.2), we get

$$
\begin{aligned}
x_p'' + \omega_0^2 x_p &= \left[2\omega_0(-a \sin \omega_0 t + b \cos \omega_0 t) + t\omega_0^2(-a \cos \omega_0 t - b \sin \omega_0 t)\right] \\
&\quad + \omega_0^2 t(a \cos \omega_0 t + b \sin \omega_0 t) \\
&= 2\omega_0(-a \sin \omega_0 t + b \cos \omega_0 t).
\end{aligned}
$$

Setting this equal to $A \cos \omega_0 t$ and equating the coefficients of the trigonometric functions, we see that we have a solution provided that

$$b = \frac{A}{2\omega_0} \quad \text{and} \quad a = 0.$$

The particular solution is

$$x_p = \frac{A}{2\omega_0} t \sin \omega_0 t,$$

and the general solution to the inhomogeneous equation is

$$x(t) = x_h(t) + x_p(t) = C_1 \cos \omega_0 t + C_2 \sin \omega_0 t + \frac{A}{2\omega_0} t \sin \omega_0 t.$$

In the special case where $x(0) = C_1 = 0$, $x'(0) = C_2 = 0$, $A = 8$, and $\omega_0 = 4$, this reduces to

$$x(t) = t \sin 4t.$$

The graph of this function is shown in Figure 4. Notice how the solution grows with time. This growth is due to the fact that the frequency of the forcing term

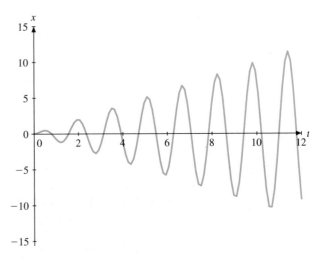

Figure 4. Forced, undamped, harmonic motion where the driving frequency equals the natural frequency.

equals the natural vibrating frequency of the spring. The force pulls and pushes at a frequency equal to the natural frequency of the spring. Thus the amplitude of the oscillatory motion of the mass increases. This type of behavior is called **resonance**, and engineers try to eliminate it in the design of mechanical systems, since oscillations that grow in amplitude can eventually cause the system to break.

In reality, mechanical systems always have some damping (friction) that will keep the solution from getting infinitely large. However, resonance can still cause mechanical motions to become large enough to cause major structural damage. A common example of the destructiveness of resonance is the shattering of a glass by the voice of a singer. The singer has to sing a note at a frequency that is very close to the natural frequency of the glass in order to cause it to resonate and eventually shatter. Another example involves the behavior of a troop of soldiers marching over a bridge. If they were to continue to march in perfect step, they might be marching at the natural frequency of some component of the bridge and cause it to resonate and eventually break. Consequently, since time immemorial troops of soldiers have broken ranks when marching over a bridge.

On the other hand, resonance is exploited in electrical systems that are governed by the same model as (7.2) in order to tune radios, for example.

Forced damped harmonic motion

If we add a damping term to the system, interesting things happen. Now we are dealing with the equation in (7.1),

$$x'' + 2cx' + \omega_0^2 x = A \cos \omega t. \tag{7.10}$$

The associated homogeneous equation is

$$x'' + 2cx' + \omega_0^2 x = 0. \tag{7.11}$$

The characteristic roots are

$$\lambda = -c \pm \sqrt{c^2 - \omega_0^2}.$$

Let's suppose that we are in the underdamped case, where $c < \omega_0$. We found in Section 4.4 that the general solution is

$$x_h = e^{-ct} \left(C_1 \cos(\eta t) + C_2 \sin(\eta t) \right),$$

where

$$\eta = \sqrt{\omega_0^2 - c^2}.$$

To find a particular solution to the inhomogeneous equation, we use undetermined coefficients, but this time it is easier to use the complex method. This means we look for a solution $z(t) = ae^{i\omega t}$ to the equation

$$z'' + 2cz' + \omega_0^2 z = Ae^{i\omega t} \tag{7.12}$$

and then set $x_p = \text{Re}(z)$. Let's substitute $z(t)$ into the left side of (7.12). We get

$$z'' + 2cz' + \omega_0^2 z = \left[(i\omega)^2 + 2c(i\omega) + \omega_0^2 \right] ae^{i\omega t} = P(i\omega)ae^{i\omega t},$$

where the polynomial

$$P(\lambda) = \lambda^2 + 2c\lambda + \omega_0^2$$

is the characteristic polynomial of the differential equation in (7.12) or (7.11). Thus equation (7.12) becomes

$$P(i\omega)ae^{i\omega t} = Ae^{i\omega t}.$$

Solving for a, we get $a = A/P(i\omega)$. Hence

$$z(t) = ae^{i\omega t} = \frac{A}{P(i\omega)}e^{i\omega t} = H(i\omega)Ae^{i\omega t}, \qquad (7.13)$$

where we have set

$$H(i\omega) = \frac{1}{P(i\omega)}.$$

The function $H(i\omega)$ is called the **_transfer function_**. To see what it is like, we look at its reciprocal

$$P(i\omega) = (i\omega)^2 + 2c(i\omega) + \omega_0^2 = (\omega_0^2 - \omega^2) + 2ic\omega.$$

We want to write this complex number in polar form, which means that we want to find a positive number R and an angle ϕ such that

$$P(i\omega) = Re^{i\phi} = R[\cos\phi + i\sin\phi].$$

We need

$$R\cos\phi = \omega_0^2 - \omega^2 \quad \text{and} \quad R\sin\phi = 2c\omega.$$

This is just polar coordinates, so

$$R = \sqrt{(\omega_0^2 - \omega^2)^2 + 4c^2\omega^2},$$

while the angle ϕ is defined by the pair of equations

$$\cos\phi = \frac{\omega_0^2 - \omega^2}{\sqrt{(\omega_0^2 - \omega^2)^2 + 4c^2\omega^2}} \quad \text{and} \quad \sin\phi = \frac{2c\omega}{\sqrt{(\omega_0^2 - \omega^2)^2 + 4c^2\omega^2}}.$$

(See the discussion of amplitude and phase in Section 4.4.) Notice that $2c\omega > 0$, so $\sin\phi > 0$. This means that the phase always satisfies $0 < \phi < \pi$. Furthermore, we have

$$\cot\phi = \frac{\omega_0^2 - \omega^2}{2c\omega} \quad \text{or} \quad \phi = \phi(\omega) = \operatorname{arccot}\left(\frac{\omega_0^2 - \omega^2}{2c\omega}\right). \qquad (7.14)$$

The last equation defines ϕ uniquely because $0 < \phi < \pi$.

Now we see that the transfer function can be written as

$$H(i\omega) = \frac{1}{P(i\omega)} = \frac{1}{R}e^{-i\phi}.$$

We will define the **_gain_** G by

$$G(\omega) = \frac{1}{R} = \frac{1}{\sqrt{(\omega_0^2 - \omega^2)^2 + 4c^2\omega^2}}, \qquad (7.15)$$

and then the transfer function is

$$H(i\omega) = G(\omega)e^{-i\phi(\omega)}. \qquad (7.16)$$

From (7.13) we see that the solution to (7.12) is

$$z(t) = H(i\omega)Ae^{i\omega t} = G(\omega)Ae^{i(\omega t - \phi)}.$$

Finally, the solution to (7.10) is

$$x_p(t) = \operatorname{Re} z(t) = G(\omega)A\cos(\omega t - \phi). \tag{7.17}$$

This form makes it clear that x_p is sinusoidal with the same frequency as the driving force. It also shows that the amplitude of x_p is the product of the gain $G(\omega)$ times the amplitude of the driving force. In addition, x_p is out of phase with the driving force by the amount $\phi = \phi(\omega)$, given in (7.14).

The general solution to the inhomogeneous equation is

$$\begin{aligned} x &= x_h + x_p \\ &= e^{-ct}(C_1\cos(\eta t) + C_2\sin(\eta t)) + G(\omega)A\cos(\omega t - \phi). \end{aligned} \tag{7.18}$$

Notice that x_h has the factor e^{-ct}, which quickly decays to 0 as $t \to \infty$. For this reason, this term is called the **transient** term. The rate of the decay of the transient term is governed by the exponential factor e^{-ct}. After the passage of time equal to $T_c = 1/c$, this decreases to $e^{-1} \approx 0.3679$, and the amplitude of the transient term has decreased to e^{-1} times its original value. T_c is called the **time constant**, and it is defined by this property. Notice that after the passage of time equal to $4T_c$ the amplitude of the transient is reduced to $e^{-4} \approx 0.0183$ times its original value, after which the effect of the transient term is usually negligible.

The particular solution x_p in (7.17) does not decay. It is therefore called the **steady-state** term. It is this term that is driven by the external force $A\cos\omega t$. After the passage of time equal to a few time constants, it is the only term that is important.

Example 7.19   Consider a forced vibrating spring where $m = 5\text{kg}$, $\mu = 7\text{kg/s}$, and $k = 3\text{kg/s}^2$, with a forcing term $2\cos 4t$. Suppose the initial conditions are such that the constants in (7.18) are $C_1 = 0$ and $C_2 = 1$. Find the amplitude and the phase of the steady-state solution. Plot the displacement of the resulting oscillation versus time. Add a plot of the steady-state oscillation.

The equation for the vibrating spring with the given parameters is

$$5x'' + 7x' + 3x = 2\cos 4t.$$

Dividing by the mass 5, we get

$$x'' + \frac{7}{5}x' + \frac{3}{5}x = \frac{2}{5}\cos 4t.$$

This is the form of the equation for the forced harmonic oscillator. The natural frequency is $\omega_0 = \sqrt{3/5} \approx 0.7746$. In addition, $c = 7/10$. With the assumptions of the example, the transient solution is

$$x_h(t) = e^{-ct}\sin\eta t,$$

where $\eta = \sqrt{\omega_0^2 - c^2} \approx 0.3317$. The gain is computed from (7.15) to be $G(4) \approx 0.0610$. Hence the amplitude of the steady-state solution is $G(4) \times (2/5) \approx 0.0244$. Finally, we compute the phase from (7.14) and get $\phi = \phi(4) = 2.7928$. A plot of the solution together with its steady-state term is given in Figure 5. Note how the solution x converges to the steady-state solution as t gets large. We have $T_c = 1/c = 10/7 = 1.428$. Notice in Figure 5, that the actual displacement is very close to the steady-state solution for $t \geq 4T_c = 5.714$.   ●

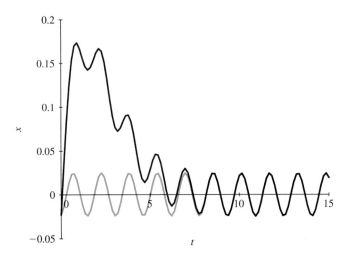

Figure 5. The motion of a forced spring. The displacement is in black, and the steady-state solution is in blue.

Transient solutions can be quite large in comparison to the steady-state solution, as we have seen in Example 7.19, as shown in Figure 5. Such large transient currents in electrical circuits can be destructive. It is almost certain that you have experienced this. Transients arise whenever an electrical circuit is turned on or off. How often have you turned on a light and one of the bulbs burned out immediately? This is caused by a large transient current flowing through an already weakened bulb. How often have you turned on a light and one of the bulbs was burned out, although it had been working fine the last time you used the light? This time the bulb most likely burned out when the lights were turned out, and again it was a large transient caused by the shutdown that did the job.

The destructive effect of transients is the reason why it is often a good idea to leave electrical equipment running even when it is not being used. For example, it is highly recommended that computers be always left on, unless it is known that they will not be used for a couple of days at the minimum. Transients are particularly harmful to hard discs.

We now examine the steady-state term (7.17) more closely. We will concentrate on the amplitude. We will examine how the gain G depends on the driving frequency ω. There are too many terms in the formula (7.15) for G to easily understand what will happen for all possible values of the parameters. We remedy this by *lumping parameters*. First, we will want to see how G changes as the driving frequency changes relative to the natural frequency, so we will introduce $s = \omega/\omega_0$ by setting $\omega = s\omega_0$ and consider G as a function of s. Then we introduce $D = 2c/\omega_0$, by substituting $c = D\omega_0/2$. The new constant D measures the effect of the damping force. When these changes are introduced, the gain takes the simpler form

$$G = \frac{1}{\omega_0^2\sqrt{(1-s^2)^2 + D^2s^2}}, \quad \text{where} \quad s = \frac{\omega}{\omega_0} \quad \text{and} \quad D = \frac{2c}{\omega_0}.$$

While still somewhat complicated, this expression for the gain is much easier to understand than (7.15). It allows us to easily see how the gain varies as the quotient $s = \omega/\omega_0$ varies. Because the natural frequency is fixed, let's look at the quantity

$$\omega_0^2 G = \frac{1}{\sqrt{(1-s^2)^2 + D^2s^2}}.$$

The behavior of $\omega_0^2 G$ as a function of s and D is shown in Figure 6. Since the natural frequency ω_0 is fixed, $D = 2c/\omega_0$ is proportional to the damping constant. Notice that for small damping constants, the gain has a significant maximum for a driving frequency that is close to the natural frequency (when $s = \omega/\omega_0$ is near 1). The maximum gets larger as the damping constant decreases. This is another example of resonance. As we see in Figure 6, the resonance increases as the damping constant decreases.

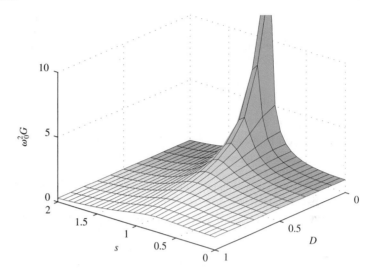

Figure 6. The gain of a forced harmonic oscillator with damping. $s = \omega/\omega_0$ and $D = 2c/\omega_0$.

Notice from Figure 6 that the maximum gain is at a frequency that is slightly smaller than the natural frequency. We will leave it as an exercise to determine the precise frequency at which the maximum gain occurs.

EXERCISES

1. In the narrative (Case 1), the substitution $x_p = a \cos \omega t + b \sin \omega t$ produced

$$x_p = \frac{A}{\omega_0^2 - \omega^2} \cos \omega t$$

as a particular solution of $x'' + \omega_0^2 = A \cos \omega t$, when $\omega \neq \omega_0$.

(a) Use the substitution $x_p = a \cos \omega t$ to produce the same result.

(b) Use the substitution $x_p = ae^{i\omega t}$ to produce the same result.

2. The function $x = \cos 6t - \cos 7t$ has mean frequency $\overline{\omega} = 13/2$ and half difference $\delta = 1/2$. Thus,

$$\cos 6t - \cos 7t = \cos\left(\frac{13}{2} - \frac{1}{2}\right)t - \cos\left(\frac{13}{2} + \frac{1}{2}\right)t,$$

$$= 2 \sin \frac{1}{2}t \sin \frac{13}{2}t.$$

Plot the graph of x, and superimpose the "envelope" of the beats, which is the slow frequency oscillation $y(t) =$

$\pm 2 \sin(1/2)t$. Use different line styles or colors to differentiate the curves.

In Exercises 3–6, plot the given function on an appropriate time interval. Use the technique of Exercise 2 to superimpose the plot of the envelope of the beats in a different line style and/or color.

3. $\cos 10t - \cos 11t$ **4.** $\cos 9t - \cos 10t$

5. $\sin 12t - \sin 11t$ **6.** $\sin 11t - \sin 10t$

7. Let $\omega_0 = 11$. Use a computer to plot the graph of the solution

$$x(t) = \frac{\cos \omega t - \cos \omega_0 t}{\omega_0^2 - \omega^2} \tag{7.20}$$

for $\omega = 9, 10, 10.5, 10.9$, and 10.99 on the time interval $[0, 24]$. Explain how these solutions approach the resonance solution as $\omega \to \omega_0$. *Hint:* Put equation (7.20) in the form $x(t) = A \sin \delta t \sin \overline{\omega} t$, and use this result to justify your conclusion.

8. If the system doesn't start from equilibrium, the beats might not be as pronounced, but they are still there. Use a numerical solver to plot the solution of

$$y'' + 144y = \cos 11t, \quad y(0) = y_0, \quad y'(0) = 0,$$

for each $y_0 = 0, 0.1, 0.2, 0.3, 0.4,$ and 0.5. Plot the solutions on the same time interval $[0, 4\pi]$ and compare the plots. Are the "slow" and "fast" frequencies still present?

9. A 1-kg mass is attached to a spring ($k = 4 \, \text{kg/s}^2$) and the system is allowed to come to rest. The spring-mass system is attached to a machine that supplies an external driving force $f(t) = 4 \cos \omega t$ Newtons. The system is started from equilibrium, the mass having no initial displacement nor velocity. Ignore any damping forces.

   (a) Find the position of the mass as a function of time.

   (b) Place your answer in the form $x(t) = A \sin \delta t \sin \overline{\omega} t$. Select an ω near the natural frequency of the system to demonstrate the "beating" of the system. Sketch a plot that shows the "beats" and include the envelope of the beating motion in your plot (see Exercise 2).

10. An undamped spring-mass system with external driving force is modeled with

$$x'' + 25x = 4 \cos 5t.$$

The parameters of this equation are "tuned" so that the frequency of the driving force equals the natural frequency of the undriven system. Suppose that the mass is displaced one positive unit and released from rest.

   (a) Find the position of the mass as a function of time. What part of the solution guarantees that this solution resonates, rather than showing the "beating" character of previous exercises?

   (b) Sketch the solution found in part (a).

11. An inductor (1 H) and a capacitor (0.25 F) are connected in series with a signal generator that provides an emf $E(t) = 12 \cos \omega t$. Assume that the system is started from equilibrium (no initial charge on the capacitor, no initial current) and ignore any damping effects.

   (a) Find the current in the system as a function of time. Plot a sample solution assuming that the signal generator provides a driving force at am frequency near the resonant frequency.

   (b) Find the current in the system as a function of time, this time assuming that the signal generator provides a driving force at resonant frequency. Plot your solution.

In Exercises 12–15, place the transfer function in the form

$$H(i\omega) = \frac{1}{R} e^{-i\phi}.$$

Use this result to find the steady-state solution of the given equation.

12. $x'' + x' + 4x = 3 \cos 2t$

13. $x'' + 2x' + 2x = 3 \sin 4t$

14. $x'' + 2x' + 4x = 2 \sin 2\pi t$

15. $x'' + 4x' + 8x = 3 \cos 2\pi t$

In Exercises 16–19, find a particular solution to the differential equation using undetermined coefficients as in Examples 5.8 or 5.12. Find and plot the solution of the initial value problem. Superimpose the plots of the transient response and the steady-state solution. Use different line styles or colors to differentiate the curves.

16. $x'' + 5x' + 4x = 2 \sin 2t, \quad x(0) = 1, \quad x'(0) = 0$

17. $x'' + 7x' + 10x = 3 \cos 3t, \quad x(0) = -1, \quad x'(0) = 0$

18. $x'' + 2x' + 2x = \cos 2t, \quad x(0) = 0, \quad x'(0) = 2$

19. $x'' + 4x' + 5x = 3 \sin t, \quad x(0) = 0, \quad x'(0) = -3$

For each equation in Exercises 20–23, calculate the time constant T_c. Plot the transient response over $[0, 4T_c]$, showing that this response dies out in $4T_c$, as advertised in the narrative.

20. The equation in Exercise 16.

21. The equation in Exercise 17.

22. The equation in Exercise 18.

23. The equation in Exercise 19.

An underdamped, driven, spring-mass system is modeled with the equations in Exercises 24–25.

   (i) Find the steady-state solution and place your answer in the form $x_p(t) = A \cos(t - \phi)$. Use a computer to plot this solution.

   (ii) Use a numerical solver to plot solutions of the model with initial conditions $(x(0), x'(0)) = (-2, 0), (-1, 0), (0, 0), (1, 0),$ and $(2, 0)$. Select a time interval that shows all of these solutions approaching the steady-state solution as $t \to \infty$.

24. $x'' + 2x' + 4x = 3 \cos t$

25. $x'' + 2x' + 5x = 2 \cos 3t$

Consider the equations in Exercises 26–27. Find the transfer function, the gain, the phase, and the steady-state response. Plot the driving function $\cos t$ and the steady-state response on the same graph. Explain how one can read the gain and phase from this graph.

26. $x'' + 0.4x' + x = \cos t$ 27. $x'' + 0.4x' + 2x = \cos t$

For the equations in Exercises 28–29, use a numerical solver to plot the solution with the initial conditions $x(0) = 0$ and $x'(0) = 0$. Being mindful of the time constant, select a time interval where the transient response has died out and superimpose the graph of the forcing function on this interval. Estimate the gain and phase from the resulting graph.

28. $x'' + 0.2x' + 1.44x = \cos t$

29. $x'' + 0.4x' + 1.69x = \cos t$

30. In Figure 7, the solution of $x'' + 2cx' + \omega_0^2 = \cos t$ is drawn in blue. The driving force, $\cos t$ is drawn in black. The initial conditions are unimportant, as you have seen that all solutions eventually approach the steady-state response when $c^2 < \omega_0^2$. (Assume that $c^2 < \omega_0^2$.) Estimate the gain and the phase, then use equations (7.15) and (7.14) to calculate the values of c and ω_0^2. Use these computed values of c and ω and your numerical solver to reproduce the image in Figure 7 (use $x(0) = 3$ and $x'(0) = 0$).

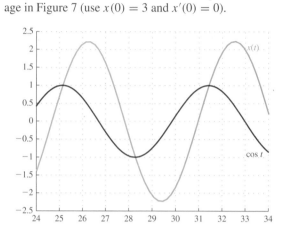

Figure 7. The response to $\cos t$.

Find the gain for each equation in Exercises 31–34 as a function of ω. Plot the graph of the gain versus the driving frequency ω and use the graph to estimate the maximum gain and the frequency at which it occurs.

31. $y'' + 0.01y' + 49y = A \cos \omega t$

32. $y'' + 0.5y' + 4y = A \sin \omega t$

33. $y'' + 0.05y' + 25y = A \sin \omega t$

34. $y'' + 0.25y' + y = A \cos \omega t$

Find the gain of

$$x'' + 2cx' + \omega_0^2 = A \cos \omega t$$

as a function of ω. If $\omega_0^2 > 2c^2$, show that the maximum gain occurs at

$$\omega_{\text{res}} = \sqrt{\omega_0^2 - 2c^2}, \qquad (7.21)$$

known as the **resonant frequency** of the driven oscillator. Use the formula to calculate the resonant frequency and compare with the estimate found from the plot in

35. Exercise 31   **36.** Exercise 32

37. Exercise 33   **38.** Exercise 34

In Exercises 39–42 plot the gain versus the frequency ω and estimate the resonant frequency from the graph. Use equation (7.21) to verify this estimate. Use a numerical solver to plot solutions of the given oscillator for selections of the driving frequency ω both near and far from the resonant frequency. Write a short paragraph describing what you learned from this exercise.

39. $y'' + 0.1y' + 25y = \cos \omega t, \quad y(0) = 0, \quad y'(0) = 0$

40. $y'' + 0.2y' + 49y = \cos \omega t, \quad y(0) = 0, \quad y'(0) = 0$

41. $y'' + 0.2y' + 49y = \cos \omega t, \quad y(0) = 0, \quad y'(0) = 0$

42. $y'' + 0.2y' + 9y = \cos \omega t, \quad y(0) = 0, \quad y'(0) = 0$

43. A driven LRC circuit is modeled by the equation

$$LI' + RI + \frac{1}{C}Q = A \cos \omega t.$$

Assume the underdamped case.

(a) Show that the charge on the capacitor (once transients have died) will achieve a maximum when the driving frequency is

$$\omega = \sqrt{\frac{1}{LC} - \frac{R^2}{2L^2}}.$$

(b) Show that the current will achieve a maximum (again, after transients have disappeared) when the driving frequency is

$$\omega = \frac{1}{\sqrt{LC}}.$$

44. An inductor (0.1 H), a resistor (100 Ω), and a capacitor (10^{-3} F) are connected in series with an emf $E(t) = 100 \sin 120\pi t$ volts. At time $t = 0$, there is no charge on the capacitor nor any current in the system. Find the current in the system as a function of time.

45. A 50-g mass stretches a spring 10 cm. As the system moves through the air, a resistive force is supplied that is proportional to, but opposite the velocity, with magnitude $0.01v$. The system is hooked to a machine that applies a driving force to the mass that is equal to $F(t) = 5 \cos 4.4t$ Newtons. If the system is started from equilibrium (no displacement, no velocity), find the position of the mass as a function of time. *Hint:* Remember that $1000 \, \text{g} = 1 \, \text{kg}$ and $100 \, \text{cm} = 1 \, \text{m}$.

PROJECT 4.8 **Nonlinear Oscillators**

Hooke's law (see equation (1.6) in Section 4.1) says that the magnitude of the restoring force of a spring is proportional to the displacement. While this is often valid, it is not always so, and even when it is, it is only valid for small values of the displacement. Here we will discuss springs that do not obey Hooke's law.

To be precise, we will study oscillators that are modeled by the equation

$$x'' + cx' + kx + lx^3 = f(t). \qquad (8.1)$$

The implication of this equation is that the restoring force of the oscillator has the form

$$R(x) = -kx - lx^3,$$

where x is the displacement from the equilibrium of the oscillator. Once we give up Hooke's law, there is a wide variety of possible restoring forces. This is just one of many possible choices. Notice that Hooke's law is being assumed if $l = 0$. If $l > 0$ the oscillator is said to be *hard* and if $l < 0$ it is *soft*. With $l \neq 0$, equation (8.1) is called ***Duffing's equation***.

If an oscillator is hard or soft, it is said to be nonlinear, simply because in these cases the differential equation (8.1) is nonlinear. The purpose of this project is to discover differences between the behaviors of nonlinear and linear oscillators. The nonlinear equation (8.1) cannot be solved explicitly, so we will have to rely on numerical solutions.

The period of undamped unforced oscillation
For a linear oscillator, modeled by equation (8.1) with $l = 0$, the natural frequency is $\omega_0 = \sqrt{k}$. This is the frequency of undamped, unforced oscillations. The period of the oscillation is $T = 2\pi/\omega_0$. Notice that the period is completely independent of the amplitude of the oscillation. Is the same true for a hard oscillator?

1. Use the parameters $k = 3$ and $l = 1$. Since the motion is undamped, we have $c = 0$, and since it is unforced, $f(t) = 0$. For several values of the amplitude A between 0.1 and 10, compute the solution to (8.1) with initial conditions $x(0) = A$ and $x'(0) = 0$. In each case, discover the period of the oscillation. Make a graph of the period T versus the amplitude A.

2. Write a short paragraph summarizing your results, emphasizing how the motion of the nonlinear oscillator differs from that of the linear oscillator. In particular describe what is happening to the period as the amplitude approaches 0, and as it gets very large.

Some comments are in order about how to perform this numerical experiment. Perhaps the easiest way to determine the period is to estimate it from the graph of the solution. You can use the graph of the displacement to do this, but then you will have to estimate where on the graph of the displacement a maximum occurs. If you use the velocity $v = x'$ instead, then it is necessary to estimate the time of a zero crossing, since $v = x' = 0$ at a maximum. It is much easier to be precise about the location of a zero crossing than that of a maximum. However, $x' = v = 0$ at a minimum as well, so care will have to be taken to ensure that the measurement is done at a maximum.

If the period is determined by graphical estimation, there is a way to decrease the measurement error. Instead of measuring one period, measure the time required for a number of periods, say 3 or 4. Then divide by the number of periods to find the period itself.

Depending on what software you are using, it might be possible to have the computer compute the period. If your computer has a routine that computes the zeros of functions, you might be able to define a function, using your differential equation solver, that returns the velocity $v(t)$ as a function of t for the oscillation. Of course, this function has to be defined in a format that is compatible with the zero finding routine, but this is possible in many mathematical software systems.

Frequency response of a hard oscillator
We examined the frequency response of a linear harmonic oscillator in Section 4.7. We discovered that the total response was the sum of a transient response and a steady-state response. The steady-state response was affected only by the driving force, and in no way depended on the initial conditions. (See equation (7.18) and the discussion that follows it.) We discovered in particular that the amplitude of the steady-state response has the form $G(\omega)A$, where A is the amplitude of the driving force and $G(\omega)$ is the gain given in (7.15). It is the graph of the gain $G(\omega)$ versus ω that we call the frequency response. Here we will see if the frequency response in the nonlinear case has behavior similar to the linear case.

1. Use the parameters $c = 1$, $k = 1$, and $l = 3$ in (8.1). We will use the sinusoidal forcing term $f(t) = 20\cos\omega t$. Experiment by solving the differential equation numerically with a number of values of ω and various initial conditions until you are convinced that the motion settles into a steady-state solution after some time. It is important for what follows that you take notice of about how long it takes to reach steady-state.[4] Submit the plot of one solution that shows the emergence of the steady-state solution.

2. Use the initial conditions $x(0) = 0$ and $v(0) = x'(0) = 0$ for several different driving frequencies ω between 2 and 6. Be sure that $\omega = 4$ is one of those values. In each case find the amplitude of the steady-state solution. Plot the amplitude versus the frequency.

3. Do the same for the initial conditions $x(0) = 6$ and $v(0) = x'(0) = 0$.

4. Plot the frequency response curves from steps 2 and 3 together on one figure, using different line types or colors to distinguish them. The phenomenon that you see illustrated is called ***Duffing's hysteresis***.

5. Write a short paragraph describing what you have discovered. Emphasize how the frequency response for the nonlinear oscillator is different from that of a linear oscillator.

The amplitude of the steady-state solution can be estimated from a plot of the solution, or by examining the numerical data. However, in many computing circumstances, it will be possible to have the computer do this for you. This can be done by removing all values of the computed displacement that might still have a significant transient content and then finding the remaining value with the largest magnitude.

---

[4] Although the steady-state solution is periodic, it will be more complicated than a simple trigonometric function. In general, it is not possible to compute the steady-state solution exactly.

5

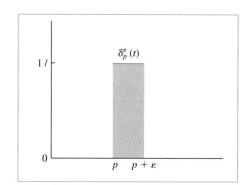

The Laplace Transform

The Laplace transform[1] offers another technique for solving linear differential equations with constant coefficients. It is a particularly useful technique when the right side of a differential equation (the inhomogeneous term) is a discontinuous function. Such functions often arise in applications in mechanics and electric circuits where the forcing term may be an impulse function that takes effect for some short time duration, or it may be a voltage that is turned off and on.

The general idea is as follows. The Laplace transform turns a differential equation into an algebraic equation. After solving the algebraic equation, the solution is found by applying the inverse Laplace transform. This process can be visualized as shown in Figure 1.

---

[1] The Laplace transform is named after Pierre-Simon Laplace. Laplace was born in Normandy in 1749. His parents were comfortably situated, although there was no evidence of academic achievement in his family. Laplace attended a priory school until he was 16 and then entered the University of Caen. It was there that he first discovered his mathematical talents and his love for the subject. At the age of 19 he left the university without a degree to go to Paris and study with Jean d'Alembert, a leading mathematician of his day. D'Alembert recognized Laplace's talents, found him a job as professor of mathematics at the Ecole Militaire, and began to direct his mathematical studies.

Almost immediately Laplace began to produce high-quality research. During a long life, he produced important works in a variety of areas. In what may be his most important work, *Traité du Mécanique Céleste*, he is credited with proving the stability of the solar system. In his *Théorie Analytique des Probabilités* he made the theory of probability an established part of mathematics.

Laplace did not have strong political beliefs. Perhaps this is why he was able to survive the French Revolution and the Napoleonic era without too much personal trouble. During most of these times he was prominent in several areas. He was a professor of mathematics and member of first the Académie des Sciences and then the Institut National des Sciences et des Arts. He also served as head of the Bureau des Longitudes. Under Napoleon, Laplace was a member, then chancellor, of the Senate and received the Legion of Honour in 1805.

After his long and productive life, Laplace died in Paris in 1827.

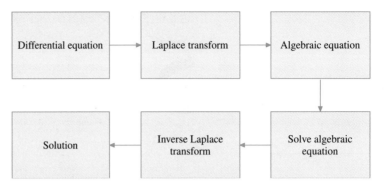

Figure 1. The use of Laplace transforms to solve differential equations.

This chapter starts with the definition of the Laplace transform along with some examples. Section 5.2 describes the various properties of the Laplace transform. The inverse Laplace transform is presented in Section 5.3. In Section 5.4, we solve differential equations using the Laplace transform. We focus on differential equations involving discontinuous forcing terms in Section 5.5. A particular type of discontinuous "function," called the Dirac delta function or impulse function, is discussed in Section 5.6. Finally, the convolution and its interplay with the Laplace transform are presented in Section 5.7.

5.1 The Definition of the Laplace Transform

The Laplace transform operates on a function, yielding another function.

DEFINITION 1.1

Suppose $f(t)$ is a function of t defined for $0 < t < \infty$. The **Laplace transform** of f is the function

$$\mathcal{L}(f)(s) = F(s) = \int_0^\infty f(t)e^{-st}\,dt \quad \text{for } s > 0. \tag{1.2}$$

Notice that the Laplace transform $F(s)$ is a function of the variable s. In (1.2) the integration is with respect to the variable t, and the resulting expression still depends on the variable s and only on s. We will often designate a function with a lowercase letter and then use the uppercase letter to designate the Laplace transform of that function. For example, the Laplace transform of f is given the label F. If we want to emphasize the original function f, we will use the notation $\mathcal{L}(f)$ for the Laplace transform of f. For an explicit function, such as $f(t) = t\cos t$, we will denote the Laplace transform by $\mathcal{L}\{t\cos t\}$.

The integral in the definition of the Laplace transform is an **improper** integral, because the upper limit is ∞. It is defined by taking the limit,

$$F(s) = \int_0^\infty f(t)e^{-st}\,dt = \lim_{T\to\infty}\int_0^T f(t)e^{-st}\,dt. \tag{1.3}$$

Let's compute some examples. We will collect the most important examples in Table 1 on page 204.

Example 1.4   Compute the Laplace transform of $f(t) = e^{at}$, where a is a constant.

Inserting $f(t) = e^{at}$ into the definition of the Laplace transform, we obtain

$$F(s) = \int_0^\infty e^{at} e^{-st}\, dt = \int_0^\infty e^{-(s-a)t}\, dt.$$

When $s = a$, the integrand is 1 and $\int_0^\infty (1)\, dt = \infty$. When $s \neq a$, we have

$$F(s) = \lim_{T \to \infty} \int_0^T e^{-(s-a)t}\, dt$$

$$= \lim_{T \to \infty} \frac{-e^{-(s-a)t}}{s-a} \bigg|_{t=0}^{T}$$

$$= \lim_{T \to \infty} \left(\frac{-e^{-(s-a)T}}{s-a} + \frac{1}{s-a} \right).$$

When $s < a$, the exponent is positive and so the term involving $e^{-(s-a)T}$ approaches infinity as $T \to \infty$. Thus, the Laplace transform of $f(t) = e^{at}$ is undefined for $s \le a$. When $s > a$, the term involving $e^{-(s-a)T}$ converges to zero as $T \to \infty$, and therefore

$$\mathcal{L}\{e^{at}\}(s) = F(s) = \frac{1}{s-a} \qquad \text{for } s > a. \qquad\bullet$$

As Example 1.4 shows, sometimes we have to restrict the domain of the Laplace transform of a function. However, the domain will always be a semi-infinite interval of the form (a, ∞).

When $a = 0$ we have $e^{0t} = 1$ for all t. Hence setting $a = 0$ in Example 1.4, we get

$$\mathcal{L}\{1\}(s) = F(s) = \frac{1}{s} \qquad \text{for } s > 0. \tag{1.5}$$

The method of integration by parts is often useful when computing Laplace transforms. We remind you that it says

$$\int u\, dv = uv - \int v\, du, \tag{1.6}$$

where u and v are functions. If they are functions of t, then $du = u'(t)\, dt$ and $dv = v'(t)\, dt$.

Example 1.7   Compute the Laplace transform of $f(t) = t$.

Using (1.3) with $f(t) = t$, we have

$$F(s) = \int_0^\infty t e^{-st}\, dt.$$

To prepare for the use of integration by parts, we write $e^{-st}\, dt = (-1/s)\, d(e^{-st})$. Then we use (1.6) with $u = t$ and $v = e^{-st}$, getting

$$\int t e^{-st}\, dt = -\frac{1}{s} \int t\, d(e^{-st})$$

$$= -\frac{1}{s} \left(t e^{-st} - \int e^{-st}\, dt \right)$$

$$= \frac{-t e^{-st}}{s} - \frac{e^{-st}}{s^2}.$$

Therefore,

$$F(s) = \lim_{T \to \infty} \left(\frac{-te^{-st}}{s} - \frac{e^{-st}}{s^2} \right)\Big|_{t=0}^{T} = \lim_{T \to \infty} \left(\frac{-Te^{-sT}}{s} - \frac{e^{-sT}}{s^2} + \frac{1}{s^2} \right).$$

As T gets large, the exponential term, e^{-sT}, tends to 0 much faster than the growth of the term T. Thus

$$\lim_{T \to \infty} \frac{-Te^{-sT}}{s} = 0.$$

(This can be verified using l'Hôpital's rule.) Therefore, the first two terms converge to zero as $T \to \infty$. Only the third term remains in the limit, so the Laplace transform of $f(t) = t$ is

$$\mathcal{L}\{t\}(s) = F(s) = \frac{1}{s^2}.$$

⬤

The process used in Example 1.7 can be used to compute the Laplace transform of any power t^n. Repeated integration by parts establishes that

$$\mathcal{L}\{t^n\}(s) = \frac{n!}{s^{n+1}}. \tag{1.8}$$

Example 1.9   Find the Laplace transform of $f(t) = \sin at$.

Using the definition of the Laplace transform with $f(t) = \sin at$, we have

$$F(s) = \int_0^\infty e^{-st} \sin at \, dt.$$

The integral can be computed using two integrations by parts. First we write

$$\int e^{-st} \sin at \, dt = -\frac{1}{a} \int e^{-st} \, d(\cos at).$$

Then we use integration by parts with $u = e^{-st}$ and $v = \cos at$, getting

$$\int e^{-st} \sin at \, dt = -\frac{1}{a} \left(e^{-st} \cos at - \int \cos at \, d(e^{-st}) \right)$$

$$= -\frac{1}{a} e^{-st} \cos at - \frac{s}{a} \int e^{-st} \cos at \, dt.$$

We will compute the integral on the right by parts. First we write

$$\int e^{-st} \sin at \, dt = -\frac{1}{a} e^{-st} \cos at - \frac{s}{a^2} \int e^{-st} \, d(\sin at).$$

Then we integrate by parts to get

$$\int e^{-st} \sin at \, dt = -\frac{e^{-st} \cos at}{a} - \frac{s}{a^2} \left(e^{-st} \sin at + s \int e^{-st} \sin at \, dt \right)$$

$$= -\frac{e^{-st} \cos at}{a} - \frac{se^{-st} \sin at}{a^2} - \frac{s^2}{a^2} \int e^{-st} \sin at \, dt.$$

The integral on the right is of the same form as the one we started with. Transferring the last term on the right to the left, we obtain

$$\frac{s^2 + a^2}{a^2} \int e^{-st} \sin at \, dt = -\frac{e^{-st} \cos at}{a} - \frac{se^{-st} \sin at}{a^2},$$

or

$$\int e^{-st} \sin at \, dt = -\frac{ae^{-st} \cos at}{s^2 + a^2} - \frac{se^{-st} \sin at}{s^2 + a^2}.$$

Inserting the limits 0 and T, we obtain

$$F(s) = \lim_{T \to \infty} \int_0^T e^{-st} \sin at \, dt$$

$$= \lim_{T \to \infty} \left(-\frac{ae^{-sT} \cos aT}{s^2 + a^2} - \frac{se^{-sT} \sin aT}{s^2 + a^2} \right) + \frac{a}{s^2 + a^2}.$$

As $T \to \infty$, the terms involving e^{-sT} converge to zero. Therefore,

$$\mathcal{L}\{\sin at\}(s) = F(s) = \frac{a}{s^2 + a^2}. \qquad \bullet$$

A computation similar to that in Example 1.9 shows that

$$\mathcal{L}\{\cos at\}(s) = \frac{s}{s^2 + a^2}. \tag{1.10}$$

Piecewise continuous functions

We now present some examples of the Laplace transform of discontinuous functions.

Example 1.11   Compute the Laplace transform of the step function

$$g(t) = \begin{cases} 1, & \text{for } 0 \le t < 1, \\ 0, & \text{for } t \ge 1. \end{cases}$$

The graph of g is shown in Figure 1. Since $g(t) = 0$ for $t > 1$, the limits in the integral in the Laplace transform become 0 and 1. Hence

$$G(s) = \mathcal{L}(g)(s) = \int_0^1 1 \, e^{-st} \, dt = \left. \frac{-e^{-st}}{s} \right|_{t=0}^1 = \frac{-e^{-s}}{s} + \frac{1}{s}. \qquad \bullet$$

Example 1.12   Compute the Laplace transform of the step function

$$\tilde{g}(t) = \begin{cases} 1, & \text{for } 0 \le t \le 1, \\ 0, & \text{for } t > 1. \end{cases}$$

The graph of $\tilde{g}$ is shown in Figure 2. Notice that the function g in Example 1.11 and $\tilde{g}$ are almost identical. In fact, they differ only at the point $t = 1$. As a result, the same computation as in Example 1.11 computes the Laplace transform of $\tilde{g}$, showing that g and $\tilde{g}$ have the same Laplace transform.

This illustrates the important fact that the value of a function at a finite number of points does not affect the Laplace transform. In fact, the values can be altered at an infinite set of points as long as there are only finitely many of them in any finite interval $(0, T)$. $\qquad \bullet$

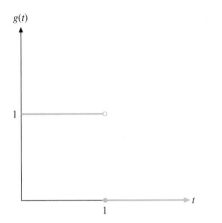

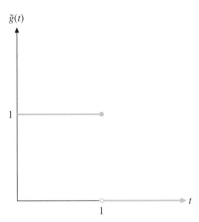

Figure 1. The step function g in Example 1.11.

Figure 2. The step function $\tilde{g}$ in Example 1.12.

We will say that a function f defined on the interval $(0, \infty)$ is *piecewise continuous* if it has only finitely many points of discontinuity over any finite interval and at every point of discontinuity the limit of f exists from both the left and the right.

Consider the function g in Example 1.11. Its only point of discontinuity is $t = 1$. For $0 \leq t < 1$, $g(t) = 1$. Hence $\lim_{t \to 1^-} g(t) = 1$. On the interval $1 \leq t < \infty$, $g(t) = 0$, so $\lim_{t \to 1^+} g(t) = 0$. Consequently, g is piecewise continuous. Of course the same is true of the function $\tilde{g}$ in Example 1.12. Since we will be interested in the Laplace transforms of piecewise continuous functions, Examples 1.10 and 1.11 show that the values of such a function at its points of discontinuity are of no importance.

The restriction that the one-sided limits exist means that a piecewise continuous function is allowed to have only jump discontinuities.

We will say that a function f is *piecewise differentiable* if it is continuous and if its derivative is piecewise continuous. An example is shown in Figure 3. Its piecewise continuous derivative is shown in Figure 4. Notice that the graph of the function in Figure 3 has corners. At these corners the function is not differentiable. Thus its derivative, shown in Figure 4, is not even defined at these points. This fact will not be material in what follows, because the value of a function at these points does not affect the Laplace transform.

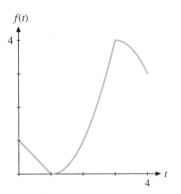

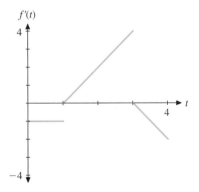

Figure 3. A piecewise differentiable function.

Figure 4. The derivative of the function in Figure 3 is piecewise continuous.

A piecewise continuous function, or even a piecewise differentiable function, is usually given by different formulas in different intervals. This was true for the function g in Example 1.11. It is also true of the function depicted in Figure 3. The Laplace transform of a piecewise defined function is computed by breaking apart the integral into the separate intervals according to the definition of the original function. Here is another example.

Example 1.13   Compute the Laplace transform of $f(t)$, where $f(t)$ is given by

$$f(t) = \begin{cases} t, & 0 \le t \le 1, \\ 1, & 1 < t < \infty. \end{cases}$$

The graph of f is shown in Figure 5.

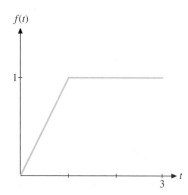

Figure 5. The piecewise differentiable function in Example 1.13.

We break up the integral of the Laplace transform into the two intervals $0 \le t \le 1$ and $1 < t < \infty$ appearing in the definition of f.

$$F(s) = \int_0^\infty e^{-st} f(t)\, dt$$

$$= \int_0^1 te^{-st}\, dt + \int_1^\infty (1)e^{-st}\, dt \qquad (1.14)$$

The first integral in (1.14) can be integrated by parts to obtain

$$\int_0^1 te^{-st}\, dt = \left.\frac{-te^{-st}}{s}\right|_{t=0}^1 + \int_0^1 \frac{e^{-st}}{s}\, dt$$

$$= \frac{-e^{-s}}{s} - \left(\left.\frac{e^{-st}}{s^2}\right|_{t=0}^1\right)$$

$$= \frac{-e^{-s}}{s} - \left(\frac{e^{-s}}{s^2} - \frac{1}{s^2}\right).$$

The second integral in (1.14) is

$$\int_1^\infty (1)e^{-st}\, dt = \lim_{T\to\infty} \left.\frac{-e^{-st}}{s}\right|_{t=1}^T = \frac{e^{-s}}{s}.$$

Combining the two integrals, we obtain

$$F(s) = \int_0^1 te^{-st}\, dt + \int_1^\infty (1)e^{-st}\, dt$$

$$= \left(\frac{-e^{-s}}{s} - \frac{e^{-s}}{s^2} + \frac{1}{s^2}\right) + \left(\frac{e^{-s}}{s}\right)$$

$$= \frac{1}{s^2} - \frac{e^{-s}}{s^2}.$$

We are almost ready to state a theorem about when a function has a Laplace transform. We need one more definition.

DEFINITION 1.15 | A function $f(t)$ is of **exponential order** if there are constants C and a such that
$$|f(t)| \le Ce^{at} \quad \text{for all } t > 0.$$

THEOREM 1.16 Suppose f is a piecewise continuous function defined on $[0, \infty)$, which is of exponential order. Then the Laplace transform $\mathcal{L}(f)(s)$ exists for large values of s. Specifically, if $|f(t)| \le Ce^{at}$, then $\mathcal{L}(f)(s)$ exists at least for $s > a$. ∎

Theorem 1.16 says that the Laplace transform exists for a wide variety of functions. However, there are functions for which the Laplace transform does not exist. For example, the function $f(t) = e^{t^2}$ is not of exponential growth, and it can be shown that it does not have a Laplace transform.

EXERCISES

In Exercises 1–10, use Definition 1.1 of the Laplace transform to find the Laplace transform of each of the following functions defined for $t > 0$.

1. $f(t) = 3$

2. $f(t) = -2$

3. $f(t) = e^{-2t}$

4. $f(t) = e^{3t}$

5. $f(t) = \cos 2t$

6. $f(t) = \sin 3t$

7. $f(t) = te^{2t}$

8. $f(t) = te^{-3t}$

9. $f(t) = e^{2t} \cos 3t$

10. $f(t) = e^{-3t} \sin 2t$

11. In the narrative, we used Definition 1.1 to show that the Laplace transform of the function defined by $f(t) = t$ is $F(s) = 1/s^2$.

(a) Use Definition 1.1 to show that the Laplace transform of the function defined by $g(t) = t^2$ is $G(s) = 2!/s^3$.

(b) Use Definition 1.1 to show that the Laplace transform of the function defined by $h(t) = t^3$ is $H(s) = 3!/s^4$.

(c) Use Definition 1.1 and mathematical induction to prove that the Laplace transform of the function defined by $f(t) = t^n$ is
$$F(s) = \frac{n!}{t^{n+1}}.$$

12. Use Definition 1.1 to show that the Laplace transform of the function defined by $f(t) = \cos \omega t$ is
$$F(s) = \frac{s}{s^2 + \omega^2}.$$

13. Use Definition 1.1 to show that the Laplace transform of the function defined by $f(t) = e^{at} \cos \omega t$ is
$$F(s) = \frac{s - a}{(s - a)^2 + \omega^2}.$$

14. Use Definition 1.1 to show that the Laplace transform of the function defined by $f(t) = e^{at} \sin \omega t$ is
$$F(s) = \frac{\omega}{(s - a)^2 + \omega^2}.$$

Each time an engineer calculates the Laplace transform of a function, the result is placed in a table of transforms for later use. For example, in Example 1.4 we generated the following transform pair:
$$f(t) = e^{at} \quad \Longleftrightarrow \quad F(s) = \frac{1}{s - a}, \text{ provided } s > a.$$

Such a table of transforms can be used in a manner similar to a table of integrals. For instance, the above transform pair easily allows the calculation of the Laplace transform of $f(t) = e^{-3t}$ as
$$F(s) = \frac{1}{s + 3}, \text{ provided } s > -3.$$

Use the Laplace transforms in Table 1 on page 204 to generate the Laplace transform of the functions in Exercises 15–24. Check each result with the result generated earlier using Definition 1.1 of the Laplace transform.

15. The function in Exercise 1

16. The function in Exercise 2

17. The function in Exercise 3

18. The function in Exercise 4

19. The function in Exercise 5

20. The function in Exercise 6

21. The function in Exercise 7

22. The function in Exercise 8

23. The function in Exercise 9

24. The function in Exercise 10

In Exercises 25–28, find the Laplace transform of the given piecewise defined function.

25. $f(t) = \begin{cases} 0, & 0 \le t < 2 \\ 2, & t \ge 2 \end{cases}$

26. $f(t) = \begin{cases} 0, & 0 \le t < 4 \\ 5, & t \ge 4 \end{cases}$

27. $f(t) = \begin{cases} t, & 0 \le t < 2 \\ 2, & t \ge 2 \end{cases}$

28. $f(t) = \begin{cases} t, & 0 \le t < 3 \\ 3, & t \ge 3 \end{cases}$

29. Engineers frequently use the **Heaviside function**, defined by

$$H(t) = \begin{cases} 0, & \text{if } t < 0, \\ 1, & \text{if } t \geq 0, \end{cases}$$

to emulate turning on a switch at a certain instant in time. Sketch the graph of

$$y(t) = H(t - 3)e^{0.2t}.$$

Calculate its Laplace transform.

30. This exercise discusses some ideas that make Theorem 1.16 plausible. Suppose that the function f is continuous on the interval $[0, \infty)$ and of exponential order; that is, there exists constants C and a such that $|f(t)| \leq Ce^{at}$ for all $t > 0$.

    (a) Show that

    $$\left| f(t)e^{-st} \right| \leq Ce^{-(s-a)t} \quad \text{for all } t > 0.$$

    (b) Since it is true that

    $$\left| \int_0^\infty f(t)e^{-st} \, dt \right| \leq \int_0^\infty \left| f(t)e^{-st} \right| \, dt$$
    $$\leq \int_0^\infty Ce^{-(s-a)t} \, dt,$$

show that the integral $\int_0^\infty f(t)e^{-st} \, dt$ is finite (convergent) for all $s > a$.

31. Suppose that the function f is continuous on the interval $[0, \infty)$ and of exponential order; that is, there exist constants C and a such that $|f(t)| \leq Ce^{at}$ for all $t > 0$.

    (a) Show that

    $$|F(s)| = \left| \int_0^\infty f(t)e^{-st} \, dt \right| \leq \frac{C}{s - a},$$

    provided that $s > a$. *Hint*: See the inequality in Exercise 30.

    (b) Use the result in part (a) to show that $\lim_{s \to \infty} F(s) = 0$. Why can't

    $$F(s) = \frac{s + 4}{s - 2}$$

    be the Laplace transform of any function?

32. It is claimed in the narrative that the function $f(t) = e^{t^2}$ is not of exponential order. Prove this claim by assuming that there exist constants C and a such that $|e^{t^2}| \leq Ce^{at}$ for all $t > 0$ and deriving a contradiction. *Hint*: $e^{t^2} \leq |e^{t^2}|$.

5.2 Basic Properties of the Laplace Transform

This section will discuss the most important properties of the Laplace transform. We will state each property in a proposition.

The Laplace transform of derivatives

The most important property for our purposes is the relationship between the Laplace transform and the derivative. As we will see, the next proposition is the key tool when using the Laplace transform to solve differential equations.

PROPOSITION 2.1 Suppose y is a piecewise differentiable function of exponential order. Suppose also that y' is of exponential order. Then for large values of s,

$$\mathcal{L}(y')(s) = s\,\mathcal{L}(y)(s) - y(0) = sY(s) - y(0), \tag{2.2}$$

where $Y(s)$ is the Laplace transform of y.

Proof Equation (2.2) is derived as follows. Since y' is of exponential order, it has a Laplace transform

$$\mathcal{L}(y')(s) = \int_0^\infty y'(t)e^{-st} \, dt = \lim_{T \to \infty} \int_0^T y'(t)e^{-st} \, dt.$$

This integral is begging us to use integration by parts. We get

$$\mathcal{L}(y')(s) = \lim_{T \to \infty} \left[e^{-st} y(t) \Big|_{t=0}^T + s \int_0^T y(t)e^{-st} \, dt \right]$$
$$= \lim_{T \to \infty} e^{-sT} y(T) - y(0) + s\,\mathcal{L}(y)(s).$$

Since y is of exponential order, there are constants C and a such that $|y(t)| \le Ce^{at}$. Consequently, the first term can be estimated by

$$e^{-sT}|y(T)| \le Ce^{-(s-a)T},$$

which converges to 0 for $s > a$ as $T \to \infty$. Therefore,

$$\mathcal{L}(y')(s) = s\,\mathcal{L}(y)(s) - y(0),$$

as claimed. ●

The effect of Proposition 2.1 is that the relatively difficult operation of finding the derivative of a function is transformed under the Laplace transform to the very easy operation of multiplication by s (ignoring $y(0)$ for the moment).

Example 2.3 Verify that Proposition 2.1 holds for the function $y(t) = t$ and its derivative $y'(t) = 1$.

As computed in Example 1.7, the Laplace transform of $y(t) = t$ is $1/s^2$. From Proposition 2.1,

$$\mathcal{L}(1)(s) = \mathcal{L}(y')(s) = s\,\mathcal{L}(y)(s) - y(0) = s\left(\frac{1}{s^2}\right) - 0 = \frac{1}{s}.$$

This agrees with the calculation $\mathcal{L}(1)(s) = 1/s$ computed in Example 1.5. ●

The Laplace transforms of higher derivatives can be computed just as easily. They follow a pattern similar to that in Proposition 2.1.

PROPOSITION 2.4 Suppose that y and y' are piecewise differentiable and continuous and that y'' is piecewise continuous. Suppose that all three are of exponential order. Then

$$\begin{aligned}
\mathcal{L}(y'')(s) &= s^2\,\mathcal{L}(y)(s) - sy(0) - y'(0) \\
&= s^2 Y(s) - sy(0) - y'(0),
\end{aligned} \tag{2.5}$$

where $Y(s)$ is the Laplace transform of y. More generally, if y and all of its derivatives up to order $k-1$ are piecewise differentiable and continuous, and $y^{(k)}$ is piecewise continuous, and all of them have exponential order, then

$$\begin{aligned}
\mathcal{L}(y^{(k)})(s) &= s^k\,\mathcal{L}(y)(s) - s^{k-1}y(0) - \cdots - sy^{(k-2)}(0) - y^{(k-1)}(0) \\
&= s^k Y(s) - s^{k-1}y(0) - \cdots - sy^{(k-2)}(0) - y^{(k-1)}(0).
\end{aligned} \tag{2.6}$$

Proof To derive these equations, Proposition 2.1 is applied repeatedly. For example,

$$\begin{aligned}
\mathcal{L}(y'')(s) &= s\,\mathcal{L}(y')(s) - y'(0) \\
&= s\,(s\,\mathcal{L}(y)(s) - y(0)) - y'(0) \\
&= s^2\,\mathcal{L}(y)(s) - sy(0) - y'(0).
\end{aligned}$$

The formula for higher derivatives is handled similarly. ●

The Laplace transform is linear

This basic property is described in the following proposition. It greatly simplifies the computation of Laplace transforms.

PROPOSITION 2.7 Suppose f and g are piecewise continuous functions of exponential order, and α and β are constants. Then

$$L\{\alpha f(t) + \beta g(t)\}(s) = \alpha \, L\{f(t)\}(s) + \beta \, L\{g(t)\}(s).$$

The point is that the Laplace transform of a linear combination of functions can be computed by taking the Laplace transform of each term separately and then adding up the result.

Proof The proposition follows because the integral is a linear operator. Hence,

$$L\{\alpha f(t) + \beta g(t)\}(s) = \int_0^\infty (\alpha f(t) + \beta g(t)) \, e^{-st} \, dt$$

$$= \alpha \int_0^\infty f(t) e^{-st} \, dt + \beta \int_0^\infty g(t) e^{-st} \, dt$$

$$= \alpha \, L\{f(t)\}(s) + \beta \, L\{g(t)\}(s).$$ ●

Example 2.8 Find the Laplace transform of

$$f(t) = 3 \sin 2t - 4t + 5e^{3t}.$$

Using Proposition 2.7 together with Examples 1.7, 1.9, and 1.4:

$$L\{3 \sin 2t - 4t + 5e^{3t}\}(s) = 3 \, L\{\sin 2t\}(s) - 4 \, L\{t\}(s) + 5 \, L\{e^{3t}\}(s)$$

$$= 3 \left(\frac{2}{4 + s^2} \right) - 4 \left(\frac{1}{s^2} \right) + 5 \left(\frac{1}{s - 3} \right).$$ ●

Example 2.9 Use Propositions 2.4 and 2.7 to transform the initial value problem

$$y'' - y = e^{2t} \quad \text{with} \quad y(0) = 0, \quad \text{and} \quad y'(0) = 1$$

into an algebraic equation involving $L(y)$. Solve the resulting equation for the Laplace transform of y.

Using first Proposition 2.7, and then Proposition 2.4, the Laplace transform of the left-hand side is

$$L(y'' - y)(s) = L(y'')(s) - L(y)(s)$$

$$= s^2 \, L(y)(s) - s y(0) - y'(0) - L(y)(s). \tag{2.10}$$

For the right-hand side we have

$$L\{e^{2t}\}(s) = \frac{1}{s - 2}. \tag{2.11}$$

Let's set $Y(s) = L(y)(s)$. Then substituting the initial conditions $y(0) = 0$, $y'(0) = 1$ into (2.10), and equating (2.10) and (2.11), we obtain

$$s^2 Y(s) - 1 - Y(s) = \frac{1}{s - 2}.$$

Since no derivatives appear, this is an algebraic equation for $Y(s)$. Solving for $Y(s)$, we get

$$Y(s) = \frac{1}{s^2 - 1}\left[\frac{1}{s - 2} + 1\right] = \frac{1}{(s + 1)(s - 2)}. \quad \bullet$$

As Example 2.9 shows, solving for $Y(s) = \mathcal{L}(y)(s)$, the Laplace transform of the solution to the initial value problem, is a straightforward algebraic process. We could now write down the solution to the initial value problem if we knew the function $y(t)$ that had $Y(s)$ as its Laplace transform. We will say that y is the *inverse Laplace transform* of Y. We will address the issue of computing the inverse Laplace transform in the next section. First we will present two useful properties of the Laplace transform.

The Laplace transform of the product of an exponential with a function

The result is a translation in the Laplace transform.

PROPOSITION 2.12   Suppose f is a piecewise continuous function of exponential order. Let $F(s)$ be the Laplace transform of f, and let c be any constant. Then

$$\mathcal{L}\{e^{ct} f(t)\}(s) = F(s - c).$$

Proof   To prove this result we unravel the left side.

$$\mathcal{L}\{e^{ct} f(t)\}(s) = \int_0^\infty e^{ct} f(t) e^{-st}\, dt = \int_0^\infty f(t) e^{-(s-c)t}\, dt$$

The expression on the right is the definition of the Laplace transform of f with s replaced by $s - c$, or $F(s - c)$. $\quad \bullet$

Proposition 2.12 can be summed up by saying that multiplication by an exponential factor in the time (t) domain results in a translation in the s domain.

Example 2.13   Compute the Laplace transform of the function

$$g(t) = e^{2t} \sin 3t.$$

The Laplace transform of the function $f(t) = \sin 3t$ is $F(s) = 3/(s^2 + 9)$ (see Example 1.9 or Table 1 on page 204). Using Proposition 2.12 with $c = 2$, the Laplace transform of g is

$$\mathcal{L}\{e^{2t} \sin 3t\}(s) = F(s - 2) = \frac{3}{(s - 2)^2 + 9} = \frac{3}{s^2 - 4s + 13}. \quad \bullet$$

The derivative of a Laplace transform

The following proposition describes the relationship between the derivative of the Laplace transform of a function and the Laplace transform of the function itself.

PROPOSITION 2.14 Suppose f is a piecewise continuous function of exponential order, and let $F(s)$ be its Laplace transform. Then

$$\mathcal{L}\{tf(t)\}(s) = -F'(s). \tag{2.15}$$

More generally, if n is any positive integer, then

$$\mathcal{L}\{t^n f(t)\}(s) = (-1)^n F^{(n)}(s). \tag{2.16}$$

Proof To prove this theorem, we start with the definition of the Laplace transform,

$$F(s) = \int_0^\infty f(t)e^{-st}\,dt.$$

We then differentiate both sides with respect to s. Taking this derivative under the integral sign on the right and applying it to the term e^{-st} gives

$$F'(s) = \frac{d}{ds}\int_0^\infty f(t)e^{-st}\,dt$$

$$= \int_0^\infty f(t)\frac{\partial}{\partial s}\{e^{-st}\}\,dt$$

$$= -\int_0^\infty tf(t)e^{-st}\,dt$$

$$= -\mathcal{L}\{tf(t)\}(s),$$

which is equivalent to equation (2.15). If we repeat this procedure, we see that each time an s-derivative is computed, a factor of $-t$ appears as the result of differentiating e^{-st} with respect to s. This establishes equation (2.16). ●

Example 2.17 Compute the Laplace transform of

$$t^2 e^{3t}.$$

Let $f(t) = e^{3t}$. From Example 1.4, its Laplace transform is $F(s) = 1/(s-3)$. Using Proposition 2.14 with $n = 2$, we obtain

$$\mathcal{L}\{t^2 e^{3t}\}(s) = (-1)^2 F''(s) = \frac{2}{(s-3)^3}.$$ ●

EXERCISES

Use the linearity of the Laplace transform (Proposition 2.7) and Table 1 of Laplace transforms on page 204 to find the Laplace transform of the functions defined on the time domain in Exercises 1–6.

1. $y(t) = 3t^2$

2. $y(t) = 5e^{3t}$

3. $y(t) = t^2 + 4t + 5$

4. $y(t) = 3 - 5t - 11t^3$

5. $y(t) = -2\cos t + 4\sin 3t$

6. $y(t) = 2\sin 3t + 3\cos 5t$

7. Use the linearity of the Laplace transform (Proposition 2.7) and Table 1 of Laplace transforms on page 204 to show that

$$\mathcal{L}\{\cosh \omega t\}(s) = \frac{s}{s^2 - \omega^2}$$

and

$$\mathcal{L}\{\sinh \omega t\}(s) = \frac{\omega}{s^2 - \omega^2}.$$

Exercises 8–13 are designed to test the validity of Proposition 2.1. In each exercise,
 (i) compute $\mathcal{L}(y')(s)$ for the given function, and
 (ii) compute $s\,\mathcal{L}(y)(s) - y(0)$ for the given function. Compare this result to that found in part (i) to verify that $\mathcal{L}(y')(s) = s\,\mathcal{L}(y)(s) - y(0)$.

8. $y(t) = t^2$ **9.** $y(t) = t^3$

10. $y(t) = e^{2t}$ **11.** $y(t) = e^{-3t}$

12. $y(t) = \cos 3t$ **13.** $y(t) = \sin 5t$

In a manner similar to that proposed in Exercises 8–13, verify the result of Proposition 2.4 for the functions defined in Exercises 14–17.

14. $y(t) = t^4$ **15.** $y(t) = e^{-2t}$

16. $y(t) = \sin 2t$ **17.** $y(t) = t^2 + 3t + 5$

In Exercises 18–25, use Propositions 2.1, 2.4, and 2.7 to transform the given initial value problem into an algebraic equation involving $\mathcal{L}(y)$. Solve the resulting equation for the Laplace transform of y.

18. $y' + 3y = t^2$, $y(0) = -1$

19. $y' - 5y = e^{-2t}$, $y(0) = 1$

20. $y' + 5y = t^2 + 2t + 3$, $y(0) = 0$

21. $y' - 4y = \cos 2t$, $y(0) = -2$

22. $y'' + y = \sin 4t$, $y(0) = 0$, $y'(0) = 1$

23. $y'' + 2y' + 2y = \cos 2t$, $y(0) = 1$, $y'(0) = 0$

24. $y'' + y' + 2y = \cos 2t + \sin 3t$, $y(0) = -1$, $y'(0) = 1$

25. $y'' + 3y' + 5y = t + e^{-t}$, $y(0) = -1$, $y'(0) = 0$

In Exercises 26–29, use Proposition 2.12 to find the Laplace transform of the given function.

26. $y(t) = e^{-t} \sin 3t$ **27.** $y(t) = e^{2t} \cos 2t$

28. $y(t) = e^{-2t}(2t + 3)$ **29.** $y(t) = e^{-t}(t^2 + 3t + 4)$

In Exercises 30–33, use Proposition 2.14 to find the Laplace transform of the given function.

30. $y(t) = t \sin 3t$ **31.** $y(t) = te^{-t}$

32. $y(t) = t^2 \cos 2t$ **33.** $y(t) = t^2 e^{2t}$

In Exercises 34–41, use the propositions in Section 2 to transform the given initial value problem into an algebraic equation involving $\mathcal{L}(y)$. Solve the resulting equation for the Laplace transform of y.

34. $y' + 2y = t \sin t$, $y(0) = 1$

35. $y' - y = t^2 e^{-2t}$, $y(0) = 0$

36. $y' + y = e^{-t} \sin 3t$, $y(0) = 0$

37. $y' - 2y = e^{2t} \cos t$, $y(0) = -2$

38. $y'' + 4y = t^2 \sin 4t$, $y(0) = 0$, $y'(0) = -1$

39. $y'' + y' + 2y = e^{-t} \cos 2t$, $y(0) = 1$, $y'(0) = -1$

40. $y'' + 2y' + 5y = t^2 e^{-t}$, $y(0) = 1$, $y'(0) = -2$

41. $y'' + 5y = 3e^{-t} \cos 4t$, $y(0) = -1$, $y'(0) = 2$

42. Here is an interesting way to compute the Laplace transform of $\cos \omega t$.

 (a) Using only Definition 1.1, show that $\mathcal{L}\{\sin t\}(s) = 1/(s^2 + 1)$.

 (b) Suppose that $f(t)$ has Laplace transform $F(s)$. Show that
$$\mathcal{L}\{f(at)\}(s) = \frac{1}{a} F\left(\frac{s}{a}\right).$$
 Use this property and the result found in part (a) to show that
$$\mathcal{L}\{\sin \omega t\}(s) = \frac{\omega}{s^2 + \omega^2}.$$

 (c) If $f(t) = \sin \omega t$, then $f'(t) = \omega \cos \omega t$. Use Proposition 2.1 to show that
$$\mathcal{L}\{\omega \cos \omega t\}(s) = \frac{s\omega}{s^2 + \omega^2},$$
 thus ensuring
$$\mathcal{L}\{\cos \omega t\}(s) = \frac{s}{s^2 + \omega^2}.$$

43. The **gamma function** is defined by
$$\Gamma(\alpha) = \int_0^\infty e^{-t} t^{\alpha-1}\, dt, \quad (\alpha > 0).$$

 (a) Prove that $\Gamma(1) = 1$.

 (b) Prove that $\Gamma(\alpha+1) = \alpha \Gamma(\alpha)$. In fact, if n is a positive integer, show that $\Gamma(n + 1) = n!$.

 (c) Show that
$$\mathcal{L}\{t^\alpha\}(s) = \frac{\Gamma(\alpha + 1)}{s^{\alpha+1}}.$$
 Indeed, if n is a positive integer, use this result to show that $\mathcal{L}\{t^n\}(s) = n!/s^{n+1}$.

44. Suppose that f is a continuous function for $t \geq 0$ and is of exponential order.

 (a) If $f(t) \leftrightarrow F(s)$ is a transform pair, prove that
$$\mathcal{L}\left\{\int_0^t f(\tau)\, d\tau\right\}(s) = \frac{F(s)}{s}.$$

 Hint: Let $g(t) = \int_0^t f(\tau)\, d\tau$. Then note that $g'(t) = f(t)$ and use Proposition 2.1 to compute $\mathcal{L}\{g'(t)\}(s)$.

 (b) Use the technique suggested in part (a) to find
$$\mathcal{L}^{-1}\left\{\frac{1}{s(s^2 + 1)}\right\}.$$

5.3 The Inverse Laplace Transform

As seen in Example 2.9, the Laplace transform turns a differential equation into an algebraic equation, which allows us to find the Laplace transform of the solution. We could then find the solution to the original differential equation if we knew how to find the function with the given Laplace transform. This is the problem of finding the ***inverse Laplace transform***.

In order to make sense out of the inverse Laplace transform, we need to know that essentially[1] only one function will have a given function as its Laplace transform. We will depend on the following result.

THEOREM 3.1   Suppose that f and g are continuous functions and that $\mathcal{L}(f)(s) = \mathcal{L}(g)(s)$ for $s > a$. Then $f(t) = g(t)$ for all $t > 0$.   ∎

The proof of this result goes beyond the scope of this book. Theorem 3.1 says that a function is uniquely determined by its Laplace transform. This result is the key to what follows. It makes possible the next definition.

DEFINITION 3.2   If f is a continuous function of exponential order and $\mathcal{L}(f)(s) = F(s)$, then we call f the ***inverse Laplace transform*** of F, and write

$$f = \mathcal{L}^{-1}(F).$$

This procedure can be viewed schematically in Figure 1. The result can be stated quite easily algebraically. It says

$$F = \mathcal{L}(f) \quad \Longleftrightarrow \quad f = \mathcal{L}^{-1}(F). \tag{3.3}$$

Although there is a general integral formula for the inverse Laplace transform, it requires a contour integral in the complex plane. We will be able to get by using the linearity of the inverse Laplace transform, and the knowledge we continue to accumulate about the transforms of specific functions. The linearity of the inverse Laplace transform is our next result.

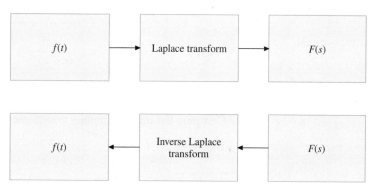

Figure 1. The Laplace transform and its inverse.

PROPOSITION 3.4   The inverse Laplace transform is linear. Suppose that $\mathcal{L}^{-1}(F) = f$ and $\mathcal{L}^{-1}(G) = g$. Then for any constants a and b,

$$\mathcal{L}^{-1}(aF + bG) = a\,\mathcal{L}^{-1}(F) + b\,\mathcal{L}^{-1}(G) = af + bg.$$

---

[1] We do not worry too much about functions as similar as those in Examples 1.11 and 1.12. To avoid examples like this we limit ourselves to continuous functions in Theorem 3.1.

Proof   By the linearity of the Laplace transform, we have

$$\mathcal{L}(af + bg) = a\,\mathcal{L}(f) + b\,\mathcal{L}(g) = aF + bG.$$

The proposition follows using (3.3).   ●

Table 1 summarizes the Laplace transforms that we have computed thus far. Because of Theorem 3.1, this table of Laplace transforms can be used to compute inverse Laplace transforms. The inverse Laplace transform of any function of s in the third column is the corresponding function of t in the second column. As we will see, the functions in the third column are the building blocks of which many Laplace transforms are constructed using linearity. We can use this information together with the linearity of the inverse Laplace transform and the properties of the Laplace transform to compute many inverse Laplace transforms.

Table I   A small table of Laplace transforms

No.	$f(t)$	$\mathcal{L}(f)(s) = F(s)$	REFERENCE
1.	1	$\dfrac{1}{s}, \quad s > 0$	Equation (1.5)
2.	t^n	$\dfrac{n!}{s^{n+1}}, \quad s > 0$	Equation (1.8)
3.	$\sin at$	$\dfrac{a}{s^2 + a^2}, \quad s > 0$	Example 1.9
4.	$\cos at$	$\dfrac{s}{s^2 + a^2}, \quad s > 0$	Equation (1.10)
5.	e^{at}	$\dfrac{1}{s - a}, \quad s > a$	Example 1.4
6.	$e^{at} \sin bt$	$\dfrac{b}{(s - a)^2 + b^2}, \quad s > a$	Prop. 2.12 with $f = \sin bt$
7.	$e^{at} \cos bt$	$\dfrac{s - a}{(s - a)^2 + b^2}, \quad s > a$	Prop. 2.12 with $f = \cos bt$
8.	$t^n e^{at}$	$\dfrac{n!}{(s - a)^{n+1}}, \quad s > a$	Prop. 2.14 with $f = e^{at}$

Example 3.5   Compute the inverse Laplace transform of

$$F(s) = \frac{1}{s - 2} - \frac{16}{s^2 + 4}.$$

Look at the entries in Table 1. We see that line 5 with $a = 2$ applies to the first summand in F. We get

$$\mathcal{L}^{-1}\left\{\frac{1}{s - 2}\right\} = e^{2t}.$$

Similarly, we see that line 3 in Table 1 applies to the second summand, giving

$$\mathcal{L}^{-1}\left\{\frac{2}{s^2 + 4}\right\} = \sin 2t.$$

Hence, by the linearity of the inverse Laplace transform,

$$\mathcal{L}^{-1}(F)(t) = \mathcal{L}^{-1}\left\{\frac{1}{s-2} - 8\frac{2}{s^2+4}\right\}$$
$$= e^{2t} - 8\sin 2t.$$

Using partial fractions

You will notice that all of the Laplace transforms in Table 1 are rational functions. In fact, most of the Laplace transforms that arise in the study of differential equations are rational functions. This means that the method of partial fractions will frequently be useful in computing inverse Laplace transforms. We will therefore illustrate some techniques for computing partial fractions.

Example 3.6 Compute the partial fraction decomposition for the rational function

$$\frac{s+9}{s^2-2s-3}.$$

The denominator factors into $s^2 - 2s - 3 = (s+1)(s-3)$, so we want to find A and B such that

$$\frac{s+9}{s^2-2s-3} = \frac{A}{s+1} + \frac{B}{s-3}. \tag{3.7}$$

We simplify the right side to get

$$\frac{s+9}{s^2-2s-3} = \frac{A(s-3) + B(s+1)}{s^2-2s-3}.$$

To have equality, the numerators must be equal, so we need to find A and B so that

$$s+9 = A(s-3) + B(s+1). \tag{3.8}$$

At this point, there are two ways to proceed. The first way is the *coefficient method*. We multiply out the right-hand side of (3.8) to get $s + 9 = (A + B)s + (B - 3A)$. To have the coefficients of the powers of s equal, we need

$$A + B = 1$$
$$-3A + B = 9.$$

This system is easily solved to get $A = -2$ and $B = 3$.

The second way is the *substitution method*. We substitute specific values for s into (3.8). The values to substitute are those which simplify the right-hand side (i.e., $s = 3$ and $s = -1$). Notice that these are the roots of the quadratic polynomial in the denominators in equation (3.7). For $s = 3$ we get $12 = 4B$ or $B = 3$, while for $s = -1$ we get $8 = -4A$, or $A = -2$.

The coefficient method has the advantage that it is straightforward and always works for any partial fraction problem. However, the substitution method is easier for Example 3.6. The substitution method will work, and it will be easier than the coefficient method when the denominator of the rational function factors into simple linear factors, as in Example 3.6. The disadvantage of the substitution method is that when the denominator does not factor into linear terms, there might not be enough obvious values of s to substitute.

Example 3.9 Compute the partial fraction decomposition for the rational function

$$\frac{2s^2 + s + 13}{(s - 1)\left((s + 1)^2 + 4\right)}.$$

The denominator is already factored as far as possible, since $(s + 1)^2 + 4$ is irreducible. The partial fraction decomposition will have the form

$$\frac{2s^2 + s + 13}{(s - 1)\left((s + 1)^2 + 4\right)} = \frac{A}{s - 1} + \frac{Bs + C}{(s + 1)^2 + 4}$$

$$= \frac{A((s + 1)^2 + 4) + (Bs + C)(s - 1)}{(s - 1)\left((s + 1)^2 + 4\right)}.$$

To use the coefficient method, we multiply out the right-hand side and set the numerators equal,

$$2s^2 + s + 13 = (A + B)s^2 + (2A - B + C)s + (5A - C).$$

Equating the coefficients, we get three equations:

$$A + B = 2,$$
$$2A - B + C = 1,$$
$$5A - C = 13.$$

By solving these equations, we find that $A = 2$, $B = 0$, and $C = -3$.

For the substitution method, we do not multiply out the right-hand side. We simply set the numerators equal,

$$2s^2 + s + 13 = A\left((s + 1)^2 + 4\right) + (Bs + C)(s - 1).$$

An obvious substitution is $s = 1$, since $s - 1$ is a factor of the denominator. This gives us $16 = 8A$, or $A = 2$. So far so good, but there is no systematic way to choose other points at which to evaluate the denominators. The only guideline is that our choice should make the computations as simple as possible. For that reason $s = 0$ is always a good choice. Substituting $s = 0$, we get $13 = 5A - C = 10 - C$. Hence $C = -3$. For a third choice we choose $s = -1$ and get $14 = 4A - 2(C - B) = 8 + 6 + 2B$. Hence $B = 0$.

The final answer is

$$\frac{2s^2 + s + 13}{(s - 1)\left((s + 1)^2 + 4\right)} = \frac{2}{s - 1} - \frac{3}{(s + 1)^2 + 4}.$$

Which method should be used? That's up to you. The key idea is that we need to compute a number of coefficients, say N. To do so we need N equations. We get very simple equations if we substitute the roots of the denominator. These enable us to compute some coefficients directly. If this is not enough, we look for other equations. These can be obtained by substitution or by equating coefficients in the numerators. You should use whatever method works best for you.

The inverse Laplace transform of rational functions

The method of partial fractions together with the Laplace transforms we have computed in Table 1 on page 204 enable us to compute the inverse Laplace transforms of most rational functions.

Example 3.10 Compute the inverse Laplace transform of the function

$$F(s) = \frac{1}{s^2 - 2s - 3}, \quad s > 3.$$

The denominator factors as $s^2 - 2s - 3 = (s - 3)(s + 1)$. Hence the partial fraction decomposition has the form

$$F(s) = \frac{1}{(s - 3)(s + 1)} = \frac{A}{s - 3} + \frac{B}{s + 1}.$$

Recombining the expression on the right, we have

$$\frac{1}{(s - 3)(s + 1)} = \frac{A(s + 1) + B(s - 3)}{(s - 3)(s + 1)}.$$

By setting the numerators equal and substituting the roots of the denominator, we get

$$s = 3 \quad \Longrightarrow \quad 1 = 4A \quad \text{or} \quad A = 1/4,$$
$$s = -1 \quad \Longrightarrow \quad 1 = -4B \quad \text{or} \quad B = -1/4.$$

Thus

$$\frac{1}{(s - 3)(s + 1)} = \frac{1}{4}\left(\frac{1}{s - 3} - \frac{1}{s + 1}\right).$$

Using Table 1 and the linearity of the inverse Laplace transform, we obtain the answer:

$$\mathcal{L}^{-1}(F)(t) = \frac{1}{4}\left(\mathcal{L}^{-1}\left\{\frac{1}{s - 3}\right\} - \mathcal{L}^{-1}\left\{\frac{1}{s + 1}\right\}\right) = \frac{e^{3t} - e^{-t}}{4}. \qquad \bullet$$

Example 3.11 Compute the inverse Laplace transform of

$$F(s) = \frac{1}{s^2 + 4s + 13}.$$

We complete the square in the denominator to obtain

$$F(s) = \frac{1}{(s^2 + 4s + 4) + 9} = \frac{1}{(s + 2)^2 + 3^2}.$$

Using line 6 in Table 1, with $a = -2$ and $b = 3$, the inverse Laplace transform is

$$\mathcal{L}^{-1}(F)(t) = \mathcal{L}^{-1}\left\{\frac{1}{3}\frac{3}{(s + 2)^2 + 3^2}\right\} = \frac{1}{3}e^{-2t}\sin(3t). \qquad \bullet$$

Example 3.12 Compute the inverse Laplace transform of

$$\frac{2s^2 + s + 13}{(s - 1)((s + 1)^2 + 4)}.$$

In Example 3.9, we computed that

$$\frac{2s^2 + s + 13}{(s-1)((s+1)^2 + 4)} = \frac{2}{s-1} - \frac{3}{(s+1)^2 + 4}.$$

Using line 5 of Table 1, we find that

$$\mathcal{L}^{-1}\left\{\frac{1}{s-1}\right\} = e^t.$$

Using line 6 of Table 1, we find that

$$\mathcal{L}^{-1}\left\{\frac{2}{(s+1)^2 + 4}\right\} = e^{-t}\sin 2t.$$

Hence

$$\mathcal{L}^{-1}\left\{\frac{2s^2 + s + 13}{(s-1)((s+1)^2 + 4)}\right\} = 2\mathcal{L}^{-1}\left\{\frac{1}{s-1}\right\} - \frac{3}{2}\mathcal{L}^{-1}\left\{\frac{2}{(s+1)^2 + 4}\right\}$$

$$= 2e^t - \frac{3}{2}e^{-t}\sin 2t.$$

EXERCISES

Using a table of Laplace transforms, much like using a table of integrals in a calculus class, is not as straightforward as it would seem. Often, one has to make adjustments to the given function in order to match a form in the table. It is the linearity of the Laplace transform and its inverse that makes such adjustments possible. For example, the form $Y(s) = 1/(2s - 3)$ is not available in Table 1, but if we make the adjustment

$$Y(s) = \frac{1}{2} \cdot \frac{1}{s - \frac{3}{2}},$$

then, by linearity,

$$y(t) = \mathcal{L}^{-1}\left\{\frac{1}{2} \cdot \frac{1}{s - \frac{3}{2}}\right\} = \frac{1}{2}\mathcal{L}^{-1}\left\{\frac{1}{s - \frac{3}{2}}\right\} = \frac{1}{2}e^{(3/2)t}.$$

Use this technique to find the inverse Laplace transforms of the functions provided in Exercises 1–6.

1. $Y(s) = \dfrac{1}{3s + 2}$

2. $Y(s) = \dfrac{2}{3 - 5s}$

3. $Y(s) = \dfrac{1}{s^2 + 4}$

4. $Y(s) = \dfrac{5s}{s^2 + 9}$

5. $Y(s) = \dfrac{3}{s^2}$

6. $Y(s) = \dfrac{2}{3s^4}$

In Exercises 1–6, we used the fact that $\mathcal{L}^{-1}(\alpha Y) = \alpha \mathcal{L}^{-1}(Y)$. However, linearity in its more general form demands that $\mathcal{L}^{-1}(\alpha X + \beta Y) = \alpha \mathcal{L}^{-1}(X) + \beta \mathcal{L}^{-1}(Y)$. The form $Y(s) = (2s + 5)/(s^2 + 4)$ is not available in Table 1, but if we make the adjustment

$$Y(s) = 2 \cdot \frac{s}{s^2 + 4} + \frac{5}{2} \cdot \frac{2}{s^2 + 4},$$

then, by linearity,

$$y(t) = 2\mathcal{L}^{-1}\left\{\frac{s}{s^2 + 4}\right\} + \frac{5}{2}\mathcal{L}^{-1}\left\{\frac{2}{s^2 + 4}\right\}$$

$$= 2\cos 2t + \frac{5}{2}\sin 2t.$$

Use this technique to find the inverse Laplace transforms of the functions given in Exercises 7–10.

7. $Y(s) = \dfrac{3s + 2}{s^2 + 25}$

8. $Y(s) = \dfrac{2 - 5s}{s^2 + 9}$

9. $Y(s) = \dfrac{1}{3 - 4s} + \dfrac{3 - 2s}{s^2 + 49}$

10. $Y(s) = \dfrac{1}{s^5} - \dfrac{5 - 9s}{s^2 + 100}$

The terminology **transform pair** is popular with engineers, and notation such as

$$y(t) \leftrightarrow Y(s)$$

is used to denote a transform pair. For example, $e^{at} \leftrightarrow 1/(s - a)$. Using this notation, if $y(t) \leftrightarrow Y(s)$ is a transform pair, then Proposition 2.12 tells us that $e^{at}y(t) \leftrightarrow Y(s - a)$ is a transform pair. For example, because

$$\cos 2t \leftrightarrow \frac{s}{s^2 + 4},$$

Proposition 2.12 tells us that

$$e^{3t}\cos 2t \leftrightarrow \frac{s - 3}{(s - 3)^2 + 4}.$$

Use this technique to find the inverse Laplace transform of the functions given in Exercises 11–14.

11. $Y(s) = \dfrac{5}{(s+2)^3}$

12. $Y(s) = \dfrac{1}{(s-1)^6}$

13. $Y(s) = \dfrac{3}{(s+2)^2 + 25}$

14. $Y(s) = \dfrac{4(s-1)}{(s-1)^2 + 4}$

Use the technique developed in Exercises 11–14 to help find the inverse Laplace transform of the functions in Exercises 15–18.

15. $Y(s) = \dfrac{2s-3}{(s-1)^2 + 5}$

16. $Y(s) = \dfrac{s}{(s+2)^2 + 4}$

17. $Y(s) = \dfrac{3s+2}{s^2 + 4s + 29}$

18. $Y(s) = \dfrac{5 - 2s}{s^2 - 2s + 5}$

In Exercises 19–36, perform the appropriate partial fraction decomposition, and then use the result to find the inverse Laplace transform of the given function.

19. $Y(s) = \dfrac{1}{(s+2)(s-1)}$

20. $Y(s) = \dfrac{1}{(s+3)(s-4)}$

21. $Y(s) = \dfrac{2s-1}{(s+1)(s-2)}$

22. $Y(s) = \dfrac{2s-2}{(s-4)(s+2)}$

23. $Y(s) = \dfrac{7s+13}{s^2 + 2s - 3}$

24. $Y(s) = \dfrac{7 - s}{s^2 + s - 2}$

25. $Y(s) = \dfrac{13s - 5}{2s^2 - s}$

26. $Y(s) = \dfrac{4s + 15}{2s^2 + 3s}$

27. $Y(s) = \dfrac{7s^2 + 3s + 16}{(s+1)(s^2 + 4)}$

28. $Y(s) = \dfrac{3s^2 + s + 1}{(s-2)(s^2 + 1)}$

29. $Y(s) = \dfrac{2s^2 + 9s + 11}{(s+1)(s^2 + 4s + 5)}$

30. $Y(s) = \dfrac{7s^2 + 20s + 53}{(s-1)(s^2 + 2s + 5)}$

31. $Y(s) = \dfrac{1}{(s-2)^2(s+1)^3}$

32. $Y(s) = \dfrac{1}{(s-1)^2(s^2 + 4)}$

33. $Y(s) = \dfrac{1}{(s+2)^2(s^2 + 9)}$

34. $Y(s) = \dfrac{1}{(s-1)^2(s^2 - 9)}$

35. $Y(s) = \dfrac{1}{(s+1)^2(s^2 - 4)}$

36. $Y(s) = \dfrac{s}{(s+2)^2(s^2 + 9)}$

5.4 Using the Laplace Transform to Solve Differential Equations

Let's finish the job of finding the solution to the initial value problem in Example 2.9.

Example 4.1 Use the Laplace transform to find the solution to the initial value problem

$$y'' - y = e^{2t} \quad \text{with} \quad y(0) = 0, \quad \text{and} \quad y'(0) = 1.$$

Even though we did this in Example 2.9, let's take the Laplace transform of the equation. Using Proposition 2.4 and the linearity of the Laplace transform, we get

$$s^2 Y(s) - sy(0) - y'(0) - Y(s) = \frac{1}{s-2}.$$

Substituting the initial conditions and solving for Y, we get

$$Y(s) = \frac{1}{(s+1)(s-2)}.$$

The denominator has two linear factors, so the substitution method can be used. The partial fraction decomposition has the form

$$\frac{1}{(s+1)(s-2)} = \frac{A}{s+1} + \frac{B}{s-2}.$$

Recombining the terms on the right, the two numerators become

$$1 = A(s-2) + B(s+1).$$

Substituting $s = -1$, this reduces to $1 = -3A$, so $A = -1/3$. Similarly, substituting $s = 2$ reveals that $B = 1/3$. Hence

$$Y(s) = \frac{1}{3}\left[\frac{1}{s-2} - \frac{1}{s+1}\right].$$

Then using the linearity of the Laplace transform, we get

$$y(t) = \mathcal{L}^{-1}(Y)$$

$$= \frac{1}{3}\mathcal{L}^{-1}\left[\frac{1}{s-2} - \frac{1}{s+1}\right] \tag{4.2}$$

$$= \frac{1}{3}\left[e^{2t} - e^{-t}\right]. \qquad \bullet$$

This example illustrates the general method by which the Laplace transform is used to solve initial value problems. In this section we will look at more examples of the method.

Homogeneous equations

First let's look at a homogeneous differential equation.

Example 4.3   Use the Laplace transform to solve the initial value problem

$$y'' - 2y' - 3y = 0 \quad \text{with} \quad y(0) = 1 \quad \text{and} \quad y'(0) = 0.$$

Using Propositions 2.4, and 2.7, we compute the Laplace transform of the equation. We will set $Y = \mathcal{L}(y)$.

$$\mathcal{L}(y'' - 2y' - 3y) = s^2 Y(s) - sy(0) - y'(0) - 2(sY(s) - y(0)) - 3Y(s)$$

$$= (s^2 - 2s - 3)Y(s) - y(0)(s - 2) - y'(0)$$

Inserting the initial conditions $y(0) = 1$ and $y'(0) = 0$ and then solving for $Y(s)$, we obtain

$$(s^2 - 2s - 3)Y(s) = s - 2 \quad \text{or} \quad Y(s) = \frac{s-2}{s^2 - 2s - 3}.$$

Notice that the denominator is the characteristic polynomial of the differential equation. We will show later that this is always the case.

To find the solution y, we must compute the inverse Laplace transform of $Y(s)$. We use partial fractions to obtain

$$Y(s) = \frac{s-2}{s^2 - 2s - 3} = \frac{1/4}{s-3} + \frac{3/4}{s+1}.$$

Using Table 1, the inverse Laplace transform of the right side is

$$y(t) = \frac{1}{4}e^{3t} + \frac{3}{4}e^{-t}.$$

Note that the initial conditions, $y(0) = 1$ and $y'(0) = 0$, are satisfied. $\bullet$

The solution to the previous example could have been easily found using the techniques of Chapter 4. In fact, since the characteristic polynomial is $s^2 - 2s - 3 = (s-3)(s+1)$, we know that e^{3t} and e^{-t} are a fundamental set of solutions. Later, we will see examples where the Laplace transform offers an easier method for finding solutions than the techniques of Chapter 4.

Inhomogeneous equations

Using Laplace transforms, inhomogeneous equations are handled in the same way as homogeneous equations. The Laplace transform is applied to both sides of the equation. It is not necessary to find the general solution to the homogeneous equation first and then find a particular solution to the inhomogeneous equation. Here is an example.

Example 4.4 Use the Laplace transform to solve the initial value problem

$$y'' + 2y' + 2y = \cos 2t \quad \text{with} \quad y(0) = 0 \quad \text{and} \quad y'(0) = 1.$$

We compute that

$$\mathcal{L}(y'' + 2y' + 2y) = s^2 Y(s) - sy(0) - y'(0) + 2(sY(s) - y(0)) + 2Y(s),$$

and

$$\mathcal{L}(\cos 2t) = \frac{s}{s^2 + 4}.$$

Setting these two expressions equal and substituting the initial conditions, we get

$$(s^2 + 2s + 2)Y(s) - 1 = \frac{s}{s^2 + 4},$$

or

$$Y(s) = \frac{1}{s^2 + 2s + 2} + \frac{s}{(s^2 + 2s + 2)(s^2 + 4)}. \tag{4.5}$$

The inverse of the Laplace transform of the first term on the right can be computed easily, so let's deal with the second term. Since $s^2 + 2s + 2 = (s + 1)^2 + 1$ cannot be factored, the partial fraction decomposition of this term has the form

$$\frac{s}{(s^2 + 2s + 2)(s^2 + 4)} = \frac{As + B}{s^2 + 2s + 2} + \frac{Cs + D}{s^2 + 4}. \tag{4.6}$$

If we combine the terms on the right of (4.6) and multiply it out completely, we find that the numerators are

$$s = (A + C)s^3 + (B + 2C + D)s^2 + (4A + 2C + 2D)s + (4B + 2D).$$

Equating the coefficients of the powers, we get four equations:

$$A + \quad C = 0,$$
$$B + 2C + \quad D = 0,$$
$$4A + 2C + 2D = 1,$$
$$4B + 2D = 0.$$

Solving this system, we get $A = 1/10$, $B = -1/5$, $C = -1/10$, and $D = 2/5$. Thus the second term on the right in (4.5) is

$$\frac{1}{10}\frac{s - 2}{s^2 + 2s + 2} - \frac{1}{10}\frac{s - 4}{s^2 + 4},$$

and (4.5) becomes

$$Y(s) = \frac{1}{s^2 + 2s + 2} + \frac{1}{10}\frac{s - 2}{s^2 + 2s + 2} - \frac{1}{10}\frac{s - 4}{s^2 + 4}.$$

To better match the entries in Table 1, we complete the square to get $s^2 + 2s + 2 = (s+1)^2 + 1$ and then rewrite the above equation as

$$Y(s) = \frac{1}{(s+1)^2 + 1} + \frac{1}{10}\frac{s+1}{(s+1)^2 + 1}$$
$$- \frac{3}{10}\frac{1}{(s+1)^2 + 1} - \frac{1}{10}\frac{s}{s^2 + 4} + \frac{2}{10}\frac{2}{s^2 + 4}.$$

We can read the inverse Laplace transforms of the summands directly from Table 1 to find that

$$y(t) = e^{-t}\sin t + \frac{1}{10}e^{-t}\cos t - \frac{3}{10}e^{-t}\sin t - \frac{1}{10}\cos 2t + \frac{2}{10}\sin 2t$$
$$= \frac{1}{10}\left\{e^{-t}(\cos t + 7\sin t) + 2\sin 2t - \cos 2t\right\}. \qquad \bullet$$

Higher-order equations

Higher-order equations are handled in essentially the same way.

Example 4.7 Find the solution to the initial value problem

$$y^{(4)} - y = 0 \quad \text{with} \quad y(0) = 0,\ y'(0) = 1,\ y''(0) = 0,\ \text{and}\ y'''(0) = 0.$$

Taking the Laplace transform of both sides of the equation $y^{(4)} - y = 0$ gives

$$\left(s^4 Y(s) - s^3 y(0) - s^2 y'(0) - sy''(0) - y'''(0)\right) - Y(s) = 0.$$

Substituting the initial conditions and then rearranging, we obtain

$$(s^4 - 1)Y(s) - s^2 = 0 \quad \text{or} \quad Y(s) = \frac{s^2}{s^4 - 1} = \frac{s^2}{(s-1)(s+1)(s^2+1)}.$$

Again we see that the denominator $s^4 - 1$ is the characteristic polynomial of the differential equation. It factors as $s^4 - 1 = (s-1)(s+1)(s^2+1)$. Since $s^2 + 1$ cannot be factored, the partial fraction decomposition of the right-hand side has the form

$$\frac{s^2}{s^4 - 1} = \frac{A}{s-1} + \frac{B}{s+1} + \frac{Cs + D}{s^2 + 1}$$
$$= \frac{A(s+1)(s^2+1) + B(s-1)(s^2+1) + (Cs+D)(s^2-1)}{(s-1)(s+1)(s^2+1)}.$$

Matching numerators, we get

$$s^2 = A(s+1)(s^2+1) + B(s-1)(s^2+1) + (Cs+D)(s^2-1).$$

We will use the substitution method. We start by substituting the roots of the characteristic polynomial.

$$s = 1 \implies 1 = 4A \quad \text{or} \quad A = 1/4 \quad \text{and}$$
$$s = -1 \implies 1 = -4B \quad \text{or} \quad B = -1/4$$

The substitution $s = 0$ is always a good choice. This gives us $0 = A - B - D = 1/4 + 1/4 - D$ or $D = 1/2$.

We need one more equation. We could choose any number for s that we have not already used, but let's try something different. The coefficient of the highest power of s that occurs is usually easy to compute. In this case the highest power is 3, so we look for the coefficient of s^3. We get $0 = A + B + C = 1/4 - 1/4 + C$, or $C = 0$. We therefore have

$$Y(s) = \frac{1/4}{s-1} - \frac{1/4}{s+1} + \frac{1/2}{s^2+1}.$$

Using Table 1 to find the inverse transforms, we obtain

$$y(t) = \frac{1}{4}e^t - \frac{1}{4}e^{-t} + \frac{1}{2}\sin t. \qquad \bullet$$

An overview of the method

If we limit ourselves to second-order equations, we are looking at initial value problems of the sort

$$ay'' + by' + cy = f(t) \quad \text{with} \quad y(0) = y_0 \quad \text{and} \quad y'(0) = y_1. \qquad (4.8)$$

Let's apply the Laplace transform to this, with $Y(s) = \mathcal{L}(y)(s)$. Using Propositions 2.1 and 2.4, we get

$$\begin{aligned}
\mathcal{L}\left(ay'' + by' + cy\right) &= a\,\mathcal{L}(y'') + b\,\mathcal{L}(y') + c\,\mathcal{L}(y) \\
&= a\left(s^2 Y(s) - sy(0) - y'(0)\right) \\
&\quad + b\left(sY(s) - y(0)\right) \\
&\quad + cY(s) \\
&= \left(as^2 + bs + c\right)Y(s) - y_0(as + b) - ay_1.
\end{aligned}$$

If y is a solution to the initial value problem in (4.8), then $\mathcal{L}\left(ay'' + by' + cy\right) = \mathcal{L}(f) = F$. Substituting into the above equation, we can solve for Y. The final result is as follows:

$$ay'' + by' + cy = f(t) \quad \text{with} \quad y(0) = y_0 \quad \text{and} \quad y'(0) = y_1$$
$$\Longleftrightarrow \quad \mathcal{L}(y) = Y, \; \mathcal{L}(f) = F, \quad \text{and} \qquad (4.9)$$
$$Y(s) = \frac{F(s)}{as^2 + bs + c} + \frac{y_0(as + b) + ay_1}{as^2 + bs + c}.$$

Notice that the denominator in the last equation in (4.9) is the characteristic polynomial of the differential equation. We have already pointed this out in several examples.

Let's define y_s to be the solution of

$$ay_s'' + by_s' + cy_s = f(t) \quad \text{with} \quad y_s(0) = y_s'(0) = 0, \qquad (4.10)$$

and y_i to be the solution of

$$ay_i'' + by_i' + cy_i = 0 \quad \text{with} \quad y_i(0) = y_0 \quad \text{and} \quad y_i'(0) = y_1. \qquad (4.11)$$

Notice that the initial value problem for y_s in (4.10) has zero initial conditions but the same forcing term as does that for y in (4.8). It is referred to as the ***state-free solution***. On the other hand, the initial value problem for y_i in (4.11) has y_i being the solution of the homogeneous equation, with the same initial conditions as does y in (4.8). It is referred to as the **input-free solution**. According to (4.9), we have

$$Y_s(s) = \mathcal{L}(y_s)(s) = \frac{F(s)}{as^2 + bs + c} \quad \text{and} \tag{4.12}$$

$$Y_i(s) = \mathcal{L}(y_i)(s) = \frac{y_0(as + b) + ay_1}{as^2 + bs + c}. \tag{4.13}$$

It follows from (4.9) that $Y = Y_s + Y_i$, and from this it follows that $y = y_s + y_i$. Thus, the solution to any initial value problem can be written as the sum of a state-free solution and an input-free solution.

We will return to this in a later section.

EXERCISES

Use the Laplace transform to solve the first-order initial value problems in Exercises 1–10.

1. $y' + 3y = e^{2t}$, $y(0) = -1$

2. $y' + 9y = e^{-t}$, $y(0) = 0$

3. $y' + 4y = \cos t$, $y(0) = 0$

4. $y' + 16y = \sin 3t$, $y(0) = 1$

5. $y' + 6y = 2t + 3$, $y(0) = 1$

6. $y' + 8y = t^2$, $y(0) = -1$

7. $y' + 8y = e^{-2t} \sin t$, $y(0) = 0$

8. $y' - 2y = e^{-t} \cos t$, $y(0) = -2$

9. $y' + y = te^t$, $y(0) = -2$

10. $y' - 4y = e^{-2t}t^2$, $y(0) = 1$

Use the Laplace transform to solve the second-order initial value problems in Exercises 11–26.

11. $y'' - 4y = e^{-t}$, $y(0) = -1$, $y'(0) = 0$

12. $y'' - 9y = -2e^t$, $y(0) = 0$, $y'(0) = 1$

13. $y'' + 4y = \cos t$, $y(0) = 1$, $y'(0) = 0$

14. $y'' + 9y = 2\sin 2t$, $y(0) = 0$, $y'(0) = -1$

15. $y'' - y = 2t$, $y(0) = 0$, $y'(0) = -1$

16. $y'' + 3y' = -3t$, $y(0) = -1$, $y'(0) = 1$

17. $y'' + y' = te^{-t}$, $y(0) = -2$, $y'(0) = 0$

18. $y'' - y' - 2y = t^2 e^{2t}$, $y(0) = 0$, $y'(0) = -1$

19. $y'' - 4y' - 5y = e^{2t}$, $y(0) = -1$, $y'(0) = 0$

20. $y'' - 2y' - 3y = e^{4t}$, $y(0) = 1$, $y'(0) = -1$

21. $y'' - y' - 2y = e^{2t}$, $y(0) = -1$, $y'(0) = 0$

22. $y'' - 3y' - 4y = e^{-t}$, $y(0) = 0$, $y'(0) = -1$

23. $y'' + 2y' + 2y = \cos t$, $y(0) = -1$, $y'(0) = 0$

24. $y'' + 4y' + 5y = -3\sin 2t$, $y(0) = 1$, $y'(0) = -1$

25. $y'' + 4y = e^{-2t} \cos t$, $y(0) = 1$, $y'(0) = 0$

26. $y'' + 9y = e^{-t} \sin t$, $y(0) = -1$, $y'(0) = 1$

27. Consider the second-order initial value problem

$$y'' - 4y' + 3y = 0, \quad y(0) = 1, \quad y'(0) = -1. \tag{4.14}$$

(a) Use the technique of Section 4.3 in Chapter 4 to show that the general solution is $y(t) = C_1 e^{3t} + C_2 e^t$. Use the initial conditions to find the values of C_1 and C_2 that produce the solution of the initial value problem (4.14).

(b) Use the Laplace transform technique to find the solution of the initial value problem (4.14) and compare with the solution found in part (a).

Use both solution techniques suggested in parts (a) and (b) in Exercise 27 to find and compare solutions of the initial value problems in Exercises 28–31.

28. $y'' - 3y' - 4y = 0$, $y(0) = 1$, $y'(0) = -1$

29. $y'' - 3y' + 2y = 0$, $y(0) = 0$, $y'(0) = -1$

30. $y'' + 4y = 0$, $y(0) = 1$, $y'(0) = -1$

31. $y'' + 2y' + 2y = 0$, $y(0) = -1$, $y'(0) = 1$

32. Consider the second-order initial value problem

$$y'' + 4y = 2\cos 3t, \quad y(0) = 1, \quad y'(0) = -1. \tag{4.15}$$

(a) Use the technique of Section 4.3, Chapter 4, to find the homogeneous solution y_h of the homogenous equation $y'' + 4y = 0$.

(b) Use the method of undetermined coefficients in Section 4.5 to find a particular solution y_p of $y'' + 4y = 2\cos 3t$.

(c) Form the general solution $y = y_h + y_p$ and use the initial conditions to find the solution of the initial value problem (4.15).

(d) Use the Laplace transform to find the solution of the initial value problem (4.15) and compare with the result found in part (c).

Use both solution techniques suggested in Exercise 32 to find and compare solutions of the initial value problems in Exercises 33–36.

33. $y'' + y = -2\cos 2t, \quad y(0) = 1, \quad y'(0) = -1$

34. $y'' + 9y = 3\sin 2t, \quad y(0) = 0, \quad y'(0) = -1$

35. $y'' + 4y = \cos 2t, \quad y(0) = 1, \quad y'(0) = -1$

36. $y'' + y = -2\sin t, \quad y(0) = -1, \quad y'(0) = 1$

Note: Exercises 35 and 36 involve resonance. See Section 4.7 for assistance.

In Exercises 37–41 we will redo some of the analysis of harmonic oscillators done in Section 4.4 of Chapter 4.

37. Consider the undamped, unforced oscillator, which is modeled by the initial value problem

$$y'' + \omega_0^2 y = 0, \quad y(0) = y_0, \ y'(0) = v_0, \quad (4.16)$$

where y_0 and v_0 are the initial displacement and velocity of the mass, respectively.

(a) Use the Laplace transform to show that the solution of equation (4.16) is

$$y(t) = y_0 \cos \omega_0 t + \frac{v_0}{\omega_0} \sin \omega_0 t.$$

(b) Show that the solution in part (a) is equivalent to

$$y(t) = A\cos(\omega_0 t - \phi),$$

where $A = \sqrt{y_0^2 + v_0^2/\omega_0^2}$ and $\tan \phi = v_0/(y_0\omega_0)$.

38. When there is damping, but still no forcing term, the initial value problem becomes

$$y'' + 2cy' + \omega_0^2 y = 0, \quad y(0) = y_0, \ y'(0) = v_0, \quad (4.17)$$

where y_0 and v_0 are the initial displacement and velocity of the mass, respectively. Show that the Laplace transform of the solution can be written

$$Y(s) = \frac{y_0 s + v_0 + 2cy_0}{(s+c)^2 + (\omega_0^2 - c^2)}. \quad (4.18)$$

A moment's reflection will reveal that taking the inverse transform of $Y(s)$ will depend upon the sign of $\omega_0^2 - c^2$. In Exercises 39–41, we will examine three cases. Notice that each has a counterpart in Section 4.4.

39. Case 1: $\omega_0^2 - c^2 > 0$. Show that this is the **underdamped** case of Section 4.4. Compute the inverse Laplace transform of $Y(s)$ to show that the solution in this case is given by

$$y(t) = y_0 e^{-ct} \cos\left(t\sqrt{\omega_0^2 - c^2}\right)$$
$$+ \frac{y_0 c + v_0}{\sqrt{\omega_0^2 - c^2}} e^{-ct} \sin\left(t\sqrt{\omega_0^2 - c^2}\right).$$

40. Case 2: $\omega_0^2 - c^2 = 0$. Show that this is the **critically damped** case of Section 4.4. Compute the inverse Laplace transform of $Y(s)$ to show that the solution in this case is given by

$$y(t) = y_0 e^{-ct} + (v_0 + by_0)te^{-ct}.$$

41. Case 3: $\omega_0^2 - c^2 < 0$. Show that this is the **overdamped** case of Section 4.4. To simplify calculations, let's set the intitial displacement as $y(0) = y_0 = 0$. Show that with this assumption the transform in (4.18) becomes

$$Y(s) = \frac{v_0}{(s+c)^2 - (c^2 - \omega_0^2)},$$
$$\quad (4.19)$$
$$= \frac{v_0}{\left(s + c + \sqrt{c^2 - \omega_0^2}\right)\left(s + c - \sqrt{c^2 - \omega_0^2}\right)}.$$

Use the technique of partial fractions to decompose (4.19), and then find the solution $y(t)$.

42. Use Laplace transform techniques to show that the solution of the undamped forced oscillator,

$$x'' + \omega_0^2 x = F_0 \sin \omega_0 t, \quad x(0) = x'(0) = 0,$$

is given by

$$x(t) = \frac{F_0}{2\omega_0^2}(\sin \omega_0 t - \omega_0 t \cos \omega_0 t).$$

5.5 Discontinuous Forcing Terms

In Chapter 4 we analyzed harmonic motion, which is modeled by the equation

$$y'' + 2cy' + \omega_0^2 y = F(t).$$

Remember that applications of harmonic motion include vibrating springs and electrical circuits.

In Section 4.7, we considered sinusoidal forcing terms involving sines and cosines. This is a reasonable model for a mechanical system involving an alternating external force or a circuit whose external voltage arises from an alternating current. In this section, we discuss how the Laplace transform is used to handle an external forcing term which is piecewise continuous. We have seen examples in Figures 1, 2, 3, 4, and 5 of Section 5.1. We will see several more in this section.

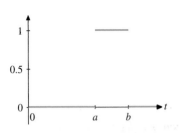

Figure 1. The function H_{ab}.

Another example is the ***interval function***

$$H_{ab}(t) = \begin{cases} 0, & \text{for } t < a; \\ 1, & \text{for } a \le t < b; \\ 0, & \text{for } b \le t < \infty; \end{cases} \tag{5.1}$$

where we assume that $a < b$. The graph of H_{ab} is shown in Figure 1. For a mechanical or electrical system, this corresponds to a force that is turned on at time $t = a$ and then turned off again at time $t = b$.

The Heaviside function

Instead of finding the Laplace transform of the interval function directly, we will use a basic building block called the ***Heaviside function***. This function is defined as

$$H(t) = \begin{cases} 0, & \text{for } t < 0; \\ 1, & \text{for } t \ge 0. \end{cases}$$

The Heaviside function represents a force that is turned on at time $t = 0$, and left on thereafter. The graph of H is shown in Figure 2.

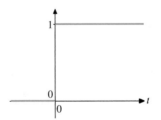

Figure 2. The Heaviside function $H(t)$.

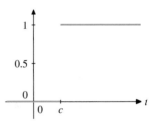

Figure 3. The shifted Heaviside function $H_c(t)$.

We will want forces that turn on at times other than 0. These are modeled by translates of H. For a real number c, we define

$$H_c(t) = H(t - c) = \begin{cases} 0, & \text{for } t < c; \\ 1, & \text{for } t \ge c. \end{cases}$$

If $c > 0$, then the graph of H_c is a translation by c units to the right of the graph of H as illustrated in Figure 3. If $c < 0$, then the graph of H_c is a translation to the left of the graph of H.

The interval function H_{ab} defined for $a < b$ in (5.1) can be described using the translates of the Heaviside function,

$$H_{ab}(t) = H_a(t) - H_b(t) = H(t - a) - H(t - b).$$

Indeed for $t < a$, both $H_b(t)$ and $H_a(t)$ are zero; for $a \le t < b$, $H_a(t) = 1$ and $H_b(t) = 0$, giving 1 for the right side; and for $b \le t$, both $H_a(t)$ and $H_b(t)$ are 1, giving 0 for the right side.

The Heaviside function H, its translates H_a, and the interval functions H_{ab} can be used to describe piecewise continuous functions in a most convenient manner.

Example 5.2   Describe the function

$$g(t) = \begin{cases} 2t, & \text{for } 0 \le t < 1; \\ 2, & \text{for } 1 \le t < \infty \end{cases}$$

using the Heaviside function.

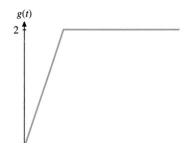

Since g is a piecewise differentiable function defined on the intervals $0 \le t < 1$ and $1 \le t$, we use the interval function $H_{01}(t)$, which is equal to 1 precisely on the interval $0 \le t < 1$, and the translated Heaviside function $H_1(t)$, which is equal to 1 for $1 \le t$. Therefore,

$$\begin{aligned} g(t) &= 2t\,H_{01}(t) + 2H_1(t) \\ &= 2t\,[H(t) - H(t-1)] + 2H(t-1) \\ &= 2t\,H(t) - 2(t-1)H(t-1). \end{aligned}$$ ●

Figure 4. The function in Example 5.2.

Example 5.3   Describe the function

$$f(t) = \begin{cases} 3, & \text{for } 0 \le t < 4; \\ -5, & \text{for } 4 \le t < 6; \\ e^{7-t}, & \text{for } 6 \le t < \infty \end{cases}$$

in terms of the Heaviside function.

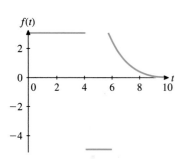

The function f is a piecewise continuous function defined on the intervals $0 \le t < 4$, $4 \le t < 6$, and $6 \le t$, so we use the interval functions H_{04}, H_{46}, and the translated Heaviside function H_6.

$$\begin{aligned} f(t) &= 3H_{04}(t) - 5H_{46}(t) + e^{7-t}H_6(t) \\ &= 3\,[H(t) - H(t-4)] - 5\,[H(t-4) - H(t-6)] + e^{7-t}H(t-6) \\ &= 3H(t) - 8H(t-4) + 5H(t-6) + e^{7-t}H(t-6) \end{aligned}$$ ●

Figure 5. The function in Example 5.3.

The Laplace transform of the Heaviside function

We just compute

$$\mathcal{L}(H_c)(s) = \int_0^\infty H_c(t)e^{-st}\,dt = \int_c^\infty e^{-st}\,dt = \frac{e^{-cs}}{s}. \tag{5.4}$$

Notice in particular that $\mathcal{L}(H)(s) = 1/s$. This is the same as the transform of the function $f(t) = 1$, which we computed in Example 1.5. This should not be surprising, since $f(t) = H(t) = 1$ for $t > 0$, and the Laplace transform only looks at the values of a function for $t > 0$.

For the interval function H_{ab}, we have $H_{ab} = H_a - H_b$. Consequently, the linearity of the Laplace transform gives us

$$\mathcal{L}(H_{ab})(s) = \mathcal{L}(H_a)(s) - \mathcal{L}(H_b)(s) = \frac{e^{-as} - e^{-bs}}{s}. \tag{5.5}$$

The Laplace transform of a translate of a function

In Examples 5.2 and 5.3, the functions $g(t)$ and $f(t)$ involve translations of the Heaviside function H. Thus, we digress and discuss the relationship between the Laplace transform of a general function f and the Laplace transform of a translate of f.

First we should clear up a point. The Laplace transform only takes into account the values of $f(t)$ for $t > 0$. We have not said anything about the values of function for negative values of t, because it just wasn't important. However, now it will be, and everything makes more sense if we assume that all functions we deal with are defined for $t \leq 0$ and have the value 0 there. Then, when we shift the time variable by c, the translate of the function $f(t)$ is the same as $H(t-c)f(t-c)$. The two are compared in Figures 6 and 7.

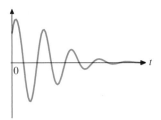

Figure 6. The graph of $f(t)$.

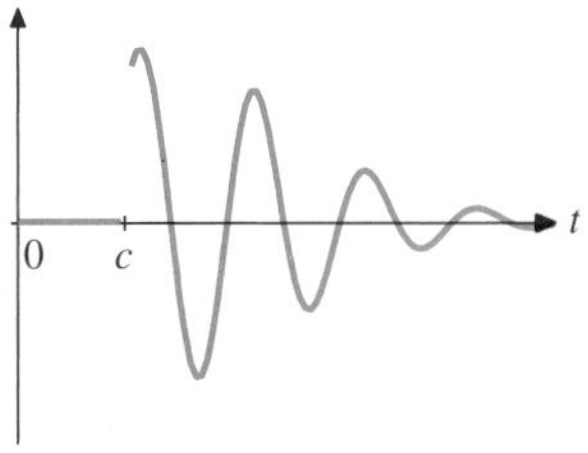

Figure 7. The graph of $H(t-c)f(t-c)$.

PROPOSITION 5.6 Suppose $f(t)$ is piecewise continuous and is of exponential order. Let $F(s)$ be the Laplace transform of f. Then, for $c \geq 0$, the Laplace transform of $H(t-c)f(t-c)$ is given by

$$\mathcal{L}\{H(t-c)f(t-c)\}(s) = e^{-cs}F(s). \qquad (5.7)$$

Notice that we restrict the amount of the shift in time, c, to be nonnegative. Proposition 5.6 states that a translation in time by the amount c results in multiplication of the Laplace transform by e^{-cs}.

Proof To prove this proposition, we use the definition of the Laplace transform on the left-hand side:

$$\mathcal{L}\{H(t-c)f(t-c)\}(s) = \int_0^\infty H(t-c)f(t-c)e^{-st}\,dt$$

$$= \int_c^\infty f(t-c)e^{-st}\,dt,$$

where the last equation follows from the fact that $H(t-c) = 1$ for $t \geq c$ and 0 for $t < c$. We will change the variable to $\tau = t - c$. Then $t = \tau + c$ and $dt = d\tau$ (since c is a constant). We obtain

$$\int_c^\infty f(t-c)e^{-st}\,dt = \int_0^\infty f(\tau)e^{-s(\tau+c)}\,d\tau.$$

Note that the limits of integration have changed from $c \leq t < \infty$ to $0 \leq \tau < \infty$

(since $\tau = t - c$). Thus,

$$\mathcal{L}\{H(t - c)f(t - c)\}(s) = \int_0^\infty f(\tau)e^{-s(\tau+c)}\, d\tau$$

$$= e^{-cs} \int_0^\infty f(\tau)e^{-s\tau}\, d\tau$$

$$= e^{-cs} F(s)$$

as desired. ●

Example 5.8   Find the Laplace transform of the function

$$g(t) = \begin{cases} 2t, & \text{for } 0 \leq t < 1; \\ 2, & \text{for } 1 \leq t < \infty. \end{cases}$$

As computed in Example 5.2,

$$g(t) = 2t\,H(t) - 2(t - 1)H(t - 1).$$

By Table 1 on page 204, $\mathcal{L}\{t\} = 1/s^2$. Using (5.7) with $c = 0$ and $c = 1$, we have

$$\mathcal{L}\{g\}(s) = 2\,\mathcal{L}\{t\,H(t)\}(s) - 2\,\mathcal{L}\{(t - 1)H(t - 1)\}(s) = \frac{2}{s^2} - \frac{2e^{-s}}{s^2}.$$ ●

Example 5.9   Find the Laplace transform of the function

$$g(t) = \begin{cases} 3, & \text{for } 0 \leq t < 4; \\ -5, & \text{for } 4 \leq t < 6; \\ e^{7-t}, & \text{for } 6 \leq t < \infty. \end{cases}$$

As computed in Example 5.3,

$$g(t) = 3H(t) - 8H(t - 4) + 5H(t - 6) + e^{7-t}H(t - 6). \tag{5.10}$$

The first three terms can be handled using (5.4). The last term can be written as

$$e^{7-t}H(t - 6) = H(t - 6)e^{-(t-6)} \times e,$$

so it too can be handled using Proposition 5.6 and Table 1. Then

$$\mathcal{L}(g)(s) = \mathcal{L}\{3H(t) - 8H(t - 4) + 5H(t - 6) + e \times e^{-(t-6)}H(t - 6)\}(s)$$

$$= \frac{3}{s} - \frac{8e^{-4s}}{s} + \frac{5e^{-6s}}{s} + e\frac{e^{-6s}}{s + 1}$$

$$= \frac{3}{s} - \frac{8e^{-4s}}{s} + \frac{5e^{-6s}}{s} + \frac{e^{1-6s}}{s + 1}.$$ ●

Example 5.11   Find the Laplace transform of

$$H(t - \pi/4)\sin t.$$

We need to express $\sin t$ in terms of $t - \pi/4$. To do so, we use the addition formula for $\sin t$ to write

$$\sin t = \sin(t - \pi/4 + \pi/4)$$
$$= \sin(t - \pi/4)\cos\pi/4 + \cos(t - \pi/4)\sin\pi/4$$
$$= \frac{1}{\sqrt{2}}\left[\sin(t - \pi/4) + \cos(t - \pi/4)\right].$$

Now we can use Proposition 5.6 and Table 1 to compute the transform.

$$\mathcal{L}\{H(t - \pi/4)\sin t\}(s) = \frac{1}{\sqrt{2}}\mathcal{L}\{H(t - \pi/4)(\sin(t - \pi/4) + \cos(t - \pi/4))\}(s)$$

$$= \frac{e^{-\pi s/4}}{\sqrt{2}}(\mathcal{L}\{\sin t\}(s) + \mathcal{L}\{\cos t\}(s))$$

$$= \frac{e^{-\pi s/4}}{\sqrt{2}}\frac{1 + s}{1 + s^2} \qquad \bullet$$

Every formula for computing the Laplace transform gives a formula for computing the inverse Laplace transform. The following proposition corresponds to Proposition 5.6.

PROPOSITION 5.12   Suppose that $f(t)$ is piecewise continuous and is of exponential type. Suppose that $F(s) = \mathcal{L}(f)(s)$. Then

$$\mathcal{L}^{-1}\{e^{-cs}F(s)\}(t) = H(t - c)f(t - c). \qquad (5.13)$$
$$\bullet$$

Here, we see that multiplication by an exponential factor in the s-domain results in a translation in the t-domain.

Example 5.14   Find the inverse Laplace transform of the function

$$\frac{e^{-2s}}{s(s^2 + 9)}.$$

We can apply Proposition 5.12 with $F(s) = 1/\{s(s^2 + 9)\}$. However, we need to compute $f = \mathcal{L}^{-1}(F)$. We use partial fractions on F to get

$$F(s) = \frac{1}{9}\left(\frac{1}{s} - \frac{s}{s^2 + 9}\right).$$

From Table 1, the inverse Laplace transform of $1/s$ is the function 1. Likewise, the inverse Laplace transform of $s/(s^2 + 9)$ is $\cos 3t$. Therefore,

$$f(t) = \mathcal{L}^{-1}(F)(t) = \frac{1}{9}(1 - \cos 3t).$$

Putting everything together, we obtain the final answer:

$$\mathcal{L}^{-1}\left\{\frac{e^{-2s}}{s(s^2 + 9)}\right\} = H(t - 2)f(t - 2)$$

$$= H(t - 2)(1 - \cos 3(t - 2))/9$$

$$= \begin{cases} 0, & \text{for } t < 2; \\ (1 - \cos 3(t - 2))/9, & \text{for } 2 \le t < \infty. \end{cases} \qquad \bullet$$

Solving initial value problems with piecewise defined forcing functions

Such forcing functions are routinely used by engineers to analyze systems. Let's look at an example.

Example 5.15   Find the solution to the initial value problem

$$y'' + y = g(t), \quad \text{with} \quad y(0) = 0 \quad \text{and} \quad y'(0) = 1,$$

where

$$g(t) = \begin{cases} 2t, & \text{for } 0 \le t < 1; \\ 2, & \text{for } 1 \le t < \infty. \end{cases}$$

Using Proposition 2.4 and the initial conditions, the Laplace transform of the left-hand side is

$$\mathcal{L}(y'' + y) = \left(s^2 Y(s) - s y(0) - y'(0)\right) + Y(s) = (s^2 + 1)Y(s) - 1.$$

From Example 5.8, $\mathcal{L}(g)(s) = G(s) = (2 - 2e^{-s})/s^2$. Hence

$$(s^2 + 1)Y(s) - 1 = \frac{2 - 2e^s}{s^2}, \quad \text{or} \quad Y(s) = \frac{2 - 2e^{-s}}{s^2(s^2 + 1)} + \frac{1}{s^2 + 1}. \qquad (5.16)$$

The solution is $y(t) = \mathcal{L}^{-1}(Y)(t)$. To compute it, we first use partial fractions to set

$$\frac{1}{s^2(s^2 + 1)} = \frac{1}{s^2} - \frac{1}{s^2 + 1}.$$

Then equation (5.16) becomes

$$Y(s) = \left(\frac{2 - 2e^{-s}}{s^2} - \frac{2 - 2e^{-s}}{s^2 + 1}\right) + \frac{1}{s^2 + 1}$$

$$= \frac{2}{s^2} - \frac{2e^{-s}}{s^2} + \frac{2e^{-s}}{s^2 + 1} - \frac{1}{s^2 + 1}.$$

From Table 1, we see that the inverse Laplace transform of $1/s^2$ is the function t, and that of $1/(s^2+1)$ is $\sin t$. Therefore, using Proposition 5.12, the inverse Laplace transform of $Y(s)$ is

$$y(t) = 2t - 2(t - 1)H(t - 1) + 2H(t - 1)\sin(t - 1) - \sin t.$$

The function y can also be written as

$$y(t) = \begin{cases} 2t - \sin t, & \text{for } 0 \le t < 1; \\ 2 + 2\sin(t - 1) - \sin t, & \text{for } 1 \le t < \infty, \end{cases}$$

since $H(t - 1) = 0$ for $0 < t < 1$ and $H(t - 1) = 1$ for $t \ge 1$. Note that the initial conditions are satisfied. Since $y(t) = 2t - \sin t$ for $0 \le t < 1$, $y(0) = 0$ and $y'(0) = 1$. A plot of $y(t)$ is given in Figure 8.   ●

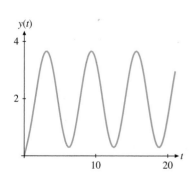

Figure 8. The solution to the initial value problem in Example 5.15.

The solution in the previous example is a piecewise differentiable function. The two pieces join together at $t = 1$ in a continuous way (since the function $2t - \sin t$ agrees with $2 + 2\sin(t - 1) - \sin t$ at $t = 1$). The first derivatives of the two pieces also match at $t = 1$. Even the second derivatives match at $t = 1$, as you can check. However, the third derivatives do not match at $t = 1$.

Periodic functions

We have dealt with periodic functions before, but let's remind ourselves of some things. First, a periodic function is one whose graph has a repetitive pattern. Formally, a function f is **periodic** with period T (or T-periodic) if $f(t + T) = f(t)$ for all t.

For example, the sine and cosine functions are 2π-periodic because $\sin(t + 2\pi) = \sin(t)$ and $\cos(t + 2\pi) = \cos(t)$. The **square wave** in Figure 9 is 2-periodic.

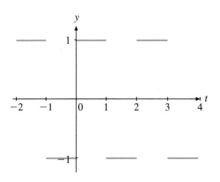

Figure 9. The square wave is periodic with period 2.

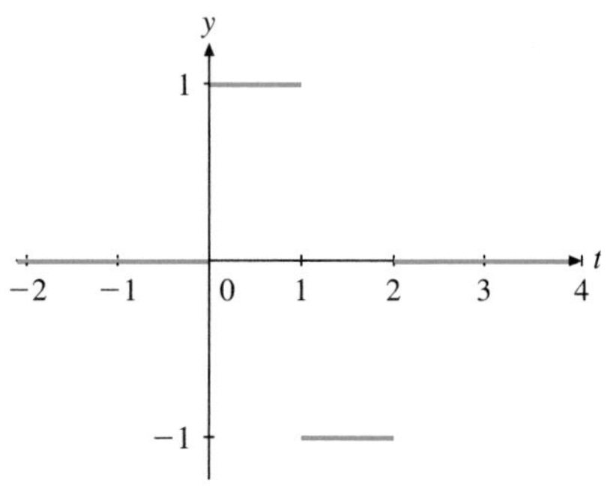

Figure 10. The window of the square wave.

Geometrically, the graph of a T-periodic function, f, repeats every interval of length T. For such a function, it is useful to define the **window** of f as

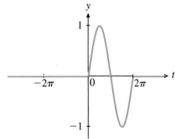

Figure 11. The window for the sine function.

$$f_T(t) = \begin{cases} f(t), & 0 \le t < T \\ 0, & T \le t < \infty, \text{ and } t < 0. \end{cases}$$

The window of the sine function is graphed in Figure 11. The window of the square wave is given in Figure 10.

The window of a periodic function is the basic building block of the function. The entire function consists of translates of the window. The following proposition uses this fact to give a useful relationship between the Laplace transform of a periodic function and the Laplace transform of its window.

PROPOSITION 5.17 Suppose f is periodic with period T and piecewise continuous. Let $F_T(s)$ be the Laplace transform of its window f_T. Then

$$\mathcal{L}(f)(s) = \frac{F_T(s)}{1 - e^{-Ts}} = \frac{\int_0^T f(t)e^{-st}\, dt}{1 - e^{-Ts}}. \tag{5.18}$$

The utility of this proposition is that it reduces the computation of the Laplace transform for a periodic function to that of its window, which is nonzero only on the finite interval, $0 \le t \le T$.

Proof For the proof, we first show that f can be written as a sum of its translates,

$$f(t) = f_T(t) + f_T(t - T) + f_T(t - 2T) + \cdots. \tag{5.19}$$

A typical periodic function f is illustrated in Figures 12 and 13 along with its window, $f_T(t)$, and the translate, $f_T(t - T)$, of its window. Graphically, we see that the translate $f_T(t - T)$ accounts for the part of $f(t)$ between T and $2T$. In general, the translate $f_T(t - kT)$ accounts for the part of $f(t)$ between kT and $(k + 1)T$. When

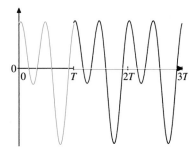

Figure 12. A periodic function f, with its window f_T plotted in blue.

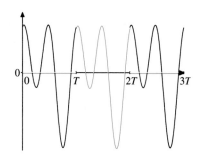

Figure 13. The same periodic function, with the translate of its window, $f_T(t - T)$, plotted in blue.

$f_T(t)$ and its translates are added together, the result is the entire periodic function $f(t)$, as claimed in equation (5.19).

Now, $H(t - kT)$ is 1 for $t \geq kT$ and zero otherwise. Hence for each integer k, $H(t - kT)f_T(t - kT) = f_T(t - kT)$ for all values of t. Therefore, from equation (5.19)

$$f(t) = f_T(t)H(t) + f_T(t - T)H(t - T) + f_T(t - 2T)H(t - 2T) + \cdots.$$

Taking the Laplace transform of each term using Proposition 5.6, we obtain

$$\mathcal{L}(f)(s) = \mathcal{L}\{f_T(t)H(t)\}(s) + \mathcal{L}\{f_T(t - T)H(t - T)\}(s)$$
$$+ \mathcal{L}\{f_T(t - 2T)H(t - 2T)\}(s) + \cdots$$
$$= F_T(s) + F_T(s)e^{-sT} + F_T(s)e^{-s2T} + \cdots$$
$$= F_T(s)\left(1 + e^{-sT} + e^{-s2T} + \cdots\right).$$

The series on the right is the geometric series $1 + r + r^2 + \cdots$, with $r = e^{-sT} < 1$. This series converges to $1/(1 - r) = 1/(1 - e^{-sT})$. Therefore,

$$\mathcal{L}(f)(s) = \frac{F_T(s)}{1 - e^{-Ts}}$$

as desired. ●

Example 5.20 Compute the Laplace transform of the square wave, f, given in Figure 9.

The square wave in Figure 9 has period $T = 2$. The window of the square wave is shown in Figure 10. Its formula is

$$f_T(t) = \begin{cases} 1, & 0 \leq t < 1; \\ -1, & 1 \leq t < 2; \\ 0, & \text{otherwise.} \end{cases}$$

In terms of the Heaviside function we can write f_T as

$$f_T(t) = H(t) - 2H_1(t) + H_2(t).$$

Its Laplace transform F_T can be computed using (5.4),

$$F_T(s) = \frac{1}{s} - 2\frac{e^{-s}}{s} + \frac{e^{-2s}}{s} = \frac{(1 - e^{-s})^2}{s}.$$

Therefore, from Proposition 5.17,

$$\mathcal{L}(f)(s) = \frac{(1 - e^{-s})^2}{s(1 - e^{-2s})} = \frac{1 - e^{-s}}{s(1 + e^{-s})}.$$ ●

Example 5.21   Solve the initial value problem

$$y'' + y = f(t), \quad \text{with} \quad y(0) = 0, \; y'(0) = 0,$$

where f is the square wave.

Using Proposition 2.4, we find that the Laplace transform of the left-hand side of the differential equation is

$$\mathcal{L}(y'' + y) = s^2 Y(s) + Y(s),$$

where $Y(s)$ is the Laplace transform of y. From Example 5.20, we must have

$$Y(s)(s^2 + 1) = \frac{1 - e^{-s}}{s(1 + e^{-s})}.$$

Solving for Y, we obtain

$$Y(s) = \frac{1}{s(s^2 + 1)} \frac{1 - e^{-s}}{1 + e^{-s}} = \frac{1}{s(s^2 + 1)} \left(\frac{2}{1 + e^{-s}} - 1 \right). \qquad (5.22)$$

Using partial fractions, the first factor on the right side of equation (5.22) becomes

$$\frac{1}{s(s^2 + 1)} = \frac{1}{s} - \frac{s}{s^2 + 1}.$$

In the second factor on the right side of equation (5.22), we can substitute the geometric series,

$$\frac{1}{1 + e^{-s}} = \frac{1}{1 - (-e^{-s})} = 1 - e^{-s} + e^{-2s} - \cdots + (-1)^n e^{-ns} + \cdots.$$

Combining these two equations with (5.22), we obtain

$$Y(s) = \left(\frac{1}{s} - \frac{s}{s^2 + 1} \right) \left(1 - 2e^{-s} + 2e^{-2s} - \cdots + (-1)^n 2e^{-ns} + \cdots \right).$$

From Table 1, we find that

$$\mathcal{L}^{-1} \left\{ \frac{1}{s} - \frac{s}{s^2 + 1} \right\} = 1 - \cos t.$$

From Proposition 5.6, we see that each factor of e^{-ns} translates this function by n and inserts a factor of $H(t - n)$. Therefore, the inverse Laplace transform of Y is

$$y(t) = (1 - \cos t)H(t) - 2(1 - \cos(t - 1))H(t - 1)$$
$$+ 2(1 - \cos(t - 2))H(t - 2) - \cdots$$
$$+ (-1)^n 2(1 - \cos(t - n))H(t - n) + \cdots.$$

Although the solution involves an infinite series, note that for any finite interval only a finite number of terms are nonzero, because $H(t - n)$ is zero for $0 \le t < n$. For example, on the interval $0 \le t < 10$,

$$y(t) = 1 - \cos t + 2 \sum_{n=1}^{9} (-1)^n (1 - \cos(t - n))H(t - n).$$

A graph of the solution $y(t)$ for $0 \le t < 10$ is given in Figure 14.

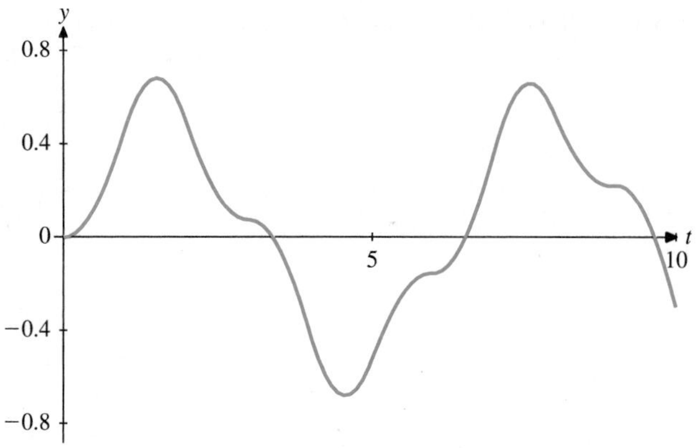

Figure 14. The solution to the initial value problem in Example 5.21.

EXERCISES

In Exercises 1–8, use Proposition 5.6 to find the Laplace transform of the given function.

1. $H(t-2)(t-2)$

2. $H(t-1)e^{2(t-1)}$

3. $H(t-\pi/4)\sin 3(t-\pi/4)$ **4.** $H(t-\pi/6)\sin(t-\pi/6)$

5. $H(t-1)t^2$

6. $H(t-2)e^{-t}$

7. $H(t-\pi/6)\sin 2t$

8. $H(t-\pi/2)\cos 3t$

9. In this exercise you will examine the effect of shifts in the time domain on the Laplace transform.

  (a) Sketch the graph of $f(t) = \sin t$ in the time domain. Find the Laplace transform $F(s) = \mathcal{L}\{f(t)\}(s)$. Sketch the graph of F in the s-domain on the interval $[0, 2]$.

  (b) Sketch the graph of $g(t) = H(t-1)\sin(t-1)$ in the time domain. Find the Laplace transform $G(s) = \mathcal{L}\{g(t)\}(s)$. Sketch the graph of G in the s-domain on the interval $[0, 2]$ on the same axes used to sketch the graph of F.

  (c) Repeat the directions in part (b) for $g(t) = H(t-2)\sin(t-2)$. Explain why engineers like to say that "a shift in the time domain leads to an attenuation (scaling) in the s-domain."

Use the Heaviside function to redefine each piecewise function in Exercises 10–15. Then use Proposition 5.6 to find its Laplace transform.

10. $f(t) = \begin{cases} 0, & \text{if } t < 0; \\ 3, & \text{if } t \geq 0. \end{cases}$

11. $f(t) = \begin{cases} 5, & \text{if } 2 \leq t < 4; \\ 0, & \text{otherwise.} \end{cases}$

12. $f(t) = \begin{cases} 0, & \text{if } t < 0; \\ t, & \text{if } 0 \leq t < 3; \\ 3, & \text{if } t \geq 3. \end{cases}$

13. $f(t) = \begin{cases} 0, & \text{if } t < 0; \\ t^2, & \text{if } 0 \leq t < 2; \\ 4, & \text{if } t \geq 2. \end{cases}$

14. $f(t) = \begin{cases} 3, & \text{if } 0 \leq t < 1; \\ 2, & \text{if } 1 \leq t < 2; \\ 1, & \text{if } 2 \leq t < 3; \\ 0, & \text{otherwise.} \end{cases}$

15. $f(t) = \begin{cases} t, & \text{if } 0 \leq t < 1; \\ 2-t, & \text{if } 1 \leq t < 3; \\ t-4, & \text{if } 3 \leq t < 4; \\ 0, & \text{otherwise.} \end{cases}$

Find the inverse Laplace transform of each function in Exercises 16–25. Create a piecewise definition for your solution that doesn't use the Heaviside function.

16. $F(s) = \dfrac{e^{-2s}}{s+3}$

17. $F(s) = \dfrac{e^{-s}}{s-2}$

18. $F(s) = \dfrac{1-e^{-s}}{s^2}$

19. $F(s) = \dfrac{2+e^{-2s}}{s^3}$

20. $F(s) = \dfrac{e^{-s}}{s^2+4}$

21. $F(s) = \dfrac{se^{-3s}}{s^2+4}$

22. $F(s) = \dfrac{1-e^{-s}}{s(s+2)}$

23. $F(s) = \dfrac{e^{-2s}}{s^2-2s-3}$

24. $F(s) = \dfrac{e^{-s}}{s(s-2)^2}$

25. $F(s) = \dfrac{2-e^{-2s}}{s^2+2s+2}$

Consider the driven, undamped oscillator

$$y'' + 4y = f(t), \quad y(0) = 0, \quad y'(0) = 0,$$

with driving force $f(t)$ pictured in Exercises 26–29. In each exercise, determine the solution, or response, of the differential equation to the input $f(t)$. Sketch the graph of the solution in the time domain.

26.

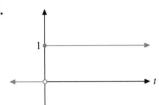

27.

28.

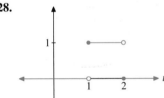

29.

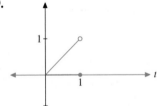

30. The following image shows the graph of a periodic function $f(t)$.

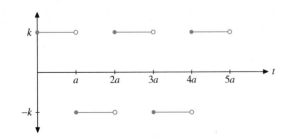

Show that the Laplace transform of $f(t)$ is

$$\mathcal{L}\{f(t)\}(s) = \frac{k}{s} \tanh \frac{1}{2} as.$$

31. The following image shows the graph of a periodic function $f(t)$.

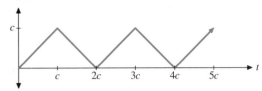

Show that the Laplace transform of $f(t)$ is

$$\mathcal{L}\{f(t)\}(s) = \frac{1}{s^2} \tanh \frac{1}{2} cs.$$

32. The following image shows the graph of a periodic function $f(t)$.

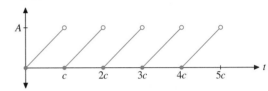

Show that the Laplace transform of $f(t)$ is

$$\mathcal{L}\{f(t)\}(s) = \frac{A}{cs^2} - \frac{A}{s(e^{cs} - 1)}.$$

33. In mathematics, the symbolism $[t]$ calls for the greatest integer not exceeding the number t. For example, $[3.1] = 3$, $[3] = 3$, and $[-1.2] = -2$. Define

$$f(t) = \begin{cases} [t], & t \geq 0 \\ 0, & \text{otherwise.} \end{cases}$$

(a) Sketch the graph of the function f. Remark why this function is often called the *staircase function*.

(b) Show that the Laplace transform is given by

$$\mathcal{L}\{f(t)\}(s) = \frac{1}{s(1 - e^{-s})}.$$

Hint: First show that $f(t) = t - h(t)$, where $h(t)$ is the function pictured in Exercise 32, using $A = 1$ and $c = 1$.

34. Functions possessing the property

$$f(t + T/2) = -f(t), \quad \text{for all } t, \tag{5.23}$$

are sometimes called **antiperiodic**.

(a) Sketch the graph of a function possessing property (5.23).

(b) Show that a function possessing property (5.23) is periodic with period T.

(c) Let f be a function possessing property (5.23). Show that the Laplace transform of f is

$$F(s) = \frac{G(s)}{1 + e^{-sT/2}},$$

where

$$G(s) = \int_0^{T/2} f(t)e^{-st} \, dt.$$

(d) An antiperiodic functions that is nonnegative on its first half-period, $0 \leq t \leq T/2$, is "rectified" by defining $f_R(t) = f(t)$ on $0 \leq t \leq T/2$, and then requiring that $f_R(t + T/2) = f_R(t)$, for all t. Find an example of an antiperiodic function that is positive on its first half-period. Sketch a graph of this function and its rectification.

(e) Show that the Laplace transform of the rectification of an antiperiodic function f is provided by

$$\mathcal{L}\{f_R(t)\}(s) = \mathcal{L} \, f(t)(s) \coth \frac{sT}{4}.$$

(f) Use parts (d) and (e) to compute the Laplace transform of $g(t) = |\sin t|$.

35. Reread Proposition 2.1, then consider the function defined by

$$y(t) = \begin{cases} 0 & t < 0; \\ t, & 0 \leq t < 1; \\ t - 1, & t \geq 1. \end{cases}$$

(a) Sketch this function. Show further that $y(t) = tH(t) - H(t - 1)$. Use the technique of this section to show that $s \, \mathcal{L}\{y(t)\}(s) - y(0) = 1/s - e^{-s}$.

(b) Calculate $y'(t)$ and provide a piecewise definition. Sketch the graph of $y'(t)$. Use the definition of the Laplace transform to compute that $\mathcal{L}\{y'(t)\}(s) = 1/s$.

(c) Note that parts (a) and (b) imply that $\mathcal{L}\{y'(t)\}(s) \neq s \, \mathcal{L}\{y(t)\}(s) - y(0)$. Why doesn't this contradict Proposition 2.1?

(d) Show, in this case, that

$$\mathcal{L}\{y'(t)\}(s) = s \, \mathcal{L}\{y(t)\}(s) - \left[y(1^+) - y(1^-)\right]e^{-s} - y(0).$$

36. Suppose that a fish population in a lake grows according to the Malthusian model $dp/dt = rp$, where $p(t)$ is the number of pounds of fish in the lake after t months. Suppose that fish are harvested from the lake at a constant rate h (pounds per month), but only for the first month of the year. This harvesting strategy is easily modeled with the Heaviside function.

$$\frac{dp}{dt} = rp - h \left(H(t) - H(t - 1)\right), \quad p(0) = p_0 \quad (5.24)$$

Note that we assume an initial fish population $p(0) = p_0$.

(a) Use the Laplace transform to show that the solution of equation (5.24) is

$$p(t) = \left(p_0 - \frac{h}{r}\right)e^{rt} + \frac{h}{r}\left(H(t) - H(t - 1)\right)$$
$$+ \frac{h}{r}H(t - 1)e^{r(t-1)}.$$

(b) Let $r = 0.05$, $p_0 = 50$. Use your numerical solver and equation (5.24) to estimate the value of h that separates extinction from survival of the fish population in the lake.

(c) Use the result in part (a) to determine the value of h (in terms of p_0 and r) that allows extinction of the fish population to occur precisely at $t = 1$. Does this result correctly predict the approximation found in part (b)? Use equation (5.24) to argue that this population will remain extinct for all time.

5.6 The Delta Function

In this section, we discuss the delta function, which is a model for a force that concentrates a large amount of energy over a short time interval. A typical example is a blow struck with a hammer.

Impulse functions

In order to motivate the definition of the delta function, we digress and discuss the impulse of a force. We make the following definition.

DEFINITION 6.1

> Suppose $F(t)$ represents a force applied to an object m at time t. Then the **impulse** of F over the time interval $a \leq t \leq b$ is defined as
>
> $$\text{impulse} = \int_a^b F(t) \, dt.$$

Geometrically, the impulse of $F(t)$ is the area under the curve $y = F(t)$, for $a \leq t \leq b$. From the point of view of mechanics, the impulse is the change in momentum of a mass as the force is applied to it over the time interval $a \le t \le b$. To see this, remember that the momentum is the product of the mass with the velocity, and by Newton's second law, the force is $F(t) = ma = m\, dv/dt$. Therefore,

$$\text{impulse} = \int_a^b F(t)\, dt = \int_a^b m\frac{dv}{dt}\, dt$$

$$= mv(t)\Big|_{t=a}^{b}$$

$$= mv(b) - mv(a),$$

which is the change of the momentum from $t = a$ to $t = b$.

Now let's consider a force of unit impulse acting over a short time interval starting at time $t = p$. An example is the force δ_p^ϵ shown in Figure 1. The formula for this function is

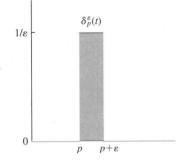

$$\delta_p^\epsilon(t) = \begin{cases} \dfrac{1}{\epsilon}, & \text{for } p \le t < p + \epsilon; \\[2mm] 0, & \text{for } t < p \text{ or } t \ge p + \epsilon \end{cases}$$

$$= \frac{1}{\epsilon}\left(H_p(t) - H_{p+\epsilon}(t)\right),$$

Figure 1. A unit impulse force over a small interval.

where H_p is the translation by p units of the Heaviside function. The force δ_p^ϵ is applied over the time interval $p \le t < p + \epsilon$ and has intensity equal to $1/\epsilon$. The area under the graph of $y = \delta_p^\epsilon(t)$ can be easily computed because the region under its graph is a rectangle of width ϵ and height $1/\epsilon$. Thus, the impulse of this force is 1.

The graphs of the function δ_p^ϵ for $\epsilon = 1, 1/2, 1/4$, and $1/8$ are given in Figure 2. Note that as ϵ gets smaller, the region under the graph of δ_p^ϵ becomes a tall, thin rectangle beginning at the point p. However, no matter what the value of ϵ, the area of this rectangle is 1.

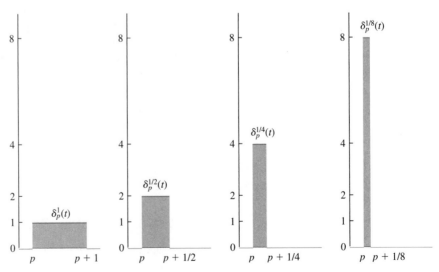

Figure 2. The functions $\delta_p^\epsilon(t)$ with $\epsilon = 1, 1/2, 1/4$, and $1/8$. In each case the area under the graph is equal to 1.

A reasonable model for a sharp, instantaneous force at time $t = p$ (e.g., a hammer blow) is obtained by taking the limit of δ_p^ϵ as $\epsilon \to 0$. We make this the definition of the delta function centered at p.

DEFINITION 6.2

> The ***delta function*** centered at $t = p$ is the limit
>
> $$\delta_p(t) = \lim_{\epsilon \to 0} \delta_p^\epsilon(t).$$
>
> When $p = 0$, we will set $\delta = \delta_0$.

Since the rectangles in Figure 2 get higher and higher as ϵ gets smaller, the delta function cannot be a function in the traditional sense. In the limit we would have $\delta_p(p) = \infty$ and $\delta_p(t) = 0$ for $t \neq p$. From the traditional view of functions, it is difficult to see how this is different from the function that is identically zero.

In fact, the delta "function" is not a function at all. It is what mathematicians call a *generalized function* or a *distribution*. We must make precise the meaning of the limit in Definition 6.2. It may seem strange, but the most useful way of viewing the delta function is to examine how it acts when integrated against another function. If ϕ is a continuous function, then

$$\int_0^\infty \delta_p(t)\phi(t)\,dt = \lim_{\epsilon \to 0} \int_0^\infty \delta_p^\epsilon(t)\phi(t)\,dt = \lim_{\epsilon \to 0} \frac{1}{\epsilon} \int_p^{p+\epsilon} \phi(t)\,dt.$$

The situation might be clearer to the reader if we make the change of variables $t = p + \epsilon\tau$ in the last integral. Since $dt = \epsilon\,d\tau$, we get

$$\int_0^\infty \delta_p(t)\phi(t)\,dt = \lim_{\epsilon \to 0} \int_0^1 \phi(p + \epsilon\tau)\,d\tau.$$

Figure 3 shows a typical continuous function $\phi(t)$. In Figure 4 we show the same function plotted over the small interval $p \leq t \leq p + \epsilon$, when ϵ is small. Over this interval $\phi(t) = \phi(p + \epsilon\tau)$ differs little from $\phi(p)$. Therefore, we can make the approximation

$$\int_0^1 \phi(p + \epsilon\tau)\,d\tau \approx \int_0^1 \phi(p)\,d\tau = \phi(p).$$

This approximation gets better as ϵ gets smaller.

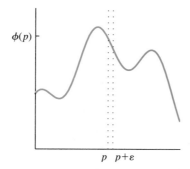

Figure 3. A continuous function $\phi(t)$.

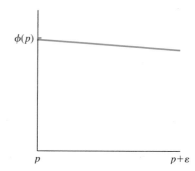

Figure 4. The function $\phi(t)$ over the small interval $[p, p + \epsilon]$.

We summarize this discussion in the following theorem.

THEOREM 6.3 Suppose $p \geq 0$ is any fixed point and let ϕ be any function that is continuous at $t = p$. Then

$$\int_0^\infty \delta_p(t)\phi(t)\,dt = \phi(p).$$ ∎

In particular, this theorem can be used to compute the Laplace transform of the delta function centered at p.

THEOREM 6.4 For $p \geq 0$, the Laplace transform of δ_p is given by

$$\mathcal{L}(\delta_p)(s) = e^{-sp}.$$

Proof Indeed, with $\phi(t) = e^{-st}$, Theorem 6.3 gives

$$\mathcal{L}(\delta_p)(s) = \int_0^\infty \delta_p(t)e^{-st}\,dt = e^{-sp},$$

which establishes Theorem 6.4. ∎

The case when $p = 0$ is especially important, so we state it as a corollary.

COROLLARY 6.5 $\mathcal{L}(\delta_0)(s) = \mathcal{L}(\delta)(s) = 1.$

Impulse response functions

Having computed the Laplace transform of the delta function, we can now solve differential equations involving the delta function. In what follows, we will be looking at a number of differential equations of the form

$$ay'' + by' + cy = f(t), \quad \text{with} \quad y(0) = y_0, \quad \text{and} \quad y'(0) = y_1, \tag{6.6}$$

with different forcing functions f and different initial conditions y_0 and y_1.

An engineer would arrive at the differential equation (6.6) as the model of a physical *system*, which might be an *RLC* circuit, a vibrating spring, or a variety of other systems. Because of this, equation (6.6) is called a ***system***. The solution y is called the ***response*** of the system to the forcing function f and the initial conditions. In what remains of this chapter we will adopt this terminology.

First we will discuss the response of our system to a very special forcing function.

DEFINITION 6.7 The solution $e(t)$ to the initial value problem

$$ae'' + be' + ce = \delta_0(t), \quad \text{with} \quad e(0) = 0, \quad \text{and} \quad e'(0) = 0$$

is called the ***unit impulse response function*** to the system modeled by the differential equation.

The unit impulse response function may be viewed as the response of a mechanical spring system that is governed by the differential equation in (6.6) and set in motion at time $t = 0$, by a hammer blow of impulse 1, represented by the delta function.

Example 6.8   Solve the following initial value problem:

$$e'' + 2e' + 2e = \delta_0(t) = \delta(t), \quad \text{with} \quad e(0) = 0, \quad \text{and} \quad e'(0) = 0.$$

That is, find the unit impulse response function for the system.

With the given initial conditions, the Laplace transform of the left-hand side of the differential equations is

$$\mathcal{L}(e'' + 2e' + 2e) = E(s)(s^2 + 2s + 2),$$

where $E(s) = \mathcal{L}(e)(s)$. Since the Laplace transform of the delta function δ_0 is 1, we get

$$E(s)(s^2 + 2s + 2) = 1.$$

Solving for E and then completing the square gives

$$E(s) = \frac{1}{s^2 + 2s + 2} = \frac{1}{(s+1)^2 + 1}. \tag{6.9}$$

Proposition 2.12 (or Table 1) now gives the solution

$$e(t) = e^{-t} \sin t.$$

Note that the initial condition $e(0) = 0$ is satisfied. It appears, however, that the other initial condition $e'(0) = 0$ is not satisfied since $e'(t) = e^{-t}(\cos t - \sin t)$, and $e'(0) = 1$. However, when using the Laplace transform, the assumption is that the function (in this case $e'(t)$) satisfies $\lim_{t \to 0^-} e'(t) = 0$. A better way of defining our initial condition in this case is to require that $e(t) = 0$ for $t < 0$. Therefore, a complete description of the solution is

$$e(t) = \begin{cases} 0, & \text{for } t < 0; \\ e^{-t} \sin t, & \text{for } t \geq 0. \end{cases}$$

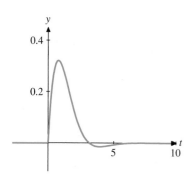

Figure 5. The unit impulse response function computed in Example 6.8.

The graph of $e(t)$ is shown in Figure 5.   ●

We see that the forcing term δ_0 causes an abrupt change in direction of the solution at $t = 0$. More precisely, we have

$$\lim_{t \to 0^-} e'(t) = 0, \quad \text{and} \quad \lim_{t \to 0^+} e'(t) = 1.$$

Technically, the derivative, e', does not exist at $t = 0$ in view of the corner on its graph, but the left limit of e' at $t = 0$ is 0. In this sense, the initial condition $e'(0) = 0$ is satisfied.

In Exercise 9, the solution to this problem is computed as a limit of solutions to the equations

$$y'' + 2y' + 2y = \delta_0^\epsilon(t)$$

as $\epsilon \to 0$.

Let's look back at Equation (6.9). It says that the Laplace transform of the unit impulse function is the reciprocal of the polynomial $s^2 + 2s + 2$. Notice that this is the characteristic polynomial of the differential equation being solved in Example 6.8. This is not an isolated phenomenon, and it is important enough to make it a theorem.

THEOREM 6.10 Let $e(t)$ be the unit impulse response function for the system modeled by the equation

$$ay'' + by' + cy = f(t).$$

The Laplace transform of e is the reciprocal of the characteristic polynomial $P(s) = as^2 + bs + c$,

$$\mathcal{L}(e)(s) = E(s) = \frac{1}{P(s)} = \frac{1}{as^2 + bs + c} \quad \text{for } s \geq 0.$$

Proof We only need to carry out the computation at the beginning of Example 6.8 for the general differential operator. We need to solve the initial value problem

$$a\frac{d^2e}{dt^2} + b\frac{de}{dt} + ce = \delta, \quad \text{with} \quad e(0) = 0, \quad \text{and} \quad e'(0) = 0.$$

With these initial conditions, and with $E(s) = \mathcal{L}(e)(s)$, the Laplace transform of the left-hand side is

$$\mathcal{L}\left(a\frac{d^2e}{dt^2} + b\frac{de}{dt} + ce\right)(s) = E(s)(as^2 + bs + c) = E(s)P(s).$$

Since the Laplace transform of the delta function $\delta = \delta_0$ is the function 1, we must have

$$E(s)P(s) = 1.$$

Hence $E(s) = 1/P(s)$, as claimed. ∎

EXERCISES

1. Recall that

$$\delta_p^\epsilon(t) = \frac{1}{\epsilon}\left(H_p(t) - H_{p+\epsilon}(t)\right).$$

(a) Show that the Laplace transform of $\delta_p^\epsilon(t)$ is given by

$$\mathcal{L}\left\{\delta_p^\epsilon(t)\right\} = e^{-sp}\frac{1 - e^{-s\epsilon}}{s\epsilon}.$$

· (b) Use l'Hôpital's rule to take the limit of the result in part (a) as $\epsilon \to 0$. How does this result agree with Theorem 6.4?

In Exercises 2–7, find the unit impulse response to the given system. Assume $y(0) = y'(0) = 0$.

2. $y'' - 4y' + 3y = \delta(t)$ **3.** $y'' - 4y' - 5y = \delta(t)$

4. $y'' + 4y = \delta(t)$ **5.** $y'' - 9y = \delta(t)$

6. $y'' + 4y' + 5y = \delta(t)$ **7.** $y'' + 2y' + 2y = \delta(t)$

8. (a) Use the fact that $\mathcal{L}\{\delta(t)\}(s) = 1$ to show that the solution of

$$x'' = 2\delta(t), \quad x(0) = x'(0) = 0$$

is $x(t) = 2tH(t)$, where H is the Heaviside function. Sketch the graph of this solution.

(b) Next, consider the equation

$$x'' = 2\delta_0^\epsilon(t), \quad x(0) = x'(0) = 0.$$

Show that the solution of this equation is given by

$$x_\epsilon(t) = \begin{cases} \frac{1}{\epsilon}t^2, & 0 \leq t < \epsilon; \\ 2t - \epsilon, & t \geq \epsilon. \end{cases}$$

Sketch the graph of this solution.

(c) Does the solution in part (b) approach the solution in part (a) as $\epsilon \to 0$, at least for $t \geq 0$?

9. Consider the equation

$$x'' + 2x' + 2x = \delta(t), \quad x(0) = x'(0) = 0. \quad (6.11)$$

(a) Use the fact that $\mathcal{L}\{\delta(t)\}(s) = 1$ to show that the solution of equation (6.11) is $x(t) = e^{-t}\sin t$.

(b) Show that the solution of

$$x'' + 2x' + 2x = \delta_0^\epsilon(t), \quad x(0) = x'(0) = 0$$

is

$$x_\epsilon(t) = \frac{1}{2\epsilon} \begin{cases} 1 - e^{-t}(\cos t + \sin t), & 0 \le t < \epsilon; \\ -e^{-t}(\cos t + \sin t) \\ \quad + e^{-(t-\epsilon)}(\cos(t - \epsilon) \\ \quad + \sin(t - \epsilon)), & t \ge \epsilon. \end{cases}$$

(c) Use l'Hôpital's rule to argue that the solution of part (b) approaches that of part (a) as $\epsilon \to 0$, at least for $t > 0$.

10. Shown is a defining graph of the function, $H_p^\epsilon(t)$. Without being too precise about things, we could argue that $H_p^\epsilon(t) \to H_p(t)$ as $\epsilon \to 0$, where $H_p(t) = H(t - p)$ and H is the Heaviside function. Sketch the graph of the derivative of $H_p^\epsilon(t)$. Compare the result with the graph of $\delta_p^\epsilon(t)$. Argue that $H_p'(t) = \delta_p(t)$.

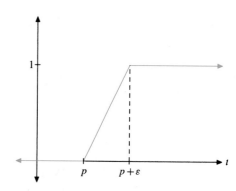

11. Use the fact that $\mathcal{L}\{\delta_p(t)\}(s) = e^{-ps}$ to show that the solution of the equation

$$x' = \delta_p(t), \quad x(0) = 0$$

is $x(t) = H_p(t)$, giving further credence to the argument in Exercise 10 that the "derivative of a unit step is a unit impulse," as engineers like to say.

5.7 Convolutions

In our overview of the use of the Laplace transform to solve initial value problems in equation (4.9) at the end of Section 5.4, we found that the solution to the initial value problem

$$ay'' + by' + cy = f(t), \quad \text{with} \quad y(0) = y'(0) = 0 \tag{7.1}$$

has Laplace transform

$$Y(s) = \frac{F(s)}{as^2 + bs + c}, \tag{7.2}$$

where $\mathcal{L}(f) = F$. The denominator on the right in (7.2) is the characteristic polynomial of the differential equation. Thus, we can use Theorem 6.10 to rewrite equation (7.2) as

$$Y(s) = E(s)F(s), \tag{7.3}$$

where $E(s)$ is the Laplace transform for the unit impulse response function $e(t)$ for the system modeled by the differential equation (7.1).

The expression on the right in (7.3) is the product of the Laplace transforms of the unit impulse response function and the forcing function. We can find the solution to (7.1) by taking the inverse Laplace transform of (7.3),

$$y(t) = \mathcal{L}^{-1}\{Y(s)\}(t) = \mathcal{L}^{-1}\{E(s)F(s)\}(t).$$

Thus we see that the product of Laplace transforms arises naturally in the solution of initial value problems. That is the topic of this section.

The convolution product

We will compute the inverse Laplace transform of a product in our first result.

PROPOSITION 7.4 Suppose that f and g are functions of exponential order. Suppose that $\mathcal{L}(f)(s) = F(s)$ and $\mathcal{L}(g)(s) = G(s)$. Then

$$\mathcal{L}^{-1}\{F(s)G(s)\}(t) = \int_0^t f(u)g(t - u) \, du. \qquad \bullet$$

Note that the variable u on the right is the variable of integration, so the resulting expression on the right is a function of t only. Of course, there is nothing special about u. The variable of integration can be anything.

The importance of this result in the solution of differential equations requires that we make a definition.

DEFINITION 7.5

The **convolution** of two piecewise continuous functions f and g is the function $f * g$ defined by

$$f * g(t) = \int_0^t f(u)g(t - u)\,du.$$

With this definition, we can write the result in Proposition 7.4 as

$$\mathcal{L}^{-1}\{\mathcal{L}(f)(s)\,\mathcal{L}(g)(s)\} = f * g.$$

If we apply the Laplace transform to this, we can state the result of Proposition 7.4 in an equivalent form, which is frequently more convenient.

THEOREM 7.6   Suppose f and g are functions of exponential order. Then

$$\mathcal{L}(f * g)(s) = \mathcal{L}(f)(s)\,\mathcal{L}(g)(s). \tag{7.7}$$

We should say a word or two about notation. The convolution product is an operation that takes priority over functional evaluation. That means that in the expression $f * g(t)$ the convolution product is computed first, before the evaluation at t. If we were really careful, we would write it as

$$f * g(t) = (f * g)(t),$$

but if we remember the priority of the operations, there is no need.

Let's prove Theorem 7.6.

Proof   To prove this theorem, we unravel the left side of Equation (7.7).

$$\begin{aligned}
\mathcal{L}(f * g)(s) &= \int_0^\infty f * g(t)e^{-st}\,dt \\
&= \int_0^\infty \left(\int_0^t f(u)g(t - u)\,du \right) e^{-st}\,dt \tag{7.8} \\
&= \int_0^\infty \int_0^t f(u)g(t - u)e^{-st}\,du\,dt
\end{aligned}$$

We want to interchange the order of integration in (7.8). In doing so we have to be careful with the limits of integration. In the inner integral we have $0 \le u \le t$, and in the outer integral $0 \le t < \infty$. We can put both of these inequalities in the one chain of inequalities

$$0 \le u \le t < \infty.$$

These limits must be maintained when the order of integration is reversed. The inner integral will be with respect to t, and the indicated limits on t are $u \le t < \infty$. The limits on u are $0 \le u < \infty$. Hence we get

$$\mathcal{L}(f * g)(s) = \int_0^\infty \int_u^\infty f(u)g(t - u)e^{-st}\,dt\,du.$$

At this point, it is natural to replace t, the variable of integration in the inner integral, by $v = t - u$. Then $t = u + v$ and $dv = dt$, so we get

$$\mathcal{L}(f * g)(s) = \int_0^\infty \int_0^\infty f(u)g(v)e^{-s(u+v)}\, dv\, du.$$

Finally, we use the exponential formula $e^{-s(u+v)} = e^{-su}e^{-sv}$ to write the multiple integral as the product of two integrals.

$$\mathcal{L}(f * g)(s) = \int_0^\infty \int_0^\infty f(u)g(v)e^{-su}e^{-sv}\, dv\, du$$

$$= \int_0^\infty f(u)e^{-su}\, du \int_0^\infty g(v)e^{-sv}\, dv$$

We recognize this as the product of two Laplace transforms, and we have

$$\mathcal{L}(f * g)(s) = \mathcal{L}(f)(s)\, \mathcal{L}(g)(s),$$

as we wanted to show. ∎

We emphasize that the convolution product of two functions is not the same as the ordinary product of those functions. This is shown by our first example.

Example 7.9   Let $f(t) = t^2 - 2t$ and $g(t) = t$. Compute $f * g(t)$.

We have

$$f * g(t) = \int_0^t f(u)g(t - u)\, du = \int_0^t (u^2 - 2u)(t - u)\, du$$

$$= \int_0^t (tu^2 - 2tu - u^3 + 2u^2)\, du = \frac{t^4}{12} - \frac{t^3}{3}.$$ ●

Since $f(t)g(t) = t^3 - 2t^2$, this example shows clearly that the convolution product is different from the ordinary product. However, as the following theorem shows, the convolution product does have many of the same properties as an ordinary product.

THEOREM 7.10   Suppose f, g, and h are piecewise continuous functions. Then

1. $f * g = g * f$
2. $f * (g + h) = f * g + f * h$
3. $(f * g) * h = f * (g * h)$
4. $f * 0 = 0$

Proof   To prove (1), we write out the definition of convolution,

$$f * g(t) = \int_0^t f(u)g(t - u)\, du,$$

and then replace the variable of integration u with $v = t - u$. Then $dv = -du$ since t is a constant in the integral. The limits of integration also change. For $u = 0$ we have $v = t$, and for $u = t$ we have $v = 0$. Thus

$$f * g(t) = -\int_t^0 g(v)f(t - v)\, dv = \int_0^t g(v)f(t - v)\, dv.$$

The expression on the right is $g * f(t)$. The fact that v is the variable of integration instead of u is not important.

The other properties are proved in the exercises. ∎

Example 7.11   Let $f(t) = \sin t$, and $g(t) = t$. Compute the convolution $f * g$ in two ways. First do it directly from Definition 7.5. Next, compute $F = \mathcal{L}(f)$ and $G = \mathcal{L}(g)$, and then compute $f * g = \mathcal{L}^{-1}\{\mathcal{L}(f)\,\mathcal{L}(g)\}$, thereby confirming Theorem 7.6.

We use the formula and then integrate by parts to get

$$f * g(t) = \int_0^t f(u)g(t - u)\,du$$

$$= \int_0^t \sin u \cdot (t - u)\,du$$

$$= -\int_0^t (t - u)\,d(\cos u)$$

$$= -(t - u)\cos u \Big|_0^t - \int_0^t \cos u\,du$$

$$= t - \sin t.$$

From Table 1 on page 204 we get $F(s) = \mathcal{L}(f)(s) = 1/(s^2 + 1)$, and $G(s) = \mathcal{L}(g)(s) = 1/s^2$. Hence,

$$F(s)G(s) = \frac{1}{s^2(s^2 + 1)} = \frac{1}{s^2} - \frac{1}{s^2 + 1}.$$

Then

$$\mathcal{L}^{-1}(FG)(t) = \mathcal{L}^{-1}\left\{\frac{1}{s^2}\right\}(t) - \mathcal{L}^{-1}\left\{\frac{1}{s^2 + 1}\right\}(t) = t - \sin t = f * g(t). \quad \bullet$$

The solution to the general initial value problem

We will now derive a formula for the solution to the general initial value problem

$$ay'' + by' + cy = f(t), \quad \text{with} \quad y(0) = y_0, \quad \text{and} \quad y'(0) = y_1. \qquad (7.12)$$

The method that we will use is illustrated in the following example.

Example 7.13   Compute the solution to the equation

$$y'' + y = g(t), \quad \text{with} \quad y(0) = 0, \quad \text{and} \quad y'(0) = 0,$$

where g is any piecewise continuous function with exponential order. In particular, compute the response when $g(t) = t$.

We refer the reader to the end of Section 5.4. In the terminology we introduced there, we are looking for a state-free solution to the differential equation.

The characteristic polynomial of the differential equation is $P(s) = s^2 + 1$. By Theorem 6.10, the Laplace transform of the unit impulse response function $e(t)$ is

$$E(s) = \frac{1}{P(s)} = \frac{1}{s^2 + 1}.$$

The initial value problem in this example is a special case of that in (7.1), so according to equation (7.3), we have

$$Y(s) = E(s)G(s).$$

By Proposition 7.4 we have

$$y(t) = e * g(t).$$

From Table 1 we see that $1/(s^2 + 1)$ is the Laplace transform of $\sin t$, so $e(t) = \sin t$. Consequently, our solution is

$$y(t) = e * g(t) = \int_0^t (\sin u)\, g(t - u)\, du = \int_0^t \sin(t - u) g(u)\, du.$$

If $g(t) = t$, the response is given by $y = e * g$, but since $e(t) = \sin t$, this was computed in Example 7.11. Thus,

$$y(t) = t - \sin t. \qquad \bullet$$

With this example in mind, let's return to the discussion of the general initial value problem in (7.12). At the end of Section 5.4, we showed that the solution to the initial value problem in (7.12) can be written as

$$y(t) = y_s(t) + y_i(t),$$

where y_s is the state-free solution and y_i is the input-free solution. The state-free solution y_s is the solution of the initial value problem

$$ay_s'' + by_s' + cy_s = f(t), \quad \text{with} \quad y_s(0) = y_s'(0) = 0.$$

The initial data $y(0)$ and $y'(0)$ give a measure of the energy in the system at time $t = 0$. Thus the state-free solution is the response of the system to the driving force with no energy in the system at the starting point. According to equation (7.3), we have

$$\mathcal{L}(y_s)(s) = E(s)F(s),$$

where $E(s)$ is the Laplace transform of the unit impulse function $e(t)$, and $F(s)$ is the Laplace transform of the forcing term $f(t)$. Thus, by Proposition 7.4, the state-free solution is given by the convolution

$$y_s(t) = e * f(t). \qquad (7.14)$$

The input-free solution y_i is the solution of

$$ay_i'' + by_i' + cy_i = 0, \quad \text{with} \quad y_i(0) = y_0, \quad \text{and} \quad y_i'(0) = y_1.$$

Thus, the input-free solution is the response to the system with no driving force. It is due entirely to the energy that is in the system at time $t = 0$. Using equation (4.13) and Theorem 6.10, we see that the input-free solution satisfies

$$\mathcal{L}(y_i)(s) = E(s)\,(y_0(as + b) + ay_1)$$
$$= ay_0 s E(s) + (ay_1 + by_0)E(s).$$

We can compute the inverse Laplace transform of the last expression using Proposition 2.1. It says that $\mathcal{L}\{e'\}(s) = sE(s) - e(0) = sE(s)$. Thus we get

$$y_i(t) = ay_0 e'(t) + (ay_1 + by_0)e(t). \qquad (7.15)$$

We will summarize our discussions in a theorem.

THEOREM 7.16   The solution to the initial value problem

$$ay'' + by' + cy = f(t), \quad \text{with} \quad y(0) = y_0, \quad \text{and} \quad y'(0) = y_1$$

can be written as

$$y(t) = y_s(t) + y_i(t).$$

If $e(t)$ is the unit impulse response function for the system, then the state-free response is

$$y_s(t) = e * f(t),$$

and the input-free response is

$$y_i(t) = ay_0 e'(t) + (ay_1 + by_0)e(t).$$ ■

Theorem 7.16 highlights the importance of the unit impulse response function. Once it is known, we have a formula for the solution of any initial value problem. We will illustrate this with some examples.

Example 7.17   Compute the solution to the initial value problem

$$y'' + y = g(t), \quad \text{with} \quad y(0) = 1, \quad \text{and} \quad y'(0) = -2,$$

where g is a piecewise continuous function.

In Example 7.13, we have already discovered that the unit impulse response function for this equation is

$$e(t) = \sin t,$$

and that the state-free solution is

$$y_s(t) = e * g(t) = \int_0^t \sin(t - u)g(u)\, du.$$

According to Theorem 7.16, the input-free solution is

$$y_i(t) = y_0 e'(t) + y_1 e(t) = \cos t - 2 \sin t.$$

Hence the solution is

$$y(t) = \cos t - 2 \sin t + \int_0^t \sin(t - u)g(u)\, du.$$ ●

Example 7.18   Compute the solution to the initial value problem

$$y'' + 4y' + 13y = g(t), \quad \text{with} \quad y(0) = -5, \quad \text{and} \quad y'(0) = 2,$$

where g is a piecewise continuous function.

We start by computing the unit impulse response function $e(t)$. The characteristic polynomial for this equation is $P(s) = s^2 + 4s + 13 = (s + 2)^2 + 9$. By Theorem 6.10, the Laplace transform $E(s)$ of $e(t)$ is the reciprocal of $P(s)$, so $E(s) = 1/((s + 2)^2 + 9)$. From Table 1, we find that

$$e(t) = \frac{1}{3} e^{-2t} \sin 3t.$$

We compute the derivative

$$e'(t) = \frac{1}{3}e^{-2t}\left(3\cos 3t - 2\sin 3t\right).$$

Then from Theorem 7.16, we know that the solution is

$$y(t) = e * g(t) + y_0 e'(t) + (y_1 + 4y_0)e(t)$$

$$= \frac{1}{3}\int_0^t e^{-2(t-u)}\sin 3(t-u)\, g(u)\, du$$

$$- \frac{5}{3}e^{-2t}\left(3\cos 3t - 2\sin 3t\right) - 6e^{-2t}\sin 3t$$

$$= \frac{1}{3}\int_0^t e^{-2(t-u)}\sin 3(t-u)\, g(u)\, du - \frac{1}{3}e^{-2t}\left(8\sin 3t + 15\cos 3t\right). \quad \bullet$$

Other properties of the convolution

One difference between an ordinary product and the convolution product is that $f * 1$ is *not* equal to f. This means that the function 1 is not the identity for the convolution product. However, as the next theorem shows, there is an identity for the convolution product.

THEOREM 7.19   Suppose f is a piecewise continuous function. Then

$$f * \delta_0 = f = \delta_0 * f.$$

Proof   Technically speaking, we have not defined the convolution of f with the delta function, since the delta function is not a piecewise continuous function. However, the delta function is defined as the limit of the functions δ_0^ϵ, where δ_0^ϵ is the function that is $1/\epsilon$ on $0 \le t < \epsilon$ and zero otherwise (see Figure 2 in Section 5.6). So

$$\delta_0 * f(t) = \lim_{\epsilon \to 0} \delta_0^\epsilon * f(t).$$

With this definition, Theorem 7.19 is a reformulation of Theorem 6.3. Indeed, the same argument we used in the proof of Theorem 6.3 shows that

$$\delta_0 * f(t) = \lim_{\epsilon \to 0}\int_0^t \delta_0^\epsilon(u) f(t-u)\, du$$

$$= \lim_{\epsilon \to 0}\frac{1}{\epsilon}\int_0^\epsilon f(t-u)\, du$$

$$= f(t). \qquad \blacksquare$$

It is interesting to verify the result of Theorem 7.19 by taking the Laplace transform. We have, by Theorem 7.6,

$$\mathcal{L}(\delta_0 * f)(s) = \mathcal{L}(\delta_0)(s)\,\mathcal{L}(f)(s) = 1 \cdot \mathcal{L}(f)(s) = \mathcal{L}(f)(s).$$

Taking the inverse Laplace transform, we get the result of Theorem 7.19.

As the next theorem shows, derivatives behave nicely with convolutions.

THEOREM 7.20   Suppose f and g are piecewise differentiable functions. Then

$$\frac{d}{dt}(f * g) = f' * g + f(0)g(t).$$

Proof   To prove this theorem, we compute the derivative of the convolution.

$$\frac{d}{dt}\{(f * g)(t)\} = \frac{d}{dt}\left\{\int_0^t f(t - u)g(u)\,du\right\}.$$

Notice that t occurs in the upper limit of the integral as well as in the integrand. Using the multivariable chain rule we compute that

$$\frac{d}{dt}\{(f * g)(t)\} = f(0)g(t) + \int_0^t \frac{\partial}{\partial t}\{f(t - u)g(u)\}\,du.$$

The only term in the integrand that depends on t is $f(t - u)$. The t-derivative of $f(t - u)$ can be computed using the chain rule. Since u is a constant (as far as t is concerned), we have

$$\frac{\partial}{\partial t}\{f(t - u)\} = f'(t - u).$$

Therefore,

$$\frac{d}{dt}\{(f * g)(t)\} = f(0)g(t) + \int_0^t f'(t - u)g(u)\,du$$
$$= f(0)g(t) + f' * g(t). \qquad \blacksquare$$

Example 7.21   Suppose that $e(t)$ is the unit impulse response function for the system modeled by

$$ay'' + by' + cy = f, \quad \text{with} \quad y(0) = y_0 \quad \text{and} \quad y'(0) = y_1.$$

Argue formally using Theorems 7.19 and 7.20 to show that $y = e * f$ is the state-free response to the system.

By Definition 6.7, $e(t)$ is the solution to the initial value problem

$$ae'' + be' + ce = \delta_0(t), \quad \text{with} \quad e(0) = 0, \quad \text{and} \quad e'(0) = 0.$$

If we set $y = e * f$, then by Theorem 7.20, $y' = e' * f$ and $y'' = e'' * f$. Hence

$$ay'' + by' + cy = a(e * f)'' + b(e * f)' + c(e * f)$$
$$= ae'' * f + be' * f + ce * f$$
$$= (ae'' + be' + ce) * f$$
$$= \delta_0 * f$$
$$= f.$$

The last in this chain of equalities follows from Theorem 7.19. The initial conditions can be checked using the initial conditions satisfied by e and Theorem 7.20.   ●

EXERCISES

1. Prove part (2) of Theorem 7.10.

2. Prove part (3) of Theorem 7.10.

3. Prove part (4) of Theorem 7.10.

In Exercises 4–9, use Definition 7.5 to calculate the convolution of the given functions.

4. $f(t) = e^{3t}$, $g(t) = e^{-2t}$ 5. $f(t) = e^t$, $g(t) = e^{2t}$

6. $f(t) = t - 1$, $g(t) = t - 2$ 7. $f(t) = t$, $g(t) = 3 - t$

8. $f(t) = t$, $g(t) = e^t$ 9. $f(t) = t^2$, $g(t) = e^{-t}$

10. Let $f(t) = e^t$ and $g(t) = e^{2t}$. Let $F(s)$ and $G(s)$ be the Laplace transforms of $f(t)$ and $g(t)$, respectively.

    (a) Compute $F(s)G(s)$.

    (b) Compute $\mathcal{L}\{f(t)g(t)\}$ and compare with the result found in part (a). What important point is made by this example?

Exercises 11–16 are designed to test the veracity of Theorem 7.6. In each exercise, perform each of the following tasks:

(i) Use Definition 7.5 to find the convolution of the given functions.

(ii) Use any technique, other than Theorem 7.6, to find the Laplace transform of the convolution found in part (i).

(iii) Compute $F(s) = \mathcal{L} f(s)$ and $G(s) = \mathcal{L} g(s)$, and the product $F(s)G(s)$. Compare this result with that found in part (ii).

11. $f(t) = \sin t$, $g(t) = t$ 12. $f(t) = \cos t$, $g(t) = t^2$

13. $f(t) = e^{-2t}$, $g(t) = t$ 14. $f(t) = e^{3t}$, $g(t) = t^2$

15. $f(t) = e^{-t}$, $g(t) = \cos t$ 16. $f(t) = e^{-2t}$, $g(t) = \sin t$

Use Theorem 7.6 (*do not use partial fraction decompositions*) to find the inverse Laplace transform of the s-domain functions given in Exercises 17–24. For example, $1/(s(s + 1))$ is the product of $F(s) = 1/s$ and $G(s) = 1/(s + 1)$. Consequently, by Theorem 7.6, the inverse Laplace transform of $1/(s(s + 1))$ is the convolution of $f(t) = 1$ and $g(t) = e^{-t}$.

17. $\dfrac{1}{s(s + 1)}$

18. $\dfrac{1}{s^2 - 3s}$

19. $\dfrac{1}{(s + 1)(s - 2)}$

20. $\dfrac{1}{s^2 - 2s - 3}$

21. $\dfrac{s}{(s - 1)(s^2 + 1)}$

22. $\dfrac{1}{(s + 1)(s^2 + 4)}$

23. $\dfrac{1}{(s^2 + 1)^2}$

24. $\dfrac{s^2}{(s^2 + 9)^2}$

25. Consider the question of finding the inverse transform of

$$\frac{s\omega}{(s^2 + \omega_0^2)(s^2 + \omega^2)} = \frac{s}{s^2 + \omega_0^2} \cdot \frac{\omega}{s^2 + \omega^2}.$$

One could proceed with a partial fraction decomposition. However, this approach would not work in the case when $\omega = \omega_0$, so let's use the convolution theorem instead. We have transform pairs

$$F(s) = \frac{s}{s^2 + \omega_0^2} \leftrightarrow f(t) = \cos \omega_0 t$$

$$G(s) = \frac{\omega}{s^2 + \omega^2} \leftrightarrow g(t) = \sin \omega t.$$

Consequently, the inverse Laplace transform of $F(s)G(s)$ is $f * g(t)$. Show that

$$f * g(t) = \frac{1}{2} \int_0^t [\sin((\omega - \omega_0)u + \omega_0 t) + \sin((\omega + \omega_0)u - \omega_0 t)] \, du.$$

Hint: You might find the trigonometric identity

$$\sin A \cos B = \frac{1}{2}[\sin(A + B) + \sin(A - B)]$$

helpful.

(a) If $\omega = \omega_0$, show that

$$f * g(t) = \frac{1}{2} t \sin \omega_0 t.$$

(b) If $\omega \neq \omega_0$, show that

$$f * g(t) = \frac{\omega}{\omega^2 - \omega_0^2}(\cos \omega_0 t - \cos \omega t).$$

In Exercises 26–31, use the technique demonstrated in Examples 7.17 and 7.18 to find the solution of the given initial value problem.

26. $y'' + 4y = g(t)$, $y(0) = 1$, $y'(0) = 1$

27. $y'' + 9y = g(t)$, $y(0) = -1$, $y'(0) = 2$

28. $y'' + 5y' + 4y = g(t)$, $y(0) = 1$, $y'(0) = 0$

29. $y'' + 4y' + 3y = g(t)$, $y(0) = -1$, $y'(0) = 1$

30. $y'' + 2y' + 5y = g(t)$, $y(0) = 1$, $y'(0) = -1$

31. $y'' + 4y' + 29y = g(t)$, $y(0) = -1$, $y'(0) = 2$

5.8 Summary   Below is a summary of the Laplace transform and its properties.

1. Definition of the Laplace transform of f (Definition 1.1):

$$L(f)(s) = F(s) = \int_0^\infty f(t)e^{-st}\, dt.$$

2. The Laplace transform is linear (Proposition 2.7):

$$L\{\alpha f(t) + \beta g(t)\}(s) = \alpha\, L\{f(t)\}(s) + \beta\, L\{g(t)\}(s).$$

3. The Laplace transform of an exponential multiplied by f is a translation of the Laplace transform of f (Proposition 2.12):

$$L\{e^{ct} f(t)\}(s) = F(s - c).$$

4. The Laplace transform of a derivative (Proposition 2.1):

$$L(y')(s) = s\, L(y)(s) - y(0) = sY(s) - y(0).$$

For a second derivative (Proposition 2.4):

$$L(y'')(s) = s^2\, L(y)(s) - sy(0) - y'(0)$$
$$= s^2 Y(s) - sy(0) - y'(0).$$

More generally (Proposition 2.4):

$$L(y^{(k)})(s) = s^k\, L(y)(s) - s^{k-1}y(0) - \cdots - sy^{(k-2)}(0) - y^{(k-1)}(0)$$
$$= s^k Y(s) - s^{k-1}y(0) - \cdots - sy^{(k-2)}(0) - y^{(k-1)}(0).$$

5. The Laplace transform of $t^n f$ is related to the nth derivative of the Laplace transform (Proposition 2.14):

$$L\{t^n f(t)\}(s) = (-1)^n F^{(n)}(s).$$

6. The Laplace transform of a translation (Proposition 5.6):

$$L\{H(t - c)f(t - c)\}(s) = e^{-cs} F(s)$$

where $H(t)$ is the Heaviside function.

7. The Laplace transform of a periodic function f can be computed in terms of the Laplace transform of its window f_T (Theorem 5.17):

$$L\{f\}(s) = \frac{F_T(s)}{1 - e^{-Ts}}.$$

8. The Laplace transform of the delta function is (Theorem 6.4)

$$L\{\delta_p\}(s) = e^{-sp}.$$

9. The unit impulse response function to the system modeled by the differential equation

$$ay'' + by' + cy = f(t)$$

is the solution $e(t)$ to the initial value problem (Definition 6.7):

$$ae'' + be' + ce = \delta_0(t), \quad \text{with} \quad e(0) = 0, \quad \text{and} \quad e'(0) = 0.$$

10. The convolution of two functions is defined as (Definition 7.5):

$$f * g(t) = \int_0^t f(u)g(t-u)\, du.$$

11. The Laplace transform of a convolution is the product of the Laplace transforms (Theorem 7.6):

$$\mathcal{L}\{f * g\}(s) = F(s)G(s).$$

12. The solution to the initial value problem

$$ay'' + by' + cy = f(t), \quad \text{with} \quad y(0) = y_0, \quad \text{and} \quad y'(0) = y_1$$

can be written as

$$y(t) = e * f(t) + ay_0 e'(t) + (ay_1 + by_0)e(t),$$

where $e(t)$ is the unit impulse response function for the system. (Theorem 7.16).

PROJECT 5.9 Forced Harmonic Oscillators

In Section 4.7, we used the method of undetermined coefficients to analyze the motion of a forced harmonic oscillator. Here we will outline the procedure to do the same analysis using the Laplace transform. We will use the standard procedure for solving differential equations using the Laplace transform that we developed in Section 5.4.

Recall from Section 4.7 (equation (7.10)) that we are looking for the solutions to the equation

$$x'' + 2cx' + \omega_0^2 x = A \cos \omega t. \tag{9.1}$$

We will use the fact that $\cos \omega t = \operatorname{Re} e^{i\omega t}$ to simplify the analysis. Instead of solving (9.1), we will solve

$$x'' + 2cx' + \omega_0^2 x = A e^{i\omega t}. \tag{9.2}$$

In doing so, we will be following the lead of electrical engineers who have learned the efficacy of using complex forcing functions.

1. The characteristic polynomial for the differential equation in (9.1) or (9.2) is $P(s) = s^2 + 2cs + \omega_0^2$. Show that in the underdamped case where $c < \omega_0$, we can write this as

$$P(s) = (s+c)^2 + \eta^2, \quad \text{where} \quad \eta^2 = \omega_0^2 - c^2.$$

2. Show that

$$\mathcal{L}\{e^{i\omega t}\}(s) = \frac{1}{s - i\omega}.$$

The integral of a complex valued function $f(t) = g(t) + ih(t)$ is defined to be

$$\int f(t)\, dt = \int g(t)\, dt + i \int h(t)\, dt.$$

However, as you will learn, in most cases the formulas are unchanged. For example, in this case the derivation of the result of Example 1.4 goes through unchanged.

3. Set $X(s) = \mathcal{L}\{x(t)\}(s)$, where $x(t)$ is a solution to (9.2). Show that

$$X(s) = \frac{A}{P(s)(s - i\omega)} - \frac{x_0 s + v_0 + 2cx_0}{P(s)}, \tag{9.3}$$

where the initial conditions on the solution are $x(0) = x_0$ and $x'(0) = v_0$.

4. Using partial fractions, the first term on the right in (9.3) can be written as

$$\frac{A}{P(s)(s - i\omega)} = \frac{\alpha}{s - i\omega} + \frac{\beta s + \gamma}{P(s)},$$

where α, β, and γ are constants. Show that $\alpha = A/P(i\omega)$. Do not bother to determine β and γ precisely. Use this result to show that

$$X(s) = \frac{A}{P(i\omega)} \cdot \frac{1}{s - i\omega} + \frac{as + b}{P(i\omega)}$$

for some constants a and b, which we will not evaluate exactly.

5. Set

$$X_0(s) = \frac{A}{P(i\omega)} \cdot \frac{1}{s - i\omega} \quad \text{and} \quad X_1(s) = \frac{as + b}{P(i\omega)}.$$

Show that

$$x_1(t) = \mathcal{L}^{-1}\{X_1\}(t) = e^{-ct}(C_1 \cos \eta t + C_2 \sin \eta t)$$

for some constants C_1 and C_2. Show that $x_1(t) \to 0$ as $t \to \infty$. In addition, show that x_1 is a solution to the homogeneous equation $x_1'' + 2cx_1' + \omega_0^2 x_1 = 0$.

6. Set $x_0(t) = \mathcal{L}^{-1}\{X_0\}(t)$. Show that

$$x_0(t) = \frac{A}{P(i\omega)} \cdot e^{i\omega t} = H(i\omega) \cdot Ae^{i\omega t}, \qquad (9.4)$$

where $H(i\omega) = 1/P(i\omega)$ is the ***transfer function*** defined in Section 4.7.

7. Show that x_0 is the steady-state response and x_1 is the transient response.

8. Follow the material in Section 4.7 to show that if $P(i\omega) = R(\omega)e^{i\Phi(\omega)}$, then

$$x_0(t) = G(\omega)Ae^{i(\omega t - \Phi(\omega))}, \qquad (9.5)$$

where the gain is $G(\omega) = 1/R(\omega)$.

Equation (9.5) shows the frequency response of the system. The gain $G(\omega)$ measures the change in the amplitude, and the function $\Phi(\omega)$ measures the phase shift.

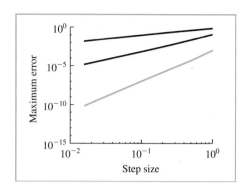

CHAPTER

6

Numerical Methods

We have been using numerical solutions of ODEs since Chapter 1. It is time to explain how they are found. First, a numerical "solution" is not a solution at all. Rather, it is a discrete approximation to a solution. That it is an approximation means that we are purposely making an error. This is a bargain with the devil. We agree to make an error, and in return we can compute an approximate solution. It is very important to understand that error.

That the solution is discrete needs some explanation. Consider the initial value problem

$$y' = f(t, y), \quad \text{with} \quad y(a) = y_0. \tag{1}$$

We are interested in the solution on the interval $a \leq t \leq b$. Let's suppose that a solution exists and denote it by $y(t)$. A *numerical solution method*, or *numerical solver*, will choose a discrete set of points $a = t_0 < t_1 < t_2 < \ldots < t_N = b$, and values $y_0, y_1, y_2, \ldots,$ and y_N, such that each y_j is approximately equal to $y(t_j)$, for $j = 1, \ldots, N$. The initial condition in (1) starts the process by giving us $y_0 = y(t_0)$.

Thus the solver only tries to approximate the solution at the discrete set of values of the independent variable t. This situation is illustrated in Figure 1, where the exact solution is the blue curve and the discrete approximate solution is plotted as the sequence of black points. The length of each vertical line is the error at that point.

In this chapter, we will explore the workings of several numerical solvers. This area of research, called the numerical analysis of ordinary differential equations, has seen enormous progress, especially since the development of the digital computer. It continues to be an active area of research. Consequently we will only be able to scratch the surface. We will first present Euler's method, which is perhaps the

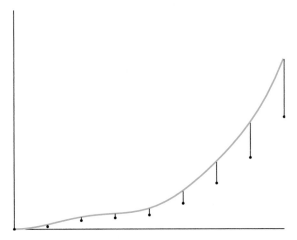

Figure 1. Exact and numerical solutions to $y' = y + \sin \pi t$, with $y(0) = 0$.

simplest solver. Euler's method cannot be considered an adequate solver in most applications. However, it is easy to explain, and it illustrates important features that are common to all solvers. We will describe the more advanced Runge-Kutta solvers in much less detail. The Runge-Kutta solvers are adequate solvers and can be used in applications. Finally, we will spend some time talking about how to use the modern solvers that are widely available today.

One other point needs to be made. We will continue to try to solve the initial value problem in (1), which involves a single first-order equation. This is simply a matter of convenience. Everything that we have said, or will say, applies equally well to first-order systems of equations. Often it will only be necessary to change our notation slightly so that our equations reflect the use of vectors instead of single equations. If you are having difficulty, refer to Section 8.1 for an explanation of the notation. There will be exercises in which the solution of systems will be required.

6.1 Euler's Method

Euler's method is an example of a *fixed-step* solver. This means that we choose the discrete set of values of the independent variable so that they divide the interval of interest into N equal subintervals. We do this by setting the *step size* $h = (b-a)/N$. Then we set $t_0 = a$, $t_1 = t_0 + h = a + h$, and, in general, $t_j = t_{j-1} + h = a + jh$. The last point of the discretization is $t_N = a + Nh = b$.

The values of the dependent variable y will be chosen iteratively. At each step, the key mathematical idea is the approximation of the graph of the unknown function $y(t)$ by its tangent line. The tangent line at the point $(t_0, y(t_0))$ is the graph of the function

$$t \to y(t_0) + y'(t_0)(t - t_0). \tag{1.1}$$

Our initial condition is $y(t_0) = y_0$, and our differential equation tells us that $y'(t_0) = f(t_0, y(t_0)) = f(t_0, y_0)$. Hence (1.1) becomes

$$t \to y_0 + f(t_0, y_0)(t - t_0). \tag{1.2}$$

Notice that although we do not know what the solution $y(t)$ is, everything on the right-hand side of (1.2) is computable. In particular, for $t = t_1 = t_0 + h$, we get

$$t_1 \to y_0 + f(t_0, y_0)\, h. \tag{1.3}$$

We now define y_1 to be the right-hand side of (1.3),

$$y_1 = y_0 + f(t_0, y_0)\, h. \tag{1.4}$$

The situation is shown in Figure 1.

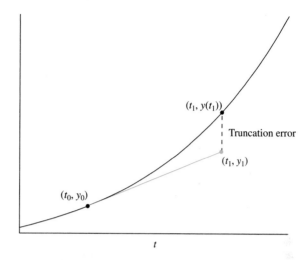

Figure 1. One step in Euler's method.

Assuming that we have already computed y_1 and t_1, we compute y_2 and t_2 in exactly the same way we computed y_1 and t_1 in (1.3) and (1.4), but starting with y_1 and t_1 instead of y_0 and t_0. We get

$$y_2 = y_1 + f(t_1, y_1)h \quad \text{and}$$
$$t_2 = t_1 + h.$$

We are now ready to define the other values $y_3, \ldots,$ and y_N inductively. We will state the method as computer pseudocode.

$$
\begin{aligned}
&\text{input } t_0 \text{ and } y_0 \\
&\text{for k = 1 to N} \\
&\quad y_k = y_{k-1} + f(t_{k-1}, y_{k-1})\, h \\
&\quad t_k = t_{k-1} + h
\end{aligned}
\tag{1.5}
$$

Notice that y_k depends only on t_{k-1} and y_{k-1}. A solver that has this property is called a ***single-step*** solver.

E x a m p l e   1 . 6   Compute the first four steps in the Euler's method approximation to the solution of $y' = f(t, y) = y - t$, with $y(1) = 1$, using the step size $h = 0.1$. Compare the results with the actual solution to the initial value problem.

We have $t_0 = 1$ and $y_0 = 1$. Then, according to (1.5), the first step is

$$y_1 = y_0 + (y_0 - t_0)\, h = 1 + (1 - 1) \times 0.1 = 1 \quad \text{and}$$
$$t_1 = t_0 + h = 1 + 0.1 = 1.1.$$

The second step is

$$y_2 = y_1 + (y_1 - t_1)\, h = 1 + (1 - 1.1) \times 0.1 = 0.99 \quad \text{and}$$
$$t_2 = t_1 + h = 1.1 + 0.1 = 1.2.$$

The third is

$$y_3 = y_2 + (y_2 - t_2)\,h = 0.99 + (0.99 - 1.2) \times 0.1 = 0.969 \quad \text{and}$$
$$t_3 = t_2 + h = 1.2 + 0.1 = 1.3.$$

Finally, the fourth and last step is

$$y_4 = y_3 + (y_3 - t_3)\,h = 0.969 + (0.969 - 1.3) \times 0.1 = 0.9359 \quad \text{and}$$
$$t_4 = t_3 + h = 1.3 + 0.1 = 1.4.$$

Having found the approximate values, let's find the exact solution. The equation $y' = y - t$ is linear, and the exact solution $y(t) = 1 + t - e^{t-1}$ is easily found. When we compute the solution at the t_ks, we get the results in Table 1.

Table 1   A few steps using Euler's method

t_k	App. sol. y_k	Exact sol. $y(t_k)$	Error $y(t_k) - y_k$
1.0	1.0000	1.0000	0.0000
1.1	1.0000	0.9948	−0.0052
1.2	0.9900	0.9786	−0.0114
1.3	0.9690	0.9501	−0.0189
1.4	0.9359	0.9082	−0.0277

Notice that the magnitude of the error increases with each step. While this is often the case, it is not always so. ●

The process carried out by hand in Exercise 1.6 can be done in more detail if you use a computer. Of course, it will be necessary to have Euler's method programmed into the computer. Routines implementing Euler's method are available as part of most mathematical computer systems. It is very easy to program. Writing a program to implement Euler's method is a good programming exercise.

Example 1.7   Consider the initial value problem $y' = y + t$, with $y(0) = 1$, on the interval $[0, 1]$. Using a computer, find and plot the Euler's method approximation to the solution for step sizes $h = 0.2$, 0.1, and 0.05. In addition, plot the exact solution and the approximate values in order to make a visual comparison.

Once again the equation is linear and we easily find the exact solution, $y(t) = 2e^t - 1 - t$. This solution is the solid black curve in Figure 2. The approximation with $h = 0.2$ is shown by the small circles, $h = 0.1$ by the black dots, and $h = 0.05$ by blue dots. ●

The error in Euler's method

Examples 1.6 and 1.7 give us some idea of the effectiveness of Euler's method. From what we see in Figure 2, it seems that as the step size decreases, the error made in the approximation decreases as well, and tends to zero as the step size does. This is indeed true, if certain minimal assumptions are valid. To understand this more fully, we need to examine the error made in Euler's method.

There are two different types of error involved. ***Truncation error*** is part of the approximation process, and ***round-off error*** occurs when any numerical computation is done on a computer or a calculator. Let's talk about round-off error first.

Round-off error is best explained with an example. Suppose you are calculating and keeping 4 decimal places in the process. You compute $1/3$ and get 0.3333. You

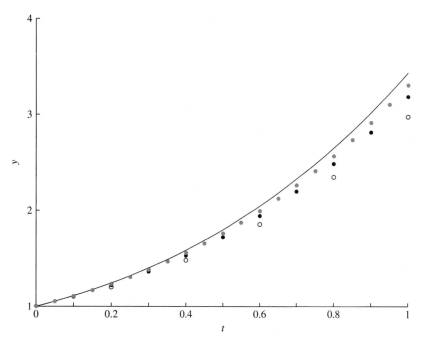

Figure 2. Exact and numerical solutions to $y' = y + t$, $y(0) = 1$.

round down to 0.3333 in this process. Next you compute 2/3 and get 0.6667. This time you round up. But now you compute

$$\frac{2}{3} - 2 \cdot \frac{1}{3} \quad \text{and get} \quad 0.6667 - 0.6666 = 0.0001,$$

even though those two numbers should be equal. The 0.0001 is an example of round-off error. Of course a computer or a calculator is computing using the binary number system, and it is computing to a much greater accuracy, but the principle is the same. Every calculation has a certain probability of producing an error in the last place.

If a lot of computations are being made, the round-off error could accumulate. However, modern computers do individual computations to such a high accuracy that round-off error is usually negligible in comparison to other errors, such as truncation error. This is almost always true for the numerical solutions of ordinary differential equations. For this reason we will not consider round-off error in our future deliberations.

The key to understanding the truncation error in any numerical method for solving ODEs is Taylor's formula. If y is a function with two continuous derivatives, Taylor's formula of order 1 says

$$y(t_0 + h) = y(t_0) + y'(t_0)h + R(h). \tag{1.8}$$

The term $R(h)$ is the Taylor remainder, and all we need to know about it is that it satisfies the inequality

$$|R(h)| \leq M h^2, \tag{1.9}$$

where M is a constant that depends on the second derivative of y.

Suppose that $y(t)$ is the solution to the initial value problem

$$y' = f(t, y) \quad \text{with} \quad y(t_0) = y_0.$$

Let's substitute the initial condition $y(t_0) = y_0$ and the formula for the derivative $y'(t_0) = f(t_0, y(t_0)) = f(t_0, y_0)$ into (1.8), getting

$$y(t_0 + h) = y(t_0) + f(t_0, y_0)h + R(h). \tag{1.10}$$

Then we can use the definition of the first approximation in Euler's method, formula (1.4), to make this

$$y(t_1) = y_1 + R(h), \quad \text{or} \quad y(t_1) - y_1 = R(h). \tag{1.11}$$

Thus the error made at the first step is simply the remainder in Taylor's formula. Putting this together with (1.9), we get

$$|y(t_1) - y_1| \le Mh^2. \tag{1.12}$$

Of course, a similar error is made at every step. This error is called the ***truncation error***. It is illustrated in Figure 1, and again in Figure 3, which shows two steps of Euler's method. Formula (1.12) tells us that the truncation error in Euler's method is bounded by a constant times the square of the step size. Since the step size is $h = (b-a)/N$, there are $N = (b-a)/h$ steps involved in covering the entire interval $[a, b]$. From the fact that at each step there is a truncation error bounded by Mh^2 (from (1.9)) and that there are $N = (b - a)/h$ steps, a naive computation seems to show that the maximum or accumulated error after N steps can be no larger than

$$N \times Mh^2 = \frac{b-a}{h}Mh^2 = M(b-a)h. \tag{1.13}$$

However, this analysis is faulty. It assumes that the truncation error made at each step stays fixed. In fact, it propagates and can get exponentially larger as it does so. To see what this means, notice that each step in the algorithm in (1.5) is exactly the same. The solution with initial condition $y(t_{k-1}) = y_{k-1}$ is approximated by the tangent line to the solution curve at the point (t_{k-1}, y_{k-1}).

This is illustrated by Figure 3, which shows two steps in Euler's method. Notice that the second step begins at the point (t_1, y_1) and uses the tangent line to the solution curve corresponding to the solution with initial value $y(t_1) = y_1$. This solution is indicated by the dashed curve in Figure 3. It is clear that this is a different solution from the one we started out to approximate. The total error at the second step has two parts. The first is the truncation error made in approximating the dashed solution, and the second is the error made at the first step, but now propagated along the solution curves. We will call this ***propagated truncation error***. The example in Figure 3 shows that the propagated truncation error can be much larger than the original truncation error.

The increase in the truncation error as it is propagated is shown more dramatically in Figure 4, which shows four steps in Euler's method. Now we see that the total error after the fourth step is the sum of the propagated truncation error of the first three steps plus the truncation error of the fourth step itself. It can be seen in Figure 4 that the propagated errors are much larger than the truncation errors themselves. In fact, the truncation error can propagate exponentially.

Despite this fact, it is possible to analyze the total error. The result is an error bound of the form

$$\text{maximum error} \le \frac{M}{L}\left(e^{L(b-a)} - 1\right)h. \tag{1.14}$$

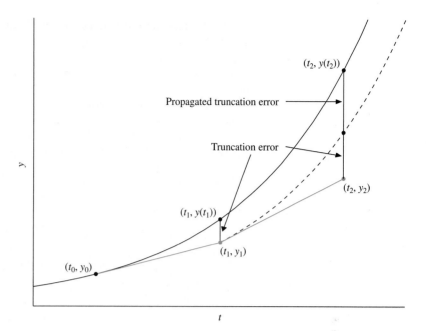

Figure 3. Two steps in Euler's method.

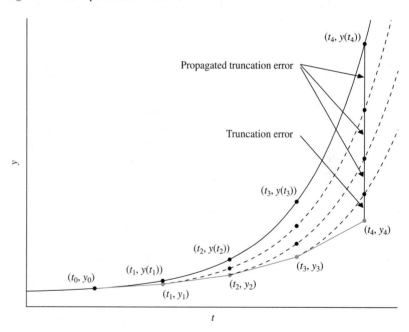

Figure 4. Four steps in Euler's method.

In (1.14), the constants M and L depend only on the function $f(t, y)$. More precisely they are

$$L = \max_{(t,y)\in R} \left| \frac{\partial f}{\partial y}(t, y) \right|, \tag{1.15}$$

and

$$M = \frac{1}{2} \max_{(t,y)\in R} \left| \frac{\partial f}{\partial t} + f\frac{\partial f}{\partial y} \right|,$$

where R is a rectangle that contains the solution curve.

Comparing the correct error bound in (1.14) with the naive estimate in (1.13) shows that the exponential propagation of truncation error has a real effect.

The inequality in (1.14) contains good news and bad news. The good news is that the right-hand side has the step size h as a factor. Thus, as h decreases we can be sure that the maximum error also decreases. We can make the error as small as we wish by using suitably small step sizes. The bad news is the dependence of the right-hand side on $b - a$, the length of the interval over which we want to find the solution. As the length of the interval gets large, the right-hand side of (1.14) grows exponentially. This means that over large intervals, Euler's method could become useless. Unfortunately, this is a feature of all numerical solvers.

Systems and Euler's method

All solution methods can be easily applied to systems of differential equations. We will demonstrate this by applying Euler's method to an example.

Consider a spring-mass system oscillating according to the model $my'' + \mu y' + ky = 0$, as introduced in Section 4.1. If there is no damping ($\mu = 0$) and $m = k$, the equation of motion becomes

$$y'' + y = 0. \tag{1.16}$$

We will use the initial conditions $y(0) = -1$ and $y'(0) = -2$.

As explained in Section 4.2, the equation in (1.16) has an equivalent first-order system. We find it by introducing the velocity $v = y'$. Then the system is

$$\begin{array}{lll} y' = f(t, y, v) = v, & & y(0) = -1 \\ & \text{with} & \\ v' = g(t, y, v) = -y, & & v(0) = -2. \end{array} \tag{1.17}$$

To apply Euler's technique, simply change the pseudocode presented earlier as needed to reflect the fact that we are solving an initial value problem for a system.

$$\begin{aligned} &\text{input } t_0, \ y_0, \text{ and } v_0 \\ &\text{for k = 1 to N} \\ &\qquad y_k = y_{k-1} + f(t_{k-1}, y_{k-1}, v_{k-1})h \\ &\qquad v_k = v_{k-1} + g(t_{k-1}, y_{k-1}, v_{k-1})h \\ &\qquad t_k = t_{k-1} + h \end{aligned} \tag{1.18}$$

Let's apply this to the system in (1.17) with step size $h = 0.1$. Our initial conditions are $t_0 = 0$, $y_0 = -1$, and $v_0 = -2$.

$$\begin{aligned} y_1 &= y_0 + v_0 h = -1 - 0.2 = -1.2 \\ v_1 &= v_0 - y_0 h = -2 + 0.1 = -1.9 \\ t_1 &= t_0 + h = 0 + 0.1 = 0.1 \end{aligned}$$

The second step is found by iterating the algorithm in (1.17).

$$\begin{aligned} y_2 &= y_1 + v_1 h = -1.2 - 0.19 = -1.39 \\ v_2 &= v_1 - y_1 h = -1.9 + 0.12 = -1.78 \\ t_2 &= t_1 + h = 0.1 + 0.1 = 0.2 \end{aligned}$$

In Section 4.1, we saw that the natural frequency of the motion is $\omega = \sqrt{k/m} = 1$, and the period of the motion is $T = 2\pi/\omega = 2\pi$. Let's iterate Euler's method over one complete period, the time interval $[0, 2\pi]$, with step size $h = 0.1$. The plot of the computed v versus y is shown in blue in Figure 5.

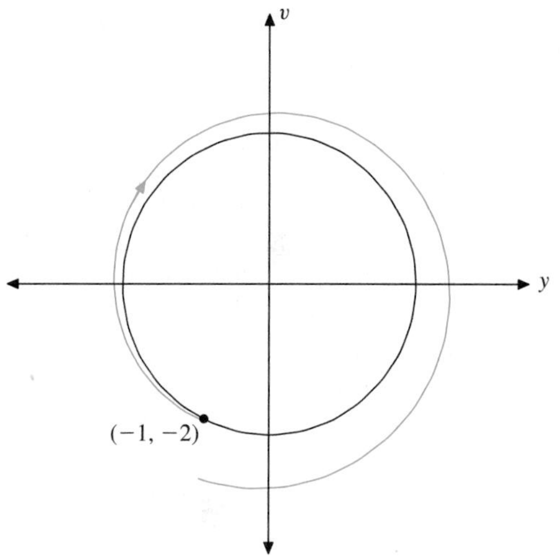

Figure 5. The error in Euler's method.

Compare the plot of the numerical solution with the exact solution ($y = -2 \sin t - \cos t$, $v = -2 \cos t + \sin t$), shown in black in Figure 5. Periodicity requires that the orbit should return to the initial condition, but the numerical solution does not return as it should. This plot effectively reveals the error involved in Euler's method. Figure 5 strikingly reflects the bound

$$\text{maximum error} \leq \frac{M}{L} \left(e^{L(b-a)} - 1 \right) h.$$

Indeed, we can reduce the error by reducing the step size h (try it!), but the length of the interval $[a, b] = [0, 2\pi]$ causes the error to grow exponentially. This is clearly visible in Figure 5.

EXERCISES

1. Consider the initial value problem

$$y' = t + y, \quad y(0) = 1.$$

When computing a solution by hand using Euler's method, it is beneficial to arrange your work in a table. Table 2 shows the computation of the first step of Euler's method, using step size $h = 0.1$. Continue with Euler's method and complete all missing entries of the table.

Table 2

k	t_k	y_k	$f(t_k, y_k) = t_k + y_k$	h	$f(t_k, y_k)h$
0	0.0	1.0	1.0	0.1	0.1
1	0.1	1.1			
2	0.2				
3	0.3				
4	0.4				
5	0.5				

For each initial value problem presented in Exercises 2–5, hand-calculate the first five iterations of Euler's method with step size $h = 0.1$. Arrange your results in the tabular form presented in Exercise 1.

2. $y' = y$, $y(0) = 1$

3. $y' = ty$, $y(0) = 1$

4. $z' = 5 - z$, $z(0) = 0$

5. $z' = x - 2z$, $z(0) = 1$

For each initial value problem presented in Exercises 6–9, perform each of the following tasks.

(i) Use a computer and Euler's method to calculate three separate approximate solutions on the interval $[0, 1]$, one with step size $h = 0.2$, a second with step size $h = 0.1$, and a third with step size $h = 0.05$.

(ii) Use the appropriate analytic method to compute the exact solution.

(iii) Plot the exact solution found in part (ii). On the same axes, plot the approximate solutions found in part (i) as discrete points, in a manner similar to that demonstrated in Figure 2.

6. $y' = y - t$, $y(0) = 0$

7. $y' + 2xy = x$, $y(0) = 8$

8. $z' - 2z = xe^{2x}$, $z(0) = 1$

9. $z' = (1 + t)z$, $z(0) = -1$

10. Consider the initial value problem

$$y' = 12y(4 - y), \quad y(0) = 1.$$

(a) Since this equation is autonomous, you can use the qualitative analysis of Section 2.9 to analyze the solution. Use the graph of $f(y) = 12y(4 - y)$ to construct the phase line. Sketch the equilibrium solutions in the ty-plane. Then sketch a rough approximation of the solution of the given initial value problem.

(b) Use Euler's method with step size $h = 0.04$ to sketch an approximate solution on the interval $[0, 2]$. What fundamental flaw does this solution exhibit?

11. Euler's method is used to solve the initial value problem

$$y' = y, \quad y(0) = 1,$$

over the time interval $[0, 2]$, generating the data in Table 3. The first column is the step size used.[1] The second column contains the Euler approximation of $y(2)$, while the third column contains the exact solution $y(2)$. The fourth column is the magnitude of the error at $t = 2$, calculated by taking the absolute value of the difference of the entries in columns two and three.

(a) Prepare a discrete plot of the error versus the step size to show that the error is proportional to the step size (i.e., $E_h = \lambda h$).

(b) Estimate the variation constant λ in the equation $E_h = \lambda h$ and use this result to calculate the step size required to assure an error no greater than 0.001 when approximating $y(2)$ with Euler's method. How many iterations will this calculation require?

(c) Use Euler's method to verify that the step size calculated in part (b) produces an error no greater in magnitude than 0.001.

For each initial value problem in Exercises 12–15, perform the following tasks.

(i) Generate a table of data similar to that in Exercise 11, where the second column contains an Euler approximation of $y(2)$ using the corresponding step size in the first column.

(ii) Use an appropriate analytic method to find the exact value of $y(2)$. Place your calculation of the magnitude of the error in the approximation of $y(2)$ in the fourth column.

(iii) Prepare a discrete plot of the error versus the step size to show that the error varies directly as the step size; i.e., $E_h = \lambda h$.

(iv) Estimate the variation constant λ in the equation $E_h = \lambda h$ and use this result to calculate the step size required to assure an error no greater in magnitude than 0.01 when approximating $y(2)$ with Euler's method. How many iterations will this calculation require?

(v) Use Euler's method to verify that the step size calculated in part (iv) produces an error no greater in magnitude than 0.01.

12. $y' = t - y, \quad y(0) = 1$

13. $y' = y + \sin t, \quad y(0) = 1$

14. $y' = ty, \quad y(0) = 1$

15. $y' = -(1/3)y^2, \quad y(0) = -1$

16. In the preceding exercises, you've seen that the error in Euler's method varies directly as the first power[1] of the step size (i.e., $E_h \approx \lambda h$). This makes Euler's method a *first-order* solver. What must be done to the step size in order to halve the error? How does this affect the number of required iterations?

17. Consider the planar system

$$x' = y,$$
$$y' = -x,$$

with initial conditions $x(0) = 1$ and $y(0) = 0$. The first few iterations of Euler's method, with step size $h = 0.1$, are shown in Table 4. Complete the remaining entries in the table.

For each initial value problem presented in Exercises 18–21, hand-calculate the first five iterations of Euler's method with step size $h = 0.1$. Arrange your results in the tabular form presented in Exercise 17.

18. $x' = y, y' = -x, x(0) = 1, y(0) = -1$

19. $x' = -2y, y' = x, x(0) = 0, y(0) = -1$

Table 3

STEP SIZE h	EULER APPROX.	TRUE VALUE	ERROR E_h
0.0625000000	6.9586667572	7.3890560989	0.4303893417
0.0312500000	7.1662761528	7.3890560989	0.2227799461
0.0156250000	7.2756697931	7.3890560989	0.1133863058
0.0078125000	7.3318505987	7.3890560989	0.0572055002
0.0039062500	7.3603235533	7.3890560989	0.0287325457
0.0019531250	7.3746571603	7.3890560989	0.0143989386
0.0009765625	7.3818484359	7.3890560989	0.0072076631

[1] We will learn that the limits on truncation error grow as a power of the step size. Hence multiplying the step size by $1/2$ should result in a decrease in the truncation error by the corresponding power of $1/2$.

Table 4				
t_k	x_k	y_k	$f(t_k, x_k, y_k)h = y_k h$	$g(t_k, x_k, y_k)h = -x_k h$
0.0	1.00	0.00	0.00	−0.10
0.1	1.00	−0.10	−0.01	−0.10
0.2	0.99	−0.20		
0.3				
0.4				
0.5				

20. $x' = x + y$, $y' = x - y$, $x(0) = 1$, $y(0) = 1$

21. $x' = -y$, $y' = x + y$, $x(0) = 1$, $y(0) = -1$

In Exercises 22–25, use a computer and Euler's method to find a solution of the given initial value problem on the time interval $[0, 2\pi]$. In each exercise, provide three plots, one of x versus t, a second of y versus t, and a third of y versus x.

22. The system in Exercise 18

23. The system in Exercise 19

24. The system in Exercise 20

25. The system in Exercise 21

In Exercises 26–29, use the substitution $v = y'$ to change the harmonic oscillator into a planar system having form

$$y' = v,$$
$$v' = f(y, v).$$

Use Euler's method to provide an approximate solution over the given time interval using the given step sizes. Provide a plot of v versus y for each step size.

26. $y'' + 4y = 0$, $y(0) = 4$, $y'(0) = 0$, $[0, 2\pi]$, $h = 0.1, 0.01, 0.001$

27. $y'' + 9y = 0$, $y(0) = 0$, $y'(0) = -4$, $[0, 2\pi]$, $h = 0.1, 0.01, 0.001$

28. $y'' + y' + 25y = 0$, $y(0) = 4$, $y'(0) = 0$, $[0, 2\pi]$, $h = 0.1, 0.01, 0.001$

29. $y'' + y' + 36y = 0$, $y(0) = -5$, $y'(0) = 0$, $[0, 2\pi]$, $h = 0.1, 0.01, 0.001$

30. Consider the harmonic oscillator

$$y'' + 4y = 0,$$

with initial conditions $y(0) = 1$ and $y'(0) = 0$. Show that this oscillator is equivalent to the planar system

$$y' = v,$$
$$v' = -4y,$$

with initial conditions $y(0) = 1$ and $v(0) = 0$.

(a) Show that $y(t) = \cos 2t$ and $v(t) = -2 \sin 2t$ is a solution of the planar system and plot the graph of v versus y on the time interval $[0, \pi]$.

(b) Use Euler's method to find approximate solutions of the planar system on the interval $[0, \pi]$ using step sizes $h = 0.2$, $h = 0.1$, and $h = 0.05$. For each step size, superimpose the graph of v versus y on the plot in part (a). Plot each of these approximate solutions as a sequence of discrete points, using a different color and/or marker. What can you say about the approximate solutions as the step size decreases?

6.2 Runge-Kutta Methods

In this section we will discuss two solvers that were invented independently around 1900 by the mathematicians Runge and Kutta. Although more modern and more effective methods have been devised, the Runge-Kutta solvers are probably the most commonly used routines for solving systems of differential equations.

Like Euler's method, the Runge-Kutta methods are fixed-step solvers. The discrete set of values of the independent variable is chosen by setting the **step size** $h = (b - a)/N$. Then we set $t_0 = a$, $t_1 = t_0 + h = a + h$, and, in general $t_j = t_{j-1} + h = a + jh$. The final value is $t_N = a + Nh = b$. The values of the dependent variable y will be chosen iteratively as in Euler's method. However, it is not as easy to find a geometric interpretation for the Runge-Kutta methods as it is for Euler's method.

The second-order Runge-Kutta method

The **second-order** Runge-Kutta method is also known as the **improved Euler's method**. Starting from the initial point (t_0, y_0), we compute two slopes:

$$s_1 = f(t_0, y_0),$$
$$s_2 = f(t_0 + h, y_0 + hs_1). \tag{2.1}$$

With these slopes, we define the next value of the dependent variable to be

$$y_1 = y_0 + h\,\frac{s_1 + s_2}{2}. \tag{2.2}$$

If we compare (2.2) with (1.4), we see that the slope $f(t_0, y_0)$ is replaced by the average of $s_1 = f(t_0, y_0)$ and s_2. It is not at all obvious why it should be so, but an analysis using Taylor's theorem reveals that there is an improvement in the estimate for the truncation error. For the second-order Runge-Kutta method, we have

$$|y(t_1) - y_1| \le M\,h^3. \tag{2.3}$$

The constant M depends on the function $f(t, y)$. Its exact value is not important. When we compare (2.3) with (1.12), we notice that the truncation error for the second-order Runge-Kutta method is controlled by the cube of the step size instead of the square.

We can now define the remaining values $y_2, y_3, \ldots,$ and y_N inductively. Once again we present the method as computer pseudocode.

input t_0 and y_0
for k = 1 to N
$$s_1 = f(t_{k-1}, y_{k-1})$$
$$s_2 = f(t_{k-1} + h, y_{k-1} + hs_1)$$
$$y_k = y_{k-1} + h\,\frac{s_1 + s_2}{2}$$
$$t_k = t_{k-1} + h$$
$$\tag{2.4}$$

When we compute the maximum error that can occur over the interval $[a, b]$, the third power in (2.3) gets reduced by one, as happened for Euler's method. We get

$$\text{maximum error} \le \frac{M}{L}\left(e^{L(b-a)} - 1\right)h^2. \tag{2.5}$$

The constant L is the same as that which occurs in (1.15) for Euler's method. The constant M is not the same as that used in (1.14), but again it depends only on $f(t, y)$. Its exact formula is not too useful.

The power of the step size h that occurs in the error estimates like (1.14) and (2.5) is called the **order** of the solver (thus the name *second-order Runge-Kutta method*). In contrast, Euler's method is a first-order method.

Again we have good news and bad news. This time the good news is better because the maximum error is bounded by the square of the step size. Since the step size is going to be small, the square will be even smaller, and we can get smaller errors with larger step sizes using the second-order Runge-Kutta method. The bad news is just as bad for this method as it was for Euler's method. The exponential dependence on $b - a$, the length of the computation interval, means that the method may become unreliable over long intervals.

Example 2.6   Compute the first four steps in the second-order Runge-Kutta approximation to the solution of $y' = f(t, y) = y - t$, with $y(1) = 1$, using the step size $h = 0.1$. Compare the results with the actual solution to the initial value problem.

We have $t_0 = 1$ and $y_0 = 1$. Following the algorithm in (2.4), we get

$$s_1 = f(t_0, y_0) = y_0 - t_0 = 1 - 1 = 0$$
$$s_2 = f(t_0 + h, y_0 + hs_1) = (y_0 + hs_1) - (t_0 + h) = 1 - 1.1 = -0.1$$
$$y_1 = y_0 + h\frac{s_1 + s_2}{2} = 1 + 0.1\left(\frac{0 - 0.1}{2}\right) = 0.995$$
$$t_1 = t_0 + h = 1 + 0.1 = 1.1.$$

A second iteration produces

$$s_1 = f(t_1, y_1) = 0.995 - 1.1 = -0.105,$$
$$s_2 = f(t_1 + h, y_1 + hs_1) = (0.995 - 0.0105) - 1.2 = -0.2155,$$
$$y_2 = y_1 + h\frac{s_1 + s_2}{2} = 0.995 + 0.1\left(\frac{-0.105 - 0.2155}{2}\right) = 0.978975,$$
$$t_2 = t_1 + h = 1.1 + 0.1 = 1.2.$$

Continue in this manner to produce the next two terms, y_3 and y_4.

Once again, the exact solution is $y(t) = 1 + t - e^{t-1}$. When we compute the solution at all of the time steps, we get Table 1, below. Comparing the results with Table 1 in Section 6.1, we notice that the errors in the fourth column have decreased significantly. ●

Table 1   A few steps using the second-order Runge-Kutta method

t_k	App. sol. y_k	Exact sol. $y(t_k)$	Error $y(t_k) - y_k$
1.0	1.0000000	1.000000000	0
1.1	0.9950000	0.994829081	−0.000170918
1.2	0.9789750	0.978597241	−0.000377758
1.3	0.9507673	0.950141192	−0.000626182
1.4	0.9090979	0.908175302	−0.000922647

The fourth-order Runge-Kutta method

Our next solution method is also due to Runge and Kutta. This is the one that most people think of as *the* Runge-Kutta method. It is probably the most commonly used solution algorithm. For most equations and systems it is suitably fast and accurate.

For this method we use four slopes. Starting with the initial point (t_0, y_0), we compute

$$s_1 = f(t_0, y_0),$$
$$s_2 = f\left(t_0 + \frac{h}{2}, y_0 + \frac{h}{2}s_1\right),$$
$$s_3 = f\left(t_0 + \frac{h}{2}, y_0 + \frac{h}{2}s_2\right),$$
$$s_4 = f(t_0 + h, y_0 + h s_3).$$

$$(2.7)$$

With these slopes we define the next value of the dependent variable:

$$y_1 = y_0 + h \frac{s_1 + 2s_2 + 2s_3 + s_4}{6}. \tag{2.8}$$

We compute the rest of the values of the independent variable iteratively, as we did before. We will again state the algorithm using computer pseudocode.

$$
\begin{aligned}
&\text{input } t_0 \text{ and } y_0 \\
&\text{for k = 1 to N} \\
&\quad s_1 = f(t_{k-1}, \; y_{k-1}) \\
&\quad s_2 = f\left(t_{k-1} + \frac{h}{2}, \; y_{k-1} + \frac{h}{2}s_1\right) \\
&\quad s_3 = f\left(t_{k-1} + \frac{h}{2}, \; y_{k-1} + \frac{h}{2}s_2\right) \\
&\quad s_4 = f(t_{k-1} + h, \; y_{k-1} + hs_3) \\
&\quad y_k = y_{k-1} + h\frac{s_1 + 2s_2 + 2s_3 + s_4}{6} \\
&\quad t_k = t_{k-1} + h
\end{aligned}
\tag{2.9}
$$

If we compare the algorithms in (1.5), (2.4), and (2.9), we see that in all three cases, y_k is obtained from y_{k-1} by adding the product of the step size and a weighted average of slopes. For Euler's method, this average reduces to one slope, the slope of the tangent line. For the Runge-Kutta methods, the average is more complicated.

The fourth-order Runge-Kutta method is more costly, since it requires us to compute the right-hand side $f(t, y)$ four times. We are compensated for this, however, by an improvement in the truncation error. Again a (very difficult) analysis using Taylor's theorem shows that

$$|y(t_1) - y_1| \le Mh^5. \tag{2.10}$$

Again the constant M depends on $f(t, y)$. Its exact value is not important. When we use this to compute the bound on the maximum error, we get a corresponding improvement,

$$\text{maximum error} \le \frac{M}{L}\left(e^{L(b-a)} - 1\right)h^4. \tag{2.11}$$

The constant L is as defined in (1.15). The constant M is not the same as that which occurs in (1.14) or (2.5). However, once again it depends only on $f(t, y)$. Because of the fourth power of the step size that occurs in (2.11), this method is called the **fourth-order Runge-Kutta** method.

E x a m p l e   2.12   Compute the first four steps in the fourth-order Runge-Kutta approximation to the solution of $y' = f(t, y) = y - t$, with $y(1) = 1$, using the step size $h = 0.1$. Compare the results with the actual solution to the initial value problem.

We have $t_0 = 1$ and $y_0 = 1$. Following the algorithm in (2.9), we get

$$s_1 = f(t_0, y_0) = 1 - 1 = 0$$

$$s_2 = f\left(t_0 + \frac{h}{2}, y_0 + \frac{h}{2}s_1\right) = 1 - 1.05 = -0.05$$

$$s_3 = f\left(t_0 + \frac{h}{2}, y_0 + \frac{h}{2}s_2\right) = 0.9975 - 1.05 = -0.0525$$

$$s_4 = f(t_0 + h, y_0 + hs_3) = 0.99475 - 1.1 = -0.10525$$

$$y_1 = y_0 + h\frac{s_1 + 2s_2 + 2s_3 + s_4}{6}$$

$$= 1 + 0.1\left(\frac{0 + 2(-0.05) + 2(-0.0525) - 0.10525}{6}\right)$$

$$= 0.99482916666667$$

$$t_1 = t_0 + h = 1 + 0.1 = 1.1.$$

Continue in this manner to produce the next three terms, y_2, y_3, and y_4.

Once again, the exact solution is $y(t) = 1 + t - e^{t-1}$. When we compute the solution at the t_ks, we get the results in Table 2.

Table 2 A few steps using the fourth-order Runge-Kutta method

t_k	App. sol. y_k	Exact sol. $y(t_k)$	Error $y(t_k) - y_k$
1.0	1.000000000	1.000000000	0
1.1	0.994829167	0.994829081	−0.000000084
1.2	0.978597429	0.978597241	−0.000000187
1.3	0.950141502	0.950141192	−0.000000310
1.4	0.908175759	0.908175302	−0.000000457

Comparing the results in this table with those in Table 1 in Section 6.1 for Euler's method and Table 1 on page 257 for the second-order Runge-Kutta, we notice that the fourth-order Runge-Kutta method produces the best results of all. ●

EXERCISES

1. Consider the initial value problem

$$y' = t + y, \quad y(0) = 1.$$

When computing a solution using the second-order Runge-Kutta method by hand, it is beneficial to arrange your work in a table. Table 3 shows the computation of the first step of the solution using the second-order Runge-Kutta method and step size $h = 0.1$. Continue with the method and complete all missing entries of the table.

For each initial value problem presented in Exercises 2–5, compute the first five iterations using the second-order Runge-Kutta method with step size $h = 0.1$. Arrange your results in the tabular form presented in Exercise 1.

2. $y' = y, \quad y(0) = 1$
3. $y' = ty, \quad y(0) = 1$
4. $z' = 5 - z, \quad z(0) = 0$
5. $z' = x - 2z, \quad z(0) = 1$

Table 3

k	t_k	y_k	s_1	s_2	h	$h(s_1 + s_2)/2$
0	0.0	1.0	1.0	1.2	0.1	0.11
1	0.1	1.11				
2	0.2					
3	0.3					
4	0.4					
5	0.5					

For each initial value problem presented in Exercises 6–9, perform each of the following tasks.

(i) Use a computer and the second-order Runge-Kutta method to calculate three separate approximate solutions on the interval $[0, 1]$, one with step size $h = 0.2$, a second with step size $h = 0.1$, and a third with step size $h = 0.05$.

(ii) Use the appropriate analytic method to compute the exact solution.

(iii) Plot the exact solution found in part (ii). On the same axes, plot the approximate solutions found in part (i) as discrete points, in a manner similar to that demonstrated in Figure 2.

6. $y' = -2ty^2$, $\quad y(0) = 1$

7. $z' + z = \cos x$, $\quad z(0) = 1$

8. $x' = t/x$, $\quad x(0) = 1$

9. $w' + w = x^2$, $\quad w(0) = 1/2$

A function defined by the equation

$$y = ax^b, \quad \text{for } 0 < x < \infty, \tag{2.13}$$

where $a > 0$, is called a **power function**. Taking the logarithm of both sides of this equation produces

$$\ln y = b \ln x + \ln a. \tag{2.14}$$

If you plot the graph of y versus x, then you get the graph of the power function defined by equation (2.13). However, if you plot $\ln y$ versus $\ln x$, then, according to equation (2.14), the resulting graph is a straight line with slope b and intercept $\ln a$. In Exercises 10–13, first plot y versus x on the interval $[0, 1]$ to produce the graph of the given power function. Then plot $\ln y$ versus $\ln x$. What is the slope of the resulting line?

10. $y = 3x^2$

11. $y = 100x^{-3}$

12. $y = 20x^{-4}$

13. $y = 5x^3$

14. If you have data points (x, y) that produce a line when you plot $\ln y$ versus $\ln x$, then you should suspect that the data points satisfy the equation of some *power function* (see Exercises 10–13).

The initial value problem $y' = y$, $y(0) = 1$, has the exact solution $y = e^x$. Step sizes of the form $1/2^k$, $k = 4, 5, \ldots, 10$, shown in the first column of Table 4, require 2^{k+1} iterations of the second-order Runge-Kutta method on the interval $[0, 2]$ to produce the corresponding approximation of $y(2)$, shown in the second column. The third column contains the exact solution $y(2)$, while the fourth column is the magnitude of the error at $t = 2$, calculated by taking the absolute value of the difference of the entries in columns two and three.

(a) Prepare a discrete plot of $\ln E_h$ versus $\ln h$. You should see a linear relationship.

(b) Use a calculator or computer to do a linear regression to find a linear approximation to the plot in part (a). Express the relationship between $\ln E_h$ and $\ln h$ in the form $\ln E_h \approx p \ln h + K$.

(c) Place the equation developed in part (b) in the form $E_h \approx Ch^p$.

(d) Use the result from part (c) to estimate the step size required to assure an error no greater than 0.001 when approximating $y(2)$ with the second-order Runge-Kutta method. How many iterations will this calculation require?

(e) Use the second-order Runge-Kutta method to check if the step size determined in part (b) produces an error no greater in magnitude than 0.001.

For each initial value problem in Exercises 15–18, prepare a data table similar to that in Exercise 14, portraying an analysis of the solution on the interval $[0, 2]$. Find an equation modeling the error versus the step size, in the form $E_h \approx Ch^p$. In each case, what appears to be the approximate value of p?

15. $y' = -y + \sin(t)$, $\quad y(0) = 1$

16. $y' = -t^2y$, $\quad y(0) = 1$

17. $y' = t/y^2$, $\quad y(0) = 1$

18. $y' = t - 3y$, $\quad y(0) = 1$

19. In the preceding exercises, you've seen that the error in the second-order Runge-Kutta method is approximately proportional to the second power of the step size (i.e., $E_h \approx \lambda h^2$). This makes the second-order Runge-Kutta method a **second-order** solver. What must be done to the step size in order to halve the error? How does this affect the number of required iterations?

For each initial value problem presented in Exercises 20–23, compute by hand the first three iterations using the fourth-order Runge-Kutta method with step size $h = 0.1$. Arrange your results in a tabular form similar to that presented in Exercise 1.

20. $y' = y$, $\quad y(0) = 1$

21. $y' = t + y$, $\quad y(0) = 1$

22. $z' = 5 - z$, $\quad z(0) = 0$

23. $z' = x - 2z$, $\quad z(0) = 1$

24. If you have data points (x, y) that produce a line when you plot $\ln y$ versus $\ln x$, then you should suspect that the data points satisfy the equation of some *power function* (see Exercises 10–13).

Table 4

Step size h	RK2 approx.	True value	Error E_h
0.0625000000	7.3798803664	7.3890560989	0.0091757326
0.0312500000	7.3867068504	7.3890560989	0.0023492486
0.0156250000	7.3884618026	7.3890560989	0.0005942963
0.0078125000	7.3889066478	7.3890560989	0.0001494511
0.0039062500	7.3890186263	7.3890560989	0.0000374727
0.0019531250	7.3890467170	7.3890560989	0.0000093819
0.0009765625	7.3890537517	7.3890560989	0.0000023472

Table 5

STEP SIZE h	RK4 APPROX.	TRUE VALUE	ERROR E_h
1.0000000	7.33506944444	7.38905609893	0.05398665449
0.5000000	7.38397032395	7.38905609893	0.00508577498
0.2500000	7.38866527357	7.38905609893	0.00039082536
0.1250000	7.38902900289	7.38905609893	0.00002709604
0.0625000	7.38905431509	7.38905609893	0.00000178384
0.0312500	7.38905598450	7.38905609893	0.00000011443
0.0156250	7.38905609169	7.38905609893	0.00000000725
0.0078125	7.38905609847	7.38905609893	0.00000000046

The initial value problem $y' = y$, $y(0) = 1$, has exact solution $y = e^x$. Step sizes of the form $1/2^k$, $k = 0, 1, \ldots, 7$, shown in the first column of Table 5, require 2^{k+1} iterations of the fourth-order Runge-Kutta method on the interval $[0, 2]$ to produce the corresponding approximation of $y(2)$, shown in the second column. The third column contains the exact solution $y(2)$, while the fourth column is the magnitude of the error at $t = 2$, calculated by taking the absolute value of the difference of the entries in columns two and three.

(a) Prepare a discrete plot of $\ln E_h$ versus $\ln h$. You should see a linear relationship.

(b) Use a calculator or computer to do a linear regression to find a linear approximation to the plot in part (a). Express the relationship between $\ln E_h$ and $\ln h$ in the form $\ln E_h \approx p \ln h + K$.

(c) Place the equation developed in part (b) in the form $E_h \approx Ch^p$.

(d) Use the result from part (c) to estimate the step size required to assure an error no greater than 0.00001 when approximating $y(2)$ with the fourth-order Runge-

Kutta method. How many iterations will this calculation require?

(e) Use the fourth-order Runge-Kutta method to check whether the step size calculated in part (d) produces an error no greater in magnitude than 0.000001.

For each initial value problem in Exercises 25–28, use the fourth-order Runge-Kutta method to prepare a data table similar to that in Exercise 24, portraying an analysis of the solution on the interval $[0, 2]$. Find an equation modeling the error versus the step size, in the form $E_h \approx Ch^p$. In each case, what appears to be the approximate value of p?

25. $y' = -y + \sin(t)$, $\quad y(0) = 1$

26. $y' = -t^2 y$, $\quad y(0) = 1$

27. $y' = t/y^2$, $\quad y(0) = 1$ **28.** $y' = t - 3y$, $\quad y(0) = 1$

29. In the preceding exercises, you've seen that the error in the fourth-order Runge-Kutta method is approximately proportional to the fourth power of the step size (i.e., $E_h \approx \lambda h^4$). This makes the fourth-order Runge-Kutta method a **fourth-order** solver. If you halve the step size, what affect will this have on the error?

6.3 Numerical Error Comparisons

The differences between the solvers described in the previous sections can best be evaluated by doing some comparisons on actual calculations. We will do that here. If you have access to a computer and programs implementing these algorithms, you are encouraged to duplicate the process. Such algorithms are available in MATLAB®, Maple, *Mathematica*, and in many other mathematical software packages. The actual results may differ somewhat because of differences in implementation.

The point will be made just as well with an easy equation as with one that is difficult, and we have to know the exact solution in order to compute the actual error being made. A good test is provided by the initial value problem

$$y' = \frac{\cos t}{1 + y}, \quad \text{with} \quad y(0) = 1. \tag{3.1}$$

This equation is separable, and with a little work we find that the solution is

$$y(t) = \sqrt{4 + 2 \sin t} - 1. \tag{3.2}$$

We will solve the equation over the interval $(0, 10)$, using various step sizes. The results using Euler's method are shown in Figure 1. What is clear from this figure is that the error gets smaller as we use smaller step sizes.

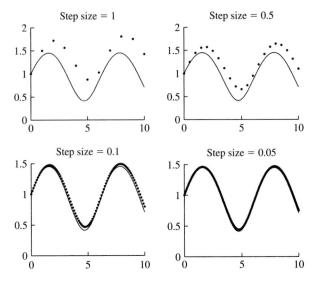

Figure 1. Results of computation using Euler's method to solve the problem in (3.1).

Next we do the same computation using the second- and fourth-order Runge-Kutta methods. The results are shown in Figures 2 and 3. Again we notice that the error gets smaller as we use smaller step sizes. In addition, by comparing Figures 1, 2, and 3 for the same step size, we see that the second-order Runge-Kutta method is more accurate than Euler's method for a given step size, and that the fourth-order Runge-Kutta method is the most accurate of the three.

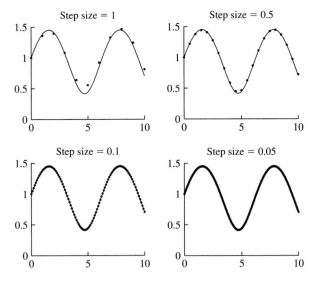

Figure 2. Results of computation using the second-order Runge-Kutta method to solve (3.1).

Notice that even with the relatively large step size of $h = 1$, the results for the fourth-order Runge-Kutta method are quite good, and with $h = 0.5$ it is difficult to see any difference between the approximation and the exact solution.

This last point is made even more evident in Figure 4, where we compare the results of the three solvers at the same step size ($h = 1$). The circles show the results using Euler's method, the black dots correspond to the second-order Runge-

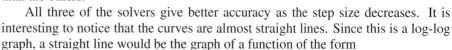

Figure 3. Results of computation using the fourth-order
Runge-Kutta method to solve (3.1).

Kutta method, and the blue dots to the fourth-order Runge-Kutta. While Euler's method and the second-order Runge-Kutta method clearly show errors, the fourth-order Runge-Kutta seems to give very good accuracy.

The two main points of this section, and indeed of the chapter, are that solvers get better as step size is decreased, and that higher-order solvers do a better job than those of lower order. Both of these points are made clearly in Figure 5. Here for a variety of step sizes we compute the maximum of the errors made by the three methods and plot them versus the step size in a log-log graph. This means we are plotting the logarithm of the maximum error against the logarithm of the step size. You will notice that for a given step size the first-order Euler's method always has the largest error, with the second-order Runge-Kutta method somewhat better. The fourth-order Runge-Kutta method is always the most accurate. Indeed, the accuracy of the fourth-order Runge-Kutta method is orders of magnitude better than the others.

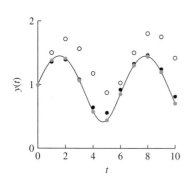

Figure 4. Comparison of the three solution methods.

All three of the solvers give better accuracy as the step size decreases. It is interesting to notice that the curves are almost straight lines. Since this is a log-log graph, a straight line would be the graph of a function of the form

$$\log(\text{maximum error}) = A \log(\text{step size}) + B,$$

where A and B are constants. The constant A is the slope of the straight line. If we exponentiate this formula, it becomes the power relationship

$$\text{maximum error} = C(\text{step size})^A,$$

where the constant $C = e^B$. Notice that A is now the power. This expression invites comparison with the inequalities in (1.14), (2.5), and (2.11). The comparison leads us to believe that the constant A is comparable to the order of the method. Thus the slopes of the graphs in Figure 5 are likely to be very close to the orders of the solvers. We will leave the checking of this to the exercises.

One final point should be made about the errors made in solving differential equations numerically. We pointed out in Section 6.1 that there were two sources of error, truncation error and round-off error. Of these two, the truncation error is the most serious on modern computers, so we have said nothing about round-off

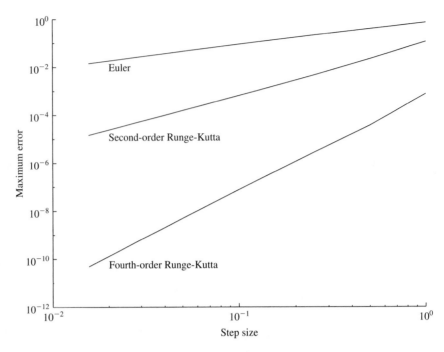

Figure 5. Comparison of the solvers at a range of step sizes.

error. In Figure 5 we see that the fourth-order Runge-Kutta methods at the smallest step sizes make errors on the order of 10^{-11}. Even for such a small error it is the truncation error that dominates. However, if we were to use step sizes an order of magnitude smaller than that used in Figure 5, the round-off error would become important. From a practical point of view, it is almost never necessary to require that errors be as small as this.

EXERCISES

In Exercises 1–6, perform each of the following tasks on the given time interval.

(i) Find the exact solution of the given initial value problem.

(ii) Use Euler's method and the given step sizes to create a plot of four graphs similar to that shown in Figure 1.

(iii) Use the second-order Runge-Kutta method and the given step sizes to create a plot of four graphs similar to that shown in Figure 2.

(iv) Use the fourth-order Runge-Kutta method and the given step sizes to create a plot of four graphs similar to that shown in Figure 3.

1. $x' = x \sin t, x(0) = 1, [0, 10], h = 1, 0.5, 0.1,$ and 0.05

2. $x' = x \cos 2t, x(0) = 1, [0, 4], h = 1, 0.5, 0.1,$ and 0.05

3. $x' = -x + \sin 2t, x(0) = 1, [0, 4], h = 1, 0.5, 0.1,$ and 0.05

4. $x' = -x + \cos t, x(0) = 1, [0, 10], h = 1, 0.5, 0.1,$ and 0.05

5. $x' = (1 + x) \cos t, x(0) = 0, [0, 6], h = 1, 0.5, 0.1,$ and 0.05

6. $x' = x^2 \cos 2t, x(0) = 1, [0, 6], h = 1, 0.5, 0.1,$ and 0.05

In Exercises 7–10, find the exact solution of the given initial value problem. Then use Euler's method, the second-order Runge-Kutta method, and the fourth-order Runge-Kutta method to determine approximate solutions at the given step size over the given time interval. Arrange the graphs of each solution on one plot, as shown in Figure 4.

7. $x' = x, \quad x(0) = 1, \quad [0, 4], \quad h = 0.5$

8. $x' = -tx, \quad x(0) = 4, \quad [0, 4], \quad h = 0.5$

9. $x' = \sin t/(x + 1), \quad x(0) = 0, \quad [0, 12], \quad h = 1$

10. $x' = (1 - x) \sin t, \quad x(0) = 0, \quad [0, 12], \quad h = 1$

11. Consider the initial value problem

$$x' = x, \quad x(0) = 1,$$

which has exact solution $x(t) = e^t$. Thus, the actual solution at $t = 1$ is $x(1) = e$. Use Euler's method, the second-order Runge-Kutta method, and the fourth-order Runge-Kutta method to approximate $x(1)$, in each case using the step size $h = 0.1$. Check your approximations with those shown in the first row of Table 1. Complete the remainder of the table.

(a) The error made in approximating $x(1)$ using Euler's method is calculated by subtracting the actual value ($x(1) = e$) from each entry in the Euler column. Plot the log of the magnitude of the error versus the log of the step size. *Note*: To get good results, work with the internal precision of the numbers in your computer or calculator, not their truncated representations in the table.

Table 1			
h	EULER	RK2	RK4
0.10000	2.59374	2.71408	2.71828
0.05000			
0.02500			
0.01250			
0.00625			

(b) Repeat the procedure in part (a) for the second-order Runge-Kutta and fourth-order Runge-Kutta data in Table 1, superimposing the plots on the graph drawn in part (a). This should give you an image similar to that in Figure 5.

(c) Use a computer or calculator to calculate the slope of the line of best fit for each method. Otherwise, use the old fashioned method: draw a line of best fit with a pencil, and use two points on the line to estimate its slope. Relate the slope of each line to the order of the method defined by the inequalities (1.14), (2.5), and (2.11).

In Exercises 12–13, repeat the procedure outlined in Exercise 11 for the given initial value problem. The analysis is to be performed on the error made by each method at the endpoint of the given time interval.

12. $x' = -tx$, $x(0) = 4$, $[0, 4]$

13. $x' = -tx^2$, $x(0) = 3$, $[0, 2]$

14. Consider the planar system

$$x' = y,$$
$$y' = -x,$$

with initial conditions $x(0) = 1$ and $y(0) = 0$.

(a) Show, by direct substitution, that $x(t) = \cos t$ and $y(t) = -\sin t$ satisfy both differential equations. Show that these solutions also satisfy the initial conditions. Plot the solution in the xy phase plane, using the time interval $[0, 2\pi]$.

(b) Use Euler's method, the second-order Runge-Kutta method, and the fourth-order Runge-Kutta method to compute approximate solutions of the initial value problem on the interval $[0, 2\pi]$. Use a step size $h = 0.5$. Superimpose the plot of each of these solutions on the plot containing the exact solution in the xy phase plane.

6.4 Practical Use of Solvers

In the previous sections, we have described three solvers. These are described as single-step solvers because the solution at each step is computed entirely in terms of the previous point. There are many other algorithms that have been discovered for the numerical solution of differential equations, but the single-step methods are probably the most commonly used. Indeed, the fourth-order Runge-Kutta method is an excellent method for most applications. In this section we will briefly describe some other methods and give some advice about the choice of algorithm.

Variable-step solvers

There are single-step methods that go beyond the fourth-order Runge-Kutta method. These add features to the basic algorithm. The most common such feature is variable-step size. With such an algorithm, the user specifies an ***error tolerance*** instead of a step size, and the algorithm chooses the step size at each step to achieve an error smaller than the tolerance specifies. Usually specifying an error tolerance is easier than specifying a good step size.

Kinky graphs and dense output

There is an annoying feature of many higher-order algorithms, especially those that use a variable step size. The high accuracy of these algorithms allows us to use step sizes so large that the graphs of the output are kinky, and do not reveal the true nature of the solution. The graphical output is not very accurate between the steps.

An example computed by the MATLAB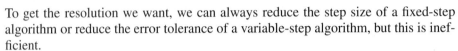® solver `ode45` is shown by the black curve in Figure 1. The initial value problem being solved is

$$y' = \frac{\cos t}{2y - 2}, \quad \text{with} \quad y(0) = 3. \tag{4.1}$$

To get the resolution we want, we can always reduce the step size of a fixed-step algorithm or reduce the error tolerance of a variable-step algorithm, but this is inefficient.

To counter this failing, modern variable-step algorithms have a feature called **dense output**. This feature provides interpolated values between the step points. The interpolated values are computed using knowledge of the differential equation, so they are highly accurate. It is usually possible to specify the number of interpolated values. Amazingly, the computation of these extra values occurs at the cost of little additional computing time. Dense output is a clever solution to the kinky curve problem.

The blue curve in Figure 1 shows the solution computed by `ode45` with 10 values interpolated between the steps. Without interpolated values, the solver computes the solution only at the vertices of the black, polygonal curve. Figure 1 shows these values to be accurate, and also shows what adding the interpolated points can do.

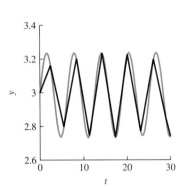

Figure 1. Solutions computed with and without interpolated values.

Stiff equations and stiff solvers

It should be clear from the preceding paragraphs that a modern, variable, single-step routine can be easy to use and highly effective. Indeed, such an algorithm is a good choice for solving any system of differential equations where no additional information is available. However, there are equations and systems of equations for which single-step solvers do not do well or even fail completely.

Typical of such systems are **stiff** systems. Stiff systems involve phenomena that have widely varying reaction rates. For example, the second-order equation

$$y'' + py' + qy = 0$$

has the general solution

$$y(t) = Ae^{\lambda_1 t} + Be^{\lambda_2 t},$$

where

$$\lambda_1, \lambda_2 = \left(-p \pm \sqrt{p^2 - 4q}\,\right)/2.$$

In the overdamped case, when $p^2 - 4q > 0$, both roots are real and negative. They can be quite different. For example, if $p = 10.1$ and $q = 1$, the roots are $\lambda_1 = -10$ and $\lambda_2 = -1/10$. With initial values $y(0) = y'(0) = 1$, the solution is

$$y(t) = (10e^{-t/10} - e^{-10t})/9.$$

In this case the term $e^{-10t}/9$ dies out much faster than the other term. Nevertheless, if we try to use a Runge-Kutta, single-step algorithm to compute the solution, we find that we must use very small step sizes or the solution is highly erratic (the technical term is unstable) and not anywhere near being accurate. In other words, the step size must be chosen to accommodate the fast reaction rate, even though that part of the solution dies out quickly.

The example in the previous paragraph is linear, and we would not be likely to want to solve such an equation numerically. However, the same situation arises with nonlinear equations. An example is the stiff van der Pol system

$$x' = y,$$
$$y' = -x + \mu(1 - x^2)y, \tag{4.2}$$

where μ is large, say $\mu = 1000$. If a single-step algorithm is used to solve this system, a very small step size must be used—so small that the solution over even a relatively small interval takes an extremely long time.

It is possible for a single equation of first order to be stiff. Consider

$$y' = e^t \cos y. \tag{4.3}$$

In this case, $y(t) = \pi/2$ is a solution. Other solutions with initial values at $t = 0$ between $-\pi/2$ and $3\pi/2$ tend to this solution as $t \to \infty$. However, because of the term e^t, the reaction rate varies wildly for y near $\pi/2$ as t gets large. A single-step solver requires increasingly small step sizes as t gets large. Indeed, the required step size is so small that the solution will take a very long time, even though to a very high accuracy the solution is almost equal to the constant solution $\pi/2$.

Fortunately, there are solvers that are designed specifically to solve stiff equations and systems. These stiff solvers take a different approach from the single-step solvers we have discussed in this chapter. While stiff solvers do not do nearly as well as single-step solvers for most equations or systems, they do a marvelous job for stiff systems.

What solver to use and how to proceed

If we have a system of differential equations about which we know very little, which solver should we use? Is there any way to tell that the system is stiff so that a stiff solver can be chosen? Unfortunately, there are no easy answers to these questions. Sometimes the presence of a large constant such as appears in the van der Pol system leads us to suspect that the system is stiff. Sometimes knowledge of what is being modeled can make us suspect that there are very different reaction rates in the system. In this case the ODEs are going to be stiff. However, occasionally it is just not possible to tell.

In the absence of an indication that an equation or a system is stiff, it is best to start with a single-step algorithm. The fourth-order Runge-Kutta method is solid and useful. However, if you have access to a variable-step method, it is usually preferable. A fixed, single-step method asked to solve a stiff system usually becomes unstable, indicated by wild oscillations of the computed solution. A variable-step algorithm trying to solve a stiff system will take such small steps that the solution bogs down, and the computation takes a long time, sometimes seeming to go on forever. If these things occur while solving a system, change to a stiff solver.

Checking your work

It is essential that you check your computational results. This can be done by comparing the computed results with qualitative information that you have discovered about the equation. Compare the results with your physical intuition about how the solution should behave. However, sometimes your physical intuition will be wrong. One of the reasons to model and compute is to discover if and when that is true.

In addition, always compute the solution twice with different step sizes. After the first run that seems satisfactory, halve the step size and run it again. If the two solutions do not agree, halve the step size once more. Continue in this way until you get agreement. If you are using a variable-step solver, reduce the tolerance by a factor of 10 and recompute the solution. Iterate this procedure until you find good agreement between successive solutions.

Working with equations that are not smooth

One more word of caution: Most solvers have been designed to work with equations where the right-hand sides are smooth.[2] If your equations do not have this property, you can expect that the solver will not work well. Usually discontinuities in the right-hand sides appear in the t-variable. When solving such an equation numerically, it is essential that you arrange that all discontinuities in t are among the points where the solution will be computed. This will prevent the solver from stepping across a discontinuity and thereby ignoring it completely. When this happens computational error always follows.

A cautionary tale

We want to instill in you a sense of caution about using ODE solvers. Sometimes, even for apparently nice equations, your solver will fail to give a correct answer. This does not contradict the results presented in Sections 6.2 and 6.3. In fact, for each solution method, we pointed to the bad news contained in the error estimates. You will notice that a common factor on the right-hand side of each of these inequalities in (1.14), (2.5), and (2.11) is

$$e^{L(b-a)} - 1. \tag{4.4}$$

We are solving the system over the interval $[a, b]$. For large intervals, the exponential term in (4.4) can be quite large. The impact is that over large computation intervals the errors can be large, even for small step sizes. It is true that the error bounds are worst-case situations, and usually the errors stay pretty small. However, worst-case situations do arise.

Consider the initial value problem

$$x' = t(x - 1) \quad x(-a) = 0, \quad \text{for } t \in [-a, a]. \tag{4.5}$$

This is a simple linear differential equation. We would not suspect that it would cause any difficulty. Since it is linear, we can find the solution,

$$x(t) = 1 - e^{(t^2 - a^2)/2}.$$

Notice in particular that $x(a) = 0$, and, in general, $x(-t) = x(t)$. Thus x is an even function of t.[3]

However, when we try to solve the equation numerically, even for modest values of a like $a = 6$, we notice a problem at step sizes like 0.1. For $a = 10$, there are problems at all step sizes. We fail to get convergence of the computed value at $a = 10$ to the correct value of $y(10) = 0$ as the step size decreases. Figure 2 shows a computed approximate solution for $a = 10$, using the fourth-order Runge-Kutta method with a step size of $h = 0.001$. This is a small step size, yet as Figure 2 shows, the result is far from perfect. Since x is even, we would expect that the solution curve would be symmetric with respect to the x-axis. The actual solution satisfies $x(10) = 0$. The computed value is larger than 0.2, so the accuracy is not too good.

---

[2] *Smooth* means that one or more derivatives are continuous.

[3] The authors are indebted to Professor John Hubbard for this simple example of how computations can go bad.

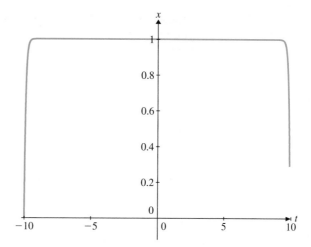

Figure 2. An approximate solution to the initial value problem in (4.5).

It is not difficult to understand why there is a problem with the initial value problem in (4.5). Notice first that $x_1(t) = 1$ is a solution to the differential equation in (4.5). Then notice that for the solution to the initial value problem (with $a = 10$) we have

$$x(0) = 1 - e^{-50}.$$

Since $e^{-50} \approx 1.93 \times 10^{-22}$, the difference between $x(0)$ and $x_1(0) = 1$ is so small that a computer operating at double precision cannot distinguish the two. Of course, any solution method is making truncation errors that are usually much larger than the double precision round-off error. Using a fourth-order Runge-Kutta routine, or any other numerical method, to approximate the solution $x(t)$ could cause the computed solution to move to any solution sufficiently close to $x(t)$, including $x_1(t)$. We cannot expect the computer to approximate accurately the solution with any solver or any solution algorithm.

This example should make it clear that it is unwise to depend on a solver without exercising a high degree of scepticism. As we said in the previous section, it is imperative that we check numerical solutions to be sure they are giving us sufficiently accurate answers.

EXERCISES

Consider the initial value problem

$$x' = e^t \cos x \quad \text{with} \quad x(0) = 2. \qquad (4.6)$$

Qualitative analysis reveals that $x(t) \to \pi/2$ as t increases. This is illustrated in Figure 3. Nevertheless, the solution becomes less and less amenable to computation as t increases. Exercises 1–3 will explore this difficulty and how the use of a stiff solver can help. In each of these you will solve (4.6) on the interval $[0, b]$ for the four cases $b = 4,\ 6,\ 8,$ and 10.

1. For each case find the largest step size of the form $h = 10^{-k}$ for $k = 1,\ 2,\ \ldots$ for which the solution computed using the fourth-order Runge-Kutta method displays the behavior shown in Figure 3. How long does the computation take with the required step size?

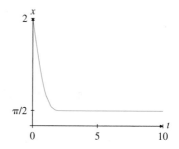

Figure 3. The solution to 4.6.

2. Solve each case using a variable-step solver. What were the largest and smallest step sizes used during the calculation? How long did the computation take?

3. Solve each case using a stiff solver. What were the largest and smallest step sizes used during the calculation? How long did the computation take?

The van der Pol system in (4.2) becomes increasingly stiff as the parameter μ increases. In Exercises 4–6 we will consider the computational solution to the van der Pol system for a variety of values of μ, and for different solvers. We will use the initial conditions $x(0) = 1$ and $y(0) = 1$. With $\mu = 2$ the solution is shown plotted in Figure 4. You will notice that the solution converges to a periodic function (called a **limit cycle**). This is characteristic of the van der Pol system regardless of the value of μ, and is what you should look for in your computed solutions. In each of the exercises you will solve the system over the interval I in the three cases ($\mu = 10$, $I = [0, 20]$), ($\mu = 50$, $I = [0, 100]$), and ($\mu = 100$, $I = [0, 200]$).

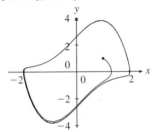

Figure 4. The solution to the van der Pol equation.

4. For each case find the largest step size of the form $h = 10^{-k}$ for $k = 1, 2, \ldots$ for which the solution computed using the fourth-order Runge-Kutta method displays the characteristic limiting behavior shown in Figure 4. How long did the computation take at the required step size?

5. Solve each case using a variable-step solver. What were the largest and smallest step sizes used during the calculation? How long did the computation take?

6. Solve each case using a stiff solver. What were the largest and smallest step sizes used during the calculation? How long does the computation take? Now use your stiff solver to find the solution with $\mu = 1000$ over the interval $[0, 2000]$. How long did this take?

7. It is the system, not the solution, that gives rise to stiffness. Show by direct substitution that $x(t) = 2e^{-t} + \sin t$ and $y(t) = 2e^{-t} + \cos t$ are solutions of the initial value problem

$$x' = -2x + y + 2\sin t, \qquad x(0) = 2$$
$$y' = x - 2y + 2(\cos t - \sin t). \qquad y(0) = 3$$

Show that they also satisfy the second initial value problem

$$x' = -2x + y + 2\sin t, \qquad x(0) = 2$$
$$y' = 998x - 999y + 999(\cos t - \sin t). \qquad y(0) = 3$$

(a) Use a variable-step solver to compute numerical solutions of the first system on the time interval $[0, 10]$, and record the time the computation took. Draw the graphs of the exact solutions x and y on $[0, 10]$. Superimpose the numerical solutions as a set of discrete points.

(b) Repeat the instructions in part (a) for the second system.

(c) Repeat the instructions in parts (a) and (b) using a stiff solver. Write a short paragraph noting any differences you see. Which system is stiff?

8. This is an exercise in choosing the correct step size or tolerance for a computation. The damped pendulum with a large forcing term leads to motions that are chaotic, and provide a computational challenge. The motion is modeled by the differential equation

$$\theta'' + \mu\theta' + \frac{g}{L}\sin\theta = A\cos\omega t,$$

where θ measures the angular displacement of the pendulum from vertical, L is the length of the pendulum, and μ is a damping constant. The model is discussed in Section 10.6. The forcing term has amplitude A and frequency ω.

Use the parameters $L = 1$, $\mu = 2$, $\omega = 2$, $g = 9.8$, and $A = 20$, and the initial conditions $\theta(0) = \pi/6$ and $\theta'(0) = 0$. The object is to find an almost optimal step size or tolerance for solving the equation over a given interval. If you are using a fixed step size solver, then for each step size h compute the solution with step size h, and then again with step size $h/2$. Plot the displacements from the two solutions in different colors on the same figure. A step size will be considered to be acceptable if there is no visual difference between the two curves. Start with $h = 0.1$ and repeat the procedure, dividing h by 2 each time, until you find the first acceptable step size.

If you are using a solver with a variable step size, then perform the same exercise by changing the tolerance each time. Start with a tolerance of 10^{-4}, and find the first acceptable tolerance. You are forewarned that some variable-step solvers use more than one tolerance. If this is true in your case you will have to be sure to reduce all of them in size in order to get a true test at the prescribed tolerance.

Perform the exercise for the intervals $0 \le t \le t_f$, where $t_f = 25$, 50, 75, and 100. Of course, your starting value for a larger interval need be no larger than the first acceptable value for a smaller interval. Based on your data for smaller intervals, what do you think would be the result $t_f = 200$? If you have a very fast computer, you might try and see if there is an acceptable step size or tolerance for this interval.

9. Use the fourth-order Runge-Kutta method to try to solve the initial value problem in (4.5) with a variety of step sizes. Write a paragraph describing your results.

10. The system

$$x' = y$$
$$y' = y^2 - x$$

is surprisingly difficult to solve accurately. Solutions with initial values of the form $x(0) = x_0 > 0$ and $y(0) = 0$ have solution curves in the phase plane that are closed. In other words, the solutions are periodic. Use your favorite solver to try to solve the system with $x_0 > 10$, and see if you get closed curves in the phase plane. More precisely, for a range of values of x_0 starting at $x_0 = 10$, record the step size or tolerance required to compute solution curves that appear to be closed. For what value of x_0 does your system completely fail for any step size or tolerance?

PROJECT 6.5 **Numerical Error Comparison**

Consider the following initial value problems.

(i) $x' = x$, with $x(0) = 1$ on the interval $[0, 1]$, with $h_0 = 0.1$

(ii) $x' = x \sin 3t$, with $x(0) = 1$ on the interval $[0, 4]$, with $h_0 = 0.4$

(iii) $y' = (1 + y^2) \cos t$, with $y(0) = 0$ on the interval $[0, 6]$, with $h_0 = 1$

(iv) $y' = y^2 \cos 2t$, with $y(0) = 1$ on the interval $[0, 6]$, with $h_0 = 0.5$

For each of these problems, carry out the following steps.

(a) Find the exact solution to the differential equation.

(b) For eight step sizes starting with h_0, and halving each time, find the approximate solution using Euler's method. Make a figure that graphically compares the computed solutions with the exact solutions for a selection of the step sizes.

A figure like Figure 2 in Section 6.1 would do, but suit yourself.

(c) Complete the tasks in (b) for the Runge-Kutta methods of orders 2 and 4.

(d) Make a figure that graphically compares the computed solutions for each of the three methods at one of the step sizes with the exact solution. A figure like Figure 4 in Section 6.3 would be suitable.

(e) For each method and each step size, find and record the maximum error over the indicated interval. It would be best to record the maximum error electronically, if possible. Make a figure plotting the maximum errors in the three methods versus step size. A figure like Figure 5 in Section 6.3 would be suitable. If it is difficult or impossible to find the maximum error, then compute the error at the end of the computation interval.

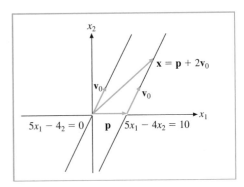

CHAPTER

7

Matrix Algebra

Our next topic in ordinary differential equations will be linear, first-order systems. We will study that subject in the next two chapters, but first we will need to understand some algebra, and that will be the topic of this chapter. Although in the end we will use this algebra to help us understand differential equations, much of what we will do here is motivated by the desire to be able to solve a system of m linear algebraic equations for n unknown numbers. Perhaps this topic seems peripheral to the study of differential equations, but in fact we will find it extremely useful in what follows.

We are interested in solving systems of equations like

$$3u + 2v - 5w = 5,$$
$$4u - v + 5w = 0.$$

Part of the difficulty in dealing with systems like this lies in the complexity of the notation. We will start by introducing new notation that will be useful in systematizing the solution process. We will then find a systematic way to solve linear systems of equations.

7.1 Vectors and Matrices

For a system like

$$3u + 2v - 5w = 5,$$
$$4u - v + 5w = 0,$$

$$(1.1)$$

we isolate the array of coefficients

$$C = \begin{pmatrix} 3 & 2 & -5 \\ 4 & -1 & 5 \end{pmatrix}$$

$$(1.2)$$

and refer to it as the ***coefficient matrix***. In general, a ***matrix*** is a rectangular array of numbers. The ***entries*** or ***components*** of a matrix are the numbers that appear in the array.

A ***column vector*** is a matrix with only one column. Thus

$$\mathbf{x} = \begin{pmatrix} u \\ v \\ w \end{pmatrix} \quad \text{and} \quad \mathbf{b} = \begin{pmatrix} 5 \\ 0 \end{pmatrix} \tag{1.3}$$

are column vectors; $\mathbf{x}$ is the ***vector of unknowns*** in the system (1.1), and $\mathbf{b}$ is the ***right-hand side*** of (1.1). In Definition 1.9 we will define the product of a matrix and a vector in such a way that the system in (1.1) can be written in the compact form

$$C\mathbf{x} = \mathbf{b}. \tag{1.4}$$

The array C in (1.2) is an example of a matrix with two rows and three columns. The general matrix with m rows and n columns has the form

$$A = \begin{pmatrix} a_{11} & a_{12} & \cdots & a_{1n} \\ a_{21} & a_{22} & \cdots & a_{2n} \\ \vdots & \vdots & \ddots & \vdots \\ a_{m1} & a_{m2} & \cdots & a_{mn} \end{pmatrix}. \tag{1.5}$$

We define the ***size*** of A to be (m, n), and A is called an $m \times n$ matrix. Notice that the element in the ith row and the jth column of A is a_{ij}. Thus the first subscript refers to the row and the second to the column.

A column vector is a matrix with only one column. Similarly, a ***row*** vector is one with only one row. Thus $\mathbf{x}$ and $\mathbf{b}$ in (1.3) are examples of column vectors, while

$$\mathbf{y} = (1, \ -3, \ 3) \quad \text{and} \quad \mathbf{z} = (-1, \ 2, \ 0, \ -3, \ 5)$$

are examples of row vectors. We will sometimes need to refer to the number of elements in a vector. For example, we will say that $\mathbf{b}$ is a 2-vector, $\mathbf{x}$ and $\mathbf{y}$ are 3-vectors, and $\mathbf{z}$ is a 5-vector.

The word *vector* is one of the most over used terms in mathematics and science. With its original meaning in physics and geometry, it refers to a quantity having direction as well as magnitude, usually illustrated by an arrow drawn from its original to its final position. Examples of this would be a tangent vector to a curve in space or the velocity of a particle. In other parts of mathematics and in many areas of application, it refers to an ordered set of two or more numbers. More broadly, a vector is an element of a vector space, and in that generality a vector can be any mathematical quantity for which there is a rule of addition and a rule of multiplication by numbers. In this very general context, a vector can even be a polynomial, a function, or something more complicated like a tensor.[1]

---

[1] The term *vector* has been taken up in other areas, where it has meanings that can be seen to be related to the original meaning (although sometimes a little thought is required). For example, in aeronautics it is a course taken by an aircraft or steered by a pilot. In computing, it refers to a sequence of consecutive locations in memory or a series of items occupying such a sequence. In medicine and biology, a vector is a person, animal, or plant that carries a pathogenic agent and acts as a potential source of infection for members of another species, such as the mosquito that carries malaria. A related meaning in genetics is a bacteriophage, which transfers genetic material from one bacterium to another or a phage or plasmid used to transfer extraneous DNA into a cell. In particle physics, it is used as an adjective to designate particles with a spin of 1, such as a vector boson.

With such a variety of meanings, it is important to determine precisely the meaning at hand. In this chapter, a vector will always be an ordered list of numbers. However, we will emphasize the geometric interpretation of the concept as well.

The 2×3 matrix C in (1.2) can be looked at as consisting of the two row vectors $(3, 2, -5)$ and $(4, -1, 5)$, or as consisting of the three column vectors $\begin{pmatrix} 3 \\ 4 \end{pmatrix}$, $\begin{pmatrix} 2 \\ -1 \end{pmatrix}$, and $\begin{pmatrix} -5 \\ 5 \end{pmatrix}$. Each interpretation is useful in different circumstances.

Algebraic properties of matrices and vectors

If two matrices A and B have the same size, we define the ***matrix sum*** $A + B$ to be the matrix whose components are the sums of the corresponding components of A and B. Thus

$$\begin{pmatrix} 2 & 3 & -1 \\ 1 & 0 & 5 \end{pmatrix} + \begin{pmatrix} 0 & 8 & 10 \\ 9 & -10 & 4 \end{pmatrix} = \begin{pmatrix} 2 & 11 & 9 \\ 10 & -10 & 9 \end{pmatrix}.$$

The set of all column vectors with n real entries will be denoted by $\mathbf{R}^n$. With the operation defined in the previous paragraph, if $\mathbf{x}, \mathbf{y} \in \mathbf{R}^n$, then $\mathbf{x} + \mathbf{y} \in \mathbf{R}^n$. For example,

$$\mathbf{x} = \begin{pmatrix} 3 \\ 1 \end{pmatrix} \quad \text{and} \quad \mathbf{y} = \begin{pmatrix} -2 \\ 1 \end{pmatrix}$$

are vectors in $\mathbf{R}^2$ and their sum is

$$\mathbf{z} = \mathbf{x} + \mathbf{y} = \begin{pmatrix} 1 \\ 2 \end{pmatrix},$$

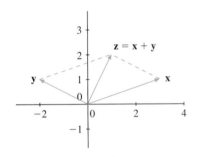

Figure 1. The sum of two vectors.

another vector in $\mathbf{R}^2$.

In Figure 1, we display these vectors as arrows with base at the origin and the arrow heads at the corresponding point in $\mathbf{R}^2$. This is to emphasize the fact that a vector possesses both a length and a direction. In this case, the direction is pointing from the origin to the point with the coordinates of the vector.

Figure 1 illustrates the geometric meaning of vector addition. The two vectors $\mathbf{x}$ and $\mathbf{y}$, together with the origin $\begin{pmatrix} 0 \\ 0 \end{pmatrix}$, define a parallelogram, and the fourth vertex of that parallelogram is the sum $\mathbf{x} + \mathbf{y}$.

Although vectors are commonly depicted as arrows with their bases at the origin, for many applications it is convenient to picture them with their bases elsewhere. This is useful when we want to emphasize the geometric notion of a vector as an object with magnitude and direction. For this purpose, we will consider all arrows that are parallel translates of each other to be equivalent. Thus the two vectors in Figure 2 are considered to be the same vector.

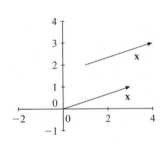

Figure 2. Parallel vectors of the same length are considered to be equivalent.

Vectors based at the origin are useful for designating the position of points. Vectors with their bases elsewhere are useful, for example, for representing the tangent vectors to curves. These "floating" vectors also allow for a new interpretation of the sum of two vectors. To add $\mathbf{x}$ and $\mathbf{y}$, we simply move the vector $\mathbf{y}$ without changing its length or its direction so that its base coincides with the head of $\mathbf{x}$. Then the vector with base at the base of $\mathbf{x}$ and head at the head of the translated $\mathbf{y}$ gives the sum $\mathbf{x} + \mathbf{y}$. See Figure 3, and compare with Figure 1. This graphical representation of vector addition is especially useful if we are adding more than one vector, as shown in Figure 4.

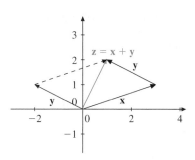

Figure 3. Another interpretation of the sum of two vectors.

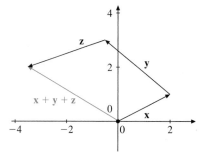

Figure 4. The sum of three vectors.

We will also define the operation of ***multiplication of a matrix by a number***. If A is a matrix, and $a \in \mathbf{R}$, then aA is the matrix in which each component of A is multiplied by a. For example,

$$\frac{1}{3} \cdot \begin{pmatrix} 0 & 1 & 3 \\ 3 & 0 & 1 \end{pmatrix} = \begin{pmatrix} 0 & 1/3 & 1 \\ 1 & 0 & 1/3 \end{pmatrix}.$$

If $\mathbf{x} \in \mathbf{R}^n$ and $c \in \mathbf{R}$, then $c\mathbf{x} \in \mathbf{R}^n$ is the vector whose components are c times those of $\mathbf{x}$. For example,

$$3 \begin{pmatrix} 5 \\ 4 \\ 2 \end{pmatrix} = \begin{pmatrix} 3 \cdot 5 \\ 3 \cdot 4 \\ 3 \cdot 2 \end{pmatrix} = \begin{pmatrix} 15 \\ 12 \\ 6 \end{pmatrix}.$$

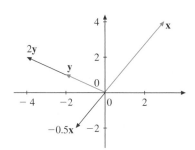

Figure 5. The product of a number and a vector.

The effect of multiplying a vector by a number is illustrated in Figure 5. Notice in particular that a positive multiple of a vector points in the same direction, but a negative multiple points in the opposite direction.

With a little imagination, the fingers on one hand can serve as a "vector laboratory." See Figure 6. The index finger can be considered as a single vector, and it can point anywhere in three dimensions. If you want to consider two different vectors, the thumb and the index finger will suffice. The thumb and the first two fingers will serve if you need three vectors. Using this laboratory can help you understand many of the mysteries of vectors.

Linear combinations of vectors and systems of equations

Suppose that $\mathbf{x}$ and $\mathbf{y}$ are two vectors in $\mathbf{R}^n$. A ***linear combination*** of $\mathbf{x}$ and $\mathbf{y}$ is any vector of the form $a\mathbf{x} + b\mathbf{y}$ where a and b are real numbers. Examples for $n = 2$

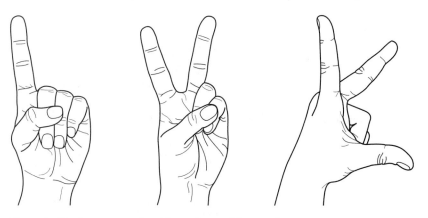

Figure 6. The fingers on one hand form a "vector laboratory."

are shown in Figure 7. It seems clear from Figure 7 that when $n = 2$, and $\mathbf{x}$ and $\mathbf{y}$ do not point in the same or opposite directions, then any vector in $\mathbf{R}^2$ can be written as linear combinations of $\mathbf{x}$ and $\mathbf{y}$. We will have more to say about this later in this chapter.

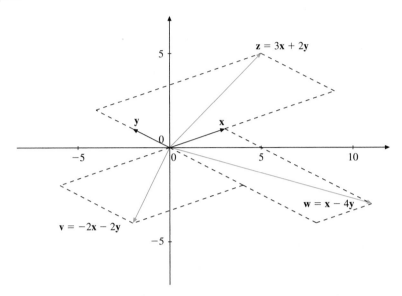

Figure 7. Linear combinations of vectors.

Of course, the concept of linear combination can be expanded to more than two vectors. Suppose that $\mathbf{v}_1, \mathbf{v}_2, \ldots, \mathbf{v}_p$ are all vectors in $\mathbf{R}^n$, and $a_1, a_2, \ldots, a_p$ are numbers. Then

$$a_1\mathbf{v}_1 + a_2\mathbf{v}_2 + \cdots + a_p\mathbf{v}_p$$

is called a linear combination of $\mathbf{v}_1, \mathbf{v}_2, \ldots, \mathbf{v}_p$.

The notion of linear combination gives a different view of a system of linear equations. Consider again the system in (1.1). We can write the system as the vector equation

$$\begin{pmatrix} 3u + 2v - 5w \\ 4u - v + 5w \end{pmatrix} = \begin{pmatrix} 5 \\ 0 \end{pmatrix}.$$

The vector on the left can be written as a linear combination, making it

$$u \begin{pmatrix} 3 \\ 4 \end{pmatrix} + v \begin{pmatrix} 2 \\ -1 \end{pmatrix} + w \begin{pmatrix} -5 \\ 5 \end{pmatrix} = \begin{pmatrix} 5 \\ 0 \end{pmatrix}. \tag{1.6}$$

Thus finding a solution to the equations in (1.1) is equivalent to finding out how to represent the right-hand side $\mathbf{b} = \begin{pmatrix} 5 \\ 0 \end{pmatrix}$ as a linear combination of the column vectors as indicated in (1.6). Notice that the vectors on the left of (1.6) are the columns of the coefficient matrix C in (1.2). We use this important observation to motivate the definition of the product of a matrix and a vector. In Definition 1.9 we will define the product of the matrix C and the vector of unknowns $\mathbf{x}$ in (1.3) as

$$C\mathbf{x} = \begin{pmatrix} 3 & 2 & -5 \\ 4 & -1 & 5 \end{pmatrix} \begin{pmatrix} u \\ v \\ w \end{pmatrix} = u \begin{pmatrix} 3 \\ 4 \end{pmatrix} + v \begin{pmatrix} 2 \\ -1 \end{pmatrix} + w \begin{pmatrix} -5 \\ 5 \end{pmatrix}.$$

With this definition, the system can be written as

$$C\mathbf{x} = \mathbf{b}.$$

Matrix multiplication and systems of equations

Consider the matrix

$$A = \begin{pmatrix} a_{11} & a_{12} & \cdots & a_{1n} \\ a_{21} & a_{22} & \cdots & a_{2n} \\ \vdots & \vdots & \ddots & \vdots \\ a_{m1} & a_{m2} & \cdots & a_{mn} \end{pmatrix}. \tag{1.7}$$

A is a matrix with m rows and n columns. We will let

$$\mathbf{a}_1 = \begin{pmatrix} a_{11} \\ a_{21} \\ \vdots \\ a_{m1} \end{pmatrix}, \quad \mathbf{a}_2 = \begin{pmatrix} a_{12} \\ a_{22} \\ \vdots \\ a_{m2} \end{pmatrix}, \quad \ldots, \quad \mathbf{a}_n = \begin{pmatrix} a_{1n} \\ a_{2n} \\ \vdots \\ a_{mn} \end{pmatrix}$$

denote the column vectors in the matrix A. We will indicate that A has columns $\mathbf{a}_1, \mathbf{a}_2, \ldots, \mathbf{a}_n$ by writing

$$A = [\mathbf{a}_1, \mathbf{a}_2, \ldots, \mathbf{a}_n].$$

Now let

$$\mathbf{x} = \begin{pmatrix} x_1 \\ x_2 \\ \vdots \\ x_n \end{pmatrix} \tag{1.8}$$

be a vector in $\mathbf{R}^n$.

DEFINITION 1.9

We define the ***product*** $A\mathbf{x}$ of the matrix A in (1.7) and the vector $\mathbf{x}$ in (1.8) to be the linear combination of the column vectors of A with coefficients from the vector $\mathbf{x}$,

$$A\mathbf{x} = x_1\mathbf{a}_1 + x_2\mathbf{a}_2 + \cdots + x_n\mathbf{a}_n.$$

The general system of m equations in n unknowns can be written as

$$\begin{aligned} a_{11}x_1 + a_{12}x_2 + \cdots + a_{1n}x_n &= b_1 \\ a_{21}x_1 + a_{22}x_2 + \cdots + a_{2n}x_n &= b_2 \\ &\vdots \\ a_{m1}x_1 + a_{m2}x_2 + \cdots + a_{mn}x_n &= b_m. \end{aligned} \tag{1.10}$$

The coefficient matrix for this system is the matrix A in (1.7). The vector of unknowns is the vector $\mathbf{x}$ in (1.8). The right-hand side is the vector

$$\mathbf{b} = \begin{pmatrix} b_1 \\ b_2 \\ \vdots \\ b_m \end{pmatrix}.$$

Notice that if we put all of the expressions on the left-hand side of (1.10) into a column vector, the result is

$$x_1\mathbf{a}_1 + x_2\mathbf{a}_2 + \cdots + x_n\mathbf{a}_n,$$

which according to Definition 1.9 is equal to $A\mathbf{x}$. Thus the general system in (1.10) can be written compactly in the matrix equation

$$A\mathbf{x} = \mathbf{b}. \qquad (1.11)$$

Notice that for the product $A\mathbf{x}$ to be defined, the number of columns in the matrix A must equal the number of rows in $\mathbf{x}$, which is just the number of elements in $\mathbf{x}$.

Computing matrix products

Definition 1.9 emphasizes one of the most important properties of the product of a matrix and a vector. There will be many times in this book when we will talk about finding the linear combination of a set of vectors with a given property. It will be important in those cases that we recognize the linear combination as a matrix product. However, there are easier ways to compute the product of a matrix and a vector than Definition 1.9.

Let's look first at the case when the matrix $A = (a_1, a_2, \cdots, a_n)$ is a row vector, or a $1 \times n$ matrix. Then the columns of A are vectors of length 1, or simply numbers. The product $A\mathbf{x}$ is the linear combination

$$A\mathbf{x} = a_1 x_1 + a_2 x_2 + \cdots + a_n x_n, \qquad (1.12)$$

and in this case it is just a sum of the products $a_j x_j$. For example,

$$\begin{pmatrix} 5 & -2 & 4 \end{pmatrix} \cdot \begin{pmatrix} -1 \\ 3 \\ 0 \end{pmatrix} = 5 \cdot (-1) + (-2) \cdot 3 + 4 \cdot 0 = -11.$$

Now suppose that A is the $m \times n$ matrix in (1.7) and $\mathbf{x}$ is the column n-vector in (1.8). Let $\mathbf{y} = A\mathbf{x}$. Then $\mathbf{y}$ is a column m-vector,

$$\mathbf{y} = \begin{pmatrix} y_1 \\ y_2 \\ \vdots \\ y_m \end{pmatrix}. \qquad (1.13)$$

It follows from Definition 1.9 that y_i, the ith component of $\mathbf{y}$, is the ith component of the linear combination $x_1 \mathbf{a}_1 + x_2 \mathbf{a}_2 + \cdots + x_n \mathbf{a}_n$. Thus we have

$$\begin{aligned} y_i &= x_1 a_{i1} + x_2 a_{i2} + \cdots + x_n a_{in} \\ &= a_{i1} x_1 + a_{i2} x_2 + \cdots + a_{in} x_n \\ &= \sum_{j=1}^{n} a_{ij} x_j, \end{aligned} \qquad (1.14)$$

for $1 \le i \le m$.

Set $\mathbf{a}_i = (a_{i1}, a_{i2}, \cdots, a_{in})$. Notice that $\mathbf{a}_i$ is the ith *row* vector in the matrix A. Furthermore, from (1.12) we recognize the sum in (1.14) to be the product of the row vector $\mathbf{a}_i$ and the column vector $\mathbf{x}$. Thus $y_i = \mathbf{a}_i \cdot \mathbf{x}$, and we see that the ith component of $\mathbf{y} = A\mathbf{x}$ is the product of the ith row vector in A and the column vector $\mathbf{x}$.

Example 1.15   Compute the product $A\mathbf{x}$, where

$$A = \begin{pmatrix} 1 & 3 & 5 \\ 7 & -2 & 4 \end{pmatrix} \quad \text{and} \quad \mathbf{x} = \begin{pmatrix} -1 \\ 2 \\ 3 \end{pmatrix}.$$

Multiplying each row in A by $\mathbf{x}$, we get

$$A\mathbf{x} = \begin{pmatrix} 1 & 3 & 5 \\ 7 & -2 & 4 \end{pmatrix} \cdot \begin{pmatrix} -1 \\ 2 \\ 3 \end{pmatrix} = \begin{pmatrix} 1 \cdot (-1) + 3 \cdot 2 + 5 \cdot 3 \\ 7 \cdot (-1) + (-2) \cdot 2 + 4 \cdot 3 \end{pmatrix} = \begin{pmatrix} 20 \\ 1 \end{pmatrix}. \quad \bullet$$

Properties of matrix multiplication

There are some important algebraic properties of the product of a matrix and a vector that we will formalize in a theorem.

THEOREM 1.16   Suppose A is an $m \times n$ matrix, $\mathbf{x}$ and $\mathbf{y}$ are in $\mathbf{R}^n$, and a is a number. Then

1. $A(a\mathbf{x}) = aA\mathbf{x}$
2. $A(\mathbf{x} + \mathbf{y}) = A\mathbf{x} + A\mathbf{y}$

■

Notice that we can put these two together and conclude that if b is another number, then

$$A(a\mathbf{x} + b\mathbf{y}) = aA\mathbf{x} + bA\mathbf{y}.$$

This can be carried even further to conclude that

$$A(a_1\mathbf{x}_1 + a_2\mathbf{x}_2 + \cdots + a_p\mathbf{x}_p) = a_1A\mathbf{x}_1 + a_2A\mathbf{x}_2 + \cdots + a_pA\mathbf{x}_p,$$

for any collection of numbers $a_1, \ldots,$ and a_p, and vectors $\mathbf{x}_1, \ldots,$ and $\mathbf{x}_p$.

We will summarize the properties of multiplication by A in Theorem 1.16 by saying that multiplication by A is a **linear** operation. No proof of Theorem 1.16 will be given. It can easily be proved by closely following the definitions. Examples will be given in the exercises.

We now extend the definition of a matrix with a vector to the product of two matrices A and B. We will assume that A is the $m \times n$ matrix in (1.7) and B is the $n \times p$ matrix with entries b_{jk}. We will regard B as a list of its column vectors,

$$B = [\mathbf{b}_1, \mathbf{b}_2, \ldots, \mathbf{b}_p],$$

where the vector $\mathbf{b}_j$ is the jth column of the matrix B,

$$\mathbf{b}_j = \begin{pmatrix} b_{1j} \\ b_{2j} \\ \vdots \\ b_{nj} \end{pmatrix}. \tag{1.17}$$

DEFINITION 1.18

The product of the two matrices A and B is defined to be

$$AB = [A\mathbf{b}_1, A\mathbf{b}_2, \ldots, A\mathbf{b}_p].$$

If we set $C = AB = (c_{ik})$, then c_{ik} (the component of C in the ith row and kth column) is the ith component of $A\mathbf{b}_k$, the kth column vector of C. Hence, using (1.14) we see that c_{ik} is the product of the ith row vector of A and the kth column vector of B. In symbols,

$$c_{ik} = a_{i1}b_{1k} + a_{i2}b_{2k} + \cdots + a_{in}b_{nk} = \sum_{j=1}^{n} a_{ij}b_{jk}. \tag{1.19}$$

Notice that for the product to be defined, the number of columns in A and the number of rows in B must be equal.

Example 1.20   Suppose

$$A = \begin{pmatrix} -2 & 3 & 1 \\ 9 & 8 & -2 \end{pmatrix} \quad \text{and} \quad B = \begin{pmatrix} 1 & 5 \\ 9 & 0 \\ 0 & -2 \end{pmatrix}.$$

Compute the product AB.

We set $B = [\mathbf{b}_1, \ \mathbf{b}_2]$, where

$$\mathbf{b}_1 = \begin{pmatrix} 1 \\ 9 \\ 0 \end{pmatrix} \quad \text{and} \quad \mathbf{b}_2 = \begin{pmatrix} 5 \\ 0 \\ -2 \end{pmatrix}.$$

Using Definition 1.9, we compute

$$A\mathbf{b}_1 = 1\begin{pmatrix} -2 \\ 9 \end{pmatrix} + 9\begin{pmatrix} 3 \\ 8 \end{pmatrix} + 0\begin{pmatrix} 1 \\ -2 \end{pmatrix} = \begin{pmatrix} 25 \\ 81 \end{pmatrix},$$

and

$$A\mathbf{b}_2 = 5\begin{pmatrix} -2 \\ 9 \end{pmatrix} + 0\begin{pmatrix} 3 \\ 8 \end{pmatrix} - 2\begin{pmatrix} 1 \\ -2 \end{pmatrix} = \begin{pmatrix} -12 \\ 49 \end{pmatrix}.$$

Consequently,

$$AB = [A\mathbf{b}_1, \ A\mathbf{b}_2] = \begin{pmatrix} 25 & -12 \\ 81 & 49 \end{pmatrix}.$$

We leave it to you to compute the product using (1.19) instead of the definition. Which do you think is easier?   ●

There are two important algebraic properties of matrices, which we will state in the next theorem.

THEOREM 1.21

Suppose that A, B, and C are matrices. Then, assuming that the sizes of A, B, and C allow the products to be defined, we have the following:

1. Multiplication is associative:

$$A(BC) = (AB)C.$$

2. Multiplication is distributive:

$$A(B + C) = AB + AC$$
$$(B + C)A = BA + CA.$$ ∎

These identities follow in a straightforward manner from Theorem 1.16 and Definition 1.18. Examples will be given in the exercises. Theorem 1.21 states that matrix multiplication is associative and distributive. However, matrix multiplication is rarely commutative. Almost any example shows this. Suppose

$$A = \begin{pmatrix} 0 & 1 \\ 9 & 2 \end{pmatrix} \quad \text{and} \quad B = \begin{pmatrix} -3 & -1 \\ 1 & 2 \end{pmatrix}.$$

Then

$$AB = \begin{pmatrix} 1 & 2 \\ -25 & -5 \end{pmatrix} \quad \text{and} \quad BA = \begin{pmatrix} -9 & -5 \\ 18 & 5 \end{pmatrix},$$

so $AB \neq BA$, and we see that the matrices A and B do not commute.

For every integer n, we have the $n \times n$ **identity** matrix

$$I = \begin{pmatrix} 1 & 0 & 0 & \cdots & 0 \\ 0 & 1 & 0 & \cdots & 0 \\ 0 & 0 & 1 & \cdots & 0 \\ \vdots & \vdots & \vdots & \ddots & \vdots \\ 0 & 0 & 0 & \cdots & 1 \end{pmatrix},$$

which has ones on the diagonal and zeros elsewhere. The name comes from the easily verified fact that

$$I\mathbf{x} = \mathbf{x} \quad \text{for all vectors } \mathbf{x}.$$

From this, it follows that

$$IA = A \quad \text{and} \quad BI = B$$

provided that A has n rows and B has n columns. Notice that the identity matrix commutes with every square matrix (i.e., $AI = IA$).

The transpose of a matrix

We have one more definition for this section.

DEFINITION 1.22

Suppose the matrix A is given by (1.7). Then the **transpose** of A is the matrix

$$A^T = \begin{pmatrix} a_{11} & a_{21} & \cdots & a_{m1} \\ a_{12} & a_{22} & \cdots & a_{m2} \\ \vdots & \vdots & \ddots & \vdots \\ a_{1n} & a_{2n} & \cdots & a_{mn} \end{pmatrix}.$$

You will notice that the transpose of a matrix is obtained by flipping the matrix along its main diagonal. For example, the transpose of the matrix

$$A = \begin{pmatrix} 0 & 1 & -2 \\ 9 & 2 & 5 \end{pmatrix} \quad \text{is the matrix} \quad A^T = \begin{pmatrix} 0 & 9 \\ 1 & 2 \\ -2 & 5 \end{pmatrix}.$$

Thus the transpose of a matrix with n rows and m columns will have m rows and n columns. In particular, the transpose of a column vector is a row vector and vice versa.

We will make use of this feature to improve the appearance of the text in this book. Even a column 2-vector in a line of text, such as $\begin{pmatrix} -2 \\ 4 \end{pmatrix}$, is not too appealing because it causes the lines of text to be too far apart. We will therefore use $(-2, 4)^T$ instead. This will allow us to refer to fairly large column vectors such as $(9, 21, 34, -5, 0)^T$ without the disruption that displaying the vector as a column vector would cause.

EXERCISES

In Exercises 1 and 2, use a sheet of graph paper and the technique presented in Figure 1 to draw the sum of the given vectors. Verify your result analytically.

1. $\mathbf{x} = \begin{pmatrix} 1 \\ 2 \end{pmatrix}$, $\mathbf{y} = \begin{pmatrix} -3 \\ 1 \end{pmatrix}$ **2.** $\mathbf{x} = \begin{pmatrix} 4 \\ -2 \end{pmatrix}$, $\mathbf{y} = \begin{pmatrix} -3 \\ -5 \end{pmatrix}$

In Exercises 3 and 4, use a sheet of graph paper and the technique presented in Figure 5 to draw the scalar product of the given scalar (number) and vector. Verify your result analytically.

3. $c = 2$, $\mathbf{x} = \begin{pmatrix} 1 \\ -3 \end{pmatrix}$ **4.** $c = -3$, $\mathbf{x} = \begin{pmatrix} -2 \\ 2 \end{pmatrix}$

In Exercises 5–8, use a sheet of graph paper and the technique presented in Figure 7 to find a linear combination of $\mathbf{x}$ and $\mathbf{y}$ that equals $\mathbf{z}$, if possible. Verify your result analytically.

5. $\mathbf{x} = \begin{pmatrix} 1 \\ 2 \end{pmatrix}$, $\mathbf{y} = \begin{pmatrix} -3 \\ 4 \end{pmatrix}$, $\mathbf{z} = \begin{pmatrix} -1 \\ 8 \end{pmatrix}$

6. $\mathbf{x} = \begin{pmatrix} 1 \\ 1 \end{pmatrix}$, $\mathbf{y} = \begin{pmatrix} 1 \\ -1 \end{pmatrix}$, $\mathbf{z} = \begin{pmatrix} 5 \\ -1 \end{pmatrix}$

7. $\mathbf{x} = \begin{pmatrix} 1 \\ 3 \end{pmatrix}$, $\mathbf{y} = \begin{pmatrix} 4 \\ 1 \end{pmatrix}$, $\mathbf{z} = \begin{pmatrix} 6 \\ -4 \end{pmatrix}$

8. $\mathbf{x} = \begin{pmatrix} -1 \\ 2 \end{pmatrix}$, $\mathbf{y} = \begin{pmatrix} 2 \\ -4 \end{pmatrix}$, $\mathbf{z} = \begin{pmatrix} 0 \\ 4 \end{pmatrix}$

9. Suppose that

$$A = \begin{pmatrix} -1 & 2 & 4 \\ 0 & 5 & 2 \\ -1 & -2 & 4 \end{pmatrix}.$$

You want to double the first column of A, triple the second column, and multiply each entry in the third column by 4. What matrix B will provide this solution via the matrix product AB?

Verify both parts of Theorem 1.16 for the scalar, vectors, and matrix given in Exercises 10 and 11.

10. $a = 2$, $A = \begin{pmatrix} -1 & 2 \\ 3 & 0 \end{pmatrix}$, $\mathbf{x} = \begin{pmatrix} 1 \\ -2 \end{pmatrix}$, $\mathbf{y} = \begin{pmatrix} 0 \\ 4 \end{pmatrix}$

11. $a = -3$, $A = \begin{pmatrix} -1 & 2 & 0 \\ 3 & 0 & -1 \\ 2 & 2 & 1 \end{pmatrix}$, $\mathbf{x} = \begin{pmatrix} 1 \\ -2 \\ 0 \end{pmatrix}$, $\mathbf{y} = \begin{pmatrix} 0 \\ -1 \\ 4 \end{pmatrix}$

12. Prove each of the following properties. None of these are hard, and some are completely trivial.

(a) For each $\mathbf{x}, \mathbf{y} \in \mathbf{R}^n$, $\mathbf{x} + \mathbf{y} \in \mathbf{R}^n$.

(b) For each $\alpha \in \mathbf{R}$, $\mathbf{x} \in \mathbf{R}^n$, $\alpha\mathbf{x} \in \mathbf{R}^n$.

(c) For each $\mathbf{x}, \mathbf{y} \in \mathbf{R}^n$, $\mathbf{x} + \mathbf{y} = \mathbf{y} + \mathbf{x}$.

(d) For each $\mathbf{x}, \mathbf{y}, \mathbf{z} \in \mathbf{R}^n$, $(\mathbf{x} + \mathbf{y}) + \mathbf{z} = \mathbf{x} + (\mathbf{y} + \mathbf{z})$.

(e) For each $\mathbf{x} \in \mathbf{R}^n$, $\mathbf{x} + \mathbf{0} = \mathbf{x}$, where each component of the vector $\mathbf{0} \in \mathbf{R}^n$ is equal to 0.

(f) For each $\mathbf{x} \in \mathbf{R}^n$, $\mathbf{x} + (-\mathbf{x}) = \mathbf{0}$, where $-\mathbf{x} = -1 \cdot \mathbf{x}$.

(g) For each $\alpha \in \mathbf{R}$, $\mathbf{x}, \mathbf{y} \in \mathbf{R}^n$, $\alpha(\mathbf{x} + \mathbf{y}) = \alpha\mathbf{x} + \alpha\mathbf{y}$.

(h) For each $\alpha, \beta \in \mathbf{R}$, $\mathbf{x} \in \mathbf{R}^n$, $(\alpha + \beta)\mathbf{x} = \alpha\mathbf{x} + \beta\mathbf{x}$.

(i) For each $\alpha, \beta \in \mathbf{R}$, $\mathbf{x} \in \mathbf{R}^n$, $(\alpha\beta)\mathbf{x} = \alpha(\beta\mathbf{x})$.

(j) For each $\mathbf{x} \in \mathbf{R}^n$, $1 \cdot \mathbf{x} = \mathbf{x}$.

13. Use Definition 1.9 and the properties proven in Exercise 12 to prove Theorem 1.16.

In Exercises 14–22, show that the matrices and scalars

$$\alpha = 3, \quad \beta = -2,$$

$$A = \begin{pmatrix} 1 & 2 \\ -1 & 0 \end{pmatrix}, \quad B = \begin{pmatrix} -1 & 1 \\ 1 & 1 \end{pmatrix}, \quad C = \begin{pmatrix} 2 & -3 \\ 1 & 1 \end{pmatrix}$$

satisfy the given identity.

14. $A + B = B + A$

15. $(A+B)+C = A+(B+C)$

16. $(AB)C = A(BC)$

17. $A(B + C) = AB + AC$

18. $(A + B)C = AC + BC$

19. $(\alpha\beta)A = \alpha(\beta A)$

20. $\alpha(AB) = (\alpha A)B = A(\alpha B)$

21. $(\alpha + \beta)A = \alpha A + \beta A$

22. $\alpha(A + B) = \alpha A + \alpha B$

In Exercises 23–26, show that the scalar and matrices

$$\alpha = -3, \qquad A = \begin{pmatrix} -2 & 4 \\ 4 & 0 \end{pmatrix}, \quad \text{and} \quad B = \begin{pmatrix} 0 & -5 \\ -2 & 3 \end{pmatrix}$$

satisfy the given identity.

23. $(A^T)^T = A$

24. $(\alpha A)^T = \alpha A^T$

25. $(A + B)^T = A^T + B^T$

26. $(AB)^T = B^T A^T$

We will use the following vectors and matrices in Exercises 27–46.

$$\mathbf{x}_1 = \begin{pmatrix} 9 \\ 5 \end{pmatrix}, \ \mathbf{x}_2 = \begin{pmatrix} -6 \\ -1 \end{pmatrix}, \ \mathbf{x}_3 = \begin{pmatrix} -1 \\ 8 \end{pmatrix}, \ \mathbf{x}_4 = \begin{pmatrix} 7 \\ -9 \end{pmatrix},$$

$$\mathbf{v}_1 = \begin{pmatrix} 10 \\ -5 \\ 3 \end{pmatrix}, \ \mathbf{v}_2 = \begin{pmatrix} 0 \\ 8 \\ 6 \end{pmatrix}, \ \mathbf{v}_3 = \begin{pmatrix} 0 \\ -9 \\ 7 \end{pmatrix}, \ \mathbf{v}_4 = \begin{pmatrix} -1 \\ 3 \\ 6 \end{pmatrix},$$

$$\mathbf{y} = \begin{pmatrix} 3 \\ -2 \end{pmatrix}, \ \mathbf{z} = \begin{pmatrix} -1 \\ 0 \\ 2 \end{pmatrix}, \ \mathbf{w} = \begin{pmatrix} 2 \\ 0 \\ -3 \end{pmatrix}, \ \mathbf{u} = \begin{pmatrix} -1 \\ 3 \\ -2 \\ 4 \end{pmatrix},$$

$$A = \begin{pmatrix} 9 & -6 \\ 5 & -1 \end{pmatrix}, \ B = \begin{pmatrix} -6 & -1 & 7 \\ -1 & 8 & -9 \end{pmatrix},$$

$$C = \begin{pmatrix} 10 & 0 & -1 \\ -5 & 8 & 3 \\ 3 & 6 & 6 \end{pmatrix}, \ D = \begin{pmatrix} 10 & 0 & 0 & -1 \\ -5 & 8 & -9 & 3 \\ 3 & 6 & 7 & 6 \end{pmatrix}$$

For Exercises 27–32,

(i) Simplify the indicated linear combination.

(ii) Write the indicated linear combination as a matrix product.

(iii) Compute the matrix product found in (ii) using the method in (1.14) and compare your result with the linear combination computed in (i).

27. $2\mathbf{x}_1 + 3\mathbf{x}_2$

28. $-\mathbf{x}_3 + 5\mathbf{x}_4$

29. $4\mathbf{x}_2 - 7\mathbf{x}_4 - 3\mathbf{x}_1$

30. $-2\mathbf{v}_2 + 4\mathbf{v}_3 + 5\mathbf{x}_4$

31. $4\mathbf{v}_1 - 3\mathbf{v}_4 + 3\mathbf{v}_2 + 4\mathbf{v}_3$

32. $\mathbf{v}_2 - 5\mathbf{v}_1 - 3\mathbf{v}_4 - 2\mathbf{v}_3$

For Exercises 33–40, compute the indicated matrix product.

33. $A\mathbf{y}$ **34.** $A\mathbf{x}_3$ **35.** $B\mathbf{z}$ **36.** $B\mathbf{w}$

37. $C\mathbf{z}$ **38.** $C\mathbf{w}$ **39.** $C\mathbf{v}_3$ **40.** $D\mathbf{u}$

41. What is the transpose of $\mathbf{y}$?

42. What is the transpose of $\mathbf{u}$?

43. What is the transpose of A?

44. What is the transpose of B?

45. What is the transpose of C?

46. What is the transpose of D?

For Exercises 47–54, write the indicated system as a matrix equation.

47. $3x + 4y = 7$

48. $-x + 4y = 3$

49. $\begin{aligned} 3x + 4y &= 7 \\ -x + 3y &= 2 \end{aligned}$

50. $\begin{aligned} x - 3y &= 5 \\ -2x + 3y &= -2 \end{aligned}$

51. $\begin{aligned} -x_1 + x_3 &= 0 \\ 2x_1 + 3x_2 &= 3 \end{aligned}$

52. $\begin{aligned} x_1 + 3x_2 - x_3 &= 2 \\ -2x_1 + 3x_2 - 2x_3 &= -3 \end{aligned}$

53. $\begin{aligned} -x_1 + x_3 &= 0 \\ 2x_1 + 3x_2 &= 3 \\ x_2 - x_3 &= 4 \end{aligned}$

54. $\begin{aligned} x_1 + 3x_2 - x_3 &= 2 \\ -2x_1 + 3x_2 - 2x_3 &= -3 \\ 2x_1 - x_3 &= 0 \end{aligned}$

7.2 Systems of Linear Equations with Two or Three Variables

Before proceeding to the solution of general systems of linear equations, we will look at systems in two or three variables. We will use what we know about vectors and matrices to systematize and simplify our solution method. We will also use what we know about geometry to get a better understanding of solution sets. In the process we will learn how to express solution sets parametrically.

One linear equation in two unknowns

Consider the equation

$$3x + 4y = 9. \tag{2.1}$$

The set of all vectors $(x, y)^T$ that solve this equation will be referred to as the **solution set** of (2.1). In this case, the solution set is a line in the plane (see Figure 1). If we solve (2.1) for x, getting $x = (9 - 4y)/3$, then the vectors in the solution set have the form

$$\begin{pmatrix} x \\ y \end{pmatrix} = \begin{pmatrix} (9 - 4y)/3 \\ y \end{pmatrix} = \begin{pmatrix} 3 \\ 0 \end{pmatrix} + y \begin{pmatrix} -4/3 \\ 1 \end{pmatrix}. \tag{2.2}$$

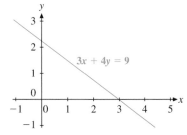

Figure 1. The solution set for a linear equation is a line.

Equation (2.2) is a ***parametric representation*** for the line defined by (2.1). This representation makes it clear that y is a ***free parameter,*** since any value of y yields a point on the line. When $y = 0$ in (2.2), we see that $\mathbf{p} = (3, 0)^T$ is a point on the line. The other points on the line are obtained by starting at $\mathbf{p}$ and adding arbitrary multiples of the vector $\mathbf{v} = (-4/3, 1)^T$. This is illustrated in Figure 2.

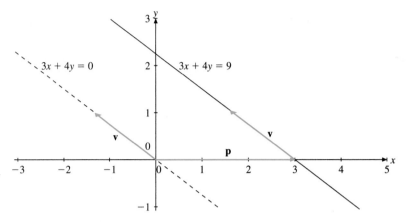

Figure 2. The solution sets for $3x + 4y = 9$ and for the homogeneous equation $3x + 4y = 0$.

A system of equations $A\mathbf{x} = \mathbf{b}$ is said to be ***homogeneous*** if the right-hand side $\mathbf{b}$ is equal to the zero vector $\mathbf{0}$. If $\mathbf{b} \neq \mathbf{0}$, the system is ***inhomogeneous***. It is interesting to compare the solution set of the inhomogeneous equation in (2.1) and that of the corresponding homogeneous equation

$$3x + 4y = 0. \tag{2.3}$$

It is easy to see that the solution set of (2.3) is the set

$$\{t\mathbf{v} \mid t \in \mathbf{R}\}, \tag{2.4}$$

where $\mathbf{v} = (-4/3, 1)^T$. By (2.2), the solution set to (2.1) is the set

$$\{\mathbf{x} = \mathbf{p} + t\mathbf{v} \mid t \in \mathbf{R}\}, \tag{2.5}$$

where $\mathbf{p} = (3, 0)^T$. Thus the general solution of the inhomogeneous equation (2.1) is the sum of the particular solution $\mathbf{p}$ plus the general solution to the corresponding homogeneous equation (2.3). This, too, is illustrated in Figure 2.

Notice that the solution set in (2.5) makes sense as long as the vectors $\mathbf{p}$ and $\mathbf{v}$ have the same number of entries.

DEFINITION 2.6

A ***line*** in $\mathbf{R}^n$ is a set of the form

$$\{\mathbf{x} = \mathbf{p} + t\mathbf{v} \mid t \in \mathbf{R}\}, \tag{2.7}$$

where $\mathbf{p}$ and $\mathbf{v} \neq \mathbf{0}$ are vectors in $\mathbf{R}^n$. Equation (2.7) is called a ***parametric equation*** for the line.

Notice that the vector $\mathbf{p}$ is the point on the line in (2.7) corresponding to $t = 0$. Any other point on the line differs from $\mathbf{p}$ by a multiple of the vector $\mathbf{v}$. Thus $\mathbf{p}$ is a point on the line and $\mathbf{v}$ gives the direction of the line.

Two equations in two unknowns

The next simplest case is two linear equations in two unknowns, such as

$$\begin{aligned} x - y &= 0 \\ 3x + 4y &= 9. \end{aligned} \tag{2.8}$$

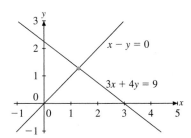

Figure 3. The solution set of (2.4) is a point.

The solution set for each of these equations is a line, as shown in Figure 3. Since we want a solution to both equations, we are looking for the intersection of two lines. Geometry tells us that two distinct lines in the plane intersect in one point unless they are parallel, in which case they do not intersect at all.

One way to solve a system like that in (2.8) is to use simple substitution. We solve the first equation for x, getting $x = y$. When we substitute this result into the second equation, we get

$$3y + 4y = 9 \quad \text{or} \quad 7y = 9.$$

Thus we are led to the system of equations

$$\begin{aligned} x - y &= 0 \\ 7y &= 9, \end{aligned} \tag{2.9}$$

which is equivalent to (2.8) in the sense that both systems have the same solutions.

Notice that the same result can be reached by the easier operation of adding -3 times the first equation in (2.8) to the second. This operation is called ***elimination*** because its goal is to eliminate one of the variables from the second equation.

By either method we obtain system (2.9), which is very easy to solve. The second equation requires $y = 9/7$, and then the first requires $x = y = 9/7$. Hence the only solution to the system in (2.8) is

$$\begin{pmatrix} x \\ y \end{pmatrix} = \begin{pmatrix} 9/7 \\ 9/7 \end{pmatrix}.$$

The method of solving the system in (2.9) by solving the last equation first and then proceeding to the first is called ***back-solving***. Unfortunately, this easy solution method is only possible with systems like that in (2.9) in which the equations involve fewer of the unknowns as we go through the system. Our strategy will be to use elimination to transform any system into a form like that in (2.9), and then to rely on back-solving.

Matrix notation

It will not be hard to convince ourselves that the method of elimination and back-solving works in general. First, however, we want to simplify our notation by getting rid of explicit reference to the unknowns. Using matrix notation, we can write (2.8) as

$$\begin{pmatrix} 1 & -1 \\ 3 & 4 \end{pmatrix} \begin{pmatrix} x \\ y \end{pmatrix} = \begin{pmatrix} 0 \\ 9 \end{pmatrix}, \tag{2.10}$$

or

$$A \begin{pmatrix} x \\ y \end{pmatrix} = \mathbf{b},$$

where A is the coefficient matrix and $\mathbf{b}$ is the right-hand side.

Notice that all information about the system in (2.8) or (2.10) is contained in the ***augmented matrix***

$$M = \begin{pmatrix} 1 & -1 & 0 \\ 3 & 4 & 9 \end{pmatrix} = [A, \mathbf{b}] \tag{2.11}$$

obtained by adjoining the right-hand side column vector $\mathbf{b}$ to the coefficient matrix A. In comparing the augmented matrix M in (2.11) and the system of equations in (2.8), it is important to notice that each row of M contains the coefficients in one of the equations. All of the information for one of the equations in the system is contained in one row of the matrix.

The effect of adding -3 times the first equation to the second is to add -3 times the first row of M to the second row to get

$$\begin{pmatrix} 1 & -1 & 0 \\ 0 & 7 & 9 \end{pmatrix}, \tag{2.12}$$

which is the augmented matrix corresponding to the system in (2.9). Using this technique, we deal with the augmented matrix instead of dealing with systems of equations. The benefit is that we do not have to continually refer to the variables x and y. The importance of this is not so obvious with two unknowns, but we will soon see that it is very effective in more complicated situations.

Other types of solution sets in two variables

In the case of (2.8), the only solution is $(9/7, 9/7)^T$, so the two lines intersect in a point, as indicated in Figure 3. This is not the only possibility. Other possibilities are illustrated by the two systems

$$\begin{aligned} 3x + 4y &= 9 \\ 6x + 8y &= 18 \end{aligned} \tag{2.13}$$

and

$$\begin{aligned} 3x + 4y &= 9 \\ 6x + 8y &= 10. \end{aligned} \tag{2.14}$$

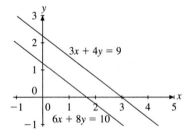

Figure 4. Equations (2.14) have no solution.

In (2.13), the second equation is simply a multiple of the first. Consequently, they have the same solution set, namely the line shown in Figure 1. In (2.14), the solution sets for the two equations are parallel lines (see Figure 4). There is no intersection and there are no solutions to (2.14). In cases like this, we say that the system of equations is ***inconsistent***.

To summarize, we have discovered that there are precisely three possibilities for the solution set of two equations in two unknowns:

1. one point
2. a line
3. no solutions

The first case is what happens most of the time, but it is not the only interesting case. Because they happen less often, the other two cases are called ***degenerate*** cases.

The case of homogeneous equations deserves special consideration. The systems in (2.1), (2.8), (2.13), and (2.14) are inhomogeneous, but the single equation

$$3x + 4y = 0 \tag{2.15}$$

is homogeneous. Notice that the zero vector $(0, 0)^T$ is always a solution to a homogeneous system in two variables. Consequently, any homogeneous system such as

$$3x + 4y = 0$$
$$x - \ y = 0$$

has the origin $(0, 0)^T$ as a solution. Therefore, homogeneous systems are always consistent, and the third of the three possibilities listed earlier cannot occur for homogeneous systems. The solution set for a homogeneous system of two equations in two variables is either one point, which must be the origin $(0, 0)^T$, or it is a line through the origin.

Solution sets in three dimensions

Examination of systems of linear equations in three variables reveals similar phenomena, although there are more possibilities. To start with, the solution set of a single equation, such as

$$6x - y + 4z = 2,$$

is a plane lying in three-dimensional space. If we solve this equation for x, we get $x = (2 + y - 4z)/6$. The points in the solution set have the form

$$\begin{pmatrix} x \\ y \\ z \end{pmatrix} = \begin{pmatrix} (2 + y - 4z)/6 \\ y \\ z \end{pmatrix}$$

$$= \begin{pmatrix} 1/3 \\ 0 \\ 0 \end{pmatrix} + y \begin{pmatrix} 1/6 \\ 1 \\ 0 \end{pmatrix} + z \begin{pmatrix} -2/3 \\ 0 \\ 1 \end{pmatrix}.$$

This is a ***parametric representation for a plane.*** In this representation, y and z are both free parameters. Any values of y and z yield a point on the plane. When both are 0, we see that $\mathbf{p} = (1/3, 0, 0)^T$ is a point on the plane. The other points are obtained by starting at $\mathbf{p}$ and adding arbitrary linear combinations of the vectors $\mathbf{v}_1 = (1/6, 1, 0)^T$ and $\mathbf{v}_2 = (-2/3, 0, 1)^T$. Using this vector notation, we can describe the solution set as

$$\{\mathbf{v} = \mathbf{p} + y\mathbf{v}_1 + z\mathbf{v}_2 \mid y, z \in \mathbf{R}\}. \tag{2.16}$$

Again we can have planes in any dimension.

DEFINITION 2.17

A ***plane*** in $\mathbf{R}^n$ is a set of the form

$$\{\mathbf{v} = \mathbf{p} + y\mathbf{v}_1 + z\mathbf{v}_2 \mid y, z \in \mathbf{R}\}, \tag{2.18}$$

where $\mathbf{p}$, $\mathbf{v}_1$, and $\mathbf{v}_2$ are vectors in $\mathbf{R}^n$ such that $\mathbf{v}_1$ and $\mathbf{v}_2$ are not multiples of each other. Equation (2.18) is called a ***parametric equation*** for the plane.

Compare this definition with that of a line in Definition 2.6. Again $\mathbf{p}$ is a point on the plane, corresponding to $y = z = 0$. Any other point on the plane differs from $\mathbf{p}$ by a linear combination of the vectors $\mathbf{v}_1$ and $\mathbf{v}_2$. Thus $\mathbf{p}$ is a point on the plane and $\mathbf{v}_1$ and $\mathbf{v}_2$ are directions in the plane.

Two equations in three unknowns

A system of two equations, such as

$$
\begin{aligned}
u - 4v + w &= -2 \\
-2u + 10v - 3w &= 4,
\end{aligned}
\tag{2.19}
$$

has a solution set that is the intersection of two planes. From geometry, we know that there are precisely three possibilities:

1. The two planes intersect in a line.
2. The two planes are the same plane and consequently the solution set is a plane.
3. The two planes are parallel, in which case there are no solutions.

To find out which is the case here we have to solve the system in (2.19). We will use matrix notation to do so.

Example 2.20   Solve the system of equations in (2.19).

The augmented matrix corresponding to (2.19) is

$$
\begin{pmatrix}
1 & -4 & 1 & -2 \\
-2 & 10 & -3 & 4
\end{pmatrix}.
\tag{2.21}
$$

We eliminate the first element in the second row by adding twice the first row to the second. In doing so, the equations and the corresponding augmented matrix become

$$
\begin{aligned}
u - 4v + w &= -2 \\
2v - w &= 0.
\end{aligned}
\qquad
\begin{pmatrix}
1 & -4 & 1 & -2 \\
0 & 2 & -1 & 0
\end{pmatrix}.
\tag{2.22}
$$

Since (2.22) was obtained from (2.19) by the process of adding a multiple of one row to another, we know that the solution sets of the systems in (2.19) and (2.22) are the same.

Notice that (2.22) can be solved by back-solving. If we assign any value to w, then the equations in (2.22) allow us to solve for u and v. We will call w a *free variable*, and set $w = t$, to remind us that w can be any number. Then we solve the last equation in (2.22) for v as a function of $w = t$,

$$
v = w/2 = t/2.
$$

Next we solve the first equation for u,

$$
\begin{aligned}
u &= -2 + 4v - w \\
&= -2 + 4 \cdot (t/2) - t \\
&= -2 + t.
\end{aligned}
$$

Thus, for any value of t, the vector

$$
\begin{pmatrix} u \\ v \\ w \end{pmatrix}
=
\begin{pmatrix} -2 + t \\ t/2 \\ t \end{pmatrix}
=
\begin{pmatrix} -2 \\ 0 \\ 0 \end{pmatrix}
+ t
\begin{pmatrix} 1 \\ 1/2 \\ 1 \end{pmatrix}
\tag{2.23}
$$

is a solution, and furthermore, every solution is of this form. Consequently, the solution set for (2.22), and therefore (2.19) as well, is completely described by (2.23). We see that when $t = 0$, the point $(-2, 0, 0)^T$ is in the solution set. The other points in the solution set are obtained by adding an arbitrary multiple of the vector $(1, 1/2, 1)^T$ to $(-2, 0, 0)^T$. Comparing with Definition 2.6, we see that this is a parametric representation of a line in $\mathbf{R}^3$. The solution set is shown in Figure 5. ●

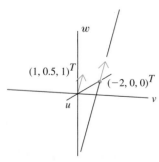

Figure 5. The solution set in Example 2.20 is a line through the point $(-2, 0, 0)^T$.

Three equations in three unknowns

Let's start with an example.

Example 2.24   Find the solution set for the system

$$
\begin{aligned}
x \quad\quad\quad + 3z &= -2 \\
-3x + 2y - 5z &= 2 \\
2z - 4y + z &= 1.
\end{aligned}
\tag{2.25}
$$

The augmented matrix for the system is

$$
\begin{pmatrix}
1 & 0 & 3 & -2 \\
-3 & 2 & -5 & 2 \\
2 & -4 & 1 & 1
\end{pmatrix}
\tag{2.26}
$$

To prepare for back-solving, we want to make the first entries in rows two and three equal to zero. In our new terminology, we want to eliminate them. We do this by adding multiples of the first row to each row. Remember that this is equivalent to adding multiples of the first equation in the system (2.25) to the second and third equations, and this does not change the solution set. Clearly, we want to add 3 times the first row of the augmented matrix to the second and -2 times the first row to the third. The resulting system and augmented matrix are

$$
\begin{aligned}
x \quad\quad\quad + 3z &= -2 \\
2y + 4z &= -4 \\
-4y - 5z &= 5
\end{aligned}
\qquad
\begin{pmatrix}
1 & 0 & 3 & -2 \\
0 & 2 & 4 & -4 \\
0 & -4 & -5 & 5
\end{pmatrix}.
\tag{2.27}
$$

We are not through. We can simplify the system still more by eliminating the coefficient of y in the third equation. We do this by adding 2 times the second row to the third (or 2 times the second equation to the third). The result is

$$
\begin{aligned}
x \quad\quad + 3z &= -2 \\
2y + 4z &= -4 \\
3z &= -3
\end{aligned}
\qquad
\begin{pmatrix}
1 & 0 & 3 & -2 \\
0 & 2 & 4 & -4 \\
0 & 0 & 3 & -3
\end{pmatrix}.
\tag{2.28}
$$

We are now ready to back-solve. The last equation, $3z = -3$, demands that $z = -1$. Using this fact, the second equation gives us

$$
y = (-4 - 4z)/2 = (-4 + 4)/2 = 0.
$$

Finally, the first equation gives us

$$
x = -2 - 3z = -2 + 3 = 1.
$$

Thus the only solution to the system is

$$
\begin{pmatrix}
x \\
y \\
z
\end{pmatrix}
=
\begin{pmatrix}
1 \\
0 \\
-1
\end{pmatrix}.
$$

Solution sets in three dimensions

The solution set in Example 2.24 consisted of a single point. What other kinds of sets can appear as solution sets? For three equations in three unknowns, we are looking at the intersection of three planes. We can most easily discover the possibilities by restricting our view to the plane that is the solution set of one of the equations. Consider first the case when each of the other two planes intersects our plane in a line. If these two lines intersect in a point, the solution set consists of that one point. This was the case in Example 2.24. If these two lines are actually the same line, the solution set is that line. If the two lines are parallel, the solution set is empty, so the system is inconsistent.

There are other possibilities. The three equations could all define the same plane, so the solution set is that plane. Two of the planes could be the same and the third intersects that plane in a line—the solution set is that line. Finally, if two of the planes are parallel, there are no solutions, and the equations are inconsistent.

To summarize, for three equations in three unknowns, there are precisely four possibilities for the solution set:

1. one point
2. a line
3. a plane
4. no solutions

Again the first case is what usually happens, but the other cases are interesting as well. Finally, we notice that if the equations are homogeneous, then the origin $(0, 0, 0)^T$ is always a solution. Thus the fourth possibility cannot occur for homogeneous equations, and the three possibilities are

1. the origin
2. a line through the origin
3. a plane through the origin

Solution sets in higher dimensions

Our intuition fails most of us when we have to visualize objects in dimensions greater than 3. However, we can usually get some intuition by thinking of analogies in dimensions 2 and 3. Let's think about how we might describe the solution sets that we have found in 2 and 3 dimensions.

Clearly, a single point in any dimension will look similar to what it does in dimensions 2 and 3. The other examples we have are lines in the plane and lines and planes in 3-space. These clearly have some features in common. First, if we can move in a particular direction in one of these, we can move arbitrarily far in that direction. We can summarize this by saying that the set is ***infinitely long***. Second, if there are two directions in which we can move, as in a plane, then we can move in any linear combination of those directions. We can summarize this by saying that the solution set is ***flat***. Finally, a line in the plane, or a line or plane in 3-space, does not take up much room. Both have negligible extent in most directions. We can summarize that by saying that the solution set is ***thin***.

We might expect that solution sets in higher dimension would have the three characteristics we see in 2 and 3 dimensions of being long, flat, and thin. We will look for those characteristics as we move to consider solution sets in higher dimension.

EXERCISES

Use Definition 2.6 to place the solution set of each of the equations in Exercises 1–4 in parametric form. In each exercise, sketch the line represented by the given equation using the parametric technique shown in Figure 2.

1. $3x - 4y = 12$

2. $x + 2y = 2$

3. $2x - y = 2$

4. $2x + 3y = 6$

For the systems in Exercises 5–8, perform each of the following tasks.

(i) Sketch the system of lines on graph paper and estimate the solution by approximating the point of intersection.

(ii) Set up an augmented matrix for the system. Use elimination to eliminate x in the second equation, then use back-solving to complete the solution of the given system. Compare this exact solution with the approximate solution found in (i).

5. $\begin{aligned} x + 2y &= 6 \\ 2x - 3y &= 6 \end{aligned}$

6. $\begin{aligned} x - 2y &= 4 \\ 3x + 2y &= 6 \end{aligned}$

7. $\begin{aligned} x + y &= 3 \\ 2x - 3y &= 6 \end{aligned}$

8. $\begin{aligned} x - y &= 4 \\ x + 3y &= 3 \end{aligned}$

For the systems in Exercises 9–12, perform each of the following tasks.

(i) Sketch the system of lines on graph paper and describe the solution set.

(ii) Is the system consistent or inconsistent?

9. $\begin{aligned} x + y &= 3 \\ 2x + 2y &= -4 \end{aligned}$

10. $\begin{aligned} x - 2y &= 4 \\ 2x - 4y &= 8 \end{aligned}$

11. $\begin{aligned} x + 2y &= 4 \\ 3x + 6y &= 12 \end{aligned}$

12. $\begin{aligned} x - 3y &= 6 \\ 2x - 6y &= 12 \end{aligned}$

Each of the equations in Exercises 13–16 represents a plane in three dimensional space. Place the solution set of the given equation in parametric form as shown in Definition 2.17.

13. $x + 2y - 3z = 12$

14. $2x + y - 3z = 6$

15. $3x - 4y + z = 12$

16. $2x + 3y + 4z = 6$

The systems in Exercises 17–20 each have two equations with three unknowns. Set up an augmented matrix for the system, eliminate x in the second equation, let the free variable z equal t, and then use back-solving to solve for x and y. Place your final answer in parametric form.

17. $\begin{aligned} x + 2y - 3z &= 6 \\ 2x + 5y + 8z &= 40 \end{aligned}$

18. $\begin{aligned} x - 3y + 4z &= 12 \\ -3x + 10y + 8z &= 40 \end{aligned}$

19. $\begin{aligned} 2x - 4y + 5z &= 40 \\ -4x + 10y - 4z &= 20 \end{aligned}$

20. $\begin{aligned} 3x - 4y + 5z &= 60 \\ -6x + 9y - 4z &= -150 \end{aligned}$

Use the technique shown in Example 2.24 to solve each system in Exercises 21–24.

21. $\begin{aligned} x + y + z &= 3 \\ 2x - y + z &= 4 \\ -x + 2y + 2z &= 6 \end{aligned}$

22. $\begin{aligned} x - y + 2z &= 4 \\ -2x + y - z &= 6 \\ 3x + y - 3z &= 7 \end{aligned}$

23. $\begin{aligned} x - 2y + 5z &= 10 \\ -2x + y + z &= 12 \\ 2x - y - z &= 6 \end{aligned}$

24. $\begin{aligned} x + y - 2z &= 4 \\ -3x + y - z &= 6 \\ 2x + 6y + 2z &= 4 \end{aligned}$

25. Can a circle in $\mathbf{R}^2$ be the solution set of a system of linear equations?

26. The set S in $\mathbf{R}^2$ contains a point $\mathbf{x}_0$ and every point that is within a distance of 2 from $\mathbf{x}_0$. Is S the solution set of a system of linear equations?

27. Consider $S = \left\{ \begin{pmatrix} t \\ 0 \end{pmatrix} \middle| t > 0 \right\}$. Is S the solution set of a system of linear equations?

28. Consider $S = \left\{ \begin{pmatrix} t \\ s \end{pmatrix} \middle| t > 0 \right\}$. Is S the solution set of a system of linear equations?

29. Consider the line L in $\mathbf{R}^2$ with the parametric representation

$$\mathbf{y} = t \begin{pmatrix} 2 \\ -3 \end{pmatrix}.$$

Is L the solution set for a system of linear equations?

30. Consider the line L in $\mathbf{R}^2$ with the parametric representation

$$\mathbf{y} = t \begin{pmatrix} -1 \\ 4 \end{pmatrix}.$$

Is L the solution set for a system of linear equations?

31. Consider the line L in $\mathbf{R}^2$ with the parametric representation

$$\mathbf{y} = \begin{pmatrix} 0 \\ 1 \end{pmatrix} + t \begin{pmatrix} 2 \\ -3 \end{pmatrix}.$$

Is L the solution set for a system of linear equations?

32. Consider the line L in $\mathbf{R}^2$ with the parametric representation

$$\mathbf{y} = \begin{pmatrix} -2 \\ 3 \end{pmatrix} + t \begin{pmatrix} 0 \\ -3 \end{pmatrix}.$$

Is L the solution set for a system of linear equations?

Roughly speaking, the **dimension** of a solution set is the minimum number of directions we must go from a fixed point in order to reach every point in the solution set.

33. What are the dimensions of the possible solution sets in $\mathbf{R}^2$?

34. What are the dimensions of the possible solution sets in $\mathbf{R}^3$?

35. Argue by analogy to decide the dimensions of the possible solution sets in $\mathbf{R}^4$.

36. Find a parametric representation for the solution set of the equation

$$x_1 + 2x_2 - 2x_3 + x_4 = 2.$$

What is its dimension? For your information, a set such as this is called a **hyperplane** in $\mathbf{R}^4$.

37. Find a parametric representation for the solution set of the system of equations

$$x_1 + 2x_2 - 2x_3 + x_4 = 2$$
$$x_2 - 3x_3 - x_4 = 3.$$

What is its dimension? How would you describe the solution set?

38. Find a parametric representation for the solution set of the system of equations

$$x_1 + 2x_2 - 2x_3 + x_4 = 2$$
$$x_2 - 3x_3 - x_4 = 3$$
$$x_3 - x_4 = 0.$$

What is its dimension? How would you describe the solution set?

7.3 Solving Systems of Equations

In Section 7.2 we learned how to solve a few systems of linear equations. Examples 2.20 and 2.24 were especially illuminating. In this section we will find a systematic way to solve systems of equations, a way that always works and that we could easily program on a computer. Sophistication is not important here. We are only interested in finding a way that always works. Furthermore, we want a method that finds all solutions when there is more than one.

As we already indicated in Section 7.2, the method will use the augmented matrix. We will eliminate coefficients to put the system into an equivalent form that can be easily solved by the method of back-solving. Our first task will be to describe the form of equations that can be solved easily in this way. We will then describe in more detail how to go about the process of elimination. Finally, we will discover some more useful facts about process of back-solving.

Row echelon form of a matrix — the goal of elimination

Look back at the augmented matrices of the systems which we solved by back-solving in Section 7.2. These are in equations (2.12), (2.22), and (2.28). To describe how these matrices are different from the others, we will introduce some terminology.

DEFINITION 3.1

The *pivot* of a row vector in a matrix is the first nonzero element of that row.

To give us a specific example to talk about, consider the system

$$2x_2 + 4x_3 = 2$$
$$x_1 \qquad - 2x_3 = -1 \qquad (3.2)$$
$$-2x_1 + 2x_2 + 8x_3 = 4.$$

This system can be written as $A\mathbf{x} = \mathbf{b}$, where

$$A = \begin{pmatrix} 0 & 2 & 4 \\ 1 & 0 & -2 \\ -2 & 2 & 8 \end{pmatrix}, \quad \mathbf{x} = \begin{pmatrix} x_1 \\ x_2 \\ x_3 \end{pmatrix}, \quad \text{and} \quad \mathbf{b} = \begin{pmatrix} 2 \\ -1 \\ 4 \end{pmatrix}. \qquad (3.3)$$

The augmented matrix for the system is

$$M = [A, \mathbf{b}] = \begin{pmatrix} 0 & 2 & 4 & 2 \\ 1 & 0 & -2 & -1 \\ -2 & 2 & 8 & 4 \end{pmatrix}. \qquad (3.4)$$

In this matrix the pivots of each row are printed in blue.

DEFINITION 3.5

> A matrix is in **row echelon form** (or just echelon form) if, in each row that contains a pivot, the pivot lies to the right of the pivot in the preceding row. Any rows that contain only zeros must be at the bottom of the matrix.

The matrices in equations (2.12), (2.22), and (2.28) are in row echelon form. Other matrices in Section 7.2 are not in row echelon form. The matrix M in (3.4) is not in row echelon form, since, for example, the pivot in row three lies below and not to the right of the pivot in row two. The matrix

$$\begin{pmatrix} P & * & * & * & * & * & * & * & * & * \\ 0 & P & * & * & * & * & * & * & * & * \\ 0 & 0 & 0 & P & * & * & * & * & * & * \\ 0 & 0 & 0 & 0 & 0 & 0 & P & * & * & * \\ 0 & 0 & 0 & 0 & 0 & 0 & 0 & P & * & * \\ 0 & 0 & 0 & 0 & 0 & 0 & 0 & 0 & 0 & 0 \\ 0 & 0 & 0 & 0 & 0 & 0 & 0 & 0 & 0 & 0 \end{pmatrix}, \tag{3.6}$$

where the Ps refer to pivots and must therefore be nonzero numbers and the asterisks refer to arbitrary numbers, is in row echelon form. This matrix illustrates some of the possibilities of row echelon form that we have not seen before. Matrices in echelon form are set up for solution by the process of back-solving, so the goal of elimination is to put the augmented matrix into row echelon form.

Row operations and elimination

In all of our examples to this point we have achieved row echelon form by the operation of adding a multiple of one row to another. There are other permissible operations. One of them is illustrated by the matrix M in (3.4). For a matrix that has a nonzero entry in the first column to be in row echelon form, it is necessary that the entry in the upper left-hand corner be a pivot—a nonzero number. The easiest way to achieve this for M is to interchange rows 1 and 2 to obtain

$$\begin{pmatrix} 1 & 0 & -2 & -1 \\ 0 & 2 & 4 & 2 \\ -2 & 2 & 8 & 4 \end{pmatrix}. \tag{3.7}$$

Notice that this is the augmented matrix for the system

$$\begin{aligned} x_1 && - 2x_3 &= -1 \\ 2x_2 + 4x_3 &= 2 \\ -2x_1 + 2x_2 + 8x_3 &= 4. \end{aligned} \tag{3.8}$$

Thus the operation on the rows of M is equivalent to interchanging the first two equations in the system in (3.2). This does not change the solutions, so it is a permissible operation.

Similarly, if we multiply an equation by a nonzero number, the solutions are not affected. When dealing with the augmented matrix this is accomplished by multiplying a row by a nonzero number.

To summarize, we have three allowed operations on the rows of a matrix (called **row operations**):

R1: Add a multiple of one row to a different row.

R2: Interchange two rows.

R3: Multiply a row by a nonzero number.

Example 3.9   Find the solution set for the system in (3.2).

We have already used R2 to reduce the problem to finding the solution set for the equivalent system in (3.8), with augmented matrix given in (3.7). Notice that one effect of this use of R2 was to raise the pivot from the second row to the first. This is a typical use of R2.

The next step is to use R1 to eliminate the first entry in the third row. We add 2 times the first row to the third, getting the system and augmented matrix

$$
\begin{array}{rcl}
x_1 & - 2x_3 &= -1 \\
2x_2 + 4x_3 &=& 2 \\
2x_2 + 4x_3 &=& 2
\end{array}
\qquad
\begin{pmatrix}
1 & 0 & -2 & -1 \\
0 & 2 & 4 & 2 \\
0 & 2 & 4 & 2
\end{pmatrix}.
\tag{3.10}
$$

By comparing the matrices in (3.4) and (3.10), we see that one of the effects of eliminating the coefficient in the third row is to move the pivot in the third row to the right. Notice that the last two equations are the same. We could stop here, but to go all the way to row echelon form we should eliminate the last row of the augmented matrix. We do this by adding -1 times the second row to the third to get

$$
\begin{array}{rcl}
x_1 & - 2x_3 &= -1 \\
2x_2 + 4x_3 &=& 2 \\
0 &=& 0
\end{array}
\qquad
\begin{pmatrix}
1 & 0 & -2 & -1 \\
0 & 2 & 4 & 2 \\
0 & 0 & 0 & 0
\end{pmatrix}.
\tag{3.11}
$$

Notice that the third row in the new augmented matrix has a common factor of 2. We could use R3 to remove this factor, but that is optional.

Because it is in row echelon form, we can back-solve the system in (3.11). The last equation is always true, so we start with the second, $2x_2 + 4x_3 = 2$. We can assign an arbitrary value to one of the variables x_2 and x_3. We choose x_3 and set $x_3 = t$ to indicate that. Then, using the second equation, we solve for

$$
x_2 = (2 - 4x_3)/2 = (2 - 4t)/2 = 1 - 2t.
$$

Finally, we solve the first equation for $x_1 = -1 + 2x_3 = -1 + 2t$. The general solution is

$$
\mathbf{x} =
\begin{pmatrix} x_1 \\ x_2 \\ x_3 \end{pmatrix}
=
\begin{pmatrix} -1 + 2t \\ 1 - 2t \\ t \end{pmatrix}
=
\begin{pmatrix} -1 \\ 1 \\ 0 \end{pmatrix}
+ t
\begin{pmatrix} 2 \\ -2 \\ 1 \end{pmatrix}.
$$

Looking back at Definition 2.6, we see that this is a parametric equation for a line in $\mathbf{R}^3$.

Each of the row operations is reversible. Subtracting the same multiple reverses R1. Interchanging the same two rows reverses the effect of R2. Multiplying by the reciprocal of the nonzero number reverses R3. This reflects the fact that the row operations replace one system of equations with another that has the same solution set.

Implementing row operations, especially for large matrices, involves a lot of arithmetic. It is very easy to make frustrating mistakes. This is just the type of thing that computers are good at. There are a number of mathematical computer programs, such as MATLAB®, Maple, and *Mathematica*, in which it is easy to manipulate matrices and to perform the row operations. After you have learned the basics using hand calculations, we encourage you to use one of these to solve larger systems to help you understand the fine points of how the solution method we have described works.

THEOREM 3.12 With operations R1, R2, and R3, any matrix can be transformed into row echelon form. ∎

Any attempt at a proof of this result will only confuse the issue. Working through a few examples is more illuminating and also suggests how a proof might be constructed.

For the application we have in mind, the original matrix will be the augmented matrix associated with a system of linear equations. The key fact is that we can use row operations to transform the matrix into a matrix in row echelon form, and this new matrix is the augmented matrix associated with a system of equations that has the same solution set as the original system. The advantage of the row echelon form is that this system can be easily solved by back-solving.

Let's look at a more complicated example.

Example 3.13 Find the general solution to the system

$$\begin{aligned}
x_2 + 3x_3 + 2x_4 + 2x_5 &= 1 \\
x_1 + 2x_2 + 3x_3 + 5x_4 + 7x_5 &= 8 \\
2x_1 + 4x_2 + 6x_3 + 9x_4 + 15x_5 &= 2.
\end{aligned} \tag{3.14}$$

First we write this as a matrix equation

$$\begin{pmatrix} 0 & 1 & 3 & 2 & 2 \\ 1 & 2 & 3 & 5 & 7 \\ 2 & 4 & 6 & 9 & 15 \end{pmatrix} \begin{pmatrix} x_1 \\ x_2 \\ x_3 \\ x_4 \\ x_5 \end{pmatrix} = \begin{pmatrix} 1 \\ 8 \\ 2 \end{pmatrix}.$$

The augmented matrix is

$$\begin{pmatrix} 0 & 1 & 3 & 2 & 2 & 1 \\ 1 & 2 & 3 & 5 & 7 & 8 \\ 2 & 4 & 6 & 9 & 15 & 2 \end{pmatrix}.$$

Again the pivots are printed in blue. The next step is to use row operations to put this into row echelon form. We need a pivot in the upper left hand corner. To achieve this we

1. interchange rows 1 and 2.

The system and the corresponding augmented matrix become

$$\begin{aligned}
x_1 + 2x_2 + 3x_3 + 5x_4 + 7x_5 &= 8 \\
x_2 + 3x_3 + 2x_4 + 2x_5 &= 1 \\
2x_1 + 4x_2 + 6x_3 + 9x_4 + 15x_5 &= 2
\end{aligned} \qquad \begin{pmatrix} 1 & 2 & 3 & 5 & 7 & 8 \\ 0 & 1 & 3 & 2 & 2 & 1 \\ 2 & 4 & 6 & 9 & 15 & 2 \end{pmatrix}.$$

The next step is to eliminate the 2 in the first column. We

2. add -2 times row 1 to row 3.

This yields

$$\begin{aligned}
x_1 + 2x_2 + 3x_3 + 5x_4 + 7x_5 &= 8 \\
x_2 + 3x_3 + 2x_4 + 2x_5 &= 1 \\
-x_4 + x_5 &= -14
\end{aligned} \qquad \begin{pmatrix} 1 & 2 & 3 & 5 & 7 & 8 \\ 0 & 1 & 3 & 2 & 2 & 1 \\ 0 & 0 & 0 & -1 & 1 & -14 \end{pmatrix}. \tag{3.15}$$

This is in row echelon form, ready for back-solving.

We have to solve the system of equations associated with the augmented matrix in (3.15). First we notice that there are pivots in the first, second, and fourth columns. We can use the method of back-solving to solve for the corresponding variables x_1, x_2, and x_4 in terms of the variables x_3 and x_5. In doing so, we can assign arbitrary values to x_3 and x_5, which means that these are free variables. Notice that there are no pivots in columns 3 and 5, which correspond to the free variables x_3 and x_5. Hence we start by assigning $x_3 = s$ and $x_5 = t$. Then the last row in (3.15) is used to solve for x_4,

$$x_4 = 14 + x_5 = 14 + t.$$

Solving the second row in (3.15) for x_2, we get

$$\begin{aligned} x_2 &= 1 - 3x_3 - 2x_4 - 2x_5 \\ &= 1 - 3s - 2(14 + t) - 2t \\ &= -27 - 3s - 4t. \end{aligned}$$

Finally we use the first row to solve for x_1:

$$\begin{aligned} x_1 &= 8 - 2x_2 - 3x_3 - 5x_4 - 7x_5 \\ &= 8 - 2(-27 - 3s - 4t) - 3s - 5(14 + t) - 7t \\ &= -8 + 3s - 4t. \end{aligned}$$

Consequently, our solution is

$$\mathbf{x} = \begin{pmatrix} x_1 \\ x_2 \\ x_3 \\ x_4 \\ x_5 \end{pmatrix} = \begin{pmatrix} -8 + 3s - 4t \\ -27 - 3s - 4t \\ s \\ 14 + t \\ t \end{pmatrix} = \begin{pmatrix} -8 \\ -27 \\ 0 \\ 14 \\ 0 \end{pmatrix} + s \begin{pmatrix} 3 \\ -3 \\ 1 \\ 0 \\ 0 \end{pmatrix} + t \begin{pmatrix} -4 \\ -4 \\ 0 \\ 1 \\ 1 \end{pmatrix},$$

(3.16)

where s and t are arbitrary numbers. ●

One solution is the vector $\mathbf{p} = (-8, -27, 0, 14, 0)^T$, as we see from (3.16) when s and t are both 0. Any other point in the solution set is obtained from this point by adding an arbitrary linear combination of the vectors $\mathbf{v}_1 = (3, -3, 1, 0, 0)^T$ and $\mathbf{v}_2 = (-4, -4, 0, 1, 1)^T$. Since $\mathbf{v}_1$ and $\mathbf{v}_2$ are not multiples of each other, there are two different directions we can move in the solution set. Looking back at Definition 2.17, we see that (3.16) is a parametric equation for a plane in $\mathbf{R}^5$. Notice that the solution set can be described as long and flat. It is also thin, since there are relatively few points in $\mathbf{R}^5$ that are solutions to the system.

Pivot variables and free variables

We should probably say a little more about the process of back-solving. First, we emphasize that it requires that the augmented matrix of the system of equations should be in row echelon form, like (3.15) or (3.6). If a matrix is in row echelon form, there can be at most one pivot in each column. Each column consists of the coefficients of one of the unknowns. We will say that a column is a *pivot column* if it contains a pivot, and we will call the associated variable a *pivot variable*. A variable associated to a column that does not contain a pivot can be assigned an arbitrary value and is therefore a *free variable*. We will therefore call a column that does not contain a pivot a *free column*.

We see that if each of the free variables is assigned a value, then each of the pivot variables is uniquely determined by the system of equations. This, then, is the strategy to be used in general in the process of back-solving. This method has been illustrated in our examples, in particular in Example 3.13.

> To summarize, the linear system $A\mathbf{x} = \mathbf{b}$ can be solved using the four steps:
>
> 1. Form the augmented matrix $M = [A, \mathbf{b}]$.
> 2. Use row operations to eliminate coefficients and reduce M to row echelon form.
> 3. Write down the simplified system.
> 4. Solve the simplified system by assigning arbitrary values to the free variables and back-solving for the pivot variables.

This method works in complete generality. It is very easily programmed to work on a computer. What we have described is the basis for almost every algorithm used by modern computers. In most mathematical computer programs, there are commands that will make the solution of systems of linear equations quite easy.

Consistency

We noticed in Section 7.2 that some systems of equations had no solutions at all—the solution set was empty. You will remember that a system is consistent if it has at least one solution. Our question here is, How do we know whether a system has any solutions at all?

According to our solution method, we need only answer this for systems in row echelon form. Look at the two augmented matrices

$$\begin{pmatrix} 1 & 2 & 3 \\ 0 & 1 & 2 \end{pmatrix} \quad \text{and} \quad \begin{pmatrix} 1 & 2 & 3 \\ 0 & 0 & 2 \end{pmatrix}.$$

Each is in row echelon form. The first system can be readily solved. The second equation in the second system is the equation

$$0 \cdot x + 0 \cdot y = 2, \quad \text{or} \quad 0 = 2.$$

This equation has no solution, so the second system is inconsistent. The first system does not have a pivot in the last column and has a solution. The second has a pivot in the last column and is inconsistent. A little reflection shows that this result is true in general.

PROPOSITION 3.17   A system of equations is consistent if and only if the augmented matrix of an equivalent system in row echelon form has no pivot in the last column.   ●

Let's see how this works out in an example.

Example 3.18   Find the solution set for the system $A\mathbf{x} = \mathbf{b}$, where

$$A = \begin{pmatrix} 0 & 2 & 4 \\ 1 & 0 & -2 \\ -2 & 2 & 8 \end{pmatrix}, \quad \text{and} \quad \mathbf{b} = \begin{pmatrix} 2 \\ -1 \\ 2 \end{pmatrix}. \tag{3.19}$$

Notice that the coefficient matrix is the same as for the system in (3.2) and the right-hand side is changed slightly. The augmented matrix of the system is

$$M = [A, \mathbf{b}] = \begin{pmatrix} 0 & 2 & 4 & 2 \\ 1 & 0 & -2 & -1 \\ -2 & 2 & 8 & 2 \end{pmatrix}.$$

We can reduce this matrix to row echelon form using the same row operations we used in Example 3.9:

1. Interchange rows 1 and 2.
2. Add 2 times the first row to the third.
3. Add -1 times the second row to the third.

The result is the matrix

$$\begin{pmatrix} 1 & 0 & -2 & -1 \\ 0 & 2 & 4 & 2 \\ 0 & 0 & 0 & -2 \end{pmatrix}.$$

The presence of the pivot in the last column means that the system is inconsistent and has no solutions.

Reduced row echelon form

It is possible to carry row reduction further than simply reaching row echelon form. We can use R3 to make all of the pivots equal to 1. In addition, we can eliminate nonzero entries above the pivots. In this way the matrix in (3.6) would be transformed to

$$\begin{pmatrix} 1 & 0 & * & 0 & * & * & 0 & 0 & * & * \\ 0 & 1 & * & 0 & * & * & 0 & 0 & * & * \\ 0 & 0 & 0 & 1 & * & * & 0 & 0 & * & * \\ 0 & 0 & 0 & 0 & 0 & 0 & 1 & 0 & * & * \\ 0 & 0 & 0 & 0 & 0 & 0 & 0 & 1 & * & * \\ 0 & 0 & 0 & 0 & 0 & 0 & 0 & 0 & 0 & 0 \\ 0 & 0 & 0 & 0 & 0 & 0 & 0 & 0 & 0 & 0 \end{pmatrix}.$$

Such a matrix is said to be in ***reduced row echelon form***.

In the process of solving equations, it is not always of great value to go to the reduced row echelon form before back-solving. Simple row echelon form is easier to get to, and there are a lot of extra row operations required to reach reduced row echelon form. On the other hand, solving a reduced row echelon form is extremely easy. It is largely a matter of personal taste. However, if your computer will compute the reduced row echelon form directly, then that is the way to go.

Let's look at one example. For the system in (3.14), we reached the row echelon form in (3.15). To transform this into reduced row echelon form, we need four additional row operations:

3. Add -2 times row 2 to row 1.
4. Add row 3 to row 1.
5. Add 2 times row 3 to row 2.
6. Multiply row 3 by -1.

This results in the reduced row echelon form

$$x_1 - 3x_3 + 4x_5 = -8$$
$$x_2 + 3x_3 + 4x_5 = -27$$
$$x_4 - x_5 = 14$$

$$\begin{pmatrix} 1 & 0 & -3 & 0 & 4 & -8 \\ 0 & 1 & 3 & 0 & 4 & -27 \\ 0 & 0 & 0 & 1 & -1 & 14 \end{pmatrix}. \qquad (3.20)$$

Those extra four row operations took some effort, but now we can read the solutions from (3.20). As before, we start with the free variables $x_5 = t$ and $x_3 = s$. Then the third equation immediately gives $x_4 = 14 + t$, the second gives $x_2 = -27 - 3s - 4t$, and the first gives $x_1 = -8 + 3s - 4t$. Thus solving is easy from (3.20), but does this make up for the extra work in reaching this form? Computer scientists actually count the operations needed to find the most efficient method. For large matrices it is certainly more efficient to use row echelon form, and not go on to reduced row echelon form. However, for an individual small matrix, reduced row echelon form might be the way to go.

EXERCISES

For each system in Exercises 1–10, perform each of the following tasks. All work is to be done by hand (pencil-and-paper calculations only).

(i) Set up the augmented matrix for the system; then place the augmented matrix in row echelon form.

(ii) If the system is inconsistent, so state, and explain why. Otherwise, proceed to the next item.

(iii) Use back-solving to find the solution. Place the final solution in parametric form.

1. $x_1 + x_2 + x_3 = 4$
$2x_1 - x_2 - x_3 = 6$
$4x_1 + x_2 + x_3 = 14$

2. $x_1 + 2x_2 - x_3 = 4$
$2x_1 - x_2 + 2x_3 = 2$
$3x_1 + x_2 + x_3 = 5$

3. $x_1 + 2x_2 + 2x_3 = 2$
$-x_1 - x_2 + 2x_3 = 4$
$x_1 + 3x_2 + 6x_3 = 7$

4. $x_1 + x_2 - 2x_3 = 2$
$3x_1 - x_2 - x_3 = 3$
$5x_1 + x_2 - 5x_3 = 7$

5. $x_1 + 2x_2 - 2x_3 = 6$
$2x_1 + 4x_2 - 4x_3 = 12$
$3x_1 + 6x_2 - 6x_3 = 18$

6. $x_1 - 2x_2 - 3x_3 = 6$
$4x_1 - 8x_2 - 12x_3 = 24$
$2x_1 - 4x_2 - 6x_3 = 12$

7. $x_2 - 2x_3 = 4$
$x_1 + 2x_2 - 2x_3 = 6$
$x_1 + 4x_2 - 6x_3 = 14$

8. $x_2 - 4x_3 = 6$
$x_1 + 2x_2 - x_3 = 2$
$2x_1 + x_2 - 2x_3 = 4$

9. $x_1 + 2x_2 - 3x_3 + x_4 = 6$
$2x_1 + x_2 - 2x_3 - x_4 = 4$
$6x_2 + 4x_3 - x_4 = 4$

10. $x_1 + 2x_2 - 3x_3 + x_4 = 6$
$x_1 - 3x_2 + x_3 - 2x_4 = 6$
$- 5x_2 + 4x_3 - 3x_4 = 0$

Using hand calculations only, place each of the augmented matrices in Exercises 11–18 in reduced row echelon form.

11. $\begin{pmatrix} 1 & 2 & -3 & 1 \\ 2 & 5 & -1 & 0 \end{pmatrix}$

12. $\begin{pmatrix} 1 & -2 & 2 & 4 \\ -2 & 5 & 2 & 0 \end{pmatrix}$

13. $\begin{pmatrix} 1 & -2 & 4 & 0 \\ -2 & 1 & 1 & 4 \end{pmatrix}$

14. $\begin{pmatrix} 1 & 0 & -2 & -2 \\ 3 & -1 & 3 & -6 \end{pmatrix}$

15. $\begin{pmatrix} 1 & 1 & -1 & 2 & 3 \\ 2 & 2 & -3 & 4 & 5 \end{pmatrix}$

16. $\begin{pmatrix} 1 & -2 & 2 & -2 & -4 \\ -5 & 10 & -10 & 12 & 8 \end{pmatrix}$

17. $\begin{pmatrix} 1 & 2 & 2 & -2 & 3 \\ 1 & 2 & -2 & 4 & 2 \\ -1 & -3 & 0 & 0 & 2 \end{pmatrix}$

18. $\begin{pmatrix} 1 & -1 & 2 & 3 & -1 \\ 2 & -2 & 2 & 0 & 0 \\ 0 & 0 & -1 & 2 & 2 \end{pmatrix}$

In Exercises 19–22, you are given an augmented matrix in reduced row echelon form. Write the system (in terms of $x_1, x_2, \ldots, x_n$) represented by the augmented matrix; then write the solution to the system in parametric form.

19. $\begin{pmatrix} 1 & 0 & -2 & 2 & -2 \\ 0 & 1 & 1 & -1 & 1 \\ 0 & 0 & 0 & 0 & 0 \end{pmatrix}$

20. $\begin{pmatrix} 1 & -1 & 0 & 0 & 1 \\ 0 & 0 & 1 & -1 & 0 \\ 0 & 0 & 0 & 0 & 0 \end{pmatrix}$

21. $\begin{pmatrix} 1 & -1 & 1 & 0 & -1 & 1 \\ 0 & 0 & 0 & 1 & 1 & 0 \\ 0 & 0 & 0 & 0 & 0 & 0 \\ 0 & 0 & 0 & 0 & 0 & 0 \end{pmatrix}$

22. $\begin{pmatrix} 1 & -1 & 0 & 0 & 1 & 1 \\ 0 & 0 & 1 & -1 & -1 & 1 \\ 0 & 0 & 0 & 0 & 0 & 0 \\ 0 & 0 & 0 & 0 & 0 & 0 \end{pmatrix}$

If you have a computer or calculator that will place an augmented matrix in reduced row echelon form, use it to help find the solution of each system $A\mathbf{y} = \mathbf{b}$ given in Exercises 23–36. Otherwise you'll have to do the calculations by hand.

23. $A = \begin{pmatrix} -6 & 8 & 0 \\ 4 & 8 & 8 \\ -2 & 2 & 7 \end{pmatrix}$ and $\mathbf{b} = \begin{pmatrix} 2 \\ 20 \\ 7 \end{pmatrix}$

24. $A = \begin{pmatrix} 3 & -3 & 1 \\ 7 & -4 & 5 \\ 4 & -3 & -3 \end{pmatrix}$ and $\mathbf{b} = \begin{pmatrix} 0 \\ 3 \\ 1 \end{pmatrix}$

25. $A = \begin{pmatrix} 4 & 2 & -5 \\ -14 & -8 & 18 \\ -3 & -2 & 4 \end{pmatrix}$ and $\mathbf{b} = \begin{pmatrix} -5 \\ 16 \\ 3 \end{pmatrix}$

26. $A = \begin{pmatrix} -4 & 10 & -6 \\ 0 & -4 & 4 \\ 2 & -10 & 8 \end{pmatrix}$ and $\mathbf{b} = \begin{pmatrix} -14 \\ 4 \\ 12 \end{pmatrix}$

27. $A = \begin{pmatrix} -3 & -3 & 1 \\ 8 & 7 & -2 \\ 8 & 6 & -1 \end{pmatrix}$ and $\mathbf{b} = \begin{pmatrix} 4 \\ -8 \\ -5 \end{pmatrix}$

28. $A = \begin{pmatrix} 5 & 9 & 2 \\ -2 & -3 & -1 \\ 0 & -2 & 1 \end{pmatrix}$ and $\mathbf{b} = \begin{pmatrix} 8 \\ -2 \\ -4 \end{pmatrix}$

29. $A = \begin{pmatrix} -12 & 12 & -8 \\ -16 & 16 & -10 \\ -3 & 3 & -1 \end{pmatrix}$ and $\mathbf{b} = \begin{pmatrix} -8 \\ -10 \\ -1 \end{pmatrix}$

30. $A = \begin{pmatrix} 0 & 4 & 6 & -7 \\ -4 & 10 & 4 & -8 \end{pmatrix}$ and $\mathbf{b} = \begin{pmatrix} 4 \\ 6 \end{pmatrix}$

31. $A = \begin{pmatrix} -5 & -4 & 4 & 3 \\ 7 & 6 & -5 & 3 \end{pmatrix}$ and $\mathbf{b} = \begin{pmatrix} 17 \\ 12 \end{pmatrix}$

32. $A = \begin{pmatrix} 2 & -3 & -2 & 2 \\ -4 & 4 & 0 & 3 \\ 8 & -8 & -1 & -7 \end{pmatrix}$ and $\mathbf{b} = \begin{pmatrix} -4 \\ -7 \\ 13 \end{pmatrix}$

33. $A = \begin{pmatrix} -7 & 7 & -8 & -3 \\ 9 & -5 & 8 & -2 \\ 5 & 0 & 2 & 8 \end{pmatrix}$ and $\mathbf{b} = \begin{pmatrix} 37 \\ -35 \\ -9 \end{pmatrix}$

34. $A = \begin{pmatrix} -7 & -4 & -5 & -9 \\ 1 & 10 & 7 & 10 \\ -2 & 0 & -3 & 0 \end{pmatrix}$ and $\mathbf{b} = \begin{pmatrix} 31 \\ -18 \\ -2 \end{pmatrix}$

35. $A = \begin{pmatrix} 8 & -6 & 9 & 8 & -1 \\ -9 & 5 & -7 & 9 & 0 \\ 1 & -4 & 1 & -3 & -7 \end{pmatrix}$ and

$\mathbf{b} = \begin{pmatrix} 15 \\ -30 \\ 9 \end{pmatrix}$

36. $A = \begin{pmatrix} -2 & 3 & 6 & -7 & -1 \\ -1 & -6 & 1 & -6 & -8 \\ 0 & -9 & -5 & -1 & 9 \end{pmatrix}$ and

$\mathbf{b} = \begin{pmatrix} 3 \\ 1 \\ 16 \end{pmatrix}$

7.4 Homogeneous and Inhomogeneous Systems

In every example of a consistent system we have seen so far, the solution set has been given by a parametric equation.[2] So far these have been points, lines, or planes; however, higher-dimensional sets do arise. In fact, any solution set can be presented parametrically, and in this section we will begin the process of understanding solution sets and their parametric representations better. We will start by looking closely at homogeneous systems. With a better understanding of these, we will study the relationship between the solution set of an inhomogeneous system and that of the related homogeneous system.

It is highly informative to compare our results here to the results for linear differential equations in Chapters 2 and 4.

Homogeneous systems

Remember that a homogeneous system has the form $A\mathbf{x} = \mathbf{0}$, where the right-hand side is $\mathbf{0}$, the zero vector. Since $A\mathbf{0} = \mathbf{0}$, a homogeneous system always has the solution $\mathbf{x} = \mathbf{0}$, and is therefore consistent. Sometimes this is the only solution, but there may be others as well. How can we tell if there are solutions other than $\mathbf{0}$? This question has important application to differential equations. We will call a solution to the homogeneous system $A\mathbf{x} = \mathbf{0}$ that is different from $\mathbf{0}$ a **nontrivial solution**. To be nontrivial, a vector must have at least one nonzero component.

Let's look at a couple of examples.

---

[2] This is true even when the solution set was a single point. For example, in Example 2.25 the solution in vector form is $\mathbf{x} = (1, 0, -1)^T$. This is a parametric equation with no parameters.

Example 4.1   Consider the system $A\mathbf{x} = \mathbf{0}$, where

$$A = \begin{pmatrix} -3 & -2 & 4 \\ 14 & 8 & -18 \\ 4 & 2 & -5 \end{pmatrix}.$$

Find the general solution.

The augmented matrix[3] for this homogeneous system is

$$M = \begin{pmatrix} -3 & -2 & 4 & 0 \\ 14 & 8 & -18 & 0 \\ 4 & 2 & -5 & 0 \end{pmatrix}.$$

Since the system is homogeneous, the last column of M consists of all zeros. We reduce M to echelon form with the following row operations.

1. Add 14/3 times row 1 to row 2.
2. Add 4/3 times row 1 to row 3.
3. Multiply rows 2 and 3 by 3.

This gives us

$$\begin{pmatrix} -3 & -2 & 4 & 0 \\ 0 & -4 & 2 & 0 \\ 0 & -2 & 1 & 0 \end{pmatrix}.$$

The final row operation is

4. Add $-1/2$ times row 2 to row 3.

This results in the echelon form

$$\begin{pmatrix} -3 & -2 & 4 & 0 \\ 0 & -4 & 2 & 0 \\ 0 & 0 & 0 & 0 \end{pmatrix}. \tag{4.2}$$

The corresponding system of equations is

$$\begin{aligned} -3x_1 - 2x_2 + 4x_3 &= 0 \\ -4x_2 + 2x_3 &= 0. \end{aligned} \tag{4.3}$$

There are pivots in the first and second columns of (4.2), so x_1 and x_2 are pivot variables, while x_3 is a free variable. We set $x_3 = t$. The second equation tells us that $4x_2 = 2x_3 = 2t$, so $x_2 = t/2$. Finally, the first equation gives $-3x_1 = 2x_2 - 4x_3 = t - 4t$, from which we get $x_1 = t$. Thus our solution set consists of all vectors of the form

$$\mathbf{x} = \begin{pmatrix} x_1 \\ x_2 \\ x_3 \end{pmatrix} = \begin{pmatrix} t \\ t/2 \\ t \end{pmatrix} = t \begin{pmatrix} 1 \\ 1/2 \\ 1 \end{pmatrix},$$

where t is arbitrary. If we set $\mathbf{v} = (1, 1/2, 1)^T$, we can write this as $\mathbf{x} = \mathbf{0} + t\mathbf{v} = t\mathbf{v}$. This is the parametric equation of a line through the origin in $\mathbf{R}^3$. ●

---

[3] Notice that we are no longer printing the pivots in blue.

Notice that in the previous example, the row operations on the augmented matrix left the last column of zeros intact. It is not difficult to see that this will always be the case for the augmented matrix of a homogeneous system. For this reason, it is not really necessary to augment the coefficient matrix when we are dealing with homogeneous equations. We can easily remember that the right-hand side is the zero vector. Consequently, we will not augment the matrix in future examples dealing with homogeneous systems.

Example 4.4 Find the solution set for the system $B\mathbf{x} = \mathbf{0}$, where

$$B = \begin{pmatrix} -3 & -2 & 4 \\ 14 & 8 & -18 \\ 4 & 2 & -4 \end{pmatrix}.$$

Notice how closely B resembles the matrix A in Example 4.1. The same row operations reduce B to the echelon form

$$\begin{pmatrix} -3 & -2 & 4 \\ 0 & -4 & 2 \\ 0 & 0 & 3 \end{pmatrix}. \tag{4.5}$$

From this, we see easily that the solution set for the system $B\mathbf{x} = \mathbf{0}$ consists of only the zero vector $\mathbf{0} = (0, 0, 0)^T$. ●

Examples 4.1 and 4.4 illustrate the possibilities for a homogeneous system. In the final step of the solution process, we have to solve a homogeneous system in row echelon form. If there is at least one free variable, as in (4.2), we can set these variables to be nonzero and solve for the pivot variables. Thus we get a nontrivial solution if there is a free variable. On the other hand, if there are no free variables, as is the case with (4.5), there are no solutions except the zero vector. Let's state this formally.

PROPOSITION 4.6 The homogeneous system $A\mathbf{x} = \mathbf{0}$ has a nontrivial solution if and only if a matrix in row echelon form that is equivalent to A has a free column. ●

We can gain information about a system from some basic numerology. Suppose that the system $A\mathbf{x} = \mathbf{b}$ consists of m equations in n unknowns. Then A is a matrix with m rows and n columns. When we transform the system to row echelon form, suppose there are p pivots and f free variables. Since there is at most one pivot in each row, we have

$$p \leq m. \tag{4.7}$$

Each column is either a pivot column or a free column, so

$$n = p + f. \tag{4.8}$$

As our first application of these formulas, let's look at the case when there are fewer equations than unknowns. This means that $m < n$. From (4.7) we see that $p \leq m < n$. Then from (4.8), $f = n - p > 0$. Hence there must be at least one free column. By Proposition 4.6, there must be a nontrivial solution. Let's state this formally.

PROPOSITION 4.9 Any homogeneous linear system with fewer equations than unknowns has a nontrivial solution. ●

Structure of the solution set to an inhomogeneous system

Let's look at a simple example.

Example 4.10 Find the solution set for the equation

$$5x_1 - 4x_2 = 10. \tag{4.11}$$

We easily see that the solution is given by $x_1 = (10 + 4x_2)/5$. In vector notation, the general solution is all vectors $\mathbf{x}$ of the form

$$\mathbf{x} = \begin{pmatrix} x_1 \\ x_2 \end{pmatrix} = \begin{pmatrix} (10 + 4x_2)/5 \\ x_2 \end{pmatrix}$$
$$= \begin{pmatrix} 2 \\ 0 \end{pmatrix} + x_2 \begin{pmatrix} 4/5 \\ 1 \end{pmatrix}.$$

If we set $\mathbf{p} = (2, 0)^T$ and $\mathbf{v}_0 = (4/5, 1)^T$, we see that the solution set for the inhomogeneous equation (4.11) is given by

$$\mathbf{x} = \mathbf{p} + x_2 \mathbf{v}_0, \quad x_2 \in \mathbf{R}. \tag{4.12}$$

This is the parametric equation for a line in the plane. Notice that $x_2 = 0$ gives the particular solution $\mathbf{p} = (2, 0)^T$.

On the other hand, we see that the solution set for the corresponding homogeneous equation

$$5x_1 - 4x_2 = 0 \tag{4.13}$$

is the line given parametrically by

$$\mathbf{v} = x_2 \mathbf{v}_0, \quad x_2 \in \mathbf{R}. \tag{4.14}$$

In summary, we see that the solution set (4.12) for the inhomogeneous equation (4.11) has the form $\mathbf{p} + \mathbf{v}$ where $\mathbf{v}$ is any vector in the solution set (4.14) for the corresponding homogeneous equation (4.13). This relationship is exhibited in Figure 1. ●

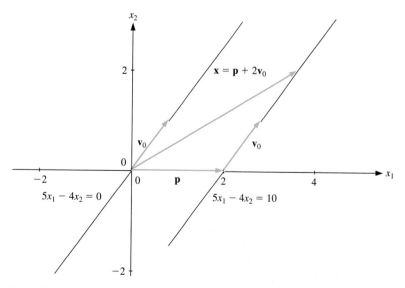

Figure 1. Relationship between the solution sets of an inhomogeneous equation and the associated homogeneous equation.

The relationship found in Example 4.10 is true in general. Consider the general system of equations

$$Ax = b,$$

which we first considered in (1.10). If the right-hand side $b = 0$, the system is homogeneous, but if $b \neq 0$, then the system is inhomogeneous. The next theorem provides a relationship between solutions of an inhomogeneous system and solutions of its associated homogeneous system.

THEOREM 4.15 Suppose p is a particular solution to the inhomogeneous system $Ax = b$. Then the solution set for $Ax = b$ has the form

$$x = p + v, \quad \text{where} \quad Av = 0.$$

Proof First, suppose that v satisfies $Av = 0$. Set $x = p + v$. Then, since $Ap = b$,

$$Ax = A(p + v) = Ap + Av = b + 0 = b,$$

so x is in the solution set for $Ax = b$.

Now suppose x is in the solution set for $Ax = b$. Let $v = x - p$. Then

$$Av = A(x - p) = Ax - Ap = b - b = 0.$$

Therefore, v is a solution to the homogeneous equation and $x = p + v$. ■

Let's look at a more complicated example.

Example 4.16 In Example 3.13 we found the solution to the system $Ax = b$, where

$$A = \begin{pmatrix} 0 & 1 & 3 & 2 & 2 \\ 1 & 2 & 3 & 5 & 7 \\ 2 & 4 & 6 & 9 & 15 \end{pmatrix} \quad \text{and} \quad b = \begin{pmatrix} 1 \\ 8 \\ 2 \end{pmatrix}.$$

Find the solution set of the homogeneous system $Ax = 0$.

In Example 3.13 we found that the solution set to the inhomogeneous system $Ax = b$ is given by the parametric equation

$$x = p + sv_1 + tv_2, \quad \text{where} \quad s, t \in \mathbf{R}, \tag{4.17}$$

$p = (-8, -27, 0, 14, 0)^T$, $v_1 = (3, -3, 1, 0, 0)^T$, and $v_2 = (-4, -4, 0, 1, 1)^T$. Notice that with the choices $s = t = 0$, the vector p satisfies $Ap = b$. That is, p is a particular solution to the inhomogeneous system. By Theorem 4.15, we know that the solution set has the form

$$x = p + v, \quad \text{where} \quad Av = 0. \tag{4.18}$$

Comparing (4.17) and (4.18), we see that the solution set of the homogeneous system is given by the parametric equation

$$v = sv_1 + tv_2, \quad \text{where} \quad s, t \in \mathbf{R}. \qquad \bullet$$

Thus, by Theorem 4.15, to find the general solution to an inhomogeneous system, it is necessary to find a particular solution to the inhomogeneous system and the general solution to the homogeneous system. If you look back at the examples in this chapter, you will see this pattern in every case.

There is another interpretation of Theorem 4.15 as exhibited in Example 4.16. Suppose we find a parametric representation for the solution set to an inhomogeneous system, like we did in (4.17) for Example 3.13. Then the part of the representation containing only the parameters is a parametric representation for the solution set for the corresponding homogeneous system.

Nullspaces

Theorem 4.15 illustrates the importance of the solution set for a homogeneous system. We will see other uses when we apply matrix algebra to differential equations. To emphasize their importance, and because we will soon get tired of using the long phrase "the solution set for a homogeneous system," we will give it a special name.

DEFINITION 4.19

> The **nullspace** of a matrix A is the set of all solutions to the homogeneous system of linear equations $A\mathbf{x} = \mathbf{0}$. The nullspace of A is denoted by null(A).

Nullspaces of matrices will be used repeatedly in our applications to differential equations. Notice that we can restate Theorem 4.15 succinctly using the terminology in Definition 4.19.

THEOREM 4.20   The solution set of the inhomogeneous system $A\mathbf{x} = \mathbf{b}$ has the form

$$\{\mathbf{x} = \mathbf{p} + \mathbf{v} \mid \mathbf{v} \in \text{null}(A)\},$$

where $\mathbf{p}$ is any particular solution to the system $A\mathbf{x} = \mathbf{b}$. ∎

Theorem 4.20 is illustrated in Examples 4.10 and 4.16. For the matrix $(5, -4)$ in Example 4.10, the nullspace consists of all multiples of the vector $\mathbf{v}_0 = (4/5, 1)^T$. For the matrix A in Example 4.16 we have

$$\text{null}(A) = \{\mathbf{v} = s\mathbf{v}_1 + t\mathbf{v}_2 \mid s, t \in \mathbf{R}\},$$

where $\mathbf{v}_1 = (3, -3, 1, 0, 0)^T$, and $\mathbf{v}_2 = (-4, -4, 0, 1, 1)^T$. This means that the nullspace consists of the set of all linear combinations of $\mathbf{v}_1$ and $\mathbf{v}_2$.

Properties of nullspaces

Nullspaces have two important properties that follow easily from Theorem 1.16.

PROPOSITION 4.21   Let A be an $n \times m$ matrix.

1. Suppose $\mathbf{x}$ and $\mathbf{y}$ are vectors in null(A). Then $\mathbf{x} + \mathbf{y}$ is also in null(A).
2. Suppose $\mathbf{x}$ is in null(A) and a is a number. Then $a\mathbf{x}$ is also in null(A).

Proof   To prove (1), notice that if $\mathbf{x}$ and $\mathbf{y}$ are vectors in null(A), then $A\mathbf{x} = \mathbf{0}$ and $A\mathbf{y} = \mathbf{0}$. By Theorem 1.16, this means that $A(\mathbf{x} + \mathbf{y}) = A\mathbf{x} + A\mathbf{y} = \mathbf{0}$. Of course this means that $\mathbf{x} + \mathbf{y}$ is in null(A). The proof of (2) is just as easy. ●

Let's look at some other examples.

Example 4.22   Find the nullspace of the matrix

$$A = \begin{pmatrix} 2 & 1 \\ 4 & 2 \end{pmatrix}. \tag{4.23}$$

To find the nullspace, we reduce A to row echelon form: Add -2 times row 1 to row 2. We get

$$\begin{pmatrix} 2 & 1 \\ 0 & 0 \end{pmatrix}.$$

Thus x_2 is a free variable, which we give an arbitrary value $x_2 = t$. Solving for x_1, we get $2x_1 = -t$. Hence every element of null(A) is of the form

$$\mathbf{x} = \begin{pmatrix} x_1 \\ x_2 \end{pmatrix} = \begin{pmatrix} -t/2 \\ t \end{pmatrix} = t \begin{pmatrix} -1/2 \\ 1 \end{pmatrix}, \tag{4.24}$$

where t is an arbitrary number. This is the parametric equation for a line through the origin in $\mathbf{R}^2$, shown in blue in Figure 2.

However, in light of Proposition 4.21, any multiple of $\mathbf{x}$ will also lie in null(A). In particular, $-2\mathbf{x}$, which equals $(1, -2)^T$, lies in null(A), which is evident in Figure 2. Thus, the nullspace of matrix A could also be described parametrically with

$$\mathbf{x} = t \begin{pmatrix} 1 \\ -2 \end{pmatrix}.$$

Indeed, any multiple of $(-1/2, 1)^T$ will do. Many software packages will choose a multiple that is a vector having length one, called a ***unit vector***.   ●

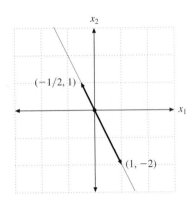

Figure 2. The nullspace in Example 4.22 is a line through the origin.

Example 4.25   Find the nullspace of the 1×3 matrix

$$B = (1, 3, -2). \tag{4.26}$$

A row vector like this is already in row echelon form, and the last two variables are free. Hence to solve $B\mathbf{x} = 0$, we can set $x_2 = s$, and $x_3 = t$. Then $x_1 + 3s - 2t = 0$, or $x_1 = -3s + 2t$, and our solution is

$$\mathbf{x} = \begin{pmatrix} x_1 \\ x_2 \\ x_3 \end{pmatrix} = \begin{pmatrix} -3s + 2t \\ s \\ t \end{pmatrix} = s \begin{pmatrix} -3 \\ 1 \\ 0 \end{pmatrix} + t \begin{pmatrix} 2 \\ 0 \\ 1 \end{pmatrix}, \tag{4.27}$$

where s and t are arbitrary. Thus the nullspace of B consists of all linear combinations of the vectors $(-3, 1, 0)^T$ and $(2, 0, 1)^T$. Equation (4.27) is the parametric equation for a plane through the origin in $\mathbf{R}^n$.   ●

Notice that we could also say that the nullspace of the matrix A in Example 4.22 consists of all linear combinations of the single vector $(-1/2, 1)^T$. This pattern is also seen if we look back at the examples in the previous section.

EXERCISES

1. Check the solution of Example 4.1 by directly multiplying

$$\begin{pmatrix} -3 & -2 & 4 \\ 14 & 8 & -18 \\ 4 & 2 & -5 \end{pmatrix} \begin{pmatrix} t \begin{pmatrix} 1 \\ 1/2 \\ 1 \end{pmatrix} \end{pmatrix}.$$

2. Check the solution of Example 4.25 by directly multiplying

$$\begin{pmatrix} 1 & 3 & -2 \end{pmatrix} \begin{bmatrix} s \begin{pmatrix} -3 \\ 1 \\ 0 \end{pmatrix} + t \begin{pmatrix} 2 \\ 0 \\ 1 \end{pmatrix} \end{bmatrix}.$$

Find all solutions of $A\mathbf{x} = 0$ for the matrices given in Exercises 3–10. Express your answer in parametric form.

3. $A = \begin{pmatrix} 1 & 0 & -2 \\ 0 & 1 & 3 \\ 0 & 0 & 0 \end{pmatrix}$ **4.** $A = \begin{pmatrix} 1 & 0 & -5 \\ 0 & 1 & 2 \\ 0 & 0 & 0 \end{pmatrix}$

5. $A = \begin{pmatrix} 1 & 0 & 2 & -2 \\ 0 & 1 & 3 & -1 \end{pmatrix}$ **6.** $A = \begin{pmatrix} 1 & 1 & 0 & 4 \\ 0 & 0 & 1 & 2 \end{pmatrix}$

7. $A = \begin{pmatrix} 1 & 0 & -1 & 0 & 3 \\ 0 & 1 & 2 & 0 & -5 \\ 0 & 0 & 0 & 1 & 2 \end{pmatrix}$

8. $A = \begin{pmatrix} 1 & 2 & 0 & 0 & -5 \\ 0 & 0 & 1 & 0 & -4 \\ 0 & 0 & 0 & 1 & 3 \end{pmatrix}$

9. $A = \begin{pmatrix} 1 & 0 & -2 & 4 \end{pmatrix}$ **10.** $A = \begin{pmatrix} 1 & 2 & -3 & 4 \end{pmatrix}$

The matrix A and the solution of $A\mathbf{x} = \mathbf{0}$ follow.

$$A = \begin{pmatrix} 1 & 0 & -4 & 2 \\ 0 & 1 & 2 & -3 \\ 0 & 0 & 0 & 0 \end{pmatrix} \qquad \mathbf{x} = s\begin{pmatrix} 4 \\ -2 \\ 1 \\ 0 \end{pmatrix} + t\begin{pmatrix} -2 \\ 3 \\ 0 \\ 1 \end{pmatrix}$$

The vectors $(4, -2, 1, 0)^T$ and $(-2, 3, 0, 1)^T$ are "special" be-cause all solutions of $A\mathbf{x} = \mathbf{0}$ can be written as a linear combi-nation of these two vectors. In Exercises 11–14, perform each of the following tasks.

(i) Use a computer or calculator to place the given matrix A in reduced row echelon form. How many free variables does the reduced row echelon form have?

(ii) Write the solution to $A\mathbf{x} = \mathbf{0}$ in parametric form. How many "special" vectors are there?

11. $A = \begin{pmatrix} 2 & -1 & 1 & -1 \\ 2 & 0 & 0 & -2 \\ 3 & -4 & 4 & 1 \end{pmatrix}$

12. $A = \begin{pmatrix} -3 & -5 & -1 & -5 \\ -2 & -3 & -1 & -3 \\ 0 & 3 & -3 & 3 \end{pmatrix}$

13. $A = \begin{pmatrix} 0 & -2 & -2 & 0 & 2 \\ -3 & 1 & 4 & 0 & -1 \\ 1 & 0 & -1 & 0 & 0 \end{pmatrix}$

14. $A = \begin{pmatrix} -1 & -1 & -2 & 2 & -3 \\ 3 & 3 & 1 & -1 & 4 \\ 0 & 0 & 2 & -2 & 2 \end{pmatrix}$

15. Consider the system

$$\begin{aligned} -x_1 + x_2 + x_3 - x_4 \quad\quad - x_6 &= 0 \\ x_1 - 4x_2 + 2x_3 + 4x_4 - 3x_5 - 2x_6 &= 0 \\ 3x_1 - 3x_2 - 3x_3 + 3x_4 \quad\quad + 3x_6 &= 0. \end{aligned}$$

(a) This system must possess at least how many free vari-ables? Answer this question without doing any calcu-lations and explain your reasoning.

(b) Could the system have more free variables than the number proposed in part (a)? Set up and reduce the augmented matrix to find out.

16. Consider the system

$$\begin{aligned} 3x_1 \quad\quad + 3x_3 - 3x_4 \quad\quad - 3x_6 + 3x_7 &= 0 \\ -x_1 + 3x_2 + 2x_3 + x_4 + 3x_5 + 4x_6 + 2x_7 &= 0 \\ x_1 - 2x_2 - x_3 - x_4 - 2x_5 - 3x_6 - x_7 &= 0 \\ -x_1 + x_2 \quad\quad + x_4 + x_5 + 2x_6 \quad\quad &= 0. \end{aligned}$$

(a) This system must possess at least how many free vari-ables? Answer this question without doing any calcu-lations and explain your reasoning.

(b) Could the system have more free variables than the number proposed in part (a)? Set up and reduce the augmented matrix to find out.

17. Prove part (2) of Proposition 4.21.

In Exercises 18–21, the matrix A and the vector $\mathbf{b}$ of the sys-tem $A\mathbf{x} = \mathbf{b}$ are given. Perform each of the following tasks for each exercise.

(i) As in Example 4.16, find the solution to the system in the form $\mathbf{x} = \mathbf{p} + \mathbf{v}$, where $\mathbf{p}$ is a particular solution and $\mathbf{v}$ is given in parametric form.

(ii) Show that the $\mathbf{v}$ is in the nullspace of the given matrix A by showing directly that $A\mathbf{v} = \mathbf{0}$. *Hint:* See Exercises 1 and 2.

18. $A = \begin{pmatrix} 2 & -2 & 1 \\ 1 & -1 & -1 \\ 0 & 0 & 3 \end{pmatrix} \qquad \mathbf{b} = \begin{pmatrix} -1 \\ 1 \\ -3 \end{pmatrix}$

19. $A = \begin{pmatrix} 5 & -2 & -5 \\ -3 & 0 & 3 \\ 0 & -3 & 0 \end{pmatrix} \qquad \mathbf{b} = \begin{pmatrix} 2 \\ 0 \\ 3 \end{pmatrix}$

20. $A = \begin{pmatrix} -3 & 3 & 3 & 0 \\ -4 & 2 & 4 & -2 \\ 1 & 0 & -1 & 1 \end{pmatrix} \qquad \mathbf{b} = \begin{pmatrix} 3 \\ 4 \\ -1 \end{pmatrix}$

21. $A = \begin{pmatrix} 2 & 0 & 0 & -4 \\ -1 & 1 & -1 & 1 \\ -1 & 1 & -1 & 1 \end{pmatrix} \qquad \mathbf{b} = \begin{pmatrix} 0 \\ 1 \\ 1 \end{pmatrix}$

In Exercises 22–25, sketch the nullspace of the given matrix in $\mathbf{R}^2$.

22. $A = \begin{pmatrix} 2 & -1 \\ 4 & -2 \end{pmatrix}$

23. $A = \begin{pmatrix} 1 & 2 \\ -2 & -4 \end{pmatrix}$

24. $A = \begin{pmatrix} 2 & 3 \\ 4 & 6 \end{pmatrix}$

25. $A = \begin{pmatrix} 3 & -2 \\ 6 & -4 \end{pmatrix}$

In Exercises 26–29, describe the nullspace of the given matrix, both parametrically and geometrically in $\mathbf{R}^3$.

26. $\begin{pmatrix} 3 & -1 & 0 \\ -1 & 0 & 2 \\ 4 & 0 & -2 \end{pmatrix}$

27. $\begin{pmatrix} 2 & 1 & -1 \\ 1 & -1 & -2 \\ 2 & 0 & -2 \end{pmatrix}$

28. $\begin{pmatrix} 4 & -4 & -4 \\ -5 & 5 & 5 \\ 3 & -3 & -3 \end{pmatrix}$

29. $\begin{pmatrix} 1 & -1 & -1 \\ 2 & -2 & -2 \\ -4 & 4 & 4 \end{pmatrix}$

7.5 Bases of a Subspace

We have discovered that the nullspace of a matrix consists of the set of *all* linear combinations of a few vectors. In Definition 5.1, we will define this as the span of those vectors, and then proceed to study this structure in more detail. We will discover the importance of the concept of linear independence. This will lead us to the notion of a basis. Finally, this will allow us to define the dimension of a nullspace.

The span of a set of vectors

We have seen the importance of the set of *all* linear combinations of a few vectors. Let's introduce a word that expresses this idea.

DEFINITION 5.1

If $\mathbf{x}_1, \mathbf{x}_2, \ldots, \mathbf{x}_k$ are vectors in $\mathbf{R}^n$, we define the *span* of $\mathbf{x}_1, \ldots, \mathbf{x}_k$ to be the set of all linear combinations of $\mathbf{x}_1, \ldots, \mathbf{x}_k$, and we will denote this set by $\text{span}(\mathbf{x}_1, \ldots, \mathbf{x}_k)$.

In Examples 4.22 and 4.25, we can now write

$$\text{null}(A) = \text{span}\left((-1/2, 1)^T\right), \tag{5.2}$$

and

$$\text{null}(B) = \text{span}\left((-3, 1, 0)^T, (2, 0, 1)^T\right). \tag{5.3}$$

In all of the cases we have seen the nullspace of a matrix is the span of a few vectors. Indeed, spans have the same basic properties that we proved for nullspaces in Proposition 4.21.

PROPOSITION 5.4   Suppose that $\mathbf{x}_1, \ldots, \mathbf{x}_k$ are vectors in $\mathbf{R}^n$, and set $V = \text{span}(\mathbf{x}_1, \ldots, \mathbf{x}_k)$.

1. If $\mathbf{x}$ and $\mathbf{y}$ are vectors in V, then $\mathbf{x} + \mathbf{y}$ is also in V.
2. If $\mathbf{x}$ is in V and a is a number, then $a\mathbf{x}$ is also in V.

Proof   Suppose $\mathbf{x}, \mathbf{y} \in V$. Then there are numbers $b_1, \ldots, b_k, c_1, \ldots, c_k$ such that

$$\mathbf{x} = b_1\mathbf{x}_1 + b_2\mathbf{x}_2 + \cdots + b_k\mathbf{x}_k$$
$$\mathbf{y} = c_1\mathbf{x}_1 + c_2\mathbf{x}_2 + \cdots + c_k\mathbf{x}_k.$$

Consequently,

$$\mathbf{x} + \mathbf{y} = (b_1 + c_1)\mathbf{x}_1 + (b_2 + c_2)\mathbf{x}_2 + \cdots + (b_k + c_k)\mathbf{x}_k$$

is a linear combination of $\mathbf{x}_1, \ldots, \mathbf{x}_k$, and consequently belongs to $\text{span}(\mathbf{x}_1, \ldots, \mathbf{x}_k)$. This verifies 1. Next, if a is a number, then

$$a\mathbf{x} = ab_1\mathbf{x}_1 + ab_2\mathbf{x}_2 + \cdots + ab_k\mathbf{x}_k$$

is also a linear combination of $\mathbf{x}_1, \ldots, \mathbf{x}_k$. Hence $a\mathbf{x}$ belongs to $\text{span}(\mathbf{x}_1, \ldots, \mathbf{x}_k)$, which verifies 2. ●

Subspaces

Both nullspaces and spans have the properties in Proposition 5.4. We will give a special name to sets of vectors that have those two properties.

DEFINITION 5.5

A nonempty subset V of $\mathbf{R}^n$ that has the following two properties is called a *subspace* of $\mathbf{R}^n$.

1. If $\mathbf{x}$ and $\mathbf{y}$ are vectors in V, then $\mathbf{x} + \mathbf{y}$ is also in V.
2. If $\mathbf{x}$ is in V and a is a number, then $a\mathbf{x}$ is also in V.

It is important to realize the full implication of Definition 5.5. Suppose that a and b are numbers, and $\mathbf{x}$ and $\mathbf{y}$ are vectors in a subspace V. Then by (2), $a\mathbf{x}$ and $b\mathbf{y}$ are in V, and by (1), $a\mathbf{x} + b\mathbf{y}$ is in V. Of course, this can be carried on to an arbitrary collection of numbers and of vectors in V. Thus

COROLLARY 5.6   Any linear combination of vectors in a subspace V is also in V.   ●

Notice that the zero vector $\mathbf{0}$ is an element of every subspace. The easiest way to see this is probably to set $a = 0$ in part (2) of Definition 5.5. In fact, the set with the single vector $\mathbf{0}$ satisfies the conditions in Definition 5.5. In a small abuse of our notation we will denote this subspace by $\mathbf{0}$. Next notice that the total space, $\mathbf{R}^n$, is also a subspace. These two subspaces will be referred to as the *trivial subspaces*.

Proposition 5.4 says that the span of a set of vectors forms a subspace. The converse is also true, but we will not prove it.

PROPOSITION 5.7   Suppose that V is a subspace of $\mathbf{R}^n$, and that $V \neq \mathbf{0}$. Then there are vectors $\mathbf{x}_1, \mathbf{x}_2,$ $\ldots, \mathbf{x}_k$ such that $V = \text{span}(\mathbf{x}_1, \mathbf{x}_2, \ldots, \mathbf{x}_k)$.   ●

Properties of spans

If $V = \text{span}(\mathbf{x}_1, \mathbf{x}_2, \ldots, \mathbf{x}_k)$, we will say that V is *spanned* by $\{\mathbf{x}_1, \mathbf{x}_2, \ldots, \mathbf{x}_k\}$, that the set of vectors $\{\mathbf{x}_1, \mathbf{x}_2, \ldots, \mathbf{x}_k\}$ *spans* V, or that $\{\mathbf{x}_1, \mathbf{x}_2, \ldots, \mathbf{x}_k\}$ is a *spanning set* for V. However we say it, it means that V is the set of all linear combinations

$$\mathbf{x} = a_1\mathbf{x}_1 + a_2\mathbf{x}_2 + \cdots + a_k\mathbf{x}_k. \tag{5.8}$$

The coefficients a_j can be any numbers. If we consider the coefficients to be parameters, equation (5.8) is a parametric equation for the subspace $V = \text{span}(\mathbf{x}_1, \mathbf{x}_2, \ldots, \mathbf{x}_k)$.

If we are given a set of vectors $\{\mathbf{x}_1, \mathbf{x}_2, \ldots, \mathbf{x}_k\}$, and another vector $\mathbf{x}$, how do we know if $\mathbf{x}$ is in $\text{span}(\mathbf{x}_1, \mathbf{x}_2, \ldots, \mathbf{x}_k)$? By Definition 5.1, we are asking if $\mathbf{x}$ is a linear combination of $\mathbf{x}_1, \mathbf{x}_2, \ldots, \mathbf{x}_k$, or if we can find coefficients $a_1, a_2, \ldots, a_k$ such that $\mathbf{x}$ can be written as in equation (5.8). According the definition of matrix-vector product (Definition 1.9), equation (5.8) is equivalent to the matrix equation $\mathbf{x} = X\mathbf{a}$, where $X = [\mathbf{x}_1, \mathbf{x}_2, \ldots, \mathbf{x}_k]$ is the matrix with columns $\mathbf{x}_1, \mathbf{x}_2, \ldots, \mathbf{x}_k$ and $\mathbf{a} = (a_1, a_2, \ldots, a_k)^T$. Thus we have the following method.

> We can determine if $\mathbf{x}$ is in $\text{span}(\mathbf{x}_1, \mathbf{x}_2, \ldots, \mathbf{x}_k)$ as follows:
>
> **1.** Form the matrix $X = [\mathbf{x}_1, \mathbf{x}_2, \ldots, \mathbf{x}_k]$.
> **2.** Solve the system $X\mathbf{a} = \mathbf{x}$.
>
>   (a) If there are no solutions, $\mathbf{x}$ is NOT in $\text{span}(\mathbf{x}_1, \mathbf{x}_2, \ldots, \mathbf{x}_k)$.
>   (b) If $\mathbf{a} = (a_1, a_2, \ldots, a_k)^T$ is a solution, then
>
> $$\mathbf{x} = a_1\mathbf{x}_1 + a_2\mathbf{x}_2 + \cdots + a_k\mathbf{x}_k$$
>
> is in $\text{span}(\mathbf{x}_1, \mathbf{x}_2, \ldots, \mathbf{x}_k)$.

Example 5.9   Consider the vectors $\mathbf{v}_1 = (-1, 2)^T$, $\mathbf{v}_2 = (1, 1)^T$, $\mathbf{v}_3 = (2, 2)^T$, and $\mathbf{v}_4 = (2, -1)^T$ shown in Figure 1. Show that $\text{span}(\mathbf{v}_1, \mathbf{v}_2) = \mathbf{R}^2$. Compute $\text{span}(\mathbf{v}_1, \mathbf{v}_3)$, $\text{span}(\mathbf{v}_2, \mathbf{v}_3)$, and $\text{span}(\mathbf{v}_1, \mathbf{v}_2, \mathbf{v}_4)$.

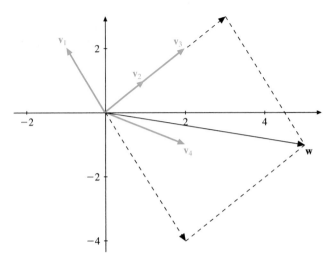

Figure 1. The vectors in Example 5.9.

To compute the span of $\mathbf{v}_1$ and $\mathbf{v}_2$, we use the matrix

$$V = [\mathbf{v}_1, \mathbf{v}_2] = \begin{pmatrix} -1 & 1 \\ 2 & 1 \end{pmatrix}.$$

To show that a vector $\mathbf{w}$ is in span($\mathbf{v}_1, \mathbf{v}_2$), we have to find a solution to the system $V\mathbf{a} = \mathbf{w}$. For example, if $\mathbf{w} = (5, -1)^T$, we look at the augmented matrix $[V, \mathbf{w}]$ and use row operations to put it into reduced row echelon form:

$$[V, \mathbf{w}] = \begin{pmatrix} -1 & 1 & 5 \\ 2 & 1 & -1 \end{pmatrix} \xrightarrow{\text{rref}} \begin{pmatrix} 1 & 0 & -2 \\ 0 & 1 & 3 \end{pmatrix}.$$

A solution to $V\mathbf{a} = \mathbf{w}$ is $\mathbf{a} = (-2, 3)^T$, and we see that $\mathbf{w} = -2\mathbf{v}_1 + 3\mathbf{v}_2$ is in span($\mathbf{v}_1, \mathbf{v}_2$). For the general vector $\mathbf{w} = (w_1, w_2)^T$ we do the same thing. The same row operations show that

$$[V, \mathbf{w}] = \begin{pmatrix} -1 & 1 & w_1 \\ 2 & 1 & w_2 \end{pmatrix} \xrightarrow{\text{rref}} \begin{pmatrix} 1 & 0 & (w_2 - w_1)/3 \\ 0 & 1 & (2w_1 + w_2)/3 \end{pmatrix}.$$

Thus we can solve $V\mathbf{a} = \mathbf{w}$ for any vector $\mathbf{w} \in \mathbf{R}^2$, so span($\mathbf{v}_1, \mathbf{v}_2$) $= \mathbf{R}^2$. That span($\mathbf{v}_1, \mathbf{v}_3$) $= \mathbf{R}^2$ is proved in the same way.

However, when we compute span($\mathbf{v}_2, \mathbf{v}_3$) we run into a complication. Forming $V = [\mathbf{v}_2, \mathbf{v}_3]$ and using $\mathbf{w} = (5, -1)^T$, we get

$$[V, \mathbf{w}] = \begin{pmatrix} 1 & 2 & 5 \\ 1 & 2 & -1 \end{pmatrix} \xrightarrow{\text{rref}} \begin{pmatrix} 1 & 2 & 0 \\ 0 & 0 & 1 \end{pmatrix}.$$

This is the augmented matrix for a system that has no solutions, so $\mathbf{w}$ is not in span($\mathbf{v}_2, \mathbf{v}_3$). Therefore, span($\mathbf{v}_2, \mathbf{v}_3$) is not equal to $\mathbf{R}^2$. The problem is that $\mathbf{v}_3 = 2\mathbf{v}_2$. Hence if $\mathbf{v} = a\mathbf{v}_2 + b\mathbf{v}_3$ is in span($\mathbf{v}_2, \mathbf{v}_3$), then

$$\mathbf{v} = a\mathbf{v}_2 + b\mathbf{v}_3 = a\mathbf{v}_2 + b \cdot 2\mathbf{v}_2 = (a + 2b)\mathbf{v}_2. \tag{5.10}$$

Thus $\mathbf{v}$ is a multiple of $\mathbf{v}_2$. Therefore,

$$\text{span}(\mathbf{v}_2, \mathbf{v}_3) = \text{span } \mathbf{v}_2 = \{t\mathbf{v}_2 \mid t \in \mathbf{R}\}.$$

We run into a similar situation when computing $span(\mathbf{v}_1, \mathbf{v}_2, \mathbf{v}_4)$. Notice that $\mathbf{v}_4 = \mathbf{v}_2 - \mathbf{v}_1$. Hence if $\mathbf{v} = a\mathbf{v}_1 + b\mathbf{v}_2 + c\mathbf{v}_4$ is a vector in $span(\mathbf{v}_1, \mathbf{v}_2, \mathbf{v}_4)$, we have

$$
\begin{aligned}
\mathbf{v} &= a\mathbf{v}_1 + b\mathbf{v}_2 + c\mathbf{v}_4 \\
&= a\mathbf{v}_1 + b\mathbf{v}_2 + c[\mathbf{v}_2 - \mathbf{v}_1] \\
&= [a - c]\mathbf{v}_1 + [b + c]\mathbf{v}_2.
\end{aligned}
\tag{5.11}
$$

Hence $\mathbf{v}$ is in $span(\mathbf{v}_1, \mathbf{v}_2)$. Thus $span(\mathbf{v}_1, \mathbf{v}_2, \mathbf{v}_4) = span(\mathbf{v}_1, \mathbf{v}_2) = \mathbf{R}^2$.   ●

For another example, consider the vectors $\mathbf{e}_1 = (1, 0, 0)^T$, $\mathbf{e}_2 = (0, 1, 0)^T$, and $\mathbf{e}_3 = (0, 0, 1)^T$ in $\mathbf{R}^3$. The vector $\mathbf{x} = (x_1, x_2, x_3)^T$ in $\mathbf{R}^3$ is a linear combination of $\mathbf{e}_1, \mathbf{e}_2$, and $\mathbf{e}_3$ in a natural way:

$$
\mathbf{x} = \begin{pmatrix} x_1 \\ x_2 \\ x_3 \end{pmatrix} = x_1 \begin{pmatrix} 1 \\ 0 \\ 0 \end{pmatrix} + x_2 \begin{pmatrix} 0 \\ 1 \\ 0 \end{pmatrix} + x_3 \begin{pmatrix} 0 \\ 0 \\ 1 \end{pmatrix} = x_1\mathbf{e}_1 + x_2\mathbf{e}_2 + x_3\mathbf{e}_3.
$$

Thus $\mathbf{R}^3 = span(\mathbf{e}_1, \mathbf{e}_2, \mathbf{e}_3)$. In a similar manner, in $\mathbf{R}^n$ we can define $\mathbf{e}_j$ to be the vector that has all zeros for entries except in the jth spot, where there is a 1. Then we see that

$$
\mathbf{R}^n = span(\mathbf{e}_1, \mathbf{e}_2, \dots, \mathbf{e}_n).
\tag{5.12}
$$

Linear dependence and independence

In Example 5.9 we discovered that $span(\mathbf{v}_2, \mathbf{v}_3) = span(\mathbf{v}_2)$ and $span(\mathbf{v}_1, \mathbf{v}_2, \mathbf{v}_4) = span(\mathbf{v}_1, \mathbf{v}_2)$. This shows that spanning sets for a subspace are not unique and do not even have to have the same number of vectors. We want to find a way to eliminate unneeded vectors from a spanning set. This will lead us to the notion of linear dependence and independence. Roughly speaking, a set of vectors is linearly dependent if one or more of them is unneeded to express the span of the set of vectors. We will build up to the general definition by examining easy subcases.

First of all if we have two vectors, then the argument leading up to equation (5.10) shows that one of them is unneeded if one is a multiple of the other. Thus we can say that two vectors are linearly dependent if one of them is a multiple of the other.

If we have three vectors the situation is more complicated. We noticed in Example 5.9 that $span(\mathbf{v}_1, \mathbf{v}_2, \mathbf{v}_4) = span(\mathbf{v}_1, \mathbf{v}_2)$. Thus $\mathbf{v}_1, \mathbf{v}_2$, and $\mathbf{v}_4$ are linearly dependent. Looking back at the argument leading up to equation (5.11), we see that this is so because $\mathbf{v}_4 = \mathbf{v}_2 - \mathbf{v}_1$ is a linear combination of $\mathbf{v}_1$ and $\mathbf{v}_2$. We are lead to say that a set of three vectors is linearly dependent if one of them is a linear combination of the other two. We would prefer a definition that treated all of the vectors the same. Notice that[4]

$$
\mathbf{v}_4 = \mathbf{v}_2 - \mathbf{v}_1 \quad \Leftrightarrow \quad \mathbf{v}_1 - \mathbf{v}_2 + \mathbf{v}_4 = \mathbf{0}.
$$

The equivalent expression on the right is a linear combination of $\mathbf{v}_1, \mathbf{v}_2$, and $\mathbf{v}_4$ that is equal to the zero vector. It is ***nontrivial*** because at least one of the coefficients is different from 0. Thus we can say that three vectors are linearly dependent if there is a nontrivial linear combination of them that is equal to the zero vector. This definition treats all of the vectors the same. Furthermore, it is readily generalizable to any number of vectors.

In our final step toward the definition, we realize that we would prefer to define linear independence instead of dependence. The result is the following:

---

[4] We will occasionally use the symbols $\Leftrightarrow$ for "is equivalent to" and $\Rightarrow$ for "implies."

DEFINITION 5.13

The vectors $x_1, x_2, \ldots, x_k$ are *linearly independent* if the only linear combination of them that is equal to the zero vector is the trivial one where all of the coefficients are equal to 0. In symbols,

$$\mathbf{0} = c_1 x_1 + c_2 x_2 + \cdots + c_k x_k \Rightarrow c_1 = c_2 = \cdots = c_k = 0.$$

Example 5.14

Which pairs of the vectors v_1, v_2, v_3, and v_4 in Example 5.9 are linearly independent?

A pair of vectors is linearly independent if neither is a multiple of the other. Since $v_3 = 2v_2$, these vectors are linearly dependent. All other pairs are linearly independent, since none are multiples of the other. However, any three of them are linearly dependent. (Why?) ●

Example 5.15

Consider the three vectors

$$v_1 = \begin{pmatrix} -3 \\ 0 \\ 1 \end{pmatrix}, \quad v_2 = \begin{pmatrix} 2 \\ 1 \\ 0 \end{pmatrix}, \quad \text{and} \quad v_3 = \begin{pmatrix} -1 \\ 1 \\ 1 \end{pmatrix}.$$

Since they are not multiples of each other, v_1 and v_2 are linearly independent. The same is true for the pair of vectors $\{v_1, v_3\}$, and for the pair $\{v_2, v_3\}$. However, it is easily verified that $v_1 + v_2 - v_3 = \mathbf{0}$, so the three vectors v_1, v_2, and v_3 are linearly dependent. ●

As Example 5.15 illustrates, it is easy to look at two vectors and decide if they are linearly dependent or independent. It is not so easy to decide if three or more vectors are linearly independent. However, there is an algebraic process for doing so. Remember that according to the definition of matrix-vector product (Definition 1.9), if $X = [x_1, x_2, \ldots, x_k]$ is the matrix with columns $x_1, x_2, \ldots, x_k$, and $c = (c_1, c_2, \ldots, c_k)^T$, then

$$Xc = c_1 x_1 + c_2 x_2 + \cdots + c_k x_k.$$

Thus we will have a nontrivial linear combination of $x_1, x_2, \ldots, x_k$ equal to $\mathbf{0}$ if and only if there is a vector $c \neq \mathbf{0}$ for which $Xc = \mathbf{0}$. Such a vector is in the nullspace of X.

Suppose that $x_1, x_2, \ldots, x_k$ are vectors in $\mathbf{R}^n$. To determine if they are linearly dependent or independent,

1. Form the matrix $X = [x_1, x_2, \ldots, x_k]$ with columns $x_1, x_2, \ldots, x_k$.
2. Find the nullspace $\text{null}(X)$.

   (a) If $\text{null}(X) = \mathbf{0}$, then $x_1, x_2, \ldots, x_k$ are linearly independent.
   (b) If $c = (c_1, c_2, \ldots, c_k)^T$ is a nonzero vector in $\text{null}(X)$, then

$$c_1 x_1 + c_2 x_2 + \cdots + c_k x_k = \mathbf{0},$$

   and $x_1, x_2, \ldots, x_k$ are linearly dependent.

Example 5.16   Consider the following vectors:

$$\mathbf{v}_1 = \begin{pmatrix} 0 \\ -2 \\ 2 \end{pmatrix} \quad \mathbf{v}_2 = \begin{pmatrix} -2 \\ -1 \\ 2 \end{pmatrix} \quad \mathbf{v}_3 = \begin{pmatrix} -2 \\ -3 \\ 4 \end{pmatrix}.$$

Either prove that they are linearly independent or find a nontrivial linear combination that is equal to **0**.

We must examine null(V), where

$$V = [\mathbf{v}_1, \mathbf{v}_2, \mathbf{v}_3] = \begin{pmatrix} 0 & -2 & -2 \\ -2 & -1 & -3 \\ 2 & 2 & 4 \end{pmatrix}.$$

To do so, we reduce the matrix to row echelon form. This can be accomplished by the row operations:

1. Interchange rows 1 and 2.
2. Add row 1 to row 3.
3. Add 1/2 times row 2 to row 3.

We get

$$\begin{pmatrix} -2 & -1 & -3 \\ 0 & -2 & -2 \\ 0 & 0 & 0 \end{pmatrix}.$$

From this, we can compute that the nullspace is spanned by the vector $\mathbf{c} = (-1, -1, 1)^T$. In particular $V\mathbf{c} = \mathbf{0}$. According to the definition of the product of a matrix and a vector, this means that

$$-\mathbf{v}_1 - \mathbf{v}_2 + \mathbf{v}_3 = \mathbf{0}.$$

Consequently, the vectors are linearly dependent and the previous formula gives a nontrivial linear combination that is equal to **0**.   ●

Example 5.17   Consider the following vectors:

$$\mathbf{v}_1 = \begin{pmatrix} 1 \\ -2 \\ 2 \end{pmatrix}, \quad \mathbf{v}_2 = \begin{pmatrix} -2 \\ -1 \\ 2 \end{pmatrix}, \quad \mathbf{v}_3 = \begin{pmatrix} -2 \\ -3 \\ 4 \end{pmatrix}.$$

Either prove that they are linearly independent or find a linear combination that is equal to **0**.

Again we examine null(V), where

$$V = [\mathbf{v}_1, \mathbf{v}_2, \mathbf{v}_3] = \begin{pmatrix} 1 & -2 & -2 \\ -2 & -1 & -3 \\ 2 & 2 & 4 \end{pmatrix}.$$

We can reduce V to row echelon form using the following row operations:

1. Add 2 times row 1 to row 2.
2. Add -2 times row 1 to row 3.
3. Add 6/5 times row 2 to row 3.

After these row operations, V becomes

$$\begin{pmatrix} 1 & -2 & -2 \\ 0 & -5 & -7 \\ 0 & 0 & -2/5 \end{pmatrix}.$$

From this, we see that the nullspace of V contains only the zero vector $\mathbf{0}$. Hence $\mathbf{v}_1$, $\mathbf{v}_2$, and $\mathbf{v}_3$ are linearly independent. ●

Example 5.18 Are the vectors

$$\mathbf{v}_1 = \begin{pmatrix} \ln(2) \\ \pi \end{pmatrix}, \quad \mathbf{v}_2 = \begin{pmatrix} e \\ \cos(3) \end{pmatrix}, \quad \text{and} \quad \mathbf{v}_3 = \begin{pmatrix} \tan(4) \\ e^3 \end{pmatrix}$$

linearly dependent or independent?

Here we have 3 vectors in $\mathbf{R}^2$, with very complicated entries. It is rather difficult to imagine there being a linear combination equal to 0. However, according to our procedure, we are looking for a solution to $V\mathbf{c} = \mathbf{0}$ where $V = [\mathbf{v}_1, \mathbf{v}_2, \mathbf{v}_3]$. This homogeneous system involves two equations in three unknowns, and according to Proposition 4.9, when there are more unknowns than equations in a homogeneous system there is always a nonzero solution. Hence the vectors are linearly dependent. Indeed, any three vectors in $\mathbf{R}^2$ are linearly dependent. By using the full force of Proposition 4.9, we see that if we have m vectors in $\mathbf{R}^n$, and $m > n$, then they are linearly dependent. ●

Bases of a subspace

A minimal spanning set will be called a basis. We will make this more precise in the following definition.

DEFINITION 5.19

> A set of vectors $\{\mathbf{x}_1, \mathbf{x}_2, \ldots, \mathbf{x}_k\}$ in a subspace V is a ***basis*** for V if it has the following properties.
>
> **1.** The vectors span V.
> **2.** The vectors are linearly independent.

The definition is illustrated in Figure 1 and Example 5.9. The vectors $\mathbf{v}_1$ and $\mathbf{v}_2$ are linearly independent. They form a basis for $\mathbf{R}^2$. On the other hand $\mathbf{v}_2$ and $\mathbf{v}_3$ are linearly dependent and $\text{span}(\mathbf{v}_2, \mathbf{v}_3) = \text{span}(\mathbf{v}_2)$. They do not form a basis for $\mathbf{R}^2$ because they do not span $\mathbf{R}^2$. They do not form a basis of $\text{span}(\mathbf{v}_2)$ because they are not linearly independent.

The next result ensures us that bases always exist and gives us a very important property of bases.

PROPOSITION 5.20 Suppose that V is a subspace of $\mathbf{R}^n$.

1. V has a basis.
2. Any two bases for V have the same number of elements. ●

The proof is not outrageously difficult, but it is also not particularly appealing, so we will not present it. Proposition 5.20 enables us to define the ***dimension*** of a subspace V to be the number of elements in any basis for V. The easiest example is $\mathbf{R}^n$. We have already shown that $\mathbf{R}^n = \text{span}\{\mathbf{e}_1, \ldots, \mathbf{e}_n\}$, where $\mathbf{e}_j \in \mathbf{R}^n$ has entries that are all zero, except for a 1 in the jth spot. However, it is easy to see that $\{\mathbf{e}_1, \ldots, \mathbf{e}_n\}$ are linearly independent as well since

$$\mathbf{0} = c_1 \mathbf{e}_1 + \cdots + c_n \mathbf{e}_n = \begin{pmatrix} c_1 \\ \vdots \\ c_n \end{pmatrix}$$

clearly implies $c_1 = \cdots = c_n = 0$. Thus the dimension of $\mathbf{R}^n$ is n. We will write this as dim $\mathbf{R}^n = n$.

One point needs to be emphasized immediately: *Bases are not unique.* There are many ways to choose a basis. We need only look back at Example 5.9 and Figure 1 for several bases of $\mathbf{R}^2$. The pairs $\{\mathbf{v}_1, \mathbf{v}_2\}$, $\{\mathbf{v}_1, \mathbf{v}_3\}$, $\{\mathbf{v}_1, \mathbf{v}_4\}$, $\{\mathbf{v}_2, \mathbf{v}_4\}$, and $\{\mathbf{v}_3, \mathbf{v}_4\}$ are all bases for $\mathbf{R}^2$.

Finding a basis for a nullspace

In Example 4.22, we calculated that the nullspace of the matrix A in (4.23) consists of the line in $\mathbf{R}^2$ parameterized by $t(1, -2)^T$. Thus, the nullspace is the span of $(1, -2)^T$, and since a single nonzero vector is always linearly independent, we see that the single vector $(1, -2)^T$ is a basis for the nullspace of A. Thus we can say that $\text{null}(A)$ is the subspace of $\mathbf{R}^2$ with basis $(1, -2)^T$.

In Example 4.25 we found that the nullspace of the matrix B in equation (4.26) is the plane parameterized by $s(-3, 1, 0)^T + t(2, 0, 1)^T$. This is another way of saying that $\text{null}(B) = \text{span}((-3, 1, 0)^T, (2, 0, 1)^T)$. The vectors $(-3, 1, 0)^T$ and $(2, 0, 1)^T$ are not multiples of each other, so they are linearly independent. We can therefore say that $\text{null}(B)$ is the subspace of $\mathbf{R}^3$ with basis $(-3, 1, 0)^T$ and $(2, 0, 1)^T$.

We have now reached our goal of finding a good way of describing a subspace of $\mathbf{R}^n$. We simply provide a basis. Examples 4.22 and 4.25 illustrate the general process of finding a basis for the nullspace of a matrix. We find a parameterization for the nullspace following the procedure developed in Section 7.3. The parameterization expresses the nullspace as the span of a few vectors, one for each free variable. These vectors are automatically linearly independent, and form a basis for the nullspace. In particular, we see that the dimension of the nullspace is equal to the number of free variables. Let's look at another example.

Example 5.21   Find the nullspace of the matrix

$$C = \begin{pmatrix} 1 & -1 & 0 & 2 & 0 \\ 0 & 0 & 1 & 2 & -1 \end{pmatrix}.$$

The matrix C is already in reduced row echelon form. The variables x_2, x_4, and x_5 are free. We set $x_2 = s$, $x_4 = t$, and $x_5 = u$ and solve the first equation (row) for the pivot variable x_1 and the second equation (row) for the pivot variable x_3. We find that $\text{null}(C)$ is parameterized by $s\mathbf{v}_1 + t\mathbf{v}_2 + u\mathbf{v}_3$, where

$$\mathbf{v}_1 = \begin{pmatrix} 1 \\ 1 \\ 0 \\ 0 \\ 0 \end{pmatrix}, \quad \mathbf{v}_2 = \begin{pmatrix} -2 \\ 0 \\ -2 \\ 1 \\ 0 \end{pmatrix}, \quad \text{and} \quad \mathbf{v}_3 = \begin{pmatrix} 0 \\ 0 \\ 1 \\ 0 \\ 1 \end{pmatrix}.$$

Thus $\text{null}(C) = \text{span}(\mathbf{v}_1, \mathbf{v}_2, \mathbf{v}_3)$.

To show that $\mathbf{v}_1$, $\mathbf{v}_2$, and $\mathbf{v}_3$ are linearly independent, consider the linear combination

$$c_1\mathbf{v}_1 + c_2\mathbf{v}_2 + c_3\mathbf{v}_3 = \begin{pmatrix} c_1 - 2c_2 \\ c_1 \\ -2c_2 + c_3 \\ c_2 \\ c_3 \end{pmatrix}.$$

If this linear combination is equal to the zero vector, $\mathbf{0}$, then all of the entries have to be equal to 0. When we focus on the entries corresponding to the free variables x_2, x_4, and x_5, we see that $c_1 = c_2 = c_3 = 0$. Thus the linear combination is the trivial one, and we conclude that $\mathbf{v}_1$, $\mathbf{v}_2$, and $\mathbf{v}_3$ are linearly independent. Therefore, $\text{null}(C)$ is the subspace of $\mathbf{R}^5$ with basis $\mathbf{v}_1$, $\mathbf{v}_2$, and $\mathbf{v}_3$. This argument for proving linear independence of the vectors spanning a nullspace works in general.

Since there are three vectors in the basis, the dimension of $\text{null}(C)$ is 3. ●

Finding a basis for a span

If we know a spanning set for a subspace, we can always find a basis by eliminating unneeded vectors from the spanning set. The following example is typical.

Example 5.22 Find a basis for $V = \text{span}(\mathbf{v}_1, \mathbf{v}_2, \mathbf{v}_3)$ where

$$\mathbf{v}_1 = \begin{pmatrix} 0 \\ -2 \\ 2 \end{pmatrix} \quad \mathbf{v}_2 = \begin{pmatrix} -2 \\ -1 \\ 2 \end{pmatrix} \quad \mathbf{v}_3 = \begin{pmatrix} -2 \\ -3 \\ 4 \end{pmatrix}.$$

These are the vectors from Example 5.16. We discovered there that $-\mathbf{v}_1 - \mathbf{v}_2 + \mathbf{v}_3 = \mathbf{0}$. Hence the three vectors are linearly dependent and cannot be a basis. We can use the fact that $\mathbf{v}_3 = \mathbf{v}_1 + \mathbf{v}_2$ to eliminate $\mathbf{v}_3$ from the spanning set. To see how this is done, suppose $\mathbf{v} \in V$. Then there are constants a_1, a_2, and a_3 such that $\mathbf{v} = a_1\mathbf{v}_1 + a_2\mathbf{v}_2 + a_3\mathbf{v}_3$. Hence

$$\begin{aligned} \mathbf{v} &= a_1\mathbf{v}_1 + a_2\mathbf{v}_2 + a_3\mathbf{v}_3 \\ &= a_1\mathbf{v}_1 + a_2\mathbf{v}_2 + a_3(\mathbf{v}_1 + \mathbf{v}_2) \\ &= (a_1 + a_3)\mathbf{v}_1 + (a_2 + a_3)\mathbf{v}_2. \end{aligned}$$

This implies that $\mathbf{v} \in \text{span}(\mathbf{v}_1, \mathbf{v}_2)$, and so $V = \text{span}(\mathbf{v}_1, \mathbf{v}_2)$. Since $\mathbf{v}_1$ and $\mathbf{v}_2$ are not multiples of each other, they are linearly independent. Therefore, $\{\mathbf{v}_1, \mathbf{v}_2\}$ is a basis for V, and we find that the dimension of V is 2. ●

EXERCISES

Consider the following collection of vectors, which you are to use in Exercises 1–8.

$\mathbf{u}_1 = (1, -2)^T$, $\quad \mathbf{u}_2 = (3, 0)^T$, $\quad \mathbf{u}_3 = (2, -4)^T$

$\mathbf{v}_1 = (1, -4, 4)^T$, $\quad \mathbf{v}_2 = (0, -2, 1)^T$, $\quad \mathbf{v}_3 = (1, -2, 3)^T$

In each exercise, if the given vector $\mathbf{w}$ lies in the span, provide a specific linear combination of the spanning vectors that equals the given vector; otherwise, provide a specific numerical argument why the given vector does not lie in the span.

1. Is the vector $\mathbf{w} = (5, -2)^T$ in the span$\{\mathbf{u}_1, \mathbf{u}_2\}$?
2. Is the vector $\mathbf{w} = (3, -6)^T$ in the span$\{\mathbf{u}_1, \mathbf{u}_3\}$?
3. Is the vector $\mathbf{w} = (3, -3)^T$ in the span$\{\mathbf{u}_1, \mathbf{u}_3\}$?
4. Is the vector $\mathbf{w} = (-3, 2, 7)^T$ in the span$\{\mathbf{v}_1, \mathbf{v}_2\}$?
5. Is the vector $\mathbf{w} = (1, 4, 1)^T$ in the span$\{\mathbf{v}_1, \mathbf{v}_3\}$?
6. Is the vector $\mathbf{w} = (1, 1, 1)^T$ in the span$\{\mathbf{v}_1, \mathbf{v}_2, \mathbf{v}_3\}$?
7. Is the vector $\mathbf{w} = (1, 0, 2)^T$ in the span$\{\mathbf{v}_1, \mathbf{v}_2, \mathbf{v}_3\}$?

8. Is the vector $\mathbf{w} = (-7, 22, -25)^T$ in the span$\{\mathbf{v}_1, \mathbf{v}_2, \mathbf{v}_3\}$?

9. Let $\mathbf{v}_1 = (1, -2)^T$ and $\mathbf{v}_2 = (2, 3)^T$. Show that span$\{\mathbf{v}_1, \mathbf{v}_2\} = \mathbf{R}^2$ by showing that any vector $\mathbf{w} = (w_1, w_2)^T$ can be written as a linear combination of $\mathbf{v}_1$ and $\mathbf{v}_2$. *Note:* Find a specific linear combination (in terms of w_1 and w_2) of $\mathbf{v}_1$ and $\mathbf{v}_2$ that equals $\mathbf{w}$.

10. Let $\mathbf{v}_1 = (0, -1, -2)^T$, $\mathbf{v}_2 = (-2, 1, -4)^T$, and $\mathbf{v}_3 = (-2, -2, 0)^T$. Show that span$\{\mathbf{v}_1, \mathbf{v}_2, \mathbf{v}_3\} = \mathbf{R}^3$ by showing that any vector $\mathbf{w} = (w_1, w_2, w_3)^T$ can be written as a linear combination of $\mathbf{v}_1, \mathbf{v}_2$, and $\mathbf{v}_3$. *Note:* Find a specific linear combination (in terms of w_1, w_2, and w_3) of $\mathbf{v}_1, \mathbf{v}_2$, and $\mathbf{v}_3$ that equals $\mathbf{w}$.

In Exercises 11–16, each set of vectors presented is linearly dependent. Use the technique of Example 5.16 to find a nontrivial linear combination of the given vectors that equal the zero vector. Check your solution.

11. $\mathbf{v}_1 = (1, 1, -2)^T$, $\mathbf{v}_2 = (1, 2, 2)^T$, and $\mathbf{v}_3 = (3, 4, -2)^T$

12. $\mathbf{v}_1 = (2, -3, 3)^T$, $\mathbf{v}_2 = (5, -2, 5)^T$, and $\mathbf{v}_3 = (-3, -1, -2)^T$

13. $\mathbf{v}_1 = (1, -2, -2)^T$, $\mathbf{v}_2 = (0, 1, 5)^T$, and $\mathbf{v}_3 = (2, -1, 11)^T$

14. $\mathbf{v}_1 = (-3, -2, -1)^T$, $\mathbf{v}_2 = (2, 0, 2)^T$, and $\mathbf{v}_3 = (7, 2, 5)^T$

15. $\mathbf{v}_1 = (1, 2, -2, 0)^T$, $\mathbf{v}_2 = (2, 0, 2, 3)^T$, $\mathbf{v}_3 = (-2, 4, -8, -6)^T$, and $\mathbf{v}_4 = (6, -4, 12, 12)^T$

16. $\mathbf{v}_1 = (1, 1, 2, -1)^T$, $\mathbf{v}_2 = (3, 1, 0, -1)^T$, $\mathbf{v}_3 = (2, 0, -2, 0)^T$, and $\mathbf{v}_4 = (1, 3, 8, -3)^T$

For each of the sets of vectors in Exercises 17–24, use the technique demonstrated in Examples 5.16 and 5.17 either to show that they are linearly independent or find a nontrivial linear combination that is equal to $\mathbf{0}$.

17. $\mathbf{v}_1 = (1, 2)^T$ and $\mathbf{v}_2 = (-1, 3)^T$

18. $\mathbf{v}_1 = (-2, 3)^T$ and $\mathbf{v}_2 = (2, -6)^T$

19. $\mathbf{v}_1 = (-1, 7, 7)^T$ and $\mathbf{v}_2 = (-3, 7, -4)^T$

20. $\mathbf{v}_1 = (-8, 9, -6)^T$ and $\mathbf{v}_2 = (-2, 0, 7)^T$

21. $\mathbf{v}_1 = (-1, 7, 7)^T$, $\mathbf{v}_2 = (-3, 7, -4)^T$, and $\mathbf{v}_3 = (-4, -14, 23)^T$

22. $\mathbf{v}_1 = (-8, 9, -6)^T$, $\mathbf{v}_2 = (-2, 0, 7)^T$, and $\mathbf{v}_3 = (8, -18, 40)^T$

23. $\mathbf{v}_1 = (-1, 7, 7)^T$, $\mathbf{v}_2 = (-3, 8, -4)^T$, and $\mathbf{v}_3 = (-4, -14, 23)^T$

24. $\mathbf{v}_1 = (-8, 9, -6)^T$, $\mathbf{v}_2 = (-2, -1, 7)^T$, and $\mathbf{v}_3 = (8, -18, 40)^T$

Use the technique of Example 5.21 to find a basis for the nullspace of the matrices given in Exercises 25–32.

25. $\begin{pmatrix} 2 & -1 \end{pmatrix}$ **26.** $\begin{pmatrix} -3 & 5 \end{pmatrix}$

27. $\begin{pmatrix} 4 & 4 \\ -2 & -2 \end{pmatrix}$ **28.** $\begin{pmatrix} 4 & 4 \\ -2 & -1 \end{pmatrix}$

29. $\begin{pmatrix} 1 & 1 & 1 \\ -5 & -2 & -5 \\ 1 & 0 & 1 \end{pmatrix}$ **30.** $\begin{pmatrix} -3 & 8 & -11 \\ -4 & 10 & -14 \\ -2 & 5 & -7 \end{pmatrix}$

31. $\begin{pmatrix} 2 & -1 & 0 & 1 \\ -1 & 1 & 1 & 0 \\ 1 & 1 & 3 & 2 \\ -3 & 3 & 3 & 0 \end{pmatrix}$ **32.** $\begin{pmatrix} -8 & 14 & -24 & 14 \\ 4 & -10 & 18 & -10 \\ 4 & -8 & 14 & -8 \\ -2 & 5 & -9 & 5 \end{pmatrix}$

Use the technique shown in Example 5.22 to help find a basis for the span of the sets of vectors in Exercises 33–40. What is the dimension of the span?

33. The set in Exercise 17   **34.** The set in Exercise 18

35. The set in Exercise 19   **36.** The set in Exercise 20

37. The set in Exercise 21   **38.** The set in Exercise 22

39. The set in Exercise 23   **40.** The set in Exercise 24

In Exercises 41–44, give a geometric description of the span of the given vectors in the given space.

41. $\mathbf{v}_1 = (1, -2)^T$, $\mathbf{v}_2 = (2, -4)^T$, in $\mathbf{R}^2$

42. $\mathbf{v}_1 = (2, -2)^T$, $\mathbf{v}_2 = (2, -4)^T$, in $\mathbf{R}^2$

43. $\mathbf{v}_1 = (-1, 3, 3)^T$, $\mathbf{v}_2 = (-3, -2, 2)^T$, and $\mathbf{v}_3 = (4, -1, -5)^T$ in $\mathbf{R}^3$

44. $\mathbf{v}_1 = (1, 2, 3)^T$, $\mathbf{v}_2 = (2, 4, 6)^T$, and $\mathbf{v}_3 = (-5, -10, -15)^T$ in $\mathbf{R}^3$

7.6 Square Matrices

In this section we will focus on square matrices. These arise as the coefficient matrices for systems of n equations in n unknowns. For purposes of application to differential equations square matrices are the most important.

We will discuss two properties of square matrices. We define a nonsingular matrix in Definition 6.1 and an invertible matrix in Definition 6.10. These definitions are quite different, yet one of the important results of this section is that they are equivalent (Proposition 6.11).

We will examine these properties of matrices using the reduction to row echelon form described in Section 7.3. The key observation is that any system of equations can be reduced to an equivalent system that is in row echelon form.

Singular and nonsingular matrices

Our first task is discovering when a system of n equations in n unknowns can be solved for every choice of the right-hand side.

DEFINITION 6.1

> The $n \times n$ matrix A is said to be ***nonsingular*** if we can solve the system $A\mathbf{x} = \mathbf{b}$ for any choice of the vector $\mathbf{b}$ in $\mathbf{R}^n$. Otherwise the matrix is ***singular***.

The augmented matrix for the system $A\mathbf{x} = \mathbf{b}$ is $M = [A, \mathbf{b}]$. When we transform this to row echelon form, we get a matrix $Q = [R, \mathbf{v}]$, which is the augmented matrix for the equivalent system $R\mathbf{x} = \mathbf{v}$. Notice that R is a row echelon transform of the matrix A.

Since the vector $\mathbf{b}$ is arbitrary, so is the vector $\mathbf{v}$. In fact, since each row operation can be reversed, we can start with the equation $R\mathbf{x} = \mathbf{v}$ and then apply the inverse of the row operations we used to go from M to Q in the opposite order to get a vector $\mathbf{b}$ that will be transformed into $\mathbf{v}$ by those row operations.

Thus the matrix A will be nonsingular if and only if the row echelon matrix R is nonsingular. For this to be true, we must be able to solve the equation $R\mathbf{x} = \mathbf{v}$ for any right-hand side. Choose the vector $\mathbf{v}$ with its last component $v_n \neq 0$. If the bottom row of R is all zeros, we get the equation $0 = v_n$, which, according to Proposition 3.17, cannot be solved. Consequently, there must be a pivot on the bottom row of R. Since R is in row echelon form, there must therefore be a pivot on each row of R. Since R has n rows and n columns, and there is at most one pivot in each column of the row echelon matrix R, each column must contain a pivot. Since each pivot must lie to the right of those above it, all of the pivots must lie on the diagonal of R, and all of the diagonal elements must be nonzero. If this is true, there are no free variables and no matter what the right-hand side $\mathbf{v}$, there is a solution, and furthermore, since there are no free variables, the solution is unique. We have discovered a characterization of nonsingularity, which we will state as a proposition.

PROPOSITION 6.2

An $n \times n$ matrix A is nonsingular if and only if an equivalent matrix in row echelon form has only nonzero entries along the diagonal. ●

The characterization of nonsingularity in Proposition 6.2 is useful because a determination can be made after some calculation that is not too difficult. However, Proposition 6.2 is a preliminary result. We will improve upon it in Theorem 7.2.

In the proof of Proposition 6.2, we concluded that if A is nonsingular, then when put into row echelon form, there are no free variables. Thus not only does a solution exist, but since there are no free variables, the solution is unique. We will state this as another proposition.

PROPOSITION 6.3

If A is nonsingular, then the equation $A\mathbf{x} = \mathbf{b}$ has a unique solution $\mathbf{x}$ for every choice of the vector $\mathbf{b}$. ●

Let's look at some examples.

Example 6.4 Is the matrix

$$A = \begin{pmatrix} -3 & 6 & 8 \\ -1 & 2 & 1 \\ 0 & 0 & 1 \end{pmatrix}$$

singular or nonsingular?

If we first add $-1/3$ times the first row to the second, and then add $3/5$ times the second row to the third, we get an equivalent matrix in row echelon form

$$\begin{pmatrix} -3 & 6 & 8 \\ 0 & 0 & -5/3 \\ 0 & 0 & 0 \end{pmatrix}.$$

The presence of zeros on the diagonal means that A is singular.   ●

Example 6.5   Is the matrix

$$A = \begin{pmatrix} 3 & -4 & -8 \\ 2 & -3 & -10 \\ 0 & 0 & 2 \end{pmatrix}$$

singular or nonsingular?

If we add $-2/3$ times the first row to the second, we get the equivalent matrix in row echelon form:

$$\begin{pmatrix} 3 & -4 & -8 \\ 0 & -1/3 & -14/3 \\ 0 & 0 & 2 \end{pmatrix}.$$

Since all of the entries on the diagonal are nonzero, A is nonsingular.   ●

Let's return to the homogeneous equation $A\mathbf{x} = \mathbf{0}$ for an $n \times n$ matrix. From our previous discussion, we know that there is a nonzero solution if and only if there is a free variable associated with a row reduced transform of the matrix A. If there is a free variable, then the diagonal entry in that column must be zero. Then by Proposition 6.2, this can happen only if the matrix is singular. This result is a proposition that will be very important in the application of linear algebra to differential equations.

PROPOSITION 6.6   If A is an $n \times n$ matrix, then the homogeneous equation $A\mathbf{x} = \mathbf{0}$ has a nonzero solution if and only if the matrix is singular.   ●

It is interesting to look at the reduced row echelon form for nonsingular $n \times n$ matrices. For these, it is particularly true that the reduced row echelon form makes it quite easy to find the solution of $A\mathbf{x} = \mathbf{b}$. The reduced row echelon form of the augmented matrix $M = [A, \mathbf{b}]$ is $Q = [R, \mathbf{r}]$, where R is the reduced row echelon form of the nonsingular matrix A. By Proposition 6.2, R must have pivots along the diagonal and since R is reduced, these must be ones. Thus $R = I$, the identity matrix, and the equation $A\mathbf{x} = \mathbf{b}$ is equivalent to $I\mathbf{x} = \mathbf{r}$, or $\mathbf{x} = \mathbf{r}$. Thus the last column vector in the reduced row echelon form of M is the solution to the equation. This is shown by the following example.

Example 6.7   Solve the system of equations $A\mathbf{x} = \mathbf{b}$, where

$$A = \begin{pmatrix} -11 & 2 & 20 & -11 \\ -6 & -3 & 6 & 3 \\ -1 & 19 & 10 & -10 \\ -8 & 26 & 8 & 8 \end{pmatrix} \quad \text{and} \quad \mathbf{b} = \begin{pmatrix} 84 \\ 9 \\ 51 \\ 0 \end{pmatrix}. \tag{6.8}$$

The reduced row echelon form of the augmented matrix

$$\begin{pmatrix} -11 & 2 & 20 & -11 & 84 \\ -6 & -3 & 6 & 3 & 9 \\ -1 & 19 & 10 & -10 & 51 \\ -8 & 26 & 8 & 8 & 0 \end{pmatrix}$$

is

$$
\begin{pmatrix}
1 & 0 & 0 & 0 & -1 \\
0 & 1 & 0 & 0 & 0 \\
0 & 0 & 1 & 0 & 2 \\
0 & 0 & 0 & 1 & -3
\end{pmatrix}. \tag{6.9}
$$

Hence the solution is the last column in (6.9),

$$
\mathbf{x} = \begin{pmatrix} -1 \\ 0 \\ 2 \\ -3 \end{pmatrix}.
$$

Invertible matrices

The process illustrated in Example 6.7 has one very useful application. First we give a definition.

DEFINITION 6.10

An $n \times n$ matrix A is **invertible** if there is an $n \times n$ matrix B such that $AB = I$ and $BA = I$. A matrix B with this property is called an **inverse** of A.

Suppose that both B_1 and B_2 are inverses of A. Then

$$
B_1 = B_1 I = B_1(AB_2) = (B_1 A)B_2 = I B_2 = B_2.
$$

Consequently, the matrix A has at most one inverse. If it exists, it will be denoted by A^{-1}.

PROPOSITION 6.11   An $n \times n$ matrix A is invertible if and only if it is nonsingular.

Proof   First suppose that A is invertible. Given any vector $\mathbf{b}$, the vector $\mathbf{x} = A^{-1}\mathbf{b}$ satisfies $A\mathbf{x} = A(A^{-1}\mathbf{b}) = (AA^{-1})\mathbf{b} = I\mathbf{b} = \mathbf{b}$. Hence A is nonsingular.

Now suppose that A is nonsingular. The identity matrix is $I = [\mathbf{e}_1, \mathbf{e}_2, \ldots, \mathbf{e}_n]$, where $\mathbf{e}_j$ is the column vector with entries that are all zeros except for a 1 in the jth position. Since A is nonsingular, we can find vectors $\mathbf{y}_j$ such that $A\mathbf{y}_j = \mathbf{e}_j$ for $j = 1, 2, \ldots, n$. Then the matrix $B = [\mathbf{y}_1, \mathbf{y}_2, \ldots, \mathbf{y}_n]$ satisfies

$$
AB = A[\mathbf{y}_1, \mathbf{y}_2, \ldots, \mathbf{y}_n] = [A\mathbf{y}_1, A\mathbf{y}_2, \ldots, A\mathbf{y}_n] = [\mathbf{e}_1, \mathbf{e}_2, \ldots, \mathbf{e}_n] = I.
$$

We must also show that $BA = I$. This requires a little work that is not too interesting, so we will not present it.

We can use the last half of this proof to describe a way of computing A^{-1}. We must find vectors $\mathbf{y}_j$ such that $A\mathbf{y}_j = \mathbf{e}_j$ for $j = 1, 2, \ldots, n$. Then $A^{-1} = [\mathbf{y}_1, \mathbf{y}_2, \ldots, \mathbf{y}_n]$. We can calculate all of the solutions $\mathbf{y}_j$ at once. Instead of augmenting A with just one column, we augment it with n columns. Set $M = [A, I]$. Then the reduced row echelon form of M is $[I, A^{-1}]$. For example, to find the inverse of the matrix

$$
A = \begin{pmatrix}
0 & 0 & -3 & -5 \\
1 & -2 & -6 & -7 \\
-1 & 3 & 7 & 7 \\
0 & -1 & -3 & -3
\end{pmatrix},
$$

we form the augmented matrix

$$M = [A, I] = \begin{pmatrix} 0 & 0 & -3 & -5 & 1 & 0 & 0 & 0 \\ 1 & -2 & -6 & -7 & 0 & 1 & 0 & 0 \\ -1 & 3 & 7 & 7 & 0 & 0 & 1 & 0 \\ 0 & -1 & -3 & -3 & 0 & 0 & 0 & 1 \end{pmatrix}.$$

The reduced row echelon form of M is

$$\begin{pmatrix} 1 & 0 & 0 & 0 & -2 & 4 & 3 & 1 \\ 0 & 1 & 0 & 0 & -3 & 6 & 6 & 5 \\ 0 & 0 & 1 & 0 & 3 & -5 & -5 & -5 \\ 0 & 0 & 0 & 1 & -2 & 3 & 3 & 3 \end{pmatrix}.$$

Hence

$$A^{-1} = \begin{pmatrix} -2 & 4 & 3 & 1 \\ -3 & 6 & 6 & 5 \\ 3 & -5 & -5 & -5 \\ -2 & 3 & 3 & 3 \end{pmatrix}.$$

EXERCISES

1. Show that the system

$$\begin{pmatrix} 1 & -2 \\ 2 & -4 \end{pmatrix} \begin{pmatrix} x_1 \\ x_2 \end{pmatrix} = \begin{pmatrix} b_1 \\ b_2 \end{pmatrix}$$

has solutions only if $\mathbf{b} = (b_1, b_2)^T$ lies on the line $-2b_1 + b_2 = 0$. Is the coefficient matrix singular or non-singular? Explain.

2. Show that the system

$$\begin{pmatrix} 1 & 2 \\ -2 & 3 \end{pmatrix} \begin{pmatrix} x_1 \\ x_2 \end{pmatrix} = \begin{pmatrix} b_1 \\ b_2 \end{pmatrix}$$

has solutions for all values of b_1 and b_2. Is the coefficient matrix singular or nonsingular? Explain.

3. Show that the system

$$\begin{pmatrix} 3 & 3 & -3 \\ -1 & -1 & 1 \\ 3 & 5 & -1 \end{pmatrix} \begin{pmatrix} x_1 \\ x_2 \\ x_3 \end{pmatrix} = \begin{pmatrix} b_1 \\ b_2 \\ b_3 \end{pmatrix}$$

has solutions only if $\mathbf{b} = (b_1, b_2, b_3)^T$ lies on the plane $b_1 + 3b_2 = 0$. Is the coefficient matrix singular or nonsingular? Explain.

Using hand calculations only (no computers or calculators), use the technique demonstrated in Examples 6.4 and 6.5 (predicated upon Proposition 6.2) to determine if matrices given in Exercises 4–11 are singular or nonsingular.

4. $\begin{pmatrix} 2 & -1 \\ 4 & -2 \end{pmatrix}$

5. $\begin{pmatrix} 1 & 2 \\ 3 & -4 \end{pmatrix}$

6. $\begin{pmatrix} 1 & 0 & 1 \\ 0 & 3 & 3 \\ -2 & 3 & 1 \end{pmatrix}$

7. $\begin{pmatrix} 1 & 0 & -1 \\ -2 & 3 & 3 \\ -2 & 3 & 1 \end{pmatrix}$

8. $\begin{pmatrix} 2 & 0 & 2 \\ -1 & 1 & 3 \\ 1 & -4 & 3 \end{pmatrix}$

9. $\begin{pmatrix} 2 & 1 & -1 \\ -1 & -3 & -2 \\ -3 & -2 & 1 \end{pmatrix}$

10. $\begin{pmatrix} -2 & 1 & 0 \\ -2 & 4 & 2 \\ 2 & -2 & 1 \end{pmatrix}$

11. $\begin{pmatrix} -1 & -1 & 1 \\ 0 & -2 & 4 \\ 3 & 0 & 3 \end{pmatrix}$

In Exercises 12–19, find all solutions of the homogeneous system $A\mathbf{x} = \mathbf{0}$ for the given coefficient matrix. Does the system have solutions other than the zero vector? Use Proposition 6.6 to determine whether the matrix A is singular or nonsingular.

12. $A = \begin{pmatrix} 1 & 2 \\ 1 & 1 \end{pmatrix}$

13. $A = \begin{pmatrix} 1 & -2 \\ 2 & -4 \end{pmatrix}$

14. $A = \begin{pmatrix} 1 & 1 & 1 \\ 1 & 1 & 0 \\ 1 & 0 & 0 \end{pmatrix}$

15. $A = \begin{pmatrix} 1 & 1 & 2 \\ 1 & 0 & 1 \\ 1 & 1 & 2 \end{pmatrix}$

16. $A = \begin{pmatrix} 1 & 1 & 1 \\ 0 & 1 & 1 \\ 0 & 0 & 0 \end{pmatrix}$

17. $A = \begin{pmatrix} 0 & -1 & -2 \\ -5 & 2 & -1 \\ -4 & 2 & 0 \end{pmatrix}$

18. $A = \begin{pmatrix} -2 & -2 & -2 & 0 \\ -1 & -4 & 2 & -3 \\ 3 & 1 & 5 & -2 \\ 3 & 3 & 3 & 0 \end{pmatrix}$

19. $A = \begin{pmatrix} 0 & 1 & 3 & 0 \\ -2 & 1 & -3 & 1 \\ -1 & -2 & 3 & 0 \\ 2 & -1 & -1 & 2 \end{pmatrix}$

In Exercises 20–27, which of the matrices are singular? If a matrix is nonsingular, find its inverse.

20. $A = \begin{pmatrix} 0 & 0 \\ 1 & 1 \end{pmatrix}$

21. $A = \begin{pmatrix} 0 & -4 \\ -1 & 2 \end{pmatrix}$

22. $A = \begin{pmatrix} 1 & 0 & 1 \\ 0 & 1 & 1 \\ 0 & 0 & 0 \end{pmatrix}$

23. $A = \begin{pmatrix} 1 & 1 & 1 \\ 0 & 1 & 1 \\ 0 & 0 & 1 \end{pmatrix}$

24. $A = \begin{pmatrix} 1 & 2 & 0 \\ 0 & 0 & 1 \\ 0 & 2 & 1 \end{pmatrix}$

25. $A = \begin{pmatrix} 1 & 2 & -3 \\ 0 & 0 & 0 \\ 0 & 1 & 1 \end{pmatrix}$

26. $A = \begin{pmatrix} 0 & -3 & -1 & 2 \\ -3 & 0 & 0 & 0 \\ 2 & 1 & -2 & -2 \\ -3 & -1 & 3 & 4 \end{pmatrix}$

27. $A = \begin{pmatrix} 3 & -1 & -3 & -1 \\ 0 & -3 & -4 & 1 \\ -2 & 1 & 2 & 1 \\ 1 & 1 & 3 & -3 \end{pmatrix}$

30. $x_1 \begin{pmatrix} -1 \\ 2 \end{pmatrix} + x_2 \begin{pmatrix} -3 \\ 6 \end{pmatrix} = \begin{pmatrix} 4 \\ -8 \end{pmatrix}$

31. $x_1 \begin{pmatrix} 1 \\ 2 \end{pmatrix} + x_2 \begin{pmatrix} 2 \\ 1 \end{pmatrix} = \begin{pmatrix} 1 \\ 0 \end{pmatrix}$

32. $\begin{pmatrix} 1 & 1 & 2 \\ 1 & 1 & -1 \\ 1 & -2 & 2 \end{pmatrix} \begin{pmatrix} x_1 \\ x_2 \\ x_3 \end{pmatrix} = \begin{pmatrix} 1 \\ 1 \\ 1 \end{pmatrix}$

33. $\begin{pmatrix} 1 & 0 & 3 \\ -1 & 1 & -1 \\ 0 & 2 & 4 \end{pmatrix} \begin{pmatrix} x_1 \\ x_2 \\ x_3 \end{pmatrix} = \begin{pmatrix} 1 \\ 1 \\ 1 \end{pmatrix}$

34. For which values of x is the matrix

$$\begin{pmatrix} 1 & 3 & -2 \\ 2 & 8 & x \\ 0 & 8 & 5 \end{pmatrix}$$

invertible?

In Exercises 28–33, without actually solving, which systems have unique solutions? Explain.

28. $\begin{aligned} x_1 + 2x_2 &= 4 \\ x_1 - x_2 &= 6 \end{aligned}$

29. $\begin{aligned} x_1 + 2x_2 &= 4 \\ 2x_1 + 4x_2 &= 8 \end{aligned}$

35. List as many properties as you can of an invertible matrix.

36. List as many properties as you can of a nonsingular matrix.

7.7 Determinants

There is a question that is still unanswered. Given a matrix A, is there an easy way to tell if its nullspace is nontrivial? For a square matrix, we found a partial answer in Section 7.6. According to Proposition 6.6, a square matrix has a nontrivial nullspace if and only if it is singular. However, we do not as yet have an easy way to tell if a matrix is singular or nonsingular. Proposition 6.2 tells us that A is nonsingular if and only if when it is transformed into row echelon form, all of the diagonal entries are nonzero, but this is not an adequate answer. A reasonable answer is provided by the determinant.

Let's look at the 2×2 case

$$A = \begin{pmatrix} a & b \\ c & d \end{pmatrix}.$$

If we assume that $a \neq 0$ and put A into row echelon form with a row operation, we get

$$\begin{pmatrix} a & b \\ 0 & d - bc/a \end{pmatrix}.$$

The diagonal entries will all be nonzero if and only if their product $a \cdot (d - bc/a) = ad - bc \neq 0$. You will recognize that $ad - bc$ is the determinant of A. Thus A is nonsingular if and only if $\det(A) \neq 0$.

Carrying through this calculation for the 3×3 matrix

$$A = \begin{pmatrix} a_{11} & a_{12} & a_{13} \\ a_{21} & a_{22} & a_{23} \\ a_{31} & a_{32} & a_{33} \end{pmatrix}$$

is tedious, but we again end up with the result that A is nonsingular if and only if $\det(A) \neq 0$. However, now the determinant has the more complicated form

$$\det(A) = a_{11}a_{22}a_{33} - a_{11}a_{23}a_{32} - a_{12}a_{21}a_{33} \\ + a_{12}a_{23}a_{31} - a_{13}a_{22}a_{31} + a_{13}a_{21}a_{32}. \tag{7.1}$$

In what follows, we will define the determinant of an arbitrary square matrix. For our purposes, the most important property of the determinant is the following result, which we have already stated for 2×2 and 3×3 matrices.

THEOREM 7.2   An $n \times n$ matrix A is nonsingular if and only if $\det(A) \neq 0$.   ∎

We will be even more specific about the application we need by stating a corollary that is related to Proposition 6.6.

COROLLARY 7.3   If A is an $n \times n$ matrix, then the homogeneous equation $A\mathbf{x} = \mathbf{0}$ has a nonzero solution if and only if $\det(A) = 0$.   ●

Using the terminology we have defined in Section 7.4, the matrix A has a non-trivial nullspace if and only if $\det(A) = 0$.

A geometric interpretation of the determinant

Suppose we have two 2-vectors $\mathbf{v}_1$ and $\mathbf{v}_2$ and we let

$$A = [\mathbf{v}_1, \ \mathbf{v}_2]$$

be the matrix with $\mathbf{v}_1$ and $\mathbf{v}_2$ as its column vectors. The vectors $\mathbf{v}_1$ and $\mathbf{v}_2$ determine the parallelogram P shown in Figure 1. It can be shown that the area of P is equal to $|\det(A)|$. Notice that A is singular if and only if the vectors $\mathbf{v}_1$ and $\mathbf{v}_2$ are linearly dependent. If they are, the vectors point in the same direction and the parallelogram degenerates to one that has area equal to 0.

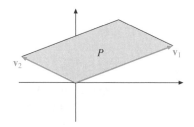

Figure 1. The area of the parallelogram P is equal to $|\det([\mathbf{v}_1, \mathbf{v}_2])|$.

There is a geometric interpretation in dimension 3 as well. Suppose that we have three 3-vectors, $\mathbf{v}_1$, $\mathbf{v}_2$, and $\mathbf{v}_3$. These generate the parallelepiped P shown in Figure 2. Let

$$A = [\mathbf{v}_1, \ \mathbf{v}_2, \ \mathbf{v}_3]$$

be the matrix with columns $\mathbf{v}_1$, $\mathbf{v}_2$, and $\mathbf{v}_3$. It can be shown that the volume of P is equal to $|\det(A)|$. The vectors $\mathbf{v}_1$, $\mathbf{v}_2$, and $\mathbf{v}_3$ are linearly dependent if and only if the matrix A is singular. Again, if the vectors are linearly dependent, the parallelepiped degenerates into one with volume 0.

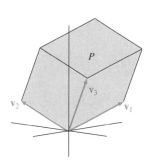

Figure 2. The volume of the parallelepiped P is equal to $|\det([\mathbf{v}_1, \mathbf{v}_2, \mathbf{v}_3])|$.

Definition of the determinant

We need to know more about the determinant and how to compute it. Since the proofs of these properties are not illuminating, we will not include them. First we need to give a definition of the determinant for larger matrices.

To motivate the definition, consider the formula in (7.1). In dimension 3, the determinant is the sum of terms, each of which is a product of three entries in the matrix. Let's look at one of those products, say the last one, $a_{13}a_{21}a_{32}$. Each term in the product has two subscripts. The first subscripts in the order of appearance are $(1, 2, 3)$ and the second subscripts are $(3, 1, 2)$. Thus the list of the first subscripts is simply a list of the numbers from 1 to 3. The list of the second subscripts is a list of the same numbers, but rearranged. Notice that the other products in (7.1) have the same feature.

The determinant in higher dimensions has the form we detected in (7.1). We need some notation. A ***permutation*** of the first n integers is a list σ of these integers in any order. For example, $\sigma = (3, 1, 2)$ is a permutation of the first three integers.

Thus a permutation is simply a vector whose entries are the integers from 1 to n in an arbitrary order, without repetitions. With $\sigma = (3, 1, 2)$, we have

$$a_{13}a_{21}a_{32} = a_{1\sigma_1}a_{2\sigma_2}a_{3\sigma_3}.$$

Let's ask how many permutations there are for three numbers. A permutation is just a list of the numbers, so we can pick the first number in the list in three different ways. Since we cannot choose the same number twice, we have only two choices for the second number. Finally, for the third number there is only one left. Thus there are $3 \times 2 = 6$ permutations for $n = 3$. They are $(1, 2, 3)$, $(1, 3, 2)$, $(2, 1, 3)$, $(2, 3, 1)$, $(3, 2, 1)$, and $(3, 1, 2)$. If we check formula (7.1) more closely, we see that every permutation of the first three integers occurs in the products. Thus the determinant in dimension 3 is just the sum of products $a_{1\sigma_1}a_{2\sigma_2}a_{3\sigma_3}$ over all permutations, with some minus signs thrown in.

To generalize the definition of the determinant, we need to know where the minus signs occur. It is clear that any permutation can be transformed into the standard, ordered list of integers by a series of interchanges of two numbers at a time. For example,

$$(3, 1, 2) \rightarrow (1, 3, 2) \rightarrow (1, 2, 3),$$

and

$$(4, 2, 1, 3) \rightarrow (1, 2, 4, 3) \rightarrow (1, 2, 3, 4).$$

This can be done in many different ways, but the remarkable thing is that the number of interchanges needed is always even or always odd. We will say that a permutation is *even* if this number is even, and *odd* if it is odd. Thus $(4, 2, 1, 3)$ is even, while $(4, 2, 3, 1)$ is odd.

Notice that for $n = 3$, the permutations $(1, 2, 3)$, $(2, 3, 1)$, and $(3, 1, 2)$ are even, and the rest are odd. In formula (7.1), it is the odd permutations that have minus signs. This will be true in all dimensions.

For a permutation σ, we will define

$$(-1)^\sigma = \begin{cases} 1 & \text{if } \sigma \text{ is even,} \\ -1 & \text{if } \sigma \text{ is odd.} \end{cases}$$

DEFINITION 7.4

The ***determinant*** of the matrix

$$A = \begin{pmatrix} a_{11} & a_{12} & \cdots & a_{1n} \\ a_{21} & a_{22} & \cdots & a_{2n} \\ \vdots & \vdots & \ddots & \vdots \\ a_{n1} & a_{n2} & \cdots & a_{nn} \end{pmatrix}$$

is defined to be

$$\det(A) = \sum_\sigma (-1)^\sigma a_{1\sigma_1} \cdot a_{2\sigma_2} \cdot \ldots \cdot a_{n\sigma_n}, \tag{7.5}$$

where the sum is over all permutations of the first n integers.

There are $n!$ permutations of the first n integers. Consequently, there are $n!$ terms in (7.5), the definition of the determinant. For $n = 3$, there are six terms as we have already seen. For $n = 4$, there are 24 terms, which is close to being unacceptable for purposes of computation. Clearly the definition is not going to be of much help in computing determinants. In fact, it is safe to say that with one important exception, which we will get to in Proposition 7.6, we will never use the definition to compute a determinant. We are going to have to find other ways of doing that.

There is one feature of (7.5) that should be noticed.

> Each summand in the definition of the determinant of A is the product of n entries from A, one entry from each row and from each column.

There is one important case where we can use the definition to compute the determinant. Consider an upper triangular matrix like

$$A = \begin{pmatrix} 1 & 5 & 9 & 3 \\ 0 & 8 & -2 & -1 \\ 0 & 0 & -5 & 6 \\ 0 & 0 & 0 & -3 \end{pmatrix}.$$

There is only one nonzero entry in the last row, $a_{44} = -3$. Hence only products containing this term can occur in the sum. But this term is also from the last column, so no other factor can be from that column. Now look at the second to last row. There are only two nonzero entries in this row, in the last and second to last columns. Since we already have a factor from the last column, we have to choose $a_{33} = -5$. Proceeding in this way, we see that the sum reduces to just one term, and $\det(A) = a_{11}a_{22}a_{33}a_{44} = 1 \cdot 8 \cdot -5 \cdot -3 = 120$.

Although done for the special case, the preceding argument works for any triangular matrix. For an $n \times n$ triangular matrix A, we get $\det(A) = a_{11}a_{22} \cdot \ldots \cdot a_{nn}$. Hence we have proved the next result.

PROPOSITION 7.6   The determinant of a triangular matrix is the product of the diagonal entries.   ●

The fact that the determinants of triangular matrices are so easy to compute will motivate much of what follows.

Determinants and row operations

Since any matrix can be reduced to row echelon form using row operations, and a matrix in row echelon form is upper triangular, it would be useful to know how row operations on matrices affect the determinant in order to make good use of Proposition 7.6. Here are the facts.

PROPOSITION 7.7   Let A be an $n \times n$ matrix.

1. If the matrix B is obtained from A by adding a multiple of one row to another, then $\det(B) = \det(A)$.
2. If the matrix B is obtained from A by interchanging two rows, then $\det(B) = -\det(A)$.
3. If the matrix B is obtained from A by multiplying a row by a constant c, then $\det(B) = c\det(A)$.   ●

Example 7.8   Compute the determinant of

$$A = \begin{pmatrix} 2 & 3 & 5 \\ 0 & 0 & 1 \\ 4 & 8 & 9 \end{pmatrix}.$$

A can be transformed to

$$R = \begin{pmatrix} 2 & 3 & 5 \\ 0 & 2 & -1 \\ 0 & 0 & 1 \end{pmatrix}$$

with two row operations:

1. Add -2 times row 1 to row 3.
2. Interchange rows 2 and 3.

The first operation leaves the determinant unchanged, and the second causes it to be multiplied by -1. Hence $\det(A) = -\det(R) = -4$. ●

Example 7.9 Compute the determinant of

$$A = \begin{pmatrix} 0 & 2 & 3 & 4 \\ 2 & 0 & 3 & 1 \\ 1 & 1 & 1 & 3 \\ 0 & 4 & -3 & 0 \end{pmatrix}.$$

We use row operations to reduce the matrix to upper triangular form. In doing so, we keep careful track of the factors introduced by row interchanges and by multiplying rows by constants. These are parts (2) and (3) of Proposition 7.7. There are many different ways to get to upper triangular form. No matter how we do it, however, we end up with $\det(A) = -82$. ●

If a matrix has two equal rows, then interchanging these rows leaves the matrix unchanged. On the other hand, according to part (2) of Proposition 7.7, the determinant is multiplied by -1. There is only one way this can happen.

COROLLARY 7.10 If a matrix A has two equal rows, then $\det(A) = 0$. ●

Determinant of the transpose

A useful property, and one that is devilishly hard to prove, has to do with the determinant of the transpose, A^T.

PROPOSITION 7.11 Let A be an $n \times n$ matrix. Then

$$\det(A^T) = \det(A).$$
●

The result immediately allows us to transfer the results about row operations to column operations.

PROPOSITION 7.12 Let A be an $n \times n$ matrix.

1. If the matrix B is obtained from A by adding a multiple of one column to another, then $\det(B) = \det(A)$.
2. If the matrix B is obtained from A by interchanging two columns, then $\det(B) = -\det(A)$.
3. If the matrix B is obtained from A by multiplying a column by a constant c, then $\det(B) = c\det(A)$. ●

COROLLARY 7.13 If a matrix A has two equal columns, then $\det(A) = 0$. ●

Expansion of a determinant by a row or a column

This is one of the most useful techniques for computing determinants. Given an $n \times n$ matrix A, we define the *ij-minor* of A to be the $(n-1) \times (n-1)$ matrix A_{ij} obtained from A by deleting the ith row and the jth column from A.

PROPOSITION 7.14 Let A be an $n \times n$ matrix.

1. For any i with $1 \le i \le n$, we have

$$\det(A) = \sum_{j=1}^{n}(-1)^{i+j}a_{ij}\det(A_{ij}).$$

2. For any j with $1 \le j \le n$, we have

$$\det(A) = \sum_{i=1}^{n}(-1)^{i+j}a_{ij}\det(A_{ij}).$$

Part (1) is called **expansion by the ith row** of the determinant of A. Part (2) is called **expansion by the jth column**. In each case, the calculation of an $n \times n$ determinant is reduced to computing n determinants of $(n-1) \times (n-1)$ matrices. This may not seem like a big help, but if the matrix has lots of zeros it can be very effective.

Example 7.15 Compute the determinant of

$$A = \begin{pmatrix} -3 & 2 & 8 \\ 0 & 4 & 0 \\ 3 & 2 & 1 \end{pmatrix}.$$

We expand by the second row to take advantage of the zeros. We get

$$\det(A) = (-1)^{2+2}4 \cdot \det\begin{pmatrix} -3 & 8 \\ 3 & 1 \end{pmatrix} = -108.$$

Example 7.16 Compute the determinant of

$$A = \begin{pmatrix} 5 & 1 & 4 & 3 \\ 0 & 2 & 3 & 0 \\ 1 & 0 & 4 & 0 \\ 0 & -1 & 0 & 1 \end{pmatrix}.$$

We expand by the second row to get

$$\det(A) = (-1)^{2+2}2\det\begin{pmatrix} 5 & 4 & 3 \\ 1 & 4 & 0 \\ 0 & 0 & 1 \end{pmatrix} + (-1)^{2+3}3\det\begin{pmatrix} 5 & 1 & 3 \\ 1 & 0 & 0 \\ 0 & -1 & 1 \end{pmatrix}.$$

Each of these determinants can be calculated by expanding by the rows containing two zeros to get

$$\det(A) = 2\left[1\det\begin{pmatrix} 5 & 4 \\ 1 & 4 \end{pmatrix}\right] - 3\left[-1\det\begin{pmatrix} 1 & 3 \\ -1 & 1 \end{pmatrix}\right]$$
$$= 2 \cdot 16 + 3 \cdot 4$$
$$= 44.$$

More properties of the determinant

There are other useful properties of the determinant that we will list here. First we give a result that might be a little surprising.

PROPOSITION 7.17   Suppose A and B are $n \times n$ matrices. Then

$$\det(AB) = \det(A)\det(B).$$

This proposition has a nice corollary.

COROLLARY 7.18   Suppose that A is a nonsingular matrix. Then

$$\det(A^{-1}) = 1/\det(A).$$

We cannot end without presenting one more application of the determinant.

PROPOSITION 7.19   A collection of n vectors $\mathbf{x}_1, \mathbf{x}_2, \ldots, \mathbf{x}_n$ in $\mathbf{R}^n$ is a basis for $\mathbf{R}^n$ if and only if the matrix $X = [\mathbf{x}_1 \ \mathbf{x}_2 \ \ldots \ \mathbf{x}_n]$, whose column vectors are the indicated vectors, has a nonzero determinant.

Proof   Notice that the equation $X\mathbf{c} = \mathbf{x}$ is equivalent to

$$c_1\mathbf{x}_1 + c_2\mathbf{x}_2 + \cdots + c_n\mathbf{x}_n = \mathbf{x}. \tag{7.20}$$

If $\det(X) \neq 0$, then X is nonsingular. Hence for each $\mathbf{x} \in \mathbf{R}^n$, there is a unique vector $\mathbf{c}$ such that (7.20) is true. This means in particular that the $\mathbf{x}$'s span $\mathbf{R}^n$. For $\mathbf{x} = 0$, the only vector $\mathbf{c}$ satisfying (7.20) is the zero vector. This means that the $\mathbf{x}$'s are linearly independent. Thus they are a basis for $\mathbf{R}^n$.

On the other hand, if $\det(X) = 0$, then there is a nonzero vector $\mathbf{c}$ such that $X\mathbf{c} = 0$, and therefore (7.20) is true with $\mathbf{x} = 0$. This means that the $\mathbf{x}$'s are linearly dependent and therefore they cannot be a basis.

We have already seen implications of Proposition 7.19 in dimension 2 and 3. Look back at Figures 1 and 2 on page 323.

EXERCISES

1. Use the following image to show that the area of the triangle spanned by the vectors $\mathbf{x}_1 = (x_1, y_1)^T$ and $\mathbf{x}_2 = (x_2, y_2)^T$ is equal to 1/2 the absolute value of the determinant of $X = [\mathbf{x}_1, \mathbf{x}_2]$, thus proving that the area of the parallelogram spanned by $\mathbf{x}_1$ and $\mathbf{x}_2$ is equal in absolute value to the determinant of $X = [\mathbf{x}_1, \mathbf{x}_2]$.

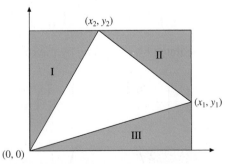

In Exercises 2–5, sketch the parallelogram spanned by the vectors $\mathbf{v}_1$ and $\mathbf{v}_2$ on graph paper. Estimate the area of your parallelogram using your sketch. Finally, compute the determinant of the matrix $[\mathbf{v}_1, \mathbf{v}_2]$ and compare with your estimate.

2. $\mathbf{v}_1 = \begin{pmatrix} 9 \\ 1 \end{pmatrix}, \mathbf{v}_2 = \begin{pmatrix} 1 \\ 9 \end{pmatrix}$ 3. $\mathbf{v}_1 = \begin{pmatrix} 1 \\ 4 \end{pmatrix}, \mathbf{v}_2 = \begin{pmatrix} 6 \\ 1 \end{pmatrix}$

4. $\mathbf{v}_1 = \begin{pmatrix} -2 \\ 5 \end{pmatrix}, \mathbf{v}_2 = \begin{pmatrix} 4 \\ 3 \end{pmatrix}$ 5. $\mathbf{v}_1 = \begin{pmatrix} 5 \\ 5 \end{pmatrix}, \mathbf{v}_2 = \begin{pmatrix} -2 \\ 6 \end{pmatrix}$

6. Prove each part of Proposition 7.7, where A is the arbitrary 2×2 matrix

$$A = \begin{pmatrix} a & b \\ c & d \end{pmatrix}.$$

In Exercises 7–12, compute the determinant of the matrix by using elementary row operations to first place the matrix in upper triangular form. Use hand calculations only. No technology is allowed.

7. $\begin{pmatrix} 3 & 0 & 0 \\ 3 & 6 & 3 \\ -18 & -18 & -9 \end{pmatrix}$ 8. $\begin{pmatrix} 5 & 6 & 4 \\ -4 & -9 & -8 \\ 4 & 6 & 5 \end{pmatrix}$

9. $\begin{pmatrix} 1 & 0 & 4 \\ -3 & 3 & -2 \\ 4 & -1 & -2 \end{pmatrix}$ 10. $\begin{pmatrix} 1 & 2 & -3 \\ 0 & 6 & -2 \\ -2 & 3 & 2 \end{pmatrix}$

11. $\begin{pmatrix} 2 & -1 & 3 & 4 \\ 0 & 2 & -2 & 0 \\ -1 & 2 & 0 & 0 \\ -1 & 3 & 1 & 2 \end{pmatrix}$ 12. $\begin{pmatrix} 3 & -3 & -2 & -1 \\ 2 & 0 & -2 & -1 \\ 1 & -2 & 0 & 0 \\ 4 & -1 & -4 & -2 \end{pmatrix}$

13. Let A be an arbitrary $n \times n$ matrix.

    (a) If the ith row of A is a scalar multiple of the jth row, prove that the determinant of A is zero. State and prove a similar statement about the columns of A.

    (b) Without computing the determinant, explain why each of the following matrices has a zero determinant.

$$\begin{pmatrix} 1 & 2 & 3 \\ -1 & 1 & 4 \\ 0 & 0 & 0 \end{pmatrix} \qquad \begin{pmatrix} -1 & 2 & 0 \\ 3 & 4 & 0 \\ 5 & 2 & 0 \end{pmatrix}$$

$$\begin{pmatrix} 1 & 2 & 3 \\ 2 & 4 & 6 \\ 5 & 1 & 2 \end{pmatrix} \qquad \begin{pmatrix} -1 & -2 & 3 \\ 1 & 2 & 1 \\ 2 & 4 & 1 \end{pmatrix}$$

14. Let A be an arbitrary $n \times n$ matrix.

    (a) If row i is a linear combination of the preceding rows, prove that the determinant of A is zero. State and prove a similar statement about the columns of A.

    (b) Without computing the determinant, explain why each of the following matrices has a zero determinant.

$$\begin{pmatrix} 1 & 2 & 3 \\ -1 & 1 & 1 \\ 0 & 3 & 4 \end{pmatrix} \qquad \begin{pmatrix} 1 & 2 & 3 \\ 3 & 0 & 3 \\ -1 & 1 & 0 \end{pmatrix}$$

$$\begin{pmatrix} 1 & 1 & 0 \\ -1 & 1 & 1 \\ 1 & 3 & 1 \end{pmatrix} \qquad \begin{pmatrix} 1 & 1 & 5 \\ -1 & 1 & 1 \\ 1 & 0 & 2 \end{pmatrix}$$

15. Suppose that matrix A is $n \times n$ and is **lower triangular** (i.e., $a_{ij} = 0$ if $i < j$). Explain why the determinant is still equal to the product of the diagonal elements. Calculate the determinant of

$$A = \begin{pmatrix} 1 & 0 & 0 & 0 \\ 1 & 2 & 0 & 0 \\ 1 & 1 & 3 & 0 \\ 1 & 1 & 1 & 4 \end{pmatrix}.$$

In Exercises 16–21, expand the matrix by row or column to calculate the determinant.

16. Matrix A in Exercise 7   17. Matrix A in Exercise 8

18. Matrix A in Exercise 9   19. Matrix A in Exercise 10

20. Matrix A in Exercise 11   21. Matrix A in Exercise 12

In Exercises 22–29, calculate the determinant of the given matrix. Determine if the matrix has a nontrivial nullspace, and if it does find a basis for the nullspace. Determine if the column vectors in the matrix are linearly independent.

22. $\begin{pmatrix} 1 & -2 \\ 2 & 3 \end{pmatrix}$ 23. $\begin{pmatrix} -1 & 2 \\ 2 & -4 \end{pmatrix}$

24. $\begin{pmatrix} 1 & 3 \\ -1 & -3 \end{pmatrix}$ 25. $\begin{pmatrix} 2 & -1 \\ 1 & 0 \end{pmatrix}$

26. $\begin{pmatrix} 1 & 1 & 1 \\ 1 & 1 & 0 \\ 1 & 0 & 0 \end{pmatrix}$ 27. $\begin{pmatrix} 1 & -2 & -4 \\ 2 & 1 & 2 \\ 3 & 0 & 0 \end{pmatrix}$

28. $\begin{pmatrix} -1 & 0 & -1 \\ 1 & 1 & 2 \\ 2 & 1 & 3 \end{pmatrix}$ 29. $\begin{pmatrix} 1 & 1 & 2 \\ -1 & 1 & 5 \\ 1 & 0 & -1 \end{pmatrix}$

In Exercises 30–39, find the values of x for which the indicated matrix has a nontrivial nullspace.

30. $\begin{pmatrix} 2 & x \\ 3 & -2 \end{pmatrix}$ 31. $\begin{pmatrix} 2 & x \\ x & 3 \end{pmatrix}$

32. $\begin{pmatrix} x & 4 \\ 3 & -2 \end{pmatrix}$ 33. $\begin{pmatrix} -1 & x \\ -x & 4 \end{pmatrix}$

34. $\begin{pmatrix} 2-x & 1 \\ 0 & -1-x \end{pmatrix}$ 35. $\begin{pmatrix} -1-x & 0 \\ 3 & 2-x \end{pmatrix}$

36. $\begin{pmatrix} -1-x & 5 & 2 \\ 0 & -x & -1 \\ 0 & 6 & -5-x \end{pmatrix}$

37. $\begin{pmatrix} 2-x & 0 & 0 \\ -1 & -x & 2 \\ 0 & -2 & 5-x \end{pmatrix}$

38. $\begin{pmatrix} 2-x & 0 & 1 \\ -3 & -1-x & -1 \\ -2 & 0 & -1-x \end{pmatrix}$

39. $\begin{pmatrix} -1-x & 2 & 2 \\ 0 & -2-x & 0 \\ -1 & 4 & 2-x \end{pmatrix}$

In Exercises 40–49, compute the determinant of the matrix. In each case, decide if there is a nonzero vector in the nullspace.

40. $\begin{pmatrix} 2 & 3 \\ -1 & 0 \end{pmatrix}$ 41. $\begin{pmatrix} -2 & 3 \\ -2 & 4 \end{pmatrix}$

42. $\begin{pmatrix} -1 & -9 & 10 \\ -7 & -19 & 26 \\ -2 & -10 & 12 \end{pmatrix}$ 43. $\begin{pmatrix} 1 & 0 & 4 \\ -3 & 3 & -2 \\ 4 & 0 & -2 \end{pmatrix}$

44. $\begin{pmatrix} 1 & 2 & -3 \\ 0 & 6 & -2 \\ -2 & 3 & 2 \end{pmatrix}$ 45. $\begin{pmatrix} 0 & 1 & 2 \\ 2 & 0 & -2 \\ -1 & 0 & 3 \end{pmatrix}$

46. $\begin{pmatrix} 10 & -1 & -3 & 9 \\ 3 & 2 & -3 & 3 \\ 3 & 1 & -2 & 3 \\ -10 & 2 & 2 & -9 \end{pmatrix}$

47. $\begin{pmatrix} 3 & 0 & 20 & -8 \\ 2 & 3 & -2 & 0 \\ 6 & 4 & 17 & -8 \\ 16 & 10 & 50 & -23 \end{pmatrix}$

48. $\begin{pmatrix} -120 & 60 & -79 & 52 & 68 & 123 \\ -262 & 162 & -216 & 124 & 184 & 262 \\ -142 & 78 & -100 & 64 & 92 & 142 \\ -262 & 162 & -216 & 124 & 184 & 262 \\ -10 & -4 & 6 & -2 & 2 & 10 \\ -112 & 60 & -79 & 52 & 68 & 115 \end{pmatrix}$

49. $\begin{pmatrix} 556 & 65 & -91 & 52 & 416 & -143 \\ 550 & 60 & -90 & 50 & 410 & -140 \\ -169 & -22 & 26 & -18 & -131 & 38 \\ -96 & -13 & 14 & -7 & -69 & 27 \\ 550 & 60 & -90 & 50 & 410 & -140 \\ 825 & 97 & -137 & 80 & 617 & -211 \end{pmatrix}$

50. Suppose that U is a 4×4 matrix with $\det(U) = -3$.

  (a) What is the value of $\det(-2U)$?

  (b) What is the value of $\det(U^3)$?

  (c) What is the value of $\det(U^{-1})$?

51. Prove or disprove: If A and B are $n \times n$ matrices, then $\det(A + B) = \det(A) + \det(B)$.

52. Two matrices A and B are *similar matrices* if there exists a nonsingular matrix S such that $A = S^{-1}BS$.

  (a) Prove: If A and B are similar, then $\det(A) = \det(B)$.

  (b) Prove: If A and B are similar, then $\det(A - \lambda I) = \det(B - \lambda I)$.

53. Make a list of as many properties as you can find that are equivalent to a matrix having a nonzero determinant.

54. Make a list of as many properties as you can find that are equivalent to a matrix having a zero determinant.

8

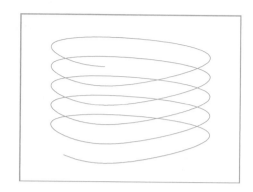

An Introduction to Systems

We are ready to move on to systems of differential equations. We will start in Section 8.1 with definitions and examples. Naturally we will be able to model many more applications using systems. In Section 8.2, we will examine ways to visualize solutions and make geometrical interpretations of a system. Section 8.3 will be devoted to qualitative analysis. This method of analysis becomes even more important when we look at systems, because relatively few systems can be solved explicitly. We will do some very basic qualitative analysis in Section 8.3, but we will have to wait for the following chapters to really get into this subject. Linear systems will be introduced in Section 8.4, together with a couple of applications. In Section 8.5, we will look at special facts that are true for linear systems.

8.1 Definitions and Examples

A system of differential equations is a set of one or more equations, involving one or more unknown functions. In many applications there are several interrelated quantities changing with respect to an independent variable, often time. Modeling such an application leads to a system of differential equations. Here is an example.

The SIR model of an epidemic

The use of systems greatly expands our ability to model applications. Let's start by modeling an epidemic. Suppose we have a population of N individuals that is subject to a communicable disease. We will assume the following facts about the disease:

- The disease is of short duration and rarely fatal.
- The disease spreads through contact between individuals.

- Individuals who have recovered from the disease are immune.

These features are present in measles, the mumps, and in the common cold.

The listed features allow us to construct a model. We divide the population into three groups. The *susceptible*, $S(t)$, are those individuals who have never had the disease. The *infected*, $I(t)$, are those who are currently ill with the disease. Finally, the *recovered*, $R(t)$, are those who have had the disease and are now immune. The total population is the sum of these three, $N = S + I + R$. We must compute the rate of change for each of these subpopulations.

Since the disease is of short duration and rarely fatal, we may ignore deaths and births. This means that the total population N is a constant. It also implies that $S(t)$, the number of susceptibles, changes only because some of them catch the disease and pass into the infected population. Since the disease spreads through contacts between susceptible and infected individuals, the rate of change is proportional to the number of contacts. Assuming that the two populations are randomly distributed over area, the number of contacts is proportional to the product SI of the two populations. Thus there is a positive constant a such that

$$\frac{dS}{dt} = -aSI.$$

The number of infected individuals, $I(t)$, changes in two ways. First, it increases as susceptible individuals get sick. We have already determined that rate to be aSI. In addition, infected individuals get well and then pass into the recovered population. Assuming that there is a fairly standard time in which a recovery takes place, the rate of recovery is proportional to the number of infected. So there is a positive constant b such that the rate of recoveries is bI. Putting these together, we see that

$$\frac{dI}{dt} = aSI - bI.$$

Finally, the number of recovered individuals increases as those who are infected are cured. We have already determined that this happens at the rate bI, so

$$\frac{dR}{dt} = bI.$$

Putting everything together, we get the system of three equations

$$\begin{aligned} S' &= -aSI, \\ I' &= aSI - bI, \\ R' &= bI. \end{aligned} \tag{1.1}$$

This is the **SIR model**. It is **nonlinear** because the right-hand side of the first and second equations contains the product SI. It is **autonomous** because the right-hand sides of the equations do not depend explicitly on the independent variable. A solution to the SIR model is a triple of functions $S(t)$, $I(t)$, and $R(t)$, which satisfy the equations in (1.1).

One good way to display the solution to a system is to plot all of the components of the solution. Suppose we look at the SIR system with $a = b = 1$:

$$\begin{aligned} S' &= -SI, \\ I' &= SI - I, \\ R' &= I, \end{aligned} \tag{1.2}$$

with initial values $S(0) = 4$, $I(0) = 0.1$, and $R(0) = 0$. The system cannot be solved explicitly, but we can compute numerical solutions. The three components of the solution are shown plotted in Figure 1.

From the figure, it is easily possible to trace the history of the disease. The number of infected individuals starts small, rises quickly, and then falls off. Interestingly, almost the entire population eventually gets the disease.

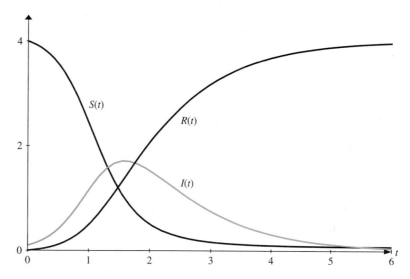

Figure 1. The components of a solution to an SIR model.

Although the SIR model in (1.1) involves three equations and three unknowns, the last equation is not really needed. Remember that part of our model requires that the total population $N = S + I + R$ is constant. If we add the equations in (1.1), we get $dN/dt = 0$, which serves as verification. In addition, the first two equations in (1.1) form a planar, autonomous system. If we were to solve only these equations for S and I, we could compute the number of recovered individuals from the equation $R = N - S - I$. In this way, we get the system

$$
\begin{aligned}
S' &= -aSI, \\
I' &= aSI - bI,
\end{aligned}
\tag{1.3}
$$

which is also referred to as the SIR model.

Vector notation

We will find it advantageous to use vector notation when dealing with first-order systems. To do so with the SIR system, for example, we relabel the unknown functions as follows:

$$
u_1(t) = S(t), \quad u_2(t) = I(t), \quad \text{and} \quad u_3(t) = R(t).
$$

We can rewrite (1.1) using this notation:

$$
\begin{aligned}
u_1' &= -au_1u_2, \\
u_2' &= au_1u_2 - bu_2, \\
u_3' &= bu_2.
\end{aligned}
\tag{1.4}
$$

Finally, we introduce the vector-valued function $\mathbf{u}(t) = (u_1(t), u_2(t), u_3(t))^T = (S(t), I(t), R(t))^T$. According to (1.4), the derivative of $\mathbf{u}$ is

$$\mathbf{u}' = \begin{pmatrix} u_1' \\ u_2' \\ u_3' \end{pmatrix} = \begin{pmatrix} -au_1u_2 \\ au_1u_2 - bu_2 \\ bu_2 \end{pmatrix}. \tag{1.5}$$

There are two things to notice about equation (1.5). First, to differentiate a vector valued function $\mathbf{u}(t)$, we simply differentiate the components,

$$\mathbf{u}' = \begin{pmatrix} u_1' \\ u_2' \\ u_3' \end{pmatrix}.$$

Second, the right-hand side of (1.5) is a vector-valued function of the vector $\mathbf{u}$. Hence if we set

$$\mathbf{f}(\mathbf{u}) = \begin{pmatrix} -au_1u_2 \\ au_1u_2 - bu_2 \\ bu_2 \end{pmatrix}, \tag{1.6}$$

then (1.5) becomes

$$\mathbf{u}' = \mathbf{f}(\mathbf{u}). \tag{1.7}$$

The initial value problem

If we consider the SIR model from a biological point of view, we realize that we should have a solution starting with any values of the three populations S, I, and R. The mathematical theory agrees with our biological intuition. Thus, the correct formulation of the *initial value problem* for the SIR system with initial time t_0 is

$$\begin{aligned} S' &= -aSI, & S(t_0) &= S_0, \\ I' &= aSI - bI, \quad \text{with} & I(t_0) &= I_0, \\ R' &= bI, & R(t_0) &= R_0. \end{aligned}$$

Using vector notation, with $\mathbf{u} = (u_1(t), u_2(t), u_3(t))^T = (S(t), I(t), R(t))^T$, as we defined it earlier, and $\mathbf{f}$ given by (1.6), this becomes

$$\mathbf{u}' = \mathbf{f}(\mathbf{u}), \quad \text{with} \quad \mathbf{u}(t_0) = \mathbf{u}_0 = (S_0, I_0, R_0)^T.$$

General first-order systems

The SIR model involves only first-order derivatives of the unknown functions S, I, and R. For that reason it is called a *first-order system*. The *order* of a system of differential equations is the highest derivative that occurs in the system.

The general first-order system of two equations has the form

$$\begin{aligned} x' &= f(t, x, y), \\ y' &= g(t, x, y), \end{aligned} \tag{1.8}$$

where f and g are functions of the three variables t, x, and y. A *solution* to the system is a pair of functions $x(t)$ and $y(t)$ that satisfies

$$\begin{aligned} x'(t) &= f(t, x(t), y(t)), \\ y'(t) &= g(t, x(t), y(t)), \end{aligned} \tag{1.9}$$

for t in some interval.

Example 1.10   Show that the pair $x(t) = \sin t$ and $y(t) = \cos t$ form a solution of the system

$$x' = y(x^2 + y^2), \tag{1.11}$$
$$y' = -x(x^2 + y^2). \tag{1.12}$$

A solution to the system is a pair of functions x and y that satisfies the equations (1.11) and (1.12) when substituted into them. After substitution of $x(t) = \sin t$ and $y(t) = \cos t$, the left-hand side of (1.11) is $x'(t) = \cos t$, and since $x^2 + y^2 = 1$, the right-hand side is $y(t)(x^2(t) + y^2(t)) = y(t) = \cos t$. Hence (1.11) is satisfied. A similar calculation shows that (1.12) is also satisfied. Hence the pair $x(t) = \sin t$ and $y(t) = \cos t$ forms a solution to the system.   ●

We will consider systems of n equations with n unknowns. If the unknown functions are $x_1(t), x_2(t), \ldots,$ and $x_n(t)$, then the system has the form

$$\begin{aligned}
x_1' &= f_1(t, x_1, x_2, \ldots, x_n), \\
x_2' &= f_2(t, x_1, x_2, \ldots, x_n), \\
&\cdots \\
x_n' &= f_n(t, x_1, x_2, \ldots, x_n),
\end{aligned} \tag{1.13}$$

where $f_1, f_2, \ldots,$ and f_n are functions of the $n + 1$ variables $t, x_1, x_2, \ldots,$ and x_n. A solution to the system (1.13) would be an n-tuple of functions $x_1(t), x_2(t), \ldots,$ and $x_n(t)$ that satisfies the equations in (1.13).

We will always require that the number of equations is equal to the number of unknowns, and this number is called the ***dimension*** of the system. Thus the system in (1.9) has dimension 2, while the general system in (1.13) has dimension n. A system of dimension 2 is called a ***planar*** system. The SIR model in (1.1) has dimension 3, but in the form of (1.3) it is a planar system.

The general system of dimension n is displayed in equation (1.13). If we set

$$\mathbf{x}(t) = (x_1, x_2, \ldots, x_n)^T, \quad \text{and}$$
$$\mathbf{f}(t, \mathbf{x}) = (f_1(t, \mathbf{x}), f_2(t, \mathbf{x}), \ldots, f_n(t, \mathbf{x}))^T,$$

the system can be written as

$$\mathbf{x}' = \mathbf{f}(t, \mathbf{x}). \tag{1.14}$$

Here we see one great advantage of vector notation. It allows us to write a system of arbitrary dimension almost as conveniently as a single first-order equation. Of course the interpretation is quite different.

The SIR example indicates that we need to specify an initial value for every unknown function. If we write the general system in vector form $\mathbf{x}' = \mathbf{f}(t, \mathbf{x})$, then we need to specify an initial value for every component of the vector $\mathbf{x}$. Thus the initial value problem for any system has the form

$$\mathbf{x}' = \mathbf{f}(t, \mathbf{x}) \quad \text{with} \quad \mathbf{x}(t_0) = \mathbf{x}_0.$$

Reduction of higher-order equations and systems to first-order systems

We have stated a couple of times that there is a system of first-order equations that is equivalent to any system of higher-order equations, in the sense that a solution to one leads easily to a solution to the other. As a result of this, every model of an application involving a higher-order equation has an equivalent model using a first-order system.

This equivalence is useful for two additional reasons. First, it allows us to spend most of our time studying first-order systems, since we know that higher-order systems will be included in all of our results. The second and equally important reason has to do with the numerical computation of solutions. Most numerical solvers are written to solve first-order systems. To solve a higher-order equation, it is necessary to use the equivalent first-order system.

One example will illustrate how to find a first-order system equivalent to a higher order equation.

E x a m p l e   1.15   Find a first-order system equivalent to the third-order, nonlinear equation

$$x''' + xx'' = \cos t. \tag{1.16}$$

The basic idea is to introduce new dependent variables for the unknown function and each derivative up to one less than the order of the equation. Hence, we introduces $u_1 = x$, $u_2 = x'$, and $u_3 = x''$. Then, using the definitions of the new variables and (1.16), we have the first-order system

$$
\begin{aligned}
u_1' &= u_2, \\
u_2' &= u_3, \\
u_3' &= -u_1 u_3 + \cos t.
\end{aligned}
\tag{1.17}
$$

Clearly, if x solves (1.16), then the set of functions $u_1 = x$, $u_2 = x'$, and $u_3 = x''$ solves (1.17).

To show that the converse is also true, suppose u_1, u_2, and u_3 solve (1.17). Set $x = u_1$. Then by the first two equations in (1.17), $x' = u_1' = u_2$ and $x'' = u_2' = u_3$. Differentiating again, we get $x''' = u_3'$, so by the last equation in (1.17),

$$x''' + xx'' = u_3' + u_1 u_3 = \cos t,$$

from which we see that x is a solution to (1.16).   ●

EXERCISES

In Exercises 1–6, indicate the dimension and whether or not the given system is autonomous. Assume that the independent variable in each example is t.

1. $x' = v$
$v' = -x - 0.02v + 2\cos t$

2. $\theta' = \omega$
$\omega' = -\dfrac{g}{L}\sin\theta + \dfrac{k}{m}\omega$

3. $x' = -ax + ay$
$y' = rx - y - xz$
$z' = -bz + xy$

4. $x' = -y - z$
$y' = x + ay$
$z' = b - z(c - x)$

5. $u_1' = u_2$
$u_2' = -\dfrac{1}{2}u_1 + \dfrac{1}{2}u_3$
$u_3' = u_4$
$u_4' = \dfrac{3}{2}u_1 + \dfrac{1}{2}u_3$

6. $u_1' = u_2$
$u_2' = -\dfrac{k_1 + k_2}{m_1}u_1 + \dfrac{k_2}{m_1}u_3$
$u_3' = u_4$
$u_4' = \dfrac{k_2}{m_2}u_1 - \dfrac{k_2 + k_3}{m_2}u_3 + \cos t$

7. Show that the functions $x(t) = 2e^{2t} - 2e^{-t}$ and $y(t) = -e^{-t} + 2e^{2t}$ are solutions of the system

$$
\begin{aligned}
x' &= -4x + 6y, \\
y' &= -3x + 5y,
\end{aligned}
$$

satisfying the initial conditions $x(0) = 0$ and $y(0) = 1$.

8. Show that the functions $x(t) = (1 + t)e^{-t}$ and $y(t) = -te^{-t}$ are solutions of the system

$$x' = y,$$
$$y' = -x - 2y,$$

satisfying the initial conditions $x(0) = 1$ and $y(0) = 0$.

9. Show that the functions $x(t) = e^{-t}(-\cos t - \sin t)$ and $v(t) = 2e^{-t} \sin t$ are solutions of the system

$$x' = v,$$
$$v' = -2x - 2v,$$

satisfying the initial conditions $x(0) = -1$ and $v(0) = 0$.

10. Show that the functions $x(t) = e^t$ and $y(t) = e^{-t}$ are solutions of the system

$$x' = x^2 y,$$
$$y' = -xy^2,$$

satisfying the initial conditions $x(0) = 1$ and $y(0) = 1$.

Write each initial value problem in Exercises 11–16 as a system of first-order equations using vector notation.

11. $y'' + 2y' + 4y = 3 \cos 2t$, $\quad y(0) = 1$, $\quad y'(0) = 0$

12. $mx'' + \mu x' + kx = F_0 \cos \omega t$, $\quad x(0) = x_0$, $\quad x'(0) = v_0$

13. $x'' + \delta x' - x + x^3 = \gamma \cos \omega t$, $\quad x(0) = x_0$, $\quad x'(0) = v_0$

14. $x'' + \mu(x^2 - 1)x' + x = 0$, $\quad x(0) = x_0$, $\quad x'(0) = v_0$

15. $\omega''' = \omega$, $\quad \omega(0) = \omega_0$, $\quad \omega'(0) = \alpha_0$, $\quad \omega''(0) = \gamma_0$

16. $y''' + y'y'' = \sin \omega t$, $\quad y(0) = \alpha$, $\quad y'(0) = \beta$, $\quad y''(0) = \gamma$

Which of the systems in Exercises 17–22 are autonomous? Assume the independent variable is t in each exercise.

17. $u' = v$ and $v' = -3u - 2v + 5 \cos t$

18. $u' = u(u^2 + v^2)$ and $v' = -v(u^2 + v^2)$

19. $u' = v \cos(u)$ and $v' = tv$

20. $u' = v \cos(u)$ and $v' = u^2 e^v$

21. $u' = v + \cos(u)$, $\quad v' = v - tw$, and $w' = 5u - 9v + 8w$

22. $u' = w + \cos(u) - 2v$, $\quad v' = u^2 e^v$, and $w' = u + v + w$

23. Write the system in Exercise 17 in vector form.

24. Write the system in Exercise 18 in vector form.

25. Write the system in Exercise 19 in vector form.

26. Write the system in Exercise 20 in vector form.

27. Write the system in Exercise 21 in vector form.

28. Write the system in Exercise 22 in vector form.

29. Use your numerical solver to produce the component solutions for the SIR model pictured in Figure 1.

30. It is sometimes useful to use normalized quantities in the SIR model. To do so, introduce the variables $s = S/N$, $i = I/N$, and $r = R/N$. Each of the new variables represents the fraction of the total population that is in the particular category. Start with the system in (1.1) and derive the system that is satisfied by s, i, and r.

31. A fisherman is located on the opposite bank of a northward flowing river, as shown in Figure 2. The man continually points the nose of his boat toward his destination (the origin at $(0, 0)$), but the current of the river pushes him downstream. Let b be the speed of the boat relative to the water and let a be the speed of the current, both measured in miles per hour. Let $x(t)$ and $y(t)$ denote the x and y position of the boat at time t. Show that the boat obeys the following equations of motion.

$$\frac{dx}{dt} = -\frac{bx}{\sqrt{x^2 + y^2}}$$

$$\frac{dy}{dt} = a - \frac{by}{\sqrt{x^2 + y^2}}$$

Hint: Break the velocity of the boat into its horizontal and vertical components, find dx/dt and dy/dt in terms of a, b, and θ, and then eliminate θ from the resulting equations.

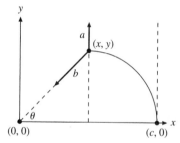

Figure 2. Fighting the current.

32. Use a numerical solver to plot the approximate path of the boat in Exercise 31. Use a number of different values for the parameters a, b, and c. Does the boat always reach its destination (the origin)?

Warning. When the boat reaches its destination, the system of differential equations becomes singular, since the right-hand sides of the equations are discontinuous when x and y are both equal to 0. This can result in your computer exhibiting strange behavior. For example, a variable step algorithm may use smaller and smaller steps, never reaching the end of the computation. Fixed step solvers will reach the end of the computation, but the results will be strange. For this reason, it is best to use final times that underestimate the time it takes to reach the destination. Start with the final time equal to c/b and do the computation a number of times with increasing final times until you are close to the destination. Before even starting this exercise, you should learn how to stop a computation if that becomes necessary.

8.2 Geometric Interpretation of Solutions

To properly interpret a solution to a system of differential equations, we need to display it graphically. Here we will discuss some ways to do that. We will also explore some geometric ways to interpret a system and its solutions. We have already done this for single first-order equations (systems of dimension 1) in Chapter 2, and for second-order equations in Chapter 4. It might be a good idea to review that material while reading this section.

We will start by deriving a model that we will follow throughout this section.

Predator–prey systems

Consider two species that exist together and interact. We will suppose that one of these species is a predator, which depends in an essential way on the other, the prey, for its food supply, and therefore for its existence. There are many examples of such situations, such as wolves and deer existing together, or sharks and food fish. It was the latter example that motivated the Italian mathematician Vito Volterra[1] to formulate the model we will present here.

We assume the existence of a prey population, denoted by $F(t)$, and of a population of predators, denoted by $S(t)$. You can think of these as food fish and sharks, respectively, but keep in mind that S and F represent any predator–prey pair. Following the discussion of population models in Chapter 3, for each of these we have a reproductive rate, which we denote by r_F for the prey and r_S for the predator. Hence, according to the discussion in Chapter 3, we have

$$
\begin{aligned}
F' &= r_F F, \\
S' &= r_S S.
\end{aligned}
\tag{2.1}
$$

It is the modeler's task to figure out what the reproductive rates are. Let's start with r_F, the reproductive rate for the prey population. We will assume that there are sufficient food resources so that in the absence of predators, the prey population would follow the Malthusian model with a positive reproductive rate. That means

$$r_F = a > 0, \quad \text{if } S = 0.$$

If there are predators, then $S > 0$ and each encounter with a predator has a certain probability of resulting in the capture and death of the prey. Consequently, the reproductive rate will decrease because of an increasing death rate resulting from predation. To a first approximation, the decrease would be proportional to the number of encounters with predators. Assuming random distribution of both predator and prey, the number of encounters for an individual prey would be proportional to S, the number of predators. Hence the decrease in the reproductive rate would also be proportional to S. Therefore,

$$r_F = a - bS, \quad a, b > 0,
\tag{2.2}$$

for some constants a and b.

---

[1] Vito Volterra was born in Ancona (then part of the Papal States) in 1860. He came from a poor family, but after attending lectures at Florence he was able to proceed to Pisa, where he graduated as a Doctor of Physics in 1882. He became Professor of Mechanics at Pisa in 1883 and later he occupied the Chair of Mathematical Physics. After being appointed to the Chair of Mechanics at Turin, he was appointed to the Chair of Mathematical Physics at Rome in 1900. Volterra's research was very wide ranging, including many aspects of mathematical physics, and later mathematical biology.

In 1922 Fascism seized Italy and Volterra fought against it in the Italian Parliament. However, by 1930 the Parliament was abolished, and when Volterra refused to take an oath of allegiance to the Fascist government in 1931, he was forced to leave the University of Rome. From the following year, he lived mostly abroad, mainly in Paris but also in Spain and other countries. He died in 1940 in Rome.

The analysis of r_S, the reproductive rate for the predators, proceeds similarly. In the absence of prey, the predator population would be Malthusian, but with a negative reproductive rate, since the predators depend on the prey for their sustenance. Hence,

$$r_S = -c < 0 \quad \text{if } F = 0.$$

The presence of the prey would increase the reproductive rate, and as before, this increase would be proportional to the size of the prey population. To a first approximation, the reproductive rate is given by

$$r_S = -c + dF \quad c, d > 0. \tag{2.3}$$

Putting everything together, the system in (2.1) becomes

$$\begin{aligned} F' &= (a - bS)F, \\ S' &= (-c + dF)S. \end{aligned} \tag{2.4}$$

This is the Volterra model of predator–prey populations. The equations are usually referred to as the Lotka-Volterra equations because the American mathematician A. J. Lotka derived the same equations.

The Lotka-Volterra model is ***nonlinear*** because the right-hand side contains the product FS. It is ***autonomous*** because the right-hand side does not depend explicitly on the independent variable.

Plots of the components of a solution

Let's look at the following special case of the Lotka-Volterra system.

$$\begin{aligned} F' &= (0.4 - 0.01S)F \\ S' &= (-0.3 + 0.005F)S \end{aligned} \tag{2.5}$$

We will use the initial conditions $F(0) = 40$ and $S(0) = 20$. Again we will use numerical solutions.

One way to represent the solutions graphically is simply to plot all of the components of the solution, as we did for the SIR system in Figure 1 of the previous section. For our predator–prey example, this is done in Figure 1 below. The graphs in Figure 1 seem to indicate that the populations vary periodically.

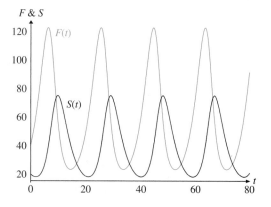

Figure 1. The components of the solution to the Lotka-Volterra model.

Parametric curves

We get another view of the solution when we look at a ***parametric plot***. We will use vector notation, setting $\mathbf{u}(t) = (F(t), S(t))^T$. Then we plot $t \to \mathbf{u}(t) = (F(t), S(t))^T$. The image is a set in $\mathbf{R}^2$; the curve is shown in Figure 2. This is an example of a solution curve in the ***phase plane***.

The graph in Figure 2 shows how the two populations interact. In particular, we notice that the curve is closed, so that it tracks over itself as t increases. This reinforces our speculation that the solution is periodic.

In Figure 1, we plotted the components of the same solution as shown in the phase plane plot in Figure 2. Notice that it is impossible to gather any precise time information from the phase plane plot in Figure 2. It is just not available there. The time information is a key element of the representation in Figure 1, however. It is possible to take the information in Figure 1 and construct the curve seen in the phase plane, Figure 2. Going in the other direction and reconstructing Figure 1 from the phase plane plot in Figure 2 is possible only approximately.

If we want to discover the direction of the curve in Figure 2 as t increases at a point $\mathbf{u}_0 = \mathbf{u}(t_0)$ on the solution curve, we can choose $h > 0$, set $\mathbf{u}_1 = \mathbf{u}(t_0 + h)$, and look at the vector $\mathbf{u}_1 - \mathbf{u}_0$. A typical example of the point $\mathbf{u}_0$ is plotted in Figure 2. Both points $\mathbf{u}_0$ and $\mathbf{u}_1$ are shown in Figure 3, together with the vector $\mathbf{u}_1 - \mathbf{u}_0$. Since it connects the points $\mathbf{u}_0$ and $\mathbf{u}_1$ on the curve, it is called a ***secant vector***.

If we let h get smaller, the secant vector also gets smaller. To prevent it from disappearing entirely as h decreases, we look at the difference quotients

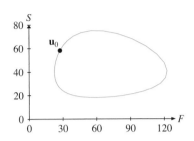

Figure 2. The phase plane plot of the predator and prey populations.

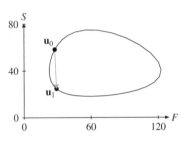

Figure 3. A secant vector.

$$\frac{1}{h}(\mathbf{u}(t_0 + h) - \mathbf{u}(t_0)) = \frac{1}{h}\begin{pmatrix} F(t_0 + h) - F(t_0) \\ S(t_0 + h) - S(t_0) \end{pmatrix}. \tag{2.6}$$

Figure 4 shows three such difference quotients for smaller values of h. Clearly, the difference quotients approach a vector that is tangent to the curve at $\mathbf{u}_0$.

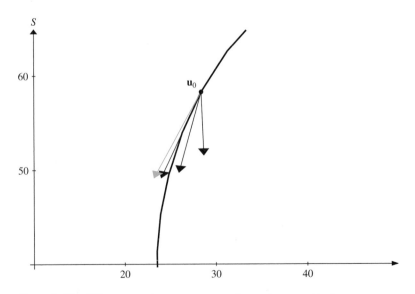

Figure 4. The difference quotient vectors approach a vector tangent to the curve.

Now let's look at what happens to the difference quotients in (2.6) as $h \to 0$.

$$\lim_{h \to 0} \frac{1}{h}(\mathbf{u}(t_0 + h) - \mathbf{u}(t_0)) = \lim_{h \to 0} \left(\begin{array}{c} (F(t_0 + h) - F(t_0))/h \\ (S(t_0 + h) - S(t_0))/h \end{array} \right)$$

$$= \left(\begin{array}{c} F'(t_0) \\ S'(t_0) \end{array} \right) = \mathbf{u}'(t_0)$$

Therefore, the derivative $\mathbf{u}'(t_0)$ is a vector that is tangent to the curve at the point $\mathbf{u}_0 = \mathbf{u}(t_0)$.

Figure 5 shows our curve with the tangent vector $\mathbf{u}'$ plotted at several points. Actually, we are cheating to some extent. At their true length, the tangent vectors are too short in comparison to the rest of the figure, so the tangent vectors are shown at twice their true length. It is frequently necessary to indulge in this kind of graphical fibbing. This is demonstrated in Figure 4, where the tangent vector is shown at its true length. It was necessary to magnify the area around $\mathbf{u}_0$ in order to show the limiting process because the vectors are too short to be seen otherwise.

The analysis that we have just done for curves in the plane also works for curves in higher dimension. In particular, if $t \to \mathbf{y}(t)$ is the parameterization for a curve in $\mathbf{R}^n$, then $\mathbf{y}'(t)$ is a vector that is tangent to the curve at the point $\mathbf{y}(t)$. Unfortunately, the ease of visualization is gone, especially if n is larger than 3. As is often the case, the best we can do for curves in higher dimension is to use the two-dimensional analogy.

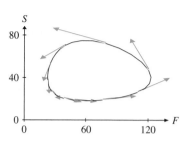

Figure 5. The tangent vectors at several points of the curve.

The phase space and the phase plane

Suppose in general that we have a planar system

$$y_1' = f(t, y_1, y_2),$$
$$y_2' = g(t, y_1, y_2).$$

If $\mathbf{y}(t) = (y_1(t), y_2(t))^T$ is a solution, then we can look at the solution curve $t \to \mathbf{y}(t)$ in the plane. In this case, the $y_1 y_2$-plane is called the **phase plane**, and the solution curve is called a **phase plane plot**, or a **solution curve** in the phase plane.

For a general system of dimension n, $\mathbf{x}' = \mathbf{f}(t, \mathbf{x})$, the space consisting only of the $\mathbf{x}$-coordinates is called the **phase space**. The phase plane is the two-dimensional version of phase space, just as the phase line is the one-dimensional version. A plot of a curve $t \to \mathbf{x}(t)$ is called a **phase space plot.** Phase space plots are especially useful in dimension 2 and sometimes in dimension 3. They become even more important for autonomous systems, but they are occasionally used for nonautonomous systems as well.

An **autonomous system** is a system in which the right-hand side does not depend explicitly on the independent variable. A planar autonomous system has the form

$$x' = f(x, y),$$
$$y' = g(x, y). \tag{2.7}$$

Suppose f and g are defined in a rectangle R in the xy-plane. Consider a solution $(x(t), y(t))$ to (2.7), and the corresponding curve $t \to (x(t), y(t))$ in the xy-plane. At each point $(x(t), y(t))$ on the solution curve, the vector

$$\left(x'(t), y'(t) \right) = \left(f(x(t), y(t)), g(x(t), y(t)) \right) \tag{2.8}$$

is tangent to the curve. The assignment of the vector $(f(x, y), g(x, y))$ on the right-hand side of (2.8) to each point (x, y) in R is called the **vector field** of the system in (2.7). Notice that the vector field can be computed from the system itself. We do not need to know solutions to find the vector field.

Let's look again at the Lotka-Volterra system in (2.5),

$$F' = (0.4 - 0.01S)F$$
$$S' = (-0.3 + 0.005F)S. \tag{2.9}$$

The vector field for this system assigns the vector

$$((0.4 - 0.01S)F, \ (-0.3 + 0.005F)S) \tag{2.10}$$

to the point (F, S) in the plane. Thus, in particular, at the point $(100, 40)$, the value of the vector field is

$$((0.4 - 0.01 \times 40) \times 100, (-0.3 + 0.005 \times 100) \times 40) = (0, 8).$$

This vector field can be visualized if we plot its value at a large number of points as we see in Figure 6.[2] Here we see the vector field at 400 points. We can easily interpolate mentally between these points to visualize the vector field at any point in the rectangle.

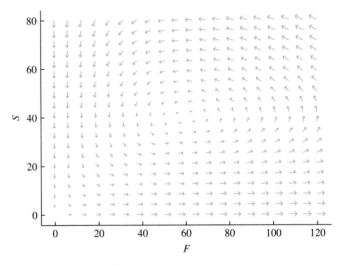

Figure 6. The phase plane for the predator–prey system.

Suppose that $(F(t), S(t))$ is a solution to the Lotka-Volterra system in (2.9), and consider the solution curve in the phase plane parameterized by

$$t \to (F(t), S(t)).$$

An example is shown in Figure 7. The tangent vectors to this curve are given by the vector field in (2.10). Any solution curve for the system (2.9) must have tangents given by the vector field (2.10). This is easily visualized using the phase plane.

It is useful to examine the solution plotted in Figure 7. The initial conditions for the solution were $F(0) = 40$ and $S(0) = 20$. Thus, the curve starts at the point

[2] Actually, in most cases drawing the arrows to full length is graphically displeasing. A better way is to show all of the arrows with a length that is proportional to the actual length. Even this seems to give too much emphasis to the longer arrows, or not enough to the shorter arrows. In Figure 6, the length of the drawn arrows is proportional to the cube root of the length of the actual arrows.

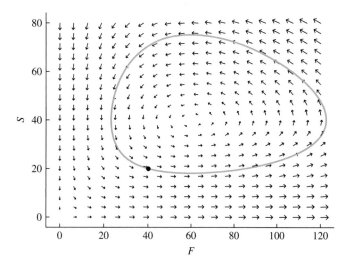

Figure 7. The phase plane for the predator–prey equations with a solution curve plotted.

(40, 20). The vector field shows that the curves proceed first to the right, and then counterclockwise around the curve. At first the prey population, $F(t)$, increases while the predator population, $S(t)$, stays relatively constant. However, the increase in the prey population provides more food for the predators, so they begin to thrive. As they thrive, they eat more of the prey and the prey population begins to decrease. At some point the prey decreases to the point where there is not enough food to sustain the predators, and that population decreases as well. Ultimately, the curve returns to the starting point, illustrating the periodic behavior of the curve.

In Figure 8, we show the phase plane plot for the SIR model we discussed earlier. We are solving only the first two equations in (1.3),

$$S' = -SI,$$
$$I' = SI - I,$$

as a planar autonomous system.

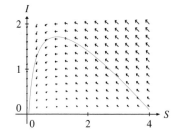

Figure 8. The phase plane for the SIR model.

The direction field

Let's start with a general planar system,

$$x' = f(t, x, y),$$
$$y' = g(t, x, y). \tag{2.11}$$

We will suppose that f and g are defined in a three-dimensional box R in (t, x, y) space with limits

$$a \le t \le b \qquad c \le x \le d \qquad e \le y \le f.$$

Now the vector field is $(f(t, x, y), g(t, x, y))$. Notice that it depends on the independent variable t, as well as on x and y. It makes little sense to look at this vector field in the phase plane because its direction is changing as t changes.

If the phase plane and the vector field are not useful, we can only go back and emulate what we did for a single equation. You will remember that for a single equation we looked at the direction field, which was an object in two dimensions. Now we must look to three dimensions. Suppose $x(t)$ and $y(t)$ are solutions to

(2.11), and consider the curve parameterized by $t \to (t, x(t), y(t))$ in the box R. At each point of this curve, there is a tangent vector, which we get by differentiating. It is

$$\left(1, x'(t), y'(t)\right), \tag{2.12}$$

and using the equations (2.11), this can be written

$$\left(1, f(t, x(t), y(t)), g(t, x(t), y(t))\right). \tag{2.13}$$

Notice that the vector in (2.13), or really the vector

$$\left(1, f(t, x, y), g(t, x, y)\right) \tag{2.14}$$

at the point (t, x, y) in $\mathbf{R}^3$, can be computed without knowing the solution to the system. Thus the system (2.11) can be looked at as an assignment of the vector (2.14) to each point in $\mathbf{R}^3$. This is completely analogous to the definition of the direction field for a single equation, and we will again call this assignment a ***direction field***. It is more difficult to visualize, because now we are working in three dimensions. Probably the best that can be done is to think of the analogy to the dimension 1 case. We will not try to display an example of the direction field.

The geometric interpretation of a solution curve is the same as for a single equation. We start at an initial point $(t_0, x_0, y_0) \in \mathbf{R}^3$. This means we want the solution to the system in (2.11) with initial values

$$x(t_0) = x_0 \quad \text{and} \quad y(t_0) = y_0.$$

The solution curve $t \to (t, x(t), y(t))$ starts at the initial point (t_0, x_0, y_0) in the direction $\left(1, f(t_0, x_0, y_0), g(t_0, x_0, y_0)\right)$, and at every point $(t, x(t), y(t))$ on the curve, the curve must be tangent to the direction field at that point.

While it is difficult to visualize the direction field in dimensions greater than 1, it is possible to visualize a solution curve. An example of such a three-dimensional plot of a solution to the predator–prey system in (2.5) is shown in Figure 9. Some of the tangent vectors to the solution curve are shown. Sometimes a three-dimensional curve is difficult to understand. In such cases, it is perhaps best to think of this by analogy to the dimension 1 case.

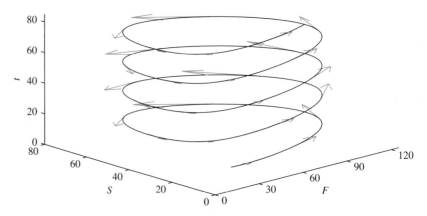

Figure 9. The solution to the predator–prey system in (2.9) starting at $F(0) = 40$ and $S(0) = 20$.

The composite graph

We have discussed three ways to visualize the solution to a planar system. The composite graph brings all three together, and shows the connections between them. Look at Figure 10. This features the solution to the predator–prey system given in (2.5) and illustrated in Figures 1, 2, and 9. The curve in blue will be recognized from Figure 9 as the three-dimensional plot of the solution. The curve on the bottom can be seen from Figure 2 to be the phase plane plot. This curve is the projection of the three-dimensional curve onto the bottom axes plane. The curves on the back and to the side are the plots of the individual components $F(t)$ and $S(t)$ plotted against t, as seen from Figure 1. These curves are the projections of the blue three-dimensional curve onto the respective axes planes.

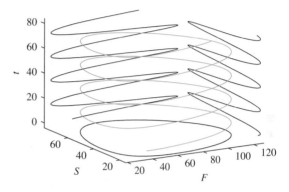

Figure 10. The composite graph of the solution to (2.9).

Clearly, there are a lot of ways to visualize the solutions to systems. Is there always one best way? The answer to that is an emphatic "No!" It is always a good idea to consider a solution from all points of view, since each adds something to our understanding.

Higher dimensions

In this section, we have spent most of our time looking at two-dimensional examples. What do we do in higher dimensions? What we do not do is become daunted by the seeming impossibility of visualizing in dimensions greater than 3. Instead we try anything that seems like it will be helpful. A simple plot of all components of the solution, like that in Figure 1, is always possible and in many cases suffices. The higher-dimensional phase space can be understood only by analogy to the two-dimensional case. However, this analogy is often enough to get useful information.

EXERCISES

In Exercises 1–6, vector-valued functions of time are presented in the form $\mathbf{x}(t) = (x_1(t), x_2(t))^T$.

(i) Use a computer or calculator to prepare a plot of both components versus time (i.e., place the plot of $t \to x_1(t)$ and $t \to x_2(t)$ on the same plot).

(ii) Use a computer or calculator to prepare a plot of $t \to (x_1(t), x_2(t))^T$ in the x_1x_2 phase plane.

1. $\mathbf{x}(t) = (2e^t - e^{-t}, e^{-t})^T$

2. $\mathbf{x}(t) = (9e^{-t} - 4e^{-2t}, 4e^{-2t})^T$

3. $\mathbf{x}(t) = (\cos t, \sin t)^T$

4. $\mathbf{x}(t) = (\cos t, -2 \sin t)^T$

5. $\mathbf{x}(t) = (e^{-t} \cos t, e^{-t} \sin t)^T$

6. $\mathbf{x}(t) = (2e^t \cos 2t, e^t \sin 2t)^T$

In Exercises 7–12, perform each of the following tasks for the given vector-valued function.

(i) Find the derivative of $\mathbf{x}(t)$.

(ii) Sketch the curve $t \to (x_1(t), x_2(t))^T$ in the x_1x_2 phase plane. At incremental points along the curve, use the derivative of $\mathbf{x}(t)$ to calculate and plot the tangent vector to the curve.

7. Use $\mathbf{x}(t)$ from Exercise 1.   **8.** Use $\mathbf{x}(t)$ from Exercise 2.

9. Use $\mathbf{x}(t)$ from Exercise 3.   **10.** Use $\mathbf{x}(t)$ from Exercise 4.

11. Use $\mathbf{x}(t)$ from Exercise 5.   **12.** Use $\mathbf{x}(t)$ from Exercise 6.

In Exercises 13–16, use your numerical solver to draw the direction field for the given planar, autonomous system. Superimpose solution trajectories for several initial conditions of your choice.

13. $\theta' = \omega$ and $\omega' = -\sin\theta$

14. $\theta' = \omega$ and $\omega' = -\sin\theta - 0.5\omega$

15. $x' = (0.4 - 0.01y)x$ and $y' = (0.005x - 0.3)y$

16. $x' = v$ and $v' = -3x - 0.2v$

In Exercises 17–20, use a numerical solver to compute the solution of the initial value problem over the indicated interval. Then prepare the following visualizations of the solution.

(i) Plot the components of the solution as functions of t.

(ii) Make a phase plane plot of the solution.

(iii) If your software has the capability, prepare a composite plot of the solution.

17. $x' = -6x + 10y$, $y' = -5x + 4y$, with $x(0) = 5$ and $y(0) = 1$ over the interval $0 \le t \le 8$

18. $x' = x^2 y$, $y' = -xy^2$, with $x(0) = 1$ and $y(0) = 1$ over the interval $0 \le t \le 4$

19. $x' = 2x - y - x^3$, $y' = x$, with $x(0) = 4$ and $y(0) = 3$ over the interval $0 \le t \le 25$

20. $x' = y$, $y' = x - x^3$, with $x(0) = 0$ and $y(0) = 0.5$ over the interval $0 \le t \le 30$

21. In the images that follow, match the $t \rightarrow (t, x(t))$ and $t \rightarrow (t, y(t))$ plots in the first column with the corresponding $t \rightarrow (x(t), y(t))$ plot in the second column.

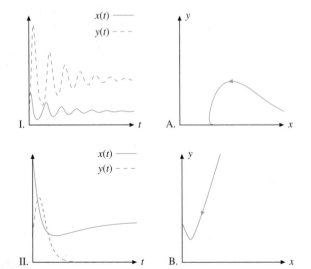

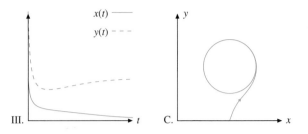

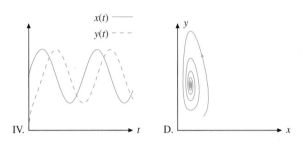

III.   C.

IV.   D.

Each figure in Exercises 22–25 contains the curves $t \rightarrow (t, x(t))$ and $t \rightarrow (t, y(t))$, solutions of the system

$$x' = f(x, y),$$
$$y' = g(x, y).$$

Sketch the solution in the phase plane; that is, plot $t \rightarrow (x(t), y(t))$ in the xy-plane.

22.

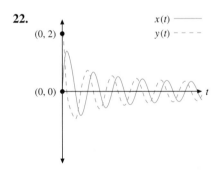

23.

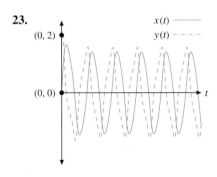

24.

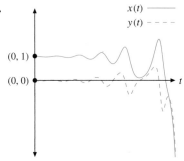

25.

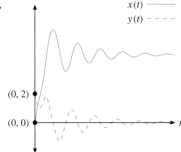

26. Consider the predator–prey system

$$F' = 0.2F - 0.1FS,$$
$$S' = -0.3S + 0.1FS.$$

(a) Use your phase plane program to compute and plot several solutions. Describe the behavior that you observe.

(b) What happens to the solution that starts with $F(0) = 3$ and $S(0) = 2$?

27. Suppose that in the absence of predators, the prey population follows a logistic model, and the predators still follow a Malthusian model. Show that when the interaction is modeled, the resulting equations are of the form

$$F' = aF - eF^2 - bFS$$
$$S' = -cS + dFS.$$

28. Consider the model in the previous exercise with the following parameters.

$$F' = 0.4F - 0.001F^2 - 0.01FS,$$
$$S' = -0.3S + 0.005FS$$

(a) Use your phase plane program to compute and plot several solutions. Describe the behavior that you observe.

(b) What happens to the solution that starts with $F(0) = 60$ and $S(0) = 34$?

29. Consider the predator–prey system

$$F' = aF - bFS,$$
$$S' = -cS + dFS.$$

(a) Provided dF/dt is neither zero nor undefined, show that

$$\frac{dS}{dF} = \frac{(-c + dF)S}{(a - bS)F}.$$

(b) Use the fact that the equation in part (a) is separable to show that its solution is given implicitly by

$$a \ln S - bS + c \ln F - dF = C,$$

where C is some arbitrary constant.

(c) If $a = 0.2$, $b = 0.1$, $c = 0.3$, and $d = 0.1$, use an implicit function plotter to sketch the solution found in part (b) for several values of C.

8.3 Qualitative Analysis

Since systems of equations can rarely be solved exactly, qualitative analysis takes on a greater importance. Here we will discuss existence and uniqueness, but for most other aspects of qualitative analysis we will have to wait. In the next chapter, we will learn how to solve linear systems with constant coefficients explicitly. That is interesting in its own right, but it is equally interesting because it provides us with one of the basic tools for doing qualitative analysis of nonlinear systems. As this indicates, qualitative analysis will be a big part of what we do in the remainder of the book.

Existence and uniqueness

Existence and uniqueness results for systems mirror those we discussed in Chapter 2 for single equations. Only the notation needs to be changed to reflect the higher dimensionality. Compare the results here with the corresponding results in Chapter 2.

Consider a system

$$\mathbf{x}' = \mathbf{f}(t, \mathbf{x}) \tag{3.1}$$

of dimension n. The vector-valued function $\mathbf{x}(t)$ has values in $\mathbf{R}^n$. We will assume that $\mathbf{f}(t, \mathbf{x})$ is defined for t in an interval $I = (a, b)$ and $\mathbf{x}$ in an open set $U \subset \mathbf{R}^n$. Let

$$R = I \times U = \{(t, \mathbf{x}) \mid a < t < b \text{ and } \mathbf{x} \in U\}.$$

With this notation, we can state our existence and uniqueness theorem.

THEOREM 3.2   Suppose the function $\mathbf{f}(t, \mathbf{x})$ is defined and continuous in the region R and that the first partial derivatives of $\mathbf{f}$ are also continuous in R. Then given any point $(t_0, \mathbf{x}_0) \in R$, the initial value problem

$$\mathbf{x}' = \mathbf{f}(t, \mathbf{x}) \quad \text{with} \quad \mathbf{x}(t_0) = \mathbf{x}_0$$

has a unique solution defined in an interval containing t_0. Furthermore, the solution will be defined at least until the solution curve $t \to (t, \mathbf{x}(t))$ leaves the region R. ∎

Notice that to save time we have put the existence and uniqueness results into one theorem. We could have stated different theorems for each, with hypotheses similar to those in the one-dimensional case, but Theorem 3.2 will suffice for our needs. You will also notice that the results are not much different than those in Chapter 2. The only difference is that in this theorem, the functions are vector valued. The geometric interpretation might seem harder or perhaps impossible because the dimension is not within your experience. However, the similarity of the multi-dimensional case to the one-dimensional case allows us to use that case and argue by analogy.

This is certainly true for the existence part. For the uniqueness part, it is better to understand the two-dimensional case and to use that to visualize higher dimensions. In Chapter 2, we made much of the fact that solution curves for single equations were plotted in the plane, and a solution curve divided the plane into two pieces—a second solution curve was not able to cross the first and so would have to stay in one of the two regions defined by the first. However, if the dimension of the system is 2, there are two equations, and solution curves are now plotted in three-dimensional space. It is no longer true that a curve divides space into two pieces, and it is always possible for a second curve to get around the first.

The situation is the same in all dimensions larger than 2. There is simply more room in higher dimensions.

Uniqueness in phase space

That being said, there is a new way of interpreting uniqueness in the case of autonomous systems in dimension 2 or higher. This has to do with solution curves in phase space. Suppose we have an autonomous system $\mathbf{x}' = \mathbf{f}(\mathbf{x})$, where the vector-valued function $\mathbf{f}(\mathbf{x})$ is continuous and has continuous derivatives, so that the uniqueness theorem applies. Suppose in addition that there are two functions $\mathbf{x}(t)$ and $\mathbf{y}(t)$ that are solutions to the system. Then $\mathbf{x}' = \mathbf{f}(\mathbf{x})$, and $\mathbf{y}' = \mathbf{f}(\mathbf{y})$. Suppose too that the solution curves meet at some point in phase space. This means that there are two (possibly different) values of the independent variable t_1 and t_2 such that

$$\mathbf{x}(t_1) = \mathbf{y}(t_2).$$

Consider the function

$$\mathbf{z}(t) = \mathbf{y}(t - t_1 + t_2).$$

We have

$$\mathbf{z}'(t) = \mathbf{y}'(t - t_1 + t_2) = \mathbf{f}(\mathbf{y}(t - t_1 + t_2)) = \mathbf{f}(\mathbf{z}(t)),$$

so $\mathbf{z}(t)$ is also a solution to the differential equation. Furthermore, we have

$$\mathbf{z}(t_1) = \mathbf{y}(t_2) = \mathbf{x}(t_1).$$

Thus $\mathbf{x}$ and $\mathbf{z}$ are solutions to the same initial value problem. By the uniqueness theorem, they must be equal for all t, so

$$\mathbf{x}(t) = \mathbf{z}(t) = \mathbf{y}(t - t_1 + t_2) \quad \text{for all } t.$$

This means that the solution curves [i.e., the curves parameterized by $t \rightarrow \mathbf{x}(t)$ and $t \rightarrow \mathbf{y}(t)$] actually coincide as sets. Only the value of the parameter is different when each of the two curves go through a point. Thus we have discovered an important geometric fact:

> Two solution curves in phase space for an autonomous system cannot meet at a point unless the curves coincide.

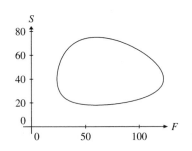

Figure 1. A solution to a predator–prey model.

This geometric fact is especially important in dimension 2. Then the solution curves are in the phase plane, and a curve divides the plane into two pieces. Thus one solution curve provides a lot of information about all others. This is one reason why the phase plane is so useful. Following is an example of how this information might be used. (We should make it clear that this uniqueness result in phase space is a property only of autonomous systems. It does not apply to nonautonomous systems.)

Example 3.3   A Lotka-Volterra model of a predator–prey system has the solution shown in Figure 1. A different solution has initial conditions $F(0) = 60$, and $S(0) = 60$. Is it possible that for some later time t we have $S(t) \geq 100$?

The point $(F(0), S(0)) = (60, 60)$ is inside the closed solution curve plotted in Figure 1. By the uniqueness theorem, the entire solution curve starting at $(F(0), S(0)) = (60, 60)$ must stay inside the closed curve, since to get out, it has to cross this solution curve. However, nowhere on the plotted solution curve is S bigger than 80, so for any solution curve starting inside the plotted curve, we must have $S(t) \leq 80$ for all t.   ●

Existence and uniqueness for linear systems

In Theorem 3.2, we discussed uniqueness for the initial value problem for general first-order systems. There is a stronger theorem if we look only at linear systems.

THEOREM 3.4   Suppose that $A = A(t)$ is an $n \times n$ matrix and $\mathbf{f}(t)$ is a column vector and that the components of both are continuous functions of t in an interval (α, β). Then, for any $t_0 \in (\alpha, \beta)$, and for any $\mathbf{y}_0 \in \mathbf{R}^n$, the inhomogeneous system

$$\mathbf{y}'(t) = A\mathbf{y}(t) + \mathbf{f}(t),$$

with initial condition

$$\mathbf{y}(t_0) = \mathbf{y}_0,$$

has a unique solution defined for all $t \in (\alpha, \beta)$.   ■

In comparison with Theorem 3.2, you will notice that we need only put a condition on the t-dependence of the right-hand side. This is because the condition of linearity ensures that the **y**-dependence is very simple, clearly simple enough to satisfy the conditions posed in Theorem 3.2. As is true with Theorem 3.2, Theorem 3.4 has two parts. The first states that an initial value problem for a linear inhomogeneous system has a solution. The conclusion here is stronger than that of Theorem 3.2, since it says that the solution exists everywhere in the interval (α, β). This is not true in general, so it is a special result for linear systems. It means that the interval of existence of the solution is as large as it possibly can be. It should be pointed out that if the coefficients are constants, then the domain of definition of the system is the entire real line, and constant functions are certainly continuous. Consequently, solutions to linear systems with constant coefficients exist on all of **R**.

The second part of Theorem 3.4 states that the solution is unique. The uniqueness part means that if **y** and **x** are two solutions to the system and satisfy the same initial condition, then $\mathbf{y}(t) = \mathbf{x}(t)$ for all t. This fact will be very important in what follows.

Equilibrium points and solutions

There are some special solutions of differential equations that can be found easily. For example, the autonomous differential equation $y' = y(4 - y)$ has constant solutions $y(t) = 0$ and $y(t) = 4$, which are found by setting the right-hand side of the differential equation to zero and solving for y. These are the equilibrium points and corresponding equilibrium solutions for the equation.

A similar analysis is possible with systems of autonomous equations. For example, let's look back at the predator–prey system in equation (2.4),

$$F' = (a - bS)F,$$
$$S' = (-c + dF)S. \tag{3.5}$$

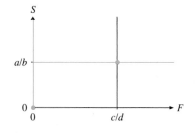

Figure 2. Nullclines and equilibrium points for the predator–prey system.

Now we have two right-hand sides, and we must set both equal to zero. The first, $(a - bS)F = 0$, is solved if either $F = 0$ or $S = a/b$. The solution set is the union of two lines in the phase plane. It is called the F-**nullcline**. It is displayed in blue in Figure 2. The solution set of the second, $(-c + dF)S = 0$, is the union of the two lines $S = 0$ and $F = c/d$. This solution set is called the S-nullcline. It is displayed in black in Figure 2.

An **equilibrium point** is where both right-hand sides are equal to zero. These will be the points that are in the intersection of the two nullclines. The system in (3.5) has two equilibrium points, $(0, 0)^T$ and $(c/d, a/b)^T$. These are marked in Figure 2.

Corresponding to each equilibrium point there is an **equilibrium solution**. For the predator–prey system, these are the constant functions

$$\begin{aligned} F(t) &= 0, \\ S(t) &= 0, \end{aligned} \quad \text{and} \quad \begin{aligned} F(t) &= c/d, \\ S(t) &= a/b. \end{aligned} \tag{3.6}$$

In each case, the constant functions F and S have derivatives equal to zero, and therefore they are solutions. Notice that the solution curve in the phase plane for a constant function consists of a single point. Thus the plots of equilibrium points are also the plots of equilibrium solutions. Figure 3 shows the equilibrium points plotted, together with a number of other solution curves for the system

$$F' = (0.4 - 0.01S)F$$
$$S' = (-0.3 + 0.005F)S. \tag{3.7}$$

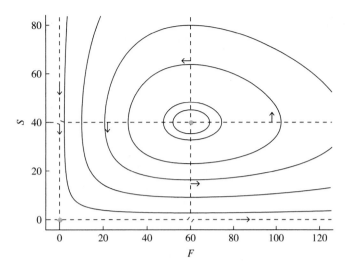

Figure 3. Equilibrium points and solutions for the predator–prey model.

For an arbitrary autonomous system $\mathbf{x}' = \mathbf{f}(\mathbf{x})$, a vector $\mathbf{x}_0$ for which the right-hand side vanishes (i.e., $\mathbf{f}(\mathbf{x}_0) = \mathbf{0}$) is called an ***equilibrium point***. The function $\mathbf{x}(t) = \mathbf{x}_0$ satisfies the equation and is called an ***equilibrium solution***.

Perhaps an inkling of the importance of equilibrium solutions can be seen in Figure 3. Note how solution trajectories tend to "veer away" from the equilibrium point at $(0, 0)$, but the equilibrium point at $(60, 40)$ acts as a "center" about which solutions seem to spiral. Already we see that the types of equilibrium points that occur in dimension 2 are more varied than we saw in dimension 1.

Let's look at another example. Consider the system

$$
\begin{aligned}
x' &= (1 - x - y)x \\
y' &= (4 - 2x - 7y)y.
\end{aligned}
\tag{3.8}
$$

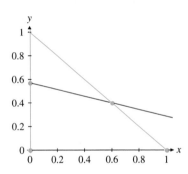

Figure 4. Nullclines and equilibrium points for the system in (3.8).

This system represents two species that are competing for resources. The derivation of the model is in Exercise 13. The nullclines and equilibrium points for this system are shown in Figure 4. The blue lines form the x-nullcline and the black lines are the y-nullcline. They intersect in the four indicated equilibrium points.

The equilibrium points are shown in Figure 5, together with several other solutions. Once more, we see a variety of behavior in the solutions near the various equilibrium points. Again we see that the types of equilibrium points become quite varied in dimension 2. We will look into this in the following two chapters.

Lest you be misled by the simplicity of the nullclines in the previous two examples, we provide the system

$$
\begin{aligned}
x' &= 2x - y + 3(x^2 - y^2) + 2xy, \\
y' &= x - 3y - 3(x^2 - y^2) + 3xy.
\end{aligned}
$$

This system does not model anything, but it does have nice properties. Its nullclines are shown in Figure 6. Clearly, the nullclines are much more complicated than the straight lines in the other examples. You are encouraged to look at this and other examples on your phase plane program to get some idea of the possibilities that can occur.

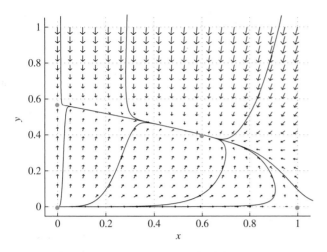

Figure 5. Several solutions for the system in (3.8).

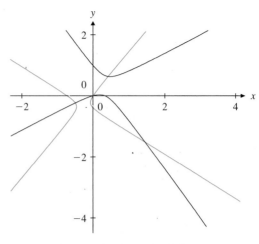

Figure 6. An example of curved nullclines.

EXERCISES

Perform each of the following tasks for the systems in Exercises 1–6.

(i) Plot the nullclines for each equation in the given system of differential equations. Use different colors or linestyles for the x-nullcline and the y-nullcline.

(ii) Calculate the coordinates of the equilibrium points. Plot each equilibrium point in your sketch from part (i) and label it with its coordinates.

1. $x' = 0.2x - 0.04xy$
   $y' = -0.1y + 0.005xy$

2. $x' = 4x - 2x^2 - xy$
   $y' = 4y - xy - 2y^2$

3. $x' = x - y - x^3$
   $y' = x$

4. $x' = 2x - y$
   $y' = -4x + 2y$

5. $x' = y$
   $y' = -\sin x - y$

6. $x' = x + y^2$
   $y' = x + y$

7. Consider the system

$$x' = 1 - (y - \sin x)\cos x$$
$$y' = \cos x - y + \sin x.$$

(a) Show that $x(t) = t$, $y(t) = \sin t$ is a solution.

(b) Plot the solution found in part (a) in the phase plane.

(c) Consider the solution to the system with the initial conditions $x(0) = \pi/2$ and $y(0) = 0$. Show that $y(t) < \sin x(t)$ for all t.

8. Consider the system

$$x' = 1 - e^x(y - e^x)$$
$$y' = 2e^x - y.$$

(a) Show that $x(t) = t$, $y(t) = e^t$ is a solution.

(b) Plot the solution found in part (a) in the phase plane.

(c) Consider the solution to the system with the initial conditions $x(0) = 0$ and $y(0) = 2$. Show that $y(t) > e^{x(t)}$ for all t.

9. Consider the system

$$x' = -y - x(x^2 - y^2)$$
$$y' = -x - y(x^2 - y^2).$$

(a) Show that $x(t) = e^t$, $y(t) = -e^t$ is a solution.

(b) Show that $x(t) = e^{-t}$, $y(t) = e^{-t}$ is a solution.

(c) Plot the solutions found in parts (a) and (b) in the phase plane.

(d) Consider the solution to the system with the initial conditions $x(0) = 1$ and $y(0) = 0$. Show that $-x(t) < y(t) < x(t)$ for all t.

10. Consider the system

$$x' = y - x(x^2 + y^2 - 1)$$
$$y' = -x - y(x^2 + y^2 - 1).$$

(a) Show that $x(t) = \sin t$, $y(t) = \cos t$ is a solution.

(b) Plot the solution found in part (a) in the phase plane.

(c) Consider the solution to the system with the initial conditions $x(0) = 0.5$ and $y(0) = 0$. Show that $x^2(t) + y^2(t) < 1$ for all t.

11. Consider the predator–prey system

$$x' = ax - bxy,$$
$$y' = -cy + dxy,$$

where $x(t)$ and $y(t)$ represent the quantity of prey and predator at time t, respectively. In (3.6), we showed that this system has two equilibrium points, one at $(0, 0)$, a second at $(c/d, a/b)$, and solution trajectories were seen to spiral about this equilibrium point in a periodic fashion. Suppose that we introduce an element of harvesting, detrimental to the rate of growth of both prey and predator, in the following manner, where all constants are assumed to be positive numbers.

$$x' = ax - bxy - \epsilon x$$
$$y' = -cy + dxy - \epsilon y \qquad (3.9)$$

(a) Assuming $\epsilon < a$, plot the nullclines and label all equilibrium points with their coordinates.

(b) Which population is least affected by the harvesting strategy in (3.9), prey or predator? Why?

(c) Adjust the example in (3.7) to

$$x' = (0.4 - 0.01y)x - 0.12x,$$
$$y' = (-0.3 + 0.005x)y - 0.12y, \qquad (3.10)$$

but keep the same initial conditions, namely, $x(0) = 40$ and $y(0) = 20$. Use your numerical solver to prepare separate plots in the xy phase plane of the solution of the original system in (3.7) and the solution of (3.10). Do your plots agree with your findings from part (b)? Why?

12. Let's make a relatively minor change in the predator–prey equations. That is, let's assume that in the absence of predators, the prey grow according to the logistic equation. Then

$$x' = ax\left(1 - \frac{x}{K}\right) - bxy,$$
$$y' = -cy + dxy,$$

where all constants are assumed to be positive numbers.

(a) Assuming $(c/d) < K$, plot the nullclines and label all equilibrium points with their coordinates.

(b) Use your numerical solver to plot the solution of

$$x' = 0.4x\left(1 - \frac{x}{100}\right) - 0.01xy,$$
$$y' = -0.3y + 0.005xy, \qquad (3.11)$$

with initial conditions $x(0) = 40$ and $y(0) = 20$. What appears to be the eventual fate of both the predator and prey populations?

(c) Use your numerical solver to complete a "phase portrait" for system (3.11). A phase portrait is a phase plane that shows a number of solution trajectories with different initial conditions, enough to portray the overall behavior of the system. When you compare solutions to the system in (3.11) with the normal predator–prey system in (3.7), what are the fundamental differences?

13. Suppose that $x(t)$ and $y(t)$ represent populations living in an environment with limited resources. Assume that in the absence of the other population each is governed by the logistic model. Under the assumption that the two populations compete for the resources, show that the two populations are governed by the model

$$x' = ax - bx^2 - cxy$$
$$y' = Ay - By^2 - Cxy,$$

where a, b, c, A, B, and C are positive constants. Notice that the system in (3.8) is a special case of this.

14. In this chapter, we have seen a number of planar (two-dimensional) autonomous systems. How many different kinds of long-term behavior can you identify?

15. Consider again the Lorenz system

$$x' = \sigma(y - x),$$
$$y' = \rho x - y - xz,$$
$$z' = -\beta z + xy.$$

If $\rho > 1$, show that the Lorenz system has three equilibrium points and find their coordinates in terms of σ, ρ, and β.

8.4 Linear Systems

We will say that the unknown functions in a system ***appear linearly*** if there are no products, powers, or higher-order functions involving the unknowns. A system where the unknown functions appear linearly is called a ***linear system***.

A few examples will help to illustrate the difference. The systems

$$\begin{array}{ccc} x' = 3x - 5y & x_1' = -2x_2 & u_1' = -\cos(t)u_2 \\ y' = -2x & x_2' = 3x_1 + 4 & u_2' = u_1 - \sin(t) \end{array}$$

and

are linear. While the independent variable may appear in a nonlinear manner, the unknowns do not. The systems

$$\begin{array}{ccc} x' = 3xy - 5y & x_1' = -2x_2 & u_1' = -\cos(tu_1) \\ y' = -2x & x_2' = 3x_1^2 + 4 & u_2' = u_1 - \sin(t) \end{array}$$

and

are nonlinear, since in each there is a nonlinear expression involving one or more of the unknowns.

Let's formulate precisely what a linear system looks like.

DEFINITION 4.1

A *linear system* of differential equations is any set of differential equations having the following form:

$$
\begin{aligned}
x_1'(t) &= a_{11}(t)x_1(t) + \cdots + a_{1n}(t)x_n(t) + f_1(t) \\
x_2'(t) &= a_{21}(t)x_1(t) + \cdots + a_{2n}(t)x_n(t) + f_2(t) \\
&\ \ \vdots \qquad\qquad\qquad \vdots \\
x_n'(t) &= a_{n1}(t)x_1(t) + \cdots + a_{nn}(t)x_n(t) + f_n(t),
\end{aligned}
\tag{4.2}
$$

where $x_1, \ldots,$ and x_n are the unknown functions. The *coefficients* $a_{ij}(t)$ and $f_i(t)$ are known functions of the independent variable, t, all defined for $t \in I$, where $I = (a, b)$ is an interval in **R**.

A system presented as in (4.2) is said to be in **standard form**. The coefficients can be constants or arbitrary functions of t, but they cannot depend on the unknown functions, $x_1, \ldots,$ and x_n. Notice that there are n equations and the same number of unknown functions. Hence the dimension of the system is n.

If all of the $f_i(t) = 0$, the system is said to be **homogeneous**. Otherwise it is **inhomogeneous**. The functions $f_1, \ldots,$ and f_n are therefore called the **inhomogeneous parts**. The inhomogeneous part is sometimes called the **forcing term**, since in physical systems inhomogeneities usually arise as external forces. In this chapter and the next, we will deal mostly with homogeneous systems.

Example 4.3 Show that the pair of functions $x_1(t) = e^{-t}$ and $x_2(t) = -e^{-t}$ is a solution to the linear system

$$
\begin{aligned}
x_1' &= x_1 + 2x_2, \\
x_2' &= 2x_1 + x_2.
\end{aligned}
\tag{4.4}
$$

Then, repeat the procedure to show that the functions $y_1(t) = e^{3t}$ and $y_2(t) = e^{3t}$ form another solution set.

Note that

$$
x_1' = -e^{-t} \quad \text{and} \quad x_1 + 2x_2 = e^{-t} - 2e^{-t} = -e^{-t},
$$

so the functions $x_1(t) = e^{-t}$ and $x_2(t) = -e^{-t}$ satisfy the first equation of system (4.4). Similarly, simple substitution shows that the given functions also satisfy the second equation of system (4.4).

In like manner, substitution will show that $y_1(t) = e^{3t}$ and $y_2(t) = e^{3t}$ form a second solution of system (4.4). We will show how to find solutions of systems like this in the next chapter. ●

Matrix notation for linear systems

Except for the inhomogeneous terms, the right-hand side of the system of differential equations in (4.2) is really a matrix product. We can utilize this to put the system into a simpler form using vector and matrix notation.

$$
\begin{pmatrix}
x_1'(t) \\
x_2'(t) \\
\vdots \\
x_n'(t)
\end{pmatrix}
=
\begin{pmatrix}
a_{11}(t) & a_{12}(t) & \cdots & a_{1n}(t) \\
a_{21}(t) & a_{22}(t) & \cdots & a_{2n}(t) \\
\vdots & \vdots & \ddots & \vdots \\
a_{n1}(t) & a_{n2}(t) & \cdots & a_{nn(t)}
\end{pmatrix}
\begin{pmatrix}
x_1(t) \\
x_2(t) \\
\vdots \\
x_n(t)
\end{pmatrix}
+
\begin{pmatrix}
f_1(t) \\
f_2(t) \\
\vdots \\
f_n(t)
\end{pmatrix}
$$

We can write this in an even more succinct form as

$$\mathbf{x}'(t) = A(t)\mathbf{x}(t) + \mathbf{f}(t), \tag{4.5}$$

where $A = A(t)$ is the $n \times n$ matrix of the coefficients $a_{ij} = a_{ij}(t)$, $\mathbf{x} = \mathbf{x}(t) = (x_1(t), \ldots, x_n(t))^T$, and $\mathbf{f} = \mathbf{f}(t) = (f_1(t), \ldots, f_n(t))^T$. We will frequently suppress explicit reference to the independent variable and write (4.5) as

$$\mathbf{x}' = A\mathbf{x} + \mathbf{f}.$$

Example 4.6   Write the system in Example 4.3 in matrix form. Write the solutions as vectors and verify using matrix multiplication that they are solutions.

For the system in (4.4), the matrix A is given by

$$A = \begin{pmatrix} 1 & 2 \\ 2 & 1 \end{pmatrix}.$$

The system is homogeneous, so $\mathbf{f} = 0$. As vectors, the solution sets given in Example 4.3 become

$$\mathbf{x}(t) = \begin{pmatrix} x_1(t) \\ x_2(t) \end{pmatrix} = \begin{pmatrix} e^{-t} \\ -e^{-t} \end{pmatrix} \quad \text{and} \quad \mathbf{y}(t) = \begin{pmatrix} y_1(t) \\ y_2(t) \end{pmatrix} = \begin{pmatrix} e^{3t} \\ e^{3t} \end{pmatrix}. \tag{4.7}$$

Substituting $\mathbf{x}$ into the system, we get

$$\mathbf{x}' = \begin{pmatrix} -e^{-t} \\ e^{-t} \end{pmatrix}, \quad \text{and} \quad A\mathbf{x} = \begin{pmatrix} 1 & 2 \\ 2 & 1 \end{pmatrix} \begin{pmatrix} e^{-t} \\ -e^{-t} \end{pmatrix} = \begin{pmatrix} -e^{-t} \\ e^{-t} \end{pmatrix}.$$

Hence $\mathbf{x}' = A\mathbf{x}$, so we have a solution. That $\mathbf{y}$ is a solution is verified in the same way.   ●

In this example, the matrix consists of constants. In a general linear system, the coefficients could be functions depending on the independent variable t. Almost everything we say in this chapter will be true for systems with variable coefficients.

Applications of linear systems

Perhaps the most important application of linear systems is in analyzing equilibrium points for nonlinear systems. We will explore this in Chapter 10. However, there are significant applications which can be modeled by linear systems themselves. Since there is a first-order linear system equivalent to any higher order linear equation, any of the applications of Chapter 4 lead to linear systems. We will present three other examples.

Multiple springs

In Chapter 4 we discovered that the motion of a vibrating spring can be modeled by a second-order differential equation, and therefore by a first-order linear system of dimension 2. A mechanical system containing more than one spring can also be modeled by a first-order linear system.

Example 4.8 Consider the coupled spring-mass system consisting of two masses linked by three springs, as shown in Figure 1. Assume there is no damping and there are no external forces. First find a system of second-order differential equations that governs the positions of masses m_1 and m_2. Then find an equivalent system of first-order differential equations.

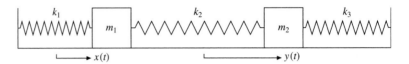

Figure 1. The three springs and two masses.

Each of the masses has its own equilibrium position where the entire system is motionless. Let $x(t)$ and $y(t)$ denote the displacements of mass m_1 and mass m_2, respectively, from these equilibrium positions. We will assume that $x(t)$ and $y(t)$ are positive to the right of their respective equilibrium positions.

First let's examine the forces acting on m_1. There are two springs connected to m_1 and the effects of both must be taken into account. The first spring exerts a force of $F_1 = -k_1 x$ on mass m_1, where k_1 is the spring constant of the first spring. The second spring exerts a force of $F_2 = k_2(y - x)$. Note the sign. If y is greater than x (so $y - x > 0$), then the second spring is stretched and the force it exerts on the first mass is directed to the right. Newton's second law tells us that

$$m_1 x'' = F_1 + F_2 = -k_1 x + k_2(y - x).$$

There are also two springs acting on the second mass. The second spring exerts the force $-F_2 = -k_2(y - x)$ (the opposite of what the second spring exerts on the first mass). The third spring exerts the force $F_3 = -k_3 y$. The resulting differential equation for y is

$$m_2 y'' = -F_2 + F_3 = -k_2(y - x) - k_3 y.$$

We rewrite the equations as

$$x'' = \frac{-(k_1 + k_2)}{m_1} x + \frac{k_2}{m_1} y,$$

$$y'' = \frac{k_2}{m_2} x - \frac{(k_2 + k_3)}{m_2} y.$$

This is a system of two second-order equations. To find an equivalent first-order system, we proceed as we did in Example 1.15. We let $\mathbf{u} = (u_1, u_2, u_3, u_4)^T$, with

$$u_1(t) = x(t), \quad u_2(t) = x'(t), \quad u_3(t) = y(t), \quad \text{and} \quad u_4(t) = y'(t).$$

Note that

$$u_2' = x'' = \frac{-(k_1 + k_2)}{m_1} x + \frac{k_2}{m_1} y = \frac{-(k_1 + k_2)}{m_1} u_1 + \frac{k_2}{m_1} u_3.$$

Likewise

$$u_4' = y'' = \frac{k_2}{m_2} x - \frac{(k_2 + k_3)}{m_2} y = \frac{k_2}{m_2} u_1 - \frac{(k_2 + k_3)}{m_2} u_3.$$

Together with the equations $u_1' = x' = u_2$ and $u_3' = y' = u_4$, we obtain the system

$$u_1' = u_2,$$

$$u_2' = \frac{-(k_1 + k_2)}{m_1} u_1 + \frac{k_2}{m_1} u_3,$$

$$u_3' = u_4,$$

$$u_4' = \frac{k_2}{m_2} u_1 - \frac{(k_2 + k_3)}{m_2} u_3.$$

This is a linear, autonomous system of dimension 4. The right-hand side can be expressed as a matrix product, and then the system is

$$\begin{pmatrix} u_1' \\ u_2' \\ u_3' \\ u_4' \end{pmatrix} = \begin{pmatrix} 0 & 1 & 0 & 0 \\ -(k_1+k_2)/m_1 & 0 & k_2/m_1 & 0 \\ 0 & 0 & 0 & 1 \\ k_2/m_2 & 0 & -(k_2+k_3)/m_2 & 0 \end{pmatrix} \begin{pmatrix} u_1 \\ u_2 \\ u_3 \\ u_4 \end{pmatrix}.$$

The initial conditions for this system involve the initial position and velocity of both masses,

$$x(0) = a_1, \ x'(0) = b_1, \quad \text{and} \quad y(0) = a_2, \ y'(0) = b_2.$$

The initial conditions for the first-order system can be stated in one vector equation as $\mathbf{u}(0) = (a_1, b_1, a_2, b_2)^T$. ●

Electrical circuits

We discussed electrical circuits in Section 3.4, and we found that all of the component laws are expressible in terms of first derivatives. It follows that any circuit involving resistors, capacitors, and inductors can be modeled by a first-order system.

Example 4.9 Find a first-order system that models the circuit in Figure 2.

The circuit contains a voltage source E, a resistor R, a capacitor C, and an inductor L. However, the capacitor and the inductor are in parallel, so this circuit is quite different from those we considered in Section 3.4.

First we use Kirchoff's current law at the junction to the right of the resistor. With the currents as indicated in Figure 2, we get

$$I = I_1 + I_2. \tag{4.10}$$

Next, Kirchoff's voltage law applied to the loop containing the source and the inductor tells us

$$E = RI + LI_1'.$$

By solving this for I_1' and using (4.10), we get the equation

$$I_1' = \frac{1}{L}[E - R(I_1 + I_2)]. \tag{4.11}$$

Kirchoff's voltage law applied to the loop containing the source, resistor, and the capacitor yields

$$E = RI + \frac{1}{C}Q,$$

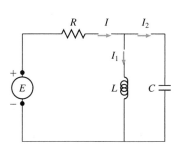

Figure 2. An electrical circuit with a capacitor and an inductor in parallel.

where Q is the charge on the capacitor. When we differentiate this equation, using $Q' = I_2$ and (4.10), this becomes

$$E' = R(I_1' + I_2') + \frac{1}{C}I_2.$$

We solve this equation for I_2' and use the substitution in (4.11) to get

$$
\begin{aligned}
I_2' &= -I_1' + \frac{1}{R}\left[E' - \frac{1}{C}I_2\right] \\
&= -\frac{1}{L}[E - R(I_1 + I_2)] + \frac{1}{R}\left[E' - \frac{1}{C}I_2\right].
\end{aligned}
\tag{4.12}
$$

Equations (4.11) and (4.12) form an inhomogeneous linear system for I_1 and I_2. This is a little easier to see if we rewrite the system as

$$
\begin{aligned}
I_1' &= -\frac{R}{L}I_1 - \frac{R}{L}I_2 + \frac{E}{L}, \\
I_2' &= \frac{R}{L}I_1 + \left(\frac{R}{L} - \frac{1}{RC}\right)I_2 + \left(\frac{E'}{R} - \frac{E}{L}\right).
\end{aligned}
\tag{4.13}
$$

To write the system in (4.13) in matrix notation, we introduce the vector $\mathbf{I} = (I_1, I_2)^T$. Then with

$$A = \begin{pmatrix} -R/L & -R/L \\ R/L & R/L - 1/RC \end{pmatrix} \quad \text{and} \quad \mathbf{F} = \begin{pmatrix} E/L \\ E'/R - E/L \end{pmatrix},$$

the system can be written as $\mathbf{I}' = A\mathbf{I} + \mathbf{F}$.　●

Mixing problems

We have seen in Section 2.5 of Chapter 2 that mixing problems lead to linear equations. There we limited ourselves mostly to mixtures in one tank, although in Example 5.4 in Section 2.5 we found that a mixing problem involving two tanks leads to two first-order linear differential equations—a linear system. That system was simple enough that we could solve it without difficulty. More complicated mixing situations lead to first-order linear systems that are not so easy to solve.

Example 4.14　As an example, consider the two tanks in Figure 3. We have two tanks, connected by two pipes. Each tank contains 500 gallons of a salt solution. Through one pipe solution is pumped from the first tank to the second at 1 gal/min. Through the other, solution is pumped at the same rate from the second tank to the first. We want to know how the salt content in each tank varies with time.

Let $x_1(t)$ and $x_2(t)$ represent the salt content in the two tanks, measured in pounds. For the tank on the left, we have

$$\text{Rate out} = 1 \text{ gal/min} \times \frac{x_1}{500} \text{ lb/gal} = \frac{x_1}{500}\text{lb/min}, \quad \text{and}$$

$$\text{Rate in} = 1 \text{ gal/min} \times \frac{x_2}{500} \text{ lb/gal} = \frac{x_2}{500}\text{lb/min}.$$

Since $dx_1/dt = \text{Rate in} - \text{Rate out}$, we have

$$\frac{dx_1}{dt} = -\frac{x_1}{500} + \frac{x_2}{500}.
\tag{4.15}$$

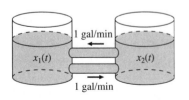

Figure 3. Two tanks feeding each other.

In a similar manner we find that

$$\frac{dx_2}{dt} = \frac{x_1}{500} - \frac{x_2}{500}. \tag{4.16}$$

Equations (4.15) and (4.16) form a first-order linear system for the unknowns x_1 and x_2. If we set $\mathbf{x} = (x_1, x_2)^T$, we can write the system using vector notation as

$$\mathbf{x}' = A\mathbf{x}, \quad \text{where} \quad A = \begin{pmatrix} -1/500 & 1/500 \\ 1/500 & -1/500 \end{pmatrix}. \tag{4.17}$$

EXERCISES

Use Definition 4.1 to determine which of the systems in Exercises 1–6 are linear systems. For those systems that are linear, which are homogeneous? Which are inhomogeneous?

1. $x_1' = -x_2$ and $x_2' = -x_1 - 2x_2 + 5\sin t$

2. $x_1' = -2x_1 + x_1 x_2$ and $x_2' = -3x_1 - x_2$

3. $x_1' = -x_2$ and $x_2' = \sin x_1$

4. $x_1' = x_1 + (\sin t)x_2$ and $x_2' = 2tx_1 - x_2$

5. $x_1' = x_2, x_2' = x_3$, and $x_3' = -x_1 - x_2 + \sin t$.

6. $x_1' = -x_1 + tx_2 - t^2 x_3, x_2' = -2x_1 + 2tx_2$, and $x_3' = -t^2 x_1 + (\sin t)x_2 + \sin x_3$

In Exercises 7–10, use direct substitution, as in Example 4.3, to show that the given pair of functions $x_1(t)$ and $x_2(t)$ is a solution of the given system.

7. $x_1' = 3x_1 + 2x_2, x_2' = -4x_1 - 3x_2, x_1(t) = e^t, x_2(t) = -e^t$.

8. $x_1' = -5x_1 - 3x_2, x_2' = 6x_1 + 4x_2, x_1(t) = e^{-2t}, x_2(t) = -e^{-2t}$.

9. $x_1' = -2x_1 - x_2, x_2' = -x_2, x_1(t) = 2e^{-t} - e^{-2t}, x_2(t) = -2e^{-t}$.

10. $x_1' = 3x_1 - 5x_2, x_2' = -2x_2, x_1(t) = -2e^{-2t} + 3e^{3t}, x_2(t) = -2e^{-2t}$.

In Exercises 11–16 place the given differential system in the form

$$\mathbf{x}' = A(t)\mathbf{x} + \mathbf{f}(t)$$

if possible. If it is not possible, explain why.

11. $x_1' = -2x_1 + 3x_2$
$x_2' = x_1 - 4x_2$

12. $(1/t)x_1' = x_1 + (\cos t)x_2$
$x_2' = (\sin t)x_1 + 3x_2$

13. $x_1' = -2x_1 + x_2^2$
$x_2' = 3x_1 - x_2$

14. $x_1' = -2x_1 + 3tx_2 + \cos t$
$tx_2' = x_1 - 4tx_2 + \sin t$

15. $t^2 x_1' = -2tx_1 x_2 + 3x_2$
$(1/t)x_2' = tx_1 - (4/t)x_2$

16. $x_1' = -2tx_1 + 3x_2 - (\cos t)x_3$
$x_2' = x_1 - 4t^2 x_3$
$x_3' = -(3/t)x_2 - \cos t$

17. Which of the systems in Exercises 11–16 are linear? Which are linear and homogeneous?

In Exercises 18–21, write the given system of equations in matrix-vector form, then use the technique of Example 4.6 to show that the given vector-valued function is a solution of the system.

18. $x_1' = -3x_1, x_2' = -2x_1 - x_2, \quad \mathbf{x} = (e^{-3t}, e^{-3t})^T$

19. $x_1' = 8x_1 - 10x_2, x_2' = 5x_1 - 7x_2, \quad \mathbf{x} = (2e^{3t}, e^{3t})^T$

20. $x_1' = -3x_1 + x_2, x_2' = -2x_1, \quad \mathbf{x} = (-e^{-2t} + e^{-t}, -e^{-2t} + 2e^{-t})^T$

21. $x_1' = -x_1 + 4x_2, x_2' = 3x_2, \quad \mathbf{x} = (e^{3t} - e^{-t}, e^{3t})^T$

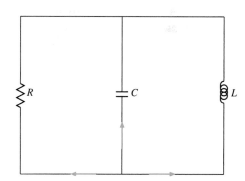

Figure 4. A parallel circuit with capacitor, resistor, and inductor.

22. Consider the parallel circuit pictured in Figure 4. Let V represent the voltage drop across the capacitor and I represent the current across the indcutor. Show that

$$V' = -\frac{V}{RC} - \frac{I}{C} \tag{4.18}$$

$$I' = \frac{V}{L}. \tag{4.19}$$

Assume the current flows in the directions indicated.

23. Consider the circuit pictured in Figure 5. Let I represent the current through the inductor and let V represent the voltage drop across the capacitor. Assume current flows in the given directions. Show that

$$LI' = -R_1 I + V \tag{4.20}$$

$$CV' = -I - \frac{V}{R_2}. \tag{4.21}$$

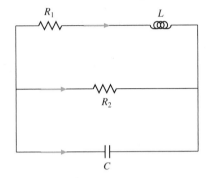

Figure 5. A more complicated example.

24. Consider the circuit pictured in Figure 6. Let I_1 and I_2 represent the current flow across the inductors L_1 and L_2, respectively. Show that the circuit is modeled by the system

$$L_1 I_1' = -R_1 I_1 - R_1 I_2 + E \tag{4.22}$$
$$L_2 I_2' = -R_1 I_1 - (R_1 + R_2) I_2 + E. \tag{4.23}$$

Use the current flow indicated in Figure 6.

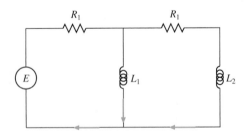

Figure 6. Two inductors, two resistors, and a voltage source E.

25. Pictured in Figure 7 are two tanks, each containing 100 gallons of a salt solution. Pure water flows into the upper tank at a rate of 4 gal/min. Salt solution drains from the upper tank into the lower tank at a rate of 4 gal/min. Finally, salt solution drains from the lower tank at a rate of 4 gal/min, effectively keeping the volume of solution in each tank at a constant 100 gal. If the initial salt content of the upper and lower tanks is 10 and 20 pounds, respectively, set up, but do not solve, an initial value problem that models the amount of salt in each tank over time. Write your model in matrix-vector form. Is the system homogeneous or inhomogeneous?

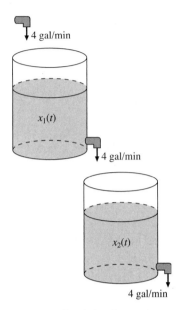

Figure 7. Coupled tanks.

26. Pictured in Figure 8 are two tanks, each containing 200 gallons of pure water. Salt solution enters the upper tank at a rate of 5 gal/min having salt concentration 0.2 lb/gal. Solution drains from the upper tank directly into the lower tank at a rate of 5 gal/min. Finally, solution leaves the lower tank at a rate of 5 gal/min, effectively maintaining the volume of solution in each tank constant. Suppose that there is initially 10 pounds of salt in each tank. Set up, but do not solve, an initial value problem that models the amount of salt in each tank over time. Place your solution in matrix-vector form. Is the system homogeneous or inhomogeneous?

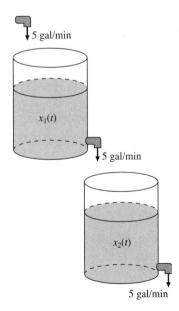

Figure 8. Coupled tanks with forcing.

27. Consider the cascading tanks in Figure 9.

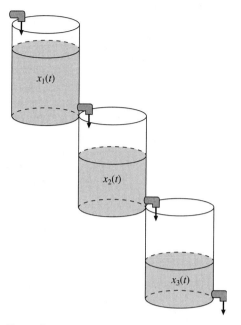

Figure 9. Three cascading tanks.

Initially, the first tank contains 100 gallons of pure water, the second 80 gallons of pure water, and the third 60 gallons of pure water. Salt solution containing 2 pounds of salt per gallon of water pours into the first tank at a rate of 5 gallons per minute. Salt solution from the first tank drains into the second tank at a rate of 5 gal/min. Finally,

salt solution drains from the second tank into the third tank at a rate of 5 gal/min, effectively keeping the volume of solution in each tank constant. Set up, but do not solve, an initial value problem that models the amount of salt in each tank over time. Place your solution in matrix-vector form. Is the system homogeneous or inhomogeneous?

28. Two masses on a frictionless tabletop are connected with a spring having spring constant k_2. The first mass is connected to a vertical support with a spring having spring constant k_1, as shown in Figure 10. Finally, the second mass is shaken harmonically via a force equaling $F = A \cos \omega t$, also shown in Figure 10. Let $x(t)$ and $y(t)$ measure the displacements of the masses m_1 and m_2, respectively, from their equilibrium positions as a function of time. If both masses start from rest at their equilibrium positions at time $t = 0$, set up, but do not solve, an initial value problem that models the position of the masses over time. Write your solution in matrix-vector form. Is the system homogeneous or inhomogeneous?

29. Three identical masses slide on a frictionless tabletop, as pictured in Figure 11.

The masses are connected with springs (and to vertical supports at each end) having identical spring constants k, also pictured in Figure 11. Let $x(t)$, $y(t)$, and $z(t)$ represent the displacement of the masses from their equilibrium positions. Suppose that each mass is initially displaced 10 cm to the right and released from rest. Set up, but do not solve, an initial value problem that models the position of the masses over time. Place your solution in matrix-vector form. Is the system homogeneous or inhomogeneous?

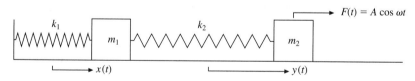

Figure 10. Coupled springs with driving force F.

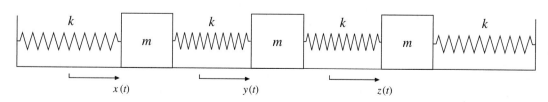

Figure 11. Three identical masses connected by identical springs.

8.5 Properties of Linear Systems

In this section, we look at some of the special properties of linear systems. This will prepare us for the next chapter, where we will explicitly solve linear systems with constant coefficients. While reading this section, you should notice that the subject of linear systems of differential equations is somewhat analogous to the study of systems of linear equations in linear algebra. It is true that many of the tools from linear algebra will be used in the process of finding solutions to systems of differential equations. However, the similarity goes beyond this.

This section is somewhat technical, but the results make it worthwhile. We will end up with a strategy that inform our search for complete solutions to linear systems in Chapter 9.

Properties of homogeneous systems

The next theorem sums up the most important effect of linearity. In mathematics courses where linearity is studied for its own sake, this is the defining property.

THEOREM 5.1 Suppose $\mathbf{x}_1$ and $\mathbf{x}_2$ are solutions to the homogeneous linear system

$$\mathbf{x}' = A\mathbf{x}. \tag{5.2}$$

If C_1 and C_2 are any constants, then $\mathbf{x} = C_1\mathbf{x}_1 + C_2\mathbf{x}_2$ is also a solution to (5.2).

Proof We just compute, using the linearity of matrix multiplication.[3]

$$\begin{aligned}
\mathbf{x}' &= (C_1\mathbf{x}_1 + C_2\mathbf{x}_2)' \\
&= C_1\mathbf{x}_1' + C_2\mathbf{x}_2' \\
&= C_1 A\mathbf{x}_1 + C_2 A\mathbf{x}_2 \\
&= A(C_1\mathbf{x}_1 + C_2\mathbf{x}_2) \\
&= A\mathbf{x}
\end{aligned}$$

■

Theorem 5.1 does *not* hold if the system is nonlinear or if the system is inhomogeneous. The following examples illustrate this.

Example 5.3 Consider the system

$$\begin{aligned}
x_1' &= x_1^2, \\
x_2' &= x_1 x_2.
\end{aligned}$$

This system is nonlinear. Consider the vector-valued function $\mathbf{x}(t) = (-1/t, 0)^T$, with components $x_1(t) = -1/t$ and $x_2(t) = 0$. Then

$$x_1'(t) = 1/t^2 = x_1(t)^2 \quad \text{and} \quad x_2(t)' = 0 = x_1(t)x_2(t),$$

so $\mathbf{x}$ is a solution. However, $\mathbf{y}(t) = 2\mathbf{x}(t) = (-2/t, 0)^T$ is not a solution, as we can see by checking the first equation in the system. Since $y_1'(t) = (-2/t)' = 2/t^2$, while $y_1^2(t) = (-2/t)^2 = 4/t^2$, this equation is not satisfied. ●

---

[3] We encourage you to prove this result for a system of dimension 2 without using matrix notation. Nothing is more convincing of the efficiency provided by matrix notation.

Example 5.4   Consider the inhomogeneous linear system

$$\begin{pmatrix} x_1' \\ x_2' \end{pmatrix} = \begin{pmatrix} 1 & 2 \\ 2 & 1 \end{pmatrix} \begin{pmatrix} x_1 \\ x_2 \end{pmatrix} + \begin{pmatrix} -6 \\ 3 + 9t \end{pmatrix}.$$

Consider $\mathbf{x}(t) = (-6t, 3t)^T$. Then $\mathbf{x}'(t) = (-6, 3)^T$. On the other hand,

$$\begin{pmatrix} 1 & 2 \\ 2 & 1 \end{pmatrix} \begin{pmatrix} -6t \\ 3t \end{pmatrix} + \begin{pmatrix} -6 \\ 3 + 9t \end{pmatrix} = \begin{pmatrix} 0 \\ -9t \end{pmatrix} + \begin{pmatrix} -6 \\ 3 + 9t \end{pmatrix} = \begin{pmatrix} -6 \\ 3 \end{pmatrix},$$

so $\mathbf{x}$ is a solution. In the same way, we can show that $\mathbf{y}(t) = (e^{-t} - 6t, -e^{-t} + 3t)^T$ is a solution.

However, $\mathbf{x} + \mathbf{y} = (e^{-t} - 12t, -e^{-t} + 6t)^T$ is not. To see this, we first compute

$$(\mathbf{x} + \mathbf{y})'(t) = \begin{pmatrix} -e^{-t} - 12 \\ e^{-t} + 6 \end{pmatrix}.$$

Next we compute

$$\begin{pmatrix} 1 & 2 \\ 2 & 1 \end{pmatrix} \begin{pmatrix} e^{-t} - 12t \\ -e^{-t} + 6t \end{pmatrix} + \begin{pmatrix} -6 \\ 3 + 9t \end{pmatrix} = \begin{pmatrix} -e^{-t} \\ e^{-t} - 18t \end{pmatrix} + \begin{pmatrix} -6 \\ 3 + 9t \end{pmatrix}$$

$$= \begin{pmatrix} -e^{-t} - 6 \\ e^{-t} + 3 - 9t \end{pmatrix}.$$

So we see that the equation is not satisfied.   ●

Theorem 5.1 shows that an arbitrary linear combination of two solutions to a linear, homogeneous system is also a solution. There is nothing special about the number two. Essentially the same proof proves the same result for a linear combination of an arbitrary number of solutions. We will state this formally.

THEOREM 5.5   Suppose that $\mathbf{x}_1, \mathbf{x}_2, \ldots,$ and $\mathbf{x}_k$ are all solutions to the homogeneous linear system $\mathbf{x}' = A\mathbf{x}$. Then any linear combination of $\mathbf{x}_1, \mathbf{x}_2, \ldots,$ and $\mathbf{x}_k$ is also a solution. Thus for any constants $C_1, C_2, \ldots,$ and C_k, the function

$$\mathbf{x}(t) = C_1\mathbf{x}_1(t) + C_2\mathbf{x}_2(t) + \cdots + C_k\mathbf{x}_k(t)$$

is a solution to $\mathbf{x}' = A\mathbf{x}$.   ■

An important example

We know that if $\mathbf{x}_1, \mathbf{x}_2, \ldots,$ and $\mathbf{x}_k$ are solutions to a linear homogeneous system, then so is any linear combination. The key question is the following: Can all solutions to a linear homogeneous system be expressed as a linear combination of certain special solutions? The following example is illuminating.

Example 5.6   In Examples 4.3 and 4.6, we pointed out that

$$\mathbf{x}_1(t) = \begin{pmatrix} e^{-t} \\ -e^{-t} \end{pmatrix} = e^{-t} \begin{pmatrix} 1 \\ -1 \end{pmatrix} \quad \text{and} \quad \mathbf{x}_2(t) = \begin{pmatrix} e^{3t} \\ e^{3t} \end{pmatrix} = e^{3t} \begin{pmatrix} 1 \\ 1 \end{pmatrix}$$

are solutions to the homogeneous system

$$\mathbf{x}'(t) = \begin{pmatrix} 1 & 2 \\ 2 & 1 \end{pmatrix} \mathbf{x}(t). \tag{5.7}$$

By Theorem 5.1, we know that any linear combination

$$\mathbf{x}(t) = C_1\mathbf{x}_1(t) + C_2\mathbf{x}_2(t)$$

is also a solution. Show that all solutions to this system can be expressed as linear combinations of $\mathbf{x}_1$ and $\mathbf{x}_2$.

Let $\mathbf{x}$ be any solution to this system. We want to show that $\mathbf{x}$ can be expressed as

$$\mathbf{x}(t) = C_1\mathbf{x}_1(t) + C_2\mathbf{x}_2(t) \tag{5.8}$$

for some choice of constants C_1 and C_2. Finding the constants so that this equation is satisfied for one value of t is difficult enough, but we want to do it for all t. We will start small and make sure that equation (5.8) is satisfied at one point, say at $t = 0$. Accordingly, we try to find C_1 and C_2 so that

$$\mathbf{x}(0) = C_1\mathbf{x}_1(0) + C_2\mathbf{x}_2(0). \tag{5.9}$$

Substituting the values of $\mathbf{x}_1(0)$ and $\mathbf{x}_2(0)$, this becomes

$$\mathbf{x}(0) = C_1 \begin{pmatrix} 1 \\ -1 \end{pmatrix} + C_2 \begin{pmatrix} 1 \\ 1 \end{pmatrix} = \begin{pmatrix} 1 & 1 \\ -1 & 1 \end{pmatrix} \begin{pmatrix} C_1 \\ C_2 \end{pmatrix}.$$

Since the function $\mathbf{x}(t)$ is known, the value $\mathbf{x}(0)$ is also known. Therefore, this is a linear system of equations for the unknown constants C_1 and C_2. From our study of matrix algebra, we know that the system can be solved provided the matrix

$$\begin{pmatrix} 1 & 1 \\ -1 & 1 \end{pmatrix} = [\mathbf{x}_1(0), \mathbf{x}_2(0)] \tag{5.10}$$

is nonsingular. This matrix will be nonsingular if the vectors $\mathbf{x}_1(0)$ and $\mathbf{x}_2(0)$ are linearly independent. In this case, perhaps the easiest way to find that out is to compute that the determinant is nonzero. Since the determinant is 2, the matrix is nonsingular and the equation

$$\mathbf{x}(0) = \begin{pmatrix} 1 & 1 \\ -1 & 1 \end{pmatrix} \begin{pmatrix} C_1 \\ C_2 \end{pmatrix} \tag{5.11}$$

can be solved for C_1 and C_2, no matter what the vector $\mathbf{x}(0)$ is. For example, if the given initial condition is

$$\mathbf{x}(0) = \begin{pmatrix} 3 \\ -1 \end{pmatrix},$$

then $C_1 = 2$ and $C_2 = 1$, as can be seen easily with a little algebra.

The functions $\mathbf{x}(t)$ and $C_1\mathbf{x}_1(t) + C_2\mathbf{x}_2(t)$ are both solutions to the system in equation (5.7). In addition, we have found constants C_1 and C_2 so that equation (5.9) is valid. This means that $\mathbf{x}(t)$ and $C_1\mathbf{x}_1(t) + C_2\mathbf{x}_2(t)$ have the same initial value at $t = 0$. According to the existence and uniqueness theorem (Theorem 3.4), $\mathbf{x}(t)$ and $C_1\mathbf{x}_1(t) + C_2\mathbf{x}_2(t)$ must agree for all t. Thus we have shown that (5.8) is true for all t.   ●

Let's go over this argument. We wanted to find the constants C_1 and C_2 so that equation (5.8) is satisfied for all t. We used algebra in (5.11) to solve for the constants that would make equation (5.8) true at $t = 0$. Then the uniqueness theorem implied that equation (5.8) is satisfied for all t.

Linear independence and dependence

Looking back on this example, we see that what allowed us to solve for the constants was the fact that the matrix in (5.11) and (5.10) is nonsingular. This is true if and only if the vectors $\mathbf{x}_1(0)$ and $\mathbf{x}_2(0)$ are linearly independent. Why were we able to use linear independence at only one value of t? We will see why in the following proposition.

PROPOSITION 5.12   Suppose that $\mathbf{y}_1(t)$, $\mathbf{y}_2(t)$, ..., and $\mathbf{y}_k(t)$ are solutions to the n-dimensional system $\mathbf{y}' = A\mathbf{y}$ defined on the interval $I = (\alpha, \beta)$.

1. If the vectors $\mathbf{y}_1(t_0)$, $\mathbf{y}_2(t_0)$, ..., and $\mathbf{y}_k(t_0)$ are linearly dependent for some $t_0 \in I$, then there are constants $C_1, C_2, \ldots$, and C_k, not all zero, such that $C_1\mathbf{y}_1(t) + C_2\mathbf{y}_2(t) + \cdots + C_k\mathbf{y}_k(t) = \mathbf{0}$ for all $t \in I$. In particular, $\mathbf{y}_1(t)$, $\mathbf{y}_2(t)$, ..., and $\mathbf{y}_k(t)$ are linearly dependent for all $t \in I$.

2. If for some $t_0 \in I$ the vectors $\mathbf{y}_1(t_0)$, $\mathbf{y}_2(t_0)$, ..., and $\mathbf{y}_k(t_0)$ are linearly independent, then $\mathbf{y}_1(t)$, $\mathbf{y}_2(t)$, ..., and $\mathbf{y}_k(t)$ are linearly independent for all $t \in I$.

Proof   If the vectors $\mathbf{y}_1(t_0)$, $\mathbf{y}_2(t_0)$, ..., and $\mathbf{y}_k(t_0)$ are linearly dependent for some $t_0 \in I$, then there are constants $C_1, C_2, \ldots$, and C_k, not all zero, such that

$$C_1\mathbf{y}_1(t_0) + C_2\mathbf{y}_2(t_0) + \cdots + C_k\mathbf{y}_k(t_0) = \mathbf{0}.$$

Define the function

$$\mathbf{y}(t) = C_1\mathbf{y}_1(t) + C_2\mathbf{y}_2(t) + \cdots + C_k\mathbf{y}_k(t).$$

The coefficients $C_1, C_2, \ldots$, and C_k were chosen so that $\mathbf{y}(t_0) = \mathbf{0}$. According to Theorem 5.5, $\mathbf{y}$ is a solution to $\mathbf{y}' = A\mathbf{y}$. However, the constant function $\mathbf{x}(t) = \mathbf{0}$ also satisfies $\mathbf{x}' = A\mathbf{x}$ and $\mathbf{x}(t_0) = \mathbf{0}$. By the uniqueness part of Theorem 3.4, this can be true only if $\mathbf{y}(t) = C_1\mathbf{y}_1(t) + C_2\mathbf{y}_2(t) + \cdots + C_k\mathbf{y}_k(t) = \mathbf{x}(t) = \mathbf{0}$ for all $t \in I$. This proves part 1.

Part 2 follows easily from part 1 by contradiction. Suppose that the vectors $\mathbf{y}_1(t_0)$, $\mathbf{y}_2(t_0)$, ..., and $\mathbf{y}_k(t_0)$ are linearly independent, but 2 is not true. If 2 is not true, there is a $t_1 \in I$ such that $\mathbf{y}_1(t_1)$, $\mathbf{y}_2(t_1)$, ..., and $\mathbf{y}_k(t_1)$ are linearly dependent. By part 1, this means that $\mathbf{y}_1(t)$, $\mathbf{y}_2(t)$, ..., and $\mathbf{y}_k(t)$ are linearly dependent for all $t \in I$, including at $t = t_0$. This contradicts our hypothesis that 2 is not true. ●

As a result of Proposition 5.12, we can confidently make the following definition.

DEFINITION 5.13

> A set of k solutions to the linear system $\mathbf{y}' = A\mathbf{y}$ is linearly independent if it is linearly independent for any one value of t.

It follows from Proposition 5.12 that if a set of solutions is linearly independent, then it is linearly independent for all values of t.

Structure of the set of solutions

Now we are ready to follow up on the line of thought that enabled us to analyze the example in (5.7).

THEOREM 5.14   Suppose $\mathbf{y}_1, \ldots$, and $\mathbf{y}_n$ are linearly independent solutions to the n-dimensional linear system

$$\mathbf{y}'(t) = A\mathbf{y}(t).$$

Then any solution $\mathbf{y}$ can be expressed as a linear combination of $\mathbf{y}_1, \ldots$, and $\mathbf{y}_n$. That is, there are constants $C_1, \ldots$, and C_n such that

$$\mathbf{y}(t) = C_1\mathbf{y}_1(t) + \cdots + C_n\mathbf{y}_n(t) \quad \text{for all } t.$$

Proof Our proof is almost word for word a repeat of the solution to Example 5.6 starting on page 363.

Let $\mathbf{y}(t)$ be an arbitrary solution to the linear system. Let t_0 be an arbitrary point in the interval of definition of the system. Then $\mathbf{y}_0 = \mathbf{y}(t_0)$ is a vector in $\mathbf{R}^n$. Since the given solutions are linearly independent, their values at t_0, $\mathbf{y}_1(t_0)$, $\mathbf{y}_2(t_0)$, ... , and $\mathbf{y}_n(t_0)$ are also linearly independent. Since there are n of them, these vectors form a basis of $\mathbf{R}^n$. Consequently, there are constants $C_1, \ldots$, and C_n such that

$$\mathbf{y}(t_0) = C_1\mathbf{y}_1(t_0) + \cdots + C_n\mathbf{y}_n(t_0) = [\mathbf{y}_1(t_0), \ldots, \mathbf{y}_n(t_0)]\mathbf{C}, \qquad (5.15)$$

where $\mathbf{C} = (C_1, \ldots, C_n)^T$. Let

$$\mathbf{x}(t) = C_1\mathbf{y}_1(t) + \cdots + C_n\mathbf{y}_n(t).$$

We have chosen the constants so that $\mathbf{x}(t_0) = \mathbf{y}(t_0)$ [see equation (5.15)]. By Theorem 5.1, $\mathbf{x}(t)$ is a solution to our system, as is $\mathbf{y}(t)$. By the uniqueness theorem, we have

$$\mathbf{y}(t) = \mathbf{x}(t) = C_1\mathbf{y}_1(t) + \cdots + C_n\mathbf{y}_n(t)$$

for all t. This completes the proof. ∎

Thus the general solution to a homogeneous linear system can be expressed as $\mathbf{y} = C_1\mathbf{y}_1(t)+\cdots+C_n\mathbf{y}_n(t)$, when $\mathbf{y}_1, \ldots$, and $\mathbf{y}_n$ are linearly independent. We will say that a set of n linearly independent solutions to a homogeneous, linear system is a ***fundamental set of solutions***.

Solution strategy

Theorem 5.14 provides us with a strategy for finding general solutions for the homogeneous system $\mathbf{y}' = A\mathbf{y}$:

> Find n linearly independent solutions $\mathbf{y}_1, \mathbf{y}_2, \ldots, \mathbf{y}_n$. The general solution is the set of all linear combinations,
>
> $$\mathbf{y} = C_1\mathbf{y}_1 + C_2\mathbf{y}_2 + \cdots + C_n\mathbf{y}_n,$$
>
> where $C_1, C_2, \ldots$, and C_n are arbitrary constants.

We will show in Chapter 9 how this can be done if the system has constant coefficients. We have to show that the n solutions are linearly independent. According to Proposition 5.12, we only have to show this for one value of t. One way to do this is to use determinants. We need to show that for one value of t

$$W(t) = \det([\mathbf{y}_1(t), \mathbf{y}_2(t), \ldots, \mathbf{y}_n(t)]) \neq 0.$$

The function $W(t)$ is called the ***Wronskian*** of $\mathbf{y}_1, \mathbf{y}_2, \ldots, \mathbf{y}_n$.

The proof of Theorem 5.14 also tells us how to find the solution to an initial value problem, assuming we have already found a fundamental set of solutions. The initial condition has the form

$$\mathbf{y}(t_0) = \mathbf{y}_0$$

for a given vector $\mathbf{y}_0 \in \mathbf{R}^n$ and some t_0. According to the theorem, our solution must have the form

$$\mathbf{y}(t) = C_1\mathbf{y}_1(t) + \cdots + C_n\mathbf{y}_n(t)$$

for all t. The constants $C_1, \dots,$ and C_n are as yet unknown, and are determined by the initial condition, as shown in (5.15). This is a set of n linear equations for the n constants, and since the vectors $\mathbf{y}_1, \mathbf{y}_2, \dots,$ and $\mathbf{y}_n$ are linearly independent, there is a unique solution. Finding the solution of this set of linear algebraic equations is a matter we discussed in Chapter 7.

Example 5.16 Consider the system of homogeneous equations

$$\mathbf{y}'(t) = \begin{pmatrix} 0 & 1 \\ -2 & 2 \end{pmatrix} \mathbf{y}(t). \tag{5.17}$$

We can easily show that

$$\mathbf{y}_1(t) = \begin{pmatrix} e^t \cos t \\ e^t (\cos t - \sin t) \end{pmatrix} \quad \text{and} \quad \mathbf{y}_2(t) = \begin{pmatrix} e^t \sin t \\ e^t (\cos t + \sin t) \end{pmatrix} \tag{5.18}$$

are solutions to the system (5.17) (just insert $\mathbf{y}_1$ and $\mathbf{y}_2$ into the system and make sure both sides of the equation are equal). Is $\{\mathbf{y}_1, \mathbf{y}_2\}$ a fundamental set of solutions for this system?

To answer this question, we need to show that $\mathbf{y}_1$ and $\mathbf{y}_2$ are linearly independent. According to Proposition 5.12, we only need to check this at one point. At $t = 0$, we have

$$\mathbf{y}_1(0) = \begin{pmatrix} 1 \\ 1 \end{pmatrix} \quad \text{and} \quad \mathbf{y}_2(t) = \begin{pmatrix} 0 \\ 1 \end{pmatrix}.$$

It is easy to see that these vectors are linearly independent, so we are through. ●

Example 5.19 Find the solution to the system in (5.17) with the initial conditions

$$\mathbf{y}(0) = \begin{pmatrix} 2 \\ 3 \end{pmatrix}.$$

The system we are solving is the same as in the previous example. There we found that a fundamental set of solutions is given by the functions in (5.18). Thus our solution is a linear combination of $\mathbf{y}_1$ and $\mathbf{y}_2$,

$$\mathbf{y}(t) = C_1 \mathbf{y}_1(t) + C_2 \mathbf{y}_2(t).$$

To find the constants, we use the initial value $\mathbf{y}(0) = (2, 3)^T$.

$$\begin{pmatrix} 2 \\ 3 \end{pmatrix} = \mathbf{y}(0)$$
$$= C_1 \mathbf{y}_1(0) + C_2 \mathbf{y}_2(0)$$
$$= [\mathbf{y}_1(0), \mathbf{y}_2(0)] \begin{pmatrix} C_1 \\ C_2 \end{pmatrix}$$
$$= \begin{pmatrix} 1 & 0 \\ 1 & 1 \end{pmatrix} \begin{pmatrix} C_1 \\ C_2 \end{pmatrix}$$

Solving this system yields $C_1 = 2$ and $C_2 = 1$. Hence our solution is

$$\mathbf{y}(t) = 2\mathbf{y}_1(t) + \mathbf{y}_2(t)$$
$$= 2 \begin{pmatrix} e^t \cos t \\ e^t (\cos t - \sin t) \end{pmatrix} + \begin{pmatrix} e^t \sin t \\ e^t (\cos t + \sin t) \end{pmatrix}$$
$$= \begin{pmatrix} e^t (2\cos t + \sin t) \\ e^t (3\cos t - \sin t) \end{pmatrix}.$$

You might check to see that this function is actually a solution and satisfies the initial condition. ●

EXERCISES

In Exercises 1–6, rewrite the system using matrix notation.

1. $x_1' = -x_1 + 3x_2$
$x_2' = 2x_2$

2. $x_1' = 6x_1 + 4x_2$
$x_2' = -8x_1 - 6x_2$

3. $x_1' = x_1 + x_2$
$x_2' = -x_1 + x_2$

4. $x_1' = -x_2$
$x_2' = x_1$

5. $x_1' = x_1 + x_2$
$x_2' = -x_1 + x_2 + e^t$

6. $x_1' = -x_2 + \sin t$
$x_2' = x_1$

In Exercises 7–10 show that the given functions are solutions to the system in the indicated Exercise. Verify that any linear combination is also a solution.

7. $\mathbf{x}(t) = \begin{pmatrix} e^{-t} \\ 0 \end{pmatrix}$, $\mathbf{y}(t) = \begin{pmatrix} e^{2t} \\ e^{2t} \end{pmatrix}$. Exercise 1.

8. $\mathbf{x}(t) = \begin{pmatrix} -e^{2t} \\ e^{2t} \end{pmatrix}$, $\mathbf{y}(t) = \begin{pmatrix} -e^{-2t} \\ 2e^{-2t} \end{pmatrix}$. Exercise 2.

9. $\mathbf{x}(t) = \begin{pmatrix} e^t \cos t \\ -e^t \sin t \end{pmatrix}$, $\mathbf{y}(t) = \begin{pmatrix} e^t \sin t \\ e^t \cos t \end{pmatrix}$. Exercise 3.

10. $\mathbf{x}(t) = \begin{pmatrix} \cos t \\ \sin t \end{pmatrix}$, $\mathbf{y}(t) = \begin{pmatrix} \sin t \\ -\cos t \end{pmatrix}$. Exercise 4.

11. Show that

$$\mathbf{x}_p(t) = \begin{pmatrix} e^t \\ 0 \end{pmatrix}$$

is a solution to the system in Exercise 5. Show that any function of the form $\mathbf{z}(t) = \mathbf{x}_p(t) + C_1\mathbf{x}(t) + C_2\mathbf{y}(t)$ is also a solution, where $\mathbf{x}$ and $\mathbf{y}$ are the functions defined in Exercise 9.

12. Show that

$$\mathbf{x}_p(t) = \frac{1}{2} \begin{pmatrix} t \sin t - \cos t \\ -t \cos t \end{pmatrix}$$

is a solution to the system in Exercise 6. Show that any function of the form $\mathbf{z}(t) = \mathbf{x}_p(t) + C_1\mathbf{x}(t) + C_2\mathbf{y}(t)$ is also a solution, where $\mathbf{x}$ and $\mathbf{y}$ are the functions defined in Exercise 10.

13. In Exercise 7 you found solutions $\mathbf{x}$ and $\mathbf{y}$ to the system in Exercise 1. Show that these solutions are linearly independent. Find the solution $\mathbf{z}(t)$ to the system in Exercise 1 that satisfies the initial condition $\mathbf{z}(0) = (0, 1)^T$.

14. In Exercise 8 you found solutions $\mathbf{x}$ and $\mathbf{y}$ to the system in Exercise 2. Show that these solutions are linearly independent. Find the solution $\mathbf{z}(t)$ to the system in Exercise 2 that satisfies the initial condition $\mathbf{z}(0) = (1, -4)^T$.

15. In Exercise 9 you found solutions $\mathbf{x}$ and $\mathbf{y}$ to the system in Exercise 3. Show that these solutions are linearly independent. Find the solution $\mathbf{z}(t)$ to the system in Exercise 3 that satisfies the initial condition $\mathbf{z}(0) = (-2, 3)^T$.

16. In Exercise 10 you found solutions $\mathbf{x}$ and $\mathbf{y}$ to the system in Exercise 4. Show that these solutions are linearly independent. Find the solution $\mathbf{z}(t)$ to the system in Exercise 4 that satisfies the initial condition $\mathbf{z}(0) = (3, 2)^T$.

17. Consider the system

$$x_1' = x_1 x_2, \qquad (5.20)$$
$$x_2' = x_2.$$

(a) Show, by direct substitution, that

$$\mathbf{x} = \begin{pmatrix} x_1 \\ x_2 \end{pmatrix} = \begin{pmatrix} 0 \\ e^t \end{pmatrix} \quad \text{and} \quad \mathbf{y} = \begin{pmatrix} y_1 \\ y_2 \end{pmatrix} = \begin{pmatrix} 1 \\ 0 \end{pmatrix}$$

are solutions of system (5.20).

(b) Show that $\mathbf{x}(t) + \mathbf{y}(t)$ is **not** a solution of system (5.20). Why doesn't this contradict Theorem 5.1?

18. Suppose that

$$\mathbf{y}_1(t) = \begin{pmatrix} 2e^{-t} \\ e^{-t} \end{pmatrix} \quad \text{and} \quad \mathbf{y}_2(t) = \begin{pmatrix} e^{2t} \\ e^{2t} \end{pmatrix}$$

are solutions of

$$\mathbf{y}' = \begin{pmatrix} -4 & 6 \\ -3 & 5 \end{pmatrix} \mathbf{y}. \qquad (5.21)$$

Furthermore, suppose that $\mathbf{x}(t)$ is also a solution of system (5.21) with initial condition

$$\mathbf{x}(0) = \begin{pmatrix} 1 \\ -1 \end{pmatrix}.$$

Argue, as in Example 5.3, that $\mathbf{x}(t)$ can be written as a linear combination of $\mathbf{y}_1(t)$ and $\mathbf{y}_2(t)$. Find this linear combination.

Each set of vectors in Exercises 19–22 is a set of solutions of the system $\mathbf{y}' = A\mathbf{y}$, for some $A = A(t)$. Use Proposition 5.12 to establish the linear dependence or independence of each set of vectors.

19. $\mathbf{y}_1(t) = \begin{pmatrix} -e^{-t} \\ -e^{-t} \\ e^{-t} \end{pmatrix}$, $\mathbf{y}_2(t) = \begin{pmatrix} 0 \\ -e^t \\ 2e^t \end{pmatrix}$,

$\mathbf{y}_3(t) = \begin{pmatrix} e^{2t} \\ 0 \\ 2e^{2t} \end{pmatrix}$

20. $\mathbf{y}_1(t) = \begin{pmatrix} e^{-t} \\ e^{-t} \\ 2e^{-t} \end{pmatrix}$, $\mathbf{y}_2(t) = \begin{pmatrix} -e^t \\ 3e^t \\ 0 \end{pmatrix}$,

$\mathbf{y}_3(t) = \begin{pmatrix} 2e^{-t} - 3e^t \\ 2e^{-t} + 9e^t \\ 4e^{-t} \end{pmatrix}$

21. $\mathbf{y}_1(t) = \begin{pmatrix} -e^{-t} + e^{2t} \\ -e^{-t} + e^{2t} \\ e^{2t} \end{pmatrix}$, $\mathbf{y}_2(t) = \begin{pmatrix} -e^t + e^{-t} \\ e^{-t} \\ 0 \end{pmatrix}$,

$\mathbf{y}_3(t) = \begin{pmatrix} e^t \\ 0 \\ 0 \end{pmatrix}$

22. $\mathbf{y}_1(t) = \begin{pmatrix} e^{-t} \\ 2e^{-t} \\ 2e^{-t} \end{pmatrix}$, $\quad \mathbf{y}_2(t) = \begin{pmatrix} e^t \\ e^t \\ 0 \end{pmatrix}$,

$\mathbf{y}_3(t) = \begin{pmatrix} 0 \\ 0 \\ e^t \end{pmatrix}$

In Exercises 23–26, verify by direct substitution that $\mathbf{y}_1(t)$ and $\mathbf{y}_2(t)$ are solutions of the given homogeneous equation. Show also that the solutions $\mathbf{y}_1(t)$ and $\mathbf{y}_2(t)$ are linearly independent. Find the solution of the given homogeneous equation with the initial condition $\mathbf{y}(0) = \mathbf{y}_0$.

23. $\mathbf{y}_1(t) = \begin{pmatrix} -e^{2t} \\ 2e^{2t} \end{pmatrix}$, $\quad \mathbf{y}_2(t) = \begin{pmatrix} -e^{-2t} \\ e^{-2t} \end{pmatrix}$,

$\mathbf{y}' = \begin{pmatrix} -6 & -4 \\ 8 & 6 \end{pmatrix} \mathbf{y}$, and $\mathbf{y}(0) = \begin{pmatrix} -5 \\ 8 \end{pmatrix}$

24. $\mathbf{y}_1(t) = \begin{pmatrix} e^{-2t} \\ 0 \end{pmatrix}$, $\quad \mathbf{y}_2(t) = \begin{pmatrix} e^{-4t} \\ e^{-4t} \end{pmatrix}$,

$\mathbf{y}' = \begin{pmatrix} -2 & -2 \\ 0 & -4 \end{pmatrix} \mathbf{y}$, and $\mathbf{y}(0) = \begin{pmatrix} 3 \\ 1 \end{pmatrix}$

25. $\mathbf{y}_1(t) = \begin{pmatrix} e^{2t} \\ e^{2t} \end{pmatrix}$, $\quad \mathbf{y}_2(t) = \begin{pmatrix} e^{2t}(t+2) \\ e^{2t}(t+1) \end{pmatrix}$,

$\mathbf{y}' = \begin{pmatrix} 3 & -1 \\ 1 & 1 \end{pmatrix} \mathbf{y}$, and $\mathbf{y}(0) = \begin{pmatrix} 0 \\ 1 \end{pmatrix}$

26. $\mathbf{y}_1(t) = \begin{pmatrix} \frac{1}{2}\cos t - \frac{1}{2}\sin t \\ \cos t \end{pmatrix}$, $\mathbf{y}_2(t) = \begin{pmatrix} \frac{1}{2}\sin t + \frac{1}{2}\cos t \\ \sin t \end{pmatrix}$,

$\mathbf{y}' = \begin{pmatrix} 1 & -1 \\ 2 & -1 \end{pmatrix} \mathbf{y}$, and $\mathbf{y}(0) = \begin{pmatrix} 1 \\ 0 \end{pmatrix}$

27. Two tanks each hold 3 liters of salt water and are connected by two pipes (see Figure 1). The salt water in each tank is kept well stirred. Pure water flows into tank A at the rate of 5 liters per minute and the salt water mixture exits tank B at the same rate. Salt water flows from tank A to tank B at the rate of 9 liters per minute and it flows from tank B to tank A at the rate of 4 liters per minute. If tank A initially contains 1 kilogram of salt and tank B contains no salt, then set up the system of differential equations that govern the mass of the salt in each tank $t \geq 0$ minutes later.

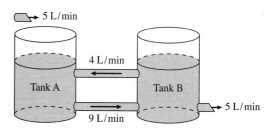

Figure 1. The two tanks in Exercise 27.

28. Suppose that we add an extra tank to the arrangement in Exercise 27, as shown in Figure 2. Further, suppose that rates of flow are defined as follows:

$$r_{in} = 5 \text{ gal/min} \qquad r_{BC} = 9 \text{ gal/min}$$
$$r_{AB} = 9 \text{ gal/min} \qquad r_{CB} = 4 \text{ gal/min}$$
$$r_{BA} = 4 \text{ gal/min} \qquad r_{out} = 5 \text{ gal/min}$$

Again, suppose that pure water is pumped into tank A at a rate of r_{in} gal/min. Suppose that each tank initially holds 200 gal of salt solution. Let $x_A(t)$, $x_B(t)$, and $x_C(t)$ represent the amount of salt (in pounds) in tank A, tank B, and tank C, respectively. Set up a system of equations modeling the salt content in each tank over time.

29. Suppose that each tank in Exercise 28 initially contains 40 pounds of salt. Use a numeric solver to plot the salt content of each tank over time.

30. Three tanks containing salt solutions are arranged in a cascade, as shown in Figure 3. Let $x(t)$, $y(t)$, and $z(t)$ represent the amount of salt, in pounds, contained in the upper, middle, and lower tanks, respectively. Pure water enters the upper tank at a rate of 10 gal/min and salt solution leaves the upper tank at the same rate, spilling into the middle tank at 10 gal/min. Salt solution leaves the middle tank at the same rate, spilling into the lower tank at a rate of 10 gal/min. Finally, solution drains from the lower tank at a rate of 10 gal/min. Initially, each tank contains 200 gallons of salt solution. Set up a system of three first-order equations that models the salt content of each tank over time.

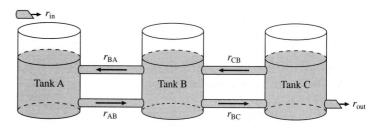

Figure 2. Lots of tanks.

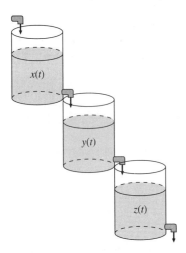

Figure 3. A cascade of tanks.

31. Consider the model of Exercise 30. Suppose that each tank initially holds 40 pounds of salt. Use a numeric solver to plot the salt content of each tank over time.

32. Consider the spring-mass system shown in Figure 4. Suppose that $k_1 = k_2 = k_3 = 1$ dyne/cm and $m_1 = m_2 = 40$ g. Further, suppose that mass m_1 is displaced 4 cm to the left and mass m_2 is displaced 3 cm to the right and both masses are released from rest. Use a numerical solver to plot the position of each mass versus time.

33. Suppose that a particle of mass m moves in the xy-plane according to a **central force law**. That is, the force on the particle is always radially directed, either toward or away from the origin. Let us assume that the force on the particle is always directed toward the origin with magnitude proportional to the mass of the particle and inversely proportional to the square of the particle's distance from the origin; that is,

$$F = \frac{km}{r^2},$$

where r is the particle's distance from the origin (see Figure 5). Let F_x and F_y represent the magnitudes of the force in the x- and y-directions, respectively. According to Newton's second law, $F_x = ma_x$ and $F_y = ma_y$, where a_x and a_y are the accelerations of the particle in the x- and y-directions, respectively. Show that

$$\frac{d^2x}{dt^2} = -\frac{kx}{r^3},$$
$$\frac{d^2y}{dt^2} = -\frac{ky}{r^3},$$

and use these results to develop a system of four first-order differential equations that model the motion of the particle.

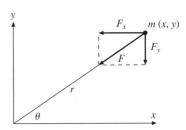

Figure 5. A particle in a central force field.

34. Use the fourth-order Runge-Kutta method to find and plot a numerical approximation of the orbit of the particle in Exercise 33 with initial conditions $x(0) = 2$, $y(0) = -0.1$, $x'(0) = 2$, $y'(0) = 0.1$. Use $k = 1$. Experiment with smaller and smaller step sizes.

35. As in Exercise 33, suppose again that the force on a particle is always directed radially inward toward the origin. If the magnitude of this force is proportional to the mass but inversely proportional to the **cube** of its distance from the origin, develop a system of four first-order differential equations that model the motion of the particle.

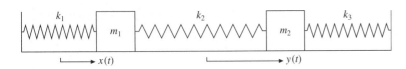

Figure 4. The three springs and two masses.

PROJECT 8.6 Long-Term Behavior of Solutions

This project has two goals. First, it will test your ability to use all of the numerical methods at your disposal. Second, you will learn experimental facts about the long-term behavior of solutions to autonomous systems in dimensions $D = 1, 2$, and 3.

Let's say something about the second goal. You will be looking at an autonomous system of the form $\mathbf{y}' = \mathbf{f}(\mathbf{y})$. We will assume that $\mathbf{f}$ is defined for all $\mathbf{y}$. One thing that can happen to a solution $\mathbf{y}(t)$ is that it tends to ∞ as $t \to \infty$. It can go in any direction, it can spiral out, or it can go straight out. None of these things matters here, since we are not interested in solutions that tend to ∞. If the solution does not "escape" in this manner, then it can be shown that the solution curve in phase space (including the phase line or the phase plane in the appropriate dimension) will approach a set. This set is called the *limit set* of the solution curve. We will refer to such a set as an *attractor* if it is the limit set for all solutions that have their initial points near the set. The second goal is to compile an experimental list of different types of limit sets. We will do this by looking at examples using numerical methods.

The list of limit sets will get larger as the dimension increases. We will start with dimension 1, where the answer is quite simple.

Dimension $D = 1$

Here we are looking at a single equation $y' = f(y)$. If you think about what kinds of limit sets you have already seen, you will probably realize that the only set that can be a limit set in a phase line is a single point. Furthermore, this point has to be an equilibrium point. Submit a plot of a solution to an autonomous equation showing this behavior.

Dimension $D = 2$

Here we get some complication. Examine each of the following examples in a phase plane computer program. For each, write a short paragraph describing the limit set that you see. Plot more than one solution curve starting at different points, and submit this plot.

(a) The damped vibrating spring.

$$y' = v,$$
$$v' = -4y - 2v$$

Compare the limit set you see with what happens in dimension 1.

(b) The undamped vibrating spring.

$$y' = v,$$
$$v' = -4y$$

You know that an undamped spring has periodic solutions. The long-term behavior here is quite different from the previous case and dimension 1. To get you started with the answer, the limit set is the closed curve consisting of the periodic solution. What happens if the initial position is the origin?

(c) The undamped pendulum.

$$\theta' = \omega,$$
$$\omega' = -\sin\theta$$

Compare with the previous two examples.

(d) The damped pendulum.

$$\theta' = \omega,$$
$$\omega' = -\sin\theta - \frac{1}{2}\omega$$

Compare with the previous examples.

(e) Competing species.

$$x' = (1 - x - y)x,$$
$$y' = (4 - 2x - 7y)y$$

Compare with the previous examples.

(f) Van der Pol's equation.

$$x' = 2x - y - x^3,$$
$$y' = x$$

Describe the limit set you see. Is the limit set a closed curve? How are the solution curves different from the periodic solutions you saw in previous examples? If you can, eliminate the first part of a solution, so that you are only looking at the limiting behavior. A plot of this would be close to being a plot of the limit set. Submit an extra plot of the results of this.

The limit set you see in this example is called a *limit cycle*.

Dimension $D = 3$

We will examine only the possibilities that arise in the Lorenz system. The system is

$$x' = -\sigma x + \sigma y$$
$$y' = \rho x - y - xz$$
$$z' = -\beta z + xy.$$

Use $\sigma = 10$ and $\beta = 8/3$. The constant ρ will vary. Use a numerical solver to solve the system. It will be sufficient to compute the solution for $0 \le t \le 50$. Even so, you may find that it takes a long time to plot the results. Be patient.

Here are some techniques that might be useful.

- Plot the three components of the solution versus t.
- Plot the solution curve in three-dimensional phase space.
- Eliminate any transient behavior so that you will only be observing the solution for large t. You can accomplish this by solving over the interval $0 \le t \le 50$ and only plotting the part with $35 \le t \le 50$.

For each case, submit a short paragraph describing the limit set. Submit one plot that best shows the phenomenon.

(a) $\rho = 1/2$. The results should remind you of phenomena you saw in dimensions 1 and 2.

(b) $\rho = 15$. You have seen this type of limit set before.

(c) $\rho = 250$. Again, you have seen this type of limit set before.

(d) $\rho = 160$. You have seen this type of limit set before, but it is a little more complicated this time.

(e) $\rho = 28$. This is new. Describe what you see as best you can. For this case, you might solve over a larger interval such as $0 \le t \le 200$.

9

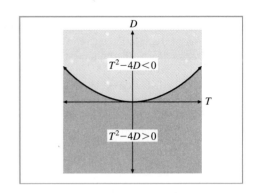

Linear Systems with Constant Coefficients

We introduced systems in Chapter 8, derived several important properties, and gave several examples. We are now ready to turn to the task of finding the general solution to linear, homogeneous equations and systems. Remember the strategy we developed in the previous chapter. For a system of dimension n, we need to find a fundamental set of solutions, which is a set of n linearly independent solutions. The general solution is a linear combination of these solutions.

We will also find the general solution for higher-order equations. However, this will be an easy job, since we need only use the results for the associated first-order system.

9.1 Overview of the Technique

We are looking for solutions to the system

$$\mathbf{y}' = A\mathbf{y}, \tag{1.1}$$

where A is a matrix with constant entries. For motivation, let's look at some systems we already understand. The simplest example is a system of dimension 1. This reduces to a single, first-order, homogeneous equation with constant coefficients and has the form

$$y' = ay. \tag{1.2}$$

In Chapter 2, we saw that the solution to this equation is the exponential function $y(t) = Ce^{at}$, where C is any constant.

Since it works in dimension 1 it is reasonable to look for solutions to $\mathbf{y}' = A\mathbf{y}$ that have an exponential character. In analogy to the solution found in (1.2), let's look for solutions of the form

$$\mathbf{y}(t) = e^{\lambda t}\mathbf{v}, \tag{1.3}$$

where $\mathbf{v}$ is a vector with constants for entries. The entries of $\mathbf{v}$ and the constant λ are as yet unknown and can be chosen to allow (1.3) to be a solution to (1.1). Let's substitute (1.3) into (1.1). The left-hand side becomes

$$\mathbf{y}' = \lambda e^{\lambda t} \mathbf{v},$$

and on the right-hand side we get

$$A\mathbf{y} = A(e^{\lambda t}\mathbf{v}) = e^{\lambda t}A\mathbf{v}.$$

Since $e^{\lambda t} \neq 0$, we will have $\mathbf{y}' = A\mathbf{y}$, provided that

$$A\mathbf{v} = \lambda\mathbf{v}.$$

The number λ and the associated vector $\mathbf{v}$ that satisfy this equation have special names.

DEFINITION 1.4

> Suppose A is an $n \times n$ matrix. A number λ is called an ***eigenvalue*** of A if there is a nonzero vector $\mathbf{v}$ such that
>
> $$A\mathbf{v} = \lambda\mathbf{v}. \tag{1.5}$$
>
> If λ is an eigenvalue, then any vector $\mathbf{v}$ satisfying (1.5) is called an ***eigenvector*** associated with the eigenvalue λ.

The requirement that $\mathbf{v}$ be nonzero is necessary, since the equation $A\mathbf{v} = \lambda\mathbf{v}$ always holds for $\mathbf{v} = \mathbf{0}$. Another observation is that if $\mathbf{v}$ is an eigenvector associated to λ, then so is any multiple of $\mathbf{v}$. Indeed, multiplying equation (1.5) by r, we get

$$A(r\mathbf{v}) = \lambda(r\mathbf{v}),$$

which shows that $r\mathbf{v}$ is also an eigenvector. We will show that the set of eigenvectors associated to a particular eigenvalue is a subspace of $\mathbf{R}^n$.

We have discovered that we can always find some exponential solutions. Since the exploitation of this idea is going to be the theme of this chapter, let's state it formally.

THEOREM 1.6 Suppose that λ is an eigenvalue of the matrix A and $\mathbf{v}$ is an associated eigenvector. Then

$$\mathbf{x}(t) = e^{\lambda t}\mathbf{v}$$

is a solution to the system $\mathbf{x}' = A\mathbf{x}$ and satisfies the initial condition $\mathbf{x}(0) = \mathbf{v}$. ∎

Finding eigenvalues

Since finding eigenvalues and eigenvectors is so important, let's discuss techniques for computing them. First, let's rewrite (1.5) as

$$\mathbf{0} = A\mathbf{v} - \lambda\mathbf{v} = A\mathbf{v} - \lambda I\mathbf{v} = [A - \lambda I]\mathbf{v}. \tag{1.7}$$

Notice that we inserted the identity matrix I into the equation. Without doing so, we could not distribute $\mathbf{v}$ out of the equation, since the products $A\mathbf{v}$ and $\lambda\mathbf{v}$ are two fundamentally different operations, and, correspondingly, the expression $A - \lambda$ makes no sense.

As a result of (1.7), we see that the defining property (1.5) has an equivalent formulation:

$$A\mathbf{v} = \lambda\mathbf{v} \iff [A - \lambda I]\mathbf{v} = \mathbf{0}. \tag{1.8}$$

The formulation on the right in equation (1.8) has a different character than (1.5). Since $\mathbf{v}$ is a nonzero vector, it means that the matrix $A - \lambda I$ has a nontrivial nullspace. By Corollary 7.3 in Section 7.7, this can happen if and only if

$$\det(A - \lambda I) = 0. \tag{1.9}$$

If we look closely at the matrix $A - \lambda I$, we notice that the unknown λ appears in each of the n diagonal terms and nowhere else. You will remember that the determinant of a matrix is a sum of products of entries of the matrix, with one from each row and one from each column. Hence, when we take the determinant in (1.9), we get a polynomial in λ, and the highest power that can occur is n. Therefore, the function on the left in (1.9) is a polynomial of degree n.

DEFINITION 1.10

> If A is an $n \times n$ matrix, the polynomial
>
> $$p(\lambda) = (-1)^n \det(A - \lambda I) = \det(\lambda I - A)$$
>
> is called the ***characteristic polynomial*** of A, and the equation
>
> $$p(\lambda) = (-1)^n \det(A - \lambda I) = 0$$
>
> is called the ***characteristic equation***.

The factor $(-1)^n$ is chosen to make the leading term in the characteristic polynomial λ^n instead of $(-1)^n \lambda^n$. We have discovered an important fact. Let's state it formally.

PROPOSITION 1.11   The eigenvalues of an $n \times n$ matrix A are the roots of its characteristic polynomial. ●

Example 1.12   Find the eigenvalues of the matrix

$$A = \begin{pmatrix} -4 & 6 \\ -3 & 5 \end{pmatrix}.$$

The characteristic polynomial is

$$
\begin{aligned}
p(\lambda) &= (-1)^2 \det(A - \lambda I) = \det(A - \lambda I) \\
&= \det \begin{pmatrix} -4 - \lambda & 6 \\ -3 & 5 - \lambda \end{pmatrix} \\
&= (-4 - \lambda)(5 - \lambda) + 18 \\
&= \lambda^2 - \lambda - 2 \\
&= (\lambda - 2)(\lambda + 1).
\end{aligned}
$$

Thus, the eigenvalues of A are 2 and -1. ●

Since we know that the eigenvalues of A are the roots of the characteristic polynomial, we have made a significant step forward. To find the eigenvalues, we only have to find the roots of the characteristic polynomial. That is not always easy, but at least it is straightforward, and in difficult cases a computer can help. Notice also that to find the eigenvalues in this way, we do not at the same time have to find associated eigenvectors, as Definition 1.4 would seem to require.

Finding eigenvectors

To find the associated eigenvectors, we go back to equation (1.8). From this, we see that the eigenvectors are simply the vectors in the nullspace of $A - \lambda I$. Thus, the set of eigenvectors is a subspace; it is the nullspace of $A - \lambda I$. The subspace consisting of all eigenvectors for a given eigenvalue λ is called the **eigenspace** of λ. Let's state this fact formally.

PROPOSITION 1.13 Let A be an $n \times n$ matrix, and let λ be an eigenvalue of A. The set of all eigenvectors associated with λ is equal to the nullspace of $A - \lambda I$. Hence, the eigenspace of λ is a subspace of $\mathbf{R}^n$. ●

Example 1.14 Find the eigenvectors for the matrix in Example 1.12.

For the eigenvalue 2, the eigenspace is the nullspace of

$$A - 2I = \begin{pmatrix} -6 & 6 \\ -3 & 3 \end{pmatrix}.$$

It is easily seen that the eigenspace is generated by the single vector $\mathbf{v}_1 = (1, 1)^T$. For the eigenvalue -1, the eigenspace is the nullspace of

$$A + I = \begin{pmatrix} -3 & 6 \\ -3 & 6 \end{pmatrix}.$$

It is easily seen that the eigenspace is generated by the vector $\mathbf{v}_2 = (2, 1)^T$. ●

Finding solutions

Let's take this example back to our original problem, solving systems of differential equations.

Example 1.15 Find a fundamental set of solutions for the system $\mathbf{y}' = A\mathbf{y}$ for the matrix A in Example 1.12.

We have done almost all of the work in Examples 1.12 and 1.14, where we computed the eigenvalues of A and the associated eigenvectors. We need only apply Theorem 1.6 to find the solutions. Using the previous two examples, we see that we have solutions

$$\mathbf{y}_1(t) = e^{2t} \begin{pmatrix} 1 \\ 1 \end{pmatrix} \quad \text{and} \quad \mathbf{y}_2(t) = e^{-t} \begin{pmatrix} 2 \\ 1 \end{pmatrix}.$$

These two solutions are linearly independent, since they are linearly independent for $t = 0$ (see Proposition 5.12 in Chapter 8). Consequently, they form a fundamental set of solutions. ●

Summary

Let us examine how far this takes us to the completion of our strategy of finding n linearly independent solutions. For a system of dimension n, the characteristic polynomial is of degree n. In general, a polynomial of degree n has n roots. Each root λ is an eigenvalue, and for each we can find an eigenvector $\mathbf{v}$. From these, we can form the exponential solution $\mathbf{y}(t) = e^{\lambda t}\mathbf{v}$. That's n solutions. If these are linearly independent (and, as we shall see, this is always the case if the eigenvalues are all different), we should be through. However, there are some complications that we will look into in the following sections.

1. **Distinct real roots:** We are essentially done here. The situation we examined in Example 1.15 is what happens in general.

2. **Complex roots:** If an eigenvalue is complex, then the exponential solution is complex valued. Since we will usually want real-valued solutions, this is a complication that we will have to deal with.

3. **Repeated roots:** Sometimes the roots of a polynomial are not distinct. If that polynomial is a characteristic polynomial, then there are fewer distinct eigenvalues than n. For each eigenvalue, we are guaranteed only one solution. Hence, our method will give us fewer solutions than we are looking for. This complication will also be dealt with in later sections.

EXERCISES

Use hand calculations to find the characteristic polynomial and eigenvalues for each of the matrices in Exercises 1–12.

1. $A = \begin{pmatrix} 12 & 14 \\ -7 & -9 \end{pmatrix}$
2. $A = \begin{pmatrix} 2 & 0 \\ 0 & 2 \end{pmatrix}$

3. $A = \begin{pmatrix} -2 & 3 \\ 0 & -5 \end{pmatrix}$
4. $A = \begin{pmatrix} -4 & 1 \\ -2 & 1 \end{pmatrix}$

5. $A = \begin{pmatrix} 5 & 3 \\ -6 & -4 \end{pmatrix}$
6. $A = \begin{pmatrix} -2 & 5 \\ 0 & 2 \end{pmatrix}$

7. $A = \begin{pmatrix} -3 & 0 \\ 0 & -3 \end{pmatrix}$
8. $A = \begin{pmatrix} 6 & 10 \\ -5 & -9 \end{pmatrix}$

9. $A = \begin{pmatrix} 1 & 2 & 3 \\ 0 & 0 & 2 \\ 0 & 3 & 1 \end{pmatrix}$
10. $A = \begin{pmatrix} 1 & 0 & 0 \\ 4 & 3 & 2 \\ -8 & -4 & -3 \end{pmatrix}$

11. $A = \begin{pmatrix} -1 & -4 & -2 \\ 0 & 1 & 1 \\ -6 & -12 & 2 \end{pmatrix}$
12. $A = \begin{pmatrix} 1 & 0 & -1 \\ -2 & -1 & 3 \\ -4 & 0 & 4 \end{pmatrix}$

13. Use a computer to find and plot the graph of the characteristic polynomials of

$$A = \begin{pmatrix} 5 & -8 & 8 \\ 4 & -7 & 16 \\ 0 & 0 & 5 \end{pmatrix} \text{ and } B = \begin{pmatrix} 11 & 0 & -8 \\ 4 & -1 & 0 \\ 16 & 0 & -13 \end{pmatrix}.$$

Use the computer to find the eigenvalues of matrices A and B and describe the relationship between the eigenvalues and the graphs of the characteristic polynomials.

14. The characteristic polynomial of

$$A = \begin{pmatrix} 5 & 4 \\ -8 & -7 \end{pmatrix}$$

is $p(\lambda) = \lambda^2 + 2\lambda - 3$. Use hand calculations to show that the matrix A satisfies the equation $p(A) = 0$ (i.e., show that $A^2 + 2A - 3I$ equals the zero matrix, where I is the 2×2 identity matrix). This result is known as the Cayley-Hamilton theorem.

15. Use a computer to verify that each of the matrices in Exercises 1–5 satisfies the Cayley–Hamilton theorem demonstrated in Exercise 14.

Use hand calculations to find a fundamental set of solutions for the system $\mathbf{y}' = A\mathbf{y}$, where A is the matrix given in Exercises 16–27.

16. $A = \begin{pmatrix} 2 & 0 \\ -4 & -2 \end{pmatrix}$
17. $A = \begin{pmatrix} 6 & -8 \\ 0 & -2 \end{pmatrix}$

18. $A = \begin{pmatrix} -3 & -4 \\ 2 & 3 \end{pmatrix}$
19. $A = \begin{pmatrix} -1 & 0 \\ 0 & -1 \end{pmatrix}$

20. $A = \begin{pmatrix} 3 & -2 \\ 4 & -3 \end{pmatrix}$
21. $A = \begin{pmatrix} 7 & 10 \\ -5 & -8 \end{pmatrix}$

22. $A = \begin{pmatrix} -3 & 14 \\ 0 & 4 \end{pmatrix}$
23. $A = \begin{pmatrix} 5 & -4 \\ 8 & -7 \end{pmatrix}$

24. $A = \begin{pmatrix} -5 & 0 & -6 \\ 26 & -3 & 38 \\ 4 & 0 & 5 \end{pmatrix}$
25. $A = \begin{pmatrix} -1 & 0 & 0 \\ 2 & -5 & -6 \\ -2 & 3 & 4 \end{pmatrix}$

26. $A = \begin{pmatrix} -1 & 2 & 0 \\ -19 & 14 & 18 \\ 17 & -11 & -17 \end{pmatrix}$

27. $A = \begin{pmatrix} -3 & 0 & 2 \\ 6 & 3 & -12 \\ 2 & 2 & -6 \end{pmatrix}$

Use a computer to find the eigenvalues and eigenvectors for the matrices in Exercises 28–37.

28. $A = \begin{pmatrix} -6 & 4 & 4 \\ -4 & 2 & 4 \\ -10 & 8 & 4 \end{pmatrix}$

29. $A = \begin{pmatrix} -7 & 2 & 10 \\ 0 & 1 & 0 \\ -5 & 2 & 8 \end{pmatrix}$

30. $A = \begin{pmatrix} 1 & -2 & 4 \\ 7 & -8 & 10 \\ 2 & -2 & 1 \end{pmatrix}$

31. $A = \begin{pmatrix} -11 & -7 & 1 \\ 20 & 13 & -2 \\ -18 & -9 & 2 \end{pmatrix}$

32. $A = \begin{pmatrix} -6 & -13 & -11 \\ 4 & 1 & 1 \\ -4 & 2 & 2 \end{pmatrix}$

33. $A = \begin{pmatrix} 5 & 0 & 4 \\ -10 & 3 & -8 \\ -14 & 2 & -11 \end{pmatrix}$

34. $A = \begin{pmatrix} -2 & -3 & 1 & 1 \\ -4 & -5 & 0 & 4 \\ 7 & 9 & -4 & -9 \\ -7 & -9 & 1 & 6 \end{pmatrix}$

35. $A = \begin{pmatrix} -4 & -7 & 0 & 7 \\ 2 & -5 & 2 & 1 \\ -1 & -11 & 1 & 8 \\ 2 & -8 & 2 & 4 \end{pmatrix}$

36. $A = \begin{pmatrix} -6 & 5 & -9 & 10 \\ 10 & -7 & 13 & -16 \\ 4 & -4 & 8 & -8 \\ -5 & 3 & -5 & 7 \end{pmatrix}$

37. $A = \begin{pmatrix} 1 & 1 & 2 & 2 \\ 2 & 0 & 2 & 2 \\ 1 & 3 & 0 & 2 \\ 4 & -8 & 4 & 2 \end{pmatrix}$

Use a computer to find a fundamental set of solutions for the system $\mathbf{y}' = A\mathbf{y}$, where A is the matrix given in Exercises 38–47.

38. $A = \begin{pmatrix} -2 & -6 & 12 \\ 0 & 3 & -2 \\ 0 & 1 & 0 \end{pmatrix}$

39. $A = \begin{pmatrix} 20 & -34 & -10 \\ 12 & -21 & -5 \\ -2 & 4 & -2 \end{pmatrix}$

40. $A = \begin{pmatrix} 0 & -2 & 2 \\ 12 & -6 & 8 \\ 9 & -1 & 3 \end{pmatrix}$

41. $A = \begin{pmatrix} -3 & -10 & 0 \\ 0 & 2 & 0 \\ 2 & 4 & -1 \end{pmatrix}$

42. $A = \begin{pmatrix} -5 & 6 & 4 \\ -18 & 16 & 8 \\ 72 & -48 & -13 \end{pmatrix}$

43. $A = \begin{pmatrix} 7 & 7 & 4 \\ -6 & -10 & -8 \\ 6 & 7 & 5 \end{pmatrix}$

44. $A = \begin{pmatrix} 3 & 0 & 0 & 0 \\ 3 & 0 & 2 & -1 \\ -4 & -2 & 5 & -8 \\ -2 & -2 & 4 & -7 \end{pmatrix}$

45. $A = \begin{pmatrix} 5 & 7 & -2 & 8 \\ -6 & -8 & 2 & -14 \\ 6 & 6 & -4 & 6 \\ 0 & 0 & 0 & 2 \end{pmatrix}$

46. $A = \begin{pmatrix} -16 & 24 & -48 & 7 \\ 10 & -7 & 23 & -5 \\ 10 & -10 & 26 & -5 \\ 6 & 4 & 8 & -5 \end{pmatrix}$

47. $A = \begin{pmatrix} 6 & 4 & -8 & 0 \\ 14 & 2 & 22 & -18 \\ 11 & 4 & 5 & -9 \\ 22 & 8 & 2 & -14 \end{pmatrix}$

48. Use Definition 1.4 to show that if $\mathbf{v}$ and $\mathbf{w}$ are eigenvectors of A associated to the eigenvalue λ, then $a\mathbf{v} + b\mathbf{w}$ is also an eigenvector associated to λ for any scalars a and b.

49. Use a computer to find the eigenvalues and determinant of each of the following matrices:

$$A = \begin{pmatrix} 6 & -8 \\ 4 & -6 \end{pmatrix}, \quad B = \begin{pmatrix} -11 & -16 \\ 8 & 13 \end{pmatrix},$$

$$\text{and} \quad C = \begin{pmatrix} 7 & -21 & -11 \\ 5 & -13 & -5 \\ -5 & 9 & 1 \end{pmatrix}.$$

Describe any relationship you see between the eigenvalues and the determinant.

50. The *trace* of a matrix is the sum of the elements on its main diagonal ($\text{tr}(A) = a_{11} + a_{22} + \cdots + a_{nn}$). For each matrix in Exercise 49, describe a relationship between the eigenvalues and the trace of the matrix.

51. The following matrices are upper triangular. Use a computer to find the eigenvalues of each matrix.

$$A = \begin{pmatrix} 2 & 3 \\ 0 & -4 \end{pmatrix}, \quad B = \begin{pmatrix} 1 & 2 & 3 \\ 0 & -1 & 4 \\ 0 & 0 & 5 \end{pmatrix},$$

$$C = \begin{pmatrix} 2 & -1 & 1 & 1 \\ 0 & 3 & -1 & 0 \\ 0 & 0 & -4 & 1 \\ 0 & 0 & 0 & 2 \end{pmatrix}$$

Describe any relationship between the eigenvalues and the upper triangular structure of these matrices. Describe a similar relationship if the matrix is lower triangular and give a detailed example.

52. In Exercise 51, you made a conjecture about the eigenvalues of a triangular matrix. Prove this conjecture.

53. The matrix

$$A = \begin{pmatrix} 3 & 10 \\ 0 & -2 \end{pmatrix}$$

has the following eigenvalue–eigenvector pairs:

$$\lambda_1 = -2 \leftrightarrow \mathbf{v}_1 = \begin{pmatrix} -2 \\ 1 \end{pmatrix} \quad \text{and} \quad \lambda_2 = 3 \leftrightarrow \mathbf{v}_2 = \begin{pmatrix} 1 \\ 0 \end{pmatrix}.$$

Form the matrices

$$V = \begin{pmatrix} -2 & 1 \\ 1 & 0 \end{pmatrix} \quad \text{and} \quad D = \begin{pmatrix} -2 & 0 \\ 0 & 3 \end{pmatrix}.$$

Note the column correspondence between the eigenvectors in V and the eigenvalues in D. Use hand calculations to show that $A = VDV^{-1}$. The change of basis represented by the eigenvectors in the matrix V is said to "diagonalize" the matrix A.

Use the technique in Exercise 53 to diagonalize the matrices in Exercises 54 and 55.

54. $A = \begin{pmatrix} 6 & 0 \\ 8 & -2 \end{pmatrix}$ **55.** $A = \begin{pmatrix} -1 & -2 \\ 4 & -7 \end{pmatrix}$

56. Why does the technique of Exercise 53 fail to diagonalize the matrix

$$A = \begin{pmatrix} 5 & 1 \\ -1 & 3 \end{pmatrix}?$$

9.2 Planar Systems

Before going to higher-dimensional systems, let's look carefully at linear systems of dimension 2. These are also called *planar systems*. In this section, we will do the algebra that will enable us to solve the system

$$\mathbf{y}' = A\mathbf{y},$$

where

$$A = \begin{pmatrix} a_{11} & a_{12} \\ a_{21} & a_{22} \end{pmatrix} \quad \text{and} \quad \mathbf{y}(t) = \begin{pmatrix} y_1(t) \\ y_2(t) \end{pmatrix}. \tag{2.1}$$

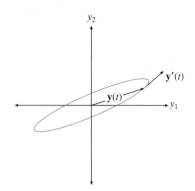

Figure 1. A solution curve in the phase plane.

We will find that some of this algebra applies equally well to higher-dimensional systems, and when this is so, we will state the general result. This will prepare us for our examination of higher-dimensional systems in Section 9.5.

Notice that as t varies, the solution curve $t \to \mathbf{y}(t)$ is a parametrically defined curve in the phase plane $\mathbf{R}^2$. An example is shown in Figure 1. We have discussed this in some detail in Section 8.2 of Chapter 8. In the next section, we look at the solution curves in the phase plane and carefully categorize the possible behaviors of solutions to planar systems.

According to the method discovered in the previous section, we want to look for exponential solutions, those of the form $\mathbf{y}(t) = e^{\lambda t} \mathbf{v}$, where λ is an eigenvalue of A and $\mathbf{v}$ is an associated eigenvector. The eigenvalues are solutions of the characteristic equation $\det(A - \lambda I) = 0$. Let's expand this in terms of the entries of A,

$$\det(A - \lambda I) = \det \begin{pmatrix} a_{11} - \lambda & a_{12} \\ a_{21} & a_{22} - \lambda \end{pmatrix}$$
$$= (a_{11} - \lambda)(a_{22} - \lambda) - a_{12}a_{21}$$
$$= \lambda^2 - (a_{11} + a_{22})\lambda + (a_{11}a_{22} - a_{12}a_{21}).$$

This is a quadratic polynomial. The constant term will be recognized as the determinant of A. We will denote this by

$$D = \det(A) = a_{11}a_{22} - a_{12}a_{21}.$$

For the coefficient of λ, we will set $T = a_{11} + a_{22}$. Notice that T is the sum of the diagonal entries of the matrix A. For a matrix A of any dimension, the **_trace_** is defined to be the sum of the diagonal elements, and it is denoted by $\text{tr}(A)$. Thus, with A as given in (2.1), we have

$$T = \text{tr}(A) = a_{11} + a_{22}.$$

With these definitions, the characteristic equation of the planar system becomes

$$\lambda^2 - T\lambda + D = 0. \qquad (2.2)$$

The eigenvalues of A are the roots of the characteristic polynomial and are given by

$$\lambda = \left(T \pm \sqrt{T^2 - 4D}\right)/2. \qquad (2.3)$$

There are three cases we must consider:

1. two distinct real roots (when $T^2 - 4D > 0$)
2. two complex conjugate roots (when $T^2 - 4D < 0$)
3. one real root of multiplicity 2 (when $T^2 - 4D = 0$)

Linear independence of solutions

Before examining these cases, let's prove a result that will make proving the independence of solutions very easy.

PROPOSITION 2.4   Suppose λ_1 and λ_2 are eigenvalues of an $n \times n$ matrix A. Suppose $\mathbf{v}_1 \neq \mathbf{0}$ is an eigenvector for λ_1 and $\mathbf{v}_2 \neq \mathbf{0}$ is an eigenvector for λ_2. If $\lambda_1 \neq \lambda_2$, then $\mathbf{v}_1$ and $\mathbf{v}_2$ are linearly independent.

Proof   Suppose there are constants c_1 and c_2 such that

$$c_1\mathbf{v}_1 + c_2\mathbf{v}_2 = \mathbf{0}. \qquad (2.5)$$

To show that $\mathbf{v}_1$ and $\mathbf{v}_2$ are linearly independent, we need to show that c_1 and c_2 are both equal to 0.

Multiply (2.5) by the matrix A. Since $A\mathbf{v}_1 = \lambda_1\mathbf{v}_1$ and $A\mathbf{v}_2 = \lambda_2\mathbf{v}_2$, we get

$$c_1\lambda_1\mathbf{v}_1 + c_2\lambda_2\mathbf{v}_2 = \mathbf{0}. \qquad (2.6)$$

Multiply (2.5) by λ_2 and subtract from (2.6) to get

$$c_1(\lambda_1 - \lambda_2)\mathbf{v}_1 = \mathbf{0}. \qquad (2.7)$$

It is our assumption that $\lambda_1 \neq \lambda_2$, so $\lambda_1 - \lambda_2 \neq 0$. In addition, the eigenvector $\mathbf{v}_1 \neq \mathbf{0}$. Hence, we must have $c_1 = 0$. We can prove that $c_2 = 0$ in a similar manner.
●

Our main application of Proposition 2.4 will be to show that solutions to a linear system of differential equations are linearly independent. Let's state this as a corollary.

COROLLARY 2.8 Suppose λ_1 and λ_2 are eigenvalues of an $n \times n$ matrix A. Suppose $\mathbf{v}_1 \neq \mathbf{0}$ is an eigenvector for λ_1 and $\mathbf{v}_2 \neq \mathbf{0}$ is an eigenvector for λ_2. If $\lambda_1 \neq \lambda_2$, then the solutions $\mathbf{y}_1(t) = e^{\lambda_1 t}\mathbf{v}_1$ and $\mathbf{y}_2(t) = e^{\lambda_2 t}\mathbf{v}_2$ are linearly independent on $\mathbf{R}$.

Proof According to Proposition 5.12 in Chapter 8, it is only necessary to show that $\mathbf{y}_1$ and $\mathbf{y}_2$ are linearly independent at one point. By Proposition 2.4, $\mathbf{y}_1(0) = \mathbf{v}_1$ and $\mathbf{y}_2(0) = \mathbf{v}_2$ are linearly independent, so we are finished. ●

Note that it is not necessary that the eigenvalues and the eigenvectors be real in Proposition 2.4 or in Corollary 2.8. If one or more is complex, the results are still true.

Distinct real eigenvalues

This is true if $T^2 - 4D > 0$. The solutions of the characteristic equation (2.2) are

$$\lambda_1 = \frac{T - \sqrt{T^2 - 4D}}{2} \quad \text{and} \quad \lambda_2 = \frac{T + \sqrt{T^2 - 4D}}{2}.$$

Then $\lambda_1 < \lambda_2$, and both are real eigenvalues of A. Notice that Example 1.15 in Section 9.1 is of this type.

Let $\mathbf{v}_1$ and $\mathbf{v}_2$ be associated eigenvectors. Then we have two exponential solutions

$$\mathbf{y}_1(t) = e^{\lambda_1 t}\mathbf{v}_1 \quad \text{and} \quad \mathbf{y}_2(t) = e^{\lambda_2 t}\mathbf{v}_2.$$

According to Corollary 2.8, the solutions $\mathbf{y}_1$ and $\mathbf{y}_2$ are linearly independent and form a fundamental set of solutions. The general solution is

$$\mathbf{y}(t) = C_1\mathbf{y}_1(t) + C_2\mathbf{y}_2(t).$$

Let's summarize this in a theorem.

THEOREM 2.9 Suppose that A is a 2×2 matrix with real eigenvalues $\lambda_1 \neq \lambda_2$. Suppose that $\mathbf{v}_1$ and $\mathbf{v}_2$ are eigenvectors associated with the eigenvalues. Then the general solution to the system $\mathbf{y}' = A\mathbf{y}$ is

$$\mathbf{y}(t) = C_1\mathbf{y}_1(t) + C_2\mathbf{y}_2(t) = C_1 e^{\lambda_1 t}\mathbf{v}_1 + C_2 e^{\lambda_2 t}\mathbf{v}_2,$$

where C_1 and C_2 are arbitrary constants. ■

Example 2.10 In Figure 2 we have two tanks connected by two pipes. Each tank contains 500 gallons of a salt solution. Through one pipe solution is pumped from the first tank to the second at 1 gal/min. Through the other solution is pumped at the same rate from the second tank to the first. Suppose that at time $t = 0$ there is no salt in the tank on the right and 100 lb in the tank on the left. Find the salt content in each tank as a function of time.

We considered this configuration in Example 4.14 in Chapter 8. There we found that if $\mathbf{x} = (x_1, x_2)^T$, where x_1 and x_2 represent the amount of salt in the two tanks, then

$$\mathbf{x}' = A\mathbf{x}, \quad \text{where} \quad A = \begin{pmatrix} -1/500 & 1/500 \\ 1/500 & -1/500 \end{pmatrix}. \tag{2.11}$$

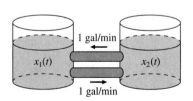

Figure 2. The two tanks in Example 2.10.

Since $T = \text{tr}(A) = -1/250$ and $D = \det(A) = 0$, the characteristic polynomial is $\lambda^2 + \lambda/250$. Thus the eigenvalues are $-1/250$ and 0. To find an eigenvector for the eigenvalue 0, we find the nullspace of $A - 0I = A$. The vector $(1, 1)^T$ is a nonzero eigenvector, so we have the solution

$$\mathbf{x}_1(t) = e^{0t} \begin{pmatrix} 1 \\ 1 \end{pmatrix} = \begin{pmatrix} 1 \\ 1 \end{pmatrix}.$$

For the eigenvalue $-1/250$, we look for the nullspace of

$$A - (-1/250)I = \begin{pmatrix} 1/500 & 1/500 \\ 1/500 & 1/500 \end{pmatrix}.$$

This time $(1, -1)^T$ works, so

$$\mathbf{x}_2(t) = e^{-t/250} \begin{pmatrix} 1 \\ -1 \end{pmatrix}$$

is a solution. According to Theorem 2.9, these solutions are linearly independent, and $\mathbf{x}_1$ and $\mathbf{x}_2$ are a fundamental set of solutions. The general solution is

$$\mathbf{x}(t) = C_1 \begin{pmatrix} 1 \\ 1 \end{pmatrix} + C_2 e^{-t/250} \begin{pmatrix} 1 \\ -1 \end{pmatrix} = \begin{pmatrix} C_1 + C_2 e^{-t/250} \\ C_1 - C_2 e^{-t/250} \end{pmatrix}. \tag{2.12}$$

At $t = 0$ we have

$$\begin{pmatrix} 100 \\ 0 \end{pmatrix} = \mathbf{x}(0) = \begin{pmatrix} C_1 + C_2 \\ C_1 - C_2 \end{pmatrix}.$$

The equations $C_1 + C_2 = 100$ and $C_1 - C_2 = 0$ have solution $C_1 = C_2 = 50$, so the solution is

$$\mathbf{x}(t) = \begin{pmatrix} x_1(t) \\ x_2(t) \end{pmatrix} = \begin{pmatrix} 50 + 50 e^{-t/250} \\ 50 - 50 e^{-t/250} \end{pmatrix}.$$

The components of the solution are plotted in Figure 3. ●

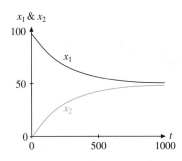

Figure 3. The salt content of the two tanks in Exercise 2.10.

Complex eigenvalues

In this case, $T^2 - 4D < 0$. The roots of the characteristic equation are the complex conjugates

$$\lambda = \frac{T + i\sqrt{4D - T^2}}{2} \quad \text{and} \quad \bar{\lambda} = \frac{T - i\sqrt{4D - T^2}}{2}. \tag{2.13}$$

This means that the matrix $A - \lambda I$ has complex entries. The properties of complex matrices are discussed in the Appendix at the end of the book. If the subject is new to you or you need to refresh your memory, please read the appendix first.

Let's look at an example.

Example 2.14 Find the eigenvalues and eigenvectors for the matrix

$$A = \begin{pmatrix} 0 & 1 \\ -2 & 2 \end{pmatrix}.$$

The eigenvalues are solutions to the characteristic equation

$$0 = \det(A - \lambda I) = \lambda^2 - T\lambda + D = \lambda^2 - 2\lambda + 2.$$

By the quadratic formula, the roots of the equation $\lambda^2 - 2\lambda + 2 = 0$ are the complex conjugates $\lambda = 1 + i$ and $\overline{\lambda} = 1 - i$.

To find an eigenvector for $\lambda = 1 + i$, we look for vectors in the nullspace of the complex matrix

$$A - \lambda I = A - (1+i)I = \begin{pmatrix} -(1+i) & 1 \\ -2 & 2 - (1+i) \end{pmatrix} = \begin{pmatrix} -1 - i & 1 \\ -2 & 1 - i \end{pmatrix}.$$

The standard methods of finding the nullspace of a matrix work, even though some of the coefficients are complex. Remember, we chose the eigenvalue λ to make this matrix singular, so do not be put off because this matrix does not *look* singular.. Let's have a little faith in our work up to now and try to find a vector in the nullspace by finding a vector that is killed by the first row of the matrix. We want a vector $\mathbf{w} = (w_1, w_2)^T$ such that $-(1 + i)w_1 + w_2 = 0$. Clearly,

$$\mathbf{w} = \begin{pmatrix} 1 \\ 1 + i \end{pmatrix} \tag{2.15}$$

is such a vector. To reassure yourself, you might show directly that

$$A\mathbf{w} = \lambda\mathbf{w}. \tag{2.16}$$

Thus, although it has complex entries, $\mathbf{w}$ is an eigenvector associated to $\lambda = 1 + i$.

If we conjugate equation (2.16), we get

$$\overline{A\mathbf{w}} = \overline{\lambda\mathbf{w}}. \tag{2.17}$$

However, since A is a real matrix, we have $\overline{A\mathbf{w}} = \overline{A}\,\overline{\mathbf{w}} = A\overline{\mathbf{w}}$. For the right-hand side of (2.17), we have $\overline{\lambda\mathbf{w}} = \overline{\lambda}\,\overline{\mathbf{w}}$. (See the appendix if these facts are new to you.) Thus, (2.17) becomes

$$A\overline{\mathbf{w}} = \overline{\lambda}\,\overline{\mathbf{w}}. \tag{2.18}$$

Thus, the complex conjugate of $\mathbf{w}$,

$$\overline{\mathbf{w}} = \begin{pmatrix} 1 \\ 1 - i \end{pmatrix},$$

is an eigenvector associated to $\overline{\lambda} = 1 - i$. ●

Let's continue with the matrix in Example 2.14. We get solutions to $\mathbf{y}' = A\mathbf{y}$ in the way indicated in Theorem 1.6. Corresponding to the eigenvalue $\lambda = 1 + i$, we have the solution

$$\mathbf{z}(t) = e^{\lambda t}\mathbf{w} = e^{(1+i)t} \begin{pmatrix} 1 \\ 1 + i \end{pmatrix}. \tag{2.19}$$

Since $\overline{\mathbf{w}}$ is an eigenvector associated with the eigenvalue $\overline{\lambda}$, we get the another complex exponential solution $e^{\overline{\lambda}t}\,\overline{\mathbf{w}}$. However,

$$\overline{\mathbf{z}}(t) = \overline{e^{\lambda t}\mathbf{w}} = e^{\overline{\lambda}t}\,\overline{\mathbf{w}}.$$

Thus, the solution corresponding to $\bar{\lambda}$ is the complex conjugate of the solution corresponding to λ,

$$\bar{\mathbf{z}}(t) = e^{\bar{\lambda}t}\,\bar{\mathbf{w}} = e^{(1-i)t}\begin{pmatrix} 1 \\ 1-i \end{pmatrix}. \tag{2.20}$$

The method we used to find the solutions in (2.19) and (2.20) is completely general. Suppose we have a matrix A with real entries, and complex conjugate eigenvalues λ and $\bar{\lambda}$, as in (2.13). Suppose that $\mathbf{w}$ is an eigenvector associated with λ. Then its complex conjugate $\bar{\mathbf{w}}$ is an eigenvector corresponding to $\bar{\lambda}$. The argument proving this in general was given in going from (2.16) to (2.18).

Notice also that $\mathbf{w}$ and $\bar{\mathbf{w}}$ are eigenvectors associated with different eigenvalues. By Corollary 2.8, $\mathbf{z}$ and $\bar{\mathbf{z}}$ form a fundamental system of solutions. This argument applies in the general situation, so we have proved the following theorem.

THEOREM 2.21　Suppose that A is a 2×2 matrix with complex conjugate eigenvalues λ and $\bar{\lambda}$. Suppose that $\mathbf{w}$ is an eigenvector associated with λ. Then the general solution to the system $\mathbf{y}' = A\mathbf{y}$ is

$$\mathbf{y}(t) = C_1 e^{\lambda t}\mathbf{w} + C_2 e^{\bar{\lambda}t}\,\bar{\mathbf{w}},$$

where C_1 and C_2 are arbitrary constants.　■

The solutions in Theorem 2.21 are complex valued. Complex-valued solutions are preferred in some situations (for example, in electrical engineering and physics). However, in other situations, it is important to find real-valued solutions. Fortunately, the real and imaginary parts of a complex solution provide the needed fundamental set of solutions.

PROPOSITION 2.22　Suppose A is an $n \times n$ matrix with real coefficients, and suppose that $\mathbf{z}(t) = \mathbf{x}(t) + i\mathbf{y}(t)$ is a solution to the system

$$\mathbf{z}' = A\mathbf{z}. \tag{2.23}$$

(a) The complex conjugate $\bar{\mathbf{z}} = \mathbf{x} - i\mathbf{y}$ is also a solution to (2.23).

(b) The real and imaginary parts $\mathbf{x}$ and $\mathbf{y}$ are also solutions to (2.23). Furthermore, if $\mathbf{z}$ and $\bar{\mathbf{z}}$ are linearly independent, so are $\mathbf{x}$ and $\mathbf{y}$.

Proof　To prove (a), we just conjugate (2.23) and remember that since A has real entries, $\bar{A} = A$. Hence

$$\bar{\mathbf{z}}' = \overline{\mathbf{z}'} = \overline{A\mathbf{z}} = \bar{A}\,\bar{\mathbf{z}} = A\bar{\mathbf{z}}.$$

The proof of part (b) is revealing. If we look at the sum of $\mathbf{z}$ and $\bar{\mathbf{z}}$ and then at their difference, we get

$$\mathbf{x} = \frac{1}{2}(\mathbf{z} + \bar{\mathbf{z}}) \quad \text{and} \quad \mathbf{y} = \frac{1}{2i}(\mathbf{z} - \bar{\mathbf{z}}). \tag{2.24}$$

Thus, $\mathbf{x}$ and $\mathbf{y}$ are linear combinations of $\mathbf{z}$ and $\bar{\mathbf{z}}$. According to Theorem 5.1 in Chapter 8, they are solutions to (2.23).

To show that $\mathbf{x}$ and $\mathbf{y}$ are linearly independent, suppose the contrary, that they are dependent. Then there is a constant c such that $\mathbf{y} = c\mathbf{x}$, so $\mathbf{z} = (1 + ic)\mathbf{x}$ and $\bar{\mathbf{z}} = (1 - ic)\mathbf{x}$. This means that $(1 - ic)\mathbf{z} - (1 + ic)\bar{\mathbf{z}} = 0$, which implies that $\mathbf{z}$ and $\bar{\mathbf{z}}$ are linearly dependent, contradicting our assumption.　●

In the case of our 2×2 matrix A with complex eigenvalue $\lambda = \alpha + i\beta$ and associated eigenvector $\mathbf{w} = \mathbf{v}_1 + i\mathbf{v}_2$, we have the solution $\mathbf{z}(t) = e^{\lambda t}\mathbf{w}$. A little algebra using Euler's formula finds the real and imaginary parts.

$$\mathbf{z}(t) = e^{\lambda t}\mathbf{w}$$
$$= e^{(\alpha + i\beta)t}(\mathbf{v}_1 + i\mathbf{v}_2)$$
$$= e^{\alpha t}(\cos\beta t + i\sin\beta t)(\mathbf{v}_1 + i\mathbf{v}_2)$$
$$= e^{\alpha t}(\cos\beta t\,\mathbf{v}_1 - \sin\beta t\,\mathbf{v}_2) + ie^{\alpha t}(\sin\beta t\,\mathbf{v}_1 + \cos\beta t\,\mathbf{v}_2)$$

Together with part (b) of Proposition 2.22, this leads immediately to the following theorem.

THEOREM 2.25 Suppose that A is a 2×2 matrix with a complex eigenvalue $\lambda = \alpha + i\beta$ and associated eigenvector $\mathbf{w} = \mathbf{v}_1 + i\mathbf{v}_2$. Then the general solution of the system $\mathbf{y}' = A\mathbf{y}$ is

$$\mathbf{y}(t) = C_1 e^{\alpha t}(\cos\beta t\,\mathbf{v}_1 - \sin\beta t\,\mathbf{v}_2) + C_2 e^{\alpha t}(\sin\beta t\,\mathbf{v}_1 + \cos\beta t\,\mathbf{v}_2),$$

where C_1 and C_2 are arbitrary constants. ∎

Let's return to the system in Example 2.14.

Example 2.26 Find a fundamental set of real solutions for the system $\mathbf{y}' = A\mathbf{y}$, where A is the matrix defined in Example 2.14.

We have already determined the two complex-valued solutions in (2.19) and (2.20). Since $\mathbf{z}(0) = \mathbf{w}$ and $\bar{\mathbf{z}}(0) = \bar{\mathbf{w}}$, which are eigenvectors corresponding to the eigenvalues λ and $\bar{\lambda}$, respectively. Since $\lambda \neq \bar{\lambda}$, we know that $\mathbf{z}$ and $\bar{\mathbf{z}}$ are linearly independent. Motivated by Proposition 2.22, let's break $\mathbf{z}(t)$ into its real and imaginary parts.

Using Euler's formula,

$$\mathbf{z}(t) = e^{(1+i)t}\begin{pmatrix} 1 \\ 1+i \end{pmatrix}$$
$$= e^t[\cos t + i\sin t]\left[\begin{pmatrix} 1 \\ 1 \end{pmatrix} + i\begin{pmatrix} 0 \\ 1 \end{pmatrix}\right]$$
$$= e^t\left\{\left[\cos t\begin{pmatrix} 1 \\ 1 \end{pmatrix} - \sin t\begin{pmatrix} 0 \\ 1 \end{pmatrix}\right] + i\left[\cos t\begin{pmatrix} 0 \\ 1 \end{pmatrix} + \sin t\begin{pmatrix} 1 \\ 1 \end{pmatrix}\right]\right\} \qquad (2.27)$$
$$= e^t\begin{pmatrix} \cos t \\ \cos t - \sin t \end{pmatrix} + ie^t\begin{pmatrix} \sin t \\ \cos t + \sin t \end{pmatrix}.$$

According to Proposition 2.22, the real and imaginary parts of $\mathbf{z}$ are also solutions. From (2.27), we see that these are

$$\mathbf{x}(t) = e^t\begin{pmatrix} \cos t \\ \cos t - \sin t \end{pmatrix} \quad \text{and} \quad \mathbf{y}(t) = e^t\begin{pmatrix} \sin t \\ \cos t + \sin t \end{pmatrix}.$$

Again according to Proposition 2.22, these solutions are linearly independent, since $\mathbf{z}$ and $\bar{\mathbf{z}}$ are linearly independent. ●

We could have found the answer in Example 2.26 by plugging into the formula in Theorem 2.25, but it is usually better to remember how to use Euler's formula and complex arithmetic to find the answers. The formula in Theorem 2.25 is useful primarily for understanding the form the general solution takes. Here is another example.

Example 2.28   Find the currents I_1 and I_2 for the circuit in Figure 4, where $R = 1$ ohm, $L = 1$ henry, and $C = 5/4$ farad. Assume that $I_1(0) = 5$ amperes and $I_2(0) = 1$ ampere.

This is the circuit in Example 4.9 in Chapter 8, except that there is no voltage source. Using the same analysis as in that example (or by simply setting $E = 0$), we find that $\mathbf{I} = (I_1, I_2)^T$ satisfies $\mathbf{I}' = A\mathbf{I}$, where

$$A = \begin{pmatrix} -1 & -1 \\ 1 & 1/5 \end{pmatrix}.$$

Since $T = \mathrm{tr}(A) = -4/5$ and $D = \det(A) = 4/5$, the characteristic polynomial is $\lambda^2 + 4\lambda/5 + 4/5$. Using the quadratic formula we see that the roots are $(-2 \pm 4i)/5$. We find an eigenvector associated with the eigenvalue $\lambda = (-2 + 4i)/5$ by finding the nullspace of the matrix

$$A - \lambda I = \begin{pmatrix} -3/5 - 4i/5 & -1 \\ 1 & 3/5 - 4i/5 \end{pmatrix}.$$

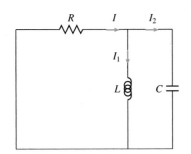

R   I I_2

I_1

L   C

Figure 4. An electrical circuit with a resistor, a capacitor and an inductor in parallel.

Proceeding as we did in Example 2.14, we find that $\mathbf{w} = (-5, 3 + 4i)^T$ is an eigenvector. The resulting complex-valued solution is $\mathbf{z}(t) = e^{\lambda t}\mathbf{w}$. To find the real and complex parts of $\mathbf{z}$, we compute

$$\begin{aligned}
\mathbf{z}(t) &= e^{\lambda t}\mathbf{w} \\
&= e^{(-2+4i)t/5} \begin{pmatrix} -5 \\ 3 + 4i \end{pmatrix} \\
&= e^{-2t/5}[\cos(4t/5) + i\sin(4t/5)]\left[\begin{pmatrix} -5 \\ 3 \end{pmatrix} + i\begin{pmatrix} 0 \\ 4 \end{pmatrix} \right] \\
&= e^{-2t/5}\left[\cos(4t/5)\begin{pmatrix} -5 \\ 3 \end{pmatrix} - \sin(4t/5)\begin{pmatrix} 0 \\ 4 \end{pmatrix} \right] \\
&\quad + ie^{-2t/5}\left[\cos(4t/5)\begin{pmatrix} 0 \\ 4 \end{pmatrix} + \sin(4t/5)\begin{pmatrix} -5 \\ 3 \end{pmatrix} \right].
\end{aligned}$$

Therefore, a fundamental set of solutions is

$$\mathbf{y}_1(t) = e^{-2t/5}\begin{pmatrix} -5\cos(4t/5) \\ 3\cos(4t/5) - 4\sin(4t/5) \end{pmatrix} \quad \text{and}$$

$$\mathbf{y}_2(t) = e^{-2t/5}\begin{pmatrix} -5\sin(4t/5) \\ 4\cos(4t/5) + 3\sin(4t/5) \end{pmatrix}.$$

The general solution is $\mathbf{I}(t) = C_1\mathbf{y}_1(t) + C_2\mathbf{y}_2(t)$.

At $t = 0$ we have

$$\begin{pmatrix} 5 \\ 1 \end{pmatrix} = \mathbf{I}(0) = \begin{pmatrix} I_1(0) \\ I_2(0) \end{pmatrix} = \begin{pmatrix} -5C_1 \\ 3C_1 + 4C_2 \end{pmatrix}.$$

Hence $C_1 = -1$ and $C_2 = 1$, so the solution is

$$\mathbf{I}(t) = \begin{pmatrix} I_1(t) \\ I_2(t) \end{pmatrix} = e^{-2t/5}\begin{pmatrix} 5\cos(4t/5) - 5\sin(4t/5) \\ \cos(4t/5) + 7\sin(4t/5) \end{pmatrix}.$$

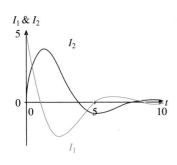

I_1 & I_2

I_2

I_1

Figure 5. The currents in Example 2.28.

The components of the solution are plotted in Figure 5.

One real eigenvalue of multiplicity 2—the easy case

This is the degenerate case when $T^2 = 4D$. The characteristic polynomial is

$$p(\lambda) = \lambda^2 - T\lambda + D = \lambda^2 - T\lambda + T^2/4 = (\lambda - T/2)^2 = (\lambda - \lambda_1)^2, \quad (2.29)$$

where $\lambda_1 = T/2$ is the only eigenvalue and it has multiplicity 2. There are two subcases, depending on the dimension of the eigenspace of λ. Since the eigenspace is a subspace of $\mathbf{R}^2$, it can have dimension 1 or 2.

If the eigenspace has dimension 2, it must be equal to all of $\mathbf{R}^2$. In this case, every vector is an eigenvector, so $A\mathbf{v} = \lambda_1\mathbf{v}$ for all $\mathbf{v} \in \mathbf{R}^2$. Suppose that

$$A = \begin{pmatrix} a & b \\ c & d \end{pmatrix}, \quad \mathbf{e}_1 = \begin{pmatrix} 1 \\ 0 \end{pmatrix}, \quad \text{and} \quad \mathbf{e}_2 = \begin{pmatrix} 0 \\ 1 \end{pmatrix}.$$

Since $A\mathbf{e}_1 = (a, c)^T$, $\lambda_1\mathbf{e}_1 = (\lambda_1, 0)^T$, and $A\mathbf{e}_1 = \lambda_1\mathbf{e}_1$, we see that $a = \lambda_1$ and $c = 0$. Similarly, since $A\mathbf{e}_2 = (b, d)^T$, $\lambda_1\mathbf{e}_2 = (0, \lambda_1)^T$, and $A\mathbf{e}_2 = \lambda_1\mathbf{e}_2$, we find that $b = 0$ and $d = \lambda_1$. Hence $A = \lambda_1 I$, and the system has the very simple form

$$x_1' = \lambda_1 x_1,$$
$$x_2' = \lambda_1 x_2,$$

or $\mathbf{x}' = \lambda_1\mathbf{x}$. For any vector $\mathbf{v}$, the solution with initial value $\mathbf{x}(0) = \mathbf{v}$ is $\mathbf{x}(t) = e^{\lambda_1 t}\mathbf{v}$.

One real eigenvalue of multiplicity 2—the interesting case

The more interesting case is when the dimension of the eigenspace is 1. Let's illustrate the problem with an example.

Example 2.30 Consider the electrical circuit in Figure 4, where $R = 1$ ohm, $L = 1$ henry, and $C = 1/4$ farad. Find all exponential solutions for the current in the circuit.

As we did in Example 2.28, we find that vector of currents, $\mathbf{I} = (I_1, I_2)^T$, satisfies $\mathbf{I}' = A\mathbf{I}$, where
$$A = \begin{pmatrix} -1 & -1 \\ 1 & -3 \end{pmatrix}.$$

The characteristic polynomial is

$$p(\lambda) = \lambda^2 - T\lambda + D = \lambda^2 + 4\lambda + 4 = (\lambda + 2)^2.$$

The only root, and therefore the only eigenvalue of A, is $\lambda_1 = -2$. We have

$$A - \lambda_1 I = A + 2I = \begin{pmatrix} 1 & -1 \\ 1 & -1 \end{pmatrix}.$$

Any vector in the nullspace of $A - \lambda_1 I$ is an eigenvector. A natural choice is $\mathbf{v}_1 = (1, 1)^T$. The corresponding solution is

$$\mathbf{x}_1(t) = e^{\lambda_1 t}\mathbf{v}_1 = e^{-2t}\begin{pmatrix} 1 \\ 1 \end{pmatrix}.$$

Since the dimension of the eigenspace of λ_1 is 1, this exhausts the possibilities for exponential solutions. Any exponential solution must be a constant multiple of $\mathbf{x}_1$.

Since we can find only one linearly independent exponential solution to the system in Example 2.30, we must try something else to get a second solution that is not a multiple of $\mathbf{x}_1(t)$. In Proposition 3.18 in Chapter 4, we found that the general solution to a second-order equation that had a characteristic equation with a double root λ has the form $y(t) = (C_1 + C_2 t)e^{\lambda t}$. It was necessary to introduce a term with a factor of t. Let's try to find a solution for our system $\mathbf{x}' = A\mathbf{x}$ of the form

$$\mathbf{x}(t) = e^{\lambda_1 t}[\mathbf{v}_2 + t\mathbf{v}_1],$$

where $\mathbf{v}_2$ and $\mathbf{v}_1$ are undetermined vectors. Differentiating, we get

$$\mathbf{x}'(t) = \lambda_1 e^{\lambda_1 t}[\mathbf{v}_2 + t\mathbf{v}_1] + e^{\lambda_1 t}\mathbf{v}_1 = e^{\lambda_1 t}[(\lambda_1\mathbf{v}_2 + \mathbf{v}_1) + \lambda_1 t\mathbf{v}_1]. \tag{2.31}$$

On the other hand,

$$A\mathbf{x}(t) = e^{\lambda_1 t}[A\mathbf{v}_2 + tA\mathbf{v}_1]. \tag{2.32}$$

We will have a solution to $\mathbf{x}' = A\mathbf{x}$ provided that the right-hand sides of (2.31) and (2.32) are equal. Since the exponential $e^{\lambda_1 t}$ is never zero, we must have

$$(\lambda_1\mathbf{v}_2 + \mathbf{v}_1) + \lambda_1 t\mathbf{v}_1 = A\mathbf{v}_2 + tA\mathbf{v}_1.$$

If this equation is to be true for all t, we must have

$$A\mathbf{v}_2 = \lambda_1\mathbf{v}_2 + \mathbf{v}_1 \quad \text{and} \tag{2.33}$$

$$A\mathbf{v}_1 = \lambda_1\mathbf{v}_1. \tag{2.34}$$

We know how to solve equation (2.34). It requires that λ_1 is an eigenvalue of A and $\mathbf{v}_1$ is an associated eigenvector. This agrees with the notation we have been using. However, at first glance, solving equation (2.33) seems problematic. After all, it is equivalent to

$$(A - \lambda_1 I)\mathbf{v}_2 = \mathbf{v}_1. \tag{2.35}$$

We know that this equation would have a solution for any right-hand side if the matrix $A - \lambda_1 I$ were nonsingular, but we chose the eigenvalue λ_1 precisely to make this matrix singular.

The situation is saved by the special nature of the matrix A. Let's look back at the case in Example 2.30. We can compute that, for the matrix A in Example 2.30,

$$(A - \lambda_1 I)^2 = \begin{pmatrix} 1 & -1 \\ 1 & -1 \end{pmatrix} \begin{pmatrix} 1 & -1 \\ 1 & -1 \end{pmatrix} = \begin{pmatrix} 0 & 0 \\ 0 & 0 \end{pmatrix}. \tag{2.36}$$

Thus, $(A - \lambda_1 I)^2 = 0I$, the matrix with all entries equal to 0. This is the key to finding a solution to equation (2.33) and to finding a second solution to the system. The Cayley-Hamilton theorem says that any matrix with characteristic polynomial $p(\lambda)$ satisfies its characteristic equation, meaning that $p(A) = 0I$. To compute $p(A)$, we systematically replace the number λ by the matrix A, including λ^2 by A^2 and $1 = \lambda^0$ by $A^0 = I$. Thus $\lambda_1 = \lambda_1 \cdot 1$ is replaced by $\lambda_1 I$. We will only use this in the case at hand, in which we have a 2×2 matrix A with a single eigenvalue of multiplicity 2. The characteristic polynomial for such a matrix was found in (2.29) to be $p(\lambda) = (\lambda - \lambda_1)^2$. Hence,

$$p(A) = (A - \lambda_1 I)^2 = 0I. \tag{2.37}$$

For the matrix in Example 2.30, we verified this in (2.36).

It follows from (2.37) that, for any vector $\mathbf{w}$ in $\mathbf{R}^2$,

$$(A - \lambda_1 I)[(A - \lambda_1 I)\mathbf{w}] = (A - \lambda_1 I)^2 \mathbf{w} = \mathbf{0}.$$

Thus, for any vector $\mathbf{w}$ in $\mathbf{R}^2$, $(A - \lambda_1 I)\mathbf{w}$ is in the nullspace of $A - \lambda_1 I$, which means that $(A - \lambda_1 I)\mathbf{w}$ is an eigenvector associated to the eigenvalue λ_1. If the eigenspace for λ_1 has dimension 1, and $\mathbf{v}_1$ is a nonzero eigenvector, then it follows that $(A - \lambda_1 I)\mathbf{w}$ is a multiple of $\mathbf{v}_1$. Hence, for any vector $\mathbf{w}$ in $\mathbf{R}^2$, there is a constant a such that

$$(A - \lambda_1 I)\mathbf{w} = a\mathbf{v}_1. \tag{2.38}$$

Furthermore, if $\mathbf{w}$ is not a multiple of $\mathbf{v}_1$, then $\mathbf{w}$ is not an eigenvector, so $(A - \lambda_1 I)\mathbf{w} \neq \mathbf{0}$. Thus the constant a in (2.38) is nonzero. If we set

$$\mathbf{v}_2 = (1/a)\mathbf{w}, \tag{2.39}$$

then

$$(A - \lambda_1 I)\mathbf{v}_2 = \frac{1}{a}(A - \lambda_1 I)\mathbf{w} = \mathbf{v}_1,$$

and we have a solution to (2.35), and therefore to (2.33).

Since our vectors $\mathbf{v}_1$ and $\mathbf{v}_2$ satisfy equations (2.33) and (2.34), we know that the functions

$$\mathbf{x}_1(t) = e^{\lambda_1 t}\mathbf{v}_1 \quad \text{and}$$
$$\mathbf{x}_2(t) = e^{\lambda_1 t}[\mathbf{v}_2 + t\mathbf{v}_1]$$

are solutions to the system $\mathbf{x}' = A\mathbf{x}$. Furthermore, since we chose $\mathbf{v}_2$ so that it was not a multiple of $\mathbf{v}_1$, $\mathbf{x}_1(0) = \mathbf{v}_1$ and $\mathbf{x}_2(0) = \mathbf{v}_2$ are linearly independent. Hence, the solutions are linearly independent, and $\mathbf{x}_1$ and $\mathbf{x}_2$ form a fundamental set of solutions.

Let's summarize the method for finding a fundamental set of solutions in a theorem.

THEOREM 2.40   Suppose that A is a 2×2 matrix with one eigenvalue λ of multiplicity 2, and suppose that the eigenspace of λ has dimension 1. Let $\mathbf{v}_1$ be a nonzero eigenvector, and choose $\mathbf{v}_2$ such that $(A - \lambda I)\mathbf{v}_2 = \mathbf{v}_1$. Then

$$\mathbf{x}_1(t) = e^{\lambda t}\mathbf{v}_1 \quad \text{and}$$
$$\mathbf{x}_2(t) = e^{\lambda t}[\mathbf{v}_2 + t\mathbf{v}_1] \tag{2.41}$$

form a fundamental set of solutions to the system $\mathbf{x}' = A\mathbf{x}$. ∎

Given the fundamental set of solutions $\mathbf{x}_1$ and $\mathbf{x}_2$, we can write the general solution as

$$\mathbf{x}(t) = C_1\mathbf{x}_1(t) + C_2\mathbf{x}_2(t)$$
$$= C_1 e^{\lambda t}\mathbf{v}_1 + C_2 e^{\lambda t}[\mathbf{v}_2 + t\mathbf{v}_1].$$

For use in later work, we will find it convenient to rearrange this solution as

$$\mathbf{x}(t) = e^{\lambda t}[(C_1 + C_2 t)\mathbf{v}_1 + C_2\mathbf{v}_2]. \tag{2.42}$$

Example 2.43   Find the currents in Example 2.30 with $I_1(0) = 5$ amperes and $I_2(0) = 1$ ampere.

In Example 2.30 we found that the vector of currents $\mathbf{I} = (I_1, I_2)^T$, satisfies $\mathbf{I}' = A\mathbf{I}$, where

$$A = \begin{pmatrix} -1 & -1 \\ 1 & -3 \end{pmatrix}.$$

We also found that A has a single eigenvalue $\lambda_1 = -2$ and the eigenvector $\mathbf{v}_1 = (1, 1)^T$. The corresponding solution is

$$\mathbf{x}_1(t) = e^{\lambda t}\mathbf{v}_1 = e^{-2t}\begin{pmatrix} 1 \\ 1 \end{pmatrix}.$$

According to Theorem 2.40, we need to find a vector $\mathbf{v}_2$ which satisfies $(A - \lambda_1 I)\mathbf{v}_2 = \mathbf{v}_1$. The process for doing this is covered in the paragraph containing formula (2.39). We can start with literally any vector $\mathbf{w}$ that is not a multiple of $\mathbf{v}_1$. Given this flexibility, we should choose a vector that contains as many zeros as possible to facilitate computation. For example, if we choose a simple vector like $\mathbf{w} = (0, 1)^T$, we proceed as follows. Compute

$$(A - \lambda_1 I)\mathbf{w} = (A + 2I)\mathbf{w} = \begin{pmatrix} 1 & -1 \\ 1 & -1 \end{pmatrix}\begin{pmatrix} 0 \\ 1 \end{pmatrix} = \begin{pmatrix} -1 \\ -1 \end{pmatrix} = -1\mathbf{v}_1.$$

Hence, we take $\mathbf{v}_2 = (-1)\mathbf{w} = (0, -1)^T$.

Our fundamental set of solutions is

$$\mathbf{x}_1(t) = e^{\lambda t}\mathbf{v}_1 = e^{-2t}\begin{pmatrix} 1 \\ 1 \end{pmatrix} \quad \text{and}$$

$$\mathbf{x}_2(t) = e^{\lambda t}(\mathbf{v}_2 + t\mathbf{v}_1) = e^{-2t}\begin{pmatrix} t \\ t-1 \end{pmatrix}. \tag{2.44}$$

The general solution is

$$\mathbf{I}(t) = C_1\mathbf{x}_1(t) + C_2\mathbf{x}_2(t) \tag{2.45}$$

$$= e^{-2t}\left(C_1\begin{pmatrix} 1 \\ 1 \end{pmatrix} + C_2\begin{pmatrix} t \\ t-1 \end{pmatrix}\right). \tag{2.46}$$

At $t = 0$ we have

$$\begin{pmatrix} 5 \\ 1 \end{pmatrix} = \mathbf{I}(0) = \begin{pmatrix} C_1 \\ C_1 - C_2 \end{pmatrix}.$$

Hence, $C_1 = 5$ and $C_2 = 4$. The solution with these initial values is

$$\mathbf{I}(t) = e^{-2t}\begin{pmatrix} 5 + 4t \\ 1 + 4t \end{pmatrix}.$$

The components of the solution are plotted in Figure 6.

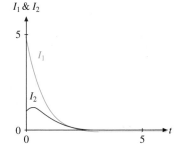

Figure 6. The currents in Example 2.43.

EXERCISES

In Exercises 1–6, the matrix A has real eigenvalues. Find the general solution of the system $\mathbf{y}' = A\mathbf{y}$.

1. $A = \begin{pmatrix} 2 & -6 \\ 0 & -1 \end{pmatrix}$

2. $A = \begin{pmatrix} -1 & 6 \\ -3 & 8 \end{pmatrix}$

3. $A = \begin{pmatrix} -5 & 1 \\ -2 & -2 \end{pmatrix}$

4. $A = \begin{pmatrix} -3 & -6 \\ 0 & -1 \end{pmatrix}$

5. $A = \begin{pmatrix} 1 & 2 \\ -1 & 4 \end{pmatrix}$

6. $A = \begin{pmatrix} -1 & 1 \\ 1 & -1 \end{pmatrix}$

In Exercises 7–12, find the solution of the initial-value problem for system $\mathbf{y}' = A\mathbf{y}$ with the given matrix A and the given initial value.

7. The matrix in Exercise 1 with $\mathbf{y}(0) = (0, 1)^T$

8. The matrix in Exercise 2 with $\mathbf{y}(0) = (1, -2)^T$

9. The matrix in Exercise 3 with $\mathbf{y}(0) = (0, -1)^T$

10. The matrix in Exercise 4 with $\mathbf{y}(0) = (1, 1)^T$

11. The matrix in Exercise 5 with $\mathbf{y}(0) = (3, 2)^T$

12. The matrix in Exercise 6 with $\mathbf{y}(0) = (1, 5)^T$

In Exercises 13 and 14, a complex vector valued function $\mathbf{z}(t)$ is given. Find the real and imaginary parts of $\mathbf{z}(t)$.

13. $\mathbf{z}(t) = e^{2it} \begin{pmatrix} 1 \\ 1+i \end{pmatrix}$ **14.** $\mathbf{z}(t) = e^{(1+i)t} \begin{pmatrix} -1+i \\ 2 \end{pmatrix}$

15. The system

$$\mathbf{y}' = \begin{pmatrix} 3 & 3 \\ -6 & -3 \end{pmatrix} \mathbf{y} \qquad (2.47)$$

has complex solution

$$\mathbf{z}(t) = e^{3it} \begin{pmatrix} -1-i \\ 2 \end{pmatrix}.$$

Verify, by direct substitution, that the real and imaginary parts of this solution are solutions of system (2.47). Then use Proposition 5.2 in Section 8.5 to verify that they are linearly independent solutions.

In Exercises 16–21, the matrix A has complex eigenvalues. Find a fundamental set of real solutions of the system $\mathbf{y}' = A\mathbf{y}$.

16. $A = \begin{pmatrix} -4 & -8 \\ 4 & 4 \end{pmatrix}$ **17.** $A = \begin{pmatrix} -1 & -2 \\ 4 & 3 \end{pmatrix}$

18. $A = \begin{pmatrix} -1 & 1 \\ -5 & -5 \end{pmatrix}$ **19.** $A = \begin{pmatrix} 0 & 4 \\ -2 & -4 \end{pmatrix}$

20. $A = \begin{pmatrix} -1 & 3 \\ -3 & -1 \end{pmatrix}$ **21.** $A = \begin{pmatrix} 3 & -6 \\ 3 & 5 \end{pmatrix}$

In Exercises 22–27, find the solution of the initial value problem for system $\mathbf{y}' = A\mathbf{y}$ with the given matrix A and the given initial value.

22. The matrix in Exercise 16 with $\mathbf{y}(0) = (0, 2)^T$

23. The matrix in Exercise 17 with $\mathbf{y}(0) = (0, 1)^T$

24. The matrix in Exercise 18 with $\mathbf{y}(0) = (1, -5)^T$

25. The matrix in Exercise 19 with $\mathbf{y}(0) = (-1, 2)^T$

26. The matrix in Exercise 20 with $\mathbf{y}(0) = (3, 2)^T$

27. The matrix in Exercise 21 with $\mathbf{y}(0) = (1, 3)^T$

28. Suppose that A is a real 2×2 matrix with one eigenvalue λ of multiplicity two. Show that the solution to the initial value problem $\mathbf{y}' = A\mathbf{y}$ with $\mathbf{y}(0) = \mathbf{v}$ is given by

$$\mathbf{y}(t) = e^{\lambda t}[\mathbf{v} + t(A - \lambda I)\mathbf{v}].$$

Hint: Verify the result by direct substitution. Remember that $(A - \lambda I)^2 = 0I$, so $A(A - \lambda I) = \lambda(A - \lambda I)$.

In Exercises 29–34 the matrix A has one real eigenvalue of multiplicity two. Find the general solution of the system $\mathbf{y}' = A\mathbf{y}$.

29. $A = \begin{pmatrix} -2 & 0 \\ 0 & -2 \end{pmatrix}$ **30.** $A = \begin{pmatrix} -3 & 1 \\ -1 & -1 \end{pmatrix}$

31. $A = \begin{pmatrix} 3 & -1 \\ 1 & 1 \end{pmatrix}$ **32.** $A = \begin{pmatrix} -2 & -1 \\ 4 & 2 \end{pmatrix}$

33. $A = \begin{pmatrix} -2 & 1 \\ -9 & 4 \end{pmatrix}$ **34.** $A = \begin{pmatrix} 5 & 1 \\ -4 & 1 \end{pmatrix}$

In Exercises 35–40, find the solution of the initial value problem for system $\mathbf{y}' = A\mathbf{y}$ with the given matrix A and the given initial value.

35. The matrix in Exercise 29 with $\mathbf{y}(0) = (3, -2)^T$

36. The matrix in Exercise 30 with $\mathbf{y}(0) = (0, -3)^T$

37. The matrix in Exercise 31 with $\mathbf{y}(0) = (2, -1)^T$

38. The matrix in Exercise 32 with $\mathbf{y}(0) = (1, 1)^T$

39. The matrix in Exercise 33 with $\mathbf{y}(0) = (5, 3)^T$

40. The matrix in Exercise 34 with $\mathbf{y}(0) = (0, 2)^T$

In Exercises 41–48, find the general solution of the system $\mathbf{y}' = A\mathbf{y}$ for the given matrix A.

41. $A = \begin{pmatrix} 2 & 4 \\ -1 & 6 \end{pmatrix}$ **42.** $A = \begin{pmatrix} -8 & -10 \\ 5 & 7 \end{pmatrix}$

43. $A = \begin{pmatrix} 5 & 12 \\ -4 & -9 \end{pmatrix}$ **44.** $A = \begin{pmatrix} -6 & 1 \\ 0 & -6 \end{pmatrix}$

45. $A = \begin{pmatrix} -4 & -5 \\ 2 & 2 \end{pmatrix}$ **46.** $A = \begin{pmatrix} -6 & 4 \\ -8 & 2 \end{pmatrix}$

47. $A = \begin{pmatrix} -10 & 4 \\ -12 & 4 \end{pmatrix}$ **48.** $A = \begin{pmatrix} -1 & 5 \\ -5 & -1 \end{pmatrix}$

In Exercises 49–56, find the solution of the initial value problem for system $\mathbf{y}' = A\mathbf{y}$ with the given matrix A and the given initial value.

49. The matrix in Exercise 41 with $\mathbf{y}(0) = (3, 1)^T$

50. The matrix in Exercise 42 with $\mathbf{y}(0) = (3, 1)^T$

51. The matrix in Exercise 43 with $\mathbf{y}(0) = (1, 0)^T$

52. The matrix in Exercise 44 with $\mathbf{y}(0) = (1, 0)^T$

53. The matrix in Exercise 45 with $\mathbf{y}(0) = (-3, 2)^T$

54. The matrix in Exercise 46 with $\mathbf{y}(0) = (4, 0)^T$

55. The matrix in Exercise 47 with $\mathbf{y}(0) = (2, 1)^T$

56. The matrix in Exercise 48 with $\mathbf{y}(0) = (5, 5)^T$

57. The Cayley-Hamilton theorem is one of the most important results in linear algebra. The proof in general is quite difficult, but for the case of a 2 matrix with a single eigenvalue λ of multiplicity 2, the proof is not so bad. We need to show that $(A - \lambda I)^2 = 0I$.

  (a) Show that it is enough to show that $(A - \lambda I)^2 \mathbf{v} = \mathbf{0}$ for every vector $\mathbf{v} \in \mathbf{R}^2$.

(b) Choose a nonzero eigenvector $\mathbf{v}_1$. Show that $(A - \lambda I)^2\mathbf{v} = \mathbf{0}$ for every vector $\mathbf{v}$ that is a multiple of $\mathbf{v}_1$.

(c) Suppose that $\mathbf{v}$ is not a multiple of $\mathbf{v}_1$. Show that $\mathbf{v}$ and $\mathbf{v}_1$ form a basis of $\mathbf{R}^2$.

(d) Set $\mathbf{w} = (A - \lambda I)\mathbf{v}$. Show that there are numbers a and b such that $\mathbf{w} = a\mathbf{v}_1 + b\mathbf{v}$.

(e) Show that $(A - \lambda I)\mathbf{w} = b\mathbf{w}$, and conclude from the fact that λ is the only eigenvalue that $b = 0$.

(f) Conclude that $(A - \lambda I)^2\mathbf{v} = \mathbf{0}$.

58. Figure 7 shows two tanks, each containing 500 gallons of a salt solution. Pure water pours into the top tank at a rate of 5 gal/s. Salt solution pours out of the bottom of the tank and into the tank below at a rate of 5 gal/s. There is a drain at the bottom of the second tank, out of which salt solution flows at a rate of 5 gal/s. As a result, the amount of solution in each tank remains constant at 500 gallons. Initially (time $t = 0$) there is 100 pounds of salt present in the first tank, and zero pounds of salt present in the tank immediately below.

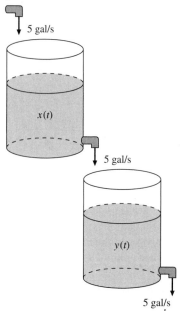

Figure 7. Two cascaded tanks.

(a) Set up, in matrix-vector form, an initial value problem that models the salt content in each tank over time.

(b) Find the eigenvalues and eigenvectors of the coefficient matrix in part (a), then find the general solution in vector form. Find the solution that satisfies the initial conditions posed in part (a).

(c) Plot each component of your solution in part (b) over a period of four time constants (see Section 4.7 or Section 2.2, Exercise 29) $[0, 4T_c]$. What is the eventual salt content in each tank? Why? Give both a physical and a mathematical reason for your answer.

59. Figure 8 shows two tanks, each containing 360 liters of a salt solution. Pure water pours into tank A at a rate of

5 L/min. There are two pipes connecting tank A to tank B. The first pumps salt solution from tank B into tank A at a rate of 4 L/min. The second pumps salt solution from tank A into tank B at a rate of 9 L/min. Finally, there is a drain on tank B from which salt solution drains at a rate of 5 L/min. Thus, each tank maintains a constant volume of 360 liters of salt solution. Initially, there are 60 kg of salt present in tank A, but tank B contains pure water.

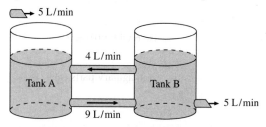

Figure 8. Two interconnected tanks.

(a) Set up, in matrix-vector form, an initial value problem that models the salt content in each tank over time.

(b) Find the eigenvalues and eigenvectors of the coefficient matrix in part (a), then find the general solution in vector form. Find the solution that satisfies the initial conditions posed in part (a).

(c) Plot each component of your solution in part (b) over a period of four time constants (see Section 4.7 or Section 2.2, Exercise 29) $[0, 4T_c]$. What is the eventual salt content in each tank? Why? Give both a physical and a mathematical reason for your answer.

60. In Exercise 22 of Section 8.4, you were given the circuit in Figure 9 and asked to show that the voltage V across the capacitor and the current I across the inductor satisfied the system

$$V' = -\frac{V}{RC} - \frac{I}{C}$$
$$I' = \frac{V}{L}.$$

Suppose that the resistance is $R = 1/2$ ohm, the capacitance is $C = 1$ farad, and the inductance is $L = 1/2$ henry. If the initial voltage across the capacitor is $V(0) = 10$ volts and there is no initial current across the inductor, solve the system to determine the voltage and current as a function of time. Plot the voltage and current as a function of time. Assume current flows in the directions indicated.

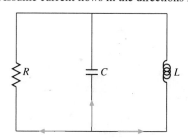

Figure 9. A parallel circuit with capacitor, resistor, and inductor.

61. Show that the voltage V across the capacitor and the current I through the inductor in Figure 10 satisfy the system

$$I' = -\frac{R_1}{L} I + \frac{1}{L} V$$

$$V' = -\frac{1}{C} I - \frac{1}{R_2 C} V.$$

Suppose that the capacitance is $C = 1$ farad, the inductance is $L = 1$ henry, the leftmost resistor has resistance $R_2 = 1$ ohm, and the rightmost resistor has resistance $R_1 = 5$ ohms. If the initial voltage across the capacitor is 12 volts and the initial current through the inductor is zero, determine the voltage V across the capacitor and the current I through the inductor as functions of time. Plot

the voltage and current as functions of time. Assume current flows in the directions indicated.

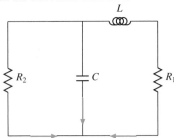

Figure 10. A more complicated circuit.

9.3 Phase Plane Portraits

Now that we know how to solve linear planar systems with constant coefficients, let's find out what the solutions look like. There is a variety of different cases, with different behaviors. We will examine the six most important cases here. There are several more. Some of them will be the subject of exercises, and a complete classification will be left as a project.

To set the stage, we will be considering the system

$$\mathbf{y}' = A\mathbf{y}, \tag{3.1}$$

where

$$A = \begin{pmatrix} a_{11} & a_{12} \\ a_{21} & a_{22} \end{pmatrix} \quad \text{and} \quad \mathbf{y}(t) = \begin{pmatrix} y_1(t) \\ y_2(t) \end{pmatrix}.$$

The characteristic polynomial is

$$\lambda^2 - T\lambda + D = 0,$$

where $D = \det(A) = a_{11}a_{22} - a_{12}a_{21}$, and $T = \operatorname{tr}(A) = a_{11} + a_{22}$.

Since the system in (3.1) is autonomous, it is natural to use the phase plane to visualize the solutions. Review Section 8.3 of Chapter 8, if necessary. In particular, remember that the uniqueness theorem implies that solution curves cannot intersect.

We will find that the solutions have strong connections to the equilibrium points of the system. For the linear homogeneous system in (3.1), the equilibrium points are those points $\mathbf{v} \in \mathbf{R}^2$ where $A\mathbf{v} = \mathbf{0}$. Thus, the set of equilibrium points is just the nullspace of A. In most of the cases that we will examine, A is nonsingular, so the origin $\mathbf{0}$ is the only equilibrium point. There will be times when A is singular, in which case the nullspace is a line in $\mathbf{R}^2$, or all of $\mathbf{R}^2$, and every point in the nullspace is an equilibrium point.

Each of the cases will have a descriptive name, like *saddle point* or *center*. This name is meant to refer to the equilibrium point at the origin. Thus, for example, we will say that the origin is a saddle point.

Real eigenvalues

In the first three cases that we will consider, the eigenvalues of A will be real and distinct. For this to be true, we must have $T^2 - 4D > 0$. The eigenvalues are

$$\lambda_1 = \frac{T - \sqrt{T^2 - 4D}}{2} \quad \text{and} \quad \lambda_2 = \frac{T + \sqrt{T^2 - 4D}}{2}.$$

Notice that $\lambda_1 < \lambda_2$.

According to Theorem 2.9, the general solution is

$$\mathbf{y}(t) = C_1 e^{\lambda_1 t}\mathbf{v}_1 + C_2 e^{\lambda_2 t}\mathbf{v}_2, \tag{3.2}$$

where C_1 and C_2 are arbitrary constants and $\mathbf{v}_1$ and $\mathbf{v}_2$ are eigenvectors associated with λ_1 and λ_2, respectively. Particular solutions are captured by assigning various values to the constants C_1 and C_2.

Exponential solutions

Two particular solutions are especially noteworthy. These occur when one, of the constants C_1 and C_2 is equal to zero. These are the solutions $C_1 e^{\lambda_1 t}\mathbf{v}_1$ and $C_2 e^{\lambda_2 t}\mathbf{v}_2$, the so-called *exponential solutions*. To simplify the notation, let's suppose that λ is an eigenvalue of the matrix A, and $\mathbf{v}$ is an associated eigenvector. Then the exponential solutions have the form

$$\mathbf{y}(t) = C e^{\lambda t}\mathbf{v}.$$

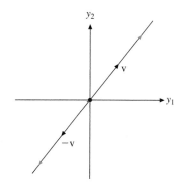

Figure 1. For a positive eigenvalue there are two unstable exponential solution curves.

Notice that as t varies, $\mathbf{y}(t)$ is always a multiple of $\mathbf{v}$. In fact, since the exponential $e^{\lambda t}$ is always positive, $\mathbf{y}(t)$ is always a positive multiple of $C\mathbf{v}$. If $C > 0$, $\mathbf{y}(t)$ is a positive multiple of $\mathbf{v}$, and if $C < 0$, $\mathbf{y}(t)$ is a positive multiple of $-\mathbf{v}$. If $\lambda > 0$, the exponential $e^{\lambda t}$ increases from 0 to ∞ as t increases from $-\infty$ to ∞, while if $\lambda < 0$ it decreases from ∞ to $-\infty$. In either case, $\mathbf{y}(t)$ traces out the half-line consisting of positive multiples of $C\mathbf{v}$. Thus there are precisely two solution curves in the phase plane depending on the sign of the constant C. Since these are half-lines, we will sometimes refer to exponential solutions as *half-line solutions.*

When $\lambda > 0$, $e^{\lambda t}$ increases. The exponential solutions tend away from the equilibrium point at the origin as t increases and tend to the origin as t decreases to $-\infty$. Solutions with this property are called *unstable* solutions. Thus, corresponding to a positive eigenvalue there are two unstable half-line solution curves. This situation is displayed in Figure 1. In this figure the black arrows indicate the vectors $\pm\mathbf{v}$, while the blue arrows indicate the direction of the flow as t increases.

On the other hand, when $\lambda < 0$, the exponential $e^{\lambda t}$ decreases to 0 as $t \to \infty$. Thus the exponential solution $\mathbf{y}(t) = C e^{\lambda t}\mathbf{v}$ approaches the equilibrium point at the origin as $t \to \infty$. Such solutions are said to be *stable*. Consequently, corresponding to a negative eigenvalue there are two stable half-line solution curves. This situation is displayed in Figure 2.

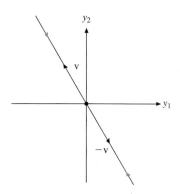

Figure 2. For a negative eigenvalue there are two stable exponential solution curves.

Saddle point

Suppose that the eigenvalues are real and have different signs, so $\lambda_1 < 0 < \lambda_2$. If we let $C_2 = 0$ in equation (3.2), then $\mathbf{y}(t) = C_1 e^{\lambda_1 t}\mathbf{v}_1$ is an exponential solution. This solution is illustrated in Figure 3 for the two cases when $C_1 = 0.8$ and $C_1 = -0.8$. As indicated there, we get two stable half-line solutions, since $\lambda_1 < 0$.

On the other hand, if we let $C_1 = 0$ in equation (3.2), then $\mathbf{y}(t) = C_2 e^{\lambda_2 t}\mathbf{v}_2$, giving us a second exponential solution. However, because $\lambda_2 > 0$, we now get two unstable half-line solution curves. This is illustrated in Figure 3 for the two cases when $C_2 = 1.2$ and $C_2 = -1.2$.

In the remaining, more interesting case, both constants in the general solution (3.2) are nonzero. In this case, the general solution,

$$\mathbf{y}(t) = C_1 e^{\lambda_1 t}\mathbf{v}_1 + C_2 e^{\lambda_2 t}\mathbf{v}_2, \tag{3.3}$$

is a superposition of the two exponential solutions. Several solutions are illustrated in Figure 4. As $t \to \infty$, the first term in (3.3), $C_1 e^{\lambda_1 t}\mathbf{v}_1$, tends to $\mathbf{0}$, and the solution

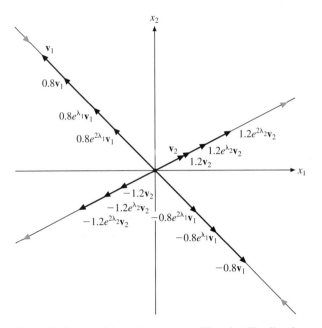

Figure 3. Exponential solutions near a saddle point. The direction of flow is indicated by the blue arrows. The black arrows indicate the position of the solution for various values of t.

gets closer to the second term, $C_2 e^{\lambda_2 t} \mathbf{v}_2$. This means that the solution curve goes to ∞, and in the process is asymptotic to the half-line generated by $C_2 \mathbf{v}_2$. On the other hand, as $t \to -\infty$, $C_2 e^{\lambda_2 t} \mathbf{v}_2 \to \mathbf{0}$, so the solution gets closer to $C_1 e^{\lambda_1 t} \mathbf{v}_1$. Geometrically this means that the solution curve goes to ∞, asymptotic to the half-line generated by $C_1 \mathbf{v}_1$. This phenomenon is illustrated in Figure 4 for specific values of the constants C_1 and C_2. It is also illustrated by the following example.

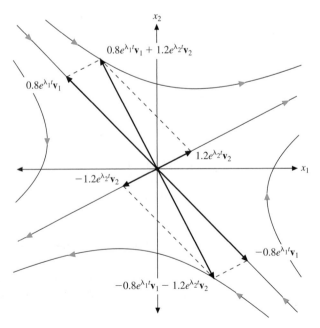

Figure 4. As $t \to \infty$, the solution $\mathbf{y}(t) = 0.8 e^{\lambda_1 t} \mathbf{v}_1 + 1.2 e^{\lambda_2 t} \mathbf{v}_2$ moves near $1.2 e^{\lambda_2 t} \mathbf{v}_2$. As $t \to -\infty$, the solution moves near $0.8 e^{\lambda_1 t} \mathbf{v}_1$.

E x a m p l e 3 . 4 The system

$$\mathbf{y}' = \begin{pmatrix} 1 & 4 \\ 2 & -1 \end{pmatrix} \mathbf{y} \tag{3.5}$$

has eigenvalues -3 and 3, with corresponding eigenvectors $(-1, 1)^T$ and $(2, 1)^T$, respectively. The eigenvalues are distinct, making

$$\mathbf{y}_1(t) = e^{-3t} \begin{pmatrix} -1 \\ 1 \end{pmatrix} \quad \text{and} \quad \mathbf{y}_2(t) = e^{3t} \begin{pmatrix} 2 \\ 1 \end{pmatrix}$$

a fundamental set of solutions. Consequently, the general solution of system (3.5) is

$$\mathbf{y}(t) = C_1 e^{-3t} \begin{pmatrix} -1 \\ 1 \end{pmatrix} + C_2 e^{3t} \begin{pmatrix} 2 \\ 1 \end{pmatrix}. \tag{3.6}$$

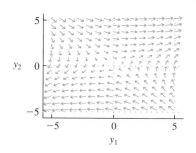

Figure 5. The arrows of the vector field indicate the presence of half-line solutions along $y_2 = -y_1$ and $y_2 = (1/2)y_1$.

We've used our numerical solver to sketch the vector field in Figure 5 associated with system (3.5).

If $C_1 = 0$, then the solution is $\mathbf{y}(t) = C_2 e^{3t}(2, 1)^T$, which traces out half-lines emanating from the origin along the line $y_2 = y_1/2$. Note how the arrows of the vector field in Figure 5 clearly indicate that these half-line solutions move away from the origin, as is expected with a growth coefficient e^{3t}. In a similar manner, if $C_2 = 0$, then the solution is $\mathbf{y}(t) = C_1 e^{-3t}(-1, 1)^T$, which traces out half-line solutions decaying into the origin along the line $y_2 = -y_1$. This time, the arrows in the vector field in Figure 5 indicate that these half-line solutions move toward the origin, as is expected with a decay coefficient e^{-3t}.

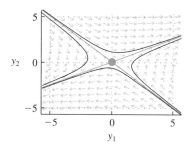

Figure 6. As $t \to \infty$, solution curves approach the half-line generated by $C_2(2, 1)^T$.

In Figure 6, we've used our numerical solver to draw the half-line solutions.[1] Remember, solution trajectories cannot cross one another, so the half-line solutions "separate" the phase plane into four distinct regions. We've plotted trajectories with initial conditions in each of these four regions. Each of these trajectories depicts the typical case where neither C_1 nor C_2 equal zero. As time moves forward, $C_1 e^{-3t}(-1, 1)^T$ decays to $\mathbf{0}$, so solutions tend toward $C_2 e^{3t}(2, 1)^T$, as is clearly evidenced in Figure 6. As $t \to -\infty$, $C_2 e^{3t}(2, 1)^T$ approaches $\mathbf{0}$. Consequently, as we move backward in time, solutions move toward $C_1 e^{-3t}(-1, 1)^T$, as is also evidenced in Figure 6. ●

As Figures 4 and 6 show, the exponential, or half-line, solutions separate the plane into four regions in which the solution curves have different behavior. For this reason, these curves (half-lines) are called **separatrices**.

The distinguishing characteristic of a system of the type that we are considering here is the existence of the separatrices. Two of these are stable solution curves that approach the equilibrium point as $t \to \infty$. Two others are unstable solutions that approach the equilibrium point as $t \to -\infty$. Any equilibrium point of a planar system (linear or nonlinear) that has this property is called a **saddle point**.

Finally, notice that if the solution curves were the altitude lines on a topographic map, then the surface would have the shape of a saddle. This is the reason for the name *saddle point*.

---

[1] Our solver draws both forward and backward solutions. So drawing half-line solutions in this case is a simple matter of drawing trajectories with initial conditions at $(2, 1)$, $(-2, -1)$, $(-1, 1)$, and $(1, -1)$.

Nodal sink

Next, suppose that both eigenvalues are negative, $\lambda_1 < \lambda_2 < 0$. Again the solution is given by (3.2). If either of the constants is zero, we get an exponential solution. However, now both eigenvalues are negative, so in both cases, the exponential or half-line solutions are all stable. For example, the exponential solutions $\mathbf{y}(t) = 0.7e^{\lambda_1 t}\mathbf{v}_1$ and $\mathbf{y}(t) = 0.5e^{\lambda_2 t}\mathbf{v}_2$ in Figure 7 decay to the origin along the half-lines generated by $0.7\mathbf{v}_1$ and $0.5\mathbf{v}_2$, respectively.

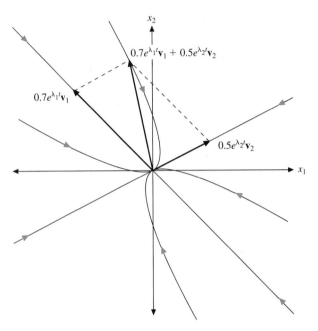

Figure 7. As $t \to \infty$, the solution $\mathbf{y}(t) = 0.7e^{\lambda_1 t}\mathbf{v}_1 + 0.5e^{\lambda_2 t}\mathbf{v}_2$ decays to the origin in a direction parallel to $0.5\mathbf{v}_2$. As $t \to -\infty$, the solution moves to infinity in a direction parallel to $0.7\mathbf{v}_1$.

In the case where both C_1 and C_2 are nonzero, the general solution (3.2) is again a superposition of the exponential solutions. To examine what happens as $t \to \infty$, we rewrite (3.2) as

$$\mathbf{y}(t) = C_1 e^{\lambda_1 t}\mathbf{v}_1 + C_2 e^{\lambda_2 t}\mathbf{v}_2 = e^{\lambda_2 t}\left[C_1 e^{(\lambda_1 - \lambda_2)t}\mathbf{v}_1 + C_2\mathbf{v}_2\right]. \qquad (3.7)$$

Look at the two factors in the last line of (3.7). The first factor, the exponential term $e^{\lambda_2 t}$, tends toward 0 as $t \to \infty$, since $\lambda_2 < 0$. On the other hand, since $\lambda_1 - \lambda_2 < 0$, the exponential in the bracketed term also goes to 0. Thus, the bracketed factor converges to $C_2\mathbf{v}_2$. Consequently, the product, $\mathbf{y}(t)$, converges to $\mathbf{0}$, but in the process, its direction gets closer to that of $C_2\mathbf{v}_2$. This means that as $\mathbf{y}(t) \to \mathbf{0}$, the solution curve becomes tangent to the half-line generated by $C_2\mathbf{v}_2$.[2]

To examine what happens as $t \to -\infty$, we rewrite (3.2) as

$$\mathbf{y}(t) = C_1 e^{\lambda_1 t}\mathbf{v}_1 + C_2 e^{\lambda_2 t}\mathbf{v}_2 = e^{\lambda_1 t}\left[C_1\mathbf{v}_1 + C_2 e^{(\lambda_2 - \lambda_1)t}\mathbf{v}_2\right].$$

Now the exponential term $e^{\lambda_1 t}$ goes to ∞, since $\lambda_1 < 0$. Since $\lambda_2 - \lambda_1 > 0$, the exponential term in the bracketed factor converges to $\mathbf{0}$. Thus, the bracketed factor

---

[2] There is a nice way to remember this behavior. Because $\lambda_1 < \lambda_2 < 0$, as $t \to \infty$, the decay of the exponential solution $\mathbf{y}(t) = C_1 e^{\lambda_1 t}\mathbf{v}_1$ is more rapid ("faster") than that of the exponential solution, $\mathbf{y}(t) = C_2 e^{\lambda_2 t}\mathbf{v}_2$. The general solution decays to the origin along a direction eventually paralleling that of the "slower" exponential solution, $\mathbf{y}(t) = C_2 e^{\lambda_2 t}\mathbf{v}_2$.

converges to $C_1\mathbf{v}_1$. Therefore, as $t \to -\infty$, the product $\mathbf{y}(t)$ gets infinitely large, but in the process, its direction approaches that of $C_1\mathbf{v}_1$.[3] This behavior is shown in Figure 7 for several choices of C_1 and C_2.

A distinguishing characteristic of a planar linear system with two negative eigenvalues is that all solution curves approach the origin as $t \to \infty$ with a well-defined tangent line. For the exponential solution $e^{\lambda_1 t}\mathbf{v}_1$, the solution approaches the origin along the line generated by $\mathbf{v}_1$. All other solutions approach the origin tangent to the line generated by $\mathbf{v}_2$. Any equilibrium point for a planar system, linear or nonlinear, that has the property that all solution curves approach the equilibrium point as $t \to \infty$ with a well-defined tangent is called a ***nodal sink***. If all solution curves approach the equilibrium point as $t \to -\infty$ with a well-defined tangent, the equilibrium point is called a ***nodal source***. Thus, we see that a planar linear system with two negative eigenvalues has a nodal sink at the origin.

Let's look at an example of a nodal sink.

Example 3.8   The system

$$\mathbf{y}' = \begin{pmatrix} -3 & -1 \\ -1 & -3 \end{pmatrix} \mathbf{y} \tag{3.9}$$

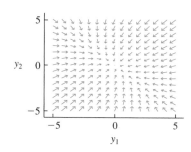

y_2

has eigenvalues -2 and -4, with corresponding eigenvectors $(-1, 1)^T$ and $(1, 1)^T$, respectively. The eigenvalues are distinct, making

$$\mathbf{y}_1(t) = e^{-2t} \begin{pmatrix} -1 \\ 1 \end{pmatrix} \quad \text{and} \quad \mathbf{y}_2(t) = e^{-4t} \begin{pmatrix} 1 \\ 1 \end{pmatrix}$$

a fundamental set of solutions. Consequently, the general solution of system (3.9) is

$$\mathbf{y}(t) = C_1 e^{-2t} \begin{pmatrix} -1 \\ 1 \end{pmatrix} + C_2 e^{-4t} \begin{pmatrix} 1 \\ 1 \end{pmatrix}. \tag{3.10}$$

Figure 8. The arrows of the vector field indicate the presence of half-line solutions along the lines $y_2 = -y_1$ and $y_2 = y_1$.

We've used our numerical solver to sketch the vector field shown in Figure 8 associated with system (3.10).

The exponential solution $\mathbf{y}(t) = C_1 e^{-2t}(-1, 1)^T$ traces out half-line solutions that decay to the origin along the line $y_2 = -y_1$. Note how the arrows of the vector field in Figure 8 clearly indicate that these solutions move toward the origin, as is expected with the decay coefficient e^{-2t}. The second exponential solution, $\mathbf{y}(t) = C_2 e^{-4t}(1, 1)^T$, is also evident in Figure 8, where the arrows indicate half-line solutions decaying to the origin along the line $y_2 = y_1$.

In the phase portrait shown in Figure 9, we've drawn the half-line solutions and four solutions where neither C_1 nor C_2 equals zero. As $t \to \infty$, note how each of these latter trajectories decays to the origin in a direction parallel to $C_1 e^{-2t}(-1, 1)^T$ (the "slower" exponential solution). As $t \to -\infty$, these solutions move toward infinity in a direction eventually paralleling the vector $C_2 e^{-4t}(1, 1)^T$ (the "faster" exponential solution).   ●

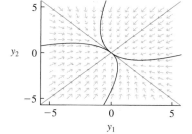

y_2

Figure 9. As $t \to \infty$, solution curves approach the origin tangent to the half-line generated by $C_1(-1, 1)^T$.

Nodal source

Next, suppose both eigenvalues are positive, $0 < \lambda_1 < \lambda_2$. The argument in this case parallels that in the description of a nodal sink, only with time reversed. Let's look at an example.

---

[3] As $t \to -\infty$, the trajectory goes to infinity, eventually turning parallel to the "faster" exponential solution, $\mathbf{y}(t) = e^{\lambda_1 t}\mathbf{v}_1$.

Example 3.11 The system

$$\mathbf{y}' = \begin{pmatrix} 3 & 1 \\ 1 & 3 \end{pmatrix} \mathbf{y} \tag{3.12}$$

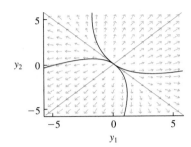

has eigenvalues 2 and 4, with corresponding eigenvectors $(-1, 1)^T$ and $(1, 1)^T$, respectively. The general solution is

$$\mathbf{y}(t) = C_1 e^{2t} \begin{pmatrix} -1 \\ 1 \end{pmatrix} + C_2 e^{4t} \begin{pmatrix} 1 \\ 1 \end{pmatrix}. \tag{3.13}$$

Note that equation (3.13) is identical to equation (3.10), with $-t$ replaced with t. Consequently, it should come as no surprise when this system's phase portrait in Figure 10 duplicates the image in Figure 9, only with time reversed.

Indeed, since all solution curves approach the origin as $t \to -\infty$ with a definite tangent, the origin is a nodal source. ●

Figure 10. As $t \to \infty$, the solutions move to infinity in a direction parallel to $C_2(1, 1)^T$.

Complex eigenvalues

The next three cases will study situations where the eigenvalues are complex. In Theorem 2.25, we saw that a complex eigenvalue $\lambda = \alpha + i\beta$ and its associated eigenvector $\mathbf{w} = \mathbf{v}_1 + i\mathbf{v}_2$ lead to the general solution

$$\mathbf{y}(t) = C_1 e^{\alpha t} (\cos \beta t \, \mathbf{v}_1 - \sin \beta t \, \mathbf{v}_2) + C_2 e^{\alpha t} (\sin \beta t \, \mathbf{v}_1 + \cos \beta t \, \mathbf{v}_2). \tag{3.14}$$

Center

Suppose the eigenvalues are purely imaginary. Then $\alpha = 0$, and equation (3.14) becomes

$$\mathbf{y}(t) = C_1 (\cos \beta t \, \mathbf{v}_1 - \sin \beta t \, \mathbf{v}_2) + C_2 (\sin \beta t \, \mathbf{v}_1 + \cos \beta t \, \mathbf{v}_2). \tag{3.15}$$

The trigonometric functions $\cos \beta t$ and $\sin \beta t$ are both periodic with period $T = 2\pi/|\beta|$. Consequently, the vector-valued function $\mathbf{y}(t)$ has the same property. This means that the solution trajectory is a *closed* curve, orbiting about the origin with period $T = 2\pi/|\beta|$.

Example 3.16 The system

$$\mathbf{y}' = \begin{pmatrix} 0 & 2 \\ -2 & 0 \end{pmatrix} \mathbf{y} \tag{3.17}$$

has an eigenvalue-eigenvector pair $\lambda = 2i$, $\mathbf{w} = (1, i)^T$. Therefore, we have the complex-valued exponential solution $\mathbf{z}(t) = e^{\lambda t} \mathbf{w}$. To find real-valued solutions we compute the real and imaginary parts of $\mathbf{z}(t)$. Using Euler's formula, we get

$$\mathbf{z}(t) = e^{2it} \begin{pmatrix} 1 \\ i \end{pmatrix}$$

$$= [\cos 2t + i \sin 2t] \left[\begin{pmatrix} 1 \\ 0 \end{pmatrix} + i \begin{pmatrix} 0 \\ 1 \end{pmatrix} \right]$$

$$= \left[\cos 2t \cdot \begin{pmatrix} 1 \\ 0 \end{pmatrix} - \sin 2t \cdot \begin{pmatrix} 0 \\ 1 \end{pmatrix} \right] + i \left[\cos 2t \cdot \begin{pmatrix} 0 \\ 1 \end{pmatrix} + \sin 2t \cdot \begin{pmatrix} 1 \\ 0 \end{pmatrix} \right]$$

$$= \begin{pmatrix} \cos 2t \\ -\sin 2t \end{pmatrix} + i \begin{pmatrix} \sin 2t \\ \cos 2t \end{pmatrix}.$$

The real and imaginary parts of $\mathbf{z}$,

$$\mathbf{y}_1(t) = \begin{pmatrix} \cos 2t \\ -\sin 2t \end{pmatrix} \quad \text{and} \quad \mathbf{y}_2(t) = \begin{pmatrix} \sin 2t \\ \cos 2t \end{pmatrix},$$

form a fundamental set of solutions, so the general solution is

$$\mathbf{y}(t) = C_1 \begin{pmatrix} \cos 2t \\ -\sin 2t \end{pmatrix} + C_2 \begin{pmatrix} \sin 2t \\ \cos 2t \end{pmatrix}. \tag{3.18}$$

We can show that the solution curves are circles in this case by showing that the components of $\mathbf{y}(t) = (y_1(t), y_2(t))^T$ satisfy $y_1^2 + y_2^2 = $ constant. We do this by showing that the derivative of $y_1(t)^2 + y_2(t)^2$ is zero. First,

$$(y_1^2 + y_2^2)' = 2y_1 y_1' + 2y_2 y_2'.$$

Now we use the system (3.17), which says that $y_1' = 2y_2$ and $y_2' = -2y_1$. Substituting, we get

$$(y_1^2 + y_2^2)' = 2y_1(2y_2) + 2y_2(-2y_1) = 0.$$

Thus the solution curves are circles centered at the origin. This is evident when we use a numerical solver to generate solution trajectories of system (3.17), as we have done in Figure 11. ●

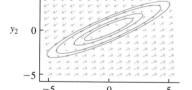

Figure 11. Circular solutions spiral about the center at the origin.

The distinguishing characteristic of this type of equilibrium point is that it is surrounded by closed solution curves. An equilibrium point for any planar system, linear or nonlinear, that has this property is called a ***center***. Thus, planar linear systems with purely imaginary eigenvalues have centers at the origin.

Not all centers have solution curves that are circles. This is not even true for linear systems. We will prove later that, for a linear center, the orbits are similar ellipses centered at the origin. For example, the system

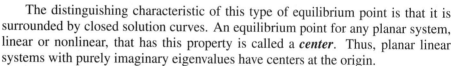

$$\mathbf{y}' = \begin{pmatrix} 4 & -10 \\ 2 & -4 \end{pmatrix} \mathbf{y} \tag{3.19}$$

Figure 12. Elliptical solutions to the system in (3.19) circle the center at the origin.

has eigenvalue $2i$ and associated eigenvector $(2+i, 1)^T$, so the equilibrium point is still a center, but the solution curves are ellipses, as shown in Figure 12.

Spiral sink

Now suppose that the real part of the eigenvalue is negative. The solution is again provided by equation (3.14),

$$\begin{aligned} \mathbf{y}(t) &= C_1 e^{\alpha t} (\cos \beta t\, \mathbf{v}_1 - \sin \beta t\, \mathbf{v}_2) + C_2 e^{\alpha t} (\sin \beta t\, \mathbf{v}_1 + \cos \beta t\, \mathbf{v}_2) \\ &= e^{\alpha t} \{C_1 (\cos \beta t\, \mathbf{v}_1 - \sin \beta t\, \mathbf{v}_2) + C_2 (\sin \beta t\, \mathbf{v}_1 + \cos \beta t\, \mathbf{v}_2)\}. \end{aligned} \tag{3.20}$$

The term inside the brackets in (3.20) is just what we see in (3.15). As we have seen, these terms are periodic with period $T = 2\pi/|\beta|$, and by themselves parameterize ellipses centered at the origin. However, these are modified by the factor $e^{\alpha t}$. Since α, the real part of the complex eigenvalue, is negative, $e^{\alpha t} \to 0$ as $t \to \infty$. Thus, while the solution curve circles the origin, it is being drawn toward it at the same time, resulting in a spiral motion. Since all solution curves spiral to the equilibrium point at the origin, all solutions are stable.

The fact that all solutions spiral around the equilibrium point while at the same time approaching it characterizes a ***spiral sink***. The term is used for nonlinear systems as well. Thus, any linear system with complex eigenvalues having negative real parts has a spiral sink at the origin. Let's look at an example.

Example 3.21   The system

$$\mathbf{y}' = \begin{pmatrix} 1 & -4 \\ 2 & -3 \end{pmatrix} \mathbf{y} \tag{3.22}$$

has an eigenvalue-eigenvector pair $\lambda = -1 + 2i$, $\mathbf{w} = (2, 1 - i)^T$. Because $\alpha = \text{Re}(\lambda) = -1 < 0$, we expect solutions to decay to the origin. In addition, because $\beta = \text{Im}(\lambda) = 2$, the natural frequency is 2, and the time to complete one revolution is $T = 2\pi/2 = \pi$. These claims are evident in Figure 13, where we've used our numerical solver to start a trajectory at a particular initial condition but restricted the time of computation to the interval $[0, \pi]$. ●

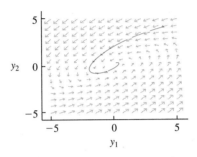
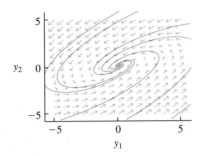

Figure 13. The solution discussed in Example 3.21.

Figure 14. Solutions spiral and decay to the origin.

In the phase portrait shown in Figure 14, a number of solution trajectories swirl about the origin, decaying to $\mathbf{0}$ as $t \to \infty$. This illustrates the behavior we expect near a spiral sink.

Spiral source

Suppose the real part of the eigenvalue is positive. Again, the general solution is given by (3.20), but now $\alpha > 0$. Therefore, the amplitude of oscillation will increase as solutions spiral about the origin, since $e^{\alpha t} \to \infty$ as $t \to \infty$. This behavior characterizes a ***spiral source.*** Thus, a linear system with complex eigenvalues having a positive real part has a spiral source at the origin.

For example, the system

$$\mathbf{y}' = \begin{pmatrix} 2 & -1 \\ 2 & 0 \end{pmatrix} \mathbf{y} \tag{3.23}$$

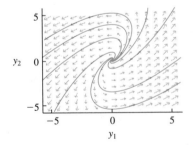

Figure 15. Solutions spiral and grow away from the origin.

has eigenvalue $\lambda = 1 + i$. A phase portrait for system (3.23) appears in Figure 15. The behavior seen in Figure 15 illustrates what is expected near a ***spiral source***.

The direction of rotation

If you don't have your numerical solver handy and you want a quick sketch of the solutions to a planar system with complex eigenvalues, an immediate problem arises. In the previous example the real part of the eigenvalue $\lambda = 1 + i$ is positive, indicating a spiral source, but is the motion clockwise or counterclockwise? One quick way to determine the rotation direction is simply to compute one vector in the vector field determined by the right-hand side of system (3.23). It is usually easier to compute the vector at a point like $(1, 0)^T$ where the computation will be easiest. We get

$$\begin{pmatrix} 2 & -1 \\ 2 & 0 \end{pmatrix} \begin{pmatrix} 1 \\ 0 \end{pmatrix} = \begin{pmatrix} 2 \\ 2 \end{pmatrix}.$$

Sketch the vector $(2, 2)^T$ at the point $(1, 0)$. Because the solution trajectory must be tangent to the vector $(2, 2)^T$ at the point $(1, 0)^T$, it is clear that the rotation is counterclockwise. This is shown in Figure 16.

More generally, if we have a matrix

$$A = \begin{pmatrix} a_{11} & a_{12} \\ a_{21} & a_{22} \end{pmatrix}$$

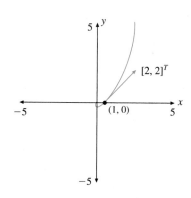

Figure 16. Often, the computation of a single vector determines the direction of rotation.

that has complex eigenvalues, then we can compute the vector field at $(1, 0)^T$. This gives

$$A \begin{pmatrix} 1 \\ 0 \end{pmatrix} = \begin{pmatrix} a_{11} & a_{12} \\ a_{21} & a_{22} \end{pmatrix} \begin{pmatrix} 1 \\ 0 \end{pmatrix} = \begin{pmatrix} a_{11} \\ a_{21} \end{pmatrix}.$$

If $a_{21} > 0$, this vector points into the upper half-plane, indicating that the rotation is counterclockwise. On the other hand, if $a_{21} < 0$, this vector points into the lower half-plane, indicating that the rotation is clockwise.

EXERCISES

For the 2×2 matrices in Exercises 1 and 2, use $p(\lambda) = \lambda^2 - T\lambda + D$, where $T = \text{tr}(A)$ and $D = \det(A)$, to compute the characteristic polynomial. Then, use $p(\lambda) = \det(A - \lambda I)$ to calculate the characteristic polynomial a second time and compare the results.

1. $A = \begin{pmatrix} -10 & -25 \\ 5 & 10 \end{pmatrix}$ **2.** $A = \begin{pmatrix} 0 & 5 \\ -1 & 4 \end{pmatrix}$

For each exponential solution in Exercises 3–6, sketch the half-line generated by the solution on a sheet of graph paper. Use a calculator to compute the vectors generated at times $t = 0$, $t = 1$, and $t = 2$, and then plot them on your half-line. Use a different color arrow to indicate the direction of motion, as shown in Figure 3.

3. $\mathbf{y}(t) = 1.2e^{0.6t} \begin{pmatrix} 1 \\ 1 \end{pmatrix}$ **4.** $\mathbf{y}(t) = -0.8e^{0.6t} \begin{pmatrix} -2 \\ 1 \end{pmatrix}$

5. $\mathbf{y}(t) = 0.8e^{-0.6t} \begin{pmatrix} 4 \\ 4 \end{pmatrix}$ **6.** $\mathbf{y}(t) = -1.2e^{-0.6t} \begin{pmatrix} 4 \\ -4 \end{pmatrix}$

In Exercises 7–9, you are given a solution $\mathbf{y}(t)$ to a planar system. On a sheet of graph paper, sketch the half-lines generated by each exponential term of this solution. Use a calculator to compute $\mathbf{y}(t)$ for $t = -5, -4, \ldots, 5$, and then plot the result on your graph paper. Draw a smooth trajectory through your solution points and use arrows to indicate the direction of motion along the solution.

7. $\mathbf{y}(t) = 0.8e^{0.4t} \begin{pmatrix} -2 \\ 1 \end{pmatrix} + 1.2e^{-0.2t} \begin{pmatrix} 4 \\ 4 \end{pmatrix}$

8. $\mathbf{y}(t) = 0.4e^{0.1t} \begin{pmatrix} -2 \\ 1 \end{pmatrix} - 0.4e^{0.4t} \begin{pmatrix} 4 \\ 4 \end{pmatrix}$

9. $\mathbf{y}(t) = -0.4e^{-0.4t} \begin{pmatrix} -2 \\ 1 \end{pmatrix} + 0.4e^{-0.2t} \begin{pmatrix} 4 \\ 4 \end{pmatrix}$

Each of Exercises 10–15 provides a general solution of $\mathbf{y}' = A\mathbf{y}$, for some A. Without the help of a computer or a calculator, sketch the half-line solutions generated by each exponential term of the solution. Then, sketch a rough approximation of a solution in each region determined by the half-line solutions. Use arrows to indicate the direction of motion on all solutions. Classify the equilibrium point as a saddle, a nodal sink, or a nodal source.

10. $\mathbf{y}(t) = C_1 e^{-t} \begin{pmatrix} 2 \\ 1 \end{pmatrix} + C_2 e^{-2t} \begin{pmatrix} -1 \\ 1 \end{pmatrix}$

11. $\mathbf{y}(t) = C_1 e^{t} \begin{pmatrix} -1 \\ -2 \end{pmatrix} + C_2 e^{2t} \begin{pmatrix} 3 \\ -1 \end{pmatrix}$

12. $\mathbf{y}(t) = C_1 e^{t} \begin{pmatrix} 1 \\ 1 \end{pmatrix} + C_2 e^{-2t} \begin{pmatrix} 1 \\ -1 \end{pmatrix}$

13. $\mathbf{y}(t) = C_1 e^{-3t} \begin{pmatrix} -4 \\ 1 \end{pmatrix} + C_2 e^{-t} \begin{pmatrix} 1 \\ 2 \end{pmatrix}$

14. $\mathbf{y}(t) = C_1 e^{-t} \begin{pmatrix} -5 \\ 2 \end{pmatrix} + C_2 e^{2t} \begin{pmatrix} -1 \\ 4 \end{pmatrix}$

15. $\mathbf{y}(t) = C_1 e^{3t} \begin{pmatrix} 4 \\ 1 \end{pmatrix} + C_2 e^{t} \begin{pmatrix} 1 \\ 5 \end{pmatrix}$

In Exercises 16–19, verify that the equilibrium point at the origin is a center by showing that the real parts of the system's complex eigenvalues are zero. In each case, calculate and sketch the vector generated by the right-hand side of the system at the point $(1, 0)$. Use this to help sketch the elliptic solution trajectory for the system passing through the point

(1, 0). Draw arrows on the solution, indicating the direction of motion. Use your numerical solver to check your result.

16. $\mathbf{y}' = \begin{pmatrix} -4 & 8 \\ -4 & 4 \end{pmatrix} \mathbf{y}$ **17.** $\mathbf{y}' = \begin{pmatrix} 0 & 3 \\ -3 & 0 \end{pmatrix} \mathbf{y}$

18. $\mathbf{y}' = \begin{pmatrix} 2 & 2 \\ -4 & -2 \end{pmatrix} \mathbf{y}$ **19.** $\mathbf{y}' = \begin{pmatrix} 0 & 1 \\ -4 & 0 \end{pmatrix} \mathbf{y}$

In Exercises 20–23, calculate the eigenvalues to determine whether the equilibrium point is a spiral sink or a source. Cal-

culate and sketch the vector generated by the right-hand side of the system at the point (1, 0). Use this to help sketch the solution trajectory for the system passing through the point (1, 0). Draw arrows on the solution, indicating the direction of motion. Use your numerical solver to check your result.

20. $\mathbf{y}' = \begin{pmatrix} -2 & 2 \\ -1 & 0 \end{pmatrix} \mathbf{y}$ **21.** $\mathbf{y}' = \begin{pmatrix} -1 & 1 \\ -5 & 3 \end{pmatrix} \mathbf{y}$

22. $\mathbf{y}' = \begin{pmatrix} 7 & -10 \\ 4 & -5 \end{pmatrix} \mathbf{y}$ **23.** $\mathbf{y}' = \begin{pmatrix} -3 & 2 \\ -4 & 1 \end{pmatrix} \mathbf{y}$

9.4 The Trace-Determinant Plane

In the previous section we examined six different types of equilibrium points that can occur for a linear, planar system. Up to this point, our analysis of equilibrium points has not been systematic. In this section we will correct that. Our goal will be a classification of all of the different types of equilibria that can occur, although we will leave the final classification as a project.

Eigenvalues and trace-determinant pairs

In Section 9.2, we showed that the characteristic polynomial of the matrix

$$A = \begin{pmatrix} a_{11} & a_{12} \\ a_{21} & a_{22} \end{pmatrix}$$

is

$$p(\lambda) = \lambda^2 - T\lambda + D, \tag{4.1}$$

where $T = \text{tr}(A) = a_{11} + a_{22}$ and $D = \det(A) = a_{11}a_{22} - a_{21}a_{12}$. The roots of the characteristic polynomial, which are the eigenvalues of the matrix A, are determined by T and D using the quadratic formula,

$$\lambda_1, \lambda_2 = \frac{T \pm \sqrt{T^2 - 4D}}{2}. \tag{4.2}$$

The fundamental theorem of algebra dictates that the characteristic polynomial must factor as

$$p(\lambda) = (\lambda - \lambda_1)(\lambda - \lambda_2),$$

where λ_1 and λ_2 are the eigenvalues of the matrix. If we multiply out the right-hand side and collect powers of λ, the characteristic polynomial becomes

$$p(\lambda) = \lambda^2 - (\lambda_1 + \lambda_2)\lambda + \lambda_1\lambda_2. \tag{4.3}$$

Comparing the coefficients of the powers of λ in (4.1) and (4.3), we discover the important relationships,

$$D = \det(A) = \lambda_1\lambda_2 \quad \text{and} \quad T = \text{tr}(A) = \lambda_1 + \lambda_2. \tag{4.4}$$

Thus, the eigenvalues determine the trace and the determinant as well, and formulas (4.2) and (4.4) complement each other.

Classification of equilibria

This duality between pairs of eigenvalues and trace-determinant pairs enables us to systematize our analysis of possible equilibrium points in terms of the trace and the determinant. Since both the trace and the determinant are real, we can use the **_trace-determinant plane_** to provide visual assistance to our efforts. This is simply a coordinate plane with coordinates T and D. (See Figures 1 and 2.)

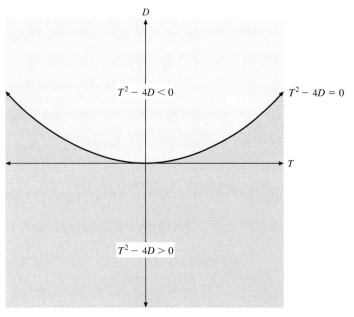

Figure 1. The trace-determinant plane.

The quadratic formula (4.2) for the eigenvalues shows that the discriminant, $T^2 - 4D$, plays a special role. If $T^2 - 4D > 0$, there are two real eigenvalues. If

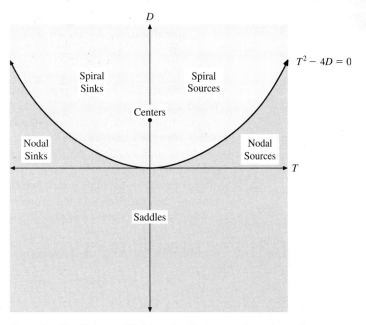

Figure 2. Classifying equilibrium points in the trace-determinant plane.

$T^2 - 4D = 0$, we have one repeated eigenvalue. Finally, if $T^2 - 4D < 0$, the eigenvalues are complex conjugates. Clearly, the sign of the discriminant is important, so we begin by sketching the graph of $T^2 - 4D = 0$ in the trace-determinant plane, as shown in Figure 1. The parabola $T^2 - 4D = 0$ divides the trace-determinant plane into two separate regions.

In the region above the parabola, we have $T^2 - 4D < 0$.[4] Therefore, matrices having trace T and determinant D, where (T, D) lies above the parabola, have complex eigenvalues. In this case, the type of the equilibrium point the system has at the origin is determined by the real part of the eigenvalue. According to equation (4.2), the real part of the eigenvalue is $T/2$. Consequently, if $T < 0$, we have a spiral sink; if $T = 0$, we have a center; and if $T > 0$, we have a spiral source. These regions are labeled in Figure 2.

In the region below the parabola, we have $T^2 - 4D > 0$, so the eigenvalues are real. Let's focus on the determinant $D = \lambda_1 \lambda_2$. If $D < 0$, the eigenvalues must have opposite signs. Hence, the equilibrium point is a saddle point. Thus, the half-plane below the T-axis corresponds to saddle points. On the other hand, if (T, D) is below the parabola but above the T-axis, then $D = \lambda_1 \lambda_2 > 0$. Therefore, the eigenvalues have the same sign. The sign is determined by the trace $T = \lambda_1 + \lambda_2$. If T is positive, both eigenvalues are positive, and the equilibrium point is a nodal source. However, if T is negative, both eigenvalues are negative and the equilibrium point is a nodal sink.

All of our findings are shown in Figure 2.

Generic and nongeneric equilibria

You will notice that we have analyzed the equilibrium points for almost every point in the trace-determinant plane. Let's look specifically at the following five types:

- **Saddle point.** A has two real eigenvalues, one positive and one negative.
- **Nodal sink.** A has two real and negative eigenvalues.
- **Nodal source.** A has two real and positive eigenvalues.
- **Spiral sink.** A has two complex conjugate eigenvalues with a negative real part.
- **Spiral source.** A has two complex conjugate eigenvalues with a positive real part.

Each of these types corresponds to a large open subset of the trace-determinant plane (Figure 2). For that reason, we will call each of these five types ***generic***. While almost all points in the trace-determinant plane correspond to one of the five generic types, there are exceptions. All of these exceptional types of equilibrium points are called ***nongeneric***.

Most important among the nongeneric equilibrium points is the center. This fact will be important when we get to the analysis of nonlinear systems. It remains to analyze the nongeneric cases that lie on the T-axis or on the parabola. We will examine these scenarios in the exercises.

Let's look at some examples of generic systems.

Example 4.5   Consider the system $\mathbf{y}' = A\mathbf{y}$, where

$$A = \begin{pmatrix} 4 & -3 \\ 15 & -8 \end{pmatrix}.$$

---

[4] This is checked easily . Note that the point $(0, 1)$ satisfies the inequality $T^2 - 4D < 0$ (substitute $T = 0$ and $D = 1$). This indicates that the region above the parabola satisfies $T^2 - 4D < 0$.

The trace is $T = -4$ and the determinant is $D = 13$. Hence, $T^2 - 4D = -36 < 0$. Consequently, this puts us in the region of the trace-determinant plane inhabited by spiral sinks. This is exactly the behavior produced by our numerical solver in Figure 3.

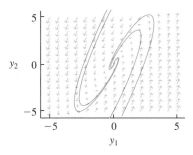

Figure 3. The spiral sink in Example 4.5.

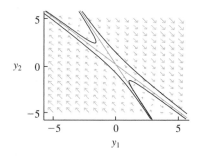

Figure 4. The saddle point in Example 4.6.

E x a m p l e 4 . 6   Consider the system $\mathbf{y}' = A\mathbf{y}$, where

$$A = \begin{pmatrix} 8 & 5 \\ -10 & -7 \end{pmatrix}.$$

The trace is $T = 1$ and the determinant is $D = -6$. This places us in the region below the T-axis. Consequently, the equilibrium point at the origin is a saddle point. This is exactly the behavior produced by our numerical solver in Figure 4.

E x a m p l e 4 . 7   Consider the system $\mathbf{y}' = A\mathbf{y}$, where

$$A = \begin{pmatrix} -2 & 0 \\ 1 & -1 \end{pmatrix}.$$

The trace is $T = -3$ and the determinant is $D = 2$. Moreover, $T^2 - 4D = 1$. This places us in the region of nodal sinks. This analysis is supported by the phase portrait produced by our numerical solver in Figure 5.

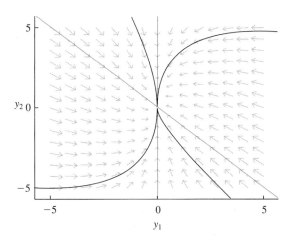

Figure 5. The nodal sink in Example 4.7.

EXERCISES

In Exercises 1–12, classify the equilibrium point of the system $\mathbf{y}' = A\mathbf{y}$ based on the position of (T, D) in the trace-determinant plane. Sketch the phase portrait by hand. Verify your result by creating a phase portrait with your numerical solver.

1. $A = \begin{pmatrix} 8 & 20 \\ -4 & -8 \end{pmatrix}$

2. $A = \begin{pmatrix} -16 & 9 \\ -18 & 11 \end{pmatrix}$

3. $A = \begin{pmatrix} 2 & -4 \\ 8 & -6 \end{pmatrix}$

4. $A = \begin{pmatrix} 8 & 3 \\ -6 & -1 \end{pmatrix}$

5. $A = \begin{pmatrix} -11 & -5 \\ 10 & 4 \end{pmatrix}$

6. $A = \begin{pmatrix} 6 & -5 \\ 10 & -4 \end{pmatrix}$

7. $A = \begin{pmatrix} -7 & 10 \\ -5 & 8 \end{pmatrix}$

8. $A = \begin{pmatrix} 4 & 3 \\ -15 & -8 \end{pmatrix}$

9. $A = \begin{pmatrix} 3 & 2 \\ -4 & -1 \end{pmatrix}$

10. $A = \begin{pmatrix} -5 & 2 \\ -6 & 2 \end{pmatrix}$

11. $A = \begin{pmatrix} -4 & 10 \\ -2 & 4 \end{pmatrix}$

12. $A = \begin{pmatrix} -2 & -6 \\ 4 & 8 \end{pmatrix}$

Degenerate nodes. Degenerate nodal sinks are equilibrium points characterized by the fact that all solutions tend toward the equilibrium point as $t \to \infty$, all tangent to the same line. The solutions of degenerate nodal sources tend toward the equilibrium point as $t \to -\infty$, all tangent to the same line. Exercises 13–19 discuss degenerate nodes.

13. The system
$$\mathbf{y}' = \begin{pmatrix} 1 & 4 \\ -1 & -3 \end{pmatrix} \mathbf{y}$$
has a repeated eigenvalue, $\lambda = -1$, but only one eigenvector, $\mathbf{v}_1 = (2, -1)^T$.

(a) Explain why this system lies on the boundary separating nodal sinks from spiral sinks in the trace-determinant plane.

(b) The general solution (see (2.42) in Section 9.2) can be written
$$\mathbf{y}(t) = e^{-t}\left((C_1 + C_2 t) \begin{pmatrix} 2 \\ -1 \end{pmatrix} + C_2 \begin{pmatrix} 0 \\ 1/2 \end{pmatrix} \right).$$

Predict the behavior of the solution in the phase plane as $t \to \infty$ and as $t \to -\infty$.

(c) Use your numerical solver to sketch the half-line solutions. Then sketch exactly one solution in each region separated by the half-line solutions. Explain how the behavior you see agrees with your findings in part (b).

In Exercises 14–16, let A be a real 2×2 matrix with one eigenvalue λ, and suppose that the eigenspace of λ has dimension 1. The matrix in Exercise 13 is an example. According to Theorem 2.40 in Section 9.2, a fundamental set of solutions for the system $\mathbf{y}' = A\mathbf{y}$ is given by
$$\mathbf{y}_1(t) = e^{\lambda t}\mathbf{v}_1 \quad \text{and}$$
$$\mathbf{y}_2(t) = e^{\lambda t}[\mathbf{v}_2 + t\mathbf{v}_1],$$

where $\mathbf{v}_1$ is a nonzero eigenvector and $\mathbf{v}_2$ satifies $(A - \lambda I)\mathbf{v}_2 = \mathbf{v}_1$. By (2.42), the general solution can be written as
$$\mathbf{y}(t) = e^{\lambda t}[(C_1 + C_2 t)\mathbf{v}_1 + C_2 \mathbf{v}_2].$$

14. Assume that the eigenvalue λ is negative.

(a) Describe the exponential solutions.

(b) Describe the behavior of the general solution as $t \to \infty$.

(c) Describe the behavior of the general solution as $t \to -\infty$.

(d) Is the equilibrium point at the origin a degenerate nodal sink or source?

15. Redo Exercise 14 under the assumption that the eigenvalue λ is positive.

16. Where do linear degenerate sources and sinks fit on the trace-determinant plane?

In Exercises 17–18, find the general solution of the given system. Write your solution in the form
$$\mathbf{y}(t) = e^{\lambda t}[(C_1 + C_2 t)\mathbf{v}_1 + C_2 \mathbf{v}_2],$$

where $\mathbf{v}_1$ is an eigenvector and $\mathbf{v}_2$ satisfies $(A - \lambda I)\mathbf{v}_2 = \mathbf{v}_1$. Without the use of a computer or a calculator, sketch the half-line solutions. Sketch exactly one solution in each region separated by the half-line solutions. Use a numerical solver to verify your result when finished. *Hint*: The solutions in this case want desperately to spiral but are prevented from doing so by the half-line solutions (solutions cannot cross). However, the suggestions regarding clockwise or counterclockwise rotation in the subsection on spiral sources apply nicely in this situation.

17. $\mathbf{y}' = \begin{pmatrix} 6 & 4 \\ -1 & 2 \end{pmatrix} \mathbf{y}$

18. $\mathbf{y}' = \begin{pmatrix} -4 & -4 \\ 1 & 0 \end{pmatrix} \mathbf{y}$

19. Consider the system
$$x' = x + ay,$$
$$y' = x + y.$$

(a) Show that the equilibrium point is a nodal source for all $0 < a < 1$.

(b) In the case $0 < a < 1$, what are the equations of the half-line solutions? Explain what happens to the half-line solutions as $a \to 0$.

(c) What happens to the system when $a = 0$? When $a < 0$?

20. **Star nodes** A star sink is an equilibrium point that has the property that there are solution curves tending to it as $t \to \infty$ tangent to any direction. A star source is defined similarly. In Section 9.2 you saw that matrices having the form

$$A = \begin{pmatrix} a & 0 \\ 0 & a \end{pmatrix}$$

have a repeated eigenvalue, $\lambda = a$. Moreover, all vectors in the $\mathbf{R}^2$ are eigenvectors of A associated with $\lambda = a$. Consider the matrices

$$B = \begin{pmatrix} 2 & 0 \\ 0 & 2 \end{pmatrix} \quad \text{and} \quad C = \begin{pmatrix} -2 & 0 \\ 0 & -2 \end{pmatrix}.$$

(a) Sketch the phase portraits for the systems $\mathbf{y}' = B\mathbf{y}$ and $\mathbf{y}' = C\mathbf{y}$ without the aid of a computer or a calculator.

(b) Is the equilibrium point at the origin a star source or a sink for the systems $\mathbf{y}' = B\mathbf{y}$ and $\mathbf{y}' = C\mathbf{y}$?

(c) Where in the trace-determinant plane will you find the systems $\mathbf{y}' = B\mathbf{y}$ and $\mathbf{y}' = C\mathbf{y}$?

(d) Where does the system $\mathbf{y}' = A\mathbf{y}$ fit in the trace-determinat plane? When is it a star sink or a star source?

21. Let A be a real 2×2 matrix. Prove that $\det(A) = 0$ if and only if one of the eigenvalues of A is equal to zero.

For each of the matrices in Exercises 22 and 23, perform the following activities.

(i) Determine where in the trace-determinate plane the system $\mathbf{y}' = A\mathbf{y}$ fits.

(ii) Find all of the equilibrium points for the system $\mathbf{y}' = A\mathbf{y}$, and plot them in the phase plane.

(iii) Find the general solution and use the result to create a phase portrait for this system, taking care to sketch enough solution curves to show what happens in every part of the phase plane.

22. $A = \begin{pmatrix} 2 & 1 \\ -10 & -5 \end{pmatrix}$ **23.** $A = \begin{pmatrix} 8 & 4 \\ -10 & -5 \end{pmatrix}$

24. How do the phase portraits of the systems in Exercises 22 and 23 differ?

25. Consider the parallel LRC circuit with current direction as shown in Figure 6. We've named the currents across the resistor, inductor, and capacitor as i_R, i_L, and i_C, respectively, to make it a bit easier to identify each current with the element it passes through.

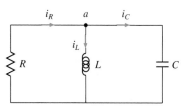

Figure 6. A parallel LRC circuit.

First, show that the current passing through the inductor and the voltage across the capacitor satisfy the system of first-order equations

$$
\begin{aligned}
v_C' &= -\frac{1}{RC} v_C - \frac{1}{C} i_L \\
i_L' &= \frac{1}{L} v_C.
\end{aligned}
\tag{4.8}
$$

It is easy to show that the system has an equilibrium point $(i_L, v_C) = (0, 0)$. Use qualitative analysis to discuss the *type* of equilibrium point. Provide a relationship amongst the parameters L, R, and C that determines when the system is *underdamped*, *critically damped*, and *overdamped* (see Section 4.4). Provide a numerical example of each case and a phase plane plot of v_C versus i_L for each numerical example.

26. Pure water enters Tank A in Figure 7 at a rate r_I gallons per minute. Two pipes connect the tanks. Through the top pipe, salt solution enters Tank A from Tank B at a rate r_B gal/min. Through the bottom pipe, salt solution enters Tank B from Tank A at a rate r_A gal/min. Finally, there is a drain on Tank B that drains salt solution from Tank B at a rate of r_D gal/min, so as to keep the level of solution in each tank at a constant volume V.

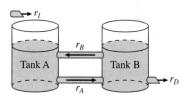

Figure 7. Two interconnected tanks.

(a) Set up a system of equations that models the salt content $x_A(t)$ and $x_B(t)$ in each tank over time.

(b) Use qualitative analysis to show that the equilibrium point of the system in part (a) is either a nodal sink or a degenerate sink, regardless of the values r_A, r_B, r_I, and r_D. That is, show that the salt content in each tank has no chance of oscillation as the salt content goes to zero.

9.5
Higher-Dimensional Systems

In the previous four sections we completely resolved the questions of solving systems of dimension 2. Much of what we did applies to higher-dimensional systems as well. For example, Theorem 1.6 is still valid and will be one of our most important tools. To remind you, it says that if λ is an eigenvalue of a matrix A and $\mathbf{v}$ is an associated eigenvector, then $\mathbf{x}(t) = e^{\lambda t}\mathbf{v}$ is the solution to $\mathbf{x}' = A\mathbf{x}$ with $\mathbf{x}(0) = \mathbf{v}$.

We continue to have the problem of showing that solutions are linearly independent. Proposition 2.4 can be generalized to help us.

PROPOSITION 5.1   Suppose $\lambda_1, \ldots, \lambda_k$ are distinct eigenvalues for an $n \times n$ matrix A, and suppose that $\mathbf{v}_i \neq \mathbf{0}$ is an eigenvector associated to λ_i for $1 \leq i \leq k$.

1. The vectors $\mathbf{v}_1, \ldots,$ and $\mathbf{v}_k$ are linearly independent.
2. The functions $\mathbf{y}_i(t) = e^{\lambda_i t} \mathbf{v}_i$, $1 \leq i \leq k$, are linearly independent solutions for the system $\mathbf{y}' = A\mathbf{y}$.

Proof   The idea for the proof of part 1 is illustrated in the proof of Proposition 2.4. We will not give the details. For part 2, we simply notice that $\mathbf{y}_i(0) = \mathbf{v}_i$. Since these vectors are linearly independent by part 1, it follows from Proposition 5.12 in Section 8.5 that the solutions are linearly independent.   ●

Real, distinct eigenvalues

Theorem 1.6 together with Proposition 5.1 allow us to solve any system with real, distinct eigenvalues.

THEOREM 5.2   Suppose the $n \times n$ matrix A has n distinct eigenvalues $\lambda_1, \ldots,$ and λ_n. Suppose that for $1 \leq i \leq n$, $\mathbf{v}_i$ is an eigenvector associated with λ_i. Then the n exponential solutions $\mathbf{y}_i(t) = e^{\lambda_i t} \mathbf{v}_i$ form a fundamental set of solutions for the system $\mathbf{y}' = A\mathbf{y}$.

Proof   There are n solutions and by part 2 of Proposition 5.1 they are linearly independent. Hence they form a fundamental set of solutions.   ■

The main application of Theorem 5.2 is when the eigenvalues are real. In this case, the exponential solutions are also real. Hence, the general solution has the form

$$\mathbf{y}(t) = C_1 e^{\lambda_1 t} \mathbf{v}_1 + C_2 e^{\lambda_2 t} \mathbf{v}_2 + \cdots + C_n e^{\lambda_n t} \mathbf{v}_n, \tag{5.3}$$

where $C_1, C_2, \ldots, C_n$ are arbitrary constants.

Example 5.4   Find the general solution to the three-dimensional system

$$\mathbf{y}' = A\mathbf{y}, \quad \text{where} \quad A = \begin{pmatrix} -9 & -3 & -7 \\ 3 & 1 & 3 \\ 11 & 3 & 9 \end{pmatrix}.$$

The eigenvalues are the solutions to the characteristic equation

$$0 = (-1)^3 \det(A - \lambda I)$$
$$= -\det \begin{pmatrix} -9 - \lambda & -3 & -7 \\ 3 & 1 - \lambda & 3 \\ 11 & 3 & 9 - \lambda \end{pmatrix}$$
$$= \lambda^3 - \lambda^2 - 4\lambda + 4.$$

The product of the roots must be equal to the constant term, 4. By checking the factors of 4 we find that the roots are $\lambda = 1, 2,$ and -2.

Now we find the associated eigenvectors. When $\lambda = 1$, the matrix $A - \lambda I$ becomes

$$\begin{pmatrix} -9 - \lambda & -3 & -7 \\ 3 & 1 - \lambda & 3 \\ 11 & 3 & 9 - \lambda \end{pmatrix} = \begin{pmatrix} -10 & -3 & -7 \\ 3 & 0 & 3 \\ 11 & 3 & 8 \end{pmatrix}.$$

The eigenvectors are vectors in the nullspace of this matrix. Using the methods we discussed in Chapter 7, we find that the nullspace is generated by the vector

$$\mathbf{v}_1 = \begin{pmatrix} -1 \\ 1 \\ 1 \end{pmatrix}.$$

By a similar procedure, we find that eigenvectors corresponding to $\lambda_2 = 2$ and $\lambda_3 = -2$, respectively, are

$$\mathbf{v}_2 = \begin{pmatrix} 4 \\ -3 \\ -5 \end{pmatrix}, \quad \text{and} \quad \mathbf{v}_3 = \begin{pmatrix} -1 \\ 0 \\ 1 \end{pmatrix}.$$

The general solution is

$$\mathbf{y}(t) = C_1 e^{\lambda_1 t} \mathbf{v}_1 + C_2 e^{\lambda_2 t} \mathbf{v}_2 + C_3 e^{\lambda_3 t} \mathbf{v}_3$$

$$= C_1 e^t \begin{pmatrix} -1 \\ 1 \\ 1 \end{pmatrix} + C_2 e^{2t} \begin{pmatrix} 4 \\ -3 \\ -5 \end{pmatrix} + C_3 e^{-2t} \begin{pmatrix} -1 \\ 0 \\ 1 \end{pmatrix}. \qquad \bullet$$

Complex eigenvalues

Theorem 5.2 applies if some or all of the eigenvalues are complex, as long as they are distinct. However, as we discovered in Section 9.2, the exponential solutions corresponding to complex eigenvalues are complex valued. Here we will show what to do to get real-valued solutions.

We are looking at a system in which the matrix A has real entries. Hence the characteristic polynomial has real coefficients. Since the eigenvalues are the roots of this polynomial, they are either real or they appear in complex conjugate pairs. Suppose that λ and $\bar{\lambda}$ are a complex conjugate pair of eigenvalues of A. Then, as we discovered in Section 9.2, we have complex conjugate eigenvectors $\mathbf{w}$ and $\bar{\mathbf{w}}$ associated with λ and $\bar{\lambda}$, respectively. These lead to two complex conjugate solutions

$$\mathbf{z}(t) = e^{\lambda t} \mathbf{w} \quad \text{and} \quad \bar{\mathbf{z}}(t) = e^{\bar{\lambda} t} \bar{\mathbf{w}}.$$

According to Proposition 2.22, the real and imaginary parts of $\mathbf{z}$ are also solutions. These solutions can be used instead of the complex solutions in the fundamental set of solutions. We proved in Proposition 2.22 that if the complex conjugate pair of solutions $\mathbf{z}$ and $\bar{\mathbf{z}}$ are linearly independent, then the same is true for the real and imaginary parts $\mathbf{x} = \operatorname{Re} \mathbf{z}$ and $\mathbf{y} = \operatorname{Im} \mathbf{z}$.

Example 5.5   Find a fundamental set of solutions for the system

$$\mathbf{x}' = A\mathbf{x}, \quad \text{where} \quad A = \begin{pmatrix} 5 & -2 & -2 \\ 7 & -4 & -2 \\ 3 & 1 & -1 \end{pmatrix}.$$

The first step is to find the eigenvalues. For matrices as complicated as these, it is best to use a computer. Our computer tells us that the eigenvalues are -2 and the complex conjugate pair $1 \pm 2i$.

For the real eigenvalue, $\lambda = -2$, we use the method in Theorem 1.6. We look for an eigenvector, which is a vector in the nullspace of

$$A - \lambda I = A + 2I = \begin{pmatrix} 7 & -2 & -2 \\ 7 & -2 & -2 \\ 3 & 1 & 1 \end{pmatrix}.$$

By observation, or by using our computer, we see that a reasonable choice of an eigenvector is $\mathbf{v} = (0, 1, -1)^T$. Our first solution is

$$\mathbf{x}_1(t) = e^{-2t} \begin{pmatrix} 0 \\ 1 \\ -1 \end{pmatrix}.$$

For the complex conjugate pair, we follow the same technique. We look for an eigenvector associated with the eigenvalue $\lambda = 1 + 2i$. Our computer tells us that $\mathbf{w} = (1 + i, 1 + i, 2)^T$ is a reasonable choice. Thus, our complex-valued solutions are

$$\mathbf{z}(t) = e^{(1+2i)t} \begin{pmatrix} 1+i \\ 1+i \\ 2 \end{pmatrix} \quad \text{and} \quad \bar{\mathbf{z}}(t) = e^{(1-2i)t} \begin{pmatrix} 1-i \\ 1-i \\ 2 \end{pmatrix}.$$

Using Euler's formula, we can find the real and imaginary parts of $\mathbf{z}$.

$$\mathbf{z}(t) = e^t e^{2it} \begin{pmatrix} 1+i \\ 1+i \\ 2 \end{pmatrix}$$

$$= e^t [\cos 2t + i \sin 2t] \left[\begin{pmatrix} 1 \\ 1 \\ 2 \end{pmatrix} + i \begin{pmatrix} 1 \\ 1 \\ 0 \end{pmatrix} \right]$$

$$= e^t \left\{ \begin{pmatrix} \cos 2t - \sin 2t \\ \cos 2t - \sin 2t \\ 2\cos 2t \end{pmatrix} + i \begin{pmatrix} \cos 2t + \sin 2t \\ \cos 2t + \sin 2t \\ 2\sin 2t \end{pmatrix} \right\}$$

We know that the real and imaginary parts of $\mathbf{z}$ are also solutions, so our two new solutions are

$$\mathbf{x}_2(t) = e^t \begin{pmatrix} \cos 2t - \sin 2t \\ \cos 2t - \sin 2t \\ 2\cos 2t \end{pmatrix} \quad \text{and} \quad \mathbf{x}_3(t) = e^t \begin{pmatrix} \cos 2t + \sin 2t \\ \cos 2t + \sin 2t \\ 2\sin 2t \end{pmatrix}.$$

To discover whether these solutions are linearly independent, we consider the matrix

$$[\mathbf{x}_1(0), \mathbf{x}_2(0), \mathbf{x}_3(0)] = \begin{pmatrix} 0 & 1 & 1 \\ 1 & 1 & 1 \\ -1 & 2 & 0 \end{pmatrix}.$$

The determinant of this matrix is 2. Since it is nonzero, the vectors $\mathbf{x}_1(0)$, $\mathbf{x}_2(0)$, and $\mathbf{x}_3(0)$ are linearly independent. Hence by Proposition 5.12 in Section 8.5 the solutions are linearly independent as well. ●

The method used in the example works in general to find real solutions from complex conjugate pairs of eigenvalues. When there is a total of n distinct eigenvalues, we get a fundamental set of solutions in this way. Here is another example.

Example 5.6 Consider the coupled spring-mass system in Figure 1. Assume that the masses are $m_1 = 3$ kg and $m_2 = 1$ kg, and the spring constants are $k_1 = k_2 = 3$ N/m and $k_3 = 1$ N/m. Find the modes of oscillation of the system.

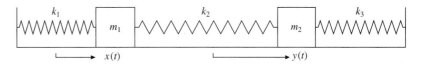

Figure 1. The three springs and two masses.

We discussed this system in Example 4.8 in Section 8.4. If $\mathbf{u} = (x, x', y, y')^T$, then $\mathbf{u}' = A\mathbf{u}$, where

$$A = \begin{pmatrix} 0 & 1 & 0 & 0 \\ -2 & 0 & 1 & 0 \\ 0 & 0 & 0 & 1 \\ 3 & 0 & -4 & 0 \end{pmatrix}$$

with the given data. Since so many of the entries of A are equal to 0, it is not too difficult to find that the characteristic polynomial of A is $\lambda^4 + 6\lambda^2 + 5$. Using the quadratic formula we see that $\lambda^2 = -1$ or -5. Therefore, the eigenvalues are $\pm i$ and $\pm i\sqrt{5}$.

Finding the eigenvectors is more difficult and we used a computer algebra program. Corresponding to $\lambda = i$ we found the eigenvector $(-i, 1, -i, 1)^T$. The complex solution is

$$\mathbf{z}_1(t) = e^{it} \begin{pmatrix} -i \\ 1 \\ -i \\ 1 \end{pmatrix} = \begin{pmatrix} \sin t \\ \cos t \\ \sin t \\ \cos t \end{pmatrix} + i \begin{pmatrix} -\cos t \\ \sin t \\ -\cos t \\ \sin t \end{pmatrix}.$$

The real and imaginary parts of this complex solution are real solutions.

Corresponding to the eigenvalue $i\sqrt{5}$ we found the eigenvector $(i\sqrt{5}, -5, -i3\sqrt{5}, 15)^T$. The complex solution is

$$z_2(t) = e^{i\sqrt{5}t} \begin{pmatrix} i\sqrt{5} \\ -5 \\ -i3\sqrt{5} \\ 15 \end{pmatrix} = \begin{pmatrix} -\sqrt{5}\sin(\sqrt{5}t) \\ -5\cos(\sqrt{5}t) \\ 3\sqrt{5}\sin(\sqrt{5}t) \\ 15\cos(\sqrt{5}t) \end{pmatrix} + i \begin{pmatrix} \sqrt{5}\cos(\sqrt{5}t) \\ -5\sin(\sqrt{5}t) \\ -3\sqrt{5}\cos(\sqrt{5}t) \\ 15\sin(\sqrt{5}t) \end{pmatrix}.$$

Again the real and imaginary parts of $\mathbf{z}_2$ are real solutions.

The general solution is a linear combination of the real and imaginary parts of $\mathbf{z}_1$ and $\mathbf{z}_2$:

$$\mathbf{u}(t) = C_1 \begin{pmatrix} \sin t \\ \cos t \\ \sin t \\ \cos t \end{pmatrix} + C_2 \begin{pmatrix} -\cos t \\ \sin t \\ -\cos t \\ \sin t \end{pmatrix}$$

$$+ C_3 \begin{pmatrix} -\sqrt{5}\sin(\sqrt{5}t) \\ -5\cos(\sqrt{5}t) \\ 3\sqrt{5}\sin(\sqrt{5}t) \\ 15\cos(\sqrt{5}t) \end{pmatrix} + C_4 \begin{pmatrix} \sqrt{5}\cos(\sqrt{5}t) \\ -5\sin(\sqrt{5}t) \\ -3\sqrt{5}\cos(\sqrt{5}t) \\ 15\sin(\sqrt{5}t) \end{pmatrix}.$$

Let's concentrate on the displacements

$$x(t) = u_1(t) = C_1 \sin t - C_2 \cos t + \sqrt{5}\left[-C_3 \sin(\sqrt{5}t) + C_4 \cos(\sqrt{5}t)\right]$$
$$y(t) = u_3(t) = C_1 \sin t - C_2 \cos t + 3\sqrt{5}\left[C_3 \sin(\sqrt{5}t) - C_4 \cos(\sqrt{5}t)\right].$$

We can find amplitudes A_1 and A_2 and phases ϕ_1 and ϕ_2 such that $C_1 \sin t - C_2 \cos t = A_1 \cos(t - \phi_1)$, and $\sqrt{5}\left[-C_3 \sin(\sqrt{5}t) + C_4 \cos(\sqrt{5}t)\right] = A_2 \cos(\sqrt{5}t - \phi_2)$. Then the displacements are

$$x(t) = A_1 \cos(t - \phi_1) + A_2 \cos(\sqrt{5}t - \phi_2)$$
$$y(t) = A_1 \cos(t - \phi_1) - 3A_2 \cos(\sqrt{5}t - \phi_2).$$

From these formulas we see that the displacements have two components. In the first the masses are oscillating in phase with frequency 1, while in the second they are oscillating in opposite directions with frequency $\sqrt{5}$, and with the amplitude of m_2 three times that of m_1.   ●

Repeated eigenvalues

If the eigenvalues are distinct, we now know how to find a fundamental set of real solutions. We have seen in Section 9.2 that repeated eigenvalues need special consideration, even in dimension 2. We expect that this situation will only get worse in higher dimensions. However, in some cases, the problem does not arise, as the next example shows.

Example 5.7   Find a fundamental set of solutions for the system

$$\mathbf{x}' = A\mathbf{x}, \quad \text{where} \quad A = \begin{pmatrix} 2 & 2 & -4 \\ 2 & -1 & -2 \\ 4 & 2 & -6 \end{pmatrix}.$$

Using our computer, we find that the characteristic polynomial of A is

$$p(\lambda) = \lambda^3 + 5\lambda^2 + 8\lambda + 4 = (\lambda + 1)(\lambda + 2)^2.$$

Hence, the eigenvalues are -1 and -2, and -2 is a repeated eigenvalue. For the eigenvalue -1, we find the eigenvector $(2, 1, 2)^T$. Hence, we have the solution

$$\mathbf{x}_1(t) = e^{-t} \begin{pmatrix} 2 \\ 1 \\ 2 \end{pmatrix}.$$

For the eigenvalue -2, we find that the eigenspace (the nullspace of $A + 2I$) has dimension 2. A basis is $(1, -2, 0)^T$ and $(1, 0, 1)^T$. For the eigenvalue $\lambda = -2$ and each of these eigenvectors, we get a solution,

$$\mathbf{x}_2(t) = e^{-2t} \begin{pmatrix} 1 \\ -2 \\ 0 \end{pmatrix} \quad \text{and} \quad \mathbf{x}_3(t) = e^{-2t} \begin{pmatrix} 1 \\ 0 \\ 1 \end{pmatrix}.$$

At $t = 0$, we have

$$[\mathbf{x}_1(0), \mathbf{x}_2(0), \mathbf{x}_3(0)] = \begin{pmatrix} 2 & 1 & 1 \\ 1 & -2 & 0 \\ 2 & 0 & 1 \end{pmatrix}.$$

Since the determinant of this matrix is -1, the three vectors are linearly independent. Consequently, the functions $\mathbf{x}_1$, $\mathbf{x}_2$, and $\mathbf{x}_3$ are linearly independent and form a fundamental set of solutions. ●

Example 5.7 shows that repeated eigenvalues do not necessarily cause trouble. Example 2.43 of Section 9.2, on the other hand, is a case where the repeated eigenvalue needed special handling. Here is another.

Example 5.8　Find a fundamental set of solutions for the system

$$\mathbf{x}' = A\mathbf{x}, \quad \text{where} \quad A = \begin{pmatrix} -1 & 2 & 1 \\ 0 & -1 & 0 \\ -1 & -3 & -3 \end{pmatrix}.$$

Again, we find that the characteristic polynomial of A is

$$p(\lambda) = \lambda^3 + 5\lambda^2 + 8\lambda + 4 = (\lambda + 1)(\lambda + 2)^2.$$

Hence, the eigenvalues are -1 and -2, and -2 is a repeated eigenvalue. The eigenvalue -1 has $(1, 1, -2)^T$ as an eigenvector, so

$$\mathbf{x}_1(t) = e^{-t} \begin{pmatrix} 1 \\ 1 \\ -2 \end{pmatrix}$$

is a solution.

This time, the eigenspace for the eigenvalue -2 has dimension 1, and it is spanned by $(1, 0, -1)^T$. Thus, we have the solution

$$\mathbf{x}_2(t) = e^{-2t} \begin{pmatrix} 1 \\ 0 \\ -1 \end{pmatrix},$$

but here our resources end. At present, we do not have a way to find the needed third solution. ●

The resolution of Example 5.8 will have to wait until after we develop a new tool in the next section. However, on the basis of the examples, we can speculate about the source of the problem. To help the discussion, let's make a couple of definitions.

Suppose A is an $n \times n$ matrix with real entries. Let $\lambda_1, \lambda_2, \ldots, \lambda_k$ be a list of the *distinct* eigenvalues of A. (For example, in Examples 5.7 and 5.8, we would have $\lambda_1 = -1$ and $\lambda_2 = -2$.) In general, the characteristic polynomial of A factors into

$$p(\lambda) = (\lambda - \lambda_1)^{q_1}(\lambda - \lambda_2)^{q_2} \cdots (\lambda - \lambda_k)^{q_k}.$$

The powers of the factors are at least 1 and satisfy $q_1 + q_2 + \cdots + q_k = n$. In the previous two examples, $q_1 = 1$ and $q_2 = 2$. We define the ***algebraic multiplicity*** of λ_j to be q_j. On the other hand, the ***geometric multiplicity*** of λ_j is d_j, the dimension of the eigenspace of λ_j.

In both Examples 5.7 and 5.8, the characteristic polynomial is $(\lambda + 1)(\lambda + 2)^2$. In Example 5.7, we have $d_1 = 1 = q_1$ and $d_2 = 2 = q_2$. However, in Example 5.8, we have $d_1 = 1 = q_1$ and $d_2 = 1 < q_2$. It is always true that $1 \leq d_j \leq q_j$. The method we used in Example 5.7 will enable us to find d_j independent solutions corresponding to the eigenvalue λ_j. It is only necessary to choose a basis for the eigenspace and use the corresponding exponential solutions. If $d_j = q_j$ for all j, then we can find a total of

$$d_1 + d_2 + \cdots + d_k = q_1 + q_2 + \cdots + q_k = n$$

solutions in this way. We would then have a fundamental set of solutions.

Consequently, we do not really need distinct eigenvalues. We can find a fundamental set of solutions provided that the geometric multiplicity of each eigenvalue is equal to its algebraic multiplicity. We will next learn how to proceed when the two multiplicities are unequal.

EXERCISES

For matrix A in Exercises 1–2, find the characteristic polynomial and the eigenvalues. Sketch the characteristic polynomial and explain the relationship between the graph of the characteristic polynomial and the eigenvalues of matrix A.

1. $A = \begin{pmatrix} 2 & 1 & 0 \\ 0 & 1 & 0 \\ 6 & 10 & -1 \end{pmatrix}$ **2.** $A = \begin{pmatrix} -1 & 6 & 2 \\ 0 & -1 & 0 \\ -1 & 11 & 2 \end{pmatrix}$

Proposition 5.1 guarantees that eigenvectors associated with distinct eigenvalues are linearly independent. In Exercises 3–6, each matrix has distinct eigenvalues; find them and their associated eigenvectors. Verify that the eigenvectors are linearly independent.

3. $A = \begin{pmatrix} 2 & 0 & 0 \\ -6 & 1 & -4 \\ -3 & 0 & -1 \end{pmatrix}$ **4.** $A = \begin{pmatrix} 1 & 0 & 0 \\ 3 & -2 & 1 \\ 5 & -5 & 2 \end{pmatrix}$

5. $A = \begin{pmatrix} -4 & 0 & 2 \\ 12 & 2 & -6 \\ -6 & 0 & 3 \end{pmatrix}$ **6.** $A = \begin{pmatrix} -5 & -2 & 0 \\ 4 & 1 & 0 \\ -3 & -1 & -2 \end{pmatrix}$

Find the general solution of the systems in Exercises 7–12.

7. $x' = 4x - 5y + 4z$
$y' = -y + 4z$
$z' = z$

8. $\mathbf{y}' = \begin{pmatrix} -3 & 0 & -1 \\ 3 & 2 & 3 \\ 2 & 0 & 0 \end{pmatrix} \mathbf{y}$

9. $x' = -3x$
$y' = -5x + 6y - 4z$
$z' = -5x + 2y$

10. $\mathbf{y}' = \begin{pmatrix} -3 & -6 & -2 \\ 0 & 1 & 0 \\ 0 & -2 & -1 \end{pmatrix} \mathbf{y}$

11. $\mathbf{y}' = \begin{pmatrix} -3 & 4 & 8 \\ -2 & 3 & 2 \\ 0 & 0 & 2 \end{pmatrix} \mathbf{y}$

12. $x_1' = 2x_1 + 4x_2 - 4x_4$
$x_2' = 3x_1 - 2x_3 - x_4$
$x_3' = x_3$
$x_4' = 3x_1 + 2x_2 - 2x_3 - 3x_4$

Find the solution of the given initial-value problems in Exercises 13–18.

13. Exercise 7 with $x(0) = 1$, $y(0) = -1$, and $z(0) = 2$

14. Exercise 8 with $\mathbf{y}(0) = (1, -1, 2)^T$

15. Exercise 9 with $x(0) = -2$, $y(0) = 0$, and $z(0) = 2$

16. Exercise 10 with $\mathbf{y}(0) = (-3, -3, 0)^T$

17. Exercise 11 with $\mathbf{y}(0) = (1, -2, 1)^T$

18. Exercise 12 with $x_1(0) = 1$, $x_2(0) = -1$, $x_3(0) = 0$, and $x_4(0) = 2$

Use Euler's formula to find the real and imaginary parts of the given complex solutions in Exercises 19–20.

19. $\mathbf{y}(t) = e^{2it} \begin{pmatrix} 1 \\ 1 + 2i \\ -3i \end{pmatrix}$ **20.** $\mathbf{y}(t) = e^{(1+i)t} \begin{pmatrix} 1 \\ 1 + i \\ 1 - i \\ 0 \end{pmatrix}$

Find the general solution of each system in Exercises 21–26.

21. $x' = -4x + 8y + 8z$
$y' = -4x + 4y + 2z$
$z' = 2z$

22. $\mathbf{y}' = \begin{pmatrix} 2 & 4 & 4 \\ 1 & 2 & 3 \\ -3 & -4 & -5 \end{pmatrix} \mathbf{y}$

23. $x' = 6x - 4z$
$y' = 8x - 2y$
$z' = 8x - 2z$

24. $\mathbf{y}' = \begin{pmatrix} -1 & 0 & 0 \\ -52 & -11 & 26 \\ -20 & -4 & 9 \end{pmatrix} \mathbf{y}$

25. $\mathbf{y}' = \begin{pmatrix} -7 & -13 & 0 \\ 2 & 3 & 0 \\ 3 & 8 & -2 \end{pmatrix} \mathbf{y}$

26. $\mathbf{y}' = \begin{pmatrix} -2 & 0 & 0 \\ 10 & 5 & -10 \\ 1 & 2 & -3 \end{pmatrix} \mathbf{y}$

In each of Exercises 27–32, find the solution of the given initial-value problems using the general solution found in the indicated exercise.

27. Exercise 21 with $x(0) = 1$, $y(0) = 0$, and $z(0) = 0$

28. Exercise 22 with $\mathbf{y}(0) = (1, -1, 0)^T$

29. Exercise 23 with $x(0) = -2$, $y(0) = -1$, and $z(0) = 0$

30. Exercise 24 with $\mathbf{y}(0) = (-2, 4, -2)^T$

31. Exercise 25 with $\mathbf{y}(0) = (-1, 1, 1)^T$

32. Exercise 26 with $\mathbf{y}(0) = (-1, 1, -1)^T$

For the systems in Exercises 33–36, find the eigenvalues and their associated eigenvectors. State the algebraic and geometric multiplicity of each eigenvalue. In the case where there are enough independent eigenvectors, state the general solution of the given system.

33. $x' = x$
$y' = x + y$
$z' = -10x + 8y + 5z$

34. $\mathbf{y}' = \begin{pmatrix} 2 & 0 & 0 \\ -6 & 2 & 3 \\ 6 & 0 & -1 \end{pmatrix} \mathbf{y}$

35. $x' = 4x$
$y' = -6x - 2y$
$z' = 7x + y - 2z$

36. $\mathbf{y}' = \begin{pmatrix} 6 & -5 & 10 \\ -1 & 2 & -2 \\ -1 & 1 & -1 \end{pmatrix} \mathbf{y}$

For the matrices in Exercises 37–44, use a computer to help find a fundamental set of solutions to the system $\mathbf{y}' = A\mathbf{y}$.

37. $A = \begin{pmatrix} -6 & 2 & -3 \\ -1 & -1 & -1 \\ 4 & -2 & 1 \end{pmatrix}$

38. $A = \begin{pmatrix} -7 & 2 & -4 \\ 42 & -11 & 18 \\ 38 & -10 & 18 \end{pmatrix}$

39. $A = \begin{pmatrix} 8 & 12 & -4 \\ -9 & -13 & 4 \\ -1 & -3 & 0 \end{pmatrix}$

40. $A = \begin{pmatrix} -1 & -2 & 4 \\ -1 & 0 & -4 \\ -1 & 2 & -6 \end{pmatrix}$

41. $A = \begin{pmatrix} -18 & -18 & 10 \\ 18 & 17 & -10 \\ 10 & 10 & -7 \end{pmatrix}$

42. $A = \begin{pmatrix} -6 & 6 & 8 \\ -12 & 16 & 24 \\ 8 & -12 & -18 \end{pmatrix}$

43. $A = \begin{pmatrix} 1 & 4 & 1 & -5 \\ -6 & -10 & -2 & 10 \\ 3 & 4 & -1 & -5 \\ -3 & -4 & -1 & 3 \end{pmatrix}$

44. $A = \begin{pmatrix} 6 & -6 & -6 & -8 \\ 8 & -8 & -6 & -8 \\ -1 & 7 & -10 & -9 \\ -8 & 6 & 6 & 6 \end{pmatrix}$

In Exercises 45–52, use a computer to help find the solution to the system $\mathbf{y}' = A\mathbf{y}$, with the given matrix A and the given intial value.

45. The matrix A in Exercise 37 with $\mathbf{y}(0) = (-6, 2, 9)^T$

46. The matrix A in Exercise 38 with $\mathbf{y}(0) = (-2, 2, 5)^T$

47. The matrix A in Exercise 39 with $\mathbf{y}(0) = (0, 8, 5)^T$

48. The matrix A in Exercise 40 with $\mathbf{y}(0) = (1, 0, 0)^T$

49. The matrix A in Exercise 41 with $\mathbf{y}(0) = (-1, 7, 3)^T$

50. The matrix A in Exercise 42 with $\mathbf{y}(0) = (-1, -4, 1)^T$

51. The matrix A in Exercise 43 with $\mathbf{y}(0) = (-1, 5, 2, 4)^T$

52. The matrix A in Exercise 44 with $\mathbf{y}(0) = (-2, -1, 6, -5)^T$

53. Consider the three cascaded tanks in Figure 2. The topmost contains 80 gallons, the second 60 gallons, and the third 40 gallons of salt solution. Let $x_1(t)$, $x_2(t)$, and $x_3(t)$ represent the salt content (in pounds) of the top tank, the second tank, and the third tank, respectively. Pure water is poured into the topmost tank at a rate of 4 gal/min.

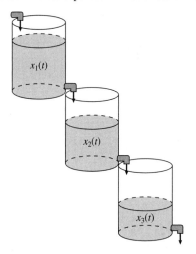

Figure 2. Three cascaded tanks.

Salt solution leaves the first tank at the same rate (4 gal/min) and pours immediately into the second tank in the cascade. Salt solution leaves the second tank at the same rate (4 gal/min) and pours immediately into the last tank in the cascade. There is a drain in the bottom tank that drains salt solution from the bottom tank at a rate of 4 gal/min. The effect of this assembly is to keep the volume of salt solution in each tank constant. If the initial salt content in the top tank is 4 pounds, the second tank 2 pounds, and the third tank 1 pound, find the salt content in each tank as a function of time. Sketch the salt content in each tank over time.

54. Two masses m_1 and m_2 slide on a frictionless table, as shown in Figure 3. The left mass is attached to a vertical support with a spring having spring constant k_1. A second spring with spring constant k_2 connects the two masses and a third spring with spring constant k_3 connects the second mass to a vertical support, as shown in Figure 3. Suppose that $m_1 = m_2 = 1$ kg, $k_1 = k_3 = 1$ N/m, and $k_2 = 2$ N/m. If the first mass is displaced 2 meters to the left, the second mass 2 meters to the right, plot the position of each mass versus time if both masses are released from rest.

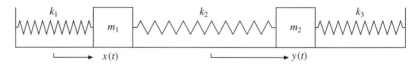

Figure 3. Coupled oscillators.

9.6 The Exponential of a Matrix

Our strategy, discussed in Section 9.1, of looking for exponential solutions in order to find a fundamental system of solutions has hit a snag in some cases where an eigenvalue has geometric multiplicity strictly smaller than its algebraic multiplicity. It is time to reexamine that strategy.

Let's go back to the beginning. When $n = 1$, we have the initial value problem

$$x' = ax, \quad x(0) = x_0,$$

which we know has the solution

$$x(t) = x_0 e^{at}. \tag{6.1}$$

We were motivated by this to look for exponential solutions to the system $\mathbf{x}' = A\mathbf{x}$ of the form

$$\mathbf{x}(t) = e^{\lambda t}\mathbf{v}. \tag{6.2}$$

We discovered that this worked, provided that λ was an eigenvalue of A and $\mathbf{v}$ was a corresponding eigenvector. The transition from (6.1) to (6.2), involving the replacement of a with the eigenvalue λ, has a certain appeal, but it has let us down in the cases we are now considering.

The use of eigenvalues in the transition from (6.1) to (6.2) does not use all of the information about the matrix A. This is made clearer in Examples 5.7 and 5.8 in Section 9.5. These matrices have the same eigenvalues, yet our method works for one but not for the other. Perhaps the most direct and complete analog of (6.1) is

$$\mathbf{x}(t) = e^{tA}\mathbf{v}. \tag{6.3}$$

At the moment, the term e^{tA}, the exponential of a matrix, makes no sense. Our task, therefore, is to make sense of the exponential of a matrix, to learn as much about it as we can, to show that (6.3) actually gives us a solution to the equation $\mathbf{x}' = A\mathbf{x}$, and to learn how to compute it.

The exponential of a matrix and solutions to differential equations

The correct way to define the exponential of a matrix is not at all obvious. We will do so using a power series, in analogy to the power series for the exponential function,

$$e^a = 1 + a + \frac{1}{2!}a^2 + \frac{1}{3!}a^3 + \cdots = \sum_{k=0}^{\infty} \frac{1}{k!}a^k. \tag{6.4}$$

DEFINITION 6.5

The *exponential of the matrix* A is defined to be

$$e^A = I + A + \frac{1}{2!}A^2 + \frac{1}{3!}A^3 + \cdots = \sum_{k=0}^{\infty} \frac{1}{k!}A^k. \qquad (6.6)$$

In this formula, $A^2 = AA$, $A^3 = AAA$, etc., all of the products being matrix products of the $n \times n$ matrix A with itself. By convention, $A^0 = I$. Consequently, all of the terms make good sense. They are all $n \times n$ matrices, so e^A is also an $n \times n$ matrix, provided that the series converges.

Convergence of this infinite series with matrix terms means that we consider the partial sum matrices

$$S_N = \sum_{k=0}^{N} \frac{1}{k!}A^k.$$

The components of S_N are very complicated expressions involving the entries of A. Convergence of the infinite series means that each component of the partial sum matrices converges. It is a fact, although we will not prove it, that the series in (6.6) converges for every matrix A. Furthermore, the convergence is rapid enough that all operations done on this series in what follows are justified.

Example 6.7   Show that the exponential of the diagonal matrix $A = \begin{pmatrix} r_1 & 0 \\ 0 & r_2 \end{pmatrix}$ is the diagonal matrix $e^A = \begin{pmatrix} e^{r_1} & 0 \\ 0 & e^{r_2} \end{pmatrix}$.

Since A is a diagonal matrix, the powers of A are easy to compute:

$$A^2 = A \cdot A = \begin{pmatrix} r_1 & 0 \\ 0 & r_2 \end{pmatrix} \cdot \begin{pmatrix} r_1 & 0 \\ 0 & r_2 \end{pmatrix} = \begin{pmatrix} r_1^2 & 0 \\ 0 & r_2^2 \end{pmatrix},$$

$$A^3 = A^2 \cdot A = \begin{pmatrix} r_1^2 & 0 \\ 0 & r_2^2 \end{pmatrix} \cdot \begin{pmatrix} r_1 & 0 \\ 0 & r_2 \end{pmatrix} = \begin{pmatrix} r_1^3 & 0 \\ 0 & r_2^3 \end{pmatrix},$$

and so forth. Therefore,

$$\begin{aligned} e^A &= I + A + \frac{A^2}{2!} + \frac{A^3}{3!} + \cdots \\ &= \begin{pmatrix} 1 & 0 \\ 0 & 1 \end{pmatrix} + \begin{pmatrix} r_1 & 0 \\ 0 & r_2 \end{pmatrix} + \frac{1}{2!}\begin{pmatrix} r_1^2 & 0 \\ 0 & r_2^2 \end{pmatrix} + \frac{1}{3!}\begin{pmatrix} r_1^3 & 0 \\ 0 & r_2^3 \end{pmatrix} + \cdots \\ &= \begin{pmatrix} 1 + r_1 + r_1^2/2! + r_1^3/3! + \cdots & 0 \\ 0 & 1 + r_2 + r_2^2/2! + r_2^3/3! + \cdots \end{pmatrix} \\ &= \begin{pmatrix} e^{r_1} & 0 \\ 0 & e^{r_2} \end{pmatrix}. \end{aligned}$$

Obviously, there is nothing special about the dimension 2 in this example. In general, the exponential of a diagonal matrix is the diagonal matrix containing the exponentials of the diagonal entries. In particular, we have

$$e^{rI} = e^r I. \qquad (6.8)$$

When $r = 0$, we have the important special case

$$e^{0I} = e^0 I = I. \qquad (6.9)$$

Unfortunately, there is no easy formula for the exponential of a matrix that is not diagonal.

Solution to the initial value problem

We will be looking at

$$e^{tA} = I + tA + \frac{t^2}{2!}A^2 + \frac{t^3}{3!}A^3 + \cdots . \qquad (6.10)$$

for a fixed $n \times n$ matrix A and a real number t. Consider e^{tA} as a function of t with values that are $n \times n$ matrices. If $\mathbf{v} \in \mathbf{R}^n$, we will also be looking at the function

$$e^{tA}\mathbf{v} = \mathbf{v} + tA\mathbf{v} + \frac{t^2}{2!}A^2\mathbf{v} + \frac{t^3}{3!}A^3\mathbf{v} + \cdots , \qquad (6.11)$$

which has values in $\mathbf{R}^n$. Remember that it is our goal to compute $e^{tA}\mathbf{v}$ for every $\mathbf{v}$ in a basis of $\mathbf{R}^n$.

Let's prove immediately that the exponential can be used to solve the initial value problem.

PROPOSITION 6.12 Suppose A is an $n \times n$ matrix.

1. Then

$$\frac{d}{dt}e^{tA} = Ae^{tA}.$$

2. If $\mathbf{v} \in \mathbf{R}^n$, the function $\mathbf{x}(t) = e^{tA}\mathbf{v}$ is the solution to the initial value problem

$$\mathbf{x}' = A\mathbf{x} \quad \text{with} \quad \mathbf{x}(0) = \mathbf{v}.$$

Proof Let's prove part (2) first, since it easily follows from part (1). First, by part (1),

$$\frac{d}{dt}\mathbf{x}(t) = \frac{d}{dt}\left(e^{tA}\mathbf{v}\right) = \frac{d}{dt}\left(e^{tA}\right)\mathbf{v} = Ae^{tA}\mathbf{v} = A\mathbf{x}(t).$$

To finish, we use (6.9) to show that $\mathbf{x}(0) = e^{0A}\mathbf{v} = I\mathbf{v} = \mathbf{v}$.

To prove part (1), we differentiate the series (6.10) term by term. This is possible since the series for the exponential converges so fast. We get

$$\frac{d}{dt}e^{tA} = \frac{d}{dt}\left(I + tA + \frac{t^2}{2!}A^2 + \frac{t^3}{3!}A^3 + \cdots\right)$$

$$= A + \frac{t}{1!}A^2 + \frac{t^2}{2!}A^3 + \cdots$$

$$= A\left(I + tA + \frac{t^2}{2!}A^2 + \cdots\right)$$

$$= Ae^{tA}.$$

Part (2) of Proposition 6.12 shows that we can solve the initial value problem if we can compute $e^{tA}\mathbf{v}$. This will be our emphasis from now on. We will start modestly.

Truncation

Notice that if most of the terms in the series in (6.11) are equal to the zero matrix, the infinite series becomes a finite sum. For example, if $A^2\mathbf{v} = \mathbf{0}$ and if $p > 2$, then $A^p\mathbf{v} = A^{p-2}A^2\mathbf{v} = A^{p-2}\mathbf{0} = \mathbf{0}$. Therefore, using (6.11), we have

$$e^{tA}\mathbf{v} = \mathbf{v} + tA\mathbf{v} + \frac{t^2}{2!}A^2\mathbf{v} + \frac{t^3}{3!}A^3\mathbf{v} + \cdots$$
$$= \mathbf{v} + tA\mathbf{v}.$$

When this happens we will say that the series for $e^{tA}\mathbf{v}$ *truncates*. Since we will refer to it often, we will state the result as a proposition.

PROPOSITION 6.13   Suppose A is an $n \times n$ matrix and $\mathbf{v}$ is an n-vector.

1. If $A\mathbf{v} = \mathbf{0}$, then $e^{tA}\mathbf{v} = \mathbf{v}$ for all t.
2. If $A^2\mathbf{v} = \mathbf{0}$, then $e^{tA}\mathbf{v} = \mathbf{v} + tA\mathbf{v}$ for all t.
3. More generally, if $A^k\mathbf{v} = \mathbf{0}$, then

$$e^{tA}\mathbf{v} = \mathbf{v} + tA\mathbf{v} + \cdots + \frac{t^{k-1}}{(k-1)!}A^{k-1}\mathbf{v} \quad \text{for all } t. \qquad \bullet$$

According to Proposition 6.13, we can compute $e^{tA}\mathbf{v}$ whenever the vector $\mathbf{v}$ is in the nullspace of A or in the nullspace of a power of A. Let's use the result to compute an example.

E x a m p l e   6.14   Consider

$$A = \begin{pmatrix} -4 & -2 & 1 \\ 4 & 2 & -2 \\ 8 & 4 & 0 \end{pmatrix}, \quad \mathbf{v} = \begin{pmatrix} -1 \\ 2 \\ 0 \end{pmatrix}, \quad \text{and} \quad \mathbf{w} = \begin{pmatrix} 0 \\ 0 \\ 1 \end{pmatrix}.$$

Compute $e^{tA}\mathbf{v}$ and $e^{tA}\mathbf{w}$.

We compute that

$$A\mathbf{v} = \begin{pmatrix} -4 & -2 & 1 \\ 4 & 2 & -2 \\ 8 & 4 & 0 \end{pmatrix} \begin{pmatrix} -1 \\ 2 \\ 0 \end{pmatrix} = \begin{pmatrix} 0 \\ 0 \\ 0 \end{pmatrix}.$$

Hence, by part (1) of Proposition 6.13, $e^{tA}\mathbf{v} = \mathbf{v}$. On the other hand,

$$A\mathbf{w} = \begin{pmatrix} -4 & -2 & 1 \\ 4 & 2 & -2 \\ 8 & 4 & 0 \end{pmatrix} \begin{pmatrix} 0 \\ 0 \\ 1 \end{pmatrix} = \begin{pmatrix} 1 \\ -2 \\ 0 \end{pmatrix} = -\mathbf{v},$$

so $A^2\mathbf{w} = \mathbf{0}$. Therefore, by part (2) of Proposition 6.13,

$$e^{tA}\mathbf{w} = \mathbf{w} + tA\mathbf{w}$$
$$= \mathbf{w} - t\mathbf{v}$$
$$= \begin{pmatrix} t \\ -2t \\ 1 \end{pmatrix}. \qquad \bullet$$

Proposition 6.13 provides a very modest beginning to the computations of $e^{tA}\mathbf{v}$. However, this modest beginning will bear fruit when we learn some properties of the exponential.

The law of exponents

The key property of the ordinary exponential function is the law of exponents,

$$e^{a+b} = e^a e^b.$$

It's time for us to explore the extent to which this property remains true for the exponential of a matrix.

PROPOSITION 6.15   **1.** If A and B are $n \times n$ matrices, then

$$e^{A+B} = e^A e^B \tag{6.16}$$

if and only if $AB = BA$.

2. If A is an $n \times n$ matrix, then e^A is a nonsingular matrix whose inverse is e^{-A}.

Proof   Part (1) is the most interesting property. If $AB = BA$ we say that A and B **commute.** Then part (1) says that the usual law of exponents does not apply to matrices unless the two matrices commute. To prove it, we first compute $e^A e^B$ and regroup the terms as follows.

$$
\begin{aligned}
e^A e^B &= \left(I + A + \frac{1}{2!}A^2 + \frac{1}{3!}A^3 + \cdots \right) \cdot \left(I + B + \frac{1}{2!}B^2 + \frac{1}{3!}B^3 + \cdots \right) \\
&= I + (A + B) + \frac{1}{2!}(A^2 + 2AB + B^2) \\
&\quad + \frac{1}{3!}(A^3 + 3A^2B + 3AB^2 + B^3) + \cdots
\end{aligned}
\tag{6.17}
$$

On the other hand, e^{A+B} involves powers of $A + B$. We have

$$(A + B)^2 = (A + B)(A + B) = A^2 + AB + BA + B^2.$$

Since $AB = BA$ (by assumption), $AB + BA = 2AB$. Therefore,

$$(A + B)^2 = A^2 + 2AB + B^2.$$

This is analogous to the familiar rule of squaring the sum of two numbers. Likewise, since A and B commute, we have

$$(A + B)^3 = A^3 + 3A^2B + 3AB^2 + B^3.$$

With this information, we compute

$$
\begin{aligned}
e^{A+B} &= I + (A + B) + \frac{1}{2!}(A + B)^2 + \frac{1}{3!}(A + B)^3 + \cdots \\
&= I + (A + B) + \frac{1}{2!}\left(A^2 + 2AB + B^2\right) \\
&\quad + \frac{1}{3!}\left(A^3 + 3A^2B + 3AB^2 + B^3\right) + \cdots .
\end{aligned}
$$

The expression on the right is the same as the right side of (6.17). Thus, $e^{A+B} = e^A e^B$.

Part (2) follows easily from part (1). Since A and $-A$ commute,

$$
\begin{aligned}
e^A e^{-A} &= e^{A-A} \quad \text{from (6.16)} \\
&= e^{0I} \\
&= I \quad \text{from (6.9)}.
\end{aligned}
$$

Thus, e^{-A} is the inverse of e^A.   ●

We can use part (1) of Proposition 6.15 together with Proposition 6.13 to greatly extend our capability of computing $e^{tA}\mathbf{v}$. Notice that if λ is a number, then

$$tA = \lambda t I + t[A - \lambda I]. \tag{6.18}$$

In addition, since the identity matrix commutes with every other matrix, the two summands in (6.18) commute. It follows from part (1) of Proposition 6.15 that

$$e^{tA} = e^{\lambda t I + t[A - \lambda I]} = e^{\lambda t I}e^{t[A-\lambda I]}. \tag{6.19}$$

Since the matrix $\lambda t I$ in the first factor in (6.19) is a diagonal matrix, we can compute that $e^{\lambda t I} = e^{\lambda t}I$, as we did in equation (6.8). Thus (6.19) becomes

$$e^{tA} = e^{\lambda t}I e^{t[A-\lambda I]} = e^{\lambda t}e^{t[A-\lambda I]}. \tag{6.20}$$

When we apply (6.20) to a vector $\mathbf{v}$, we get

$$e^{tA}\mathbf{v} = e^{\lambda t}e^{t[A-\lambda I]}\mathbf{v}. \tag{6.21}$$

Therefore, we can compute $e^{tA}\mathbf{v}$ if we can compute $e^{t[A-\lambda I]}\mathbf{v}$ for some number λ. In particular, according to Proposition 6.13, we can compute $e^{tA}\mathbf{v}$ if $\mathbf{v}$ is in the nullspace of a power of $A - \lambda I$ for some number λ. We can use equation (6.21) and Proposition 6.13 to prove the following:

PROPOSITION 6.22   Suppose A is an $n \times n$ matrix, λ is a number, and $\mathbf{v}$ is an n-vector.

1. If $[A - \lambda I]\mathbf{v} = \mathbf{0}$, then $e^{tA}\mathbf{v} = e^{\lambda t}\mathbf{v}$ for all t.
2. If $[A - \lambda I]^2\mathbf{v} = \mathbf{0}$, then $e^{tA}\mathbf{v} = e^{\lambda t}(\mathbf{v} + t[A - \lambda I]\mathbf{v})$ for all t.
3. More generally, if k is a positive integer and $[A - \lambda I]^k\mathbf{v} = \mathbf{0}$, then

$$e^{tA}\mathbf{v} = e^{\lambda t}\left(\mathbf{v} + t[A - \lambda I]\mathbf{v} + \cdots + \frac{t^{k-1}}{(k-1)!}[A - \lambda I]^{k-1}\mathbf{v}\right) \quad \text{for all } t. \;\bullet$$

Look carefully at part (1) of Proposition 6.22. Remember that $[A - \lambda I]\mathbf{v} = \mathbf{0}$ if and only if λ is an eigenvalue and $\mathbf{v}$ is an associated eigenvector. In this case we get the solution

$$\mathbf{y}(t) = e^{tA}\mathbf{v} = e^{\lambda t}\mathbf{v}.$$

This agrees with what we learned in Section 9.1. The other parts of Proposition 6.22 will enable us to go further.

Generalized eigenvectors and the corresponding solutions

Let's return to Example 5.8 in Section 9.5, where we were unable to find a fundamental set of solutions, to see if Proposition 6.22 will help us. Remember that we only need to compute $e^{tA}\mathbf{v}$ for enough vectors $\mathbf{v}$ to provide a fundamental system of solutions.

Example 6.23   Find a fundamental set of solutions for the system

$$\mathbf{x}' = A\mathbf{x}, \quad \text{where} \quad A = \begin{pmatrix} -1 & 2 & 1 \\ 0 & -1 & 0 \\ -1 & -3 & -3 \end{pmatrix}.$$

We have already discovered that the eigenvalue $\lambda_1 = -1$ has multiplicity 1, and the vector $\mathbf{v}_1 = (1, 1, -2)^T$ is an eigenvector. The eigenvalue $\lambda_2 = -2$ has algebraic multiplicity 2 and geometric multiplicity 1. Its eigenspace is generated by $\mathbf{v}_2 = (1, 0, -1)^T$. With this information, we have computed the exponential solutions

$$\mathbf{x}_1(t) = e^{tA}\mathbf{v}_1 = e^{-t} \begin{pmatrix} 1 \\ 1 \\ -2 \end{pmatrix} \quad \text{and} \quad \mathbf{x}_2(t) = e^{tA}\mathbf{v}_2 = e^{-2t} \begin{pmatrix} 1 \\ 0 \\ -1 \end{pmatrix}.$$

We need a third solution, and since $\lambda_2 = -2$ has algebraic multiplicity 2, let's try to find it using part (2) of Proposition 6.22. If we can find a vector $\mathbf{v}$ such that $[A - \lambda I]^2\mathbf{v} = \mathbf{0}$, then

$$e^{tA}\mathbf{v} = e^{\lambda t} \left(\mathbf{v} + t[A - \lambda I]\mathbf{v} \right). \tag{6.24}$$

We compute that

$$[A + 2I]^2 = \begin{pmatrix} 1 & 2 & 1 \\ 0 & 1 & 0 \\ -1 & -3 & -1 \end{pmatrix} \begin{pmatrix} 1 & 2 & 1 \\ 0 & 1 & 0 \\ -1 & -3 & -1 \end{pmatrix} = \begin{pmatrix} 0 & 5 & 0 \\ 0 & 1 & 0 \\ 0 & -8 & 0 \end{pmatrix}.$$

The nullspace of this matrix has dimension 2, and it has a basis $(1, 0, 0)^T$ and $(0, 0, 1)^T$. However, we have already computed the solution $\mathbf{x}_2$ using the initial value $\mathbf{v}_2 = (1, 0, -1)^T$. Notice that $\mathbf{v}_2$ is in the nullspace of $[A + 2I]^2$ since $[A + 2I]\mathbf{v}_2 = \mathbf{0}$. We want to choose a basis for the nullspace of $[A + 2I]^2$ that includes $\mathbf{v}_2$, so we need a second vector in the nullspace of $[A + 2I]^2$ that is not a multiple of $\mathbf{v}_2$. There are lots of choices, but let's pick $\mathbf{v}_3 = (1, 0, 0)^T$. Then, using (6.24), our third solution is

$$\mathbf{x}_3(t) = e^{tA}\mathbf{v}_3 = e^{-2t} \left(\mathbf{v}_3 + t[A + 2I]\mathbf{v}_3 \right)$$

$$= e^{-2t} \left[\begin{pmatrix} 1 \\ 0 \\ 0 \end{pmatrix} + t \begin{pmatrix} 1 & 2 & 1 \\ 0 & 1 & 0 \\ -1 & -3 & -1 \end{pmatrix} \begin{pmatrix} 1 \\ 0 \\ 0 \end{pmatrix} \right]$$

$$= e^{-2t} \begin{pmatrix} 1 + t \\ 0 \\ -t \end{pmatrix}.$$

Since the vectors $\mathbf{x}_1(0) = \mathbf{v}_1$, $\mathbf{x}_2(0) = \mathbf{v}_2$, $\mathbf{x}_3(0) = \mathbf{v}_3$ are linearly independent, we have a fundamental set of solutions. ●

As this example indicates, when eigenvalues have algebraic multiplicity greater than 1, we can compute extra solutions by looking for vectors in the nullspace of $[A - \lambda I]^p$ for $p > 1$. If λ is an eigenvector of A, and $[A - \lambda I]^p\mathbf{v} = 0$ for some integer $p \geq 1$, we will call $\mathbf{v}$ a **generalized eigenvector**. In fact, generalized eigenvectors provide all of the solutions we need because of the following result from linear algebra.

THEOREM 6.25 Suppose λ is an eigenvector of A with algebraic multiplicity q. Then there is an integer $p \leq q$ such that the dimension of the nullspace of $[A - \lambda I]^p$ is equal to q.
■

The solution procedure

As a result of Proposition 6.22 and Theorem 6.25, we have a general procedure for solving linear systems. Suppose the matrix A has distinct eigenvalues $\lambda_1, \lambda_2, \ldots, \lambda_k$, and that λ_j has algebraic multiplicity q_j. Since $q_1 + q_2 + \cdots + q_k = n$, we need to find q_j linearly independent solutions corresponding to the eigenvalue λ_j.

We do this with the following three-step procedure to find q linearly independent solutions corresponding to an eigenvalue λ of algebraic multiplicity q:

1. Find the smallest integer p such that the nullspace of $[A - \lambda I]^p$ has dimension q.
2. Find a basis $\{\mathbf{v}_1, \mathbf{v}_2, \ldots, \mathbf{v}_q\}$ of the nullspace of $[A - \lambda I]^p$.
3. For each $\mathbf{v}_j$, $1 \le j \le q$, we have the solution

$$\mathbf{x}_j(t) = e^{tA}\mathbf{v}_j$$
$$= e^{\lambda t}\left(\mathbf{v}_j + t[A - \lambda I]\mathbf{v}_j + \cdots + \frac{t^{p-1}}{(p-1)!}[A - \lambda I]^{p-1}\mathbf{v}_j\right).$$

We state without proof that if the solutions chosen for each eigenvalue are linearly independent, then the n solutions will also be linearly independent.

The three-step procedure works for complex eigenvalues as well as real. However, it is better to treat a complex conjugate pair of eigenvalues together. Suppose $\lambda = \alpha + i\beta$ is a complex eigenvalue of algebraic multiplicity q. Then the same is true for the complex conjugate $\bar{\lambda} = \alpha - i\beta$. The solutions corresponding to $\bar{\lambda}$ are complex conjugates of solutions corresponding to λ. Therefore, use the three-step procedure to find q linearly independent complex solutions $\mathbf{z}_1(t), \mathbf{z}_2(t), \ldots, \mathbf{z}_q(t)$ associated with λ. Then the conjugates $\bar{\mathbf{z}}_1(t), \bar{\mathbf{z}}_2(t), \ldots, \bar{\mathbf{z}}_q(t)$ are solutions associated with $\bar{\lambda}$. Thus we get $2q$ complex solutions associated with the λ and $\bar{\lambda}$. To get real-valued solutions, simply set $\mathbf{x}_j(t) = \operatorname{Re} \mathbf{z}_j(t)$ and $\mathbf{y}_j(t) = \operatorname{Im} \mathbf{z}_j(t)$. This will yield $2q$ real solutions for the conjugate pair of eigenvalues λ and $\bar{\lambda}$.

We should work out a few examples to see how the procedure works.

Example 6.26   Find a fundamental set of solutions to the system

$$\mathbf{y}' = A\mathbf{y}, \quad \text{where} \quad A = \begin{pmatrix} -1 & -2 & 1 \\ 0 & -4 & 3 \\ 0 & -6 & 5 \end{pmatrix}.$$

The characteristic equation is

$$\lambda^3 - 3\lambda - 2 = (\lambda + 1)^2(\lambda - 2) = 0,$$

so the eigenvalues are -1 and 2, with -1 having algebraic multiplicity 2.

The eigenspace for the eigenvalue $\lambda = 2$ is spanned by $\mathbf{v}_1 = (0, 1, 2)^T$. Hence,

$$\mathbf{y}_1(t) = e^{tA}\mathbf{v}_1 = e^{2t}\begin{pmatrix} 0 \\ 1 \\ 2 \end{pmatrix}$$

is our first solution.

For the eigenvalue $\lambda = -1$, we have

$$A - \lambda I = A + I = \begin{pmatrix} 0 & -2 & 1 \\ 0 & -3 & 3 \\ 0 & -6 & 6 \end{pmatrix} \quad \text{and} \quad (A + I)^2 = \begin{pmatrix} 0 & 0 & 0 \\ 0 & -9 & 9 \\ 0 & -18 & 18 \end{pmatrix}.$$

The eigenspace corresponding to $\lambda = -1$ is the nullspace of $A + I$, and it has dimension 1. Thus, the eigenvalue $\lambda = -1$ has geometric multiplicity 1. The nullspace of $(A + I)^2$ has dimension 2, which is the algebraic multiplicity of $\lambda = -1$. Therefore, we need to compute $e^{tA}\mathbf{v}$ for a basis of the nullspace of $(A + I)^2$. Notice that the nullspace of $A + I$ is a subset of the nullspace of $(A + I)^2$, so an eigenvector is a generalized eigenvector. It is usually a good idea to choose an eigenvector as part of our basis of the nullspace of $(A + I)^2$, since the solution corresponding to an eigenvector is so easy to compute. Hence, we select $\mathbf{v}_2 = (1, 0, 0)^T$. Since $\mathbf{v}_2$ is an eigenvector, the corresponding solution is

$$\mathbf{y}_2(t) = e^{tA}\mathbf{v}_2 = e^{-t}\begin{pmatrix} 1 \\ 0 \\ 0 \end{pmatrix}.$$

We need our third vector $\mathbf{v}_3$ to be in the nullspace of $(A + I)^2$, but linearly independent of $\mathbf{v}_2$. Perhaps the simplest example is $\mathbf{v}_3 = (0, 1, 1)^T$. The corresponding solution is

$$\begin{aligned} \mathbf{y}_3(t) &= e^{tA}\mathbf{v}_3 \\ &= e^{-t}\left(\mathbf{v}_3 + t(A + I)\mathbf{v}_3\right) \\ &= e^{-t}\left(\begin{pmatrix} 0 \\ 1 \\ 1 \end{pmatrix} + t\begin{pmatrix} -1 \\ 0 \\ 0 \end{pmatrix}\right) \\ &= e^{-t}\begin{pmatrix} -t \\ 1 \\ 1 \end{pmatrix}. \end{aligned}$$

For $n \times n$ matrices with $n \leq 3$, it is possible to do all of the needed computations by hand. For larger matrices, the work gets tedious, and it does not increase our understanding. For the larger examples that follow, we encourage you to use a computer. The computations can be made quickly, and the concepts can be easily explored.

The next example will show some different features.

Example 6.27 Find a fundamental set of solutions for $\mathbf{y}' = A\mathbf{y}$, where

$$A = \begin{pmatrix} 7 & 5 & -3 & 2 \\ 0 & 1 & 0 & 0 \\ 12 & 10 & -5 & 4 \\ -4 & -4 & 2 & -1 \end{pmatrix}.$$

Using a computer, we find that the characteristic polynomial is

$$\lambda^4 - 2\lambda^3 + 2\lambda - 1 = (\lambda + 1)(\lambda - 1)^3.$$

Hence, the eigenvalues are $\lambda = -1$, which has algebraic multiplicity 1, and $\lambda = 1$, which has algebraic multiplicity 3.

A computer also tells us that the eigenspace for $\lambda = -1$ is generated by the vector $\mathbf{v}_1 = (1, 0, 2, -1)^T$. Since this is an eigenvector, the corresponding solution is

$$\mathbf{y}_1(t) = e^{tA}\mathbf{v}_1 = e^{-t}\begin{pmatrix} 1 \\ 0 \\ 2 \\ -1 \end{pmatrix}.$$

Again, a computer tells us that the nullspace of

$$A - I = \begin{pmatrix} 6 & 5 & -3 & 2 \\ 0 & 0 & 0 & 0 \\ 12 & 10 & -6 & 4 \\ -4 & -4 & 2 & -2 \end{pmatrix}$$

has dimension 2. Thus, the geometric multiplicity of $\lambda = 1$ is 2. Our computer tells us that the eigenspace is spanned by $\mathbf{v}_2 = (1, 0, 2, 0)^T$ and $\mathbf{v}_3 = (1, -2, 0, 2)^T$. Since $\mathbf{v}_2$ and $\mathbf{v}_3$ are eigenvectors, the corresponding solutions are easily computed. They are

$$\mathbf{y}_2(t) = e^{tA}\mathbf{v}_2 = e^t \begin{pmatrix} 1 \\ 0 \\ 2 \\ 0 \end{pmatrix} \quad \text{and} \quad \mathbf{y}_3(t) = e^{tA}\mathbf{v}_3 = e^t \begin{pmatrix} 1 \\ -2 \\ 0 \\ 2 \end{pmatrix}.$$

Now we compute that the nullspace of

$$(A - I)^2 = \begin{pmatrix} -8 & -8 & 4 & -4 \\ 0 & 0 & 0 & 0 \\ -16 & -16 & 8 & -8 \\ 8 & 8 & -4 & 4 \end{pmatrix}$$

has dimension 3. Hence, we can find a third solution associated to $\lambda = 1$ by finding a vector $\mathbf{v}_4$ in the nullspace of $(A - I)^2$, which is linearly independent of $\mathbf{v}_2$ and $\mathbf{v}_3$. We will choose $\mathbf{v}_4 = (0, 0, 1, 1)^T$ because it has lots of zero entries. It is left to you to check that $\mathbf{v}_2$, $\mathbf{v}_3$, and $\mathbf{v}_4$ are linearly independent. Since $(A - I)^2\mathbf{v}_4 = \mathbf{0}$, the corresponding solution is

$$\mathbf{y}_4(t) = e^{tA}\mathbf{v}_4 = e^t \left(\mathbf{v}_4 + t(A - I)\mathbf{v}_4 \right)$$

$$= e^t \left\{ \begin{pmatrix} 0 \\ 0 \\ 1 \\ 1 \end{pmatrix} + t \begin{pmatrix} -1 \\ 0 \\ -2 \\ 0 \end{pmatrix} \right\} = e^t \begin{pmatrix} -t \\ 0 \\ 1 - 2t \\ 1 \end{pmatrix}.$$

Since we chose $\mathbf{v}_4$ to be independent of $\mathbf{v}_2$ and $\mathbf{v}_3$, our solutions are independent and we have a fundamental set. ●

We need an example involving complex eigenvalues.

Example 6.28 Find a fundamental system of solutions for the system $\mathbf{x}' = A\mathbf{x}$, where

$$A = \begin{pmatrix} 6 & 6 & -3 & 2 \\ -4 & -4 & 2 & 0 \\ 8 & 7 & -4 & 4 \\ 1 & 0 & -1 & -2 \end{pmatrix}.$$

Using a computer, we find that the eigenvalues of A are $-1 \pm i$, each with algebraic multiplicity 2. We compute that

$$A - (-1 + i)I = \begin{pmatrix} 7 - i & 6 & -3 & 2 \\ -4 & -3 - i & 2 & 0 \\ 8 & 7 & -3 - i & 4 \\ 1 & 0 & -1 & -1 - i \end{pmatrix}.$$

This matrix has a nullspace of dimension 1, so the eigenvalue $\lambda = -1 + i$ has geometric multiplicity 1. Our computer tells us that $\mathbf{w}_1 = (1 + i, 0, 2 + 2i, -1)^T$ is an eigenvector. The corresponding complex-valued solution is

$$\mathbf{z}_1(t) = e^{tA}\mathbf{w}_1 = e^{(-1+i)t}\mathbf{w}_1 = e^{(-1+i)t}\begin{pmatrix} 1+i \\ 0 \\ 2+2i \\ -1 \end{pmatrix}.$$

To find another complex-valued solution, we compute

$$(A - (-1+i)I)^2 = \begin{pmatrix} 2-14i & 3-12i & -2+6i & -4i \\ 8i & -2+6i & -4i & 0 \\ 8-16i & 6-14i & -6+6i & -8i \\ -2-2i & -1 & 1+2i & -2+2i \end{pmatrix}.$$

We look for a vector $\mathbf{w}_2$ in this nullspace that is not an eigenvector and therefore not a multiple of $\mathbf{w}_1$. One choice is $\mathbf{w}_2 = (1 + 2i, -2 - 2i, 0, 2)^T$. Since

$$(A - (-1+i)I)^2\mathbf{w}_2 = \mathbf{0},$$

the corresponding solution is

$$\mathbf{z}_2(t) = e^{tA}\mathbf{w}_2 = e^{(-1+i)t}\left(\mathbf{w}_2 + t(A - (-1+i)I)\mathbf{w}_2\right)$$

$$= e^{(-1+i)t}\begin{pmatrix} (1+t) + i(2+t) \\ -2-2i \\ 2t+2it \\ 2-t \end{pmatrix}.$$

Corresponding to the complex conjugate eigenvalue $-1 - i$, we have the conjugate generalized eigenvectors $\overline{\mathbf{w}}_1$ and $\overline{\mathbf{w}}_2$ and the corresponding conjugate solutions $\overline{\mathbf{z}}_1$ and $\overline{\mathbf{z}}_2$. These are the required four solutions. To find real solutions, we use the real and imaginary parts of the complex solutions. If $\mathbf{z}_1 = \mathbf{x}_1 + i\mathbf{y}_1$ and $\mathbf{z}_2 = \mathbf{x}_2 + i\mathbf{y}_2$, then $\mathbf{x}_1, \mathbf{x}_2, \mathbf{y}_1$, and $\mathbf{y}_2$ are the four needed real solutions. ●

EXERCISES

Use Definition 6.5 to calculate e^A for the matrices in Exercises 1–4.

1. $A = \begin{pmatrix} -2 & -4 \\ 1 & 2 \end{pmatrix}$ **2.** $A = \begin{pmatrix} 1 & 1 \\ -1 & -1 \end{pmatrix}$

3. $A = \begin{pmatrix} 1 & -1 & 0 \\ 1 & -1 & 0 \\ 0 & 0 & 0 \end{pmatrix}$ **4.** $A = \begin{pmatrix} -2 & 1 & -3 \\ -1 & 1 & -1 \\ 1 & -1 & 1 \end{pmatrix}$

5. Suppose that the matrix A satisfies $A^2 = \alpha A$, where $\alpha \neq 0$.

(a) Use Definition 6.5 to show that

$$e^{tA} = I + \frac{e^{\alpha t} - 1}{\alpha}A.$$

(b) Use part (a) to compute e^{tA} for

$$A = \begin{pmatrix} 1 & 1 & 1 \\ 1 & 1 & 1 \\ 1 & 1 & 1 \end{pmatrix}.$$

6. There are many important series in mathematics, such as the exponential series. For example,

$$\cos t = \sum_{k=0}^{\infty}(-1)^k\frac{t^{2k}}{(2k)!} = 1 - \frac{t^2}{2!} - \frac{t^4}{4!} + \cdots \quad \text{and}$$

$$\sin t = \sum_{k=0}^{\infty}(-1)^k\frac{t^{2k+1}}{(2k+1)!} = t - \frac{t^3}{3!} + \frac{t^5}{5!} - \cdots .$$

Use these infinite series together with Definition 6.5 to show that

$$e^{t\begin{pmatrix} 0 & -1 \\ 1 & 0 \end{pmatrix}} = \begin{pmatrix} \cos t & -\sin t \\ \sin t & \cos t \end{pmatrix}.$$

7. Use the result of Exercise 6 to show that if

$$A = \begin{pmatrix} a & -b \\ b & a \end{pmatrix},$$

then

$$e^{tA} = e^{at} \begin{pmatrix} \cos bt & -\sin bt \\ \sin bt & \cos bt \end{pmatrix}.$$

Hint: $A = aI + b \begin{pmatrix} 0 & -1 \\ 1 & 0 \end{pmatrix}.$

8. If

$$A = \begin{pmatrix} a & b \\ 0 & a \end{pmatrix},$$

find e^{tA}. *Hint:* See the hint for Exercise 7.

9. Let

$$A = \begin{pmatrix} 0 & -2 \\ 0 & 0 \end{pmatrix} \quad \text{and} \quad B = \begin{pmatrix} 0 & 0 \\ 2 & 0 \end{pmatrix}.$$

(a) Show that $AB \neq BA$.

(b) Evaluate e^{A+B}. *Hint:* This is a simple computation if you use Exercise 7.

(c) Use Definition 6.5 to evaluate e^A and e^B. Use these results to compute $e^A e^B$ and compare this with the result found in part (b). What have you learned from this exercise?

10. If $A = PDP^{-1}$, prove that $e^{tA} = Pe^{tD}P^{-1}$.

Use the results of Exercise 53 of Section 9.1 and Exercise 10 to calculate e^{tA} for each matrix in Exercises 11–12.

11. $A = \begin{pmatrix} -2 & 6 \\ 0 & -1 \end{pmatrix}$ **12.** $A = \begin{pmatrix} -2 & 0 \\ -3 & -3 \end{pmatrix}$

13. Let A be a 2×2 matrix with a single eigenvalue λ of algebraic multiplicity 2 and geometric multiplicity 1. Prove that

$$e^{At} = e^{\lambda t} [I + (A - \lambda I)t].$$

In Exercises 14–17, each matrix has an eigenvalue of algebraic multiplicity 2 but geometric multiplicity 1. Use the technique of Exercise 13 to compute e^{tA}.

14. $A = \begin{pmatrix} -2 & 1 \\ -1 & 0 \end{pmatrix}$ **15.** $A = \begin{pmatrix} -1 & 0 \\ 1 & -1 \end{pmatrix}$

16. $A = \begin{pmatrix} 0 & 1 \\ -1 & -2 \end{pmatrix}$ **17.** $A = \begin{pmatrix} -3 & -1 \\ 4 & 1 \end{pmatrix}$

Each of the matrices in Exercises 18–25 has only one eigenvalue λ. In each exercise, determine the smallest k such that $(A - \lambda I)^k = 0$. The use the fact that

$$e^{tA} = e^{\lambda t} \left[I + t(A - \lambda I) + \frac{t^2}{2!}(A - \lambda I)^2 + \cdots \right]$$

to compute e^{tA}.

18. $A = \begin{pmatrix} -1 & 0 & 0 \\ -1 & 1 & -1 \\ -2 & 4 & -3 \end{pmatrix}$ **19.** $A = \begin{pmatrix} -1 & -1 & 0 \\ -1 & 0 & -1 \\ -1 & 2 & -2 \end{pmatrix}$

20. $A = \begin{pmatrix} -2 & -1 & 0 \\ 0 & 0 & 1 \\ 0 & -4 & -4 \end{pmatrix}$ **21.** $A = \begin{pmatrix} -2 & 0 & 0 \\ 0 & -2 & 0 \\ -1 & 1 & -2 \end{pmatrix}$

22. $A = \begin{pmatrix} 1 & -1 & 2 & 0 \\ 0 & 1 & 0 & 0 \\ 0 & 0 & 1 & 0 \\ 0 & -1 & 2 & 1 \end{pmatrix}$

23. $A = \begin{pmatrix} -5 & 0 & -1 & 4 \\ -4 & 0 & 1 & 5 \\ 4 & -4 & -5 & -4 \\ 0 & -1 & -1 & -2 \end{pmatrix}$

24. $A = \begin{pmatrix} 0 & 4 & 5 & -2 \\ 1 & -5 & -7 & 3 \\ 0 & 2 & 3 & -1 \\ 3 & -10 & -13 & 6 \end{pmatrix}$

25. $A = \begin{pmatrix} 1 & 0 & 0 & 0 \\ -9 & 4 & 1 & 4 \\ 13 & -3 & -1 & -5 \\ 2 & -1 & 0 & 0 \end{pmatrix}$

Do the following for each of the matrices in Exercises 26–33. Exercises 26–29 can be done by hand, but you should use a computer for the rest.

(i) Find the eigenvalues.

(ii) For each eigenvalue, find the algebraic and the geometric multiplicities.

(iii) For each eigenvalue λ, find the smallest integer k such that the dimension of the nullspace of $(A - \lambda I)^k$ is equal to the algebraic multiplicity.

(iv) For each eigenvalue λ, find q linearly independent generalized eigenvectors, where q is the algebraic multiplicity of λ.

(v) Verify that the collection of the generalized eigenvectors you find in part (iv) for all of the eigenvalues is linearly independent.

(vi) Find a fundamental set of solutions for the system $\mathbf{y}' = A\mathbf{y}$.

26. $A = \begin{pmatrix} -2 & 1 & -1 \\ 1 & -3 & 0 \\ 3 & -5 & 0 \end{pmatrix}$ **27.** $A = \begin{pmatrix} 1 & 0 & 1 \\ 2 & 2 & -2 \\ 0 & 0 & 2 \end{pmatrix}$

28. $A = \begin{pmatrix} 0 & 1 & 0 \\ -4 & 4 & 0 \\ -2 & 0 & 1 \end{pmatrix}$ **29.** $A = \begin{pmatrix} -1 & 0 & 0 \\ 2 & -5 & -1 \\ 0 & 4 & -1 \end{pmatrix}$

30. $A = \begin{pmatrix} 11 & -42 & 4 & 28 \\ -12 & 39 & -4 & -28 \\ 0 & 0 & -1 & 0 \\ -24 & 81 & -8 & -57 \end{pmatrix}$

31. $A = \begin{pmatrix} 18 & -7 & 24 & 24 \\ 15 & -8 & 20 & 16 \\ 0 & 0 & -1 & 0 \\ -12 & 4 & -15 & -17 \end{pmatrix}$

32. $A = \begin{pmatrix} 0 & -30 & -42 & 40 & -48 & 14 \\ 1 & 7 & 9 & -9 & 10 & -2 \\ -1 & 5 & 8 & -6 & 6 & -2 \\ 2 & 45 & 64 & -60 & 72 & -20 \\ 2 & 33 & 47 & -45 & 55 & -15 \\ 0 & 7 & 11 & -10 & 10 & -1 \end{pmatrix}$

33. $A = \begin{pmatrix} 2 & 0 & 0 & 0 & 0 & 1 \\ -14 & -2 & -7 & 11 & -9 & -8 \\ -9 & -3 & -3 & 7 & -6 & -4 \\ -19 & -5 & -9 & 17 & -12 & -9 \\ -29 & -7 & -13 & 23 & -16 & -15 \\ 19 & 5 & 9 & -15 & 12 & 11 \end{pmatrix}$

34. Consider the system

$$\mathbf{y}' = \begin{pmatrix} -2 & 2 & -1 \\ -4 & 3 & 0 \\ 0 & -1 & 3 \end{pmatrix} \mathbf{y}.$$

A student computes the fundamental solution set

$$\mathbf{y}_1(t) = e^{2t} \begin{pmatrix} 1 \\ 4 \\ 4 \end{pmatrix}, \quad \mathbf{y}_2(t) = e^t \begin{pmatrix} 1-t \\ 1-2t \\ -t \end{pmatrix},$$

$$\mathbf{y}_3(t) = e^t \begin{pmatrix} -1+2t \\ 4t \\ 1+2t \end{pmatrix}.$$

Her study partner computes a different fundamental solution set,

$$\mathbf{y}_1(t) = e^{2t} \begin{pmatrix} 1 \\ 4 \\ 4 \end{pmatrix}, \quad \mathbf{y}_2(t) = e^t \begin{pmatrix} 3-5t \\ 1-10t \\ -2-5t \end{pmatrix},$$

$$\mathbf{y}_3(t) = e^t \begin{pmatrix} 3-7t \\ -1-14t \\ -4-7t \end{pmatrix}.$$

Are they both correct? Both wrong? One correct, the other wrong? Why?

In Exercises 35–45, find a fundamental set of solutions for the given system. Exercises 35–40 can be done by hand, but use a computer for the rest.

35. $\mathbf{y}' = \begin{pmatrix} 6 & 0 & -4 \\ -2 & 4 & 5 \\ 1 & 0 & 2 \end{pmatrix} \mathbf{y}$

36. $\mathbf{y}' = \begin{pmatrix} 8 & 3 & 2 \\ 0 & 4 & 0 \\ -8 & -6 & 0 \end{pmatrix} \mathbf{y}$

37. $x' = -2x - 4y + 13z$
$y' = 5y - 4z$
$z' = y + z$

38. $x' = -x + 5y + 3z$
$y' = y + z$
$z' = -2y - 2z$

39. $\mathbf{x}' = \begin{pmatrix} 5 & -1 & 0 & 2 \\ 0 & 3 & 0 & 4 \\ 1 & 1 & -1 & -3 \\ 0 & -1 & 0 & 7 \end{pmatrix} \mathbf{x}$

40. $\mathbf{x}' = \begin{pmatrix} -12 & -1 & 8 & 10 \\ -8 & 0 & -1 & 9 \\ 0 & 0 & 5 & 0 \\ -17 & -1 & 8 & 15 \end{pmatrix} \mathbf{x}$

41. $\mathbf{x}' = \begin{pmatrix} -1 & 0 & 0 & 2 \\ -6 & 13 & 0 & -42 \\ 0 & -6 & -2 & 13 \\ -2 & 5 & 0 & -16 \end{pmatrix} \mathbf{x}$

42. $\mathbf{x}' = \begin{pmatrix} -8 & -2 & 3 & 12 \\ -3 & -2 & 2 & 6 \\ 2 & 0 & -3 & -4 \\ -4 & -1 & 2 & 6 \end{pmatrix} \mathbf{x}$

43. $\mathbf{x}' = \begin{pmatrix} -2 & 2 & -2 & 0 & -3 \\ -1 & 0 & -1 & 0 & -3 \\ 15 & -16 & -1 & 10 & 33 \\ 12 & -13 & 1 & 6 & 26 \\ -5 & 5 & 0 & -3 & -12 \end{pmatrix} \mathbf{x}$

44. $\mathbf{x}' = \begin{pmatrix} -4 & 3 & 6 & 4 & 2 \\ 0 & -8 & -10 & -8 & 2 \\ -1 & 7 & 10 & 9 & -1 \\ 1 & -4 & -7 & -7 & 0 \\ -1 & -1 & -1 & -1 & 0 \end{pmatrix} \mathbf{x}$

45. $x_1' = 5x_1 + 7x_2 + x_3 + x_4 + 8x_5$
$x_2' = 3x_1 + 6x_2 + 5x_3 + 4x_4 + 5x_5$
$x_3' = -3x_1 - 8x_2 - 2x_3 - 5x_4 - 12x_5$
$x_4' = 3x_1 + 14x_2 + 8x_3 + 10x_4 + 18x_5$
$x_5' = -4x_1 - 9x_2 - 6x_3 - 5x_4 - 9x_5$

46. Pictured in Figure 1 are two tanks, each containing 100 gallons of a salt solution. Pure water flows into the upper tank at a rate of 4 gal/min. Salt solution drains from the upper tank into the lower tank at a rate of 4 gal/min. Finally, salt solution drains from the lower tank at a rate of 4 gal/min, effectively keeping the volume of solution in each tank at a constant 100 gal. If the initial salt content of the upper and lower tanks is 10 and 20 pounds, respectively, use an exponential matrix approach to find the salt content of each tank as a function of time.

47. Pure water pours into the topmost tank shown in Figure 2 at a rate of 4 gal/min. Salt solution drains from the topmost tank into the second tank at a rate of 4 gal/min. There is a drain at the bottom of the third tank from which salt solution drains at a rate of 4 gal/min. The effect of this cascading of tanks is to keep the volume of solution in each tank at a constant 100 gallons. Initially, the first tank holds 10 pounds, the second 8 pounds, and the third 4 pounds of salt. Set up an initial value problem that models the salt content $x_1(t)$, $x_2(t)$, and $x_3(t)$ in the three cascading tanks, then solve using the technique of Example 6.27. Finally, sketch the graph of the salt content in each tank over time and predict the salt content in each tank after 8 minutes.

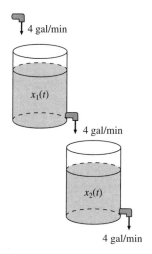

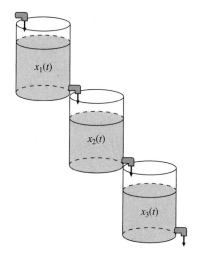

Figure 1. Two cascading tanks.

Figure 2. Three cascading tanks.

9.7 Qualitative Analysis of Linear Systems

In Section 9.3, we were able to do a systematic study of the types of equilibrium points that occur in two dimensions. If we tried to do this in three dimensions, there would be a much larger variety of cases. Fortunately, it is not usually necessary to have a classification as precise as the two-dimensional one. Instead, we will settle for a simpler classification based on the notion of stability, which means that we will focus on the behavior of solutions to linear systems as $t \to \infty$. We have computed the general solutions for arbitrary linear systems, so it is just a matter of going back to previous sections and finding the t dependence of these solutions.

Form of the solutions

We now know from the solution procedure found in Section 9.6 that every solution to a linear system is a linear combination of solutions corresponding to generalized eigenvectors. First, let's look at the easiest case. This is the case of an eigenvalue λ (real or complex) whose geometric and algebraic multiplicities coincide. Then every generalized eigenvector is an eigenvector, so every solution is of the form $e^{tA}\mathbf{v} = e^{\lambda t}\mathbf{v}$. Every component of the solution is a constant multiple of $e^{\lambda t}$.

Next let's look at what happens if the geometric multiplicity is smaller than the algebraic multiplicity. By the results of the previous section, the solution with initial value equal to a generalized eigenvector $\mathbf{v}$ has the form

$$\mathbf{x}(t) = e^{tA}\mathbf{v} = e^{\lambda t}\left(\mathbf{v} + t(A - \lambda I)\mathbf{v} + \cdots + \frac{t^{p-1}}{(p-1)!}(A - \lambda I)^{p-1}\mathbf{v}\right). \quad (7.1)$$

Every component in the vector-valued function $\mathbf{x}(t)$ is of the form $e^{\lambda t}P(t)$, where $P(t)$ is a polynomial.

Suppose first that λ is a real eigenvalue. Then every component of $\mathbf{x}(t)$ has the form

$$e^{\lambda t}P(t), \quad (7.2)$$

where $P(t)$ is a real-valued polynomial.

If $\lambda = \alpha + i\beta$ is complex with algebraic multiplicity q, then the generalized eigenvector $\mathbf{v}$ is complex as well. Then every component of the solution $\mathbf{x}(t)$ in (7.1) has the form $e^{\lambda t}P(t)$, where now the polynomial $P(t)$ is complex valued. Using

Euler's formula $e^{\lambda t} = e^{(\alpha+i\beta)t} = e^{\alpha t}(\cos \beta t + i \sin \beta t)$, and writing $P(t) = Q(t) + iR(t)$, we see that every component of $\mathbf{x}(t)$ has the form

$$e^{\lambda t} P(t) = e^{\alpha t}(\cos \beta t + i \sin \beta t)(Q(t) + iR(t))$$
$$= e^{\alpha t}[(Q(t)\cos \beta t - R(t)\sin \beta t) + i(R(t)\cos \beta t + Q(t)\sin \beta t)].$$

Consequently,

$$\mathrm{Re}\left(e^{\lambda t} P(t)\right) = e^{\alpha t}(Q(t)\cos \beta t - R(t)\sin \beta t) \quad \text{and}$$
$$\mathrm{Im}\left(e^{\lambda t} P(t)\right) = e^{\alpha t}(R(t)\cos \beta t + Q(t)\sin \beta t).$$

Putting everything together, we realize that every component of every real solution is the sum of functions of the form

$$e^{\alpha t}(p(t)\cos \beta t + q(t)\sin \beta t), \tag{7.3}$$

where p and q are polynomials and α is the real part of an eigenvalue.

Using these facts, we can easily prove Theorem 7.4.

THEOREM 7.4 Let A be an $n \times n$ matrix.

1. Suppose that the real part of every eigenvalue of A is negative. Then every solution to the system $\mathbf{x}' = A\mathbf{x}$ tends toward the equilibrium point at the origin as $t \to \infty$.

2. Suppose that A has at least one eigenvalue with a positive real part. Then there are solutions to the system $\mathbf{x}' = A\mathbf{x}$ starting arbitrarily close to the equilibrium point at the origin that get arbitrarily large as $t \to \infty$.

Since the real part of a real eigenvalue is the eigenvalue itself, the hypotheses in both parts of the theorem refer to both the real and the complex eigenvalues of A.

Proof According to what we discovered in our survey, every entry in every solution corresponding to a generalized eigenvector has the form (7.3), where the eigenvalue is $\lambda = \alpha + i\beta$. To prove part (1), it suffices to prove that every function of this kind tends toward 0 as $t \to \infty$, when $\alpha < 0$.

By breaking the function in (7.3) into its summands, it suffices to prove that

$$e^{\alpha t} t^k \cos \beta t \to 0 \quad \text{and} \quad e^{\alpha t} t^k \sin \beta t \to 0 \quad \text{as } t \to \infty$$

for any integer k. Since the trigonometric functions oscillate between -1 and 1, it suffices to prove that

$$e^{\alpha t} t^k = \frac{t^k}{e^{-\alpha t}} \to 0 \quad \text{as } t \to \infty$$

for any integer k. This follows by applying l'Hôpital's rule to the fraction k times. Remember that $-\alpha > 0$. The exponential in the denominator grows much faster than the power in the numerator.

We will prove part (2) only in the case when A has a real, positive eigenvalue λ. The proof in the case where A has a complex eigenvector with a positive real part is similar. Let $\mathbf{v}$ be an eigenvector. The solution $\mathbf{x}(t) = \epsilon e^{\lambda t}\mathbf{v}$ satisfies $\mathbf{x}(0) = \epsilon \mathbf{v}$. For small ϵ, this vector can be made arbitrarily close to the origin $\mathbf{0}$. Since $\lambda > 0$, $e^{\lambda t} \to \infty$ as $t \to \infty$. Thus, $\mathbf{x}(t)$ gets arbitrarily large. ∎

Stability

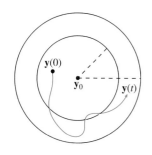

Figure 1. A stable equilibrium point.

We need some language to describe the phenomena described in Theorem 7.4. We give the definitions for arbitrary autonomous systems, linear or nonlinear. Suppose that $\mathbf{y}' = \mathbf{f}(\mathbf{y})$ is an autonomous system and that $\mathbf{y}_0$ is an equilibrium point.

We will say that $\mathbf{y}_0$ is **stable** if, for every $\epsilon > 0$, there is a $\delta > 0$ such that if $\mathbf{y}(t)$ is a solution that satisfies $|\mathbf{y}(0) - \mathbf{y}_0| < \delta$, then $|\mathbf{y}(t) - \mathbf{y}_0| < \epsilon$ for all $t > 0$. The relationship between ϵ and δ is shown in Figure 1. Roughly speaking, $\mathbf{y}_0$ is stable if every solution that starts close to $\mathbf{y}_0$ stays close as t increases.

We will say that an equilibrium point is **asymptotically stable** if it is stable and there is an $\eta > 0$ such that every solution $\mathbf{y}(t)$ satisfying $|\mathbf{y}(0) - \mathbf{y}_0| < \eta$ approaches $\mathbf{y}_0$ as $t \to \infty$. This means that every solution that starts reasonably close to $\mathbf{y}_0$ approaches $\mathbf{y}_0$ as t increases. Figure 2 illustrates an asymptotically stable equilibrium point. Thus, a spiral sink or a nodal sink is asymptotically stable. We will call any asymptotically stable equilibrium point a **sink**. An example of a stable equilibrium point that is not asymptotically stable is a dimension 2 linear system that has its center at the origin. The solutions are all periodic. They neither move away from nor approach the origin. Of course, any equilibrium point that is asymptotically stable is stable.

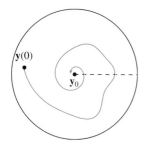

Figure 2. An asymptotically stable equilibrium point.

An equilibrium point $\mathbf{y}_0$ is said to be **unstable** if it is not stable. This requires that there are solution curves starting arbitrarily close to $\mathbf{y}_0$ that move far away as t increases. To be precise, $\mathbf{y}_0$ is unstable if there is an $\epsilon > 0$ such that for any $\delta > 0$ there is a solution $\mathbf{y}(t)$ with $|\mathbf{y}(0) - \mathbf{y}_0| < \delta$, but for which there are values of t with $|\mathbf{y}(t) - \mathbf{y}_0| > \epsilon$.

Examples of unstable equilibrium points include any spiral source or any nodal source. In these cases, every solution moves away as t increases. Another example is any saddle point. Remember that although a saddle has two stable solutions that approach the equilibrium point as $t \to \infty$, most solutions approach ∞.

Finally, we want to give a name to equilibrium points that are the opposite of asymptotically stable. A **source** is an equilibrium point at which *any* solution starting arbitrarily close to the equilibrium point moves away as t increases. Thus, any nodal source or spiral source is a source.

Clearly, a source is unstable, but not every unstable equilibrium point is a source. For example, a saddle point is unstable, but it is not a source, since the stable orbits approach the equilibrium point.

Using this language to reinterpret Theorem 7.4, we see that, for the linear system with constant coefficients $\mathbf{x}' = A\mathbf{x}$, the origin $\mathbf{0}$ is asymptotically stable if the real part of every eigenvalue of A is negative. On the other hand, if there is at least one eigenvalue with a positive real part, then $\mathbf{0}$ is unstable.

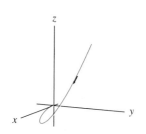

Figure 3. A solution to the system in Example 7.5.

Example 7.5 Consider the system $\mathbf{x}' = A\mathbf{x}$, where

$$A = \begin{pmatrix} -5/2 & -1 & 2 \\ -7/3 & -5/3 & 8/3 \\ -8/3 & -4/3 & 7/3 \end{pmatrix}.$$

Using a computer, we find that the eigenvalues are -1, $-1/2$, and $-1/3$. Since all the eigenvalues are negative, the origin is an asymptotically stable equilibrium point, or a sink. An example of a solution is shown plotted in phase space in Figure 3. ●

Example 7.6   Consider the system $\mathbf{x}' = A\mathbf{x}$, where

$$A = \begin{pmatrix} -8.1 & 0 & -4 \\ 18.2 & 1 & 8 \\ 20 & 0 & 7.9 \end{pmatrix}.$$

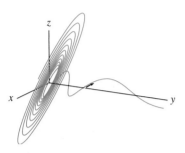

Figure 4. A solution to the system in Example 7.6.

The characteristic polynomial of A is $\lambda^3 - 0.8\lambda^2 + 15.81\lambda - 16.01$. It is easily checked that $\lambda = 1$ is an eigenvalue. The existence of a positive eigenvalue means that the origin is an unstable equilibrium point. The other eigenvalues are the complex conjugate pair $-0.1 \pm 4i$. These have a negative real part, but this fact cannot overcome the one positive eigenvalue. An example of a solution is shown plotted in phase space in Figure 4. The vector $\mathbf{v}_1 = (0, 1, 0)^T$ is an eigenvector for the eigenvalue 1, so the corresponding exponential solution is $e^t\mathbf{v}_1$. As Figure 4 shows, the spiral part of the solution coming from the complex eigenvalues dies out in time, and the solution approaches the y-axis.   ●

Notice that there is a wide class of systems that Theorem 7.4 does not address. Any system for which all eigenvalues have nonpositive real parts and that has at least one eigenvalue with a zero real part does not meet the hypotheses of either of the parts. Such systems have to be examined individually.

EXERCISES

In Exercises 1–8, use Theorem 7.4 to classify the equilibrium point at the origin as unstable, stable, or asymptotically stable. Verify your findings by sketching a phase portrait of the system.

1. $x' = -0.2x + 2.0y$
$y' = -2.0x - 0.2y$

2. $x' = 4x$
$y' = 3x + y$

3. $x' = -6x - 15y$
$y' = 3x + 6y$

4. $x' = 2x$
$y' = -3x - y$

5. $\mathbf{y}' = \begin{pmatrix} 0.1 & 2.0 \\ -2.0 & 0.1 \end{pmatrix} \mathbf{y}$

6. $\mathbf{y}' = \begin{pmatrix} -0.2 & 0.0 \\ -0.1 & -0.1 \end{pmatrix} \mathbf{y}$

7. $\mathbf{y}' = \begin{pmatrix} 1 & -4 \\ 1 & -3 \end{pmatrix} \mathbf{y}$

8. $\mathbf{y}' = \begin{pmatrix} 2 & -1 \\ 1 & 0 \end{pmatrix} \mathbf{y}$

In Exercises 9–16, use Theorem 7.4 to classify the equilibrium point at the origin as unstable, stable, or asymptotically stable. Compute and plot three solutions with randomly chosen initial values to check your result.

9. $\mathbf{y}' = \begin{pmatrix} -3 & -4 & 2 \\ -2 & -7 & 4 \\ -3 & -8 & 4 \end{pmatrix} \mathbf{y}$

10. $\mathbf{y}' = \begin{pmatrix} -3 & -1 & 0 \\ 2 & 0 & 0 \\ -6 & -1 & 3 \end{pmatrix} \mathbf{y}$

11. $x' = -x + 3y + 4z$
$y' = y + 6z$
$z' = -3y - 5z$

12. $x' = 2x + y$
$y' = -2x$
$z' = -4x - 6y - 2z$

13. $\mathbf{y}' = \begin{pmatrix} 0 & 0 & -1 \\ -1 & 0 & 0 \\ 4 & -2 & -3 \end{pmatrix} \mathbf{y}$

14. $\mathbf{y}' = \begin{pmatrix} 3 & -3 & -5 \\ 0 & 1 & 0 \\ 0 & -3 & -2 \end{pmatrix} \mathbf{y}$

15. $\mathbf{y}' = \begin{pmatrix} 3 & -2 & -5 & 3 \\ 16 & -6 & -17 & 9 \\ -14 & 5 & 15 & -8 \\ -19 & 8 & 23 & -13 \end{pmatrix} \mathbf{y}$

16. $\mathbf{y}' = \begin{pmatrix} -3 & 3 & 0 & -4 \\ 4 & -7 & 0 & 8 \\ 0 & -1 & -3 & 2 \\ 4 & -6 & 0 & 7 \end{pmatrix} \mathbf{y}$

17. Consider the system

$$x' = -3x,$$
$$y' = -2x - y,$$
$$z' = -2z.$$

(a) Find the eigenvalues and eigenvectors. Use your numerical solver to sketch the half-lines generated by the exponential solutions $\mathbf{y}(t) = C_i e^{\lambda_i t}\mathbf{v}_i$ for $i = 1, 2$, and 3.

(b) The six half-line solutions found in part (a) come in pairs that form straight lines. These three lines, taken two at a time, generate three planes. In turn these planes divide phase space into eight octants. Use your numerical solver to add solution trajectories with initial conditions in each of the eight octants.

(c) Based on your phase portrait, what would be an appropriate name for the equilibrium point in this case?

18. Consider the system

$$\mathbf{y}' = \begin{pmatrix} 1 & -1 & 0 \\ 0 & 2 & 0 \\ 0 & 0 & 3 \end{pmatrix} \mathbf{y}.$$

(a) Find the eigenvalues and eigenvectors. Use your numerical solver to sketch the half-lines generated by the exponential solutions $\mathbf{y}(t) = C_i e^{\lambda_i t} \mathbf{v}_i$ for $i = 1, 2,$ and 3.

(b) The six half-line solutions found in part (a) come in pairs that form straight lines. These three lines, taken two at a time, generate three planes. In turn these planes divide phase space into eight octants. Use your numerical solver to add solution trajectories with initial conditions in each of the eight octants.

(c) Based on your phase portrait, what would be an appropriate name for the equilibrium point in this case?

19. Consider the system

$$\mathbf{y}' = \begin{pmatrix} -1 & -10 & 0 \\ 10 & -1 & 0 \\ 0 & 0 & -1 \end{pmatrix} \mathbf{y}.$$

(a) Find the eigenvalues and eigenvectors and a fundamental set of solutions.

(b) What happens to solutions with initial condition on the z-axis? Why? Use your numerical solver to start a phase portrait with trajectories having initial conditions $\mathbf{y}(0) = (0, 0, 1)^T$ and $\mathbf{y}(0) = (0, 0, -1)^T$.

(c) What happens to solutions with initial conditions in the xy-plane? Why? Use your numerical solver to append the solution trajectory with initial condition $\mathbf{y}(0) = (1, 1, 0)^T$ to your phase portrait.

(d) What happens to solutions with initial conditions above the xy-plane? Below the xy-plane? Why? Use your numerical solver to append solutions with initial conditions $\mathbf{y}(0) = (1, 1, 1)^T$ and $\mathbf{y}(0) = (-1, -1, -1)^T$ to your phase portrait.

9.8 Higher-Order Linear Equations

A *linear equation* of order n is of the form

$$y^{(n)} + a_1(t)y^{(n-1)} + \cdots + a_{n-1}(t)y' + a_n(t)y = F(t). \tag{8.1}$$

An example is the equation of the vibrating spring that we derived in Chapter 4,

$$y'' + \frac{\mu}{m}y' + \frac{k}{m}y = \frac{1}{m}F(t), \tag{8.2}$$

where m is the mass of the spring, μ is the damping constant, k is the spring constant, and $F(t)$ is the external driving force.

The equivalent system

In Section 8.1, we learned that we can replace the single higher-order equation (8.1) with an equivalent first-order system of dimension n. To do so, we introduce new variables

$$x_1 = y, \ x_2 = y', \ \ldots, \ x_n = y^{(n-1)}. \tag{8.3}$$

Following the procedure from Section 8.1, we find that these variables satisfy the system of equations

$$
\begin{aligned}
x_1' &= x_2 \\
x_2' &= x_3 \\
&\ \vdots \\
x_{n-1}' &= x_n \\
x_n' &= -a_1(t)x_n - \cdots - a_{n-1}(t)x_2 - a_n(t)x_1 + F(t).
\end{aligned}
\tag{8.4}
$$

The system in (8.4) is a system of n equations involving the n unknown functions $x_1, x_2, \cdots, x_n$, which is equivalent to the single equation in (8.1) in the sense that the function $y(t)$ is a solution to (8.1) if and only if the vector-valued function

$$\mathbf{x}(t) = \begin{pmatrix} x_1(t) \\ x_2(t) \\ \vdots \\ x_n(t) \end{pmatrix} = \begin{pmatrix} y(t) \\ y'(t) \\ \vdots \\ y^{(n-1)}(t) \end{pmatrix} \tag{8.5}$$

is a solution to the system in (8.4). Notice, in particular, that if $\mathbf{x}(t)$ is a solution to the system in (8.4), then it is the first component of $\mathbf{x}$, $y(t) = x_1(t)$, which is the solution to the higher-order equation in (8.1).

Using matrix notation, (8.4) becomes

$$\mathbf{x}' = A\mathbf{x} + \mathbf{f}, \tag{8.6}$$

where

$$A = \begin{pmatrix} 0 & 1 & 0 & \cdots & 0 \\ 0 & 0 & 1 & \cdots & 0 \\ \vdots & \vdots & \vdots & \ddots & \vdots \\ 0 & 0 & 0 & \cdots & 1 \\ -a_n(t) & -a_{n-1}(t) & -a_{n-2}(t) & \cdots & -a_1(t) \end{pmatrix} \tag{8.7}$$

and $\mathbf{f}(t) = (0, 0, \ldots, F(t))^T$. The linear equation in (8.1) is homogeneous if $F(t) = 0$. Thus, the equation is homogeneous if and only if the corresponding system (8.6) is homogeneous.

The intimate connection between higher-order linear equations and first-order linear systems will enable us to translate every result we have obtained for first-order systems into a result for higher-order linear equations. Our first example of this will be the formulation of the initial value problem. Notice that, for the system (8.6), it is necessary to specify the vector $\mathbf{x}(t_0)$ as the initial condition. Given the definition of $\mathbf{x}$ in (8.5), the initial value problem for the higher-order equation in (8.1) requires us to specify the values of the unknown function and all of its derivatives up to order $n - 1$. Thus, the initial value problem for a higher-order equation is properly stated as

$$y^{(n)} + a_1(t)y^{(n-1)} + \cdots + a_{n-1}(t)y' + a_n(t)y = F(t), \quad \text{with}$$
$$y(t_0) = y_0, \ y'(t_0) = y_1, \ \ldots, \text{ and } y^{(n-1)}(t_0) = y_{n-1}. \tag{8.8}$$

Existence and uniqueness

Next, we can use the results about systems to find an appropriate existence and uniqueness theorem for a higher-order linear equation. The appropriate theorem follows immediately from Theorem 3.4 in Section 8.3, applied to the equivalent system.

THEOREM 8.9 Suppose the coefficients of the equation

$$\frac{d^n y}{dt^n} + a_1(t)\frac{d^{n-1}y}{dt^{n-1}} + \cdots + a_{n-1}(t)\frac{dy}{dt} + a_n(t)y = F(t) \tag{8.10}$$

are continuous functions of t in an interval (α, β). Then, for any $t_0 \in (\alpha, \beta)$, and for any constants $y_0, y_1, \cdots, y_{n-1}$, equation (8.10), together with the initial conditions

$$y(t_0) = y_0, \ y'(t_0) = y_1, \ \ldots, \text{ and } y^{(n-1)}(t_0) = y_{n-1},$$

has a unique solution defined for all $t \in (\alpha, \beta)$. ■

The linear structure of the space of solutions

We know that $y(t)$ is a solution to the homogeneous, higher-order equation

$$y^{(n)} + a_1(t)y^{(n-1)} + \cdots + a_{n-1}(t)y' + a_n(t)y = 0 \qquad (8.11)$$

if and only if $\mathbf{x}(t) = (y(t), y'(t), \ldots, y^{(n-1)})^T$ is a solution to the corresponding system

$$\mathbf{x}' = A\mathbf{x}, \qquad (8.12)$$

where A is the matrix defined in (8.7). Notice, in particular, that the first component of the vector $\mathbf{x}$ is the solution y to (8.11).

THEOREM 8.13   Suppose that $y_1(t), y_2(t), \ldots,$ and $y_k(t)$ are all solutions to (8.11). Then any linear combination of these functions is also a solution.

Proof   Let

$$\mathbf{x}_j(t) = \begin{pmatrix} y_j(t) \\ y_j'(t) \\ \vdots \\ y_j^{(n-1)}(t) \end{pmatrix} \quad \text{for } j = 1, \ldots, k. \qquad (8.14)$$

For each j, $\mathbf{x}_j$ is a solution to the system (8.12). According to Theorem 5.5 in Chapter 8, any linear combination of the vectors $\mathbf{x}_j$ is also a solution to (8.12). Let $C_1, \ldots,$ and C_k be constants. Then $\mathbf{x} = C_1\mathbf{x}_1 + \cdots + C_k\mathbf{x}_k$ is a solution to (8.12). This means that the first component of $\mathbf{x}$ is a solution to (8.11). Since the first component of $\mathbf{x}$ is $y = C_1y_1 + \cdots + C_ky_k$, the theorem is proved. ∎

Theorem 8.13 can also be proved directly. You are encouraged to prove it yourself.

Linear independence and dependence of functions

Before going further, we need to discuss the notion of linear dependence and independence for scalar-valued functions.

DEFINITION 8.15   Suppose the functions $y_1(t), y_2(t), \ldots,$ and $y_n(t)$ are all defined on the interval (α, β). The functions are *linearly dependent* if there are constants $c_1, c_2, \ldots,$ and c_n, not all of them equal to 0, such that

$$c_1 y_1(t) + c_2 y_2(t) + \cdots + c_n y_n(t) = 0,$$

for all $t \in (\alpha, \beta)$. The functions are *linearly independent* if they are not linearly dependent.

E x a m p l e 8 . 1 6   Show that the functions $\sin t$ and $\cos t$ are linearly independent.

To show this, suppose they are not. Then there are constants a and b, with at least one of them not equal to 0, such that $a \cos t + b \sin t = 0$ for all t. Evaluate this equation, first at $t = 0$, and then at $t = \pi/2$. The first gives us

$$0 = a \cos 0 + b \sin 0 = a.$$

In a similar manner, the second yields $b = 0$. This contradiction proves that the functions are linearly independent. $\bullet$

Example 8.17 Let ϕ be a fixed number and show that the functions $\sin t$, $\cos t$, and $\cos(t + \phi)$ are linearly dependent.

The addition law for the cosine says that

$$\cos(t + \phi) = \cos \phi \cos t - \sin \phi \sin t$$

or

$$\cos \phi \cos t - \sin \phi \sin t - \cos(t + \phi) = 0.$$

Since the coefficient of $\cos(t + \phi)$ is nonzero, this is a nontrivial linear combination that is equal to 0 for all t. Therefore, the functions are linearly dependent. $\bullet$

It is fairly easy to decide if two functions are linearly independent or dependent. If y_1 and y_2 are dependent, there are constants C_1 and C_2, not both of which are equal to 0, such that $C_1 y_1(t) + C_2 y_2(t) = 0$. Suppose, without loss of generality, that $C_1 \neq 0$ and solve for y_1 to get

$$y_1(t) = -\frac{C_2}{C_1} y_2(t).$$

Thus, y_1 is a constant multiple of y_2. Consequently, two functions are linearly dependent if and only if one is a constant multiple of the other. For example, $e^t = e^{-t} \cdot e^{2t}$. While e^t is a multiple of e^{2t}, the factor e^{-t} is not a constant. Therefore, e^t and e^{2t} are independent. We used this idea in Section 4.1, where we discussed second-order linear equations.

We need a quick way to decide when n functions are linearly independent for $n \geq 3$. This is not always easy, but when they are all solutions to the same linear, homogeneous equation like (8.11), it is easier. There is a way that will always work, although it does require some computation. Once more, this method relates to the fact that if y is a solution to (8.11), then the vector-valued function $\mathbf{x} = (y, y', \ldots, y^{(n-1)})^T$ is a solution to the equivalent system (8.12).

Let $y_1(t), y_2(t), \ldots, y_n(t)$ be n solutions to the higher-order equation (8.11) and let $\mathbf{x}_1, \mathbf{x}_2, \ldots, \mathbf{x}_n$ be the corresponding solutions to the first-order system (8.12). Remember that $\mathbf{x}_j = (y_j, y'_j, \ldots, y_j^{(n-1)})^T$. If the functions $y_1(t), y_2(t), \ldots,$ and $y_n(t)$ are linearly dependent, there are constants $c_1, c_2, \ldots,$ and c_n, not all of them equal to 0, such that

$$c_1 y_1(t) + c_2 y_2(t) + \cdots + c_n y_n(t) = 0$$

for all t. If we differentiate this formula repeatedly, we see that

$$c_1 y_1^{(j)}(t) + c_2 y_2^{(j)}(t) + \cdots + c_n y_n^{(j)}(t) = 0$$

for every j. Of course, this implies that

$$c_1 \mathbf{x}_1(t) + c_2 \mathbf{x}_2(t) + \cdots + c_n \mathbf{x}_n(t) = 0$$

for all t, so $\mathbf{x}_1(t)$, $\mathbf{x}_2(t)$, ..., and $\mathbf{x}_n(t)$ are linearly dependent for all t.

Conversely, suppose that the solutions $\mathbf{x}_1(t)$, $\mathbf{x}_2(t)$, ..., and $\mathbf{x}_n(t)$ are linearly dependent for some t_0. According to Proposition 5.12 in Chapter 8, there are constants $c_1, c_2, \ldots$, and c_n, not all of them equal to 0, such that

$$c_1\mathbf{x}_1(t) + c_2\mathbf{x}_2(t) + \cdots + c_n\mathbf{x}_n(t) = 0$$

for all t. If we look at the first component of this vector equation, we get

$$c_1 y_1(t) + c_2 y_2(t) + \cdots + c_n y_n(t) = 0$$

for all t, so $y_1, y_2, \ldots$, and y_n are linearly dependent.

Thus, $y_1, y_2, \ldots$, and y_n are linearly independent if and only if the corresponding vector-valued functions $\mathbf{x}_1, \mathbf{x}_2, \ldots$, and $\mathbf{x}_n$ are. One way to check this is to look at the Wronskian. The **Wronskian** of $y_1, y_2, \ldots$, and y_n is defined to be the Wronskian of $\mathbf{x}_1, \mathbf{x}_2, \ldots$, and $\mathbf{x}_n$,

$$W(t) = \det \begin{pmatrix} y_1(t) & y_2(t) & \cdots & y_n(t) \\ y_1'(t) & y_2'(t) & \cdots & y_n'(t) \\ \vdots & \vdots & \ddots & \vdots \\ y_1^{(n-1)}(t) & y_2^{(n-1)}(t) & \cdots & y_n^{(n-1)}(t) \end{pmatrix}. \tag{8.18}$$

If the Wronskian is nonzero at one point, $\mathbf{x}_1, \mathbf{x}_2, \ldots$, and $\mathbf{x}_n$ are linearly independent at that point. By Proposition 5.12 in Chapter 8, they are linearly independent as functions. We will summarize our discussion with Proposition 8.19.

PROPOSITION 8.19 The solutions $y_1(t)$, $y_2(t)$, ..., $y_n(t)$ to equation (8.11) are linearly independent if and only if the corresponding solutions to the system (8.12) are linearly independent. This, in turn, is equivalent to $W(t_0) \neq 0$ for some t_0. ●

Example 8.20 It can be verified by substitution that $y_1(t) = e^{-2t}\cos 2t$, $y_2(t) = e^{-2t}\sin 2t$, and $y_3(t) = e^{-3t}$ are all solutions to

$$y''' + 7y'' + 20y' + 24y = 0.$$

Are they linearly independent?

Since the equation has dimension 3, the best way to answer this question is to use the Wronskian,

$$W(t) = \det \begin{pmatrix} y_1 & y_2 & y_3 \\ y_1' & y_2' & y_3' \\ y_1'' & y_2'' & y_3'' \end{pmatrix}$$

$$= \det \begin{pmatrix} e^{-2t}\cos 2t & e^{-2t}\sin 2t & e^{-3t} \\ -2e^{-2t}(\cos 2t + \sin 2t) & 2e^{-2t}(\cos 2t - \sin 2t) & -3e^{-3t} \\ 8e^{-2t}\sin 2t & -8e^{-2t}\cos 2t & 9e^{-3t} \end{pmatrix}.$$

At $t = 0$,

$$W(0) = \det \begin{pmatrix} 1 & 0 & 1 \\ -2 & 2 & -3 \\ 0 & -8 & 9 \end{pmatrix}.$$

We compute that $W(0) = 10$. Since this is not zero, the solutions are linearly independent. ●

Structure of the general solution

Finally, the correspondence between the higher-order equation in (8.11) and the corresponding system in (8.12) allows us to deduce the next theorem, which describes the structure of the space of solutions to (8.11).

THEOREM 8.21 Suppose that $y_1(t)$, $y_2(t)$, ..., and $y_n(t)$ are linearly independent solutions to equation (8.11). Then every solution to (8.11) is a linear combination of $y_1(t)$, $y_2(t)$, ..., and $y_n(t)$. ∎

The solution strategy for a higher-order equation is almost the same as that for a system. To find the general solution to the nth-order linear, homogeneous equation (8.11), we need to find a ***fundamental set of solutions***; that is, a set of n solutions $y_1(t)$, $y_2(t)$, ..., and $y_n(t)$ that are linearly independent. Then the general solution is a linear combination

$$y(t) = C_1 y_1(t) + C_2 y_2(t) + \cdots + C_n y_n(t). \tag{8.22}$$

To find the particular solution to (8.11) that satisfies the initial conditions $y(t_0) = y_0$, $y'(t_0) = y_1$, ..., and $y^{(n-1)} = y_{n-1}$, we substitute the initial conditions into (8.22). This gives us

$$C_1 y_1(t_0) + C_2 y_2(t_0) + \cdots + C_n y_n(t_0) = y_0,$$
$$C_1 y_1'(t_0) + C_2 y_2'(t_0) + \cdots + C_n y_n'(t_0) = y_1,$$
$$\vdots$$
$$C_1 y_1^{(n-1)}(t_0) + C_2 y_2^{(n-1)}(t_0) + \cdots + C_n y_n^{(n-1)}(t_0) = y_{n-1},$$

a system of n linear equations for the n unknown constants. Since the solutions are a fundamental set, these equations have a unique solution, which can be found by the methods we discussed in Chapter 7.

Example 8.23 Show that $y(t) = C_1 \cos t + C_2 \sin t + C_3 e^t$ is the general solution to

$$y''' - y'' + y' - y = 0.$$

Find the solution that satisfies the initial conditions $y(0) = 2$, $y'(0) = 1$, and $y''(0) = 0$.

Let $y_1(t) = \cos t$, $y_2(t) = \sin t$ and $y_3(t) = e^t$. Each of these functions is easily shown by substitution to be a solution to $y''' - y'' + y' - y = 0$. To see that they are linearly independent, we compute the Wronskian:

$$W(t) = \det \begin{pmatrix} y_1 & y_2 & y_3 \\ y_1' & y_2' & y_3' \\ y_1'' & y_2'' & y_3'' \end{pmatrix} = \det \begin{pmatrix} \cos t & \sin t & e^t \\ -\sin t & \cos t & e^t \\ -\cos t & -\sin t & e^t \end{pmatrix}.$$

It suffices to check the Wronskian at one point. At $t = 0$, we have

$$W(0) = \det \begin{pmatrix} 1 & 0 & 1 \\ 0 & 1 & 1 \\ -1 & 0 & 1 \end{pmatrix} = 2.$$

Since $W(0) \neq 0$, $y_1(t)$, $y_2(t)$, and $y_3(t)$ are linearly independent, the general solution is

$$y(t) = C_1 y_1(t) + C_2 y_2(t) + C_3 y_3(t) = C_1 \cos t + C_2 \sin t + C_3 e^t. \tag{8.24}$$

To solve the initial-value problem, we must find C_1, C_2, and C_3 so that the conditions are satisfied. Differentiating (8.24), we get

$$y'(t) = -C_1 \sin t + C_2 \cos t + C_3 e^t$$
$$y''(t) = -C_1 \cos t - C_2 \sin t + C_3 e^t.$$

Evaluating at $t = 0$, we get

$$2 = y(0) = C_1 + C_3,$$
$$1 = y'(0) = C_2 + C_3,$$
$$0 = y''(0) = -C_1 + C_3.$$

Solving these equations, we find that $C_1 = 1$, $C_2 = 0$, and $C_3 = 1$. Hence, the solution is

$$y(t) = \cos t + e^t.$$

Finding a fundamental set of solutions

Since we know that y is a solution to

$$\frac{d^n y}{dt^n} + a_1 \frac{d^{n-1} y}{dt^{n-1}} + \cdots + a_{n-1} \frac{dy}{dt} + a_n y = 0 \tag{8.25}$$

if and only if $\mathbf{x} = (y, y', \ldots, y^{(n-1)})^T$ is a solution to the system

$$\mathbf{x}' = A\mathbf{x}, \tag{8.26}$$

where A is the matrix in (8.7), we could just find a fundamental set of solutions for the system in (8.26) and use that set to find the fundamental set for (8.25). However, there is an easier procedure, and we will use the information gained from the solutions to the system to find it.

Recall that we solved systems by looking for exponential solutions. Suppose that $\mathbf{x}(t) = e^{\lambda t} \mathbf{v}$ is a solution of (8.26). Then the first component of $\mathbf{x}(t)$ is a solution to (8.25), and it is a constant multiple of $e^{\lambda t}$. So let's substitute $y(t) = e^{\lambda t}$ into (8.25). Since $y^{(k)}(t) = \lambda^k e^{\lambda t}$ for every integer k, we get

$$e^{\lambda t} \left(\lambda^n + a_1 \lambda^{n-1} + \cdots + a_{n-1} \lambda + a_n \right) = 0.$$

Since $e^{\lambda t} \neq 0$, we can cancel that factor, and we see that λ must be a root of the polynomial equation

$$p(\lambda) = \lambda^n + a_1 \lambda^{n-1} + \cdots + a_{n-1} \lambda + a_n = 0. \tag{8.27}$$

It is not hard to see that the characteristic polynomial of the system $\mathbf{x}' = A\mathbf{x}$ in (8.26) is $(-1)^n p(\lambda)$. We will call the polynomial $p(\lambda)$ in (8.27) the **characteristic polynomial** of the differential equation in (8.25). Equation (8.27) is called the **characteristic equation**. Thus, the roots of the characteristic equation of (8.25) are equal to the eigenvalues of the matrix A in system (8.26).

We have to give some consideration to the linear independence of the solutions we find. If we have n solutions y_1, y_2, ..., and y_n to (8.25), these give rise to n solutions $\mathbf{x}_1$, $\mathbf{x}_2$, ..., and $\mathbf{x}_n$ to the system in (8.25). According to Proposition 8.19, the functions y_1, y_2, ..., and y_n are linearly independent if and only if the functions $\mathbf{x}_1$, $\mathbf{x}_2$, ..., and $\mathbf{x}_n$ are linearly independent. But, according to Theorem 5.2, the functions $\mathbf{x}_1$, $\mathbf{x}_2$, ..., and $\mathbf{x}_n$ are linearly independent if the eigenvalues of A are distinct. Since the eigenvalues of A are the roots of the characteristic equation of (8.25), we see that if the characteristic equation has n distinct roots, then the exponential solutions are linearly independent.

Real roots

Corresponding to the real root λ, we have the solution $y(t) = e^{\lambda t}$. If the roots are distinct, this is all we need to know.

Example 8.28 Find a fundamental set of solutions to the equation

$$y''' - y' = 0.$$

The characteristic equation is $\lambda^3 - \lambda = 0$. The roots are $-1, 0,$ and 1. Thus,

$$y_1(t) = e^{-t}, \quad y_2(t) = e^{0t} = 1, \quad \text{and} \quad y_3(t) = e^t$$

are solutions. Since the roots of the characteristic polynomial are distinct, these solutions are linearly independent. Hence, they form a fundamental set of solutions. ●

Suppose λ is a root of the characteristic polynomial of algebraic multiplicity q. According to Section 9.6, we can find q linearly independent solutions of the system of the form

$$\mathbf{x}(t) = e^{\lambda t} \left(\mathbf{v} + t(A - \lambda I)\mathbf{v} + \cdots + \frac{t^{q-1}}{(q-1)!}(A - \lambda I)^{q-1}\mathbf{v} \right).$$

The first component of $\mathbf{x}$ is a solution to (8.25) and has the form $P(t)e^{\lambda t}$, where $P(t)$ is a polynomial of at most degree $q - 1$. Thus, we can find q polynomials $P_j(t)$, $j = 1, 2, \ldots, q$, of degree $q - 1$ or less, such that the functions $y_j(t) = P_j(t)e^{\lambda t}$ are solutions and are linearly independent.

This leads us to wonder if the functions

$$y(t) = t^k e^{\lambda t}, \quad \text{where} \quad k = 0, 1, \ldots, q - 1,$$

are solutions. After all, there are precisely q functions of this type. It turns out that these q functions are indeed linearly independent solutions of (8.25). To see this, we use some of the linear algebra we learned in Chapter 7. We will give the proof, but we will leave the verification of some details to the Exercises.

For $\mathbf{a} = (a_1, a_2, \ldots, a_q)^T \in \mathbf{R}^q$, we define

$$y_{\mathbf{a}}(t) = \left(a_1 + a_2 t + a_3 t^2 + \cdots + a_q t^{q-1} \right) e^{\lambda t}. \tag{8.29}$$

In other words, we assign to the vector $\mathbf{a}$, the function $y_{\mathbf{a}}(t) = p(t)e^{\lambda t}$, where $p(t)$ is the polynomial whose coefficients are the elements of $\mathbf{a}$. We define the subset $V \subset \mathbf{R}^q$ by saying that $\mathbf{a} \in V$ if $y_{\mathbf{a}}$ is a solution to (8.25). It is not difficult to verify that if $\mathbf{a}$ and $\mathbf{b}$ are vectors and α and β are numbers, then

$$y_{\alpha \mathbf{a} + \beta \mathbf{b}} = \alpha y_{\mathbf{a}} + \beta y_{\mathbf{b}}. \tag{8.30}$$

Because of the linear nature of the set of solutions to (8.25), Theorem 8.13, and (8.30), we can verify that

$$V \text{ is a subspace of } \mathbf{R}^q. \tag{8.31}$$

Now, consider the q linearly independent solutions of $y_j(t) = P_j(t)e^{\lambda t}$. For each j, let $\mathbf{a}_j$ be the vector of coefficients of the polynomial $P_j(t)$, so that $y_j = y_{\mathbf{a}_j}$. Then $\mathbf{a}_j \in V$ for $j = 1, \ldots, k$. It is easily verified that the vectors

$$\mathbf{a}_1, \mathbf{a}_2, \ldots, \mathbf{a}_k \text{ are linearly independent.} \tag{8.32}$$

Since V contains q linearly independent vectors, the dimension of V is at least q. But, since V is a subspace of $\mathbf{R}^q$, this means that $V = \mathbf{R}^q$.

Finally, we focus on the standard basis of $V = \mathbf{R}^q$, which consists of the vectors $\mathbf{e}_j$, $j = 1, 2, \dots, q$, where all of the entries in $\mathbf{e}_j$ are equal to 0 except for the jth, which is equal to 1. The corresponding functions

$$y_{\mathbf{e}_j}(t) = t^{j-1}e^{\lambda t}$$

are solutions to (8.25) and are linearly independent.

Let's summarize what we have learned about the solutions associated with the real roots.

THEOREM 8.33 If λ is a real root to the characteristic polynomial of algebraic multiplicity q, then

$$y_1(t) = e^{\lambda t}, \quad y_2(t) = te^{\lambda t}, \quad \dots, \quad \text{and} \quad y_q(t) = t^{q-1}e^{\lambda t}$$

are q linearly independent solutions. ■

Example 8.34 Find a fundamental set of solutions to the equation

$$y'''' - 2y''' + 2y' - 1 = 0.$$

The characteristic polynomial is

$$\lambda^4 - 2\lambda^3 + 2\lambda - 1 = (\lambda - 1)^3(\lambda + 1).$$

Thus, the roots are ± 1, and 1 has multiplicity 3. From the root $\lambda = -1$, we get the solution $y_1(t) = e^{-t}$. From the root $\lambda = 1$, we get three linearly independent solutions $y_2(t) = e^t$, $y_3(t) = te^t$, and $y_4(t) = t^2e^t$. ●

Complex roots

Since the equation in (8.25) has real coefficients, the characteristic polynomial does as well. Hence, complex roots come in complex conjugate pairs, $\lambda = \alpha + i\beta$ and $\overline{\lambda} = \alpha - i\beta$. We have the corresponding solutions

$$z(t) = e^{\lambda t} = e^{\alpha t}(\cos \beta t + i \sin \beta t) \quad \text{and} \quad \overline{z}(t) = e^{\overline{\lambda} t} = e^{\alpha t}(\cos \beta t - i \sin \beta t).$$

To find the real solutions, we take the real and imaginary parts of $z(t)$ and $\overline{z}(t)$. Thus,

$$x(t) = e^{\alpha t}\cos \beta t \quad \text{and} \quad y(t) = e^{\alpha t}\sin \beta t$$

are real solutions.

Example 8.35 Find a fundamental set of solutions to

$$y''' + 7y'' + 19y' + 13y = 0.$$

The characteristic polynomial is

$$\lambda^3 + 7\lambda^2 + 19\lambda + 13 = (\lambda + 1)(\lambda^2 + 6\lambda + 13).$$

Thus, the roots are -1, and $-3 \pm 2i$. The real root -1 leads us to the solution $y_1(t) = e^{-t}$. The complex conjugate pair $-3 \pm 2i$ yields the real solutions

$$y_2(t) = e^{-3t}\cos 2t \quad \text{and} \quad y_3(t) = e^{-3t}\sin 2t.$$
●

Now suppose that $\lambda = \alpha + i\beta$ is a complex root of multiplicity q. Then $\bar{\lambda} = \alpha - i\beta$ is also a complex root of multiplicity q. By the same reasoning we used in the real case, we get complex conjugate pairs of solutions

$$z(t) = t^k e^{\lambda t} \quad \text{and} \quad \bar{z}(t) = t^k e^{\bar{\lambda} t}, \quad \text{where} \quad k = 0, 1, \ldots, q - 1.$$

To find the real solutions, we take the real and imaginary parts,

$$x(t) = t^k e^{\alpha t} \cos \beta t \quad \text{and} \quad y(t) = t^k e^{\alpha t} \sin \beta t, \quad \text{where} \quad k = 0, 1, \ldots, q - 1.$$

Let's summarize the facts about the solutions associated with the complex roots.

THEOREM 8.36   If $\lambda = \alpha + i\beta$ is a complex root of the characteristic polynomial with multiplicity q, then so is $\bar{\lambda} = \alpha - i\beta$. In addition,

$$x_1(t) = e^{\alpha t} \cos \beta t, \ x_2(t) = t e^{\alpha t} \cos \beta t, \ \ldots,$$
$$\text{and} \ x_q(t) = t^{q-1} e^{\alpha t} \cos \beta t$$
$$y_1(t) = e^{\alpha t} \sin \beta t, \ y_2(t) = t e^{\alpha t} \sin \beta t, \ \ldots,$$
$$\text{and} \ y_q(t) = t^{q-1} e^{\alpha t} \sin \beta t$$

are $2q$ linearly independent solutions.   ■

Example 8.37   Find a fundamental set of solutions to

$$y'''' + 4y''' + 14y'' + 20y' + 25y = 0.$$

The characteristic polynomial is

$$\lambda^4 + 4\lambda^3 + 14\lambda^2 + 20\lambda + 25 = (\lambda^2 + 2\lambda + 5)^2.$$

Consequently, we have roots $-1 \pm 2i$, each of multiplicity 2. Thus, we have solutions

$$y_1(t) = e^{-t} \cos 2t, \quad y_2(t) = e^{-t} \sin 2t, \quad y_3(t) = t e^{-t} \cos 2t,$$
$$\text{and} \quad y_4(t) = t e^{-t} \sin 2t.$$ ●

EXERCISES

1. The function $y(t)$ is a solution of the homogeneous equation $y'' - 2y' - 3y = 0$ if and only if

$$\mathbf{x}(t) = \begin{pmatrix} x_1(t) \\ x_2(t) \end{pmatrix} = \begin{pmatrix} y(t) \\ y'(t) \end{pmatrix}$$

is a solution of

$$\mathbf{x}' = \begin{pmatrix} 0 & 1 \\ 3 & 2 \end{pmatrix} \mathbf{x}.$$

(a) Use direct substitution to show that

$$\mathbf{x}_1(t) = \begin{pmatrix} e^{3t} \\ 3e^{3t} \end{pmatrix} \quad \text{and} \quad \mathbf{x}_2(t) = \begin{pmatrix} e^{-t} \\ -e^{-t} \end{pmatrix}$$

are solutions of system (1). Show that $\mathbf{x}_1(t)$ and $\mathbf{x}_2(t)$ are linearly independent.

(b) Use direct substitution to show that the first component of the general solution $\mathbf{x}(t) = C_1\mathbf{x}_1(t) + C_2\mathbf{x}_2(t)$ is a solution of $y'' - 2y' - 3y = 0$.

2. The function $y(t)$ is a solution of the homogeneous equation $y'' + 4y = 0$ if and only if

$$\mathbf{x}(t) = \begin{pmatrix} x_1(t) \\ x_2(t) \end{pmatrix} = \begin{pmatrix} y(t) \\ y'(t) \end{pmatrix}$$

is a solution of

$$\mathbf{x}' = \begin{pmatrix} 0 & 1 \\ -4 & 0 \end{pmatrix} \mathbf{x}.$$

(a) Use direct substitution to show that

$$\mathbf{x}_1(t) = \begin{pmatrix} \sin 2t \\ 2\cos 2t \end{pmatrix} \quad \text{and} \quad \mathbf{x}_2(t) = \begin{pmatrix} \cos 2t \\ -2\sin 2t \end{pmatrix}$$

are solutions of system (2). Show that $\mathbf{x}_1(t)$ and $\mathbf{x}_2(t)$ are linearly independent.

(b) Use direct substitution to show that the first component of the general solution $\mathbf{x}(t) = C_1\mathbf{x}_1(t) + C_2\mathbf{x}_2(t)$ is a solution of $y'' + 4y = 0$.

Use Definition 8.15 and the technique of Example 8.16 to show that each set of functions in Exercises 3–6 is linearly independent.

3. $y_1(t) = e^t$ and $y_2(t) = e^{2t}$

4. $y_1(t) = e^t \cos t$ and $y_2(t) = e^t \sin t$

5. $y_1(t) = \cos t$, $y_2(t) = \sin t$, and $y_3(t) = e^t$

6. $y_1(t) = e^t$, $y_2(t) = te^t$, and $y_3(t) = t^2 e^t$

In Exercises 7–12, use Proposition 8.19 and the technique of Example 8.20 to show that the given solutions are linearly independent and form a fundamental set of solutions of the given equation.

7. The equation $y'' + 9y = 0$ has solutions $y_1(t) = \cos 3t$ and $y_2(t) = \sin 3t$.

8. The equation $y'' + 9y' - 10y = 0$ has solutions $y_1(t) = e^{-10t}$ and $y_2(t) = e^t$.

9. The equation $y'' - 4y' + 4y = 0$ has solutions $y_1(t) = e^{2t}$ and $y_2(t) = te^{2t}$.

10. The equation $y''' - 3y'' + 9y' - 27y = 0$ has solutions $y_1(t) = \cos 3t$, $y_2(t) = \sin 3t$, and $y_3(t) = e^{3t}$.

11. The equation $y''' - 3y'' + 3y' - y = 0$ has solutions $y_1(t) = e^t$, $y_2(t) = te^t$, and $y_3(t) = t^2 e^t$.

12. The equation $y^{(4)} - 13y'' + 36y = 0$ has solutions $y_1(t) = \cos 3t$, $y_2(t) = \sin 3t$, $y_3(t) = \cos 2t$, and $y_4(t) = \sin 2t$.

13. Consider the equation

$$y''' + ay'' + by' + cy = 0. \tag{8.38}$$

(a) If $e^{\lambda t}$ is a solution of equation (8.38), provide details showing that $\lambda^3 + a\lambda^2 + b\lambda + c = 0$.

(b) Write the third-order equation (8.38) as a system of first-order equations, placing your answer in the form $\mathbf{x}' = A\mathbf{x}$. Calculate the characteristic polynomial of system $\mathbf{x}' = A\mathbf{x}$ and compare it with the characteristic polynomial of the equation in (8.38).

Each equation in Exercises 14–21 has a characteristic equation possessing distinct real roots. Find the general solution of each equation.

14. $y''' - 2y'' - y' + 2y = 0$ **15.** $y''' - 3y'' = 4y' - 12y$

16. $y^{(4)} - 5y'' + 4y = 0$ **17.** $y^{(4)} + 36y = 13y''$

18. $y''' + 2y'' - 5y' - 6y = 0$ **19.** $y''' + 30y = 4y'' + 11y'$

20. $y^{(5)} + 3y^{(4)} - 5y''' - 15y'' + 4y' + 12y = 0$

21. $y^{(5)} - 4y^{(4)} - 13y''' + 52y'' + 36y' - 144y = 0$

Each equation in Exercises 22–27 has a characteristic equation possessing real roots of various multiplicities. Find the general solution of each equation.

22. $y''' - 3y' + 2y = 0$ **23.** $y''' + y'' = 8y' + 12y$

24. $y''' + 6y'' + 12y' + 8y = 0$

25. $y''' + 3y'' + 3y' + y = 0$

26. $y^{(5)} + 3y^{(4)} - 6y''' - 10y'' + 21y' - 9y = 0$

27. $y^{(5)} - y^{(4)} - 6y''' + 14y'' - 11y' + 3y = 0$

Each equation in Exercises 28–33 has a characteristic equation possessing some complex zeros, some of which are repeated. Find the general solution of each equation.

28. $y''' - y'' + 4y' - 4y = 0$ **29.** $y''' + 2y = y''$

30. $y^{(4)} + 17y'' + 16y = 0$ **31.** $y^{(4)} + y = -2y''$

32. $y^{(5)} - 9y^{(4)} + 34y''' - 66y'' + 65y' - 25y = 0$

33. $y^{(6)} + 3y^{(4)} + 3y'' + y = 0$

Find the solution of each initial-value problem presented in Exercises 34–43.

34. $y'' - 2y' - 3y = 0$, with $y(0) = 4$ and $y'(0) = 0$

35. $y'' + 2y' + 5y = 0$, with $y(0) = 2$ and $y'(0) = 0$

36. $y'' + 4y' + 4y = 0$, with $y(0) = 2$ and $y'(0) = -1$

37. $y'' - 2y' + y = 0$, with $y(0) = 1$ and $y'(0) = 0$

38. $y''' - 4y'' - 7y' + 10y = 0$, with $y(0) = 1$, $y'(0) = 0$, and $y''(0) = -1$

39. $y''' - 7y'' + 11y' - 5y = 0$, with $y(0) = -1$, $y'(0) = 1$, and $y''(0) = 0$

40. $y''' - 2y' + 4y = 0$, $y(0) = 1$, with $y'(0) = -1$, and $y''(0) = 0$

41. $y''' - 6y'' + 12y' - 8y = 0$, with $y(0) = -2$, $y'(0) = 0$, and $y''(0) = 2$

42. $y''' - 3y' + 52y = 0$, $y(0) = 0$, with $y'(0) = -1$, and $y''(0) = 2$

43. $y^{(4)} + 8y'' + 16y = 0$, with $y(0) = 0$, $y'(0) = -1$, $y''(0) = 2$, and $y'''(0) = 0$

In Exercises 44–46, we will verify some of the steps leading to the proof of Theorem 8.33.

44. Verfiy (8.30).

45. Verfiy (8.31). See the definition of a subspace in Definition 5.5 in Section 7.5.

46. Verfiy (8.32). *Hint*: Suppose that there are constants $c_1, \ldots, c_k$ such that $c_1\mathbf{a}_1 + \cdots + c_k\mathbf{a}_k = \mathbf{0}$. Show that $c_1y_1(t) + \cdots + c_k y_k(t) = 0$. Conclude that all of the constants must be equal to zero.

9.9 Inhomogeneous Linear Systems

Inhomogeneous linear systems are systems of the form

$$\mathbf{y}' = A(t)\mathbf{y} + \mathbf{f}(t), \tag{9.1}$$

where A is an $n \times n$ matrix, $\mathbf{y}$ is the column vector of unknown functions, and $\mathbf{f}$ is a column vector of known functions. Remember that $\mathbf{f}$ is called the inhomogeneous, or forcing, term. Our solution strategy comes from understanding the structure of the general solution that is contained in Theorem 9.2.

THEOREM 9.2 Suppose that $\mathbf{y}_p$ is a particular solution to the inhomogeneous equation (9.1) and that $\mathbf{y}_1, \mathbf{y}_2, \ldots,$ and $\mathbf{y}_n$ form a fundamental set of solutions to the associated homogeneous equation

$$\mathbf{y}' = A(t)\mathbf{y}. \tag{9.3}$$

Then the general solution to the inhomogeneous equation (9.1) is given by

$$\mathbf{y} = \mathbf{y}_p + C_1\mathbf{y}_1 + C_2\mathbf{y}_2 + \cdots + C_n\mathbf{y}_n,$$

where C_1, C_2, and C_n are arbitrary constants.

Notice that the general solution can be written as $\mathbf{y} = \mathbf{y}_p + \mathbf{y}_h$, where $\mathbf{y}_h = C_1\mathbf{y}_1 + C_2\mathbf{y}_2 + \cdots + C_n\mathbf{y}_n$ is the general solution to the corresponding homogeneous system. Thus, to find the general solution to the inhomogeneous system, we first find the general solution, $\mathbf{y}_h$, to the corresponding homogeneous system (9.3). Next, we find a particular solution, $\mathbf{y}_p$, to the inhomogeneous equation. The general solution to the inhomogeneous equation is then $\mathbf{y} = \mathbf{y}_h + \mathbf{y}_p$.

Proof Suppose that $\mathbf{y}$ is a solution to (9.1). We are given that $\mathbf{y}_p$ is also a solution, so we have the two equations

$$\mathbf{y}' = A\mathbf{y} + \mathbf{f}, \quad \text{and}$$
$$\mathbf{y}'_p = A\mathbf{y}_p + \mathbf{f}.$$

Subtracting and using the linearity of matrix multiplication and differentiation, we get

$$(\mathbf{y} - \mathbf{y}_p)' = A(\mathbf{y} - \mathbf{y}_p).$$

Therefore, $\mathbf{y} - \mathbf{y}_p$ is a solution to the associated homogeneous equation (9.3). Since $\mathbf{y}_1, \mathbf{y}_2, \ldots,$ and $\mathbf{y}_n$ form a fundamental set of solutions, there are constants $C_1, C_2, \ldots,$ and C_n such that $\mathbf{y} - \mathbf{y}_p = C_1\mathbf{y}_1 + C_2\mathbf{y}_2 + \cdots + C_n\mathbf{y}_n$. Consequently,

$$\mathbf{y} = \mathbf{y}_p + C_1\mathbf{y}_1 + C_2\mathbf{y}_2 + \cdots + C_n\mathbf{y}_n,$$

as promised. ■

For systems with constant coefficients, we already know how to solve the homogeneous equation, so in this section we will concentrate on finding one particular solution to an inhomogeneous system.

Fundamental matrices

We start with a fundamental set of solutions, $\mathbf{y}_1, \mathbf{y}_2, \ldots,$ and $\mathbf{y}_n$ to the associated homogeneous system $\mathbf{y}' = A\mathbf{y}$. Let $Y(t)$ be the $n \times n$ matrix whose ith column is $\mathbf{y}_i(t)$, for $1 \leq i \leq n$. Thus,

$$Y = [\mathbf{y}_1, \mathbf{y}_2, \ldots, \mathbf{y}_n].$$

A matrix like Y, whose columns form a fundamental set of solutions for the system $\mathbf{y}' = A\mathbf{y}$, is called a ***fundamental matrix.***

Since $\mathbf{y}'_i = A\mathbf{y}_i$ for each i, we have

$$Y' = [\mathbf{y}'_1, \mathbf{y}'_2, \ldots, \mathbf{y}'_n] = [A\mathbf{y}_1, A\mathbf{y}_2, \ldots, A\mathbf{y}_n] = AY, \tag{9.4}$$

where the product AY is matrix multiplication.

Notice also that $Y(t)$ is invertible for all t, since its column vectors are linearly independent, a property of any fundamental set of solutions. We can now state a proposition that allows us to see fundamental matrices in a slightly different light.

PROPOSITION 9.5   A matrix-valued function $Y(t)$ is a fundamental matrix for the system $\mathbf{y}' = A\mathbf{y}$ if and only if

$$Y' = AY \quad \text{and} \quad Y_0 = Y(t_0) \text{ is invertible for some } t_0.$$

Proof   We have already shown that a fundamental matrix has these two properties. To prove the converse, notice that $Y' = AY$ means that $\mathbf{y}'_j = A\mathbf{y}_j$ for every column vector in Y. Thus, the n columns of Y are solutions to the homogeneous system. Since $Y(t_0)$ is invertible, these solutions are linearly independent at t_0. Thus, by Proposition 5.12 of Chapter 8, the column vectors are linearly independent solutions and form a fundamental set of solutions.   ●

Remember from Proposition 6.12 that the exponential e^{tA} satisfies $de^{tA}/dt = Ae^{tA}$. In addition, if we evaluate e^{tA} at $t = 0$, we get $e^{0A} = I$, which is certainly invertible. Thus, the exponential e^{tA} is an example of a fundamental matrix for the system $\mathbf{y}' = A\mathbf{y}$.

Variation of parameters for systems

To find a particular solution to the inhomogeneous equation in (9.1) by variation of parameters, we look for a solution of the form

$$\mathbf{y}_p(t) = Y(t)\mathbf{v}(t), \tag{9.6}$$

where $\mathbf{v}(t) = (v_1(t), \ldots, v_n(t))^T$ is a column vector of functions to be determined, and $Y = [\mathbf{y}_1, \mathbf{y}_2, \ldots, \mathbf{y}_n]$ is a fundamental matrix. Using matrix multiplication,

$$\mathbf{y}_p = [\mathbf{y}_1, \mathbf{y}_2, \ldots, \mathbf{y}_n] \begin{pmatrix} v_1 \\ \vdots \\ v_n \end{pmatrix} = v_1 \mathbf{y}_1 + \cdots + v_n \mathbf{y}_n. \tag{9.7}$$

Written in this way, we see that $\mathbf{y}_p$ is a linear combination of the column vectors of Y, a fundamental set of solutions to the homogeneous system, with coefficients that are the unknown functions $v_1(t), \ldots,$ and $v_n(t)$. Notice the similarity of this expression with that used in Section 4.6 of Chapter 4 for a single equation.

We want $\mathbf{y}_p$ to be a solution to the inhomogeneous equation $\mathbf{y}'_p = A\mathbf{y}_p + \mathbf{f}$, so we compute both sides. If we differentiate (9.6) and use (9.4), we get

$$\mathbf{y}'_p = Y\mathbf{v}' + Y'\mathbf{v} = Y\mathbf{v}' + (AY)\mathbf{v}. \tag{9.8}$$

On the other hand,

$$A\mathbf{y}_p + \mathbf{f} = A(Y\mathbf{v}) + \mathbf{f} = (AY)\mathbf{v} + \mathbf{f}. \tag{9.9}$$

The function $\mathbf{y}_p$ will be a solution to $\mathbf{y}'_p = A\mathbf{y}_p + \mathbf{f}$ provided that the right-hand sides of (9.8) and (9.9) are equal,

$$Y\mathbf{v}' + (AY)\mathbf{v} = (AY)\mathbf{v} + \mathbf{f},$$

or

$$Y\mathbf{v}' = \mathbf{f}.$$

The columns of Y are a fundamental set of solutions to the homogeneous system, so they are linearly independent. Therefore, $Y(t)$ is a nonsingular matrix for all t, and its inverse $Y(t)^{-1}$ exists. We can use it to solve the above equation for $\mathbf{v}'$, obtaining

$$\mathbf{v}' = Y^{-1}\mathbf{f}.$$

To find $\mathbf{v}$, we integrate and get

$$\mathbf{v}(t) = \int Y(t)^{-1}\mathbf{f}(t)\,dt.$$

This formula is analogous to the formula found in Section 4.6 of Chapter 4 for a single second-order differential equation. Inserting $\mathbf{v}$ into (9.6), we obtain

$$\mathbf{y}_p(t) = Y(t)\mathbf{v}(t) = Y(t)\left(\int Y(t)^{-1}\mathbf{f}(t)\,dt\right). \tag{9.10}$$

In many circumstances, it is useful to use the particular solution that is equal to $\mathbf{0}$ at some initial point t_0 (perhaps at $t_0 = 0$). This solution is found by using a definite integral to integrate $\mathbf{v}' = Y^{-1}\mathbf{f}$ instead of the indefinite integral used in (9.10). Doing so, we get

$$\mathbf{y}_p(t) = Y(t)\left(\int_{t_0}^{t} Y(s)^{-1}\mathbf{f}(s)\,ds\right). \tag{9.11}$$

The general solution to the inhomogeneous equation is $\mathbf{y} = \mathbf{y}_h + \mathbf{y}_p$, where $\mathbf{y}_h$ is the general solution to the homogeneous equation. We write $\mathbf{y}_h$ as

$$\mathbf{y}_h(t) = c_1\mathbf{y}_1(t) + \cdots + c_n\mathbf{y}_n(t)$$

$$= [\mathbf{y}_1(t), \mathbf{y}_2(t), \ldots, \mathbf{y}_n(t)]\begin{pmatrix} c_1 \\ \vdots \\ c_n \end{pmatrix}$$

$$= Y(t)\mathbf{c},$$

where $\mathbf{c} = (c_1, \ldots, c_n)^T$ is a column vector of constants. Using this together with (9.11), we obtain

$$\mathbf{y}(t) = Y(t)\mathbf{c} + Y(t)\left(\int_{t_0}^{t} Y(s)^{-1}\mathbf{f}(s)\,ds\right). \tag{9.12}$$

If initial conditions, $\mathbf{y}(t_0) = \mathbf{y}_0$, are given, then inserting $t = t_0$ into (9.12) gives

$$\mathbf{y}_0 = \mathbf{y}(t_0) = Y(t_0)\mathbf{c},$$

since the integral from t_0 to t_0 is zero. Therefore,

$$\mathbf{c} = Y(t_0)^{-1}\mathbf{y}_0.$$

Inserting $\mathbf{c}$ into (9.12) gives

$$\mathbf{y}(t) = Y(t)\left(Y(t_0)^{-1}\mathbf{y}_0 + \int_{t_0}^{t} Y(s)^{-1}\mathbf{f}(s)\,ds\right). \tag{9.13}$$

Equation (9.13) is the variation of parameters formula for the solution to the inhomogeneous system with a given initial condition. This is important enough to state as a theorem.

THEOREM 9.14 Suppose that A is a real $n \times n$ matrix and that $Y(t)$ is a fundamental matrix for the system $\mathbf{y}' = A\mathbf{y}$. Let $\mathbf{f}(t)$ be a vector-valued function. Then the solution to the initial value problem

$$\mathbf{y}' = A\mathbf{y} + \mathbf{f} \quad \text{with} \quad \mathbf{y}(t_0) = \mathbf{y}_0$$

is given by (9.13). ∎

Example 9.15 Find the general solution to the inhomogeneous system $\mathbf{y}' = A\mathbf{y} + \mathbf{f}$, where

$$A = \begin{pmatrix} 1 & 2 \\ 2 & 1 \end{pmatrix} \quad \text{and} \quad \mathbf{f}(t) = \begin{pmatrix} e^{-t} \\ 0 \end{pmatrix}.$$

Find the solution with initial condition $\mathbf{y}(0) = (0, 0)^T$.

Proceeding as we did in Section 9.2, we find that the eigenvalues for A are $\lambda_1 = -1$ and $\lambda_2 = 3$ with corresponding eigenvectors

$$\mathbf{v}_1 = \begin{pmatrix} -1 \\ 1 \end{pmatrix} \quad \text{and} \quad \mathbf{v}_2 = \begin{pmatrix} 1 \\ 1 \end{pmatrix}.$$

The solution to the homogeneous system $\mathbf{y}' = A\mathbf{y}$ is $y_h = C_1\mathbf{y}_1 + C_2\mathbf{y}_2$, where

$$\mathbf{y}_1(t) = e^{-t}\mathbf{v}_1 = \begin{pmatrix} -e^{-t} \\ e^{-t} \end{pmatrix} \quad \text{and} \quad \mathbf{y}_2(t) = e^{3t}\mathbf{v}_2 = \begin{pmatrix} e^{3t} \\ e^{3t} \end{pmatrix}.$$

The fundamental matrix is the 2×2 matrix whose columns are $\mathbf{y}_1$ and $\mathbf{y}_2$. Thus,

$$Y(t) = [\mathbf{y}_1(t), \mathbf{y}_2(t)] = \begin{pmatrix} -e^{-t} & e^{3t} \\ e^{-t} & e^{3t} \end{pmatrix}.$$

The inverse of Y is found in the way described in Section 7.6 in Chapter 7. We apply Gaussian elimination to the matrix

$$\begin{pmatrix} -e^{-t} & e^{3t} & 1 & 0 \\ e^{-t} & e^{3t} & 0 & 1 \end{pmatrix}$$

to bring it into reduced row echelon form. The fact that the entries are functions changes nothing. This process ends with

$$\begin{pmatrix} 1 & 0 & -e^{t}/2 & e^{t}/2 \\ 0 & 1 & e^{-3t}/2 & e^{-3t}/2 \end{pmatrix}.$$

The inverse of Y is the right-hand 2×2 block of this matrix,

$$Y^{-1}(t) = \begin{pmatrix} -e^t/2 & e^t/2 \\ e^{-3t}/2 & e^{-3t}/2 \end{pmatrix}.$$

We will use formula (9.10) to find our solution. In preparation, we multiply $\mathbf{f}$ by Y^{-1} to get

$$Y^{-1}(t)\mathbf{f}(t) = \begin{pmatrix} -e^t/2 & e^t/2 \\ e^{-3t}/2 & e^{-3t}/2 \end{pmatrix} \begin{pmatrix} e^{-t} \\ 0 \end{pmatrix} = \begin{pmatrix} -1/2 \\ e^{-4t}/2 \end{pmatrix}.$$

Next, we integrate this vector-valued function by integrating each component separately. This results in

$$\int Y^{-1}(t)\mathbf{f}(t)\, dt = \begin{pmatrix} -t/2 \\ -e^{-4t}/8 \end{pmatrix}.$$

We then multiply by $Y(t)$ to get the final answer,

$$\mathbf{y}_p(t) = Y(t)\left(\int Y(t)^{-1}\mathbf{f}(t)\, dt\right)$$

$$= \begin{pmatrix} -e^{-t} & e^{3t} \\ e^{-t} & e^{3t} \end{pmatrix} \begin{pmatrix} -t/2 \\ -e^{-4t}/8 \end{pmatrix}$$

$$= \begin{pmatrix} te^{-t}/2 - e^{-t}/8 \\ -te^{-t}/2 - e^{-t}/8 \end{pmatrix}.$$

The general solution is

$$\mathbf{y} = \mathbf{y}_h + \mathbf{y}_p$$
$$= C_1\mathbf{y}_1 + C_2\mathbf{y}_2 + \mathbf{y}_p$$
$$= C_1\begin{pmatrix} -e^{-t} \\ e^{-t} \end{pmatrix} + C_2\begin{pmatrix} e^{3t} \\ e^{3t} \end{pmatrix} + \begin{pmatrix} te^{-t}/2 - e^{-t}/8 \\ -te^{-t}/2 - e^{-t}/8 \end{pmatrix}.$$

To find the solution $\mathbf{y}$ with initial value $\mathbf{y}(0) = (0, 0)^T$, we could use the previous formula and solve for C_1 and C_2. However, it is easier to use (9.11).

$$\mathbf{y}(t) = \begin{pmatrix} -e^{-t} & e^{3t} \\ e^{-t} & e^{3t} \end{pmatrix} \int_0^t \begin{pmatrix} -1/2 \\ e^{-4s}/2 \end{pmatrix} ds$$

$$= \begin{pmatrix} -e^{-t} & e^{3t} \\ e^{-t} & e^{3t} \end{pmatrix} \begin{pmatrix} -t/2 \\ (1 - e^{-4t})/8 \end{pmatrix}$$

$$= \frac{1}{8}\begin{pmatrix} (-1 + 4t)e^{-t} + e^{3t} \\ (-1 - 4t)e^{-t} + e^{3t} \end{pmatrix}$$

Undetermined Coefficients

We used this method with some success in Section 4.5 of Chapter 4 to solve inhomogeneous second-order equations. It can also be used for systems.

Example 9.16 For the circuit in Figure 1, suppose that $R = 1$ ohm, $L = 1$ henry, and $C = 5/4$ farad. Suppose that the electromotive force is provided by a battery and $E = 5$ volts. Assume that $I_1(0) = 5$ amperes and $I_2(0) = 1$ ampere, and find the currents I_1 and I_2 as functions of time t.

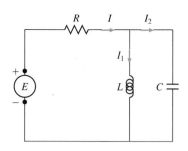

Figure 1. An electrical circuit with a capacitor and an inductor in parallel.

We considered this circuit in Example 4.9 in Chapter 8. Using the result there and the data given, the vector of currents $\mathbf{I} = (I_1, I_2)^T$ satisfies $\mathbf{I}' = A\mathbf{I} + \mathbf{F}$, where

$$A = \begin{pmatrix} -1 & -1 \\ 1 & 1/5 \end{pmatrix} \quad \text{and} \quad \mathbf{F} = \begin{pmatrix} 5 \\ -5 \end{pmatrix}.$$

We considered the homogeneous system in Example 2.28 of this chapter, where we found that a fundamental set of solutions to the homogeneous system is

$$\mathbf{y}_1(t) = e^{-2t/5} \begin{pmatrix} -5\cos(4t/5) \\ 3\cos(4t/5) - 4\sin(4t/5) \end{pmatrix} \quad \text{and}$$

$$\mathbf{y}_2(t) = e^{-2t/5} \begin{pmatrix} -5\sin(4t/5) \\ 4\cos(4t/5) + 3\sin(4t/5) \end{pmatrix}.$$

Thus,

$$Y(t) = e^{-2t/5} \begin{pmatrix} -5\cos(4t/5) & -5\sin(4t/5) \\ 3\cos(4t/5) - 4\sin(4t/5) & 4\cos(4t/5) + 3\sin(4t/5) \end{pmatrix}$$

is a fundamental matrix.

The general solution to $\mathbf{I}' = A\mathbf{I} + \mathbf{F}$ has the form $\mathbf{I} = \mathbf{P} + \mathbf{Q}$, where $\mathbf{P}$ is a particular solution, and $\mathbf{Q} = a_1\mathbf{y}_1 + a_2\mathbf{y}_2 = Y\mathbf{a}$ is a solution to the homogeneous equation. Because the right-hand side, $\mathbf{F} = (5, -5)^T$, is a vector of constants, it is natural to look for a particular solution $\mathbf{P} = (p_1, p_2)^T$, which is also a vector of constants. $\mathbf{P}$ must satisfy $\mathbf{0} = \mathbf{P}' = A\mathbf{P} + \mathbf{F}$. Hence we must have $A\mathbf{P} = -\mathbf{F}$, or

$$\begin{pmatrix} -1 & -1 \\ 1 & 1/5 \end{pmatrix} \begin{pmatrix} p_1 \\ p_2 \end{pmatrix} = \begin{pmatrix} -5 \\ 5 \end{pmatrix}.$$

We solve this system to find that $\mathbf{P} = (5, 0)^T$ is a solution.

The constants $\mathbf{a} = (a_1, a_2)^T$ are chosen so that $\mathbf{I} = \mathbf{P} + \mathbf{Q}$ satisfies the initial conditions,

$$\begin{pmatrix} 5 \\ 1 \end{pmatrix} = \mathbf{I}(0) = \mathbf{P} + \mathbf{Q}(0) = \begin{pmatrix} 5 \\ 0 \end{pmatrix} + \mathbf{Q}(0).$$

This requires $\mathbf{Q}(0) = Y(0)\mathbf{a} = (0, 1)^T$. Hence we must solve

$$\begin{pmatrix} -5 & 0 \\ 3 & 4 \end{pmatrix} \begin{pmatrix} a_1 \\ a_2 \end{pmatrix} = \begin{pmatrix} 0 \\ 1 \end{pmatrix}.$$

From this we conclude that $a_1 = 0$ and $a_2 = 1/4$, so $\mathbf{Q} = \mathbf{y}_2/4$. Therefore, the solution we are seeking is

$$\mathbf{I}(t) = \mathbf{P} + \mathbf{Q}(t) = \begin{pmatrix} 5 - 5e^{-2t/5}\sin(4t/5)/4 \\ e^{-2t/5}[\cos(4t/5) + 3\sin(4t/5)/4] \end{pmatrix}.$$

Computing the exponential of a matrix

You will have noticed that in Section 9.6 we defined the exponential of a matrix, but except for special cases, we did not explain how the exponential could be computed. We did not explain it because we were able to achieve our goal of finding a fundamental set of solutions to a system of differential equations without doing so. Nevertheless, we have left an obvious question unanswered. Here, we will rectify that situation.

There is a reversal of roles taking place here. We originally defined the exponential of a matrix to help us find a fundamental set of solutions to the homogeneous system. We were able to do that without actually computing the exponential. Now we will use a fundamental set of solutions to help us find the exponential.

Suppose A is an $n \times n$ matrix. Let $\mathbf{y}_1, \mathbf{y}_2, \ldots,$ and $\mathbf{y}_n$ be a fundamental set of solutions for the system $\mathbf{y}' = A\mathbf{y}$. The matrix-valued function

$$Y(t) = [\mathbf{y}_1(t), \mathbf{y}_2(t), \ldots, \mathbf{y}_n(t)] \tag{9.17}$$

is a fundamental matrix for the system $\mathbf{y}' = A\mathbf{y}$. Set

$$Y_0 = Y(0) = [\mathbf{y}_1(0), \mathbf{y}_2(0), \ldots, \mathbf{y}_n(0)]. \tag{9.18}$$

Since $\mathbf{y}_1, \mathbf{y}_2, \ldots, \mathbf{y}_n$ form a fundamental set of solutions, we know that the vectors $\mathbf{y}_1(0), \mathbf{y}_2(0), \ldots, \mathbf{y}_n(0)$ are linearly independent. Hence, $\det(Y_0) \neq 0$, and Y_0 is invertible.

Now let's set $E(t) = Y(t)Y_0^{-1}$. From (9.18), we see that

$$E(0) = Y(0)Y_0^{-1} = I.$$

Using Proposition 9.5, we get

$$\frac{d}{dt}E(t) = \frac{d}{dt}Y(t)Y_0^{-1} = AY(t)Y_0^{-1} = AE(t).$$

Thus, $E(t)$ is a solution to the initial value problem

$$\frac{d}{dt}E(t) = AE(t) \quad \text{and} \quad E(0) = I.$$

On the other hand, using Proposition 6.12 for the first part, we have

$$\frac{d}{dt}e^{tA} = Ae^{tA} \quad \text{and} \quad e^{0I} = I.$$

Consequently, the uniqueness theorem implies that $E(t) = e^{tA}$. Let's state this formally.

PROPOSITION 9.19 Suppose that Y is a fundamental matrix for the system $\mathbf{y}' = A\mathbf{y}$. Then the exponential e^{tA} can be computed as

$$e^{tA} = Y(t)Y_0^{-1}, \tag{9.20}$$

where $Y_0 = Y(0)$.
●

Thus, if we know a fundamental set of solutions of $\mathbf{y}' = A\mathbf{y}$, we can compute the exponential e^{tA}. Let's look at some examples.

Example 9.21 Compute e^{tA}, where
$$A = \begin{pmatrix} -4 & 6 \\ -3 & 5 \end{pmatrix}.$$

The matrix A has eigenvalues -1 and 2, with eigenvectors $(2, 1)^T$ and $(1, 1)^T$, respectively. Hence,
$$\mathbf{y}_1(t) = e^{-t} \begin{pmatrix} 2 \\ 1 \end{pmatrix} \quad \text{and} \quad \mathbf{y}_2(t) = e^{2t} \begin{pmatrix} 1 \\ 1 \end{pmatrix}$$

make up a fundamental set of solutions. We set
$$Y(t) = [\mathbf{y}_1(t), \mathbf{y}_2(t)] = \begin{pmatrix} 2e^{-t} & e^{2t} \\ e^{-t} & e^{2t} \end{pmatrix}.$$

At $t = 0$, we have
$$Y_0 = Y(0) = \begin{pmatrix} 2 & 1 \\ 1 & 1 \end{pmatrix}.$$

We compute that
$$Y_0^{-1} = \begin{pmatrix} 1 & -1 \\ -1 & 2 \end{pmatrix}.$$

Then
$$e^{tA} = Y(t)Y_0^{-1} = \begin{pmatrix} 2e^{-t} - e^{2t} & 2e^{2t} - 2e^{-t} \\ e^{-t} - e^{2t} & 2e^{2t} - e^{-t} \end{pmatrix}.$$

If we want to find the exponential of A, we simply evaluate at $t = 1$:
$$e^A = \begin{pmatrix} 2e^{-1} - e^2 & 2e^2 - 2e^{-1} \\ e^{-1} - e^2 & 2e^2 - e^{-1} \end{pmatrix}.$$

●

Example 9.22 Compute e^{tA}, where
$$A = \begin{pmatrix} -4 & 5 \\ -2 & 2 \end{pmatrix}.$$

This time, A has complex eigenvalues $-1 \pm i$. An eigenvector associated with $-1 + i$ is $(3 - i, 2)^T$. The corresponding complex solution is
$$\mathbf{z}(t) = e^{(-1+i)t} \begin{pmatrix} 3 - i \\ 2 \end{pmatrix}$$
$$= e^{-t} \left[(\cos t + i \sin t) \left(\begin{pmatrix} 3 \\ 2 \end{pmatrix} - i \begin{pmatrix} 1 \\ 0 \end{pmatrix} \right) \right]$$
$$= e^{-t} \left[\begin{pmatrix} 3\cos t + \sin t \\ 2\cos t \end{pmatrix} + i \begin{pmatrix} 3\sin t - \cos t \\ 2\sin t \end{pmatrix} \right].$$

Thus,
$$\mathbf{y}_1(t) = e^{-t} \begin{pmatrix} 3\cos t + \sin t \\ 2\cos t \end{pmatrix} \quad \text{and} \quad \mathbf{y}_2(t) = e^{-t} \begin{pmatrix} 3\sin t - \cos t \\ 2\sin t \end{pmatrix}$$

make up a fundamental set of solutions. We set
$$Y(t) = [\mathbf{y}_1(t), \mathbf{y}_2(t)] = e^{-t} \begin{pmatrix} 3\cos t + \sin t & 3\sin t - \cos t \\ 2\cos t & 2\sin t \end{pmatrix}.$$

Then

$$Y_0 = Y(0) = \begin{pmatrix} 3 & -1 \\ 2 & 0 \end{pmatrix}.$$

We compute that

$$Y_0^{-1} = \begin{pmatrix} 0 & 1/2 \\ -1 & 3/2 \end{pmatrix}.$$

Finally,

$$e^{tA} = Y(t)Y_0^{-1} = e^{-t} \begin{pmatrix} \cos t - 3\sin t & 5\sin t \\ -2\sin t & \cos t + 3\sin t \end{pmatrix}.$$

●

EXERCISES

In Exercises 1–10, use the variation of parameters technique demonstrated in Example 9.15 to find the general solution to $\mathbf{y}' = A\mathbf{y} + \mathbf{f}$.

1. $A = \begin{pmatrix} 5 & 6 \\ -2 & -2 \end{pmatrix}$ and $\mathbf{f} = \begin{pmatrix} e^t \\ e^t \end{pmatrix}$

2. $A = \begin{pmatrix} 3 & -4 \\ 2 & -3 \end{pmatrix}$ and $\mathbf{f} = \begin{pmatrix} e^{-t} \\ e^t \end{pmatrix}$

3. $A = \begin{pmatrix} -3 & 6 \\ -2 & 4 \end{pmatrix}$ and $\mathbf{f} = \begin{pmatrix} 3 \\ 4 \end{pmatrix}$

4. $A = \begin{pmatrix} -3 & 10 \\ -3 & 8 \end{pmatrix}$ and $\mathbf{f} = \begin{pmatrix} e^{-t} \\ e^{2t} \end{pmatrix}$

5. $A = \begin{pmatrix} 3 & 2 \\ -1 & 1 \end{pmatrix}$ and $\mathbf{f} = \begin{pmatrix} 0 \\ e^{2t} \end{pmatrix}$

6. $A = \begin{pmatrix} 4 & 2 \\ -1 & 2 \end{pmatrix}$ and $\mathbf{f} = \begin{pmatrix} t \\ e^{3t} \end{pmatrix}$

7. $A = \begin{pmatrix} -4 & -2 & 6 \\ 6 & 3 & -6 \\ -2 & -1 & 4 \end{pmatrix}$ and $\mathbf{f} = \begin{pmatrix} -\sin t \\ 2\sin t \\ 0 \end{pmatrix}$

8. $A = \begin{pmatrix} 1 & -18 & 8 \\ 0 & 14 & -6 \\ 0 & 35 & -15 \end{pmatrix}$ and $\mathbf{f} = \begin{pmatrix} 0 \\ 1 \\ 0 \end{pmatrix}$

9. $A = \begin{pmatrix} -6 & -6 & 18 \\ 10 & 13 & -42 \\ 2 & 3 & -10 \end{pmatrix}$ and $\mathbf{f} = \begin{pmatrix} 0 \\ 2 \\ 0 \end{pmatrix}$

10. $A = \begin{pmatrix} 11 & -7 & -4 \\ -6 & 4 & 2 \\ 42 & -27 & -15 \end{pmatrix}$ and $\mathbf{f} = \begin{pmatrix} 1 \\ 0 \\ 0 \end{pmatrix}$

11. In Example 9.16, we used the method of *undetermined coefficients* to find a particular solution. Use the variation of parameters method to find a particular solution for the circuit in Example 9.16. (*Note:* You will quickly find that this is a very difficult problem to complete using hand calculations only. If you have a computer algebra system (CAS) on your computer, use it to assist with some of the computations.)

Use the technique of *undetermined coefficients* demonstrated

in Example 9.16 to find particular solutions for the system $\mathbf{y}' = A\mathbf{y} + \mathbf{f}$ in Exercises 12–13.

12. $A = \begin{pmatrix} -1 & 0 \\ 3 & 2 \end{pmatrix}$, $\mathbf{f} = \begin{pmatrix} 2 \\ -5 \end{pmatrix}$

13. $A = \begin{pmatrix} -6 & 8 \\ -4 & 6 \end{pmatrix}$, $\mathbf{f} = \begin{pmatrix} 2 \\ 3 \end{pmatrix}$

14. Use the method of *undetermined coefficients* and the "natural choice" $\mathbf{y}_p = \mathbf{a}e^t$, where $\mathbf{a} = (a_1, a_2)^T$, to find a particular solution of the system $\mathbf{y}' = A\mathbf{y} + \mathbf{f}$, where

$$A = \begin{pmatrix} -1 & 0 \\ 1 & -2 \end{pmatrix} \quad \text{and} \quad \mathbf{f} = \begin{pmatrix} 2e^t \\ -e^t \end{pmatrix}.$$

15. Consider the system $\mathbf{y} = A\mathbf{y} + \mathbf{f}$, where

$$A = \begin{pmatrix} -2 & 0 \\ 1 & -1 \end{pmatrix} \quad \text{and} \quad \mathbf{f} = \begin{pmatrix} 0 \\ \sin t \end{pmatrix}.$$

Because the first derivative of the sine function is a cosine function, we might hazard a guess that involves both sines and cosines. Use the method of *undetermined coefficients* and the guess $\mathbf{y}_p = \mathbf{a}\cos t + \mathbf{b}\sin t$, where $\mathbf{a} = (a_1, a_2)^T$ and $\mathbf{b} = (b_1, b_2)^T$ to find a particular solution of the system.

16. Use the method of *undetermined coefficients* to find a particular solution of the system in Exercise 3. Unfortunately, the first natural guess, namely $\mathbf{y}_p = \mathbf{a}$ is also a term in the homogeneous solution for this system. So, instead, try $\mathbf{y}_p = \mathbf{a}t + \mathbf{b}$, where $\mathbf{a} = (a_1, a_2)^T$ and $\mathbf{b} = (b_1, b_2)^T$. This will lead to a system of equations in a_1, a_2, b_1, and b_2 with one resulting free variable. Make a choice for the free variable to help find a particular solution for the system. Check your solution.

17. Use the method of *undetermined coefficients* to find a particular solution of the system in Example 9.15. Unfortunately, the natural guess, namely $\mathbf{y}_p = \mathbf{a}e^{-t}$ is also a term of the homogeneous solution for this system. So, instead, set $\mathbf{y}_p = (\mathbf{a}t + \mathbf{b})e^{-t}$ in the equation $\mathbf{y}' = A\mathbf{y} + \mathbf{f}$, where $\mathbf{a} = (a_1, a_2)^T$ and $\mathbf{b} = (b_1, b_2)^T$. This will lead to a system of equations in a_1, a_2, b_1, and b_2 with one resulting free variable. Find an appropriate choice of the free variable that leads to the same particular solution found in the Example 9.15.

18. Consider the circuit shown in Figure 2.

Assume all currents are initially zero. Find the current through each inductor as a function of time t.

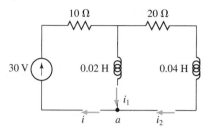

Figure 2. Another driven circuit.

19. Consider the circuit shown in Figure 3. Assume that the initial current through each inductor is zero amps. Find the current through each inductor as a function of time

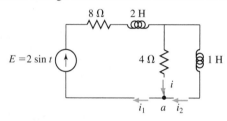

Figure 3. A driven circuit.

(a) using *undetermined coefficients* to find a particular solution (*Hint:* See Exercise 15), and

(b) solve the exercise a second time using *variation of parameters* to find a particular solution.

In Exercises 20–25, use the technique of Examples 9.21 and 9.22 to find e^{tA} for the given matrix A.

20. $A = \begin{pmatrix} 3 & 0 \\ -4 & -1 \end{pmatrix}$ 　　**21.** $A = \begin{pmatrix} -1 & 4 \\ -2 & 5 \end{pmatrix}$

22. $A = \begin{pmatrix} -2 & 1 \\ -5 & 2 \end{pmatrix}$ 　　**23.** $A = \begin{pmatrix} 2 & -2 \\ 1 & 0 \end{pmatrix}$

24. $A = \begin{pmatrix} -4 & -1 \\ 1 & -2 \end{pmatrix}$ 　　**25.** $A = \begin{pmatrix} 0 & 1 \\ -4 & 4 \end{pmatrix}$

In Exercises 26–29, use the technique of Examples 9.21 and 9.22 to find e^{tA}. Then use Proposition 6.12 to find the solution of the system $\mathbf{y}' = A\mathbf{y}$ for the given initial condition.

26. $A = \begin{pmatrix} 5 & 3 \\ -6 & -4 \end{pmatrix}$, $\mathbf{y}_0 = \begin{pmatrix} 1 \\ 1 \end{pmatrix}$

27. $A = \begin{pmatrix} -7 & -3 \\ 6 & 2 \end{pmatrix}$, $\mathbf{y}_0 = \begin{pmatrix} 1 \\ 0 \end{pmatrix}$

28. $A = \begin{pmatrix} 2 & -2 \\ 4 & -2 \end{pmatrix}$, $\mathbf{y}_0 = \begin{pmatrix} 1 \\ 1 \end{pmatrix}$

29. $A = \begin{pmatrix} -1 & 1 \\ -4 & -5 \end{pmatrix}$, $\mathbf{y}_0 = \begin{pmatrix} 2 \\ -1 \end{pmatrix}$

30. Pictured in Figure 4 are two tanks, each containing 200 gallons of water. Salt solution enters the upper tank at a rate of 5 gal/min having salt concentration 0.2 lb/gal. Solution drains from the upper tank directly into the lower tank at a rate of 5 gal/min. Finally, solution leaves the lower tank at a rate of 5 gal/min, effectively maintaining the volume of solution in each tank constant. Initially, each tank contains pure water. Use an exponential matrix approach to find the salt content in each tank as a function of time t.

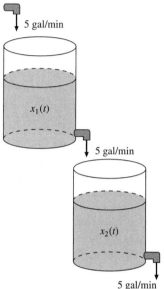

Figure 4. Coupled tanks with forcing.

Notice that $\mathbf{y}(t) = e^{(t-t_0)A}\mathbf{v}$ is a solution of the system $\mathbf{y}' = A\mathbf{y}$ satisfying the initial condition $\mathbf{y}(t_0) = \mathbf{v}$. Use this result in Exercises 31–34 to solve $\mathbf{y}' = A\mathbf{y}$ with the given initial condition.

31. $A = \begin{pmatrix} 5 & 0 \\ 6 & -1 \end{pmatrix}$, $\mathbf{y}(1) = \begin{pmatrix} -1 \\ 3 \end{pmatrix}$

32. $A = \begin{pmatrix} 0 & 1 \\ -2 & -3 \end{pmatrix}$, $\mathbf{y}(-1) = \begin{pmatrix} 1 \\ 0 \end{pmatrix}$

33. $A = \begin{pmatrix} 6 & -3 \\ 15 & -6 \end{pmatrix}$, $\mathbf{y}(1) = \begin{pmatrix} -1 \\ 0 \end{pmatrix}$

34. $A = \begin{pmatrix} -4 & 0 \\ -1 & -4 \end{pmatrix}$, $\mathbf{y}(2) = \begin{pmatrix} -2 \\ 2 \end{pmatrix}$

In Exercises 35–38, use the technique of Examples 9.21 and 9.22 to find e^{tA} for the given matrix A. You may want to use a computer to help with some calculations.

35. $A = \begin{pmatrix} 5 & 0 & 4 \\ 3 & -2 & 6 \\ -2 & 0 & -1 \end{pmatrix}$

36. $A = \begin{pmatrix} 1 & -2 & 1 \\ 0 & 1 & 0 \\ -2 & 4 & -1 \end{pmatrix}$

37. $A = \begin{pmatrix} 2 & 0 & 3 & 0 \\ 5 & 1 & 3 & -3 \\ 0 & 0 & -1 & 0 \\ 4 & 0 & 4 & -2 \end{pmatrix}$

38. $A = \begin{pmatrix} 0 & 0 & 0 & 2 \\ -1 & -2 & 0 & -1 \\ 2 & 2 & -3 & 2 \\ 0 & 1 & 0 & -2 \end{pmatrix}$

39. Show that if A is a real $n \times n$ matrix then the solution to the initial value problem

$$\mathbf{y}' = A\mathbf{y} + \mathbf{f} \quad \text{with} \quad \mathbf{y}(0) = \mathbf{y}_0$$

is given by

$$\mathbf{y}(t) = e^{tA}\mathbf{y}_0 + \int_0^t e^{(t-s)A}\mathbf{f}(s)\,ds.$$

PROJECT 9.10 Phase Plane Portraits

The goal of this project is to acquire knowledge of all possible planar, linear systems and their phase portraits. This is meant to be a discovery process. You have to discover all of the types of systems we have not yet discovered. You will recall that six types were discussed in Section 9.3; it's up to you to discover the remaining types. Some of the types are given names in this book. Please use those names. When a type has not been named, make up a *reasonable* name yourself. You should read the Exercises for Sections 9.3 and 9.4.

We will not give a formal definition of what is meant by *type*. We will say that the types of systems are differentiated by the formulas for the general solution and the appearance of the solution curves in the phase plane. These in turn are affected by the properties of the eigenvalues, including whether they are real or complex, positive or negative, as well as by their multiplicities.

How many types are there? That's up to you to discover. You will have to defend your answer. Let's just say that, including the six types discussed in Section 9.3, there are more than 10 and less than 20.

You are to do the following:

1. Make a chart of the trace-determinant plane on a single sheet of paper, with all of the types listed and with references to the phase portrait, for an example of that type of system. The chart should be modeled after Figure 2 in Section 9.3. In fact, this figure provides the start for your inquiries.

2. For each type that you discover, find an example and submit a phase portrait produced either by hand or with the aid of a computer. Be sure to add labels and refer to them

in your trace-determinant plane in the analysis called for in part (3) and in the statements called for in parts (4) and (5). It will be up to you to find a system of the given type. One key to doing this is to keep things simple. Diagonal matrices are just fine, as are upper triangular matrices and matrices used in examples and exercises in this book.

3. Do an analysis of the behavior of all solutions to each system. This should be along the same lines as is done in Section 9.3 for the six types discussed there. Be sure to find all of the equilibrium points of the system and figure out how they affect the solutions. In addition, find all exponential solutions and the general solution. It is not necessary to repeat any part of the discussion in Section 9.3. This includes the analysis for the six types discussed there. Instead, it will suffice to refer to that discussion. However, do so in a way that makes it easy for a reader to follow your discussion.

4. Prepare a short statement in which you argue that you have indeed found *all* possible types of systems. In other words, show that every linear, planar system is represented by one of the types on your list.

5. While you are working on this project, you may notice things about linear planar systems that pique your curiosity—things that you think are interesting, but perhaps do not understand completely. Write up a short statement about these things. If you have time to explore them, include the results of your exploration. If not, simply describe your observations and state your questions and conjectures. Make this statement clearly distinguishable from the rest of your project submission.

PROJECT 9.11 Oscillations of Linear Molecules

Molecules are made up of atoms, which are bound together by bonding forces between the atoms. The atoms move within the molecule subject to these forces. In general the atomic bonds can stretch or compress, they can bend, or they can rotate. In addition molecules move through space under the influence of external forces, or, in the absence of external forces, in a purely inertial manner. For any particular molecule these motions can

be modeled with a system of differential equations. As you can see, the combination of all of the motions can be quite complicated, and the resulting system of differential equations must be just as complicated.

In this project[5] we will examine the internal motion of linear molecules. These are molecules such as oxygen (O_2, see Figure 1), which has two atoms, and carbon dioxide (CO_2, see

---

[5] This project originated as a project in the course *Computation in the Natural Sciences* at Rice University. The author of the project was John Hutchinson, a professor of chemistry.

Figure 3), which has three. We will assume that there are no external forces on the molecule. Hence the only forces acting are the internal bonding forces between adjacent atoms. Even simple molecules like these can bend or rotate. To simplify things, we will look only at the stretch or compression of the linear bonds between adjoining atoms. Our goal will be to find what are called the ***normal modes*** of oscillation, a term that we will define later.

A diatomic molecule

We will illustrate the method by giving many of the details of the analysis of a diatomic molecule, and then we will leave it to you to analyze more complicated molecules. Consider a molecule such as oxygen (O_2), which has two atoms. We will assume that this diatomic molecule and its individual atoms are constrained to move on a line, as is indicated in Figure 1. Let x_1 and x_2 be the coordinates of the two atoms with respect to a fixed, but otherwise arbitrary origin. We are concerned with oscillations that are much smaller than the usual distance between the atoms, so we may assume that $x_1 < x_2$ at all times.

Figure 1. The oxygen molecule, containing two atoms of oxygen.

We will also assume that the atoms in the molecule act like two particles which are connected by a spring. A spring has an equilibrium position, but this is not true for the individual atoms in a molecule. Instead there is an equilibrium distance d between the atoms called the ***equilibrium bond length***. When $x_2 - x_1 = d$, the force between them is zero. For small displacements, the force is proportional to the amount that the distance between the atoms differs from the equilibrium distance. Thus, the force on the first atom with displacement x_1 is

$$F_1 = k(x_2 - x_1 - d),$$

where k is the ***bond constant***. If $x_2 - x_1 > d$, the force is positive, which draws the first atom to the right. if $x_2 - x_1 < d$, the force is negative and the first atom is being forced to the left.

Similarly, the force acting on the second atom is

$$F_2 = k(x_1 - x_2 + d).$$

Notice that $F_2 = -F_1$. This means that the two atoms are attracted to each other when the distance between them is greater than d, and they are repelled from each other when their separation is less than d. In addition, the sum of the forces is 0, so the total force on the molecule is 0. The result is that the motion of the molecule as a whole is purely inertial. According to Newton's laws, it moves along the line with constant velocity.

If we assume that the atoms are of equal mass, then according to Newton's second law, the motion of the atoms is governed by the following system of second-order differential equations:

$$mx_1'' = F_1 = k(x_2 - x_1 - d)$$
$$mx_2'' = F_2 = k(x_1 - x_2 + d). \tag{11.1}$$

The differential equations in (11.1) are linear, second-order equations. The system would be homogeneous except for the presence of d. We can take care of that by making a change of variables. Introduce the modified displacements

$$y_1 = x_1 \quad \text{and} \quad y_2 = x_2 - d \quad \text{or}$$
$$x_1 = y_1 \quad \text{and} \quad x_2 = y_2 + d. \tag{11.2}$$

If we also divide each equation by m, then the system (11.1) becomes the homogeneous system

$$y_1'' = \frac{k}{m}(y_2 - y_1)$$
$$y_2'' = \frac{k}{m}(y_1 - y_2). \tag{11.3}$$

We can write this in matrix form. If we introduce the matrix

$$K = \begin{pmatrix} -k/m & k/m \\ k/m & -k/m \end{pmatrix} \tag{11.4}$$

and let $\mathbf{y} = (y_1, y_2)^T$ be the vector of displacements, the system becomes

$$\mathbf{y}'' = K\mathbf{y}. \tag{11.5}$$

This is a second-order system. We can write this as a more familiar first-order system if we let $v_1 = x_1' = y_1'$ and $v_2 = x_2' = y_2'$ denote the velocities of the two atoms. Then the second-order system in (11.3) is equivalent to the first-order system

$$y_1' = v_1 \qquad\qquad v_1' = \frac{k}{m}(y_2 - y_1)$$
$$\text{and}$$
$$y_2' = v_2 \qquad\qquad v_2' = \frac{k}{m}(y_1 - y_2). \tag{11.6}$$

The first-order system can be written in matrix form in the standard way. Let

$$\mathbf{u} = \begin{pmatrix} y_1 \\ y_2 \\ v_1 \\ v_2 \end{pmatrix} \quad \text{and} \quad A = \begin{pmatrix} 0 & 0 & 1 & 0 \\ 0 & 0 & 0 & 1 \\ -k/m & k/m & 0 & 0 \\ k/m & -k/m & 0 & 0 \end{pmatrix}.$$

Then the system (11.6) can be written as

$$\mathbf{u}' = A\mathbf{u}. \tag{11.7}$$

Of course, the systems in (11.5) and (11.7) are equivalent. A solution $\mathbf{y}(t)$ to (11.5) leads to a solution $\mathbf{u}(t)$ to (11.7) in the way outlined in the previous paragraph. On the other hand, a solution $\mathbf{u}(t)$ to (11.7) leads to a solution to (11.5) by taking $\mathbf{y}(t) = (u_1(t), u_2(t))^T$. We could, therefore, solve the second-order system (11.5) by solving the more familiar first-order system (11.7). However, a better way is to use what we know about first-order systems to motivate a solution procedure for the second-order system.

We have learned to solve first-order systems like (11.7) by looking for exponential solutions. However, if $\mathbf{u}$ is exponential, then the vector, $\mathbf{y}(t) = (u_1(t), u_2(t))^T$, consisting of the first two components of $\mathbf{u}$ is also exponential. So let's look for an exponential solution, $\mathbf{y}(t) = e^{\lambda t}\mathbf{w}$, to (11.5). Substituting into the left- and right-hand sides of (11.5), we get

$$\mathbf{y}'' = e^{\lambda t}\lambda^2\mathbf{w} \quad \text{and} \quad K\mathbf{y} = e^{\lambda t}K\mathbf{w}.$$

Thus the exponential function $\mathbf{y}(t) = e^{\lambda t}\mathbf{w}$ will satisfy $\mathbf{y}'' = K\mathbf{y}$ provided $K\mathbf{w} = \lambda^2\mathbf{w}$. This means that we need λ^2 to be an eigenvalue of K with $\mathbf{w}$ an associated eigenvector.

Let's turn that around. Suppose that σ is an eigenvector of K, and $\mathbf{w}$ is an associated eigenvector. If $\sigma \neq 0$, then σ has two square roots, $\pm\sqrt{\sigma}$. Consequently, the functions

$$\mathbf{y}_1(t) = e^{\sqrt{\sigma}t}\mathbf{w} \quad \text{and} \quad \mathbf{y}_2(t) = e^{-\sqrt{\sigma}t}\mathbf{w}$$

are solutions to (11.5).

The case when the eigenvalue $\sigma = 0$ is special, since 0 has only one square root. However, we would still expect there to be two corresponding solutions. Notice that the eigenspace of $\sigma = 0$ is just the nullspace of K, so an eigenvector $\mathbf{w}$ satisfies $K\mathbf{w} = 0$. If $f(t)$ is any function of t, and we set $\mathbf{y}(t) = f(t)\mathbf{w}$, then $K\mathbf{y} = 0$. The function $\mathbf{y}(t)$ will be a solution of equation (11.5) provided that $\mathbf{y}'' = 0$. Notice that $\mathbf{y}'' = f''(t)\mathbf{w}$, so we need the function f to satisfy $f'' = 0$. This is true for any function of the form $f(t) = C_1 + C_2t$. To sum up, if $\sigma = 0$ is an eigenvalue of K, and $\mathbf{w}$ is in the nullspace of K, then any function of the form $\mathbf{y}(t) = (C_1 + C_2t)\mathbf{w}$, where C_1 and C_2 are arbitrary constants is a solution to (11.5).

We are now ready to analyze the oxygen molecule.

1. Show that the characteristic equation for the matrix K in (11.4) is $p(\sigma) = \sigma(\sigma + 2k/m)$, so the eigenvalues of K are $\sigma_1 = -\omega^2$, where $\omega = \sqrt{2k/m}$, and $\sigma_2 = 0$.

2. Show that the vector $\mathbf{w}_2 = (1, 1)^T$ is an eigenvector associated with $\sigma_2 = 0$, and that the corresponding motion of the molecule is

$$\mathbf{y}(t) = (y_0 + v_0t)\begin{pmatrix} 1 \\ 1 \end{pmatrix},$$

where y_0 and v_0 are arbitrary constants. Show that in terms of the original displacements this motion is described by $x_1(t) = y_0 + v_0t$ and $x_2(t) = (y_0 + d) + v_0t$. Thus the molecule moves rigidly along the line to which it is constrained with constant velocity v_0. This is the **inertial mode** of the motion.

3. Show that the vector $\mathbf{w}_1 = (1, -1)^T$ is an eigenvector of K associated with $\sigma_1 = -\omega^2$. Show, using Euler's formula and the information on amplitude and phase in Section 4.4, that the corresponding motion of the molecule can be written as

$$\mathbf{y}(t) = C\cos(\omega t - \phi)\begin{pmatrix} 1 \\ -1 \end{pmatrix}.$$

The vector

$$C\begin{pmatrix} 1 \\ -1 \end{pmatrix}$$

is called the **vector of amplitudes** for the oscillation and ϕ is called the **phase**. Describe this motion. Oscillatory motions such as this are called **normal modes** of oscillation of the molecule.

The displacements x_1 and x_2 are plotted over two periods ($T = 2\pi/\omega$) of oscillation in Figure 2. For convenience we have set the phase $\phi = 0$, and we have chosen the vector of amplitudes small enough to ensure that $x_1(t) < x_2(t)$. This plot displays the essential features of the normal mode.

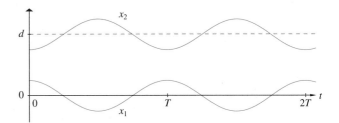

Figure 2. The stretching mode for an oxygen molecule.

A linear triatomic molecule

Admittedly there are easier ways to find the oscillations of a diatomic molecule. However the method we have used pays off when there are three or more atoms in the molecule. We will set up the equations for a linear triatomic molecule (such as CO_2). See Figure 3. Numbering the atoms from one to three, starting from the left, we have three position coordinates x_1, x_2, and x_3. We will assume that the oscillations are very small, so that we always have $x_1 < x_2 < x_3$.

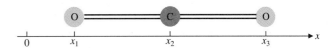

Figure 3. A molecule of carbon dioxide.

The binding forces between the atoms are similar in nature to what we found for the diatomic molecule. Thus, there is an equilibrium bond length d_{12} between atoms 1 and 2, and an equilibrium bond length d_{23} between atoms 2 and 3. We allow for the possibility that each atom is different, and therefore each has a different mass. We also assume that the bond constants for the binding forces are different. Let k_{12} be the bond constant for the force between atoms 1 and 2, and k_{23} that for the force between atoms 2 and 3.

4. Show that the equations of motion for the three atoms are

$$m_1x_1'' = k_{12}[x_2 - x_1 - d_{12}]$$
$$m_2x_2'' = -k_{12}[x_2 - x_1 - d_{12}] + k_{23}[x_3 - x_2 - d_{23}]$$
$$m_3x_3'' = -k_{23}[x_3 - x_2 - d_{23}].$$

Show that we can reduce this system to a homogeneous system by choosing the modified displacements $y_1 = x_1$,

$y_2 = x_2 - d_{12}$, and $y_3 = x_3 - d_{12} - d_{23}$, and that the resulting equations are

$$y_1'' = \frac{k_{12}}{m_1}[y_2 - y_1]$$

$$y_2'' = -\frac{k_{12}}{m_2}[y_2 - y_1] + \frac{k_{23}}{m_2}[y_3 - y_2] \qquad (11.8)$$

$$y_3'' = -\frac{k_{23}}{m_3}[y_3 - y_2].$$

Find the matrix K for which the system in (11.8) can be written as $\mathbf{y}'' = K\mathbf{y}$.

From this point on the analysis of the modes of motion of a carbon dioxide molecule can be done either analytically or numerically. For either approach you should notice that the two oxygen atoms have the same mass, so $m_3 = m_1$, and the two bond constants are the same so $k_{12} = k_{23} = k$. With these simplifications the modes can be found algebraically, but the computation cannot be called trivial.

To facilitate the numerical approach we need to give you the data on the masses and bonding constants. First, however, we need to worry about units. Any consistent set of units could be used, but for problems involving atoms it is not usually convenient to use the International System. The set of "atomic units" is most appropriate. The atomic unit of mass is the mass of the electron. It is denoted by m_e, and $1 \ m_e = 9.1091 \times 10^{-31}$ kg. The unit of length is the radius of the orbit of the lowest energy electron in the Bohr model of the hydrogen atom. It is called the **bohr** or the **bohr radius**, it is denoted by a_o, and $1 \ a_o = 5.29167 \times 10^{-11}$ m. Finally, the atomic unit of time is the period of the the lowest energy electron in the Bohr model of the hydrogen atom. It is denoted by au, and 1 au $= 2.41888 \times 10^{-17}$ s.

In these units, the mass of a carbon atom is $m_2 = 21,876$ m_e and the mass of an oxygen atom is $m_1 = m_3 = 29,168$ m_e. The carbon-oxygen binding force constant in CO_2 is $k_{12} = k_{23} = k = 0.4758$ in atomic units.

5. Calculate the eigenvalues and eigenvectors of K.

6. Describe the inertial mode for a CO_2 molecule.

7. There are two normal modes of oscillation for the CO_2 molecule. Describe the motion of each in terms of its frequency and vector of amplitudes. One mode might be called the symmetric stretch, and the other the antisymmetric stretch. Figure out which is which.

8. For each normal mode plot the displacements x_1, x_2, and x_3 on a single graph versus time. You need to know that the equilibrium bond length of each CO in CO_2 is 2.2 a_o. Choose the multiplicative factor in the vector of amplitudes small enough to ensure that $x_1(t) < x_2(t) < x_3(t)$, and large enough to be easily visible, and set the phase equal to 0. Use the same time interval for both modes. Choose it to span two periods for the slowest oscillation. These graphs display the nature of the modes quite nicely.

A more complicated example

Larger molecules have more oscillatory normal modes. For example, cyanoacetylene, HCCCN, is a linear molecule with five atoms (see Figure 4). The normal modes can be found using the method used for O_2 and CO_2. Since HCCCN has five molecules, we expect that there will be four normal modes of oscillation in addition to the inertial mode. Reasonable choices of the force constants in atomic units are: $k_{HC_1} = 0.378$, $k_{C_1C_2} = 1.087$, $k_{C_2C_3} = 0.441$, and $k_{C_3N} = 1.031$. The mass of a hydrogen atom is $1,823 \ m_e$ and the mass of a nitrogen atom is $25,522 \ m_e$. For HCCCN, the bond lengths are from left to right: HC_1 2.0281 a_o, C_1C_2 2.2716 a_o, C_2C_3 2.6664 a_o, and C_3N 2.1776 a_o. Notice that the two CC lengths and the two CC force constants are not close to being equal. This reflects the fact that the C_1C_2 bond is a triple bond, while the C_2C_3 bond is a single bond, as indicated in Figure 4.

9. Use a computer to calculate the inertial mode and the normal modes for the stretching motions in cyanoacetylene. Plot the motions of the atoms in each normal mode.

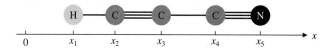

Figure 4. A molecule of cyanoacetylene, HCCCN.

10

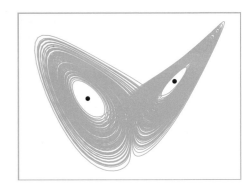

Nonlinear Systems

We are now ready to apply all of the machinery we have developed to the analysis of autonomous, nonlinear systems. The most important question to be answered for any system is, "What happens to solutions over the long term?" Although sometimes we are interested in an individual solution, most of the time we will want to know what happens to all solutions.

We will focus first on equilibrium points and equilibrium solutions. Solutions with initial conditions near an equilibrium point can frequently be analyzed by considering the linearization of the nonlinear system. There is some similarity between our equilibrium point analysis and what we did in dimension 1 in Section 2.9. However, the higher dimensions that we deal with here require new methods that are more difficult than the method we used in Chapter 2.

The results of the equilibrium point analysis will give us local results. These tell us what happens to solutions that start near an equilibrium point. Global methods are those that enable us to talk about all solutions, even those that start relatively far from any equilibrium point. Such methods are more difficult to find and have limited application. Nonetheless, we will describe some of these.

In addition, we will present a variety of applications of the methods.

10.1 The Linearization of a Nonlinear System

In Section 1.1, we saw that the algebraic definition of the derivative says that a differentiable function f can be approximated by a linear function. More precisely,

$$f(y_0 + u) = f(y_0) + f'(y_0)u + R(u), \tag{1.1}$$

where $R(u)$ is the remainder term and satisfies

$$\frac{R(u)}{u} \to 0 \quad \text{as } u \to 0. \tag{1.2}$$

This means that the function $f(y_0 + u)$ can be approximated near the point $u = 0$ by the linear function $L(u) = f(y_0) + f'(y_0)u$.

The whole of the differential calculus, as well as much of mathematical analysis, can be considered to be the application of this idea. The properties of linear functions are easily understood. Using the approximation in (1.1), we argue that the nonlinear function has the same or similar properties. This idea can be carried over to other circumstances. In the calculus of variations, this idea is used to study the extrema of functions defined on infinite dimensional spaces, such as the space of "all" curves or of "all" surfaces. In this section we will approximate a nonlinear system of differential equations by a linear system. The properties of the linear system will then be used to discover properties of the nonlinear system.

The one-dimensional case

We will begin by looking closely at the single first-order equation

$$y' = f(y) \tag{1.3}$$

as a motivation for what we will do for a higher-dimensional system. Suppose that y_0 is an equilibrium point, so that $f(y_0) = 0$. In Theorem 9.8 of Section 2.9, we proved that if $f'(y_0) < 0$, then the equilibrium point y_0 is stable, while if $f'(y_0) > 0$, y_0 is unstable. The proof in Section 2.9 was a one-variable proof that does not extend to higher dimensions. We will look at the situation again from a different point of view. We will not get any better results, but this new method is applicable to higher-dimensional systems.

The algebraic definition of the derivative is formulated in (1.1). It is important to notice that (1.2) says that the remainder $R(u)$ is much smaller than u for small values of u. In our case, y_0 is an equilibrium point, so $f(y_0) = 0$. Therefore, (1.1) becomes

$$f(y_0 + u) = f'(y_0)u + R(u), \tag{1.4}$$

where the remainder $R(u)$ is so much smaller than u that

$$\frac{R(u)}{u} \to 0 \quad \text{as } u \to 0.$$

We will assume that $f'(y_0) \neq 0$. It then follows that

$$\frac{R(u)}{f'(y_0)u} = \frac{1}{f'(y_0)} \cdot \frac{R(u)}{u} \to 0 \quad \text{as } u \to 0. \tag{1.5}$$

Let's change variables in (1.3) by setting $y = y_0 + u$. The new variable u measures the displacement of y from the equilibrium point y_0. Notice first that $y' = (y_0 + u)' = u'$. Then, using (1.3), we get $u' = y' = f(y) = f(y_0 + u)$. Finally, using (1.4), this becomes

$$u' = f'(y_0)u + R(u). \tag{1.6}$$

Remember that we are interested in what happens for $y = y_0 + u$ near the equilibrium point y_0, which means that we are interested in u near 0. Of course, 0 is the equilibrium point for the differential equation in (1.6) that corresponds to the equilibrium point y_0 for equation (1.3).

It follows from (1.5) that for small u, $|R(u)|$ is much smaller than $|f'(y_0)u|$. For such values of u, it is interesting to compare the differential equation in (1.6) to the equation

$$\tilde{u}' = f'(y_0)\tilde{u}, \tag{1.7}$$

where we have ignored the remainder term. Of course, this is a different equation, but it has the advantage that it is linear, homogeneous, and has constant coefficients.

We can write down the solutions to (1.7). If, for convenience, we set $a = f'(y_0)$, the solutions have the form $\tilde{u}(t) = Ce^{at}$. We are only interested in the qualitative behavior of $\tilde{u}$ as $t \to \infty$. Notice that if $a = f'(y_0) < 0$, then all solutions to (1.7) tend to 0 as $t \to \infty$. Thus, the equilibrium point of (1.7) at 0 is stable. If $a = f'(y_0) > 0$, then all solutions to (1.7) tend to $\pm\infty$ as $t \to \infty$, so 0 is unstable.

According to Theorem 9.8 of Section 2.9, the equilibrium point y_0 for the differential equation (1.3) is stable if $f'(y_0) < 0$ and unstable if $f'(y_0) > 0$. Thus, the behavior of the solutions to (1.3) for $y = y_0 + u$ near the equilibrium point y_0 is the same as the behavior of the solutions of (1.7) for $\tilde{u}$ near its equilibrium point at 0. A proof of Theorem 9.8 can be constructed starting with the considerations that led from equation (1.3) to equation (1.7). More important, it provides a starting point for analyzing what happens in higher dimensions.

The linear equation in (1.7) is called the *linearization of (1.3) at the equilibrium point* y_0. The key observation is that, in many circumstances, the solutions to the linearization reflect the behavior of the solutions to the original equation.

The two-dimensional case

Let's look at a planar, autonomous system

$$x' = f(x, y),$$
$$y' = g(x, y). \tag{1.8}$$

Again we will look at solutions that start near an equilibrium point (x_0, y_0).

The algebraic definition of the derivative works in several variables as it does in one, but the formulas are more complicated. If f and g are differentiable at (x_0, y_0), we have

$$f(x_0 + u, y_0 + v) = f(x_0, y_0) + \frac{\partial f}{\partial x}(x_0, y_0)u + \frac{\partial f}{\partial y}(x_0, y_0)v + R_f(u, v),$$
$$\tag{1.9}$$
$$g(x_0 + u, y_0 + v) = g(x_0, y_0) + \frac{\partial g}{\partial x}(x_0, y_0)u + \frac{\partial g}{\partial y}(x_0, y_0)v + R_g(u, v).$$

The remainders $R_f(u, v)$ and $R_g(u, v)$ get very small as the pair (u, v) gets small. Precisely, we have

$$\frac{R_f(u, v)}{\sqrt{u^2 + v^2}} \to 0 \quad \text{and} \quad \frac{R_g(u, v)}{\sqrt{u^2 + v^2}} \to 0 \quad \text{as} \quad \sqrt{u^2 + v^2} \to 0.$$

Thus the remainder terms get very small, even in comparison to $\sqrt{u^2 + v^2}$, which measures the distance from (x_0, y_0) to $(x_0 + u, y_0 + v)$.

Since (x_0, y_0) is an equilibrium point for the system (1.8), we have

$$f(x_0, y_0) = g(x_0, y_0) = 0.$$

Consequently, (1.9) becomes

$$f(x_0 + u, y_0 + v) = \frac{\partial f}{\partial x}(x_0, y_0)u + \frac{\partial f}{\partial y}(x_0, y_0)v + R_f(u, v),$$
$$\tag{1.10}$$
$$g(x_0 + u, y_0 + v) = \frac{\partial g}{\partial x}(x_0, y_0)u + \frac{\partial g}{\partial y}(x_0, y_0)v. + R_g(u, v).$$

Let's introduce u and v as new variables in the system (1.8) by setting $x = x_0 + u$ and $y = y_0 + v$. Notice that $x' = (x_0 + u)' = u'$ and $y' = (y_0 + v)' = v'$. From (1.8), we get

$$u' = x' = f(x, y) = f(x_0 + u, y_0 + v) \quad \text{and}$$
$$v' = y' = g(x, y) = g(x_0 + u, y_0 + v).$$

Using (1.10), this becomes

$$u' = \frac{\partial f}{\partial x}(x_0, y_0)u + \frac{\partial f}{\partial y}(x_0, y_0)v + R_f(u, v),$$

$$v' = \frac{\partial g}{\partial x}(x_0, y_0)u + \frac{\partial g}{\partial y}(x_0, y_0)v + R_g(u, v). \tag{1.11}$$

Again, we are interested in what happens to solutions to (1.8) for $(x, y) = (x_0 + u, y_0 + v)$ near the equilibrium point (x_0, y_0). This translates into solutions of (1.11) near the equilibrium point at the origin, $(0, 0)$. For values of (u, v) near the origin, the remainders $R_f(u, v)$ and $R_g(u, v)$ are very small, so it is interesting to compare the system in (1.11) to the system where we ignore the remainders,

$$\tilde{u}' = \frac{\partial f}{\partial x}(x_0, y_0)\tilde{u} + \frac{\partial f}{\partial y}(x_0, y_0)\tilde{v},$$

$$\tilde{v}' = \frac{\partial g}{\partial x}(x_0, y_0)\tilde{u} + \frac{\partial g}{\partial y}(x_0, y_0)\tilde{v}. \tag{1.12}$$

The system in (1.12) is called the **linearization of (1.8) at the equilibrium point** (x_0, y_0). The linearization is linear, homogeneous, and has constant coefficients. Thus we can find the solutions explicitly using the techniques of Chapter 9. It should be emphasized that the systems in (1.11) and (1.12) are not the same. They do not have the same solutions. The best we can expect is that there will be qualitative similarity between the solutions to the two systems near their respective equilibrium points. The solutions of the linearization can be found explicitly. If there is a qualitative similarity, it can be used to discover qualitative information about the solutions to the nonlinear system. This is the approach that we will take in what follows.

If we introduce the vector $\mathbf{u} = (\tilde{u}, \tilde{v})^T$, we can write the linearization in vector form as

$$\mathbf{u}' = J\mathbf{u}, \tag{1.13}$$

where the matrix J is

$$J = J(x_0, y_0) = \begin{pmatrix} \dfrac{\partial f}{\partial x}(x_0, y_0) & \dfrac{\partial f}{\partial y}(x_0, y_0) \\[2ex] \dfrac{\partial g}{\partial x}(x_0, y_0) & \dfrac{\partial g}{\partial y}(x_0, y_0) \end{pmatrix}. \tag{1.14}$$

The matrix $J(x_0, y_0)$ in (1.14) is called the **Jacobian** of $(f, g)^T$ at the point $(x_0, y_0)^T$.

It should be mentioned that to do the analysis leading to the results that will follow, we will need the hypothesis that the matrix $J(x_0, y_0)$ is nonsingular. It is only with this condition that the remainder terms in (1.11) are smaller than the other terms on the right-hand sides of these equations. This assumption for dimension 2 is similar to assuming that $f'(y_0) \neq 0$ in dimension 1, which leads to (1.5).

Interacting species

We will pause to derive a model of the interaction between species. Variants of this model will provide examples for us as we proceed through this chapter. In Section 3.1, we studied single populations, and in Section 8.2, we derived the Lotka-Volterra model of the interaction between a prey population and its predator. We will enlarge on these examples to model a variety of interactions between species. You are encouraged to review Sections 3.1 and 8.2, since the more general model is derived using the same principles used there.

Let's start with two species with populations $x_1(t)$ and $x_2(t)$. Let r_1 and r_2 be the reproductive rates. Then our basic model of the evolution of the two species is

$$\begin{aligned} x_1' &= r_1 x_1, \\ x_2' &= r_2 x_2. \end{aligned} \tag{1.15}$$

To complete the model, we must derive the form of the reproductive rates.

Let's look at r_1 and suppose first that $x_2 = 0$. In the absence of another population, we are looking at the growth of a single population. We discussed this in Section 3.1. There we derived two models. In the Malthusian model, $r_1 = a_1$ is constant, and in the logistic model, $r_1 = a_1 - b_1 x_1$. We can model both of these with

$$r_1 = a_1 - b_1 x_1 \quad \text{if} \quad x_2 = 0. \tag{1.16}$$

If $b_1 = 0$, this is the Malthusian model, while if $b_1 > 0$, it is the logistic model. If $b_1 > 0$, the population has a finite limit as t increases. In this case, we will sometimes say that the population x_1 has a *logistic limit*.

If $x_2 > 0$, we have to account for the interactions between the two species. When we discussed the predator–prey model in Section 8.2, we pointed out that the reproductive rate r_1 measured the probability of one individual surviving and reproducing. To a first approximation, the effect of interactions with the second population on the reproductive rate is to increase or decrease the rate shown in (1.16) by an amount proportional to the number of encounters per unit time. If we assume that the populations are well distributed in space, then the number of encounters is proportional to x_2. Hence, to a first approximation, the reproductive rate has the form

$$r_1 = a_1 - b_1 x_1 + c_1 x_2, \tag{1.17}$$

where a_1, b_1, and c_1 are all constants.

Similarly, the reproductive rate for x_2 is

$$r_2 = a_2 - b_2 x_2 + c_2 x_1.$$

The constants in this formula have an interpretation similar to those in (1.17). Consequently, from (1.15), the model of the interaction between the species is

$$\begin{aligned} x_1' &= (a_1 - b_1 x_1 + c_1 x_2) x_1, \\ x_2' &= (a_2 - b_2 x_2 + c_2 x_1) x_2. \end{aligned} \tag{1.18}$$

Let's review the role of the constants in (1.18). The constant a_i is the reproductive rate for x_i when both populations are very small. It can be positive or negative. Next, the nonnegative constant b_i measures the interaction of the population x_i with itself. If $b_i = 0$, we are assuming a Malthusian growth rate for x_i. If it is positive, then the growth is modeled by the logistic equation, and there is a logistic limit on the size of the species.

The constants c_1 and c_2 measure the interactions between the two species. For example, c_1 measures the effect of the presence of x_2 on the species x_1. If the presence of x_2 has a positive effect on x_1, then c_1 is positive. This would be the case, for example, if the two populations cooperated in the use of resources or if x_1 preyed upon x_2. On the other hand, c_1 is negative if the presence of x_2 has a deleterious effect on x_1. This would be the case if the two species competed for resources or if x_2 preyed upon x_1. The effect of x_1 on x_2 is measured by c_2. Its interpretation is exactly the same as for c_1 with the roles of the two species reversed.

Let's look at an example. Suppose we have two populations that compete for resources. An example would be cattle and sheep, which competed for food and space in some western states during the late nineteenth century. Both species thrive if the populations are small, so a_1 and a_2 are positive. In the absence of the other species, they are governed by the logistic model, so b_1 and b_2 are both positive. Since they compete for resources, each has a negative effect of the growth of the other. Hence c_1 and c_2 are negative.

Example 1.19   Consider the competing species system

$$x' = (1 - x - y)x,$$
$$y' = (4 - 7x - 3y)y. \tag{1.20}$$

Find and analyze the linearization at the equilibrium point in the positive quadrant.

To find the equilibrium points, we solve the system

$$(1 - x - y)x = 0,$$
$$(4 - 7x - 3y)y = 0.$$

The solutions to the first equation, $(1 - x - y)x = 0$, form the x-nullcline, which is the union of the two lines defined by $x = 0$ and $x + y = 1$. The y-nullcline consists of the solutions to the second equation, $(4 - 7x - 3y)y = 0$. It is the union of the two lines defined by $y = 0$ and $7x + 3y = 4$. The equilibrium points are the intersections of the nullclines. There are four, $(0, 0)$, $(0, 4/3)$, $(1, 0)$, and $(1/4, 3/4)$. Only the last of these is in the positive quadrant.

Next we compute the Jacobian

$$J(x, y) = \begin{pmatrix} \dfrac{\partial f}{\partial x} & \dfrac{\partial f}{\partial y} \\[2mm] \dfrac{\partial g}{\partial x} & \dfrac{\partial g}{\partial y} \end{pmatrix} = \begin{pmatrix} 1 - 2x - y & -x \\ -7y & 4 - 7x - 6y \end{pmatrix}.$$

At the equilibrium point $(1/4, 3/4)$, this becomes

$$J(1/4, 3/4) = \begin{pmatrix} -1/4 & -1/4 \\ -21/4 & -9/4 \end{pmatrix}.$$

The linearization is the system $\mathbf{u}' = J\mathbf{u}$, with J as its matrix of coefficients. Since $\det(J) = -3/4$ is negative, the linearization has a saddle point at the origin.   ●

The phase plane for the nonlinear system in (1.20) is shown in Figure 1, including the equilibrium points and several solutions. There are two solutions that end at the equilibrium point at $(1/4, 3/4)$ as $t \to \infty$, and two that end there as $t \to -\infty$. These solutions are plotted in blue.

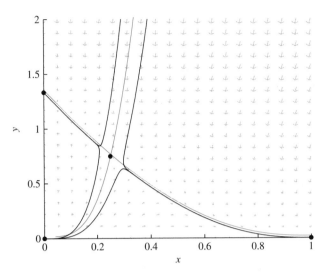

Figure 1. The phase portrait for the competing species system in Example 1.19.

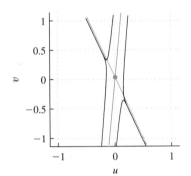

Figure 2. The phase portrait for the linearization of the competing species system in Example 1.19.

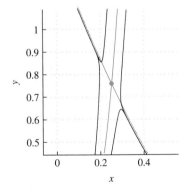

Figure 3. A closer look at a neighborhood of the equilibrium point in Figure 1.

The phase portrait for the linearization at $(1/4, 3/4)$ is shown in Figure 2. The half-line solutions are in blue. While there are some remarkable similarities between the two phase planes, there are also some significant differences. In particular, the phase plane for the linearization (Figure 2) shows only one equilibrium point. It is a feature of the linearization that it focuses on only one of the equilibrium points of the nonlinear system. It is only when we get far from the equilibrium point $(1/4, 3/4)$ in the phase plane for the nonlinear system (Figure 1) that we see differences from the linear system.

If we zoom in on this equilibrium point in Figure 1 and at the same time make the scales along the axes the same for both, we get Figure 3 for the nonlinear system. You will notice that Figures 2 and 3 are very similar. In fact, the only real difference is in the labels along the axes. These reflect the fact that the equilibrium point for the nonlinear system is $(1/4, 3/4)$, while that for the linearization is $(0, 0)$.

Example 1.19 demonstrates how closely the behavior of solutions to the linearization can mimic the solutions of the original nonlinear system, at least if we limit our observation to a neighborhood of the equilibrium point. Is this always true? Unfortunately not, but there are important cases where it does happen.

Characterization of equilibrium points

In Section 9.3, we examined a number of different phase portraits for a linear, planar system

$$\mathbf{y}' = A\mathbf{y}.$$

Recall that we found the following generic types:

- **Saddle point.** A has two real eigenvalues, one positive and one negative.
- **Nodal sink.** A has two negative real eigenvalues.
- **Nodal source.** A has two positive real eigenvalues.
- **Spiral sink.** A has two complex-conjugate eigenvalues with negative real parts.
- **Spiral source.** A has two complex-conjugate eigenvalues with positive real parts.

If we refer back to the phase-determinant plane in Figure 2 in Section 9.4, we see that each of these types corresponds to a large open subset of the trace-determinant plane. This is why we called these types generic. While almost all points in the trace-determinant plane correspond to one of the five generic types, there are exceptions. All of these exceptional types of equilibrium points are called nongeneric. For example, a center is nongeneric. It is when the linearization has a generic equilibrium point that the behavior of the linearization echoes the behavior of the nonlinear system.

THEOREM 1.21   Consider the planar system

$$
\begin{aligned}
x' &= f(x, y), \\
y' &= g(x, y),
\end{aligned}
\tag{1.22}
$$

where the functions f and g are continuously differentiable. Suppose that (x_0, y_0) is an equilibrium point. If the linearization of (1.22) at (x_0, y_0) has a generic equilibrium point at the origin, then the equilibrium point for the system in (1.22) at (x_0, y_0) is of the same type. ∎

For the competing species system in Example 1.19, we saw that the linearization had a saddle point. Since a saddle point is generic, Theorem 1.21 applies and predicts that the nonlinear system will have a saddle point at $(1/4, 3/4)$. Saddle points are characterized by having exactly two stable solutions that tend to the point as $t \to \infty$ and two unstable solutions that tend to the point as $t \to -\infty$. These curves are plotted in blue in Figure 1. Notice that the same behavior is shown by the linearization in Figure 2.

Example 1.23   Consider the following model of predator–prey populations, with a logistic limit on the prey population,

$$
\begin{aligned}
F' &= (0.4 - 0.002F - 0.01S)F \\
S' &= (-0.3 + 0.005F)S.
\end{aligned}
\tag{1.24}
$$

Analyze the equilibrium point in the positive quadrant.

The system has three equilibrium points, but only $(60, 28)$ is in the positive quadrant. Let's compute the Jacobian,

$$
J(F, S) = \begin{pmatrix} 0.4 - 0.004F - 0.01S & -0.01F \\ 0.005S & -0.3 + 0.005F \end{pmatrix}.
$$

At the equilibrium point $(60, 28)$ this becomes

$$
J = \begin{pmatrix} -0.12 & -0.6 \\ 0.14 & 0 \end{pmatrix}.
$$

We compute that $D = \det(J) = 0.084$ and $T = \operatorname{tr}(J) = -0.12$. Since $D > 0$, $T < 0$, and $T^2 - 4D = -0.3216 < 0$, the linearization has a spiral sink at the origin. Since a spiral sink is generic, the predator–prey system has a spiral sink at $(60, 28)$.

The phase plane for our system with one solution drawn is shown in Figure 4. The phase plane for the linearization is shown in Figure 5. Notice the similarity. ●

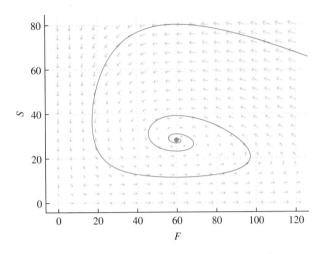

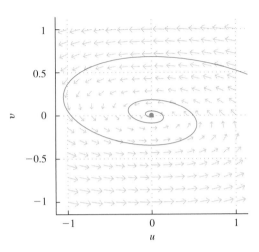

Figure 4. The phase portrait for the predator–prey system in Example 1.23

Figure 5. The phase portrait for the linearization of the predator–prey system in Example 1.23.

Centers

A center is nongeneric, so if the linearization of a planar system has a center, we cannot apply Theorem 1.21. Unfortunately, if the linearization has a center, the equilibrium point for the nonlinear system may fail to be a center. The solution curves can have very different behavior from the closed orbits of the linearization. This is demonstrated by the next example.

Example 1.25 Consider the system

$$x' = y + \alpha x(x^2 + y^2),$$
$$y' = -x + \alpha y(x^2 + y^2). \tag{1.26}$$

For $\alpha = 5$ and $\alpha = -5$, find the linearization at the origin. Compare the behavior of the solutions of (1.26) with that of the linearization.

The Jacobian is

$$J(x, y) = \begin{pmatrix} 3\alpha x^2 + \alpha y^2 & 1 + 2\alpha xy \\ -1 + 2\alpha xy & \alpha x^2 + 3\alpha y^2 \end{pmatrix}.$$

At the origin, this becomes

$$J(0, 0) = \begin{pmatrix} 0 & 1 \\ -1 & 0 \end{pmatrix}.$$

Hence, the linearization at the origin is

$$u' = v,$$
$$v' = -u.$$

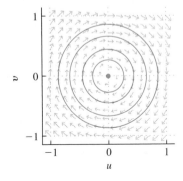

Figure 6. The linearization of the system in Example 1.25.

The interesting thing is that there is no dependence on α. In fact, except for the change of notation, the linearization is the same system as (1.26) with $\alpha = 0$.

The determinant of $J(0, 0)$ is 1 and the trace is 0, so the linearization has a center at the origin. Thus the solution curves for the linearization are closed curves surrounding the origin. Some solutions are shown in Figure 6.

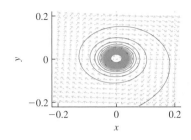

Figure 7. Solutions to (1.26) with $\alpha = 5$.

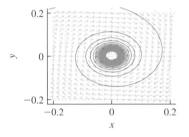

Figure 8. Solutions to (1.26) with $\alpha = -5$.

The situation with the nonlinear system is quite different. A solution curve for $\alpha = 5$ is shown in Figure 7. The curve spirals out from the origin as t increases. As $t \to -\infty$, the curve spirals in. The numerically computed curve seems to stop before it actually gets to the origin, but we need to examine that more closely. On the other hand, Figure 8 shows a solution curve to the system with $\alpha = -5$. Here we seem to have the opposite behavior. Now the solution curve spirals in as $t \to \infty$ and seems to stop short of the origin.

To see what actually happens near the origin, suppose that $x(t)$, $y(t)$ is a solution to the nonlinear system, and consider the function

$$r(t) = \sqrt{x(t)^2 + y(t)^2}.$$

Since $r(t)$ is the distance from the point $(x(t), y(t))$ to the origin, we are only interested in where $r > 0$. We can use the system (1.26) to compute the derivative of r. It is easier to compute the derivative of $r^2 = x(t)^2 + y(t)^2$. First we have $(r^2)' = 2rr'$. Hence, using the system in (1.26),

$$
\begin{aligned}
2rr' &= (r^2)' \\
&= \left(x(t)^2 + y(t)^2\right)' \\
&= 2xx' + 2yy' \\
&= 2x(y + \alpha x(x^2 + y^2)) + 2y(-x + \alpha y(x^2 + y^2)) \\
&= 2\alpha(x^2 + y^2)^2 \\
&= 2\alpha r^4.
\end{aligned}
$$

Thus, for $r \neq 0$,

$$r' = \alpha r^3. \tag{1.27}$$

This autonomous equation in (1.27) can be solved,[1] but qualitative analysis of the solutions will be enough for our purposes. If $\alpha > 0$, the equation in (1.27) has an unstable equilibrium point at $r = 0$. Therefore, $r(t)$ is increasing as t increases, and $r(t) \to 0$ as $t \to -\infty$. Since $r(t)$ is the distance from $(x(t), y(t))$ to the origin, this means that the solution curve in Figure 7 moves away from the origin as t increases and approaches the origin as $t \to -\infty$. Consequently, the nonlinear system has a spiral source at the origin, although its linearization has a center.

By a similar argument, we see that if $\alpha < 0$, the solution tends to the origin as $t \to \infty$. In this case, the nonlinear system has a spiral sink at the origin, while its linearization has a center. Thus we have two nonlinear systems (equation (1.26) with $\alpha < 0$ and $\alpha > 0$) having the same linearization with a center at the origin. Nevertheless, one has a spiral sink and the other has a spiral source. The computer-generated curves are misleading in this case. Actually, the curves spiral so rapidly and approach the origin so slowly that it is difficult to get the computer to show this behavior. ●

As the example shows, linear analysis is inconclusive if the linearization has a center at the origin. Remember that an equilibrium point for a planar autonomous system is a center if it is surrounded by closed periodic solution curves. Unfortunately, an equilibrium point for a nonlinear system can fail to be a center even though the linearization at that point has a center.

In the five generic cases covered by Theorem 1.21, the linearization gives very precise information. As we pointed out, the five generic cases correspond to large

---

[1] The solution is $r(t) = 1/\sqrt{C - 2\alpha t}$.

open parts of the trace-determinant plane, as is shown in Figure 2 in Section 9.4. Nongeneric types of equilibrium points correspond to the boundaries of these regions in the trace-determinant plane. This is not an accident. Nonlinear systems can be considered as perturbations of linear systems. These perturbations can cause slight movements of the system as represented in the trace-determinant plane. If the system is generic and corresponds to a point in one of the large open areas, the small perturbation cannot move that point out of that area. On the other hand, if the system is nongeneric, corresponding to a point on one of the boundaries, a small movement can change the system dramatically. Example 1.25 shows that clearly.

EXERCISES

For each of the systems in Exercises 1–8, perform each of the following tasks.

(i) Sketch the nullclines for each equation. Use a distinctive marking for each nullcline so they can be distinguished.

(ii) Use analysis to find the equilibrium points for the system. Label each equilibrium point on your sketch with its coordinates.

(iii) Use the Jacobian to classify each equilibrium point (spiral source, nodal sink, etc.).

1. $x' = x(6 - 3x) - 2xy$
$y' = y(5 - y) - xy$

2. $x' = x(6 - 2x - 3y)$
$y' = y(1 - x - y)$

3. $x' = 2x - 2x^2 - xy$
$y' = 2y - xy - 2y^2$

4. $x' = 2x(1 - x/2) - xy$
$y' = 4y(1 - y/4) - 4xy$

5. $x' = x(4y - 5)$
$y' = y(3 - x)$

6. $x' = 1.2x - xy$
$y' = -0.5y + xy$

7. $x' = y$
$y' = -\sin x - y$

8. $x' = y$
$y' = -\cos x - 0.5y$

For Exercises 9–16, use your numerical solver to compare the phase portrait of the nonlinear system with that of its linearization near the indicated equilibrium point.

9. Exercise 1, $(2, 0)$

10. Exercise 2, $(3, 0)$

11. Exercise 3, $(2/3, 2/3)$

12. Exercise 4, $(2/3, 4/3)$

13. Exercise 5, $(3, 5/4)$

14. Exercise 6, $(1/2, 6/5)$

15. Exercise 7, $(2\pi, 0)$

16. Exercise 8, $(-\pi/2, 0)$

17. Consider the competing-species model of Example 1.19.

$$x' = (1 - x - y)x$$
$$y' = (4 - 7x - 3y)y$$

(a) Use your numerical solver to sketch the phase portrait of this system on the domain $D = \{(x, y) : -1 \leq x \leq 2, -1 \leq y \leq 2\}$. Label the equilibrium points with their coordinates.

(b) Show that the substitutions $x = 1/4 + u$ and $y = 3/4 + v$ transform the competing-species model to

$$u' = -\frac{1}{4}u - \frac{1}{4}v - u^2 - uv,$$

$$v' = -\frac{21}{4}u - \frac{9}{4}v - 3v^2 - 7uv.$$

Sketch the phase portrait of this system on D and label the equilibrium points with their coordinates. Explain the effect of the transformations $x = 1/4 + u$ and $y = 3/4 + v$.

(c) If you throw away the terms of the system in part (b) having degree 2 or higher, what relation does the resulting system have with linearization provided by the Jacobian in Example 1.19?

18. Consider the predator–prey model of Example 1.23.

$$F' = (0.4 - 0.002F - 0.01S)F$$
$$S' = (-0.3 + 0.005F)S$$

(a) Use your numerical solver to sketch the phase portrait of this system on the domain $D = \{(F, S) : -100 \leq F \leq 250, -80 \leq S \leq 80\}$. Label the equilibrium points with their coordinates.

(b) Show that the substitutions $F = 60 + u$ and $S = 28 + v$ transform the competing-species model to

$$u' = -\frac{3}{25}u - \frac{3}{5}v - \frac{1}{500}u^2 - \frac{1}{100}uv,$$

$$v' = \frac{7}{50}u + \frac{1}{200}uv.$$

Sketch the phase portrait of this system on D and label the equilibrium points with their coordinates. Explain the effect of the transformations $F = 60 + u$ and $S = 28 + v$.

(c) If you throw away the terms of the system in part (b) having degree 2 or higher, what relation does the resulting system have with linearization provided by the Jacobian in Example 1.23?

19. Transforming a system into polar coordinates can sometimes aid in the analysis of the behavior of the system near an equilibrium point. In Figure 9, the Cartesian coordinates of the point P are (x, y), but the polar coordinates of the point P are (r, θ), where r is the radial length and θ is the angle made with the positive x-axis (counterclockwise rotations assumed positive). It is easily seen that $x = r \cos \theta$, $y = r \sin \theta$,

$$r^2 = x^2 + y^2 \quad \text{and} \quad \tan \theta = \frac{y}{x}.$$

(a) Use these last two relations to prove that

$$r\frac{dr}{dt} = x\frac{dx}{dt} + y\frac{dy}{dt} \quad \text{and} \quad \frac{d\theta}{dt} = \frac{1}{r^2}\left(x\frac{dy}{dt} - \frac{dx}{dt}y\right).$$

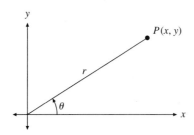

Figure 9. Polar coordinate transformations.

(b) Consider again the system of Example 1.25,

$$\frac{dx}{dt} = y + \alpha x(x^2 + y^2),$$

$$\frac{dy}{dt} = -x + \alpha y(x^2 + y^2).$$

Use the transformations developed in part (a) to show that the equivalent system, in polar coordinates, is given by

$$\frac{dr}{dt} = \alpha r^3,$$

$$\frac{d\theta}{dt} = -1.$$

(c) Use the polar form of the system found in part (b) to argue that the equilibrium point $(0, 0)$ of the original system is a spiral source if $\alpha = 5$ and a spiral sink if $\alpha = -5$.

20. The system

$$\frac{dx}{dt} = -y - x^3$$

$$\frac{dy}{dt} = x \tag{1.28}$$

has an isolated equilibrium point at $(0, 0)$.

(a) Use the Jacobian to produce a linearization of the system near the equilibrium point $(0, 0)$. What type of equilibrium point is predicted by this linearization?

(b) Use your numerical solver to compare the phase portraits of (1.28) and its linearization near $(0, 0)$.

(c) Use the polar-Cartesian transformations in part (a) of Exercise 19 and (1.28) to show that $dr/dt = -x^4/r$.

(d) Show that $d\theta/dt = 1 + x^3y/r^2$; then show $x^3y/r^2 \to 0$ as $r \to 0$. *Hint:* Use polar coordinates.

(e) Use the results of parts (c) and (d) to explain the behavior of solution trajectories of (1.28) near $(0, 0)$.

In Exercises 21–25, find the signs of the coefficients in the model in (1.18) required to model the pair of interacting species described. Use logistic limits unless told otherwise, or unless the reproductive rate when the populations are small is negative. (Possible signs are positive, negative, or zero.)

21. The first species is a parasite and the second is the host. The presence of the parasite is beneficial to the host. (If you are are puzzled, look up the definition of parasite.)

22. The first species is a parasite and the second is the host. The presence of the parasite is harmful to the host.

23. Two species are dependent on each other for their continued existence, and they cooperate to better their situation.

24. The first species is a predator preying on the second but is able to survive in the absence of the prey.

25. Two species are healthy on their own, and they cooperate to create better circumstances for both.

26. Create a model for a food chain consisting of three kinds of fish, where the second preys on the first and the third preys on the second. The first species thrives on its own but has limited resources. The second species will survive without the first but has limited resources. The third will die out if the second is not present.

27. Create a general model for three intracting species along the lines of the model for two in (1.18). Include all possible interactions in your model. Interpret the signs of all of the constants.

10.2 Long-Term Behavior of Solutions

In the previous section, we made a start on the problem of deciding the answer to one of the most important questions about systems of differential equations, "What happens to all solutions as $t \to \infty$?" Let's remind ourselves of some terminology we introduced earlier.

Stability

In Section 9.7, we defined the terms *stable*, *unstable*, and *asymptotically stable* and applied them to linear systems. The main result was Theorem 7.4 of that section. Now we will find analogous results for nonlinear systems. Refer back to Section 9.7 for the terminology used.

Linearization of higher-dimensional systems

Suppose we have an autonomous nonlinear system of dimension n,

$$\mathbf{y}' = \mathbf{f}(\mathbf{y}). \tag{2.1}$$

The solution $\mathbf{y}$ is a vector-valued function, $\mathbf{y}(t) = (y_1(t), y_2(t), \ldots, y_n(t))^T$. The right-hand side is also vector valued, $\mathbf{f}(\mathbf{y}) = (f_1(\mathbf{y}), f_2(\mathbf{y}), \ldots, f_n(\mathbf{y}))^T$.

We want to explore the behavior of solutions near an equilibrium point $\mathbf{y}_0$. The algebraic definition of the derivative has in general an interpretation like that in (1.10), but now we can state it more succinctly as

$$\mathbf{f}(\mathbf{y}_0 + \mathbf{u}) = \mathbf{f}(\mathbf{y}_0) + J(\mathbf{y}_0)\mathbf{u} + \mathbf{R}(\mathbf{u}),$$

where the J is the Jacobian matrix of $\mathbf{f}$ at $\mathbf{y}_0$,

$$J = \begin{pmatrix} \dfrac{\partial f_1}{\partial y_1} & \dfrac{\partial f_1}{\partial y_2} & \cdots & \dfrac{\partial f_1}{\partial y_n} \\[2mm] \dfrac{\partial f_2}{\partial y_1} & \dfrac{\partial f_2}{\partial y_2} & \cdots & \dfrac{\partial f_2}{\partial y_n} \\[2mm] \vdots & \vdots & \ddots & \vdots \\[2mm] \dfrac{\partial f_n}{\partial y_1} & \dfrac{\partial f_n}{\partial y_2} & \cdots & \dfrac{\partial f_n}{\partial y_n} \end{pmatrix}, \tag{2.2}$$

and the remainder $\mathbf{R}(\mathbf{u})$ satisfies

$$\frac{\mathbf{R}(\mathbf{u})}{|\mathbf{u}|} \to 0 \quad \text{as} \quad \mathbf{u} \to \mathbf{0}.$$

Proceeding as we did in the two-dimensional case, we introduce the ***linearization*** at the equilibrium point $\mathbf{y}_0$ by

$$\mathbf{u}' = J\mathbf{u}.$$

Given our success in two dimensions, we expect that we will be able to say something productive about the equilibrium point at $\mathbf{y}_0$ by analyzing the linearization. Indeed this is true, but we cannot expect the specificity of Theorem 1.21 in the general case.

THEOREM 2.3   Suppose that $\mathbf{y}_0$ is an equilibrium point for the autonomous system $\mathbf{y}' = \mathbf{f}(\mathbf{y})$. Let J be the Jacobian of $\mathbf{f}$ at $\mathbf{y}_0$.

1. Suppose that the real part of every eigenvalue of J is negative. Then $\mathbf{y}_0$ is an asymptotically stable equilibrium point.

2. Suppose that J has at least one eigenvalue with positive real part. Then $\mathbf{y}_0$ is an unstable equilibrium point. ■

Refer back to Section 9.7, where we analyzed the stability of linear systems. In particular, Theorem 7.4 in that section is the corresponding theorem for linear systems. Indeed that theorem is the starting point of the difficult proof of Theorem 2.3, which we will omit.

Example 2.4 Consider the system

$$x' = -4x + y - xy + x^2$$
$$y' = -x - 2y + y^2.$$

Analyze the equilibrium point at the origin.

We compute that the Jacobian at the origin is

$$J = \begin{pmatrix} -4 & 1 \\ -1 & -2 \end{pmatrix}.$$

We have $D = \det(J) = 9$, $T = \text{tr}(J) = -6$, and $T^2 - 4D = 0$. Consequently, (T, D) lies on the parabola $T^2 - 4D = 0$, and J has only one eigenvalue $\lambda = -3$ with multiplicity 2. Therefore, the equilibrium point at the origin is nongeneric. Hence, Theorem 1.21 does not apply. However, since both eigenvalues are negative, Theorem 2.3 does apply and we know that the origin is an asymptotically stable equilibrium point. The phase portrait for this system is shown in Figure 1. ●

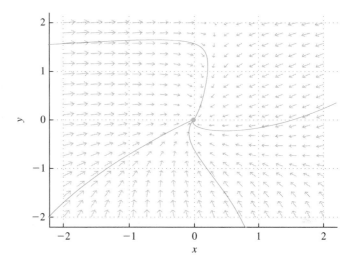

Figure 1. The phase plane portrait for the system in Example 2.4. The origin is a stable equilibrium point, although it is degenerate.

The example shows that although Theorem 1.21 gives more precise information when it is applicable, Theorem 2.3 sometimes provides information when Theorem 1.21 does not.

The Lorenz system

In 1963, the meteorologist and mathematician E. N. Lorenz discovered the very interesting three-dimensional system

$$x' = -ax + ay$$
$$y' = rx - y - xz \qquad (2.5)$$
$$z' = -bz + xy.$$

The system is known as the **Lorenz** system. It represents a simplified model for atmospheric turbulence beneath a thunderhead. The parameters a, b, and r are positive constants.

Let's find the equilibrium points. We must solve

$$-ax + ay = 0,$$
$$rx - y - xz = 0,$$
$$-bz + xy = 0.$$

From the first equation, we see that $x = y$. Setting $y = x$ in the second equation, we conclude that $x(r - 1 - z) = 0$, so $x = 0$ or $z = r - 1$. Again setting $y = x$, the third equation becomes $bz = x^2$. If $x = 0$, we see that $y = z = 0$ as well. Thus, the origin $(0, 0, 0)$ is an equilibrium point. If $z = r - 1$, then $x = y = \pm\sqrt{b(r-1)}$. These will be real only if $r \geq 1$. Consequently, there are three equilibrium points if $r > 1$ and one if $r \leq 1$.

We compute that the Jacobian is

$$J(x, y, z) = \begin{pmatrix} -a & a & 0 \\ r - z & -1 & -x \\ y & x & -b \end{pmatrix}. \tag{2.6}$$

At the origin, this becomes

$$J(0, 0, 0) = \begin{pmatrix} -a & a & 0 \\ r & -1 & 0 \\ 0 & 0 & -b \end{pmatrix}.$$

Since $J = J(0, 0, 0)$ has two zeros in the bottom row, we can compute the characteristic polynomial by expanding along the bottom row. We get

$$\det(J - \lambda I) = (-b - \lambda)\left((-a - \lambda)(-1 - \lambda) - ar\right)$$
$$= -(\lambda + b)(\lambda^2 + (a + 1)\lambda + a(1 - r)).$$

Thus, the eigenvalues are $-b$ and the roots of $\lambda^2 + (a + 1)\lambda + a(1 - r) = 0$. By the quadratic formula, these are

$$\lambda = \frac{1}{2}\left[-(a + 1) \pm \sqrt{(a + 1)^2 - 4a(1 - r)}\right]. \tag{2.7}$$

The discriminant

$$(a + 1)^2 - 4a(1 - r) = (a - 1)^2 + 4ar \tag{2.8}$$

is always positive, so the roots in (2.7) are real. The root with a plus sign in (2.7) will be negative if the discriminant (2.8) is smaller than $(a + 1)^2$. This happens if and only if $r < 1$. Therefore, for $r < 1$, all of the eigenvalues of the Jacobian are negative, and by Theorem 2.3, the Lorenz system has an asymptotically stable equilibrium point at the origin. On the other hand, if $r > 1$, the eigenvalue with the plus sign in (2.7) is positive, so the origin is unstable.

Example 2.9 With the parameters $a = 10$, $b = 8/3$, and $r = 1/2$, compute and plot the solution to the Lorenz system with initial conditions $x(0) = -2$, $y(0) = -1$, and $z(0) = 1$.

The required solution is shown in Figure 2. All three components converge to 0 as t increases, so this solution converges to the sink at the origin. ●

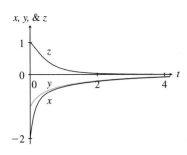

Figure 2. The solution to the Lorenz system in Example 2.9.

For $r > 1$, the Lorenz system has the two additional equilibrium points

$$\mathbf{c}^+ = \begin{pmatrix} \sqrt{b(r-1)} \\ \sqrt{b(r-1)} \\ r-1 \end{pmatrix} \quad \text{and} \quad \mathbf{c}^- = \begin{pmatrix} -\sqrt{b(r-1)} \\ -\sqrt{b(r-1)} \\ r-1 \end{pmatrix}.$$

The stability of these points is somewhat harder to establish, but it can be done. In what follows, we will assume that $a = 10$ and $b = 8/3$. These are the values that Lorenz used in his study. We will continue to allow r to vary.

It is possible, although difficult, to compute the eigenvalues of $J(\mathbf{c}^{\pm})$ and conclude using Theorem 2.3 that $\mathbf{c}^+$ and $\mathbf{c}^-$ are asymptotically stable if $1 < r < 470/19 \approx 24.74$, and unstable if $r > 470/19$. Thus when $r > 470/19$, all three equilibrium points are unstable.

Example 2.10   While it is difficult to analyze the equilibrium points $\mathbf{c}^+$ and $\mathbf{c}^-$ in general, it is relatively easy to determine the stability for specific values of r using the formula for the Jacobian in (2.6). With the parameters $a = 10$, $b = 8/3$, and $r = 15$, determine the stability of $\mathbf{c}^+$ and $\mathbf{c}^-$. Compute and plot the solution with initial conditions $x(0) = -2$, $y(0) = -1$, and $z(0) = 1$.

Using (2.6) and the parameter values, we find that

$$J(\mathbf{c}^+) = \begin{pmatrix} -10 & 10 & 0 \\ 1 & -1 & -6.11 \\ 6.11 & 6.11 & -8/3 \end{pmatrix}.$$

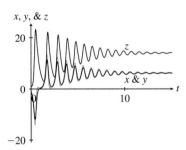

Figure 3. The solution to the Lorenz system in Example 2.10.

Using a computer, we find that the eigenvalues are approximately -12.9663 and $-0.3502 \pm 7.5804i$. Since the real parts are all negative, we see that $\mathbf{c}^+$ is asymptotically stable by Theorem 2.3. The stability of $\mathbf{c}^-$ is proved in the same way.

The solution is plotted in Figure 3. The x- and z-components are plotted in black and the y-component in blue. You will notice that the solution converges to $\mathbf{c}^+$.   ●

The behavior of the Lorenz system when all three equilibrium points are unstable is what was most interesting to Lorenz and to modern researchers as well. Here we can only touch on the subject. We will look at the case when $r = 28$. This was the case that Lorenz focused on. It can be shown that the solution curves for any initial conditions remain bounded. (See Exercise 28 in Section 10.7.) However, since $r = 28 > 470/19$, all of the equilibrium points are now unstable, so the solutions cannot converge to them. What does happen to the solutions is very interesting and came as quite a surprise to researchers in the field at the time.

An example is illustrated in Figures 4 and 5. Here we are computing and plotting the solution with initial conditions $x(0) = -2$, $y(0) = -1$, and $z(0) = 1$. In Figure 4, we plot the components of the solution with the y component shown in blue. There is a movement away from the initial conditions to a motion that seems to be centered around $\mathbf{c}^+$ (where x and y are positive), and then later around $\mathbf{c}^-$ (where x and y are negative). However, that motion does not show any regularity. This is true of any solution to the Lorenz system with these parameters.

In Figure 5, we present a three-dimensional plot of the solution. Actually, the plot shows the solution starting where the plot in Figure 4 ends. The point of this is to show where almost all solutions end up after the transients involved with the initial conditions are out of the system. The butterfly-shaped object is called the ***Lorenz attractor***. Indeed, almost every solution ends up approaching this set. The two wings of the butterfly are located near the two equilibrium points $\mathbf{c}^+$ and $\mathbf{c}^-$,

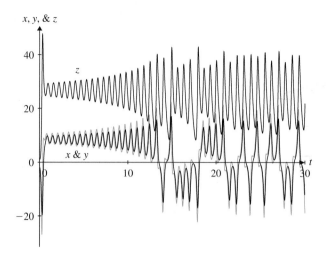

Figure 4. A solution to the Lorenz system with $r = 28$.

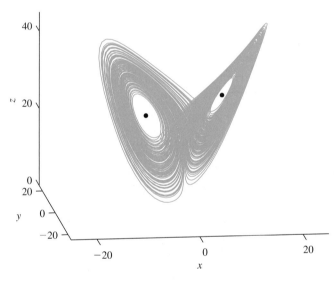

Figure 5. The Lorenz attractor.

which are plotted in black in Figure 5. Since these are unstable, the solutions do not converge to them. Notice that the solution rotates around one of these points and then moves over to a neighborhood of the other. This pattern repeats infinitely often.

We will have more to say about the Lorenz system in later sections.

EXERCISES

For each of the systems in Exercises 1–8, find the equilibrium points and analyze their stability.

1. $x' = x(1 - y)$
$\quad y' = x^2 - y$

2. $x' = x(2 - y)$
$\quad y' = y(-3 + x)$

3. $x' = (x + 9y)(1 - y)$
$\quad y' = -x - 5y$

4. $x' = x + y$
$\quad y' = y(1 - x^2)$

5. $x' = -x + 3z$
$\quad y' = -y + 2z$
$\quad z' = x^2 - 2z$

6. $x' = -x + z$
$\quad y' = -y + z$
$\quad z' = y^2 - 2z$

7. $x' = y - z$
$\quad y' = x - z$
$\quad z' = x^2 + y^2 - 2z$

8. $x' = y - z$
$\quad y' = x + z$
$\quad z' = y^2 - 2z$

For each of the systems in Exercises 9–16, show that the origin is an equilibrium point and analyze its stability.

9. $x' = -x + 4y + 4z + 2xy - yz$
$y' = -3y - 2z + x^2 + y^2$
$z' = 4y + 3z$

10. $x' = y + 1 - \cos x$
$y' = -x - 2y - z + \sin z$
$z' = -2x - 2y$

11. $x' = -5x + 6y + x^2 y$
$y' = -3x + y + e^z - 1$
$z' = -3x + 2y - \sin z$

12. $x' = -1 + e^x + 6y + 6z$
$y' = e^y - 1 + 3xz$
$z' = -3x - 15y - 5z + x^3$

13. $x' = 6x - 4y + x^3 - y^2$
$y' = 12x - 8y - 1 + \cos z^2$
$z' = 9x - 5y - 3z$

14. $x' = -5x - y + \sin z$
$y' = 2x - 2y + 1 - e^z$
$z' = 28x + 18y + z$

15. $x_1' = x_1 + 3x_2 - 3x_3$
$x_2' = -2\sin x_2 + x_3 + \ln(x_4 + 1)$
$x_3' = -x_3 + x_1(x_2 - x_4)$
$x_4' = -9x_1 - 9x_2 + 11x_3 - 2x_4$

16. $x_1' = -\sin x_1 + 1 - \cos x_2$
$x_2' = 2x_1 - x_2 - 2e^{x_4} + 2$
$x_3' = 6x_1 + 9x_2 - 2x_3 - 18x_4$
$x_4' = x_1 - 2x_4 + x_2^2 + x_3^2$

17. Find and analyze the equilibrium points for the Lorenz system when $a = 10$, $b = 8/3$, and $r = 7$. Numerically compute and plot a solution that starts near each of the equilibrium points.

18. Find and analyze the equilibrium points for the Lorenz system when $a = 10$, $b = 8/3$, and $r = 18$. Numerically compute and plot a solution that starts near each of the equilibrium points.

19. Find and analyze the equilibrium points for the Lorenz system when $a = 10$, $b = 8/3$, and $r = 28$. Numerically compute and plot a solution that starts near each of the equilibrium points.

20. In the text, we say that almost all solution curves for the Lorenz system tend to the Lorenz attractor as t increases. For the parameters $a = 10$, $b = 8/3$, and $r = 28$, find three different solution curves that do not tend to the attractor as t increases.

21. Consider a food chain consisting of three species with populations $x_1(t)$, $x_2(t)$, and $x_3(t)$, where x_1 preys upon x_2, which in turn preys upon x_3. Assume that both predator populations will die out in the absence of their prey. Assume a logistic limit for x_3 only.

(a) Show that the food chain is modeled by

$$x_1' = (-a_1 + b_1 x_2)x_1,$$
$$x_2' = (-a_2 - c_2 x_1 + b_2 x_3)x_2,$$
$$x_3' = (a_3 - b_3 x_2 - d x_3)x_3,$$

where all of the constants are positive.

(b) Consider the special case

$$x_1' = (-1 + 2x_2)x_1,$$
$$x_2' = (-0.5 - 2x_1 + 4x_3)x_2,$$
$$x_3' = (2 - x_2 - 2x_3)x_3.$$

Find and analyze the equilibrium point with all populations positive.

10.3 Invariant Sets and the Use of Nullclines

The analysis of equilibrium points in the previous two sections provides valuable information about solutions to autonomous systems. However, it is of limited use. Each of the results that we stated is *local* in the sense that it is valid only for solutions that have initial values in a neighborhood of an equilibrium point. Nothing is said about what happens to solutions that start far away from any equilibrium point. To analyze such solutions, we need methods that apply to solutions that start anywhere. We will call such methods *global methods*.

Global methods are harder to find, and typically they only apply to a few situations. It is necessary to find a combination of them that works for each particular system. However, there are some methods that can be used often, even though none of them is a panacea. In this section, we begin to develop these methods.

Invariant sets

An *invariant set* for a system $\mathbf{y}' = \mathbf{f}(\mathbf{y})$ of dimension n is a set $S \subset \mathbf{R}^n$ with the property that if $\mathbf{y}(t)$ is a solution to $\mathbf{y}' = \mathbf{f}(\mathbf{y})$, with its initial value $\mathbf{y}(0) \in S$, then

$\mathbf{y}(t) \in S$ for all $t \geq 0$. Thus, if S contains the initial value, it must contain the entire solution curve from that point on.

The preceding definition more correctly defines a set that is positively invariant, since it is only required that S contain the solution curve for $t \geq 0$. A set is negatively invariant if it contains the solution curve for $t \leq 0$. However, we will seldom need to discuss negative invariance, so when we say *invariant*, we will mean "positively invariant." When we need to discuss negative invariance, we will mention that specifically.

Let's look first at some easily understood examples of invariant sets. First of all, a set consisting of one equilibrium point is invariant. Next, a single solution curve is clearly invariant. Then we see that any set that is the union of equilibrium points and solution curves is invariant.

Example 3.1   Analyze all equilibrium points for the competing species system

$$\begin{aligned} x' &= (1 - x - y)x \\ y' &= (4 - 7x - 3y)y. \end{aligned} \qquad (3.2)$$

Show that the x- and y-axes are invariant. Show that the positive quadrant is invariant under this system.

We examined this system in Example 1.19. There we found four equilibrium points, $(0, 0)$, $(0, 4/3)$, $(1, 0)$, and $(1/4, 3/4)$. We analyzed $(1/4, 3/4)$ and discovered that it is a saddle. We also computed the Jacobian,

$$J(x, y) = \begin{pmatrix} 1 - 2x - y & -x \\ -7y & 4 - 7x - 6y \end{pmatrix}.$$

Let's analyze the other equilibrium points, starting with $(0, 0)$. Here the Jacobian is

$$J(0, 0) = \begin{pmatrix} 1 & 0 \\ 0 & 4 \end{pmatrix}.$$

We can read the eigenvalues from the diagonal of this diagonal matrix and discover that $(0, 0)$ is a nodal source. At $(0, 4/3)$, we have

$$J(0, 4/3) = \begin{pmatrix} -1/3 & 0 \\ -28/3 & -4 \end{pmatrix}.$$

Again, we can read the eigenvalues from the diagonal of a triangular matrix like this one, and we see that this is a nodal sink. At $(1, 0)$, the Jacobian is

$$J(1, 0) = \begin{pmatrix} -1 & -1 \\ 0 & -3 \end{pmatrix}.$$

This too is a nodal sink.

Let's look at the x-axis first. Suppose $(x_0, 0)$ is a point on the axis. Let $x(t)$ be the solution to $x' = (1 - x)x$, with $x(0) = x_0$, and let $y(t) = 0$. By direct substitution, we see that $(x(t), y(t))$ is a solution to the system (3.2). Thus, by the uniqueness theorem, any solution starting in the x-axis stays there in both the positive and negative time direction. The same argument shows that the y-axis is positively and negatively invariant.

To show that the positive quadrant is invariant, notice that a solution curve can leave the quadrant only by crossing one of the axes. By the uniqueness theorem, this is impossible, since to do so it has to cross a solution curve that is contained in the axis.

Some solution curves to (3.2) are shown in Figure 1. Some of these solution curves approach the nodal sinks at $(0, 4/3)$ and $(1, 0)$. However, by the uniqueness theorem, they never reach these points in a finite amount of time.   ●

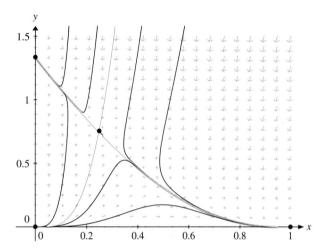

Figure 1. Solutions to the competing species model in Example 3.1.

The careful reader will have noticed that we have actually shown that the positive quadrant is both positively and negatively invariant. Since the system in (3.2) models populations, we would expect the solutions to be positive. The fact that the solutions to the model stay in the positive quadrant serves as a reality check for the model. If the solutions did not remain in the positive quadrant, we might suspect that the model is incorrect.

Our next example will be a set that is only positively invariant.

Example 3.3   Show that the blue rectangle R in Figure 2, defined by $0 < x < 1$ and $0 < y < 3/2$, is invariant for the system in (3.2).

The reason this is true is clearly illustrated in Figure 2. If we look at the portions of the boundary of R along the lines $x = 1$ and $y = 3/2$, all of the vector field arrows point into the rectangle R. Thus, as t increases, all solution curves starting

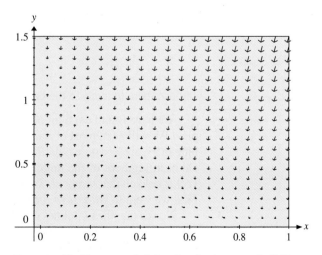

Figure 2. The blue rectangle is invariant for the system in (3.2).

on these portions of the boundary of R cross into R, and none come out. By the previous example, solution curves cannot cross the axes, so they are trapped in R.

These geometric considerations must be confirmed by analytic arguments. Along the line $x = 1$, the first equation in (3.2) is $x' = (1 - x - y)x = -y < 0$. Hence, the x component of the solution is decreasing, and the solution curve is moving into R. Similarly, along $y = 3/2$, the second equation in (3.2) is

$$y' = (4 - 7x - 3y)y = -3(1/2 + 7x)/2 < 0,$$

since $0 \leq x \leq 1$. This time the y-component of the solution is decreasing, so again the solution curve is moving into R. $\qquad\bullet$

Figure 1 shows the phase plane for the system in Examples 3.1 and 3.3. The locations of the equilibrium points are shown, and the separatrices for the saddle at $(1/4, 3/4)$ are plotted in blue. We know that the equilibrium points at $(0, 0)$ and $(1/4, 3/4)$ are unstable, while the other two are nodal sinks. From our analysis, we know that any solution curve starting near one of the sinks is attracted to it as t increases. This is clear from Figure 1. In fact, Figure 1 shows more. It indicates that almost all solution curves in the positive quadrant approach one of the sinks. The only exceptions are the two stable solution curves that approach the saddle point $(1/4, 3/4)$. If we look closely, we see that these curves separate the initial values whose solution curves lead to the two sinks.

All of these statements are based on looking at Figure 1. Can we verify them with analytic arguments? In Example 3.3, we showed that any solution curve starting in the rectangle R stays there as t increases. We know that solutions that start near one of the nodal sinks at $(0, 4/3)$ and $(1, 0)$ are attracted to the sinks. However, this is far from showing what seems to be true from observations of Figure 1. We need a little more precision, and we will learn how to provide it next.

Use of nullclines for global analysis

Nullclines were introduced in Section 8.3. Now we will examine how they can be used to decide the global nature of the solutions to a nonlinear system. Let's return to the system we considered in Examples 1.19, 3.1, and 3.3. We will show that certain regions bounded by the nullclines are invariant, and that will help us further analyze the system.

Example 3.4 Find the nullclines for the system

$$x' = (1 - x - y)x$$
$$y' = (4 - 7x - 3y)y,$$

and use them to analyze the behavior of all solutions to the system.

We have found out a lot of information about this system in the previous examples. We will need most of it in what follows. The nullcline information is shown in blue in Figure 3. The x nullcline is shown dashed, and the y-nullcline is dot-dashed. On each segment of the nullclines, there are arrows that show the direction of the vector field on that segment. We will show how to accumulate that information later. First we want to show how it can be used.

Notice that the nullclines divide the positive quadrant into four pieces, which we have labeled with Roman numerals. We will examine what can happen to solution curves in each of these regions. Let's start with region II. This region is a triangle. One of its sides is contained in the x-axis, which we know is invariant. Thus, no solution curve can cross it. Notice that the arrows on the other two sides of the triangle point into the region. This means that every solution curve that crosses these sides goes into region II. Hence, no solution curve can escape from region II, so region II is invariant. Furthermore, within the region, the direction of the curves is between the directions on the boundary. Hence, the curves must move in the "southeasterly" direction, staying in region II. There is no alternative but that they are ultimately attracted to the sink at $(1, 0)$. Similar arguments apply in region IV. It, too, is invariant, and every curve in region IV is attracted to the sink at $(0, 4/3)$.

Let's concentrate on region I. Here, the direction of the curves is "southwest," because the direction must mediate between the directions shown on the bounding nullclines. Included among the solution curves in region I is a stable solution curve for the saddle point at $(1/4, 3/4)$. This curve, shown plotted in black in Figure 3, approaches the saddle point as $t \to \infty$. Every other curve in region I either approaches one of the sinks at $(0, 4/3)$ and $(1, 0)$ directly, or is forced into region II or region IV and then onto one of the two sinks. Similar reasoning shows that the same thing happens to solutions starting in region III. There is one stable solution curve that approaches the saddle at $(1/4, 3/4)$, and all other curves are forced into regions II and IV and then onto one of the two sinks.

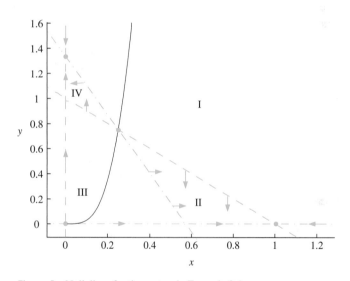

Figure 3. Nullclines for the system in Example 3.4.

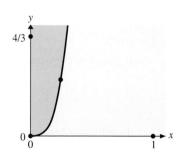

Figure 4. The basins of attraction of the sinks in Example 3.4.

Thus, except for the two stable solution curves for the saddle point, all solutions tend to one of the nodal sinks as t increases. How do we decide which sink a curve approaches? By the uniqueness theorem, any solution curve that starts to the left of a stable solution curve must stay there. Therefore, it is clear that any solution curve starting to the left of the stable curves ends up at $(0, 4/3)$, while any curve starting to the right of them ends up at $(1, 0)$.

We define the ***basin of attraction*** of a sink $\mathbf{y}_0$ to be all of those points $\mathbf{y}$ with the property that the solution curve starting at $\mathbf{y}$ approaches the sink $\mathbf{y}_0$ as $t \to \infty$. With this definition, we see that in this example the basin of attraction of the sink at $(0, 4/3)$ is the region to the left of the stable curves, while the basin of attraction of $(1, 0)$ is all points to the right of these curves. The portions of the basins of attraction in the first quadrant are shown in Figure 4.

It remains to show how to discover the nullcline information. One way to do so is to have a computer program that does it for us. One was used to generate Figure 3. However, in this case, it is not difficult to develop the nullcline information without technical assistance.

We start by finding the nullclines. The x-nullcline consists of the solutions to

$$(1 - x - y)x = 0.$$

There are two pieces, corresponding to the two factors, each of them a straight line,

$$x = 0 \quad \text{and} \quad x + y = 1.$$

At any point on the x nullcline, the vector field has, by definition, a zero in the first component. Thus, the vector field arrow points up or down. We need to discover which is the case. The actual value of the vector field on the x nullcline is

$$((1 - x - y)x, (4 - 7x - 3y)y) = (0, (4 - 7x - 3y)y).$$

Thus, the arrow points up if $(4 - 7x - 3y)y > 0$ and down otherwise. Since we are looking only where y is positive, the arrows point up if $4 - 7x - 3y > 0$ and down otherwise. Finding and analyzing the y-nullcline will help us resolve the question of the direction of the arrows on the x-nullcline.

The y-nullcline is defined by

$$(4 - 7x - 3y)y = 0.$$

Again, there are two straight lines, corresponding to the two factors,

$$y = 0 \quad \text{and} \quad 7x + 3y = 4.$$

Along the y-nullcline, the vector field has a zero in the second component. Thus, the vector field arrow points either left or right.

However, the y-nullcline also separates regions in which the second component of the vector field is positive or negative. We can use this information to discover on which portions of the x nullcline the arrow points up. We have $4 - 7x - 3y = 0$ on the y nullcline and in particular on the line $7x + 3y = 4$. By substituting some point above and to the right of this line, say $x = y = 1$, we discover that $4 - 7x - 3y < 0$ above and to the right of the line and $4 - 7x - 3y > 0$ below and to the left of the line. Thus, field arrows point up on the portion of the x-nullcline below and to the left of the line $7x + 3y = 4$, and down above and to the right of it. Similar considerations allow us to determine the directions of the arrows on the y-nullcline.

EXERCISES

In Exercises 1–4, show that the x- and y-axes are invariant sets for the given system. Then explain why each of the four quadrants is invariant for the given system.

1. $x' = (2 - x - y)x$
     $y' = (3 - 3x - y)y$

2. $x' = 4x(1 - x) - xy$
     $y' = y(3 - y) - xy$

3. $x' = (1 - x - y)x$
     $y' = (6 - 2x - 3y)y$

4. $x' = 4x(3 - x) - 3xy$
     $y' = y(1 - y) - xy$

Use the technique of Example 3.3 to answer the questions posed in Exercises 5–8.

5. Show that the set $S = \{(x, y) : 0 \le x \le 3, 0 \le y \le 3\}$ is invariant for the system of Exercise 1.

6. Show that the set $S = \{(x, y) : 0 \le x \le 3, 0 \le y \le 4\}$ is

invariant for the system of Exercise 2.

7. Show that the set $S = \{(x, y) : 0 \le x \le 4, 0 \le y \le 3\}$ is invariant for the system of Exercise 3.

8. Show that the set $S = \{(x, y) : 0 \le x \le 4, 0 \le y \le 5\}$ is invariant for the system of Exercise 4.

In Exercises 9–12, perform each of the following tasks for the given system.

(i) Set up an xy phase plane on graph paper. Sketch the nullclines. Then use algebraic analysis to find and label each of the equilibrium points on your plot.

(ii) Use the Jacobian to classify each equilibrium point: nodal sink, saddle, and so on.

(iii) Use the analysis demonstrated at the end of Example 3.4 to indicate the flow of solution trajectories across each nullcline, as pictured in Figure 3.

(iv) Finally, use the information provided in (i)–(iii) to sketch the phase portrait for the given system in the first quadrant. Be sure to check your hand-drawn results with your numerical solver.

9. The system of Exercise 1   10. The system of Exercise 2

11. The system of Exercise 3   12. The system of Exercise 4

The graph of a nullcline is not always a line, and we do not have to restrict ourselves to the first quadrant when sketching phase portraits. In Exercises 13–16, perform each of the tasks (i)–(iv) required in Exercises 9–12, only this time, sketch the phase portrait in all four quadrants of the xy phase plane.

13. $x' = 1 - y$
    $y' = y - x^2$

14. $x' = y - x$
    $y' = x - y^2$

15. $x' = y - x^3$
    $y' = x - y$

16. $x' = 1 - x^2 - y^2$
    $y' = x - y$

17. Let u and v represent two populations whose interaction is beneficial to the growth of each species, and which are individually subject to a logistic limit. (See Section 10.1 for a description of interacting species.) Suppose their interaction is modeled by the system

$$u' = u(1 - u + av),$$
$$v' = rv(1 - v + bu),$$

where r, a, and b are positive constants.

(a) Show that the system has equilibrium points $(0, 0)$, $(1, 0)$, $(0, 1)$, and

$$u = \frac{1 + a}{1 - ab} \quad \text{and} \quad v = \frac{1 + b}{1 - ab}.$$

(b) In the case that $ab > 1$, provide a complete analysis of the system addressing each task (i)–(iv) assigned in Exercises 9–12. Limit your analysis to the first quadrant (there is no such thing as a negative population). Discuss the eventual fate of each population.

(c) In the case that $ab < 1$, provide a complete analysis of the system addressing each task (i)–(iv) assigned in Exercises 9–12. Limit your analysis to the first quadrant. Discuss the eventual fate of each population.

18. A chemostat[2] is a device used to maintain a constant supply of bacteria for biological studies in the laboratory.

Think of a chamber containing bacteria and nutrients essential for the bacteria's growth. The bacteria eat the nutrient and multiply. A nutrient reservoir replenishes lost nutrients, and the experimentalist harvests available bacteria as needed for research. Let N and C represent the number of bacteria in the chamber and the concentration of the nutrient, respectively. One possible model for the chemostat (using dimensionless variables) is given by the system

$$\frac{dN}{dt} = \alpha_1 \left(\frac{C}{1 + C} \right) N - N,$$

$$\frac{dC}{dt} = -\left(\frac{C}{1 + C} \right) N - C + \alpha_2,$$

where $\alpha_1, \alpha_2 > 0$.

(a) Show that the system has two equilibrium points,

$$(N_1, C_1) = \left(\alpha_1 \left(\alpha_2 - \frac{1}{\alpha_1 - 1} \right), \frac{1}{\alpha_1 - 1} \right)$$

and $(N_2, C_2) = (0, \alpha_2)$.

(b) If we assume that $\alpha_1 > 1$ and $\alpha_2 > 1/(\alpha_1 - 1)$, then the equilibrium point (N_1, C_1) lies in the first quadrant. Show that the Jacobian, evaluated at the first equilibrium point, is given by

$$J(N_1, C_1) = \begin{pmatrix} 0 & \alpha_1 A \\ -1/\alpha_1 & -(A + 1) \end{pmatrix},$$

where $A = N_1/(1 + C_1)^2$. Use this result to show that the equilibrium point (N_1, C_1) is stable.

(c) Show that

$$J(N_2, C_2) = \begin{pmatrix} \alpha_1 B - 1 & 0 \\ -B & -1 \end{pmatrix},$$

where $B = \alpha_2/(1 + \alpha_2)$. Assuming the same conditions as in part (b), show that the equilibrium point at (N_2, C_2) is a saddle.

(d) Choose numerical values for α_1 and α_2 satisfying $\alpha_1 > 1$ and $\alpha_2 > 1/(\alpha_1 - 1)$. Sketch the nullclines in the NC phase plane, label equilibrium points with their coordinates, and indicate the direction of solution trajectories across each nullcline. Then use all of this information to sketch the phase portrait in the first quadrant for the chemostat model. Check your results with your numerical solver.

10.4 Long-Term Behavior of Solutions to Planar Systems

Is it possible to give a short list of all possible behaviors of solutions to a system? The short answer is "yes" for a system of dimension 2 and "no" for higher dimensions. We will provide the list for dimension 2 in this section.

---

[2] For a full development of the chemostat model, see Leah Edelstein-Keshet's *Mathematical Models in Biology*, McGraw Hill Companies, 1988.

Suppose we have a system defined in a set U and a solution $\mathbf{y}(t)$ that starts at a point $\mathbf{y}(0) = \mathbf{y}_0$ in U. The solution can move out toward the boundary of U or to infinity. However, we are more interested in what happens to solution curves that stay in a bounded subset of U for all $t \geq 0$. So far we have seen two possibilities. We have seen many examples of solution curves that approach a point, and we have seen closed solution curves, corresponding to periodic solutions.

We need some terminology. We define the **(forward) limit set** of the solution $\mathbf{y}(t)$ that starts at $\mathbf{y}(0) = \mathbf{y}_0$ to be the set of all limit points of the solution curve. This set will be denoted by $\omega(\mathbf{y}_0)$. A point $\mathbf{x}$ is in $\omega(\mathbf{y}_0)$ if there is a sequence $t_1 < t_2 < \cdots$ that approaches ∞ such that $\mathbf{y}(t_k) \to \mathbf{x}$. Sometimes a set is the forward limit set for all solution curves starting near a set. In this case, we will call the set an **attractor**.

If a solution curve approaches a single point, then its forward limit set consists of that one point. Any sink is the limit set of every solution curve staring near it. Thus, a sink is an example of an attractor.

If the solution is periodic and the solution curve is closed, then every point on the curve is in the limit set. To see this, suppose $x(t)$ is the solution and T is the period. Let t_0 be arbitrary. Then $x(t_0) = x(t_0 + T) = x(t_0 + 2T) = \cdots$. The sequence $t_0 + kT$ approaches ∞, but the points $x(t_0 + kT)$ are all equal to $x(t_0)$. Hence the point $x(t_0)$ is in the limit set of the periodic solution curve. Since t_0 is arbitrary, we see that the limit set of a periodic solution is the solution curve itself.

We need to know if there are any other possibilities besides a point and a periodic solution curve. In the plane, there is a very short list, which we will give in this section. In higher dimensions, the situation is quite different. The complete answer is still unknown. We will discuss this briefly at the end of this section.

The next theorem provides some important information about limit sets.

THEOREM 4.1 Suppose the system $\mathbf{y}' = \mathbf{f}(\mathbf{y})$ is defined in the set U.

1. If the solution curve starting at $\mathbf{y}_0$ stays in a bounded subset of U, then the limit set $\omega(\mathbf{y}_0)$ is not empty.
2. Any limit set is positively and negatively invariant. ∎

Let's examine some implications of Theorem 4.1. Suppose that $\omega(\mathbf{y}_0)$ consists of a single point $\mathbf{x}_0$. According to part (2) of Theorem 4.1, this singleton must be invariant. This means that the solution curve starting at $\mathbf{x}_0$ must stay at $\mathbf{x}_0$. In other words, the point $\mathbf{x}_0$ is an equilibrium point. Thus, if a solution curve converges to a point, that point must be an equilibrium point.

Every set that is positively and negatively invariant contains the full solution curve starting at any point. Hence, such a set is the union of solution curves. Therefore, another implication of part (2) of Theorem 4.1 is that every limit set is the union of solution curves.

Limit cycles

Here is an example to think about. It exhibits a new type of attractor.

Example 4.2 Consider the system

$$\begin{aligned} x' &= -y + x(1 - x^2 - y^2), \\ y' &= x + y(1 - x^2 - y^2). \end{aligned} \tag{4.3}$$

Express the system in polar coordinates and describe all solutions.

If r and θ are the polar coordinates, we have

$$x = r \cos \theta \quad \text{and} \quad y = r \sin \theta.$$

We can solve for r and θ, getting

$$r^2 = x^2 + y^2 \quad \text{and} \quad \tan \theta = \frac{y}{x}.$$

Differentiating the first equation and using our system (4.3), we get

$$
\begin{aligned}
2rr' &= \frac{d}{dt} r^2 \\
&= 2(xx' + yy') \\
&= 2[x(-y + x(1 - x^2 - y^2)) + y(x + y(1 - x^2 - y^2))] \\
&= 2(x^2 + y^2)(1 - x^2 - y^2) \\
&= 2r^2(1 - r^2).
\end{aligned}
$$

Hence, for $r \neq 0$,

$$r' = r(1 - r^2). \tag{4.4}$$

Similarly, when we differentiate $\tan \theta$, we get

$$
\begin{aligned}
\sec^2(\theta)\, \theta' &= \frac{d}{dt} \tan \theta \\
&= (xy' - yx')/x^2 \\
&= [x(x + y(1 - x^2 - y^2)) - y(-y + x(1 - x^2 - y^2))]/x^2 \\
&= (x^2 + y^2)/x^2.
\end{aligned}
$$

Since $\sec^2 \theta = r^2/x^2 = (x^2 + y^2)/x^2$, we get

$$\theta' = 1. \tag{4.5}$$

Thus, in polar coordinates, our system becomes the simple pair of equations in (4.4) and (4.5). From (4.5), we see that $\theta = t + C$. Therefore, θ is steadily increasing with speed 1. This means that all solution curves spiral around the origin in the counterclockwise direction.

Qualitative analysis of equation (4.4) for the radius r discloses equilibrium points at 0 and 1. For $0 < r < 1$, r is increasing, and hence it increases to 1. For $r > 1$, r is decreasing, and it decreases to 1. Thus, $r = 1$ is an asymptotically stable equilibrium point for (4.4). For the original system, this means that the unit circle in the phase plane is a solution curve, and all other solution curves starting away from the origin spiral toward it. Hence, the unit circle is the limit set for any point in $\mathbf{R}^2$, except for the origin. This situation is shown in Figure 1.   ●

A closed solution curve toward which other solution curves spiral is called a *limit cycle*. If the solution curves spiral toward the limit cycle as $t \to \infty$, it is an attractor and it is called an *attracting limit cycle*. If the solution curves spiral away from the limit cycle as t increases and move toward the limit cycle as $t \to -\infty$, it is called a *repelling limit cycle*. In Example 4.2, the unit circle is an attracting limit cycle. It is also possible that the solution curves will spiral toward the limit cycle on one side and away from it on the other.

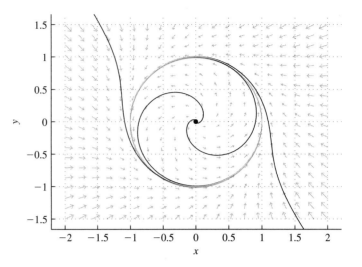

Figure 1. All solutions to the system (4.3) converge to the unit circle.

Limiting graphs

We have almost completed our list of the possible long-term behaviors for a solution of a planar system. There is one more possibility, which is illustrated by the next example.

Example 4.6 Consider the system

$$x' = (y + x/5)(1 - x^2),$$
$$y' = -x(1 - y^2).$$
(4.7)

Find and analyze the equilibrium points. With a numerical solver, compute and plot one solution curve starting somewhere inside the unit square.

We will not give all of the details. There are seven equilibrium points, $(0, 0)$, $(1, 1)$, $(1, -1)$, $(-1, 1)$, $(-1, -1)$, $(-5, 1)$, and $(5, -1)$. The Jacobian is

$$J = \begin{pmatrix} (1 - x^2)/5 - 2x(y + x/5) & 1 - x^2 \\ y^2 - 1 & 2xy \end{pmatrix}.$$

Computing J at each of the equilibrium points, we find that the origin is a spiral source; $(1, 1)$, $(1, -1)$, $(-1, 1)$, and $(-1, -1)$ are all saddle points; and $(-5, 1)$ and $(5, -1)$ are nodal sinks.

In Figure 2, we show a solution curve inside the unit square. As $t \to -\infty$, the curve spirals into the spiral source at the origin, as we might expect. However, as $t \to \infty$ the curve spirals outward, approaching the entire boundary of the unit square. Thus, the boundary of the unit square is the limit set for this solution. Since the boundary of the unit square is not a limit cycle, this is a new type of limit set. ●

Let's look more closely at the limit set in Example 4.6, starting with the line $x = 1$. Suppose we start a solution curve at $(1, 0)$. We define $y(t)$ to be the solution of $y' = y^2 - 1$, with the initial condition $y(0) = 0$,[3] and we set $x(t) = 1$. Then by direct substitution, we see that x and y are a solution to the system (4.7) (notice how the system simplifies when $x = 1$). Since $y' = y^2 - 1$ and $y(0) = 0$, qualitative analysis shows that $y(t)$ is decreasing, approaching -1 as $t \to \infty$, and approaching

---
[3] The solution is $y(t) = -\tanh(t) = (e^{-2t} - 1)/(e^{-2t} + 1)$.

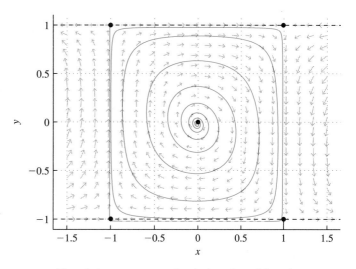

Figure 2. The solution curve approaches the boundary of the unit square, which is a cyclic graph.

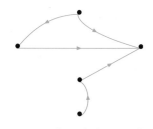

Figure 3. A directed planar graph.

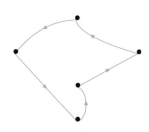

Figure 4. A cyclic graph.

1 as $t \to -\infty$. Thus, we have a solution curve $(1, y(t))$ for (4.7), which approaches the saddle at $(1, 1)$ as t decreases to $-\infty$, and the saddle at $(1, -1)$ as t increases to ∞. Consequently, this curve must be a stable separatrix for $(1, -1)$ and an unstable separatrix for $(1, 1)$.

In the same way, we find that the lines $x = \pm 1$ and $y = \pm 1$ contain all of the separatrices for the four saddle points. In particular, the boundary of the unit square consists entirely of separatrices and the four saddle points.

We need some terminology to describe the kind of limit set we see in Example 4.6. A ***planar graph*** is a collection of points in the plane, called ***vertices***, and nonintersecting curves, called ***edges***, which connect the vertices. If the edges each have an associated direction, the graph is said to be ***directed***. Examples are shown in Figures 3 and 4. A ***cyclic graph*** is a directed planar graph which forms a closed loop. The graph in Figure 4 is a cyclic graph, while that in Figure 3 is not.

The boundary of the unit square in Example 4.6 is a cyclic graph. The equilibrium points at the corners are the vertices, and separatrices on the sides are the edges. Since every solution curve has a direction, each of the separatrices has a natural direction, which makes the boundary of the square a cyclic graph. The solution curve in Example 4.6 approaches this graph as t increases, so the boundary is the limit set of the solution.

We can now make a list of all possible limit sets in the plane.

THEOREM 4.8   Suppose that the planar system

$$x' = f(x, y)$$
$$y' = g(x, y) \tag{4.9}$$

is defined for $(x, y) \in U \subset \mathbf{R}^2$. Suppose that f and g have continuous derivatives in U and that the system has finitely many equilibrium points. If S is the limit set for a solution to (4.9), then S is one of the following:

- An equilibrium point
- A closed solution curve
- A cyclic graph with vertices that are equilibrium points and edges that are solution curves ∎

Theorem 4.8 is called Bendixson's theorem, and the three alternatives therein are referred to as the **Bendixson alternatives**. They are sometimes called the **Poincaré-Bendixson alternatives**. The next theorem is called the Poincaré-Bendixson theorem. It is easily derived from Bendixson's theorem and we will leave the proof as an exercise.

THEOREM 4.10 Let R be a closed and bounded planar region that is positively invariant for (4.9). If R contains no equilibrium points, then there is a closed solution curve in R. ∎

It should be mentioned that the results of this section are true if we systematically look at what happens as $t \to -\infty$. In that case, we look at backward limit sets, which are defined in an analogous way to forward limit sets. Everything else is pretty much the same.

The van der Pol system

The system

$$
\begin{aligned}
x' &= 2x - y - x^3 \\
y' &= x
\end{aligned}
\tag{4.11}
$$

is a variant of a system that was derived by Balthazar van der Pol to model the behavior of an electric circuit. The vector field for the system is shown in Figure 5. We wish to analyze the solutions to this system, and we will need to use Theorems 4.1, 4.8, and 4.10. The key step in showing that Theorem 4.1 applies is contained in the next example.

Example 4.12 Show that the blue region S in Figure 5 is invariant under the van der Pol system.

The boundary of the region S in Figure 5 consists of two pieces. The outer boundary is a hexagon, and the inner boundary is the unit circle. To prove the invariance, we will show that solution curves along each piece of the boundary enter S along that portion. To that effect, we need only show that at every point of the

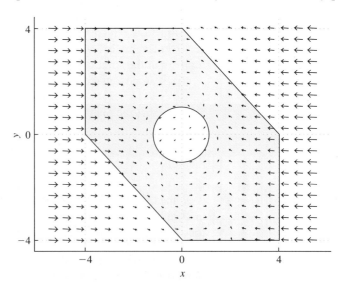

Figure 5. The blue region S is invariant under the van der Pol system in (4.11).

boundary of S the vector field points into S. This seems likely from the directions of the vector field in Figure 5, but we want to show this analytically. First consider the segment $x = 4$, $-4 \leq y \leq 0$. Here $x' = 2x - y - x^3 = -56 - y \leq -52$, so x is decreasing and the field arrows point into S. Similar arguments can be made along the other vertical and horizontal segments of the boundary.

The slanted portions of the outer boundary are segments of the lines defined by $x + y = \pm 4$. Everywhere in S we have $-4 < x + y < 4$. Thus, it suffices to show that the quantity $x + y$ is decreasing along a solution curve at any point on the upper slanted portion (defined by $x + y = 4$) and increasing along any solution curve on the lower portion.

The upper portion is defined by $x + y = 4$ and $0 \leq x \leq 4$. If x and y are components of a solution and $x + y = 4$, we have

$$
\begin{aligned}
(x + y)' &= x' + y' \\
&= (2x - y - x^3) + x \quad \text{by the differential equations} \\
&= 3x - x^3 - y \\
&= 4x - x^3 - 4 \quad \text{since } y = 4 - x.
\end{aligned}
$$

By differentiating, we find that the maximum of $4x - x^3 - 4$, for $0 \leq x \leq 4$ is at $x = 2/\sqrt{3}$, and its value there is $8/\sqrt{3} - 8/\sqrt{27} - 4 \approx -0.9208$. Hence, along this segment $(x + y)' = 4x - x^3 - 4 \leq 8/\sqrt{3} - 8/\sqrt{27} - 4 < 0$. Since $x + y$ is decreasing along a solution curve at any point on the upper slanted portion of the boundary, such a solution curve must be moving into S. A similar argument shows that $x + y$ is increasing along the lower slanted part of the boundary of S and a solution curve starting on this portion of the boundary is also moving into S.

The inner portion of the boundary of S is the unit circle, defined by $r = \sqrt{x^2 + y^2} = 1$. To show that all solution curves enter S along the unit circle, we have to show that $r' \geq 0$ on the unit circle. We have

$$
rr' = xx' + yy' = x(2x - y - x^3) + yx = x^2(2 - x^2).
$$

If $r^2 = x^2 + y^2 = 1$, then $0 \leq x^2 \leq 1$. In particular, we see that $r' \geq 0$ on the unit circle.

Thus, at every point of the boundary of S, solution curves are moving into S. Consequently, S is invariant. ●

Since S is invariant, every solution curve that starts in S stays there. Since S is a bounded set, all of the conditions in Theorem 4.1 are satisfied. Therefore, every solution curve starting in S has a nonempty forward limit set.

You may have noticed that the only equilibrium point for the van der Pol system is at the origin. Hence, S is invariant and contains no equilibrium points. According to the Poincaré-Bendixson theorem (Theorem 4.10), there is a closed solution curve in S. A solution curve starting at a point on the boundary of S cannot be closed since it cannot return to the point on the boundary. Its limit set must be a closed solution curve in S, and this limit set must be a limit cycle. The situation is illustrated in Figure 6. Several solution curves that spiral to the solution curve in blue are shown, which is the limit cycle predicted by Theorem 4.10.

It is possible that S contains more than one limit cycle, but the numerical data plotted in Figure 6 shows that there is only one. Since solution curves spiral toward the limit cycle both from outside and inside as $t \to \infty$, this is an attracting limit cycle.

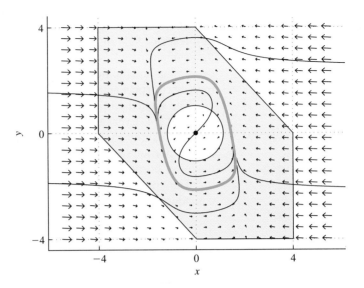

Figure 6. The phase plane for the van der Pol system in (4.11), showing the limit cycle.

Long-term behavior of solutions in higher dimensions

Bendixson's theorem (4.8) gives a satisfactory answer to our search for a description of all possible limit sets in two dimensions. The situation is not at all resolved in dimensions greater than two. In fact, this is an area of much current research.

The best that we can say here is that the three alternatives in Theorem 4.8 occur in higher dimensions and that there are other alternatives as well. We saw one new possibility in Section 10.2, when we discussed the Lorenz system. In particular, in Figure 5 of Section 10.2, we displayed the Lorenz attractor. You will recall that we mentioned that almost every solution to the Lorenz system is attracted to the Lorenz attractor as t increases. It is clearly not one of the three Bendixson alternatives.

The keys to the proof of Bendixson's theorem are the uniqueness theorem and the fact that a solution curve in the plane separates the plane into two distinct sets. Of course, the uniqueness theorem is true in higher dimensions, but in dimensions higher than three, curves do not separate space. This fact makes the analysis much harder and also makes possible limit sets like the Lorenz attractor.

Like all limit sets, the Lorenz attractor is the union of solution curves. However, the behavior of these solutions is highly chaotic. Suffice it to say that the set is not well understood. Furthermore, it is not just a mathematical curiosity. The origin of the Lorenz attractor in meteorology leads us to believe that the strangeness we find in it reflects strangeness and unpredictability that exist in nature.

EXERCISES

In Exercises 1–4, use the polar coordinate transformations shown in Example 4.2 to find a limit cycle of the system. Is the stability of the limit cycle attracting or repelling or neither? Sketch a phase portrait without the aid of a technology. Verify your results with your numerical solver.

1. $x' = y + x\left(1 - \sqrt{x^2 + y^2}\right)$

    $y' = -x + y\left(1 - \sqrt{x^2 + y^2}\right)$

2. $x' = -y + x\left(\sqrt{x^2 + y^2} - 3\right)$

    $y' = x + y\left(\sqrt{x^2 + y^2} - 3\right)$

3. $x' = -3y + x(4 - x^2 - y^2)$

    $y' = 3x + y(4 - x^2 - y^2)$

4. $x' = 5y + x(x^2 + y^2 - 9)$

    $y' = -5x + y(x^2 + y^2 - 9)$

5. Use polar coordinates to show that the system

$$x' = -y + x\left(\sqrt{x^2 + y^2} - 3 + \frac{2}{\sqrt{x^2 + y^2}}\right),$$

$$y' = x + y\left(\sqrt{x^2 + y^2} - 3 + \frac{2}{\sqrt{x^2 + y^2}}\right),$$

has *two* limit cycles. Discuss the stability of each. Verify your results with your numerical solver.

6. Consider the system

$$x' = -y + x(x^2 + y^2)\sin\frac{\pi}{\sqrt{x^2 + y^2}},$$

$$y' = x + y(x^2 + y^2)\sin\frac{\pi}{\sqrt{x^2 + y^2}}.$$

Use polar coordinates to show that the circles $r = 1/n$, $n = 1, 2, 3, \ldots$ are limit cycles. Which are attracting? Repelling? Why? Verify your results with your numerical solver.

7. Use polar coordinates to show that the system

$$x' = y + x(x^2 + y^2 - 1)^2,$$

$$y' = -x + y(x^2 + y^2 - 1)^2$$

has an **unstable** limit cycle, attractive on one side, repelling on the other. Verify your results with your numerical solver.

8. Consider the system

$$x' = -x + 2y$$

$$y' = -x + y + (x - 2y)(x^2 - 2xy + 2y^2 - 1).$$

Find all forward and backward limit sets for solution trajectories in the plane. Use a numerical solver to sketch a phase portrait of the system. *Hint*: Consider the function $F(x, y) = x^2 - 2xy + 2y^2$. Show that the sets $F(x, y) = C$ are ellipses. Compute the derivative of $F(x(t), y(t))$, where $x(t)$ and $y(t)$ are solutions to the system.

9. Consider the system

$$x' = x + y - x(x^2 + 3y^2)$$

$$y' = -x + 2y - 2y^3.$$

Find all forward and backward limit sets for solution trajectories in the plane. Use a numerical solver to sketch a phase portrait of the system. *Hint*: Calculate r' where $r = \sqrt{x^2 + y^2}$.

10. Use Bendixson's theorem (Theorem 4.8) to prove the Poincaré-Bendixson theorem (Theorem 4.10).

11. Consider the system

$$x' = y,$$

$$y' = -x + y(1 - 3x^2 - 2y^2).$$

(a) Show that if $r = \sqrt{x^2 + y^2}$, then

$$rr' = y^2(1 - 3x^2 - 2y^2).$$

(b) Show that the annular set $R = \{(x, y) \mid 1/2 \leq \sqrt{x^2 + y^2} \leq 1\}$ is positively invariant for the system. *Hint*: Show that $2(x^2 + y^2) \leq 3x^2 + 2y^2 \leq 3(x^2 + y^2)$.

(c) Use the Poincaré-Bendixson theorem (Theorem 4.10) to show that there is a limit cycle in the set R.

(d) Use your numerical solver to sketch the vector field associated with the system. Superimpose a plot of the boundaries of the annulus, and show that solutions entering the annulus move toward a limit cycle.

12. Consider the system

$$x' = x + y - x(x^2 + 3y^2)$$

$$y' = -x + y - 2y^3.$$

Use the Poincaré-Bendixson theorem (Theorem 4.10) to show that there is a limit cycle. *Hint*: Calculate r', where $r = \sqrt{x^2 + y^2}$. Follow the steps in Exercise 11 by finding an annular set that is negatively invariant.

13. Consider the system

$$x' = x - y - x(3x^2 + y^2)$$

$$y' = x + y - y(2x^3 + y^2).$$

Use the Poincaré-Bendixson theorem (Theorem 4.10) to show that there is a limit cycle. *Hint*: Calculate r', where $r = \sqrt{x^2 + y^2}$. Follow the steps in Exercise 11 by finding an annular set that is invariant.

14. Let $E(x, y)$ be a function of two variables. Consider the system

$$x' = E_y - E_x E,$$
$$y' = -E_x - E_y E, \tag{4.13}$$

where $E_x = \partial E/\partial x$ and $E_y = \partial E/\partial y$. If $(x(t), y(t))$ is a solution of system (4.13), we set $E(t) = E(x(t), y(t))$. Show that

$$\frac{dE}{dt} = -(E_x^2 + E_y^2)E$$

along this solution curve.[4] Use this result to argue that along any solution curve $E \to 0$ as $t \to \infty$. Therefore, the zero set, $E(x, y) = 0$, attracts solutions to it.

In Exercises 15–22, use the given $E = E(x, y)$ to construct the system in (4.13). Use your numerical solver to locate solutions that have planar graphs or limit cycles as limit sets. If your system supports implicit function plotting, superimpose the zero set $E(x, y) = 0$ on your plot and note that it contains the limit set.

---

[4] *Hint*: $dE/dt = (\partial E/\partial x)(dx/dt) + (\partial E/\partial y)(dy/dt)$. This form of the chain rule is usually studied in multivariable calculus.

15. $E(x, y) = x^2 - 2xy + 2y^2 - 1$

16. $E(x, y) = x^4 + x^3 - x^2 + y^4 + y^2 - 1$

17. $E(x, y) = xy(x - 1)(y - 1)$

18. $E(x, y) = xy(3 - x - y)$

19. $E(x, y) = x^2(1 + x) - y^2$

20. $E(x, y) = x(1 - x^2 - y^2)$

21. $E(x, y) = (y - 1)(y - x^2)$

22. $E(x, y) = x^2(1 - x^2) - y^2(1 - y^2)$

23. Consider the system

$$x' = y,$$
$$y' = -x + y - y^3. \qquad (4.14)$$

(a) Show that $(0, 0)$ is the only critical value and use the Jacobian to show that it is a spiral source.

(b) Use your numerical solver to construct the vector field associated with system (4.14). Superimpose the polygonal path ABCDA, where A $= (-3, 0)$, B $= (-3, 3)$, C $= (3, 0)$, and D $= (3, -3)$. Does a cursory glance seem to indicate that the vector field points *inward* everywhere along this boundary?

(c) Use analysis similar to that demonstrated in Example 4.12 to show that the vector field points inward everywhere along the boundary ABCDA. For example, along the segment BC, show that $x + 2y = 3$ and $0 \le y \le 3$. Now show that $(x + 2y)' < 0$, thereby ensuring that solution curves are moving into the region along the segment CD.

(d) Explain why parts (a)–(c) guarantee the existence of a limit cycle in the region bounded by the path ABCDA. Use your numerical solver to verify this conclusion.

In Exercises 24–25, perform each task requested in parts (a)–(d) of Exercise 23. This time it is up to you to construct your

own boundary and show that the region contained therein is invariant for the given system. Argue that the region contains a limit cycle, then verify this claim with your numerical solver.

24. $x' = 3x - y + x^3$
    $y' = x$

25. $x' = -y$
    $y' = y + 3x - y^3$

26. Consider the system

$$x' = a - x + x^2 y,$$
$$y' = b - x^2 y, \qquad (4.15)$$

where a and b are positive constants.

(a) Show that the system has exactly one equilibrium point (x_0, y_0); then evaluate the Jacobian at this point.

(b) There is a region containing (x_0, y_0) that is invariant for system (4.15) (not easy to prove). Assuming this, show that a limit cycle will exist if

$$b - a > (a + b)^3.$$

If you have an implicit function plotter, sketch the curve $b - a = (a + b)^3$ in the first quadrant $(a, b > 0)$. Determine the region in the first quadrant where $b - a > (a + b)^3$; then use your numerical solver to verify the existence of a limit cycle for system (4.15) for several (a, b) pairs selected from this region.

27. Suppose S is a directed planar graph that is a limit set for the system

$$x' = f(x, y)$$
$$y' = g(x, y).$$

Suppose that S contains an equilibrium point (x_0, y_0) and that the Jacobian at (x_0, y_0) is nonsingular. Show that (x_0, y_0) must be a saddle point. Show that any edge in S that connects to (x_0, y_0) is a separatrix.

10.5 Conserved Quantities

In Figure 1, we see a portion of a solution curve to the planar system

$$x' = f(x, y)$$
$$y' = g(x, y) \qquad (5.1)$$

in the phase plane. Notice that between the points labeled A and B, the curve is the graph of y as a function of x. Can we discover what that function is from the differential equation?

At the point (x_0, y_0) in Figure 1, we have plotted the vector field, showing its components $f(x_0, y_0)$ and $g(x_0, y_0)$. Since the vector is tangent to the solution curve at (x_0, y_0), the slope of the curve is equal to the slope of its tangent vector. Assuming that the slope is finite, we have

$$\frac{dy}{dx} = \frac{g(x, y)}{f(x, y)}. \qquad (5.2)$$

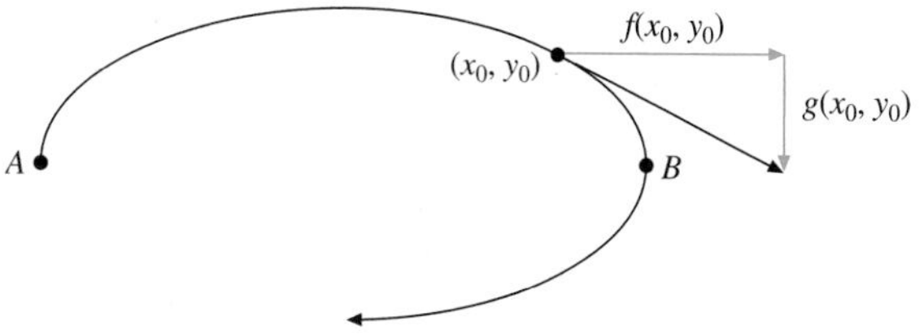

Figure 1. The part of the solution curve between A and B is the graph of a function.

We can also find this result using the chain rule and (5.1). We have[5]

$$\frac{dy}{dx} = \frac{dy}{dt}\frac{dt}{dx} = \frac{dy}{dt}\bigg/\frac{dx}{dt} = \frac{g(x, y)}{f(x, y)}.$$

Thus, the system in (5.1) leads to the single first-order equation (5.2) for y as a function of x. Sometimes this equation can be solved, perhaps only with an implicit relationship between x and y. Even this can provide very useful information. Let's look at an elementary example.

Example 5.3 Using the velocity $v = y'$, the equation $my'' + ky = 0$ for the undamped spring can be written as the system

$$\begin{aligned} y' &= v, \\ v' &= -\frac{k}{m}y. \end{aligned} \tag{5.4}$$

Find the relationship between the components y and v of the solution.

From (5.2), we get

$$\frac{dv}{dy} = \frac{-ky}{mv}.$$

When we separate variables and solve, we get the implicit relationship

$$\frac{1}{2}mv^2 + \frac{1}{2}ky^2 = C, \tag{5.5}$$

where C is an arbitrary constant. It is better to write this equation as

$$\frac{1}{2}mv(t)^2 + \frac{1}{2}ky(t)^2 = C$$

to emphasize that this is true for all values of t.

Equation (5.5) has physical significance. The term

$$K = \frac{1}{2}mv^2 = \frac{1}{2}m\left(\frac{dy}{dt}\right)^2$$

is the **kinetic energy** of the spring-mass system, while

$$U = \frac{1}{2}ky^2 = \int_0^y k\eta\, d\eta$$

---

[5] In this derivation, we use the fact that

$$\frac{dt}{dx} = \frac{1}{\dfrac{dx}{dt}},$$

a consequence of the inverse function theorem.

is the *potential energy*. Thus, the quantity

$$E = K + U = \frac{1}{2}mv^2 + \frac{1}{2}ky^2$$

is the *total energy* of the system. Equation (5.5) says that the total energy of the spring-mass system is conserved during its motion. ●

Thus, we have shown that the displacement $y(t)$ and the velocity $v(t)$ of an undamped harmonic oscillator satisfy the relationship in (5.5). The set of pairs (y, v) that satisfy (5.5) form an ellipse centered at the origin. Examples for different values of the constant C are shown in Figure 2, where we have used the parameters $m = 1$ and $k = 3$. The curves correspond to $C = 1/2, \ 2, \ 9/2, \ 8,$ and $25/2$.

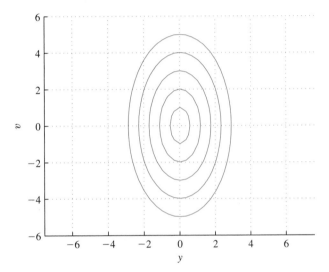

Figure 2. Level curves in the phase plane for the total energy, $E = \frac{1}{2}mv^2 + \frac{1}{2}ky^2$.

In Figure 3, we show a graph of the total energy $E(y, v) = \frac{1}{2}mv^2 + \frac{1}{2}ky^2$ as a function of y and v. This is a surface in three dimensions. On the (y, v) plane beneath the surface, some of the ellipses defined by $\frac{1}{2}mv^2 + \frac{1}{2}ky^2 = C$ are shown for various values of the constant C. These curves include those pairs (y, v) where the surface is at the same height, $E(y, v) = C$, above the (y, v) plane. For this reason, such curves are called the *level curves* of the function E. Thus, the solution curves for the undamped vibrating springs are the level curves of the total energy.

If we show that the total energy remains unchanged along a solution curve, this will mean that the solution curve is contained in a level curve of the energy. This can be proved directly. It is only necessary to show that

$$\frac{d}{dt}\left(\frac{1}{2}mv(t)^2 + \frac{1}{2}ky(t)^2\right) = 0.$$

This can be seen using the system (5.4), as follows:

$$\frac{d}{dt}\left(\frac{1}{2}mv^2 + \frac{1}{2}ky^2\right) = mvv' + kyy'$$

$$= mv\left(-\frac{k}{m}y\right) + kyv$$

$$= 0.$$

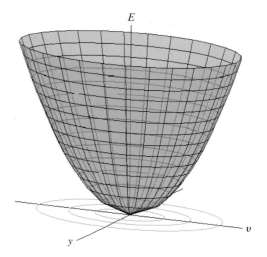

Figure 3. The total energy $E(y, v) = \frac{1}{2}mv^2 + \frac{1}{2}ky^2$ for the vibrating spring in Example 5.3.

A quantity, like the energy $E = \frac{1}{2}mv^2 + \frac{1}{2}ky^2$ for the undamped vibrating spring, that is constant along solution curves is called a ***conserved quantity*** of the system. Conserved quantities can sometimes be found by solving equation (5.2), as we did for the vibrating spring.

The effect of damping

In a damped vibrating spring, the damping slowly decreases the total energy in the system.

Example 5.6 Consider the system for the damped vibrating spring,

$$y' = v$$
$$v' = -\frac{k}{m}y - \frac{\mu}{m}v. \tag{5.7}$$

Show that the total energy $E = \frac{1}{2}mv^2 + \frac{1}{2}ky^2$ decreases along a solution curve.

This can be shown by proceeding as we did earlier. Along a solution curve $t \to (y(t), v(t))$, E becomes a function of t,

$$E(t) = \frac{1}{2}mv(t)^2 + \frac{1}{2}ky(t)^2.$$

We can show that E is decreasing by differentiating it.

$$\frac{dE}{dt} = \frac{d}{dt}\left(\frac{1}{2}mv^2 + \frac{1}{2}ky^2\right)$$
$$= mvv' + kyy'$$
$$= mv\left(-\frac{k}{m}y - \frac{\mu}{m}v\right) + kyv \tag{5.8}$$
$$= -\mu v^2$$
$$\le 0.$$

If the damping constant $\mu > 0$, the derivative is negative except where the velocity $v = 0$. However, when this is true, we see from (5.7) that $v' \neq 0$, unless $y = 0$ as well. This means that the solution curve passes through the points where $E' = 0$, and then E continues to decrease, unless we have $v = y = 0$. At this point $E = 0$, its minimum value. ●

It is interesting to see what this means in terms of the surface of E shown in Figure 3. To see the relationship clearly, it is illuminating to *lift* a solution curve to the surface. If $(y(t), v(t))$ is a solution, we consider the curve $t \rightarrow (y(t), v(t), E(t))$. This curve lies in the graph of E, as is shown in Figures 4 and 5. The blue curves on the surface are the lifts of the corresponding blue solution curves in the (y, v) plane below the surface.

If there is no damping, so $\mu = 0$, then according to (5.8), $E(t)$ is not changing. Hence, the lifted curve stays at the same level on the surface. This case is shown in Figure 4. On the other hand, if $\mu > 0$, there is damping, and then $E(t)$ is decreasing. Consequently, the lifted curve is always moving lower on the surface, the graph of $E(y, v) = \frac{1}{2}mv^2 + \frac{1}{2}ky^2$. This is illustrated in Figure 5. Under these circumstances, the curve must approach the lowest point on the surface, the minimum of $E(y, v) = \frac{1}{2}mv^2 + \frac{1}{2}ky^2$, which occurs at the origin.

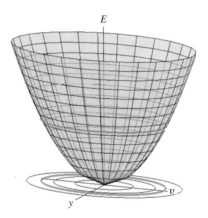

Figure 4. A lifted solution curve for the undamped vibrating spring.

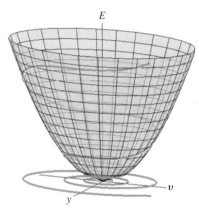

Figure 5. A lifted solution curve for the damped vibrating spring.

What we see in this simple example will occur in other more complicated examples to follow.

EXERCISES

Use the technique of Example 5.3, motivated by equation (5.2), to find a conserved quantity for each of the systems in Exercises 1–10. Verify that your solution is actually conserved on solution trajectories of the given system.

1. $y' = v$
$v' = -2y$

2. $y' = v$
$v' = 3 - y$

3. $y' = v$
$v' = -y + y^2$

4. $y' = v$
$v' = 2y - y^2$

5. $y' = v$
$v' = y - y^3$

6. $y' = v$
$v' = -2y + y^3$

7. $x' = e^y$
$y' = e^x$

8. $x' = x(1 - 2y)$
$y' = y(-1 + 2x)$

9. $x' = y$
$y' = -\sin x$

10. $x' = 1/(1 + y^2)$
$y' = e^x$

For the given systems in Exercises 11–20, perform each of the following tasks.

(i) Use the technique of Example 5.3 to find a conserved quantity.

(ii) Use your numerical solver to sketch the vector field associated with each system, then superimpose level curves of the conserved quantity on the vector field.

11. The system in Exercise 1

12. The system in Exercise 2

13. The system in Exercise 3

14. The system in Exercise 4

15. The system in Exercise 5

16. The system in Exercise 6

17. The system in Exercise 7

18. The system in Exercise 8

19. The system in Exercise 9

20. The system in Exercise 10

For Exercises 21–24, use the technique in Example 5.6.

21. Let $E(y, v)$ be the conserved quantity for the system in Exercise 1. Compute E' along the solution curves to the system

$$y' = v$$
$$v' = -v - 2y.$$

22. Let $E(y, v)$ be the conserved quantity for the system in Exercise 2. Compute E' along the solution curves to the system

$$y' = v$$
$$v' = 3 - v - y.$$

23. Let $E(y, v)$ be the conserved quantity for the system in Exercise 6. Compute E' along the solution curves to the

system

$$y' = v$$
$$v' = -v - 2y + y^3.$$

24. Let $E(x, y)$ be the conserved quantity for the system in Exercise 9. Compute E' along the solution curves to the system

$$x' = y$$
$$y' = -y - \sin x.$$

25. Consider the system

$$\frac{dx}{dt} = y,$$

$$\frac{dy}{dt} = -x + x^3. \qquad (5.9)$$

(a) Show that the quantity $H(x, y) = x^2 - (1/2)x^4 + y^2$ is conserved on solution trajectories of system (5.9).

(b) The level curve $H(x, y) = 1/2$ is the union of six distinct solutions of system (5.9). Plot the level curve $H(x, y) = 1/2$. Then use your numerical solver to superimpose the plots of the six solutions contained in the level curve.

10.6 Nonlinear Mechanics

We can easily generalize our analysis of the spring in the previous section to nonlinear examples of one-dimensional mechanics. Let's suppose that we have a mechanical system that is modeled by the equation

$$y'' = f(y, y'). \qquad (6.1)$$

If y is the displacement of a particle of mass m, then by Newton's second law, $my'' = F$, where F is the force. Assuming that the force depends only on y and y', we have $y'' = F(y, y')/m$, so we have an equation of the type in (6.1) with $f = F/m$. By writing our equation in the form in (6.1), we are allowing for more general situations. For example, y could be the angular displacement of a pendulum.

We are now familiar with the fact that if we introduce the velocity $v = y'$ as an unknown, then the second-order equation in (6.1) is equivalent to the first-order system

$$y' = v,$$
$$v' = f(y, v). \qquad (6.2)$$

Conservative systems

In the case when the force does not depend on the velocity, the system in (6.2) simplifies to

$$y' = v,$$
$$v' = f(y). \qquad (6.3)$$

We can find a conserved quantity for this system using the methods of the previous section. From (6.3) and (5.2), we get the equation

$$\frac{dv}{dy} = \frac{f(y)}{v}.$$

Separating variables and solving this equation yields

$$\frac{1}{2}v^2 = \int f(y)\,dy + C. \tag{6.4}$$

The terms in equation (6.4) have physical significance. We will refer to $K(v) = \frac{1}{2}v^2$ as the kinetic energy,[6] and $U(y) = -\int f(y)\,dy$ as the potential energy.[7] Thus, (6.4) says that the total energy

$$E(y, v) = K(v) + U(y) = \frac{1}{2}v^2 - \int f(y)\,dy$$

is conserved during the motion of this system. Systems like (6.3), where the force depends only on the displacement, are called **conservative systems** because the total energy is conserved.

As we saw in the previous section, the undamped vibrating spring is a conservative system. When damping is present, energy is dissipated, and the system is not conservative.

The fact that the energy is conserved means that any solution curve is contained in a level curve of the energy. This means that the solution curves in the phase plane are defined implicitly by equation (6.4), which we will write in the slightly modified form

$$\frac{1}{2}v^2 + U(y) = E, \tag{6.5}$$

where E is a constant.

The phase plane curves for a conservative system

A one-dimensional, conservative physical system like (6.3) can be qualitatively understood by analyzing the potential $U(y)$. There are several important considerations that allow us to do this. During this process, we will refer frequently to Figure 1. The upper graph in that figure is a graph of a potential energy U. The lower graph is the phase plane information that is implied by the shape of the graph of U.

The following considerations allow us to analyze the physics:

- The equilibrium points of (6.3) are where $v = 0$ and $f(y) = -U'(y) = 0$. Hence, they are completely determined by the critical points of U. (Remember that the critical points are where $U'(y) = 0$.)

---

[6] If y is a length, then the actual kinetic energy is $\frac{1}{2}mv^2$, a constant multiple of our definition. In other cases, where, for example, y is an angle, the kinetic energy will again be a constant multiple of $\frac{1}{2}v^2$. In every case the actual potential energy will be the same constant multiple of $U = -\int_{y_0}^y f(y)\,dy$.

[7] Notice that the potential energy is defined as an indefinite integral and is only defined up to an arbitrary additive constant. Physicists define the potential energy as $U(y) = -\int_{y_0}^y f(\eta)\,d\eta$, where y_0 is a fixed reference displacement. However, the reference displacement is completely arbitrary and is reflected in the arbitrary constant in the indefinite integral that we use. We encourage you to determine that this constant does not affect what follows in any significant way.

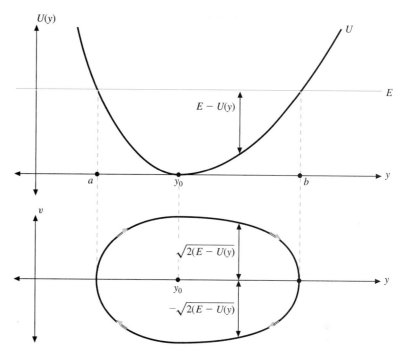

Figure 1. Analyzing motion using the potential energy. The upper graph is the graph of U. The lower one is the resulting phase plane diagram.

- The solution curves in the phase plane are defined implicitly by equation (6.5). Each solution curve has two branches defined by

$$v = \pm\sqrt{2(E - U(y))}. \tag{6.6}$$

Since the two branches are negatives of each other, the solution curves in the phase plane are symmetric with respect to the y-axis.

- The implicit-function theorem from multivariable calculus tells us that the solution curves will be smooth except possibly where $\partial E/\partial v = v = 0$ and $\partial E/\partial y = -f(y) = 0$. Notice that these are at the equilibrium points of (6.3).

- We notice that the velocity v is positive on the upper half of the solution curve, and because $y' = v$, the displacement y is increasing there. Thus, the vector field points to the left where $v < 0$ and to the right where $v > 0$.

- Since $\frac{1}{2}v^2 \geq 0$, it follows from (6.5) that on any solution curve we have

$$U(y) \leq E. \tag{6.7}$$

This restricts the range of the displacement y over a solution curve. An example is shown in Figure 1. This is an example of what is called a **potential well**. More examples are shown in Figure 2. You will find it helpful to imagine a ball rolling along the graph of U, with velocity in the y-direction given by (6.6). When this ball reaches an endpoint where $U(y) = E$, the ball's velocity is 0. At this point, the ball turns and goes the other way. Thus, the ball is constrained to remain in the potential well.

Let's first look at a local minimum of U, such as the point y_0 in the upper graph in Figure 1. For an energy level E slightly larger than $U(y_0)$, such as that shown in Figure 1, the set of y defined by (6.7) is a small interval containing y_0, such as (a, b) in the figure. Therefore, the set in the phase plane defined by (6.5) is the closed curve indicated in the lower half of Figure 1. As this illustrates, the solution curve corresponding to any energy level slightly larger than the local minimum is a closed curve. This means that the equilibrium point defined by a local minimum of the potential energy is a center.

A more complicated potential energy is analyzed in Figure 2. This potential energy has two local minima, y_1 and y_2. Several energy levels are examined and the corresponding solution curves have been drawn.

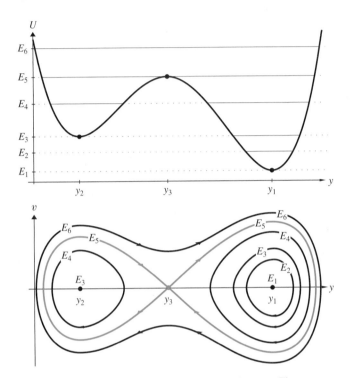

Figure 2. Analyzing a more complicated potential energy. The upper graph is graph of U. The lower one is the resulting phase plane diagram.

Next let's look at a local maximum of U, such as y_3 in Figure 2. We know that $(y_3, 0)$ is an equilibrium point. Consequently, the solutions that approach this point must take forever to do so. These are stable solutions for the equilibrium point. Similarly, the solutions that correspond to the ball rolling away from this point must be unstable solutions, and we see that the equilibrium point defined by a local maximum of the potential energy is a saddle point.

The pendulum

We now have all of the mathematical tools that we need to analyze the motion of a pendulum. The pendulum consists of a bob with mass m connected by a solid rod of length L to a fixed pivot (see Figure 3). The rod is considered to be massless. If the bob is moved from its equilibrium position directly below the pivot, it will move on a circle of radius L with center at the pivot. Let θ be the angle between the rod and true vertical. The displacement of the bob from equilibrium is measured along the circle and is therefore $L\theta$. Its velocity is the derivative of this, $L\theta'$.

The bob is affected by the force of gravity, which points directly down and has magnitude mg. However, the portion of the force that is parallel to the rod is balanced by the rod itself. The unbalanced portion of the force is tangential to the circle and perpendicular to the rod. Hence, this tangential force is $-mg \sin \theta$. The

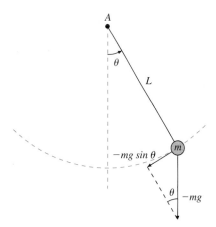

Figure 3. A pendulum under the force of gravity.

minus sign indicates that the force is always acting to decrease the magnitude of the angular displacement. If there is a damping force that is proportional to the velocity, it has the form $-\mu L\theta'$, where μ is a constant. Since the acceleration of the pendulum is $L\theta''$, Newton's second law is

$$mL\theta'' = -mg\sin\theta - \mu L\theta'.$$

If we divide by mL, we get

$$\theta'' = -\frac{g}{L}\sin\theta - \frac{\mu}{m}\theta'.$$

This equation has the same form as equation (6.1), but the displacement is now θ, the angular displacement.

We can find the equivalent first-order system by introducing the angular velocity $\omega = \theta'$. In the usual way, we find that the dynamics of the pendulum are governed by the nonlinear system

$$\theta' = \omega,$$
$$\omega' = -\frac{g}{L}\sin\theta - \frac{\mu}{m}\omega.$$

We will set $a = g/L$ and $b = \mu/m$. We will refer to b as the damping constant. Then the system for the pendulum is

$$\begin{aligned} \theta' &= \omega, \\ \omega' &= -a\sin\theta - b\omega. \end{aligned} \tag{6.8}$$

The undamped pendulum

Notice that the system in (6.8) has the same form as that in (6.2), with y replaced by θ and v by ω. The forcing term $f(\theta, \omega) = -a\sin\theta - b\omega$ is, in general, a function of both θ and ω. The preceding analysis shows that the pendulum system is conservative if there is no damping (i.e., if $b = 0$). Let's assume this is true for the moment. The system for the undamped pendulum reduces to

$$\begin{aligned} \theta' &= \omega, \\ \omega' &= -a\sin\theta. \end{aligned} \tag{6.9}$$

In this case, the potential energy is

$$U(\theta) = -\int f(\theta)\,d\theta = \int a\sin\theta\,d\theta = -a\cos\theta.$$

The graph of U is shown in Figure 4.

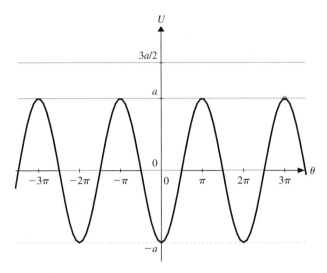

Figure 4. The potential energy of the pendulum.

Notice that U is periodic with period 2π, reflecting the fact that θ measures the angle at which the pendulum is displaced. The graph of the potential energy in Figure 4 leads to the phase plane plot in Figure 5. It is also periodic with period 2π. The physical situation of the pendulum itself is periodic. It would be a good idea to imagine the graphs in Figures 4 and 5 rolled up so that the lines $\theta = -\pi$ and $\theta = \pi$ were identified. Then the potential energy would be plotted on a cylinder, and instead of a phase plane, we would have a phase cylinder.

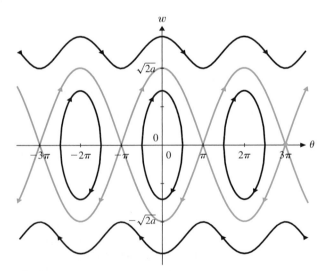

Figure 5. Level curves for the energy of the undamped pendulum. The blue curve consists of the separatrices.

For the pendulum the potential energy has a minimum value of $-a$ when $\theta = 2k\pi$ for any integer k. Therefore, all of the equilibrium points $(2k\pi, 0)$ are centers.

Of course, they all correspond to the one physical equilibrium where the bob hangs motionless directly below the pivot. If the total energy

$$E = \frac{1}{2}\omega^2 - a\cos\theta$$

lies between $-a$ and a, the solution curves are closed. These correspond to what might be called the normal small-amplitude oscillations of the undamped pendulum.

The potential energy achieves its maximum value of a at $\theta = (2k+1)\pi$ for any integer k. Thus, all of the equilibrium points $((2k+1)\pi, 0)$ are saddle points. They correspond to the unstable physical equilibrium where the pendulum bob is balanced motionless directly above the pivot. The stable and unstable solution curves for the undamped pendulum in this case are shown in blue in Figure 5. Notice that the blue level set consists of a series of solution curves, each starting at an unstable equilibrium point and ending at an adjacent one. They are all separatrices for the saddle points.

If the total energy is larger than a, then the pendulum has enough energy to pass over the top of the pivot. The motion continues in one direction around the pivot, slowing as the bob reaches the position directly over the pivot and then accelerating as it moves downward.

The damped pendulum

We have found that the undamped pendulum has equilibrium points when $\theta = k\pi$ for some integer k and when $\omega = 0$. If we examine the right-hand sides in (6.8), we see that the damped pendulum has the same equilibrium points. Let's analyze them. The Jacobian is

$$J(\theta, \omega) = \begin{pmatrix} 0 & 1 \\ -a\cos\theta & -b \end{pmatrix}.$$

If $\theta = 2k\pi$, then $\cos\theta = 1$, so

$$J = \begin{pmatrix} 0 & 1 \\ -a & -b \end{pmatrix}.$$

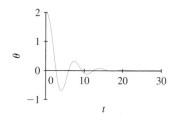

Then $D = \det(J) = a > 0$ and $T = \operatorname{tr}(J) = -b < 0$. We see that the equilibrium point is a sink, but the actual type depends on the damping constant b.

To be more precise in our evaluation, we compute the discriminant $T^2 - 4D = b^2 - 4a$. If $b^2 < 4a$, the discriminant is negative and the equilibrium point is a spiral sink. This is the underdamped case. An example of the motion of an underdamped pendulum is shown in Figure 6. If $b^2 > 4a$, the discriminant is positive and the equilibrium point is a nodal sink. This is the overdamped case. A typical graph of the displacement is shown in Figure 7. If $b^2 = 4a$, the damping is critical, and we cannot say anything more than that the equilibrium point is a sink.

If θ is an odd multiple of π, then $\cos\theta = -1$ and the Jacobian is

$$J = \begin{pmatrix} 0 & 1 \\ a & -b \end{pmatrix}.$$

Figure 6. The angular displacement of an underdamped pendulum.

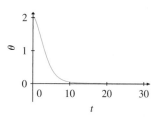

Figure 7. The angular displacement of an overdamped pendulum.

Now $D = \det(J) = -a < 0$, so we know this is a saddle point. Mathematically, these saddle points are unstable, reflecting the physical instability of the pendulum bob balanced above the pivot. It is interesting to reflect on the behavior of the separatrices. A stable separatrix would be a solution that ends up (after an infinitely long time) balanced above the pivot. If we were to try to achieve this with a pendulum, it is highly unlikely that we would succeed. Some of the time the bob would stop short, and the rest of the time it would pass over the pivot and continue down the other side. Theoretically, there must be a possibility between these two, where the bob just comes to a stop above the pivot. This solution is a stable separatrix. On the other hand, an unstable separatrix starts at the equilibrium point and then proceeds away from it.

The phase plane for the underdamped pendulum is shown in Figure 8. Notice that the stable solution curves come to rest at the saddles, but the unstable solution curves spiral into the sinks. Each of the sinks has a basin of attraction, which is readily seen to be bounded by the stable solution curves of the neighboring saddles.

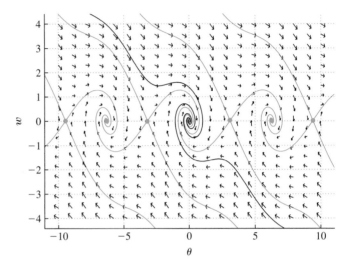

Figure 8. The phase plane for a damped pendulum.

Energy and the damped pendulum

Let's look at what happens to the energy along the solution curve $(\theta(t), \omega(t))$ to the system (6.8) for the pendulum. Along the curve, the energy is

$$E(t) = E(\theta(t), \omega(t)) = \frac{1}{2}\omega^2(t) - a\cos\theta(t).$$

Differentiating and using the system (6.8), we see that

$$E' = \omega\omega' + a\sin\theta\ \theta' = \omega(-a\sin\theta - b\omega) + a\omega\sin\theta = -b\omega^2.$$

If there is no damping, $b = 0$, so $E' = 0$, and the energy is conserved along the solution curves, confirming what we already know. However, if there is damping, then $b > 0$ and $E' = -b\omega^2 \leq 0$. Thus, the energy is decreasing except where the angular velocity $\omega = 0$. At such a point, however, the second equation in (6.8) becomes $\omega' = -a\sin\theta$, which is nonzero unless $\theta = k\pi$. The combination $\omega = 0$ and $\theta = k\pi$ occurs only at an equilibrium point. Thus, if our solution is not an equilibrium solution, at every point where $E' = -b\omega^2 \neq 0$, the angular velocity is changing. Therefore, such points are isolated, and overall the energy continues to decrease along the solution curve.

Where can this process end? We refer you to Figure 8. The stable solution curves for the saddles end at the saddles, of course. With these exceptions, all solution curves end at the stable equilibrium points, which are the minimums of the energy function.

Hamiltonian systems

There is a special type of system that is defined by a conserved quantity. Let $H(x, y)$ be a function of two variables that is twice continuously differentiable. The **Hamiltonian system** defined by H is the system

$$x' = \frac{\partial H}{\partial y},$$

$$y' = -\frac{\partial H}{\partial x}. \tag{6.10}$$

The function H is called the **Hamiltonian**.

For example, any conservative physical system

$$y' = v,$$
$$v' = f(y)$$

is Hamiltonian. If $U(y) = -\int f(y)\,dy$ is a potential function, we take the total energy $H(y, v) = \frac{1}{2}v^2 + U(y)$ as our Hamiltonian. Then

$$y' = v = \frac{\partial H}{\partial v},$$

$$v' = f(y) = -U'(y) = -\frac{\partial H}{\partial y},$$

so the system is Hamiltonian.

Like the energy for a conservative system, the Hamiltonian is always a conserved quantity for the corresponding Hamiltonian system. To see this, suppose $(x(t), y(t))$ is a solution to (6.10). We compute using the multivariable chain rule,

$$\frac{d}{dt} H(x(t), y(t)) = \frac{\partial H}{\partial x} x' + \frac{\partial H}{\partial y} y'$$

$$= \frac{\partial H}{\partial x} \frac{\partial H}{\partial y} + \frac{\partial H}{\partial y} \left(-\frac{\partial H}{\partial x} \right)$$

$$= 0.$$

Thus, $H(x(t), y(t))$ is constant along any solution curve, so H is a conserved quantity. The level curve $H(x, y) = C$ consists of one or more solution curves.

Suppose we are given a system

$$x' = f(x, y),$$
$$y' = g(x, y).$$

Is there a way to tell whether the system is Hamiltonian? If it is Hamiltonian, there is a function H such that

$$f = \frac{\partial H}{\partial y} \quad \text{and}$$

$$g = -\frac{\partial H}{\partial x}. \tag{6.11}$$

Consequently, we are really asking if and when we can solve the system of partial differential equations in (6.11). The answer is given in detail by the next theorem.

THEOREM 6.12 Consider the system

$$x' = f(x, y),$$
$$y' = g(x, y),$$

(6.13)

where f and g are continuously differentiable.

1. If the system is Hamiltonian, then

$$\frac{\partial f}{\partial x} = -\frac{\partial g}{\partial y}.$$

(6.14)

2. If equation (6.14) is true in a rectangle R, then the system is Hamiltonian in R. A Hamiltonian for the system is given by

$$H(x, y) = \int f(x, y)\, dy + \phi(x),$$

(6.15)

where the function ϕ satisfies

$$\phi'(x) = -\frac{\partial}{\partial x} \int f(x, y)\, dy - g(x, y).$$

(6.16)

Proof   Suppose first that the system is Hamiltonian, with H as its Hamiltonian. Then f and g are given by (6.11). From this, we get

$$\frac{\partial f}{\partial x} = \frac{\partial}{\partial x}\left(\frac{\partial H}{\partial y}\right) = \frac{\partial^2 H}{\partial x\, \partial y} \quad \text{and} \quad \frac{\partial g}{\partial y} = \frac{\partial}{\partial y}\left(-\frac{\partial H}{\partial x}\right) = -\frac{\partial^2 H}{\partial y\, \partial x}.$$

Since H is twice continuously differentiable, the mixed second partials are equal, so (6.14) follows.

Now suppose that (6.14) is true in a rectangle R. We need to find a function H that satisfies (6.11). Let's start by integrating the first equation in (6.11) with respect to y. By the fundamental theorem of calculus, we have

$$H(x, y) = \int f(x, y)\, dy + \phi(x).$$

This is equation (6.15). Since we are integrating with respect to y, the constant of integration $\phi(x)$ can still depend on x.

To discover what ϕ is, we differentiate H as given in (6.15) with respect to x.

$$\frac{\partial H}{\partial x} = \frac{\partial}{\partial x} \int f(x, y)\, dy + \phi'(x)$$

According to the second equation in (6.11), we want $\partial H / \partial x = -g$, so ϕ must satisfy

$$\phi'(x) = -\frac{\partial}{\partial x} \int f(x, y)\, dy - g(x, y).$$

This is equation (6.16). It can be solved by integration if the right-hand side is a function of x only. The hypothesis of part (2) of the theorem guarantees that this is the case. To see this, we show that the derivative of the right-hand side with respect to y is 0. We have

$$\frac{\partial}{\partial y}\left(\frac{\partial}{\partial x}\int f(x, y)\, dy + g(x, y)\right) = \frac{\partial}{\partial y}\left(\frac{\partial}{\partial x}\int f(x, y)\, dy\right) + \frac{\partial g}{\partial y}$$

$$= \frac{\partial}{\partial x}\left(\frac{\partial}{\partial y}\int f(x, y)\, dy\right) + \frac{\partial g}{\partial y}$$

$$= \frac{\partial f}{\partial x} + \frac{\partial g}{\partial y}$$

$$= 0$$

by (6.14). ∎

Example 6.17   Consider the system

$$\begin{aligned} x' &= y, \\ y' &= x(x - 1). \end{aligned} \qquad (6.18)$$

Show that the system is Hamiltonian and find a Hamiltonian function.

With $f(x, y) = y$ and $g(x.y) = x(x - 1)$, we have

$$\frac{\partial f}{\partial x} = 0 = -\frac{\partial g}{\partial y}.$$

Therefore, the system is Hamiltonian.

To find the Hamiltonian function, we follow the steps in the proof. First,

$$H(x, y) = \int f(x, y)\, dy + \phi(x) = \int y\, dy + \phi(x) = \frac{y^2}{2} + \phi(x).$$

Then, to find ϕ, we use

$$-g(x, y) = \frac{\partial H}{\partial x}(x, y) = \phi'(x).$$

or

$$\phi'(x) = -x(x - 1).$$

From this, we choose $\phi(x) = \frac{1}{2}x^2 - \frac{1}{3}x^3$. Finally,

$$H(x, y) = \frac{1}{2}(x^2 + y^2) - \frac{1}{3}x^3.$$

●

The equilibrium points of Hamiltonian systems

We saw earlier that conservative mechanical systems can have equilibrium points that are saddle points or centers. In fact, these are the only types of equilibrium points that can occur. We will show that this is true for any Hamiltonian system. To see this, suppose we have a Hamiltonian system

$$x' = \frac{\partial H}{\partial y},$$

$$y' = -\frac{\partial H}{\partial x}.$$

We compute the Jacobian

$$
J = \begin{pmatrix} \dfrac{\partial^2 H}{\partial x\, \partial y} & \dfrac{\partial^2 H}{\partial y^2} \\[2ex] -\dfrac{\partial^2 H}{\partial x^2} & -\dfrac{\partial^2 H}{\partial y\, \partial x} \end{pmatrix}.
$$

If $\det(J) < 0$, the equilibrium point is a saddle. If $\det(J) > 0$, we must be a little more careful. The trace of the Jacobian is

$$
\frac{\partial^2 H}{\partial x\, \partial y} - \frac{\partial^2 H}{\partial y\, \partial x} = 0,
$$

since the mixed second partials of H are equal. In particular, the trace of the Jacobian is equal to 0 at any equilibrium point. Therefore, if $\det(J) > 0$, the linearization has a center at the origin, but this does not necessarily imply that the nonlinear Hamiltonian system has a center.

Notice that any equilibrium point for the Hamiltonian system is a critical point[8] for the Hamiltonian H. Let's apply the second derivative test from multivariable calculus to determine whether the critical point is an extremum or not. This requires that we look at the determinant of the matrix of second partial derivatives of H. However,

$$
\det \begin{pmatrix} \dfrac{\partial^2 H}{\partial x^2} & \dfrac{\partial^2 H}{\partial y\, \partial x} \\[2ex] \dfrac{\partial^2 H}{\partial x\, \partial y} & \dfrac{\partial^2 H}{\partial y^2} \end{pmatrix} = \frac{\partial^2 H}{\partial x^2}\frac{\partial^2 H}{\partial y^2} - \left(\frac{\partial^2 H}{\partial x\, \partial y} \right)^2 = \det(J) > 0.
$$

By the second derivative test for functions of two variables, if $\det(J) > 0$, the critical point of H must be either a local maximum or a local minimum. In either case, the level sets of H (which are the solution curves of the Hamiltonian system) near the critical value are closed curves. Thus, it follows that the equilibrium point is a center.

Linear Hamiltonian systems

Consider the planar, linear system

$$
\mathbf{x}' = A\mathbf{x}, \quad \text{where} \quad A = \begin{pmatrix} a & b \\ c & d \end{pmatrix}. \tag{6.19}
$$

In terms of the coordinates of $\mathbf{x} = (x_1, x_2)^T$, the system is

$$
\begin{aligned}
x_1' &= f(x_1, x_2) = ax_1 + bx_2, \\
x_2' &= g(x_1, x_2) = cx_1 + dx_2.
\end{aligned}
$$

According to Theorem 6.12, this system is Hamiltonian if and only if

$$
\frac{\partial f}{\partial x_1} = -\frac{\partial g}{\partial x_2}, \quad \text{or} \quad a = -d.
$$

---

[8] Remember that a critical point for a function H of several variables is a point where all first partial derivatives of H are equal to 0.

Thus, the linear system (6.19) is Hamiltonian if and only if

$$A = \begin{pmatrix} a & b \\ c & -a \end{pmatrix}.$$

Saying this in another way, we see that a planar, linear system is Hamiltonian if and only if its trace is 0.

Let's find a Hamiltonian function. It must be

$$H(x_1, x_2) = \int (ax_1 + bx_2)\, dx_2 = ax_1 x_2 + \frac{1}{2} bx_2^2 + \phi(x_1),$$

where $\phi(x_1)$ is yet to be determined. The function ϕ is found by differentiating and noticing that

$$\frac{\partial H}{\partial x_1} = -g(x_1, x_2) = -(cx_1 - ax_2).$$

This becomes

$$ax_2 + \phi'(x_1) = -(cx_1 - ax_2),$$

so $\phi(x_1) = -cx_1^2/2$. Thus, a Hamiltonian for our linear system is the quadratic polynomial

$$H(x_1, x_2) = \frac{1}{2} bx_2^2 - \frac{1}{2} cx_1^2 + ax_1 x_2. \tag{6.20}$$

We know that the equilibrium point at the origin is either a saddle point or a center. The exact type of the equilibrium point is decided by the determinant, $\det(A) = -a^2 - bc$. The nature of the level sets of the polynomial H are determined by its discriminant, which is

$$a^2 - 4\left(-\frac{c}{2}\right)\left(\frac{b}{2}\right) = a^2 + bc = -\det(A).$$

We will consider two cases:

- If $\det(A) < 0$, or, equivalently, if $-bc < a^2$, then the equilibrium point is a saddle point. This same condition means that the discriminant of H is positive, which implies that the level sets of the quadratic Hamiltonian (6.20) are hyperbolas.
- If $\det(A) > 0$, or, equivalently, if $-bc > a^2$, then the equilibrium point is a center. This same condition means that the discriminant of H is negative, which implies that the level sets of our Hamiltonian (6.20) are ellipses centered at the origin. Since linear centers are characterized by $T = \mathrm{tr}(A) = 0$ and $D > 0$, we see that every linear center is surrounded by solution curves that are ellipses centered at the origin.

EXERCISES

Consider the system

$$y' = v,$$
$$v' = f(y).$$

In Exercises 1–6, a potential function $U(y) = -\int f(y)\,dy$ is drawn. Make a copy of the potential function; then align the yv-phase plane below the potential axes, as shown in Figure 1. Use the potential function U to create the corresponding phase portrait in the yv-phase plane.

1.

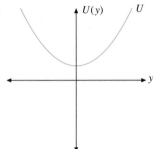

2.

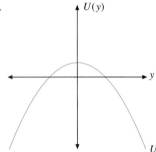

3.

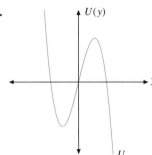

4.

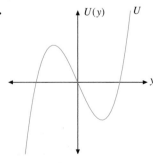

5.

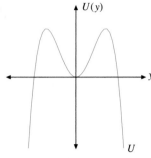

6.

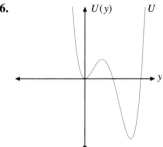

Consider the system

$$y' = v,$$
$$v' = f(y).$$

In Exercises 7–12, you are given an equation for the force $f(y)$. In each exercise, perform the following tasks:

(i) Find the equilibrium points for the system.

(ii) Find a potential function $U(y)$ for the system.

(iii) As in Figure 1, align the potential axes with that of the yv phase plane. Plot the potential function and the equilibrium points, then construct a phase portrait in the yv-phase plane with this information.

7. $f(y) = 1 - 2y$ **8.** $f(y) = y$

9. $f(y) = 4y - y^2$ **10.** $f(y) = y^2 - 9y$

11. $f(y) = \sin y$ **12.** $f(y) = -\cos y$

Consider the system

$$y' = v,$$
$$v' = f(y).$$

In Exercises 13–16, use the given $f(y)$ to aid in completing each of the following tasks:

(i) Use your numerical solver to sketch the phase portrait of the system. Label and classify each equilibrium point: center, saddle, and so on.

(ii) Use the relation

$$\frac{1}{2}v^2 + U(y) = E$$

to find the equation of the separatrices. Sketch these separatrices on the phase portrait developed in part (i).

13. $f(y) = y^3 - y$

14. $f(y) = y^2(y^2 - 1)$

15. $f(y) = -\sin y$

16. $f(y) = \cos y$

17. If the restoring force of a spring is a linear function of the displacement, as in $f(y) = -ky$, then the model $my'' + f(y) = 0$ becomes $my'' + ky = 0$, which yields simple harmonic motion. But what if the restoring force of the spring is nonlinear? For example, let the restoring force be defined by $f(y) = -(\alpha y + \beta y^3)$, where $\alpha > 0$. Define the **stiffness** of the spring as $f'(y)$. A **soft** spring is one whose stiffness decreases as the displacement increases, but a **hard** spring is one whose stiffness increases with increased displacement. Use the graph of f' to find conditions on β such that the spring defined by f is a hard (soft) spring.

18. Consider the equation

$$y'' + \alpha y + \beta y^3 = 0,$$

which models a spring-mass system ($m = 1$) with a nonlinear restoring force.

(a) If the spring is a hard spring (see Exercise 17), show that all trajectories in the yv-phase plane ($v = y'$) are closed orbits circling an isolated equilibrium point at the origin.

(b) If the spring is a soft spring (see Exercise 17), identify the equilibrium points. Then find and use a potential function $U(y)$ to aid in sketching a phase portrait in the yv-phase plane. Find the equations of the separatrices and include them on your plot.

19. In equation (6.8), it was assumed that the damping was proportional to the angular velocity ω. The term $-b\omega$ guarantees that the damping is always *opposite* the motion. Now, let's make a different assumption about the damping (i.e., let's assume that the magnitude of the damping force is proportional to the *square* of the angular velocity). Consider

$$\begin{aligned} \theta' &= \omega, \\ \omega' &= -a\sin\theta - b\omega|\omega|, \end{aligned} \quad (6.21)$$

$a, b > 0$, and explain why the use of the absolute-value symbols is consistent with our new assumption. Next,

(a) find the equilibrium points of this new system, and

(b) use your numerical solver to create a phase portrait for the system, using $a = 1$ and $b = 0.2$. Use your phase portrait to classify the equilibrium points: centers, saddles, and so on.

20. Consider system (6.21) of Exercise 19, again using $a = 1$ and $b = 0.2$. Suppose that the pendulum is given the following initial conditions: $\theta = 0$ and $\omega = \overline{\omega}$, this initial angular velocity being provided by striking the bob with an impulsive, hammer like blow.

(a) Use your numerical solver to determine approximate bounds on $\overline{\omega}$ (both min and max) so that the bob spins over the top, going past its unstable position exactly once, eventually (after a long time) coming to rest at the equilibrium position defined by $\theta = 2\pi$, $\omega = 0$.

(b) Using the substitution $\sqrt{a}y = \omega$, show that system (6.21) is equivalent to the single equation

$$y\frac{dy}{d\theta} + by^2 = -\sin\theta,$$

for $y > 0$, or, better yet,

$$\frac{d}{d\theta}y^2 + 2by^2 = -2\sin\theta.$$

Note that this last equation is linear in y^2. Use an integrating factor to show that the general solution is given by

$$y^2 = c_1 e^{-2b\theta} + \frac{2}{1+4b^2}\cos\theta - \frac{4b}{1+4b^2}\sin\theta.$$

(c) There are saddle equilibrium points at $(n\pi, 0)$, n an odd integer. (See Exercise 19.) Use the last result of part (b) to find the equations of the stable separatrices in the upper half of the phase plane that approach the saddle equilibrium points at $(n\pi, 0)$, n an odd integer. Show that these separatrices cross the ω axis in the $\theta\omega$ phase plane at

$$\overline{\omega}^2 = \frac{2a}{1+4b^2}\left[1 + e^{2nb\pi}\right].$$

Does this equation produce the same results as that found in part (a)?

21. Suppose that the planar, linear system

$$\begin{aligned} x' &= ax + by \\ y' &= cx + dy \end{aligned}$$

has the quadratic polynomial $H(x, y) = Ax^2 + Bxy + Cy^2$ as a conserved quantity. Show that $(a+d)(ad-bc) = 0$. Thus, the matrix of the system either has determinant or trace equal to 0.

In Exercises 22–29, determine if the given system is Hamiltonian. If the system is Hamiltonian, find its Hamiltonian function.

22. $x' = y$
$y' = -x + x^3$

23. $x' = -2x - 3y^2$
$y' = -3x^2 + 2y$

24. $x' = 5x - 3y^2$
$y' = 3x^2 - 4y$

25. $x' = 3y^2$
$y' = -3x^2$

26. $x' = x - 4y$
$y' = 4x + 2y$

27. $x' = -x + 2y$
$y' = -2x + y$

28. $x' = -2y \sin(x^2 + y^2)$
$\quad\; y' = \;\;\; 2x \sin(x^2 + y^2)$

29. $x' = \cos x$
$\quad\; y' = -y \sin x + 2x$

30. A particle moves along a line according to the equation of motion

$$x'' = -2x + x^2,$$

where x marks the position of the particle on the line at time t.

(a) Change the equation to a system of first-order equations, show that the system is Hamiltonian, and find the Hamiltonian function $H(x, y)$ for the system.

(b) Find the level curve of $H(x, y)$ that contains the separatrices for the system. Plot this level curve; then shade in red the regions of the plane where the system oscillates.

(c) How many distinct solution curves are contained in the level set of part (b)?

31. The methods of this section apply to physical systems with more than one degree of freedom. Suppose there are two displacements, y_1 and y_2, with associated velocities $v_1 = y_1'$ and $v_2 = y_2'$, and suppose that the system is modeled by the system of equations $\mathbf{y}'' = \mathbf{f}(\mathbf{y})$, where $\mathbf{y} = (y_1, y_2)^T$, and the forcing term is $\mathbf{f} = (f_1, f_2)^T$. We also put the velocities into the vector $\mathbf{v} = (v_1, v_2)^T$. Then we can rewrite the second order system as the first-order system

$$\mathbf{y}' = \mathbf{v}$$
$$\mathbf{v}' = \mathbf{f}(\mathbf{y}).$$

We will say that this system is **_conservative_** if there is a function $U(\mathbf{y})$ such that $\partial U / \partial y_1 = -f_1$ and $\partial U / \partial y_2 = -f_2$. Show that if the system is conservative, then the total energy

$$E(y, v) = \frac{1}{2}|\mathbf{v}|^2 + U(\mathbf{y}) = \frac{1}{2}(v_1^2 + v_2^2) + U(\mathbf{y})$$

is a conserved quantity for the first-order system.

10.7 The Method of Lyapunov

Some of the most useful global methods of studying stability are due to Aleksandr Lyapunov.[9] The key idea is one that we have been using occasionally but that we will now explain in detail.

Variation of a function along a solution curve

Consider a planar autonomous system

$$\begin{aligned} x' &= f(x, y), \\ y' &= g(x, y). \end{aligned} \tag{7.1}$$

Suppose we have a continuously differentiable function $V(x, y)$ and we ask how V varies along a solution curve. This means that we have a solution $(x(t), y(t))$ to (7.1), and we want to know how the function $V(x(t), y(t))$ varies as t varies. In particular, is it increasing or decreasing? Of course, the way to find the answer is to differentiate. The multivariable chain rule[10] and the system of equations in (7.1) provide the answer:

$$\begin{aligned} \frac{d}{dt} V(x(t), y(t)) &= \frac{\partial V}{\partial x} \frac{dx}{dt} + \frac{\partial V}{\partial y} \frac{dy}{dt} \\ &= \frac{\partial V}{\partial x} f(x, y) + \frac{\partial V}{\partial y} g(x, y). \end{aligned} \tag{7.2}$$

Notice that the right-hand side of (7.2) can be computed without actually knowing the solution. Let's define

$$\dot{V}(x, y) = \frac{\partial V}{\partial x}(x, y) f(x, y) + \frac{\partial V}{\partial y}(x, y) g(x, y). \tag{7.3}$$

---

[9] Aleksandr Mikhailovich Lyapunov was born in Yaroslavl, Russia in 1857. He died a violent death in 1918, in Odessa, Russia. His work concentrated on the stability of equilibrium and motion of a mechanical system and the stability of a uniformly rotating fluid. He devised important methods of approximation. Lyapunov's methods, which he introduced in 1899, provide ways of determining the stability of sets of ordinary differential equations.

[10] We will be using techniques from multivariable calculus in this section.

Then (7.2) becomes

$$\frac{d}{dt} V(x(t), y(t)) = \dot{V}(x(t), y(t)).$$

Thus, we can say that V is increasing along the solution curve through a point (x, y) if $\dot{V}(x, y) > 0$, and it is decreasing if $\dot{V}(x, y) < 0$. We can do this without knowing the solution!

It's worth looking at the definition of $\dot{V}$ in (7.3) a little more closely. Remember that the gradient of V is the vector

$$\mathbf{grad}\, V = \left(\frac{\partial V}{\partial x}, \frac{\partial V}{\partial y} \right)^T.$$

If we put the system in (7.1) into vector form, the right-hand sides become the vector field $\mathbf{F} = (f, g)^T$. Notice that (7.3) defines $\dot{V}$ as the dot product of the gradient vector $\mathbf{grad}\, V$ and the field vector $\mathbf{F}$,

$$\dot{V} = \mathbf{grad}\, V \cdot \mathbf{F}.$$

Recall that for any vector $\mathbf{v}$, the dot product $\mathbf{grad}\, V \cdot \mathbf{v}$ is the derivative of V in the direction $\mathbf{v}$. Thus, $\dot{V}$ is the derivative of V in the direction of the vector field $\mathbf{F} = (f, g)^T$ defined by the system (7.1).

Let's change our point of view slightly. We want to use the information about how V varies along a solution curve to obtain information about the solution curve itself. It is useful to conceive of the xy-plane as a map of some terrain where the elevation is given by the function V. Thus, the three-dimensional view of the terrain is just the graph of V. See Figure 1 for an example. We can get some idea of what the terrain looks like by plotting the level curves of V in the xy-plane. These are the curves defined by the equation $V(x, y) = C$ for various constants C. Since we are imagining that V is the elevation, the level curves are where the terrain has the same elevation. Four such curves are shown in blue on the surface in Figure 1. If we project the level curves onto the xy-plane, we get Figure 2. This is called a **topographic** map of the terrain. The level curves of V are the lines of equal elevation and are usually called **contour lines**.

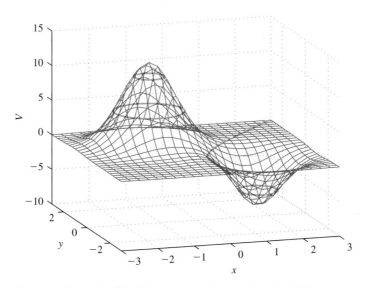

Figure 1. The graph of V with level curves at $V = -5, 0, 5,$ and 10.

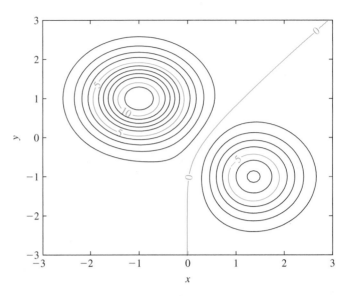

Figure 2. The level curves of the function V graphed in Figure 1.

Recall that **grad** V points in the direction in which V is increasing the fastest. In particular, this direction is orthogonal to the level curves of V. If $\dot{V}(x, y) > 0$, then we know that V is increasing in the direction of the vector field. Thus, the elevation along the solution curve through the point (x, y) is increasing. Looked at slightly differently, this means that the solution curve at (x, y) is moving into that portion of the terrain where the elevation is greater than it is at (x, y). In other words, the curve is going uphill. Similarly, if $\dot{V}(x, y) < 0$, the solution curve at (x, y) is going downhill.

The considerations of the previous paragraphs can be carried over to higher-dimensional systems, although the elevation metaphor becomes strained. Suppose we have an autonomous system

$$\mathbf{x}' = \mathbf{f}(\mathbf{x}), \tag{7.4}$$

where $\mathbf{x} = (x_1, x_2, \ldots, x_n)^T$ and $\mathbf{f}(\mathbf{x}) = (f_1(\mathbf{x}), f_2(\mathbf{x}), \ldots, f_n(\mathbf{x}))^T$. Again, suppose that we have a function $V(\mathbf{x})$, and we ask how V varies along a solution curve. This means that we have a solution $\mathbf{x}(t)$ to (7.4), and we want to know how the function $V(\mathbf{x}(t))$ behaves as a function of t. In particular, is it increasing or decreasing? Of course, the way to find the answer is to differentiate. The chain rule provides the answer:

$$
\begin{aligned}
\frac{d}{dt} V(\mathbf{x}(t)) &= \sum_{i=1}^{n} \frac{\partial V}{\partial x_i}(\mathbf{x}(t)) \frac{dx_i}{dt}(t) \\
&= \sum_{i=1}^{n} \frac{\partial V}{\partial x_i}(\mathbf{x}(t)) f_i(\mathbf{x}(t)).
\end{aligned}
\tag{7.5}
$$

Notice that the right-hand sides of (7.5) can be computed without actually knowing the solution. Let's define

$$\dot{V}(\mathbf{x}) = \mathbf{grad}\, V(\mathbf{x}) \cdot \mathbf{f}(\mathbf{x}) = \sum_{i=1}^{n} \frac{\partial V}{\partial x_i}(\mathbf{x}) f_i(\mathbf{x}).$$

Then (7.5) becomes

$$\frac{d}{dt} V(\mathbf{x}(t)) = \dot{V}(\mathbf{x}(t)).$$

Lyapunov's theorems

Suppose that the function V is defined in a set S that includes $\mathbf{x}_0$ and satisfies

$$V(\mathbf{x}_0) = 0 \quad \text{and} \quad V(\mathbf{x}) > 0 \quad \text{for all } \mathbf{x} \in S \text{ with } \mathbf{x} \neq \mathbf{x}_0. \tag{7.6}$$

Such a function will be said to be *positive definite* in the set S with a minimum at $\mathbf{x}_0$. If V is a function of two variables, its graph will look like a bowl, at least near the minimum $\mathbf{x}_0$. We will relax this definition by saying that V is *positive semidefinite* in S if the inequality in (7.6) is changed to $V(\mathbf{x}) \geq 0$ for all $x \in S$. If the inequalities are reversed, we will say that V is *negative definite* or *negative semidefinite*.

We have already seen several examples of positive definite (or semidefinite) functions.

- The function $V(x, y) = x^2 + y^2$ is positive definite in all of $\mathbf{R}^2$ with a minimum at the origin $(0, 0)$. More generally, $V(\mathbf{x}) = x_1^2 + x_2^2 + \cdots + x_n^2$ is positive definite in all of $\mathbf{R}^n$ with a minimum at $\mathbf{0}$.
- Suppose we have a conservative mechanical system with potential energy $U(y)$ and total energy $E(y, v) = \frac{1}{2}v^2 + U(y)$. If y_0 is an isolated local minimum of U, then $(y_0, 0)$ is a local minimum of E. The function $V(y, v) = E(y, v) - U(y_0)$ is positive definite in a set S containing $(y_0, 0)$.

Consider a system (either (7.1) or (7.4)) that has an equilibrium at $\mathbf{x}_0$. Suppose we can find a function V that is positive definite on a set S with a minimum at $\mathbf{x}_0$ and that has the additional property that $\dot{V}$ is negative definite on S. Then every solution curve would move downhill and seemingly would continue to do so until it ended up at $\mathbf{x}_0$. This requires that $\mathbf{x}_0$ be an asymptotically stable equilibrium point. This intuition is correct and serves as the main idea in the proof of the following theorem, which is due to Lyapunov.

THEOREM 7.7 Suppose the system $\mathbf{x}' = \mathbf{f}(\mathbf{x})$ has an equilibrium point at $\mathbf{x}_0$. Suppose there is a continuously differentiable function V defined on a neighborhood U of $\mathbf{x}_0$ that is positive definite with a minimum at $\mathbf{x}_0$.

1. If $\dot{V}$ is negative semidefinite in U, then $\mathbf{x}_0$ is a stable equilibrium point.
2. If $\dot{V}$ is negative definite in U, then $\mathbf{x}_0$ is an asymptotically stable equilibrium point. ∎

Let's review some examples where Theorem 7.7 applies.

Example 7.8 In Example 1.25, we considered the system

$$x' = y + \alpha x(x^2 + y^2),$$
$$y' = -x + \alpha y(x^2 + y^2).$$

The function $V(x, y) = x^2 + y^2$ is positive definite in $\mathbf{R}^2$, with an isolated minimum at the origin. We compute

$$\dot{V}(x, y) = \frac{\partial V}{\partial x}(x, y) f(x, y) + \frac{\partial V}{\partial y}(x, y) g(x, y)$$
$$= 2x(y + \alpha x(x^2 + y^2)) + 2y(-x + \alpha y(x^2 + y^2))$$
$$= 2\alpha(x^2 + y^2)^2.$$

If $\alpha < 0$, $\dot{V}$ is negative definite in $\mathbf{R}^2$. Consequently, Theorem 7.7 implies that the origin is an asymptotically stable equilibrium point. This is true even though the

Jacobian of the system at $(0, 0)$ has imaginary eigenvalues, which means that the linearization has a center.

If you look closely at the analysis in Example 1.25, you will find that we essentially provided a proof of Theorem 7.7 for this special case. ●

Example 7.9 Consider the vibrating spring system

$$y' = v,$$
$$v' = -\frac{k}{m}y - \frac{\mu}{m}v.$$

The total energy of the system is

$$E(y, v) = \frac{1}{2}mv^2 + \frac{1}{2}ky^2.$$

The energy is positive definite on $\mathbf{R}^2$ with an isolated minimum at the origin. We compute

$$\dot{E} = \frac{\partial E}{\partial y}y' + \frac{\partial E}{\partial v}v' = kyv + mv\left(-\frac{k}{m}y - \frac{\mu}{m}v\right) = -\mu v^2.$$

Assuming that the damping constant μ is positive, $\dot{E}$ is negative semidefinite, so by Theorem 7.7, the origin is a stable equilibrium point. We need to know more in order to conclude that the origin is asymptotically stable. We will return to this example later. ●

The argument used in Example 7.9 can be applied to the pendulum and to many other nonlinear physical systems.

It should be remarked that if the function V in Theorem 7.7 is negative definite and $\dot{V}$ is positive (semi-)definite, the conclusion still stands. We need only replace V with $-V$ to get a function that satisfies the hypotheses. Any function that satisfies the hypotheses of part (2) of Theorem 7.7 is called a ***Lyapunov*** function for the given system. It is a ***strong*** Lyapunov function if part (1) is satisfied.

It is interesting to compare Theorem 7.7 with Theorem 2.3. Both lead to the conclusion that an equilibrium point has certain stability properties. Theorem 2.3 is easier to use since we need to discover some properties of the Jacobian, which only requires some computation. On the other hand, Theorem 7.7 requires that we know a Lyapunov function. It is frequently very difficult to find a Lyapunov function. There seem to be no general ways of doing so.

Nevertheless, Theorem 7.7 is useful for two reasons. First, it can sometimes be used when Theorem 2.3 does not apply, as, for instance, in Example 7.8. The second reason is that a Lyapunov function is defined on its domain, which is usually a well-understood set. For example, in both of our examples, the domain is all of $\mathbf{R}^2$. The domain then gives us some idea of the region over which the stability of the equilibrium point extends. By contrast, we can conclude from Theorem 2.3 only the existence of some neighborhood of the equilibrium point where the stability is valid. Nothing is said about how large that neighborhood is. Thus, using a Lyapunov function and Theorem 7.7 gives a genuinely global result, while Theorem 2.3 allows only local conclusions.

The key idea in the method of Lyapunov is summarized in the next theorem. It is frequently useful all by itself.

THEOREM 7.10 Suppose the system $\mathbf{x}' = \mathbf{f}(\mathbf{x})$ is defined in a set U. Suppose that $\mathbf{x}(t)$ is a solution that has a forward limit set ω. Furthermore, suppose that the function V is continuously differentiable in U and satisfies

$$\dot{V}(\mathbf{x}(t)) \leq 0 \quad \text{for all } t \geq 0.$$

Then ω is contained in the set

$$\{x \in U \mid \dot{V}(x) = 0\}. \qquad \blacksquare$$

Example 7.11 Let's return to the damped vibrating spring in Example 7.9. We saw that $\dot{E} = -\mu v^2 \leq 0$ in all of $\mathbf{R}^2$. According to Theorem 7.10, the limit set of every solution curve is contained in the set where $\dot{E} = 0$, which is the y-axis, defined by $v = 0$. However, a limit set must be a union of solution curves, and the only solution curve contained in the y-axis is the equilibrium point at the origin. Thus, every solution curve for the damped vibrating spring is attracted to the origin. ●

Theorem 7.10 will be very useful in the next section.

EXERCISES

In Exercises 1–10, is the given function positive definite in an open neighborhood containing $(0, 0)$? Positive semidefinite? Negative definite? Negative semidefinite? None of these? Justify your answer in each case.

1. $V(x, y) = x^2 + 2y^2$ **2.** $V(x, y) = -x^2 - 2y^2$

3. $V(x, y) = x + y^2$ **4.** $V(x, y) = (x + y)^2$

5. $V(x, y) = -2y^2$ **6.** $V(x, y) = x^2$

7. $V(x, y) = x^2 + y$ **8.** $V(x, y) = 2xy - x^2 - y^2$

9. $V(x, y) = x^2 - 2xy + 3y^2$

10. $V(x, y) = x^2 - 6xy + 9y^2$

11. Prove that the quadratic polynomial $V(x, y) = ax^2 + 2bxy + cy^2$ is positive definite with a minimum at $(x, y) = (0, 0)$ if and only if $a > 0$ and $ac - b^2 > 0$.

12. Suppose that $S(x, y)$ is negative definite with a maximum at (x_0, y_0).

(a) Show that $V(x, y) = -S(x, y)$ is a Lyapunov function for the system

$$x' = \frac{\partial S}{\partial x}, \qquad y' = \frac{\partial S}{\partial y}. \qquad (7.12)$$

(b) Find a Lyapunov function for the system

$$x' = 2y - 2x, \qquad y' = 2x - 2y - 4y^3.$$

A system of the type in 7.12 is called a ***gradient system***.

Consider the system

$$\frac{dx}{dt} = y$$
$$\frac{dy}{dt} = -x - y^3. \qquad (7.13)$$

Use the chain rule (7.2) to calculate dV/dt for each function $V = V(x, y)$ in Exercises 13–16 on the solution curves of the system in (7.13).

13. $V(x, y) = x^2 + y^2$ **14.** $V(x, y) = x^3 + y^3$

15. $V(x, y) = x^2 + xy - y^2$

16. $V(x, y) = 2x^2 + 3xy - y^2$

17. Consider the system

$$\frac{dx}{dt} = -y - x^3,$$
$$\frac{dy}{dt} = x - y^3.$$

(a) By calculating the Jacobian, show that the linearization predicts that the equilibrium point at $(0, 0)$ is a center.

(b) Show that the function $V(x, y) = x^2 + y^2$ is positive definite on a neighborhood containing the origin. Show that $\dot{V}$ is negative definite on solution trajectories of the system in this same neighborhood, establishing that the equilibrium point at the origin is asymptotically stable.

(c) Sketch a phase portrait of the system that highlights the asymptotic stability of the system at the origin.

Use the positive definite function $V(x, y) = x^2 + y^2$ to argue that the systems in Exercises 18–19 have asymptotically stable equilibrium points at the origin.

18. $x' = -x - 2y^2$
$\quad\ y' = 2xy - y^3$

19. $x' = -x + y$
$\quad\ y' = -x - y$

In Exercises 20–21, use the positive definite function $V(x, y) = x^2 + y^2$ to argue that the equilibrium point at the origin is asymptotically stable. In each exercise, sketch a phase portrait highlighting the asymptotic stability of the origin. *Hint:* In each case, $\dot{V}$ will not be negative definite (or negative semidefinite) for all x and y. However, you can *restrict* the domain of V so that $\dot{V}$ is negative semidefinite *on a neighborhood* of the origin. From there, you can argue that the equilibrium point is asymptotically stable.

20. $x' = -x + y^2$
$\quad\ y' = -xy - x^2$

21. $x' = -x + xy$
$\quad\ y' = -y + xy$

22. Consider the system[11]

$$\dot{x}_1 = x_2,$$
$$\dot{x}_2 = -x_2^3 - x_1^3.$$

Show that

$$V(x_1, x_2) = \frac{1}{2}x_1^4 + x_2^2$$

is positive definite on a neighborhood of the equilibrium point at the origin. Furthermore, show that $\dot{V}$ is negative semidefinite on this neighborhood, and that the origin is stable. Finally, use Theorem 7.10 to show that the origin is asymptotically stable. Sketch a phase portrait of the system that highlights this stability of the system at the origin.

23. Consider the second-order equation

$$x'' + x' - \frac{1}{3}(x')^3 + x = 0. \tag{7.14}$$

Use the standard substitutions $x_1 = x$ and $x_2 = x'$ to write equation (7.14) as a system of first-order equations having an equilibrium point at the origin.

(a) Using the positive definite function $V(x_1, x_2) = x_1^2 + x_2^2$, show that

$$\dot{V}(x_1, x_2) = -\frac{2}{3}x_2^2(3 - x_2^2)$$

on solution trajectories of the system. Argue that $\dot{V}$ is negative semidefinite on $\{(x_1, x_2) : |x_2| < \sqrt{3}\}$ and show that the equilibrium point at the origin is stable.

(b) Given that $\dot{V}$ is negative semidefinite on $\{(x_1, x_2) : |x_2| < \sqrt{3}\}$, use Theorem 7.10 to argue that the equilibrium point at the origin is asymptotically stable. Provide a phase portrait of the system highlighting the asymptotic stability of the equilibrium point at the origin.

Show that the second-order equations in Exercises 24–25 have asymptotically stable solutions $\mathbf{x}(t) = \mathbf{0}$. *Hint:* See Exercises 22 and 23.

24. $\ddot{x} + \dot{x} + x^3 = 0$

25. $\ddot{x} + (\dot{x})^3 + x^3 = 0$

26. Let m, g, and L be positive constants.

(a) Show that

$$V(\theta, \omega) = \frac{1}{2}mL^2\omega^2 + mgL(1 - \cos\theta) \tag{7.15}$$

is positive definite in an open neighborhood containing $(\theta, \omega) = (0, 0)$.

(b) Show that $\dot{V}(\theta, \omega)$ is negative semidefinite on solution trajectories of

$$\dot{\theta} = \omega$$
$$\dot{\omega} = -\frac{\mu}{m}\omega - \frac{g}{L}\sin\theta$$

on an open neighborhood containing $(\theta, \omega) = (0, 0)$.

27. Consider the second-order equation

$$x'' + f(x)x' + g(x) = 0, \tag{7.16}$$

where f and g are continuous differentiable functions satisfying

(i) $f(x) > 0$ for all x, and

(ii) $xg(x) > 0$ for all $|x| < k$, where k is some positive, real number.

(a) Show that $g(0) = 0$ and sketch some possible graphs for the function g. Define

$$G(x) = \int_0^x g(u)\,du,$$

and show that G has an absolute minimum at $x = 0$.

(b) Show that equation (7.16) has an equivalent first-order system representation

$$x' = y,$$
$$y' = -f(x)y - g(x); \tag{7.17}$$

then show that the function

$$V(x, y) = \frac{1}{2}y^2 + G(x)$$

is positive definite in a neighborhood containing the origin.

(c) Show that the function V is negative semidefinite on solution trajectories of system (7.17). Then, use Theorem 7.10 to argue that the origin is asymptotically stable.

---

[11] It is typical in physics and engineering classes to use $\dot{x}_1$ to represent dx_1/dt, the derivative of x_1 with respect to time. A similar comment applies to $\dot{x}_2$.

28. At the end of Section 10.2 we discussed the Lorenz system

$$x' = -ax + ay$$
$$y' = rx - y - xz$$
$$z' = -bz + xy.$$

There we made the claim that almost all solution curves approached the Lorenz attractor as t increased. We are not in a position to prove that, but in this exercise we will show that every solution to the Lorenz system remains bounded. Consider the function

$$V(x, y, z) = rx^2 + ay^2 + a(z - 2r)^2.$$

(a) Show that

$$\dot{V} = -2a \left[rx^2 + y^2 + b(z - r)^2 - br^2 \right].$$

(b) Assuming the fact that $(z - r)^2 \geq (z - 2r)^2/2 - r^2$, show that

$$rx^2 + y^2 + b(z - r)^2 \geq mV - br^2,$$

where m is the smallest of the three numbers 1, $1/a$, and $b/2a$.

(c) Use parts (a) and (b) to show that

$$\dot{V} \leq -2maV + 4abr^2.$$

(d) Use part (c) to show that $\dot{V} \leq -2abr^2 < 0$ everywhere outside of the ellipsoid

$$R = \left\{ (x, y, z) | V(x, y, z) \leq 3br^2/m \right\}.$$

(e) Use part (d) to show that the ellipsoid R is invariant and that every solution curve ends up inside R.

10.8 Predator–Prey Systems

We derived the Lotka-Volterra predator–prey model in Section 8.2. Here we will analyze it in more detail, and we will look at a couple of variants that have been proposed.

The Lotka-Volterra model

If $x(t)$ is the prey population and $y(t)$ is the predator, then the Lotka-Volterra model is the system

$$\begin{aligned} x' &= x(a - by), \\ y' &= y(-c + dx), \end{aligned} \tag{8.1}$$

where all of the constants are positive.

This model was first studied by the Italian mathematician Vito Volterra. He was approached by the young biologist Umberto D'Ancona (who was engaged to his daughter) with a problem that puzzled Italian fishermen and biologists. The fishermen had noticed that during World War I, when there was less fishing than usual, the percentage of their catch that consisted of sharks and other predatory creatures went up. After the war, when fishing returned to normal, the percentage went back down to prewar levels. Neither the fishermen nor the biologists could find a reason for this. Volterra invented the predator–prey model (8.1) to explain the phenomenon.[12] We will offer his explanation after a few preparatory remarks.

We notice that the x- and y-axes are invariant. Of course, this implies that the positive quadrant is also invariant. There are two equilibrium points. The origin is a saddle point, with the x- and y-axes as separatrices. The other equilibrium point is at the point

$$(x_0, y_0) = (c/d, a/b).$$

The Jacobian is

$$J(x, y) = \begin{pmatrix} a - by & -bx \\ dy & -c + dx \end{pmatrix}.$$

At the equilibrium point,

$$J(x_0, y_0) = \begin{pmatrix} 0 & -bc/d \\ ad/b & 0 \end{pmatrix}.$$

---

[12] The American mathematician Alfred Lotka discovered the model later and independently.

Since the trace is equal to 0 and the determinant is positive, we see that the linearization has a center. One of our goals is to show that the equilibrium point (x_0, y_0) is a center for the nonlinear system (8.1).

Let's look for a conserved quantity. We have

$$\frac{dy}{dx} = \frac{y(-c + dx)}{x(a - by)}.$$

Since this equation is separable, we can solve it. We find that the solution curves are given implicitly by the equation

$$H(x, y) = by - a \ln y + dx - c \ln x = C, \tag{8.2}$$

where C is a constant. Notice that

$$H(x, y) = F(x) + G(y), \tag{8.3}$$

where $F(x) = dx - c \ln x$, and $G(y) = by - a \ln y$. These functions have the same form with different coefficients. The graph of F is shown in Figure 1. We have

$$\lim_{x \to 0^+} F(x) = \lim_{x \to \infty} F(x) = \infty. \tag{8.4}$$

G will have the same properties. In addition, F has an absolute minimum at $x = x_0 = c/d$. Its value at the minimum is $F_0 = c(1 - \ln x_0)$. Similarly, G has an absolute minimum at $y = y_0 = a/b$, and its value there is $G_0 = a(1 - \ln y_0)$. Thus H has a minimum at the equilibrium point (x_0, y_0). Its value there is $F_0 + G_0$.

We could have shown the graph of G in addition to that of F, but it is unnecessary since the shape of the graph of G is qualitatively the same as that of F shown in Figure 1. Instead, we choose to show the inverses of G in Figure 2. Notice that since the graph of G has the same parabolic shape as does that of F in Figure 1, G has the two inverses shown in Figure 2.[13] The two inverses of G are defined only for $z \geq G_0$, the minimum value of G.

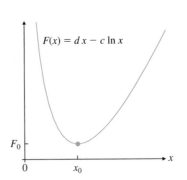

Figure 1. The graph of $F(x) = dx - c \ln(x)$.

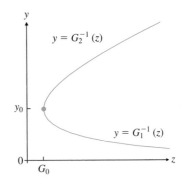

Figure 2. The graph of the inverses of $G(y) = by - a \ln(y)$.

Using the inverses of G, we can solve the implicit equation (8.2), which we rewrite as

$$H(x, y) = F(x) + G(y) = C.$$

---

[13] The situation is essentially the same as it would be if $G(y) = y^2$. In that case, we would be blessed with the formulas $G_1^{-1}(z) = -\sqrt{z}$ and $G_2^{-1}(z) = \sqrt{z}$. The only real difference is that in the current situation, no formulas are available.

There are two branches,

$$y = G_1^{-1}(C - F(x)), \quad \text{and} \quad y = G_2^{-1}(C - F(x)). \tag{8.5}$$

For these to be defined, we must have $C - F(x) \geq G_0$, or $F(x) \leq C - G_0$. Since $F(x) \geq F_0$, for there to be any solutions at all we must have $C \geq F_0 + G_0$. If this is true, then the situation is shown in the upper graph in Figure 3. Because of the parabolic graph of F, there are numbers x_1 and x_2 such that

$$F(x) \leq C - G_0 \quad \text{if and only if} \quad x_1 \leq x \leq x_2.$$

Notice that at the endpoints x_1 and x_2, we have $F(x_1) = F(x_2) = C - G_0$, or $C - F(x_1) = C - F(x_2) = G_0$. Since, as indicated in Figure 2, the two branches of G^{-1} agree at G_0, we conclude that the graphs of the two functions in (8.5) agree at x_1 and x_2. Thus the graphs of these two branches of the solution curve defined in (8.2) form a closed curve, as illustrated in Figure 3. Hence, if the constant C is larger than $F_0 + G_0$, the minimum value of H, then the solution curve defined by (8.2) must be closed. Consequently, all solutions are periodic. We had earlier conjectured this fact because of our computations, but now we know it is true. The equilibrium point for the predator–prey system is a center.

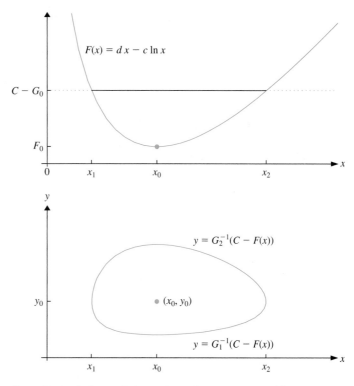

Figure 3. Analyzing a solution to the predator–prey system. The upper graph is a graph of $F(x)$, and the lower graph is a portion of the phase plane.

Volterra's explanation of the fishing phenomenon

Volterra's explanation begins by calculating the time averages of the predator and prey populations. Suppose that $(x(t), y(t))$ is a solution and that T is its period.

We want to compute the averages of x and y over the period. To do this, we notice (using (8.1)) that

$$\frac{d}{dt} \ln x(t) = \frac{x'}{x} = a - by.$$

If we integrate this over a period, we get

$$\int_0^T \frac{d}{dt} \ln x(t)\, dt = \int_0^T (a - by(t))\, dt = aT - b \int_0^T y(t)\, dt.$$

On the other hand, since x is periodic with period T, we have $x(0) = x(T)$, and

$$\int_0^T \frac{d}{dt} \ln x(t)\, dt = \ln(x(T)) - \ln(x(0)) = 0.$$

From this we see that the average value of y over a period is

$$\frac{1}{T} \int_0^T y(t)\, dt = \frac{a}{b} = y_0.$$

Similarly, the average value of x is

$$\frac{1}{T} \int_0^T x(t)\, dt = \frac{c}{d} = x_0.$$

To explain why the fisherman caught more predators during the war than they did before and after, Volterra included the effect of fishing in his model. We assume that the fishing is indiscriminant. This means that our fishing methods are as likely to catch available predators as prey. Under this assumption, the effect of fishing is to remove a common percentage of both the predators and the prey per unit time. Let e be that percentage. Then, his model with fishing is

$$\begin{aligned} x' &= x(a - by) - ex = x((a - e) - by), \\ y' &= y(-c + dx) - ey = y(-(c + e) + dx). \end{aligned} \tag{8.6}$$

Assuming that $e < a$, this is exactly the same kind of system as in (8.1). Everything we have derived for that system is true for the model with fishing. The change in the coefficients means that the equilibrium point has shifted to

$$(x_1, y_1) = \left(\frac{c + e}{d}, \frac{a - e}{b} \right),$$

and the components of this point represent the average populations. With fishing, the average of the prey population is $x_1 = (c + e)/d$, while the average of the predators is $y_1 = (a - e)/b$.

We notice that $x_1 > x_0$ and $y_1 < y_0$. Thus, the average of the prey population has increased as a result of fishing, while the average of the predators has decreased. We have discovered the counterintuitive fact that fishing can actually increase the size of the fish population, if at the same time the fishing removes the predators that kill the fish. As a result of these changes, the percent of the catch that consists of fish will go up with increased fishing, and this explains what the Italian fisherman saw.

The effect of logistic limits on population growth

One criticism of the Lotka-Volterra predator–prey model is that, in the absence of the predators, the equation for the prey alone becomes

$$x' = ax.$$

This is the Malthusian model, and it predicts that the prey grows exponentially without limit. It would be more realistic to put a logistic limit on this growth. This would make the equation

$$x' = x(a - ex),$$

where both a and e are positive constants. To integrate this into our predator–prey system, we change (8.1) to

$$\begin{aligned} x' &= x(a - ex - by), \\ y' &= y(-c + dx). \end{aligned} \tag{8.7}$$

We could put a logistic limit on the predators as well, but this is less serious, since they die out in the absence of the prey. Let's analyze the system in (8.7).

Just as for (8.1), we notice that the x- and y-axes are invariant. Again, this implies that the positive quadrant is also invariant. Since the line $by + ex = a$ has a negative slope, there are now three equilibrium points. The origin is a saddle point, with x- and y-axes as separatrices. In addition, the points

$$\mathbf{p}_1 = \left(\frac{a}{e}, 0\right) \quad \text{and} \quad \mathbf{p}_2 = (x_0, y_0) = \left(\frac{c}{d}, \frac{ad - ec}{bd}\right)$$

are equilibrium points. The location and type of these equilibrium points depend on the sign of $y_0 = (ad - ec)/bd$. There are two cases.

If $y_0 < 0$, then $\mathbf{p}_2$ is not in the positive quadrant and is of no interest, since populations are always nonnegative. The best way to analyze this case is to use nullclines. The nullcline information is shown in Figure 4. The nullclines divide the positive quadrant into three regions, labeled I, II, and III. Notice from the direction of the nullclines that solutions starting in region I move into region II, and solutions in region II move into region III, or proceed directly to $\mathbf{p}_1$. Finally, we see that

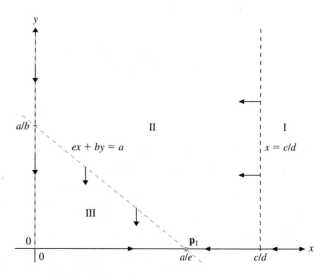

Figure 4. The nullclines for a predator–prey system where the predators die out.

region III is invariant, and all solutions are directed toward the equilibrium point $\mathbf{p}_1$. It is not necessary for the analysis, but a computation of the Jacobian reveals that $\mathbf{p}_1$ is a nodal sink. Thus, we see that if $y_0 < 0$, the predators die out over time, and the prey approaches the equilibrium population of $x = a/e$.

The case when $y_0 > 0$ is more interesting. In this case, all three equilibrium points are of interest. The nullcline information is shown in Figure 5. It tells us that solution curves move from region I to region II to region III to region IV and then back to region I. Thus, the solutions spiral around the equilibrium point $\mathbf{p}_2$. However, we cannot decide from the nullcline information whether the curves are closed or not.

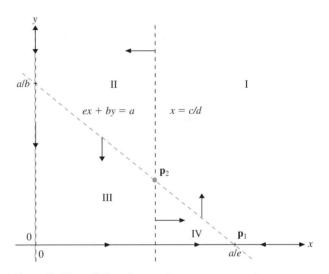

Figure 5. The nullclines for a predator–prey system where predators and prey coexist.

To determine the true behavior we look for a Lyapunov function. For motivation, we look at the conserved quantity H in (8.2), which has its minimum value at $(c/d, a/b)$. We want the minimum value to be at $\mathbf{p}_2 = (c/d, (ad - ec)/bd)$, so we define

$$W(x, y) = by - \left(\frac{ad - ec}{d}\right) \ln y + dx - c \ln x.$$

Notice that we have $W(x, y) = F(x) + G(y)$ and both F and G are of the form of the function graphed in Figure 1. Furthermore, W is defined where both x and y are positive, and W has an isolated minimum at $\mathbf{p}_2$. The function $V(x, y) = W(x, y) - W(\mathbf{p}_2)$ is positive definite in the positive quadrant with minimum at $\mathbf{p}_2$.

To use the method of Lyapunov we must compute $\dot{V} = \dot{W}$:

$$\dot{V} = \frac{\partial W}{\partial x} x' + \frac{\partial W}{\partial y} y'$$

$$= \left(d - \frac{c}{x}\right) x(a - ex - by) + \left(b - \frac{ad - ec}{dy}\right) y(-c + dx)$$

$$= -\frac{e}{d}(dx - c)^2.$$

Since $\dot{V} \leq 0$, the sets defined by $V(x, y) \leq C$ are positively invariant. Any solution $(x(t), y(t))$ stays in the bounded set defined by $V(x, y) \leq V(x(0), y(0))$. By Theorem 4.1, the solution curve has a nonempty limit set. By Theorem 7.10, the limit set is an invariant subset of the set where $\dot{V} = 0$, which is the line $x = c/d$. However, at any point on that line, except at the equilibrium point $\mathbf{p}_2$, the solution curve leaves the line. (See Figure 5.) Consequently, the only invariant subset of $x = c/d$ is the equilibrium point $\mathbf{p}_2$. As a result, we see that all solution curves tend to $\mathbf{p}_2$.

We conclude that if $y_0 > 0$, then the predator and prey populations tend to an equilibrium situation where the two populations coexist. This equilibrium is at $\mathbf{p}_2$, which we now know is asymptotically stable.

A refined predator–prey model[14]

The Lotka-Volterra model has received a significant amount of criticism. The original model, without the logistic limit on the growth of the prey population, is criticized for that fact. On the other hand, if we introduce a logistic limit, we have shown that the model predicts either that the predators will die out or that the predators and prey will converge to an equilibrium point. While almost periodic variation of predators and prey has been observed, no one has observed this type of limiting behavior.

When a model turns out to be unsatisfactory, we have to rethink our assumptions. Let's look at the prey population $x(t)$ first. In the absence of the predators, where $y = 0$, we assume that we have a logistic model, which we write as

$$x' = rx \left(1 - \frac{x}{K} \right).$$

In our derivation of the predator–prey equations, we assumed that the decrease of the prey due to the predators would be proportional to the number of contacts between predators and the prey, and therefore to the product xy. This led us to the equation

$$x' = rx \left(1 - \frac{x}{K} \right) - axy. \tag{8.8}$$

Let's rethink this assumption. In particular, would the attrition of the prey continue to be proportional to xy if x is very large? The strict proportionality assumes that the probability that a predator will attack is the same for every encounter. Is this reasonable? If there are a large number of encounters, then many of them will occur when the predator has recently eaten and is not hungry. A well-fed predator is less likely to attack than one that is hungry. When we think about things this way, we see that proportionality to the product xy is valid for small x but not for large x. The effect should become smaller than proportionality to the product xy would imply. There are a number of ways to account for this, but we will assume that the affect of predation is proportional to $xy/(x + B)$ for some positive constant B. Notice that as x gets large, this quantity becomes approximately proportional to y, and not the larger quantity xy.[15] Thus, we replace the term axy in (8.8) with $Arxy/(x + B)$. The net result is that our model for the prey population has the form

$$x' = rx \left(1 - \frac{x}{K} - \frac{Ay}{x + B} \right). \tag{8.9}$$

For the predator population, we argue similarly, and we get the equation

$$y' = sy \left(-1 + \frac{Cx}{x + D} \right). \tag{8.10}$$

---

[14] There is much more to be said about predator–prey models. The interested reader is referred to *The Theory of Evolution and Dynamical Systems* by Josef Hofbauer and Karl Sigmund, London Mathematical Society, 1996, and *Mathematical Biology* by J. D. Murray, Springer-Verlag, 1993. These books illustrate the growing field of mathematical biology.

[15] See Exercise 2 for an alternate approach to this change.

Using dimensionless quantities

We want to examine the system in equations (8.9) and (8.10) for all values of the parameters. There are seven of them, and it is difficult to keep track of them all. In cases like this, it is a good idea to choose dimensionless variables that allow us to reduce the number of the parameters and make the equations simpler as well. In this case we will compare the populations to K, the carrying capacity of the prey population in the absence of the predators. We will set

$$x = Ku \quad \text{and} \quad y = Kv/A.$$

With these changes, our equations become

$$Ku' = rKu\left(1 - u - \frac{Kv}{Ku + B}\right),$$

$$\frac{K}{A}v' = \frac{K}{A}sv\left(-1 + \frac{CKu}{Ku + D}\right).$$

If we divide each equation by its leading constant, the equations simplify to

$$u' = ru\left(1 - u - \frac{v}{u + b}\right),$$

$$v' = sv\left(-1 + \frac{cu}{u + d}\right),$$

where we have introduced the new constants

$$b = \frac{B}{K}, \quad c = C, \quad \text{and} \quad d = \frac{D}{K}.$$

We now have five parameters. We can get rid of one more by choosing a different unit of time. We do this by setting $\tau = rt$. Then

$$\frac{d}{d\tau} = \frac{1}{r}\frac{d}{dt},$$

so our equations become

$$\frac{du}{d\tau} = u\left(1 - u - \frac{v}{u + b}\right),$$

$$\frac{dv}{d\tau} = \rho v\left(-1 + \frac{cu}{u + d}\right),$$

where $\rho = s/r$. Now we have four parameters, a manageable number. For convenience, we will write this system as

$$u' = u\left(1 - u - \frac{v}{u + b}\right),$$

$$v' = \rho v\left(-1 + \frac{cu}{u + d}\right). \tag{8.11}$$

Analysis of the refined model

We start our analysis by noting that the u- and v-axes are invariant. Consequently, the positive quadrant is also invariant. Next, we look for equilibrium points. The second equation in (8.11) gives us two alternatives. For $v = 0$, the first equation requires that $u = 0$ or 1. Thus, the origin $(0, 0)$ and $(1, 0)$ are equilibrium points. If $v \neq 0$, then the second equation requires that

$$-1 + \frac{cu}{u + d} = 0 \quad \text{or} \quad u = u_0 = \frac{d}{c - 1}. \tag{8.12}$$

Since $u_0 \neq 0$, the first equation in (8.11) requires that $v = v_0$, where

$$1 - u_0 - \frac{v_0}{u_0 + b} = 0 \quad \text{or} \quad v_0 = (u_0 + b)(1 - u_0). \tag{8.13}$$

The most interesting case is when both u_0 and v_0 are positive, and that is the case we will analyze. The other cases have some interest, but we will leave them for the exercises. From (8.13), it follows that v_0 will be positive if $u_0 < 1$. Then, from (8.12), both u_0 and v_0 will be positive if

$$0 < d < c - 1. \tag{8.14}$$

We want to find a bounded invariant set that contains the equilibrium point (u_0, v_0). We will find one that is a triangle T with two sides contained in the axes (see the triangle outlined in blue in Figure 6). The third side will be a line L through the equilibrium point at $(1, 0)$ with slope $-\alpha$, where $\alpha > 0$ is yet to be chosen. This line has equation $v = \alpha(1 - u)$ for $0 \leq u \leq 1$. Since the two axes are invariant, the triangle will be invariant if all solution curves starting at a point in L enter the triangle. It is visually clear in Figure 6 that if the line L is steep enough (α large enough), then all solution curves crossing L will be moving into T. Then T will be invariant. However, we have to prove this.

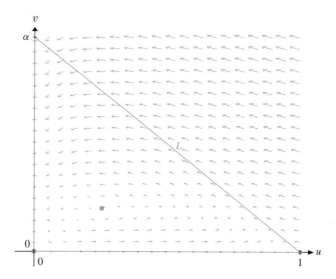

Figure 6. The invariant triangle T, vector field, and equilibrium points for the predator–prey system in equation (8.11).

To study how solution curves behave when crossing L, we use the function $W(u, v) = \alpha(u - 1) + v$. Notice that L is the set where $W(u, v) = 0$. Since $W(0, 0) = -\alpha < 0$, we know that W is negative below and to the left of L and positive above and to the right. In particular, $W(u, v) \leq 0$ in the triangle T. All solutions through L will move into T if W is decreasing along solution curves on the line L. This, in turn, will be true if $\dot{W}(u, v) \leq 0$ for $(u, v) \in L$.

We compute

$$\dot{W} = \alpha u' + v' = \alpha u \left(1 - u - \frac{v}{u + b}\right) + \rho v \left(-1 + \frac{cu}{u + d}\right).$$

If we substitute $v = \alpha(1 - u)$, we get

$$\dot{W} = \alpha u \left(1 - u - \frac{\alpha(1 - u)}{u + b}\right) + \rho \alpha(1 - u) \left(-1 + \frac{cu}{u + d}\right)$$

$$= \alpha u(1 - u) \left[1 - \frac{\alpha}{u + b} + \rho \left(\frac{c}{u + d} - \frac{1}{u}\right)\right].$$

Since $\alpha u(1 - u) \geq 0$, to ensure that $\dot{W}(u, v) \leq 0$, we need to choose α so that

$$1 - \frac{\alpha}{u + b} + \rho \left(\frac{c}{u + d} - \frac{1}{u}\right) \leq 0 \quad \text{for} \quad 0 < u \leq 1.$$

Isolating α, we see that we need

$$\alpha \geq (u + b) \left[1 + \rho \left(\frac{c}{u + d} - \frac{1}{u}\right)\right] \quad \text{for} \quad 0 < u \leq 1.$$

However, the quantity on the right is a continuous function of u for $0 < u \leq 1$, and it tends to $-\infty$ as $u \to 0^+$. Hence, it is bounded above, and we can choose α to be any upper bound. In this way, we find the invariant triangle T shown in Figure 6. Notice that we do not need to know the precise value of α. We only need to know that one exists that makes the triangle T invariant.

Next, we compute the Jacobian for the system in (8.11). With some effort, we find that

$$J(u, v) = \begin{pmatrix} 1 - 2u - \dfrac{bv}{(u + b)^2} & -\dfrac{u}{u + b} \\ \dfrac{\rho cdv}{(u + d)^2} & \rho \left(-1 + \dfrac{cu}{u + d}\right) \end{pmatrix}.$$

Substituting, we get

$$J(0, 0) = \begin{pmatrix} 1 & 0 \\ 0 & -\rho \end{pmatrix} \quad \text{and} \quad J(1, 0) = \begin{pmatrix} -1 & -\dfrac{1}{b + 1} \\ 0 & \dfrac{\rho(c - d - 1)}{d + 1} \end{pmatrix}.$$

Thus the origin is clearly a saddle point. The point $(1, 0)$ is also a saddle as a result of our assumption in (8.14).

At the third equilibrium point (u_0, v_0) we have, using (8.12),

$$J(u_0, v_0) = \begin{pmatrix} 1 - 2u_0 - \dfrac{bv_0}{(u_0 + b)^2} & -\dfrac{u_0}{u_0 + b} \\ \dfrac{\rho cdv_0}{(u_0 + d)^2} & 0 \end{pmatrix}.$$

The determinant is positive. We must examine the trace T more closely. Using (8.13), we get

$$T = 1 - 2u_0 - \frac{bv_0}{(u_0 + b)^2}$$

$$= 1 - 2u_0 - \frac{b(1 - u_0)}{u_0 + b}$$

$$= \frac{u_0(1 - 2u_0 - b)}{u_0 + b}.$$

We see that there are two cases, depending on the sign of $1 - 2u_0 - b$.

- If $1 - 2u_0 - b < 0$, the trace is negative, so (u_0, v_0) is a sink. Every solution curve that starts in our invariant triangle must have a limit set. With a little more work we can show that every solution curve tends to (u_0, v_0), and the predators and prey reach an equilibrium situation.

- If $1 - 2u_0 - b > 0$, the trace is positive, so (u_0, v_0) is a source. Again, every solution curve that starts in our invariant triangle must have a limit set, but now it cannot be the source (u_0, v_0). It cannot be one of the saddles at the origin and $(1, 0)$ either. According to Theorem 4.8, there is only one possibility. There must be a limit cycle in our invariant triangle. This case is illustrated in Figure 7.

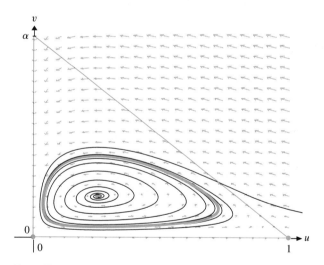

Figure 7. Some solution curves, including the limit cycle.

EXERCISES

1. Consider the system

$$x' = x(a - ex - by),$$
$$y' = y(-c + dx),$$

which we considered in (8.7). It is assumed that the parameters a, b, c, d, and e are positive constants. In the text, we showed that

$$\mathbf{p}_2 = (x_0, y_0) = \frac{ad - ec}{bd}$$

is a sink when $y_0 > 0$ without computing the Jacobian. To gain an appreciation for our use of Theorem 7.10 in this model, show, using the Jacobian, that there is a sink at $\mathbf{p}_2$ when $y_0 > 0$.

Exercises 2–5 refer to the predator–prey model in (8.9) and (8.10).

2. If we adjust the first equation of the predator–prey system in (8.7) to

$$x' = rx\left(1 - \frac{x}{K}\right) - axy, \qquad (8.15)$$

then the first term, $rx(1 - x/K)$, gives the growth rate of the prey in the absence of predators. The second term, axy, might be more intuitive if you think in terms of the **predation rate**, which is defined to be the number of prey killed per unit time per predator.

(a) In equation (8.15), show that the predation rate is ax. Your argument should include an analysis of the units involved in balancing the equation.

(b) Sketch the graph of the predation rate found in part (a) versus the prey population. Do you think the behavior of the predation rate in your graph is reasonable? Explain.

(c) What is the predation rate in equation (8.9)? Sketch the graph of this predation rate versus the prey population. Is the behavior of the predation rate in this graph more reasonable than that of part (b)? Explain. Your explanation should examine the predation rate for both low and high prey populations.

3. Present an argument to justify the model proposed by equation (8.10). Use reasoning similar to that used to develop the model in equation (8.9).

4. Sketch a plot of the nullclines of the system in (8.11) for the three cases $u_0 < 0$, $0 < u_0 < 1$, and $1 < u_0$.

5. The case when $0 < u_0 < 1$ was discussed in this section. For the other two cases in Exercise 4, use the nullcline information to find the limiting behavior of all solutions.

6. Suppose that the predator population obeys the logistic law

$$y' = sy\left(1 - \frac{y}{K}\right), \qquad (8.16)$$

where s is the reproductive rate for small populations and K is the carrying capacity. This equation has been studied extensively in this text, and it is well known that the predator population will approach the carrying capacity of the environment with the passage of time. A different way to model the effect of the presence of the prey on the predator is to allow the carrying capacity K to vary as the prey population changes.

(a) Assume that it takes N prey to support a single predator, where N is a positive constant. If the prey population is x, how many predators can the environment sustain?

(b) Adjust equation (8.16) to reflect the changing carrying capacity found in part (a).

PROJECT 10.9 Human Immune Response to Infectious Disease

Probably everyone has suffered through an infectious disease such as measles or the mumps. The only good thing about such an experience is that once cured, the patient is henceforth immune to that disease. Some people think that the common cold is another example. They postulate that there are a large number (around 100 different types) of cold viruses, and each of these requires its own immune response. It is certainly true that children are much more prone to colds than are adults, and older persons rarely get colds, all of which provides evidence in favor of this theory.

How does the human body develop an immunity to an infectious disease? The answer is still unknown, but the mathematical biologists Roy M. Anderson and Robert M. May[16] have proposed a theory and a mathematical model. The key fact is that the presence of virus cells stimulates the body to produce lymphocytes that are capable of attacking and killing virus cells. Furthermore, in the Anderson/May model, the mechanism for producing the lymphocytes is self-regulating.

Behavior of lymphocytes in the absence of the virus

Anderson and May posit the existence of two lymphocyte populations $E_1(t)$ and $E_2(t)$. In the absence of the virus, the populations of the lymphocytes are regulated by interactions between them. The key properties are as follows:

- New lymphocytes of type i are produced by bone marrow at a constant rate Λ_i.
- Lymphocytes of type i die at a per capita rate of μ_i.
- Lymphocytes of the two types proliferate due to contact with each other at a rate that saturates for large values of E_1 and E_2.

---

[16] See *Infectious Diseases of Humans* by Roy M. Anderson and Robert M. May, Oxford University Press, 1992.

These considerations lead to the differential equations

$$E_1' = \Lambda_1 - \mu_1 E_1 + a_1 E_1 E_2 / (1 + b_1 E_1 E_2),$$
$$E_2' = \Lambda_2 - \mu_2 E_2 + a_2 E_1 E_2 / (1 + b_2 E_1 E_2). \quad (9.1)$$

The first two terms on the right of each equation correspond to the first two properties. The complicated term that follows models the interaction between the two types of lymphocytes. Notice that for small values of the product $E_1 E_2$, this term accounts for a rate of increase that is roughly proportional to the product. However, as the product increases, the rate of increase goes no higher than a_i / b_i. This is what we mean when we say that the rate *saturates*.

1. Use the parameters $\Lambda_1 = \Lambda_2 = 1$, $\mu_1 = \mu_2 = 1.25$, $a_1 = a_2 = 0.252$, and $b_1 = b_2 = 0.008$. Find the equilibrium points for the system in (9.1) in the positive quadrant.

   The algebra needed to find the equilibrium points is quite difficult. The analysis can be assisted by examining the phase plane as drawn by a computer. Include the nullclines if this is possible, since they can help to find the equilibrium points.

2. Determine the type of each equilibrium point.

   This can be done using the Jacobian; however, the algebra involved here is even harder. Use any techniques that will assist you. This might be the time to use a computer algebra system if you have one available. Some phase plane programs will compute the type of an equilibrium point, and you should feel free to use them if available. In addition, qualitative analysis using the nullclines can help.

3. For each asymptotically stable equilibrium point found in the first two steps, determine the basin of attraction.

   This can be done using qualitative analysis involving the regions bounded by the nullclines. You should also plot the stable and unstable solutions corresponding to any saddle points you find. Submit a plot of the phase plane showing all of the equilibrium points, the stable and unstable solutions, and the basins of attraction.

   Assuming a baby were born with very few or no lymphocytes, what would you predict would be the number present at the age of 21? We will refer to this stable equilibrium point as the *virgin state*.

Behavior in the presence of the virus

Now we are ready to examine what happens when the body is invaded by the virus. Let $V(t)$ denote the number of virus cells in the body. The interaction between the lymphocytes and the virus has the following properties:

- The virus cells have an intrinsic growth rate r.
- Lymphocytes of type 1 kill virus cells in proportion to the numbers of contacts between them, and they proliferate because of these contacts.
- Lymphocytes of type 2 do not directly interact with the

virus, but they continue to regulate the growth of cells of type 1 as before.

As a result of these interactions, we get the system

$$E_1' = \Lambda_1 - \mu_1 E_1 + a_1 E_1 E_2 / (1 + b_1 E_1 E_2) + K V E_1,$$
$$E_2' = \Lambda_2 - \mu_2 E_2 + a_2 E_1 E_2 / (1 + b_2 E_1 E_2), \quad (9.2)$$
$$V' = rV - kV E_1.$$

The term rV represents the intrinsic growth rate of the virus, and $kV E_1$ represents the rate at which virus cells are destroyed by type-1 lymphocytes. The term $K V E_1$ represents the rate of growth of type-1 lymphocytes due to interactions with the virus.

We will use the parameters $r = 0.1$, $k = 0.01$, and $K = 0.05$ in what follows.

4. Show that

$$\begin{pmatrix} E_1 \\ E_2 \\ V \end{pmatrix} = \begin{pmatrix} 1 \\ 1 \\ 0 \end{pmatrix}, \quad \begin{pmatrix} 5 \\ 5 \\ 0 \end{pmatrix}, \quad \text{and} \quad \begin{pmatrix} 20 \\ 20 \\ 0 \end{pmatrix}$$

   are equilibrium points for (9.2). It is not so easy to show that these are the only viable equilibrium points.[17]

5. It is even more difficult to compute the Jacobian of (9.2).[18] However, at the three equilibrium points found in Step 4, the Jacobians are

$$J_1 = \begin{pmatrix} -1.002 & 0.248 & 0.05 \\ 0.248 & -1.002 & 0 \\ 0 & 0 & 0.09 \end{pmatrix},$$

$$J_5 = \begin{pmatrix} -0.375 & 0.875 & 0.25 \\ 0.875 & -0.375 & 0 \\ 0 & 0 & 0.05 \end{pmatrix}, \quad \text{and}$$

$$J_{20} = \begin{pmatrix} -0.9643 & 0.2857 & 1 \\ 0.2857 & -0.9643 & 0 \\ 0 & 0 & -0.1 \end{pmatrix},$$

   respectively. You are not required to prove this.

   Analyze the stability of the three equilibrium points. What would happen to a person whose lymphocytes were at the virgin state and a small amount of the virus were introduced? What do you think would happen to the state of the lymphocyte/virus system as $t \to \infty$? Can you prove this?

6. In the absence of firm mathematical reasoning, we will go to the computer and compute the solutions to (9.2) for initial conditions near the virgin state. Compute the solution with initial conditions $E_1 = E_2 = V = 1$. Turn in a plot of all three components.

7. Next, we will examine what happens when a person with lymphocytes in the immune state is infected. Compute the solution with initial conditions $E_1 = E_2 = 20$ and $V = 1$.

---

[17] There are two others that are not viable because they have one or more of the populations negative.

[18] Please do not allow this comment to prevent you from trying. The computation is not impossible.

Turn in a plot of all three components. Compute and plot the solution with initial conditions $E_1 = E_2 = 20$ and $V = 30$.

8. Write a paragraph or two describing what the model predicts will happen to the numbers of lymphocytes in the body during and after the first infection, and then after subsequent infections. Use all of the information gathered in steps 1 through 7. Be sure to bring in the effects of the various equilibrium points. Do more computations if you think they are necessary. Do you think this model captures significant aspects of the behavior of the immune system?

PROJECT 10.10 Analysis of Competing Species

The goal of this project is to analyze most of the possibilities for systems modeling two competitive species that are in the same location and depend upon the same resources for food.

Derivation of the model

Let $X(t)$ and $Y(t)$ be the two populations. We will briefly describe the derivation of the competing-species model, with emphasis on the population X. In the absence of the population Y, X is governed by the logistic equation

$$X' = r\left(1 - \frac{X}{K_x}\right)X,$$

where r is the reproductive rate for small populations and K_x is the carrying capacity. The reproductive rate for X is $r(1 - X/K_x)$. The introduction of the population Y will reduce this rate, and to a first approximation the reduction will be proportional to Y. Hence the reproductive rate becomes $r(1 - X/K_x) - AY$, where A is the constant of proportionality. The differential equation satisfied by X is

$$X' = \left[r\left(1 - \frac{X}{K_x}\right) - AY\right]X. \tag{10.1}$$

In exactly the same way, we find that the equation for Y is

$$Y' = \left[s\left(1 - \frac{Y}{K_y}\right) - BX\right]Y, \tag{10.2}$$

where s is the reproductive rate for Y at small populations, K_y is the carrying capacity for Y, and the constant B is the proportionality constant that measures the effect of the population X on the reproductive rate for Y.

Dimensionless variables

The system for X and Y in equations (10.1) and (10.2) has six parameters. We can reduce the number of parameters and ease the analysis in other ways by introducing the dimensionless variables

$$x = \frac{X}{K_x} \quad \text{and} \quad y = \frac{Y}{K_y}.$$

1. Show that, in terms of x and y, equations (10.1) and (10.2) for the competing species become

$$\begin{aligned} x' &= r(1 - x - ay)x, \\ y' &= s(1 - y - bx)y, \end{aligned} \tag{10.3}$$

where the new constants are

$$a = AK_y/r \quad \text{and} \quad b = BK_x/s.$$

Now we have only four parameters in the system (10.3). Another benefit of introducing the dimensionless variables is that since x and y represent fractions of the carrying capacities, we know that the interesting range of both variables is the interval [0, 1].

Analysis of the cases

The most important benefit of introducing the dimensionless variables is that all cases can be analyzed in terms of the parameters a and b. (It is frequently the case that introducing dimensionless parameters results in lumping parameters in an effective manner.) The parameters r and s turn out to have little bearing on the classification. (Please use $r = 0.4$ and $s = 0.6$ for whatever computations you do in order to have some consistency.)

There are four cases that you are to consider. For each, describe what happens to all solutions where the populations are nonnegative as $t \to \infty$. We will refer to such solutions as **viable solutions.** Submit your description and whatever computer-generated figures you wish to support your description.

2. Consider when $0 < a < 1$ and $0 < b < 1$. A typical example is $a = 3/4$ and $b = 2/3$.

3. Consider when $0 < a < 1$ and $b > 1$. A typical example is $a = 3/4$ and $b = 4$.

4. Consider when $a > 1$ and $0 < b < 1$. A typical example is $a = 4$ and $b = 3/4$. (This part is a gift. Compare it to the previous case.)

5. Consider when $a > 1$ and $b > 1$. A typical example is $a = 5$ and $b = 4$.

6. Submit a short summary paragraph explaining all of the possibilities.

To earn partial credit, it suffices to analyze the typical case, with the numerical values provided. For complete credit, you need to do the analysis algebraically in terms of the parameters a and b in the given ranges.

In your analysis, you should use any of the techniques you have learned and that apply. Most of what you need is in this chapter. In particular, on your way to answering the question, you should do any or all of the following that apply.

- Find the equilibrium points.

- Find the type of each viable equilibrium point (*viable* means that both populations are nonnegative).

- Show that the positive quadrant is invariant. Find other important invariant regions that help you to analyze the solutions.

- Find the nullclines and the direction of the vector field on portions of the nullclines.

- Use a computer phase plane program to compute and display various aspects of the systems in the typical cases. This can be used to motivate and illustrate your argu-

ments. In particular, you may find it useful to plot the separatrices associated with saddle points.

- Find the basins of attraction for sinks.

You will notice that you are not required to analyze the competing-species model when $a = 1$ or $b = 1$. These are special cases that should not arise very often in practice. If you are curious about them, look at them in a computer phase plane program.

11

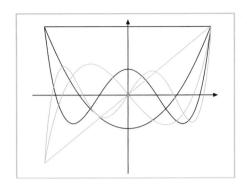

Series Solutions to Differential Equations

One of the applications of partial differential equations is the discovery of the modes of vibration of a circular drum. The displacement, $u(t, x, y)$, of the drum depends on the time t, and the point (x, y) on the drum. Analysis of the tension of the drum leads to the partial differential equation

$$\frac{\partial^2 u}{\partial t^2} = c^2 \left(\frac{\partial^2 u}{\partial x^2} + \frac{\partial^2 u}{\partial y^2} \right),$$

where c is a constant. This is a special case of the wave equation.

There are solutions of the form $u(t, x, y) = \cos(c\mu t) \cdot Y(r)$, where $r = \sqrt{x^2 + y^2}$ is the distance from the point (x, y) to the center of the drum, and μ is a constant. The function $Y(r)$ must satisfy the ordinary differential equation

$$r^2 Y'' + r Y' + \mu^2 r^2 Y = 0. \tag{1}$$

You will notice that this is a second-order, linear, homogeneous equation with variable coefficients. It is not an equation that we have solved or for which we have discovered solution methods. It is a minor variant of Bessel's equation. Our point in bringing it up is to give an example of how solving partial differential equations leads to new problems in ordinary differential equations.

Bessel's equation is one of many ordinary differential equations that arise from applications. In addition there are the equations of Legendre, Haenkel, Airy, and many others. The study of these equations and their solutions is the beginning of the topic of *special functions*. The properties of these special functions are discovered in a variety of ways. One of the most important methods involves finding power series representations for the solutions. That is the topic of this chapter.

Equation (1) presents additional difficulties because the coefficient of the highest derivative vanishes at $r = 0$, which represents the center of the drum. Such a point is called a singular point for the differential equation. The methods for studying solutions near singular points will be studied starting in Section 11.4.

11.1 Review of Power Series

In this section we will review without proofs the basic concepts of power series to prepare ourselves for the use of the technique of series solution. In this chapter, the independent variable will usually be x, reflecting the fact that in most applications the independent variable is a spatial variable instead of time.

Basic definitions

We start with the definition of a power series.

DEFINITION 1.1

A **power series about the point** x_0 is a series of the form

$$\sum_{n=0}^{\infty} a_n(x - x_0)^n = a_0 + a_1(x - x_0) + a_2(x - x_0)^2 + a_3(x - x_0)^3 + \cdots. \quad (1.2)$$

The series is said to **converge** at x if the sequence of **partial sums**

$$S_N(x) = \sum_{n=0}^{N} a_n(x - x_0)^n$$

$$= a_0 + a_1(x - x_0) + a_2(x - x_0)^2 + \cdots + a_N(x - x_0)^N$$

converges as $N \to \infty$. The **sum** of the series at the point x is defined to be the limit of the partial sums, so we write

$$\sum_{n=0}^{\infty} a_n(x - x_0)^n = \lim_{N \to \infty} S_N(x).$$

If the limit of the sequence of partial sums does not exist, then the series is said to **diverge** at x.

In many situations we will have $x_0 = 0$, in which case the power series has the simpler form

$$\sum_{n=0}^{\infty} a_n x^n = a_0 + a_1 x + a_2 x^2 + a_3 x^3 + \cdots. \quad (1.3)$$

There is no loss of generality in considering only series of the simpler form in (1.3), since the substitution $y = x - x_0$ in (1.2) reduces it to a series of the form in (1.3).

Notice that in Definition 1.1 we write the series in two forms. For example, we have

$$\sum_{n=1}^{\infty} \frac{x^n}{n^2} = x + \frac{x^2}{4} + \frac{x^3}{9} + \frac{x^4}{16} + \cdots.$$

We will call the form on the left the **summation form** and the form on the right the **open form**. The summation form is usually more compact and contains more precise information about the series since it contains all of the coefficients. However, the open form is sometimes easier to understand, and when we cannot find a closed form for the coefficients it may be the only form we can use. We will give some preference to the summation form, but we will frequently use the open form to add clarity to our presentation. You should feel free to use both. Even the most expert manipulators of power series move back and forth between the two.

A polynomial is a series with all but a finite number of the a_n equal to zero. Since it is a finite sum, it converges for all x. A power series defines a function

$$f(x) = \sum_{n=0}^{\infty} a_n (x - x_0)^n,$$

at those points x where the series converges. As you will see, such functions act very much like polynomials. They may be considered to be polynomials of infinite degree.

Example 1.4 Show that the **geometric series** $\sum_{n=0}^{\infty} x^n$ converges for $|x| < 1$ and that

$$\sum_{n=0}^{\infty} x^n = \frac{1}{1 - x} \quad \text{for } |x| < 1. \tag{1.5}$$

Show that the series diverges for $|x| \geq 1$.

The partial sums $S_N(x) = \sum_{n=0}^{N} x^n$ can be evaluated as follows. We have

$$(1 - x)S_N(x) = (1 - x)(1 + x + x^2 + \cdots + x^N)$$
$$= (1 + x + x^2 + \cdots + x^N) - (x + x^2 + \cdots + x^N + x^{N+1})$$
$$= 1 - x^{N+1}.$$

Therefore,

$$S_N(x) = \sum_{n=0}^{N} x^n = \frac{1 - x^{N+1}}{1 - x},$$

provided $x \neq 1$. If $|x| < 1$, then $x^{N+1} \to 0$ as $N \to \infty$. In this case, the partial sums converge to $1/(1 - x)$. If $|x| > 1$, then x^{N+1} diverges and therefore the power series diverges. If $x = 1$, then $S_N(1) = N + 1$, while if $x = -1$ the partial sums alternate between 1 and 0, so the series diverges in these cases. ●

Interval of convergence

The series in Example 1.4 converges only for x in the interval $-1 < x < 1$. Notice that this interval is centered at $x_0 = 0$. This behavior is typical of power series, as we see in the next theorem.

THEOREM 1.6 For any power series $\sum_{n=0}^{\infty} a_n (x - x_0)^n$ there is an R, either a nonnegative number or ∞, such that the series converges if $|x - x_0| < R$ and diverges if $|x - x_0| > R$. ■

The quantity R in Theorem 1.6 is called the **radius of convergence** of the power series. The interval $-R < x - x_0 < R$ is called the **interval of convergence**. There are three cases that deserve attention. If $R = 0$, the series converges only for $x = x_0$. If $R = \infty$, the series converges for all real numbers x. Finally, if $0 < R < \infty$, Theorem 1.6 says that the series converges on the finite, open interval $-R < x - x_0 < R$, and diverges if $|x - x_0| > R$.

The ratio test

The radius of convergence of a power series is often difficult to determine. However, the next result is sometimes useful for this purpose.

THEOREM 1.7   Suppose the terms of the series $\sum_{n=0}^{\infty} A_n$ have the property that

$$\lim_{n \to \infty} \frac{|A_{n+1}|}{|A_n|} = L$$

exists. If $L < 1$ the series converges, while if $L > 1$ the series diverges.   ∎

The test for convergence in Theorem 1.7 is called the *ratio test*. For the power series $\sum_{n=0}^{\infty} a_n (x - x_0)^n$, we have

$$L = \lim_{n \to \infty} \frac{|a_{n+1}(x - x_0)^{n+1}|}{|a_n(x - x_0)^n|} = |x - x_0| \lim_{n \to \infty} \frac{|a_{n+1}|}{|a_n|}.$$

We have $L < 1$ if and only if

$$|x - x_0| < R = \lim_{n \to \infty} \frac{|a_n|}{|a_{n+1}|}. \tag{1.8}$$

Hence, assuming the limit defining R exists, it is the radius of convergence.

Example 1.9   Find the radii of convergence for the series $\sum_{n=0}^{\infty} x^n$, $\sum_{n=1}^{\infty} x^n/n$, and $\sum_{n=1}^{\infty} x^n/n^2$. Examine the convergence at the endpoints of the interval of convergence.

You will recognize the first series as the geometric series, which we examined in Example 1.4. The radius of convergence was found to be $R = 1$. For the second series we have $a_n = 1/n$. Hence, by the ratio test the radius of convergence is

$$R = \lim_{n \to \infty} \frac{|a_n|}{|a_{n+1}|} = \lim_{n \to \infty} \frac{n + 1}{n} = \lim_{n \to \infty} \left(1 + \frac{1}{n}\right) = 1.$$

For the third series we have $a_n = 1/n^2$. Now the radius of convergence is

$$R = \lim_{n \to \infty} \frac{|a_n|}{|a_{n+1}|} = \lim_{n \to \infty} \frac{(n + 1)^2}{n^2} = \lim_{n \to \infty} \left(1 + \frac{2}{n} + \frac{1}{n^2}\right) = 1.$$

Hence all three series have radius of convergence $R = 1$. However, they have different behavior at the end points of the interval of convergence. The first series is the geometric series, which diverges at both endpoints, as we saw in Example 1.4. The second series reduces to the divergent harmonic series $\sum_{n=1}^{\infty} 1/n$ at $x = 1$, and to the convergent alternating harmonic series $\sum_{n=1}^{\infty} (-1)^n/n$ at $x = -1$. Finally, since the series $\sum_{n=1}^{\infty} 1/n^2$ converges, the series $\sum_{n=1}^{\infty} x^n/n^2$ converges at both endpoints ± 1.   ●

Notice that the interval of convergence is always open, and that Theorem 1.6 says nothing about what happens at the endpoints of the interval where $|x - x_0| = R$. As the three series in Example 1.9 show, any possible combination of convergence and divergence can be expected at the endpoints of the interval of convergence.

Example 1.10   Find the radius of convergence for the series

$$\sum_{n=0}^{\infty} \frac{2^n x^{2n}}{2n(n + 1)}.$$

Since only the even powers occur in the first of these series, we cannot use the formula in (1.8). However, we can still use the ratio test in Theorem 1.7. The terms of the series are $A_n = 2^n x^{2n}/(2n(n+1))$, so

$$\frac{|A_{n+1}|}{|A_n|} = \frac{2^{n+1} x^{2(n+1)} \cdot 2n(n+1)}{2(n+1)(n+2) \cdot 2^n x^{2n}} = x^2 \frac{2n}{n+2} \to 2x^2.$$

By the ratio test, the series converges if $2x^2 < 1$, so the radius of convergence is $R = 1/\sqrt{2}$. ●

Algebraic operations on series

The algebraic operations of addition, subtraction, and multiplication of power series are defined formally just as they are for polynomials. It is simply a matter of collecting terms of the same degree. However, unlike the case for polynomials, we have to worry about the interval of convergence. We will do that after we have completed the formal definitions.

The **sum** and **difference** of the two series

$$\sum_{n=0}^{\infty} a_n x^n = a_0 + a_1 x + a_2 x^2 + \cdots \quad \text{and} \quad \sum_{n=0}^{\infty} b_n x^n = b_0 + b_1 x + b_2 x^2 + \cdots$$

are defined by

$$\sum_{n=0}^{\infty} a_n x^n \pm \sum_{n=0}^{\infty} b_n x^n = (a_0 \pm b_0) + (a_1 \pm b_1)x + (a_2 \pm b_2)x^2 + \cdots$$

$$= \sum_{n=0}^{\infty} (a_n \pm b_n)x^n.$$

(1.11)

The formal multiplication of series is slightly more complicated, but again it is very similar to the multiplication of polynomials. We expand the product using the distributive law and then collect equal powers of x. Thus,

$$\left(\sum_{n=0}^{\infty} a_n x^n\right) \left(\sum_{m=0}^{\infty} b_m x^m\right)$$
$$= \left(a_0 + a_1 x + a_2 x^2 + \cdots\right) \left(b_0 + b_1 x + b_2 x^2 + \cdots\right)$$
$$= a_0 \left(b_0 + b_1 x + b_2 x^2 + \cdots\right)$$
$$\quad + a_1 x \left(b_0 + b_1 x + b_2 x^2 + \cdots\right)$$
$$\quad + a_2 x^2 \left(b_0 + b_1 x + b_2 x^2 + \cdots\right)$$
$$\quad + \cdots$$
$$= a_0 b_0 + (a_1 b_0 + a_0 b_1)x + (a_2 b_0 + a_1 b_1 + a_0 b_2)x^2 + \cdots.$$

Continuing this pattern, we see that the general term in the product is $c_p x^p$, where

$$c_p = a_p b_0 + a_{p-1} b_1 + \cdots + a_1 b_{p-1} + a_0 b_p$$
$$= \sum_{k=0}^{p} a_{p-k} b_k.$$

(1.12)

Thus the formal expression for the **product** of two series is

$$\left(\sum_{n=0}^{\infty} a_n x^n\right)\left(\sum_{m=0}^{\infty} b_m x^m\right) = \sum_{p=0}^{\infty} c_p x^p, \tag{1.13}$$

where the coefficients of the product are given by (1.12).

While the formal definitions of the algebraic operations are easy consequences of normal arithmetic operations, we need to determine the radius of convergence of the sum and product. The next proposition gives the best general result.

PROPOSITION 1.14   The radius of convergence of the sum, difference, or product of two power series is at least as large as the smaller of the radii of convergence of the original two series.

⬤

The coefficient of x^p in the product of two series in (1.12) can get quite complicated for large p. We are not always fortunate enough to get a nice closed form. In applications it is frequently necessary to compute only the first few terms. When p is relatively small, formula (1.12) can be used effectively to compute the coefficient.

Shifting the index

The index n in the series $\sum_{n=0}^{\infty} a_n (x - x_0)^n$ is a dummy index. It can be replaced by any other letter, and it can be shifted up or down to suit our needs as long as we make the change consistently. The following examples show how a shift in the index of a series allows us to manipulate power series.

Example 1.15   Combine the sum

$$\sum_{n=1}^{\infty} \frac{n}{n+1} x^{n-1} + \sum_{n=1}^{\infty} \frac{1}{n^2} x^n \tag{1.16}$$

into one series.

We will shift the index in the first series so that the power of x that appears is x^n, the same as the power that appears in the second series. To do this we make the substitution in the index of $p = n - 1$, or, equivalently, $n = p + 1$. Doing so in the first series, we get

$$\sum_{n=1}^{\infty} \frac{n}{n+1} x^{n-1} = \sum_{p=0}^{\infty} \frac{p+1}{p+2} x^p. \tag{1.17}$$

Notice that we have changed the lower limit of the sum to reflect the change in the index. Since the index n starts at 1, $p = n - 1$ starts at 0. Next, recognizing that the index is a dummy index, we can replace the p in (1.17) with n. With this change, (1.16) becomes

$$\sum_{n=0}^{\infty} \frac{n+1}{n+2} x^n + \sum_{n=1}^{\infty} \frac{1}{n^2} x^n.$$

We are now ready to add the two series using the formula (1.11), except for the fact that the two series start at different values of the index. We handle this by isolating the first term in the first series to make the two series start at the same point.

$$\sum_{n=0}^{\infty} \frac{n+1}{n+2} x^n + \sum_{n=1}^{\infty} \frac{1}{n^2} x^n = \frac{1}{2} + \sum_{n=1}^{\infty} \frac{n+1}{n+2} x^n + \sum_{n=1}^{\infty} \frac{1}{n^2} x^n$$

$$= \frac{1}{2} + \sum_{n=1}^{\infty} \left(\frac{n+1}{n+2} + \frac{1}{n^2} \right) x^n.$$

An application of the ratio test shows that the radius of convergence of the final series is equal to 1, as is that of each of the two original series. ●

Example 1.18 Compute the product $(1 + x^2) \sum_{n=1}^{\infty} n x^n$.

Proceeding as if both factors were polynomials, we get

$$(1 + x^2) \sum_{n=1}^{\infty} n x^n = \sum_{n=1}^{\infty} n x^n + x^2 \sum_{n=1}^{\infty} n x^n$$

$$= \sum_{n=1}^{\infty} n x^n + \sum_{n=1}^{\infty} n x^{n+2}. \tag{1.19}$$

As in Example 1.15, we have two sums with different powers of x. We will shift the index in the second sum. We set $p = n + 2$, or $n = p - 2$, and the second sum becomes $\sum_{n=1}^{\infty} n x^{n+2} = \sum_{p=3}^{\infty} (p-2) x^p$ If we remember that the index is a dummy index, we can replace the index p with n to get $\sum_{p=3}^{\infty} (p-2) x^p = \sum_{n=3}^{\infty} (n-2) x^n$. Then (1.19) becomes

$$\sum_{n=1}^{\infty} n x^n + \sum_{n=3}^{\infty} (n-2) x^n = x + 2x^2 + \sum_{n=3}^{\infty} n x^n + \sum_{n=3}^{\infty} (n-2) x^n$$

$$= x + 2x^2 + \sum_{n=3}^{\infty} (2n-2) x^n.$$

The ratio test shows that the radius of convergence of the final series is 1, as is that of the original series $\sum_{n=1}^{\infty} n x^n$. ●

Differentiating power series

Suppose that $\sum_{n=0}^{\infty} a_n (x - x_0)^n$ is a power series with radius of convergence $R > 0$. Then

$$f(x) = \sum_{n=0}^{\infty} a_n (x - x_0)^n = a_0 + a_1 (x - x_0) + a_2 (x - x_0)^2 + \cdots \tag{1.20}$$

defines a function of x in the interval $|x - x_0| < R$.

THEOREM 1.21 The function $f(x)$ defined in (1.20) is infinitely differentiable in its interval of convergence. Its derivative can be computed by differentiating the series term by term. Thus,

$$f'(x) = \frac{d}{dx} \left\{ a_0 + a_1 (x - x_0) + a_2 (x - x_0)^2 + a_3 (x - x_0)^3 + \cdots \right\}$$

$$= a_1 + 2a_2 (x - x_0) + 3a_3 (x - x_0)^2 + \cdots \tag{1.22}$$

$$= \sum_{n=1}^{\infty} n a_n (x - x_0)^{n-1}.$$

The power series for f' has the same radius of convergence as does that of f. ■

Higher derivatives can be computed in the same way. For example, the second derivative of $f(x)$ is

$$f''(x) = \frac{d^2}{dx^2} \left\{ \sum_{n=0}^{\infty} a_n (x - x_0)^n \right\} = \sum_{n=2}^{\infty} n(n-1) a_n (x - x_0)^{n-2}. \qquad (1.23)$$

Identity theorem for power series

Notice that if we evaluate the function $f(x)$ defined in (1.20) at the point x_0, all terms in the series vanish except the first, so

$$f(x_0) = a_0. \qquad (1.24)$$

Similarly, from (1.22) and (1.23), we get

$$f'(x_0) = a_1 \quad \text{and} \quad f''(x_0) = 2a_2. \qquad (1.25)$$

If we differentiate (1.23) and evaluate it at $x = x_0$, we see that

$$f'''(x_0) = 3 \cdot 2 \cdot a_3. \qquad (1.26)$$

Solving for the coefficients in (1.24), (1.25), and (1.26), we get

$$a_0 = f(x_0), \quad a_1 = f'(x_0), \quad a_2 = \frac{f''(x_0)}{2}, \quad \text{and} \quad a_3 = \frac{f'''(x_0)}{3 \cdot 2}.$$

We begin to see a pattern here, which is confirmed in the next theorem.

THEOREM 1.27   Suppose that the series $f(x) = \sum_{n=0}^{\infty} a_n (x - x_0)^n$ has a positive radius of convergence. Then

$$a_n = \frac{f^{(n)}(x_0)}{n!}. \qquad \blacksquare$$

Thus the coefficients of a power series like that in (1.20) are determined by the values of the sum $f(x)$. Theorem 1.27 is called the ***identity theorem***. The reason for this name becomes more clear when we state the next corollary.

COROLLARY 1.28   Suppose we have two power series with the same sum,

$$\sum_{n=0}^{\infty} a_n (x - x_0)^n = \sum_{n=0}^{\infty} b_n (x - x_0)^n \quad \text{for } |x - x_0| < R,$$

where $R > 0$. Then $a_n = b_n$ for all n. In particular, if $\sum_{n=0}^{\infty} a_n (x - x_0)^n = 0$ for $|x - x_0| < R$, where $R > 0$, then $a_n = 0$ for all n.   ●

We will find Corollary 1.28 very useful in the rest of this chapter.

Integrating power series

We can also integrate power series by integrating the series term by term.

THEOREM 1.29 Suppose the power series

$$f(x) = \sum_{n=0}^{\infty} a_n (x - x_0)^n = a_0 + a_1(x - x_0) + a_2(x - x_0)^2 + \cdots$$

converges for $|x - x_0| < R$, where $R > 0$. Then the power series for f can be integrated term by term, obtaining

$$\int f(x)\, dx = C + \sum_{n=0}^{\infty} a_n \frac{(x - x_0)^{n+1}}{n + 1}$$

$$= C + a_0(x - x_0) + a_1 \frac{(x - x_0)^2}{2} + a_2 \frac{(x - x_0)^3}{3} + \cdots,$$

where C is the constant of integration. The radius of convergence of the integrated series is the same as that of the original series. ∎

Analytic functions and Taylor series

We have seen that power series define functions. Now let's go in the other direction. If we are given a function $f(x)$, can we express f as a power series about a given point x_0? For example, can we express the function $f(x) = \ln(1 + x^2)$ as a power series about $x_0 = 0$?

We get some information from what we already know. If the function f is represented as a power series, say

$$f(x) = \sum_{n=0}^{\infty} a_n (x - x_0)^n$$

near the point x_0, then Theorem 1.21 tells us that the function f must be infinitely differentiable. Next, the identity theorem, Theorem 1.27, tells us that

$$a_n = \frac{f^{(n)}(x_0)}{n!} \quad \text{for all } n.$$

Not all infinitely differentiable functions can be represented by power series in this way. However, many commonly used functions can be so represented, and many solutions to differential equations can as well. We will give such functions a name.

DEFINITION 1.30

A function f is **analytic** at x_0 provided $f(x) = \sum_{n=0}^{\infty} a_n (x - x_0)^n$, where the series converges for all x in some interval containing x_0.

Clearly, all polynomials are analytic functions. The discussion preceding Definition 1.30 is summed up in the next theorem.

THEOREM 1.31 Suppose f is analytic at x_0. Then

$$f(x) = \sum_{n=0}^{\infty} \frac{f^{(n)}(x_0)}{n!} (x - x_0)^n$$

for all x in the interval of convergence of the series on the right. ∎

The series on the right is called the *Taylor series* expansion of f about x_0. Here are some examples of Taylor series (in the case $x_0 = 0$).

$$e^x = \sum_{n=0}^{\infty} \frac{x^n}{n!} = 1 + x + \frac{x^2}{2!} + \frac{x^3}{3!} + \cdots \qquad (1.32)$$

$$\cos x = \sum_{n=0}^{\infty} \frac{(-1)^n x^{2n}}{(2n)!} = 1 - \frac{x^2}{2!} + \frac{x^4}{4!} - \frac{x^6}{6!} + \cdots$$

$$\sin x = \sum_{n=0}^{\infty} \frac{(-1)^n x^{2n+1}}{(2n+1)!} = x - \frac{x^3}{3!} + \frac{x^5}{5!} - \frac{x^7}{7!} + \cdots$$

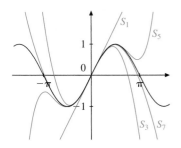

Figure 1. Partial sums of the Taylor series for $\sin(x)$.

These Taylor series are easily found by using Theorem 1.31. The convergence of the Taylor series for $\sin(x)$ is illustrated in Figure 1.

Computing Taylor series

Theorem 1.31 does not always provide the best way to find the Taylor series for a function. Since the coefficients of the series expansion of an analytic function are unique, any legitimate mathematical process used to compute these coefficients will lead to the same answer. Here is an example.

Example 1.33   Compute the series expansion of $f(x) = \ln(1 + x^2)$ about $x_0 = 0$.

The derivatives of $\ln(1 + x^2)$ get complicated very quickly, so using Theorem 1.31 is not the best way to proceed. Instead, we notice that $f(x) = g(x^2)$, where $g(y) = \ln(1 + y)$. We will find the Taylor series for $g(y)$ and then substitute $y = x^2$. We could easily find the Taylor series for g using Theorem 1.31, but we will illustrate another way to proceed.

Notice that $g'(y) = 1/(1 + y)$. From Example 1.4 we see that

$$g'(y) = \frac{1}{1+y} = \sum_{n=0}^{\infty} (-y)^n = \sum_{n=0}^{\infty} (-1)^n y^n.$$

Integrating this series, using Theorem 1.29, we get

$$g(y) = C + \sum_{n=0}^{\infty} \frac{(-1)^n}{n+1} y^{n+1} = C + \sum_{n=1}^{\infty} \frac{(-1)^{n-1}}{n} y^n.$$

We notice that $C = g(0) = \ln(1) = 0$, so

$$\ln(1 + y) = g(y) = \sum_{n=1}^{\infty} \frac{(-1)^{n-1}}{n} y^n.$$

Finally, after substituting $y = x^2$, we get

$$\ln(1 + x^2) = g(x^2) = \sum_{n=1}^{\infty} \frac{(-1)^{n-1}}{n} x^{2n}.$$

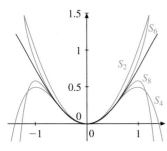

Figure 2. The partial sums for the Taylor series of $\log(1 + x^2)$.

It is an easy application of the ratio test to show that the radius of convergence of this series is 1. The first few partial sums of the Taylor series are plotted in Figure 2. This figure indicates the convergence for $|x| < 1$ and the divergence for $|x| > 1$.   ●

The reciprocal of an analytic function

The sum, difference, and product of two analytic functions are analytic. This follows from our study of the algebraic properties of power series. It is also true that the quotient of two analytic functions at x_0 is analytic at x_0 provided the denominator is nonzero at x_0.

Suppose that f is a function which is analytic at x_0. In Exercise 37 we show how to compute the coefficients of the Taylor series for $1/f$. However, in what we will do we will only need to find the radius of convergence for the Taylor series for $1/f$. There is a useful criteria for doing so that comes out of the theory of functions of a complex variable. We will state it in the following theorem for the case when f is a polynomial.

THEOREM 1.34   Suppose

$$g(x) = \frac{1}{P(x)},$$

where $P(x)$ is a polynomial. Then g is analytic at the point x_0 provided $P(x_0) \neq 0$. The radius of convergence of the power series of g about the point x_0 is $R = |x_0 - z_0|$, where z_0 is the complex root of $P(z)$ that is the closest to x_0. ∎

Example 1.35   Find the radius of convergence of the Taylor series expansion for

$$g(x) = \frac{1}{4 + x^2}$$

about the point $x_0 = 0$.

We will do this in two ways, providing in this case a verification of Theorem 1.34. The complex roots of $P(z) = 4 + z^2$ are $z = \pm 2i$, which are both 2 units away from $x_0 = 0$. Therefore, Theorem 1.34 implies that the radius of convergence of g is 2.

Another way to obtain the radius of convergence is to notice that

$$g(x) = \frac{1}{4 + x^2} = \frac{1}{4} \cdot \frac{1}{1 - (-x^2/4)} = \frac{1}{4} \sum_{n=0}^{\infty} \left(\frac{-x^2}{4} \right)^n,$$

which is a geometric series, and converges for $\left| -x^2/4 \right| < 1$, or for $|x| < 2$. ●

EXERCISES

Find the radius of convergence of each of the series in Exercises 1–12.

1. $\displaystyle\sum_{n=0}^{\infty} \frac{x^n}{n+1}$

2. $\displaystyle\sum_{n=0}^{\infty} \frac{x^n}{\sqrt{n+3}}$

3. $\displaystyle\sum_{n=0}^{\infty} nx^n$

4. $\displaystyle\sum_{n=0}^{\infty} \sqrt{n}x^{n+1}$

5. $\displaystyle\sum_{n=0}^{\infty} \frac{x^{n+1}}{n!}$

6. $\displaystyle\sum_{n=0}^{\infty} \frac{(x-2)^n}{n^2+1}$

7. $\displaystyle\sum_{n=0}^{\infty} n!(x-1)^n$

8. $\displaystyle\sum_{n=0}^{\infty} \frac{2^n}{n+1}x^n$

9. $\displaystyle\sum_{n=2}^{\infty} \frac{x^n}{\ln n}$

10. $\displaystyle\sum_{n=0}^{\infty} \frac{3^n}{n!}(x+2)^n$

11. $\displaystyle\sum_{n=1}^{\infty} \frac{nx^n}{1 \cdot 3 \cdot 5 \cdots (2n-1)}$

12. $\displaystyle\sum_{n=0}^{\infty} \frac{n!x^n}{(2n)!}$

Find the Taylor series about the given point for each of the functions in Exercises 13–18. In each case find the radius of convergence.

13. $f(x) = \cos 2x$ about $x_0 = 0$

14. $f(x) = \sin x^2$ about $x_0 = 0$

15. $f(x) = \sin x$ about $x_0 = \pi$

16. $f(x) = \ln x$ about $x_0 = 1$

17. $f(x) = \dfrac{1}{x}$ about $x_0 = 3$

18. $f(x) = \ln\left(\dfrac{1+x}{1-x}\right)$ about $x = 0$

In Exercises 19–24 find the power series $\sum_{n=0}^{\infty} a_n x^n$ for the function $f(x)$.

19. $f(x) = (1+x) \sum_{n=0}^{\infty} x^n$

20. $f(x) = (1+x^2) \sum_{n=1}^{\infty} \dfrac{x^n}{n}$

21. $f(x) = \sum_{n=0}^{\infty} x^n - \sum_{n=1}^{\infty} \dfrac{x^{n-1}}{n}$

22. $f(x) = (1+x^2) \sin x$

23. $f(x) = e^x - e^{-x}$

24. $f(x) = e^x + \sum_{n=2}^{\infty} \dfrac{x^{n-2}}{n(n+1)}$

25. Find the Taylor series of the function $f(x) = 1/(3+x)$ about $x_0 = 0$ first by using Taylor's formula and then by writing

$$f(x) = \frac{1}{3} \cdot \frac{1}{1-(-x/3)}$$

and using the formula for a geometric series: $\sum_{n=0}^{\infty} r^n = 1/(1-r)$. Find the radius of convergence for this series.

26. Find the Taylor series of the function $f(x) = 1/x$ about $x_0 = 2$ first by using Taylor's formula and then by writing

$$f(x) = \frac{1}{2} \cdot \frac{1}{1-(2-x)/2}$$

and using the formula for a geometric series: $\sum_{n=0}^{\infty} r^n = 1/(1-r)$. Find the radius of convergence for this series.

In Exercises 27–30 find the Taylor series of each of the function about $x_0 = 0$ using any technique. Find the radius of convergence R. Plot the first three different partial sums and the function f on an interval slightly larger than $[-R, R]$ if $R < \infty$, or on $[-2, 2]$ if $R = \infty$. (See Figures 1 and 2.)

27. $f(x) = 1/(4+x^2)$

28. $f(x) = \ln(1-x^2)$

29. $f(x) = e^{-2x^2}$

30. $f(x) = \sinh 3x = (e^{3x} - e^{-3x})/2$

31. Find the Taylor series for $f(x) = 1/x^2$ about $x_0 = 2$. What is the radius of convergence?

32. Find the Taylor series for $f(x) = 1/x^4$ about $x_0 = 1$. What is the radius of convergence?

33. Compute the value of the series $\sum_{n=1}^{\infty} nx^n$ by differentiating the formula for a geometric series: $\sum_{n=0}^{\infty} x^n = 1/(1-x)$. What is the radius of convergence of this series?

34. Compute the value of the series $\sum_{n=1}^{\infty} x^n/(n+1)$ by integrating the formula for a geometric series: $\sum_{n=0}^{\infty} x^n = 1/(1-x)$. What is the radius of convergence of this series?

35. Find the first four terms of the Taylor series of $f(x) = \tan x$ about $x_0 = 0$.

36. Find the first four terms of the Taylor series of $f(x) = e^{x^2} \cos x$ about $x_0 = 0$.

37. Suppose that $f(x) = \sum_{n=0}^{\infty} a_n x^n$, where $f(0) = a_0 \neq 0$. Let $g(x) = 1/f(x)$ be its reciprocal. Show that if $g(x) = \sum_{n=0}^{\infty} b_n x^n$, then the coefficients can be calculated inductively by $b_0 = 1/a_0$, and $b_n = -[a_1 b_{n-1} + \cdots + a_n b_0]/a_0$ for $n \geq 1$.

11.2 Series Solutions Near Ordinary Points

We are now ready to solve differential equations using power series. We will start with a simple example of first order and then show what happens in general. The basic idea is to start with a power series with undetermined coefficients and to use the differential equation and the initial conditions to determine the coefficients.

An example of a first-order equation

Consider the differential equation $y' - 2xy = 0$. You will recognize that this first-order equation is separable, and you will find that the general solution has the form $y(x) = Ce^{x^2}$, where C is an arbitrary constant. In our first example we will find a series solution for this equation. As a rule, it is not beneficial to find power series solutions when exact solutions are available. Our purpose is to illustrate the method in a simple case.

Example 2.1 Find a series solution for the differential equation

$$y' - 2xy = 0. \tag{2.2}$$

We will look for a solution of the form

$$y(x) = \sum_{n=0}^{\infty} a_n x^n, \tag{2.3}$$

where the coefficients a_n are as yet undetermined. We start by noticing, either by direct evaluation or by using Theorem 1.27, that

$$a_0 = y(0).$$

Hence, the first coefficient is determined by the initial condition satisfied by the solution. We will use the differential equation to determine the rest.

Differentiating y, we have

$$y'(x) = \sum_{n=0}^{\infty} n a_n x^{n-1} = \sum_{n=1}^{\infty} n a_n x^{n-1}. \tag{2.4}$$

Substituting the expressions in (2.3) and (2.4) into the differential equation (2.2) gives

$$0 = y' - 2xy$$
$$= \sum_{n=1}^{\infty} n a_n x^{n-1} - 2x \sum_{n=0}^{\infty} a_n x^n \tag{2.5}$$
$$= \sum_{n=1}^{\infty} n a_n x^{n-1} - \sum_{n=0}^{\infty} 2a_n x^{n+1}.$$

To add these two series we shift the index in the first one so that the two powers are the same. Since the power in the second series is $n + 1$, we want the new index p in the first series to satisfy $n - 1 = p + 1$. Hence we set $p = n - 2$ or $n = p + 2$. Then the first series becomes

$$\sum_{n=1}^{\infty} n a_n x^{n-1} = \sum_{p=-1}^{\infty} (p + 2) a_{p+2} x^{p+1} = a_1 + \sum_{n=0}^{\infty} (n + 2) a_{n+2} x^{n+1}.$$

To get the last expression we have first changed the (dummy) index from p to n, and then isolated the first term in the series so that both series in (2.5) start at $n = 0$. Substituting this into (2.5), we get

$$0 = a_1 + \sum_{n=0}^{\infty} (n + 2) a_{n+2} x^{n+1} - \sum_{n=0}^{\infty} 2a_n x^{n+1}$$
$$= a_1 + \sum_{n=0}^{\infty} \left[(n + 2) a_{n+2} - 2a_n \right] x^{n+1}.$$

By the corollary to the identity theorem (Corollary 1.28), this equation implies that all of the coefficients are equal to 0. From the constant term we obtain $a_1 = 0$ and from the general term we see that

$$a_{n+2} = \frac{2a_n}{n + 2}, \quad \text{for all } n \geq 0. \tag{2.6}$$

To sum up, we have discovered that $a_0 = y(0)$ and $a_1 = 0$. In addition, we have equation (2.6), which is called a **_recurrence formula_**. It will allow us to compute all of the other coefficients. Let us first look at the odd values of n. In turn, we get

$a_3 = 2a_1/3 = 0$, $a_5 = 2a_3/5 = 0$, and so forth. Inductively we see that all of the odd-numbered coefficients are equal to 0.

Next we look at the even numbered coefficients. We have

$$a_2 = \frac{2a_0}{2} = y(0) \qquad a_4 = \frac{2a_2}{4} = \frac{y(0)}{2}$$

$$a_6 = \frac{2a_4}{6} = \frac{y(0)}{2 \cdot 3} \qquad a_8 = \frac{2a_6}{8} = \frac{y(0)}{2 \cdot 3 \cdot 4}.$$

From these calculations we infer that the general pattern for the even-numbered coefficients is

$$a_{2k} = \frac{y(0)}{k!}, \quad \text{for } k \geq 0.$$

Since all of the odd-numbered coefficients are equal to 0, the series in (2.3) becomes

$$y(x) = \sum_{k=0}^{\infty} a_{2k} x^{2k} = y(0) \sum_{k=0}^{\infty} \frac{x^{2k}}{k!}.$$

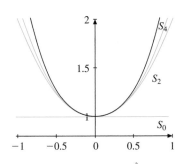

Figure 1. Partial sums for e^{x^2}.

You will recognize the series as that of e^{x^2}. Hence the solution is $y(x) = y(0)e^{x^2}$, as we already knew. A few partial sums are plotted in Figure 1. ●

If the manipulation of the summation form of the series is confusing, it is a good idea to do the computation using the open form of the series. We start with

$$y(x) = a_0 + a_1 x + a_2 x^2 + a_3 x^3 + a_4 x^4 + a_5 x^5 + \cdots \quad \text{and}$$

$$y'(x) = a_1 + 2a_2 x + 3a_3 x^2 + 4a_4 x^3 + 5a_5 x^4 + \cdots.$$

When we substitute these expressions into the differential equation (2.2) and collect coefficients of like powers of x, we get

$$
\begin{aligned}
0 &= y' - 2xy \\
&= \left(a_1 + 2a_2 x + 3a_3 x^2 + 4a_4 x^3 + \cdots\right) - \left(2a_0 x + 2a_1 x^2 + 2a_2 x^3 + \cdots\right) \\
&= a_1 + (2a_2 - 2a_0)x + (3a_3 - 2a_1)x^2 + (4a_4 - 2a_2)x^3 + \cdots.
\end{aligned}
$$

By the identity theorem, the coefficient of each power of x must be zero. Therefore,

$$a_1 = 0, \quad 2a_2 - 2a_0 = 0, \quad 3a_3 - 2a_1 = 0, \quad 4a_4 - 2a_2 = 0, \quad \ldots,$$

or

$$a_1 = 0, \quad a_2 = \frac{2}{2}a_0, \quad a_3 = \frac{2}{3}a_1, \quad a_4 = \frac{2}{4}a_2, \quad \ldots.$$

From the first of these identities we get $a_1 = 0$. From the rest we can conjecture the general pattern of the recursion formula in (2.6). The computation of the coefficients follows as before.

Second-order equations

The method we used for the first-order equation in Example 2.1 applies almost without change to higher-order equations. It is for higher-order equations that the method achieves its potential, since there are many important new functions that arise as solutions of such equations, for which there are no exact formulas in terms of elementary functions.

We will be trying to find power series solutions for linear differential equations of the form

$$L(x)y'' + M(x)y' + N(x)y = 0, \qquad (2.7)$$

where the coefficients $L(x)$, $M(x)$, and $N(x)$ are functions. It will frequently be useful to divide the equation by the leading coefficient $L(x)$, in which case the equation has the form

$$y'' + P(x)y' + Q(x)y = 0, \qquad (2.8)$$

where $P(x) = M(x)/L(x)$ and $Q(x) = N(x)/L(x)$. We will find it useful to move back and forth between representations of a differential equation in the form in (2.7), and that in (2.8). Each of these forms has its uses.

An example is **Legendre's equation**,

$$(1 - x^2)y'' - 2xy' + n(n+1)y = 0, \qquad (2.9)$$

where n is a nonnegative constant, usually an integer. Legendre's equation arises when we solve partial differential equations in regions with spherical symmetry. Another example is **Bessel's equation**,

$$x^2 y'' + xy' + (x^2 - r^2)y = 0, \qquad (2.10)$$

where r is a nonnegative constant. Bessel's equation arises when we solve partial differential equations in regions with circular symmetry, such as a disk or a cylinder. We will find the solutions to both equations later in this chapter.

Both Legendre's equation and Bessel's equation are given in the form of (2.7), but they can be easily transformed into the form of (2.8) by dividing by the leading coefficient.

Ordinary points and singular points

The method of solution for an equation of the type in (2.8) will depend on the nature of the coefficient functions $P(x)$ and $Q(x)$ near the point which is the center of our attention.

DEFINITION 2.11

If the coefficients $P(x)$ and $Q(x)$ of an equation of the form (2.8) are both analytic at the point x_0, then x_0 is called an ***ordinary point*** for the equation. A point which is not an ordinary point is called a ***singular point***.

In order to decide if an equation given in the form in (2.7) has an ordinary or a singular point at x_0, it is necessary to put it into the form in (2.8) by dividing by the leading coefficient $L(x)$. For Legendre's equation we get

$$y'' - \frac{2x}{1 - x^2}y' + \frac{n(n+1)}{1 - x^2}y = 0. \qquad (2.12)$$

The polynomial $1 - x^2$ is analytic, so its reciprocal is analytic except where $1 - x^2 = 0$, or at $x = \pm 1$. Thus the coefficients of the equation in (2.12) are analytic except at ± 1, which means that every real number except ± 1 is an ordinary point for Legendre's equation, and ± 1 are singular points. Similarly, we see that $x_0 = 0$ is the only singular point for Bessel's equation in (2.10).

We will look for power series solutions of equations of the forms (2.7) and (2.8) near a point x_0. This means we will look for a solution of the form

$$y(x) = \sum_{n=0}^{\infty} a_n(x - x_0)^n = a_0 + a_1(x - x_0) + a_2(x - x_0)^2 + a_3(x - x_0)^3 + \cdots,$$

where the coefficients a_n are undetermined. By direct evaluation, or from the identity theorem, we see that

$$a_0 = y(x_0) \quad \text{and} \quad a_1 = y'(x_0). \tag{2.13}$$

Thus the first two coefficients are determined by the initial conditions. The rest of the a_n are determined recursively by inserting the series expansion for y and the series expansions for $L(x)$, $M(x)$, and $N(x)$ into the given differential equation (in the form in (2.7)) and then comparing the coefficients of the powers of x.

We will be looking for a fundamental set of solutions. The next proposition will help us.

PROPOSITION 2.14   The solutions $y_1(x)$ and $y_2(x)$ to an equation of the form (2.7) or (2.8) with the initial conditions

$$y_1(x_0) = 1, \quad y_1'(x_0) = 0, \quad \text{and} \quad y_2(x_0) = 0, \quad y_2'(x_0) = 1$$

form a fundamental set of solutions.

Proof   The Wronskian of y_1 and y_2 at $x = x_0$ is

$$W = \det \begin{pmatrix} y_1(x_0) & y_2(x_0) \\ y_1'(x_0) & y_2'(x_0) \end{pmatrix} = \det \begin{pmatrix} 1 & 0 \\ 0 & 1 \end{pmatrix} = 1.$$

Since $W \neq 0$, we see that y_1 and y_2 are linearly independent, so these functions form a fundamental set of solutions.   ●

This choice of a fundamental set of solutions frequently has the additional feature of having simpler series expansions, as we will see. Let's look at an example that illustrates our solution procedure.

Example 2.15   Find the general series solution to the equation

$$y'' + xy' + y = 0.$$

Find the particular solution with $y(0) = 0$ and $y'(0) = 2$.

We will look for a solution of the form

$$y = \sum_{n=0}^{\infty} a_n x^n,$$

where the coefficients are undetermined. Our example has the form of (2.8) with coefficients $P(x) = x$ and $Q(x) = 1$. Since both are polynomials, $x_0 = 0$ is an ordinary point. We know that $a_0 = y(0)$ and $a_1 = y'(0)$, so the first two coefficients are determined by the initial conditions of the solution. We will use the differential equation to find the rest.

The basic idea is to substitute power series for y and its derivatives into the differential equation, and then to shift indices and add the terms to end up with one series. There are many ways to do this. However, the whole process is made easier if we recognize that there are several ways to write the power series for the derivatives of y. For example, by shifting the index we have

$$y' = \sum_{n=1}^{\infty} na_n x^{n-1} = \sum_{n=0}^{\infty} (n+1)a_{n+1}x^n, \quad \text{and} \tag{2.16}$$

$$y'' = \sum_{n=2}^{\infty} n(n-1)a_n x^{n-2} = \sum_{n=0}^{\infty} (n+2)(n+1)a_{n+2}x^n. \tag{2.17}$$

Notice, however, that for $n = 0$, $na_n = 0$, so using the first expression in (2.16), we can write

$$y' = \sum_{n=0}^{\infty} na_n x^{n-1}, \quad \text{and} \quad xy' = \sum_{n=0}^{\infty} na_n x^n.$$

When we insert the forms of the power series for y, xy', and y'' all of which have terms involving x^n into the differential equation, it becomes

$$0 = y'' + xy' + y$$

$$= \sum_{n=0}^{\infty} (n+2)(n+1)a_{n+2}x^n + \sum_{n=0}^{\infty} na_n x^n + \sum_{n=0}^{\infty} a_n x^n \tag{2.18}$$

$$= \sum_{n=0}^{\infty} \left[(n+2)(n+1)a_{n+2} + na_n + a_n \right] x^n.$$

Because of our choice of the series for the terms in the differential equation, we do not have to shift an index at this point. Furthermore, we have avoided the need to isolate a beginning term or two of the final series.

By the identity theorem, the coefficients of all powers of x in (2.18) must be zero. Therefore, setting the coefficient of x^n equal to 0, we find that

$$a_{n+2} = -\frac{n+1}{(n+2)(n+1)}a_n = -\frac{1}{n+2}a_n, \quad \text{for } n \geq 0. \tag{2.19}$$

This equation is the **recurrence formula**, and it allows us to compute all of the coefficients recursively, starting with the first two, $a_0 = y(0)$ and $a_1 = y'(0)$, which we already know.

To get a general formula for a_n in terms of a_0 and a_1, we write out enough terms to see the pattern. We start with the even-numbered coefficients. From (2.19) we have in turn

$$a_2 = -\frac{1}{2}a_0 \qquad\qquad a_4 = -\frac{1}{4}a_2 = \frac{1}{4\cdot 2}a_0$$

$$a_6 = -\frac{1}{6}a_4 = -\frac{1}{6\cdot 4\cdot 2}a_0 \qquad a_8 = -\frac{1}{8}a_6 = \frac{1}{8\cdot 6\cdot 4\cdot 2}a_0. \tag{2.20}$$

In the same manner we can use the recurrence formula (2.19) to obtain expressions for the odd-numbered coefficients.

$$a_3 = -\frac{1}{3}a_1 \qquad\qquad a_5 = -\frac{1}{5}a_3 = \frac{1}{5\cdot 3}a_1$$

$$a_7 = -\frac{1}{7}a_5 = -\frac{1}{7\cdot 5\cdot 3}a_1 \qquad a_9 = -\frac{1}{9}a_7 = \frac{1}{9\cdot 7\cdot 5\cdot 3}a_1. \tag{2.21}$$

Thus, the general solution can be written as

$$y(x) = a_0 \left[1 - \frac{1}{2} x^2 + \frac{1}{4 \cdot 2} x^4 - \frac{1}{6 \cdot 4 \cdot 2} x^6 + \cdots \right]$$
$$+ a_1 \left[x - \frac{1}{3} x^3 + \frac{1}{5 \cdot 3} x^5 - \frac{1}{7 \cdot 5 \cdot 3} x^7 + \cdots \right]. \tag{2.22}$$

If we let $y_1(x)$ be the solution with $a_0 = y_1(0) = 1$ and $a_1 = y_1'(0) = 0$, then

$$y_1(x) = 1 - \frac{1}{2} x^2 + \frac{1}{4 \cdot 2} x^4 - \frac{1}{6 \cdot 4 \cdot 2} x^6 + \cdots . \tag{2.23}$$

Similarly, if we let $y_2(x)$ be the solution with $a_0 = y_2(0) = 0$ and $a_1 = y_2'(0) = 1$,

$$y_2(x) = x - \frac{1}{3} x^3 + \frac{1}{5 \cdot 3} x^5 - \frac{1}{7 \cdot 5 \cdot 3} x^7 + \cdots . \tag{2.24}$$

By Proposition 2.14, y_1 and y_2 form a fundamental set of solutions, and by (2.22), the general solution is the linear combination

$$y(x) = a_0 y_1(x) + a_1 y_2(x).$$

For the given initial conditions we have $a_0 = y(0) = 0$ and $a_1 = y'(0) = 2$. The particular solution to this initial value problem is

$$y(x) = 2 y_2(x) = 2 \left[x - \frac{1}{3} x^3 + \frac{1}{5 \cdot 3} x^5 - \frac{1}{7 \cdot 5 \cdot 3} x^7 + \cdots \right]. \qquad \bullet$$

To put the solutions in (2.23) and (2.24) into summation form we need to find a closed form for the coefficients. The first four even-numbered coefficients were computed in (2.20). With a little thought, we see that the general pattern is

$$a_{2k} = (-1)^k \frac{1}{2k \cdot (2k - 2) \cdot \ldots \cdot 2} a_0, \quad \text{for } k \geq 1. \tag{2.25}$$

Note that the $(-1)^k$ in (2.25) causes the sign of the coefficient to alternate. It is positive when k is even (e.g., for a_4 and a_8), and negative when k is odd (e.g., for a_2 and a_6). Next, notice that

$$2k \cdot (2k - 2) \cdot \ldots \cdot 2 = 2^k k!,$$

so we can write

$$a_{2k} = (-1)^k \frac{1}{2^k k!} a_0.$$

By examination we see that this expression is valid for $k = 0$ as well as for $k \geq 1$. Hence, since we have set $a_0 = 1$, we have

$$y_1(x) = \sum_{k=0}^{\infty} (-1)^k \frac{1}{2^k k!} x^{2k}. \tag{2.26}$$

The solution y_1 and some of its partial sums are shown in Figure 2.

The first four odd-numbered coefficients were computed in (2.21). The general pattern is

$$a_{2k+1} = (-1)^k \frac{1}{(2k + 1) \cdot (2k - 1) \cdot \ldots \cdot 3} a_1, \quad \text{for all } k \geq 1. \tag{2.27}$$

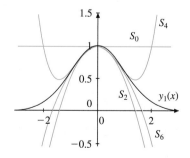

Figure 2. The solution y_1 in (2.26) and some of its partial sums.

To find a more compact expression for the denominator in (2.27), we multiply and divide it by $2k \cdot (2k - 2) \cdot \cdots \cdot 2$ to obtain

$$(2k + 1) \cdot (2k - 1) \cdot \ldots \cdot 3 = \frac{(2k + 1)!}{2k \cdot (2k - 2) \cdot \ldots \cdot 2} = \frac{(2k + 1)!}{2^k \, k!},$$

so

$$a_{2k+1} = (-1)^k \frac{2^k \, k!}{(2k + 1)!} a_1.$$

Once more we see that this expression is valid for $k = 0$ as well as for $k \geq 1$. Hence since we have set $a_1 = 1$, we have

$$y_2(x) = \sum_{k=0}^{\infty} (-1)^k \frac{2^k \, k!}{(2k + 1)!} x^{2k+1}. \tag{2.28}$$

The solution y_2 and some of its partial sums are shown in Figure 3.

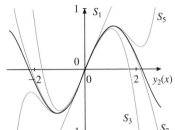

Figure 3. The solution y_2 in (2.28) and some of its partial sums.

Solutions near ordinary points

Examples 2.1 and 2.15 illustrate the general method of solving differential equations with power series. Now we will examine the method in more detail.

In Example 2.15 we computed a series solution to the given differential equation without worrying about whether or not the computed solution converges. However, a formally computed solution is no good if the end result is not a convergent series. Fortunately, the following theorem assures us that solutions do converge.

THEOREM 2.29 Suppose x_0 is an ordinary point for the equation $y'' + P(x)y' + Q(x)y = 0$, so P and Q are analytic at x_0. Then for any numbers y_0 and y_1 there is a power series solution to the initial value problem

$$y'' + P(x)y' + Q(x)y = 0, \quad \text{with} \quad y(x_0) = y_0 \quad \text{and} \quad y'(x_0) = y_1. \tag{2.30}$$

The radius of convergence of the solution about x_0 is at least as large as the minimum of the radii of convergence of the power series for P and Q about x_0. ∎

In Example 2.15 the equation is

$$y'' + xy' + y = 0.$$

Here, $P(x) = x$ and $Q(x) = 1$ are both polynomials and therefore their series expansions have infinite radii of convergence. The solution is analytic at $x_0 = 0$, and its series expansion has an infinite radius of convergence.

Example 2.31 What does Theorem 2.29 allow us to say about the radius of convergence of the power series for solutions to the differential equation

$$(1 + x^2)y'' - 2y = 0 \tag{2.32}$$

about the point $x_0 = 0$?

First, we divide by the leading coefficient $1 + x^2$ to put the differential equation in the form used in Theorem 2.29, obtaining

$$y'' - \frac{2}{1 + x^2} y = 0. \tag{2.33}$$

The coefficient functions $P(x) = 0$ and $Q(x) = -2/(1 + x^2)$. According to Theorem 1.34, Q is analytic at the origin. Since $x^2 + 1 = 0$ for $x = \pm i$, Q has radius of convergence equal to 1. Hence Theorem 2.29 implies that the radius of convergence of a solution about $x_0 = 0$ is at least 1. ●

We will not give a complete proof of Theorem 2.29, but we will find it useful to understand how the differential equation allows us to compute the coefficients of the solution. We will assume that $x_0 = 0$ in order to simplify the computations, but that assumption is of no importance. Since the coefficients P and Q of the equation in (2.30) are analytic at $x_0 = 0$, they have the power series

$$P(x) = \sum_{n=0}^{\infty} p_n x^n \quad \text{and} \quad Q(x) = \sum_{n=0}^{\infty} q_n x^n.$$

The solution will also be analytic, so we set

$$y(x) = \sum_{n=0}^{\infty} a_n x^n,$$

where the coefficients are as yet undetermined. The first two derivatives of the solution are given by

$$y'(x) = \sum_{n=1}^{\infty} n a_n x^{n-1} = \sum_{n=0}^{\infty} (n+1) a_{n+1} x^n \quad \text{and}$$

$$y''(x) = \sum_{n=2}^{\infty} n(n-1) a_n x^{n-2} = \sum_{n=0}^{\infty} (n+2)(n+1) a_{n+2} x^n.$$

According to (1.13) and (1.12), the products that appear in equation (2.30) have the series

$$P(x)y'(x) = \left(\sum_{n=0}^{\infty} p_n x^n \right) \left(\sum_{n=0}^{\infty} (n+1) a_{n+1} x^n \right)$$

$$= \sum_{n=0}^{\infty} \left(\sum_{k=0}^{n} p_{n-k}(k+1) a_{k+1} \right) x^n \quad \text{and}$$

$$Q(x)y(x) = \left(\sum_{n=0}^{\infty} q_n x^n \right) \left(\sum_{n=0}^{\infty} a_n x^n \right)$$

$$= \sum_{n=0}^{\infty} \left(\sum_{k=0}^{n} q_{n-k} a_k \right) x^n.$$

Thus the differential equation becomes

$$0 = y'' + P(x)y' + Q(x)y$$

$$= \sum_{n=0}^{\infty} \left[(n+2)(n+1) a_{n+2} + \sum_{k=0}^{n} p_{n-k}(k+1) a_{k+1} + \sum_{k=0}^{n} q_{n-k} a_k \right] x^n.$$

Setting the coefficients equal to 0 and solving for a_{n+2}, we obtain

$$a_{n+2} = -\frac{\sum_{k=0}^{n}(k+1) p_{n-k} a_{k+1} + \sum_{k=0}^{n} q_{n-k} a_k}{(n+2)(n+1)}, \quad \text{for } n \geq 0. \qquad (2.34)$$

Equation (2.34) is the general **recurrence formula**. Notice that it enables us to compute a_{n+2} provided we already know $a_0, a_1, \ldots, a_{n+1}$. We start with the initial conditions of the solution, $a_0 = y(0)$, and $a_1 = y'(0)$, and the recurrence formula allows us to compute, in order, a_2, a_3, a_4, and so on. In general, we need all of the lower-numbered coefficients in order to compute a_{n+2}. Most specific examples are much simpler. For this reason it is not usually a good idea to use (2.34) in specific cases.

Let's look at two more examples that illustrate aspects of power series solutions.

Example 2.35   Find a fundamental set of solutions for the differential equation

$$(1 - x^3)y'' + 6xy = 0. \tag{2.36}$$

The preceding discussion might suggest that we should look for a series solution using the equation in the form

$$y'' + \frac{6x}{1 - x^3}y = 0.$$

However, that would require finding and then using the power series expansion for $6x/(1 - x^3)$. It is much easier to use the equation in the form given in (2.36), since then we only have to deal with polynomials.

We start by expanding the equation to

$$0 = y'' - x^3 y'' + 6xy. \tag{2.37}$$

As we did in Example 2.15, we express the power series for y'', $x^3 y''$, and $6xy$ in terms of x^n. For y'', this is given in (2.17). For $x^3 y''$, we use the first expression in (2.17), add the terms for $n = 0$ and $n = 1$, which are equal to 0, and then shift the index, to get

$$x^3 y'' = x^3 \sum_{n=0}^{\infty} n(n-1)a_n x^{n-2} = \sum_{n=1}^{\infty} (n-1)(n-2)a_{n-1}x^n.$$

For $6xy$ we only need to shift the index. Making these substitutions, equation (2.37) becomes

$$0 = \sum_{n=0}^{\infty} (n+2)(n+1)a_{n+2}x^n - \sum_{n=1}^{\infty} (n-1)(n-2)a_{n-1}x^n + 6\sum_{n=1}^{\infty} a_{n-1}x^n$$

$$= 2a_2 + \sum_{n=1}^{\infty} \left[(n+2)(n+1)a_{n+2} - \{(n-1)(n-2) - 6\}a_{n-1} \right] x^n$$

$$= 2a_2 + \sum_{n=1}^{\infty} \left[(n+2)(n+1)a_{n+2} - (n-4)(n+1)a_{n-1} \right] x^n.$$

Since all of the coefficients have to be equal to 0, we conclude that $a_2 = 0$, and that we have the recurrence formula

$$a_{n+2} = \frac{n-4}{n+2}a_{n-1}, \quad \text{for } n \geq 1.$$

Notice that the recurrence formula relates coefficients with indices that differ by 3. Since $a_2 = 0$, it follows that $a_5 = a_8 = \cdots = 0$. Thus $a_{3k+2} = 0$ for all $k \geq 0$. Starting with an arbitrary a_0, we see that $a_3 = -a_0$, and $a_6 = 0$. It follows that $a_{3k} = 0$ for all $k \geq 2$. It remains to compute a_{3k+1}. To do so we change the recurrence formula by setting $n + 2 = 3k + 1$, or $n = 3k - 1$, to get

$$a_{3k+1} = \frac{3k-5}{3k+1}a_{3(k-1)+1}, \quad \text{for } k \geq 1.$$

The first few instances are

$$a_4 = -\frac{2}{4}a_1, \quad a_7 = \frac{1}{7}a_4, \quad a_{10} = \frac{4}{10}a_7, \quad \text{and} \quad a_{13} = \frac{7}{13}a_{10}.$$

Multiplying these equations from a_4 to a_{3k+1}, we get

$$a_4 \cdot a_7 \cdot a_{10} \cdot a_{13} \cdot \ldots \cdot a_{3k+1} = -\frac{2}{4}a_1 \cdot \frac{1}{7}a_4 \cdot \frac{4}{10}a_7 \cdot \frac{7}{13}a_{10} \cdot \ldots \cdot \frac{3k-5}{3k+1}a_{3(k-1)+1}.$$

We see that the numerators on the right cancel with the denominators two steps to the left. Hence, after all of the cancellations we get

$$a_{3k+1} = -\frac{2}{(3k-2)(3k+1)}a_1, \quad \text{for } k \geq 1.$$

Notice that this equation is valid for $k = 0$ as well.

Putting everything together, we see that the general solution is

$$y(x) = a_0(1 - x^3) - 2a_1 \sum_{k=0}^{\infty} \frac{x^{3k+1}}{(3k-2)(3k+1)}.$$

The functions

$$y_1(x) = 1 - x^3 \quad \text{and} \quad y_2(x) = -2\sum_{k=0}^{\infty} \frac{x^{3k+1}}{(3k-2)(3k+1)}$$

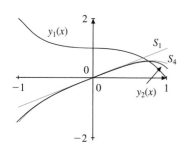

are a fundamental set of solutions. These solutions are shown in Figure 4, together with the partial sums $S_1(x) = x$ and $S_4(x) = x - x^4/2$ for y_2. Notice that by the ratio test, the radius of convergence for y_2 is $R = 1$, and that S_4 is already a good approximation for y_2 over the entire interval of convergence.   ●

Figure 4. The solutions y_1 and y_2 in Example 2.35, together with two partial sums for y_2.

Our final example shows that sometimes we cannot find an exact formula for the coefficients.

Example 2.38   Find the recurrence formula for the equation

$$(1 + x)y'' + y = 0,$$

and use it to find the first five terms of the power series for two linearly independent solutions.

Set $y(x) = \sum_{n=0}^{\infty} a_n x^n$. Notice that we can write $xy'' = \sum_{n=0}^{\infty}(n+1)na_{n+1}x^n$. Hence, substituting the power series for y'', xy'', and y expressed in terms of x^n, we get

$$0 = y'' + xy'' + y$$

$$= \sum_{n=0}^{\infty}(n+2)(n+1)a_{n+2}x^n + \sum_{n=0}^{\infty}(n+1)na_{n+1}x^n + \sum_{n=0}^{\infty}a_n x^n$$

$$= \sum_{n=0}^{\infty}[(n+2)(n+1)a_{n+2} + (n+1)na_{n+1} + a_n]x^n.$$

Since all of the coefficients must be equal to 0, we get the recurrence formula

$$a_{n+2} = -\frac{n(n+1)a_{n+1} + a_n}{(n+2)(n+1)}, \quad \text{for } n \geq 0. \tag{2.39}$$

This recurrence formula expresses a_{n+2} in terms of the previous two coefficients. In theory it allows us to compute all of the coefficients and to find two linearly independent solutions. However, this recurrence formula is much more

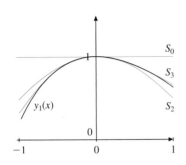

Figure 5. The solution y_1 in Example 2.38, together with some partial sums.

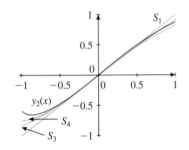

Figure 6. The solution y_2 in Example 2.38, together with some partial sums.

difficult to solve than those in Examples 2.15 and 2.35. Those were **one-step** recurrence formulas, since they expressed a_{n+2} as a multiple of one previously found coefficient. The formula in (2.39) is a **two-step** recurrence formula. We have described methods to find closed-form expressions for the coefficients when they are given by a one-step recursion, but this is a very difficult problem when the recursion is multistep.

However, we can use the recurrence formula to compute as many coefficients as we wish. For example, for the solution y_1 which satisfies $a_0 = y_1(0) = 1$, and $a_1 = y_1'(0) = 0$, we can compute in order $a_2 = -1/2$, $a_3 = 1/6$, and $a_4 = -1/24$. Thus

$$y_1(x) = 1 - \frac{x^2}{2} + \frac{x^3}{6} - \frac{x^4}{24} + \cdots .$$

Similarly, for the solution y_2, which satisfies $a_0 = y_2(0) = 0$, and $a_1 = y_2'(0) = 1$, we can compute in order $a_2 = 0$, $a_3 = -1/6$, and $a_4 = 1/12$. Thus

$$y_2(x) = x - \frac{x^3}{6} + \frac{x^4}{12} + \cdots . \tag{2.40}$$

If we write the differential equation in the form

$$y'' + \frac{1}{1+x} y = 0,$$

then using Theorem 2.29 we see that the radius of convergence of the power series solutions is at least 1. The coefficient $1/(1+x)$ leads us to believe that it is impossible to solve the equation past $x = -1$, and that the radius of convergence is exactly 1. In Figures 5 and 6 we have plotted the solutions y_1 and y_2 computed using a differential equation solver. In each case we have also plotted some partial sums. Clearly the convergence is quite rapid over the entire interval of convergence. ●

EXERCISES

In Exercises 1–14, solve the given equation exactly using a technique from a previous chapter. Then find a power series solution and verify that it is the series expansion of the exact solution.

1. $y' = 3y$

2. $y' + 2y = 0$

3. $y' = -x^3 y$

4. $y' = x^2 y$

5. $(1-x)y' + y = 0$

6. $(1+x)y' - y = 0$

7. $(x-4)y' + y = 0$

8. $(2-x)y' = y$

9. $(2-x)y' + 2y = 0$

10. $(x-3)y' + 2y = 0$

11. $y'' = y'$

12. $y'' = 9y$

13. $y'' + y = 0$

14. $y'' + 4y = 0$

In Exercises 15–20, verify that $x_0 = 0$ is an ordinary point of the given differential equation. Then find two linearly independent solutions to the differential equation valid near $x_0 = 0$. Estimate the radius of convergence of the solutions.

15. $y'' + x^2 y = 0$

16. $y'' + xy = 0$

17. $y'' + 2xy' - y = 0$

18. $y'' + xy' + 2y = 0$

19. $y'' + xy' - y = 0$

20. $(1+x^2)y'' + 2xy' - 2y = 0$

In Exercises 21–24, verify that $x_0 = 0$ is an ordinary point. Find S_4, the partial sum of order 4 for two linearly independent solutions. Estimate the radius of convergence of the solutions.

21. $(1+x)y'' + y = 0$.

22. $(1+x^2)y'' + y' - 2y = 0$.

23. $y'' - (\cos x)y = 0$.

24. $y'' + (\sin x)y' + y = 0$.

In Exercises 25–34, plot the two solutions found in the given exercise together with the first three distinct partial sums found in that exercise. Plot over the designated interval.

25. Exercise 15, $[-5, 5]$

26. Exercise 15, $[-2, 7]$

27. Exercise 17, $[-3, 3]$

28. Exercise 17, $[-4, 4]$

29. Exercise 19, $[-4, 4]$

30. Exercise 19, $[-4, 4]$

31. Exercise 21, $[-1, 1]$

32. Exercise 22, $[-1, 1]$

33. Exercise 23, $[-4, 4]$. For y_1 plot S_0 and S_4. For y_2 plot S_1 and S_4.

34. Exercise 24, $[-4, 4]$. For y_2 plot S_1 and S_3.

35. Show that $y(x) = (1+x)^p$, where p is an arbitrary constant, is the unique solution of the initial value problem

$$(1+x)y' = py, \quad y(0) = 1.$$

Find a series solution of the initial value problem, showing that

$$(1 + x)^p = 1 + px + \frac{p(p - 1)}{2!} x^2 + \cdots$$
$$+ \frac{p(p - 1) \cdots (p - n + 1)}{n!} x^n + \cdots .$$

Notice that if p is a positive integer, the series truncates at $n = p$, and the result is the binomial formula. The series for general p is called the **binomial series.** What is the radius of convergence of the binoimal series?

36. Show that the solution $y_1(x)$ to the equation in Example 2.15 has the closed form $y_1(x) = e^{-x^2/2}$.

11.3 Legendre's Equation

A good example of the results of Theorem 2.29 is Legendre's equation of order n (see equation (2.9)),

$$(1 - x^2)y'' - 2xy' + n(n + 1)y = 0, \tag{3.1}$$

where n is a parameter. Not only will the computation illustrate the theory, but it will inform us about the Legendre functions and polynomials.

We have already noticed that the origin, $x_0 = 0$, is an ordinary point for this equation, so by Theorem 2.29, we know that we can find power series solutions. We will leave it to Exercise 8 to show that the solutions have a radius of convergence of at least 1.

Since n is a parameter in Legendre's equation, we will use k as the index of summation, and look for a power series solution of the form $y(x) = \sum_{k=0}^{\infty} a_k x^k$. The expanded version of equation (3.1) is

$$y'' - x^2 y'' - 2xy' + n(n + 1)y = 0.$$

Following the method of the previous section, we want to express the power series for y'', $x^2 y''$, xy', and y in terms of x^k. Starting with the series in (2.16) and (2.17), we see that

$$y''(x) = \sum_{k=0}^{\infty}(k + 2)(k + 1)a_{k+2}x^k \qquad -x^2 y''(x) = \sum_{k=0}^{\infty} -k(k - 1)a_k x^k$$

$$-2x\, y'(x) = \sum_{k=0}^{\infty} -2ka_k x^k \qquad n(n + 1)\, y(x) = \sum_{k=0}^{\infty} n(n + 1)a_k x^k.$$

Adding these together, we get

$$0 = \sum_{k=0}^{\infty} \left[(k + 2)(k + 1)a_{k+2} + (n(n + 1) - k(k - 1) - 2k)a_k\right] x^k$$

$$= \sum_{k=0}^{\infty} \left[(k + 2)(k + 1)a_{k+2} + (n - k)(n + k + 1)a_k\right] x^k.$$

By the identity theorem, all of the coefficients must be equal to 0. Solving for a_{k+2}, we get the recurrence formula

$$a_{k+2} = -\frac{(n - k)(n + k + 1)}{(k + 2)(k + 1)}a_k, \quad \text{for } k \geq 0. \tag{3.2}$$

Notice that a_{k+2} is determined by a_k. Hence all of the even-numbered coefficients are determined by a_0, and all of the odd-numbered coefficients are determined by a_1.

The first three even-numbered terms are

$$a_2 = -\frac{n \cdot (n+1)}{2 \cdot 1} a_0$$

$$a_4 = -\frac{(n-2) \cdot (n+3)}{4 \cdot 3} a_2$$

$$= \frac{n \cdot (n-2) \cdot (n+1) \cdot (n+3)}{4!} a_0$$

$$a_6 = -\frac{(n-4) \cdot (n+5)}{6 \cdot 5} a_4$$

$$= -\frac{n \cdot (n-2) \cdot (n-4) \cdot (n+1) \cdot (n+3) \cdot (n+5)}{6!} a_0.$$

From these computations we can see that the general pattern is

$$a_{2k} = (-1)^k C_{2k} \, a_0,$$

where $C_0 = 1$ and

$$C_{2k} = \frac{n \cdot (n-2) \cdot \ldots \cdot (n-2k+2) \cdot (n+1) \cdot (n+3) \cdot \ldots \cdot (n+2k-1)}{(2k)!}$$

$$(3.3)$$

for $k \geq 1$.

The first three odd-numbered coefficients are

$$a_3 = -\frac{(n-1) \cdot (n+2)}{3 \cdot 2} a_1$$

$$a_5 = -\frac{(n-3) \cdot (n+4)}{5 \cdot 4} a_3$$

$$= \frac{(n-1) \cdot (n-3) \cdot (n+2) \cdot (n+4)}{5!} a_1$$

$$a_7 = -\frac{(n-5) \cdot (n+6)}{7 \cdot 6} a_5$$

$$= -\frac{(n-1) \cdot (n-3) \cdot (n-5) \cdot (n+2) \cdot (n+4) \cdot (n+6)}{7!} a_1.$$

The general pattern for the odd-numbered coefficient is

$$a_{2k+1} = (-1)^k C_{2k+1} \, a_1,$$

where $C_1 = 1$ and

$$C_{2k+1} = \frac{(n-1) \cdot (n-3) \cdot \ldots \cdot (n-2k+1) \cdot (n+2) \cdot (n+4) \cdot \ldots \cdot (n+2k)}{(2k+1)!}$$

$$(3.4)$$

for $k \geq 1$.

For the particular solution $y_1(x)$ with initial conditions $a_0 = y_1(0) = 1$, and $a_1 = y_1'(0) = 0$, we have

$$y_1(x) = \sum_{k=0}^{\infty} (-1)^k C_{2k} x^{2k}$$

$$(3.5)$$

$$= 1 - \frac{n \cdot (n+1)}{2!} x^2 + \frac{n \cdot (n-2) \cdot (n+1) \cdot (n+3)}{4!} x^4 - \cdots .$$

For the particular solution $y_2(x)$ with initial conditions $a_0 = y_2(0) = 0$, and $a_1 = y_2'(0) = 1$, we have

$$y_2(x) = \sum_{k=0}^{\infty}(-1)^k C_{2k+1} x^{2k+1}$$

$$= x - \frac{(n-1)\cdot(n+2)}{3!}x^3 + \frac{(n-1)\cdot(n-3)\cdot(n+2)\cdot(n+4)}{5!}x^5 - \cdots.$$

$$(3.6)$$

By Proposition 2.14, y_1 and y_2 are linearly independent. Hence the general solution is the linear combination

$$y(x) = a_0 y_1(x) + a_1 y_2(x).$$

Legendre polynomials

Suppose that the parameter n is a nonnegative integer. If $n = 2l$ is even and $k > l$, then the formula for C_{2k} in (3.3) contains the factor $n - 2l = 0$. Hence $C_{2k} = 0$ for $k > l$, and the infinite series for y_1 in (3.5) truncates to the polynomial

$$y_1(x) = \sum_{k=0}^{l}(-1)^k C_{2k} x^{2k}.$$

Notice that y_1 is a polynomial of degree $n = 2l$, and contains only even powers of x. In the same way we see that if $n = 2l + 1$ is an odd integer, then y_2 is an odd polynomial of degree $n = 2l + 1$, given by the truncated series

$$y_2(x) = \sum_{k=0}^{l}(-1)^k C_{2k+1} x^{2k+1}.$$

We set

$$P_n(x) = \begin{cases} \dfrac{y_1(x)}{y_1(1)}, & \text{for } n \text{ even,} \\[2mm] \dfrac{y_2(x)}{y_2(1)}, & \text{for } n \text{ odd.} \end{cases}$$

$$(3.7)$$

Each P_n is a polynomial of degree n and is a solution to Legendre's equation of order n. It is normalized by (3.7) so that $P_n(1) = 1$. P_n is called the **Legendre polynomial of degree** n. Notice that P_n is an even polynomial if n is even and an odd polynomial if n is odd. The first few Legendre polynomials are

$$P_0(x) = 1, \qquad\qquad\qquad P_1(x) = x,$$
$$P_2(x) = (3x^2 - 1)/2, \qquad\qquad P_3(x) = (5x^3 - 3x)/2, \qquad (3.8)$$
$$P_4(x) = (35x^4 - 30x^2 + 3)/8, \qquad P_5(x) = (63x^5 - 70x^3 + 15x)/8.$$

These polynomials are plotted in Figure 1.

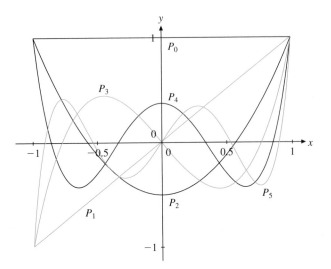

Figure 1. The first six Legendre polynomials.

Properties of Legendre polynomials

There are a number of properties of Legendre polynomials that are useful. We will list them here without proof.

- A very important fact for use in studying partial differential equations is that if $y(x)$ is a nonzero solution to Legendre's equation of order n which is bounded at both ± 1, then the order n must be an integer and y must be a multiple of the Legendre polynomial P_n.

- The series expansions for the polynomials are not very revealing. Here is one that might be more appealing;

$$P_n(x) = \frac{1}{2^n} \sum_{k=0}^{[n/2]} \frac{(-1)^k (2n - 2k)!}{k!(n-k)!(n-2k)!} x^{n-2k}.$$

  In this formula $[n/2]$ stands for the greatest integer smaller than or equal to $n/2$.

- Next we have **Rodrigues's formula**,

$$P_n(x) = \frac{1}{2^n n!} \frac{d^n}{dx^n} (x^2 - 1)^n. \tag{3.9}$$

- The **recursion formula**

$$P_n(x) = \frac{1}{n} \left[(2n-1)x P_{n-1}(x) - (n-1) P_{n-2}(x) \right] \tag{3.10}$$

  is often useful for computing the polynomials.

- There is a variety of other identities connecting the polynomials and their first derivatives. These do not have names, so we will just list them.

$$\begin{aligned}
P_n' &= n P_{n-1} + x P_{n-1}', & n P_n &= n x P_{n-1} + (x^2 - 1) P_{n-1}', \\
x P_n' - n P_n &= P_{n-1}', & (1 - x^2) P_n' + n x P_n &= n P_{n-1}.
\end{aligned} \tag{3.11}$$

- The polynomial P_n has n real roots, all of them in the interval $(-1, 1)$.

- The polynomials satisfy $|P_n(x)| \le 1$ for $-1 \le x \le 1$. They have the values $P_n(1) = 1$, and $P_n(-1) = (-1)^n$ at the endpoints, while

$$P_{2n+1}(0) = 0 \quad \text{and} \quad P_{2n}(0) = (-1)^n \frac{(2n)!}{2^{2n}(n!)^2}. \tag{3.12}$$

EXERCISES

1. In the following steps we will verify that

$$g(t, x) = \frac{1}{\sqrt{1 - 2xt + t^2}} = \sum_{n=0}^{\infty} P_n(x)t^n. \tag{3.13}$$

Because of this identity, $g(t, x)$ is called the **generating function** for the Legendre polynomials. Notice that g is an analytic function of t at $t = 0$.

(a) Show that there are functions $Q_n(x)$ such that

$$g(t, x) = \sum_{n=0}^{\infty} Q_n(x)t^n, \quad \text{for } |t| < 1. \tag{3.14}$$

Verify that $Q_n(x)$ is a polynomial. (*Hint*: Use the binomial series (see Exercise 35 in Section 2).)

(b) Show by direct substitution that $g(t, x) = (1 - 2xt + t^2)^{-1/2}$ satisfies the partial differential equation

$$\frac{\partial}{\partial x}\left[(1 - x^2)\frac{\partial g}{\partial x}\right] + t\frac{\partial^2}{\partial t^2}[tg] = 0. \tag{3.15}$$

(c) Substitute the series in (3.14) into (3.15) to show that

$$\sum_{n=0}^{\infty}\left\{\left[(1 - x^2)Q_n'(x)\right]' + n(n+1)Q_n(x)\right\}t^n = 0. \tag{3.16}$$

(d) Conclude from (3.16) that Q_n is a solution to Legendre's equation of order n.

(e) Notice that when $x = 1$ we have

$$g(t, 1) = \frac{1}{1 - t} = \sum_{n=0}^{\infty} t^n,$$

and, as a result, $Q_n(1) = 1$. Conclude that $Q_n(x) = P_n(x)$, the Legendre polynomial of order n.

2. Establish the recursion formula using the following two steps.

(a) Differentiate both sides of equation (3.13) with respect to t to show that

$$(x - t)\sum_{n=0}^{\infty} P_n(x)t^n = (1 - 2xt + t^2)\sum_{n=1}^{\infty} nP_n(x)t^{n-1}.$$

(b) Equate the coefficients of t^n in this equation to show that

$$P_1(x) = xP_0(x), \quad \text{and}$$

$$(n+1)P_{n+1}(x) = (2n+1)xP_n(x)$$
$$- nP_{n-1}(x), \quad \text{for } n \ge 1.$$

(c) Assuming that $P_0(x) = 1$, use the recurrence formulas in part (b) to derive the formulas in (3.8).

3. Show that

$$P_{2n+1}(0) = 0 \quad \text{and} \quad P_{2n}(0) = (-1)^n \frac{(2n)!}{2^{2n}(n!)^2}.$$

4. Show that $P_n'(1) = (-1)^{n+1}P_n'(-1) = n(n+1)/2$. (*Hint*: Use Legendre's equation (3.1).)

5. The differential equation $y'' + xy = 0$ is called **Airy's equation**, and its solutions are called **Airy functions**. Find series for the solutions y_1 and y_2 where $y_1(0) = 1$ and $y_1'(0) = 0$, while $y_2(0) = 0$ and $y_2'(0) = 1$. What is the radius of convergence for these two series?

6. The differential equation $(1 - x^2)y'' - xy' + p^2y = 0$ is called **Chebychev's equation**.

(a) Find series for the solutions y_1 and y_2, where $y_1(0) = 1$ and $y_1'(0) = 0$, while $y_2(0) = 0$ and $y_2'(0) = 1$. What is the radius of convergence for these two series?

(b) Show that when p is a nonnegative integer, either y_1 or y_2 is a polynomial of degree p.

(c) For integer $p > 0$, the **Chebychev polynomial of degree** p is the polynomial solution of degree p to Chebychev's equation multiplied by a suitable constant to ensure that the coefficient of x^p is 2^{p-1}. It is denoted by $T_p(x)$. By convention we set $T_0(x) = 1$. Find the first six Chebychev polynomials. Plot them on one figure.

7. The differential equation $y'' - 2xy' + py = 0$ is called **Hermite's equation**.

(a) Find series for the solutions y_1 and y_2, where $y_1(0) = 1$ and $y_1'(0) = 0$, while $y_2(0) = 0$ and $y_2'(0) = 1$. What is the radius of convergence for these two series?

(b) Show that when p is a nonnegative, even integer, either y_1 or y_2 is a polynomial of degree $p/2$.

(c) For integer $n \ge 0$, the **Hermite polynomial of degree** n is the polynomial solution of degree n to Hermite's equation with $p = 2n$, multiplied by a suitable constant to insure that the coefficient of x^n is 2^n. It is denoted by $H_n(x)$. Find the first six Hermite polynomials. Plot them on one figure.

8. Use Theorem 2.29 to show that the solutions to Legendre's equation have a radius of convergence of at least 1.

11.4 Types of Singular Points—Euler's Equation

In the previous two sections we were completely successful in finding series solutions for differential equations at ordinary points. We also introduced Bessel's equation,

$$x^2 y'' + x y' + (x^2 - r^2)y = 0, \tag{4.1}$$

where r is a nonnegative constant. The point $x = 0$ is the only singular point for Bessel's equation. Using our techniques, we could find series solutions to Bessel's equation centered at any point $x_0 \neq 0$. You might well wonder why that would not suit our purposes.

In some applications of Bessel's equation the variable x will be the distance of a point from the origin in polar coordinates. It will be very important to understand how the solution behaves when x is close to 0, and the point is close to the origin. A series solution with center at a nonzero point, say at $x_0 = 1$, could be shown to have radius of convergence equal to 1, but, as Example 1.9 shows, it would still be very difficult to say what happens as x gets close to 0, since it is an endpoint of the interval of convergence. In this section and in the next two sections we will find slightly different methods that will allow us to find series expansions for the solutions at singular points, so we will be able to study the behavior of the solutions near such points. In particular, we will be able to do so for solutions to Bessel's equation.

Types of singular points

We will continue to consider equations of the form

$$y'' + P(x)y' + Q(x)y = 0. \tag{4.2}$$

We will be looking at the equation near a singular point x_0, which means that one or both of $P(x)$ and $Q(x)$ fails to be analytic at x_0. However, the point x_0 cannot be too singular. The limitations are just those which will allow the solution method we will describe in the next section to work, and they are made precise in the next definition.

DEFINITION 4.3

A singular point x_0 for equation (4.2) is called a *regular singular point* if

$$(x - x_0)P(x) \quad \text{and} \quad (x - x_0)^2 Q(x)$$

are both analytic at x_0. A singular point that is not regular is called an *irregular singular point*.

We can write Bessel's equation (4.1) in the form of (4.2) by dividing by x^2, getting

$$y'' + \frac{1}{x}y' + \frac{x^2 - r^2}{x^2}y = 0.$$

Thus $P(x) = 1/x$ and $Q(x) = (x^2 - r^2)/x^2$, so $xP(x) = 1$ and $x^2 Q(x) = x^2 - r^2$ are both analytic at $x_0 = 0$. Thus $x_0 = 0$ is a regular singular point for Bessel's equation.

Example 4.4 Find the singular points of

$$(x^2 - 1)^2 y'' - (x - 1)y' + 3y = 0$$

and classify them as regular or irregular.

To find the singular points we rewrite the equation in the standard form of (4.2) by dividing by the leading coefficient $(x^2 - 1)^2 = (x - 1)^2(x + 1)^2$. We get

$$y'' - \frac{1}{(x - 1)(x + 1)^2} y' + \frac{3}{(x - 1)^2(x + 1)^2} y = 0.$$

In this case,

$$P(x) = -\frac{1}{(x - 1)(x + 1)^2} \quad \text{and} \quad Q(x) = \frac{3}{(x - 1)^2(x + 1)^2}.$$

Both of these functions are analytic everywhere except where the denominators are zero. Thus the singular points are $x_0 = \pm 1$. For $x_0 = 1$, we have

$$(x - 1)P(x) = -\frac{1}{(x + 1)^2} \quad \text{and} \quad (x - 1)^2 Q(x) = \frac{3}{(x + 1)^2},$$

which are analytic near $x_0 = 1$. Therefore, $x_0 = 1$ is classified as a regular singular point.

For $x_0 = -1$, we have

$$(x + 1)P(x) = -\frac{1}{(x - 1)(x + 1)} \quad \text{and} \quad (x + 1)^2 Q(x) = \frac{3}{(x - 1)^2}.$$

Note that $(x + 1)P(x)$ fails to be analytic at $x_0 = -1$ due to the presence of the factor of $x + 1$ in the denominator. Therefore, $x_0 = -1$ is an irregular singular point. There is no need to even check $(x + 1)^2 Q(x)$ (which happens to be analytic at $x_0 = -1$). ●

We can put equation (4.2) into another useful form if we multiply it by $(x - x_0)^2$, and use the notation $p(x) = (x - x_0)P(x)$ and $q(x) = (x - x_0)^2 Q(x)$. The equation becomes

$$(x - x_0)^2 y'' + (x - x_0)p(x)y' + q(x)y = 0. \tag{4.5}$$

When we put the equation into this form, we see that x_0 is a regular singular point if and only if the functions $p(x)$ and $q(x)$ are analytic at x_0. Notice that Bessel's equation in (4.1) is in this form, with $p(x) = 1$, and $q(x) = x^2 - r^2$, so in this format it is easily seen that $x_0 = 0$ is a regular singular point for Bessel's equation.

Example 4.6   We pointed out in Section 11.2 that $x_0 = \pm 1$ are singular points for Legendre's equation

$$(1 - x^2)y'' - 2xy' + n(n + 1)y = 0. \tag{4.7}$$

Are these singular points regular?

Let's look at $x_0 = 1$. We will put the equation into the form (4.5). To do so we first notice that the leading term $1 - x^2 = (1 - x)(1 + x)$. Consequently, if we divide (4.7) by $1 + x$ and multiply it by $1 - x$, we get

$$(x - 1)^2 y'' + \frac{2x}{1 + x}(x - 1)y' + n(n + 1)\frac{1 - x}{1 + x} y = 0.$$

This is in the form of (4.5) with $p(x) = 2x/(1+x)$ and $q(x) = n(n+1)(1-x)/(1+x)$. Since both of these are analytic at $x_0 = 1$, we see that $x_0 = 1$ is a regular singular point. In a similar way, we see that $x_0 = -1$ is also a regular singular point. ●

Euler's equation

The easiest example of an equation in the form of (4.5) with a regular singular point at x_0 is when both functions p and q are constants. Its solution will provide some motivation for the solution technique we describe in the next section. Without loss of generality we can let $x_0 = 0$. Then the equation becomes

$$x^2 y'' + pxy' + qy = 0. \tag{4.8}$$

This equation is called **Euler's equation**. Its distinguishing feature is that every time the function y is differentiated, it is multiplied by x. Thus we have the terms xy' and $x^2 y''$.

In thinking about possible solutions for Euler's equation, we are led to consider power functions, since if $y(x) = x^s$, then differentiating y reduces the power by one, while multiplying by x restores the power. More precisely, we have

$$xy'(x) = sx^s \quad \text{and} \quad x^2 y''(x) = s(s-1)x^s,$$

so all of the terms in Euler's equation (4.8) are multiples of x^s. Making the substitutions, we get

$$\begin{aligned} x^2 y'' + pxy' + qy &= [s(s-1) + ps + q]x^s \\ &= [s^2 + (p-1)s + q]x^s. \end{aligned} \tag{4.9}$$

Thus $y(x) = x^s$ is a solution provided that

$$I(s) = s^2 + (p-1)s + q = 0. \tag{4.10}$$

The polynomial $I(s)$ is called the **indicial polynomial**, and equation (4.10) is called the **indicial equation**. Since it is a quadratic equation, it generally has two roots, s_1 and s_2, and $I(s)$ factors into $I(s) = (s - s_1)(s - s_2)$. The roots could be complex, but here we will only treat the case when s_1 and s_2 are real. The complex case will be left to Exercise 23. Each real root leads to a solution of Euler's equation. Thus

$$y_1(x) = x^{s_1} \quad \text{and} \quad y_2(x) = x^{s_2}$$

are both solutions for $x > 0$.[1] If the roots are real and different, then we have two solutions which are linearly independent since they are not constant multiples of each other. Therefore, the general solution is

$$y(x) = C_1 y_1(x) + C_2 y_2(x) = C_1 x^{s_1} + C_2 x^{s_2}.$$

Example 4.11 Find all solutions to the differential equation

$$x^2 y'' + 4xy' + 2y = 0. \tag{4.12}$$

This is Euler's equation with $p = 4$ and $q = 2$. The indicial equation is

$$I(s) = s(s-1) + 4s + 2 = s^2 + 3s + 2 = (s+1)(s+2) = 0.$$

Since the indicial equation has roots -1 and -2, $y_1(x) = x^{-1}$ and $y_2(x) = x^{-2}$ are solutions to (4.12). Note that y_1 and y_2 are linearly independent, since neither is a constant multiple of the other. This fundamental set of solutions is plotted in Figure 1. Therefore, any solution to this differential equation on the interval $x > 0$ is given by

$$y = C_1 y_1 + C_2 y_2 = C_1 x^{-1} + C_2 x^{-2}.$$

Since any nonzero solution is unbounded at the origin, no initial conditions can be specified there. However, initial conditions can be specified at any other value of x.

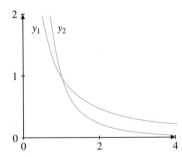

Figure 1. The fundamental set of solutions found in Example 4.11.

---

[1] Notice that x^s is not defined for $x < 0$ unless s is an integer or $s = p/q$ where q is odd. Therefore, we limit the domain of the solution to $x > 0$. For $x < 0$, the solution is $|x|^s$. See Theorem 4.20.

Operator notation

You are well aware that differentiation is an operation that takes a function $f(x)$ and yields another function, df/dx. This is an example of an **operator**. By definition an operator inputs a function and outputs another function.

Another example of an operator is multiplication by a function. Thus, if $P(x)$ is a function, then

$$f(x) \rightarrow Pf(x) = P(x) \cdot f(x)$$

is an operator. Both of the operators d/dx and P have the special property that they are **linear operators**. By definition, this means that if a and b are constants, then

$$\frac{d}{dx}(af + bg) = a\frac{df}{dx} + b\frac{dg}{dx} \quad \text{and} \quad P(af + bg) = aPf + bPg.$$

This property is easily checked.

From these two basic operators we can construct more complicated examples of linear operators. Thus if P and Q are functions, we can define the linear operator

$$L = \frac{d^2}{dx^2} + P(x)\frac{d}{dx} + Q(x). \tag{4.13}$$

When applied to a function y, we have

$$Ly = \frac{d^2y}{dx^2} + P(x)\frac{dy}{dx} + Q(x)y.$$

It is easily checked that the operator L is again linear.

Most important for our use of operator notation is that, using (4.13), equation (4.2) can be written succinctly as

$$Ly = 0.$$

We will find it convenient to use operator notation in what follows. You should look back over the beginning of this section to see how many of the equations can be rewritten using operator notation. Operator notation will relieve us from the necessity of writing out the formulas in detail.

In particular, if p and q are constants, and we introduce the operator

$$L = x^2\frac{d^2}{dx^2} + px\frac{d}{dx} + q, \tag{4.14}$$

then Euler's equation (4.8) can be written as $Ly = 0$. We can rewrite (4.9) using this operator. However, let's consider y to be a function of both x and s, so that

$$y(s, x) = x^s.$$

Using the fact that $I(s) = (s - s_1)(s - s_2)$, formula (4.9) becomes

$$Ly(s, x) = I(s)x^s = (s - s_1)(s - s_2)x^s. \tag{4.15}$$

Since $I(s_1) = I(s_2) = 0$, we can conclude immediately that $y(s_1, x)$ and $y(s_2, x)$ are solutions to Euler's equation, as we saw earlier.

When the indicial equation has a double root

If the indicial equation has a double root, then $s_2 = s_1$, so $I(s) = (s - s_1)^2$, and (4.15) becomes

$$Ly(s, x) = (s - s_1)^2 x^s. \tag{4.16}$$

We still have one solution,

$$y_1(x) = y(s_1, x) = x^{s_1}.$$

We need to find a second solution, and we do so by differentiating (4.16) with respect to s.

Notice that the operator L in (4.14) involves only the variable x. As a result, the order of applying L and $\partial/\partial s$ is unimportant. Therefore, using (4.16), we have

$$L\frac{\partial y}{\partial s} = \frac{\partial}{\partial s} Ly = \frac{\partial}{\partial s}[(s - s_1)^2 x^s].$$

We remember that $x^s = e^{s \ln x}$, so $\partial x^s/\partial s = x^s \ln x$. Therefore, the equation becomes

$$L\frac{\partial y}{\partial s} = 2(s - s_1)x^s + (s - s_1)^2 x^s \ln x. \tag{4.17}$$

Finally, we evaluate (4.17) at $s = s_1$. On the right-hand side we get 0. Therefore, if we set

$$y_2(x) = \frac{\partial y}{\partial s}(s_1, x) = \frac{\partial x^s}{\partial s}\bigg|_{s=s_1} = x^{s_1} \ln x, \tag{4.18}$$

equation (4.17) becomes

$$Ly_2 = 0.$$

Since $y_1(x) = x^{s_1}$ and $y_2(x) = x^{s_1} \ln x$ are not constant multiples of each other, they are linearly independent.

Example 4.19 Find the general solution to the equation

$$x^2 y'' + 3xy' + y = 0.$$

Find the particular solution satisfying the initial conditions $y(1) = 4$ and $y'(1) = -2$.

The indicial equation is

$$I(s) = s(s - 1) + 3s + 1 = s^2 + 2s + 1 = (s + 1)^2 = 0.$$

Since $s_1 = -1$ is double root, two linearly independent solutions are

$$y_1(x) = x^{-1} \quad \text{and} \quad y_2(x) = x^{-1} \ln x.$$

This fundamental set of solutions is shown in Figure 2. The general solution is

$$y(x) = C_1 x^{-1} + C_2 x^{-1} \ln x.$$

To find the solution satisfying the given initial conditions, we first differentiate y, obtaining

$$y'(x) = -C_1 x^{-2} - C_2 x^{-2} \ln x + C_2 x^{-2}.$$

The initial conditions, $y(1) = 4$ and $y'(1) = -2$, become

$$4 = y(1) = C_1$$
$$-2 = y'(1) = -C_1 + C_2.$$

The solutions to these equations are $C_1 = 4$ and $C_2 = 2$. The particular solution satisfying the initial conditions is

$$y(x) = 4x^{-1} + 2x^{-1} \ln x. \qquad \bullet$$

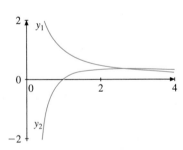

Figure 2. The fundamental set of solutions found in Example 4.19.

The general result

We are ready to summarize our results about Euler's equation. However, there is one point to clear up. In the preceding discussion we have tacitly assumed that $x > 0$ and have found the solutions accordingly. But what about negative values of x? Our theorem will include the results for negative x without proof. Exercise 18 will address the issue.

THEOREM 4.20   Consider the Euler equation

$$(x - x_0)^2 y'' + p(x - x_0)y' + qy = 0 \tag{4.21}$$

and the associated indicial equation $I(s) = s(s-1) + ps + q = s^2 + (p-1)s + q = 0$.

1. If the indicial equation has two distinct real roots s_1 and s_2, then a fundamental set of solutions defined for $x \neq x_0$ is

$$y_1(x) = |x - x_0|^{s_1} \quad \text{and} \quad y_2(x) = |x - x_0|^{s_2}.$$

2. If the indicial equation has one root s_1 of multiplicity 2, then

$$y_1(x) = |x - x_0|^{s_1} \quad \text{and} \quad y_2(x) = |x - x_0|^{s_1} \ln|x - x_0|$$

is a fundamental set of solutions defined for $x \neq x_0$. ∎

Theorem 4.20 does not discuss what happens when the indicial equation has complex roots. This will be treated in Exercise 23.

EXERCISES

In Exercises 1–9, classify each singular point of the given equation.

1. $x^2 y'' + 4xy' - 2xy = 0$

2. $x^2(1-x)y'' + (x-2)y' - 3xy = 0$

3. $x^2(2-x)^2 y'' + 4xy' + 5y = 0$

4. $(x^2 - 4)y'' - (x-2)y' + 2y = 0$

5. $(t^2 - t - 6)y'' + (t^2 - 9)y' + 2y = 0$

6. $(4 - x^2)^2 y'' + x(2-x)y' + (2+x)y = 0$

7. $(t^2 - 1)y'' + (t-2)y' + y = 0$

8. $(x - \pi)^2 y'' + (\sin x)y' + (\cos x)y = 0$

9. $(\sin^2 t)y'' + ty' - 3y = 0$

In Exercises 10–17, find the general solution to each example of Euler's equation.

10. $x^2 y'' - 3xy' - 5y = 0$ **11.** $x^2 y'' + 5xy' + 3y = 0$

12. $x^2 y'' + 3xy' + y = 0$ **13.** $-x^2 y'' + xy' - y = 0$

14. $x^2 y'' - 2xy' - 4y = 0$ **15.** $x^2 y'' + 9xy' + 16y = 0$

16. $2x^2 y'' + 10xy' + 8y = 0$ **17.** $x^2 y'' - 5xy' + 9y = 0$

18. Show that the change of variables $\tilde{x} = -x$ in Euler's equation

$$x^2 y'' + pxy' + qy = 0 \tag{4.22}$$

yields the equation

$$\tilde{x}^2 \frac{d^2 y}{d\tilde{x}^2} + p\tilde{x}\frac{dy}{d\tilde{x}} + qy = 0.$$

Use this to show that if $y(x)$ is a solution to (4.22) for $x > 0$, then $y(|x|)$ is a solution for all $x \neq 0$.

In Exercises 19–22, find the general solution. Then find the solution that satisfies the given initial conditions.

19. $x^2 y'' + 2xy' - 6y = 0$, $y(1) = 3$ and $y'(1) = 1$.

20. $x^2 y'' - 2y = 0$, $y(1) = 1$ and $y'(1) = 3$.

21. $(x-3)^2 y'' + 5(x-3)y' + 4y = 0$, $y(4) = 1$ and $y'(4) = 1$.

22. $(x-1)^2 y'' - 6y = 0$, $y(0) = 1$ and $y'(0) = 1$.

23. Suppose that the indicial equation to Euler's equation

$$x^2 y'' + pxy' + qy = 0 \tag{4.23}$$

has complex roots $s = \alpha + i\beta$ and $\bar{s} = \alpha - i\beta$. Arguing formally, a complex-valued solution to this equation is given by

$$y(x) = x^s = x^{\alpha + i\beta} = x^\alpha x^{i\beta}$$

$$= x^\alpha e^{i\beta \ln x}$$

$$= x^\alpha [\cos(\beta \ln x) + i \sin(\beta \ln x)].$$

A good guess for a fundamental set of solutions is the real and imaginary parts of the complex-valued solution, so we

set
$$y_1(x) = |x|^\alpha \cos(\beta \ln |x|)$$
and
$$y_2(x) = |x|^\alpha \sin(\beta \ln |x|).$$

Verify that y_1 and y_2 are indeed solutions to (4.23) using the following outline.

(a) Show that if $s = \alpha + i\beta$ and $\bar{s} = \alpha - i\beta$ are the roots of the indicial equation to (4.23), then

$$p = 1 - 2\alpha \quad \text{and} \quad q = \alpha^2 + \beta^2.$$

(b) Use part (a) to show that y_1 and y_2 are solutions to (4.23) by direct substitution into the differential equation.

In Exercises 24–27, find the general solution using the method in Exercise 23.

24. $x^2 y'' + 2xy' + y = 0$ **25.** $3x^2 y'' + y = 0$

26. $2x^2 y'' + xy' + y = 0$ **27.** $x^2 y'' + xy' + 4y = 0$

In Exercises 28–31, use the method in Exercise 23 to find the general solution. Then find the solution that satisfies the given initial condition.

28. $x^2 y'' + xy' + 4y = 0$, $y(1) = -2$, $y'(1) = 2$

29. $x^2 y'' + 3xy' + 2y = 0$, $y(1) = 3$, $y'(1) = -2$.

30. $x^2 y'' + 2xy' + y = 0$, $y(1) = -1$, $y'(1) = (1 + \sqrt{3})/2$.

31. $x^2 y'' - xy' + 6y = 0$ $y(1) = 2$, $y'(1) = 2 - 3\sqrt{5}$.

32. Suppose that the indical equation for Euler's equation

$$x^2 y'' + pxy' + qy = 0 \qquad (4.24)$$

has a double root at $s = s_0$. Then $y_1(x) = x^{s_0}$ is one solution. Use the method of **reduction of order** to find the second solution. Follow the outline below.

(a) The technique of reduction of order is to look for a second solution of the form $y_2(x) = u(x)y_1(x)$, where u is a function yet to be found, and y_1 is a known solution. Substitute this formula for y_2 into (4.24) and use the fact that y_1 already satisfies (4.24) to show that

$$u' = Cx^{-(p+2s_0)}$$

for some constant C.

(b) Use the fact that s_0 is a double root to the indicial equation to show that $p = 1 - 2s_0$ (*Hint:* If the indicial equation has a double root, then it factors as $(s - s_0)^2$.)

(c) Use the first and second parts of this exercise to conclude that $u' = Cx^{-1}$ and so $u = C \ln x$. Therefore, $y_2 = x^{s_0} \ln x$ is a second linearly independent solution to (4.24).

33. We can also solve Euler's equation (4.24) by converting it into a second-order equation with constant coefficients. Make the substitution $x = e^t$ and $Y(t) = y(e^t)$. Therefore, $t = \ln x$ and $y(x) = Y(\ln x)$ for $x > 0$.

(a) Show that $y(x)$ satisfies (4.24) if and only if $Y(t)$ satisfies

$$Y''(t) + (p - 1)Y'(t) + qY(t) = 0, \quad \text{for } t > 0. \tag{4.25}$$

(b) Show that the characteristic polynomial of (4.25) is equal to the indicial polynomial of (4.24).

(c) If the characteristic polynomial of (4.25) has two distinct real roots s_1 and s_2, then the general solution to (4.25) is $Y(t) = C_1 e^{s_1 t} + C_2 e^{s_2 t}$. Verify that $y(x) = Y(\ln x)$ is the general solution to Euler's equation (4.24) for $x > 0$.

(d) If the characteristic polynomial of (4.25) has one real root s_0 of multiplicity 2, then the general solution to (4.25) is $Y(t) = C_1 e^{s_0 t} + C_2 t e^{s_0 t}$. Verify that $y(x) = Y(\ln x)$ is the general solution to Euler's equation (4.24) for $x > 0$.

(e) If the characteristic polynomial of (4.25) has two complex conjugate roots $s = \alpha + i\beta$ and $\bar{s} = \alpha - i\beta$, then the general solution to (4.25) is $Y(t) = e^{\alpha t}(C_1 \cos(\beta t) + C_2 \sin(\beta t))$. Verify that $y(x) = Y(\ln x)$ is the general solution to Euler's equation (4.24) given for this case in Exercise 23.

11.5 Series Solutions Near Regular Singular Points

Consider the differential equation

$$x^2 y'' + 5xy' + 3y = 0. \tag{5.1}$$

This is an Euler equation, and using the methods of the previous section we find that the general solution is $y(x) = C_1 x^{-1} + C_2 x^{-3}$. Since every such solution except the trivial one $y = 0$ fails to be continuous at $x = 0$, the only power series solution to (5.1) is $y = 0$. Notice also that $x = 0$ is a regular singular point for (5.1). Thus to solve an equation near a regular singular point we have to expand our solution methods. We will do that in this section.

We start with an equation of the form

$$y'' + P(x)y' + Q(x)y = 0.$$

Everything is simpler if we put the equation into the form of equation (4.5) by multiplying by $(x - x_0)^2$. The equation becomes

$$(x - x_0)^2 y'' + (x - x_0) p(x) y' + q(x) y = 0, \qquad (5.2)$$

where $p(x) = (x - x_0) P(x)$ and $q(x) = (x - x_0)^2 Q(x)$. Then x_0 is a regular singular point if and only if both $p(x)$ and $q(x)$ are analytic at x_0. For simplicity, we will assume that the regular singular point is $x_0 = 0$.[2]

The method of Frobenius

We will find solutions using the ***method of Frobenius***. It is motivated by our results in the previous section on the Euler equation and our results in Section 11.2 on power series. We will be solving the equation

$$x^2 y'' + x p(x) y' + q(x) y = 0. \qquad (5.3)$$

Notice that since the coefficients $p(x)$ and $q(x)$ are analytic at 0, they have power series expansions at that point. On the other hand, the factors x^2 and x in the coefficients make equation (5.3) look like Euler's equation. Thus, equation (5.3) contains elements of an equation near an ordinary point and of Euler's equation. Instead of looking for a solution which is just a power series, as we do at an ordinary point, or just a power function, as we did for Euler's equation, we look for a solution which is the product of a power series and a power function. Thus we set

$$y(x) = x^s \sum_{n=0}^{\infty} a_n x^n = \sum_{n=0}^{\infty} a_n x^{s+n} \qquad (5.4)$$
$$= a_0 x^s + a_1 x^{s+1} + a_2 x^{s+2} + \cdots .$$

A solution of the type in (5.4) will be called a ***Frobenius solution***.

Both the power s and the coefficients a_n are as yet undetermined. However, we may assume that the coefficient a_0 of x^s is nonzero. If this is not true, we can increase s so that it is true. To see this, look at the open form of the series in (5.4). Suppose that $a_0 = a_1 = \cdots = a_{k-1} = 0$, but $a_k \neq 0$. If we replace s by $s + k$, we get a series of the type in (5.4) with the first coefficient nonzero.

Let's look at the use of the method of Frobenius in an example.

Example 5.5 Find the general solution to the equation

$$2xy'' + y' - 4y = 0 \qquad (5.6)$$

near the point $x_0 = 0$.

To put the equation into the form of equation (5.3), we must multiply the equation by $x/2$. It becomes

$$x^2 y'' + \frac{x}{2} y' - 2xy = 0. \qquad (5.7)$$

---

[2] This does not involve any loss of generality, since we can make the substitution $\tilde{x} = x - x_0$ for the independent variable. With this substitution the equation becomes

$$\tilde{x}^2 y'' + \tilde{x} p(\tilde{x} + x_0) y' + q(\tilde{x} + x_0) y = 0,$$

where $p(\tilde{x} + x_0)$ and $q(\tilde{x} + x_0)$ are analytic at $\tilde{x}_0 = 0$.

Comparing it to (5.3), we see that $p(x) = 1/2$ and $q(x) = -2x$, both of which are analytic. Hence $x_0 = 0$ is a regular singular point.

We look for a Frobenius solution of the type given in (5.4). Differentiating term by term, we get

$$y'(x) = \sum_{n=0}^{\infty} (s+n)a_n x^{s+n-1} \quad \text{and} \quad y''(x) = \sum_{n=0}^{\infty} (s+n)(s+n-1)a_n x^{s+n-2}.$$

Therefore,

$$xy'(x) = \sum_{n=0}^{\infty} (s+n)a_n x^{s+n} \quad \text{and} \quad x^2 y''(x) = \sum_{n=0}^{\infty} (s+n)(s+n-1)a_n x^{s+n}.$$

$$(5.8)$$

We will refer to the expressions y, xy', and $x^2 y''$ as the **fundamental expressions**. The series for these expressions are presented in equations in (5.4) and (5.8). Notice that each of these series contains the expression $a_n x^{s+n}$. This fact will make combining sums easier.

We substitute the series for y in (5.4) into the differential equation. We could use either (5.6) or (5.7) or another form derived from one of these. The form of the equation we use should include the fundamental expressions, and it should avoid complicated expressions. Even a fraction like the $1/2$ in equation (5.7) should be avoided if possible. We choose to use (5.6), but we multiply it by x so that it contains fundamental expressions. Thus, the form we will use is

$$2x^2 y'' + xy' - 4xy = 0.$$

In the first step, we rewrite the equation, gathering the fundamental expressions times a fixed power of x. In the case at hand, there are two groupings, which we put between square brackets, obtaining

$$0 = 2x^2 y'' + xy' - 4xy = [2x^2 y'' + xy'] - 4x \cdot [y].$$

With the equation in this form, we can easily substitute using the series for the fundamental expressions from (5.4) and (5.8). Combining the series for the terms in the bracketed expressions, we get

$$0 = \sum_{n=0}^{\infty} [2(s+n)(s+n-1) + (s+n)]a_n x^{s+n} - 4x \sum_{n=0}^{\infty} a_n x^{s+n}.$$

Notice that the coefficient of a_n in the first sum can be written as $I(s+n)$, where

$$I(s) = 2s(s-1) + s = 2s^2 - s = s(2s-1).$$

The quadratic polynomial $I(s)$ is called the **indicial polynomial**. In the second sum, we distribute the x through the sum and then shift the index so that all terms have the same power $s+n$. Finally, we combine the two sums. These steps are recorded as

$$0 = \sum_{n=0}^{\infty} I(s+n)a_n x^{s+n} - 4\sum_{n=0}^{\infty} a_n x^{s+n+1}$$

$$= \sum_{n=0}^{\infty} I(s+n)a_n x^{s+n} - 4\sum_{n=1}^{\infty} a_{n-1} x^{s+n} \qquad (5.9)$$

$$= I(s)a_0 x^s + \sum_{n=1}^{\infty} [I(s+n)a_n - 4a_{n-1}]x^{s+n}.$$

In the last step we isolated the term for $n = 0$ in order to combine the two sums.

According to the identity theorem, the series in (5.9) is equal to 0 if and only if the coefficient of each power of x is equal to 0. Since $a_0 \neq 0$, the coefficient of x^s being equal to 0 implies that

$$I(s) = s(2s - 1) = 0. \tag{5.10}$$

This is the ***indicial equation***. The roots of the indicial equation are the powers s which lead to Frobenius solutions using the series for y in (5.4). From (5.10), we see that either $s = 0$ or $s = 1/2$.

Using the factored form of the indicial polynomial $I(s) = s(2s - 1)$, the coefficient of x^{s+n} for $n \geq 1$ in (5.9) is

$$(s + n)(2s + 2n - 1)a_n - 4a_{n-1} = 0.$$

This is the ***recurrence formula***. It allows us to compute the coefficients inductively starting with a given $a_0 \neq 0$. Solving for a_n, we get

$$a_n = \frac{4}{(s + n)(2s + 2n - 1)}a_{n-1}, \quad \text{for } n \geq 1. \tag{5.11}$$

We will first look for a solution corresponding to the root $s = 0$ of the indicial equation. The recurrence formula becomes

$$a_n = \frac{4}{n(2n - 1)}a_{n-1}, \quad \text{for } n \geq 1.$$

The first few instances are

$$a_1 = \frac{4}{1 \cdot 1}a_0, \quad a_2 = \frac{4}{2 \cdot 3}a_1, \quad \text{and} \quad a_3 = \frac{4}{3 \cdot 5}a_2.$$

Multiplying the recurrence formulas for a_1 to a_n, we get

$$a_1 \cdot a_2 \cdot a_3 \cdot \ldots \cdot a_n = \frac{4}{1 \cdot 1}a_0 \cdot \frac{4}{2 \cdot 3}a_1 \cdot \frac{4}{3 \cdot 5}a_2 \cdot \ldots \cdot \frac{4}{n(2n - 1)}a_{n-1}.$$

Canceling common factors and rearranging, we get

$$a_n = \frac{4^n}{n! \cdot 1 \cdot 3 \cdot \ldots \cdot (2n - 1)}a_0, \quad \text{for } n \geq 1.$$

We can simplify this expression by noticing that

$$1 \cdot 3 \cdot \ldots \cdot (2n - 1) = \frac{(2n)!}{2 \cdot 4 \cdot \ldots \cdot 2n} = \frac{(2n)!}{2^n n!}. \tag{5.12}$$

Hence,

$$a_n = \frac{8^n}{(2n)!}a_0, \quad \text{for } n \geq 1.$$

Notice that this formula is valid for $n = 0$ as well. With $a_0 = 1$, and remembering that $s = 0$, we get the particular solution

$$y_1(x) = \sum_{n=0}^{\infty} a_n x^{s+n} = \sum_{n=0}^{\infty} \frac{8^n}{(2n)!}x^n. \tag{5.13}$$

Some of the partial sums $\sum_{n=0}^{N}(8x)^n/(2n)!$ are plotted in Figure 1 to illustrate the convergence.

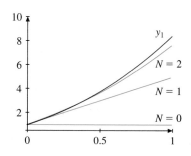

Figure 1. Convergence of the series in (5.13).

For the root $s = 1/2$ of the indicial equation, the recurrence formula (5.11) becomes

$$a_n = \frac{4}{n(2n+1)}a_{n-1}, \quad \text{for } n \geq 1.$$

The first few instances are

$$a_1 = \frac{4}{1 \cdot 3}a_0, \quad a_2 = \frac{4}{2 \cdot 5}a_1, \quad \text{and} \quad a_3 = \frac{4}{3 \cdot 7}a_2.$$

Multiplying the recurrences formulas for a_1 to a_n and canceling common terms, we get

$$a_n = \frac{4^n}{n! \cdot 3 \cdot 5 \cdot \ldots \cdot (2n+1)}a_0, \quad \text{for } n \geq 1.$$

Using a computation similar to that in (5.12), we get

$$a_n = \frac{8^n}{(2n+1)!}a_0, \quad \text{for } n \geq 1.$$

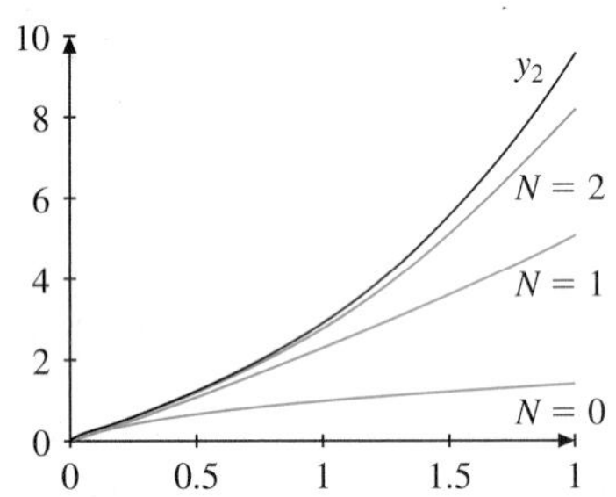

Again this equation is valid for $n = 0$.

Setting $a_0 = 1$ and remembering that $s = 1/2$, we get the particular solution

$$y_2(x) = x^s \sum_{n=0}^{\infty} a_n x^n = x^{1/2} \sum_{n=0}^{\infty} \frac{8^n}{(2n+1)!} x^n. \tag{5.14}$$

Figure 2. Convergence of the series in (5.14).

Some of the partial sums $x^{1/2} \sum_{n=0}^{N}(8x)^n/(2n+1)!$ are plotted in Figure 2 to illustrate the convergence.

The two solutions y_1 and y_2 are not multiples of each other, so they are linearly independent, and the general solution is the general linear combination of the two,

$$y(x) = C_1 y_1(x) + C_2 y_2(x). \qquad \bullet$$

Let's look at another example that illustrates some points about finding Frobenius solutions.

Example 5.15 Find a fundamental set of solutions to the equation

$$2x^2(1-x^2)y'' - xy' + y = 0 \tag{5.16}$$

near $x_0 = 0$.

To put the equation into the form of equation (5.3) we must multiply the equation by $1/[2(1-x^2)]$, getting

$$x^2 y'' - x\frac{1}{2(1-x^2)}y' + \frac{1}{2(1-x^2)}y = 0. \tag{5.17}$$

Then $p(x) = -1/[2(1-x^2)]$ and $q(x) = 1/[2(1-x^2)]$ are analytic at $x_0 = 0$. Therefore, the point $x_0 = 0$ is a regular singular point.

To solve the equation, we use equation (5.16). Notice that it contains the fundamental expressions. So does equation (5.17), but this equation also contains the coefficient $x/(1-x^2)$, which has to be expanded into a power series, greatly complicating the computations. Following the steps we used in Example 5.5, we first rewrite the equation, gathering together the fundamental expressions times a fixed power of x, getting

$$0 = 2x^2(1-x^2)y'' - xy' + y$$
$$= [2x^2 y'' - xy' + y] - 2x^2 \cdot [x^2 y''].$$

Next, we substitute the series for the fundamental expressions, obtaining

$$0 = \sum_{n=0}^{\infty} [2(s+n)(s+n-1) - (s+n) + 1] a_n x^{s+n}$$

$$- 2 \sum_{n=0}^{\infty} (s+n)(s+n-1) a_n x^{s+n+2}.$$

The coefficient of a_n in the first sum can be written as $I(s+n)$, where the indicial polynomial is

$$I(s) = 2s(s-1) - s + 1 = 2s^2 - 3s + 1 = (2s-1)(s-1). \tag{5.18}$$

We shift the index in the second sum so that the same power of x appears in both sums. Then we combine the two sums. In doing so we have to isolate the terms for $n = 0$ and $n = 1$. We get

$$0 = I(s)a_0 x^s + I(s+1)a_1 x^{s+1}$$

$$+ \sum_{n=2}^{\infty} [I(s+n)a_n - 2(s+n-2)(s+n-3)a_{n-2}]x^{s+n}. \tag{5.19}$$

The coefficient of x^s is $I(s)a_0$. Since $a_0 \neq 0$, we get the indicial equation $I(s) = (2s-1)(s-1) = 0$. Its roots are $s_1 = 1$ and $s_2 = 1/2$. In order to have a solution, we must have $s = s_1$ or $s = s_2$.

The coefficient of x^{s+1} in (5.19) is $I(s+1)a_1$, which must be equal to 0. This is a special case of the recurrence formula. Notice that $I(s+1) \neq 0$ if $s = s_1 = 1$ or $s = s_2 = 1/2$. Hence we must have $a_1 = 0$.

The general recurrence formula

$$I(s+n)a_n = 2(s+n-2)(s+n-3)a_{n-2}, \quad \text{for } n \geq 2 \tag{5.20}$$

comes from setting the coefficient of x^{s+n} in (5.19) equal to 0. Notice that $I(s+n) \neq 0$ if $s = s_1 = 1$ or $s = s_2 = 1/2$, and n is a positive integer. Hence we can use the recurrence formula to solve inductively for all of the coefficients. In particular, since $a_1 = 0$, the recurrence formula for $n = 3$ implies that $a_3 = 0$. Proceeding inductively, we see that all odd-numbered coefficients are equal to 0. Rewriting the recurrence formula in (5.20) for even integers, we get

$$I(s+2n)a_{2n} = 2(s+2n-2)(s+2n-3)a_{2n-2}, \quad \text{for } n \geq 1.$$

Using the factored form of the indicial polynomial in (5.18), this becomes

$$(2s+4n-1)(s+2n-1)a_{2n} = 2(s+2n-2)(s+2n-3)a_{2n-2}, \quad \text{for } n \geq 1. \tag{5.21}$$

Setting $s = s_1 = 1$, this formula becomes

$$2n(4n+1)a_{2n} = 2(2n-1)(2n-2)a_{2n-2}, \quad \text{for } n \geq 1.$$

Notice that for $n = 1$ the right-hand side is equal to 0. Hence $a_2 = 0$, and it follows inductively that $a_{2n} = 0$ for $n \geq 1$. Therefore, a_0 is the only nonzero coefficient. Setting $a_0 = 1$, our solution for $s = s_1 = 1$ is very simply

$$y_1(x) = a_0 x^{s_1} = x.$$

For $s = s_2 = 1/2$ things do not work out quite that easily. With a little work the recurrence formula (5.21) becomes

$$4n(4n - 1)a_{2n} = (4n - 3)(4n - 5)a_{2n-2}, \quad \text{for } n \geq 1$$

or

$$a_{2n} = \frac{(4n - 3)(4n - 5)a_{2n-2}}{4n(4n - 1)}, \quad \text{for } n \geq 1. \qquad (5.22)$$

For small n we have

$$a_2 = -\frac{a_0}{4 \cdot 1 \cdot 3} \qquad a_4 = \frac{5 \cdot 3 \cdot a_2}{4 \cdot 2 \cdot 7}$$
$$a_6 = \frac{9 \cdot 7 \cdot a_4}{4 \cdot 3 \cdot 11} \qquad a_8 = \frac{13 \cdot 11 \cdot a_6}{4 \cdot 4 \cdot 15}. \qquad (5.23)$$

The product of the first n of the recurrence formulas is

$$a_2 \cdot a_4 \cdot a_6 \cdot a_8 \cdot \ldots \cdot a_{2n}$$
$$= -\frac{a_0}{4 \cdot 1 \cdot 3} \cdot \frac{5 \cdot 3 \cdot a_2}{4 \cdot 2 \cdot 7} \cdot \frac{9 \cdot 7 \cdot a_4}{4 \cdot 3 \cdot 11} \cdot \frac{13 \cdot 11 \cdot a_6}{4 \cdot 4 \cdot 15} \qquad (5.24)$$
$$\cdot \ldots \cdot \frac{(4n - 3) \cdot (4n - 5) \cdot a_{2n-2}}{4 \cdot n \cdot (4n - 1)}.$$

After simplifying, this becomes

$$a_{2n} = -\frac{1 \cdot 5 \cdot 9 \cdot 13 \cdot \ldots \cdot (4n - 3) \cdot 1 \cdot 3 \cdot 7 \cdot 11 \cdot \ldots \cdot (4n - 5)}{4^n \cdot n! \cdot 3 \cdot 7 \cdot 11 \cdot 15 \cdot \ldots \cdot (4n - 1)} a_0$$
$$= -\frac{1 \cdot 5 \cdot 9 \cdot 13 \cdot \ldots \cdot (4n - 3)}{4^n \cdot n! \cdot (4n - 1)} a_0, \quad \text{for } n \geq 1. \qquad (5.25)$$

If we set $a_0 = 1$ and remember that $s_2 = 1/2$, our second solution is

$$y_2(x) = x^{1/2} \sum_{n=0}^{\infty} a_{2n} x^{2n}$$
$$= x^{1/2} \left[1 - \frac{1}{4 \cdot 3} x^2 - \frac{5}{4^2 \cdot 2 \cdot 7} x^4 - \cdots \right]. \qquad \bullet$$

In proceeding from the recurrence formula in (5.22) to the final answer in (5.25), there are a couple of things to keep in mind.

- Do not do any cancellation until close to the end. For example, resist the natural impulse to cancel the many common factors in the product in (5.24). It is better to see the general pattern first, as we do in the first formula in (5.25).
- In proceeding from (5.23) to (5.24), gather together the factors that have the same position in the recurrence formula (5.22).

Exceptional cases

In our two examples we were able to find two independent solutions. This will not always be possible. For example, if the indicial equation has two equal roots $s_1 = s_2$, then we cannot find a second Frobenius solution. We have already illustrated this for the Euler equation $x^2 y'' + 3xy' + y = 0$ in Example 4.19. We found one Frobenius solution, $y_1(x) = x^{-1}$. The second solution, $y_2(x) = x^{-1} \ln x$, is not a Frobenius solution.

Exceptional cases can also arise if $s_1 - s_2 = N$ is a positive integer, as the next example shows.

Example 5.26 Try to find the solutions to the equation $xy'' + y = 0$.

We first multiply the equation by x to put it into the form of equation (5.3). Then, substituting $y(x) = \sum_{n=0}^{\infty} a_n x^{s+n}$ in the usual way, we get

$$0 = x^2 y'' + xy = s(s-1)a_0 x^s + \sum_{n=1}^{\infty} \left[(s+n)(s+n-1)a_n + a_{n-1} \right] x^{s+n}.$$

Thus the indicial equation is $I(s) = s(s-1) = 0$, and the roots are $s_1 = 1$ and $s_2 = 0$. We will leave finding the Frobenius solution corresponding to the larger root $s_1 = 1$ to Exercise 26. For the smaller root $s_2 = 0$ the recurrence formula is

$$n(n-1)a_n = -a_{n-1}.$$

For $n = 1$ this becomes $0 \cdot a_1 = -a_0$, and, since $a_0 \neq 0$, this equation cannot be satisfied by any choice of a_1. So there is no Frobenius solution corresponding to the smaller root $s_2 = 0$. ●

As Example 5.26 illustrates, there is always a Frobenius solution corresponding to the larger root, but perhaps not to the smaller root. We will describe a method of finding a second solution in the exceptional cases in the next section.

A summary of the method of Frobenius

The method can be summarized as follows:

To solve an equation of the form $L(x)y'' + M(x)y' + N(x)y = 0$ near a regular singular point x_0,

1. If necessary, multiply by a factor to put the equation in the form

$$B(x)(x - x_0)^2 y'' + C(x)(x - x_0)y' + D(x)y = 0, \quad (5.27)$$

where $B(x_0) \neq 0$, and B, C, and D are analytic near x_0.
2. Substitute $y(x) = (x - x_0)^s \sum_{n=0}^{\infty} a_n (x - x_0)^n$ and its derivatives into (5.27) and rearrange into a series of the form

$$A_0(x - x_0)^s + A_1(x - x_0)^{s+1} + \cdots = 0.$$

3. Set $A_0 = 0$ to find the indicial equation and its roots $s_2 \leq s_1$.
4. For the larger root, $s = s_1$, set $A_1 = A_2 = \cdots = 0$ to find the recurrence formula and determine the coefficients $a_1, a_2, \ldots$. The first solution is $y_1(x) = (x - x_0)^{s_1} \sum_{n=0}^{\infty} a_n (x - x_0)^n$.
5. If possible, repeat the last step for $s = s_2$ to find $y_2(x)$.

The key step is step 2. This can be simplified somewhat by using the fundamental expressions in (5.4) and (5.8)

EXERCISES

1. Differential equations do not usually have power series solutions near singular points, even if they are regular. As an example, consider the equation

$$x^2 y'' + 4xy' + 2y = 0,$$

which is an Euler's equation, and has a regular singular point at $x_0 = 0$. Show that if the series $y(x) = \sum_{n=0}^{\infty} a_n x^n$ is a solution, then $a_n = 0$ for all n.

2. After we defined a regular singular point in Definition 4.3, we mentioned that the definition characterized those differential equations for which the method of Frobenius works. In this exercise we will demonstrate that to some extent.

   (a) Show that 0 is a regular singular point for $x^2 y'' - y = 0$ and an irregular singular point for $x^3 y'' - y = 0$.

   (b) Show that if $y(x) = x^s \sum_{n=0}^{\infty} a_n x^n$ is a solution to $x^3 y'' - y = 0$, then $a_n = 0$ for all n. Hence $x^3 y'' - y = 0$ has no Frobenius solutions, while $x^2 y'' - y = 0$ does.

Find the indicial equation for the differential equation given in Exercises 3–6 at the indicated singularity.

3. $x^2 y'' + 4xy' + 2y = 0$ at $x = 0$

4. $(x^2 - 1)^2 y'' + (x + 1)y' - y = 0$ at $x = -1$

5. $4x \sin x y'' - 3y = 0$ at $x = 0$

6. $(\cos x - 1)y'' + xy' + y = 0$ at $x = 0$

For the differential equations in Exercises 7–10, find the indicial polynomial for the singularity at $x = 0$. Then find the recurrence formula for the largest of the roots to the indicial equation.

7. $2x(x - 1)y'' + 3(x - 1)y' - y = 0$

8. $xy'' + y' - 2y = 0$

9. $xy'' + (1 - x)y' - y = 0$ 10. $x^2 y'' - xy' + x^2 y = 0$

In Exercises 11–25, find two Frobenius series solutions.

11. $4xy'' + 2y' + y = 0$

12. $2x(x + 1)y'' + 3(x + 1)y' - y = 0$

13. $2x^2 y'' - 3xy' + (2 + 2x)y = 0$

14. $4x^2 y'' + 4xy' - (4x^2 + 1)y = 0$

15. $x^2 y'' + xy' + (x^2 - 1/16)y = 0$

16. $3x^2 y'' + 2xy' + x^2 y = 0$

17. $2x^2 y'' - xy' + (x + 1)y = 0$

18. $3x^2 y'' + (x - x^2)y' - y = 0$

19. $2(x + x^2)y'' + y' - y = 0$

20. $2xy'' + y' + xy = 0$

21. $2xy'' + y' + y = 0$

22. $2x^2 y'' - xy' + (1 + x)y = 0$

23. $2x^2 y'' - xy' + (1 - x^2)y = 0$

24. $2x^2 y'' + 3xy' - (x^2 + 1)y = 0$

25. $2xy'' + y' + x^2 y = 0$

26. Find the Frobenius solution to the equation $xy'' + y = 0$ considered in Example 5.26 corresponding to the larger root $s_1 = 1$ of the indicial equation.

27. Bessel's equation of order 0 is $x^2 y'' + xy' + x^2 y = 0$. Show that the indicial polynomial has only one distinct root. Show that the Frobenius solution corresponding to this root is

$$y(x) = \sum_{n=0}^{\infty} \frac{(-1)^n}{2^{2n}(n!)^2} x^{2n}.$$

28. Bessel's equation of order 1 is $x^2 y'' + xy' + (x^2 - 1)y = 0$. Show that the indicial polynomial has roots $s_1 = 1$ and $s_2 = -1$. Show that

$$y(x) = x \sum_{n=0}^{\infty} \frac{(-1)^n}{2^{2n}(n + 1)! n!} x^{2n}$$

is a Frobenius solution corresponding to $s_1 = 1$. Show that there is no Frobenius solution corresponding to $s_2 = -1$.

29. Bessel's equation of order 1/2 is $x^2 y'' + xy' + (x^2 - 1/4)y = 0$. Show that the indicial polynomial has roots $s_1 = 1/2$ and $s_2 = -1/2$. Show that

$$y(x) = x^{1/2} \sum_{n=0}^{\infty} \frac{(-1)^n}{(2n + 1)!} x^{2n}$$

is a Frobenius solution corresponding to $s_1 = 1/2$. Show that there is a Frobenius solution corresponding to $s_2 = -1/2$.

30. Show that $x = 1$ and $x = -1$ are regular singular points of the **Chebychev equation**

$$(1 - x^2)y'' - xy' + p^2 y = 0,$$

where p is a real number. Find the roots of the indicial equation and compute the recurrence relations for two linearly independent solutions about the point $x = 1$. (*Hint*: It might be easier to make the substitution $t = x - 1$, or $x = t + 1$, and then to find the Frobenius solutions as functions of t near $t = 0$. Then substituting back will give the series in terms of $x - 1$.)

31. Show that $x = 0$ is a regular singular point to the *Laguerre equation:*

$$xy'' + (1 - x)y' + py = 0.$$

Show that $s = 0$ is a double root for the indicial equation and compute the recurrence relation for one series solution to this differential equation about $x = 0$. Show that if p is a positive integer, then this solution is a polynomial (called a *Laguerre polynomial*).

32. Suppose $y_1(x)$ is a solution to the differential equation

$$y'' + p(x)y' + q(x)y = 0, \quad x > 0. \tag{5.28}$$

Use the reduction of order technique[3] to show that a second solution is

$$y_2(x) = y_1(x) \int \frac{e^{-\int p(x)\, dx}}{[y_1(x)]^2}\, dx.$$

33. The *hypergeometric equation*

$$x(1 - x)y'' + [\gamma - (1 + \alpha + \beta)x]y' - \alpha\beta y = 0, \tag{5.29}$$

where α, β, and γ are constants, occurs frequently in mathematics as well as in physical applications.

(a) Show that $x = 0$ is a regular singular point for (5.29). Find the indicial polynomial and its roots.

(b) Show that $x = 1$ is a regular singular point for (5.29). Find the indicial polynomial and its roots.

(c) Suppose that $1 - \gamma$ is not a positive integer. Show that one solution of (5.29) for $0 < x < 1$ is given by

$$y_1(x) = 1 + \frac{\alpha \cdot \beta}{\gamma \cdot 1!}x$$
$$+ \frac{\alpha \cdot (\alpha + 1) \cdot \beta \cdot (\beta + 1)}{\gamma \cdot (\gamma + 1) \cdot 2!}x^2 + \cdots.$$

(d) Suppose that $1 - \gamma$ is not an integer. Show that a second solution to (5.29) for $0 < x < 1$ is given by

$$y_2(x) = x^{1-\gamma}\left[1 + \frac{(\alpha - \gamma + 1)(\beta - \gamma + 1)}{(2 - \gamma)1!}x\right.$$
$$+ \frac{(\alpha - \gamma + 1)(\alpha - \gamma + 2)(\beta - \gamma + 1)(\beta - \gamma + 2)}{(2 - \gamma)(3 - \gamma)2!}x^2$$
$$\left.+ \cdots\right].$$

11.6 Series Solutions Near Regular Singular Points—The General Case

In Examples 5.5 and 5.15 in the previous section we were able to use the method of Frobenius to find a fundamental set of solutions. However, in Example 5.26, we were able to find only one such solution. In this section we will find a second solution using a technique that is very similar to that used in Section 4 for the exceptional case of Euler's equation. To do this, we need to go through the Frobenius solution process in general.

We will consider the equation

$$x^2 y'' + xp(x)y' + q(x)y = 0, \tag{6.1}$$

where the coefficients $p(x)$ and $q(x)$ are analytic at 0 and have power series expansions

$$p(x) = \sum_{n=0}^{\infty} p_n x^n \quad \text{and} \quad q(x) = \sum_{n=0}^{\infty} q_n x^n,$$

which converge in an interval containing $x = 0$. We will use operator notation and write equation (6.1) as

$$Ly = x^2 y'' + xp(x)y' + q(x)y = 0. \tag{6.2}$$

We will look for solutions of the form

$$y(s, x) = x^s \sum_{n=0}^{\infty} a_n(s)x^n. \tag{6.3}$$

The leading coefficient a_0 will be a nonzero constant, but we will allow the coefficients $a_n(s)$ to depend on s for $n \geq 1$. We remind you that (6.3) is one of the

---

[3] See Exercise 32 in Section 11.4 on page 566.

fundamental expressions and that, using y' to denote $\partial y/\partial x$, the other two are

$$xy'(s, x) = \sum_{n=0}^{\infty}(s + n)a_n(s)x^{s+n} \quad \text{and}$$

$$x^2 y''(s, x) = \sum_{n=0}^{\infty}(s + n)(s + n - 1)a_n(s)x^{s+n}. \tag{6.4}$$

Notice that the first term in (6.1) is one of the fundamental expressions, while the second and third terms are products of fundamental expressions and the coefficient functions $p(x)$ and $q(x)$. Expanding these products using formulas (1.13) and (1.12) on page 536, we get

$$xp(x)y' = p(x) \cdot xy'$$

$$= \left(\sum_{n=0}^{\infty} p_n x^n\right) \cdot x^s \left(\sum_{n=0}^{\infty}(s + n)a_n x^n\right)$$

$$= x^s \cdot \sum_{n=0}^{\infty}\left(\sum_{k=0}^{n} p_{n-k}(s + k)a_k\right)x^n,$$

and

$$q(x)y = \left(\sum_{n=0}^{\infty} q_n x^n\right) \cdot x^s \left(\sum_{n=0}^{\infty} a_n x^n\right) = x^s \cdot \sum_{n=0}^{\infty}\left(\sum_{k=0}^{n} q_{n-k}a_k\right)x^n.$$

Adding everything together, we get

$$Ly(s, x) = x^2 y'' + xp(x)y' + q(x)y$$

$$= x^s \cdot \sum_{n=0}^{\infty} A_n(s)x^n, \tag{6.5}$$

where the coefficients are

$$A_n(s) = (s + n)(s + n - 1)a_n(s) + \sum_{k=0}^{n}\left[p_{n-k}(s + k) + q_{n-k}\right]a_k(s). \tag{6.6}$$

We get a solution to equation (6.1) provided that $A_n(s) = 0$ for $n \geq 0$.

The indicial equation

For $n = 0$, equation (6.6) becomes

$$A_0(s) = [s(s - 1) + p_0 s + q_0]a_0$$

$$= \left[s^2 + (p_0 - 1)s + q_0\right]a_0 \tag{6.7}$$

$$= I(s)a_0,$$

where

$$I(s) = s^2 + (p_0 - 1)s + q_0 \tag{6.8}$$

is the **indicial polynomial**. Since we are assuming that $a_0 \neq 0$, the coefficient $A_0(s) = 0$ only if $I(s) = 0$. This is the **indicial equation**, and its roots are the only powers s for which there can be solutions. We will only consider the case when the indicial equation has two real roots s_1 and s_2, and we will assume that they are ordered by $s_2 \leq s_1$. Then

$$A_0(s) = a_0 I(s) = a_0(s - s_1)(s - s_2). \tag{6.9}$$

Notice that we will not immediately choose s to be one of the roots s_1 or s_2.

The recurrence formula

For $n \geq 1$ we bring together the terms involving a_n in (6.6). Then

$$A_n(s) = [(s+n)(s+n-1) + p_0(s+n) + q_0]a_n(s)$$

$$+ \sum_{k=0}^{n-1}[p_{n-k}(s+k) + q_{n-k}]a_k(s)$$

$$= [(s+n)^2 + (p_0-1)(s+n) + q_0]a_n(s)$$

$$+ \sum_{k=0}^{n-1}[p_{n-k}(s+k) + q_{n-k}]a_k(s).$$

Notice that the coefficient of a_n is $I(s+n)$, where I is the indicial polynomial. Hence

$$A_n(s) = I(s+n)a_n(s) + \sum_{k=0}^{n-1}[p_{n-k}(s+k) + q_{n-k}]a_k(s). \qquad (6.10)$$

We will continue to treat s as a variable, but we will require that a_0 be a nonzero constant. We define the functions $a_n(s)$ inductively to make $A_n(s) = 0$ for $n \geq 1$. By (6.10), this means that $a_n(s)$ must satisfy the ***recurrence formula,***

$$I(s+n)a_n(s) = -\sum_{k=0}^{n-1}[p_{n-k}(s+k) + q_{n-k}]a_k(s), \quad \text{for } n \geq 1. \qquad (6.11)$$

This recurrence formula can be used to solve inductively for the coefficient functions $a_n(s)$, and these functions will be well defined except at those points s where $I(s+n) = 0$ for some positive integer n.

Since $A_n(s) = 0$ for $n \geq 1$, (6.5) becomes

$$Ly(s, x) = A_0(s)x^s = a_0 I(s)x^s = a_0(s - s_1)(s - s_2)x^s. \qquad (6.12)$$

The right-hand side of (6.12) vanishes for $s = s_1$ and $s = s_2$, so $y(s_1, x)$ and $y(s_2, x)$ are solutions, provided they are defined.

The solution for the larger root $s = s_1$

According to (6.9), $I(s) = 0$ only for $s = s_1$ and $s = s_2$. Since $s_2 \leq s_1 < s_1 + n$, $I(s_1 + n) \neq 0$ for all $n \geq 1$. Hence we can always solve the recurrence formula (6.11) to find $a_n(s)$ for s near s_1. To be precise, we pick $a_0 \neq 0$. Then

$$y_1(x) = y(s_1, x) = x^{s_1}\sum_{n=0}^{\infty}a_n(s_1)x^n = \sum_{n=0}^{\infty}a_n(s_1)x^{s_1+n} \qquad (6.13)$$

is a Frobenius solution corresponding to the root s_1.

The second solution when $s_1 - s_2$ is not a nonnegative integer

If $s_1 - s_2$ is not a nonnegative integer, $s_2 + n \neq s_1$ for any n. Therefore, $I(s_2+n) \neq 0$ for all $n \geq 1$. Hence the recurrence formula (6.11) can be used to inductively solve for $a_n(s)$ for s near s_2. Thus, the Frobenius solution corresponding to s_2 is

$$y_2(x) = y(s_2, x) = x^{s_2}\sum_{n=0}^{\infty}a_n(s_2)x^n = \sum_{n=0}^{\infty}a_n(s_2)x^{s_2+n}. \qquad (6.14)$$

Second solution when $s_2 = s_1$

Since the roots are equal, the indicial equation is now $I(s) = (s - s_1)^2 = 0$. Notice that if $n \geq 1$, then $I(s + n) \neq 0$ for s near s_1. Hence the coefficients $a_n(s)$ are well-defined functions of s near $s = s_1$. In this case, equation (6.12) becomes

$$Ly(s, x) = a_0(s - s_1)^2 x^s. \tag{6.15}$$

The right-hand side of (6.15) equals 0 when $s = s_1$, so

$$y_1(x) = y(s_1, x) = x^{s_1} \sum_{n=0}^{\infty} a_n(s_1)x^n = \sum_{n=0}^{\infty} a_n(s_1)x^{s_1+n}$$

is a solution, agreeing with what we have already found in (6.13). As we did in Section 11.4 for Euler's equation, let's differentiate (6.15) with respect to s. Since the order of differentiation is not important, we get

$$L\left(\frac{\partial y}{\partial s}\right) = \frac{\partial}{\partial s}Ly = a_0[(s - s_1)^2 x^s \ln x + 2(s - s_1)x^s].$$

Notice that because of the factor of $(s - s_1)^2$ in (6.15), there is a factor of $s - s_1$ in the derivative. Evaluating at $s = s_1$, we see that

$$L\left(\frac{\partial y}{\partial s}\bigg|_{s=s_1}\right) = \left(\frac{\partial}{\partial s}Ly\right)\bigg|_{s=s_1} = 0.$$

Consequently, we get the second solution

$$y_2(x) = \frac{\partial y}{\partial s}(s_1, x)$$

$$= x^{s_1} \ln x \sum_{n=0}^{\infty} a_n(s_1)x^n + x^{s_1} \sum_{n=0}^{\infty} a_n'(s_1)x^n \tag{6.16}$$

$$= y_1(x) \ln x + x^{s_1} \sum_{n=1}^{\infty} a_n'(s_1)x^n.$$

Notice that the sum in the last expression starts at $n = 1$ instead of at $n = 0$. This is because a_0 is a constant, so $a_0' = 0$.

The exceptional cases and Bessel's equation

We need some examples, and all of the examples in this section will be special cases of Bessel's equation of order r,

$$x^2 y'' + xy' + (x^2 - r^2)y = 0. \tag{6.17}$$

We have already verified in Section 11.4 that $x_0 = 0$ is a regular singular point. If we isolate the fundamental expressions times powers of x in (6.17), it becomes

$$[x^2 y'' + xy' - r^2 y] + x^2[y] = 0.$$

Substituting $y(s, x)$ from (6.3) and the other fundamental expressions from (6.4), we get

$$\sum_{n=0}^{\infty}[(s + n)(s + n - 1) + (s + n) - r^2]a_n(s)x^{s+n} + x^2 \sum_{n=0}^{\infty} a_n(s)x^{s+n} = 0.$$

Shifting the index in the second sum and isolating the first two terms in the first sum, this becomes

$$(s^2 - r^2)a_0 x^s + [(s+1)^2 - r^2]a_1(s)x^{s+1}$$
$$+ \sum_{n=2}^{\infty} \big[[(s+n)^2 - r^2]a_n(s) + a_{n-2}(s) \big] x^{s+n} = 0. \tag{6.18}$$

Setting the coefficient of the first term in (6.18) equal to 0, we get the indicial equation $I(s) = s^2 - r^2 = 0$. Thus,

$$\text{the roots are} \quad s_1 = r \quad \text{and} \quad s_2 = -r. \tag{6.19}$$

Since $s_1 - s_2 = 2r$, there are possible exceptional cases whenever $2r$ is an integer. Setting the coefficient of the second term in (6.18) equal to 0, we get

$$[(s+1)^2 - r^2]a_1(s) = (s+1+r)(s+1-r)a_1(s) = 0. \tag{6.20}$$

We record this here for later use. Setting the coefficient of the general term in (6.18) equal to 0, we get the recurrence formula

$$[(s+n)^2 - r^2]a_n(s) = (s+n+r)(s+n-r)a_n(s) = -a_{n-2}(s) \tag{6.21}$$

for $n \geq 2$, which we again record for later use.

Example 6.22   Find a fundamental set of solutions to Bessel's equation of order $r = 0$.

From (6.19) we see that the roots are $s_1 = s_2 = 0$. Equation (6.20) becomes $(s+1)^2 a_1(s) = 0$. Since $s+1 \neq 0$ for s near 0, this requires $a_1(s) = 0$. The recurrence formula (6.21) becomes

$$(s+n)^2 a_n(s) = -a_{n-2}(s), \quad \text{or} \quad a_n(s) = -\frac{a_{n-2}(s)}{(s+n)^2}.$$

Since $a_1(s) = 0$, we immediately conclude that $a_{2n+1}(s) = 0$ for all n, and for the even-numbered terms we have

$$a_{2n}(s) = -\frac{a_{2n-2}(s)}{(s+2n)^2}, \quad \text{for } n \geq 1.$$

When we solve inductively, we get

$$a_{2n}(s) = (-1)^n \frac{a_0}{[(s+2) \cdot \ldots \cdot (s+2n)]^2}, \quad \text{for } n \geq 1. \tag{6.23}$$

Evaluating at $s = s_1 = 0$, we have

$$a_{2n}(0) = (-1)^n \frac{a_0}{2^{2n}(n!)^2}, \quad \text{for } n \geq 1.$$

This is also valid for $n = 0$, so our first solution is

$$y(0, x) = a_0 \sum_{n=0}^{\infty} \frac{(-1)^n}{(n!)^2} \left(\frac{x}{2}\right)^{2n}. \tag{6.24}$$

For the solution y_2 in (6.16) we have to compute the derivatives of $a_{2n}(s)$. First, since a_0 is a constant, $a_0' = 0$. The somewhat daunting task of differentiating the right-hand side in (6.23) is made easier by differentiating

$$\ln |a_{2n}(s)| = \ln |a_0| - 2\sum_{k=1}^{n} \ln |s + 2k|$$

instead. This method is called *logarithmic differentiation*. It is frequently useful when it is necessary to differentiate a product. It leaves us with the easier task of differentiating a sum instead of a product. Differentiating both sides, we get

$$\frac{1}{a_{2n}(s)} a'_{2n}(s) = -2 \sum_{k=1}^{n} \frac{1}{s + 2k}.$$

Evaluating at $s = 0$, this becomes

$$a'_{2n}(0) = -a_{2n}(0) \sum_{k=1}^{n} 1/k.$$

We will use the notation

$$H(n) = \begin{cases} 0, & \text{if } n = 0, \\ \sum_{k=1}^{n} 1/k, & \text{if } n > 0 \end{cases} \tag{6.25}$$

for the partial sums of the harmonic series $\sum_{k=1}^{\infty} 1/k$. Using this,

$$a'_{2n}(0) = (-1)^{n+1} \frac{a_0}{2^{2n}(n!)^2} H(n).$$

According to (6.16), a second solution is

$$\frac{\partial y}{\partial s}(0, x) = a_0 y(0, x) \ln x + a_0 \sum_{n=1}^{\infty} \frac{(-1)^{n+1} H(n)}{(n!)^2} \left(\frac{x}{2}\right)^{2n}.$$

Choosing $a_0 = 1$, our solutions are

$$y_1(x) = \sum_{n=0}^{\infty} \frac{(-1)^n}{(n!)^2} \left(\frac{x}{2}\right)^{2n} \quad \text{and}$$

$$y_2(x) = y_1(x) \ln x + \sum_{n=1}^{\infty} \frac{(-1)^{n+1} H(n)}{(n!)^2} \left(\frac{x}{2}\right)^{2n}. \tag{6.26}$$

●

Second solution when $s_1 - s_2$ is a positive integer—the false exception

Suppose that $s_1 - s_2 = N$ is a positive integer. We can use $s = s_2$ and the recurrence formula in (6.11),

$$I(s_2 + n)a_n(s_2) = -\sum_{k=0}^{n-1} \left[p_{n-k}(s_2 + k) + q_{n-k} \right] a_k(s_2), \tag{6.27}$$

to solve for a_n for $1 \le n \le N - 1$. However, since $I(s_2 + N) = I(s_1) = 0$, for $n = N$ the equation becomes

$$0 \cdot a_N(s_2) = -\sum_{k=0}^{N-1} \left[p_{N-k}(s_2 + k) + q_{N-k} \right] a_k(s_2). \tag{6.28}$$

We cannot solve for $a_N(s_2)$ if the right-hand side is not equal to 0.

As we shall see in Example 6.29, sometimes the right-hand side of (6.28) is equal to zero. If so, we can set $a_N(s_2)$ equal to anything we want and proceed to solve for the rest of the coefficients using the recurrence formula. This case is called a *false exception*. Notice that in the case of a false exception, if we start with the smaller root $s = s_2$, then we have two arbitrary constants, a_0 and $a_N(s_2)$. As a result, we will get the general solution without doing any work with the larger root.

Example 6.29   Find a fundamental set of solutions for the Bessel equation of order $r = 1/2$.

From (6.19) we see that the roots are $s_1 = 1/2$ and $s_2 = -1/2$. Notice that $s_1 - s_2 = 1$, a positive integer. We will set $s = s_2 = -1/2$. With this choice equation (6.20) becomes $0 \cdot a_1(-1/2) = 0$. Therefore, we have a false exception and $a_1(-1/2)$ is arbitrary. With $s = s_2 = -1/2$ the recurrence formula (6.21) becomes

$$n(n-1)a_n(-1/2) + a_{n-2}(-1/2) = 0, \quad \text{or} \quad a_n(-1/2) = -\frac{a_{n-2}(-1/2)}{n(n-1)}.$$

For $n = 2k + 1$, an odd integer, this becomes

$$a_{2k+1}(-1/2) = -a_{2k-1}(-1/2)/2k(2k+1).$$

From this we conclude that

$$a_{2k+1}(-1/2) = \frac{(-1)^k a_1(-1/2)}{(2k+1)!}.$$

In the same way we discover that

$$a_{2k}(-1/2) = (-1)^k \frac{a_0}{(2k)!}.$$

Remembering that $s = s_2 = -1/2$, and setting $a_1 = a_1(-1/2)$, the general solution is

$$y(x) = a_0 x^{-1/2} \sum_{k=0}^{\infty} (-1)^k \frac{x^{2k}}{(2k)!} + a_1 x^{-1/2} \sum_{k=0}^{\infty} \frac{(-1)^k x^{2k+1}}{(2k+1)!}, \tag{6.30}$$

where the coefficients a_0 and a_1 are arbitrary.

You may recognize the series in (6.30) as the series for $\cos x$ and $\sin x$. Thus the solution is

$$y(x) = a_0 x^{-1/2} \cos x + a_1 x^{-1/2} \sin x. \qquad \bullet$$

It is important to realize that we began looking for our solution in the previous example with the smaller root $s_2 = -1/2$. In the process we discovered that $a_1 = a_1(-1/2)$ could be given an arbitrary value and as a result we were able to find the general solution without actually using the larger root. This method will always work with a false exception.

Second solution when $s_1 - s_2$ is a positive integer—the real exception

When the right-hand side in (6.28) is not equal to zero, we cannot solve for $a_N(s_2)$ using the recurrence formula. We will call this a *real exception*. Let's look at an example.

Example 6.31   Find a fundamental set of solution to Bessel's equation of order $r = 1$.

From (6.19) we see that the roots are $s_1 = 1$ and $s_2 = -1$. Notice that $s_1 - s_2 = 2$, a positive integer. For $r = 1$, equation (6.20) becomes $s(s+2)a_1(s) = 0$. Hence $a_1(s) = 0$ for $s = s_1 = 1$ or $s = s_2 = -1$. For $r = 1$ the recurrence formula (6.21) becomes

$$(s+n+1)(s+n-1)a_n(s) = -a_{n-2}(s). \tag{6.32}$$

If we set $s = s_2 = -1$, and let $n = 2$, this becomes $0 \cdot a_2(-1) = -a_0$. Since we insist that $a_0 \neq 0$, this equation is not solvable and we have a true exception.

Nevertheless, we will follow the method outlined at the beginning of this section and solve the recurrence formula for the coefficients as functions of s. Since $a_1(s) = 0$, it follows from (6.32) that all odd-numbered coefficients are equal to 0. For the even-numbered coefficients, (6.32) becomes

$$(s + 2n + 1)(s + 2n - 1)a_{2n}(s) = -a_{2n-2}(s), \quad \text{for } n \geq 1.$$

For $n = 1$,

$$a_2(s) = -\frac{a_0}{(s+3)(s+1)}, \tag{6.33}$$

and for $n \geq 1$,

$$a_{2n}(s) = \frac{(-1)^n a_0}{(s+2n+1) \cdot \ldots \cdot (s+3) \cdot (s+2n-1) \cdot \ldots \cdot (s+1)}. \tag{6.34}$$

Notice that $a_{2n}(s)$ is not defined at $s = s_2 = -1$ for $n \geq 1$ because of the factor $s + 1$ in the denominator. Consequently, the function

$$y(s, x) = x^s \sum_{n=0}^{\infty} a_{2n}(s)x^{2n} = \sum_{n=0}^{\infty} a_{2n}(s)x^{s+2n}$$

is not defined at $s = s_2 = -1$.

However, let's continue to follow the plan. The differential operator in this example is

$$Ly = x^2 y'' + xy' - (x^2 - 1)y.$$

According to (6.12), we have

$$Ly(s, x) = a_0(s - s_1)(s - s_2)x^s = a_0(s - 1)(s + 1)x^s. \tag{6.35}$$

It would appear that $Ly(-1, x) = 0$. Unfortunately, we have shown that $y(s, x)$ is not defined for $s = s_2 = -1$.

The trick to finding a solution corresponding to the smaller root is to multiply $y(s, x)$ by the factor $s - s_2 = s + 1$ to get

$$Y(s, x) = (s + 1)y(s, x) = \sum_{n=0}^{\infty}(s + 1)a_{2n}(s)x^{s+2n} = \sum_{n=0}^{\infty} b_{2n}(s)x^{s+2n},$$

where $b_{2n}(s) = (s + 1)a_{2n}(s)$. Multiplying by $s + 1$ has two good effects. To see the first one, let's compute the first two b_{2n}. Using (6.33), we get

$$b_0(s) = (s + 1)a_0 \quad \text{and} \quad b_2(s) = (s + 1)a_2(s) = -\frac{a_0}{s + 3}. \tag{6.36}$$

Notice that $b_2(s)$ is defined for $s = -1$, since multiplying by $s + 1$ cancels the same factor that appears in the denominator of $a_2(s)$. From (6.34) we get the coefficients for $n \geq 1$,

$$b_{2n}(s) = \frac{(-1)^n a_0}{(s+2n+1) \cdot \ldots \cdot (s+3) \cdot (s+2n-1) \cdot \ldots \cdot (s+3)}. \tag{6.37}$$

Again, multiplying by $s + 1$ removes that factor from the denominator, so $b_{2n}(s)$ is defined at $s = -1$ for all n. Hence the function $Y(s, x)$ is defined for $s = -1$.

The second good effect of multiplying $y(s, x)$ by $s + 1$ comes from the fact that

$$LY(s, x) = L(s + 1)y(s, x) = (s + 1)Ly(s, x) = a_0(s - 1)(s + 1)^2 x^s, \tag{6.38}$$

as we see from (6.35). Notice the factor of $(s + 1)^2$ on the right. This is what occurred in equation (6.15), and enabled us to find two solutions in the case when the roots were equal. In exactly the same way, we find that

$$L\left(Y(-1, x)\right) = 0, \quad \text{and} \quad L\left(\frac{\partial Y}{\partial s}(-1, x)\right) = 0.$$

Consequently, we get two solutions

$$Y(-1, x) = x^{-1} \sum_{n=0}^{\infty} b_{2n}(-1)x^n, \quad \text{and} \tag{6.39}$$

$$\frac{\partial Y}{\partial s}(-1, x) = Y(-1, x) \ln x + x^{-1} \sum_{n=0}^{\infty} b'_{2n}(-1)x^n. \tag{6.40}$$

Let's compute them. For $n = 0$ and $n = 1$, setting $s = s_2 = -1$ in (6.36) yields $b_0(-1) = 0$ and $b_2(-1) = -a_0/2$, while (6.37) yields

$$b_{2n}(-1) = \frac{(-1)^n a_0}{2^{2n-1} n!(n-1)!}, \quad \text{for } n \geq 1.$$

Hence, the solution in (6.39) is

$$Y(-1, x) = x^{-1} \sum_{n=0}^{\infty} b_{2n}(-1)x^{2n} = a_0 x^{-1} \sum_{n=1}^{\infty} \frac{(-1)^n x^{2n}}{2^{2n-1} n!(n-1)!}.$$

By shifting the index so the sum starts at $n = 0$ and rearranging things a little, we find that our first solution is

$$Y(-1, x) = -a_0 \sum_{n=0}^{\infty} \frac{(-1)^n}{n!(n+1)!} \left(\frac{x}{2}\right)^{2n+1}. \tag{6.41}$$

For $n = 0$ and $n = 1$ the derivatives are $b'_0(-1) = a_0$ and $b'_2(-1) = a_0/4$. We compute the derivative of the general coefficient using logarithmic differentiation of equation (6.37), obtaining

$$b'_{2n}(-1) = a_0 \frac{(-1)^{n+1}[H(n) + H(n-1)]}{2^{2n} n!(n-1)!}, \quad \text{for } n \geq 1. \tag{6.42}$$

where H is the function defined in (6.25). Therefore, the solution in (6.40) is

$$\frac{\partial Y}{\partial s}(-1, x) = Y(-1, x) \ln x + x^{-1} \sum_{n=0}^{\infty} b'_{2n}(-1)x^{2n}$$

$$= Y(-1, x) \ln x \tag{6.43}$$

$$+ a_0 x^{-1} \left[1 + \sum_{n=1}^{\infty} \frac{(-1)^{n+1}[H(n) + H(n-1)]}{2^{2n} n!(n-1)!} x^{2n} \right].$$

We choose $a_0 = -1$ and get the first solution

$$y_1(x) = Y(-1, x) = \sum_{n=0}^{\infty} \frac{(-1)^n}{n!(n+1)!} \left(\frac{x}{2}\right)^{2n+1}. \tag{6.44}$$

Next we shift the index in the sum in (6.43) and rearrange things to get the second solution

$$y_2(x) = \frac{\partial Y}{\partial s}(-1, x)$$

$$= y_1(x) \ln x - \frac{1}{x} - \frac{1}{2} \sum_{n=0}^{\infty} \frac{(-1)^n[H(n+1) + H(n)]}{n!(n+1)!} \left(\frac{x}{2}\right)^{2n+1}. \tag{6.45}$$

Consequently, we have once more found two linearly independent solutions starting from the smaller root of the indicial equation. It is not necessary to consider the larger root in the exceptional cases, unless we are mystified and want to verify that the solution y_1 in equation (6.44) is actually a solution corresponding to the larger of the two roots. However, that follows from the fact that y_1 is a solution, and the leading term is a multiple of x^{s_1}. ●

The procedure we used in Example 6.31 works in general. However, different examples of the exceptional case may still require different techniques in carrying out the details. Here are some points to keep in mind. We will not prove them. You might try to prove them yourself.

- The recurrence formula in (6.11) enables the computation of all of the coefficients $a_n(s)$. If $n < N = s_1 - s_2$, then $I(s + n) \neq 0$ at $s = s_2$. Consequently, $a_n(s)$ is defined at $s = s_2$. We have seen that this is not the case for $n \geq N$. This has the result that the coefficients $b_n(s) = (s - s_2)a_n(s)$ have a different character for $n < N$ and for $n \geq N$. In particular, we have

$$b_n(s_2) = 0 \quad \text{and} \quad b'_n(s_2) = a_n(s_2), \quad \text{for } 0 \leq n < N.$$

- Because of this, it is a good idea to split the sum in the definition of $Y(s, x)$ as

$$Y(s, x) = (s - s_2)y(s, x)$$

$$= \sum_{n=0}^{N-1} b_n(s)x^{s+n} + \sum_{n=N}^{\infty} b_n(s)x^{s+n}.$$

- For $n = N$, it turns out that $b_N(s_2) = Ba_0$, and the constant $B = 0$ if and only if we have a false exception.

- If we have a true exception, then it follows from the recurrence formula that $b_{N+n}(s_2) = Ba_n(s_1)$ for all $n \geq 0$, with the constant B from the previous bullet. It follows that $Y(s_2, x) = By(s_1, x)$.

- The technique of logarithmic differentiation works well but cannot be applied to b_n if $b_n(s_2) = 0$. However, in such a case, it must be true that $a_n(s)$ is defined at $s = s_2$, and then $b'_n(s_2) = a_n(s_2)$.

The general result

Let's sum up our results in a theorem. We will also extend the results to a general regular, singular point x_0, and to values of x on both sides of x_0.

THEOREM 6.46   Suppose that x_0 is a regular singular point for the differential equation

$$(x - x_0)^2 y'' + (x - x_0)p(x)y' + q(x)y = 0,$$

where the coefficient functions have power series expansions

$$p(x) = \sum_{n=0}^{\infty} p_n(x - x_0)^n \quad \text{and} \quad q(x) = \sum_{n=0}^{\infty} q_n(x - x_0)^n,$$

which converge in the interval $|x - x_0| < R$. Suppose that the indicial equation

$$I(s) = s(s - 1) + p_0 s + q_0 = s^2 + (p_0 - 1)s + q_0 = 0$$

has roots s_1 and s_2, where $s_2 \le s_1$. Define

$$y(s, x) = |x - x_0|^s \sum_{n=0}^{\infty} a_n(s)(x - x_0)^n, \tag{6.47}$$

where a_0 is a nonzero constant and the rest of the coefficients are defined inductively by the recurrence formula

$$I(s + n)a_n(s) = -\sum_{k=0}^{n-1} \left[p_{n-k}(s + k) + q_{n-k} \right] a_k(s), \quad \text{for } n = 1, 2, \dots. \tag{6.48}$$

1. The function $y_1(x) = y(s_1, x)$ is always a solution.
2. If $s_1 - s_2$ is not a nonnegative integer, the function $y_2(x) = y(s_2, x)$ is a second solution.
3. If $s_1 = s_2$, a second solution is given by

$$y_2(x) = y(s_1, x) \ln |x - x_0| + |x - x_0|^{s_1} \sum_{n=1}^{\infty} a_n'(s_1)(x - x_0)^n. \tag{6.49}$$

4. If $s_1 - s_2 = N$ is a positive integer, and the right-hand side of (6.48) vanishes for $n = N$ and $s = s_2$, then with a_N assigned any value, a second solution is given by $y_2(x) = y(s_2, x)$.
5. If $s_1 - s_2 = N$ is a positive integer, and the right-hand side of (6.48) does not vanish for $n = N$ and $s = s_2$, then, with $b_n(s) = (s - s_2)a_n(s)$, a second solution is provided by

$$y_2(x) = y_1(x) \ln |x - x_0|$$
$$+ |x - x_0|^{s_2} \sum_{n=0}^{\infty} b_n'(s_2)(x - x_0)^n. \tag{6.50}$$

The series in all solutions converge at least for $|x - x_0| < R$.   ∎

We end by remarking that in cases 4 and 5, when $s_1 - s_2$ is a positive integer, it is computationally easier to start with the smaller root s_2, since both solutions will turn up, as we illustrated in Examples 6.29 and 6.31.

EXERCISES

In Exercises 1–4, the indicial equation corresponding to the given differential equation has equal roots. Find a fundamental set of solutions for the given differential equation.

1. $x^2 y'' + 3xy' + (1 - 2x)y = 0$
2. $x^2 y'' - x(1 + x)y' + y = 0$
3. $xy'' + (1 - x)y' - y = 0$
4. $x^2 y'' + x(x - 3)y' + 4y = 0$

In Exercises 5–8, the indicial equation corresponding to the given differential equation has roots differing by a positive in-

teger. However, in each case, a false degeneracy occurs. Find a fundamental set of solutions for the given differential equation.

5. $xy'' - (4 + x)y' + 2y = 0$
6. $x^2 y'' + x^2 y' - 2y = 0$
7. $xy'' - (3 + x)y' + 2y = 0$
8. $x^2 y'' + (6x + x^2)y' + xy = 0$

In Exercises 9–12, the indicial equation corresponding to the given differential equation has roots differing by a positive integer. In each case, a real degeneracy occurs. Find a fundamental set of solutions for the given differential equation.

9. $xy'' + y = 0$

10. $x^2 y'' - x(6 + x)y' + 10y = 0$

11. $x^2 y'' - 3xy' + (3 + 4x)y = 0$

12. $xy'' + (3 + 2x)y' + 8y = 0$

Find a fundamental set of solutions for the differential equations in Exercises 13–22.

13. $x^2 y'' + x(2 + 3x)y' - 2y = 0$

14. $2xy'' + 6y' + y = 0$

15. $xy'' + xy' + y = 0$

16. $xy'' - (1 + x^2)y' - 2xy = 0$

17. $xy'' + y' - y = 0$

18. $4x^2 y'' + (1 - 2x)y = 0$

19. $x^2 y'' + xy' - (1 + x^2)y = 0$

20. $x^2 y'' + x(x - 3)y' + 4y = 0$

21. $x^2 y'' + 2x(x - 2)y' + 2(2 - 3x)y = 0$

22. $x^2 y'' + x(3 + x^2)y' + (1 + 3x^2)y = 0$

23. In Section 11.3 we considered

$$(1 - x^2)y'' - 2xy' + n(n + 1)y = 0,$$

which is Legendre's equation of order n.

(a) Show that the points $x = \pm 1$ are regular singular points for Legendre's equation.

(b) Use Theorem 6.46 to show that for any real number n Legendre's equation of order n has one nonzero solution which is bounded at $x_0 = 1$, and that every other solution is unbounded there.

(c) Do the same thing for $x_0 = -1$. (*Remark:* Only when n is an integer is the same solution bounded at both endpoints. These are the Legendre polynomials. However, showing this is a much more difficult problem.)

11.7 Bessel's Equation and Bessel Functions

We have already solved three instances of Bessel's equation of order r,

$$x^2 y'' + xy' + (x^2 - r^2)y = 0. \tag{7.1}$$

In this section we want to solve the other cases and put the equation into a unified context. The solutions are called **Bessel functions**, and they have very important application to the solution of partial differential equations. Notice that Bessel's equations of orders r and $-r$ are the same.

By (6.19) the roots of the indicial polynomial are $s = \pm r$. We will have exceptional cases when $2r$ is an integer. In (6.20) we showed that

$$(s + 1 + r)(s + 1 - r)a_1(s) = 0, \tag{7.2}$$

and in (6.21) we derived the recurrence formula

$$(s + n + r)(s + n - r)a_n(s) = -a_{n-2}(s), \quad \text{for } n \geq 2. \tag{7.3}$$

Solution for $s = r$ when r is not a negative integer

Notice that equation (7.2) is satisfied if we set $a_1(s) = 0$ no matter what r is. Then, if we set $a_n(s) = 0$ for all odd integers n, the recurrence formula in (7.3) is satisfied for n odd.

For the even-numbered coefficients, equation (7.3) becomes

$$(s + 2n + r)(s + 2n - r)a_{2n}(s) = -a_{2n-2}(s), \quad \text{for } n \geq 1. \tag{7.4}$$

For $s = r$ the recurrence formula becomes

$$4n(n + r)a_{2n}(r) = -a_{2n-2}(r), \quad \text{for } n \geq 1.$$

Since r is not a negative integer, the coefficient of a_{2n} is nonzero for all integers $n \geq 1$, the recurrence formula is easily solved inductively to give

$$a_{2n}(r) = (-1)^n \frac{a_0}{n!(r + 1) \cdot (r + 2) \cdot \ldots \cdot (r + n)2^{2n}}, \quad \text{for } n \geq 1. \tag{7.5}$$

Consequently, as long as r is not a negative integer, the function

$$y(r, x) = a_0 x^r \left[1 + \sum_{n=1}^{\infty} \frac{(-1)^n}{n!(r+1) \cdot (r+2) \cdot \ldots \cdot (r+n)} \left(\frac{x}{2}\right)^{2n} \right] \qquad (7.6)$$

is a solution to Bessel's equation of order r.

The gamma function

The formula for the coefficient in (7.5) is rather unwieldy. It could be made easier if we had a better way of writing the product $(r+1) \cdot (r+2) \cdot \ldots \cdot (r+n)$. If r is an integer, we can use the factorial to write

$$(r+1) \cdot (r+2) \cdot \ldots \cdot (r+n) = \frac{(r+n)!}{r!}.$$

However, if r is not an integer, $r!$ is not yet defined so this is not possible.

A convenient way of extending the factorial to numbers that are not integers is provided by the gamma function. It comes up naturally in a variety of circumstances, so we will spend some time discussing it. The ***gamma function*** is defined by

$$\Gamma(x) = \int_0^{\infty} e^{-t} t^{x-1} \, dt, \quad \text{for } x > 0. \qquad (7.7)$$

This improper integral converges at ∞ for all x. For $x < 1$ it is also improper at 0, and it converges there only for $x > 0$. We can easily evaluate

$$\Gamma(1) = \int_0^{\infty} e^{-t} \, dt = \lim_{R \to \infty} \int_0^{R} e^{-t} \, dt = 1.$$

The most important property of the gamma function is its ***functional relationship***

$$\Gamma(x + 1) = x\Gamma(x), \quad \text{for } x > 0. \qquad (7.8)$$

This can be proved using integration by parts. We have

$$\begin{aligned}
\Gamma(x + 1) &= \int_0^{\infty} e^{-t} t^x \, dt = \lim_{R \to \infty} \int_0^{R} e^{-t} t^x \, dt \\
&= \lim_{R \to \infty} -\int_0^{R} t^x \, d(e^{-t}) \\
&= \lim_{R \to \infty} \left\{ -R^x e^{-R} + x \int_0^{R} e^{-t} t^{x-1} \, dt \right\} \\
&= x\Gamma(x).
\end{aligned}$$

Starting with $\Gamma(1) = 1$ and using the functional relationship, we quickly get $\Gamma(2) = 1$, $\Gamma(3) = 1 \cdot 2$, $\Gamma(4) = 1 \cdot 2 \cdot 3$, and generally

$$\Gamma(n + 1) = n!, \quad \text{for any integer } n \geq 0. \qquad (7.9)$$

Thus the gamma function is a function defined for $x > 0$, which interpolates the factorials of the integers.

The functional relationship (7.8) can also be used to extend the domain of the gamma function to include any value of x except for 0 and the negative integers. If we rewrite the functional relationship as

$$\Gamma(x) = \frac{\Gamma(x+1)}{x},$$

we see that although the left-hand side $\Gamma(x)$ is defined by (7.7) only for $x > 0$, the right-hand side is also defined for $-1 < x < 0$. Thus we can use this formula to extend the range of definition of $\Gamma(x)$ to the interval $-1 < x < 0$.

More generally, choose any positive integer n. Then, using the functional relationship several times, we have

$$\begin{aligned} \Gamma(x) &= \frac{\Gamma(x+1)}{x} = \frac{\Gamma(x+2)}{x \cdot (x+1)} = \cdots \\ &= \frac{\Gamma(x+n)}{x \cdot (x+1) \cdot \ldots \cdot (x+n-1)}. \end{aligned} \tag{7.10}$$

Notice that $\Gamma(x + n)$ is defined by (7.7) for $x > -n$, so formula (7.10) provides a definition for $\Gamma(x)$ for all $x > -n$, except for $x = 0, -1, \cdots, -(n-1)$. The graph of the gamma function is shown in Figure 1.

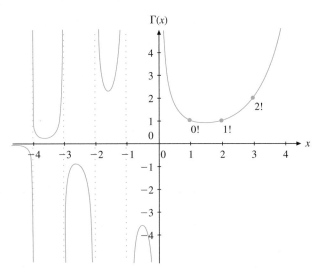

Figure I. The gamma function $\Gamma(x)$.

It will be useful to know the value of $\Gamma(x)$ when x is a half-integer. Using the function relationship, we see that if we know $\Gamma(1/2)$, we will be able to compute the rest. First, we notice that with the substitution $t = y^2$, we get

$$\Gamma(1/2) = \int_0^\infty e^{-t} t^{-1/2} \, dt = 2 \int_0^\infty e^{-y^2} \, dy.$$

The new integral can be computed using the unusual trick of replacing the single integral with a multiple integral, which is then evaluated using polar coordinates.

We have

$$\Gamma(1/2)^2 = 2 \int_0^\infty e^{-x^2}\, dx \cdot 2 \int_0^\infty e^{-y^2}\, dy$$

$$= 4 \int_0^\infty \int_0^\infty e^{-(x^2+y^2)}\, dx\, dy$$

$$= 4 \int_0^{\pi/2} \int_0^\infty e^{-r^2} r\, dr\, d\theta$$

$$= 4 \cdot \frac{\pi}{2} \cdot \left(-\frac{e^{-r^2}}{2} \right) \Big|_0^\infty$$

$$= \pi.$$

Hence

$$\Gamma(1/2) = \sqrt{\pi}. \tag{7.11}$$

Back to Bessel functions

In keeping with formula (7.9), we will define the factorial by

$$x! = \Gamma(x+1) \quad \text{for } x \text{ not a negative integer.}$$

Notice that using the functional relationship (7.8), we have

$$(r+n)! = \Gamma(r+n+1) = \Gamma(r+1) \cdot (r+1) \cdot (r+2) \cdot \ldots \cdot (r+n)$$
$$= r! \cdot (r+1) \cdot (r+2) \cdot \ldots \cdot (r+n).$$

We can now simplify the coefficient $a_{2n}(r)$ in (7.5) to

$$a_{2n}(r) = (-1)^n \frac{a_0 r!}{n!(r+n)! 2^{2n}} \quad \text{for } n \geq 1.$$

Since this formula is valid for $n = 0$ as well, the solution in (7.6) becomes

$$y(r,x) = a_0 r! 2^r \left(\frac{x}{2}\right)^r \sum_{n=0}^\infty \frac{(-1)^n}{n!(r+n)!} \left(\frac{x}{2}\right)^{2n}.$$

The coefficient a_0 can be assigned any nonzero value so we set $a_0 = 1/(r! 2^r)$ to simplify the notation. Then our solution is denoted by

$$J_r(x) = \sum_{n=0}^\infty \frac{(-1)^n}{n!(r+n)!} \left(\frac{x}{2}\right)^{2n+r} = \left(\frac{x}{2}\right)^r \sum_{n=0}^\infty \frac{(-1)^n}{n!(r+n)!} \left(\frac{x}{2}\right)^{2n}. \tag{7.12}$$

The function J_r is called the ***Bessel function of the first kind, of order r***. Notice that J_r is defined for any real number r that is not a negative integer, and it is a solution to Bessel's equation of order $|r|$. Examples of Bessel functions of the first kind are shown in Figure 2.

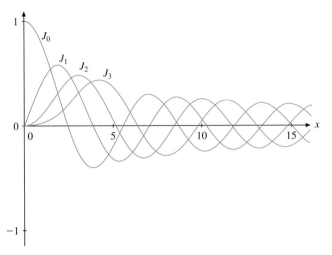

Figure 2. Four examples of Bessel functions of the first kind.

The second solution when $r < 0$ is not an integer

Since $s_2 = -r$ is not a negative integer, a second solution is

$$J_{-r}(x) = \sum_{n=0}^{\infty} \frac{(-1)^n}{n!(n-r)!} \left(\frac{x}{2}\right)^{2n-r} = \left(\frac{x}{2}\right)^{-r} \sum_{n=0}^{\infty} \frac{(-1)^n}{n!(n-r)!} \left(\frac{x}{2}\right)^{2n}. \qquad (7.13)$$

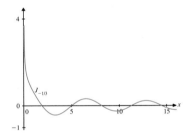

Figure 3.

Notice that for x near 0, J_{-r} acts like x^{-r} and is therefore unbounded, while J_r acts like x^r and is bounded. Since $r \neq 0$, these two solutions are linearly independent, and form a fundamental set of solutions. See Figure 3 for a plot of $J_{-1/3}$.

The second solution when $r = k + 1/2$ is a positive half-integer

In this case $2r = 2k + 1$ is an odd integer. This is an exceptional case. However, since r is not an integer, we already have found two linearly independent Frobenius solutions $J_{k+1/2}$ and $J_{-(k+1/2)}$. This indicates that when r ia a half-integer we have a false exception. By setting the odd numbered coefficients equal to zero at the very beginning, we have obviated the need to analyze this case separately. However, we refer you to Example 6.29 for the case when $r = 1/2$.

When $r = k + 1/2$ is a half-integer, the Bessel functions are closely related to the trigonometric functions. For $r = 1/2$, we leave it as an exercise to show that

$$J_{1/2}(x) = \sqrt{\frac{2}{\pi x}} \sin x \quad \text{and} \quad J_{-1/2}(x) = \sqrt{\frac{2}{\pi x}} \cos x. \qquad (7.14)$$

These functions are plotted in Figure 4.

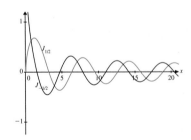

Figure 4.

The second solution when $r = 0$

This case was explored in Example 6.22. There we found the solutions

$$y_1(x) = \sum_{n=0}^{\infty} \frac{(-1)^n}{(n!)^2} \left(\frac{x}{2}\right)^{2n} \quad \text{and}$$

$$y_2(x) = y_1(x) \ln x + \sum_{n=1}^{\infty} \frac{(-1)^{n+1} H(n)}{(n!)^2} \left(\frac{x}{2}\right)^{2n}, \qquad (7.15)$$

where the function H is defined in (6.25). Examination shows that $y_1(x) = J_0(x)$. We will set

$$
\begin{aligned}
Y_0(x) &= \frac{2}{\pi} \left[y_2(x) + (\gamma - \ln 2) J_0(x) \right] \\
&= \frac{2}{\pi} \left[\left(\ln \frac{x}{2} + \gamma \right) J_0(x) + \sum_{n=1}^{\infty} \frac{(-1)^{n+1} H(n)}{(n!)^2} \left(\frac{x}{2} \right)^{2n} \right],
\end{aligned}
\tag{7.16}
$$

where γ is **Euler's constant**, defined by

$$
\begin{aligned}
\gamma &= \lim_{n \to \infty} \left[H(n) - \ln n \right] \\
&= \lim_{n \to \infty} \left[1 + \frac{1}{2} + \frac{1}{3} + \cdots + \frac{1}{n} - \ln n \right] \\
&= 0.5772156 \cdots.
\end{aligned}
\tag{7.17}
$$

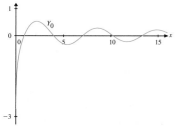

Figure 5.

Y_0 is called the **Bessel function of the second kind of order 0**. Y_0 is plotted in Figure 5.

Second solution when r is a positive integer

This time we have a truly exceptional case, and we have to use part 5 of Theorem 6.46. The case $r = 1$ was examined in Example 6.31. There we found the solutions

$$
\begin{aligned}
y_1(x) &= \sum_{n=0}^{\infty} \frac{(-1)^n}{n!(n+1)!} \left(\frac{x}{2} \right)^{2n+1}, \quad \text{and} \\
y_2(x) &= y_1(x) \ln x - \frac{1}{x} - \frac{1}{2} \sum_{n=0}^{\infty} \frac{(-1)^n [H(n+1) + H(n)]}{n!(n+1)!} \left(\frac{x}{2} \right)^{2n+1}.
\end{aligned}
\tag{7.18}
$$

Examination shows that $y_1(x) = J_1(x)$. We will set

$$
Y_1(x) = \frac{2}{\pi} \left[y_2(x) + (\gamma - \ln 2) J_1(x) \right].
\tag{7.19}
$$

This is a special case of the most common form of a second solution, called the **Bessel function of the second kind of integer order r**:

$$
\begin{aligned}
Y_r(x) = \frac{2}{\pi} &\left[\left(\ln \frac{x}{2} + \gamma \right) J_r(x) - \frac{1}{2} \sum_{n=0}^{r-1} \frac{(r-n-1)!}{n!} \left(\frac{x}{2} \right)^{2n-r} \right. \\
&\left. + \frac{1}{2} \sum_{n=0}^{\infty} \frac{(-1)^{n+1} [H(n) + H(n+r)]}{n!(r+n)!} \left(\frac{x}{2} \right)^{2n+r} \right].
\end{aligned}
\tag{7.20}
$$

It is important for applications that we notice that the second solution is unbounded near $x = 0$. We can see this because the first sum in (7.20) begins with the term x^{-r}.

Bessel functions of the second kind

Except when r is an integer, a fundamental set of solutions to Bessel's equation of order r is given by J_r and J_{-r}. The general solution is

$$a J_r(x) + b J_{-r}(x), \tag{7.21}$$

where a and b are arbitrary constants. However, it is frequently convenient to use

$$Y_r(x) = \frac{\cos(r\pi) J_r(x) - J_{-r}(x)}{\sin(r\pi)} \tag{7.22}$$

as the second solution when r is not an integer. Y_r is called the **Bessel function of the second kind of order** r.

Since Y_r is a linear combination of J_r and J_{-r}, it is a solution to Bessel's equation. Since the coefficient of J_{-r} in this linear combination is nonzero, it is linearly independent from J_r. However, it does seem like a strange choice. There are at least two reasons for this particular choice. The first reason is that, with some effort, it can be shown that

$$\lim_{r \to k} Y_r(x) = Y_k(x),$$

when k is a nonnegative integer. Notice that for integer subscripts the Bessel functions of the second kind were previously defined in (7.16) and (7.20). Thus the family Y_r is continuous in the parameter r.

The second reason comes from the asymptotic behavior of Bessel's functions as x gets very large. In Exercise 8 you are asked to show that if $y(x)$ is a solution to Bessel's equation of order r, then $u(x) = x^{1/2} y(x)$ satisfies the equation

$$u'' + \left(1 - \frac{r^2 - 1/4}{x^2}\right) u = 0. \tag{7.23}$$

The coefficient of u in (7.23) gets very close to 1 as x gets large, so the equation approaches $u'' + u = 0$. This equation has the trigonometric functions $\cos(x - \alpha)$ and $\sin(x - \alpha)$ as solutions for any choice of α. It might therefore be expected that solutions to (7.23) are closely related to these trigonometric functions. This is true, although it is not easily demonstrated. The precise facts are that

$$J_r(x) = \sqrt{\frac{2}{\pi x}} \cos(x - \alpha_r) + \frac{R_1(x)}{x^{3/2}}$$
$$Y_r(x) = \sqrt{\frac{2}{\pi x}} \sin(x - \alpha_r) + \frac{R_2(x)}{x^{3/2}}, \tag{7.24}$$

where $\alpha_r = \pi(2r + 1)/4$, and the functions $R_1(x)$ and $R_2(x)$ are bounded as $x \to \infty$. It is precisely for the linear combination in (7.22) that we have the second equation in (7.24).

From (7.24) we see that the Bessel functions are oscillatory functions that decrease in amplitude like $1/\sqrt{x}$. Since the remainder term approaches 0 at the faster rate of $1/x^{3/2}$, the agreement between the Bessel function and the damped trigonometric functions gets better as x increases. The agreement between the two is really quite remarkable. For $r = 1$ the situation is illustrated in Figure 6. The graphs of the Bessel functions are in blue, and the corresponding damped trigonometric functions from (7.24) are plotted in black.

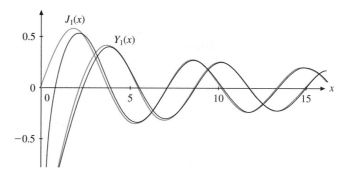

Figure 6. Asymptotic agreement between Bessel functions and damped trigonometric functions.

The zeros of the Bessel functions

It follows from (7.24) that the Bessel functions have infinitely many zeros in the interval $0 < x < \infty$. Furthermore, from (7.24) we conclude that the zeros of J_r must be close to the zeros of $\cos(x - \alpha_r)$, which have the form $x = \alpha_r + k\pi + \pi/2$, where k is any nonnegative integer. In particular, if $r = n$ is a nonnegative integer, and $\alpha_{n,k}$ is the kth zero of J_n, we would expect that $\alpha_{n,k} \approx (2n + 3)\pi/4 + k\pi$. The first few zeros of the some of the Bessel functions are shown in Table 1. We leave it to you to examine how close the approximation really is.

Table 1 The zeros $\alpha_{n,k}$ of the Bessel function J_n

n	$k = 1$	$k = 2$	$k = 3$	$k = 4$	$k = 5$	$k = 6$	$k = 7$
0	2.4048	5.5201	8.6537	11.7915	14.9309	18.0711	21.2116
1	3.8317	7.0156	10.1735	13.3237	16.4706	19.6159	22.7601
2	5.1356	8.4172	11.6198	14.7960	17.9598	21.1170	24.2701
3	6.3802	9.7610	13.0152	16.2235	19.4094	22.5827	25.7482
4	7.5883	11.0647	14.3725	17.6160	20.8269	24.0190	27.1991
5	8.7715	12.3386	15.7002	18.9801	22.2178	25.4303	28.6266

EXERCISES

1. Derive the formulas in (7.14) using the series expansion for the Bessel functions.

There are many identities relating the Bessel functions of different orders and their derivatives. In Exercises 2–7, you are asked first to derive them, and then to use them.

2. By differentiating the series term by term, prove that

$$[x^p J_p]' = x^p J_{p-1}.$$

3. By differentiating the series term by term, prove that

$$[x^{-p} J_p]' = -x^{-p} J_{p+1}.$$

4. Using Exercises 2 and 3, show that

$$J_p' = J_{p-1} - (p/x)J_p \quad \text{and} \quad J_p' = -J_{p+1} + (p/x)J_p.$$

5. Using Exercise 4, show that

$$J_{p+1}(x) = \frac{2p}{x}J_p(x) - J_{p-1}(x) \quad \text{and} \quad 2J_p' = J_{p-1} - J_{p+1}.$$

6. Use Exercises 1 and 5 to show that

$$J_{3/2}(x) = \sqrt{\frac{2}{\pi x}} \frac{\sin x - x \cos x}{x}.$$

7. Using Rolle's theorem and Exercise 4, show that between any two positive zeros of J_0 there is a zero of J_1, and that between any two positive zeros of J_1 there is a zero of J_0.

Many differential equations which arise in applications can be transformed into Bessel's equation. In Exercises 8–12 we will explore some of the possibilities.

8. Show that the substitution $y(x) = x^{-1/2}u(x)$ transforms Bessel's equation (7.1) into

$$u'' + \left(1 - \frac{r^2 - 1/4}{x^2}\right)u = 0.$$

Use this to show that $x^{-1/2} \sin x$ and $x^{-1/2} \cos x$ are solutions to Bessel's equation of order 1/2.

9. Show that the substitution $y(x) = u(\mu x)$ transforms the equation $x^2 y'' + xy' + (\mu^2 x^2 - p^2)y = 0$ into Bessel's equation of order p and the general solution is $y(x) = a J_p(\mu x) + b Y_p(\mu x)$.

10. Airy's equation is

$$\frac{d^2 y}{dx^2} + xy = 0.$$

Show that if we set $u = x^{-1/2} y$ and $t = \frac{2}{3} x^{3/2}$, Airy's equation becomes

$$t^2 \frac{d^2 u}{dt^2} + t \frac{du}{dt} + (t^2 - 1/9)u = 0.$$

Show that the general solution of Airy's equation is

$$y(x) = x^{1/2} \left[a J_{1/3}(2x^{3/2}/3) + b J_{-1/3}(2x^{3/2}/3) \right].$$

11. Exercises 8–10 are special cases of a more general transformation formula that is useful for recognizing some of the equations which can be transformed into Bessel's equation, which we write as

$$t^2 \frac{d^2 u}{dt^2} + t \frac{du}{dt} + (t^2 - r^2)u = 0.$$

Show that the substitutions $t = ax^b$ and $u = x^c y$, where a, b, and c are constants, transform this equation into

$$x^2 \frac{d^2 y}{dx^2} + \alpha x \frac{dy}{dx} + [\beta^2 x^{2b} + \gamma]y = 0,$$

where $\alpha = 2c + 1$, $\beta = ab$, and $\gamma = c^2 - r^2 b^2$.

12. An undamped spring with a spring constant that is decaying exponentially over time is modeled by the equation

$$m \frac{d^2 y}{dt^2} + ke^{-\epsilon t} y = 0. \qquad (7.25)$$

Show that with the change of variables $x = e^{-\epsilon t/2}$ this equation becomes

$$x^2 \frac{d^2 y}{dx^2} + x \frac{dy}{dx} + \frac{4k}{m\epsilon^2} x^2 y = 0.$$

With the assistance of Exercise 9, show that the general solution to equation (7.25) is

$$y(t) = a J_0(\mu e^{-\epsilon t/2}) + b Y_0(\mu e^{-\epsilon t/2}),$$

where $\mu = 2\sqrt{k/m}/\epsilon$.

13. Use the method of Section 11.6 to derive the general solution to Bessel's equation of order 1.

12

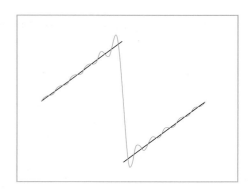

Fourier Series

In 1807, the French mathematician and physicist Joseph Fourier submitted a paper on heat conduction to the Academy of Sciences of Paris. In this paper Fourier made the claim that any function $f(x)$ can be expanded into an infinite sum of trigonometric functions,

$$f(x) = \frac{a_0}{2} + \sum_{k=1}^{\infty} [a_k \cos(kx) + b_k \sin(kx)], \quad \text{for } -\pi \le x \le \pi.$$

The paper was rejected after it was read by some of the leading mathematicians of his day. They objected to the fact that Fourier had not presented much in the way of proof for this statement, and most of them did not believe it.

In spite of its less than glorious start, Fourier's paper was the impetus for major developments in mathematics and in the application of mathematics. His ideas forced mathematicians to come to grips with the definition of a function. This, together with other metamathematical questions, caused nineteenth-century mathematicians to rethink completely the foundations of their subject, and to put it on a more rigorous foundation. Fourier's ideas gave rise to a new part of mathematics, called harmonic analysis or Fourier analysis. This, in turn, fostered the introduction at the end of the nineteenth century of a completely new theory of integration, now called the Lebesgue integral.

The applications of Fourier analysis outside of mathematics continue to multiply. One important application pertains to signal analysis. Here, $f(x)$ could represent the amplitude of a sound wave, such as a musical note, or an electrical signal from a CD player or some other device (in this case x represents time and is usually replaced by t). The Fourier series representation of a signal represents a decomposition of this signal into its various frequency components. The terms $\sin kx$ and

$\cos kx$ oscillate with numerical frequency[1] of $k/2\pi$. Signals are often corrupted by noise, which usually involves the high-frequency components (when k is large). Noise can sometimes be filtered out by setting the high-frequency coefficients (the a_k and b_k when k is large) equal to zero.

Data compression is another increasingly important problem. One way to accomplish data compression uses Fourier series. Here the goal is to be able to store or transmit the essential parts of a signal using as few bits of information as possible. The Fourier series approach to the problem is to store (or transmit) only those a_k and b_k that are larger than some specified tolerance and discard the rest. Fortunately, an important theorem (the Riemann-Lebesgue lemma, which is our Theorem 2.11) assures us that only a small number of Fourier coefficients are significant, and therefore the aforementioned approach can lead to significant data compression.

12.1 Computation of Fourier Series

The problem that we wish to address is the one faced by Fourier. Suppose that $f(x)$ is a given function on the interval $[-\pi, \pi]$. Can we find coefficients, a_n and b_n, so that

$$f(x) = \frac{a_0}{2} + \sum_{n=1}^{\infty} [a_n \cos nx + b_n \sin nx], \quad \text{for } -\pi \le x \le \pi? \quad (1.1)$$

Notice that, except for the term $a_0/2$, the series is an infinite linear combination of the basic terms $\sin nx$ and $\cos nx$ for n a positive integer. These functions are periodic with period $2\pi/n$, so their graphs trace through n periods over the interval $[-\pi, \pi]$. Figure 1 shows the graphs of $\cos x$ and $\cos 5x$, and Figure 2 shows the graphs of $\sin x$ and $\sin 5x$. Notice how the functions become more oscillatory as n increases.

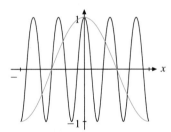

Figure 1. The graphs of $\cos x$ and $\cos 5x$.

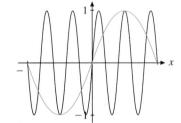

Figure 2. The graphs of $\sin x$ and $\sin 5x$.

The orthogonality relations

Our task of finding the coefficients a_n and b_n for which (1.1) is true is facilitated by the following lemma. These orthogonality relations are one of the keys to the whole theory of Fourier series.

LEMMA 1.2 Let p and q be positive integers. Then we have the following **orthogonality relations**.

$$\int_{-\pi}^{\pi} \sin px \, dx = \int_{-\pi}^{\pi} \cos px \, dx = 0 \quad (1.3)$$

$$\int_{-\pi}^{\pi} \sin px \cos qx \, dx = 0 \quad (1.4)$$

---

[1] Be sure you know the difference between angular frequency, k in this case, and numerical frequency. It is explained in Section 4.1.

$$\int_{-\pi}^{\pi} \cos px \cos qx \, dx = \begin{cases} \pi, & \text{if } p = q \\ 0, & \text{if } p \neq q \end{cases} \tag{1.5}$$

$$\int_{-\pi}^{\pi} \sin px \sin qx \, dx = \begin{cases} \pi, & \text{if } p = q \\ 0, & \text{if } p \neq q \end{cases} \tag{1.6}$$

We will leave the proof of these identities for the exercises.

Computation of the coefficients

The orthogonality relations enable us to find the coefficients a_n and b_n in (1.1). Suppose we are given a function f that can be expressed as

$$f(x) = \frac{a_0}{2} + \sum_{k=1}^{\infty} [a_k \cos kx + b_k \sin kx] \tag{1.7}$$

on the interval $[-\pi, \pi]$. To find a_0, we simply integrate the series (1.7) term by term. Using the orthogonality relation (1.3), we see that

$$\int_{-\pi}^{\pi} f(x) \, dx = a_0 \pi. \tag{1.8}$$

To find a_n for $n \geq 1$, we multiply both sides of (1.7) by $\cos nx$ and integrate term by term, getting

$$\int_{-\pi}^{\pi} f(x) \cos nx \, dx = \int_{-\pi}^{\pi} \left(\frac{a_0}{2} + \sum_{k=1}^{\infty} [a_k \cos kx + b_k \sin kx] \right) \cos nx \, dx$$

$$= \frac{a_0}{2} \int_{-\pi}^{\pi} \cos nx \, dx$$

$$+ \sum_{k=1}^{\infty} a_k \int_{-\pi}^{\pi} \cos kx \cos nx \, dx \tag{1.9}$$

$$+ \sum_{k=1}^{\infty} b_k \int_{-\pi}^{\pi} \sin kx \cos nx \, dx.$$

Using the orthogonality relations in Lemma 1.2, we see that all the terms on the right-hand side of (1.9) are equal to zero, except for

$$a_n \int_{-\pi}^{\pi} \cos nx \cos nx \, dx = a_n \cdot \pi.$$

Hence, equation (1.9) becomes

$$\int_{-\pi}^{\pi} f(x) \cos nx \, dx = a_n \cdot \pi, \quad \text{for } n \geq 1,$$

so, including equation (1.8), [2]

$$a_n = \frac{1}{\pi} \int_{-\pi}^{\pi} f(x) \cos nx \, dx, \quad \text{for } n \geq 0. \tag{1.10}$$

To find b_n, we multiply equation (1.7) by $\sin nx$ and then integrate. By reasoning similar to the computation of a_n, we obtain

$$b_n = \frac{1}{\pi} \int_{-\pi}^{\pi} f(x) \sin nx \, dx, \quad \text{for } n \geq 1. \tag{1.11}$$

---

[2] We used the expression $a_0/2$ instead of a_0 for the constant term in the Fourier series (1.7) so formulas like equation (1.10) would be true for $n = 0$ as well as for larger n.

Definition of Fourier series

We define the **right-hand and left-hand limits** of a function f at a point x_0 to be

$$f(x_0^+) = \lim_{x \to x_0^+} f(x) \quad \text{and} \quad f(x_0^-) = \lim_{x \to x_0^-} f(x).$$

The notation $x \to x_0^+$ under the first limit means that the limit is taken as x approaches x_0 from the right where $x > x_0$. The notation $x \to x_0^-$ has a similar interpretation. A function f defined on an interval I is **piecewise continuous** if it has only finitely many points of discontinuity, and if both the left- and right-hand limits exist at every point of discontinuity.

If f is a piecewise continuous function on the interval $[-\pi, \pi]$, we can compute the coefficients a_n and b_n using (1.10) and (1.11). Thus we can define the Fourier series for any such function.

DEFINITION 1.12

Suppose that f is a piecewise continuous function on the interval $[-\pi, \pi]$. With the coefficients computed using (1.10) and (1.11), we define the **Fourier series associated to** f by

$$f(x) \sim \frac{a_0}{2} + \sum_{n=1}^{\infty} [a_n \cos nx + b_n \sin nx]. \tag{1.13}$$

The finite sum

$$S_N(x) = \frac{a_0}{2} + \sum_{n=1}^{N} [a_n \cos nx + b_n \sin nx] \tag{1.14}$$

is called the **partial sum of order** N for the Fourier series in (1.13). We say that the Fourier series converges at x if the sequence of partial sums converges at x as $N \to \infty$. We use the symbol $\sim$ in (1.13) because we cannot be sure that the series converges. We will explore the question of convergence in the next section, and we will see in Theorem 2.3 that for functions that are minimally well behaved, the $\sim$ can be replaced by an equals sign for most values of x.

Example 1.15 Find the Fourier series associated with the function

$$f(x) = \begin{cases} 0, & \text{for } -\pi \le x < 0, \\ \pi - x, & \text{for } 0 \le x \le \pi. \end{cases}$$

We compute the coefficient a_0 using (1.8) or (1.10). We have

$$a_0 = \frac{1}{\pi} \int_{-\pi}^{\pi} f(x)\,dx = \frac{1}{\pi} \int_0^{\pi} (\pi - x)\,dx = \frac{\pi}{2}.$$

For $n \ge 1$, we use (1.10), and integrate by parts to get

$$a_n = \frac{1}{\pi} \int_{-\pi}^{\pi} f(x) \cos nx\,dx = \frac{1}{\pi} \int_0^{\pi} (\pi - x) \cos nx\,dx$$

$$= \frac{1}{n\pi} \int_0^{\pi} (\pi - x)\,d(\sin nx)$$

$$= \frac{1}{n\pi} (\pi - x) \sin nx \Big|_0^{\pi} + \frac{1}{n\pi} \int_0^{\pi} \sin nx\,dx$$

$$= \frac{1}{n^2\pi} (1 - \cos n\pi).$$

Thus, since $\cos n\pi = (-1)^n$, the even numbered coefficients are $a_{2n} = 0$, and the odd numbered coefficients are $a_{2n+1} = 2/[\pi(2n+1)^2]$ for $n \geq 0$.

We compute b_n using (1.11). Again we integrate by parts to get

$$
\begin{aligned}
b_n &= \frac{1}{\pi} \int_{-\pi}^{\pi} f(x) \sin nx\, dx = \frac{1}{\pi} \int_0^{\pi} (\pi - x) \sin nx\, dx \\
&= -\frac{1}{n\pi} \int_0^{\pi} (\pi - x)\, d(\cos nx) \\
&= -\frac{1}{n\pi}(\pi - x)\cos nx \Big|_0^{\pi} - \frac{1}{n\pi} \int_0^{\pi} \cos nx\, dx \\
&= \frac{1}{n}.
\end{aligned}
$$

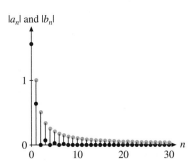

$|a_n|$ and $|b_n|$

The magnitude of the coefficients is plotted in Figure 3, with $|a_n|$ in black and $|b_n|$ in blue. Notice how the coefficients decay to 0. The Fourier series for f is

$$
f(x) \sim \frac{\pi}{4} + \frac{2}{\pi} \sum_{n=0}^{\infty} \frac{\cos(2n+1)x}{(2n+1)^2} + \sum_{n=1}^{\infty} \frac{\sin nx}{n}. \tag{1.16}
$$

Figure 3. The Fourier coefficients for the function in Example 1.15.

Let's examine the experimental evidence for convergence of the Fourier series in Example 1.15. The partial sums of orders 3, 30, and 300 for the Fourier series in Example 1.15 are shown in Figures 4, 5, and 6, respectively. In these figures the function f is plotted in black and the partial sum in blue. The evidence of these figures is that the Fourier series converges to $f(x)$, at least away from the discontinuity of the function at $x = 0$.

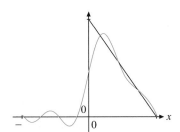

Figure 4. The partial sum of order 3 for the function in Example 1.15.

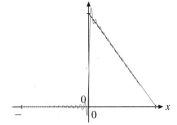

Figure 5. The partial sum of order 30 for the function in Example 1.15.

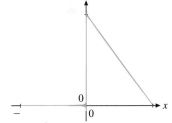

Figure 6. The partial sum of order 300 for the function in Example 1.15.

Fourier series on a more general interval

It is very natural to consider functions defined on $[-\pi, \pi]$ when studying Fourier series because in applications the argument x is frequently an angle. However, in other applications (such as heat transfer and the vibrating string) the argument represents a length. In such a case it is more natural to assume that x is in an interval of the form $[-L, L]$. It is a matter of a simple change of variable to go from $[-\pi, \pi]$ to a more general integral.

Suppose that $f(x)$ is defined for $-L \leq x \leq L$. Then the function $F(y) = f(Ly/\pi)$ is defined for $-\pi \leq y \leq \pi$. For F we have the Fourier series defined in

Definition 1.12. Using the formula $y = \pi x/L$, the coefficients a_n are given by

$$a_n = \frac{1}{\pi} \int_{-\pi}^{\pi} F(y) \cos ny \, dy$$

$$= \frac{1}{\pi} \int_{-\pi}^{\pi} f\left(\frac{Ly}{\pi}\right) \cos ny \, dy$$

$$= \frac{1}{L} \int_{-L}^{L} f(x) \cos \frac{n\pi x}{L} \, dx.$$

The formula for b_n is derived similarly. Thus equations (1.10) and (1.11) are the special case for $L = \pi$ of the following more general result.

THEOREM 1.17   If $f(x) = a_0/2 + \sum_{n=1}^{\infty}[a_n \cos(n\pi x/L) + b_n \sin(n\pi x/L)]$ for $-L \le x \le L$, then

$$a_n = \frac{1}{L} \int_{-L}^{L} f(x) \cos \frac{n\pi x}{L} \, dx, \quad \text{for } n \ge 0, \tag{1.18}$$

$$b_n = \frac{1}{L} \int_{-L}^{L} f(x) \sin \frac{n\pi x}{L} \, dx, \quad \text{for } n \ge 1. \tag{1.19}$$

∎

Keep in mind that Theorem 1.17 only shows that *if* f can be expressed as a Fourier series, then the coefficients a_n and b_n must be given by the formulas in (1.18) and (1.19). The theorem does not say that an arbitrary function can be expanded into a convergent Fourier series.

The case when $n = 0$ in (1.18) deserves special attention. Since $\cos 0 = 1$, it says

$$a_0 = \frac{1}{L} \int_{-L}^{L} f(x) \, dx.$$

Thus $a_0/2$ is the average of f over the interval $[-L, L]$.

We will also extend Definition 1.12 to functions defined on the interval $[-L, L]$.

DEFINITION 1.20   Suppose that f is a piecewise continuous function on the interval $[-L, L]$. With the coefficients computed using (1.18) and (1.19), we define the *Fourier series associated to* f by

$$f(x) \sim \frac{a_0}{2} + \sum_{n=1}^{\infty}\left[a_n \cos\left(\frac{n\pi x}{L}\right) + b_n \sin\left(\frac{n\pi x}{L}\right)\right]. \tag{1.21}$$

Even and odd functions

The computation of the Fourier coefficients can often be facilitated by taking note of the symmetries of the function f.

DEFINITION 1.22   A function $f(x)$ defined on an interval $-L \le x \le L$ is said to be *even* if $f(-x) = f(x)$ for $-L \le x \le L$, and *odd* if $f(-x) = -f(x)$ for $-L \le x \le L$.

The graph of an even function is symmetric about the y-axis as shown in Figure 7. Examples include $f(x) = x^2$ and $f(x) = \cos x$. The graph of an odd function is symmetric about the origin as shown in Figure 8. Examples include $f(x) = x^3$ and $f(x) = \sin x$.

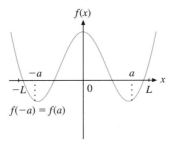

Figure 7. The graph of an even function.

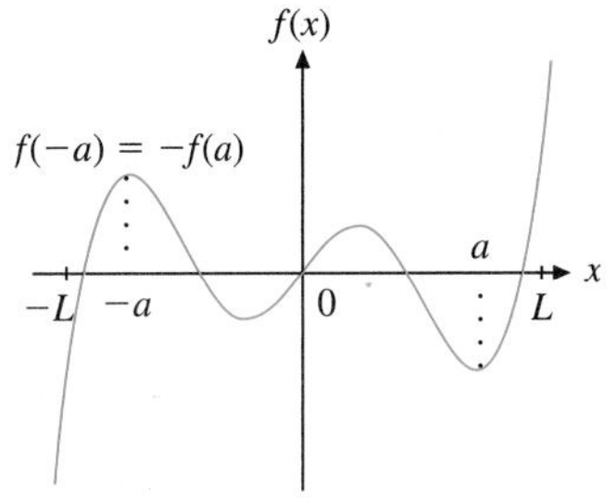

Figure 8. The graph of an odd function.

The following properties follow from the definition.

PROPOSITION 1.23   Suppose that f and g are defined on the interval $-L \le x \le L$.

1. If both f and g are even, then fg is even.
2. If both f and g are odd, then fg is even.
3. If f is even and g is odd, then fg is odd.
4. If f is even, then
$$\int_{-L}^{L} f(x)\,dx = 2\int_{0}^{L} f(x)\,dx.$$
5. If f is odd, then
$$\int_{-L}^{L} f(x)\,dx = 0.$$
●

We will leave the proof for the exercises. If we remember that the integral of f computes the algebraic area under the graph of f, parts 4 and 5 of Proposition 1.23 can be seen in Figures 7 and 8.

The Fourier coefficients of even and odd functions

Parts 4 and 5 of Proposition 1.23 simplify the computation of the Fourier coefficients of a function that is either even or odd. For example, if f is even, then, since $\sin(n\pi x/L)$ is odd, $f(x)\sin(n\pi x/L)$ is odd by part 3 of Proposition 1.23, and by part 5,
$$b_n = \frac{1}{L}\int_{-L}^{L} f(x)\sin\frac{n\pi x}{L}\,dx = 0.$$

Consequently, no computations are necessary to find b_n. Using similar reasoning, we see that $f(x)\cos(n\pi x/L)$ is even, and therefore
$$a_n = \frac{1}{L}\int_{-L}^{L} f(x)\cos\frac{n\pi x}{L}\,dx = \frac{2}{L}\int_{0}^{L} f(x)\cos\frac{n\pi x}{L}\,dx.$$

Frequently integrating from 0 to L is simpler than integrating from $-L$ to L.

Just the opposite occurs for an odd function. In this case the a_n are zero and the b_n can be expressed as an integral from 0 to L. We will leave this as an exercise. We summarize the preceding discussion in the following theorem.

THEOREM 1.24   Suppose that f is piecewise continuous on the interval $[-L, L]$.

1. If $f(x)$ is an even function, then its associated Fourier series will involve only the cosine terms. That is, $f(x) \sim a_0/2 + \sum_{n=1}^{\infty} a_n \cos(n\pi x/L)$ with

$$a_n = \frac{2}{L} \int_0^L f(x) \cos \frac{n\pi x}{L} \, dx, \quad \text{for } n \geq 0.$$

2. If $f(x)$ is an odd function, then its associated Fourier series will involve only the sine terms. That is, $f(x) \sim \sum_{n=1}^{\infty} b_n \sin(n\pi x/L)$ with

$$b_n = \frac{2}{L} \int_0^L f(x) \sin \frac{n\pi x}{L} \, dx, \quad \text{for } n \geq 1. \tag{1.25}$$

∎

Let's look at another example of a Fourier series.

Example 1.26   Find the Fourier series associated to the function $f(x) = x$ on $-\pi \leq x \leq \pi$.

The function f is odd, so according to Theorem 1.24 its Fourier series will involve only the sine terms. The coefficients are

$$b_n = \frac{2}{\pi} \int_0^{\pi} x \sin(nx) \, dx.$$

Using integration by parts, we obtain

$$b_n = \frac{2}{\pi} \left(\frac{-\pi \cos n\pi}{n} + \frac{1}{n} \int_0^{\pi} \cos n\pi x \, dx \right) = 2 \frac{(-1)^{n+1}}{n}.$$

Thus, the Fourier series of $f(x) = x$ on the interval $[-\pi, \pi]$ is

$$f(x) \sim 2 \sum_{n=1}^{\infty} \frac{(-1)^{n+1}}{n} \sin nx. \tag{1.27}$$

●

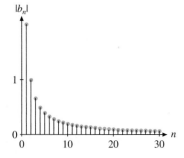

Figure 9. The Fourier coefficients for the function in Example 1.26.

The magnitude of the coefficients is plotted in Figure 9. The a_n are not shown, since they are all equal to 0. The partial sums of orders 5, 11, and 51 for the Fourier series in (1.27) are shown in Figures 10, 11, and 12 respectively. In these figures the function $f(x) = x$ is plotted in black and the partial sum in blue. These figures provide evidence that the Fourier series converges to $f(x)$, at least on the open interval $(-\pi, \pi)$. At $x = \pm\pi$ every term in the series is equal to 0. Therefore, the series converges to 0 at $\pm\pi$, but not to $f(\pm\pi) = \pm\pi$.

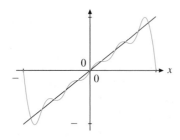

Figure 10. The partial sum of order 5 for $f(x) = x$.

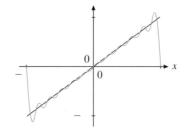

Figure 11. The partial sum of order 11 for $f(x) = x$.

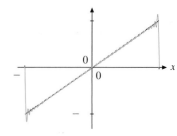

Figure 12. The partial sum of order 51 for $f(x) = x$.

Example 1.28   Compute the Fourier series for the saw-tooth wave f graphed in Figure 13 on the interval $[-1, 1]$.

The graph in Figure 13 on the interval $[-1, 1]$ consists of two lines with slope $+2$ and -2, respectively. The formula for f on the interval $-1 \leq x \leq 1$ is given by

$$f(x) = \begin{cases} 1 + 2x, & \text{if } -1 \leq x \leq 0, \\ 1 - 2x, & \text{if } 0 \leq x \leq 1. \end{cases}$$

The function f is even and is periodic with period 2.

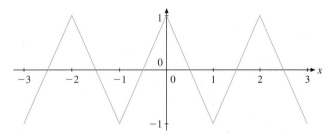

Figure 13. A saw-tooth-shaped wave.

Since f is an even function, we see using Theorem 1.24 that only the cosine terms appear in the Fourier series, and the coefficients are given by

$$a_n = 2 \int_0^1 (1 - 2x) \cos(n\pi x)\, dx, \quad \text{for } n \geq 0.$$

For $n = 0$ we can compute the integral by observation,

$$a_0 = 2 \int_0^1 (1 - 2x)\, dx = 0.$$

For $n > 0$, we use integration by parts to obtain

$$a_n = 2 \int_0^1 (1 - 2x) \cos(n\pi x)\, dx = \frac{4}{n^2\pi^2}(1 - \cos n\pi).$$

Since $\cos n\pi = (-1)^n$, we see that

$$a_{2n} = 0 \quad \text{and} \quad a_{2n+1} = \frac{8}{(2n+1)^2\pi^2}, \quad \text{for } n \geq 0.$$

Thus we have

$$f(x) \sim \frac{8}{\pi^2} \sum_{n=0}^{\infty} \frac{1}{(2n+1)^2} \cos((2n+1)\pi x). \qquad \bullet$$

The magnitude of the coefficients is plotted in Figure 14. The b_n are not shown, since they are all equal to 0. Notice how fast the coefficients decay to 0, in comparison to those in Figures 3 and 9. The graph of the partial sum of order 3,

$$S_3(x) = \frac{8}{\pi^2} \left[\cos \pi x + \frac{1}{9} \cos 3\pi x \right],$$

is shown in Figure 15. The sum of these two terms gives a pretty accurate approximation of the saw-tooth wave. This reflects the fact that the coefficients decay rapidly, as shown in Figure 14. The partial sum of order 9 is plotted in Figure 16. Notice that the poorest approximation occurs at the "corners" of the graph of the saw-tooth. This is where the function fails to be differentiable, and these facts are connected.

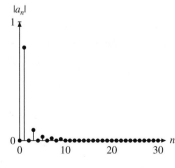

Figure 14. The Fourier coefficients for the function in Example 1.28.

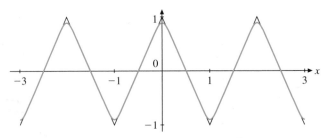

Figure 15. The partial sum of order 3 for the saw-tooth function.

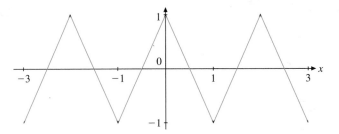

Figure 16. The partial sum of order 9 for the saw-tooth function.

Example 1.29   Find the Fourier series of the function $f(x) = \sin 3x + 2\cos 4x$ on the interval $[-\pi, \pi]$.

Since f is already given as a sum of sines and cosines, no work is needed. The Fourier series of f is just $\sin 3x + 2\cos 4x$. This example illustrates an important point. According to Theorem 1.17, the Fourier coefficients of a function are uniquely determined by the function. Thus, by inspection, $b_3 = 1$, $a_4 = 2$ and all other coefficients are equal to 0. By uniqueness, these are the same values as would have been obtained by computing the integrals in Theorem 1.17 for the a_n and b_n.●

Example 1.30   Find the Fourier series of the function $f(x) = \sin^2 x$ on the interval $[-\pi, \pi]$.

In this example, f is not explicitly given as a linear combination of sines and cosines, so there is some work to do. However, if we use the trigonometric identity

$$\sin^2 x = \frac{1}{2}(1 - \cos 2x),$$

the right side is the desired Fourier series, since it is a finite linear combination of terms of the form $\cos nx$.   ●

EXERCISES

In Exercises 1–6, expand the given function in a Fourier series valid on the interval $-\pi \le x \le \pi$. Plot the function and two partial sums of your choice over the interval $-\pi \le x \le \pi$. Plot the same partial sums over the interval $-3\pi \le x \le 3\pi$.

1. $f(x) = |\sin x|$

2. $f(x) = |x|$

3. $f(x) = \begin{cases} 0, & -\pi \le x < 0, \\ x, & 0 \le x \le \pi. \end{cases}$

4. $f(x) = \begin{cases} 0, & -\pi \le x < 0, \\ \sin x, & 0 \le x \le \pi. \end{cases}$

5. $f(x) = x \cos x$ **6.** $f(x) = x \sin x$

In Exercises 7–16, find the Fourier series for the indicated function on the indicated interval. Plot the function and two partial sums of your choice over the interval.

7. $f(x) = \begin{cases} 1 + x, & \text{for } -1 \le x \le 0, \\ 1, & \text{for } 0 < x \le 1, \end{cases}$ on $[-1, 1]$

8. $f(x) = 4 - x^2$ on $[-2, 2]$

9. $f(x) = x^3$ on $[-1, 1]$

10. $f(x) = \sin x \cos^2 x$ on $[-\pi, \pi]$

11. $f(x) = \begin{cases} 0, & \text{for } -1 \le x \le 0, \\ x^2, & \text{for } 0 < x \le 1, \end{cases}$ on $[-1, 1]$

12. $f(x) = \begin{cases} \sin \pi x / 2, & \text{for } -2 \le x \le 0, \\ 0, & \text{for } 0 < x \le 2, \end{cases}$ on $[-2, 2]$

13. $f(x) = \begin{cases} \cos \pi x, & \text{for } -1 \le x \le 0, \\ 1, & \text{for } 0 < x \le 1, \end{cases}$ on $[-1, 1]$

14. $f(x) = \begin{cases} 1 + x, & \text{for } -1 \le x \le 0, \\ 1 - x, & \text{for } 0 < x \le 1, \end{cases}$ on $[-1, 1]$

15. $f(x) = \begin{cases} 2 + x, & \text{for } -2 \le x \le 0, \\ -2 + x, & \text{for } 0 < x \le 2, \end{cases}$ on $[-2, 2]$

16. $f(x) = \begin{cases} 2, & \text{for } -2 \le x \le 0, \\ 2 - x, & \text{for } 0 < x \le 2, \end{cases}$ on $[-2, 2]$

17. Expand the function $f(x) = x^2$ in a Fourier series valid on the interval $-\pi \le x \le \pi$. Plot both f and the partial sum S_N for $N = 1, 3, 5, 7$. Observe how the graphs of the partial Fourier series approximates the graph of f. Plot the same graphs over the interval $-2\pi \le x \le 2\pi$.

18. Expand the function $f(x) = x^2$ in a Fourier series valid on the interval $-1 \le x \le 1$. Plot both f and the partial sum S_N for $N = 1, 3, 5, 7$. Observe how the graphs of the partial sums approximate the graph of f. Plot the same graphs over the interval $-2 \le x \le 2$.

In Exercises 19–22, determine if the function f is even, odd, or neither.

19. $f(x) = |\sin x|$

20. $f(x) = x + 3x^3$

21. $f(x) = e^x$

22. $f(x) = x + x^2$

23. Use the addition formulas for sin and cos to show that

$$\cos \alpha \cos \beta = \frac{1}{2}[\cos(\alpha - \beta) + \cos(\alpha + \beta)],$$

$$\sin \alpha \sin \beta = \frac{1}{2}[\cos(\alpha - \beta) - \cos(\alpha + \beta)], \quad (1.31)$$

$$\sin \alpha \cos \beta = \frac{1}{2}[\sin(\alpha - \beta) + \sin(\alpha + \beta)].$$

24. Prove Lemma 1.2. *Hint*: Use Exercise 23.

25. Complete the derivation of equation (1.11) for the coefficient b_n.

26. Prove parts 1, 2, and 3 of Proposition 1.23.

27. Prove parts 4 and 5 of Proposition 1.23.

28. Prove part 2 of Theorem 1.24.

29. From Theorem 1.24, the Fourier series of an odd function consists only of sine-terms. What additional symmetry conditions on f will imply that the sine coefficients with even indices will be zero? Give an example of a function satisfying this additional condition.

30. Suppose that f is a function which is periodic with period T and differentiable. Show that f' is also periodic with period T.

31. Suppose that f is a function defined on **R**. Show that there is an odd function f_{odd} and an even function f_{even} such that $f(x) = f_{\text{odd}}(x) + f_{\text{even}}(x)$ for all x.

12.2 Convergence of Fourier Series

Suppose that f is a piecewise continuous function on the interval $[-L, L]$, and that

$$f(x) \sim \frac{a_0}{2} + \sum_{n=1}^{\infty} \left[a_n \cos \left(\frac{n\pi x}{L} \right) + b_n \sin \left(\frac{n\pi x}{L} \right) \right] \quad (2.1)$$

is its associated Fourier series. Two questions arise immediately whenever an infinite series is encountered. The first question is, does the series converge? The second question arises if the series converges. Can we identify the sum of the series? In particular, does the Fourier series of a function f converge at x to $f(x)$ or to something else? These are the questions we will address in this section.[3]

Fourier series and periodic functions

The partial sums of the Fourier series in (2.1) have the form

$$S_N(x) = \frac{a_0}{2} + \sum_{n=1}^{N} \left[a_n \cos \left(\frac{n\pi x}{L} \right) + b_n \sin \left(\frac{n\pi x}{L} \right) \right]. \quad (2.2)$$

---

[3] Theorem 1.17 does not answer this question, since it assumes that $f(x)$ equals its Fourier series and then describes what the Fourier coefficients have to be.

The function $S_N(x)$ is a finite linear combination of the trigonometric functions $\cos(n\pi x/L)$ and $\sin(n\pi x/L)$, each of which is periodic with period $2L$.[4] Hence for every N the partial sum S_N is a function that is periodic with period $2L$. Consequently, if the partial sums converge at each point x, the limit function must also be periodic with period $2L$.

Let's consider again the function $f(x) = x$, which we treated in Example 1.26. $f(x)$ is defined for all real numbers x, and it is not periodic. We found that its Fourier series on the interval $[-\pi, \pi]$ is

$$2\sum_{n=1}^{\infty} \frac{(-1)^{n+1}}{n} \sin nx.$$

The partial sums of this series are all periodic with period 2π. Therefore, if the Fourier series converges, the limit function will be periodic with period 2π. Thus the limit cannot be equal to $f(x) = x$ everywhere. The evidence from the graphs of the partial sums in Figures 10, 11, and 12 of the previous section indicates that the series does converge to $f(x) = x$ on the interval $(-\pi, \pi)$. Since the limit must be 2π-periodic, we expect that the limit is closely related to the periodic extension of $f(x) = x$ from the interval $(-\pi, \pi)$. The situation is illustrated in Figure 1, which shows the partial sum of order 5 over 3 periods. The periodic extension of $f(x) = x$ is shown plotted in black.

Since it will appear repeatedly, let's denote the **periodic extension** of a function $f(x)$ defined on an interval $[-L, L]$ by $f_p(x)$. Usually it is easier to understand the periodic extension of a function graphically than it is to give an understandable formula for it. This is illustrated for $f(x) = x$ in Figure 1. However, for the record, the formula for the periodic extension[5] is

$$f_p(x) = f(x - 2kL) \quad \text{for } (2k-1)L < x \le (2k+1)L.$$

You will notice that f_p is periodic with period $2L$.

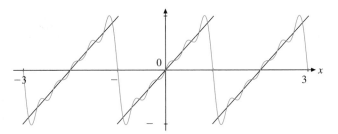

Figure 1. The partial sum of order 5 for the series in Example 1.26 over three periods.

To get a feeling for whether or not the Fourier series of $f(x) = x$ converges to its periodic extension f_p, we graph both f_p and the partial sum of order 21 in Figure 2. Note that the graph of S_{21} (the blue curve) is close to the graph of f_p except at the points of discontinuity of f_p, which occur at the odd multiples of π. The accuracy of the approximation of $f_p(x)$ by $S_{21}(x)$ gets worse as x gets closer

---

[4] We studied periodic functions in Section 5.5, but let's refresh our memory. A function $g(x)$ is **periodic** with period T if $g(x+T) = g(x)$ for all x. Notice that every integral multiple of a period is also a period. The smallest period of $\cos(n\pi x/L)$ and $\sin(n\pi x/L)$ is $2L/n$, so $n \cdot (2L/n) = 2L$ is also a period. Thus each function in the partial sum S_N in (2.2) is periodic with period $2L$.

[5] Notice we use less than ($<$) at the lower endpoint of each interval and less than or equal ($\le$) at the upper endpoint. A choice is necessary to avoid having two values at the endpoints. This is not the only possible choice, but it is as good as any.

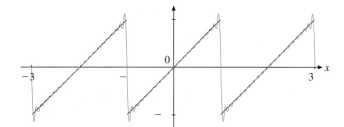

Figure 2. The partial sum of order 21 for the series in Example 1.26 over three periods.

to a point of discontinuity. This is necessary, simply because each partial sum is a continuous function, while f_p is not. Furthermore, we see that $S_N(\pi) = 0$ for all N. Hence the Fourier series converges to 0 at π, and not to $f_p(\pi) = \pi$. The same phenomenon occurs at $x = (2k + 1)\pi$ for any integer k, and these are the points of discontinuity of f_p. We will see these considerations reflected in our convergence theorem.

Piecewise continuous functions

We defined what it means for a function to be piecewise continuous on a finite interval in Section 12.1. We will say that a function defined on all of **R** is piecewise continuous if it is piecewise continuous on every finite interval.

For example, the periodic extension f_p of $f(x) = x$ on the interval $[-\pi, \pi]$ is piecewise continuous with discontinuities at all odd multiples of π. At $x_0 = \pi$ we have the one-sided limits $f_p(\pi^+) = -\pi$ and $f_p(\pi^-) = \pi$. This is clear from the graph of the function in Figures 1 and 2.

A function f is continuous at a point x_0 if and only if both one-sided limits exist and $f(x_0^+) = f(x_0^-) = f(x_0)$. For a piecewise continuous function the left- and right-hand limits exist everywhere. In fact, if f is any function that is continuous on the interval $[-L, L]$, then the periodic extension f_p is piecewise continuous on all of **R**, and its only possible points of discontinuity are the odd multiples of L.

We will say that the function f has a **_left-hand derivative_** at x_0 if the left-hand limit $f(x_0^-)$ exists and the limit

$$\lim_{x \to x_0^-} \frac{f(x) - f(x_0^-)}{x - x_0}$$

exists. Similarly, we will say that f has a **_right-hand derivative_** at x_0 if $f(x_0^+)$ exists and the limit

$$\lim_{x \to x_0^+} \frac{f(x) - f(x_0^+)}{x - x_0}$$

exists.

If f is differentiable at x_0, then f has left- and right-hand derivatives there and both are equal to $f'(x_0)$. However, a function can be differentiable from both sides at a point without being continuous there. For example, the left- and right-hand derivatives of the periodic extension f_p of $f(x) = x$ on the interval $[-\pi, \pi]$ exist at every point and are equal to 1, even at the points of discontinuity. Again this is easily seen from Figures 1 and 2.

Another example is the saw-tooth wave in Example 1.28. This function is continuous everywhere, but fails to be differentiable where x is an integer. However, both one-sided deivatives exist everywhere, as can easily be seen in Figure 16 in

Section 12.1. For example, the left-hand derivative at $x_0 = 0$ is $+1$ and the right-hand derivative is -1.

Convergence

Since a Fourier series converges to a periodic function, we may as well assume from the beginning that the function f is already periodic. If, as was the case in Example 1.26, we are given a function that is not periodic, then it is necessary to look at the periodic extension of the function.

We are now in a position to state our main theorem on the convergence of Fourier series. A proof is beyond the scope of this text, but one can be found in any advanced book on Fourier series.

THEOREM 2.3   Suppose $f(x)$ is a piecewise continuous function that is periodic with period $2L$. If the left- and right-hand derivatives of f exist at x_0 then the Fourier series for f converges at x_0 to

$$\frac{f(x_0^+) + f(x_0^-)}{2}. \tag{2.4}$$

∎

If f is continuous at x_0, then $f(x_0^+) = f(x_0^-) = f(x_0)$, so Theorem 2.3 concludes that the Fourier series converges at x_0 to $f(x_0)$.

Theorem 2.3 assumes the existence of the left- and right-hand derivatives of f *only at* the point x_0, and concludes that the Fourier series converges *only at* x_0. This indicates that it is only the smoothness of f near x_0 that affects the convergence there. However, in most of the examples that we will consider, the left- and right-hand derivatives will exist everywhere. We will consider this special case in the next corollary.

COROLLARY 2.5   Suppose $f(x)$ is a piecewise continuous function that is periodic with period $2L$. Suppose in addition that f has left- and right-hand derivatives at every point.

1. At every point x_0 where f is continuous the Fourier series for f converges to $f(x_0)$.

2. At every point x_0 where f is not continuous the Fourier series for f converges to

$$\frac{f(x_0^+) + f(x_0^-)}{2}. \tag{2.6}$$

●

Example 2.7   We have verified that the hypothesis of Corollary 2.5 holds for the periodic extension f_p of $f(x) = x$ on $-\pi \le x \le \pi$. Show that the conclusion of Corollary 2.5 holds at any point of discontinuity.

The points of discontinuity are $(2k + 1)\pi$, the odd integral multiples of π. We have $f_p([(2k + 1)\pi]^+) = -\pi$ and $f_p([(2k + 1)\pi]^-) = \pi$. Therefore,

$$\frac{f_p([(2k + 1)\pi]^+) + f_p([(2k + 1)\pi]^-)}{2} = 0.$$

We have also seen that the Fourier series of f is

$$2 \sum_{n=1}^{\infty} \frac{(-1)^{n+1}}{n} \sin nx.$$

When $x = (2k + 1)\pi$ every term in the series is equal to 0. Hence the series converges to 0, so the conclusion of Theorem 2.6 is valid at $x = (2k + 1)\pi$.   ●

E x a m p l e 2.8   Let $f(x) = x^2$ on the interval $-1 \leq x \leq 1$. Without computing the Fourier coefficients, explicitly describe the sum of the Fourier series of f for all x.[6]

Of course, $f(x) = x^2$ is not periodic. Therefore, we must consider its periodic extension, f_p, graphed in black in Figure 3. Note that f_p is continuous everywhere, and it is differentiable except at the odd integers, $x = 2k + 1$. At these points the left- and right-hand derivatives exist. Thus the left- and right-hand derivatives exist everywhere, and Corollary 2.5 implies that its Fourier series converges to $f_p(x)$ for all x.   ●

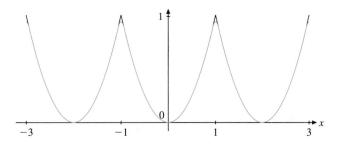

Figure 3. The partial sum $S_5(x)$ for the function in Example 2.8.

E x a m p l e 2.9   Consider the function

$$f(x) = \begin{cases} -1, & \text{for } -1 \leq x < 0, \\ 0, & \text{for } x = 0, \\ 1, & \text{for } 0 < x \leq 1. \end{cases}$$

Find the Fourier series for f, and describe the sum of its Fourier series.

Since f is only defined over the interval $[-1, 1]$, we are really looking at the Fourier series of its periodic extension f_p, shown plotted in black in Figure 4. For obvious reasons, f_p is called the **square wave**. Since f (and f_p) is an odd function, Theorem 1.24 says that only the sine terms are present in the Fourier series, and that the coefficients are

$$b_n = 2 \int_0^1 f(x) \sin n\pi x \, dx = 2 \int_0^1 \sin n\pi x \, dx = -\frac{2}{n\pi}[(-1)^n - 1].$$

Hence

$$b_{2n} = 0 \quad \text{and} \quad b_{2n+1} = \frac{4}{(2n+1)\pi}.$$

The Fourier series associated to f_p (and to f) is

$$f_p(x) \sim \frac{4}{\pi} \sum_{n=0}^{\infty} \frac{1}{2n+1} \sin(2n+1)\pi x. \tag{2.10}$$

The function f_p is piecewise continuous, with discontinuities at all of the integers. In addition, its left- and right-hand derivatives exist everywhere. Thus, the Fourier

---

[6] The computation of this series is Exercise 18 in Section 12.1.

series converges to $f_p(x)$ if x is not an integer. If $x = k$ is an integer, the series converges to

$$\frac{f_p(k^+) + f_p(k^-)}{2} = \frac{1 + (-1)}{2} = 0.$$

In fact, each term of the Fourier series in (2.10) is equal to 0 when $x = k$ is an integer. The partial sum of the Fourier series of order 11 is shown plotted in Figure 4.

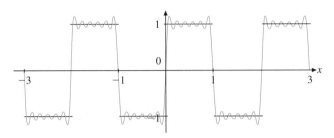

Figure 4. The partial sum $S_{11}(x)$ for the square wave in Example 2.9.

Gibb's phenomenon

Suppose that the piecewise continuous function f has a discontinuity at x_0, but that the left- and right-hand derivatives of f exist at x_0. As Theorem 2.3 points out, the Fourier series of f converges at x_0 to $[f(x_0^+) + f(x_0^-)]/2$. However, if you look closely at the graphs of the partial sums near the points of discontinuity in Figures 1, 2, and 4, we see that the graph of the partial sum overreaches the graph of the function on each side of the discontinuity. This effect is called ***Gibb's phenomenon***.

To examine Gibb's phenomenon a little more deeply, let's look at the graphs of some high-order partial sums for the square wave. The partial sums S_{301} and S_{601} are plotted in Figures 5 and 6, but only for $-0.01 \leq x \leq 0.01$, so that we may see the overrun clearly. Notice that both partial sums display the overrun that is characteristic of Gibb's phenomenon. In the two cases the amount of the overrun is approximately the same, but for the higher-order sum the duration is smaller.

It can be proved that whenever a function f satisfies the hypotheses of Theorem 2.3, but has a discontinuity at x_0, the graphs of the partial sums of the Fourier series display Gibb's phenomenon near x_0. Furthermore, the ratio of the length of the interval between the upper peak and the lower peak of the partial sum to $|f(x_0^+) - f(x_0^-)|$ is approximately 1.179 in every case.

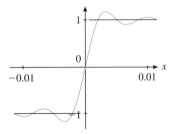

Figure 5. The partial sum $S_{301}(x)$ for the square wave.

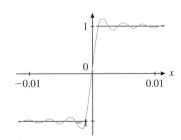

Figure 6. The partial sum $S_{601}(x)$ for the square wave.

The Riemann-Lebesgue lemma

Notice that in every example we have considered, the Fourier coefficients approach 0 as the frequency gets large. This is demonstrated, in particular, in Figures 3, 9, and 14 in Section 12.1. These examples are typical of the behavior of Fourier coefficients, as the next theorem, known as the Riemann-Lebesgue lemma, shows.

THEOREM 2.11   Suppose f is a piecewise continuous function on the interval $a \leq x \leq b$. Then

$$\lim_{k \to \infty} \int_a^b f(x) \cos kx \, dx = \lim_{k \to \infty} \int_a^b f(x) \sin kx \, dx = 0.$$ ∎

As we will see in Section 12.5, this theorem has important applications. Basically, this theorem states that any given signal can be approximated very well by a few dominant Fourier coefficients because most of the Fourier coefficients are near zero.

The intuitive reason behind this theorem is that as k gets very large, $\sin kx$ and $\cos kx$ oscillate much more rapidly than does f (see Figures 1 and 2 in Section 12.1). If k is large, $f(x)$ is nearly constant on two adjacent periods of $\sin kx$ or $\cos kx$. The integral over each period is almost zero, since the areas above and below the x-axis almost cancel.

Interpretation of the Fourier coefficients

Suppose that f is a function that satisfies the hypotheses of Corollary 2.5. Then $f(x)$ is equal to its Fourier series

$$f(x) = \frac{a_0}{2} + \sum_{n=1}^{\infty} \left[a_n \cos\left(\frac{n\pi x}{L}\right) + b_n \sin\left(\frac{n\pi x}{L}\right) \right],$$ (2.12)

except at those points where f is not continuous. Let's look more closely at the nth summand, which we can rewrite in terms of its amplitude and phase[7] as

$$f_n(x) = a_n \cos\left(\frac{n\pi x}{L}\right) + b_n \sin\left(\frac{n\pi x}{L}\right) = A_n \cos\left(\frac{n\pi x}{L} - \phi_n\right),$$ (2.13)

where $A_n = \sqrt{a_n^2 + b_n^2}$ and $\tan \phi_n = b_n/a_n$. We see that $f_n(x)$, defined in (2.13), is an oscillation with amplitude A_n and frequency[8] $\omega_n = n\pi/L$. We will call f_n the **component of f at frequency** $\omega_n = n\pi/L$. Notice that $\omega_n = n\omega_1$, where $\omega_1 = \pi/L$, so all of the frequencies are integer multiples of the **fundamental frequency** ω_1.

We can interpret Corollary 2.5 as saying that any function that satisfies its hypotheses is an infinite linear combination of oscillatory components at frequencies that are integer multiples of the fundamental frequency. The component of f at frequency ω_n has amplitude $A_n = \sqrt{a_n^2 + b_n^2}$. The amplitude is a numerical measure of the importance of the component in the Fourier expansion. By the Riemann-Lebesgue lemma, the Fourier coefficients decay to 0 as n increases, so the amplitudes A_n do as well. As a result, the components at the smaller frequencies dominate the Fourier series in (2.12). This fact is illustrated by the plots of the magnitudes of the coefficients in Figures 3, 9, and 14 in Section 12.1.

---

[7] See Section 4.4.

[8] This is an angular frequency. Remember that we are using angular frequencies instead of numerical frequencies unless otherwise stated.

Fourier coefficients for periodic functions

In this section we have been looking at periodic functions, since it is only such functions that can be the sums of Fourier series. It is worth pointing out that for a periodic function with period $2L$, the coefficients can be computed by an integral over any interval of length $2L$. More precisely, we have the following:

PROPOSITION 2.14 Suppose that f is a piecewise continuous function that is periodic with period $2L$. Then for any c the Fourier coefficients for f are given by

$$a_n = \frac{1}{L} \int_c^{c+2L} f(x) \cos \frac{n\pi x}{L} \, dx, \quad \text{for } n \geq 0,$$

$$b_n = \frac{1}{L} \int_c^{c+2L} f(x) \sin \frac{n\pi x}{L} \, dx, \quad \text{for } n \geq 1. \qquad \bullet$$

We will leave the proof to the exercises.

EXERCISES

In Exercises 1–6, determine if the function f is periodic or not. If it is periodic, find the smallest positive period.

1. $f(x) = |\sin x|$ **2.** $f(x) = \cos 3\pi x$

3. $f(x) = x$ **4.** $f(x) = \sin(x) + \cos(x/2)$

5. $f(x) = x^2$ **6.** $f(x) = e^x$

In Exercises 7–14, find the sum of the Fourier series for indicated function at every point in **R** without computing the series. Each of these is an exercise in Section 12.1. Although that is not very important, the reference is included in parentheses.

7. $f(x) = \begin{cases} 0, & -\pi \leq x < 0, \\ x, & 0 \leq x \leq \pi \end{cases}$ on $[-\pi, \pi]$
(See Exercise 3)

8. $f(x) = \begin{cases} 0, & -\pi \leq x < 0, \\ \sin x, & 0 \leq x \leq \pi \end{cases}$ on $[-\pi, \pi]$
(See Exercise 4)

9. $f(x) = \begin{cases} 1 + x, & \text{for } -1 \leq x \leq 0 \\ 1, & \text{for } 0 < x \leq 1 \end{cases}$ on $[-1, 1]$
(See Exercise 7)

10. $f(x) = 4 - x^2$ on $[-2, 2]$ (See Exercise 8)

11. $f(x) = x^3$ on $[-1, 1]$ (See Exercise 9)

12. $f(x) = \begin{cases} 0, & \text{for } -1 \leq x \leq 0, \\ x^2, & \text{for } 0 < x \leq 1 \end{cases}$ on $[-1, 1]$
(See Exercise 11)

13. $f(x) = \begin{cases} \sin \pi x/2, & \text{for } -2 \leq x \leq 0, \\ 0, & \text{for } 0 < x \leq 2 \end{cases}$ on $[-2, 2]$
(See Exercise 12)

14. $f(x) = \begin{cases} 2, & \text{for } -2 \leq x \leq 0, \\ 2 - x, & \text{for } 0 < x \leq 2 \end{cases}$ on $[-2, 2]$
(See Exercise 16)

15. Compute the Fourier series for the function $f(x) = |x|$ on the interval $[-\pi, \pi]$. (See Exercise 2 in Section 12.1.) Use the result and Theorem 2.3 to show that

$$\sum_{n=0}^{\infty} \frac{1}{(2n+1)^2} = \frac{\pi^2}{8}.$$

16. Compute the Fourier series for the function $f(x) = x^2$ on the interval $[-\pi, \pi]$. (See Example 2.8.) Use the result and Theorem 2.3 to show that

$$\sum_{n=1}^{\infty} \frac{(-1)^{n+1}}{n^2} = \frac{\pi^2}{12} \quad \text{and} \quad \sum_{n=1}^{\infty} \frac{1}{n^2} = \frac{\pi^2}{6}.$$

17. Compute the Fourier series for the function $f(x) = x^4$ on the interval $[-\pi, \pi]$. Use the result, Theorem 2.3, and Exercise 16 to show that

$$\sum_{n=1}^{\infty} \frac{(-1)^{n+1}}{n^4} = \frac{7\pi^4}{120} \quad \text{and} \quad \sum_{n=1}^{\infty} \frac{1}{n^4} = \frac{\pi^4}{90}.$$

18. Expand the function

$$f(x) = \begin{cases} 0, & -1 < x \leq -1/2, \\ 1, & -1/2 < x \leq 1/2, \\ 0, & 1/2 < x \leq 1, \end{cases}$$

in a Fourier series valid on the interval $-1 \leq x \leq 1$. Plot the graph of f and the partial sums of order N for $N = 5, 10, 20,$ and 40, as in Exercise 17 in Section 12.1. Notice how much slower the series converges to f in this example than in Exercise 17 in Section 12.1. What accounts for the slow rate of convergence in this example?

19. Expand the function $f(x) = e^{rx}$ in a Fourier series valid for $-\pi \le x \le \pi$. For the case $r = 1/2$, plot the partial sums of orders $N = 10$, 20, and 30 of the Fourier series along with the graph of f_p over the intervals $-\pi \le x \le \pi$ and $-2\pi \le x \le 2\pi$.

20. Use the previous exercise to compute the Fourier coefficients for the function $f(x) = \sinh x = (e^x - e^{-x})/2$ and $f(x) = \cosh(x) = (e^x + e^{-x})/2$ over the interval $-\pi \le x \le \pi$.

21. Use Theorem 2.3 to determine the sum of the Fourier series of the function f defined in Exercise 18 for each x in the interval $-1 \le x \le 1$.

22. Suppose that f is periodic with period T and differentiable. Show that f' is also periodic with period T.

23. Suppose that f is periodic with period T. Show that

$$\int_b^{b+T} f(x)\, dx = \int_a^{a+T} f(x)\, dx$$

for any a and b. Use this result to prove Proposition 2.14.

24. Suppose that f is periodic with period T. Define

$$F(x) = \int_0^x f(y)\, dy.$$

Show that if $\int_0^T f(y)\, dy = 0$, then F is periodic with period T. (*Hint:* Use Exercise 23.)

12.3 Fourier Cosine and Sine Series

In this section we will examine the possibility of finding Fourier series of the forms

$$f(x) = \sum_{n=1}^{\infty} b_n \sin \frac{n\pi x}{L}, \quad \text{for } 0 \le x \le L,$$

and

$$f(x) = \frac{a_0}{2} + \sum_{n=1}^{\infty} a_n \cos \frac{n\pi x}{L}, \quad \text{for } 0 \le x \le L.$$

The basic idea behind our method comes from Theorem 1.24.

Fourier cosine series

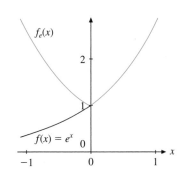

Figure 1. The even extension of $f(x) = e^x$.

According to Theorem 1.24, the Fourier series of an even function contains only cosine terms. If the function $f(x)$ is defined for $0 \le x \le L$, we can extend it to $-L \le x \le 0$ as an even function. The **even extension** of f is defined by

$$f_e(x) = \begin{cases} f(x), & \text{if } 0 \le x \le L, \\ f(-x), & \text{if } -L \le x < 0. \end{cases}$$

For the function $f(x) = e^x$ on the interval $[0, 1]$, the even extension f_e is plotted in blue in Figure 1.

Since the function f_e is an even function defined on $[-L, L]$, Theorem 1.24 tells us that its Fourier series has the form

$$f_e(x) \sim \frac{a_0}{2} + \sum_{n=1}^{\infty} a_n \cos\left(\frac{n\pi x}{L}\right), \quad \text{for } -L \le x \le L, \tag{3.1}$$

where

$$a_n = \frac{2}{L} \int_0^L f_e(x) \cos\left(\frac{n\pi x}{L}\right) dx, \quad \text{for } n \ge 0.$$

Since $f_e(x) = f(x)$ for $0 \le x \le L$, this formula becomes

$$a_n = \frac{2}{L} \int_0^L f(x) \cos\left(\frac{n\pi x}{L}\right) dx, \quad \text{for } n \ge 0. \tag{3.2}$$

Furthermore, if we restrict ourselves to the interval $[0, L]$, where $f_e(x) = f(x)$, we can write

$$f(x) \sim \frac{a_0}{2} + \sum_{n=1}^{\infty} a_n \cos\left(\frac{n\pi x}{L}\right), \quad \text{for } 0 \leq x \leq L, \tag{3.3}$$

with the coefficients given by (3.2). The series in (3.3), with the coefficients given in (3.2), is called the **Fourier cosine series** for f on the interval $[0, L]$.

Example 3.4 Find the Fourier cosine series for $f(x) = e^x$ on the interval $[0, 1]$.

The coefficients in (3.2) become

$$a_n = 2\int_0^1 e^x \cos n\pi x \, dx = \frac{2}{1 + n^2\pi^2}\left[(-1)^n e - 1\right].$$

This evaluation can be done by direct computation, by looking the integral up in an integral table, or by using a computer and a symbolic algebra program. The magnitude of these coefficients is plotted in Figure 2. Notice how quickly the coefficients decay to 0. The Fourier series is

$$e^x \sim (e - 1) + 2\sum_{n=1}^{\infty} \frac{(-1)^n e - 1}{1 + n^2\pi^2} \cos n\pi x$$

$$\sim (e - 1) - \frac{2(e + 1)}{1 + \pi^2} \cos \pi x + \frac{2(e - 1)}{1 + 4\pi^2} \cos 2\pi x + \dots \tag{3.5}$$

on the interval $[0, 1]$.

The partial sum $S_3(x)$ is plotted in blue in Figure 3. The black curve in Figure 3 is the periodic extension of the even extension of the function $f(x) = e^x$. We will call this the **even periodic extension** of f, and we will denote it by f_{ep}. Since, in this case, f_{ep} is continuous and satisfies the hypotheses of Corollary 2.5, the Fourier series converges everywhere to $f_{ep}(x)$. ●

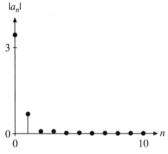

Figure 2. The Fourier cosine coefficients for $f(x) = e^x$.

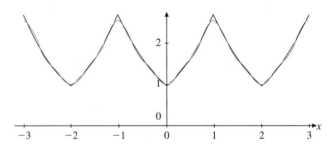

Figure 3. The partial sum S_3 of the Fourier cosine series for $f(x) = e^x$ plotted over three periods.

Fourier sine series

In a similar manner, a function f can be expanded in a series which involves only sine terms. Again motivated by Theorem 1.24, we consider the **odd extension of** f, which is defined by

$$f_o(x) = \begin{cases} f(x), & \text{if } 0 < x \leq L, \\ 0, & \text{if } x = 0, \\ -f(-x), & \text{if } -L \leq x < 0. \end{cases}$$

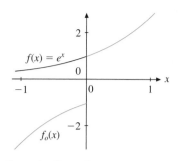

Figure 4. The odd extension of $f(x) = e^x$.

The odd extension of $f(x) = e^x$ is plotted in blue in Figure 4. Since the function f_o is an odd function defined on $[-L, L]$, Theorem 1.24 tells us that its Fourier series has only sine terms. Proceeding as before, we find that

$$f(x) \sim \sum_{n=1}^{\infty} b_n \sin\left(\frac{n\pi x}{L}\right), \quad \text{for } 0 < x \leq L, \tag{3.6}$$

where

$$b_n = \frac{2}{L} \int_0^L f(x) \sin\left(\frac{n\pi x}{L}\right) dx, \quad \text{for } n \geq 1. \tag{3.7}$$

The series in (3.6), with the coefficients given in (3.7), is called the ***Fourier sine series*** for f on the interval $[0, L]$.

Example 3.8   Find the Fourier sine series for $f(x) = e^x$ on the interval $[0, 1]$.

The coefficients in (3.7) become

$$b_n = 2 \int_0^1 e^x \sin n\pi x \, dx = \frac{2n\pi[1 - (-1)^n e]}{1 + n^2\pi^2}.$$

This evaluation can be done by direct computation, by looking the integral up in an integral table, or by using a computer and a symbolic algebra program. Thus we have

$$e^x \sim \sum_{n=1}^{\infty} \frac{2n\pi[1 - (-1)^n e]}{1 + n^2\pi^2} \sin n\pi x$$
$$\sim \frac{2\pi(e+1)}{1 + \pi^2} \sin \pi x - \frac{4\pi(e-1)}{1 + 4\pi^2} \sin 2\pi x + \frac{6\pi(e+1)}{1 + 9\pi^2} \sin 3\pi x + \dots \tag{3.9}$$

on the interval $[0, 1]$. The magnitude of the coefficients is plotted in Figure 5. Notice that the sine coefficients do not decay nearly as rapidly as do the cosine coefficients in Figure 2. The partial sum of order 3 is plotted in blue in Figure 6. It is interesting to compare this figure with Figure 3. The black curve in Figure 6 is the ***odd periodic extension*** of $f(x) = e^x$, which we denote by f_{op}. In this case f_{op} satisfies the hypotheses of Corollary 2.5 and fails to be continuous only at the integers. Consequently, the Fourier sine series converges to $f_{op}(x)$ everywhere except at the odd integers.   ●

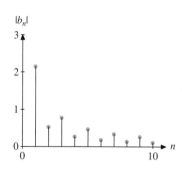

Figure 5. The Fourier sine coefficients for $f(x) = e^x$.

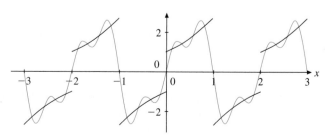

Figure 6. The partial sum S_3 of the Fourier sine series for $f(x) = e^x$ plotted over three periods.

Note that while we used the even and odd extensions of f (f_e and f_o) to help derive the cosine and sine expansions, the formulas for a_n and b_n involve only the function f on the interval $[0, L]$. This is reflected by the fact that the cosine and sine expansions converge to f only on $(0, L)$. Outside this interval the cosine and sine expansions converge to f_{ep} and f_{op}, respectively. Examples 3.4 and 3.8 illustrate these facts, but another example might help to put things into perspective.

Example 3.10   Find the complete Fourier series for $f(x) = e^x$ on the interval $[-1, 1]$.

From (1.18) we have

$$a_n = \int_{-1}^{1} e^x \cos n\pi x \, dx = (-1)^n \frac{e - 1/e}{1 + n^2\pi^2}, \quad \text{for } n \geq 0,$$

while from (1.19) we have

$$b_n = \int_{-1}^{1} e^x \sin n\pi x \, dx = (-1)^{n+1} \frac{n\pi(e - 1/e)}{1 + n^2\pi^2}, \quad \text{for } n \geq 1.$$

Hence the complete Fourier series for e^x on $[-1, 1]$ is

$$e^x = \left(e - \frac{1}{e}\right) \left\{\frac{1}{2} + \sum_{n=1}^{\infty} \frac{(-1)^n}{1 + n^2\pi^2} [\cos n\pi x - n\pi \sin n\pi x]\right\}. \tag{3.11}$$

The magnitude of the coefficients is plotted in Figure 7, with $|a_n|$ in black and $|b_n|$ in blue. The partial sum of order 3 is plotted in blue Figure 8.

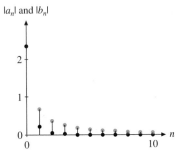

$|a_n|$ and $|b_n|$

Figure 7. The coefficients of the complete Fourier series for $f(x) = e^x$.

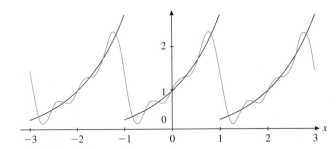

Figure 8.   The partial sum S_3 of the complete Fourier series for $f(x) = e^x$ plotted over three periods.

The periodic extension of e^x satisfies the hypotheses of Corollary 2.5 and fails to be continuous only at the odd integers. Consequently, the Fourier series converges to the periodic extension everywhere except at the odd integers.   ●

Now we have three Fourier series that converge to $f(x) = e^x$ on the interval $(0, 1)$. The first in (3.5) contains only cosine terms. The second in (3.9) contains only sine terms. The third in (3.11) contains both sine and cosine terms. It is interesting to compare graphs of the partial sums in Figures 3, 6, and 8. The difference between the three is what happens outside of the interval $(0, 1)$. The cosine series converges to f_{ep}, the even periodic extension of f. The sine series converges to f_{op}, the odd periodic extension of f, except at the odd integers. And, finally, the full Fourier series converges to f_p, the periodic extension of f, except at the odd integers.

Of course, the same three series can be considered for any piecewise continuous function defined on an interval of the form $[-L, L]$.

EXERCISES

In Exercises 1–4, give a piecewise definition of f_o, the odd extension for f as defined on the given interval. Sketch the graph of f_o. Sketch the graph of f_{op} over three periods.

1. $f(x) = 1 - x$, $[0, 2]$ **2.** $f(x) = 1 - 2x$, $[0, 1]$

3. $f(x) = x^2 - 1$, $[0, 2]$ **4.** $f(x) = x^2 - 2$, $[0, 2]$

In Exercises 5–8, give a piecewise definition of f_e, the even extension for f as defined on the given interval. Sketch the graph of f_e Sketch the graph of f_{ep} over three periods.

5. $f(x) = 1 - x$, $[0, 2]$ **6.** $f(x) = 1 - 2x$, $[0, 1]$

7. $f(x) = x^2 - 1$, $[0, 2]$ **8.** $f(x) = x^2 - 2$, $[0, 2]$

In Exercises 9–20, expand the given function in a Fourier cosine series valid on the interval $0 \le x \le \pi$. Plot the function and two partial sums of your choice over the interval $0 \le x \le \pi$. Plot the same partial sums and the function the series converges to over the interval $-3\pi \le x \le 3\pi$.

9. $f(x) = x$ **10.** $f(x) = \sin x$

11. $f(x) = \cos x$ **12.** $f(x) = 1$

13. $f(x) = \pi - x$ **14.** $f(x) = x^2$

15. $f(x) = x^3$ **16.** $f(x) = x^4$

17. $f(x) = \begin{cases} 1, & 0 \le x < \pi/2, \\ 0, & \pi/2 \le x \le \pi \end{cases}$

18. $f(x) = \begin{cases} x, & 0 \le x < \pi/2, \\ \pi/2, & \pi/2 \le x \le \pi \end{cases}$

19. $f(x) = x \cos x$ **20.** $f(x) = x \sin x$

In Exercises 21–32, expand the given function in a Fourier sine series valid on the interval $0 \le x \le \pi$. Plot the function and two partial sums of your choice over the interval $0 \le x \le \pi$.

Plot the same partial sums and the function the series converges to over the interval $-3\pi \le x \le 3\pi$.

21. Same as Exercise 9   **22.** Same as Exercise 10

23. Same as Exercise 11   **24.** Same as Exercise 12

25. Same as Exercise 13   **26.** Same as Exercise 14

27. Same as Exercise 15   **28.** Same as Exercise 16

29. Same as Exercise 17   **30.** Same as Exercise 18

31. Same as Exercise 19   **32.** Same as Exercise 20

33. Show that the functions $\cos(n\pi x/L)$, $n = 0, 1, 2, \ldots$ are orthogonal on the interval $[0, L]$. This means that

$$\int_0^L \cos(n\pi x/L) \, \cos(p\pi x/L) \, dx = 0, \quad \text{if } p \ne n.$$

Hint: Use Exercise 23 in Section 12.1.

34. Show that the functions $\sin(n\pi x/L)$, $n = 1, 2, 3, \ldots$ are orthogonal on the interval $[0, L]$. This means that

$$\int_0^L \sin(n\pi x/L) \, \sin(p\pi x/L) \, dx = 0, \quad \text{if } p \ne n.$$

Hint: Use Exercise 23 in Section 12.1.

35. Show that

$$\int_0^1 \cos(2n\pi x) \sin(2k\pi x) \, dx = 0.$$

Hint: Use Exercise 23 in Section 12.1.

36. If $f(x)$ is continuous on the interval $0 \le x \le L$, show that its even periodic extension is continuous everywhere. Does this statement hold for the odd periodic extension? What additional condition(s) is (are) necessary to ensure that the odd periodic extension is everywhere continuous?

12.4 The Complex Form of a Fourier Series

If the piecewise continuous function f is periodic with period $2L$, then its Fourier series is

$$f(x) \sim \frac{a_0}{2} + \sum_{n=1}^{\infty} \left[a_n \cos\left(\frac{n\pi x}{L}\right) + b_n \sin\left(\frac{n\pi x}{L}\right) \right], \tag{4.1}$$

where the coefficients are given by

$$\begin{aligned} a_n &= \frac{1}{L} \int_{-L}^{L} f(x) \cos\left(\frac{n\pi x}{L}\right) dx, \quad \text{for } n \ge 0, \text{ and} \\ b_n &= \frac{1}{L} \int_{-L}^{L} f(x) \sin\left(\frac{n\pi x}{L}\right) dx, \quad \text{for } n \ge 1. \end{aligned} \tag{4.2}$$

Sometimes it is useful to express the Fourier series in complex form using the complex exponentials, e^{inx} for $n = 0, \pm 1, \pm 2, \ldots$. This is possible because of the close connection between the complex exponentials and the trigonometric functions. We explored this connection in the appendix to this book and in Section 4.3. The most important facts to know about the complex exponential are Euler's formula

$$e^{iy} = \cos y + i \sin y, \tag{4.3}$$

which defines the exponential, and that all of the familiar properties of the real exponential remain true for the complex exponential.

If we write down Euler's formula with y replaced by $-y$, we get

$$e^{-iy} = \cos y - i \sin y. \tag{4.4}$$

Solving (4.3) and (4.4) for $\cos y$ and $\sin y$, we see that

$$\cos y = \frac{e^{iy} + e^{-iy}}{2} \quad \text{and} \quad \sin y = \frac{e^{iy} - e^{-iy}}{2i}. \tag{4.5}$$

Let's substitute these expressions into the Fourier series (4.1). The nth term in the sum is the component of f at the frequency $\omega_n = n\pi/L$, and it becomes

$$
\begin{aligned}
f_n(x) &= a_n \cos\left(\frac{n\pi x}{L}\right) + b_n \sin\left(\frac{n\pi x}{L}\right) \\
&= \frac{a_n}{2}\left(e^{in\pi x/L} + e^{-in\pi x/L}\right) + \frac{b_n}{2i}\left(e^{in\pi x/L} - e^{-in\pi x/L}\right) \\
&= \frac{a_n - ib_n}{2}e^{in\pi x/L} + \frac{a_n + ib_n}{2}e^{-in\pi x/L} \\
&= \alpha_n e^{in\pi x/L} + \alpha_{-n}e^{-in\pi x/L},
\end{aligned}
\tag{4.6}
$$

where we have substituted

$$\alpha_n = \frac{a_n - ib_n}{2} \quad \text{and} \quad \alpha_{-n} = \frac{a_n + ib_n}{2}, \quad \text{for } n \geq 1. \tag{4.7}$$

We will also write the constant term as $f_0(x) = \alpha_0 = a_0/2$. Separating the positive and negative terms, the Fourier series can be written as

$$f(x) \sim \sum_{n=-\infty}^{\infty} \alpha_n e^{in\pi x/L}. \tag{4.8}$$

Notice that by (4.6), the component of f at frequency $\omega_n = n\pi/L$ is given by $f_n(x) = \alpha_n e^{i\omega_n x} + \alpha_{-n}e^{-i\omega_n x}$. As a result, when we talk in terms of low frequency components we have to consider the coefficients α_n and α_{-n} for small values of n.

We can use (4.2) to express the coefficients α_n in terms of the function f. For example, for $n \geq 1$ we have

$$
\begin{aligned}
\alpha_n &= \frac{a_n - ib_n}{2} \\
&= \frac{1}{2L}\int_{-L}^{L} f(x)\left[\cos\left(\frac{n\pi x}{L}\right) - i\sin\left(\frac{n\pi x}{L}\right)\right]dx \\
&= \frac{1}{2L}\int_{-L}^{L} f(x)e^{-in\pi x/L}\,dx.
\end{aligned}
$$

The corresponding formulas for $n = 0$ and for $n < 0$ can be computed in the same way, and we discover that

$$\alpha_n = \frac{1}{2L}\int_{-L}^{L} f(x)e^{-in\pi x/L}\,dx, \quad \text{for all } n. \tag{4.9}$$

It is important to notice that while α_n is the coefficient of $e^{in\pi x/L}$ in the Fourier series (4.8), it is $e^{-in\pi x/L}$ which appears in the integral in (4.9).

The series (4.8), with the coefficients computed using (4.9), is called the **complex Fourier series** for the function f. There are several differences between the Fourier series involving cosines and sines, given in Definition 1.20, and the Fourier series using complex exponentials presented here. First, the complex Fourier series involves a sum from $n = -\infty$ to $n = \infty$, rather than a sum from $n = 0$ to $n = \infty$. Next, for the complex Fourier series, there is one succinct formula (4.9) for the Fourier coefficients, rather than the two separate formulas for a_n and b_n in (1.18) and (1.19). For this reason, and also because computations using exponentials are easier than those using trigonometric functions, many scientists and engineers prefer to use the complex version of the Fourier series.

Example 4.10 Find the complex Fourier series for the function $f(x) = e^x$ on the interval $[-1, 1]$.

This is the function we examined in Examples 3.4, 3.8, and 3.10. For this function it is much easier to compute the complex Fourier coefficients than the real ones. The nth coefficient is

$$\alpha_n = \frac{1}{2} \int_{-1}^{1} e^x e^{-in\pi x} \, dx$$

$$= \frac{1}{2} \int_{-1}^{1} e^{(1-in\pi)x} \, dx$$

$$= \frac{1}{2(1 - in\pi)} \left[e^{1-in\pi} - e^{-1+in\pi} \right]$$

$$= \frac{(-1)^n}{2(1 - in\pi)} (e - 1/e).$$

The last identity follows since $e^{in\pi} = e^{-in\pi} = (-1)^n$.

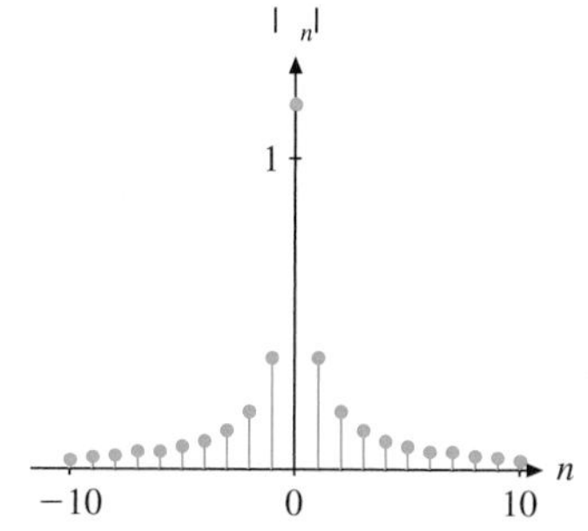

Figure 1. The coefficients of the complex Fourier series for $f(x) = e^x$.

The magnitude of the coefficients is plotted in Figure 1. Notice that we included negative indices. The complex Fourier series is

$$e^x \sim \frac{e - 1/e}{2} \sum_{n=-\infty}^{\infty} \frac{(-1)^n}{1 - in\pi} e^{in\pi x} \quad \text{for } -1 \le x \le 1.$$

Relation between the real and complex Fourier series

We derived the complex Fourier series from the real series. In doing so we found that the complex coefficients can be computed from the real coefficients using (4.7). In turn, we can solve these relationships for the real coefficients in terms of the complex coefficients, getting

$$a_0 = 2\alpha_0, \quad a_n = \alpha_n + \alpha_{-n}, \quad \text{and} \quad b_n = i(\alpha_n - \alpha_{-n}), \quad \text{for } n \ge 1. \quad (4.11)$$

These equations simplify somewhat if the function f is real valued. In that case $\overline{f(x)} = f(x)$, so

$$\overline{\alpha_n} = \overline{\frac{1}{2L} \int_{-L}^{L} f(x) e^{-in\pi x/L} \, dx} = \frac{1}{2L} \int_{-L}^{L} \overline{f(x)} \cdot \overline{e^{-in\pi x/L}} \, dx$$

$$= \frac{1}{2L} \int_{-L}^{L} f(x) e^{in\pi x/L} \, dx = \alpha_{-n}.$$

Consequently, if f is real valued,

$$a_n = \alpha_n + \overline{\alpha_n} = 2 \operatorname{Re} \alpha_n \quad \text{and} \quad b_n = i(\alpha_n - \overline{\alpha_n}) = -2 \operatorname{Im} \alpha_n. \quad (4.12)$$

E x a m p l e 4 . 1 3   Compute the coefficients of the real Fourier series for the function $f(x) = e^x$ on the interval $[-1, 1]$.

We computed the complex coefficients in Example 4.10 and found that

$$\alpha_n = \frac{(-1)^n}{2(1 - in\pi)}(e - 1/e).$$

Since the function is real valued, we can use (4.12) to find that

$$a_n = 2\,\mathrm{Re}\,\alpha_n = \frac{(-1)^n(e - 1/e)}{1 + n^2\pi^2} \quad \text{and}$$

$$b_n = -2\,\mathrm{Im}\,\alpha_n = \frac{(-1)^{n+1}n\pi(e - 1/e)}{1 + n^2\pi^2}. \qquad \bullet$$

EXERCISES

1. Show that the complex Fourier coefficients for an even, real-valued function are real.   Show that the complex Fourier coefficients for an odd, real-valued function are purely imaginary (i.e., their real parts are zero).

In Exercises 2–11, find the complex Fourier series for the given function on the interval $[-\pi, \pi]$.

2. $f(x) = x$

3. $f(x) = |x|$

4. $f(x) = \begin{cases} -1, & -\pi \leq x < 0, \\ 1, & 0 \leq x \leq \pi \end{cases}$

5. $f(x) = \begin{cases} 0, & -\pi \leq x < 0, \\ 1, & 0 \leq x \leq \pi \end{cases}$

6. $f(x) = x^2$

7. $f(x) = e^{bx}$

8. $f(x) = x^3$

9. $f(x) = \pi - x$

10. $f(x) = |\cos x|$

11. $f(x) = |\sin x|$

12. Two complex valued functions f and g are said to be orthogonal on the interval $[a, b]$ if $\int_a^b f(x)\overline{g(x)}\, dx = 0$. Show that the functions e^{ipx} and e^{iqx} are orthogonal on $[-\pi, \pi]$ if p and q are different integers.

13. Use the method of proof of Theorem 1.17 and Exercise 12 to show that if $f(x) = \sum_{-\infty}^{\infty} \alpha_n e^{inx}$ for $-\pi \leq x \leq \pi$, then

$$\alpha_n = \frac{1}{2\pi} \int_{-\pi}^{\pi} f(x)e^{-inx}\, dx.$$

12.5 The Discrete Fourier Transform and the FFT

Suppose that $f(t)$ is piecewise continuous for $0 \leq t \leq 2\pi$.[9] Then, using Proposition 2.14,

$$f(t) \sim \sum_{k=-\infty}^{\infty} \alpha_k e^{ikt}, \quad \text{where} \quad \alpha_k = \frac{1}{2\pi} \int_0^{2\pi} f(t)e^{-ikt}\, dt. \qquad (5.1)$$

The Fourier coefficients α_k are often too difficult to compute exactly. In such a case it is useful to approximate the coefficients using a numerical integration technique such as the trapezoid rule.

We remind you that for a function F defined on the interval $[0, 2\pi]$, the trapezoid rule for approximating the integral $\int_0^{2\pi} F(t)\, dt$ with step size $h = 2\pi/N$ is

$$\int_0^{2\pi} F(t)\, dt \approx h\left[\frac{1}{2}F(0) + F(h) + F(2h) + \cdots + F((N-1)h) + \frac{1}{2}F(Nh)\right].$$

If $F(t)$ is 2π-periodic, then $F(0) = F(2\pi) = F(Nh)$, and the preceding formula becomes

$$\int_0^{2\pi} F(t)\, dt \approx h\sum_{j=0}^{N-1} F(jh) = \frac{2\pi}{N}\sum_{j=0}^{N-1} F(2\pi j/N).$$

---

[9] In most applications of the material in this section, the function f represents a time-dependent signal. Consequently, we will use t instead of x as the independent variable.

Applying this formula to the integral for the Fourier coefficient in (5.1), we get

$$\alpha_k = \frac{1}{2\pi} \int_0^{2\pi} f(t)e^{-ikt}\, dt \approx \frac{1}{N}\sum_{j=0}^{N-1} f(2\pi j/N)e^{-2\pi ijk/N}. \tag{5.2}$$

Let's set

$$y_j = f(2\pi j/N) \quad \text{and} \quad w = e^{2\pi i/N}.$$

Then $\overline{w} = e^{-2\pi i/N}$, and the approximation becomes

$$\alpha_k \approx \frac{1}{N}\sum_{j=0}^{N-1} y_j \overline{w}^{jk}. \tag{5.3}$$

The sum on the right side of equation (5.3) involves the discrete values $y_j = f(2\pi j/N)$. The values of $f(t)$ for $t \neq 2\pi j/N$ are ignored. It is a common occurrence in the digital age to replace a time dependent function with such a discrete sample of that function. For example, $f(t)$ may represent a music signal that we want to transmit over the internet. The internet, or any other computer network, allows only discrete signals, so to transmit the music we replace the continuous signal $f(t)$ with the discrete sample $y_j = f(2\pi j/N)$ for $j = 0, 1, 2, \ldots, N-1$.

In many digital applications signals arise that are not represented by a continuous function at all. Instead, they arise as discrete values y_j at a discrete set of times t_j. Such a signal is illustrated in Figure 1. Here, the horizontal axis represents time, which has been divided into many small time intervals.

The discrete Fourier transform

For discrete signals such as that illustrated in Figure 1, it is often useful to consider the transform we found in equation (5.3). We will assume that the signal is an infinite sequence $y = \{y_j \mid -\infty < j < \infty\}$ that is **periodic with period** N, meaning that $y_{N+j} = y_j$ for all j.

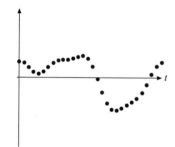

Figure 1. A discrete signal.

DEFINITION 5.4

Let $y = \{y_j\}$ be a sequence of complex numbers that is periodic with period N. The **discrete Fourier transform** of y is the sequence $\widehat{y} = \{\widehat{y}_k\}$, where

$$\widehat{y}_k = \sum_{j=0}^{N-1} y_j e^{-2\pi ikj/N} = \sum_{j=0}^{N-1} y_j \overline{w}^{jk}, \quad \text{for } -\infty < k < \infty. \tag{5.5}$$

For the last expression in (5.5) we use the notation $w = e^{2\pi i/N}$, so that $\overline{w} = e^{-2\pi i/N}$.

An important property of $w = e^{2\pi i/N}$ is that $w^N = e^{2\pi i} = 1$. Of course, it follows that $\overline{w}^N = 1$. From this we see that

$$\widehat{y}_{k+N} = \sum_{j=0}^{N-1} y_j \overline{w}^{j(k+N)} = \sum_{j=0}^{N-1} y_j \overline{w}^{jk}\overline{w}^{jN} = \sum_{j=0}^{N-1} y_j \overline{w}^{jk} = \widehat{y}_k.$$

Thus the discrete Fourier transform is also periodic of period N.

Let's look back at equation (5.3), where we used the trapezoid rule to approximate α_k, the kth Fourier coefficient of the function f. Using (5.5), we can now write (5.3) as

$$\alpha_k \approx \frac{\widehat{y}_k}{N}. \tag{5.6}$$

It follows from the Riemann-Lebesgue lemma that $\alpha_k \to 0$ as $k \to \pm\infty$. On the other hand, the sequence $\widehat{y}_k$ is periodic. This implies that (5.3) is not a good approximation for large k. In fact, The trapezoid rule algorithm used to approximate the integral in (5.2) loses accuracy as the integrand $f(t)e^{ikt}$ becomes more oscillatory as the frequency (and index) k increases. Therefore, we would expect equation (5.3) to provide a good approximation only for k that are relatively small compared to N.

There is another, related factor to consider. We have previously talked of the importance of the low frequency components of a function. When we use the complex Fourier series, this means that we include both α_k and α_{-k} for small values of k. By (5.6) and the periodicity of the sequence $\widehat{y}$,

$$\alpha_{-k} \approx \frac{\widehat{y}_{-k}}{N} = \frac{\widehat{y}_{N-k}}{N}.$$

Therefore, when considering small frequency components while using the discrete Fourier transform, we must include both $\widehat{y}_k$ and $\widehat{y}_{N-k}$ for small nonnegative values of the index k.

Example 5.7 Use the discrete Fourier transform to compute approximately the first 64 Fourier coefficients of the function

$$f(t) = e^{-t^2/10}[\sin 2t + 2\cos 4t + 0.4\sin t \sin 10t]$$

on the interval $[0, 2\pi]$.

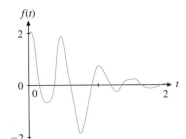

Figure 2. The function in Example 5.7.

The function f is plotted in Figure 2. Because of the terms involving $\sin 2t$ and $\cos 4t$, we would expect that the Fourier coefficients of order 2 and 4 would be large. With $N = 64$, we set $y_j = f(2\pi j/N)$ for $0 \le j \le N - 1$. Then we use the fast Fourier transform function in MATLAB® to compute the discrete Fourier transform $\widehat{y}$. The magnitude of the $\widehat{y}_k$ is plotted in Figure 3. Indeed, the coefficients corresponding to $k = 4$ and $k = 60 = N - 4$ are the largest. Notice how the coefficients with index k and $N - k$ are largest for small k. Thus the coefficients corresponding to small frequency components dominate.

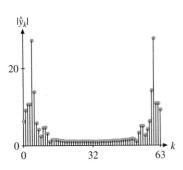

Figure 3. The discrete Fourier transform of the discretization of the function in Example 5.7.

Now let's look at (5.5) and restrict ourselves to $k = 0, 1, \ldots, N - 1$. The kth equation expresses $\widehat{y}_k$ as a linear combination of $\{y_j \mid 0 \le j \le N - 1\}$. These N equations can be expressed as the single matrix equation

$$\begin{pmatrix} \widehat{y}_0 \\ \widehat{y}_1 \\ \widehat{y}_2 \\ \vdots \\ \widehat{y}_{N-1} \end{pmatrix} = \begin{pmatrix} 1 & 1 & 1 & \cdots & 1 \\ 1 & \overline{w} & \overline{w}^2 & \cdots & \overline{w}^{N-1} \\ 1 & \overline{w}^2 & \overline{w}^4 & \cdots & \overline{w}^{2(N-1)} \\ \vdots & \vdots & \vdots & \ddots & \vdots \\ 1 & \overline{w}^{N-1} & \overline{w}^{2(N-1)} & \cdots & \overline{w}^{(N-1)^2} \end{pmatrix} \begin{pmatrix} y_0 \\ y_1 \\ y_2 \\ \vdots \\ y_{N-1} \end{pmatrix}. \tag{5.8}$$

It will be useful to use vector notation. We will set

$$\mathbf{y} = (y_0, y_1, \ldots, y_{N-1})^T, \quad \text{and}$$
$$\widehat{\mathbf{y}} = (\widehat{y}_0, \widehat{y}_1, \ldots, \widehat{y}_{N-1})^T.$$

With this notation, equation (5.8) becomes

$$\widehat{\mathbf{y}} = F\mathbf{y}, \tag{5.9}$$

where

$$F = \begin{pmatrix} 1 & 1 & 1 & \cdots & 1 \\ 1 & \overline{w} & \overline{w}^2 & \cdots & \overline{w}^{N-1} \\ 1 & \overline{w}^2 & \overline{w}^4 & \cdots & \overline{w}^{2(N-1)} \\ \vdots & \vdots & \vdots & \ddots & \vdots \\ 1 & \overline{w}^{N-1} & \overline{w}^{2(N-1)} & \cdots & \overline{w}^{(N-1)^2} \end{pmatrix}.$$ (5.10)

The inverse discrete Fourier transform

Equation (5.8) gives the formula for computing the discrete Fourier coefficients in terms of the original discrete signal. Many applications require the reverse operation, the computation of the original discrete signal, y_k, from its discrete Fourier coefficients, $\widehat{y}_k$. Therefore, we would like to solve for the y_k in equation (5.8) or, equivalently, we need to find the inverse of the matrix F in (5.10).

Computing the inverse of F is somewhat difficult, so we will simply give the result. Consider the complex conjugate of F

$$\overline{F} = \begin{pmatrix} 1 & 1 & 1 & \cdots & 1 \\ 1 & w & w^2 & \cdots & w^{N-1} \\ 1 & w^2 & w^4 & \cdots & w^{2(N-1)} \\ \vdots & \vdots & \vdots & \ddots & \vdots \\ 1 & w^{N-1} & w^{2(N-1)} & \cdots & w^{(N-1)^2} \end{pmatrix}.$$

Direct computation shows that

$$\overline{F} \cdot F = NI \quad \text{or} \quad F^{-1} = \frac{1}{N}\overline{F}.$$

The computation of $\overline{F} \cdot F$ is not too difficult. For example, when $N = 3$,

$$\overline{F} \cdot F = \begin{pmatrix} 1 & 1 & 1 \\ 1 & w & w^2 \\ 1 & w^2 & w^4 \end{pmatrix} \begin{pmatrix} 1 & 1 & 1 \\ 1 & \overline{w} & \overline{w}^2 \\ 1 & \overline{w}^2 & \overline{w}^4 \end{pmatrix}.$$

An explicit computation shows that this matrix product is $3I$. For example, the $(2, 1)$-entry of this matrix product is

$$1 + w + w^2 = \frac{1 - w^3}{1 - w} = 0,$$

since $w^3 = 1$. On the other hand, the $(2, 2)$-entry is

$$1 + |w|^2 + |w|^4 = 3.$$

We summarize this discussion in the next theorem.

THEOREM 5.11   The original signal y_j, $j = 0, \ldots, N - 1$, can be computed from its discrete Fourier transform, $\widehat{y}_k$, $k = 0, \ldots, N - 1$, using

$$y_j = \frac{1}{N} \sum_{k=0}^{N-1} \widehat{y}_k w^{jk}, \quad \text{for } -\infty < j < \infty.$$

We can write this in matrix form as

$$\mathbf{y} = \frac{1}{N}\overline{F}\widehat{\mathbf{y}}.$$ (5.12)

Noise filtering

Practical applications of this theorem are numerous. We will mention two. The first involves filtering noise from a signal. When a signal is transmitted, it is often corrupted by interference from background radiation or other sources. The corrupted part of the signal is called noise. In many applications, the noise appears with a certain frequency range that is different from the dominant frequencies of the original signal. To filter out noise, a discrete Fourier transform of the signal is computed using (5.8). Then the Fourier coefficients, $\widehat{y}_k$, corresponding to the noisy, undesirable frequencies are set equal to zero. The signal is then recomputed from the new Fourier coefficients using equation (5.12). Since the frequency components corresponding to the noise have been removed, the resulting signal should contain much less noise than the original.

Frequently, noise occurs at relatively high frequencies. Suppose that we add the noise term $N(t) = 2\sin(50t)$ to the function in Example 5.7. The resulting signal is $g(t) = f(t) + N(t)$, and it is plotted in Figure 4. It is difficult to see that the signal of interest is the function $f(t)$ plotted in Figure 2. We set $y_j = g(2\pi j/N)$ for $0 \le j \le N-1$, with $N = 256$, and take the discrete Fourier transform. The result is plotted in Figure 5. Notice the large terms at $k = 50$ and $k = 206 = N - 50$. To eliminate the high-frequency noise, we "zero out" the high-frequency coefficients by setting $\widehat{y}_k = 0$ for $13 \le k \le N - 13 = 243$, and compute the inverse transform. The resulting function is plotted in blue in Figure 6, while the original function f is plotted in black. The graphical comparison shows that we have effectively recovered the wanted signal from the noisy one. The most significant difference occurs at the two endpoints. This is a result of Gibb's phenomenon. Since f_p, the periodic extension of f, is not continuous, we have to expect this.

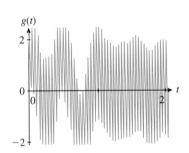

Figure 4. A signal in the presence of high-frequency noise.

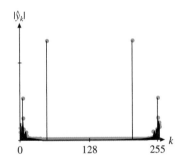

Figure 5. The discrete Fourier transform of the noisy signal.

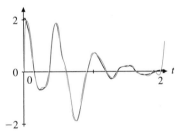

Figure 6. The result of filtering out the high frequencies.

Data compression

A second application involves data compression. The goal is to store or transmit a signal using the fewest possible bits of data. One way to accomplish this is to store or transmit only the dominant Fourier coefficients of a given signal. In view of the Riemann-Lebesgue lemma, Theorem 2.11, only a finite number of Fourier coefficients are dominant, since these coefficients get very small as the frequency gets large. Thus, a compression routine can be implemented in a three-step process. First, we compute the discrete Fourier coefficients using equation (5.8). Then we set all of the small Fourier coefficients equal to zero, storing only the dominant Fourier coefficients. Finally, to recover the compressed signal, use equation (5.12). What constitutes "small" depends on the application and the tolerance for error. There is a trade-off between the number of Fourier coefficients that are set equal to zero and the accuracy of the compressed signal. The larger the amount of compression, the more coefficients that are set equal to zero, and the greater the difference between the compressed signal and the original signal.

As an example, we set equal to 0 all coefficients for the function in Example 5.7 that were smaller than $1/10$ of the largest coefficient. This resulted in 21 nonzero coefficients. Again we computed the inverse transform, and plotted the result in blue in Figure 7. The function $f(t)$ is plotted in black. The comparison shows that there is loss, but not a great deal.

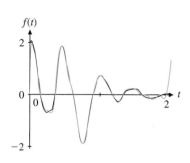

Figure 7. The result of removing small coefficients.

The fast Fourier transform

Calculating the discrete Fourier transform using equation (5.5) or (5.8) involves lots of computations. Computing each $\widehat{y}_k$ using (5.5) requires the sum of N products of two numbers. We will call the combination of a multiplication and an addition a **multiply-add**, and we will refer to it as an MA. Thus computing each $\widehat{y}_k$ requires N MAs. Computing the complete discrete Fourier transform means computing $\widehat{y}_k$ for $0 \le k \le N - 1$. This requires N^2 MAs.

The computation can be speeded up using the multiplicative nature of N. Suppose that $N = pq$, where the factors p and q are both bigger than 1. The index in the sum in (5.5) can be written as $j = \alpha p + \beta$, where $0 \le \alpha \le q - 1$ and $0 \le \beta \le p - 1$. In terms of α and β, the sum in (5.5) becomes the double sum

$$\widehat{y}_k = \sum_{\beta=0}^{p-1} \sum_{\alpha=0}^{q-1} y_{\alpha p+\beta}\, \overline{w}^{(\alpha p+\beta)k} = \sum_{\beta=0}^{p-1} \left(\sum_{\alpha=0}^{q-1} y_{\alpha p+\beta}\, \overline{w}^{\alpha p k} \right) \overline{w}^{\beta k}. \tag{5.13}$$

We will isolate the inner sum by setting

$$\widehat{y}_{\beta,k} = \sum_{\alpha=0}^{q-1} y_{\alpha p+\beta}\, \overline{w}^{\alpha p k}. \tag{5.14}$$

Then (5.13) becomes

$$\widehat{y}_k = \sum_{\beta=0}^{p-1} \widehat{y}_{\beta,k}\, \overline{w}^{\beta k}, \quad \text{for } 0 \le k \le N - 1. \tag{5.15}$$

The idea is to compute $\widehat{y}_{\beta,k}$ first using (5.14), and then compute the Fourier transform using (5.15). The savings in the computation comes from realizing that $\widehat{y}_{\beta,k}$ is periodic in k with period q. To see this, we first remember that $\overline{w}^N = 1$ and $N = pq$. Then we have

$$\widehat{y}_{\beta,k+q} = \sum_{\alpha=0}^{q-1} y_{\alpha p+\beta}\, \overline{w}^{\alpha p(k+q)} = \sum_{\alpha=0}^{q-1} y_{\alpha p+\beta}\, \overline{w}^{\alpha p k} = \widehat{y}_{\beta,k}.$$

Thus we only need to compute $\widehat{y}_{\beta,k}$ for $0 \le \beta \le p - 1$ and $0 \le k \le q - 1$. Since computing each $\widehat{y}_{\beta,k}$ requires q MAs, computing all of them requires $pq \cdot q = pq^2 = Nq$ MAs. Now computing the N components of the Fourier transform using (5.15) requires Np additional MAs, for a total of $N(p + q)$. If N, p, and q are all large numbers, the sum $p + q$ is much smaller than the product $N = pq$.

The process outlined in the previous paragraph can be iterated if N has more factors. If $N = p_1 \cdot p_2 \cdot \ldots \cdot p_n$, the number of MAs required is reduced to $N(p_1 + p_2 + \cdots + p_n)$. This algorithm for computing the discrete Fourier transform is called the **fast Fourier transform (FFT)**. Clearly the FFT works best if N has a large number of very small factors, the best being when N is a power of 2. This is the most commonly used case. When $N = 2^L$ the FFT can compute the Fourier coefficients with only about $N \cdot 2L = 2N \log_2 N$ MAs. For example, if $N = 2^{10} = 1024$, the FFT requires only about 20,000 MAs versus the one million or so that are required using (5.8). The savings get more impressive as N gets larger. A similar FFT routine exists for computing the inverse discrete Fourier transform. The mathematical computer programs MATLAB®, *Mathematica*, and Maple all have built-in commands for the FFT and inverse FFT.

EXERCISES

All of these exercises are designed to be done with a mathematical computer program such as MATLAB®, Maple, or *Mathematica*.

1. Consider the function

   $$f(t) = e^{-t^2/10} \left(\cos 2t + 2 \sin 4t + 0.4 \cos 2t \cos 40t \right).$$

   For what values of n would you expect the Fourier coefficients to be largest? Why? Compute the coefficients numerically through $n = 50$ and see if you are right. (You can use a fast Fourier transform algorithm with $N = 256$ to do this if you wish.) Plot the partial sum of the Fourier series of order $n = 6$ and compare with the plot of the original $f(x)$.

2. Consider the function

   $$g(t) = e^{-t^2/8} \left[\cos 2t + 2 \sin 4t + 0.4 \cos 2t \cos 10t \right],$$

   for $0 \le t \le 2\pi$. Compute numerically the partial sum of the Fourier series of order $N = 25$. Zero out any coefficients that have absolute value smaller than 10% of the maximum. Plot the resulting series and compare with the original function $g(t)$. Try experimenting with different tolerances (other than 10%).

3. Show that if $y = \{y_m\}$ is a sequence of real numbers that is periodic with period N and $\widehat{y}$ is the discrete Fourier transform of y, then the complex conjugate of $\widehat{y}_m$ is $\widehat{y}_{N-m}$. (As a result, when m is small relative to N, $\widehat{y}_{N-m}$ has to be considered a low-frequency coefficient, since it is equal to the conjugate of $\widehat{y}_m$, which is approximately equal to the conjugate of the mth Fourier coefficient.)

The next three problems require the use of the fast Fourier transform on a computer (e.g., the FFT routines in MATLAB® or Maple, or *Mathematica*).

4. **Filtering** Let

   $$f(t) = e^{-t^2/10} \left(\sin(2t) + 2 \cos(4t) + 0.4 \sin(t) \sin(50t) \right).$$

   Discretize f by setting $y_k = f(2k\pi/256)$, for $k = 0, \ldots, 255$. Use the fast Fourier transform to compute $\widehat{y}_k$ for $0 \le k \le 255$. According to Exercise 3, the low-frequency coefficients are $\widehat{y}_0, \ldots, \widehat{y}_m$ and $\widehat{y}_{256-m}, \ldots, \widehat{y}_{255}$ for some low value of m. Filter out the high-frequency terms by setting $\widehat{y}_k = 0$ for $m \le k \le 255 - m$ with $m = 6$. Apply the inverse fast Fourier transform to this new set of $\widehat{y}_k$ to compute the y_k (now filtered); plot the new values of y_k and compare with the original function. Experiment with other values of m.

5. **Compression** Let tol $= 0.01$. In Exercise 4, if $|\widehat{y}_k| <$ tol $\times M$, where $M = \max_{0 \le k \le 255} |\widehat{y}_k|$, set $\widehat{y}_k$ equal to zero. Apply the inverse fast Fourier transform to this new set of $\widehat{y}_k$ to compute the y_k. Plot the new values of y_k and compare with the original function. Experiment with other values of tol. Keep track of the percentage of Fourier coefficients that have been filtered out. The MATLAB® sort command is useful for finding a value for tol in order to filter out a specified percentage of coefficients.

6. Repeat the previous two exercises over the interval $0 \le t \le 1$ with the function

   $$f(t) = -52t^4 + 100t^3 - 49t^2 + 2 + N(100(t - 1/3))$$
   $$+ N(200(t - 2/3))$$

   where $N(t) = te^{-t^2}$.

13

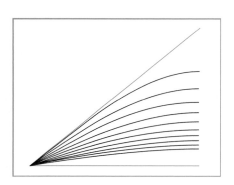

Partial Differential Equations

We will now consider differential equations that model change where there is more than one independent variable. For example, the temperature in an object changes with time and with the position within the object. The rates of change lead to partial derivatives, and the equations relating them are called partial differential equations. The applications of the subject are many, and the types of equations that arise have a great deal of variety. We will limit our study to the equations that arise most frequently in applications. These model heat flow and simple waves. The differential equation models for heat flow and the vibrating string will be derived in Sections 13.1 and 13.3, where we will also describe some of their properties. We will then systematically study each of the equations, solving them in some cases using the method of separation of variables.

13.1 Derivation of the Heat Equation

Heat is a form of energy that exists in any material. Like any other form of energy, heat is measured in joules (1 J = 1 Nm). However, it is also measured in calories (1 cal = 4.184 J) or sometimes in British thermal units (1 BTU = 252 cal = 1.054 kJ).

The amount of heat within a given volume is defined only up to an additive constant. We will assume the convention of saying that the amount of heat is equal to 0 when the temperature is equal to 0. Suppose that ΔV is a small volume in which the temperature u is almost constant. It has been found experimentally that the amount of heat ΔQ in ΔV is proportional to the temperature u and to the mass $\Delta m = \rho \, \Delta V$, where ρ is the mass density of the material. Thus the amount of heat in ΔV is given by

$$\Delta Q = c\rho u \, \Delta V. \tag{1.1}$$

The new constant c is called the **specific heat**. It measures the amount of heat required to raise 1 unit of mass of the material 1 degree of temperature. We will usually use the Celsius or Kelvin scales for temperature.

Let's consider a thin rod that is insulated along its length, as seen in Figure 1. If the length of the rod is L, the position along the rod is given by x, where $0 \leq x \leq L$. Since the rod is insulated, there is no transfer of heat from the rod except at its two ends. We may therefore assume that the temperature u depends only on x and on the time t.

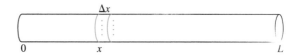

Figure 1. The variation of temperature in an insulated rod.

Consider a small section of the rod between x and $x + \Delta x$. Let S be the cross-sectional area of the rod. The volume of the section is $S\,\Delta x$, so (1.1) becomes

$$\Delta Q = c\rho u S\,\Delta x.$$

Therefore, the amount of heat at time t in the portion U of the rod defined by $a \leq x \leq b$ is given by the integral

$$Q(t) = S \int_a^b c\rho u(t, x)\,dx. \tag{1.2}$$

The specific heat and the density sometimes vary from point to point and more rarely with time as well. However, we will usually be dealing with homogeneous, time independent materials for which both the specific heat and the density are constants.

The heat equation models the flow of heat through the material. It is derived by computing the time rate of change of Q in two different ways. The first way is to differentiate (1.2). Differentiating under the integral sign, we get

$$\frac{dQ}{dt} = \frac{d}{dt} S \int_a^b c\rho u\,dx = S \int_a^b \frac{\partial}{\partial t}[c\rho u]\,dx.$$

Of course, if the specific heat and the density do not vary with time, this becomes

$$\frac{dQ}{dt} = S \int_a^b c\rho \frac{\partial u}{\partial t}\,dx. \tag{1.3}$$

The second way to compute the time rate of change of Q is to notice that, in the absence of heat sources within the rod, the quantity of heat in U can change only through the flow of heat across the boundaries of U at $x = a$ and $x = b$. The rate of heat flow through a section of the rod is called the **heat flux through the section**. Consider the section of the rod between $x = a$ and $a + \Delta x$. Experimental study of heat conduction reveals that the flow of heat across such a section has the following properties:

- Heat flows from hot positions to cold positions at a rate proportional to the difference in the temperatures on the two sides of the section. Thus the heat flux through the section is proportional to $u(a + \Delta x, t) - u(a, t)$.
- The heat flux through the section is inversely proportional to Δx, the width of the section.
- The heat flux through the section is proportional to the area of S of the boundary of the section.

Putting these three points together, we see that there is a coefficient C such that the heat flux *into* U at $x = a$ is given approximately by

$$-CS\frac{u(a + \Delta x, t) - u(a, t)}{\Delta x}. \tag{1.4}$$

The coefficient C is called the **thermal conductivity**. It is positive since, if $u(a + \Delta x, t) > u(a, t)$, then the temperature is hotter inside U than it is outside, and the heat flows *out of* U at $x = a$. The thermal conductivity is usually constant, but it may depend on the temperature u and the position x.

If we let Δx go to 0 in (1.4), the difference quotient approaches $\partial u/\partial x$, and we see that the heat flux into U at $x = a$ is

$$-CS\frac{\partial u}{\partial x}(a, t). \tag{1.5}$$

The same argument at $x = b$ shows that the heat flux into U at $x = b$ is

$$CS\frac{\partial u}{\partial x}(b, t). \tag{1.6}$$

The total time rate of change of Q is the sum of the rates at the two ends. Using the fundamental theorem of calculus, this is

$$\frac{dQ}{dt} = S\left[C\frac{\partial u}{\partial x}(b, t) - C\frac{\partial u}{\partial x}(a, t)\right] = S\int_a^b \frac{\partial}{\partial x}\left(C\frac{\partial u}{\partial x}\right) dx. \tag{1.7}$$

If the thermal conductivity C is independent of x, this becomes

$$\frac{dQ}{dt} = CS\int_a^b \frac{\partial^2 u}{\partial x^2} dx. \tag{1.8}$$

In equations (1.3) and (1.8) we have two formulas for the rate of heat flow into U. Setting them equal, we see that

$$\int_a^b c\rho\frac{\partial u}{\partial t} dx = C\int_a^b \frac{\partial^2 u}{\partial x^2} dx \quad \text{or} \quad \int_a^b \left(c\rho\frac{\partial u}{\partial t} - C\frac{\partial^2 u}{\partial x^2}\right) dx = 0.$$

This is true for all $a < b$, which can be true only if the integrand is equal to 0. Hence,

$$c\rho\frac{\partial u}{\partial t} - C\frac{\partial^2 u}{\partial x^2} = 0$$

throughout the material. If we divide by $c\rho$, and set $k = C/c\rho$, the equation becomes

$$\frac{\partial u}{\partial t} - k\frac{\partial^2 u}{\partial x^2} = 0, \quad \text{or} \quad \frac{\partial u}{\partial t} = k\frac{\partial^2 u}{\partial x^2}. \tag{1.9}$$

The constant k is called the **thermal diffusivity** of the material. The units of k are $(\text{length})^2/\text{time}$. The values of k for some common materials are listed in Table 1.

Equation (1.9) is called the **heat equation**. As we have shown, it models the flow of heat through a material and is satisfied by the temperature. It should be noticed that if we have a wall with height and width that are large in comparison to the thickness L, then the temperature in the wall away from its ends will depend only on the position within the wall. Consequently, we have a one dimensional problem, and the variation of the temperature is modeled by the heat equation in (1.9).

A similar derivation shows that the diffusion of a substance through a liquid or a gas satisfies the same equation. In this case it is the concentration u that satisfies the equation. For this reason equation (1.9) is also referred to as the **diffusion equation**.

Table 1 Thermal diffusivities of common materials			
MATERIAL	k (cm²/sec)	MATERIAL	k (cm²/sec)
Aluminum	0.84	Gold	1.18
Brick	0.0057	Granite	0.008–0.018
Cast iron	0.17	Ice	0.0104
Copper	1.12	PVC	0.0008
Concrete	0.004–0.008	Silver	1.70
Glass	0.0043	Water	0.0014

Subscript notation for derivatives

We will find it useful to abbreviate partial derivatives by using subscripts to indicate the variable of differentiation. For example, we will write

$$u_x = \frac{\partial u}{\partial x}, \quad u_y = \frac{\partial u}{\partial y}, \quad u_{yx} = \frac{\partial^2 u}{\partial x \, \partial y}, \quad \text{and} \quad u_{xx} = \frac{\partial^2 u}{\partial x^2}.$$

Using this notation, we can write the heat equation in (1.9) quite succinctly as

$$u_t = k u_{xx}.$$

The inhomogeneous heat equation

Equation (1.9) was derived under the assumption that there is no source of heat within the material. If there are heat sources, we can modify the model to accommodate them. If we look back to equation (1.7), which accounts for the rate of flow of heat into U, we see that we must modify the right-hand side to account for internal sources. We will assume that the heat source is spread throughout the material and that heat is being added at the rate of $p(u, x, t)$ thermal units per unit volume per second.

Notice that we allow the rate of heat inflow to depend on the temperature u, as well as on x and t. An example would be a rod that is not completely insulated along its length. Then heat would flow into or out of the rod along its length at a rate that is proportional to the difference between the temperature in U and the ambient temperature, so $p(u, x, t) = \alpha[u - T]$, where T is the ambient temperature.

Assuming there is a source of heat, equation (1.7) becomes

$$\frac{dQ}{dt} = CS\left[\frac{\partial u}{\partial x}(b, t) - \frac{\partial u}{\partial x}(a, t)\right] + S\int_a^b p(u, x, t)\, dx.$$

The rest of the derivation is unchanged, and in the end we get

$$c\rho\frac{\partial u}{\partial t} = C\frac{\partial^2 u}{\partial x^2} + p, \quad \text{or} \quad \frac{\partial u}{\partial t} = k\frac{\partial^2 u}{\partial x^2} + \frac{p}{c\rho}. \tag{1.10}$$

Because of the term involving p, equation (1.10) is called the **inhomogeneous heat equation**, while equation (1.9) is called the **homogeneous heat equation.**

Initial conditions

We have seen that ordinary differential equations have many solutions, and to determine a particular solution we specify initial conditions. The situation is more complicated for partial differential equations.

For example, specifying initial conditions for a temperature requires giving the temperature at each point in the material at the initial time. In the case of the rod this means that we give a function $f(x)$ defined for $0 \le x \le L$ and we look for a solution to the heat equation that also satisfies

$$u(x, 0) = f(x), \quad \text{for } 0 \le x \le L. \tag{1.11}$$

Types of boundary conditions

In addition to specifying the initial temperature, it will be necessary to specify conditions on the boundary of the material. For example, the temperature may be fixed at one endpoint of the rod as the result of the material being embedded in a source of heat kept at a constant temperature. The temperatures might well be different at the two ends of the rod. Thus if the temperature at $x = 0$ is T_0 and that at $x = L$ is T_L, then the temperature $u(x, t)$ satisfies

$$u(0, t) = T_0 \quad \text{and} \quad u(L, t) = T_L, \quad \text{for all } t. \tag{1.12}$$

Boundary conditions of the form in (1.12) specifying the value of the temperature at the boundary are called ***Dirichlet conditions***.

In other circumstances one or both ends of the rod might be insulated. This means that there is no flow of heat into or out of the rod at these points. According to the discussion leading to equation (1.5), this means that

$$\frac{\partial u}{\partial x} = 0 \tag{1.13}$$

at an insulated point. This type of boundary condition is called a ***Neumann condition***. A rod could satisfy a Dirichlet condition at one boundary point and a Neumann condition at the other.

There is a third condition that occurs, for example, when one end of the rod is poorly insulated from the exterior. According to Newton's law of cooling, the flow of heat across the insulation is proportional to the difference in the temperatures on the two sides of the insulation. If this is true at the endpoint $x = 0$, then arguing along the same lines as we did in the derivation of equation (1.5), we see that there is a positive number α such that

$$\frac{\partial u}{\partial x}(0, t) = \alpha(u(0, t) - T), \tag{1.14}$$

where T is the temperature outside the insulation and u is the temperature at the endpoint $x = 0$. Poor insulation at the endpoint $x = L$ leads in the same way to a boundary condition of the form

$$\frac{\partial u}{\partial x}(L, t) = -\beta(u(L, t) - T), \tag{1.15}$$

where $\beta > 0$. Boundary conditions of the type in (1.14) and (1.15) are called ***Robin conditions***.

Robin boundary conditions also arise when a solid wall meets a fluid or a gas. In such a case a thin boundary layer is formed, which shields the rest of the fluid or gas from the temperature in the wall. The constant β is sometimes called the ***heat transfer coefficient***.

Initial/boundary value problems

Putting everything together, we see that the temperature $u(x,t)$ in an insulated rod with Dirichlet boundary conditions must satisfy the heat equation together with initial and boundary conditions. The complete problem is to find a function $u(x,t)$ such that

$$u_t(x,t) = ku_{xx}, \quad \text{for } 0 < x < L \text{ and } t > 0,$$
$$u(0,t) = T_0, \quad \text{and} \quad u(L,t) = T_L, \quad \text{for } t > 0, \tag{1.16}$$
$$u(x,0) = f(x), \quad \text{for } 0 \le x \le L.$$

The function $f(x)$ is the initial temperature distribution. The initial/boundary value problem is illustrated in Figure 2. As we have indicated, the Dirichlet boundary condition at each endpoint in (1.16) could be replaced with a Neumann or a Robin condition.

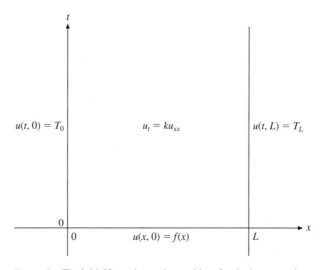

Figure 2. The initial/boundary value problem for the heat equation.

The maximum principle

One of the major tenets of the theory of heat flow is that heat flows from hot areas to colder areas. From this starting point, physical reasoning allows us to conclude that the temperature $u(t,x)$ cannot get too hot or too cold in the region where it satisfies the heat equation. To be precise, let

$$m = \min_{0 \le x \le L} f(x) \quad \text{and} \quad M = \max_{0 \le x \le L} f(x).$$

Then, if $u(t,x)$ is a solution to the initial/boundary value problem in (1.16),

$$\min\{m, T_0, T_L\} \le u(t,x) \le \max\{M, T_0, T_L\} \quad \text{for } 0 \le t \text{ and } 0 \le x \le L.$$

This result is called the ***maximum principle for the heat equation.*** In English it says that a temperature $u(t,x)$ defined for $0 \le t$ and $0 \le x \le L$ must achieve its maximum value (and its minimum value) on the boundary of the region where it is defined. Thus in Figure 2 The temperature $u(t,x)$ in the indicated half-strip must achieve its maximum and minimum values on the three lines that form its boundary.

Linearity

If u and v are functions and α and β are constants, then

$$\frac{\partial}{\partial x}(\alpha u + \beta v) = \alpha \frac{\partial u}{\partial x} + \beta \frac{\partial v}{\partial x}. \tag{1.17}$$

We will express this standard fact about $\partial/\partial x$ by saying that it is a ***linear operator***. It is an ***operator*** because it "operates" on a function u and yields another function $\partial u/\partial x$. That it is ***linear*** simply means that (1.17) is satisfied. It follows easily that more complicated differential operators, such as

$$\frac{\partial^2}{\partial x^2} \quad \text{and} \quad \frac{\partial^2}{\partial x \, \partial y}$$

are also linear. It then follows that the heat operator

$$\frac{\partial}{\partial t} - k \frac{\partial^2}{\partial x^2}$$

is linear. This implies the following theorem.

THEOREM 1.18   The homogeneous heat equation is a linear equation, meaning that if u and v satisfy

$$u_t = k u_{xx} \quad \text{and} \quad v_t = k v_{xx},$$

and α and β are constants, then the linear combination $w = \alpha u + \beta v$ satisfies $w_t = k w_{xx}$, so w is also a solution to the homogeneous heat equation.   ∎

We will make frequent use of Theorem 1.18. It will enable us to build up more complicated solutions as linear combinations of basic solutions.

EXERCISES

1. Suppose that the temperature at each point of a rod of length L is originally at $15°$. Suppose that starting at time $t = 0$ the left end is kept at $5°$ and the right end at $25°$. Write down the complete description of the initial/boundary value problem the temperature in the rod must obey.

2. Show that the temperature in the rod in Exercise 1 must satisfy $5 \le u(x, t) \le 25$ for $t \ge 0$ and $0 \le x \le L$.

3. Suppose the specific heat, density, and thermal conductivity depend on x, and are not constant. Show that the heat equation becomes

$$\frac{\partial}{\partial t}[c\rho u] = \frac{\partial}{\partial x}\left[C \frac{\partial u}{\partial x} \right].$$

4. If our rod is insulated at both ends, we would expect that the total amount of heat in the rod does not change with time. Show that this follows from equation (1.7).

5. Prove Theorem 1.18 by showing that $w_t = k w_{xx}$.

6. Solutions to the Dirichlet problem in (1.16) are unique. This means that if both u and v satisfy the conditions in (1.16), then $u(x, t) = v(x, t)$ for $t \ge 0$ and $0 \le x \le L$. Use the linearity of the heat equation and the maximum principle to prove this fact.

7. Suppose we have an insulated aluminum rod of length L. Suppose the rod is at a constant temperature of $15°$K, and that starting at time $t = 0$, the left-hand end point is kept at $20°$K and the right-hand endpoint is kept at $35°$K. Provide the initial/boundary value problem that must be satisfied by the temperature $u(t, x)$.

8. Suppose we have an insulated gold rod of length L. Suppose the rod is at a constant temperature of $15°$K, and that starting at time $t = 0$, the left-hand end point is kept at $20°$K while the right-hand endpoint is kept insulated. Provide the initial/boundary value problem that must be satisfied by the temperature $u(t, x)$.

9. Suppose we have an insulated silver rod of length L. Suppose the rod is at a constant temperature of $15°$K, and that starting at time $t = 0$, the right-hand end point is kept at $35°$K while the left-hand endpoint is only partially insulated, so heat is lost there at a rate equal to 0.0013 times the difference between the temperature of the rod at this point and the ambient temperature $T = 15°$K. Provide the initial/boundary value problem that must be satisfied by the temperature $u(t, x)$.

13.2 Separation of Variables for the Heat Equation

We will start this section by solving the initial/boundary value problem

$$u_t(x, t) = k u_{xx}(x, t), \quad \text{for } t > 0 \text{ and } 0 < x < L, \tag{2.1}$$

$$u(0, t) = T_0 \quad \text{and} \quad u(L, t) = T_L, \quad \text{for } t > 0, \tag{2.2}$$

$$u(x, 0) = f(x), \quad \text{for } 0 \le x \le L \tag{2.3}$$

that we posed in (1.16).

Steady-state temperatures

It is useful for both mathematical and physical purposes to split the problem into two parts. We first find the steady-state temperature that satisfies the boundary conditions in (2.2). A **steady-state** temperature is one that does not depend on time. Then $u_t = 0$, so the heat equation (2.1) simplifies to $u_{xx} = 0$. Hence we are looking for a function $u_s(x)$ defined for $0 \le x \le L$ such that

$$\frac{\partial^2 u_s}{\partial x^2}(x) = 0, \quad \text{for } 0 < x < L, \tag{2.4}$$

$$u_s(0, t) = T_0 \quad \text{and} \quad u_s(L, t) = T_L, \quad \text{for } t > 0.$$

The solution to this boundary value problem is easily found, since the general solution of the differential equation is $u_s(x) = Ax + B$, where A and B are arbitrary constants. Then the boundary conditions reduce to

$$u_s(0) = B = T_0 \quad \text{and} \quad u_s(L) = AL + B = T_L.$$

We conclude that $B = T_0$ and $A = (T_L - T_0)/L$, so the steady-state temperature is

$$u_s(x) = (T_L - T_0)\frac{x}{L} + T_0.$$

It remains to find $v = u - u_s$. By Theorem 1.18, it will be a solution to the heat equation, since both u and u_s are, and the heat equation is linear. The boundary and initial conditions that v satisfies can be calculated from those for u and u_s in (2.2), (2.3), and (2.4). Thus, $v = u - u_s$ must satisfy

$$v_t(x, t) = k v_{xx}(x, t), \quad \text{for } 0 < x < L \text{ and } t > 0,$$
$$v(0, t) = v(L, t) = 0, \quad \text{for } t > 0, \tag{2.5}$$
$$v(x, 0) = g(x) = f(x) - u_s(x), \quad \text{for } 0 \le x \le L.$$

The most important fact is that the boundary conditions for v are $v(0, t) = v(L, t) = 0$. When the right-hand sides are equal to 0 we say that the boundary conditions are **homogeneous**. This will make finding the solution a lot easier.

Having found the steady-state temperature u_s and the temperature v, the solution to the original problem is $u(x, t) = u_s(x) + v(x, t)$.

Solution with homogeneous boundary conditions

We will find the solution to the initial/boundary value problem with homogeneous boundary conditions in (2.5) using the technique of **separation of variables**. It should be noted that separation of variables can only be used to solve an initial/boundary value problem when the boundary conditions are homogeneous. Since this is the first time we are using the technique, and since it is a technique we will use throughout this chapter, we will go through the process slowly. The basic idea of the method of separation of variables is to hunt for solutions in the product form

$$v(x, t) = X(x)T(t), \tag{2.6}$$

where $T(t)$ is a function of t and $X(x)$ is a function of x. We will insist that the product solution v satisfies the homogeneous boundary conditions. Since $0 = v(0, t) = X(0)T(t)$ for all $t > 0$, we conclude that $X(0) = 0$. A similar argument shows that $X(L) = 0$. This leads to a two-point boundary value problem for X that we will solve. In the end we will have found enough solutions of the factored form so that we will be able to solve the initial/boundary value problem in (2.5) using an infinite linear combination of them.

There are three steps to the method.

Step 1. Separate the PDE into two ODEs. When we insert $v = X(x)T(t)$ into the heat equation $v_t = kv_{xx}$, we get

$$X(x)T'(t) = kX''(x)T(t). \tag{2.7}$$

The key step is to separate the variables by bringing everything depending on t to the left, and everything depending on x to the right. Dividing (2.7) by $kX(x)T(t)$, we get

$$\frac{T'(t)}{kT(t)} = \frac{X''(x)}{X(x)}.$$

Since x and t are independent variables, the only way that the left-hand side, a function of t, can equal the right-hand side, a function of x, is if both functions are constant. Consequently, there is a constant that we will write as $-\lambda$, such that

$$\frac{T'(t)}{kT(t)} = -\lambda \quad \text{and} \quad \frac{X''(x)}{X(x)} = -\lambda,$$

or

$$T' + \lambda kT = 0 \quad \text{and} \quad X'' + \lambda X = 0. \tag{2.8}$$

The first equation has the general solution

$$T(t) = Ce^{-\lambda kt}. \tag{2.9}$$

We have to work a little harder on the second equation.

Step 2. Set up and solve the two-point boundary value problem. Since we insist that the solution X satisfies the homogeneous boundary conditions, the complete problem to be solved in finding X is

$$X'' + \lambda X = 0 \quad \text{with } X(0) = X(L) = 0. \tag{2.10}$$

Notice that the problem in (2.10) is not the standard initial value problem we have been solving up to now. There are two conditions imposed, but instead of both being imposed at the initial point $x = 0$, there is one condition imposed at each endpoint of the interval. Accordingly, this is called a ***two-point boundary value problem***. It is also called a ***Sturm-Liouville problem***.[1]

---

[1] This is our first example of a Sturm-Liouville problem. We will study them in some detail in Sections 13.6 and 13.7.

Another point to be made is that the constant λ is still undetermined. Furthermore, as we will see, for most values of λ the only solution to (2.10) is the function that is identically 0. Solving a Sturm-Liouville problem amounts to finding the numbers λ for which there are nonzero solutions to (2.10).

DEFINITION 2.11

A number λ is called an **eigenvalue** for the Sturm-Liouville problem in (2.10) if there is a nonzero function X that solves (2.10). If λ is an eigenvalue, then any function that satisfies (2.10) is called an **eigenfunction**.[2]

The solution to a Sturm-Liouville problem like (2.10) is the list of its eigenvalues and eigenfunctions. Notice that because of the linearity of the differential equation in (2.10), any constant multiple of an eigenfunction is also an eigenfunction. We will usually choose the constant that leads to the least complicated form for the eigenfunction.

Let's return to the example in (2.10). We will first show that there are no negative eigenvalues. To see this, set $\lambda = -r^2$, where $r > 0$. The equation in (2.10) becomes $X'' - r^2 X = 0$, which has general solution $X(x) = C_1 e^{rx} + C_2 e^{-rx}$. The boundary conditions are

$$0 = X(0) = C_1 + C_2$$
$$0 = X(L) = C_1 e^{rL} + C_2 e^{-rL}.$$

From the first equation, $C_2 = -C_1$. Inserting this into the second equation, we get

$$0 = C_1(e^{rL} - e^{-rL}).$$

Since $r \neq 0$, the factor in parenthesis on the right is nonzero. Hence $C_1 = 0$, which in turn implies that $C_2 = 0$, so the only solution is $X(x) = 0$. This means that λ is not an eigenvalue. This agrees with our physical intuition about heat flow. If there were a solution X with $\lambda < 0$, then, according to (2.6) and (2.9), the product solution to the heat equation would be $v(x, t) = e^{-\lambda kt} X(x)$. If $\lambda < 0$, this solution would grow exponentially in magnitude as t increases. In fact, we notice experimentally that temperatures tend to remain stable over time in the absence of heat sources.

This argument can be repeated if $\lambda = 0$. In this case the differential equation becomes $X'' = 0$, which has the general solution $X(x) = ax + b$, where a and b are constants. The boundary conditions become $0 = X(0) = b$ and $0 = X(L) = aL + b$, from which we easily conclude that $a = b = 0$.

Next suppose that $\lambda > 0$ and set $\lambda = \omega^2$, where $\omega > 0$. Then the differential equation in (2.10) is $X'' + \omega^2 X = 0$, which has the general solution

$$X(x) = a \cos \omega x + b \sin \omega x.$$

For this solution the boundary condition $X(0) = 0$ becomes $a = 0$. Then the boundary condition $X(L) = 0$ becomes

$$b \sin \omega L = 0.$$

We are only interested in nonzero solutions, so we must have $\sin \omega L = 0$. This occurs if $\omega L = n\pi$ for some positive integer n. When this is true we have the

---

[2] You will observe that finding the eigenvalues and eigenfunctions of a Sturm-Liouville problem is similar in many ways to finding the eigenvalues and eigenvectors of a matrix. It might be useful to compare the situation here with Section 9.1.

eigenvalue $\lambda = \omega^2 = n^2\pi^2/L^2$. For any nonzero constant b, $X(x) = b\sin(n\pi x/L)$ is an eigenfunction. The simplest thing to do is to set $b = 1$.

In summary, the eigenvalues and eigenfunctions for the Sturm-Liouville problem in (2.10) are

$$\lambda_n = \frac{n^2\pi^2}{L^2} \quad \text{and} \quad X_n(x) = \sin\left(\frac{n\pi x}{L}\right), \quad \text{for } n = 1, 2, 3, \ldots. \tag{2.12}$$

Finally, by incorporating (2.6), (2.9), and (2.12), we get the product solutions,

$$v_n(x, t) = e^{-n^2\pi^2 kt/L^2}\sin\left(\frac{n\pi x}{L}\right), \quad \text{for } n = 1, 2, 3, \ldots, \tag{2.13}$$

to the heat equation, that also satisfy the boundary conditions $v_n(0, t) = v_n(L, t) = 0$.

Step 3. Satisfying the initial condition. Having found infinitely many product solutions in (2.13), we can use the linearity of the heat equation (see Theorem 1.18) to conclude that any finite linear combination of them is also a solution. Hence, if b_n is a constant for each n, then for any N the function

$$v(x, t) = \sum_{n=1}^{N} b_n v_n(x, t) = \sum_{n=1}^{N} b_n e^{-n^2\pi^2 kt/L^2}\sin\left(\frac{n\pi x}{L}\right)$$

is a solution to the heat equation that satisfies the homogeneous boundary conditions.

We are naturally led to consider the infinite series

$$v(x, t) = \sum_{n=1}^{\infty} b_n v_n(x, t) = \sum_{n=1}^{\infty} b_n e^{-n^2\pi^2 kt/L^2}\sin\left(\frac{n\pi x}{L}\right). \tag{2.14}$$

We will assume that the coefficients b_n are such that this series converges, and that the resulting function v satisfies the heat equation and the homogeneous boundary conditions. These facts are true formally.[3] They are also true in the cases that we will consider, but we will not verify this. To do so requires some lengthy mathematical arguments that would not significantly add to our understanding of the issue.

Referring back to our original initial/boundary value problem in (2.5), we see that the function v defined in (2.14) satisfies everything except the initial condition $v(x, 0) = g(x) = f(x) - u_s(x)$. However, we have yet to determine the coefficients b_n. Using the series definition for v in (2.14), the initial condition becomes

$$g(x) = v(x, 0) = \sum_{n=1}^{\infty} b_n \sin\left(\frac{n\pi x}{L}\right), \quad \text{for } 0 \leq x \leq L. \tag{2.15}$$

Equation (2.15) will be recognized as the Fourier sine expansion for the initial temperature g. According to Section 12.3, and in particular equation (3.7), the values of b_n are given by

$$b_n = \frac{2}{L}\int_0^L g(x)\sin\left(\frac{n\pi x}{L}\right)dx. \tag{2.16}$$

Substituting these values into (2.14) gives a complete solution to the homogeneous initial/boundary value problem in (2.5). As indicated previously, the function $u(x, t) = u_s(x) + v(x, t)$ satisfies the original initial/boundary value problem in equations (2.1), (2.2), and (2.3).

---

[3] *Formally* means that we ignore the mathematical niceties of verifying that we can differentiate the function v by differentiating the terms in the infinite series.

Example 2.17 Suppose a rod of length 1 meter (100 cm) is originally at 0°C. Starting at time $t = 0$, one end is kept at the constant temperature of 100°C, while the other is kept at 0°C. Find the temperature distribution in the rod as a function of time and position. Assume that the thermal diffusivity of the rod is $k = 1$ cm^2/sec.

If we use the meter as the unit of length, then $k = 0.0001$ m^2/sec. The temperature in the rod, $u(x, t)$, must solve the initial/boundary value problem

$$u_t(x, t) = 0.0001\, u_{xx}(x, t), \quad \text{for } t > 0 \text{ and } 0 < x < 1,$$
$$u(0, t) = 0 \quad \text{and} \quad u(1, t) = 100, \quad \text{for } t > 0, \tag{2.18}$$
$$u(x, 0) = 0, \quad \text{for } 0 \le x \le 1.$$

Following the discussion at the beginning of this section, we write the temperature distribution as $u = u_s + v$, where $u_s(x)$ is the steady-state temperature with the same boundary conditions as u, and v is a temperature with homogeneous boundary conditions, and the same initial condition as $u - u_s$. The steady-state temperature u_s must satisfy

$$u_s'' = 0 \quad \text{with} \quad u_s(0) = 0 \quad \text{and} \quad u_s(1) = 100.$$

We easily see that $u_s(x) = 100\,x$.

Then the temperature $v = u - u_s$ must satisfy

$$v_t(x, t) = 0.0001 v_{xx}(x, t), \quad \text{for } t > 0 \text{ and } 0 < x < 1,$$
$$v(0, t) = 0 \quad \text{and} \quad v(1, t) = 0, \quad \text{for } t > 0, \tag{2.19}$$
$$v(x, 0) = -100\,x, \quad \text{for } 0 \le x \le 1.$$

The boundary values are homogeneous, so we can use the formula for the solution in (2.14), with $k = 0.0001$ and $L = 1$, to get

$$v(x, t) = \sum_{n=1}^{\infty} b_n e^{-0.0001 n^2 \pi^2 t} \sin n\pi x. \tag{2.20}$$

The coefficients are determined by the initial condition. Setting $t = 0$ in (2.20) and using $v(x, 0) = -100\,x$, we obtain

$$-100\,x = \sum_{n=1}^{\infty} b_n \sin n\pi x.$$

Therefore, the b_n are the Fourier sine coefficients of $-100\,x$ on the interval $(0, 1)$, which by (2.16) are

$$b_n = 2 \int_0^1 (-100\,x) \sin n\pi x \, dx = -200 \int_0^1 x \sin n\pi x \, dx = (-1)^n \frac{200}{n\pi}.$$

Thus,

$$v(x, t) = \frac{200}{\pi} \sum_{n=1}^{\infty} \frac{(-1)^n}{n} e^{-0.0001 n^2 \pi^2 t} \sin n\pi x.$$

Finally, the temperature in the rod is

$$u(x, t) = u_s(x) + v(x, t) = 100\,x + \frac{200}{\pi} \sum_{n=1}^{\infty} \frac{(-1)^n}{n} e^{-0.0001\, n^2 \pi^2 t} \sin n\pi x. \tag{2.21}$$

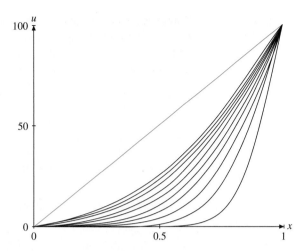

Figure 1. The temperature in the rod in Example 2.17.

The temperature is plotted in Figure 1. The initial temperature is $u(x, 0) = 0°C$. The steady-state temperature is plotted in blue. The black curves represent the temperature distribution after 200 second intervals. Notice how the temperature increases with time throughout the rod to the steady-state temperature. Heat flows from hot to cold, so to maintain the new temperature of 100°C at the right endpoint, heat must flow into the rod at this point. It then flows through the rod, raising the temperature in the process. Some heat has to flow out of the rod at the left endpoint to maintain the temperature there. Eventually the rod reaches steady state, at which point as much heat flows out of the rod at $x = 0$ as flows in at $x = 1$. ●

The rate of convergence

The general term in the infinite series in equation (2.14) is

$$b_n e^{-n^2\pi^2 kt/L^2} \sin\left(\frac{n\pi x}{L}\right). \tag{2.22}$$

Since the sine function is bounded in absolute value by 1, this term is bounded by $|b_n| e^{-n^2\pi^2 kt/L^2}$. By the Riemann-Lebesgue lemma (see Theorem 2.11 in Section 12.2), the Fourier coefficient $b_n \to 0$ as $n \to \infty$. On the other hand, the exponential term $e^{-n^2\pi^2 kt/L^2} \to 0$ extremely rapidly as $n \to \infty$, at least if the product kt is relatively large. As a result the series in equation (2.14) converges rapidly for large values of the time t. The result is that the sum of the series in (2.14) can be accurately approximated by using relatively few of the terms of the infinite series. Sometimes one term is enough.

Example 2.23 For the rod in Example 2.17, how many terms of series in (2.21) are needed to approximate the solution within one degree for $t = 10$, 100, and 1000? Estimate how long it will take before the heat in the rod is everywhere within 5° of the steady-state temperature.

The general term in the series in (2.21) is bounded by $200 e^{-0.0001 n^2 \pi^2 t}/n\pi$. We will estimate the error by computing the first omitted term.[4] Thus we want to find the smallest integer n for which $200 e^{-0.0001(n+1)^2\pi^2 t}/[(n+1)\pi] < 1$. Since we

---

[4] This is a rough estimate and is not usually a good idea. It is justified in this case because the terms are decreasing so rapidly.

cannot solve this inequality for n, we compute the left-hand side for values of n and t until we get the correct values. For $t = 10$ we discover that we need 12 terms, while for $t = 100$ we need 5, and for $t = 1000$ one term will suffice.

For the temperature of the rod to be within $5°$ of the steady-state temperature, we will certainly need the first term in the infinite series in (2.21) to be less than 5. If we solve $200e^{-0.0001\pi^2 t}/\pi = 5$, we obtain $t = 2,578$ sec. We compute that for $t = 2,578$, the second term in the series is about 0.0012, so $t = 2,578$ sec is a good estimate. However, in view of the fact that we are ignoring terms, and an estimate is not expected to have four place accuracy, $2,600$ sec might be preferable, and since $2,580$ sec is 43 minutes, that might be even better.   ●

Insulated boundary points

As mentioned in Section 13.1, if the boundary points of the rod are insulated, there is no flow of heat through the endpoints of the rod, and the correct boundary conditions are the Neumann conditions $u_x(0, t) = 0 = u_x(L, t)$. The initial/boundary value problem to be solved is now

$$
\begin{aligned}
u_t(x, t) &= ku_{xx}(x, t), \quad \text{for } t > 0 \text{ and } 0 < x < L, \\
u_x(0, t) &= 0 \quad \text{and} \quad u_x(L, t) = 0, \quad \text{for } t > 0, \\
u(x, 0) &= f(x), \quad \text{for } 0 \le x \le L.
\end{aligned}
\tag{2.24}
$$

We will use the method of separation of variables again, starting by looking for product solutions $u(x, t) = X(x)T(t)$. Notice that since the Neumann boundary conditions are homogeneous, it is not necessary to find the steady-state solution first.

Step 1. Separate the PDE into two ODEs. This first step is unchanged. The product $u(x, t) = X(x)T(t)$ is a solution only if the factors satisfy the differential equations

$$
T' + \lambda k T = 0 \quad \text{and} \quad X'' + \lambda X = 0,
\tag{2.25}
$$

where λ is a constant. The first equation has the general solution

$$
T(t) = Ce^{-\lambda k t}.
\tag{2.26}
$$

Step 2. Set up and solve the two-point boundary value problem. We will again insist that the product solution satisfy the boundary conditions. Since $0 = u_x(0, t) = X'(0)T(t)$ for all $t > 0$, we must have $X'(0) = 0$. A similar argument shows that $X'(L) = 0$, so we want to solve

$$
X'' + \lambda X = 0 \quad \text{with } X'(0) = X'(L) = 0.
\tag{2.27}
$$

This is the two-point or Sturm-Liouville boundary value problem for the Neumann problem. As before, we find that there are no negative eigenvalues (see Exercise 22). If $\lambda = 0$ the differential equation in (2.27) becomes $X'' = 0$, which has the general solution $X(x) = ax + b$. The first boundary condition is $0 = X'(0) = a$, leaving us with the constant function $X(x) = b$. This function also satisfies the second boundary $X'(L) = 0$, so $\lambda = 0$ is an eigenvalue. We will choose the simplest nonzero constant $b = 1$ and set $X_0(x) = 1$. The corresponding function in (2.26) is $T_0 = C$, which is also a constant. Once more we choose $C = 1$ so the resulting product solution to the heat equation is the constant function

$$
u_0(x, t) = X_0(x)T_0(t) = 1.
$$

For $\lambda > 0$, we set $\lambda = \omega^2$, where $\omega > 0$. Then the differential equation in (2.27) is $X'' + \omega^2 X = 0$, which has the general solution $X(x) = a \cos \omega x + b \sin \omega x$. The boundary condition $X'(0) = 0$ becomes $\omega b = 0$. Since $\omega > 0$, we have $b = 0$. Then the boundary condition $X'(L) = 0$ becomes

$$\omega a \sin \omega L = 0.$$

Since we are only interested in nonzero solutions, we must have $\sin \omega L = 0$. Therefore, $\omega L = n\pi$ for some positive integer n. When this is true we have $\lambda = \omega^2 = n^2\pi^2/L^2$, and $X(x) = a \cos(n\pi x/L)$. Again a can be any nonzero constant, and the simplest choice is $a = 1$.

In summary, the eigenvalues and eigenfunctions for the Sturm Liouville problem in (2.27) are

$$\lambda_n = \frac{n^2\pi^2}{L^2} \quad \text{and} \quad X_n(x) = \cos\left(\frac{n\pi x}{L}\right), \quad \text{for } n = 0, 1, 2, 3, \ldots . \quad (2.28)$$

Notice that in the case $n = 0$, $\lambda_0 = 0$ and $X_0(x) = 1$, as we found earlier. For every nonnegative integer n we get the product solution

$$u_n(x, t) = e^{-n^2\pi^2 kt/L^2} \cos\left(\frac{n\pi x}{L}\right) \quad (2.29)$$

to the heat equation by using (2.26). Observe that this solution also satisfies the boundary conditions

$$\frac{\partial u_n}{\partial x}(0, t) = \frac{\partial u_n}{\partial x}(L, t) = 0.$$

Step 3. Satisfy the initial conditions. Having found infinitely many product solutions in (2.29), we can use the linearity of the heat equation (see Theorem 1.18) to conclude that any linear combination of the product solutions is also a solution. Hence if a_n is a constant for each n, the function

$$\begin{aligned} u(x, t) &= \frac{a_0}{2} + \sum_{n=1}^{\infty} a_n u_n(x, t) \\ &= \frac{a_0}{2} + \sum_{n=1}^{\infty} a_n e^{-n^2\pi^2 kt/L^2} \cos\left(\frac{n\pi x}{L}\right) \end{aligned} \quad (2.30)$$

is formally a solution. Setting $u(x, 0) = f(x)$, we obtain the equation

$$f(x) = \frac{a_0}{2} + \sum_{n=0}^{\infty} a_n \cos\left(\frac{n\pi x}{L}\right). \quad (2.31)$$

This is the Fourier cosine expansion of f on the interval $0 \leq x \leq L$. From Section 12.3 of Chapter 12, and especially equation (3.2) in that section, we see that the coefficients a_n are given by

$$a_n = \frac{2}{L} \int_0^L f(x) \cos\left(\frac{n\pi x}{L}\right) dx, \quad \text{for } n \geq 0. \quad (2.32)$$

Substituting these values into (2.30) gives a complete solution to the heat equation with Neumann boundary conditions.

Notice that each term in the infinite sum in (2.30) tends to 0 as $t \to \infty$. Using this and the definition of the coefficient a_0 we see that

$$\lim_{t \to \infty} u(x, t) = \frac{a_0}{2} = \frac{1}{L} \int_0^L f(x)\, dx.$$

Thus as t increases in an insulated rod, the temperature tends to a constant equal to the average of the initial temperature.

Example 2.33 Suppose a rod of length 1 meter made from a material with thermal diffusivity $k = 1 \text{ cm}^2/\text{sec}$ is originally at steady state with its temperature maintained at $0°C$ at $x = 0$ and at $100°C$ at $x = 1$. (See Example 2.17.) Starting at time $t = 0$, both ends are insulated. Find the temperature distribution in the rod as a function of time and position. Find the constant temperature which is approached as $t \to \infty$. Estimate how long it will take for all portions of the rod to get to within $5°C$ of the final temperature.

According to our analysis in Example 2.17, the steady-state temperature is $f(x) = 100\,x$, with x measured in meters. This will be the initial temperature. With length measured in meters, $k = 0.0001 \text{ m}^2/\text{sec}$. Our new initial/boundary value problem is

$$\begin{aligned}
u_t(x, t) &= 0.0001\, u_{xx}(x, t), \quad \text{for } t > 0 \text{ and } 0 < x < 1, \\
u_x(0, t) &= u_x(1, t) = 0, \quad \text{for } t > 0, \\
u(x, 0) &= f(x) = 100\, x, \quad \text{for } 0 \le x \le 1.
\end{aligned} \qquad (2.34)$$

The solution as given in (2.30) with $k = 0.0001$ and $L = 1$ is

$$u(x, t) = \frac{a_0}{2} + \sum_{n=1}^{\infty} a_n e^{-0.0001\, n^2 \pi^2 t} \cos n\pi x. \qquad (2.35)$$

The initial condition becomes

$$u(x, 0) = 100\, x = \frac{a_0}{2} + \sum_{n=1}^{\infty} a_n \cos n\pi x.$$

The a_n are the Fourier cosine coefficients of $100\,x$ on the interval $[0, 1]$, so $a_0 = 100$, and

$$a_n = 2 \int_0^1 100\, x \cos n\pi x\, dx = \begin{cases} 0, & \text{for } n > 0 \text{ even}, \\ -\dfrac{400}{n^2 \pi^2}, & \text{for } n \text{ odd}. \end{cases}$$

Substituting into (2.35), using $n = 2p + 1$, we get the solution

$$u(x, t) = 50 - \frac{400}{\pi^2} \sum_{p=0}^{\infty} \frac{1}{(2p + 1)^2} e^{-0.0001\,(2p+1)^2 \pi^2 t} \cos(2p + 1)\pi x. \qquad (2.36)$$

Notice that each of the terms in the series, with the exception of the constant first term, includes an exponential factor that approaches 0 as $t \to \infty$. Thus the temperature in the rod approaches the constant, steady-state temperature of $50°C$ as $t \to \infty$. Notice also that $50°C$ is the average of the initial temperature over the rod. This reflects the fact that the ends are insulated, and no heat flows into or out of the rod.

We suspect that one of the exponential terms (with $p = 0$) in equation (2.36) will suffice to find how long it takes for the temperature to be within $5°$ of the constant steady-state temperature. We solve $400\, e^{-0.0001\pi^2 t}/\pi^2 = 5$ to get $t = 2,120$ sec. We check that the contribution to the temperature of the $p = 1$ term is less than 3×10^{-8}, so $2,120$ sec is a good estimate.

The temperature is shown in Figure 2. The initial temperature $f(x) = 100\,x$ and the constant steady-state temperature of $50°C$ are shown plotted in blue. The black curves are the temperature profiles plotted at time intervals of 300s.   ●

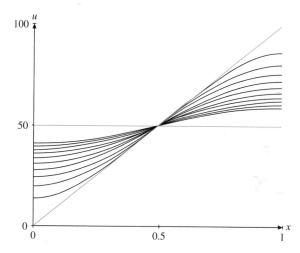

Figure 2.   The temperature for the rod in Example 2.33.

EXERCISES

1. Consider a rod 50 cm long with thermal diffusivity $k\ \mathrm{cm^2/sec}$. Originally the rod is at a constant temperature of $100°C$. Starting at time $t = 0$ the ends of the rod are immersed in an ice bath at temperature $0°C$. Show that the temperature $u(x, t)$ in the rod for $t > 0$ is given by

$$u(x, t) = \sum_{p=0}^{\infty} \frac{400}{(2p+1)\pi} e^{-k(2p+1)^2\pi^2 t/2500} \sin\left(\frac{n\pi x}{50}\right).$$

(2.37)

If the rod is made of gold, find the thermal diffusivity in Table 1 on page 630, and estimate how long it takes the temperature in the rod to decrease everywhere to less than $10°C$. How many terms in the series for u are needed to approximate the temperature within one degree at $t = 100$ sec. On one figure, plot the temperature versus x for $t = 0, 100, 200, 300, 400$.

2. Estimate how long it takes the temperature in the rod in Exercise 1 to decrease everywhere to less than $10°C$ if it is made of aluminum, silver, or PVC. For aluminum and silver, how many terms of the series in (2.37) are needed to approximate the temperature throughout the rod within $1°$ when $t = 100$ sec. For PVC, how many terms are needed to approximate the temperature throughout the rod within $1°$ when $t = 1$ day.

3. Consider a wall made of brick 10 cm thick, which separates a room in a house from the outside. The room is kept at $20°$.

(a) Originally the outside temperature is $10°C$ and the temperature in the wall has reached steady state. What is the temperature in the wall at this point?

(b) There is a sudden cold snap and the outside temperature drops to $-10°C$. Find the temperature in the wall as a function of position and time.

4. The wall of a furnace is 10 cm thick, and built from a refractory material with thermal diffusivity $k = 5 \times 10^{-5}\mathrm{cm^2/sec}$. Originally there is no fire in the furnace and the temperature of the furnace and the outside are both $20°C$. At $t = 0$, a fire is lit and the inside of the furnace is quickly raised to $420°C$. Find the temperature in the wall for $t > 0$.

In Exercises 5–8, find the temperature $u(t, x)$ in a rod modeled by the initial/boundary value problem

$$u_t(x, t) = k u_{xx}(x, t), \quad \text{for } t > 0 \text{ and } 0 < x < L,$$
$$u(0, t) = T_0 \quad \text{and} \quad u(L, t) = T_L, \quad \text{for } t > 0,$$
$$u(x, 0) = f(x), \quad \text{for } 0 \le x \le L$$

with the indicated values of the parameters.

5. $k = 4$, $L = 1$, $T_0 = 0$, $T_L = 0$, and $f(x) = x(1 - x)$

6. $k = 2$, $L = \pi$, $T_0 = 0$, $T_L = 0$, and $f(x) = \sin 2x - \sin 4x$

7. $k = 1$, $L = \pi$, $T_0 = 0$, $T_L = 0$, and $f(x) = \sin^2 x$

8. $k = 1$, $L = 1$, $T_0 = 0$, $T_L = 2$, and $f(x) = x$

In Exercises 9–12, use the temperature computed in the given exercise. Plot the initial temperature versus x and add the plots of the temperatue versus x for a number of time values like those in the text that show the significant portion of the change of the temperature. (Approximate the solution with an appropriate partial sum.) In addition, plot $y = u_x(0, t)$ and $y = u_x(L, t)$ as functions of t. Recall from (1.5) and (1.6) that these terms are proportional to the heat flux through the endpoints of the rod. Give a physical description of what is happening to the temperature as time increases. Include the information from the graphs of the flux and the graphs of the solution.

9. Exercise 5   **10.** Exercise 6

11. Exercise 7   **12.** Exercise 8

In Exercises 13–18, find the temperature $u(t, x)$ in a rod modeled by the initial/boundary value problem

$$u_t(x, t) = ku_{xx}(x, t), \quad \text{for } t > 0 \text{ and } 0 < x < L,$$
$$u_x(0, t) = u_x(L, t) = 0, \quad \text{for } t > 0,$$
$$u(x, 0) = f(x), \quad \text{for } 0 \leq x \leq L$$

with the indicated values of the parameters. Plot the solution for a number of time values like those in the text that show the significant portion of the change of the temperature. Give a physical explanation of what is happening to the solution as time progresses.

13. $k = 1$, $L = 1$, and $f(x) = \begin{cases} x, & 0 \leq x < 1/2, \\ (1 - x), & 1/2 \leq x \leq 1 \end{cases}$

14. $k = 1$, $L = 2$, and $f(x) = \begin{cases} 1, & 0 \leq x < 1 \\ 0, & 1 \leq x \leq 2 \end{cases}$

15. $k = 1$, $L = 1$, and $f(x) = \sin(\pi x)$

16. $k = 1$, $L = 1$, and $f(x) = \cos(\pi x)$

17. $k = 1$, $L = 2$, and $f(x) = \begin{cases} 0, & 0 \leq x \leq 1 \\ (x - 1), & 1 < x \leq 2 \end{cases}$

18. $k = 1/3$, $L = 2$, and $f(x) = x(2 - x)$

In Exercises 19–21, we will consider heat flow in a rod of length L, where an internal heat source, given by $p(x)$, is present. As indicated in equation (1.10), this leads to the initial/boundary value problem

$$\frac{\partial u}{\partial t} - k\frac{\partial^2 u}{\partial x^2} = \frac{p(x)}{c\rho} \quad 0 < x < L, \ t > 0 \tag{2.38}$$
$$u(0, t) = A, \quad u(L, t) = B,$$
$$u(x, 0) = f(x) \quad 0 \leq x \leq L,$$

for the inhomogeneous heat equation, where f and p are given (known) functions of x and A and B are constants.

19. The corresponding steady-state solution is the function $v(x)$ that satisfies the partial differential equation and the boundary conditions. Show that $v(x)$ satisfies

$$v''(x) = -\frac{p(x)}{C}, \quad \text{with} \quad v(0) = A \quad \text{and} \quad v(L) = B.$$

(Remember that $k = C/c\rho$, where C is the thermal conductivity.) Suppose that $u_h(x, t)$ is the solution to the initial/boundary value problem

$$\frac{\partial u_h}{\partial t} - k\frac{\partial^2 u_h}{\partial x^2} = 0 \quad 0 < x < L, \ t > 0,$$
$$u_h(0, t) = 0 = u_h(L, t) \quad t > 0,$$
$$u_h(x, 0) = f(x) - v(x) \quad 0 \leq x \leq L$$

for the homogeneous heat equation. Show that the function $u(x, t) = u_h(x, t) + v(x)$ is a solution to the initial/boundary value problem in (2.38).

20. Use Exercise 19 to find the solution to the initial/boundary value problem in (2.38) with $k = 1$, $L = 1$, $p(x)/c\rho = 6x$, $A = 0$, $B = 1$, and $f(x) = \sin \pi x$.

21. Use Exercise 19 to find the solution to the initial/boundary value problem in (2.38) with $k = 1$, $L = 1$, $p(x)/c\rho = e^{-x}$, $A = 1$, $B = -1/e$, and $f(x) = \sin 2\pi x$.

22. Show that the Sturm-Liouville problem in (2.27) has no negative eigenvalues.

13.3 The Wave Equation

We will start with the derivation of the wave equation in one space dimension. We will be modeling the vibrations of a wire or a string that is stretched between two points. A violin string is a very good example. We will also look at two techniques for solving the wave equation.

Derivation of the wave equation in one space variable

We assume the string is stretched from $x = 0$ to $x = L$. We are looking for the function $u(x, t)$ that describes the vertical displacement of the wire at position x and at time t. We assume the string is fixed at both endpoints, so $u(0, t) = u(L, t) = 0$ for all t. We will ignore the force of gravity, so at equilibrium we have $u(x, t) = 0$ for all x and t, which means that the string is in a straight line between the two fixed endpoints.

To derive the differential equation that models a vibrating string, we have to make some simplifying assumptions. In mathematical terms the assumptions amount to assuming that both $u(x, t)$, the displacement of the string, and $\partial u / \partial x$, the slope of the string, are small in comparison to L, the length of the string.

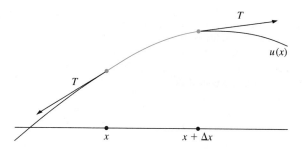

Figure 1. The forces acting on a portion of a vibrating string.

Consider the portion of the string above the small interval between x and $x + \Delta x$, as illustrated in blue in Figure 1. The forces acting on this portion come from the tension T in the string. The tension is a force that the rest of the string exerts on this particular part. For the portion in Figure 1, tension acts at the endpoints. We assume that the tension is so large that the string acts as if it were perfectly flexible and can bend without the requirement of a bending force. With that assumption, the tension acts tangentially to the string.

The tension at the point x is resolved into its horizontal and vertical components in Figure 2. We are assuming that the positive direction is upward. The vertical component is $T_u = -T \sin \theta$, and the horizontal component is $T_x = -T \cos \theta$. The slope of the graph of u at the point x is

$$\frac{\partial u}{\partial x} = \tan \theta.$$

We are assuming that the slope is very small, so θ is small. Therefore, $\cos \theta \approx 1$, and $\tan \theta \approx \sin \theta$. As a result, we have

$$T_u \approx -T \frac{\partial u}{\partial x}(x, t) \quad \text{and} \quad T_x \approx -T.$$

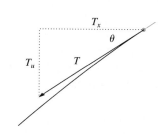

Figure 2. The resolution of the tension at the point x.

In a similar manner, we find that horizontal component of the force at $x + \Delta x$ is approximately T, which cancels the horizontal component at x. More interesting is the fact that the vertical component of the force at $x + \Delta x$ is approximately

$$T \frac{\partial u}{\partial x}(x + \Delta x, t),$$

so the total force acting in the vertical direction on the small portion of the string is

$$F \approx T \left(\frac{\partial u}{\partial x}(x + \Delta x, t) - \frac{\partial u}{\partial x}(x, t) \right).$$

The length of the segment of string is close to Δx. If the string is uniform and has linear mass density ρ, then the mass of the segment is $m = \rho \Delta x$. The acceleration of the segment in the vertical direction is $\partial^2 u / \partial t^2$. By Newton's second law, we have $ma = F$, which translates into

$$\rho \Delta x \frac{\partial^2 u}{\partial t^2} \approx T \left(\frac{\partial u}{\partial x}(x + \Delta x, t) - \frac{\partial u}{\partial x}(x, t) \right).$$

Dividing by Δx and taking the limit as Δx goes to 0, we have

$$\rho \frac{\partial^2 u}{\partial t^2} = T \lim_{\Delta x \to 0} \frac{1}{\Delta x} \left(\frac{\partial u}{\partial x}(x + \Delta x, t) - \frac{\partial u}{\partial x}(x, t) \right) = T \frac{\partial^2 u}{\partial x^2}.$$

If we set $c^2 = T/\rho$, the equation becomes

$$u_{tt} = c^2 u_{xx}. \tag{3.1}$$

This is the wave equation in one space variable. The constant c has dimensions length/time, so it is a velocity.

Notice that the homogeneous wave equation in (3.1) is linear. Once again we can build complicated solutions out of simpler ones.

Solution to the wave equation by separation of variables

Let's turn to the solution of the equation for the vibrating string. Since the wave equation is of order 2 in t, we are required to specify the initial velocity of the string as well as the initial displacement. Thus we are led to the initial/boundary value problem

$$
\begin{aligned}
u_{tt}(x, t) &= c^2 u_{xx}(x, t), \quad \text{for } 0 < x < L \text{ and } t > 0, \\
u(0, t) &= 0 \quad \text{and} \quad u(L, t) = 0, \quad \text{for } t > 0, \\
u(x, 0) &= f(x) \quad \text{and} \quad u_t(x, 0) = g(x), \quad \text{for } 0 \le x \le L.
\end{aligned}
\tag{3.2}
$$

We will find the solution using separation of variables. Since the process is similar to that used in previous examples, we will omit some of the details. Notice that the boundary conditions in (3.2) are homogeneous, so we can proceed directly with the separation of variables. The starting point is to look for product solutions of the form $u(x, t) = X(x)T(t)$.

Step 1. Separate the PDE into two ODEs. Inserting $u(x, t) = X(x)T(t)$ into the wave equation and separating variables gives

$$\frac{X''(x)}{X(x)} = \frac{T''(t)}{c^2 T(t)}.$$

Since x and t are independent variables, each side of this equation must equal a constant, which we will denote by $-\lambda$. Thus the factors must satisfy the differential equations

$$X'' + \lambda X = 0 \quad \text{and} \quad T'' + \lambda c^2 T = 0. \tag{3.3}$$

Step 2. Set up and solve the two-point boundary value problem. The first equation in (3.3) together with the boundary condition $u(0, t) = 0 = u(L, t)$ implies that X must solve the two-point boundary value problem

$$X''(x) + \lambda X(x) = 0 \quad \text{with} \quad X(0) = 0 = X(L). \tag{3.4}$$

We have seen this Sturm-Liouville problem before in (2.10). The solutions, given in (2.12), are

$$\lambda_n = \frac{n^2 \pi^2}{L^2} \quad \text{and} \quad X_n(x) = \sin\left(\frac{n\pi x}{L}\right), \quad \text{for } n = 1, 2, 3, \dots.$$

Step 3. Satisfy the initial conditions. With $\lambda_n = n^2\pi^2/L^2$, the second equation in (3.3) is

$$T'' + \left(\frac{cn\pi}{L}\right)^2 T = 0.$$

The functions $\cos(cn\pi t/L)$ and $\sin(cn\pi t/L)$ form a fundamental set of solutions. Consequently, we have found the product solutions

$$u_n(x, t) = \sin\left(\frac{n\pi x}{L}\right)\cos\left(\frac{cn\pi t}{L}\right) \quad \text{and} \quad v_n(x, t) = \sin\left(\frac{n\pi x}{L}\right)\sin\left(\frac{cn\pi t}{L}\right),$$

for $n = 1, 2, 3, \ldots$. Since the wave equation is linear, the function

$$
\begin{aligned}
u(x, t) &= \sum_{n=1}^{\infty} [a_n u_n(x, t) + b_n v_n(x, t)] \\
&= \sum_{n=1}^{\infty} \sin\left(\frac{n\pi x}{L}\right)\left[a_n \cos\left(\frac{cn\pi t}{L}\right) + b_n \sin\left(\frac{cn\pi t}{L}\right)\right]
\end{aligned}
\tag{3.5}
$$

is a solution to the wave equation for any choice of the coefficients a_n and b_n that ensures that the series will converge. Further, $u(x, t)$ also satisfies the homogeneous boundary conditions.

The first initial condition is

$$f(x) = u(x, 0) = \sum_{n=1}^{\infty} a_n \sin\frac{n\pi x}{L}.$$

To satisfy this condition, we choose the coefficients a_n to be

$$a_n = \frac{2}{L}\int_0^L f(x) \sin\frac{n\pi x}{L}\, dx, \tag{3.6}$$

the Fourier sine coefficients for f. The second initial condition involves the derivative $u_t(x, t)$. Differentiating (3.5) term by term, we see that

$$u_t(x, t) = \sum_{n=1}^{\infty} \frac{cn\pi}{L}\sin\left(\frac{n\pi x}{L}\right)\left[-a_n \sin\left(\frac{cn\pi t}{L}\right) + b_n \cos\left(\frac{cn\pi t}{L}\right)\right].$$

The second initial condition now becomes

$$g(x) = u_t(x, 0) = \sum_{n=1}^{\infty} b_n \frac{cn\pi}{L}\sin\frac{n\pi x}{L}.$$

Therefore, $b_n cn\pi/L$ should be the Fourier sine coefficients for g, or

$$b_n = \frac{2}{cn\pi}\int_0^L g(x) \sin\frac{n\pi x}{L}\, dx. \tag{3.7}$$

Inserting the values of a_n and b_n into (3.5) gives the complete solution to the wave equation.

Notice that every solution is an infinite linear combination of the product solutions

$$\sin\left(\frac{cn\pi t}{L}\right)\sin\left(\frac{n\pi x}{L}\right) \quad \text{and} \quad \cos\left(\frac{cn\pi t}{L}\right)\sin\left(\frac{n\pi x}{L}\right).$$

These solutions are periodic in time with frequency $\omega_n = nc\pi/L$. All of these frequencies are integer multiples of the *fundamental frequency* $\omega_1 = c\pi/L$. In music the contributions for $n > 1$ are referred to as *higher harmonics*. It is the fundamental frequency that our ears focus on, but the higher harmonics add body to the sound. This coupling of a fundamental frequency with the higher harmonics is thought to be accountable for the pleasing sound of a vibrating string. We will see later that the situation is different for the vibrations of a drum.

Example 3.8 Suppose that a string is stretched and fixed at $x = 0$ and $x = \pi$. The string is plucked in the middle, which means that its shape is described by[5]

$$f(x) = \begin{cases} x, & \text{if } 0 \leq x < \pi/2, \\ \pi - x, & \text{if } \pi/2 \leq x \leq \pi. \end{cases}$$

At $t = 0$ the string is released with initial velocity $g(x) = 0$. Find the displacement of the string as a function of x and t. Assume that for this string we have $c = 0.002$.

The solution is given by (3.5). We have only to find the coefficients a_n and b_n. Since $g(x) = 0$, we have $b_n = 0$. The coefficients a_n are the Fourier sine coefficients of f on the interval $(0, \pi)$, and they are given by

$$a_n = \frac{2}{\pi}\int_0^\pi f(x)\sin nx \, dx.$$

Inserting the definition of f, and evaluating the integral, we find that $a_n = 0$ if n is even, and if $n = 2k + 1$ is odd we have

$$a_{2k+1} = (-1)^k \frac{4}{\pi(2k+1)^2}.$$

Substituting into (3.5), we see that

$$u(x, t) = \sum_{k=0}^\infty (-1)^k \frac{4}{\pi(2k+1)^2}\sin(2k+1)x \cdot \cos 0.002(2k+1)t \tag{3.9}$$

is the solution. ●

The rate of convergence

The general term in the series in equation (3.5) is

$$\sin\left(\frac{n\pi x}{L}\right)\left[a_n\cos\left(\frac{cn\pi t}{L}\right) + b_n\sin\left(\frac{cn\pi t}{L}\right)\right]. \tag{3.10}$$

---

[5] The wave equation was derived under the assumption that the displacements and the slopes were small. While this is not true for this and the other examples that we will examine, it is true for smaller, more realistic initial displacements, such as $0.001 \times f(x)$. Since the wave equation is linear, the solution with this initial condition is $0.001\times$ the solution we find in Example 3.8.

The first factor, $\sin(n\pi x/L)$ is bounded in absolute value by 1. We can express the second factor in terms of its amplitude and phase,

$$a_n \cos\left(\frac{cn\pi t}{L}\right) + b_n \sin\left(\frac{cn\pi t}{L}\right) = A_n \cos\left(\frac{cn\pi t}{L} - \phi_n\right), \tag{3.11}$$

where the amplitude $A_n = \sqrt{a_n^2 + b_n^2}$. Thus, the general term in (3.10) is bounded by A_n for all $t > 0$. We can judge the convergence of the solution in equation (3.5) by the rate of convergence of $\sum_{n=1}^{\infty} A_n$. Notice that the rate of convergence of $\sum_{n=1}^{\infty} A_n$ does not change as t increases.

Example 3.12 The displacement of the string in Example 3.8 is given by the series in (3.9). How many terms must be included if we approximate the solution by the sum including all terms satisfying $A_{2k+1} > 0.01$? How many if we include all terms satisfying $A_{2k+1} > 0.001$?

We see that $A_{2k+1} = 4/[\pi(2k+1)^2]$. For any acceptable error e, we have $A_{2k+1} < e$ if $k > 1/\sqrt{\pi e} - 1/2$. Thus for an acceptable error of $e = 0.01$ we must keep all terms with $k \leq 5$, and for $e = 0.001$ terms with $k \leq 17$ are needed. ●

Comparing Examples 2.23 and 3.12, we see that many more terms are needed to get the required accuracy for solutions to the wave equation than are needed for solutions to the heat equation. The exponential decay of the terms in the solution to the heat equation makes that series converge much faster.

D'Alembert's solution

Let's examine another approach to solving the wave equation in one space variable. We start by finding all solutions to the wave equation

$$u_{tt}(x, t) = c^2 u_{xx}(x, t) \tag{3.13}$$

without worrying about initial or boundary conditions. We do this by introducing new variables $\xi = x + ct$ and $\eta = x - ct$. By the chain rule,

$$\frac{\partial u}{\partial x} = \frac{\partial u}{\partial \xi}\frac{\partial \xi}{\partial x} + \frac{\partial u}{\partial \eta}\frac{\partial \eta}{\partial x} = \frac{\partial u}{\partial \xi} + \frac{\partial u}{\partial \eta}.$$

Similarly, we have $u_t = c[u_\xi - u_\eta]$. Differentiating once more using the chain rule, we see that

$$u_{xx} = [u_\xi + u_\eta]_x = [u_\xi + u_\eta]_\xi + [u_\xi + u_\eta]_\eta = u_{\xi\xi} + 2u_{\xi\eta} + u_{\eta\eta}.$$

Similarly, $u_{tt} = c^2[u_{\xi\xi} - 2u_{\xi\eta} + u_{\eta\eta}]$. Therefore, $u_{tt} - c^2 u_{xx} = -4c^2 u_{\xi\eta}$. Consequently, in the new variables the wave equation has the form $u_{\xi\eta} = 0$.

If we read this equation as

$$\frac{\partial}{\partial \eta} u_\xi = 0,$$

we can integrate to find that

$$u_\xi(\xi, \eta) = H(\xi),$$

where $H(\xi)$ is an arbitrary[6] function of ξ. We can now integrate once more to find that

$$u(\xi, \eta) = \int H(\xi)\, d\xi + G(\eta),$$

---

[6] Although the argument used requires that H is a differentiable function, it is really true that H can be an arbitrary function. The same is true for the functions F and G that follow.

where $G(\eta)$ is an arbitrary function of η. If we set $F(\xi) = \int H(\xi)\,d\xi$, we find that

$$u(\xi, \eta) = F(\xi) + G(\eta),$$

where F and G are arbitrary functions.

   In terms of the original variables, we see that every solution to the wave equation (3.13) has the form

$$u(x, t) = F(x + ct) + G(x - ct), \tag{3.14}$$

where F and G are arbitrary functions. It is easily verified that any function of the form in (3.14) is a solution to the wave equation. The general solution to the wave equation in (3.14) is called the ***d'Alembert solution***.

Traveling waves

If we choose $F = 0$ in (3.14), we see that $u(x, t) = G(x - ct)$ is a solution to the wave equation. Let's get an idea of what this solution looks like. Figure 3 shows the graph of a function $G(x)$ that is a nonzero bump centered at $x = 0$. Figure 4 shows the graph of $G(x - ct)$, where $t > 0$ is fixed. Notice that the graph of $G(x - ct)$ is now centered at $x = ct$. From this we see that as t increases, the solution $u(x, t) = G(x - ct)$ to the wave equation has a graph versus x that is a bump moving to the right as t increases. Furthermore, since the wave has moved a distance ct in time t, it is moving to the right with speed c.

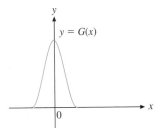

Figure 3. The graph of $G(x)$.

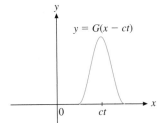

Figure 4. The graph of $G(x - ct)$ for $t > 0$.

Similarly, the solution $F(x + ct)$ represents a wave moving to the left with speed c as t increases. We will call solutions of the form $G(x - ct)$ and $F(x + ct)$ ***traveling waves***.

   As a result, we see that the d'Alembert solution in (3.14) represents the general solution to the wave equation (3.13) as the sum of two traveling waves, one moving to the right with speed c and the other moving to the left with speed c.

Solving the initial/boundary value problem

The d'Alembert solution in (3.14) can be used to find the solution to the initial/boundary value problem that we encountered in (3.2). To make the argument somewhat easier to follow, we will make the assumption that the initial velocity is 0, so the initial/boundary value problem we will solve is

$$
\begin{aligned}
&u_{tt}(x, t) = c^2 u_{xx}(x, t), \quad \text{for } 0 < x < L \text{ and } t > 0, \\
&u(0, t) = 0 \quad \text{and} \quad u(L, t) = 0, \quad \text{for } t > 0, \\
&u(x, 0) = f(x) \quad \text{and} \quad u_t(x, 0) = 0, \quad \text{for } 0 \le x \le L.
\end{aligned}
\tag{3.15}
$$

In the process we will gain additional information about the solution.

We start with a d'Alembert solution $u(x, t) = F(x+ct)+G(x-ct)$ from (3.14). We will use the initial and boundary conditions in (3.15) to find out what F and G have to be. We will assume that F and G are defined for all values of x. Observe that $u_t(x, t) = c[F'(x + ct) - G'(x - ct)]$. Therefore, the initial conditions imply that

$$f(x) = u(x, 0) = F(x) + G(x), \quad \text{and}$$
$$0 = u_t(x, 0) = c[F'(x) - G'(x)],$$

for $0 \leq x \leq L$. The second equation can be integrated to yield $F(x) - G(x) = C$, where C is a constant. Solving these two linear equations, we get $F(x) = [f(x) + C]/2$ and $G(x) = [f(x) - C]/2$ for $0 \leq x \leq L$. When we substitute into (3.14), we see that the constant cancels, so we may as well take $C = 0$. Thus we have

$$F(x) = G(x) = \frac{1}{2} f(x), \quad \text{for } 0 \leq x \leq L. \tag{3.16}$$

Next we use the boundary conditions. Setting $x = 0$ in (3.14), and using (3.16), we obtain $0 = u(0, t) = F(ct) + F(-ct)$, or $F(-ct) = -F(ct)$ for $t > 0$. Consequently, F must be an odd function. From (3.16) we get

$$F(x) = \frac{1}{2} f_o(x), \quad \text{for } -L \leq x \leq L, \tag{3.17}$$

where f_o is the odd extension of f.[7]

The second boundary condition is

$$0 = u(L, t) = F(L + ct) + F(L - ct).$$

If we set $ct = y + L$ in this formula, we get $F(y + 2L) + F(-y) = 0$. Using the fact that F is odd, this becomes

$$F(y + 2L) = -F(-y) = F(y).$$

This means that F must be periodic with period $2L$. Building on (3.17), we conclude that

$$F(x) = \frac{1}{2} f_{op}(x), \quad \text{for all } x \in \mathbf{R},$$

where f_{op} is the odd periodic extension of f to the whole real line. Thus the solution to the initial/boundary value problem in (3.15) is

$$u(x, t) = \frac{1}{2} \left[f_{op}(x + ct) + f_{op}(x - ct) \right]. \tag{3.18}$$

Example 3.19   Suppose that a string of length 1 m originally has the shape of the graph on the left in Figure 5 and has initial velocity 0. Assuming that $c = 1$m/sec, find the displacement of the string as a function of x and t.

The mathematical formula for the function f is given by

$$f(x) = \begin{cases} x - 3/8, & \text{for } 3/8 \leq x \leq 1/2 \\ 5/8 - x, & \text{for } 1/2 \leq x \leq 5/8 \\ 0, & \text{otherwise.} \end{cases}$$

According to the previous discussion, the solution is given by (3.18). The graph of the odd extension of f is given on the right in Figure 5.

Figure 6 shows the displacement of the string at several times. Notice how the initial wave splits into a forward wave and a backward wave, which then reflect when they hit the boundary points at $x = 0$ and $x = 1$.   ●

---

[7] We studied periodic extensions in Section 12.2 and odd and even extensions in Section 12.3.

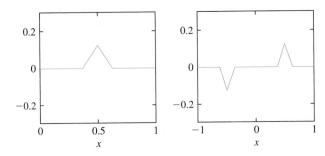

Figure 5. The initial displacement $f(x)$ for the string in Example 3.19, and its odd extension.

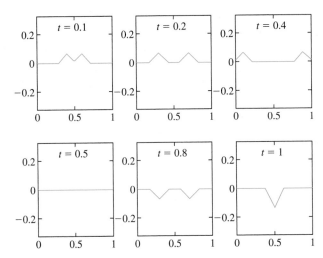

Figure 6. The displacement of the string in Example 3.19 at several times.

EXERCISES

In Exercises 1–6, use Fourier series to find the displacement $u(x, t)$ of the string of length L with fixed endpoints, initial displacement $u(x, 0) = f(x)$, and initial velocity $u_t(x, 0) = g(x)$. Assume that $c = 1$.

1. $f(x) = x(1 - x)/4$, $g(x) = 0$, and $L = 1$

2. $f(x) = \begin{cases} x/10, & \text{for } 0 \le x < 5 \\ 1 - x/10, & \text{for } 5 \le x \le 10, \end{cases}$ $g(x) = 0$, and $L = 10$

3. $f(x) = 0$, $g(x) = 1$, and $L = 1$

4. $f(x) = 0$, $g(x) = \begin{cases} -1/2, & \text{for } 0 \le x < 1/2 \\ 1/2, & \text{for } 1/2 \le x \le 1, \end{cases}$ and $L = 1$

5. $f(x) = 0$, $g(x) = \begin{cases} 1, & \text{for } 1 \le x \le 2 \\ 0, & \text{otherwise}, \end{cases}$ and $L = 3$

6. $f(x) = x(1 - x)/4$, $g(x) = -1$ and $L = 1$

In Exercises 7–8, use the d'Alembert solution (3.14) to find the displacement $u(x, t)$ of the string of length L with fixed endpoints, initial displacement $u(x, 0) = f(x)$, and initial velocity $u_t(x, 0) = 0$. Sketch the solution as a function of x, $0 \le x \le L$, for the specific values of t that are given.

7. $c = 2$, $f(x) = \sin \pi x$, and $L = 1$. Plot $u(x, t)$ as a function of x for $t = 0$, $1/8$, $1/4$, $1/2$, $3/4$, 1.

8. $c = 1$, $L = 10$, and

$$f(x) = \begin{cases} 0, & \text{for } 0 \le x \le 5, \\ x - 5, & \text{for } 5 < x \le 6, \\ 7 - x, & \text{for } 6 < x \le 7, \\ 0, & \text{for } 7 < x \le 10 \end{cases}$$

Plot $u(x, t)$ as a function of x for $t = 2$, 4, 6, 8, 10, 12.

9. Suppose that we have a string of length $L = 1$ with fixed endpoints, and $c = 1$. In this section we discussed two methods of finding the displacement $u(x, t)$ of the string with initial displacement $u(x, 0) = f(x)$, and initial velocity $u_t(x, 0) = g(x) = 0$. The first solution is

$$u_1(x, t) = \sum_{n=1}^{\infty} a_n \sin n\pi x \cos n\pi t,$$

where a_n are the Fourier sine coefficients for f (i.e., $f(x) = \sum_n a_n \sin n\pi x$). The second solution is d'Alembert's solution,

$$u_2(x, t) = \frac{1}{2} \left(f_{op}(x + t) + f_{op}(x - t) \right).$$

Show that these two solutions are the same. (*Hint:* Use the trigonometric identity $\sin A \cos B = [\sin(A + B) + \sin(A - B)]/2$ to transform u_1 into u_2.)

10. Use the method of separation of variables to find the general solution for the initial/boundary value problem

$$u_{tt}(x, t) + u_t(x, t) + u(x, t) = u_{xx}(x, t),$$
$$\text{for } 0 < x < 1 \text{ and } t > 0,$$
$$u(0, t) = 0 = u(1, t) \quad \text{for } t > 0,$$
$$u(x, 0) = f(x) \quad \text{for } 0 \le x \le 1,$$
$$u_t(x, 0) = 0 \quad \text{for } 0 \le x \le 1.$$

Express the solution in terms of the Fourier sine coefficients of the function f on the interval $0 \le x \le 1$. The differential equation in this problem is called the **telegraph equation**.

11. D'Alembert's solution can also be used to find the displacement $u(x, t)$ of a string with fixed endpoints having initial displacement $u(x, 0) = 0$ and initial velocity $u_t(x, 0) = g(x)$. Follow the derivation of the solution in (3.18) to show that the solution is given by

$$u(x, t) = \frac{1}{2c} \int_{x-ct}^{x+ct} g_{op}(s) \, ds,$$

where g_{op} is the odd periodic extension of g. (*Hint:* At some point it will be necessary to know that the derivative of an even function is odd, and vice versa.)

12. Use Exercise 11 to find the displacement $u(x, t)$ of a string of length L with fixed endpoints, where $c = 1$, $u(x, 0) = 0$ and ,

$$u_t(x, 0) = g(x) = \begin{cases} 0 & \text{for } 0 \le x < 1, \\ 2 & \text{for } 1 \le x \le 2, \\ 0 & \text{for } 2 < x \le 3. \end{cases}$$

Plot $u(x, t)$ as a function of x for $t = 0, 0.25, 0.5, 1.5, 4.5,$ 6.

13. Use Exercise 11 and the solution in (3.18) to show that the displacement $u(x, t)$ of a string of length L with fixed

endpoints having initial displacement $u(x, 0) = f(x)$ and initial velocity $u_t(x, 0) = g(x)$ is

$$u(x, t) = \frac{1}{2} \left(f_{op}(x + ct) + f_{op}(x - ct) \right)$$
$$+ \frac{1}{2c} \int_{x-ct}^{x+ct} g_{op}(s) \, ds.$$

14. The displacement of a wire or string that is stretched horizontally between two fixed endpoints actually satisfies the equation

$$u_{tt} = c^2 u_{xx} - g, \tag{3.20}$$

where g is the aceleration due to gravity. Usually the force of gravity is ignored because it is so much smaller than the tension in the string. In this exercise we will consider a string of length L, and include gravity.

(a) Find a steady-state solution v. This means that v is independent of t and satisfies equation (3.20) and the boundary conditions. If $u(x, t)$ is a solution to (3.20), what equation does $w(x, t) = u(x, t) - v(x)$ satisfy?

(b) Use separation of variables to find the solution $u(x, t)$ to (3.20) which satisfies the boundary conditions $u(0, t) = u(L, t) = 0$ and the initial conditions $u(x, 0) = 0 = u_t(x, 0)$ for $0 \le x \le L$.

15. The total energy in a vibrating string is

$$E(t) = \frac{1}{2} \int_0^L \left[\rho u_t^2 + T u_x^2 \right] dx. \tag{3.21}$$

Show that if $u(0, t) = 0 = u(L, t)$ for all $t > 0$, then $E(t)$ is constant. Thus, the energy in the string is conserved. (*Hint:* Differentiate (3.21) under the integral. Then use the wave equation and prove that $u_t(0, t) = 0 = u_t(L, t)$ for all $t > 0$.)

16. If you pluck a violin string, and then finger the string, fixing it precisely in the middle, the tone increases by one octave. In mathematical terms this means that the frequency is doubled. Explain why this happens.

17. Our derivation of the wave equation ignored any damping effects of the medium in which the string is vibrating. If damping is taken into account, the equation becomes

$$u_{tt} = c^2 u_{xx} - 2k u_t, \tag{3.22}$$

where k is a damping constant which we will assume satisfies $0 < k < \pi c/L$, where L is the length of the string.

(a) Find all product solutions $u(x, t) = X(x)T(t)$ to (3.22) which satisfy the boundary conditions $u(0, t) = u(L, t) = 0$ for $t > 0$.

(b) Find a series representation for the solution $u(x, t)$ which satisfies the boundary conditions and the initial conditions $u(x, 0) = f(x)$ and $u_t(x, 0) = g(x)$.

13.4 Laplace's Equation

So far we have considered partial differential equations where there was only one spatial dimension. Now we want to begin to study situations where there is more than one. Our discussion will be limited to the three most important examples, Laplace's equation, the heat equation, and the wave equation.

The Laplacian operator and Laplace's equation

The Laplacian operator is a part of all of the partial differential equations we will discuss. The discussion naturally begins with the gradient. In two spatial dimensions the gradient of a function $u(x, y)$ is

$$\nabla u = \left(\frac{\partial u}{\partial x}, \frac{\partial u}{\partial y} \right)^T.$$

For a function $u(x, y, z)$ of three variables we have

$$\nabla u = \left(\frac{\partial u}{\partial x}, \frac{\partial u}{\partial y}, \frac{\partial u}{\partial z} \right)^T.$$

In greater generality, for a function $u(\mathbf{x})$, where $\mathbf{x} = (x_1, x_2, \ldots, x_n)^T \in \mathbf{R}^n$, the gradient is the vector

$$\nabla u(\mathbf{x}) = \left(\frac{\partial u}{\partial x_1}(\mathbf{x}), \frac{\partial u}{\partial x_2}(\mathbf{x}), \ldots, \frac{\partial u}{\partial x_n}(\mathbf{x}) \right)^T.$$

This equation also defines the gradient as a vector valued differential operator, which we write as

$$\nabla = \left(\frac{\partial}{\partial x_1}, \frac{\partial}{\partial x_2}, \ldots, \frac{\partial}{\partial x_n} \right)^T.$$

Notice that in dimension $n = 1$, the gradient is just the ordinary derivative,

$$\nabla u = \frac{du}{dx}.$$

The *Laplacian operator* or, more simply, the *Laplacian*, is roughly the "square" of the gradient operator. It is denoted by ∇^2, and it is defined by[8]

$$\nabla^2 u = \nabla \cdot \nabla u = \frac{\partial^2 u}{\partial x_1^2} + \frac{\partial^2 u}{\partial x_2^2} + \cdots + \frac{\partial^2 u}{\partial x_n^2}.$$

Observe that the dot in $\nabla \cdot \nabla$ is the vector dot product. Thus the Laplacian operator is the square of the gradient operator if we are using the dot product. Using the subscript notation for derivatives, the notation specializes in low dimensions to

$$\nabla^2 u(x) = u_{xx}(x), \quad \text{for } n = 1,$$
$$\nabla^2 u(x, y) = u_{xx}(x, y) + u_{yy}(x, y), \quad \text{for } n = 2,$$
$$\nabla^2 u(x, y, z) = u_{xx}(x, y, z) + u_{yy}(x, y, z) + u_{zz}(x, y, z), \quad \text{for } n = 3.$$

The equation

$$\nabla^2 u(\mathbf{x}) = 0 \tag{4.1}$$

---

[8] Many mathematicians and scientists use the notation $\Delta u = \nabla^2 u$. However, we will follow the usage that we think is most common.

is called **Laplace's** equation. We have seen that in one space dimension, steady-state temperatures satisfy Laplace's equation. This is true in two or three dimensions as well. There are many other applications. For example, a conservative force $\mathbf{F}$ has a potential u, which is a function for which $\mathbf{F} = -\nabla u$. If in addition the force is divergence free, then the potential u satisfies Laplace's equation. In particular, this applies to an electric force in regions of space where there are no charges present, or to a gravitational force in regions where there is no mass.

A solution to Laplace's equation is called a **harmonic function**. Laplace's equation and harmonic functions are widely studied by mathematicians, both for their important applications and because of their intrinsic interest.

The heat equation

In one space dimension temperatures satisfy the heat equation $u_t = k u_{xx}$. If we replace u_{xx} by the Laplacian of u, the same is true in higher dimensions. Thus, if $u(\mathbf{x}, t)$ represents the temperature at a point $\mathbf{x}$ in space and at time t, then u satisfies the **heat equation**

$$\frac{\partial u}{\partial t}(\mathbf{x}, t) = k \nabla^2 u(\mathbf{x}, t), \tag{4.2}$$

where k is a constant called the **thermal diffusivity**. In low dimensions we can write the heat equation as

$$u_t = k(u_{xx} + u_{yy}), \text{ for } n = 2, \quad \text{and} \quad u_t = k(u_{xx} + u_{yy} + u_{zz}), \text{ for } n = 3.$$

A steady-state temperature is a temperature which does not depend on t. Notice that for a steady-state temperature u, the heat equation in (4.2) reduces to Laplace's equation (4.1).

The wave equation

In one space dimension the displacement of a vibrating string satisfies the wave equation $u_{tt} = c^2 u_{xx}$. Once again, if we replace u_{xx} by the Laplacian of u, then wave phenomena in higher dimensions satisfy the same equation. The wave equation is the equation

$$u_{tt} = c^2 \nabla^2 u,$$

where c is a constant that has the dimensions of velocity. The wave equation describes a variety of oscillatory behavior. For example, in two dimensions it describes the motion of a drum head. In three dimensions it describes electromagnetic waves.

Linearity

Laplace's equation, the heat equation, and the wave equation are all linear equations. We will use this to build up more and more complicated solutions as linear combinations of more basic solutions.

Boundary conditions for Laplace's equation

In this section and the next we will find solutions to Laplace's equation in a rectangle and in a disk in the plane $\mathbf{R}^2$. We could do this with any of the boundary conditions we discussed for the heat equation in Section 13.1. For example, the *Dirichlet problem* is to solve the boundary value problem

$$\nabla^2 u(x, y) = u_{xx} + u_{yy} = 0, \quad \text{for } (x, y) \in D,$$
$$u(x, y) = f(x, y), \quad \text{for } (x, y) \in \partial D, \tag{4.3}$$

where D is a region in $\mathbf{R}^2$ and ∂D is its boundary. The boundary condition $u(x, y) = f(x, y)$ is called a *Dirichlet condition*. The problem of finding a function u satisfying (4.3) for a given f defined on the boundary ∂D is called the *Dirichlet problem*. Notice that being able to solve the Dirichlet problem means that the steady-state temperature in a region D is completely determined by the temperature on the boundary ∂D.

If the boundary of the region ∂D is insulated, there is no flow of heat across the boundary. This means that the temperature is not varying in the direction normal to the boundary. Let $\mathbf{n}(x, y)$ denote the vector of length 1 at the point $(x, y) \in \partial D$, which is orthogonal to the boundary at (x, y) and points out of D. The vector $\mathbf{n}$ is called the *unit exterior normal* to the boundary of D, and $\partial u / \partial \mathbf{n} = \nabla u \cdot \mathbf{n}$ is called the *normal derivative* of u. Since the temperature is not varying in the direction $\mathbf{n}$, $\partial u / \partial \mathbf{n} = 0$. More generally, we can specify the normal derivative at each point of the boundary. Then we would have

$$\frac{\partial u}{\partial \mathbf{n}}(x, y) = g(x, y), \quad \text{for } (x, y) \in \partial D, \tag{4.4}$$

where g is a function defined on the boundary of D. This is called a *Neumann condition*. If we replace the Dirichlet condition in (4.3) with the Neumann condition, the problem is called the *Neumann problem*. We could also impose a *Robin condition*

$$\frac{\partial u}{\partial \mathbf{n}}(x, y) - \alpha(x, y) u(x, y) = h(x, y), \quad \text{for } (x, y) \in \partial D, \tag{4.5}$$

where α and h are functions defined on the boundary of D, to get the *Robin problem*.

The maximum principle for harmonic functions

Harmonic functions are solutions to Laplace's equation and therefore represent steady-state temperatures. Since heat flows from hot areas to colder areas, a steady-state temperature cannot be higher at one point than it is everywhere around it. Therefore, a solution to Laplace's equation cannot have a local maximum (or a local minimum). This fact is referred to as the *maximum principle for harmonic functions.*

If $u(x, y)$ is a solution to the Dirichlet problem (4.3) in a region D, then it follows from the maximum principle that u achieves its maximum and minimum values on the boundary ∂D. This is also sometimes called the maximum principle.

The mean value property of harmonic functions

Suppose that u is a harmonic function in a region D. Suppose also that $\mathbf{p} = (x_0, y_0)^T \in D$, and that $r > 0$ is so small that the disk U of radius r and center $\mathbf{p}$ is completely contained in D. Then *the mean value property of harmonic functions* states that the value $u(\mathbf{p})$ is the average of u over U. In other words,

$$u(\mathbf{p}) = \frac{1}{\pi r^2} \int_U u(x, y) \, dx \, dy.$$

If you think of u as a steady-state temperature, you find that the temperature at any point must be the average of the temperatures in any disk centered at that point. Clearly, this fact reflects the fact that heat flows from hot to cold.

Solution on a rectangle with Dirichlet boundary conditions

We shall consider the Dirichlet problem for the rectangle

$$D = \{(x, y) \mid 0 < x < a \text{ and } 0 < y < b\}$$

illustrated in Figure 1.

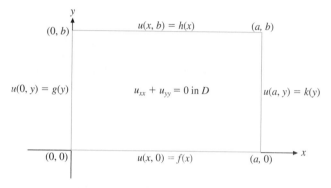

Figure 1. The Dirichlet problem for the rectangle D.

The boundary conditions specify the temperature u on each of the four sides as indicated in Figure 1. The full statement of the Dirichlet problem is

$$u_{xx}(x, y) + u_{yy}(x, y) = 0, \quad \text{for } (x, y) \in D,$$
$$u(x, 0) = f(x) \quad \text{and} \quad u(x, b) = h(x), \quad \text{for } 0 \le x \le a, \tag{4.6}$$
$$u(0, y) = g(y) \quad \text{and} \quad u(a, y) = k(y), \quad \text{for } 0 \le y \le b,$$

where f, g, h, and k are given functions.

We will reduce the problem to one that can be solved using separation of variables by imposing homogeneous boundary conditions on two opposite sides of the rectangle and the correct boundary condition from (4.6) on the remaining sides. The problem is to find u such that

$$u_{xx}(x, y) + u_{yy}(x, y) = 0, \quad \text{for } (x, y) \in D,$$
$$u(x, 0) = f(x) \quad \text{and} \quad u(x, b) = h(x), \quad \text{for } 0 \le x \le a, \tag{4.7}$$
$$u(0, y) = 0 \quad \text{and} \quad u(a, y) = 0, \quad \text{for } 0 \le y \le b.$$

There is the similar problem where homogeneous boundary conditions are imposed on the top and bottom of the rectangle, which can be solved using the same technique. Using the linearity of Laplace's equation, the sum of the two is the solution to (4.6).

We start the separation of variables by looking for product solutions of the form $u(x, y) = X(x)Y(y)$. We want u to satisfy the homogeneous boundary conditions, which means that $X(0) = X(a) = 0$. Substituting $u(x, y) = X(x)Y(y)$ into Laplace's equation, we obtain

$$X''(x)Y(y) + X(x)Y''(y) = 0.$$

Upon separating variables in the usual way, we obtain the differential equations

$$X'' + \lambda X = 0 \quad \text{and} \quad Y'' - \lambda Y = 0. \tag{4.8}$$

The function X must satisfy the homogeneous boundary conditions, so we want to solve the Sturm-Liouville problem

$$X'' + \lambda X = 0 \quad \text{with } X(0) = X(a) = 0. \tag{4.9}$$

This is the same problem that arose in our study of the heat equation with Dirichlet conditions (see (2.10)). The eigenvalues and eigenfunctions are

$$\lambda_n = \frac{n^2\pi^2}{a^2} \quad \text{and} \quad X_n(x) = \sin\left(\frac{n\pi x}{a}\right), \quad \text{for } n = 1, 2, 3, \ldots. \tag{4.10}$$

The factor Y satisfies the differential equation $Y'' - \lambda Y = 0$. We now know that λ is one of the eigenvalues, so let's write $\lambda = \omega^2$, where $\omega = n\pi/a$. Then the equation $Y'' - \omega^2 Y = 0$ has the fundamental set of solutions $e^{\omega y}$ and $e^{-\omega y}$. While these are the standard solutions, it will be convenient to use

$$\sinh \omega y = \frac{e^{\omega y} - e^{-\omega y}}{2} \quad \text{and} \quad \sinh \omega(y - b) = \frac{e^{\omega(y-b)} - e^{-\omega(y-b)}}{2}.$$

These functions are linear combinations of $e^{\omega y}$ and $e^{-\omega y}$, so they are solutions to the equation $Y'' - \omega^2 Y = 0$. They are not multiples of each other, so they are linearly independent and therefore form a fundamental set of solutions. The advantage for us is that $\sinh \omega y$ vanishes at $y = 0$, while $\sinh \omega(y - b)$ vanishes at $y = b$. This fact will facilitate finding the solution that satisfies the inhomogeneous boundary conditions.

Thus, for each positive integer n, we set $\omega = n\pi/a$ and we get two product solutions to Laplace's equation

$$u_n(x, y) = \sinh\left(\frac{n\pi y}{a}\right) \sin\left(\frac{n\pi x}{a}\right) \quad \text{and}$$
$$v_n(x, y) = \sinh\left(\frac{n\pi(y - b)}{a}\right) \sin\left(\frac{n\pi x}{a}\right) \tag{4.11}$$

that satisfy the homogeneous part of the boundary conditions.

Using the linearity of Laplace's equation, the function

$$u(x, y) = \sum_{n=1}^{\infty} a_n u_n(x, y) + \sum_{n=1}^{\infty} b_n v_n(x, y)$$

$$= \sum_{n=1}^{\infty} a_n \sinh\left(\frac{n\pi y}{a}\right) \sin\left(\frac{n\pi x}{a}\right) \tag{4.12}$$

$$+ \sum_{n=1}^{\infty} b_n \sinh\left(\frac{n\pi(y - b)}{a}\right) \sin\left(\frac{n\pi x}{a}\right)$$

is a solution to Laplace's equation for any constants a_n and b_n for which the series converges. In addition, u satisfies the homogeneous part of the boundary conditions.

The coefficients a_n and b_n are chosen to satisfy the inhomogeneous boundary conditions in (4.7). For $y = 0$ the first sum in (4.12) vanishes, so

$$f(x) = u(x, 0) = \sum_{n=1}^{\infty} b_n \sinh\left(\frac{n\pi(-b)}{a}\right) \sin\left(\frac{n\pi x}{a}\right).$$

This will be recognized as the Fourier sine expansion of $f(x)$, so using equation (3.7) of Section 12.3 in Chapter 12, we have

$$b_n \sinh\left(\frac{n\pi(-b)}{a}\right) = \frac{2}{a}\int_0^a f(x)\sin\left(\frac{n\pi x}{a}\right)dx. \qquad (4.13)$$

Using the fact that sinh is an odd function, we solve for

$$b_n = \frac{-2}{a\sinh(n\pi b/a)}\int_0^a f(x)\sin\left(\frac{n\pi x}{a}\right)dx.$$

Similarly, the boundary condition at $y = b$ requires

$$a_n = \frac{2}{a\sinh(n\pi b/a)}\int_0^a h(x)\sin\left(\frac{n\pi x}{a}\right)dx. \qquad (4.14)$$

Example 4.15 Find the steady-state temperature $u(x, y)$ in a square plate 1 m on a side where $u(x, 1) = x - x^2$ for $0 \le x \le 1$, and $u(x, y) = 0$ on the other three sides.

The square is the rectangle in Figure 1 with $a = b = 1$. The boundary temperatures are given by $f = g = k = 0$, and $h(x) = x - x^2$. The solution is given by (4.12). Since $f(x) = 0$, (4.13) implies that $b_n = 0$. To compute a_n, we use (4.14). We compute that

$$\int_0^1 (x - x^2)\sin n\pi x\, dx = \frac{2}{n^3\pi^3}(1 - \cos n\pi) = \begin{cases} \dfrac{4}{n^3\pi^3}, & \text{if } n \text{ is odd,} \\ 0, & \text{if } n \text{ is even.} \end{cases}$$

This can be accomplished by integrating by parts twice. Then, from (4.12) and (4.14), we conclude that

$$u(x, y) = \sum_{k=0}^{\infty} \frac{8}{(2k+1)^3\pi^3 \sinh(2k+1)\pi} \sin(2k+1)\pi x \cdot \sinh(2k+1)\pi y.$$

Truncating the above sum at $k = 10$, an approximate solution is graphed in Figure 2. Note that the solution agrees with the graph of $h(x) = x - x^2$ on the part of the boundary where $y = 1$. The boundary values of the other three sides are all zero, as specified in (4.7). ●

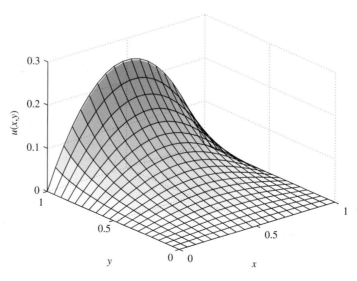

Figure 2. An approximate solution to the Dirichlet problem in Example 4.15.

EXERCISES

1. Consider a rectangular metal plate $a = 1$ m wide and $b = 2$ m long, as shown in Figure 1. Suppose that the temperature is $10°C$ on the bottom edge, and $0°C$ on the others. Find the steady-state temperature $u(x, y)$ throughout the plate.

2. Consider a rectangular metal plate $a = 10$ cm wide and $b = 25$ cm long, as shown in Figure 1. Suppose that the temperature is $u(x, 2) = 20°C$ on the top edge, $u(x, 0) = (100°C)x$ on the bottom, and $0°C$ on the others. Find the steady-state temperature $u(x, y)$ throughout the plate.

3. Show that when $f = h = 0$, the general solution of the boundary value problem in (4.6) is

$$u(x, y) =$$

$$\sum_{n=1}^{\infty} \left[a_n \sinh \frac{n\pi x}{b} + b_n \sinh \frac{n\pi(x-a)}{b} \right] \sin \frac{n\pi y}{b},$$

where

$$a_n = \frac{2}{b \sinh(n\pi a/b)} \int_0^b k(y) \sin(n\pi y/b) \, dy$$

and

$$b_n = \frac{-2}{b \sinh(n\pi a/b)} \int_0^b g(y) \sin(n\pi y/b) \, dy.$$

4. Suppose that you have a square plate for which the temperature on one side is kept at a uniform temperature of $100°$, and the other three sides are kept at $0°$. What is the temperature in the middle of the square? (*Hint:* Don't do any compuation of series. Use physical intuition, the symmetry of the square, and the linearity of the Laplacian.)

Exercises 5–10 are concerned with the boundary value problem in (4.6) in the rectangle D of width a and height b shown in Figure 1. Compute the solution for the given boundary functions $f, g, h,$ and k. Draw a hand sketch of what you think the graph of u over D should be and then compare with a computer drawn graph of the exact solution or of the first 10 terms or so of the solution.

5. $a = b = 1$, $f = g = k = 0$, and

$$h(x) = \begin{cases} x, & \text{for } 0 \le x \le 1/2, \\ 1-x, & \text{for } 1/2 < x \le 1 \end{cases}$$

6. $a = 2, b = 1, f = g = k = 0$, and

$$h(x) = \begin{cases} 1, & \text{for } 0 \le x \le 1, \\ -1, & \text{for } 1 < x \le 2 \end{cases}$$

7. $a = 1, b = 1, f(x) = \sin(2\pi x)$, and $g = h = k = 0$

8. $a = 1, b = 2, f(x) = \sin^2(\pi x)$, and $g = h = k = 0$

9. $a = 1, b = 2, f(x) = -1, h(x) = 1$, and $g = k = 0$

10. $a = b = 1, g = k = 0, f(x) = \sin(2\pi x)$, and

$$h(x) = \begin{cases} x, & \text{for } 0 \le x \le 1/2, \\ x-1, & \text{for } 1/2 < x \le 1 \end{cases}$$

11. (a) Consider the rectangle $D_L = \{(x, y) \mid 0 < x < 1 \text{ and } 0 < y < L\}$. Compute the solution of the boundary value problem

$$u_{xx}(x, y) + u_{yy}(x, y) = 0 \text{ for } (x, y) \in D$$

$$u(x, 0) = f(x) \text{ and } u(x, L) = 0 \text{ for } 0 \le x \le 1,$$

$$u(0, y) = 0 \text{ and } u(a, y) = 0 \text{ for } 0 \le y \le L,$$

where $f(x)$ is a piecewise differentiable function on the interval $0 \le x \le 1$, with the Fourier sine series $f(x) \sim \sum_{n=0}^{\infty} B_n \sin(n\pi x)$.

(b) Consider the infinite strip $D = \{(x, y) \mid 0 < x < 1 \text{ and } 0 < y < \infty\}$. Find the temperature $u(x, y)$ on D that is equal to 0 on the infinite sides and satisfies $u(x, 0) = f(x)$ for $0 \le x \le 1$. (*Hint:* Use part (a) to solve on the rectangle with bounds $0 \le x \le L$, and $u(L, y) = 0$. Then let L increase to ∞ and find the limiting temperature.)

12. Show that the steady-state temperature in a region D is completely determined by the temperature on the boundary ∂D. In other words, if $u(x, y)$ and $v(x, y)$ are two possible steady-state temperatues that satisfy $u(x, y) = v(x, y)$ for every point $(x, y) \in \partial D$, then $u(x, y) = v(x, y)$ at every point $(x, y) \in D$. (*Hint:* Consider $w = u - v$ and apply the maximum principle.)

13.5 Laplace's Equation on a Disk

Now we turn our attention to finding steady-state temperatures in regions with circular symmetry. One example is a metal disk. Another is a pipe, which has a cross-section which is a ring or annulus, described mathematically as the region between two concentric circles.

Finding a steady-state temperature involves solving the Dirichlet problem (4.3). Therefore, for a metal disk D of radius a centered at the origin, we want to find u such that

$$u_{xx} + u_{yy} = 0, \quad \text{if } x^2 + y^2 < a^2,$$
$$u(x, y) = f(x, y), \quad \text{if } x^2 + y^2 = a^2,$$

$$(5.1)$$

where f is a function defined on the boundary of the disk. The geometry is illustrated in Figure 1.

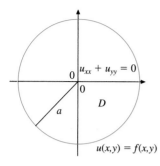

The Laplacian in other coordinate systems

Since D is a circular domain, our problem will be more easily solved if we use polar coordinates r and θ. We will derive the form of the Laplacian in polar coordinates in some detail so the derivation can be a model to be used with other coordinate systems.

The original Cartesian coordinates x and y and the polar coordinates r and θ are related by

$$x = r\cos\theta \qquad \text{and} \qquad r^2 = x^2 + y^2$$
$$y = r\sin\theta \qquad\qquad \tan\theta = y/x.$$

Differentiating $r^2 = x^2 + y^2$, we get $2r\,\partial r/\partial x = 2x$. Solving this equation for the partial derivative and then doing the same calculation for the y-derivative, we get

$$\frac{\partial r}{\partial x} = \frac{x}{r} = \cos\theta \quad \text{and} \quad \frac{\partial r}{\partial y} = \frac{y}{r} = \sin\theta.$$

In the same way, by differentiating $\tan\theta = y/x$ we find that

$$\frac{\partial\theta}{\partial x} = -\frac{y}{r^2} = -\frac{\sin\theta}{r} \quad \text{and} \quad \frac{\partial\theta}{\partial y} = \frac{x}{r^2} = \frac{\cos\theta}{r}.$$

If u is a function, then the chain rule implies that

$$u_x = u_r \cdot \frac{\partial r}{\partial x} + u_\theta \cdot \frac{\partial\theta}{\partial x} = u_r \cdot \cos\theta - u_\theta \cdot \frac{\sin\theta}{r}.$$

Differentiating once more using the chain rule, we see that

$$\begin{aligned}
u_{xx} &= \frac{\partial}{\partial r}\left[u_r \cdot \cos\theta - u_\theta \cdot \frac{\sin\theta}{r} \right]\cos\theta \\
&\quad - \frac{\partial}{\partial\theta}\left[u_r \cdot \cos\theta - u_\theta \cdot \frac{\sin\theta}{r} \right]\frac{\sin\theta}{r} \\
&= u_{rr} \cdot \cos^2\theta - 2u_{r\theta} \cdot \frac{\sin\theta\cos\theta}{r} + u_{\theta\theta} \cdot \frac{\sin^2\theta}{r^2} \\
&\quad + u_r \cdot \frac{\sin^2\theta}{r} + 2u_\theta \cdot \frac{\sin\theta\cos\theta}{r^2}.
\end{aligned}$$

In exactly the same way we compute that

$$u_{yy} = u_{rr} \cdot \sin^2\theta + 2u_{r\theta} \cdot \frac{\sin\theta\cos\theta}{r} + u_{\theta\theta} \cdot \frac{\cos^2\theta}{r^2} + u_r \cdot \frac{\cos^2\theta}{r} - 2u_\theta \cdot \frac{\sin\theta\cos\theta}{r^2}.$$

Thus

$$\nabla^2 u = u_{xx} + u_{yy} = u_{rr} + \frac{1}{r}u_r + \frac{1}{r^2}u_{\theta\theta}. \tag{5.2}$$

Using this technique, we can find the form of the Laplacian in any coordinate system, although the details can be tedious. We will state two more results. Cylindrical coordinates are defined by the relations

$$\begin{aligned}
x &= r\cos\theta & r^2 &= x^2 + y^2 \\
y &= r\sin\theta \quad\text{and}\quad & \tan\theta &= y/x \\
z &= z & z &= z.
\end{aligned}$$

In cylindrical coordinates the Laplacian operator has the form

$$\nabla^2 = \frac{\partial^2}{\partial r^2} + \frac{1}{r}\frac{\partial}{\partial r} + \frac{1}{r^2}\frac{\partial^2}{\partial \theta^2} + \frac{\partial^2}{\partial z^2}. \tag{5.3}$$

Spherical coordinates r, θ, and ϕ are illustrated in Figure 2. They are related to the Cartesian coordinates by

$$
\begin{aligned}
x &= r\cos\theta\sin\phi, \\
y &= r\sin\theta\sin\phi, \qquad \text{and} \\
z &= r\cos\phi
\end{aligned}
\qquad
\begin{aligned}
r^2 &= x^2 + y^2 + z^2, \\
\tan\theta &= y/x, \\
\tan\phi &= \sqrt{x^2+y^2}\big/z.
\end{aligned}
$$

The expression for the Laplacian in spherical coordinates is

$$\nabla^2 u = \frac{1}{r^2}(r^2 u_r)_r + \frac{1}{r^2\sin\phi}(\sin\phi \cdot u_\phi)_\phi + \frac{1}{r^2\sin^2\phi}u_{\theta\theta}. \tag{5.4}$$

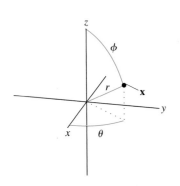

Figure 2. Spherical coordinates of a point in $\mathbf{R}^3$.

The Dirichlet problem on the disk

Using polar coordinates, the Dirichlet problem in (5.1) becomes

$$
\begin{aligned}
u_{rr} + \frac{1}{r}u_r + \frac{1}{r^2}u_{\theta\theta} &= 0, \quad \text{for } 0 < r < a, \\
u(a,\theta) &= f(\theta), \quad \text{for } 0 \le \theta \le 2\pi.
\end{aligned}
\tag{5.5}
$$

The function f is supposed to be defined on the circle of radius $r = a$. Since this circle is parameterized by $\theta \to (a\cos\theta, a\sin\theta)$, we can consider f to be a function of θ. Since it is really a function of $\sin\theta$ and $\cos\theta$, which are periodic with period 2π, the function f must be 2π-periodic.

We solve the problem using separation of variables in polar coordinates by looking for product functions of the form $u(r,\theta) = R(r)T(\theta)$, which are solutions to Laplace's equation. Just like f, T must be 2π-periodic.

When we insert the function $u(r,\theta) = R(r)T(\theta)$ into Laplace's equation, we obtain

$$\left[R''(r) + \frac{1}{r}R'(r)\right]T(\theta) + \frac{1}{r^2}R(r)T''(\theta) = 0.$$

We multiply by $r^2/[R(r)T(\theta)]$ to separate the r variable from the θ variable, obtaining

$$\frac{r^2 R''(r) + r R'(r)}{R(r)} = -\frac{T''(\theta)}{T(\theta)}.$$

Both sides must be equal to a constant, λ, so we obtain the following two equations:

$$r^2 R''(r) + r R'(r) - \lambda R(r) = 0 \quad \text{and} \quad T''(\theta) + \lambda T(\theta) = 0. \tag{5.6}$$

The function T must be 2π-periodic. Therefore, it must solve the Sturm-Liouville problem

$$T''(\theta) + \lambda T(\theta) = 0 \quad \text{with } T \ 2\pi\text{-periodic.} \tag{5.7}$$

Notice that the boundary condition is different than in previous Sturm-Liouville problems. These conditions are called **periodic boundary conditions**.

Let's first look for nonzero solutions to (5.7) with $\lambda < 0$. We write $\lambda = -s^2$, with $s > 0$. The differential equation becomes $T'' - s^2 T = 0$, which has the general solution $T(\theta) = Ae^{s\theta} + Be^{-s\theta}$. However, no function of this type is periodic, so there are no nonzero solutions for $\lambda < 0$.

If $\lambda = 0$, the differential equation in (5.7) is $T'' = 0$, which has the general solution $T(\theta) = A + B\theta$. Since T must be 2π-periodic, we conclude that $B = 0$. Thus, for $\lambda = 0$ the nonzero solutions are any constant function, that is a multiple of the function

$$c_0(\theta) = 1.$$

For $\lambda > 0$ we set $\lambda = \omega^2$, where $\omega > 0$. Then the differential equation has the form $T'' + \omega^2 T = 0$, which has the general solution

$$T(\theta) = A \cos \omega\theta + B \sin \omega\theta.$$

Since this function must be 2π-periodic, we conclude that ω must be a positive integer. Thus, any linear combination of the functions

$$c_n(\theta) = \cos n\theta \quad \text{and} \quad s_n(\theta) = \sin n\theta$$

will be a solution to the Sturm-Liouville problem in (5.7).

To sum up, the eigenvalues for the Sturm-Liouville problem in (5.7) are $\lambda_n = n^2$, for n any nonnegative integer. The corresponding eigenfunctions are the single function $c_0(\theta) = 1$ for $n = 0$, and the pair of functions $c_n(\theta) = \cos n\theta$ and $s_n(\theta) = \sin n\theta$ for $n \geq 1$.

In view of the fact that $\lambda = n^2$, where n is a nonnegative integer, the differential equation for R in (5.6) becomes

$$r^2 R'' + r R' - n^2 R = 0. \tag{5.8}$$

This is a special case of Euler's equation, which we studied in Section 11.3. A fundamental set of solutions is

$$\begin{aligned} r^0 = 1 \quad &\text{and} \quad \ln r, \quad \text{for } n = 0, \\ r^n \quad &\text{and} \quad r^{-n}, \quad \text{for } n \geq 1. \end{aligned} \tag{5.9}$$

However, there is a hidden boundary condition. We are really looking at functions defined on the disk, and the point $r = 0$ corresponds to the center of the disk. We want our solutions[9] to be bounded there, so the solutions $\ln r$ for $n = 0$ and r^{-n} for $n > 0$ are not viable. Consequently, we are led to the solution

$$R_n(r) = r^n, \quad \text{for } n \geq 0.$$

The corresponding product solutions to Laplace's equation are $u_0(r, \theta) = 1$ and

$$\begin{aligned} u_n(r, \theta) &= R_n(r)c_n(\theta) = r^n \cos n\theta \quad \text{and} \\ v_n(r, \theta) &= R_n(r)s_n(\theta) = r^n \sin n\theta \end{aligned}$$

for $n \geq 1$. Since the Laplacian is linear, the function

$$\begin{aligned} u(r, \theta) &= \frac{A_0}{2} + \sum_{n=1}^{\infty} A_n u_n(r, \theta) + B_n v_n(r, \theta) \\ &= \frac{A_0}{2} + \sum_{n=1}^{\infty} r^n (A_n \cos n\theta + B_n \sin n\theta) \end{aligned} \tag{5.10}$$

---

[9] It is good to remember that our solutions represent temperatures.

is a solution to Laplace's equation on the disk for any constants A_n and B_n for which the series converges.

The boundary condition $u(a, \theta) = f(\theta)$ now becomes

$$f(\theta) = u(a, \theta) = \frac{A_0}{2} + \sum_{n=1}^{\infty} a^n (A_n \cos n\theta + B_n \sin \theta).$$

This is the complete Fourier series for the boundary function f. According to Theorem 1.17 and Proposition 2.14 in Chapter 12, the coefficients must be

$$A_n = \frac{1}{a^n \pi} \int_0^{2\pi} f(\theta) \cos n\theta \, d\theta, \quad \text{for } n \geq 0,$$

$$B_n = \frac{1}{a^n \pi} \int_0^{2\pi} f(\theta) \sin n\theta \, d\theta, \quad \text{for } n \geq 1.$$

Inserting these values into (5.10) yields the solution to the Laplace equation (5.2).

Example 5.11 A beer can of radius 1 inch is full of beer and lies on its side, halfway submerged in the snow (see Figure 3). The snow keeps the bottom half of the beer can at $0°C$ while the sun warms the top half of the can to $1°C$. Find the steady-state temperature inside the can.

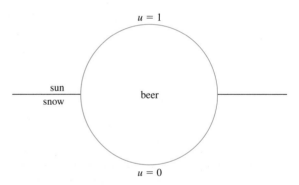

Figure 3. The can of beer in Example 5.11.

The boundary of the beer can is a circle of radius 1 inch, which in polar coordinates can be described by the equations $r = 1$ and $0 \leq \theta \leq 2\pi$. The temperature function on the boundary is given by

$$f(\theta) = \begin{cases} 1, & \text{for } 0 \leq \theta \leq \pi, \\ 0, & \text{for } \pi < \theta < 2\pi. \end{cases}$$

Thus, we wish to solve (5.1) with $a = 1$ and f as given.[10]

We need to compute the Fourier coefficients of f. First, we have

$$A_0 = \frac{1}{\pi} \int_0^{2\pi} f(\theta) \, d\theta = \frac{1}{\pi} \int_0^{\pi} 1 \, d\theta = 1.$$

Next, for $n \geq 1$,

$$A_n = \frac{1}{\pi} \int_0^{2\pi} f(\theta) \cos n\theta \, d\theta = \frac{1}{\pi} \int_0^{\pi} \cos n\theta \, d\theta = 0.$$

---

[10] The function f is defined explicitly only for $0 \leq \theta < 2\pi$. It is to be understood that we mean the period extension to **R**.

Finally,

$$B_n = \frac{1}{\pi} \int_0^{2\pi} f(\theta) \sin n\theta \, d\theta = \frac{1}{\pi} \int_0^{\pi} \sin n\theta \, d\theta = \frac{1 - (-1)^n}{n\pi}.$$

If n is even, $B_n = 0$, and if $n = 2k + 1$ is odd, $B_{2k+1} = 2/[(2k+1)\pi]$. Substituting the coefficients into (5.10), we see that the solution to (5.2) is

$$u(r, \theta) = \frac{1}{2} + \sum_{k=0}^{\infty} \frac{2r^{2k+1}}{\pi(2k+1)} \sin(2k+1)\theta.$$

Figure 4 shows the graph of the partial sum of the solution up to $k = 20$ over the unit disk.

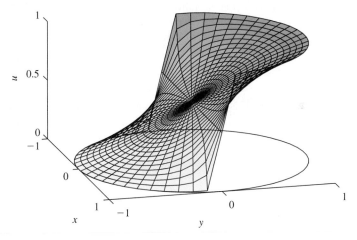

Figure 4.   The temperature of the beer in Example 5.11.

The vertical axis is the temperature of the beer in the can. Notice that the graph of the temperature on the boundary of the disk jumps from 0 to 1 halfway around the disk as is consistent with the temperature of the surface of the beer can.   ●

EXERCISES

1. Verify that the Laplacian has the form in (5.3) in cylindrical coordinates.

2. Verify that the Laplacian has the form in (5.4) in spherical coordinates.

3. Suppose that the temperature on the surface of the beer can in Example 5.11 is $f(x, y) = 1 + y$, where y is the distance above the snow. Find the steady-state temperature throughout the can. (*Hint:* This exercise is easier than it might look. Keep in mind that you want to solve the Dirichlet problem in (5.1).)

4. Suppose that the snow keeps the temperature on the bottom half of the surface of the beer can in Example 5.11 at 0°C, but this time suppose the vertical sun's rays keep the very top of the can at 1°C and that the temperature at any other point on the top half of the can is proportional to the sine of the angle between the sun's rays and the tan-

gent line to the can. Give an intuitive argument why this is a reasonable model for the temperature on the boundary. (*Hint:* Look at how the intensity of the sunlight on the can's surface depends on the angle between the sun's rays and the tangent to the can.) Compute the steady-state temperature at any point on the inside of the beer can. Draw a hand sketch of what you think is the graph of the solution (over the region $0 \le r \le 1$ and $0 \le \theta \le 2\pi$) and then compare with the graph of the first 10 terms or so of the computed solution.

In Exercises 5–8, find the steady-state temperature in a disk of radius 1, with the given temperature $f(\theta)$, for $0 \le \theta \le 2\pi$.

5. $f(\theta) = \sin^2 \theta$ **6.** $f(\theta) = \cos^2 \theta$

7. $f(\theta) = \theta(2\pi - \theta)$ **8.** $f(\theta) = \sin \theta \cos \theta$

If we are looking for the steady-state temperature in a ring-shaped plate, then the hidden boundary condition used to eliminate half of the solutions from (5.9) does not come into play. Exercises 9–11 deal with this situation.

9. Consider a plate that is ring shaped. Its boundary consists of two concentric circles with radii $a < b$. Suppose that the inner circle is kept at a uniform temperature T_1 and the outer circle at a uniform temperature of T_2. Find the steady-state temperature throughout the plate. (*Hint:* Since in polar coordinates the temperature on the boundary does not depend on θ, you can conclude that the steady-state temperature doesn't either.)

10. Suppose that the outer boundary, where $r = b$, of the ring shaped plate in Exercise 9 is insulated, and the inner boundary, where $r = a$, is kept at the uniform temperature T. Find the steady-state temperature in the plate. (*Hint:* According to (4.4), since the plate is insulated at $r = b$, the normal derivative of the temperature u is equal to 0

there. With the circular symmetry we have, this means that $u_r = 0$.)

11. Consider once more the plate in Exercise 9. Now suppose that the temperature on the inner boundary is uniformly equal to $0°$, and on the outer boundary is given by $f(x, y)$. Find the steady-state temperature throughout the plate.

12. Suppose we have a semicircular plate of radius a. Suppose that the temperature on the flat base is kept at $0°$, while on the curved portion the temperature is described by $f(x, y)$. Find the steady-state temperature throughout the plate.

13. Suppose that we have a plate that is shaped like a piece of pie. It is a segment of a circle of radius a, with angle θ_0. Suppose the temperature is fixed at $0°$ along the flat portions of the boundary, and is given by $f(\theta)$ for $0 \le \theta \le \theta_0$ along the curved portion. Find the steady-state temperature throughout the plate.

13.6 Sturm-Liouville Problems

One of the steps in the method of separation of variables is the solution of a Sturm-Liouville problem. The prototypical example appeared in (2.10). It was to find numbers λ and nonzero functions X defined on the interval $[0, L]$ for which

$$-X'' = \lambda X \quad \text{with} \quad X(0) = X(L) = 0. \tag{6.1}$$

We have rewritten the differential equation in the form we will adopt in this section. The solutions to the Sturm-Liouville problem in (6.1) were found in (2.12). They are

$$\lambda_n = \frac{n^2\pi^2}{L^2} \quad \text{and} \quad X_n(x) = \sin\left(\frac{n\pi x}{L}\right), \quad \text{for } n = 1, 2, 3, \dots .$$

We saw another example using the same differential equation, but with different boundary conditions in (2.27). Let's look at some more examples.

Example 6.2 Consider a rod of length L that is kept at a constant temperature of $0°$ at $x = 0$ and is insulated at $x = L$. The temperature $u(x, t)$ in the rod satisfies the initial/boundary value problem

$$u_t(x, t) = ku_{xx}(x, t), \quad \text{for } t > 0 \text{ and } 0 < x < L,$$
$$u(0, t) = 0 \quad \text{and} \quad u_x(L, t) = 0, \quad \text{for } t > 0,$$
$$u(x, 0) = f(x), \quad \text{for } 0 \le x \le L,$$

where $f(x)$ is the temperature in the rod at time $t = 0$. If we look for a product solution $u(x, t) = X(x)T(t)$, the function X must satisfy

$$-X'' = \lambda X \quad \text{with } X(0) = X'(L) = 0. \tag{6.3}$$

This is a Sturm-Liouville problem of a type we have not seen before. ●

Example 6.4 Suppose for the same rod that the insulation at $x = L$ is poor and slowly leaks heat (see (1.14) and (1.15)). Then the boundary condition at that point will be a

Robin condition. Assuming for simplicity that the ambient temperature is $T = 0$, the temperature satisfies the initial/boundary value problem

$$u_t(x, t) = ku_{xx}(x, t), \quad \text{for } t > 0 \text{ and } 0 < x < L,$$
$$u(0, t) = 0 \quad \text{and} \quad u_x(L, t) + \gamma u(L, t) = 0, \quad \text{for } t > 0,$$
$$u(x, 0) = f(x), \quad \text{for } 0 \le x \le L,$$

where γ is a positive constant. This time, if we look for a product solution $u(x, t) = X(x)T(t)$, the function X must satisfy

$$-X'' = \lambda X \quad \text{with } X(0) = 0 \text{ and } X'(L) + \gamma X(L) = 0. \tag{6.5}$$

Again this is a Sturm-Liouville problem of a type we have not seen before. ●

Example 6.6 Suppose that the insulation along the length of the rod in Example 6.2 is not perfect, and slowly leaks heat. In this case, by an argument similar to that in Section 13.1, we are led to the initial/boundary value problem

$$u_t(x, t) = k\,[u_{xx}(x, t) - q(x)u(x, t)], \quad \text{for } t > 0 \text{ and } 0 < x < L,$$
$$u(0, t) = 0 \quad \text{and} \quad u_x(L, t) = 0, \quad \text{for } t > 0,$$
$$u(x, 0) = f(x), \quad \text{for } 0 \le x \le L,$$

where $q(x) > 0$ is a measure of the leakiness of the insulation. Again let's look for a product solution $u(x, t) = X(x)T(t)$. We find that X must satisfy

$$-X'' + qX = \lambda X \quad \text{with } X(0) = X'(L) = 0. \tag{6.7}$$

This Sturm-Liouville problem involves a different differential operator than the previous two. ●

As these examples indicate, there is a large variety of initial/boundary value problems that might be solved using the method of separation of variables. Each of these problems leads to a Sturm-Liouville problem. In this section and the next we will study these problems in general. In addition to providing us with additional techniques to solve initial/boundary value problems, the study of Sturm-Liouville problems in general will provide us with important insights into much of what we have studied both in this chapter and in Chapter 12.

The differential operator

We will assume that the differential equation in the Sturm-Liouville problem has the form

$$-(p\phi')' + q\phi = \lambda w\phi, \tag{6.8}$$

where p, q, and w are functions of x for $a \le x \le b$. It will be convenient to use operator notation in this section. The differential operators we will deal with have the form

$$L\phi = -(p\phi')' + q\phi. \tag{6.9}$$

Using this notation, the differential equation in (6.8) can be written as

$$L\phi = \lambda w\phi. \tag{6.10}$$

The function $w(x)$ is called the **weight function**.

In equations (6.1), (6.3), and (6.5), the differential operator is $L\phi = -\phi''$. Hence $p(x) = 1$ and $q(x) = 0$. In equation (6.7) the differential operator is $L\phi = -\phi'' + q\phi$, so again $p(x) = 1$ but now $q(x) > 0$. In all four cases the differential equation is written as $L\phi = \lambda\phi$, so the weight function is $w(x) = 1$.

In this section we will only consider **nonsingular** Sturm-Liouville problems.

DEFINITION 6.11

A Sturm-Liouville problem involving the equation

$$L\phi = -(p\phi')' + q\phi = \lambda w \phi$$

is nonsingular if

- the coefficient $p(x)$ and its derivative $p'(x)$ are both continuous on $[a, b]$, and $p(x) > 0$ for $a \le x \le b$,
- the coefficient $q(x)$ is piecewise continuous on $[a, b]$, and
- the weight function w is continuous and positive on $[a, b]$.

These conditions can be relaxed, but only at the endpoints a and b. If so, that endpoint is said to be **singular**. For example, if $p(a) = 0$, the endpoint a is singular. We will discuss some singular Sturm-Liouville problems later in this chapter.

Operators of the form (6.9) are said to be **formally self-adjoint**. The most important property of formally self-adjoint operators is in the following proposition.

PROPOSITION 6.12   Let L be a differential operator of the type in (6.9). If f and g are two functions defined on (a, b) that have continuous second derivatives, then

$$\int_a^b Lf \cdot g \, dx = \int_a^b f \cdot Lg \, dx + p(fg' - f'g)\Big|_a^b. \tag{6.13}$$

We will leave the proof to Exercise 12.

The property of formally self-adjoint operators displayed in Proposition 6.12 is not true for most differential operators. This property is the main reason for limiting our consideration to formally self-adjoint operators. In Exercise 13 we will exhibit a differential operator that does not have this property.

The assumption that our operator is formally self-adjoint might seem too restrictive. We could consider the more general operator

$$M\phi = -P\phi'' - Q\phi' + R\phi.$$

However, assuming that $P(x) > 0$ for all $x \in (a, b)$, it is always possible to find a function μ so that the operator $L = \mu M$ is formally self-adjoint. To see this, notice that

$$\mu M\phi = -\mu P\phi'' - \mu Q\phi' + \mu R\phi, \quad \text{while}$$
$$L\phi = -p\phi'' - p'\phi' + q\phi.$$

For these to be equal, it is necessary to find functions $\mu(x)$, $p(x)$, and $q(x)$ so that

$$p = \mu P, \quad p' = \mu Q, \quad \text{and} \quad q = \mu R.$$

The quotient of the first two equations gives us the linear differential equation $p' = (Q/P)p$, which can be solved to find p. Then we take $\mu = p/P$ and $q = \mu R$. This process is exemplified in Exercises 6–9.

The boundary conditions

In each of the examples at the beginning of this section, we imposed two boundary conditions. We will do so in general. The most general boundary condition has the form

$$B\phi = \alpha\phi(a) + \beta\phi'(a) + \gamma\phi(b) + \delta\phi'(b) = 0.$$

Notice that the condition mixes the values of ϕ and ϕ' at the two endpoints. We will consider only one pair of boundary conditions that mix the endpoints in this way. These are

$$B_1\phi = \phi(a) - \phi(b) = 0 \quad \text{and} \quad B_2\phi = \phi'(a) - \phi'(b) = 0. \qquad (6.14)$$

We will call these the ***periodic boundary conditions*** because they are satisfied by a function ϕ that is periodic with period $b - a$.

Boundary conditions that involve only one endpoint are called ***unmixed***. We will consider pairs of unmixed boundary conditions where one of the conditions applies to each endpoint. The most general unmixed boundary conditions have the form

$$B_1\phi = \alpha_1\phi'(a) + \beta_1\phi(a) = 0 \quad \text{and} \quad B_2\phi = \alpha_2\phi'(b) + \beta_2\phi(b) = 0, \qquad (6.15)$$

where α_1, α_2, β_1, and β_2 are constants. In order that the boundary conditions be meaningful, we will insist that the vectors (α_1, β_1), and (α_2, β_2) are nonzero.

The general unmixed boundary condition in (6.15) splits into three cases, depending on the coefficients. For the endpoint a they are

1. $\phi(a) = 0$. The Dirichlet condition (if $\alpha_1 = 0$).
2. $\phi'(a) = 0$. The Neumann condition (if $\beta_1 = 0$).
3. $\phi'(a) + \gamma_1\phi(a) = 0$. The Robin condition (if neither α_1 nor β_1 is equal to 0).

You should compare these conditions with our discussion of boundary conditions in Sections 13.1 and 13.4.

The eigenvalues and eigenfunctions

The Sturm-Liouville problem for a given operator L, weight function w, and boundary conditions B_1 and B_2 on an interval (a, b) is to find all numbers λ and nonzero functions ϕ such that

$$\begin{aligned} L\phi = \lambda w\phi \quad &\text{on the interval } (a, b), \text{ and} \\ B_1\phi = B_2\phi &= 0. \end{aligned} \qquad (6.16)$$

Any number λ for which there is a nonzero function ϕ satisfying (6.16) is called an ***eigenvalue*** of the Sturm-Liouville problem. If λ is an eigenvalue, then any function ϕ that satisfies (6.16) is called an associated ***eigenfunction***.[11] Thus, our problem is to find all of the eigenvalues and eigenfunctions for a Sturm-Liouville boundary value problem. Notice that if c_1 and c_2 are constants and ϕ_1 and ϕ_2 are eigenfunctions, then, because L is linear, $L(c_1\phi_1 + c_2\phi_2) = c_1 L\phi_1 + c_2 L\phi_2 = \lambda w(c_1\phi_1 + c_2\phi_2)$, so any linear combination of eigenfunctions is also an eigenfunction. In particular, any constant multiple of an eigenfunction is also an eigenfunction. We will usually choose the constant for which the eigenfunction has the simplest algebraic form.

We have seen two examples of the solution to a Sturm-Liouville problem in (2.10) and (2.27). Let's look at another.

---

[11] You are encouraged to compare the discussion here of eigenvalues and eigenfunctions with the discussion in Section 9.1 of eigenvalues and eigenvectors of a matrix.

Example 6.17   Find the eigenvalues and eigenfunctions for the operator $L\phi = -\phi''$ on the interval $[-\pi, \pi]$ with periodic boundary conditions, and with weight $w = 1$.

We analyzed what is essentially the same problem in Section 13.5. See equation (5.7) and the following text. The eigenvalues and eigenfunctions are

$$\lambda_0 = 0 \quad \text{with} \quad c_0(x) = 1, \quad \text{and}$$

$$\lambda_n = n^2 \quad \text{with} \quad c_n(x) = \cos nx \quad \text{and} \quad s_n(x) = \sin nx, \quad \text{for } n \geq 1.$$

We will leave the details to Exercise 14.   ●

Properties of eigenvalues and eigenfunctions

The *multiplicity* of an eigenvalue is the number of linearly independent eigenfunctions associated to it. Notice that the positive eigenvalues in Example 6.17 have multiplicity 2, while all of the eigenvalues in (2.12) and (2.27) have multiplicity 1.

Suppose that λ is any number, and let's look at the differential equation $L\phi = -(p\phi')' + q\phi = \lambda w\phi$. If we write this out and rearrange it, we get

$$p\phi'' + p'\phi' + (\lambda w - q)\phi = 0. \tag{6.18}$$

This is a second order, linear differential equation. From Section 4.1 we know that it has a fundamental set of solutions consisting of two linearly independent functions $\phi_1(x)$ and $\phi_2(x)$. The general solution is the linear combination

$$\phi(x) = A\phi_1(x) + B\phi_2(x). \tag{6.19}$$

For any λ, the boundary conditions put constraints on the coefficients A and B in (6.19). As we have seen in our examples, for most values of λ the two boundary conditions will imply that both A and B are equal to zero, and the only solution is the zero function. Thus, most numbers λ are not eigenvalues.

In Example 6.17, the boundary conditions were satisfied by all solutions to the differential equation, so they did not constrain the coefficients A and B at all. The multiplicity is 2. In general, if an eigenvalue has multiplicity 2, then there are two linearly independent solutions to (6.16). These two solutions form a fundamental set of solutions to the differential equation, so every solution to the differential equation also solves the boundary conditions. The only other possibility is that the multiplicity is 1. In this case, the two boundary conditions combine to put only one nontrivial constraint on A and B.

Our examples show that eigenvalues are rare, and the next theorem shows that this is true in general.

THEOREM 6.20   The eigenvalues for a nonsingular Sturm-Liouville problem with either unmixed or periodic boundary conditions, repeated according to their multiplicity, form a sequence of real numbers

$$\lambda_1 \leq \lambda_2 \leq \lambda_3 \leq \cdots \quad \text{where} \quad \lambda_n \to \infty. \tag{6.21}$$

For each eigenvalue λ_n there is an associated eigenfunction which we will denote by $\phi_n(x)$. If we have a repeated eigenvalue, $\lambda_n = \lambda_{n+1}$, the eigenfunctions ϕ_n and ϕ_{n+1} can be chosen to be linearly independent. The eigenfunctions are all real valued. ■

The proof of Theorem 6.20 requires techniques that are beyond the scope of this book, so we will not present it.[12]

Example 6.17 shows that eigenvalues with multiplicity 2 do occur. However, it is rare, as is shown by the following result.

PROPOSITION 6.22   Suppose that one of the boundary conditions for a nonsingular Sturm Liouville problem is unmixed. Then every eigenvalue has multiplicity 1.   ●

The proof of Proposition 6.22 is left to Exercise16.   Under the hypotheses of Proposition 6.22, the eigenvalues satisfy

$$\lambda_1 < \lambda_2 < \lambda_3 < \cdots \quad \text{and} \quad \lambda_n \to \infty, \tag{6.23}$$

instead of the less restrictive inequalities in (6.21).

There is one additional fact that will frequently speed our search for eigenvalues and eigenfunctions.

PROPOSITION 6.24   Suppose that we have a nonsingular Sturm-Liouville problem

$$L\phi = -(p\phi')' + q\phi = \lambda w\phi,$$
$$B_1\phi = \alpha_1\phi'(a) + \beta_1\phi(a) = 0,$$
$$B_2\phi = \alpha_2\phi'(b) + \beta_2\phi(b) = 0,$$

where

(a) $q(x) \geq 0$ for $a \leq x \leq b$, and

(b) the boundary conditions on a function ϕ imply that $p\phi\phi'|_a^b \leq 0$.

Then all of the eigenvalues are nonnegative. If $\lambda = 0$ is an eigenvalue, the corresponding eigenfunctions are the constant functions.

Proof   Suppose that λ is an eigenvalue and ϕ is an associated eigenfunction. Consider the following computation, where the last step involves an integration by parts.

$$\lambda \int_a^b \phi^2(x)\, w(x)\, dx = \int_a^b \phi(x) \cdot [\lambda w(x)\phi(x)]\, dx = \int_a^b \phi(x) \cdot L\phi(x)\, dx$$

$$= -\int_a^b \phi(p\phi')'\, dx + \int_a^b q\phi^2\, dx \tag{6.25}$$

$$= \int_a^b p(x)\phi'(x)^2\, dx - p\phi\phi'\big|_a^b + \int_a^b q(x)\phi(x)^2\, dx$$

Since the Sturm-Liouville problem is nonsingular, the coefficient $p(x)$ is always positive. By our hypotheses, the coefficient $q(x)$ is nonnegative, and the term $-p\phi\phi'|_a^b \geq 0$. Hence all of the terms on the right-hand side of the equation are nonnegative. Consequently,

$$\lambda \int_a^b \phi^2(x)\, w(x)\, dx \geq 0,$$

and since the weight function $w(x)$ is positive, we conclude that $\lambda \geq 0$.

If $\lambda = 0$ is an eigenvalue, then all three terms on the right-hand side of (6.25) must be equal to 0. In particular, $\int_a^b p(x)\phi'(x)^2\, dx = 0$. Since $p(x) > 0$, this means that $\phi'(x) = 0$, so ϕ is a constant function.   ●

---

[12] See *Theory of Ordinary Differential Equations* by E. Coddington and N. Levinson (Krieger, New York, 1984) or *A First Course in Partial Differential Equations* by H. Weinberger (Dover, New York, 1995).

Let's end the section by finding the eigenvalues and eigenfunctions for the Sturm-Liouville problems in Examples 6.2, 6.4, and 6.6.

Example 6.26   Find the eigenvalues and eigenfunctions for the Sturm-Liouville problem in Example 6.2.

Let's rewrite equation (6.3) with $X(x)$ replaced by $\phi(x)$ to get

$$-\phi'' = \lambda\phi \quad \text{for } x \in (0, L), \text{ with } \quad \phi(0) = 0 = \phi'(L).$$

The coefficients are $p = 1$ and $q = 0$. Thus q is nonnegative, and the boundary conditions imply that $p\phi\phi' = \phi\phi'$ vanishes at each endpoint. Therefore, by Proposition 6.24, all of the eigenvalues are nonnegative. If $\lambda = 0$ is an eigenvalue, then the eigenfunction ϕ is a constant. However, $\phi(0) = 0$, so $\phi(x) = 0$. Thus $\lambda = 0$ is not an eigenvalue.

Thus all eigenvalues are positive. If we set $\lambda = \omega^2$, where $\omega > 0$, the differential equation becomes $\phi'' + \omega^2\phi = 0$. The general solution is $\phi(x) = A\cos\omega x + B\sin\omega x$. The first boundary condition says that $0 = \phi(0) = A$. Then the second boundary condition says that $\phi'(L) = \omega B\cos\omega L = 0$. Since we are looking for nonzero solutions, $B \neq 0$. Hence we must have $\cos\omega L = 0$. This is true only if $\omega L = \pi/2 + n\pi = (2n+1)\pi/2$, where n is a nonnegative integer. Thus our eigenvalues and eigenfunctions are

$$\lambda_n = \frac{(2n+1)^2\pi^2}{4L^2} \quad \text{and} \quad \phi_n(x) = \sin\frac{(2n+1)\pi x}{2L}, \quad \text{for } n = 0, 1, 2, \ldots .$$

The first five eigenfunctions are shown in Figure 1.   ●

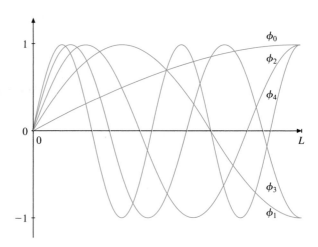

Figure 1. The first five eigenfunctions for the Sturm-Liouville problem in Example 6.26.

Example 6.27   Find the eigenvalues and eigenfunctions for the Sturm-Liouville problem in Example 6.4.

Let's rewrite (6.5) as

$$-\phi'' = \lambda\phi \quad \text{for } x \in (0, L), \text{ with } \quad \phi(0) = 0 = \phi'(L) + \gamma\phi(L).$$

Again $p = 1$ and $q = 0$. We have $p(0)\phi(0)\phi'(0) = 0$, while $p(L)\phi(L)\phi'(L) = -\gamma\phi(L)^2 \leq 0$, since $\gamma > 0$. Therefore, Proposition 6.24 shows that all of the eigenvalues are nonnegative. If $\lambda = 0$ is an eigenvalue then the eigenfunction ϕ is a constant. However, since $\phi(0) = 0$, $\phi(x) = 0$, so $\lambda = 0$ is not an eigenvalue. For $\lambda > 0$, we write $\lambda = \omega^2$, where $\omega > 0$. The differential equation has the general solution $\phi(x) = A \cos \omega x + B \sin \omega x$, where A and B are arbitrary constants. The first boundary condition implies that $0 = \phi(0) = A$. Hence $\phi(x) = B \sin \omega x$. The second boundary condition implies that

$$0 = \phi'(L) + \gamma \phi(L) = B \left[\omega \cos \omega L + \gamma \sin \omega L \right].$$

Since $\phi(x) \neq 0$, $B \neq 0$. Hence the second factor must vanish. Dividing by $\cos \omega L$ and rearranging, we get

$$\tan \omega L = -\frac{\omega}{\gamma}. \tag{6.28}$$

For those values of ω that solve equation (6.28), $\lambda = \omega^2$ is an eigenvalue and $\phi(x) = \sin \omega x$ is an associated eigenfunction.

Equation (6.28) cannot be solved exactly, but it can be solved to any desired degree of accuracy using numerical methods. To see what the solutions look like, let's first simplify the equation somewhat by setting $\theta = \omega L$, and $\alpha = 1/(\gamma L)$. Then (6.28) becomes

$$\tan \theta = -\alpha \theta. \tag{6.29}$$

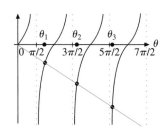

In Figure 2, we plot $f(\theta) = \tan \theta$ in black and $g(\theta) = -\alpha \theta$ in blue. The points where the two graphs intersect correspond to the values of θ that solve (6.29). From Figure 2 we see that there are infinitely many solutions to (6.29), which we will write as the increasing sequence θ_j for $j = 1, 2, 3, \ldots$. Again from Figure 2, we see that

$$(j - 1/2)\pi < \theta_j < j\pi$$

Figure 2. The solutions to $\tan \theta = -\alpha \theta$.

and that θ_j gets closer to $(j - 1/2)\pi$ as j increases. For each j, $\omega_j = \theta_j/L$. This leads to the eigenvalues and eigenfunctions

$$\lambda_j = \omega_j^2 = \frac{\theta_j^2}{L^2} \quad \text{and} \quad \phi_j(x) = \sin \omega_j x = \sin \frac{\theta_j x}{L}, \quad \text{for } j = 1, 2, 3, \ldots. \tag{6.30}$$

For the case when $L = \gamma = 1$, the first five eigenfunctions are plotted in Figure 3. ●

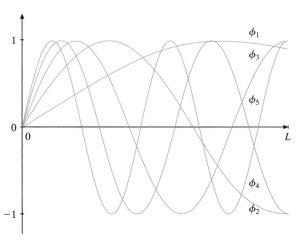

Figure 3. The first five eigenfunctions for the Sturm-Liouville problem in Example 6.27.

Example 6.31 Find the eigenvalues and eigenfunctions for the Sturm-Liouville problem in Example 6.6.

Let's rewrite (6.7) as

$$-\phi'' + q\phi = \lambda\phi, \quad \text{for } x \in (0, L), \text{ with } \phi(0) = \phi'(L) = 0. \tag{6.32}$$

If the coefficient q is not constant, we will not usually be able to find solutions explicitly, but Theorem 6.20 guarantees that they exist. However, if q is a positive constant, we can rewrite the differential equation in (6.32) as

$$-\phi'' = (\lambda - q)\phi.$$

Since $\lambda - q$ is a constant, we see that the problem is almost the same as that in Example 6.26. To be precise, $\lambda - q$ must be an eigenvalue for the Sturm-Liouville problem in Example 6.26, with the corresponding eigenfunction. Thus the eigenvalues and eigenfunctions are

$$\lambda_n = q + \frac{(2n+1)^2\pi^2}{4L^2} \quad \text{and} \quad \phi_n(x) = \sin\frac{(2n+1)\pi x}{2L}, \quad \text{for } n = 0, 1, 2, \ldots.$$

●

EXERCISES

1. Which of the following operators are formally self-adjoint?

(a) $L\phi = \phi'' + \phi'$ (b) $L\phi = x\phi'' + \phi'$

(c) $L\phi = x\phi'' + 2\phi'$ (d) $L\phi = (\cos x)\phi'' + (\sin x)\phi'$

(e) $L\phi = (\sin x)\phi'' + (\cos x)\phi'$

(f) $L\phi = (1 - x^2)\phi'' - 2x\phi'$

In Exercises 2–5, find the eigenvalues and eigenfunctions for the given Sturm-Liouville problem.

2. $-\phi'' = \lambda\phi$ with $\phi'(0) = \phi'(1) = 0$

3. $-\phi'' = \lambda\phi$ with $\phi'(0) = \phi(1) = 0$

4. $-\phi'' = \lambda\phi$ with $\phi'(0) = \phi'(1) + \phi(1) = 0$

5. $-\phi'' = \lambda\phi$ with $\phi'(0) - \phi(0) = \phi(1) = 0$

In Exercises 6–9, use the procedure given after the statement of Proposition 6.12 to transform the given differential equation into a formally self-adjoint equation.

6. $\phi'' + 4\phi' + \lambda\phi = 0$ **7.** $2x\phi'' + \lambda\phi = 0$

8. $x(x - 1)\phi'' + 2x\phi' + \lambda\phi = 0$

9. $x^2\phi'' - 2x\phi' + \lambda\phi = 0$

10. Following the lead of Example 6.27, show how to graphically find the eigenvalues for the Sturm-Liouville problem

$$-\phi'' = \lambda\phi \quad \text{with} \quad \phi'(0) - \phi(0) = \phi'(1) + \phi(1) = 0.$$

Find the eigenfunctions as well.

11. Consider the Sturm-Liouville problem

$$-\phi'' = \lambda\phi \quad \text{with} \quad \phi(0) = \phi'(1) - a\phi(1) = 0,$$

where $a > 0$.

(a) Show that this problem does not satisfy the hypotheses of Proposition 6.24.

(b) Show that all eigenvalues are positive if $0 < a < 1$, that 0 is the smallest eigenvalue if $a = 1$, and that there is one negative eigenvalue if $a > 1$.

12. Prove Proposition 6.12. (*Hint:* Start with $\int_a^b Lf \cdot g \, dx$, insert the definition of L in (6.9), and then integrate by parts twice.)

13. Not all differential operators have the property in Proposition 6.12.

(a) Use Proposition 6.12 to show that if L is a formally self-adjoint operator, then

$$\int_a^b Lf \cdot g \, dx = \int_a^b f \cdot Lg \, dx$$

for any two functions f and g that vanish at both endpoints.

(b) Consider the operator $L\phi = \phi'' + \phi'$ on the interval $[0, 1]$. Show that the integral identity in part (a) is not true for L with $f(x) = x(1-x)$ and $g(x) = x^2(1-x)$, and therefore L does not have the property in Proposition 6.12.

14. Verify that the eigenvalues and eigenfunctions for the Sturm Liouville problem in Example 6.17 are those listed there.

15. Show that if $u(x, t) = X(x)T(t)$ is a product solution of the differential equation in Example 6.6, together with the boundary conditions, then X must be a solution to the Sturm-Liouville problem in (6.7).

16. Prove Proposition 6.22. (*Hint*: Suppose that the boundary condition at $x = a$ is unmixed. Let λ be an eigenvalue and

suppose that ϕ_1 and ϕ_2 are eigenfunctions. Let W be the Wronskian of ϕ_1 and ϕ_2. Use the boundary condition to show that $W(a) = 0$. Then use Proposition 1.26 of Section 4.1.)

13.7 Orthogonality and Generalized Fourier Series

You may have noticed that the eigenfunctions in the examples of Sturm-Liouville problems in Sections 2, 3, and 4 were the bases of Fourier sine and cosine series. In addition, in Example 6.17 we have a Sturm-Liouville problem for which the eigenfunctions are the basis of complete Fourier series. You may have asked yourself if the eigenfunctions of other Sturm-Liouville problems lead to similar expansions. In this section we will carry out the derivation of such series.

Inner products, and orthogonality

The key idea in the derivation of Fourier series in Chapter 12 was the notion of orthogonality. It will also be important here, and it is time to put the idea into its proper framework. This involves the use of an inner product.

DEFINITION 7.1

Suppose that f and g are piecewise continuous functions on the interval $[a, b]$. The ***inner product*** of f and g with weight function $w(x) > 0$ is defined to be

$$(f, g)_w = \int_a^b f(x)g(x) \, w(x) \, dx. \tag{7.2}$$

If $w(x) = 1$, we will denote $(f, g)_w$ by (f, g). Thus

$$(f, g) = \int_a^b f(x)g(x) \, dx. \tag{7.3}$$

Notice that

$$(f, g)_w = (f, wg) = (wf, g). \tag{7.4}$$

Some elementary properties of the inner product are easily discovered. First, the inner product is ***symmetric***, which means that

$$(f, g)_w = (g, f)_w.$$

Second, the inner product is ***linear in each component***. For example, if a and b are constants, then

$$(af + bg, h)_w = a(f, h)_w + b(g, h)_w.$$

Finally, the inner product is ***positive***, meaning that

$$(f, f)_w > 0, \quad \text{unless} \quad f(x) = 0.$$

Our first result using this new definition will throw some light on why we write our differential operator in the formally self-adjoint form of (6.9).

PROPOSITION 7.5

Suppose that we have a nonsingular Sturm-Liouville equation

$$L\phi = -(p\phi')' + q\phi = \lambda w\phi, \tag{7.6}$$

together with the unmixed boundary conditions

$$
\begin{aligned}
B_1\phi &= \alpha_1\phi'(a) + \beta_1\phi(a) = 0,\\
B_2\phi &= \alpha_2\phi'(b) + \beta_2\phi(b) = 0.
\end{aligned}
\tag{7.7}
$$

If f and g are two functions defined on $[a, b]$ that have continuous second derivatives and satisfy the boundary conditions, then

$$
(Lf, g) = (f, Lg).
\tag{7.8}
$$

Proof According to Proposition 6.12, we have

$$
\int_a^b Lf \cdot g\, dx = \int_a^b f \cdot Lg\, dx + p(fg' - f'g)\Big|_a^b.
$$

Hence, to prove (7.8) we need to show that $p(fg' - f'g)\big|_a^b = 0$ if both f and g satisfy the boundary conditions. In fact, $fg' - f'g$ is equal to 0 at each of the endpoints. We will show this for the endpoint $x = a$. Since both f and g satisfy the boundary condition at $x = a$, we get the system of equations $\alpha_1 f'(a) + \beta_1 f(a) = 0$ and $\alpha_1 g'(a) + \beta_1 g(a) = 0$. In matrix form this can be written

$$
\begin{pmatrix} f'(a) & f(a) \\ g'(a) & g(a) \end{pmatrix} \begin{pmatrix} \alpha_1 \\ \beta_1 \end{pmatrix} = \begin{pmatrix} 0 \\ 0 \end{pmatrix}.
$$

Since the vector $(\alpha_1, \beta_1)^T$ is nonzero, the determinant of the matrix, $f'(a)g(a) - f(a)g'(a)$, must be equal to 0, as we wanted to show. ●

The property of the boundary value problem with unmixed boundary conditions expressed in (7.8) is critical to the theory that we are presenting. It is important to the proof of Theorem 6.20, as well as to the results that follow. We will say that a boundary value problem is **self-adjoint** if $(Lf, g) = (f, Lg)$ for any two functions that satisfy the boundary conditions.

DEFINITION 7.9

Two real valued functions f and g defined on the interval $[a, b]$ are said to be **orthogonal with respect to the weight** w if

$$
(f, g)_w = \int_a^b f(x)g(x)\, w(x)\, dx = 0.
$$

You will notice that this is the sense in which we used the term orthogonal for Fourier series in Chapter 12. The eigenfunctions of a Sturm-Liouville problem have orthogonality properties similar to those we discovered for the sines and cosines in Chapter 12.

PROPOSITION 7.10 Suppose that ϕ_j and ϕ_k are eigenfunctions of the Sturm-Liouville problem defined by (7.6) and (7.7) associated to different eigenvalues $\lambda_j \neq \lambda_k$. Then ϕ_j and ϕ_k are orthogonal with respect to the weight w.

Proof Since ϕ_j and ϕ_k are eigenfunctions associated to λ_j and λ_k, we have

$$
L\phi_j = \lambda_j w\phi_j \quad \text{and} \quad L\phi_k = \lambda_k w\phi_k.
$$

Hence

$$(L\phi_j, \phi_k) = \lambda_j (w\phi_j, \phi_k) = \lambda_j (\phi_j, \phi_k)_w \quad \text{and}$$
$$(L\phi_k, \phi_j) = \lambda_k (w\phi_k, \phi_j) = \lambda_k (\phi_k, \phi_j)_w.$$

Using these equations, the properties of the inner product, and Proposition 7.5, we have

$$\lambda_j (\phi_j, \phi_k)_w = (L\phi_j, \phi_k) = (\phi_j, L\phi_k) = \lambda_k (\phi_j, \phi_k)_w.$$

Thus

$$(\lambda_j - \lambda_k)(\phi_j, \phi_k)_w = 0.$$

Since $\lambda_j - \lambda_k \neq 0$, we must have $(\phi_j, \phi_k)_w = 0$, so ϕ_j and ϕ_k are orthogonal with respect to the weight w. ●

Generalized Fourier series

In Chapter 12 we saw how the orthogonality properties of the sines and cosines led to Fourier series expansions for functions. The orthogonality result in Proposition 7.10 will allow us to find an analog to Fourier series based on the eigenfunctions of a Sturm-Liouville problem. You are encouraged to observe the similarity of this development with that for Fourier series in Chapter 12.

First we assume that a function can be expressed as an infinite linear combination of eigenfunctions, and derive a formula for the coefficients.

PROPOSITION 7.11   Suppose that $\{\phi_n \mid n = 1, 2, \dots\}$ is the sequence of orthogonal eigenfunctions for a nonsingular Sturm-Liouville problem on the interval $[a, b]$. Suppose that

$$f(x) = c_1\phi_1(x) + c_2\phi_2(x) + \cdots = \sum_{n=1}^{\infty} c_n\phi_n(x) \tag{7.12}$$

for $a < x < b$. Then

$$c_n = \frac{(f, \phi_n)_w}{(\phi_n, \phi_n)_w} = \frac{\int_a^b f(x)\phi_n(x)\, w(x)\, dx}{\int_a^b \phi_n^2(x)\, w(x)\, dx}. \tag{7.13}$$

Proof   If we compute the inner product of f and ϕ_k using (7.12) and Proposition 7.10, we get[13]

$$(f, \phi_k)_w = \left(\sum_{n=1}^{\infty} c_n\phi_n, \phi_k\right)_w = \sum_{n=1}^{\infty} c_n(\phi_n, \phi_k)_w = c_k(\phi_k, \phi_k)_w,$$

from which the result follows. ●

Given a piecewise continuous function f on $[a, b]$, we can evaluate the inner products $(f, \phi_n)_w$ and $(\phi_n, \phi_n)_w$, and therefore the coefficients c_n in (7.13). Then we can write down the infinite series

$$f(x) \sim c_1\phi_1(x) + c_2\phi_2(x) + \cdots = \sum_{n=1}^{\infty} c_n\phi_n(x). \tag{7.14}$$

---

[13] We are quietly assuming that the series for f converges fast enough that we can distribute the sum out of the inner product.

DEFINITION 7.15

> The series in (7.14) with coefficients given by (7.13) is called the *generalized Fourier series* for the function f. The coefficients c_n are called the *generalized Fourier coefficients* of f.

Two questions immediately come to mind. Does the series converge? If the series converges, does it converge to the function f? The answers are almost the same as for Fourier series.

THEOREM 7.16 Suppose that $\{\phi_n \mid n = 1, 2, \ldots\}$ is the sequence of orthogonal eigenfunctions for a nonsingular Sturm-Liouville problem on the interval $[a, b]$. Suppose also that f is a piecewise continuous function on the interval $[a, b]$.

1. If the left- and right-hand derivatives of f exist at a point $x_0 \in (a, b)$, then the generalized Fourier series in (7.14) converges at x_0 to

$$\frac{f(x_0^+) + f(x_0^-)}{2}.$$

2. If the right-hand derivative of f exists at a and f satisfies the boundary condition at a, then the series converges at a to $f(a)$.

3. If the left-hand derivative of f exists at b and f satisfies the boundary condition at b, then the series converges at b to $f(b)$. ■

Notice that if f is continuous at a point $x_0 \in (a, b)$ and is differentiable there, then the generalized Fourier series converges to $f(x_0)$ at x_0 by part 1 of Theorem 7.16.

Example 7.17 Find the generalized Fourier series for the function $f(x) = 100x/L$ on the interval $[0, L]$ using the eigenfunctions of the Sturm Liouville problem in Example 6.26. Discuss the convergence properties of the series.

The eigenfunctions are $\phi_n(x) = \sin((2n+1)\pi x/(2L))$ and the weight function is $w(x) = 1$. Hence,

$$(\phi_n, \phi_n) = \int_0^L \sin^2 \frac{(2n+1)\pi x}{2L} \, dx = \frac{L}{2}.$$

Next, using integration by parts, we get

$$(f, \phi_n) = \frac{100}{L} \int_0^L x \cdot \sin \frac{(2n+1)\pi x}{2L} \, dx = (-1)^n \frac{400L}{(2n+1)^2 \pi^2}.$$

Consequently, the coefficients are

$$c_n = \frac{(f, \phi_n)}{(\phi_n, \phi_n)} = (-1)^n \frac{800}{(2n+1)^2 \pi^2}, \tag{7.18}$$

and the generalized Fourier series is

$$f(x) = \frac{100x}{L} = \sum_{n=0}^{\infty} (-1)^n \frac{800}{(2n+1)^2 \pi^2} \sin \frac{(2n+1)\pi x}{2L}.$$

Theorem 7.16 guarantees that the series will converge to $f(x)$ for x in the open interval $(0, L)$. Since f satisfies the boundary condition at $x = 0$, convergence there is also guaranteed. In fact, at $x = 0$ we have $f(0) = 0$, and each term of the series is also equal to 0, so the series converges there. The sum of the first two terms and the sum of the first eight terms of the generalized Fourier series are shown in blue in Figure 1, while the function f is shown in black. The series seems to converge very rapidly for all values of x. It appears that the series converges to $f(L) = 100$ at $x = L$, although this is not guaranteed by Theorem 7.16. Notice, however, that the error seems to be greatest at this endpoint.   ●

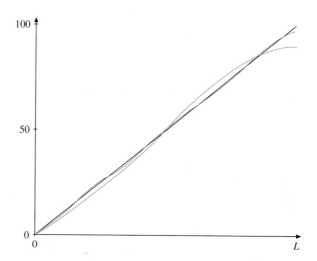

Figure 1. The partial sums of orders 2 and 8 for the generalized Fourier series in Example 7.17.

The results of Example 7.17 will enable us to solve the initial/boundary value problem for the heat equation in the next example.

Example 7.19   Suppose a rod of length L is at steady state, with the temperature maintained at $0°$ at the left-hand endpoint, and at $100°$ at the right-hand endpoint. At time $t = 0$, the heat source at the right-hand endpoint is removed and that point is insulated. Find the temperature in the rod as a function of t and x. Suppose that the thermal diffusivity is $k = 1$.

The initial temperature distribution in the rod is $100x/L$. The temperature at $x = 0$ is maintained at $0°$. The end at $x = L$ is insulated, so we have a Neumann boundary condition there. Thus we need to solve the initial/boundary value problem

$$
\begin{aligned}
u_t(x, t) &= u_{xx}(x, t), \quad \text{for } t > 0 \text{ and } 0 < x < L, \\
u(0, t) &= 0 \quad \text{and} \quad u_x(L, t) = 0, \quad \text{for } t > 0, \\
u(x, 0) &= 100x/L, \quad \text{for } 0 \le x \le L.
\end{aligned}
\tag{7.20}
$$

Notice that the boundary conditions are already homogeneous, so we do not have to find the steady-state temperature first. Substituting the product solution $u(x, t) = X(x)T(t)$ into the heat equation, and separating variables, we see that the factors must satisfy the differential equations

$$
T' + \lambda T = 0 \quad \text{and} \quad X'' + \lambda X = 0,
$$

where λ is a constant. The first equation has the general solution

$$
T(t) = Ce^{-\lambda t}.
\tag{7.21}
$$

As usual, we insist that X satisfy the boundary conditions, so we want to solve the Sturm-Liouville problem

$$X'' + \lambda X = 0 \quad \text{with } X(0) = X'(L) = 0.$$

We did this is in Example 6.26. The solutions are

$$\lambda_n = \frac{(2n+1)^2\pi^2}{4L^2} \quad \text{and} \quad X_n(x) = \sin\frac{(2n+1)\pi x}{2L}, \quad \text{for } n = 0, 1, 2, \ldots.$$

Thus, for every nonnegative integer n we get the product solution

$$u_n(x, t) = e^{-(2n+1)^2\pi^2 t/4L^2} \sin\frac{(2n+1)\pi x}{2L}$$

to the heat equation by using (7.21). This solution also satisfies the boundary conditions $u_n(0, t) = \partial u_n/\partial x(L, t) = 0$.

By the linearity of the heat equation, the function

$$u(x, t) = \sum_{n=0}^{\infty} c_n u_n(x, t) = \sum_{n=0}^{\infty} c_n e^{-(2n+1)^2\pi^2 t/4L^2} \sin\frac{(2n+1)\pi x}{2L} \tag{7.22}$$

is also a solution to the heat equation, provided the series converges. Furthermore, since each of the functions u_n satisfies the homogeneous boundary conditions, so does the linear combination u.

To satisfy the initial condition in (7.20), we must have

$$100x/L = u(x, 0) = \sum_{n=0}^{\infty} c_n \sin\frac{(2n+1)\pi x}{2L}.$$

This is the problem we solved in Example 7.17. The coefficients are those in (7.18). Hence, our solution is

$$u(x, t) = \sum_{n=0}^{\infty} (-1)^n \frac{800}{(2n+1)^2\pi^2} e^{-(2n+1)^2\pi^2 t/4L^2} \sin\frac{(2n+1)\pi x}{2L}.$$

The evolution of the temperature $u(x, t)$ is depicted in Figure 2. The initial and steady-state temperatures are plotted in blue. The black curves represent the temperature after increments of 0.1 s. Notice how the temperature steadily decreases throughout the rod to the steady-state temperature of $0°$. ●

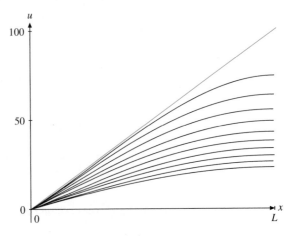

Figure 2. The temperature distribution in Example 7.19.

Example 7.23   Suppose that the rod in Example 7.19 is weakly insulated at $x = L$, and satisfies the Robin condition $u_x(L, t) + \gamma u(L, t) = 0$ posed in Example 6.4. Find the temperature in the rod as a function of t and x.

We will not go into details, since the analysis is very similar to that in the previous example. The only difference is that the Robin boundary condition leads to the Sturm-Liouville problem in Example 6.27. The eigenvalues and eigenfunctions are

$$\lambda_n = \frac{\theta_n^2}{L^2} \quad \text{and} \quad \phi_n(x) = \sin \frac{\theta_n x}{L}, \tag{7.24}$$

where θ_n is the nth positive solution of the equation $\tan \theta = -\theta/\gamma L$, which comes from (6.29). If we multiply this equation by $\gamma L \cos \theta$, we see that θ_n is the nth positive solution of the equation

$$\theta \cos \theta + \gamma L \sin \theta = 0. \tag{7.25}$$

Proposition 7.10 assures us that the eigenfunctions in (7.24) are orthogonal on $[0, L]$. Theorem 7.16 assures us that the generalized Fourier series based on these eigenvalues will converge to $f(x) = 100x/L$, at least for $0 < x < L$. This series has the form

$$\frac{100x}{L} = \sum_{n=1}^{\infty} c_n \sin \frac{\theta_n x}{L}.$$

The coefficients can be calculated using (7.13). The calculation is similar to the computation in the previous example. If equation (7.25) is used to express everything in terms of $\cos \theta_n$, the result is

$$c_n = -\frac{200(\gamma L + 1) \cos \theta_n}{\theta_n (\gamma L + \cos^2 \theta_n)}. \tag{7.26}$$

With $\gamma = L = 1$ the partial sums of order 2, 8, and 20 are shown in Figure 3.

The solution to the initial/boundary value problem is

$$u(x, t) = \sum_{n=1}^{\infty} c_n e^{-\lambda_n t} \sin \frac{\theta_n x}{L},$$

where the eigenvalues are given in (7.24), and the coefficients are given in (7.26). With $\gamma = L = 1$, the temperature is plotted in black in Figure 4 with temperatures given every 0.05 s.

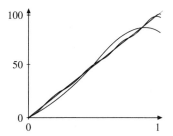

Figure 3. The partial sums of order 2, 8 and 20 for the generalized Fourier series in Example 7.23.

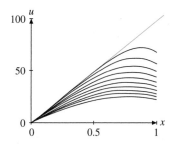

Figure 4. The temperature distribution in Example 7.23.

EXERCISES

In Exercises 1–4, find the generalized Fourier series for the indicated function using the eigenfunctions from Example 6.26 on the interval [0, 1].

1. $f(x) = 1$

2. $f(x) = \sin \pi x$ (*Hint:* Remember the trigonmetric identity $\sin \alpha \sin \beta = [\cos(\alpha - \beta) - \cos(\alpha + \beta)]/2$.)

3. $f(x) = 1 - x$ **4.** $f(x) = \sin^2 \pi x$

In Exercises 5–8, find the generalized Fourier series for referenced function using the eigenfunctions from Example 6.27 on the interval [0, 1] with $\gamma = 1$.

5. See Exercise 1   **6.** See Exercise 2

7. See Exercise 3   **8.** See Exercise 4

In Exercises 9–12, solve the heat equation $u_t = u_{xx}$ on the interval $0 < x < 1$ with the boundary conditions $u(0, t) = u_x(1, t) = 0$, and with $u(x, 0) = f(x)$ for the referenced function f.

9. See Exercise 1   **10.** See Exercise 2

11. See Exercise 3   **12.** See Exercise 4

In Exercises 13–16, solve the heat equation $u_t = u_{xx}$ on the interval $0 < x < 1$ with the boundary conditions $u(0, t) = u_x(1, t) + u(1, t) = 0$, and with $u(x, 0) = f(x)$ for the referenced function f.

13. See Exercise 5   **14.** See Exercise 6

15. See Exercise 7   **16.** See Exercise 8

17. Find the steady-state temperature u in a square plate of side length 1, where $u(x, 0) = T_1$, $u(x, 1) = T_2$, $u(0, y) = 0$, and $u_x(1, y) = 0$. (*Hint:* Look back at the methods used in Section 13.4.)

13.8 Temperatures in a Ball—Legendre Polynomials

We solved the problem of finding the steady-state temperature at points inside a disk or a rectangle in Sections 13.4 and 13.5. Now we want to do the same thing for the ball. The ball of radius a is defined to be

$$B = \left\{ (x, y, z) \,\middle|\, x^2 + y^2 + z^2 < a^2 \right\}.$$

The boundary of the ball is the sphere

$$S = \left\{ (x, y, z) \,\middle|\, x^2 + y^2 + z^2 = a^2 \right\}.$$

According to (4.3), we want to find the function $u(x, y, z, t)$ on B that satisfies

$$\nabla^2 u(x, y, z) = 0, \quad \text{for } (x, y, z) \in B,$$
$$u(x, y, z) = f(x, y, z), \quad \text{for } (x, y, z) \in S, \tag{8.1}$$

where f is a given function defined on the sphere S.

Since we have spherical symmetry, it is best to use spherical coordinates r, θ, and ϕ, which we discussed in Section 13.5. They are related to Cartesian coordinates by

$$x = r \cos \theta \sin \phi, \quad y = r \sin \theta \sin \phi, \quad z = r \cos \phi. \tag{8.2}$$

The expression for the Laplacian in spherical coordinates is

$$\nabla^2 u = \frac{1}{r^2}(r^2 u_r)_r + \frac{1}{r^2 \sin \phi}(\sin \phi \cdot u_\phi)_\phi + \frac{1}{r^2 \sin^2 \phi} u_{\theta\theta}.$$

In spherical coordinates the ball B and its boundary sphere S are described by

$$B = \left\{ (r, \theta, \phi) \,\middle|\, 0 \le r < a, \ -\pi < \theta \le \pi, \ 0 \le \phi \le \pi \right\}, \quad \text{and}$$
$$S = \left\{ (a, \theta, \phi) \,\middle|\, -\pi < \theta \le \pi, \ 0 \le \phi \le \pi \right\}.$$

Since the boundary temperature is defined on S, where $r = a$, in spherical coordinates it is given by

$$F(\theta, \phi) = f(a \cos \theta \sin \phi, a \sin \theta \sin \phi, a \cos \phi).$$

Thus the boundary condition in (8.1) can be written as $u(a, \theta, \phi) = F(\theta, \phi)$.

We will solve the problem in the easier case when the temperature u is axially symmetric, meaning that it depends only on the radius r and the polar angle ϕ, and not on the variable θ. Then the boundary value problem in (8.1) becomes

$$\frac{1}{r^2}(r^2 u_r)_r + \frac{1}{r^2 \sin\phi}(\sin\phi \cdot u_\phi)_\phi = 0, \quad \text{for } 0 \le r < a \text{ and } 0 \le \phi \le \pi,$$
$$u(a, \phi) = F(\phi), \quad \text{for } 0 \le \phi \le \pi. \tag{8.3}$$

For (x, y, z) in the boundary of the ball, we have $z = a\cos\phi$, and $\phi = \cos^{-1}(z/a)$. Thus the boundary temperature will be axially symmetric if and only if the function $f(x, y, z)$ depends only on z. Then we have $F(\phi) = f(a\cos\phi)$.

We look for product solutions of the form $u(r, \phi) = R(r) \cdot T(\phi)$. For such functions the differential equation in (8.3) becomes

$$\frac{(r^2 R')' \cdot T}{r^2} + \frac{(\sin\phi \cdot T')' \cdot R}{r^2 \sin\phi} = 0.$$

After separating the variables, we see that there is a constant λ such that

$$(r^2 R')' = \lambda R \quad \text{and} \quad -(\sin\phi \cdot T')' = \lambda \sin\phi \cdot T. \tag{8.4}$$

A singular Sturm-Liouville problem

We will solve the second equation in (8.4) first. To be specific, we want to solve

$$-(\sin\phi \cdot T')' = \lambda \sin\phi \cdot T, \quad \text{for } 0 \le \phi \le \pi. \tag{8.5}$$

Notice that the points in the ball where $\phi = 0$ are on the positive z-axis, and $\phi = \pi$ corresponds to the negative z-axis. The product function u must be well behaved[14] on the entire sphere, including along the z-axis. Therefore, the factor $T(\phi)$ must be well behaved at $\phi = 0, \pi$.

The differential equation in (8.5) becomes more familiar when we make the substitution $s = \cos\phi$. Then $\sin^2\phi = 1 - \cos^2\phi = 1 - s^2$, and by the chain rule,

$$\frac{d}{d\phi} = \frac{ds}{d\phi}\frac{d}{ds} = -\sin\phi\frac{d}{ds}.$$

Hence,

$$-\frac{d}{d\phi}\left(\sin\phi\frac{dT}{d\phi}\right) = -\sin\phi\frac{d}{ds}\left[\sin^2\phi\frac{dT}{ds}\right] = -\sin\phi\frac{d}{ds}\left[(1 - s^2)\frac{dT}{ds}\right].$$

Therefore, the differential equation in (8.5) becomes

$$LT = -\frac{d}{ds}\left[(1 - s^2)\frac{dT}{ds}\right] = \lambda T, \quad \text{for } -1 \le s \le 1.$$

Notice that since $\phi \in [0, \pi]$, $s = \cos\phi \in [-1, 1]$. The operator L, defined in this equation, is formally self-adjoint, but while the coefficient $p(s) = 1 - s^2$ is positive on $(-1, 1)$, it vanishes at both endpoints. Thus the operator L is singular at both endpoints.

---

[14] The expression *well behaved* means bounded, continuous, or continuously differentiable. However, it is best kept a little vague.

Although L is singular, from Proposition 6.12 we see that

$$(Lf, g) = (f, Lg) + (1 - s^2)(fg' - f'g)\big|_{-1}^{1} = (f, Lg),$$

for any functions f and g that have continuous second derivatives on $[-1, 1]$. Thus, no explicit boundary conditions are needed to make the operator L self-adjoint, although we do need that f and g and their first derivatives are continuous on the closed interval $[-1, 1]$. Since both the physics and the mathematics agree, we are led to pose the problem to find numbers λ and functions T such that

$$-\big((1 - s^2)T'\big)' = \lambda T, \quad \text{for } -1 \leq s \leq 1,$$

$$\text{with } T \text{ and } T' \text{ continuous on } [-1, 1]. \tag{8.6}$$

Although this is a self-adjoint Sturm-Liouville problem, it is singular, so we cannot blindly apply the results of Sections 13.6 and 13.7. Indeed, the theory of singular Sturm-Liouville problems leads to a variety of new phenomena. For example, Theorem 6.20, which states that the eigenvalues of a nonsingular Sturm-Liouville problem form a sequence that converges to ∞, is not true in general for singular problems. In fact, it can happen that every positive real number is an eigenvalue. Singular Sturm-Liouville problems are best analyzed on an ad hoc basis. When we do this for the problem in (8.6), we discover that all of the results in Sections 13.6 and 13.7 remain true.[15]

In particular, the proof of Proposition 6.24 can be easily modified for this case, and we see that all of the eigenvalues are nonnegative. Hence we can write $\lambda = n(n+1)$ where n is a nonnegative real number. Writing out the differential equation in (8.6), we get

$$(1 - s^2)T'' - 2sT' + n(n+1)T = 0.$$

This will be recognized as Legendre's equation, which we studied in Section 11.3. By (8.6), we need solutions that are bounded at both endpoints. It is a fact, not easily proven, that we get solutions to Legendre's equation that are bounded at both endpoints only if n is a nonnegative integer. Furthermore, the only solution that is bounded at both endpoints is $P_n(s)$, the Legendre polynomial of degree n (see Exercise 23 in Section 11.6 for partial results in this direction). Thus, the solution to the Sturm-Liouville problem in (8.6) is

$$\lambda_n = n(n+1) \quad \text{and} \quad P_n(s), \quad \text{for } n = 0, 1, 2, \dots. \tag{8.7}$$

From Proposition 7.10, we see that two Legendre polynomials of different degrees are orthogonal. Since the weight in equation (8.6) is $w(s) = 1$, we have

$$(P_j, P_k) = \int_{-1}^{1} P_j(s) P_k(s)\, ds = 0, \quad \text{if } j \neq k.$$

We state without proof that

$$(P_n, P_n) = \int_{-1}^{1} P_n^2(s)\, ds = \frac{2}{2n + 1}.$$

According to Theorem 7.16, if g is a piecewise continuous function on $[-1, 1]$, then it has an associated **Legendre series**

$$g(s) \sim \sum_{n=0}^{\infty} c_n P_n(s), \quad \text{with} \quad c_n = \frac{(g, P_n)}{(P_n, P_n)} = \frac{2n + 1}{2} \int_{-1}^{1} g(s) P_n(s)\, ds. \tag{8.8}$$

---

[15] See *Theory of Ordinary Differential Equations* by E. Coddington and N. Levinson (Krieger, New York, 1984) or *A First Course in Partial Differential Equations* by H. Weinberger (Dover, New York, 1995).

Solution to the boundary value problem

If we substitute $s = \cos\phi$ into (8.7), we see that the solutions to the second equation in (8.4) are

$$\lambda_n = n(n+1) \quad \text{and} \quad T_n(\phi) = P_n(\cos\phi), \quad \text{for } n = 0, 1, 2, \ldots,$$

where $P_n(s)$ is the Legendre polynomial of degree n. With $\lambda_n = n(n+1)$, the first equation in (8.4) becomes

$$r^2 R'' + 2r R' - n(n+1)R = 0.$$

This will be recognized as a special case of Euler's equation (see Section 11.3). The only solution that is bounded near $r = 0$ is $R(r) = r^n$. Hence the product solutions are of the form

$$r^n P_n(\cos\phi), \quad \text{for } n = 0, 1, 2, \ldots,$$

and we look for a solution of the form

$$u(r,\phi) = \sum_{n=0}^{\infty} c_n r^n P_n(\cos\phi). \tag{8.9}$$

By the linearity of the Laplacian, if the series converges, this function is a solution to Laplace's equation. Thus we need only show that we can find the coefficients c_n so that the boundary condition in (8.3),

$$u(a,\phi) = \sum_{n=0}^{\infty} c_n a^n P_n(\cos\phi) = F(\phi),$$

is satisfied. If we again set $s = \cos\phi$, then $F(\phi) = f(a\cos\phi) = f(as)$, so we want

$$f(as) = \sum_{n=0}^{\infty} c_n a^n P_n(s). \tag{8.10}$$

This is just the Legendre series for $f(as)$, so by (8.8) we need

$$c_n a^n = \frac{2n+1}{2} \int_{-1}^{1} f(as) P_n(s)\, ds. \tag{8.11}$$

Example 8.12   Find the steady-state temperature in a ball of radius $a = 1$, when the boundary is kept at the temperature $f(z) = 1 - z^2$.

Since the boundary temperature depends only on z, it is axially symmetric. Since $f(s) = 1 - s^2$ is a polynomial of degree 2, we expect that it is a linear combination of the first three Legendre polynomials, $P_0(s) = 1$, $P_1(s) = s$, and $P_2(s) = (3s^2 - 1)/2$. We easily see that in this case (8.10) becomes $f(s) = 2[P_0(s) - P_2(s)]/3$. Then, using (8.9), we see that the solution is

$$u(r,\phi) = \frac{2}{3}[P_0(\cos\phi) - r^2 P_2(\cos\phi)]$$

$$= \frac{2}{3}\left[1 - r^2 \frac{3\cos^2\phi - 1}{2}\right]$$

$$= \frac{2 + r^2}{3} - r^2 \cos^2\phi.$$

We can express the temperature in Cartesian coordinates using (8.2). In fact, this is quite easy, since $r^2 = x^2 + y^2 + z^2$, and $r \cos \phi = z$. We see that the steady-state temperature is given by

$$u(x, y, z) = \frac{2 + x^2 + y^2 - 2z^2}{3}.$$

●

EXERCISES

1. Find the steady-state temperature in a ball, assuming that the surface of the ball is kept at a uniform temperature of T.

2. Find the steady-state temperature in a ball of radius $a = 1$, assuming that the surface of the ball is kept at the temperature $f(z) = 1 - z$.

3. Find the steady-state temperature in a ball of radius $a = 1$, assuming that the surface of the ball is kept at the temperature $f(z) = z^3$.

4. Find the steady-state temperature in a ball of radius $a = 1$, assuming that the surface of the ball is kept at the temperature $f(z) = z^4$.

In Section 11.3 we presented the identity $x P_n'(x) - n P_n(x) = P_{n-1}'(x)$, and stated that $P_{2n+1}(0) = 0$, while $P_{2n}(0) = (-1)^n \frac{(2n)!}{2^{2n} (n!)^2}$. You will find these facts useful in Exercises 5–8.

5. Suppose a ball of radius 1 is exactly half immersed into ice, so that the bottom half of the surface is at 0°C, while the upper half is kept at 10°C. Find the first three nonzero

terms in the series expansion (8.9) of the steady-state temperature in the ball.

6. Find the complete Legendre series for the temperature in Exercise 5.

7. Suppose the surface of the ball of radius $a = 1$ is kept at the temperature

$$f(z) = \begin{cases} z & \text{if } 0 \le z \le 1, \\ 0 & \text{if } -1 \le z < 0. \end{cases}$$

Find the first three nonsero terms in the series expansion (8.9) of the steady-state temperature in the ball.

8. Find the complete Legendre series for the temperat.. Exercise 7.

9. Without doing any series computations, what is the temperature at the center of the ball in Exercise 5?

10. Suppose we have a spherical shell with inner radius 1 and outer radius 2. Suppose that the inner boundary is kept at 0°, and the outer boundary at 10°. Find the steady-state temperature throughout the shell.

13.9 The Heat and Wave Equations in Higher Dimension

We have successfully used the method of separation of variables to solve the heat equation and the wave equation when there is only one space variable. The method is also applicable when there are several space variables. In theory, the method is the same as it is with one space variable. In place of the Sturm-Liouville problem that comes up naturally in one space variable, there is the more general eigenvalue problem for the Laplacian. When the space domain is a rectangle or a sphere, the geometric symmetry allows us to solve this problem using our favorite method of separation of variables.

Heat transfer on a rectangle

As an example of the method, let's consider the rectangle D of width a and height b which we first discussed in Section 13.3. The rectangle is illustrated in Figure 1, together with the initial and boundary conditions to be satisfied by the temperature $u(t, x, y)$.

Notice that the boundary value of the temperature is described differently on each edge of the rectangle. To simplify our notation, we will define the function $g(x, y)$ on the boundary of the disk, ∂D, by

$$g(x, y) = \begin{cases} g_1(x), & \text{if } y = 0, \\ g_2(x), & \text{if } y = b, \\ g_3(y), & \text{if } x = 0, \\ g_4(y), & \text{if } x = a. \end{cases}$$

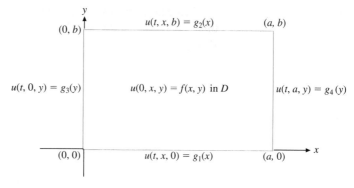

Figure 1. The Dirichlet problem for the rectangle D.

Then the initial/boundary value problem for the heat equation in the rectangle is

$$u_t(t, x, y) = k\nabla^2 u(t, x, y), \quad \text{for } (x, y) \in D \text{ and } t > 0,$$
$$u(t, x, y) = g(x, y), \quad \text{for } (x, y) \in \partial D \text{ and } t > 0, \qquad (9.1)$$
$$u(0, x, y) = f(x, y), \quad \text{for } (x, y) \in D.$$

Let's suppose that the initial temperature is constant throughout D:

$$u(0, x, y) = f(x, y) = T_1, \quad \text{for } 0 < x < a \text{ and } 0 < y < b. \qquad (9.2)$$

Let's also suppose that beginning at time $t = 0$ the boundary of the rectangle is submitted to a source of heat at the constant temperature T_2. Hence the boundary condition is

$$u(t, x, y) = g(x, y) = T_2, \quad \text{for } (x, y) \in \partial D \text{ and } t > 0. \qquad (9.3)$$

We want to discover how the temperature in D varies as t increases.

Reduction to homogeneous boundary conditions

Our first step is to reduce the problem to one with homogeneous boundary conditions, as we did in Section 13.2. We do this by finding the steady-state solution u_s that solves the boundary value problem

$$\nabla^2 u_s(x, y) = 0, \quad \text{for } (x, y) \in D,$$
$$u_s(x, y) = g(x, y) = T_2, \quad \text{for } (x, y) \in \partial D, \qquad (9.4)$$

for Laplace's equation. We showed how to solve this problem for the rectangle D in Section 13.4. However, in our case the boundary temperature is constant, so we are led to expect that $u_s(x, y) = T_2$. Substituting into (9.4) verifies that this is correct.

Having found the steady-state temperature, it remains to find $v = u - u_s$. By combining the information in (9.1) and (9.4), we see that v must solve the homogeneous initial/boundary value problem

$$v_t(t, x, y) = k\nabla^2 v(t, x, y), \quad \text{for } x, y \in D \text{ and } t > 0.$$
$$v(t, x, y) = 0, \quad \text{for } (x, y) \in \partial D \text{ and } t > 0. \qquad (9.5)$$
$$v(0, x, y) = F(x, y) = f(x, y) - u_s(x, y), \quad \text{for } (x, y) \in D.$$

In the case at hand, $F(x, y) = T_1 - T_2$. To solve the problem in (9.5), we use the method of separation of variables. It will be useful to compare what we do here with the method used in Section 13.2.

Step 1. Separate the PDE into an ODE in t and a PDE in (x, y). When we insert the product $v = T(t)\phi(x, y)$ into the heat equation $v_t = k\nabla^2 v$, we obtain $T'(t)\phi(x, y) = kT(t)\nabla^2\phi(x, y)$. Separating the variable t from the pair of variables x and y, we get the two differential equations

$$T' + \lambda kT = 0 \quad \text{and} \quad -\nabla^2\phi = \lambda\phi, \tag{9.6}$$

where λ is a constant. Notice the similarity with equation (2.8). The first equation has the solution

$$T(t) = Ce^{-\lambda kt}. \tag{9.7}$$

It is the second equation that requires our attention. This time it is a partial differential equation.

Step 2. Solve the eigenvalue problem for the Laplacian. We will insist that the product solution $v(t, x, y) = T(t)\phi(x, y)$ satisfy the homogeneous boundary condition coming from (9.5). Since this condition affects only the factor ϕ, the problem to be solved is finding λ and ϕ such that

$$-\nabla^2\phi = \lambda\phi, \quad \text{with } \phi(x, y) = 0 \text{ for } (x, y) \in \partial D. \tag{9.8}$$

This is called an ***eigenvalue problem*** for the Laplacian. Using the same terminology we used for the Sturm-Liouville problem in dimension $n = 1$, the number λ is called an ***eigenvalue***, and the function ϕ is an ***associated eigenfunction***. In our case we have the Dirichlet boundary condition $\phi(x, y) = 0$ for $(x, y) \in \partial D$, but we could have Neumann or Robin conditions.

We look for product solutions $\phi(x, y) = X(x)Y(y)$. The differential equation becomes $-X''(x)Y(y) - X(x)Y''(y) = \lambda X(x)Y(y)$. When we separate variables, we see that there must be constants μ and ν such that

$$-X'' = \mu X \quad \text{and} \quad -Y'' = \nu Y, \quad \text{with} \quad \mu + \nu = \lambda. \tag{9.9}$$

Next, look at the boundary condition. For example, we require that $\phi(0, y) = X(0)Y(y) = 0$ for $0 < y < b$. This means that we must have $X(0) = 0$. In the same way, we see that $X(a) = 0$, and $Y(0) = Y(b) = 0$. Together with the differential equations in (9.9), we see that we have Sturm-Liouville problems for both X and Y. It is essentially the same problem for both, and it is the problem we solved in Section 13.2, ending with equation (2.12). According to (2.12), we have solutions

$$\mu_i = \frac{i^2\pi^2}{a^2} \quad \text{and} \quad X_i(x) = \sin\frac{i\pi x}{a}$$

$$\nu_j = \frac{j^2\pi^2}{b^2} \quad \text{and} \quad Y_j(y) = \sin\frac{j\pi y}{b},$$

for $i, j = 1, 2, 3, \ldots$. To sum up, the eigenvalue problem for the rectangle D has solution

$$\lambda_{i,j} = \frac{i^2\pi^2}{a^2} + \frac{j^2\pi^2}{b^2} \quad \text{and} \quad \phi_{i,j}(x, y) = \sin\frac{i\pi x}{a} \cdot \sin\frac{j\pi y}{b}, \tag{9.10}$$

for $i, j = 1, 2, 3, \ldots$.

Step 3. Solving the initial/boundary value problem. The finish of the process is very much like it was in dimension $n = 1$. Notice that for each pair of positive integers i and j we have the solution $T_{i,j}(t) = e^{-\lambda_{i,j}kt}$ from (9.7). The product $T_{i,j}(t)\phi_{i,j}(x, y) = e^{-\lambda_{i,j}kt}\phi_{i,j}(x, y)$ is a solution to the heat equation and the homogeneous boundary conditions. Using the linearity of the heat equation, and assuming that there are no convergence problems, we see that any function of the form

$$v(t, x, y) = \sum_{i=1}^{\infty}\sum_{j=1}^{\infty} c_{i,j}e^{-\lambda_{i,j}kt}\phi_{i,j}(x, y) = \sum_{i=1}^{\infty}\sum_{j=1}^{\infty} c_{i,j}e^{-\lambda_{i,j}kt} \sin\frac{i\pi x}{a} \cdot \sin\frac{j\pi y}{b}$$

will be a solution to the heat equation and will also satisfy the boundary conditions. In order that the initial condition be solved, we set $t = 0$ to get

$$v(0, x, y) = F(x, y) = \sum_{i=1}^{\infty}\sum_{j=1}^{\infty} c_{i,j}\phi_{i,j}(x, y)$$

$$= \sum_{i=1}^{\infty}\sum_{j=1}^{\infty} c_{i,j} \sin\frac{i\pi x}{a} \cdot \sin\frac{j\pi y}{b}, \tag{9.11}$$

for $0 \le x \le a$ and $0 \le y \le b$. Equation (9.11) is a Fourier series in both x and y simultaneously. Using the orthogonality relations for Fourier series, we see that

$$\int_{D} \phi_{i,j}\,\phi_{i',j'}\,dx\,dy = \int_0^a\int_0^b \sin\frac{i\pi x}{a}\sin\frac{j\pi y}{b}\sin\frac{i'\pi x}{a}\sin\frac{j'\pi y}{b}\,dx\,dy$$

$$= \int_0^a \sin\frac{i\pi x}{a}\sin\frac{i'\pi x}{a}\,dx \int_0^b \sin\frac{j\pi y}{b}\sin\frac{j'\pi y}{b}\,dy \tag{9.12}$$

$$= \begin{cases} \dfrac{ab}{4}, & \text{if } i = i' \text{ and } j = j', \\ 0, & \text{otherwise.} \end{cases}$$

Therefore, if we multiply the series in (9.11) by $\phi_{i',j'}$ and integrate over the rectangle D, we get

$$\int_{D} F\,\phi_{i',j'}\,dx\,dy = \sum_{i=1}^{\infty}\sum_{j=1}^{\infty} c_{i,j} \int_{D} \phi_{i,j}\,\phi_{i',j'}\,dx\,dy = \frac{ab}{4}c_{i',j'}.$$

Consequently, the coefficients are given by

$$c_{i,j} = \frac{4}{ab}\int_{D} F(x, y)\,\phi_{i,j}(x, y)\,dx\,dy. \tag{9.13}$$

In our case, we have $F(x, y) = T_1 - T_2$, so

$$c_{i,j} = \frac{4}{ab}(T_1 - T_2)\int_0^a \sin\frac{i\pi x}{a}\,dx \int_0^b \sin\frac{j\pi y}{b}\,dy$$

$$= (T_1 - T_2)\frac{4}{ij\pi^2}[1 - \cos i\pi][1 - \cos j\pi].$$

Hence, $c_{i,j} = 0$ unless both i and j are odd, and

$$c_{2i+1,2j+1} = \frac{16(T_1 - T_2)}{(2i + 1)(2j + 1)\pi^2}.$$

Thus, the solution to the homogeneous initial/boundary value problem is

$$v(t, x, y) = \sum_{i=0}^{\infty} \sum_{j=0}^{\infty} \frac{16(T_1 - T_2)}{\pi^2(2i + 1)(2j + 1)} e^{-\lambda_{2i+1,2j+1}kt} \phi_{2i+1,2j+1}(x, y).$$

The solution to the original problem is

$$u(t, x, y) = u_s(x, y) + v(t, x, y)$$

$$= T_2 + \sum_{i=0}^{\infty} \sum_{j=0}^{\infty} \frac{16(T_1 - T_2)}{\pi^2(2i + 1)(2j + 1)} e^{-\lambda_{2i+1,2j+1}kt} \phi_{2i+1,2j+1}(x, y).$$

Vibrations of a rectangular drum

Without much more work we can analyze the modes of vibration of a rectangular drum. The displacement $u(t, x, y)$ of the drum is governed by the wave equation $u_{tt} = c^2\nabla^2 u$. The edge of the drum is fixed, so it satisfies the homogeneous boundary condition $u(t, x, y) = 0$ for $(x, y) \in \partial D$. The drum has an initial displacement $f_0(x, y)$ and initial velocity $f_1(x, y)$. Hence the displacement of the drum satisfies the initial/boundary value problem

$$\begin{aligned}
u_{tt}(t, x, y) &= c^2\nabla^2 u(t, x, y), \quad \text{for } (x, y) \in D \text{ and } t > 0, \\
u(t, x, y) &= 0, \quad \text{for } (x, y) \in \partial D \text{ and } t > 0, \\
u(0, x, y) &= f_0(x, y), \quad \text{for } (x, y) \in D, \\
u_t(0, x, y) &= f_1(x, y), \quad \text{for } (x, y) \in D.
\end{aligned}$$
(9.14)

We look for product solutions to the wave equation, so we set $u(t, x, y) = T(t)\phi(x, y)$ and substitute into the wave equation, getting

$$T''(t)\phi(x, y) = c^2T(t)\nabla^2\phi(x, y).$$

Arguing as we have before, we get the two differential equations

$$T'' + \lambda c^2 T = 0 \quad \text{and} \quad -\nabla^2\phi = \lambda\phi,$$
(9.15)

where λ is a constant. Since are looking for solutions to the second equation that vanish on the boundary of D, we have once more the eigenvalue problem in (9.8), and the solutions are those in (9.10).

It remains to solve the first equation in (9.15) with $\lambda = \lambda_{i,j}$. If we set

$$\omega_{i,j} = c\sqrt{\lambda_{i,j}} = c\pi\sqrt{\frac{i^2}{a^2} + \frac{j^2}{b^2}},$$
(9.16)

the equation becomes $T'' + \omega_{i,j}^2 T = 0$. This equation has the fundamental set of solutions $\sin\omega_{i,j}t$ and $\cos\omega_{i,j}t$. Hence for the eigenvalue $\lambda_{i,j}$ we have two linearly independent product solutions

$$\sin(\omega_{i,j}t) \cdot \phi_{i,j}(x, y) \quad \text{and} \quad \cos(\omega_{i,j}t) \cdot \phi_{i,j}(x, y),$$
(9.17)

where $\phi_{i,j}(x, y)$ is the solution found in (9.10). Every solution to the initial/boundary value problem in (9.14) is an infinite series in these product solutions. Hence if u is a solution we have

$$u(t, x, y) = \sum_{i=1}^{\infty} \sum_{j=1}^{\infty} [a_{i,j}\cos(\omega_{i,j}t) + b_{i,j}\sin(\omega_{i,j}t)]\phi_{i,j}(x, y).$$

Evaluating u and u_t at $t = 0$, we see that

$$f_0(x, y) = u(0, x, y) = \sum_{i=1}^{\infty} \sum_{j=1}^{\infty} a_{i,j} \phi_{i,j}(x, y), \quad \text{and}$$

$$f_1(x, y) = u_t(0, x, y) = \sum_{i=1}^{\infty} \sum_{j=1}^{\infty} \omega_{i,j} b_{i,j} \phi_{i,j}(x, y). \tag{9.18}$$

These are double Fourier series like that in (9.11), so the coefficients can be evaluated using (9.13). We get

$$a_{i,j} = \frac{4}{ab} \int_D f_0(x, y) \phi_{i,j}(x, y) \, dx \, dy \quad \text{and}$$

$$b_{i,j} = \frac{4}{ab\omega_{i,j}} \int_D f_1(x, y) \phi_{i,j}(x, y) \, dx \, dy.$$

Notice that the product solutions in (9.17) are periodic in time with frequency $\omega_{i,j}$ given in (9.16). Unlike the case of the vibrating string, these frequencies are not integer multiples of the lowest frequency $\omega_{1,1}$. Consequently the vibrations of a rectangular drum will not have the fine musical qualities of a violin.

EXERCISES

In Exercises 1– 6, we will further explore heat transfer with in a square plate D of side length 1. Suppose first that the plate has three sides which are kept at $0°$, while the fourth side is insulated. Then the boundary conditions for the temperature can be written as

$$u(t, x, 0) = u(t, x, 1) = 0, \quad \text{and}$$
$$u(t, 0, y) = u_x(t, 1, y) = 0. \tag{9.19}$$

Notice that the steady-state temperature in the plate is $0°$. The temperature in D satisfies the heat equation $u_t = k\nabla^2 u$.

1. Suppose that $u(t, x, y) = T(t)\phi(x, y)$ is a product solution of the heat equation, together with the boundary conditions in (9.19). Show that there is a constant λ such that

   (a) T satisfies the equation $T' + \lambda T = 0$.

   (b) ϕ satisfies $-\nabla^2\phi = \lambda\phi$, together with the boundary conditions

   $$\phi(x, 0) = \phi(x, 1) = 0, \quad \text{and}$$
   $$\phi(0, y) = \phi_x(1, y) = 0. \tag{9.20}$$

2. Solve the eigenvalue problem for the Laplacian in part (b) of Exercise 1, and show that the solutions are

   $$\lambda_{p,q} = \frac{(2p + 1)^2 \pi^2}{4} + q^2 \pi^2 \quad \text{with}$$

   $$\phi_{p,q}(x, y) = \sin\left(\frac{(2p + 1)\pi x}{2}\right) \sin q\pi y,$$

   for $p \geq 0$ and $q \geq 1$.

3. Show that

   $$\int_D \phi_{p,q}(x, y) \phi_{p',q'}(x, y) \, dx \, dy$$

   $$= \begin{cases} 1/4, & \text{if } p = p' \text{ and } q = q' \\ 0, & \text{otherwise.} \end{cases}$$

4. Suppose that the initial temperature is $u(0, x, y) = f(x, y)$. Show that the temperature is given by

   $$u(t, x, y) = \sum_{p=0}^{\infty} \sum_{q=1}^{\infty} c_{p,q} e^{\lambda_{p,q} kt} \phi_{p,q}(x, y),$$

   where the coefficients are given by

   $$c_{p,q} = 4 \int_D f(x, y)\phi_{p,q}(x, y) \, dx \, dy.$$

5. How must the solution in Exercise 4 be modified if the boundary conditions are changed to

   $$u(t, x, 0) = u(t, x, 1) = T_1,$$

   and

   $$u(t, 0, y) = u_x(t, 1, y) = 0?$$

6. How must the solution in Exercise 4 be modified if the plate is insulated on two opposite edges, so that the boundary conditions are changed to

   $$u(t, x, 0) = u(t, x, 1) = 0,$$

   and

   $$u_x(t, 0, y) = u_x(t, 1, y) = 0?$$

7. Suppose we have a square drum with side length π, and suppose that it is plucked in the middle and then released. Then its initial displacement is given by $f(x, y) = \min\{x, y, \pi - x, \pi - y\}$, while its initial velocity is 0. The graph of f is a four-sided pyramid with height $\pi/2$. Use the techniques of this section to compute the displacement as a function of both time and space. This seemingly daunting task is made easier if you follow these steps.

(a) Show that

$$f(x, y) = [F(x - y) - F(x + y)]/2,$$

where $F(z)$ is the periodic extension of $\pi - |z|$ from the interval $[-\pi, \pi]$ to the reals. (*Hint:* Just check the cases.)

(b) Compute the Fourier series for F on the interval $[-\pi, \pi]$.

(c) Use the formula in 7(a) and the addition formula for the cosine to complete the computation of the double

Fourier series for f. You will notice that the series has the form $f(x, y) = \sum_{p=1}^{\infty} a_p \sin px \sin py$. Comparing this with the series that appears in (9.18), we see that the coefficients of all of the off-diagonal terms are equal to 0.

(d) Find the displacement $u(t, x, y)$ in the way described in this section. Is the vibration of the drum with these initial conditions periodic in time?

8. Using the terminology in Exercise 7, show that the function

$$u(t, x, y) = \frac{1}{4}[F(x - y + \sqrt{2}ct)$$
$$+ F(x - y - \sqrt{2}ct)$$
$$- F(x + y + \sqrt{2}ct)$$
$$- F(x + y - \sqrt{2}ct)]$$

is a solution to the wave equation and also satisfies the initial and boundary conditions in Exercise 7.

13.10 Domains with Circular Symmetry—Bessel Functions

In this section we will analyze the vibrations of a circular drum. Let D be the disk of radius a, which we describe as

$$D = \{(x, y) \in \mathbf{R}^2 \,|\, x^2 + y^2 < a^2\}.$$

The displacement $u(t, x, y)$ of the circular drum on D satisfies the initial/boundary value problem in (9.14). Separation of variables leads us once again to the two differential equations in (9.15). Hence we are led to an eigenvalue problem for the disk D, which is to find all numbers λ and functions ϕ such that

$$-\nabla^2 \phi(x, y) = \lambda \phi(x, y), \quad \text{for } (x, y) \in D, \text{ and}$$
$$\phi(x, y) = 0, \quad \text{for } (x, y) \in \partial D. \tag{10.1}$$

As we did in Section 13.5, we will use polar coordinates (see equation (5.2)) to solve the problem in (10.1). In these coordinates the eigenvalue problem in (10.1) becomes

$$-\left[\phi_{rr} + \frac{1}{r}\phi_r + \frac{1}{r^2}u_{\theta\theta}\right](r, \theta) = \lambda \phi(r, \theta), \quad \text{for } r < a,$$
$$\phi(a, \theta) = 0, \quad \text{for } 0 \le \theta \le 2\pi. \tag{10.2}$$

When we substitute a product function of the form $\phi(r, \theta) = R(r)U(\theta)$ into the differential equation in (10.2), we get

$$R_{rr}(r)U(\theta) + \frac{1}{r}R_r(r)U(\theta) + \frac{1}{r^2}R(r)U_{\theta\theta}(\theta) + \lambda R(r)U(\theta) = 0.$$

To separate variables, we multiply by r^2/RU, obtaining

$$\frac{r^2 R_{rr} + r R_r + \lambda r^2 R}{R} + \frac{U_{\theta\theta}}{U} = 0.$$

This sum of a function of r and a function of θ can be equal to 0 only if each is constant. Hence there is a constant μ such that

$$r^2 R_{rr} + r R_r + \lambda r^2 R - \mu R = 0 \quad \text{and} \quad U_{\theta\theta} + \mu U = 0. \tag{10.3}$$

We will solve the second equation in (10.3) first. Remember that θ represents the polar angle in the disk, so the solution U must be periodic with period 2π in θ. We examined the resulting Sturm-Liouville problem in Example 6.17 in Section 13.6, and found that we must have $\mu = n^2$, where n is a nonnegative integer, and that the eigenfunctions are

$$
\begin{aligned}
&1, \quad \text{for } n = 0 \\
&\sin n\theta, \quad \text{and} \quad \cos n\theta, \quad \text{for } n \geq 1.
\end{aligned}
\tag{10.4}
$$

Bessel functions

Substituting $\mu = n^2$ into the first equation in (10.3) and then rearranging it, we get the equation

$$
r^2 \frac{d^2 R}{dr^2} + r \frac{dR}{dr} - n^2 R = -\lambda r^2 R.
\tag{10.5}
$$

After dividing by r and multiplying by -1, it becomes

$$
-(rR')' + \frac{n^2}{r} R = \lambda r R.
\tag{10.6}
$$

The operator L defined by $LR = -(rR')' + n^2 R/r$ that appears in (10.6) is formally self-adjoint. However, the coefficient $p(r) = r$ vanishes at $r = 0$, and the coefficient $q(r) = n^2/r$ has an infinite discontinuity there. Hence, the operator is singular at $r = 0$.

As we did for the Legendre equation in Section 13.8, we will require that the eigenfunctions are continuous at $r = 0$ together with their first derivatives. At the other boundary point $r = a$, the boundary condition comes from (10.1). Our Sturm-Liouville problem is

$$
\begin{aligned}
&-(rR')' + \frac{n^2}{r} R = \lambda r R, \quad \text{for } 0 < r < a, \\
&R \text{ and } R' \text{ are continuous at } r = 0, \\
&\qquad\qquad R(a) = 0.
\end{aligned}
\tag{10.7}
$$

Notice that the weight function is $w(r) = r$.

Once again we are fortunate. Even though the Sturm-Liouville problem is singular, it has all of the properties of nonsingular problems that we described in Sections 13.6 and 13.7. In particular, Proposition 6.24 remains true, and we see that all of the eigenvalues are positive. Hence, we can write $\lambda = \nu^2$, where $\nu > 0$. If we make the change of variables $s = \nu r$ in the differential equation in (10.5) and rearrange it, it becomes

$$
s^2 \frac{d^2 R}{ds^2} + s \frac{dR}{ds} + [s^2 - n^2] R = 0.
$$

(See Exercise 9 in Section 11.7.) This is Bessel's equation of order n. In Section 11.7 we discovered that a fundamental set of solutions is the pair $J_n(s)$ and $Y_n(s)$. Therefore, the general solution to the differential equation in (10.7) is $R(r) = A J_n(\nu r) + B Y_n(\nu r)$. However, since $Y_n(\nu r)$ has an infinite singularity at $r = 0$, it does not satisfy the boundary condition at $r = 0$ in (10.7). Therefore, $B = 0$. Taking $A = 1$, we have $R(r) = J_n(\nu r)$.

It remains to satisfy the boundary condition $J_n(\nu a) = 0$. We discussed the zeros of the Bessel functions in Section 11.7. There are infinitely many of them. If $\alpha_{n,k}$ is the kth zero of J_n, then we need $\nu = \nu_{n,k} = \alpha_{n,k}/a$. Consequently, the solutions to the Sturm-Liouville problem in (10.7) are

$$\lambda_k = \frac{\alpha_{n,k}^2}{a^2} \quad \text{and} \quad R_k(r) = J_n(\alpha_{n,k}r/a), \quad \text{for } k = 1, 2, \ldots . \tag{10.8}$$

From Proposition 7.10, we see that the functions R_k are orthogonal with respect to the weight $w(r) = r$. This means that

$$(R_k, R_j)_r = \int_0^a J_n(\alpha_{n,k}r/a) J_n(\alpha_{n,j}r/a)r\, dr = 0, \quad \text{if } j \neq k.$$

A rather difficult computation shows that

$$(R_k, R_k)_r = \int_0^a J_n(\alpha_{n,k}r/a)^2\, r dr = \frac{a^2}{2} J_{n+1}^2(\alpha_{n,k}).$$

If f is a piecewise continuous function on $[0, a]$, then its associated **Fourier-Bessel series** is

$$f(r) \sim \sum_{k=1}^{\infty} c_k J_n(\alpha_{n,k}r/a), \tag{10.9}$$

where the coefficients are given by

$$c_k = \frac{(f, R_k)_r}{(R_k, R_k)_r} = \frac{2}{a^2 J_{n+1}^2(\alpha_{n,k})} \int_0^a f(r) J_n(\alpha_{n,k}r/a)\, r\, dr. \tag{10.10}$$

The integral in (10.10) is difficult to compute, even for the simplest functions f. Not infrequently it is necessary to compute the integral approximately for small values of k.

Solution to the eigenvalue problem on the disk

Bringing together the results in (10.4) and (10.8), we see that the solutions to the eigenvalue problem in (10.1) for the Laplacian on the disk are

$$\lambda_{0,k} = \frac{\alpha_{0,k}^2}{a^2} \quad \text{with} \quad \phi_{0,k}(r, \theta) = J_0(\alpha_{0,k}r/a),$$
$$\text{for } n = 0 \text{ and } k = 1, 2, 3, \ldots$$

$$\lambda_{n,k} = \frac{\alpha_{n,k}^2}{a^2} \quad \text{with} \quad \begin{cases} \phi_{n,k}(r, \theta) = \cos n\theta \cdot J_n(\alpha_{n,k}r/a) \quad \text{and} \\ \psi_{n,k}(r, \theta) = \sin n\theta \cdot J_n(\alpha_{n,k}r/a), \end{cases} \tag{10.11}$$
$$\text{for } n = 1, 2, 3, \ldots \text{ and } k = 1, 2, 3, \ldots .$$

By integrating using polar coordinates, and using the orthogonality relations for the Bessel functions and the trigonometric functions, we see that the eigenfunctions $\phi_{n,k}$ and $\psi_{n,k}$ satisfy the orthogonality relations

$$\int_D \phi_{n,k}\phi_{n',k'}\, dx\, dy = \begin{cases} \pi a^2 J_{n+1}^2(\alpha_{n,k})/2, & \text{if } n' = n \text{ and } k' = k. \\ 0, & \text{otherwise,} \end{cases}$$

$$\int_D \psi_{n,k}\psi_{n',k'}\, dx\, dy = \begin{cases} \pi a^2 J_{n+1}^2(\alpha_{n,k})/2, & \text{if } n' = n \text{ and } k' = k. \\ 0, & \text{otherwise,} \end{cases} \tag{10.12}$$

$$\int_D \phi_{n,k}\psi_{n',k'}\, dx\, dy = 0, \quad \text{in all cases.}$$

The solution of the wave equation

For the time dependence of the product solution to the wave equation, we must solve the first equation in (9.15). With $\omega_{n,k}^2 = c^2 \lambda_{n,k} = (c\alpha_{n,k}/a)^2$, this equation becomes $T'' + \omega_{n,k}^2 T = 0$. The solutions are $\cos(\omega_{n,k}t)$ and $\sin(\omega_{n,k}t)$. Thus the product solutions of the wave equation are of the form

$$
\begin{aligned}
\cos(\omega_{n,k}t) \cdot \phi_{n,k}(r,\theta), & \quad \cos(\omega_{n,k}t) \cdot \psi_{n,k}(r,\theta), \\
\sin(\omega_{n,k}t) \cdot \phi_{n,k}(r,\theta), & \quad \text{and} \quad \sin(\omega_{n,k}t) \cdot \psi_{n,k}(r,\theta),
\end{aligned}
\tag{10.13}
$$

for all appropriate choices of the indices. By linearity, any function of the form

$$
\begin{aligned}
u(t,r,\theta) = & \sum_{k=1}^{\infty} J_0\left(\frac{\alpha_{0,k}r}{a}\right)\left[A_{0,k}\cos\frac{c\alpha_{0,k}t}{a} + B_{0,k}\sin\frac{c\alpha_{0,k}t}{a}\right] \\
& + \sum_{n=1}^{\infty}\sum_{k=1}^{\infty} J_n\left(\frac{\alpha_{n,k}r}{a}\right)\left[A_{n,k}\cos n\theta + B_{n,k}\sin n\theta\right]\cos\frac{c\alpha_{n,k}t}{a} \\
& + \sum_{n=1}^{\infty}\sum_{k=1}^{\infty} J_n\left(\frac{\alpha_{n,k}r}{a}\right)\left[C_{n,k}\cos n\theta + D_{n,k}\sin n\theta\right]\sin\frac{c\alpha_{n,k}t}{a}
\end{aligned}
\tag{10.14}
$$

is formally a solution to the wave equation on the disk that satisfies Dirichlet boundary conditions.

The coefficients are evaluated using the initial conditions $u(0,r,\theta) = f_0(r,\theta)$ and $u_t(0,r,\theta) = f_1(r,\theta)$. Evaluating (10.14) at $t=0$, we see that

$$
\begin{aligned}
f_0(r,\theta) = & \sum_{k=1}^{\infty} A_{0,k} J_0\left(\frac{\alpha_{0,k}r}{a}\right) \\
& + \sum_{n=1}^{\infty}\sum_{k=1}^{\infty} J_n\left(\frac{\alpha_{n,k}r}{a}\right)\left[A_{n,k}\cos n\theta + B_{n,k}\sin n\theta\right].
\end{aligned}
$$

The coefficients can be found in the usual way, using the orthogonality relations in (10.12). We get

$$
A_{n,k} = \frac{2}{\pi a^2 J_{n+1}(\alpha_{n,k})} \int_0^a \int_0^{2\pi} f_0(r,\theta) J_n\left(\frac{\alpha_{n,k}r}{a}\right)\cos n\theta \, r \, dr \, d\theta, \quad \text{and}
$$

$$
B_{n,k} = \frac{2}{\pi a^2 J_{n+1}^2(\alpha_{n,k})} \int_0^a \int_0^{2\pi} f_0(r,\theta) J_n\left(\frac{\alpha_{n,k}r}{a}\right)\sin n\theta \, r \, dr \, d\theta.
\tag{10.15}
$$

The remaining coefficients in (10.14) can be evaluated in the same way using the initial velocity. They are

$$
C_{n,k} = \frac{2}{c\pi a\alpha_{n,k} J_{n+1}^2(\alpha_{n,k})} \int_0^a \int_0^{2\pi} f_1(r,\theta) J_n\left(\frac{\alpha_{n,k}r}{a}\right)\cos n\theta \, r \, dr \, d\theta, \quad \text{and}
$$

$$
D_{n,k} = \frac{2}{c\pi a\alpha_{n,k} J_{n+1}^2(\alpha_{n,k})} \int_0^a \int_0^{2\pi} f_1(r,\theta) J_n\left(\frac{\alpha_{n,k}r}{a}\right)\sin n\theta \, r \, dr \, d\theta.
\tag{10.16}
$$

The fundamental modes of vibration of a drum

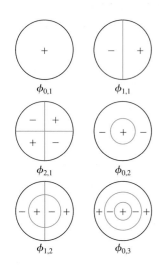

$\phi_{0,1}$ $\phi_{1,1}$

$\phi_{2,1}$ $\phi_{0,2}$

$\phi_{1,2}$ $\phi_{0,3}$

Figure 1. The nodal sets for some fundamental modes.

Notice that the product solutions in (10.12) represent vibrations of the drum with frequency $\omega_{n,k} = c\alpha_{n,k}/a$, and with an amplitude that varies over the drum like the functions $\phi_{n,k}$ and $\psi_{n,k}$. For this reason, the functions $\phi_{n,k}$ and $\psi_{n,k}$ are referred to as the *fundamental modes* of vibration for the drum.

The frequencies are proportional to the zeros of the Bessel's functions. According to Table 1 in Section 11.7, the four smallest zeros are $\alpha_{0,1} = 2.4048$, $\alpha_{1,1} = 3.8317$, $\alpha_{2,1} = 5.1356$, and $\alpha_{0,2} = 5.5201$. We see that the frequencies are clearly not integer multiples of a lowest, fundamental frequency, which is the case for the vibrating string. This explains the quite different sounds of a kettle drum and a violin.

The *nodal set* of a fundamental mode is the set where it vanishes. During a vibration in a fundamental mode, the points in the nodal set do not move. Since $\phi_{0,1}(r, \theta) = J_0(\alpha_{0,1}r/a)$ is not equal to 0 for $r < a$, its nodal set is empty. Similarly, $\phi_{1,1} = \cos\theta\, J_1(\alpha_{1,1}r/a) = 0$ only where $\cos\theta = 0$, so its nodal set is the y-axis. The nodal sets for several fundamental modes are shown in Figure 1. The $+$ and $-$ signs indicate regions where the drum head has opposite displacement during the oscillation. As n and k get large, the motion of the drum in a fundamental mode can get quite complicated.

If you strike a kettle drum in the middle, seemingly a natural place to do so, you will get a mixture of all of the frequencies as shown in (10.14). The result is a sound that is really awful. Naturally, professional tympanists avoid this. Instead, they carefully strike the drum near the edge. The result is that the lowest frequency is eliminated from the mixture. In fact, a professional tympanist gets a sound that is almost a pure $\phi_{1,1}$ mode.

EXERCISES

1. Verify the orthogonality relations in (10.12).

2. Verify the formulas in (10.15) and (10.16).

3. Plot the nodal sets for the fundamental modes $\phi_{1,3}$, $\phi_{3,2}$, and $\phi_{2,4}$.

4. Suppose that the initial displacement of the drum is a function $u(0, r, \theta) = f(r)$, that is independent of the angle θ, and the initial velocity is 0. How does the series for the solution in (10.14) simplify?

5. Suppose that the initial temperature in a disk D of radius a is $u(0, r, \theta) = f(r)$, where $f(r)$ is a function of the radius r only. It is safe to assume that $u = u(t, r)$ is also independent of the angle θ.

   (a) Show that $\nabla^2 u = u_{rr} + u_r/r$.

   (b) Assuming that the temperature vanishes on the boundary of the disk, the initial/boundary value problem in polar coordinates is

   $$u_t = k\left[u_{rr} + \frac{1}{r}u_r\right], \quad \text{for } 0 \le r < a \text{ and } t > 0,$$

   $$u(t, a) = 0, \quad \text{for } t > 0,$$

   $$u(0, r) = f(r), \quad \text{for } 0 \le r < a.$$

   Find the product solutions that satisfy the Dirichlet boundary condition.

(c) Find a series expansion for the temperature $u(t, r, \theta)$.

6. Find a series expansion for the solution to the initial/boundary value problem

$$u_t(t, x, y) = k\nabla^2 u(t, x, y), \quad \text{for } (x, y) \in D \text{ and } t > 0,$$

$$u(t, x, y) = 0, \quad \text{for } (x, y) \in \partial D \text{ and } t > 0,$$

$$u(t, x, y) = f(x, y), \quad \text{for } (x, y) \in D,$$

where D is the disk of radius a.

7. Consider the initial/boundary value problem

$$u_t(t, x, y) = k\nabla^2 u(t, x, y), \quad \text{for } (x, y) \in D \text{ and } t > 0,$$

$$\frac{\partial u}{\partial \mathbf{n}}(t, x, y) = 0, \quad \text{for } (x, y) \in \partial D \text{ and } t > 0,$$

$$u(t, x, y) = f(x, y), \quad \text{for } (x, y) \in D,$$

where D is the disk of radius 1. The problem models the temperature in a circular plate when the boundary is insulated.

   (a) What is the eigenvalue problem for the disk that arises when you solve this problem by separation of variables?

   (b) Restate the eigenvalue problem in part (a) in polar coordinates.

   (c) What are the eigenvalues?

8. Consider the cylinder described in cylindrical coordinates by

$$C = \{(r, \theta, z) \mid 0 \leq r < a, 0 \leq \theta \leq 2\pi, \text{ and } 0 < z < L\}.$$

(See Section 13.5 for a discussion of the Laplacian in cylindrical coordinates.)

(a) Suppose $u(r, \theta, z) = \phi(r, \theta)Z(z)$ is a product solution of Laplace's equation in C. Show that there is a constant λ such that

$$-\nabla^2 \phi = \lambda \phi, \quad \text{and} \quad Z'' = \lambda Z.$$

(b) Find the product solutions to Laplace's equation that vanish on the curved portion of the boundary of the cylinder, so that $\phi(a, \theta) = 0$ for $0 \leq \theta \leq 2\pi$.

(c) Find a series solution to the boundary value problem

$$\nabla^2 u(r, \theta, z) = 0, \text{ for } (r, \theta, z) \in C,$$
$$u(r, \theta, 0) = f(r, \theta), \text{ for } 0 \leq r < a \text{ and } 0 \leq \theta \leq 2\pi,$$
$$u(r, \theta, L) = 0, \text{ for } 0 \leq r < a \text{ and } 0 \leq \theta \leq 2\pi,$$
$$u(a, \theta, z) = 0, \text{ for } 0 \leq \theta \leq 2\pi \text{ and } 0 < z < L,$$

for a steady-state temperature in C.

9. Find the solutions to the eigenvalue problem for the Laplacian on the cylinder C. This means we want to find numbers λ and functions ϕ such that

$$-\nabla^2 \phi = \lambda \phi \quad \text{in } C, \text{ and} \quad \phi = 0 \quad \text{on the boundary of } C.$$

(a) First suppose that $\phi(r, \theta, z) = A(r, \theta)B(z)$ and show that there are numbers μ and ν such that $\mu + \nu = \lambda$ for which

$$-\nabla^2 A = \mu A \quad \text{for } (r, \theta) \in D$$
$$A(a, \theta) = 0 \quad \text{for } (r, \theta) \in \partial D,$$

and

$$-B'' = \nu B \quad \text{for } 0 < z < L$$
$$B(0) = B(L) = 0.$$

(b) Solve the two eigenvalue problems in part (a) to find the eigenvalues and eigenfunctions for the Laplacian on C.

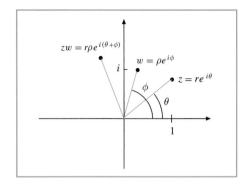

Complex Numbers and Matrices

The use of complex numbers is an essential tool in the study of differential equations. In this appendix we will gather the essential facts about the algebra and geometry of complex numbers. We will also talk briefly about matrices with complex entries.

Complex numbers

A complex number has the form $z = x + iy$, where x and y are real numbers. The number i satisfies $i^2 = -1$. In the complex number $z = x + iy$, the real number x is called the **real part** of z and is denoted by $x = \operatorname{Re} z$. The real number y is called the **imaginary part** of z and is denoted by $y = \operatorname{Im} z$.

Complex addition and multiplication satisfy the usual rules for real numbers. For example,

$$(3 + 5i) + (2 - 3i) = (3 + 2) + (5 - 3)i = 5 + 2i.$$

In multiplication it is necessary to use $i^2 = -1$. For example,

$$
\begin{aligned}
(3 + 5i) \cdot (4 - 2i) &= 3(4 - 2i) + 5i(4 - 2i) \\
&= 12 - 6i + 20i - 10i^2 \\
&= 12 + 14i + 10 \\
&= 22 + 14i.
\end{aligned}
$$

The **complex conjugate** of the complex number $z = x + iy$ is the number $\bar{z} = x - iy$. Notice that conjugation affects only the imaginary part of the complex number, replacing the imaginary part with its negative. In particular, we see that

$$\bar{z} = z \quad \text{if and only if } z \text{ is a real number.}$$

We can solve the two equations $z = x + iy$ and $\bar{z} = x - iy$ for x and y, obtaining

$$x = \operatorname{Re} z = \frac{z + \bar{z}}{2} \quad \text{and} \quad y = \operatorname{Im} z = \frac{z - \bar{z}}{2i}. \tag{A.1}$$

The process of conjugation preserves algebraic combinations. Suppose that $z = x + iy$ and $w = u + iv$ are complex numbers. Then the following facts are proved directly from the definition of the conjugate.

$$\begin{aligned}
\overline{z + w} &= \bar{z} + \bar{w} \\
\overline{z - w} &= \bar{z} - \bar{w} \\
\overline{zw} &= \bar{z} \cdot \bar{w}
\end{aligned} \tag{A.2}$$

The ***absolute value*** of a complex number $z = x + iy$ is the real number

$$|z| = \sqrt{x^2 + y^2}.$$

The absolute value of a complex number has many uses. It is also called the magnitude of the number, and for many purposes that name is highly illuminating. Notice that

$$z\bar{z} = |z|^2. \tag{A.3}$$

Formula (A.3) provides the secret to computing the reciprocal of a complex number. We have

$$\frac{1}{z} = \frac{1}{z} \cdot \frac{\bar{z}}{\bar{z}} = \frac{\bar{z}}{z\bar{z}} = \frac{\bar{z}}{|z|^2}. \tag{A.4}$$

For example,

$$\frac{1}{4 - 3i} = \frac{4 + 3i}{|4 - 3i|^2} = \frac{4 + 3i}{25}.$$

Knowing how to compute reciprocals and products, we can also compute quotients. Thus, using (A.4), we have

$$\frac{w}{z} = w \cdot \frac{1}{z} = \frac{w\bar{z}}{|z|^2}. \tag{A.5}$$

Using (A.5), together with the third formula in (A.2), we see that

$$\overline{\left(\frac{z}{w}\right)} = \frac{\bar{z}}{\bar{w}}.$$

Other important properties of the absolute value follow most easily from (A.3). If z and w are two complex numbers, then

$$|zw| = |z||w| \quad \text{and} \quad \left|\frac{z}{w}\right| = \frac{|z|}{|w|}. \tag{A.6}$$

Most people feel more comfortable with complex numbers when they are represented as points in the complex plane. We will identify the complex number $z = x + iy$ with the point in the plane having cartesian coordinates (x, y). This identification is illustrated in Figure 1. Under this identification the x-axis contains the real numbers and is therefore called the ***real axis***. Similarly, the y-axis is called the ***imaginary axis***. Figure 1 also shows the complex conjugate $\bar{z} = x - iy$. This is the reflection of z in the real axis.

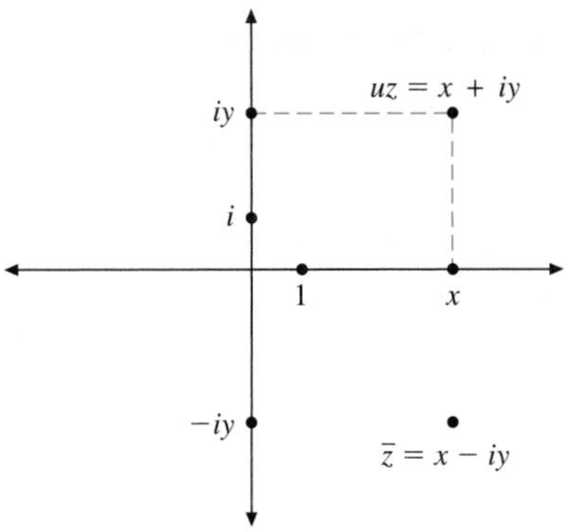

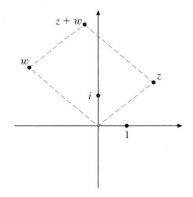

Figure 1. The complex number $z = x + iy$ and its conjugate $\bar{z} = x - iy$.

Figure 2. The sum of two complex numbers.

Figure 2 provides the geometric interpretation of complex addition. If z and w are complex numbers, then the sum $z + w$ corresponds to the fourth vertex of the parallelogram with the other three vertices at the origin 0, z, and w.

Polar coordinates have an especially appealing interpretation for complex numbers. (See Figure 3.) The complex number $z = x + iy$ has polar coordinates $r \geq 0$ and θ, with $-\pi < \theta \leq \pi$, defined by the standard equations

$$x = r \cos \theta \quad \text{and} \quad y = r \sin \theta.$$

Then

$$r = \sqrt{x^2 + y^2} = |z| \quad \text{and} \quad \tan \theta = \frac{y}{x}.$$

Hence we can identify r with $|z|$. The angle θ is called the **argument** of z. Using the polar coordinates, we can write

$$z = r \cos \theta + ir \sin \theta = r[\cos \theta + i \sin \theta]. \tag{A.7}$$

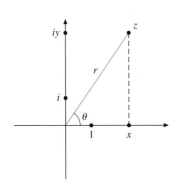

Figure 3. Polar coordinates for the complex number $z = x + iy$.

Equation (A.7) takes on a different character after we introduce the complex exponential. For a real number θ we define

$$e^{i\theta} = \cos \theta + i \sin \theta. \tag{A.8}$$

We will refer to this formula as **Euler's formula**. The definition is motivated by the Taylor series for the function

$$e^x = 1 + \frac{x}{1!} + \frac{x^2}{2!} + \frac{x^3}{3!} + \frac{x^4}{4!} + \cdots .$$

Inserting $x = i\theta$, and using $i^2 = -1$, $i^3 = -i$, $i^4 = 1$, and so forth, we obtain

$$e^{i\theta} = 1 + \frac{(i\theta)}{1!} + \frac{(i\theta)^2}{2!} + \frac{(i\theta)^3}{3!} + \frac{(i\theta)^4}{4!} + \cdots$$

$$= \left(1 - \frac{\theta^2}{2!} + \frac{\theta^4}{4!} + \cdots \right) + i \left(\theta - \frac{\theta^3}{3!} + \cdots \right).$$

The real part is the Taylor series for $\cos \theta$ and the imaginary part is the Taylor series for $\sin \theta$. Thus $e^{i\theta} = \cos \theta + i \sin \theta$.

As a result of Euler's formula, we can rewrite equation (A.7) as

$$z = r e^{i\theta}. \tag{A.9}$$

This very concise formula expresses the polar coordinates of a complex number. Its usefulness is enhanced after we learn about the full complex exponential and its properties. We define the exponential of a complex number by assuming that the addition formula for the exponential is still valid. The **exponential** of the complex number $z = x + iy$ is defined to be

$$e^z = e^{x+iy} = e^x e^{iy} = e^x (\cos y + i \sin y). \tag{A.10}$$

Thus, e^{x+iy} is the complex number with real part $e^x \cos y$ and imaginary part $e^x \sin y$.

The complex exponential satisfies all the familiar rules for real exponents. We set these rules down in the following proposition.

PROPOSITION A.11   The complex exponential satisfies the following properties:

1. $e^{z_1 + z_2} = e^{z_1} e^{z_2}$
2. $e^{z_1 - z_2} = e^{z_1} / e^{z_2}$
3. $(e^z)^r = e^{rz}$ for any real number r
4. $\overline{e^z} = e^{\bar{z}}$
5. $|e^z| = e^{\operatorname{Re} z}$
6. $\dfrac{d}{dt}\{e^{\lambda t}\} = \lambda e^{\lambda t}$ for any complex number λ

Proof   The straightforward verification of these formulas is left to the reader. We will only verify the last property in the special case when $\lambda = i$. The derivative of such a complex valued function is computed by differentiating the real and imaginary parts in the ordinary way. We have

$$\frac{d}{dt}\{e^{it}\} = \frac{d}{dt}\{\cos t + i \sin t\} \quad \text{by definition of } e^{it}$$
$$= \frac{d}{dt}\cos t + i\frac{d}{dt}\sin t$$
$$= -\sin t + i \cos t$$
$$= i(\cos t + i \sin t)$$
$$= i e^{it}.$$

Proposition A.11 allows us to make a geometric interpretation of the product of two complex numbers. Suppose that in polar coordinates we have $z = re^{i\theta}$ and $w = \rho e^{i\phi}$. Then by Part (1) of Proposition A.11, the product is

$$zw = re^{i\theta} \cdot \rho e^{i\phi} = r\rho e^{i(\theta + \phi)}.$$

We automatically get the polar coordinates of the product. We see that $|zw| = r\rho$ and that the argument of zw is $\theta + \phi$, the sum of the arguments of z and w. This means that the effect of multiplying z by $w = \rho e^{i\phi}$ is to multiply the absolute value of z by $\rho = |w|$, and to rotate z by ϕ, the argument of w. See Figure 4.

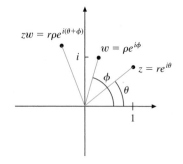

Figure 4. The product of two complex numbers.

Complex matrices

As a general statement, most operations on complex numbers transfer with little or no change to operations on complex vectors and matrices. An example of a complex matrix is

$$M = \begin{pmatrix} 4 + 5i & 3 - 2i & -3i \\ 2 + i & 4 & 0 \end{pmatrix}. \tag{A.12}$$

Any complex matrix can be split into its real and imaginary parts in exactly the same way as complex numbers are split. We have

$$M = A + iB, \quad \text{where} \quad A = \text{Re}(M) \quad \text{and} \quad B = \text{Im}(M). \tag{A.13}$$

For the example in (A.12), we have

$$A = \text{Re}(M) = \begin{pmatrix} 4 & 3 & 0 \\ 2 & 4 & 0 \end{pmatrix} \quad \text{and} \quad B = \text{Im}(M) = \begin{pmatrix} 5 & -2 & -3 \\ 1 & 0 & 0 \end{pmatrix}.$$

The complex conjugate of a complex matrix is computed component by component. Thus, for the matrix in (A.12), we have

$$\overline{M} = \begin{pmatrix} \overline{4+5i} & \overline{3-2i} & \overline{-3i} \\ \overline{2+i} & \overline{4} & \overline{0} \end{pmatrix} = \begin{pmatrix} 4-5i & 3+2i & 3i \\ 2-i & 4 & 0 \end{pmatrix}.$$

If $M = A + iB$, where $A = \text{Re}(M)$ and $B = \text{Im}(M)$, then $\overline{M} = A - iB$, just as though the matrices were complex numbers. From these considerations, we find that

$$M \text{ is a real matrix if and only if } \overline{M} = M, \tag{A.14}$$

and

$$\text{Re}(M) = \frac{1}{2}(M + \overline{M}); \quad \text{Im}(M) = \frac{1}{2i}(M - \overline{M}). \tag{A.15}$$

Finally, the operation of conjugation behaves well with the matrix operations of addition and multiplication. If M and N are complex matrices, then

$$\overline{M + N} = \overline{M} + \overline{N}. \tag{A.16}$$

If $\mathbf{z}$ is a complex vector, then

$$\overline{M\mathbf{z}} = \overline{M}\,\overline{\mathbf{z}}. \tag{A.17}$$

Answers to Odd-Numbered Problems

Chapter 1
Section 1.1

1. $y'(t) = ky(t)$ **3.** $y'(t) = ky(t)(100 - y(t))$

5. $y'(t) = -k/y(t)$ **7.** $y'(t) = k(77 - y(t))$

9. $-k(x'(t))|x'(t)| = mx''(t)$ **11.** $V(t) = kI'(t)$

Section 1.2

1. $f'(x) = 3$ **3.** $f'(x) = 15\cos 5x$

5. $f'(x) = 3e^{3x}$ **7.** $f'(x) = 1/x$

9. $f'(x) = 1 + \ln x$

11. $f'(x) = (2x\ln x - x)/(\ln x)^2$

15. $L(x) = \dfrac{\sqrt{2}}{2} - \dfrac{\sqrt{2}}{2}\left(x - \dfrac{\pi}{4}\right).$

17. $L(x) = x$

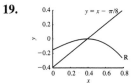

19.

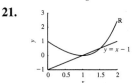

Note that both graphs approach zero as $x \to \pi/8$, but the graph of R approaches zero at a more rapid rate.

21.

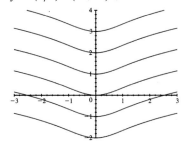

Note that both graphs approach zero as $x \to 1$, but the graph of R approaches zero at a more rapid rate.

Section 1.3

1. $y = t^2 + 3t + C$

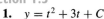

3. $y = (-1/2)\cos 2t + (2/3)\sin 3t + C$

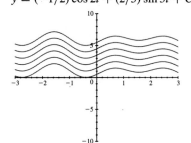

5. $y = (1/2)\ln(1 + t^2) + C$

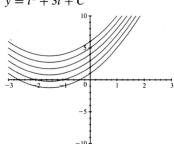

A-1

7. $y = \dfrac{t^2 e^{3t}}{3} - \dfrac{2t e^{3t}}{9} + \dfrac{2e^{3t}}{27} + C$

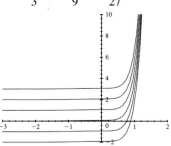

9. $y = \left(-e^{-2\omega}\cos\omega - 2\sin\omega e^{-2\omega}\right)/5 + C$

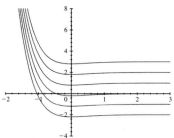

11. $x = -s^2 e^{-s} - 2se^{-s} - 2e^{-s} + C$

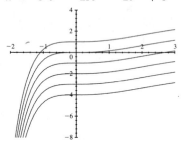

13. $r = \ln\left(\dfrac{u}{1-u}\right)$

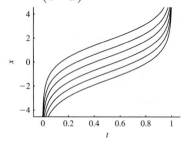

15. $y(t) = 2t^2 - 6t + 1$

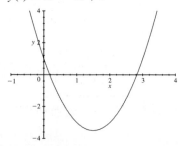

17. $x(t) = (-1/2)e^{-t^2} + (3/2)$

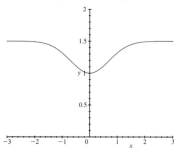

19. $s(r) = \dfrac{r^2 \sin 2r}{2} + \dfrac{r\cos 2r}{2} - \dfrac{\sin 2r}{4} + 1$

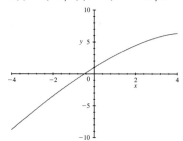

21. $x(t) = (-2/3)(4-t)^{3/2} + 19/3$

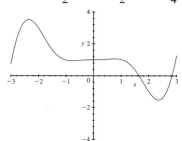

23. $y(t) = (1/4)\ln|t| + (3/4)\ln|t+4| - (3/4)\ln 3$

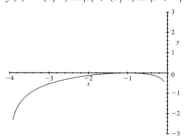

25. Velocity = 20.6 meters/sec; height = 108.9 meters.

27. The maximum height is 740.69 meters at $t = 12.24$.

Chapter 2
Section 2.1

1. $y' = -\dfrac{(1+t)y}{t^2}.$

3.

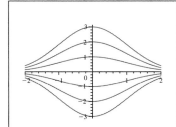

5.

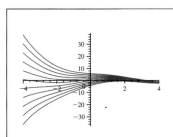

7. For $y(t) = 0$, $y'(t) = 0$ and
$y(t)(4 - y(t)) = 0(4 - 0) = 0$.

9. a. $[t^2 - 4y^2]' = 2(t - 4yy') = 0$
   b. $y = \pm(1/2)\sqrt{t^2 - C^2}$
   c. Either $-\infty < t < -C$ or $C < t < \infty$
   d.

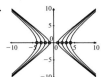

11. Interval of existence is $(-\infty, \ln(5)/4)$.

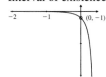

13. $y(t) = \dfrac{1}{3}t^2 + \dfrac{5}{3t}$. Interval of existence is $(0, \infty)$.

15. $y(t) = 2/(-1 + e^{-2t/3})$. Interval of existence is
$(-\ln(3)/2, \infty)$.

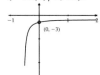

17.

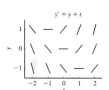

19.

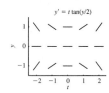

21.

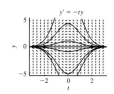

23.

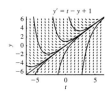

25.

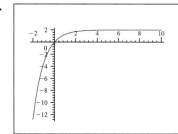

27.

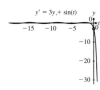

29.

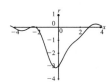

31.

33.

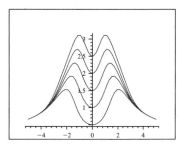

35. 124 mg   **37.** 29 days   **39.** 91 critters

Section 2.2

1. $y(x) = Ae^{x^2/2}$ **3.** $y(x) = \ln(e^x + C)$

5. $y(x) = De^{(1/2)x^2 + x}$ **7.** $y(x) = -2 \pm \sqrt{x^2 + E}$

9. $y(x) = e^{De^{\tan^{-1} x}}$

11. The solution is given implicitly by the equation
$y^4 - 8y - 2x^2 = A$.

13. $y(x) = -2x$, $(0, \infty)$

15. $y(x) = \sqrt{1 - 2\cos x}$, $(\pi/3, 5\pi/3)$

17. $y(t) = \tan(t + \pi/4)$, $(-3\pi/4, pi/4)$

19. For $y(0) = 1$, $y(x) = \sqrt{1 + x^2}$, $(-\infty, \infty)$, shown with a solid curve in the next figure. For $y(0) = -1$, $y(x) = -\sqrt{1 + x^2}$, $(-\infty, \infty)$, shown with a dashed curve in the next figure.

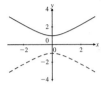

21. $y(x) = 2 + e^{-x}$, and $y(x) = 2 - e^{-x}$ on $(-\infty, \infty)$

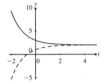

25. 12.4251 hours   **27.** 89.1537 mg

29. The time constant is $T_\lambda \approx 17.3$ hr.

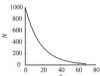

31. $T_{1/2} \approx 1582$ yr   **33.** 9:54 P.M.

35. 5.85 minutes   **37.** $y(x) = C/x$

39. $r(\theta) = C \sin^2(\theta/2)$

41. $(x, y, z) = (Cy^4, y, \sqrt{4 - C^2 y^8 - y^2/4})$, where y is the independent variable

43. $y(x) = Kx^4$

Section 2.3

1. $t = c/5g = 612,240$ seconds, and $x = gt^2/2 = 1.84 \times 10^{14}$ meters.

3. $t = 7.2438$ sec and $d = 340(8 - t) = 257.1$ m

5. 57.12 m and 3.4142 s   **7.** $a = -5.8$ ft/s^2

9. $v_{\text{term}} = -mg/r = -0.196$ m/s

11. $v_0 = 16.2665$ m/s without air resistance and $v_0 = 18.1142$ m/s with air resistance

13. 7.9010 m   **15.** 4.93 mi/s

17. 0.405 seconds   **19.** $mx'' = -mg - ke^{-ax}x'$

Section 2.4

1. $y(t) = 2 + Ce^{-t}$ **3.** $y(x) = \dfrac{\sin x + C}{x^2}$

5. $x(t) = t(t + 1)^2 + C(t + 1)^2$

7. $y(x) = \dfrac{\sin x + C}{1 + x}$ **9.** $i(t) = E/R + Ce^{-Rt/L}$

11. $y(x) = \dfrac{x - \cos x + C}{\sec x + \tan x}$

13. a. $y(x) = 1 + Ce^{-\sin x}$, C any real number

    b. We get the same solutions.

15. $y(x) = 2 - 3(x^2 + 1)^{-3/2}$

17. $x(t) = \sin t - 1 + 2e^{-\sin t}$

19. $y(x) = (1/2)(2x + 3)^{1/2} \ln(2x + 3)$ on $(-3/2, +\infty)$

21. $x(t) = \dfrac{1 + \sin t}{1 + t}$ on $(-\infty, -1)$

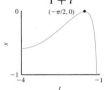

23. $y(x) = \dfrac{1}{x(C - x)}$

25. $y(x) = \pm 1/\sqrt{Cx^2 - x^4}$

27. b. $y(t) = \dfrac{1}{t} - \dfrac{2Bt}{1 + Bt^2}$

29. At approximately 11:12 P.M.

31. $y(t) = \dfrac{5}{2} + Ce^{-2t}$ **33.** $y(t) = \dfrac{4}{3}t^2 + C/t$

35. $y(t) = 2 + Ce^{x^2}$ **37.** $y(t) = (2t - 4) + 5e^{-t/2}$

39. $y(x) = x^2 - 1$ **41.** $y(t) = \dfrac{4 + 2t^2 + t^4}{4(1 + t^2)^2}$

43. a. $T_h = Fe^{-kt}$, F an arbitrary constant

Section 2.5

1. a. 9.038 lb   b. 46.2098 m   c. 20 lb

3. a. 0.25 lb/gal or 25 lb   b.

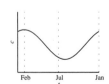

5. 21 lb

7. a. 21.5342 lb   b. $t = 8.7868$ min

9. b. Approximately 1.41 years.

11. a. The concentration varies periodically.   b. Early in February

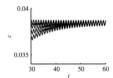

13. a. 0.0298 km^3 b. 10.18 yr

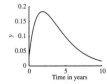

Section 2.6

1. $dF = 2y\,dx + (2x + 2y)\,dy$

3. $dF = \dfrac{x\,dx + y\,dy}{\sqrt{x^2 + y^2}}$

5. dF
$$= \frac{x^2 y\,dx + y^3\,dx - y\,dx + x^3\,dy + xy^2\,dy + x\,dy}{x^2 + y^2}$$

7. $dF = \left(\dfrac{2x}{x^2 + y^2} + \dfrac{1}{y}\right)dx + \left(\dfrac{2y}{x^2 + y^2} - \dfrac{x}{y^2}\right)dy$

9. Exact. $F(x, y) = x^2 + xy - 3y^2 = C$

11. Not exact

13. Exact. $F(x, y) = x^3 + xy - y^3 = C$

15. Exact. $F(u, v) = \dfrac{u^2 + 2vu - v^2}{2} = C$

17. Not exact

19. Exact. $F(x, t) = x \sin 2t - t^2 = C$

21. Not exact

23. $F(x, y) = \dfrac{x^2 y^2}{2} - \ln x + \ln y = C$

25. $F(x, y) = x - \frac{1}{2}\ln(x^2 + y^2) = C$

27. $\mu(x) = \dfrac{1}{x}$. $F(x, y) = xy - \ln x - \dfrac{y^2}{2} = C$

29. $\mu(y) = \dfrac{1}{y^2}$. $F(x, y) = \dfrac{yx + x^2}{y} = C$

31. Degree one   **33.** Degree one

35. $x^2 - Cx = y^2$ **37.** $F(x, y) = xy + \frac{3}{2}x^2 = C$

39. $y(x) = \dfrac{x + Cx^4}{1 - 2Cx^3}$

41. The following three graphs show the cases where $a = 1$, and $k = 1/2,\ 1,\ 3/2$.

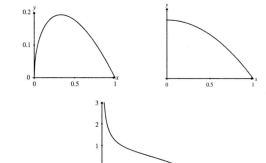

43. b. The orthogonal family is defined implicitly by

$$G(x, y) = \frac{y}{x^2 + y^2} = C.$$

The original curves are the solid curves in the following figure, and the orthogonal family is dashed.

45. $\arctan\left(\dfrac{y}{x}\right) - \dfrac{y^4}{4} = C$

47. $\arctan\left(\dfrac{y}{x}\right) - \dfrac{(y^2 + x^2)^2}{4} = C$

49. $Cxy = \dfrac{x + y}{x - y}$

Section 2.7

1. $f, \partial f/\partial y = 2y$ continuous on $\mathbf{R}^2$. Unique solution.

3. $f, \partial f/\partial y = t/(1 + y^2)$ continuous on $\mathbf{R}^2$. Unique solution.

5. $f, \partial f/\partial y = -t/(x + 1)^2$ continuous on rectangle containing $(0, 0)$ but not $x = -1$. Unique solution.

7. The general solution is $y(t) = t \sin t + Ct$.

9. The y-derivative of the right-hand side $f(t, y) = 3y^{2/3}$ is $2y^{-1/3}$, which is not continuous at $y = 0$. Hence the hypotheses of Theorem 7.16 are not satisfied.

11. The exact solution is $y(t) = -1 + \sqrt{t^2 - 3}$. The interval of solution is $(\sqrt{3}, \infty)$. The solver has trouble near $\sqrt{3}$. The point where the difficulty arises is circled in the following figure.

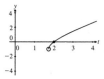

13. The exact solution is $y(t) = -1 + \sqrt{4 + 2\ln(1 - t)}$. The interval of existence is $(-\infty, 1 - e^{-2})$. The solver has trouble near $1 - e^{-2}$. The point where the difficulty arises is circled in the following figure.

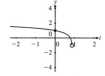

15. The solution is defined implicitly by the equation $y^3/3 + y^2 - 3y = 2t^3/3$. The solver has trouble near $(t_1, 1)$, where $t_1 = -(5/2)^{1/3} \approx -1.3572$, and also near $(t_2, -3)$, where $t_2 = (22/2)^{1/3} \approx 2.3811$. The points where the difficulty arises are circled in the following figure.

17. The computed solution is shown in the following figure.

$q(4) = 0.5851$

19. The computed solution is shown in the following figure.

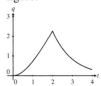

$q(4) = 2(1 + e^{-2})e^{-2} \approx 0.3073$

21. b. The right-hand side of the equation, $f(t, y) = 3y^{2/3}$, is continuous, but $\partial f/\partial y = 2y^{-1/3}$ is not continuous where $y = 0$. Hence the hypotheses of Theorem 7.16 are not satisfied.

23. A second solution is

$$\omega(t) = \begin{cases} 0, & t < 0 \\ 5t^4, & t \geq 0 \end{cases}$$

Normal form $y' = 4y/t$ reveals that $\partial/\partial y(4y/t) = 4/t$ is discontinuous at $t = 0$. The hypotheses of the Uniqueness Theorem are not satisfied.

25. No

27. This is true because of the uniqueness theorem.

29. This is true because of the uniqueness theorem.

31. This is true because of the uniqueness theorem.

Section 2.8

1. $x(0) = 0.8009$ **3.** $x(0) = 0.9596$

5. $x(0) = 0.7275$ **7.** $x(0) = 0.7290106$

9. $x(0) = -3.2314$ **11.** $x(0) = -3.2320923$

17. a. $|y(t) - x(t)| \leq |y(0) - x(0)|e^{2t}$

b. $x(t) = [2\sin t - \cos t]/5$, and $y(t) = [2\sin t - \cos t]/5 - e^{-2t}/10$. Yes.

c. The maximum predicted error is achieved for all $t < 0$.

Section 2.9

1. Autonomous. $P = 20000$ is an unstable equilibrium point.

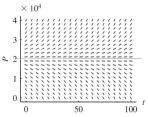

3. Not autonomous

5. $q = 2$ and $k\pi$, where k is any integer, positive or negative are equilibrium points. $q = 2$ is unstable and the stability alternates.

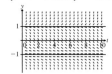

7. $y = 3$ is an unstable equilibrium point.

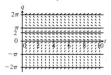

9. $y = -1$ and $y = 1$ are unstable equilibrium points.

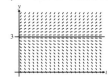

11.

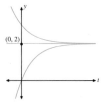

13.

15. The graph of f.   The phase line.

The solutions.

17. The graph of f.

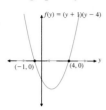

The phase line.

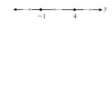

$f(y) = 6 - y$ and the phase line on the y-axis.

The solutions.

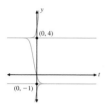

25. The graph of the right-hand side, with the phase line information on the y-axis.

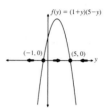

The solution is $y(t) = \dfrac{5 - e^{-6t}}{1 + e^{-6t}}$. $\lim_{t \to \infty} y(t) = 5$.

27. $x = -2$ is unstable. $x = 2$ is asymptotically stable.

29. a. $f(x) = x^2$, $f(x) = x^3$, or $f(x) = x^4$

b. $f(x) = -x^3$, $f(x) = -x^5$, or $f(x) = -x^7$

31. $c = 3$ lb/gal

19. The graph of f.

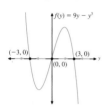

The phase line.

The solutions.

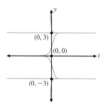

Chapter 3
Section 3.1

1. $r = \ln 3 \approx 1.0986$. 24300 cells

3. $r = (1/10) \ln 3 \approx 0.1099$.
$t = (10 \ln 2)/\ln 3 \approx 6.3093$ hr.

5. The long-time activity of the population depends entirely on the sign of $P(0) - (h/r)$. If it is negative, then the population will die out. If it is positive, the population will grow exponentially. If it is zero, then the population will remain at a constant h/r.

7. $h \geq 2500 \ln 2 \approx 1732.9$

13. $r \approx 0.0241$, $t \approx 167.6713$

17. a. The curves with initial conditions $P(0) = 410$, $P(0) = 414$, $P(0) = 415$, $P(0) = 420$, and $P(0) = 450$.

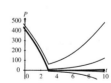

The critical population is between 414 and 415.

b. $P_0 = 414.5557$

19. a. $P' = 0.1P(1 - P/10) - 0.01P$.

b. $P = 0$ is unstable, and $P = 9$ is asymptotically stable.

c. The population tends to the carrying capacity $P = 9$.

21. The graph of f.

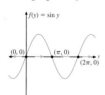

The phase line.

The solutions.

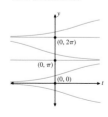

23. $y(t) = 6 - 4e^{-t}$. $\lim_{t \to \infty} y(t) = 6$. The graph of

21. The maxima yield is $rK/4$ and the carrying capacity is $K/2$ when $\gamma = r/2$

Section 3.3
1. a. $1978.50   b. 28.5 yr
3. $25,553   **5.** 7.5 yrs
7. $8,798.15   **9.** 3.5%
11. a. $P(n+1) = \left(1 + \frac{I}{m}\right) P(n)$, $P(0) = P_0$
   b. $P(n) = P_0 \left(1 + \frac{I}{m}\right)^n$
15. a. $242.66 per month   b. $243.32 per month

Section 3.4
1. $Q(t) = 10 - 10e^{-t/2}$
3. $Q(t) = (-40 \cos 2t + 10 \sin 2t + 40e^{-t/2})/17$
5. $Q = 51(1 - e^{-t/2})/5 - t/10$
7. $I(t) = 10(1 - e^{-t/10})$
9. $I(t) = \dfrac{50(-20\pi \cos 2\pi t + 20\pi e^{-t/10} + \sin 2\pi t)}{1 + 400\pi^2}$
11. $I(t) = 300 - 20t - 300e^{-t/10}$
13. $Q(t) = EC(1 - e^{-t/RC})$
15. Maximum value of $Q = 6.69$ at $t = 25 \ln 5 \approx 40.24$
16. Maximum value of $I = 2.5$ at $t = 13.86$
17. a. 23.87   b. 23.87
19. The nine solutions are plotted below.

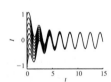

The frequency is π Hz for both.
21. The general solution to the differential equation is
$$Q(t) = \frac{EC}{1 + R^2C^2\omega^2}(\sin \omega t - RC\omega \cos \omega t) + Ae^{-t/RC},$$
where A is an arbitrary constant. The term $Ae^{-t/RC}$ dies out as t increases, so it is the transient term.
23. $Q(t) = k_1\sqrt{LC} \sin\left(\dfrac{t}{\sqrt{LC}} + k_2\right)$, and
$I(t) = k_1 \cos\left(\dfrac{t}{\sqrt{LC}} + k_2\right)$, where k_1 and k_2 are constants

Chapter 4
Section 4.1
1. linear, inhomogeneous   **3.** nonlinear
5. nonlinear   **7.** nonlinear
9. $2y'' + 39.2y = 0$, $y(0) = 0.12$, $y'(0) = 0$.
11. $2y'' + 0.05y' + 39.2y = -0.5 \cos \pi t$, $y(0) = 0.12$, $y'(0) = 0$
17. $y1/y2 = e^{-3t}$, which is nonconstant. Further, $W(t) = 3e^t$, which is never zero.
19. $y1/y2 = \cot 3t$, which is nonconstant. Further, $W(t) = +3e^{-4t}$, which is never zero.

21. There is no contradiction because Proposition 1.27 requires that both y_1 and y_2 be solutions of a differential equation of the form
$$y'' + p(t)y' + q(t)y = 0.$$

23. $y(t) = 2\cos 4t - (1/4)\sin 4t$

25. $y(t) = 2e^{-4t} + 7te^{-4t}$

27. $y(t) = C_1t + C_2t^2$ **29.** $y(t) = C_1t + C_2t^3$

Section 4.2
1. $y' = v$, $v' = -2v + 3y$

3. $y' = v$, $v' = -3v - 4y + 2\cos 2t$

5. $y' = v$, $v' = -\mu(t^2 - 1)v - y$

9. Position vs. time   Velocity vs. time

Position and velocity vs. time   Velocity vs. position
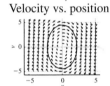

11. Position vs. time   Velocity vs. time

Position and velocity vs. time   Velocity vs. position
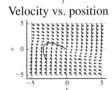

13. Position vs. time   Velocity vs. time

Position and velocity vs. time   Velocity vs. position
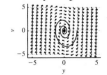

15. Position vs. time Velocity vs. time

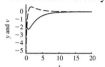

Position and velocity vs. time Velocity vs. position

17. a. The displacement of the mass is at a maximum when the velocity is zero.

b. No. The velocity has maximum magnitude when the acceleration is 0. At these points $y = -\mu v/k \neq 0$.

19. **21.**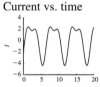

23. Charge vs. time Current vs. time

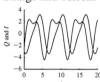 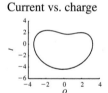

Charge and current vs. time Current vs. charge

25. Charge vs. time Current vs. time

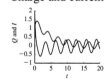

Charge and current vs. time Current vs. charge

 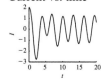

27. Charge vs. time Current vs. time

Charge and current vs. time Current vs. charge

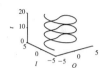 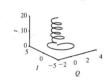

29. 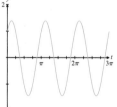 **31.**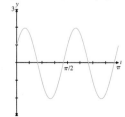

Section 4.3

1. $y(t) = C_1 e^{2t} + C_2 e^{-t}$

3. $y(t) = C_1 e^{-3t} + C_2 e^{-2t}$

5. $y(t) = C_1 e^{-(1/2)t} + C_2 e^{t}$

7. $y(t) = C_1 e^{-t/3} + C_2 e^{t}$

9. $y(t) = C_1 \cos t + C_2 \sin t$

11. $y(t) = e^{-2t}(C_1 \cos t + C_2 \sin t)$

13. $y(t) = C_1 \cos \sqrt{2}t + C_2 \sin \sqrt{2}t$

15. $y(t) = C_1 e^{t} \cos \sqrt{3}t + C_2 e^{t} \sin \sqrt{3}t$

17. $y(t) = (C_1 + C_2 t)e^{2t}$

19. $y(t) = (C_1 + C_2 t)e^{-t/2}$

21. $y(t) = (C_1 + C_2 t)e^{-t/4}$

23. $y(t) = (C_1 + C_2 t)e^{-3t/4}$

25. $y(t) = (1/3)e^{2t} - (4/3)e^{-t}$

27. $y(t) = e^{t}(-2 \cos 4t + (5/4) \sin 4t)$

29. $y(t) = (2 + 9t)e^{-5t}$

31. $y(t) = e^{-t}\left(\cos \sqrt{2}t + (\sqrt{2}/2) \sin \sqrt{2}t\right)$

33. $y(t) = (2/3)e^{(t+1)/4} + (1/3)e^{-(t+1)/2}$

35. $y(t) = (1 - t)e^{-6(t-1)}$

Section 4.4

1. $y = \sqrt{2} \cos(2t - \pi/4)$

3. $y = 2 \cos(4t - \pi/3)$

5. $y = \sqrt{0.05}\cos(2.5t - \phi)$, where $\phi = \arctan(-0.5)$

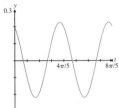

7. $y = \sqrt{2}e^{-t/2}\cos(5t - \pi/4)$

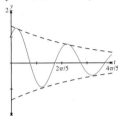

9. $y = \sqrt{0.05}e^{-0.1t}\cos(2t - \phi)$, where $\phi = \arctan 0.5$

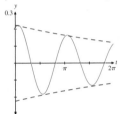

11. $y = 0.5\cos 5t$

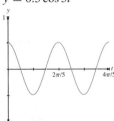

13. $k = 32/5$, $v_0 = 0$

15. a. 4.0×10^{-5} coulombs

    b. Amplitude is 4.0×10^{-5}. Frequency is
$1/\sqrt{12} \times 10^6 \approx 2.887 \times 10^5$ rad/s. Phase is 0.

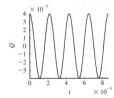

23. $I = \frac{5}{2}e^{-5t/2}\cos(5t + \pi/2)$

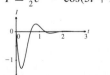

25. $Q = \frac{\sqrt{5}}{50}e^{-2t}\cos(t - \phi)$, where $\phi = \arctan 2$.

Section 4.5

1. $y = 2e^{-3t}$ **3.** $y = 3e^{-t}$

5. $y_p = -1/5\cos 3t$

7. $y_p = -(21/100)\cos 2t + (3/100)\sin 2t$

9. If $z(t) = x(t) + iy(t)$ is a solution of
$z'' + pz' + qz = Ae^{i\omega t}$, then the real and imaginary
parts of z are solutions of $x'' + px' + qx = A\cos\omega t$
and $y'' + py' + qy = A\sin\omega t$, respectively.

11. $y = (1/5)\sin 2t$

13. $y = (42/221)\cos 3t - (2/221)\sin 3t$

15. $y = (1/4)t - 9/16$

17. $y = (1/4)t^3 - (9/16)t^2 + (15/32)t - 9/128$

19. $y = -(1/14)e^{5t} - (1/2)e^{-t} + (4/7)e^{-2t}$

21. $y = e^t(-(3/5)\cos 2t - (11/20)\sin 2t) +$
$(3/5)\cos t - (3/10)\sin t$

23. $y = (-23 + 5t)e^t + t^3 + 6t^2 + 18t + 24$

25. $y_p = -(2/3)te^{-t}$ **27.** $y_p = -(1/6)t\cos 3t$

29. $y_p = (5/2)t^2e^{-3t}$

31. $y = 1 - (1/10)\cos 2t + (1/5)\sin 2t$

33. $y = (2/25)t + 3/25 + (1/10)t\sin 5t$

35. $y = -(5/13)\cos 2t + (1/13)\sin 2t$

37. $y = (1/2)t^2e^{-2t} - (1/8)\cos 2t$

39. $y = ((1/6)t + 5/36)e^{-4t}$

41. $y = (t^2 + 4t + 6)e^{-2t}$

43. $y = t(-(1/3)t^2 - t - 2)e^{-2t}$

47. $y = (2 - t)\cos t + \sin t$

Section 4.6

1. $y_p = -(1/9)\cos 3t \ln|\sec 3t + \tan 3t|$

3. $y_p = -t - 3$ **5.** $y_p = (1/2)t^2e^t$

7. $x_p = -2 + \sin t \ln|\sec t + \tan t|$

9. $x(t) = \frac{1}{2}\tan t + \frac{3}{2}(\cos t)(\ln|\sec t + \tan t|)$

11. $y_p = 1 - (1/2)t\cos t - \cos t \ln|\sec t + \tan t|$
$+ (1/4)\sin t$

13. $y(t) = C_1 t + C_2/t^3 - 1/(4t)$

Section 4.7

3. $y = 2\sin\frac{1}{2}t \sin\frac{21}{2}t$ **5.** $y = 2\sin\frac{1}{2}t \cos\frac{23}{2}t$

7. The graph of x with $\omega = 10.99$ is shown in the following figure.

9. a. $x(t) = \dfrac{4}{4 - \omega^2}(\cos \omega t - \cos 2t)$

b. With $\omega = 1.84$, $x(t) = \dfrac{2}{0.19} \sin 0.1t \sin 1.9t$. The graph of x and its envelope is presented in the following figure.

11. a. $I(t) = \dfrac{6\omega}{\omega^2 - 4}(2 \sin \omega t - \omega \sin 2t)$. With $\omega = 1.8$, we get the following graph.

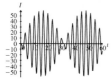

b. $I(t) = 6t \cos 2t - 3 \sin 2t$

13. $x(t) = \dfrac{3}{\sqrt{260}} \sin(4t - \phi) = \dfrac{3}{\sqrt{260}} \cos(4t - \phi - \pi/2)$, where $\phi = \operatorname{arccot}(-7/4) \approx 2.6224$.

15. $x(t) = 0.0745 \cos(2\pi t - 2.4678)$

17. $x(t) = \dfrac{3}{442}(\cos 3t + 21 \sin 3t) - \dfrac{71}{39}e^{-2t} + \dfrac{83}{102}e^{-5t}$

19. $x(t) = \dfrac{3}{8}(\sin t - \cos t) + \dfrac{1}{8}e^{-2t}[3 \cos t - 21 \sin t]$

21. Use the more slowly decaying term to determine the time constant; thus, $T_c = 1/2$. The plot of the transient solution $x_h(t) = -\dfrac{71}{39}e^{-2t} + \dfrac{83}{102}e^{-5t}$ follows.

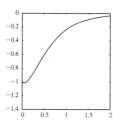

23. $T_c = 1/2$. The plot of the transient solution $x_h = e^{-2t}[3 \cos t - 21 \sin t]/8$ on $[0, 4T_c] = [0, 2]$ follows.

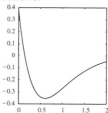

25. steady-state: $x(t) = \dfrac{1}{\sqrt{13}} \cos(3t - \phi)$, where $\cot \phi = -2/3$

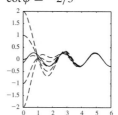

27. steady-state: $x(t) = \dfrac{1}{\sqrt{1.16}} \cos(t - \phi)$, where $\cot \phi = 2.5$; gain: $1/\sqrt{1.16} \approx 0.9285$; phase: $\phi = \operatorname{arccot} 2.5 \approx 0.3805$

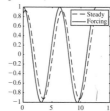
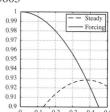

29. The solution steadies after $4T_c = 20$. Gain: 1.25, phase: 0.53.

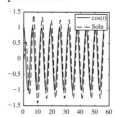
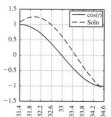

31. $G(\omega) = \dfrac{1}{R(\omega)} = \dfrac{1}{\sqrt{(49 - \omega^2)^2 + 0.0001\omega^2}}$. Maximum gain near $\omega \approx 7$.

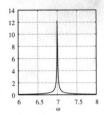

33. $G(\omega) = \frac{1}{R(\omega)} = \frac{1}{\sqrt{(25-\omega^2)^2+0.0025\omega^2}}$. Maximum gain near $\omega \approx 5$.

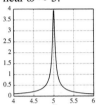

35. Maximum gain at $\omega_{res} \approx 7.0000$.

37. Maximum gain at $\omega_{res} \approx 4.9999$.

39.

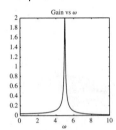
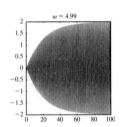

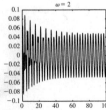

Driving frequencies near the resonant frequencies nearly double the amplitude. Driving frequencies far from the resonant frequency attenuate the signal.

41.

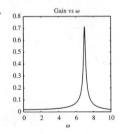
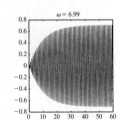

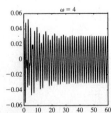

Driving frequencies near the resonant frequency attenuate the driving signal to about 70% of its amplitude. Driving frequencies far from the resonant frequency severely attenuate the signal.

45. $y(t) = e^{-0.1t}(-1.2715\cos 9.8990t - 0.0192\sin 9.8990t) + 1.2716\cos(4.4t - 0.0112)$

Chapter 5
Section 5.1

1. $3/s$, provided $s > 0$

3. $1/(s+2)$, provided $s > -2$

5. $s/(4+s^2)$, provided $s > 0$

7. $1/(s-2)^2$, provided $s > 2$

9. $(s-2)/(9+(s-2)^2)$, provided $s > 2$

15. $3/s$, provided $s > 0$

17. $1/(s+2)$, provided $s > -2$

19. $s/(s^2+4)$, provided $s > 0$

21. $1/(s-2)^2$, provided $s > 2$

23. $(s-2)/((s-2)^2+9)$, provided $s > 2$

25. $2e^{-2s}/s$, provided $s > 0$

27. $(1-e^{-2s})/s^2$, provided $s > 0$

29.

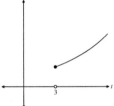

$Y(s) = e^{-3(s-0.2)}/(s-0.2)$, provided $s > 0.2$

Section 5.2

1. $6/s^3$, provided $s > 0$

3. $(2+4s+5s^2)/s^3$, provided $s > 0$

5. $(-2s^3+12s^2-18s+12)/((s^2+1)(s^2+9))$, provided $s > 0$

9. In each case, $\mathcal{L}(y')(s) = 6/s^3$.

11. In each case, $\mathcal{L}(y')(s) = -3/(s+3)$.

13. In each case, $\mathcal{L}(y')(s) = 5s/(s^2+25)$.

15. In each case, $\mathcal{L}(y'')(s) = 4/(s+2)$.

17. In each case, $\mathcal{L}(y'')(s) = 2/s$.

19. $(s+3)/((s-5)(s+2))$

21. $(-2s^2+s-8)/((s-4)(s^2+4))$

23. $(s^3+2s^2+5s+8)/((s^2+4)(s^2+2s+2))$

25. $(-s^4-4s^3-2s^2+s+1)/(s^2(s+1)(s^2+3s+5))$

27. $(s-2)/(s^2-4s+8)$

29. $(4s^2+11s+9)/(s+1)^3$ **31.** $1/(s+1)^2$

33. $2/(s-2)^3$ **35.** $2/((s-1)(s+2)^3)$

37. $(-2s^2+9s-12)/((s-2)(s^2-4s+5))$

39. $(s^3+2s^2+6s+1)/((s^2+s+2)(s^2+2s+5))$

41. $(-s^3-10s+37)/((s^2+5)(s^2+2s+17))$

Section 5.3

1. $(1/3)e^{-(2/3)t}$ **3.** $(1/2)\sin 2t$

5. $3t$ **7.** $3\cos 5t + (2/5)\sin 5t$

9. $-(1/4)e^{(3/4)t} + (3/7)\sin 7t - 2\cos 7t$

11. $(5/2)t^2 e^{-2t}$ **13.** $(3/5)e^{-2t}\sin 5t$

15. $e^t(2\cos\sqrt{5}t - (\sqrt{5}/5)\sin\sqrt{5}t)$

17. $e^{-2t}(3\cos 5t - (4/5)\sin 5t)$

19. $-(1/3)e^{-2t} + (1/3)e^t$

21. $e^{-t} + e^{2t}$ **23.** $2e^{-3t} + 5e^t$

25. $5 + (3/2)e^{(1/2)t}$ **27.** $4e^{-t} + 3\cos 2t$

29. $2e^{-t} + e^{-2t}\sin t$

31. $(1/27)e^{2t}(t-1) + (1/27)e^{-t}(1 + 2t + (3/2)t^2)$

33. $(4/169)e^{-2t} + (1/13)te^{-2t} - (4/169)\cos 3t -$
$(5/507)\sin 3t$

35. $(2/9)e^{-t} - (1/3)te^{-t} - (1/4)e^{-2t} + (1/36)e^{2t}$

Section 5.4

1. $-(6/5)e^{-3t} + (1/5)e^{2t}$

3. $-(4/17)e^{-4t} + (4/17)\cos t + (1/17)\sin t$

5. $(5/9)e^{-6t} + 4/9 + (1/3)t$

7. $(1/37)e^{-8t} - (1/37)e^{-2t}\cos t + (6/37)e^{-2t}\sin t$

9. $-(7/4)e^{-t} - (1/4)e^t + (1/2)te^t$

11. $-(5/12)e^{2t} - (1/4)e^{-2t} - (1/3)e^{-t}$

13. $(2/3)\cos 2t + (1/3)\cos t$

15. $-2t + (1/2)e^t - (1/2)e^{-t}$

17. $-1 - e^{-t} - te^{-t} - (1/2)t^2 e^{-t}$

19. $-(7/9)e^{-t} - (1/9)e^{5t} - (1/9)e^{2t}$

21. $-(5/9)e^{-t} - (4/9)e^{2t} + (1/3)te^{3t}$

23. $-(6/5)e^{-t}\cos t - (8/5)e^{-t}\sin t + (1/5)\cos t +$
$(2/5)\sin t$

25. $(58/65)\cos 2t + (9/65)\sin 2t + (7/65)e^{-2t}\cos t -$
$(4/65)e^{-2t}\sin t$

27. a. $C_1 = -1$, $C_2 = 2$.   b. $2e^t - e^{3t}$

29. $e^t - e^{-2t}$ **31.** $-e^{-t}\cos t$

33. $(1/3)\cos t - \sin t + (2/3)\cos 2t$

35. $\cos 2t - (1/2)\sin 2t + (1/4)t\sin 2t$

41. $y(t) =$
$-v_0\Big/\left(2\sqrt{c^2 - \omega_0^2}\right)\left[e^{-\left(c+\sqrt{c^2-\omega_0^2}\right)t} - e^{-\left(c-\sqrt{c^2-\omega_0^2}\right)t}\right]$

Section 5.5

1. e^{2s}/s^2 **3.** $3e^{-\pi s/4}/(s^2 + 9)$

5. $e^{-s}(2 + 2s + s^2)/s^3$

7. $(1/2)e^{-\pi s/6}(4 + \sqrt{3}s^2)/(s^2 + 4)$

9. a. The graph of f: The graph of F:

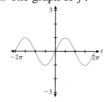   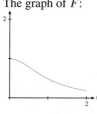

b. The graph of g: The graph of G:

    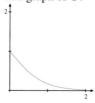

c. The graph of g: The graph of G:

    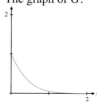

11. $(5/s)(e^{-2s} - e^{-4s})$

13. $2/s^3 - 2e^{-2s}/s^3 - 4e^{-2s}/s^2$

15. $(1/s^2)(1 - 2e^{-s} + 2e^{-3s} - e^{-4s})$

17. $f(t) = \begin{cases} 0, & \text{if } 0 \le t < 1, \\ e^{2(t-1)}, & \text{if } 1 \le t < \infty \end{cases}$

19. $f(t) = \begin{cases} t^2, & \text{if } 0 \le t < 2, \\ t^2 + (1/2)(t-2)^2, & \text{if } 2 \le t < \infty \end{cases}$

21. $f(t) = \begin{cases} 0, & \text{if } 0 \le t < 2, \\ \cos 2(t-2), & \text{if } 2 \le t < \infty \end{cases}$

23. $f(t) = \begin{cases} 0, & \text{if } 0 \le t < 2, \\ \left[e^{3(t-2)} - e^{-(t-2)}\right]/2, & \text{if } 2 \le t < \infty \end{cases}$

25. $f(t) =$
$\begin{cases} 2e^{-t}\sin t, & \text{if } 0 \le t < 2, \\ 2e^{-t}\sin t - e^{-(t-2)}\sin(t-2), & \text{if } 2 \le t < \infty \end{cases}$

27. $1/4 - (1/4)\cos 2t - H(t-1)[1/4 - (1/4)\cos 2(t-1)]$

29. $t/4 - (1/8)\sin 2t - H(t-1)[(1/4)t -$
$(1/8)\sin 2(t-1) - (1/4)\cos 2(t-1)]$

33.

35. a.

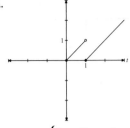

b. $y'(t) = \begin{cases} 0, & \text{if } t < 0, \\ 1, & \text{if } 0 < t < 1, \\ 1, & \text{if } t > 1. \end{cases}$

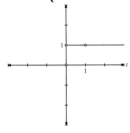

c. y is not continuous, and therefore not piecewise differentiable, so the hypotheses of Proposition 2.1 are not satisfied;.

Section 5.6

3. $e(t) = (1/6)e^{5t} - (1/6)e^{-t}$

5. $e(t) = -(1/6)e^{-3t} + (1/6)e^{3t}$

7. $e(t) = e^{-t}\sin t$

Section 5.7

5. $e^{2t} - e^{t}$ **7.** $(3/2)t^2 - (1/6)t^3$

9. $t^2 - 2t + 2 - 2e^{-t}$

11. $f * g(t) = t - \sin t$ and
$\mathcal{L}\{f * g(t)\}(s) = 1/(s^2(s^2 + 1))$

13. $f * g(t) = \frac{1}{4}e^{-2t} + \frac{1}{2}t - \frac{1}{4}$ and
$\mathcal{L}\{f * g(t)\}(s) = 1/(s^2(s + 2))$

15. $f * g(t) = -\frac{1}{2}e^{-t} + \frac{1}{2}[\cos t + \sin t]$ and
$\mathcal{L}\{f * g(t)\}(s) = s/((s + 1)(s^2 + 1))$

17. $1 - e^{-t}$ **19.** $(1/3)(e^{2t} - e^{-t})$

21. $(1/2)e^{t} - (1/2)\cos t + (1/2)\sin t$

23. $(1/2)\sin t - (1/2)t\cos t$

27. $(1/3)\int_0^t (\sin 3u)g(t - u)\,du - \cos 3t + (2/3)\sin 3t$

29. $(1/2)\int_0^t (e^{-u} - e^{-3u})g(t - u)\,du - e^{-t}$

31. $(1/5)\int_0^t e^{-2u}(\sin 5u)g(t - u)\,du - e^{-2t}\cos 5t$

Chapter 6
Section 6.1

1.

k	t_k	y_k	$f(t_k, y_k) = y_k$	h	$f(t_k, y_k)h$
0	0.0	1.0000	1.0000	0.1	0.1000
1	0.1	1.1000	1.2000	0.1	0.1200
2	0.2	1.2200	1.4200	0.1	0.1420
3	0.3	1.3620	1.6620	0.1	0.1662
4	0.4	1.5282	1.9282	0.1	0.1928
5	0.5	1.7210	2.2210	0.1	0.2221

3.

k	t_k	y_k	$f(t_k, y_k) = t_k y_k$	h	$f(t_k, y_k)h$
0	0.0	1.0000	0.0000	0.1	0.0000
1	0.1	1.0000	0.1000	0.1	0.0100
2	0.2	1.0100	0.2020	0.1	0.0202
3	0.3	1.0302	0.3091	0.1	0.0309
4	0.4	1.0611	0.4244	0.1	0.0424
5	0.5	1.1036	0.5518	0.1	0.0552

5.

k	x_k	z_k	$f(x_k, z_k) = x_k - 2z_k$	h	$f(x_k, z_k)h$
0	0.0	1.0000	-2.0000	0.1	-0.2000
1	0.1	0.8000	-1.5000	0.1	-0.1500
2	0.2	0.6500	-1.1000	0.1	-0.1100
3	0.3	0.5400	-0.7800	0.1	-0.0780
4	0.4	0.4620	-0.5240	0.1	-0.0524
5	0.5	0.4096	-0.3192	0.1	-0.0319

7. $y = 1/2 + (15/2)e^{-x^2}$

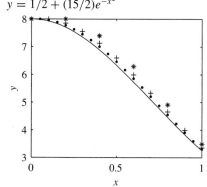

9. $z = -e^{t+(1/2)t^2}$

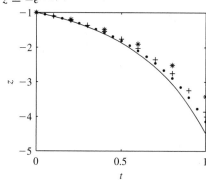

11. **a.**

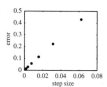

b. $\lambda \approx 1.454 \times 10^{-4}$, 13,757 iterations

13. $y = -(1/2)\sin t - (1/2)\cos t + (3/2)e^t$, $\lambda \approx 12.1321$, $h \approx 8.2426 \times 10^{-4}$, $N \approx 2{,}426$.

15. $y = 3/(t-3)$, $\lambda \approx 2.8670$, $h \approx 0.003487$, $N \approx 573$

17.

t_k	x_k	y_k	$f(t_k, x_k, y_k)h = y_k h$	$g(t_k, x_k, y_k)h = -x_k h$
0.0	1.0000	0.0000	0.0000	-0.1000
0.1	1.0000	-0.1000	-0.0100	-0.1000
0.2	0.9900	-0.2000	-0.0200	-0.0990
0.3	0.9700	-0.2990	-0.0299	-0.0970
0.4	0.9401	-0.3960	-0.0396	-0.0940
0.5	0.9005	-0.4900	-0.0490	-0.0901

19.

t_k	x_k	y_k	$f(t_k, x_k, y_k)h = -2y_k h$	$g(t_k, x_k, y_k)h = x_k h$
0.0	0.0000	-1.0000	-0.1000	0.0000
0.1	0.2000	-1.0000	-0.1000	-0.0200
0.2	0.4000	-0.9800	-0.0980	-0.0400
0.3	0.5960	-0.9400	-0.0940	-0.0596
0.4	0.7840	-0.8804	-0.0880	-0.0784
0.5	0.9601	-0.8020	-0.0802	-0.0960

21.

t_k	x_k	y_k	$f(t_k, x_k, y_k)h = -y_k h$	$g(t_k, x_k, y_k)h = (x_k + y_k)h$
0.0	1.0000	-1.0000	0.1000	0.0000
0.1	1.1000	-1.0000	0.1000	0.0100
0.2	1.2000	-0.9900	0.0990	0.0210
0.3	1.2990	-0.9690	0.0969	0.0330
0.4	1.3959	-0.9360	0.0936	0.0460
0.5	1.4895	-0.8900	0.0890	0.0599

23. Step size $h = 0.05$. x versus t is solid line, y versus t is dashed line.

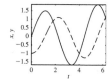

25. Step size $h = 0.05$. x versus t is solid line, y versus t is dashed line.

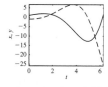
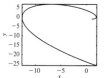

27. Step size $h = 0.1$

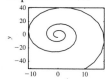

Step size $h = 0.01$

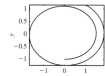

Step size $h = 0.001$

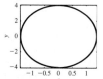

29. Step size $h = 0.1$ Step size $h = 0.01$ Step size $h = 0.001$

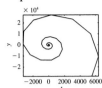

Section 6.2

1.

k	t_k	y_k	s_1	s_2	h	$h(s_1 + s_2)/2$
0	0.0	1.0000	1.0000	1.2000	0.1	0.1100
1	0.1	1.1100	1.2100	1.4310	0.1	0.1321
2	0.2	1.2421	1.4421	1.6863	0.1	0.1564
3	0.3	1.3985	1.6985	1.9683	0.1	0.1833
4	0.4	1.5818	1.9818	2.2800	0.1	0.2131
5	0.5	1.7949	2.2949	2.6244	0.1	0.2460

3.

k	t_k	y_k	s_1	s_2	h	$h(s_1 + s_2)/2$
0	0.0	1.0000	0.0000	0.1000	0.1	0.0050
1	0.1	1.0050	0.1005	0.2030	0.1	0.0152
2	0.2	1.0202	0.2040	0.3122	0.1	0.0258
3	0.3	1.0460	0.3138	0.4309	0.1	0.0372
4	0.4	1.0832	0.4333	0.5633	0.1	0.0498
5	0.5	1.1331	0.5665	0.7138	0.1	0.0640

5.

k	x_k	z_k	s_1	s_2	h	$h(s_1 + s_2)/2$
0	0.0	1.0000	−2.0000	−1.5000	0.1	−0.1750
1	0.1	0.8250	−1.5500	−1.1400	0.1	−0.1345
2	0.2	0.6905	−1.1810	−0.8448	0.1	−0.1013
3	0.3	0.5892	−0.8784	−0.6027	0.1	−0.0741
4	0.4	0.5152	−0.6303	−0.4042	0.1	−0.0517
5	0.5	0.4634	−0.4268	−0.2415	0.1	−0.0334

7. $z = (1/2) \sin x + (1/2) \cos x + (1/2)e^{-x}$

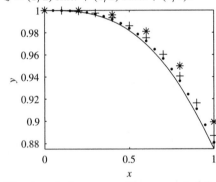

9. $w = x^2 - 2x + 2 - (3/2)e^{-x}$

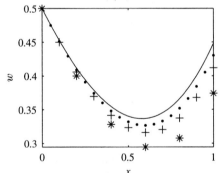

11. The slope is the same as the exponent of the power function $100x^{-3}$.

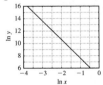

13. The slope is the same as the exponent of the power function $5x^3$.

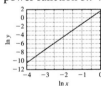

15. $E_h \approx 0.03770h^{1.9938}$

STEP SIZE h	RK2 APPROX.	TRUE VALUE	ERROR E_h
0.06250000	0.86557660	0.86572506	0.00014846
0.03125000	0.86568735	0.86572506	0.00003771
0.01562500	0.86571556	0.86572506	0.00000950
0.00781250	0.86572267	0.86572506	0.00000238
0.00390625	0.86572446	0.86572506	0.00000060
0.00195313	0.86572491	0.86572506	0.00000015
0.00097656	0.86572502	0.86572506	0.00000004

17. $E_h \approx 0.008522h^{1.958}$

STEP SIZE h	RK2 APPROX.	TRUE VALUE	ERROR E_h
0.06250000	1.91289601	1.91293118	0.00003517
0.03125000	1.91292139	1.91293118	0.00000979
0.01562500	1.91292861	1.91293118	0.00000257
0.00781250	1.91293053	1.91293118	0.00000066
0.00390625	1.91293102	1.91293118	0.00000017
0.00195313	1.91293114	1.91293118	0.00000004
0.00097656	1.91293117	1.91293118	0.00000001

19.   Dividing the step size by $\sqrt{2}$ halves the error.

21.

t_k	y_k	s_1	s_2	s_3	s_4	$h\frac{s_1+2s_2+2s_3+s_4}{6}$
0.0	1.0000	1.0000	1.1000	1.1050	1.2105	0.1103
0.1	1.1103	1.2103	1.3209	1.3264	1.4430	0.1325
0.2	1.2428	1.4428	1.5649	1.5711	1.6999	0.1569
0.3	1.3997					

23.

x_k	z_k	s_1	s_2	s_3	s_4	$h\frac{s_1+2s_2+2s_3+s_4}{6}$
0.0	1.0000	-2.0000	-1.7500	-1.7750	-1.5450	-0.1766
0.1	0.8234	-1.5468	-1.3422	-1.3626	-1.1743	-0.1355
0.2	0.6879	-1.1758	-1.0082	-1.0250	-0.8708	-0.1019
0.3	0.5860					

25. $E_h \approx 0.001138h^{4.352}$

STEP SIZE h	RK4 APPROX.	TRUE VALUE	ERROR E_h
1.0000000	0.87036095021	0.86572505654	0.00463589367
0.5000000	0.86580167967	0.86572505654	0.00007662313
0.2500000	0.86572599485	0.86572505654	0.00000093831
0.1250000	0.86572502504	0.86572505654	0.00000003150
0.0625000	0.86572505215	0.86572505654	0.00000000439
0.0312500	0.86572505620	0.86572505654	0.00000000034
0.0156250	0.86572505652	0.86572505654	0.00000000002
0.0078125	0.86572505654	0.86572505654	0.00000000000

27. $E_h \approx 0.002157h^{4.055}$

STEP SIZE h	RK4 APPROX.	TRUE VALUE	ERROR E_h
0.25000000	1.91293958	1.91293118	0.00000840
0.12500000	1.91293164	1.91293118	0.00000046
0.06250000	1.91293121	1.91293118	0.00000003
0.03125000	1.91293118	1.91293118	0.00000000
0.01562500	1.91293118	1.91293118	0.00000000
0.00781250	1.91293118	1.91293118	0.00000000
0.00390625	1.91293118	1.91293118	0.00000000

29.   Halving the step size scales the error by 1/16.

Section 6.3

1. $x = e^{1-\cos t}$.

Euler's method.

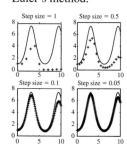

RK2

RK4

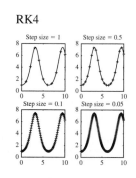

3. $x = (1/5) \sin 2t - (2/5) \cos 2t + (7/5)e^{-t}$.

Euler's method.

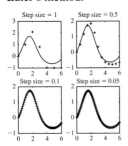

RK2

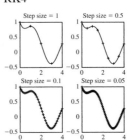

RK4

5. $x = -1 + e^{\sin t}$.

Euler's method.

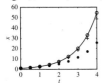

RK2

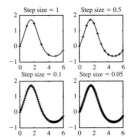

RK4

7. $x = e^t$. Exact (solid line), Euler (dots), RK2 (plus signs), RK4 (circles).

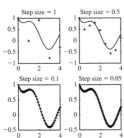

9. $x = -1 + \sqrt{3 - 2\cos t}$. Exact (solid line), Euler (dots), RK2 (plus signs), RK4 (circles).

11. a.

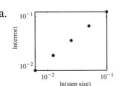

b. Euler (solid line), RK2 (dashed), RK4 (dotted).

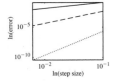

c. Using a computer, the slopes were 0.9716, 1.9755, and 3.9730.

13. $x = 6/(3t^2 + 2)$. Error slopes: Eul (1.0135), RK2 (2.0303), RK4 (4.0256). Euler (solid line), RK2 (dashed), RK4 (dotted)

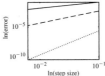

Section 6.4

1. $b = 4 \to h = 0.1, t \approx 0.11$ s. $b = 6 \to h = 0.01$, $t \approx 1.43$ s. $b = 8 \to h = 0.001, t \approx 19.12$ s. $b = 10 \to h = 0.0001, t \approx 238.6$ s. Answers run on 166 MHz machine. Answers will vary on other systems.

3. $b = 4 \to t \approx 0.72$ s, min $= 0.2420$, max $= 0.4232$. $b = 6 \to t \approx 0.22$ s, min $= 0.2422$, max $= 0.6304$. $b = 8 \to t \approx 0.28$ s, min $= 0.1571$, max $= 0.8$. $b = 10 \to t \approx 0.44$ s, min $= 4.3 \times 10^{-4}$, max $= 1$. Answers will vary on different systems.

5. $\mu = 10, I = [0, 20] \to t \approx 2.03$ s, min $= 0.0051$, max $= 0.0889$. $\mu = 50, I = [0, 100] \to t \approx 39.27$ s, min $= 0.0010$, max $= 0.0585$. $\mu = 100, I = [0, 200] \to t \approx 187.9$ s, min $= 5.3 \times 10^{-4}$, max $= 0.0442$. Answers will vary on different systems.

7. a. About 0.22 seconds on a 300-MHz PC.

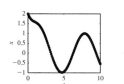

b. About 12 seconds on a 300-MHz PC.

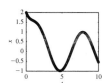

c. About 2.31 seconds on a 300-MHz PC. The second system is stiff.

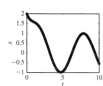

9. Solution for $h = 0.1$ Solution for $h = 0.01$

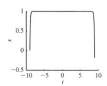

Solution for $h = 0.001$

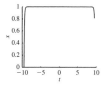

Chapter 7
Section 7.1

1.

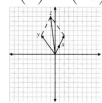

3.

5. $2 \begin{pmatrix} 1 \\ 2 \end{pmatrix} + 1 \begin{pmatrix} -3 \\ 4 \end{pmatrix} = \begin{pmatrix} -1 \\ 8 \end{pmatrix}$

7. $-2 \begin{pmatrix} 1 \\ 3 \end{pmatrix} + 2 \begin{pmatrix} 4 \\ 1 \end{pmatrix} = \begin{pmatrix} 6 \\ -4 \end{pmatrix}$

9. $B = \begin{pmatrix} 2 & 0 & 0 \\ 0 & 3 & 0 \\ 0 & 0 & 4 \end{pmatrix}$

15. $(A + B) + C = A + (B + C) = \begin{pmatrix} 2 & 0 \\ 1 & 2 \end{pmatrix}$

17. $A(B + C) = AB + AC = \begin{pmatrix} 5 & 2 \\ -1 & 2 \end{pmatrix}$

19. $(\alpha\beta)A = \alpha(\beta A) = \begin{pmatrix} -6 & -12 \\ 6 & 0 \end{pmatrix}$

21. $(\alpha + \beta)A = \alpha A + \beta A = \begin{pmatrix} 1 & 2 \\ -1 & 0 \end{pmatrix}$

23. $(A^T)^T = \begin{pmatrix} -2 & 4 \\ 4 & 0 \end{pmatrix}^T = \begin{pmatrix} -2 & 4 \\ 4 & 0 \end{pmatrix} = A$

25. $(A + B)^T = A^T + B^T = \begin{pmatrix} -2 & 2 \\ -1 & 3 \end{pmatrix}$

27. $\begin{pmatrix} 0 \\ 7 \end{pmatrix}$ **29.** $= \begin{pmatrix} -100 \\ 44 \end{pmatrix}$ **31.** $\begin{pmatrix} 43 \\ -41 \\ 40 \end{pmatrix}$

33. $\begin{pmatrix} 39 \\ 17 \end{pmatrix}$ **35.** $\begin{pmatrix} 20 \\ -17 \end{pmatrix}$ **37.** $\begin{pmatrix} -12 \\ 11 \\ 9 \end{pmatrix}$

39. $\begin{pmatrix} -7 \\ -51 \\ -12 \end{pmatrix}$ **41.** $\begin{pmatrix} 3 & -2 \end{pmatrix}$

43. $\begin{pmatrix} 9 & 5 \\ -6 & -1 \end{pmatrix}$ **45.** $\begin{pmatrix} 10 & -5 & 3 \\ 0 & 8 & 6 \\ -1 & 3 & 6 \end{pmatrix}$

47. $\begin{pmatrix} 3 & 4 \end{pmatrix} \begin{pmatrix} x \\ y \end{pmatrix} = \begin{pmatrix} 7 \end{pmatrix}$

49. $\begin{pmatrix} 3 & 4 \\ -1 & 3 \end{pmatrix} \begin{pmatrix} x \\ y \end{pmatrix} = \begin{pmatrix} 7 \\ 2 \end{pmatrix}$

51. $\begin{pmatrix} -1 & 0 & 1 \\ 2 & 3 & 0 \end{pmatrix} \begin{pmatrix} x_1 \\ x_2 \\ x_3 \end{pmatrix} = \begin{pmatrix} 0 \\ 3 \end{pmatrix}$

53. $\begin{pmatrix} -1 & 0 & 1 \\ 2 & 3 & 0 \\ 0 & 1 & -1 \end{pmatrix} \begin{pmatrix} x_1 \\ x_2 \\ x_3 \end{pmatrix} = \begin{pmatrix} 0 \\ 3 \\ 4 \end{pmatrix}.$

Section 7.2

1. $\begin{pmatrix} x \\ y \end{pmatrix} = \begin{pmatrix} 4 \\ 0 \end{pmatrix} + y \begin{pmatrix} 4/3 \\ 1 \end{pmatrix}$

3. $\begin{pmatrix} x \\ y \end{pmatrix} = \begin{pmatrix} 0 \\ -2 \end{pmatrix} + x \begin{pmatrix} 1 \\ 2 \end{pmatrix}$

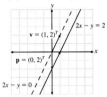

5. $(30/7, 6/7)$ **7.** $(3, 0)$

9. inconsistent   **11.** consistent

13. $\begin{pmatrix} x \\ y \\ z \end{pmatrix} = \begin{pmatrix} 12 - 2y + 3z \\ y \\ z \end{pmatrix}$

$= \begin{pmatrix} 12 \\ 0 \\ 0 \end{pmatrix} + y \begin{pmatrix} -2 \\ 1 \\ 0 \end{pmatrix} + z \begin{pmatrix} 3 \\ 0 \\ 1 \end{pmatrix}$

15. $\begin{pmatrix} x \\ y \\ z \end{pmatrix} = \begin{pmatrix} x \\ y \\ 12 - 3x + 4y \end{pmatrix}$

$= \begin{pmatrix} 0 \\ 0 \\ 12 \end{pmatrix} + x \begin{pmatrix} 1 \\ 0 \\ -3 \end{pmatrix} + y \begin{pmatrix} 0 \\ 1 \\ 4 \end{pmatrix}$

17. $\begin{pmatrix} x \\ y \\ z \end{pmatrix} = \begin{pmatrix} 31t - 50 \\ 28 - 14t \\ t \end{pmatrix} = \begin{pmatrix} -50 \\ 28 \\ 0 \end{pmatrix} + t \begin{pmatrix} 31 \\ -14 \\ 1 \end{pmatrix}$

19. $\begin{pmatrix} x \\ y \\ z \end{pmatrix} = \begin{pmatrix} 120 - (17/2)t \\ 50 - 3t \\ t \end{pmatrix} = \begin{pmatrix} 120 \\ 50 \\ 0 \end{pmatrix} + t \begin{pmatrix} -17/2 \\ -3 \\ 1 \end{pmatrix}$

21. $(x, y, z) = (0, -1/2, 7/2)$

23. $(x, y, z) = (-160/3, -230/3, -18)$

25. No. The set is not a point or a line.

27. No. The set is not a point or a line.

29. Yes. Consider $3x + 2y = 0$.

31. Yes. Consider $3x + 2y = 2$.

33. A point has zero dimension. A line is one dimensional.

35. In $\mathbf{R}^4$ we can expect dimensions 0, 1, 2, and 3.

37. $\mathbf{x} = (-4, 3, 0, 0)^T + x_3(-4, 3, 1, 0)^T + x_4(-3, 1, 0, 1)^T$. The set has dimension 2, a translated plane in $\mathbf{R}^4$.

Section 7.3

1. $\begin{pmatrix} x_1 \\ x_2 \\ x_3 \end{pmatrix} = \begin{pmatrix} 10/3 \\ -t + 2/3 \\ t \end{pmatrix} = \begin{pmatrix} 10/3 \\ 2/3 \\ 0 \end{pmatrix} + t \begin{pmatrix} 0 \\ -1 \\ 1 \end{pmatrix}$

3. Inconsistent

5. $\begin{pmatrix} x_1 \\ x_2 \\ x_3 \end{pmatrix} = \begin{pmatrix} 6 - 2s + 2t \\ s \\ t \end{pmatrix}$

$= \begin{pmatrix} 6 \\ 0 \\ 0 \end{pmatrix} + s \begin{pmatrix} -2 \\ 1 \\ 0 \end{pmatrix} + t \begin{pmatrix} 2 \\ 0 \\ 1 \end{pmatrix}$

7. $\begin{pmatrix} x_1 \\ x_2 \\ x_3 \end{pmatrix} = \begin{pmatrix} -2 - 2t \\ 4 + 2t \\ t \end{pmatrix} = \begin{pmatrix} -2 \\ 4 \\ 0 \end{pmatrix} + t \begin{pmatrix} -2 \\ 2 \\ 1 \end{pmatrix}$

9. $\begin{pmatrix} x_1 \\ x_2 \\ x_3 \end{pmatrix} = \begin{pmatrix} 1/3 + 43t/36 \\ 4/3 - 2t/9 \\ -1 + 7t/12 \\ t \end{pmatrix} = \begin{pmatrix} 1/3 \\ 4/3 \\ -1 \\ 0 \end{pmatrix} + t \begin{pmatrix} 43/36 \\ -2/9 \\ 7/12 \\ 1 \end{pmatrix}$

11. $\begin{pmatrix} 1 & 0 & -13 & 5 \\ 0 & 1 & 5 & -2 \end{pmatrix}$

13. $\begin{pmatrix} 1 & 0 & -2 & -8/3 \\ 0 & 1 & -3 & -4/3 \end{pmatrix}$

15. $\begin{pmatrix} 1 & 1 & 0 & 2 & 4 \\ 0 & 0 & 1 & 0 & 1 \end{pmatrix}$

17. $\begin{pmatrix} 1 & 0 & 0 & 3 & 23/2 \\ 0 & 1 & 0 & -1 & -9/2 \\ 0 & 0 & 1 & -3/2 & 1/4 \end{pmatrix}$

19. $\begin{pmatrix} x_1 \\ x_2 \\ x_3 \\ x_4 \end{pmatrix} = \begin{pmatrix} -2 + 2s - 2t \\ 1 - s + t \\ s \\ t \end{pmatrix}$

$= \begin{pmatrix} -2 \\ 1 \\ 0 \\ 0 \end{pmatrix} + s \begin{pmatrix} 2 \\ -1 \\ 1 \\ 0 \end{pmatrix} + t \begin{pmatrix} -2 \\ 1 \\ 0 \\ 1 \end{pmatrix}$

21.

$\begin{pmatrix} x_1 \\ x_2 \\ x_3 \\ x_4 \\ x_5 \end{pmatrix} = \begin{pmatrix} 1 + u - v + w \\ u \\ v \\ -w \\ w \end{pmatrix}$

$= \begin{pmatrix} 1 \\ 0 \\ 0 \\ 0 \\ 0 \end{pmatrix} + u \begin{pmatrix} 1 \\ 1 \\ 0 \\ 0 \\ 0 \end{pmatrix} + v \begin{pmatrix} -1 \\ 0 \\ 1 \\ 0 \\ 0 \end{pmatrix} + w \begin{pmatrix} 1 \\ 0 \\ 0 \\ -1 \\ 1 \end{pmatrix}$

23. $(1, 1, 1)^T$

25. $\mathbf{y} = (-2, 3/2, 0)^T + t(1, 1/2, 1)^T$, t free

27. $\mathbf{y} = (1, -2, 1)^T$

29. $\mathbf{x} = (0, 0, 1)^T + y(1, 1, 0)^T$, y free

31. $\mathbf{y} = (-75, 179/2, 0, 0)^T + x_3(2, -3/2, 1, 0)^T + x_4(15, -18, 0, 1)^T$, x_3, x_4 free

33. $\mathbf{y} = (-1, 2, -2, 0)^T + t(-31/4, 41/4, 123/8, 1)^T$, t free

35. $\mathbf{y} = \begin{pmatrix} (528 + 445s - 44t)/83 \\ (-165 - 82s - 173t)/83 \\ (-441 - 524s - 67t)/83 \\ s \\ t \end{pmatrix}$

$= \begin{pmatrix} 528/83 \\ -165/83 \\ -441/83 \\ 0 \\ 0 \end{pmatrix} + s \begin{pmatrix} 445/83 \\ -82/83 \\ -524/83 \\ 1 \\ 0 \end{pmatrix} + t \begin{pmatrix} -44/83 \\ -173/83 \\ -67/83 \\ 0 \\ 1 \end{pmatrix}$

Section 7.4

3. $\mathbf{c} = \begin{pmatrix} x_1 \\ x_2 \\ x_3 \end{pmatrix} = \begin{pmatrix} 2t \\ -3t \\ t \end{pmatrix} = t \begin{pmatrix} 2 \\ -3 \\ 1 \end{pmatrix}$

5. $\mathbf{x} = \begin{pmatrix} x_1 \\ x_2 \\ x_3 \\ x_4 \end{pmatrix} = \begin{pmatrix} -2s + 2t \\ -3s + t \\ s \\ t \end{pmatrix} = s \begin{pmatrix} -2 \\ -3 \\ 1 \\ 0 \end{pmatrix} + t \begin{pmatrix} 2 \\ 1 \\ 0 \\ 1 \end{pmatrix}$

7. $\mathbf{x} = \begin{pmatrix} x_1 \\ x_2 \\ x_3 \\ x_4 \\ x_5 \end{pmatrix} = \begin{pmatrix} s - 3t \\ -2s + 5t \\ s \\ -2t \\ t \end{pmatrix} = s \begin{pmatrix} 1 \\ -2 \\ 1 \\ 0 \\ 0 \end{pmatrix} + t \begin{pmatrix} -3 \\ 5 \\ 0 \\ -2 \\ 1 \end{pmatrix}$

9. $\mathbf{x} = \begin{pmatrix} x_1 \\ x_2 \\ x_3 \\ x_4 \end{pmatrix} = \begin{pmatrix} 2v - 4w \\ u \\ v \\ w \end{pmatrix}$

$= u \begin{pmatrix} 0 \\ 1 \\ 0 \\ 0 \end{pmatrix} + v \begin{pmatrix} 2 \\ 0 \\ 1 \\ 0 \end{pmatrix} + w \begin{pmatrix} -4 \\ 0 \\ 0 \\ 1 \end{pmatrix}$

11. Two special vectors (solutions are not unique): $(0, 1, 1, 0)^T$ and $(1, 1, 0, 1)^T$. Two free variables.

13. Three special vectors (solutions are not unique): $(1, -1, 1, 0, 0)^T$, $(0, 0, 0, 1, 0)^T$, and $(0, 1, 0, 0, 1)^T$. Three free variables.

15. a. At least 3 free variables.

b. Actually 4 free variables.

19. $\mathbf{x} = \begin{pmatrix} x_1 \\ x_2 \\ x_3 \end{pmatrix} = \begin{pmatrix} t \\ -1 \\ t \end{pmatrix} = \begin{pmatrix} 0 \\ -1 \\ 0 \end{pmatrix} + t \begin{pmatrix} 1 \\ 0 \\ 1 \end{pmatrix}$

21. $\mathbf{x} = \begin{pmatrix} x_1 \\ x_2 \\ x_3 \\ x_4 \end{pmatrix} = \begin{pmatrix} 2t \\ 1 + s + t \\ s \\ t \end{pmatrix} = \begin{pmatrix} 0 \\ 1 \\ 0 \\ 0 \end{pmatrix} + s \begin{pmatrix} 0 \\ 1 \\ 1 \\ 0 \end{pmatrix} + t \begin{pmatrix} 2 \\ 1 \\ 0 \\ 1 \end{pmatrix}$

23.

25.

27. $\mathbf{x} = \begin{pmatrix} x_1 \\ x_2 \\ x_3 \end{pmatrix} = \begin{pmatrix} t \\ -t \\ t \end{pmatrix} = t \begin{pmatrix} 1 \\ -1 \\ 1 \end{pmatrix}$ This is a line through the origin in $\mathbf{R}^3$.

29. $\mathbf{x} = \begin{pmatrix} x_1 \\ x_2 \\ x_3 \end{pmatrix} = \begin{pmatrix} s + t \\ s \\ t \end{pmatrix} = s \begin{pmatrix} 1 \\ 1 \\ 0 \end{pmatrix} + t \begin{pmatrix} 1 \\ 0 \\ 1 \end{pmatrix}$ This is a plane passing through the origin in $\mathbf{R}^3$.

Section 7.5

1. Yes. $\mathbf{w} = 1\mathbf{u}_1 + \frac{4}{3}\mathbf{u}_2$ **3.** No   **5.** No

7. Yes. $\mathbf{w} = 1\mathbf{v}_1 - 2\mathbf{v}_2 + 0\mathbf{v}_3$. Answers may vary.

9. $\mathbf{w} = (3w_1/7 - 2w_2/7)(1, -2)^T + (2w_1/7 + w_2/7)(2, 3)^T$

11. $-2\mathbf{v}_1 - \mathbf{v}_2 + \mathbf{v}_3 = \mathbf{0}$. Answers will vary.

13. $-2\mathbf{v}_1 - 3\mathbf{v}_2 + \mathbf{v}_3 = \mathbf{0}$. Answers will vary.

15. $-2\mathbf{v}_1 + 0\mathbf{v}_2 + 2\mathbf{v}_3 + \mathbf{v}_4 = \mathbf{0}$. Answers will vary.

17. Independent   **19.** Independent

21. Independent   **23.** Independent

25. $B = \{(1/2, 1)^T\}$ **27.** $B = \{(-1, 1)^T\}$

29. $B = \{(-1, 0, 1)^T$

31. $B = \{(-1, -2, 1, 0)^T, (-1, -1, 0, 1)^T\}$

33. $B = \{(1, 2)^T, (-1, 3)^T\}$, dimension $= 2$

35. $B = \{(-1, 7, 7)^T, (-3, 7, -4)^T\}$, dimension $= 2$

37. $B = \{(-1, 7, 7)^T, (-3, 7, -4)^T, (-4, -14, 23)^T\}$, dimension $= 3$

39. $B = \{(-1, 7, 7)^T, (-3, 8, -4)^T, (-4, -14, 23)^T\}$, dimension $= 3$

41. A line through the origin in $\mathbf{R}^2$.

43. A plane through the origin.

Section 7.6

1. Singular   **3.** Singular   **5.** Nonsingular

7. Nonsingular   **9.** Singular   **11.** Singular

13. $\mathbf{x} = x_2(2, 1)^T$, x_2 free. Singular.

15. $\mathbf{x} = x_3(-1, -1, 0)^T$, x_3 free. Singular

17. $\mathbf{x} = x_3(-1, -2, 1)^T$, x_3 free. Singular.

19. The only solution is $\mathbf{x} = (0, 0, 0, 0)^T$. Nonsingular.

21. Nonsingular, $A^{-1} = \begin{pmatrix} -1/2 & -1 \\ -1/4 & 0 \end{pmatrix}$

23. Nonsingular, $A^{-1} = \begin{pmatrix} 1 & -1 & 0 \\ 0 & 1 & -1 \\ 0 & 0 & 1 \end{pmatrix}$

25. Singular   **27.** Singular

29. No unique solution   **31.** Unique solution

33. No unique solution

35. Suppose that A is invertible. Then
- A is nonsingular.
- The only solution of the homogeneous system $A\mathbf{y} = \mathbf{0}$ is the zero vector $\mathbf{0}$.
- The equation $A\mathbf{x} = \mathbf{b}$ has a unique solution for any right-hand side $\mathbf{b}$.
- If A is put into row echelon form, then the diagonal entries of he result are nonzero.

• If A is put into reduced row echelon form, then the result is the identity matrix.

Section 7.7

3. Area $= 23$ **5.** Area $= 40$

7. 0   **9.** -44 **11.** 8

15. 24   **17.** -9 **19.** -10 **21.** 0

23. $\det = 0$, $B = \{(2, 1)^T\}$, dependent

25. $\det = 1$, $B = \{(0, 0)^T\}$, independent

27. $\det = 0$, $B = \{(0, -2, 1)^T\}$, dependent

29. $\det = 1$, $B = \{(0, 0, 0)^T\}$, independent

31. $x = \pm\sqrt{6}$

33. $x = \pm 2$ **35.** $x = -1, 2$

37. $x = 1, 2, 4$ **39.** $x = -2, 0, 1$

41. -2; no   **43.** -54; no   **45.** -4; no

47. 1; no   **49.** 0; yes   **51.** False

53. $\det A \neq 0$ is equivalent to each of the following:
   • A is nonsingular.
   • A is invertible.
   • null(A) is trivial.
   • The system $A\mathbf{x} = \mathbf{b}$ has a unique solution for every right-hand side $\mathbf{b}$.
   • If A is $n \times n$, the column vectors in A are a basis for $\mathbf{R}^n$.
   • When A is reduced to row echelon form the diagonal entices are all nonzero.
   • When A is reduced to reduced row echelon form the result is the identity matrix.

Chapter 8
Section 8.1

1. Dimension 2, nonautonomous

3. Dimension 3, autonomous

5. Dimension 4, autonomous

11. With $\mathbf{u} = (u_1, u_2)^T = (y, y')^T$ we have $u_1' = u_2$, $u_2' = -2u_2 - 4u_1 + 3\cos 2t$, with initial conditions $\mathbf{u}(0) = (1, 0)^T$.

13. With $\mathbf{u} = (u_1, u_2)^T = (x, x')^T$, we have $u_1' = u_2$, $u_2' = -\delta u_2 + u_1 - u_1^3 + \gamma \cos \omega t$, with initial conditions $\mathbf{u}(0) = (x_0, v_0)^T$.

15. With $\mathbf{u} = (u_1, u_2, u_3)^T = (\omega, \omega', \omega'')^T$, we have $u_1' = u_2, u_2' = u_3, u_3' = u_1$, with initial conditions $\mathbf{u}(0) = (\omega_0, \alpha_0, \gamma_0)^T$.

17. Nonautonomous   **19.** Nonautonomous

21. Nonautonomous

23. Let $x_1 = u$, $x_2 = v$. Then the system can be written $\mathbf{x}' = \mathbf{f}(t, \mathbf{x})$, where $\mathbf{x}' = (x_1', x_2')^T$ and $\mathbf{f}(t, \mathbf{x}) = (x_2, -3x_1 - 2x_2 + 5\cos t)^T$.

25. Let $x_1 = u$, $x_2 = v$. Then the system can be written $\mathbf{x}' = \mathbf{f}(t, \mathbf{x})$, where $\mathbf{x}' = (x_1', x_2')^T$ and $\mathbf{f}(t, \mathbf{x}) = (x_2 \cos x_1, t x_2)^T$.

27. Let $x_1 = u$, $x_2 = v$, and $x_3 = \omega$. Then the system can be written $\mathbf{x}' = \mathbf{f}(t, \mathbf{x})$, where $\mathbf{x} = (x_1', x_2', x_3')^T$ and $\mathbf{f}(t, \mathbf{x}) = (x_2 + \cos x_1, x_2 - t x_3, 5x_1 - 9x_2 + 8x_3)^T$.

Section 8.2

1. x_1 is the solid curve, x_2 is the dashed curve.

The plot of $t \to (x_1(t), x_2(t))$.

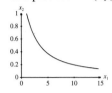

3. x_1 is the solid curve, x_2 is the dashed curve.

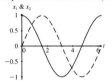

The plot of $t \to (x_1(t), x_2(t))$.

5. x_1 is the solid curve, x_2 is the dashed curve.

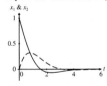

The plot of $t \to (x_1(t), x_2(t))$.

7. $\mathbf{x}'(t) = (2e^t + e^{-t}, -e^{-t})$. In the following figure, the tangent vectors are plotted at 25% of their actual length.

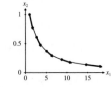

9. $\mathbf{x}'(t) = (-\sin t, \cos t)$. In the following figure, the tangent vectors are plotted at 25% of their actual length.

11. $\mathbf{x}'(t) = (e^{-t}(-\sin t - \cos t), e^{-t}(\cos t - \sin t))$. In the following figure, the tangent vectors are plotted at 25% of their actual length.

13.

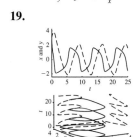

15.

17.

19.

21. I $\mapsto$ D, II $\mapsto$ A, III $\mapsto$ B, IV $\mapsto$ C

23.

25.

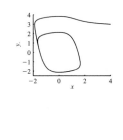

29. c. An implicit function plotter was used to sketch

$$0.2 \ln S - 0.1S + 0.3 \ln F - 0.1F = C,$$

for $C = -0.2, -0.4, -0.6, -0.8,$ and -1.

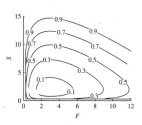

Section 8.3

1. Eq. pts: $(0, 0), (20, 5)$

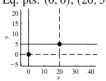

3. Eq. pts: $(0, 0)$

5. Eq. pts: $(k\pi, 0)$, k and integer.

7. b.

9. c.

11. a.

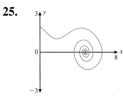

b. The prey. See eq pt in the figure of part (a).

c.

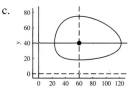
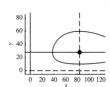

15. $(0, 0, 0), (\sqrt{\beta(\rho - 1)}, \sqrt{\beta(\rho - 1)}, \rho - 1)$, and $(-\sqrt{\beta(\rho - 1)}, -\sqrt{\beta(\rho - 1)}, \rho - 1)$

Section 8.4

1. Linear, inhomogeneous **3.** Nonlinear
5. Linear, inhomogeneous

11. $\begin{pmatrix} x_1 \\ x_2 \end{pmatrix}' = \begin{pmatrix} -2 & 3 \\ 1 & -4 \end{pmatrix} \begin{pmatrix} x_1 \\ x_2 \end{pmatrix}$

13. Nonlinear   **15.** Nonlinear

17. Exercises 11, 12, 14, and 16 are linear; 13 and 15 are nonlinear. Of the linear systems, only 11 and 12 are homogeneous.

19. $\begin{pmatrix} x_1 \\ x_2 \end{pmatrix}' = \begin{pmatrix} 8 & -10 \\ 5 & -7 \end{pmatrix} \begin{pmatrix} x_1 \\ x_2 \end{pmatrix}$

21. $\begin{pmatrix} x_1 \\ x_2 \end{pmatrix}' = \begin{pmatrix} -1 & 4 \\ 0 & 3 \end{pmatrix} \begin{pmatrix} x_1 \\ x_2 \end{pmatrix}$

25. $\mathbf{x} = \begin{pmatrix} -1/25 & 0 \\ 1/25 & -1/25 \end{pmatrix} \mathbf{x}, \ \mathbf{x}(0) = (10, 20)^T$, homogeneous

27. $\begin{pmatrix} x_1 \\ x_2 \\ x_3 \end{pmatrix}' = \begin{pmatrix} -1/20 & 0 & 0 \\ 1/20 & -1/16 & 0 \\ 0 & 1/16 & -1/12 \end{pmatrix} \begin{pmatrix} x_1 \\ x_2 \\ x_3 \end{pmatrix}$
$+ \begin{pmatrix} 10 \\ 0 \\ 0 \end{pmatrix}, \ \mathbf{x}(0) = (0, 0, 0)^T$, inhomogeneous

29.
$$\begin{pmatrix} x_1 \\ x_2 \\ x_3 \\ x_4 \\ x_5 \\ x_6 \end{pmatrix}' =$$
$$\begin{pmatrix} 0 & 1 & 0 & 0 & 0 & 0 \\ -2k/m & 0 & k/m & 0 & 0 & 0 \\ 0 & 0 & 0 & 1 & 0 & 0 \\ k/m & 0 & -2k/m & 0 & k/m & 0 \\ 0 & 0 & 0 & 0 & 0 & 1 \\ 0 & 0 & k/m & 0 & -2k/m & 0 \end{pmatrix} \begin{pmatrix} x_1 \\ x_2 \\ x_3 \\ x_4 \\ x_5 \\ x_6 \end{pmatrix},$$
$\mathbf{x}(0) = (10, 0, 10, 0, 10, 0)^T$, homogeneous

Section 8.5

1. $\begin{pmatrix} x_1 \\ x_2 \end{pmatrix}' = \begin{pmatrix} -1 & 3 \\ 0 & 2 \end{pmatrix} \begin{pmatrix} x_1 \\ x_2 \end{pmatrix}$

3. $\begin{pmatrix} x_1 \\ x_2 \end{pmatrix}' = \begin{pmatrix} 1 & 1 \\ -1 & 1 \end{pmatrix} \begin{pmatrix} x_1 \\ x_2 \end{pmatrix}$

5. $\begin{pmatrix} x_1 \\ x_2 \end{pmatrix}' = \begin{pmatrix} 1 & 1 \\ -1 & 1 \end{pmatrix} \begin{pmatrix} x_1 \\ x_2 \end{pmatrix} + \begin{pmatrix} 0 \\ e^t \end{pmatrix}$

13. $\mathbf{z}(t) = \begin{pmatrix} -e^{-t} + e^{2t} \\ e^{2t} \end{pmatrix}$

15. $\mathbf{z}(t) = \begin{pmatrix} -2e^t \cos t + 3e^t \sin t \\ 2e^t \sin t + 3e^t \cos t \end{pmatrix}$

19. Independent   **21.** Independent

23. $\mathbf{y}(t) = \begin{pmatrix} -3e^{2t} - 2e^{-2t} \\ 6e^{2t} + 2e^{-2t} \end{pmatrix}$

25. $\mathbf{y}(t) = \begin{pmatrix} -te^{2t} \\ (1-t)e^{2t} \end{pmatrix}$.

27. $dx_A/dt = \frac{4x_B}{3} - 3x_A, dx_B/dt = 3x_A - 3x_B$

29. $dx_A/dt = x_B/50 - 9x_A/200,$
$dx_B/dt = 9x_A/200 - 3x_B/200 + x_C/50,$
$dx_C/dt = 9x_B/200 - 9x_C/200$

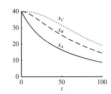

31. $dx/dt = -x/20, dy/dt = x/20 - y/20,$
$dz/dt = y/20 - z/20$

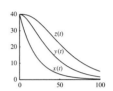

33. With $(u_1, u_2, u_3, u_4)^T = (x, x', y, y')^T, u_1' = u_2,$
$u_2' = -ku_1/(u_1^2 + u_3^2)^{3/2}, u_3' = u_4,$
$u_4' = -ku_3/(u_1^2 + u_3^2)^{3/2}$

35. As in Exercise 33, with
$(u_1, u_2, u_3, u_4)^T = (x, x', y, y')^T, u_1' = u_2,$
$u_2' = -ku_1/(u_1^2 + u_3^2)^2, u_3' = u_4,$
$u_4' = -ku_3/(u_1^2 + u_3^2)^2$

Chapter 9
Section 9.1

1. $p(\lambda) = \lambda^2 - 3\lambda - 10, \lambda_1 = 5,$ and $\lambda_2 = -2$
3. $p(\lambda) = \lambda^2 + 7\lambda + 10, \lambda_1 = -2,$ and $\lambda_2 = 5$
5. $p(\lambda) = \lambda^2 - \lambda - 2, \lambda_1 = 2,$ and $\lambda_2 = -1$
7. $p(\lambda) = \lambda^2 + 6\lambda + 9, \lambda_1 = 3,$ and $\lambda_2 = 3$
9. $p(\lambda) = (\lambda - 1)(\lambda - 3)(\lambda + 2), \lambda_1 = 1, \lambda_2 = 3,$ and $\lambda_3 = -2$
11. $p(\lambda) = (\lambda + 1)(\lambda - 1)(\lambda - 2), \lambda_1 = -1, \lambda_2 = 1,$ and $\lambda_3 = 2.$
13. The zeros of the graph are the eigenvalues.

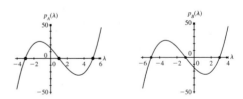

17. $\mathbf{y}_1(t) = e^{6t}(1, 0)^T, \mathbf{y}_2(t) = e^{-2t}(1, 1)^T$
19. $\mathbf{y}_1(t) = e^t(1, 0)^T, \mathbf{y}_2(t) = e^t(0, 1)^T$
21. $\mathbf{y}_1(t) = e^{-3t}(1, -1)^T, \mathbf{y}_2(t) = e^{2t}(2, -1)^T$
23. $\mathbf{y}_1(t) = e^{-3t}(1, 2)^T, \mathbf{y}_2(t) = e^t(1, 1)^T$
25. $\mathbf{y}_1(t) = e^{-t}(1, -1, 1)^T$
$\mathbf{y}_2(t) = e^{-2t}(0, -2, 1)^T$
$\mathbf{y}_3(t) = e^t(0, -1, 1)^T$
27. $\mathbf{y}_1(t) = e^{-3t}(-1, 1, 0)^T$
$\mathbf{y}_2(t) = e^{-2t}(2, 0, 1)^T$
$\mathbf{y}_3(t) = e^{-t}(1, 3/2, 1)^T$

29. $3 \to (1, 0, 1)^T$
  $1 \to (1, -1, 1)^T$
  $-2 \to (2, 0, 1)^T$

31. $3 \to (1, -2, 0)^T$
  $2 \to (1, -2, -1)^T$
  $-1 \to (1, -1, 3)^T$

33. $1 \to (-1, -1, 1)^T$
  $-3 \to (-1, 1, 2)^T$
  $-1 \to (-2, 1, 3)^T$

35. $3 \to (1, 1, 2, 2)^T$
  $-1 \to (0, 1, 3/2, 1)^T$
  $-2 \to (0, 1, 1, 1)^T$
  $-4 \to (-1, 2, 1, 2)^T$

37. $4 \to (1, 1, 1, 0)^T$
  $-2 \to (0, 0, -1, 1)^T$
  $2 \to (-1, -1, -1, 1)^T$
  $-1 \to (-1, 0, 1, 0)^T$

39. $\mathbf{y}_1(t) = e^{3t}(2, 1, 0)^T$
  $\mathbf{y}_2(t) = e^{-4t}(-1, -1, 1)^T$
  $\mathbf{y}_3(t) = e^{-2t}(2, 1, 1)^T$

41. $\mathbf{y}_1(t) = e^{-t}(0, 0, 1)^T$
  $\mathbf{y}_2(t) = e^{-3t}(-1, 0, 1)^T$
  $\mathbf{y}_3(t) = e^{2t}(-2, 1, 0)^T$

43. $\mathbf{y}_1(t) = e^{t}(1, -2, 2)^T$
  $\mathbf{y}_2(t) = e^{-3t}(1, -2, 1)^T$
  $\mathbf{y}_3(t) = e^{4t}(-1, 1, -1)^T$

45. $\mathbf{y}_1(t) = e^{-4t}(1, -1, 0, 0)^T$
  $\mathbf{y}_2(t) = e^{-2t}(-1, 1, 0, 0)^T$
  $\mathbf{y}_3(t) = e^{2t}(-1, -1, -1, 1)^T$
  $\mathbf{y}_4(t) = e^{-t}(-3/2, 1, -1, 0)^T$

47. $\mathbf{y}_1(t) = e^{-5t}(0, 2, 1, 2)^T$
  $\mathbf{y}_2(t) = e^{-2t}(1, 0, 1, 2)^T$
  $\mathbf{y}_3(t) = e^{4t}(0, 2, 1, 1)^T$
  $\mathbf{y}_4(t) = e^{2t}(1, 1, 1, 2)^T$

49. A has eigenvalues 2 and -2, and $\det(A) = -4$. B has eigenvalues -3 and 5, and $\det(B) = -15$. C has eigenvalues 2, -3, and -4, and $\det(C) = 24$. The product of the eigenvalues equals the determinant.

51. The main diagonal of an upper triangular or lower triangular matrix contains the eigenvalues.

55. $V = \begin{pmatrix} 1 & 1 \\ 2 & 1 \end{pmatrix}$ and $D = \begin{pmatrix} -5 & 0 \\ 0 & -3 \end{pmatrix}$

Section 9.2

1. $\mathbf{y}(t) = C_1 e^{2t} \begin{pmatrix} 1 \\ 0 \end{pmatrix} + C_2 e^{-t} \begin{pmatrix} 2 \\ 1 \end{pmatrix}$

3. $\mathbf{y}(t) = C_1 e^{-4t} \begin{pmatrix} 1 \\ 1 \end{pmatrix} + C_2 e^{-3t} \begin{pmatrix} 1 \\ 2 \end{pmatrix}$

5. $\mathbf{y}(t) = C_1 e^{2t} \begin{pmatrix} 2 \\ 1 \end{pmatrix} + C_2 e^{3t} \begin{pmatrix} 1 \\ 1 \end{pmatrix}$

7. $\mathbf{y}(t) = -2 e^{2t} \begin{pmatrix} 1 \\ 0 \end{pmatrix} + e^{-t} \begin{pmatrix} 2 \\ 1 \end{pmatrix}$

9. $\mathbf{y}(t) = e^{-4t} \begin{pmatrix} 1 \\ 1 \end{pmatrix} - e^{-3t} \begin{pmatrix} 1 \\ 2 \end{pmatrix}$

11. $\mathbf{y}(t) = e^{2t} \begin{pmatrix} 2 \\ 1 \end{pmatrix} + e^{3t} \begin{pmatrix} 1 \\ 1 \end{pmatrix}$

13. $\text{Re}\,(\mathbf{z}(t)) = (\cos 2t, \cos 2t - \sin 2t)^T$
  $\text{Im}\,(\mathbf{z}(t)) = (\sin 2t, \cos 2t + \sin 2t)^T$

17. $\mathbf{y}_1(t) = e^{t} \begin{pmatrix} \cos 2t \\ -\cos 2t + \sin 2t \end{pmatrix}$,
  $\mathbf{y}_2(t) = e^{t} \begin{pmatrix} \sin 2t \\ -\cos 2t - \sin 2t \end{pmatrix}$

19. $\mathbf{y}_1(t) = e^{-2t} \begin{pmatrix} -\cos 2t + \sin 2t \\ \cos 2t \end{pmatrix}$,
  $\mathbf{y}_2(t) = e^{-2t} \begin{pmatrix} -\cos 2t - \sin 2t \\ \sin 2t \end{pmatrix}$

21. $\mathbf{y}_1(t) = e^{4t} \begin{pmatrix} 6\cos \sqrt{17}t \\ -\cos \sqrt{17}t + \sqrt{17} \sin \sqrt{17}t \end{pmatrix}$,
  $\mathbf{y}_2(t) = e^{4t} \begin{pmatrix} 6\sin \sqrt{17}t \\ -\sqrt{17}\cos \sqrt{17}t - \sin \sqrt{17}t \end{pmatrix}$

23. $\mathbf{y}(t) = -e^{t} \begin{pmatrix} \sin 2t \\ -\cos 2t - \sin 2t \end{pmatrix}$

25. $\mathbf{y}(t) = e^{-2t} \begin{pmatrix} -\cos 2t + 3\sin 2t \\ 2\cos 2t - \sin 2t \end{pmatrix}$

27. $\mathbf{y}(t) = \dfrac{1}{6} e^{4t} \begin{pmatrix} 6\cos \sqrt{17}t \\ -\cos \sqrt{17}t + \sqrt{17} \sin \sqrt{17}t \end{pmatrix}$
  $- \dfrac{19\sqrt{17}}{102} \begin{pmatrix} 6\sin \sqrt{17}t \\ -\sqrt{17}\cos \sqrt{17}t - \sin \sqrt{17}t \end{pmatrix}$.

29. $\mathbf{y}(t) = C_1 e^{-2t}(1, 0)^T + C_2 e^{-2t}(0, 1)^T$

31. $\mathbf{y}(t) = e^{2t}[(C_1 + C_2 t)(1, 1)^T + C_2(1, 0)^T]$

33. $\mathbf{y}(t) = e^{t}[(C_1 + C_2 t)(1, 3)^T + C_2(-1/3, 0)^T]$

35. $\mathbf{y}(t) = e^{-2t}(3, -2)^T$

37. $\mathbf{y}(t) = e^{2t}(2 + 3t, -1 + 3t)^T$

39. $\mathbf{y}(t) = e^{t}(5 - 12t, 3 - 36t)^T$

41. $\mathbf{y}(t) = e^{4t}[(C_1 + C_2 t)(2, 1)^T + C_2(-1, 0)^T]$

43. $\mathbf{y}(t) = C_1 e^{-t}(-2, 1)^T + C_2 e^{-3t}(-3/2, 1)^T$

45. $\mathbf{y}(t) = C_1 e^{-t}(5\cos t, -3\cos t + \sin t)^T + C_2 e^{-t}(5\sin t, -\cos t - 3\sin t)^T$

47. $\mathbf{y}(t) = C_1 e^{-4t}(2, 3)^T + C_2 e^{-2t}(1, 2)^T$

49. $\mathbf{y}(t) = e^{4t}(3 - 2t, 1 - t)^T$

51. $\mathbf{y}(t) = (4e^{-t} - 3e^{-3t}, -2e^{-t} + 2e^{-3t})^T$

53. $\mathbf{y}(t) = e^{-t}(-3\cos t - \sin t, 2\cos t)^T$

55. $\mathbf{y}(t) = (6e^{-4t} - 4e^{-2t}, 9e^{-4t} - 8e^{-2t})^T$

59. a. $\begin{pmatrix} x_A \\ x_B \end{pmatrix}' = \begin{pmatrix} -1/40 & 1/90 \\ 1/40 & -1/40 \end{pmatrix} \begin{pmatrix} x_A \\ x_B \end{pmatrix}$,
  $\mathbf{x}(0) = (60, 0)^T$

  b. $\mathbf{x} = 15 e^{-t/120} \begin{pmatrix} 2 \\ 3 \end{pmatrix} - 15 e^{-t/24} \begin{pmatrix} -2 \\ 3 \end{pmatrix}$

  c.

  taxis

  The salt content in each tank goes to zero.

61. $V = (4\sqrt{3} + 6)e^{(-3+\sqrt{3})t} - (4\sqrt{3} - 6)e^{(-3-\sqrt{3})t}$
$I = 2\sqrt{3}e^{(-3+\sqrt{3})t} - 2\sqrt{3}e^{(-3-\sqrt{3})t}$.

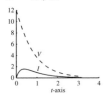

Section 9.3

1. $p(\lambda) = \lambda^2 - T\lambda + D = \lambda^2 + 25$

3.

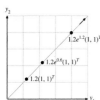

5.

7.

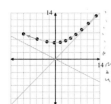

9.

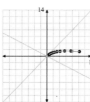

11.

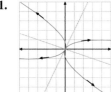

13.

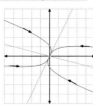

15.

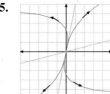

17.

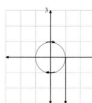

19.

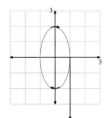

21. Spiral source

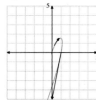

23. Spiral sink

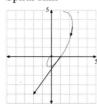

Section 9.4

1. Center with clockwise motion

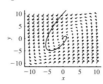

3. Spiral sink with counterclockwise motion

5. Nodal sink   **7.** Saddle

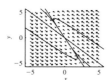

9. Spiral source with clockwise rotation.   **11.** Center with clockwise rotation.

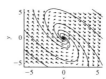
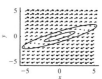

13.   a. $T^2 - 4D = 0$.   b. $\mathbf{y}(t) \to \mathbf{0}$ as $t \to \infty$

c. Nongeneric sink with clockwise rotation.

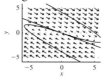

15.   a. $\to \infty$ along the half-line generated by $C_1\mathbf{v}_1$.

b. $\to \infty$, becoing parallel to the half-line generated by $C_2\mathbf{v}_1$.

c. $\to \mathbf{0}$ tangent to the half-line generated by $-C_2\mathbf{v}_1$.

d. Degenerate nodal source

17. $\mathbf{y}(t) = e^{4t}\left[(C_1 + C_2t)\begin{pmatrix} 2 \\ -1 \end{pmatrix} + C_2\begin{pmatrix} 1 \\ 0 \end{pmatrix}\right]$

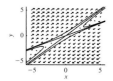

19.   a. Degenerate nodal source

b. The half-line solutions $y = \pm x/\sqrt{a}$ coalesce as $a \to 0$.

c. When $a = 0$, the half-line solutions coalesce, $T^2 - 4D = -$, and we have a degenerate (nongeneric) nodal source. When $a < 0$, then $T^2 - 4D = 4a < 0$, and we have a spiral source.

23. (i) On the T-axis.

(ii) Every point on line $y = -2x$ is an equilibrium point.

(iii) $\mathbf{y}(t) = C_1 \begin{pmatrix} 1 \\ -2 \end{pmatrix} + C_2 e^{3t} \begin{pmatrix} 4 \\ -5 \end{pmatrix}$

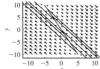

25. Overdamped if $L - 4R^2C > 0$. Critically damped if $L - 4R^2C = 0$. Underdamped if $L - 4R^2C < 0$.

Section 9.5

1. $p(\lambda) = (\lambda + 1)(\lambda - 2)(\lambda - 1)$, eigs: -1, 2, and 1, which are the horizontal intercepts of p.

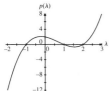

3. $2 \to (-1, 2, 1)^T$, $1 \to (0, 1, 0)^T$, $-1 \to (0, 2, 1)^T$

5. $0 \to (1, 0, 2)^T$, $2 \to (0, 1, 0)^T$, $-1 \to (-2, 2, -3)^T$

7. $C_1 e^{-t}(1, 1, 0)^T + C_2 e^{4t}(1, 0, 0)^T + C_3 e^{t}(2, 2, 1)^T$

9. $C_1 e^{4t}(0, 2, 1)^T + C_2 e^{-3t}(1, 1, 1)^T + C_3 e^{2t}(0, 1, 1)^T$

11. $C_1 e^{-t}(2, 1, 0)^T + C_2 e^{t}(1, 1, 0)^T + C_3 e^{2t}(0, -2, 1)^T$

13. $-5e^{-t}(1, 1, 0)^T + 2e^{4t}(1, 0, 0)^T + 2e^{t}(2, 2, 1)^T$

15. $-2e^{4t}(0, 2, 1)^T - 2e^{-3t}(1, 1, 1)^T + 6e^{2t}(0, 1, 1)^T$

17. $e^{-t}(2, 1, 0)^T - e^{t}(1, 1, 0)^T + e^{2t}(0, -2, 1)^T$

19. $\text{Re}(\mathbf{y}(t)) =$
$(\cos 2t, \cos 2t - 2\sin 2t, -3\cos 2t + 3\sin 2t)^T$
$\text{Im}(\mathbf{y}(t)) =$
$(\sin 2t, 2\cos 2t + \sin 2t, -3\cos 2t - 3\sin 2t)^T$

21. $C_1 e^{2t}(0, -1, 1)^T + C_2(\cos 4t + \sin 4t, \cos 4t, 0)^T + C_3(-\cos 4t + \sin 4t, \sin 4t, 0)^T$

23. $C_1 e^{-2t}(0, 1, 0)^T + C_2 e^{2t}(\cos 4t - \sin 4t, 2\cos 4t, 2\cos 4t)^T + C_3 e^{2t}(\cos 4t + \sin 4t, 2\sin 4t, 2\sin 4t)^T$

25. $C_1 e^{-2t}(0, 0, 1)^T + C_2 e^{-2t}(-5\cos t - \sin t, 2\cos t, 3\cos t + \sin t)^T + C_3 e^{-2t}(\cos t - 5\sin t, 2\sin t, -\cos t + 3\sin t)^T$

27. $(x(t), y(t), z(t))^T = (\cos 4t - \sin 4t, -\sin 4t, 0)^T$

29. $(x(t), y(t), z(t))^T = (e^{2t}(-2\cos 4t - 2\sin 4t), -e^{-2t} - 4e^{2t}\sin 4t, -4e^{2t}\sin 4t)^T$

31. $\mathbf{y}(t) = e^{-2t}(-\cos t - \sin t, \cos t + 3\sin t, 1 + 5\sin t)^T$

33. $\lambda_1 = 1$, $q_1 = 2$, $d_1 = 1$, $\lambda_2 = 5$, $q_2 = 1$, $d_2 = 1$

35. $\lambda_1 = 4$, $q_1 = 1$, $d_1 = 1$, $\lambda_2 = -2$, $q_2 = 2$, $d_2 = 1$

37. $\mathbf{y}_1(t) = e^{-2t}(1, -1, -2)^T$
$\mathbf{y}_2(t) = e^{-3t}(-1, 0, 1)^T$
$\mathbf{y}_3(t) = e^{-t}(-1, -1, 1)^T$

39. $\mathbf{y}_1(t) = e^{-t}(0, 1, 3)^T$
$\mathbf{y}_2(t) = e^{-2t}(-2\cos 2t, 2\cos 2t, \cos 2t - \sin 2t)^T$
$\mathbf{y}_3(t) = e^{-2t}(-2\sin 2t, 2\sin 2t, \cos 2t + \sin 2t)^T$

41. $\mathbf{y}_1(t) = e^{-2t}(1, -2, -2)^T$
$\mathbf{y}_2(t) = e^{-3t}(-6\cos 2t - 2\sin 2t, 8\cos 2t + \sin 2t, 5\cos 2t)^T$
$\mathbf{y}_3(t) = e^{-3t}(2\cos 2t - 6\sin 2t, -\cos 2t + 8\sin 2t, 5\sin 2t)^T$

43. $\mathbf{y}_1(t) = e^{-t}(-1, 2, -1, 1)^T$
$\mathbf{y}_2(t) = e^{-2t}(4, -3, 0, 0)^T$
$\mathbf{y}_3(t) = e^{-2t}(1, 0, -3, 0)^T$
$\mathbf{y}_4(t) = e^{-2t}(5, 0, 0, 3)^T$

45. $\mathbf{y}(t) = \begin{pmatrix} -3e^{-2t} - 2e^{-3t} - e^{-t} \\ 3e^{-2t} - e^{-t} \\ 6e^{-2t} + 2e^{-3t} + e^{-t} \end{pmatrix}$

47. $\mathbf{y}(t) = \begin{pmatrix} 38e^{-2t}\sin 2t \\ 8e^{-t} - 38e^{-2t}\sin 2t \\ 24e^{-t} - 19e^{-2t}\cos 2t - 19e^{-2t}\sin 2t \end{pmatrix}$

49. $\mathbf{y}(t) = \begin{pmatrix} 7e^{-2t} - e^{-3t}(8\cos 2t - 44\sin 2t) \\ -14e^{-2t} + e^{-3t}(21\cos 2t + 53\sin 2t) \\ -14e^{-2t} + e^{-3t}(17\cos 2t + 31\sin 2t) \end{pmatrix}$

51. $\mathbf{y}(t) = \begin{pmatrix} -e^{-t} \\ 2e^{-t} + 3e^{-2t} \\ -e^{-t} + 3e^{-2t} \\ e^{-t} + 3e^{-2t} \end{pmatrix}$

53. $x_1 = 4e^{-t/20}$
$x_2 = 12e^{-t/20} - 10e^{-t/15}$
$x_3 = 16e^{-t/20} - 20e^{-t/15} + 5e^{-t/10}$

Section 9.6

1. $e^A = \begin{pmatrix} -1 & -4 \\ 1 & 3 \end{pmatrix}$

3. $e^A = \begin{pmatrix} 2 & -1 & 0 \\ 1 & 0 & 0 \\ 0 & 0 & 1 \end{pmatrix}$

5. b. $e^{tA} = \begin{pmatrix} (e^{3t} + 2)/3 & (e^{3t} - 1)/3 & (e^{3t} - 1)/3 \\ (e^{3t} - 1)/3 & (e^{3t} + 2)/3 & (e^{3t} - 1)/3 \\ (e^{3t} - 1)/3 & (e^{3t} - 1)/3 & (e^{3t} + 2)/3 \end{pmatrix}$

9. b. $e^{A+B} = \begin{pmatrix} \cos 2 & -\sin 2 \\ \sin 2 & \cos 2 \end{pmatrix}$

c. $e^A e^B = \begin{pmatrix} -3 & -2 \\ 2 & 1 \end{pmatrix}$, so $e^{A+B} \neq e^A e^B$ in this example where $AB \neq BA$.

11. $e^{tA} = \begin{pmatrix} e^{2t} & 2e^{2t} - 2e^{-t} \\ 0 & e^{-t} \end{pmatrix}$

15. $e^{tA} = e^{-t} \begin{pmatrix} 1 & 0 \\ t & 1 \end{pmatrix}$

17. $e^{tA} = e^{-t} \begin{pmatrix} 1 - 2t & -t \\ 4t & 1 + 2t \end{pmatrix}$

19. $k = 3$,
$$e^{-tA} = e^{-t}\begin{pmatrix} 1 + t^2/2 & -t - t^2/2 & t^2/2 \\ -t & 1 + t & -t \\ -t - t^2/2 & 2t + t^2/2 & 1 - t - t^2/2 \end{pmatrix}$$

21. $k = 2$, $e^{tA} = e^{-2t}\begin{pmatrix} 1 & 0 & 0 \\ 0 & 1 & 0 \\ -t & t & 1 \end{pmatrix}$

23. $k = 2$,
$$e^{tA} = e^{-3t}\begin{pmatrix} 1 - 2t & 0 & -t & 4t \\ -4t & 1 + 3t & t & 5t \\ 4t & -4t & 1 - 2t & -4t \\ 0 & -t & -t & 1 + t \end{pmatrix}$$

25. $k = 4$, $e^{tA} = e^t\begin{pmatrix} 1 & 0 & 0 & 0 \\ -9t - 3t^2 + t^3/6 & 1 + 3t + t^2 & t + t^2/2 & 4t + 3t^2/2 \\ 13t - 9t^2/2 + t^3/6 & -3t + t^2 & 1 - 2t + t^2/2 & -5t + 3t^2/2 \\ 2t + 7t^2/2 - t^3/6 & -t - t^2 & -t^2/2 & 1 - t - 3t^2/2 \end{pmatrix}$

27. $\mathbf{y}_1(t) = e^t(-1, 2, 0)^T$, $\mathbf{y}_2(t) = e^{2t}(0, 1, 0)^T$
$\mathbf{y}_3(t) = e^{2t}(1, 0, 1)^T$

29. $\mathbf{y}_1(t) = e^{-t}(1, 0, 2)^T$, $\mathbf{y}_2(t) = e^{-3t}(0, -1, 2)^T$
$\mathbf{y}_3(t) = e^{-3t}(0, 1 - 2t, 4t)^T$

31. $\mathbf{y}_1(t) = e^{-3t}(1, 3, 0, 0)^T$
$\mathbf{y}_2(t) = e^{-3t}(-7 - 3t, -9t, 0, 6)^T$
$\mathbf{y}_3(t) = e^{-t}(-2, -2, 0, 1)^T$
$\mathbf{y}_4(t) = e^{-t}(-1 - 2t, 1 - 2t, 1, t)^T$

33. $\mathbf{y}_1(t) = e^t\begin{pmatrix} -1 - t - t^2/2 \\ 2 + t \\ 1 - t - t^2 \\ -t^2/2 \\ 2t + t^2/2 \\ t^2/2 \end{pmatrix}$

$\mathbf{y}_2(t) = e^t\begin{pmatrix} -1 - t - t^2/4 \\ 3/2 + t/2 \\ -3t/2 - t^2/2 \\ -t/2 - t^2/4 \\ 1 + 3t/2 + t^2/4 \\ t/2 + t^2/4 \end{pmatrix}$

$\mathbf{y}_3(t) = e^t\begin{pmatrix} -2 - t - 3t^2/4 \\ 5/2 + 3t/2 \\ -t/2 - 3t^2/2 \\ -1 + t/2 - 3t^2/4 \\ 10t/4 + 3t^2/4 \\ 1 - t/2 + 3t^2/4 \end{pmatrix}$

$\mathbf{y}_4(t) = e^{2t}\begin{pmatrix} -1/6 \\ -7/6 \\ 1 \\ 0 \\ 0 \\ 0 \end{pmatrix}$, $\mathbf{y}_5(t) = e^{2t}\begin{pmatrix} 5/6 \\ -1/6 \\ 0 \\ 1 \\ 0 \\ 0 \end{pmatrix}$

$\mathbf{y}_6(t) = e^{2t}\begin{pmatrix} -1/2 \\ -1/2 \\ 0 \\ 0 \\ 1 \\ 0 \end{pmatrix}$

35. $\mathbf{y}_1(t) = e^{4t}(1 + 2t, -2t + t^2/2, t)^T$,
$\mathbf{y}_2(t) = e^{4t}(0, 1, 0)^T$,
$\mathbf{y}_3(t) = e^{4t}(-4t, 5t - t^2, 1 - 2t)^T$

37. $\mathbf{y}_1(t) = e^{-2t}(1, 0, 0)^T$, $\mathbf{y}_2(t) = e^{3t}(1, 2, 1)^T$,
$\mathbf{y}_3(t) = e^{3t}(3 - 2t, -4t, 1 - 2t)^T$

39. $\mathbf{y}_1(t) = e^{-t}(0, 0, 1, 0)^T$, $\mathbf{y}_2(t) = e^{5t}(6, 0, 1, 0)^T$,
$\mathbf{y}_3(t) = e^{5t}(1, 2, 0, 1)^T$,
$\mathbf{y}_4(t) = e^{5t}(3 + 2t, 4t, 0, 1 + 2t)^T$

41. $\mathbf{y}_1(t) = e^{-2t}(0, 0, 1, 0)^T$, $\mathbf{y}_2(t) = e^{-2t}(-2, 2, t, 1)^T$,
$\mathbf{y}_3(t) =$
$e^{-t}(\cos 2t, -3\sin 2t, -2\cos 2t + \sin 2t, -\sin 2t)^T$,
$\mathbf{y}_4(t) =$
$e^{-t}(\sin 2t, 3\cos 2t, -\cos 2t - 2\sin 2t, \cos 2t)^T$

43. $\mathbf{y}_1(t) = e^{-t}(-3, -2, -2, -2, 1)^T$
$\mathbf{y}_2(t) = (1/2)e^{-2t}(8 + 12t - 5t^2 + t^3, 6 + 4t + t^2 + t^3, 24t - 5t^2 + t^3, 18t, -10t + 3t^2)^T$
$\mathbf{y}_3(t) = (-1/2)e^{-2t}(2 + 12t - 5t^2 + t^3, 4t + t^2 + t^3, -6 + 24t - 5t^2 + t^3, 18t, 10t + 3t^2)^T$
$\mathbf{y}_4(t) = (1/2)e^{-2t}(-4 + t^2, 4 + t, t^2, 6, 2t)^T$
$\mathbf{y}_5(t) = -e^{-2t}(8 + 9t - 5t^2 + t^3, t + t^2 + t^3, 21t - 5t^2 + t^3, 18t, -3 - 10t + 3t^2)^T$

45. $\mathbf{y}_1(t) = e^{-t}(1, -1, 0, 1, 0)^T$
$\mathbf{y}_2(t) = e^{-t}(1 + t, -2 - t, 1, t, 1)^T$
$\mathbf{y}_3(t) =$
$1/2e^{4t}(-2, 4t - t^2, 2 - 5t + t^2, 10t - 2t^2, 4t + t^2)^T$
$\mathbf{y}_4(t) = 1/2e^{4t}(-2, 2t - t^2, -4t + t^2, 2 + 6t - 2t^2, -2t + t^2)^T$
$\mathbf{y}_5(t) = 1/2e^{4t}(-2, -2 - t^2, -2t + t^2, 2t - 2t^2, 2 + t^2)^T$

47. $\mathbf{x}(t) = e^{-t/25}\begin{pmatrix} 10 \\ 2t/5 + 8 \\ t^2/125 + 8t/25 + 4 \end{pmatrix}$
At $t = 8$ minutes, the amount of salt in each tank is
$\mathbf{x}(8) = (7.2614\text{ lb}, 8.1328\text{ lb}, 5.1353\text{ lb})^T$

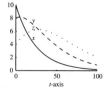

Section 9.7

1. $\lambda = -0.2 \pm 2i$, so asymptotically stable.

3. $\lambda = \pm 3i$, so stable center.

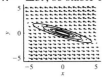

5. $\lambda = 0.1 \pm 2i$, so unstable, spiral source.

7. Single, repeated eigenvalue, $\lambda = -1$, so asymptotically stable.

9. $\lambda = -1$, $\lambda = -2$, and $\lambda = -3$, so asymptotically stable.

11. $\lambda = -1$, $\lambda = -2 \pm 3i$, so asymptotically stable.

13. $\lambda = -1$, $\lambda = -1 \pm i$, so aysmptotically stable.

15. $\lambda = 2$ and $\lambda = -1$, so unstable.

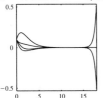

17. a. Exponential solutions: $C_1 e^{-3t}(1, 1, 0)^T$, $C_2 e^{-2t}(0, 0, 1)^T$, $C_3 e^{-t}(0, 1, 0)^T$

b. We selected initial conditions $(1, 0, 1)^T$, $(-1, 0, 1)^T$, $(1/2, 1, 1)^T$, $(-1/2, -1, 1)^T$, $(1, 0, -1)^T$, $(-1, 0, -1)^T$, $(1/2, 1, -1)^T$, and $(-1/2, -1, -1)^T$ to craft the portrait in the following figure.

c. Nodal sink

19. a. $\mathbf{y}_1(t) = e^{-t}(0, 0, 1)^T$,
$\mathbf{y}_2(t) = e^{-t}(\cos 10t, \sin 10t, 0)^T$,
$\mathbf{y}_3(t) = e^{-t}(\sin 10t, -\cos 10t, 0)^T$

b. They remain on z-axis and decay to the origin.

c. They remain in the xy-plane and spiral into the origin.

d. They spiral into the origin from above or below the xy-plane.

Section 9.8

15. $y(t) = C_1 e^{-2t} + C_2 e^{2t} + C_3 e^{3t}$

17. $y(t) = C_1 e^{-3t} + C_2 e^{-2t} + C_3 e^{2t} + C_4 e^{3t}$

19. $y(t) = C_1 e^{-3t} + C_2 e^{2t} + C_3 e^{5t}$

21. $y(t) = C_1 e^{-3t} + C_2 e^{-2t} + C_3 e^{2t} + C_4 e^{3t} + C_5 e^{4t}$

23. $y(t) = C_1 e^{-2t} + C_2 t e^{-2t} + C_3 e^{3t}$

25. $y(t) = C_1 e^{-t} + C_2 t e^{-t} + C_3 t^2 e^{-t}$

27. $y(t) = C_1 e^t + C_2 t e^t + C_3 t^2 e^t + C_4 t^3 e^t + C_5 e^{-3t}$

29. $y(t) = C_1 e^{-t} + C_2 e^t \cos t + C_3 e^t \sin t$

31. $y(t) = C_1 \cos t + C_2 t \cos t + C_3 \sin t + C_4 t \sin t$

33. $y(t) = C_1 \cos t + C_2 t \cos t + C_3 t^2 \cos t + C_4 \sin t + C_5 t \sin t + C_6 t^2 \sin t$

35. $y(t) = 2e^{-t} \cos 2t + e^{-t} \sin 2t$

37. $y(t) = e^t - te^t$

39. $y(t) = -(13/16)e^t + (11/4)te^t - (3/16)e^{5t}$

41. $y(t) = -2e^{2t} + 4te^{2t} - 3t^2 e^{2t}$

43. $y(t) = (1/2)t \cos 2t - (3/4) \sin 2t + (1/2)t \sin 2t$

Section 9.9

1. $y(t) = C_1 e^t (3, -2)^T + C_2 e^{2t} (2, -1)^T + (-9te^t - 10e^t, 6te^t + 5e^t)^T$

3. $y(t) = C_1 (2, 1)^T + C_2 (3e^t, 2e^t)^T + (-12t - 15, -6t - 10)^T$

5. $y(t) = C_1 e^{2t} \begin{pmatrix} \sin t - \cos t \\ \cos t \end{pmatrix} + C_2 e^{2t} \begin{pmatrix} -\cos t - \sin t \\ \sin t \end{pmatrix} + e^{2t} \begin{pmatrix} 2 \\ -1 \end{pmatrix}$

7. $y(t) = C_1 \begin{pmatrix} -1 \\ 2 \\ 0 \end{pmatrix} + C_2 \begin{pmatrix} e^{2t} \\ 0 \\ e^{2t} \end{pmatrix} + C_3 \begin{pmatrix} 0 \\ 3e^t \\ e^t \end{pmatrix} + \begin{pmatrix} \cos t \\ -2 \cos t \\ 0 \end{pmatrix}$

9. $y(t) = C_1 \begin{pmatrix} -1 \\ 4 \\ 1 \end{pmatrix} + C_2 \begin{pmatrix} -3e^{-2t} \\ 2e^{-2t} \\ 0 \end{pmatrix} + C_3 \begin{pmatrix} 0 \\ 3e^{-t} \\ e^{-t} \end{pmatrix} + \begin{pmatrix} 3 - 6t \\ 24t - 20 \\ 6t - 6 \end{pmatrix}$

13. $y_p = (-3, -5/2)^T$

15. $y_p = \begin{pmatrix} 0 \\ (-1/2) \cos t + (1/2) \sin t \end{pmatrix}$

19. $i_1 = (1/15)e^{-2t} + (2/195)e^{-8t} - (1/13) \cos t + (14/65) \sin t$
$i_2 = (2/15)e^{-2t} - (2/195)e^{-8t} - (8/65) \cos t + (12/65) \sin t$

21. $\begin{pmatrix} 2e^t - e^{3t} & 2e^{3t} - 2e^t \\ e^t - e^{3t} & 2e^{3t} - e^t \end{pmatrix}$

23. $\begin{pmatrix} \cos t + \sin t & -2 \sin t \\ \sin t & \cos t - \sin t \end{pmatrix}$

25. $e^{2t} \begin{pmatrix} 1 - 2t & t \\ -4t & 1 + 2t \end{pmatrix}$

27. $e^{tA} = \begin{pmatrix} 2e^{-4t} - e^{-t} & e^{-4t} - e^{-t} \\ 2e^{-t} - 2e^{-4t} & 2e^{-t} - e^{-4t} \end{pmatrix}$,
$y(t) = \begin{pmatrix} 2e^{-4t} - e^{-t} \\ 2e^{-t} - 2e^{-4t} \end{pmatrix}$

29. $e^{tA} = e^{-3t} \begin{pmatrix} 1 + 2t & t \\ -4t & 1 - 2t \end{pmatrix}$,
$y(t) = e^{-3t} \begin{pmatrix} 2 + 3t \\ -1 - 6t \end{pmatrix}$

31. $e^{tA} = \begin{pmatrix} e^{5t} & 0 \\ e^{5t} - e^{-t} & e^{-t} \end{pmatrix}$, $y(t) = \begin{pmatrix} -e^{5t-5} \\ 4e^{1-t} - e^{5t-5} \end{pmatrix}$

33. $e^{tA} = \begin{pmatrix} \cos 3t + 2 \sin 3t & -\sin 3t \\ 5 \sin 3t & \cos 3t - 2 \sin 3t \end{pmatrix}$,
$y(t) = \begin{pmatrix} -\cos 3(t - 1) - 2 \sin 3(t - 1) \\ -5 \sin 3(t - 1) \end{pmatrix}$

35. $e^{tA} = \begin{pmatrix} 2e^{3t} - e^t & 0 & 2e^{3t} - 2e^t \\ e^t - e^{-2t} & e^{-2t} & 2e^t - 2e^{-2t} \\ e^t - e^{3t} & 0 & 2e^t - e^{3t} \end{pmatrix}$

37. $e^{tA} = \begin{pmatrix} e^{2t} & 0 & e^{2t} - e^{-t} & 0 \\ 2e^{2t} - e^t - e^{-2t} & e^t & 2e^{2t} - 2e^t + e^{-t} - e^{-2t} & e^{-2t} - e^t \\ 0 & 0 & e^{-t} & 0 \\ e^{2t} - e^{-2t} & 0 & e^{2t} - e^{-2t} & e^{-2t} \end{pmatrix}$

Chapter 10
Section 10.1

1.

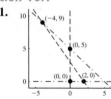

The x-nullcline is dashed and the y-nullcline is dot-dashed. $(0, 0)$ is a nodal source. $(0, 5)$ is a nodal sink. $(2, 0)$ is a saddle. $(-4, 9)$ is a saddle.

3.

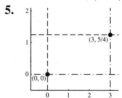

The x-nullcline is dashed and the y-nullcline is dot-dashed. $(0, 0)$ is a source. $(0, 1)$ is a saddle. $(1, 0)$ is a saddle. $(2/3, 2/3)$ is a nodal sink.

5.

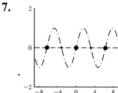

The x-nullcline is dashed and the y-nullcline is dot-dashed. $(0, 0)$ is a saddle. $(3, 5/4)$ cannot be classified.

7.

The x-nullcline is dashed and the y-nullcline is dot-dashed. $(k\pi, 0)$ is a spiral sink if k is even and a saddle if k is odd.

9. Linearization near $(2, 0)$ is at the right.

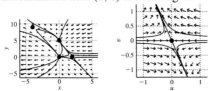

11. Linearization near $(2/3, 2/3)$ at the right.

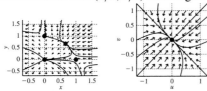

13. Linearization near $(3, 5/4)$ at the right.

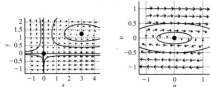

15. Linearization near $(2\pi, 0)$ at the right.

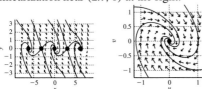

17. a.

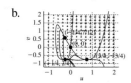

b.

c. If we throw away the terms having degree 2 or more, we get the linearization.

21. $x_1' = (a_1 - b_1x_1 + c_1x_2)x_1$
$x_2' = (a_2 - b_2x_2 + c_2x_1)x_2$
where $a_1 < 0$, $b_1 = 0$, $c_1 > 0$, $a_2 > 0$, $b_2 > 0$, and $c_2 > 0$

23. $x_1' = (a_1 - b_1x_1 + c_1x_2)x_1$
$x_2' = (a_2 - b_2x_2 + c_2x_1)x_2$
where $a_1 < 0$, $b_1 = 0$, $c_1 > 0$, $a_2 < 0$, $b_2 = 0$, and $c_2 > 0$

25. $x_1' = (a_1 - b_1x_1 + c_1x_2)x_1$
$x_2' = (a_2 - b_2x_2 + c_2x_1)x_2$
where $a_1 > 0$, $b_1 > 0$, $c_1 > 0$, $a_2 > 0$, $b_2 > 0$, and $c_2 > 0$

27. The model is

$$x_1' = (a_1 + b_{11}x_1 + b_{12}x_2 + b_{13}x_3)x_1$$
$$x_2' = (a_2 + b_{21}x_1 + b_{22}x_2 + b_{23}x_3)x_2$$
$$x_3' = (a_3 + b_{31}x_1 + b_{32}x_3 + b_{33}x_3)x_3.$$

The constant a_j is the reproductive rate of the population x_j in the absence of the others. The constant b_{jj} is negative if there is a logistic limit, and zero if not. For $i \neq j$ the constant b_{ij} measures the effect of x_j on the reproduction of x_i. If $b_{ij} > 0$ the effect is to increase the reproductive rate, and if $b_{ij} < 0$ the reproductive rate is diminished.

Section 10.2

1. $(0, 0)$ is a saddle. $(\pm 1, 1)$ are spiral sinks.

3. $(0, 0)$ is a nongeneric sink. $(-5, 1)$ is a saddle.

5. $(0, 0, 0)$ is a sink. $(2/3, 4/9, 2/9)$ is unstable.

7. $(0, 0, 0)$ is unstable. $(1, 1, 1)$ is asymptotically stable.

9. $(0, 0, 0)$ is unstable. $(1, 1, 1)$ is asymptotically stable.

11. The origin is asymptotically stable.

13. 0 is also an eigenvalue, so it is not possible to classify the equilibrium point using the Jacobian.

15. The origin is asymptotically stable.

17. **0** is unstable. $\mathbf{c}^+$ and $\mathbf{c}^-$ are asymptotically stable. Solutions starting at $(2, 1, 1)^T$ and $-2, -1, 1)^T$.

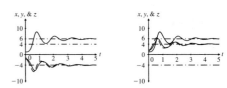

19. All three equilibrium points are unstable. Solutions starting at $(8, 8, 27)^T$ and $(-8, -8, 27)^T$.

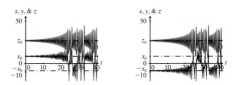

21. b. The equilibrium point is $\mathbf{x} = (5/4, 1/2, 3/4)^T$. It is asymptotically stable.

Section 10.3

9.

The origin is a nodal source. $(2, 0)$ is a nodal sink. $(0, 3)$ is also a nodal sink. $(1/2, 3/2)$ is a saddle. All solution curves flow to one of the sinks at $(2, 0)$ and $(0, 3)$, with the exception of the two stable solution curves for the saddle at $(1/2, 3/2)$.

11.

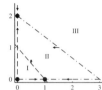

The origin is a nodal source. $(1, 0)$ is a saddle. $(0, 2)$ is a nodal sink. All solution curves in the positive quadrant flow to the sink at $(0, 2)$.

13.

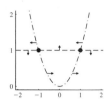

$(1, 1)$ is a saddle. $(-1, 1)$ is a spiral source.

15.

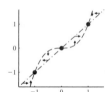

The origin is a saddle. $(1, 1)$ and $(-1, -1)$ are nodal sinks. All solutions converge to one of the two sinks, except for the two stable solutions for the saddle point.

17. b. All solutions tend to ∞.

c. All solutions tend to the equilibrium point

$$u = \frac{1+a}{1-ab} \quad \text{and} \quad v = \frac{1+b}{1-ab}.$$

Section 10.4

1. $r' = r(1 - r)$, so $r = 1$ is an attracting limit cycle.

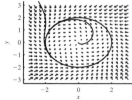

3. $r' = r(4 - r^2)$, so $r = 2$ is an attracting limit cycle.

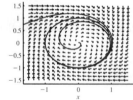

5. $r' = (r - 2)(r - 1)$, so $r = 1$ is an attracting limit cycle, while $r = 2$ is a repelling limit cycle.

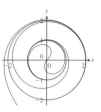

7. $r' = r(r^2 - 1)^2$, so $r = 1$ is an unstable limit cycle.

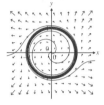

9. $r' = (x^2 + 2y^2)(1 - r^2)/r$.
$r^2\theta' = -(x^2 - xy - y^2) + xyr^2$. Notice that the unit circle is an attractive, invariant set. The equilibrium points are $(0, 0)$, a spiral source, $(1/\sqrt{2}, 1/\sqrt{2})$ and $(-1/\sqrt{2}, -1/\sqrt{2})$ which are degenerate in the sense that the Jacobian has 0 determinant. All solutions are attracted to one of these latter two equilibrium points, but they are not sinks!

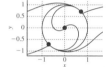

11. b. $r' > 0$ on $r = 1/2$ and $r' < 0$ on $r = 1$

c. The only equilibrium points of the system is $(0, 0)$.

d.

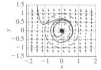

13. Compute that $rr' = r^2 - [3x^4 + 3x^2y^2 + y^4]$. Prove that $r^4 \leq 3x^4 + 3x^2y^2 + y^4 \leq 3r^4$. Then the annulus R defined by $1/3 \leq r \leq 1$ is invariant.

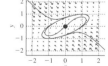

15.

17.

19.

21.

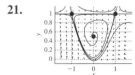

23. b. 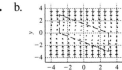   d.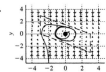

25. b.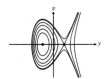

25.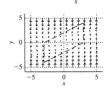

Section 10.5

1. $E(y, v) = y^2 + v^2/2$

3. $E(y, v) = y^3/3 - y^2/2 - v^2/2$

5. $E(y, v) = y^2/2 - y^4/4 - v^2/2$

7. $E(x, y) = e^x - e^y$

9. $E(x, y) = \cos x - y^2/2$

11. The conserved quantity is $E(y, v) = y^2 + v^2/2$.

13. The conserved quantity is
$E(y, v) = y^3/3 - y^2/2 - v^2/2$.

15. The conserved quantity is
$E(y, v) = y^2/2 - y^4/4 - v^2/2$.

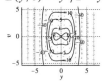

17. The conserved quantity is $E(x, y) = e^x - e^y$.

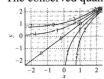

19. The conserved quantity is $E(x, y) = \cos x - y^2/2$.

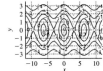

21. $E' = -v^2$ **23.** $E' = v^2$

Section 10.6

1. 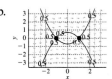   **3.**

5.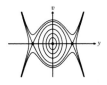

7. Min at $y = 1/2$, $U(y) = -y + y^2$

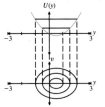

9. Min at $y = 4$, Max at $y = 0$, $U(y) = -2y^2 + (1/3)y^3$

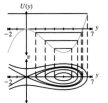

11. Minima at $y = k\pi$, k odd, maxima at $y = k\pi$, k even.
$U(y) = \cos y$

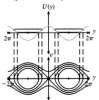

13. Separatrix $\frac{1}{2}v^2 - \frac{1}{4}y^4 + \frac{1}{2}y^2 = \frac{1}{4}$, saddles at $(-1, 0)$
and $(1, 0)$, center at $(0, 0)$

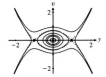

15. Separatrix $\frac{1}{2}v^2 - \cos y = 1$, saddles at $(-\pi, 0)$ and $(\pi, 0)$, centers at $(-2\pi, 0)$, $(0, 0)$, $(2\pi, 0)$ on $[-2\pi, 2\pi]$

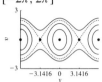

17. $\beta > 0$ implies hard spring; $\beta < 0$ yields a soft spring.

19. a. Equilibrium points: $(k\pi, 0)$, k an integer.

b. Spiral sink: $(k\pi, 0)$, k even. Saddles: $(k\pi, 0)$, k odd.

21. *Hint*: If $dH/dt = 0$, then equating coefficients leads to a system of three equations in A, B, and C.

23. $H(x, y) = -2xy - y^3 + x^3$

25. $H(x, y) = y^3 + x^3$

27. $H(x, y) = -xy + y^2 + x^2$

29. Not Hamiltonian

31. $\dot{E} = 0$, so E is conserved.

Section 10.7

1. Positive definite   **3.** None

5. Negative semidefinite   **7.** None

9. Positive definite   **13.** $dV/dt = -2y^4$

15. $dV/dt = 2y^4 - xy^3 + y^2 + 4xy - x^2$

17. c.

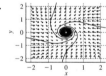

19. *Hint*: Argue that $\dot{V}(x, y) = -2(x^2 + y^2)$ is negative definite.

21. *Hint*: Show that $\dot{V}(x, y) = -2(x^2(1 - y) + y^2(1 - x))$.

23. b.

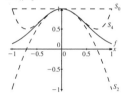

Section 10.8

5. In either of these two cases all solution curves approach the equilibrium point at $(1, 0)$.

Chapter 11
Section 11.1

1. $R = 1$ **3.** $R = 1$ **5.** $R = \infty$

7. $R = 0$ **9.** $R = 1$ **11.** $R = \infty$

13. $\sum_{n=0}^{\infty}(-1)^n(2x)^{2n}/(2n)!$

15. $\sum_{n=0}^{\infty}(-1)^{n+1}(x - \pi)^{2n+1}/(2n + 1)!$

17. $\sum_{n=0}^{\infty}(-1)^n(x - 3)^n/3^{n+1}$

19. $f(x) = 1 + 2\sum_{n=1}^{\infty} x^n$

21. $f(x) = \sum_{n=0}^{\infty}\dfrac{n}{n + 1}x^n$

23. $f(x) = 2\sum_{n=0}^{\infty}\dfrac{x^{2n}}{(2n)!}$

25. $\sum_{n=0}^{\infty}(-1)^n x^n/3^{n+1}$, $R = 3$

27. $\sum_{n=0}^{\infty}(-1)^n x^{2n}/4^{n+1}$, $R = 2$

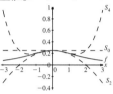

29. $\sum_{n=0}^{\infty}(-1)^n 2^n x^{2n}/n!$, $R = \infty$

31. $\sum_{n=0}^{\infty}(-1)^n(n + 1)(x - 2)^n/2^{n+2}$, $R = 2$

33. $x/(1 - x)^2$, provided $|x| < 1$

35. $f(x) \approx x + x^3/3$

Section 11.2

1. $y(x) = a_0 \sum_{n=0}^{\infty}(3x)^n/n! = a_0 e^{3x}$

3. $y(x) = a_0 \sum_{n=0}^{\infty}(-x^4/4)^n/n! = a_0 e^{-x^4/4}$

5. $y(x) = a_0(1 - x)$

7. $y(x) = a_0 \sum_{n=0}^{\infty}(x/4)^n = -4a_0/(x - 4)$

9. $y(x) = a_0(2 - x)^2/4$

11. $y(x) = a_0 + a_1 \sum_{n=1}^{\infty} x^n/n! = a_0 + a_1 e^x$

13. $y(x) = a_0 \sum_{n=0}^{\infty}(-1)^n x^{2n}/(2n)! + a_1 \sum_{n=0}^{\infty}(-1)^n x^{2n+1}/(2n + 1)! = a_0 \cos x + a_1 \sin x$

15. $y_1(x) = \displaystyle\sum_{n=0}^{\infty} \dfrac{(-1)^n x^{4n}}{4^n n![3 \cdot 7 \cdots (4n - 1)]}$

$y_2(x) = \displaystyle\sum_{n=0}^{\infty} \dfrac{(-1)^n x^{4n+1}}{4^n n![5 \cdot 9 \cdots (4n + 1)]}$

17. $y_1(x) = 1 + x^2/2 + \sum_{n=2}^{\infty}(-1)^{n+1}(3 \cdot 7 \cdot 11 \cdots (4n - 5))x^{2n}/(2n)!$,
$y_2(x) = x - x^3/6 + \sum_{n=2}^{\infty}(-1)^n(5 \cdot 9 \cdot 12 \cdots (4n - 3))x^{2n+1}/(2n + 1)!$
All solutions have radius of convergence $R = \infty$.

19. $y_1(x) = 1 + x^2/2 + \sum_{n=2}^{\infty}(-1)^{n+1}(1 \cdot 3 \cdot 5 \cdots (2n - 3))x^{2n}/(2n)!$,
$y_2(x) = x$. All solutions have radius of convergence $R = \infty$.

21. $y_1(x) = 1 - \frac{x^2}{2} + \frac{x^3}{6} + \cdots$ and $y_2(x) = x - \frac{x^3}{6} + \cdots$.
All solutions have radius of convergence $R \geq 1$.

23. $y_1(x) = 1 + \frac{x^2}{2} + \cdots$, $y_2(x) = x + \frac{x^3}{6} + \cdots$. All
solutions have infinite radius of convergence.

25.

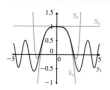

27.

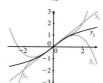

29.

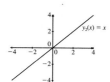

31.

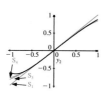

33.

35. $R = 1$

Section 11.3

5. $y_1(x) = 1 + \sum_{n=1}^{\infty} (-1)^n x^{3n}/((2 \cdot 3) \cdot (5 \cdot 6) \cdots (3n-1)(3n))$,
$y_2(x) = x + \sum_{n=1}^{\infty} (-1)^n x^{3n+1}/((3 \cdot 4) \cdot (6 \cdot 7) \cdots (3n)(3n+1))$,
$R = \infty$

7. a. $y_1(x) = 1 - \sum_{n=1}^{\infty} p(4-p)(8-p) \cdots (4n - p - 4) x^{2n}/(2n)!$,
$R = \infty$
$y_2(x) = x + \sum_{n=1}^{\infty} (2-p)(6-p) \cdots (4n - p - 2) x^{2n+1}/(2n+1)!$, $R = \infty$

c. $H_0(x) = 1$, $H_1(x) = 2x$, $H_2(x) = 4x^2 - 2$,
$H_3(x) = 8x^3 - 12x$, $H_4(x) = 16x^4 - 48x^2 + 12$,
$H_5(x) = 32x^5 - 160x^3 + 120x$

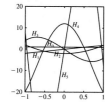

Section 11.4

1. The point $x = 0$ is a regular singular point.

3. The point $x = 2$ is an irregular singular point, $x = 0$ is a regular singular point.

5. The point $t = 3$ is a regular singular point, $t = -2$ is a regular singular point.

7. The points $t = 1$ and $t = -1$ are regular singular points.

9. The points $t = k\pi$, k an integer, $k \neq 0$, are irregular singular points. The point $k = 0$ is a regular singular point.

11. $y(x) = C_1 x^{-3} + C_2 x^{-1}$

13. $y(x) = C_1 x + C_2 x \ln x$

15. $y(x) = C_1 x^{-4} + C_2 x^{-4} \ln x$

17. $y(x) = C_1 x^3 + C_2 x^3 \ln x$

19. $y(x) = x^{-3} + 2x^2$

21. $y(x) = (x-3)^{-2} + 3(x-3)^{-2} \ln(x-3)$

25. $y(x) = C_1 x^{1/2} \cos(\sqrt{3} \ln x/6) + C_2 x^{1/2} \sin(\sqrt{3} \ln x/6)$

27. $y(x) = C_1 \cos(2 \ln x) + C_2 \sin(2 \ln x)$

29. $y(x) = 3x^{-1} \cos(\ln x) + x^{-1} \sin(\ln x)$

31. $y(x) = 2x \cos(\sqrt{5} \ln x) - 3x \sin(\sqrt{5} \ln x)$

Section 11.5

3. $s^2 + 3s + 2 = 0$ **5.** $s^2 - s - 3 = 0$

7. $a_n = (2n-3)a_{n-1}/(2n+1)$, $n \geq 1$

9. $a_n = a_{n-1}/n$, $n \geq 1$

11. $y_1(x) = \sum_{n=0}^{\infty} (-1)^n x^n/(2n)!$,
$y_2(x) = x^{1/2} \sum_{n=0}^{\infty} (-1)^n x^n/(2n+1)!$

13. $y_1(x) = x^2 [1 + \sum_{n=1}^{\infty} (-1)^n 2^n x^n/\{n!(5 \cdot 7 \cdots (2n+3))\}]$,
$y_2(x) = x^{1/2} [1 + 2x + \sum_{n=2}^{\infty} (-1)^{n+1} 2^n x^n/\{n!(1 \cdot 3 \cdots (2n-3))\}]$

15. $y_1(x) = x^{1/4} [1 + \sum_{n=1}^{\infty} (-1)^n x^{2n}/\{[5 \cdot 9 \cdots (4n+1)]n!\}]$
$y_2(x) = x^{-1/4} [1 + \sum_{n=1}^{\infty} (-1)^n x^{2n}/\{[3 \cdot 7 \cdots (4n-1)]n!\}]$

17. $y_1(x) = x^1 \sum_{n=0}^{\infty} (-1)^n 2^n x^n/(2n+1)!$,
$y_2(x) = x^{1/2} [1 + \sum_{n=1}^{\infty} (-1)^n 2^{n-1} x^n/\{n(2n-1)!\}]$

19. $y_1 = x^{1/2} [1 + \sum_{n=1}^{\infty} (-1)^n (-3) \cdot 1 \cdot 13 \cdots (4n^2 - 8n + 1) x^n/(2n+1)!]$
$y_2 = 1 + \sum_{n=1}^{\infty} (-1)^n 2^n \cdot (-1) \cdot (-1) \cdot 3 \cdot 11 \cdots (2n^2 - 6n + 3) x^n/(2n)!$

21. $y_1(x) = x^{1/2} \sum_{n=0}^{\infty} (-1)^n 2^n x^n/(2n+1)!$,
$y_2(x) = \sum_{n=0}^{\infty} (-2x)^n/(2n)!$

23. $y_1(x) = x [1 + \sum_{n=1}^{\infty} x^{2n}/\{2^n n![5 \cdot 9 \ldots (4n+1)]\}]$,
$y_2(x) = x^{1/2} [1 + \sum_{n=1}^{\infty} x^{2n}/\{2^n n![3 \cdot 7 \ldots (4n-1)]\}]$

25. $y_1(x) = x^{1/2} [1 + \sum_{n=1}^{\infty} (-1)^n x^{3n}/\{3^n n![7 \cdot 13 \ldots (6n+1)]\}]$,
$y_2(x) = 1 + \sum_{n=1}^{\infty} (-1)^n x^{3n}/\{3^n n![5 \cdot 11 \ldots (6n-1)]\}$

31. The recurrence relation is
$(s+n)^2 a_n - (s+n-1-p)a_{n-1} = 0$, $n \geq 1$.

Section 11.6

1. $y_1(x) = \sum_{n=0}^{\infty} 2^n x^{n-1}/(n!)^2$,
$y_2(x) = y_1 \ln x - \sum_{n=1}^{\infty} 2^{n+1} x^{n-1} H(n)/(n!)^2$

3. $y_1(x) = \sum_{n=0}^{\infty} x^n/n! = e^x$,
$y_2(x) = e^x \ln x - \sum_{n=1}^{\infty} H(n) x^n/n!$
$H(n) = \sum_{k=1}^{n} (1/k)$

5. $y_1(x) = 1 + x/2 + x^2/12$,
$y_2(x) = x^5 + \sum_{n=6}^{\infty} 60 x^n/[(n-2)(n-1)n(n-5)!]$

7. $y_1(x) = 1 + 2x/3 + x^2/6$,
$y_2(x) = x^4 + \sum_{n=5}^{\infty} 24(n-3)x^n/n!$

9. $y_1(x) = -x + \sum_{n=2}^{\infty} (-1)^n x^n/[n!(n-1)!]$,
$y_2(x) = y_1(x) \ln x + 1 + x +$
$\sum_{n=2}^{\infty} (-1)^{n+1}[H(n) + H(n-1)]x^n/[n!(n-1)!]$

11. $y_1(x) = 2x^3 \sum_{n=3}^{\infty} (-1)^n 4^n x^{n+1}/[n!(n-2)!]$,
$y_2(x) = y_1(x) \ln x - x - 4x^2 - x^3 -$
$\sum_{n=3}^{\infty} (-4)^n [H(n) + H(n-2) - 1]x^{n+1}/[n!(n-2)!]$

13. $y_1(x) = x^{-2} - 3x^{-1} + 9/2$,
$y_2(x) = x + 6 \sum_{n=4}^{\infty} (-3)^{n-1} x^{n-2}/n!$

15. $y_1(x) = x + \sum_{n=2}^{\infty} (-1)^{n+1} x^n/(n-1)! = xe^{-x}$,
$y_2(x) =$
$xe^{-x} \ln x - 1 + \sum_{n=2}^{\infty} (-1)^n H(n-1)x^n/(n-1)!$
$H(n) = \sum_{k=1}^{n} 1/k$

17. $y_1(x) = \sum_{n=0}^{\infty} x^n/(n!)^2$,
$y_2(x) = y_1(x) \ln x - 2 \sum_{n=1}^{\infty} H(n)x^n/(n!)^2$,
$H(n) = \sum_{k=1}^{n} 1/k$

19. $y_1(x) = x/2 + \sum_{n=2}^{\infty} x^{2n-1}/[2^{2n-1} n!(n-1)!]$,
$y_2(x) = y_1(x) \ln x + x^{-1}[1 - x^2/4 -$
$\sum_{n=2}^{\infty}[H(n) + H(n-1)]x^n/[2^{2n} n!(n-1)!]$,
$H(n) = \sum_{k=1}^{n} 1/k$

21. $y_1(x) = x - 2x^2 + 2x^3$,
$y_2(x) = x^4 + \sum_{n=4}^{\infty} 6(-2)^{n-3} x^{n+1}/n!$

Section 11.7

13. We set $b_n = (s + 2)a_n$, where $a_n(s)$ are defined by the recurrence formula, and $Y(s, x) = \sum_{n=0}^{\infty} b_n(s)x^{s+n}$. Then with $a_0 = -2$ in both cases, our solutions are

$$y_1(x) = J_2(x) \quad \text{and}$$

$$y_2(x) = J_2(x) \ln x - \frac{2}{x^2} - \frac{1}{2}$$

$$- \frac{1}{2} \sum_{n=0}^{\infty} \frac{(-1)^n [H(n) + H(n+2)]}{n!(n+2)!} \left(\frac{x}{2}\right)^{2n+2}$$

$$+ \frac{1}{2} J_2(x).$$

Chapter 12
Section 12.1

1. $f(x) = \sum_{n=1}^{\infty} (4/\pi(1 - 4n^2)) \cos 2nx$. The partial sum S_6 on $[-\pi, \pi]$ and $[-3\pi, 3\pi]$ follow.

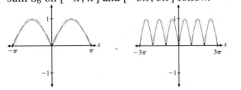

3. $f(x) = \pi/4 + \sum_{n=1}^{\infty} [((-1)^n - 1)/(\pi n^2)] \cos nx + [(-1)^{n+1}/n] \sin nx$. The partial sum S_6 on $[-\pi, \pi]$ and $[-3\pi, 3\pi]$ follow.

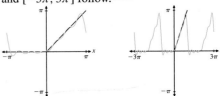

5. $f(x) = (-1/4) \sin x + \sum_{n=2}^{\infty} [2n(-1)^n/(n^2 - 1)] \sin nx$. The partial sum S_6 on $[-\pi, \pi]$ and $[-3\pi, 3\pi]$ follow.

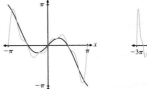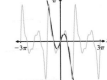

7. $f(x) \sim 3/4 + (2/\pi^2) \sum_{n=1}^{\infty} (1/(2n+1)^2) \cos((2n+1)\pi x) - (1/\pi) \sum_{n=1}^{\infty} ((-1)^n/n) \sin n\pi x$. In the following figures, the first figure displays S_3, the second S_6.

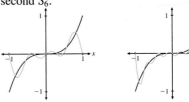

9. $\sum_{n=1}^{\infty} (2(-1)^n(6 - n^2\pi^2)/(n^3\pi^3)) \sin n\pi x$. In the following figures, the first figure displays S_3, the second S_6.

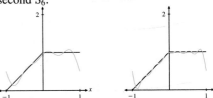

11. $1/6 + \sum_{n=1}^{\infty} [(2(-1)^n/(n^2\pi^2)) \cos n\pi x + ((2(-1)^n - n^2\pi^2(-1)^n - 2)/(n^3\pi^3)) \sin n\pi x]$. In the following figures, the first figure displays S_3, the second S_6.

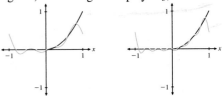

13. $f(x) \sim 1/2 + (1/2)\cos \pi x + (2/\pi)\sin \pi x + \sum_{n=2}^{\infty}\{-n[(1 + (-1)^n)/((n-1)(n+1)\pi)] + (1/(n\pi))[1 - (-1)^n]\}\sin n\pi x$. In the following figures, the first figure displays S_3, the second S_6.

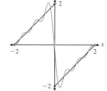

15. $\sum_{n=1}^{\infty}(-4/(n\pi))\sin(n\pi x/2)$. In the following figures, the first figure displays S_3, the second S_6.

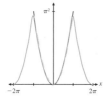

17. $f(x) = \pi^2/3 + 4\sum_{n=1}^{\infty}(-1)^n(\cos nx)/n^2$. The partial sum S_7 is displayed on $[-\pi, \pi]$ and $[-2\pi, 2\pi]$.

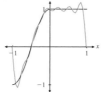

19. f is even. **21.** f is neither.

29. Consider symmetry defined by $f(L - x) = f(x)$ on $[0, L]$. For example, consider $f(x) = -x - 2$ if $-2 \le x < -1$, x if $-1 \le x < 1$, and $-x + 2$ if $1 \le x \le 2$.

31. $f_{\text{odd}}(x) = [f(x) - f(-x)]/2$ and $f_{\text{even}}(x) = [f(x) + f(-x)]/2$

Section 12.2

1. f is periodic with period π.

3. Nonperiodic   **5.** Nonperiodic

7. The Fourier series converges to $f_p(x)$ except at the odd multiples of π. At the odd multiples of π the series converges to $[\pi + 0]/2 = \pi/2$.

9. The Fourier series converges to $f_p(x)$ except at the odd integers. At the odd integers, the series converges to $[1 + 0]/2 = 1/2$.

11. The Fourier series converges to $f_p(x)$ except at the odd integers. At the odd integers, the series converges to $[1 - 1]/2 = 0$.

13. The Fourier series converges to $f_p(x)$ everywhere.

15. $|x| \sim \pi/2 - (4/\pi)\sum_{n=1}^{\infty}\cos((2n+1)x)/(2n+1)^2$

17. $x^4 \sim \pi^4/5 + 8\sum_{n=1}^{\infty}(-1)^n(\pi^2/n^2 - 6/n^4)\cos nx$

19. $f(x) = (1/\pi)\sinh(\pi/2)[2 + \sum_{n=1}^{\infty}((-1)^n/(n^2 + 1/4))\cos nx - 2\sum_{n=1}^{\infty}((-1)^n n/(n^2 + 1/4))\sin nx]$. The plots use $N = 10$ terms.

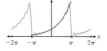

21. At the points of discontinuity, the series converges to $1/2$.

Section 12.3

1. $f_o(x) = \begin{cases} -(1 + x), & -2 \le x < 0, \\ 0, & x = 0, \\ 1 - x, & 0 < x \le 2. \end{cases}$

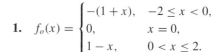
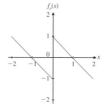
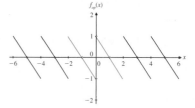

3. $f_o(x) = \begin{cases} -(x^2 - 1), & -2 \le x < 0, \\ 0, & x = 0, \\ x^2 - 1, & 0 < x \le 2. \end{cases}$

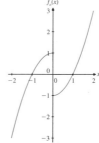

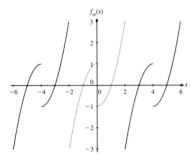

5. $f_e(x) = \begin{cases} 1 + x, & -2 \le x < 0, \\ 1 - x, & 0 \le x \le 2. \end{cases}$

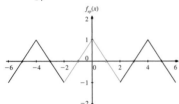

7. $f_e(x) = \begin{cases} x^2 - 1, & -2 \le x < 0, \\ x^2 - 1, & 0 \le x \le 2. \end{cases}$

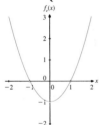

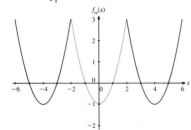

9. $f(x) = \pi/2 + \sum_{n=1}^{\infty} \left[2((-1)^n - 1)/(\pi n^2) \right] \cos nx.$
The partial sum S_6 is shown on the interval $[0, \pi]$ and $[-3\pi, 3\pi]$.

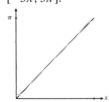
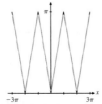

11. $f(x) = \cos x$. The partial sum S_6 is shown on the interval $[0, \pi]$ and $[-3\pi, 3\pi]$.

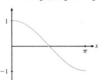
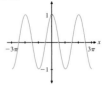

13. $f(x) = \pi/2 + \sum_{n=1}^{\infty} \left[2(1 - (-1)^n)/(\pi n^2) \right] \cos nx.$
The partial sum S_6 is shown on the interval $[0, \pi]$ and $[-3\pi, 3\pi]$.

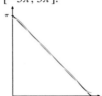
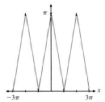

15. $f(x) = \pi^3/4 + \sum_{n=1}^{\infty} \left[6\pi(-1)^n/n^2 - 12((-1)^n - 1)/(\pi n^4) \right] \cos nx.$
The partial sum S_6 is shown on the interval $[0, \pi]$ and $[-3\pi, 3\pi]$.

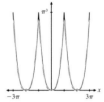

17. $f(x) = 1/2 + \sum_{n=1}^{\infty} \left[2/(n\pi) \right] \sin(n\pi/2) \cos nx$. The partial sum S_6 is shown on the interval $[0, \pi]$ and $[-3\pi, 3\pi]$.

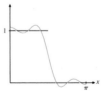
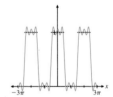

19. $f(x) = -2/\pi + (\pi/2) \cos x + \sum_{n=2}^{\infty} \left[2 \left((-1)^{n-1} - 1 \right) (n^2 + 1)/\{\pi(n - 1)^2(n + 1)^2\} \right] \cos nx$. The partial sum S_6 is shown on the interval $[0, \pi]$ and $[-3\pi, 3\pi]$.

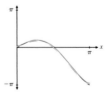
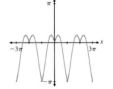

21. $f(x) = \sum_{n=1}^{\infty} \left[2(-1)^{n+1}/n \right] \sin nx$. The partial sum S_6 is shown on the interval $[0, \pi]$ and $[-3\pi, 3\pi]$.

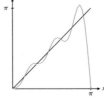
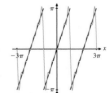

23. $f(x) =$
$\sum_{n=2}^{\infty} \left[(2((-1)^n + 1)n)/(\pi(n^2 - 1)) \right] \sin nx$. The
partial sum S_6 is shown on the interval $[0, \pi]$ and
$[-3\pi, 3\pi]$.

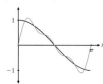
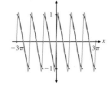

25. $f(x) = \sum_{n=1}^{\infty} (2/n) \sin nx$. The partial sum S_6 is
shown on the interval $[0, \pi]$ and $[-3\pi, 3\pi]$.

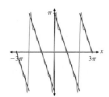

27. $f(x) = \sum_{n=1}^{\infty} \left[2(-1)^n (6 - n^2\pi^2)/n^3 \right] \sin nx$. The
partial sum S_6 is shown on the interval $[0, \pi]$ and
$[-3\pi, 3\pi]$.

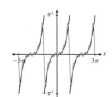

29. $f(x) = \sum_{n=1}^{\infty} \left[2(1 - \cos(n\pi/2))/n\pi \right] \sin nx$. The
partial sum S_6 is shown on the interval $[0, \pi]$ and
$[-3\pi, 3\pi]$.

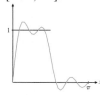
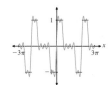

31. $f(x) = (-1/2)\sin x +$
$\sum_{n=2}^{\infty} \left[(-1)^n 2n/(n^2 - 1) \right] \sin nx$. The partial sum S_6
is shown on the interval $[0, \pi]$ and $[-3\pi, 3\pi]$.

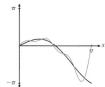
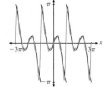

Section 12.4

3. $f(x) \sim \pi/2 - (2/\pi) \sum_{n \text{ odd}} e^{inx}/n^2$

5. $f(x) \sim 1/2 + \sum_{n \text{ odd}} e^{inx}/(\pi i n)$

7. $f(x) \sim ((\sinh b\pi)/\pi) \sum_{n=-\infty}^{\infty} (-1)^n e^{inx}/(b - in)$

9. $f(x) \sim \pi + i \sum_{n \neq 0} [(-1)^{n+1}/n] e^{inx}$

11. $2/\pi + \sum_{n \neq 0, \, n \text{ even}} (2/(\pi(n^2 - 1))) e^{inx}$.

Section 12.5

1. The plots of f, its DFT with $N = 256$, and the partial
sum of order 6.

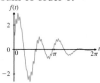

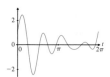

5. The plots of f, the DFT, and the result after
compression with tol $= 0.01$:

Chapter 13
Section 13.1

1. $u_t(x, t) = k u_{xx}$,   for $0 \le x \le L$ and $t > 0$,
$u(0, t) = 5$,   and $u(L, t) = 25$,   for $t > 0$,
$u(x, 0) = 15$,   for $0 \le x \le L$

7. $u_t(x, t) = 0.86 u_{xx}$,   for $0 < x < L$ and $t > 0$,
$u(0, t) = 20$,   and $u(L, t) = 35$,   for $t > 0$,
$u(x, 0) = 15$,   for $0 \le x \le L$

9. $u_t(x, t) = 1.71 u_{xx}$,   for $0 < x < L$ and $t > 0$,
$u_x(0, t) = 0.0013(u(t, 0) - 15)$,   and
$u(L, t) = 35$,   for $t > 0$,
$u(x, 0) = 15$,   for $0 \le x \le L$

Section 13.2

1. $u(x, t) = \sum_{p=0}^{\infty} \frac{400}{(2p+1)\pi} e^{-1.18 \times (2p+1)^2 \pi^2 t/2500} \sin\left(\frac{n\pi x}{50}\right)$
One term. 546 sec.

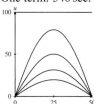

3. a. The steady-state temperature is $f(x) = 20 - x$.

b. $u(x, t) = (20 - 3x) + \sum_{n=1}^{\infty} b_n e^{-k \times n^2 \pi^2 t/100} \sin \frac{n\pi x}{10}$,
where $b_n = (-1)^{n+1} 40/n\pi$, and $k = 0.0057$.

5. $u(x, t) =$
$\frac{4}{\pi^3} \sum_{n=0}^{\infty} \frac{1}{(2n+1)^3} e^{-4(2n+1)^2\pi^2 t} \sin((2n+1)\pi x)$

7. $u(x, t) = -\frac{8}{\pi} \sum_{n=0}^{\infty} \frac{1}{(2n+1)[(2n+1)^2-4]} e^{-(2n+1)^2\pi^2 t}$
$\sin((2n+1)\pi x)$

9.

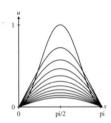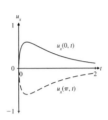

11.

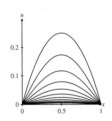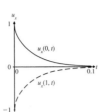

13. $u(x, t) =$
$\frac{1}{4} - \frac{8}{\pi^2} \sum_{n=0}^{\infty} \frac{1}{(4n+2)^2} e^{-(4n+2)^2\pi^2 t} \cos((4n+2)\pi x)$
The solution is plotted below at $t = 0$ and then at
intervals of 0.005.

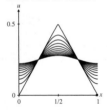

15. $u(x, t) = \frac{2}{\pi} - \frac{4}{\pi} \sum_{n=1}^{\infty} \frac{1}{4n^2-1} e^{-4n^2\pi^2 t} \cos(2n\pi x)$.
The solution is plotted in the following figure at $t = 0$
and then at intervals of 0.1

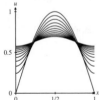

17. $u(x, t) = \frac{1}{4} + \sum_{n=1}^{\infty} a_n e^{-n^2\pi^2 t/4} \cos\left(\frac{n\pi x}{2}\right)$, where for
$n \geq 1$,

$$a_n = \frac{4}{n^2\pi^2} \times \begin{cases} 0, & \text{if } n = 4k, \\ -1, & \text{if } n = 4k \pm 1, \\ 2, & \text{if } n = 4k + 2 \end{cases}$$

The solution is plotted below at $t = 0$ and then at
intervals of 0.1.

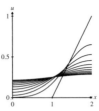

21. $u(x, t) = 1 - x - e^{-x} + e^{-\pi^2 t} \sin \pi x$
$- \sum_{n=1}^{\infty} e^{-n^2\pi^2 t} \left[\frac{2}{n\pi} - \frac{2n\pi[1-(-1)^n/e]}{1+n^2\pi^2} \right] \sin n\pi x$

Section 13.3

1. $u(x, t) = \sum_{k=0}^{\infty} \frac{2 \sin((2k+1)\pi x) \cos((2k+1)\pi t)}{(2k+1)^3\pi^3}$

3. $u(x, t) = 4 \sum_{n=0}^{\infty} \frac{\sin(2n+1)\pi x \cdot \sin(2n+1)\pi t}{(2n+1)^2\pi^2}$

5. $u(x, t) = \sum_{n=1}^{\infty} b_n \sin(n\pi x/3) \sin(n\pi t/3)$, where

$$b_n = \frac{6[\cos(n\pi/3) - \cos(2n\pi/3)]}{n^2\pi^2}$$

7. $u(x, t) = \sin \pi x \cos 2\pi t$

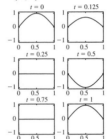

17. a. $u_n(x, t) = e^{-kt} \sin(\mu_n t) \sin\left(\frac{n\pi x}{L}\right)$ and
$v_n(x, t) = e^{-kt} \cos(\mu_n t) \sin\left(\frac{n\pi x}{L}\right)$ for $n = 1, 2, \ldots$
where $\mu_n = \sqrt{(n^2\pi^2 c^2/L^2) - k^2}$

b. $u(x, t) =$
$e^{-kt} \sum_{n=1}^{\infty} [a_n \cos(\mu_n t) + b_n \sin(\mu_n t)] \sin\left(\frac{n\pi x}{L}\right)$
where

$$a_n = \frac{2}{L} \int_0^L f(x) \sin\left(\frac{n\pi x}{L}\right) dx$$

and

$$b_n = -\frac{2}{L\mu_n} \int_0^L g(x) \sin\left(\frac{n\pi x}{L}\right) dx$$

Section 13.4

1. $u(x, y) = \sum_{k=0}^{\infty} \frac{40}{(2k+1)\pi} \frac{\sinh[(2k+1)\pi(2-y)]}{\sinh[(4k+2)\pi]} \sin(n\pi x)$

5. $u(x, y) = 4 \sum_{n=0}^{\infty} \frac{(-1)^n \sinh((2n+1)\pi y) \sin((2n+1)\pi x)}{(2n+1)^2\pi^2 \sinh((2n+1)\pi)}$

7. $u(x, y) = \frac{-\sinh(2\pi(y-2)) \sin(2\pi x)}{\sinh(4\pi)}$

9. $u(x, y) = \frac{4}{\pi} \sum_{n=0}^{\infty} \frac{\sinh((2n+1)\pi y) + \sinh((2n+1)\pi(y-2))}{(2n+1) \sinh((4n+2)\pi)}$
$\sin((2n+1)\pi x)$

11. a. $u(x, y) = -\sum_{n=0}^{\infty} B_n \frac{\sinh(n\pi(y-L))}{\sinh(n\pi L)} \sin(n\pi x)$

b. $u(x, y) = \sum_{n=1}^{\infty} B_n e^{-n\pi y} \sin(n\pi x)$

Section 13.5

3. $u(x, y) = 1 + y$

5. $u(r, \theta) = [1 - r^2 \cos 2\theta]/2$

7. $u(r, \theta) = \frac{2\pi^2}{3} - 4 \sum_{n=1}^{\infty} \frac{r^n \cos n\theta}{n^2}$

9. $u(r) = T_1 + (T_2 - T_1) \ln(r/a)/\ln(b/a)$

11. $u(r, \theta) = \frac{C_0 \ln(r/a)}{2} + \sum_{n=1}^{\infty} \left[C_n(r^{-n} - a^{-2n}r^n) \cos n\theta + D_n(r^{-n} - a^{-2n}r^n) \sin n\theta \right]$
where

$$C_0 \ln(b/a) = \frac{1}{\pi} \int_{-\pi}^{\pi} f(\theta) \, d\theta$$

$$C_n(b^{-n} - a^{-2n}b^n) = \frac{1}{\pi} \int_{-\pi}^{\pi} f(\theta) \cos n\theta \, d\theta$$

$$D_n(b^{-n} - a^{-2n}b^n) = \frac{1}{\pi} \int_{-\pi}^{\pi} f(\theta) \sin n\theta \, d\theta$$

13. $u(r, \theta) = \sum_{n=1}^{\infty} B_n r^{n\pi/\theta_0} \sin(n\theta/\theta_0)$, where
$B_n = \frac{2}{\theta_0 a^n} \int_0^{\theta_0} f(\theta) \sin(n\theta/\theta_0) \, d\theta$

Section 13.6

1. (b), (e), and (f)

3. $\lambda_n = (\pi/2 + n\pi)^2$ and $\phi_n(x) = \cos(\pi/2 + n\pi)x$ for $n = 0, 1, 2, 3, \ldots$

5. The eigenvalues and eigenfunctions are

$$\lambda_n = \omega_n^2 \quad \text{and} \quad \phi_n(x) = \omega_n \cos \omega_n x + \sin \omega_n x,$$

for $n = 1, 2, 3, \ldots$, where ω_n is the nth positive root of $\tan \omega = -1/\omega$.

7. $-\phi'' = \lambda(1/2x)\phi$ **9.** $-(x^{-2}\phi')' = \lambda x^{-4}\phi$

Section 13.7

1. $f(x) = 1 = \sum_{n=0}^{\infty} \frac{4}{(2n+1)\pi} \sin \frac{(2n+1)\pi x}{2}$

3. $f(x) = 1 - x = \sum_{n=0}^{\infty} \left[\frac{4}{(2n+1)\pi} + (-1)^{n+1} \frac{8}{(2n+1)^2\pi^2} \right] \sin \frac{(2n+1)\pi x}{2}$

5. $1 = f(x) = \sum_{n=1}^{\infty} \frac{2(1-\cos \theta_n)}{\theta_n(1+\cos^2 \theta_n)} \sin(\theta_n x)$

7. $1 - x = \sum_{n=1}^{\infty} \frac{2(1+\cos \theta_n)}{\theta_n(1+\cos^2 \theta_n)} \sin \theta_n x$

9. $u(t, x) = \sum_{n=0}^{\infty} \frac{4}{(2n+1)\pi} e^{-(2n+1)^2\pi^2 t/4} \sin \frac{(2n+1)\pi x}{2}$

11. $u(t, x) = \sum_{n=0}^{\infty} \left[\frac{4}{(2n+1)\pi} + (-1)^{n+1} \frac{8}{(2n+1)^2\pi^2} \right] \cdot e^{-(2n+1)^2\pi^2 t/4} \sin \frac{(2n+1)\pi x}{2}$

13. $u(t, x) = \sum_{n=1}^{\infty} \frac{2(1-\cos \theta_n)}{\theta_n(1+\cos^2 \theta_n)} e^{-\theta_n^2 t} \sin \theta_n x$

15. $u(t, x) = \sum_{n=1}^{\infty} \frac{2(1+\cos \theta_n)}{\theta_n(1+\cos^2 \theta_n)} e^{-\theta_n^2 t} \sin \theta_n x$

17. $u(x, y) = \sum_{n=1}^{\infty} \frac{4}{(2n+1)\pi \sinh((2n+1)\pi/2)} \sin \left(\frac{(2n+1)\pi x}{2} \right) \cdot \left[T_2 \sinh \left(\frac{(2n+1)\pi y}{2} \right) - T_1 \sinh \left(\frac{(2n+1)\pi(y-1)}{2} \right) \right]$

Section 13.8

1. $u_s(x, y) = T$

3. $u(x, y, z) = z^3 + \frac{3}{5}z(1 - x^2 - y^2 - z^2)$

5. $c_0 = 5, c_1 = 15/2, c_2 = 0, c_3 = -35/8$

7. $c_0 = 1/4, c_1 = 1/2, c_2 = 5/16, c_3 = 0$

9. $u = 5°$

Section 13.9

7. b. $F(z) = \frac{\pi}{2} + \frac{4}{\pi} \sum_{n=0}^{\infty} \frac{\cos(2n+1)z}{(2n+1)^2}$

c. $f(x, y) = \frac{4}{\pi} \sum_{n=1}^{\infty} \frac{\sin(2n+1)x \cdot \sin(2n+1)y}{(2n+1)^2}$

d. $u(t, x, y) = \frac{4}{\pi} \sum_{n=1}^{\infty} \frac{\sin(2n+1)x \cdot \sin(2n+1)y \cdot \cos \sqrt{2}c(2n+1)t}{(2n+1)^2}$
The vibration is periodic in time.

Section 13.10

3.

$\phi_{1,3}$ $\phi_{3,2}$ $\phi_{2,4}$

5. b. $e^{-kt\alpha_{0,p}^2/a^2} J_0(\alpha_{0,p}r/a)$ for $p = 1, 2, \cdots$

c. $u(t, r) = \sum_{p=1}^{\infty} c_p e^{-kt\alpha_{0,p}^2/a^2} J_0(\alpha_{0,p}r/a)$,
where $c_p = \dfrac{2}{a^2 J_1^2(\alpha_{0,p})} \int_0^a f(r) J_0(\alpha_{0,p}r/a) r \, dr$

7. b. $-\nabla^2 \phi(x, y) = \lambda \phi(x, y), \quad \text{for } (x, y) \in D$

$\dfrac{\partial \phi}{\partial \mathbf{n}}(x, y) = 0, \quad \text{for } (x, y) \in \partial D$

c. $-\left[\phi_{rr} + \frac{1}{r}\phi_r + \frac{1}{r^2}\phi_{\theta\theta} \right](r, \theta) = \lambda \phi(r, \theta)$,
for $r < 1, \phi_r(a, \theta) = 0$ for $0 \le \theta \le 2\pi$

d. $\lambda_{0,0} = 0$ and $\lambda_{n,k} = \beta_{n,k}^2$ for $n \ge 0$ and $k \ge 1$, where $\beta_{n,k}$ is the kth zero of J_n'.

9. b. The eigenvalues for the cylinder are

$$\lambda_{n,k,l} = \mu_{nk} + \nu_l = \frac{\alpha_{n,k}^2}{a^2} + \frac{l^2\pi^2}{L^2},$$

for $n = 0, 1, 2, \ldots, k = 1, 2, \ldots$, and $l = 1, 2, \ldots$. The associated eigenfunctions are

$$u_{0,k,l} = \phi_{0,k} \cdot Z_l = J_0(\alpha_{0,k}r/a) \cdot \sin \frac{l\pi z}{L},$$

for $n = 0$, and

$$u_{n,k,l} = \phi_{n,k} \cdot Z_l = \cos n\theta \cdot J_n(\alpha_{n,k}r/a) \cdot \sin \frac{l\pi z}{L},$$

$$v_{n,k,l} = \psi_{n,k} \cdot Z_l = \sin n\theta \cdot J_n(\alpha_{n,k}r/a) \cdot \sin \frac{l\pi z}{L},$$

for $n \ge 1$.

Index